LIMIT STATE DESIGN
OF REINFORCED CONCRETE

LIMIT STATE DESIGN
OF REINFORCED CONCRETE

By

Dr. B.C. PUNMIA
Formerly,
Professor and Head, Deptt. of Civil Engineering, &
Dean, Faculty of Engineering
M.B.M. Engineering College,
Jodhpur

Er. ASHOK KUMAR JAIN
Director,
Arihant Consultants,
Jodhpur

Dr. ARUN KUMAR JAIN
Assistant Professor
M.B.M. Engineering College,
Jodhpur

(AS PER IS : 456–2000)

LAXMI PUBLICATIONS (P) LTD
(An ISO 9001:2008 Company)
BENGALURU • CHENNAI • COCHIN • GUWAHATI • HYDERABAD
JALANDHAR • KOLKATA • LUCKNOW • MUMBAI • RANCHI • NEW DELHI
BOSTON (USA) • ACCRA (GHANA) • NAIROBI (KENYA)

LIMIT STATE DESIGN OF REINFORCED CONCRETE

Printed and bound in India

Typeset at : Arihant Consultants, Jodhpur.

First Edition : June 2007, *Reprints* : April 2009, 2010, July 2011, June 2012, May 2013, May 2014, July 2015

Revised Edition : 2016

ISBN : 978-81-318-0241-0

PUBLISHED IN INDIA BY

LAXMI PUBLICATIONS (P) LTD
(An ISO 9001:2008 Company)
113, GOLDEN HOUSE, DARYAGANJ, NEW DELHI-110002, INDIA
Telephone : 91-11-4353 2500, 4353 2501
Fax : 91-11-2325 2572, 4353 2528
www.laxmipublications.com info@laxmipublications.com

Branches

Branch	Phone
✆ **Bengaluru**	080-26 75 69 30
✆ **Chennai**	044-24 34 47 26
✆ **Cochin**	0484-237 70 04, 405 13 03
✆ **Guwahati**	0361-254 36 69, 251 38 81
✆ **Hyderabad**	040-27 55 53 83, 27 55 53 93
✆ **Jalandhar**	0181-222 12 72
✆ **Kolkata**	033-22 27 43 84
✆ **Lucknow**	0522-220 99 16
✆ **Mumbai**	022-24 91 54 15, 24 92 78 69
✆ **Ranchi**	0651-220 44 64

C—11415/016/02

Printed at :

Preface

Reinforced concrete occupies a leading position in modern construction of buildings, bridges, water tanks, aqueducts, bunkers, silos and other similar structures. This treatise contains upto-date information on design, analysis and construction of engineered concrete structures, based on limit state design philosophy. Its intention is to provide engineers, architects and students of Civil Engineering and Architecture with authoritative theoretical and practical design information. The book is based on the Fourth Revision of IS 456 adopted by the Bureau of Indian Standards in the year 2000.

In the present book, the subject matter has been introduced gradually, beginning with the properties and characteristics of materials used in reinforced concrete. Chapter 2 deals with various design concepts including the limit state design philosophy. Chapters 3, 4. 5 and 6 deal with the limit state of collapse in flexure, incorporating the theory and design of singly reinforced rectangular sections, doubly reinforced rectangular sections, non-rectangular sections and flanged sections respectively. Chapters 7, 8 and 9 give the limit state of collapse in shear, bond and torsion respectively while Chapter 10 deals with the limit state of collapse in deflection and cracking. Chapter 11 incorporates all the above limit states for the design of various types of beams and beam-slab construction. Chapters 12, 13 and 14 present the analysis and design of two way slabs, circular slabs and flat slabs respectively while Chapter 15 deals with various types of stair cases. Chapters 16, 17, 18 and 19 deal with the analysis and design of various cases of columns, including slender columns. Chapter 20 gives the yield line theory and design of slabs, while Chapter 21 deals with the redistribution of moments in fixed and continuous beams. The design of footings and retaining walls have been given in Chapters 22 and 23 respectively, while Chapter 24 deals with analysis of building frames. Chapter 25 deals exhaustively with the analysis and design of prestressed concrete. The design of miscellaneous structures such as deep beams, corbels, brackets etc., have been incorporated in Chapter 26, while Chapter 27 deals with the detailing of earthquake resistant R.C. structures. Lastly, Appendix-A gives the Working stress design method.

Each Chapter begins with clear statements of pertinent definitions, design theory and design principles and elaborate design procedure. The basic principles are supplemented with numerous design examples and illustrations along with detailed drawings. Various designs in this book are based on latest Indian Standards. The tables and curves giving permissible stresses and/or design coefficients, as well as basic rules for design, taken from various Indian Standards are gratefully acknowledged. Inspite of every care taken to check the numerical work, some errors may remain, and we shall be obliged for any intimation of these readers may discover

The Authors are thankful to Shri M.S. Gahlot for the fine laser type setting of the book. The Authors are also thankful to Shri R.K. Gupta, Managing Director, Laxmi Publications, for taking keen interest in the publication of First edition of the book and bring it out nicely and quickly.

Jodhpur
Akshaya Tritiya,
20th April, 2007

B.C. PUNMIA
ASHOK K. JAIN
ARUN K. JAIN

Preface

Contents

SYMBOLS

A = Total area of section.

A_e = Equivalent area of section.

A_k = Area of concrete core.

A_p = Area of pretensioning tendons.

A_{sc} = Area of steel (comp.)

A_{st} = Area of steel (tensile)

A_{stp} = Positive moment reinforcement.

A_{stn} = Negative moment reinforcement.

A_{sh} = Horizontal reinforcement.

A_l = Area of longitudinal torsional reinforcement.

A_{sv} = Total cross-sectional area of stirrup legs or bent up bars within distance s_v ; vertical reinforcement

A_w = Area of web reinforcement.

A_{φ} = Area of cross-section of one bar.

a = Lever arm.

a_c = Area of concrete.

a_{cs} = Deflection due to shrinkage.

a_v = Shear span.

B = Flange width of T-beam.

b = Width.

b_f = Effective flange width of T or L beam.

b_r = Width of rib.

C = Compressive force.

C_u = Ultimate compressive force in concrete.

c = Compressive stress in concrete.

c' = Stress in concrete surrounding compressive steel.

D = Depth.

D_f = Thickness of flange.

D_k = Diameter of conctete core.

d = Effective depth.

d_c = Cover to compressive steel.

d_s = Depth of slab.

d_t = Cover to tensile steel.

d_w = Depth of rib or web.

e = Eccentricity ; compressive steel depth factor ($= d_c / d$).

e_a = Additional eccentricity

E_c = Short term modulus of elasticity of concrete.

E_s = Modulus of elasticity of steel.

F = Shear force.

F_d = Design load

F_r = Radial shear force.

f = Stress (in general).

f_{ck} = Characteristic compressive strength of concrete.

f_k = Characteristic strength of material.

f_p = Characteristic strength of prestressing steel.

f_{pu} = Ultimate tensile stress in tendons.

f_y = Characteristic strength of steel , yield stress of steel.

f_{yd} = Design yield stress of steel = $0.87 f_y$.

f_{yh} = Characteristic strength of helical reinforcement.

H = Height.

I = Moment of inertia.

I_e = Equivalent moment of inertia of section.

j = Lever arm factor in working stress design.

K_a = Coefficient of active earth pressure.

K_p = Coefficient of passive earth pressure.

k = Neutral axis depth factor (n/d) in working stress design.

k_1, k_2, k_3 = Stress block parameters.

k_b = Distance of bottom kern level.

k_t = Distance of top kern level.

k_u = Neutral axis depth factor in limit state design.

L = Length.

L_a = Anchorage length.

L_d = Development length.

l = Effective length of column Length Bond length.

M = Bending moment ; moment.

M_a = Additional moment for long column.

M_r = Moment of resistance ; Radial bending moment.

M_t = Torsional moment.

M_u = Ultimate moment or factored moment.

$M_{u\ lim}$ = Limiting moment of resistance

M_θ = Circumferential bending moment.

m = Modular ratio.

m_u = Ultimate moment of resistance per unit length of an yield line.

n = Depth of neutral axis.

n_c = Depth of critical neutral axis.

$\Sigma 0$ = Sum of perimeter of bars.

P_a = Active earth pressure.

P_b = Axial load of column corresponding to balanced failure.

P_e = Effective prestressing force.

P_i = Initial prestressing force.
P_p = Passive earth pressure.
P_u = Ultimate Load ; ultimate axial compressive force.
P_{uc} = Factored axial compressive force.
P_{ut} = Factored axial tensile force.
P_{uz} = Axial load capacity of column under pure axial load.
p = Percentage steel.
p' = Reinforcement ratio (A_{st}/bd).
p_a = Active earth pressure intensity.
p_e = Net upward soil pressure.
p_p = Passive earth pressure intensity.
Q = Shear resistance.
q = Shear stress (due to bending).
q_o = Safe bearing capacity of soil.
q' = Shear stress due to torsion.
R = Radius ; Resistance factor $\left(=\frac{1}{2}cjk\right)$ in working stress design.
R_u = Ultimate resistance factor in limit state design.
r = Radius ; cost ratio of steel and concrete ; L/B ratio.
S_d = Standard deviation.
s = Spacing of bars ; standard deviation.
s_a = Average bond stress.
s_b = Local bond stress.
T = Tensile force ; Thickness of wall ; Torsional moment.
T_u = Torsional moment ; Ultimate tensile force in steel.
t = Tensile stress in steel.
t_c' = Compressive stress in compressive steel.
V_u = Factored shear force.
V_{uc} = Ultimate shear resisted by concrete.
V_{us} = Ultimate shear carried by web steel.
W = Point load ; Total load.
W_{cr} = Design surface crack width.
W_u = Ultimate load or factored load.
w_u = Ultimate uniformly distributed load.
X = Co-ordinate.
x_u = Depth of neutral axis (limit state design)
$x_{u,max}$ = Limiting or critical value of depth of neutral axis.
Z = Distance : Co-ordinate.
Y, y = Co-ordinate
Z_B, Z_L = Bending moment coefficients.
z = Lever arm

α = Inclination ; coefficient ; inclination of web steel.

β = Surcharge angle.

γ = Unit weight of soil.

γ' = Submerged unit weight of soil.

γ_f = Partial safety factor appropriate to the loading

γ_m = Partial safety factor appropriate to the material.

σ_{cc} = Permissible stress in concrete (direct comp).

σ_{cc}' = Direct compressive stress in concrete.

σ_{cbc} = Permissible compressive stress in concrete due to bending.

σ_{cu} = Ultimate compressive stress in concrete cubes.

σ_{sc} = Permissible compressive stress in bars.

σ_{sh} = Permissible stress in helical reinforcement.

σ_{sp} = Permissible punching shear stress.

σ_{st} = Permissible tensile stress in reinforcement.

σ_{sy} = Yield point compressive stress in steel.

μ = Coefficient of friction.

φ = Diameter of bar

ϕ = angle ; angle of internal friction

τ_{bd} = Design bond stress.

τ_c = Design of shear strength of concrete

τ_{ce} = Enhanced shear strength

$\tau_{c\,max}$ = Max. shear stress in concrete with shear reinforcement.

τ_v = Nominal shear stress.

ε_c = Strain in concrete

ε_{cs} = Ultimate shrinkage strain of concrete

ε_{cu} = Ultimate strain in concrete = 0.0035

ε_{su} = Ultimate strain in steel.

ψ_{cs} = Shrinkage curvature.

θ = Creep coefficient

ξ_c = Ultimate shear capacity of concrete.

1

Reinforced Concrete Materials

1.1. CEMENT CONCRETE

Concrete is a product obtained artificially by hardening of the mixture of cement, sand, gravel and water in predetermined proportions. When these ingredients are mixed, they form a plastic mass which can be poured in suitable moulds, called *forms*, and set on standing into hard solid mass. The chemical reaction of cement and water, in the mix, is relatively slow and requires time and favourable temperature for its completion. This time, known as *setting time* may be divided into three distinct phases. The first phase, designated as time of *initial set*, requires from 30 minutes to about 60 minutes for completion. During this phase, the mixed concrete decreases its plasticity and develops pronounced resistance to flow. The second phase, known as *final set*, may vary between 5 to 6 hours after the mixing operation. During this phase, concrete appears to be relatively soft solid without surface hardness. The third phase consists of *progressive hardening* and increase in strength. The process is rapid in the initial stage, until about one month after mixing, at which time the concrete almost attains the major portion of its potential hardness and strength.

Depending on the quality and proportions of the ingredients used in the mix, the properties of concrete vary almost as widely as different kinds of stones. Concrete has enough strength in compression, but has little strength in tension. Due to this, concrete as such is weak in bending, shear and torsion. Hence the use of *plain concrete*, described above, is limited to applications where great compressive strength and weight are the principal requirements and where tensile stresses are either totally absent or are extremely low. However, to use cement concrete for common structures such as beams, slabs, retaining structures etc, *steel bars* may be placed at tensile zones of the structure which may then be concreted. The steel bars, known as *steel reinforcement*, embedded in the concrete, take the tensile stresses. The concrete so obtained is termed as *reinforced cement concrete*, commonly abbreviated as R.C.C.

1.2. CLASSIFICATION AND COMPOSITION OF CEMENT

1. *Classification*

Cement may be classified into five groups : (1) Portland Cement, (2) High Alumina Cement, (3) Super Sulphated Cement, (4) Natural Cements and (5) Special Cements, with the following subdivisions :

1. *Portland Cements*
   - (a) Ordinary Portland Cement
   - (b) Rapid Hardening Cement
   - (c) Extra Rapid Hardening Cement
   - (d) Low Heat Portland Cement
   - (e) Portland Blast furnace Slag Cement
   - (f) Portland-Puzzolana Cement
   - (g) Sulphate Resisting Portland Cement
   - (h) White Portland Cement
   - (i) Coloured Portland Cement
2. *High Alumina Cement*
3. *Super Sulphated Cement*
4. *Natural Cements*
5. *Special Cements*
   - (a) Masonry Cement
   - (b) Trief Cement
   - (c) Expansive Cement
   - (d) Oil Well Cement

RECOMMENDATIONS OF IS 456 : 2000

The cement used for *manufacture of concrete* shall be any of the following and the type selected should be appropriate for the intended use :

(*a*) 33 Grade Ordinary Portland Cement Conforming to IS 269
(*b*) 43 Grade Ordinary Portland Cement Conforming to IS 8112
(*c*) 53 Grade Ordinary Portland Cement Conforming to IS 12269.
(*d*) Rapid Hardening Portland Cement Conforming to IS 8041
(*e*) Portland Slag Cement .Conforming to IS 455.
(*f*) Portland Pozzolana Cement (Fly ash based), conforming to IS 1489 (Part 1)
(*g*) Portland Pozzolana Cement (Calcined clay based), conforming to IS 1489 (Part 2)
(*h*) Hydrophobic Cement Conforming to IS 8043
(*i*) Low Heat Portland Cement Conforming to IS 12600
(*k*) Sulphate Resisting Portland Cement Conforming to IS 12330

Other combinations of Portland cement with mineral admixtures of quality conforming with relevant Indian Standards laid down may also be used in the manufacture of concrete provided that there are satisfactory data on their suitability, such as performance test on concrete containing them. Low heat Portland cement conforming to IS 12600 shall be used with adquate precautions with regard to the removal of form work etc. High alumina cement conforming to IS 6452 or super sulphated cement conforming to IS 6909 may be used only under special circumstances with the prior approval of the engineer-in-charge. Specialist literature may be consulted for guidance regarding the use of these types of cement.

2. *Composition of Portland Cement*

The principal raw materials used in the manufacture of cement are :

(a) Argillaceous or silicates of alumina in the form of clays and shales.

(b) Calcareous or calcium carbonate, in the form of lime stone, chalk and marl which is a mixture of clay and calcium carbonate.

The ingradients are mixed in the proportion of about two parts of calcareous material to one part of argillaceous material and then crushed and ground in ball mills in a dry state or mixed in a wet state. The dry powder or the wet slurry is then burnt in a rotary kiln at a temperature between 1400° to 1500°C. The clinker obtained from the kiln is first cooled and then passed on to ball mills where gypsum is added and it is ground to the requisite fineness according to the class of product.

The chief chemical constituents of Portland cement are as follows :

Lime (CaO)	60 to 67%
Silica (SiO_2)	17 to 25%
Alumina (Al_2O_3)	3 to 8%
Iron Oxide (Fe_2O_3)	0.5 to 6%
Magnesia (MgO)	0.1 to 4%
Sulphur Trioxide (SO_3)	1 to 3%
Soda and/or Potash (Na_2O+K_2O)	0.5 to 1.3%

The above constituents forming the raw materials undergo chemical reactions during burning and fusion, and combine to form the following compounds (called Bogue compounds) in the finished products :

	Compound	*Abbreviated designation*
1.	Tricalcium silicate ($3CaO.SiO_2$)	C_3S
2.	Dicalcium silicate ($2CaO.SiO_2$)	C_2S
3.	Tricalcium aluminate ($3CaO.Al_2O_3$)	C_3A
4.	Tetracalcium alumino-ferrite ($4CaO.Al_2O_3.Fe_2O_3$)	C_4AF

The proportions of the above four compounds vary in the various Portland cements. Tricalcium silicate and dicalcium silicates contribute most to the eventual strength. Initial setting of Portland cement is due to the tricalcium aluminate. Tricalcium silicate hydrates quickly and contributes more to the early strength. The contribution of dicalcium silicate takes place after 7 days and may continue for upto 1 year. Tricalcium aluminate hydrates quickly, generates much heat and makes only a small contribution to the strength within the first 24 hours. Tetracalcium alumino-ferrite is comparatively inactive. All the four compounds generate heat when mixed with water, the aluminate generating the maximum heat and the dicalcium silicate generating the minimum. Due to this, tricalcium aluminate is responsible for most of the undesirable properties of concrete. Cement having less C_3A will have higher ultimate strength, less generation of heat and less cracking. Table 1.1 gives the composition and percentage of the four compounds for normal, rapid hardening and low heat Portland cement.

TABLE 1.1 COMPOSITION AND COMPOUND CONTENT OF PORTLAND CEMENT (After Lea)

	Normal	*Rapid hardening*	*Low heat*
***(a)* Composition : Percent**			
Lime	63.1	64.5	60
Silica	20.6	20.7	22.5
Alumina	6.3	5.2	5.2
Iron Oxide	3.6	2.9	4.6
***(b)* Compound : Percent**			
C_3S	40	50	25
C_2S	30	21	45
C_3A	11	9	6
C_4AF	12	9	14

3. ***Ordinary Portland Cement (IS : 269)*** : The properties of various types of portland cement differ because of relative proportions of the four compounds and the fineness to which the cement clinker is ground. The *Ordinary Portland Cement* or the *Setting Cement* is the basic Portland cement and is manufactured in larger quantities than all the others. It is admirably suited for use in general concrete construction where there is no exposure to sulphates in the soil or in ground water.

4. ***Rapid Hardening Portland Cement (IS : 8041)*** : This cement is also known as *high-early strength cement*. It is similar to ordinary Portland cement except that it is ground finer, possesses more C_3S and less C_2S than the ordinary Portland Cement. The magnitude of the increase in strength is gauged from the fact that the strength developed at the age of 3 days is about the same as 7-day strength of ordinary Portland cement with the same water-cement-ratio. The main advantage of a rapid hardening cement is that shuttering may be removed much earlier, thus saving considerable time and expenses. Similarly, in the concrete products in industry, moulds can be released quicker. Rapid hardening cement is also used for road work where it is imperative to open the road traffic with the minimum delay.

5. ***Extra Rapid Hardening Cement*** : Extra rapid hardening cement is obtained by inter-grinding calcium chloride with rapid Hardening Portland cement. The normal addition of $CaCl_2$ is 2% (of the commercial 70% $CaCl_2$) by weight of the rapid hardening cement. The addition of $CaCl_2$ also imparts quick setting properties. Hence this cement should be placed and fully compacted within 30 minutes of mixing.

6. ***Low Heat Portland Cement (IS : 12600)*** : When concrete is poured in any structure, an increase in temperature occurs and a certain amount of heat is generated. This is due to the chemical reaction that takes place while the cement is setting and hardening. Low heat Portland cement is used in massive constructions like abutments, retaining walls, dams, etc., where the rate at which the heat can be lost at the surface is lower than at which the heat is initially generated. The heat generated in ordinary cement at the end of 3 days may be of the order of 80 calories per gram of cement, while in low heat cement it is 50 calories per gram. It has low percentagae of C_3A and relatively more C_2S and less C_3S than ordinary Portland cement. This is achieved by restricting the amount of calcium and increasing the silicates present in the raw materials of manufacture. Therefore, it has low rate of gain of strength, but the ultimate strength is practically the same.

7. ***Portland Blast Furnace Cement (IS : 455)*** : This cement is made by intergrinding Portland cement clinker and blast furnace slag, the proportion of the slag being not less than 25% or more than 65% by weight of cement, as prescribed by IS : 455. The slag should be granulated blast furnace slag of high lime content, which is produced by rapid quenching of molten slag obtained during the manufacture of pig iron in a blast furnace. It is usual for the Portland cement clinker to be ground with a slag, a small percentage of gypsum being added to regulate the setting time. The blending of the Portland cement clinker with the slag, by no means detracts from any desired property of cement. Indeed, it confers upon it some additional advantage. This is because the granulated slag itself possesses latent hydraulic properties which are tremendously activated when the slag is crystalised

and integrated with Portland cement clinker. In general, blast furnace cement will be found to gain strength more slowly than the ordinary Portland cement. It has less heat of hydration than ordinary Portland cement. From the point of view of a builder and the structural engineer, Portland blastfurnace slag cement may be used for all purpose for which ordinary Portland cement is used. In addition, in view of its low heat evolution, it can be used in mass concrete structures such as dams, retaining walls, foundations and bridge abutments.

8. ***Portland Pozzolana Cement (IS : 1489)*** : Portland Pozzolana cement is manufactured either by intergrinding Portland cement clinker and pozzolana or by intimately and uniformly blending Portland cement and fine pozzolana. While intergrinding presents no difficulty, *blending* tends to result in a non-uniform product and Indian Standard is specific in specifying that the latter method should be confined to factories and other such works where intimate blending can be ensured through mechanical means. As per Indian Standard, the proportion of pozzolana may vary from 10 to 25% by weight of cement. The pozzolana used in the manufacture of Portland Pozzolana cement in India is, at present, burnt clay or shale, or fly ash. Although pozzolanas have no cementing value themselves, they have the property of combining with free lime to produce a stable lime pozzolana compound which has definite cementitious properties. This cement has higher resistance to chemical agencies and to attack by sea water, because of absence of free lime. Portland Pozzolana cement also has a lower heat of evolution. Portland Pozzolana cement is frequently stated to have a lower rate of development of strength than ordinary Portland cement. However, when the pozzolana is selected with care and is refined and ground with Portland cement clinker under controlled conditions, the compressive strengths reached by Portland pozzolana cement are comparabale with those reached by ordinary Portland cement. This can be seen from the following table which compares the strength at different ages of Portland Pozzolana cement and Ordinary Portland cement manufactured at the cement works of the Associated Cement Companies (ACC) Ltd., India :

	Compressive Strength, N/mm^2	
Age in days	*Portland Pozzolana Cement*	*Ordinary Portland Cement*
3	19.6 – 21.6	18.6 – 22.6
7	25.5 – 32.4	26.5 – 31.4
28	36.3 – 47.1	35.3 – 51.0

9. ***Sulphate Resisting Cement (IS : 12330)*** : In sulphate resisting Portland cement, the quantity of tricalcium aluminate is strictly limited. They are normally ground finer than Portland cement. The action of sulphates is to form sulpho-aluminates which have expansive properties and so cause disintegration of the concrete. Sulphate resistisng cement should be allowed to harden in the air for as long as possible to allow a resistant skin to be formed through carbonation by the action of atmospheric carbon dioxide.

10. ***White and Coloured Portland Cements*** : The greyish colour of Portland cements is due to the presence of iron oxide. White Portland cement is manufactured in such a way that the percentage of iron oxide is limited to less than 1%. To achieve this, superior raw materials, such as chalk and lime stone having low percentage of iron, and white clay (China clay) are used. Sodium aluminium fluoride (cryolite) is added to act as flux in the absence of iron oxide. Oil fuel is used in place of pulverised coal in the kilning

process in order to avoid contamination by coal ash. Coloured Portland cements are usually obtained by adding strong pigments, upto 10% to the ordinary or white cement, during grinding of clinker. The essential requirements of a good pigment are that it should be permanent and should be chemically inert when mixed with cement.

11. ***High Alumina Cement*** : High Alumina cement, also known as *aluminous cement foundu* is manufactured in entirely different way from that of Portland cements. The raw materials used for its manufacture are chalk and bauxite which is a special clay of extremely high alumina content. The manufacture of this type of cement is more expansive than the Portland cements, though it has many advantages over other types of cements. High alumina cement is characterised by its dark colour, high early strength, high heat of hydration and resistance to chemical attack. It thus produces concrete of far greater strength and in considerably less time even than Rapid-Hardening Portland cement, allowing earlier removal of the formwork. Its rapid hardening properties arise from the presence of calcium aluminate, chiefly monocalcium aluminate ($Al_2O_3 \cdot CaO$), as the predominant compound in place of calcium silicates of Portland cement and after setting and hardening there is no free hydrated lime as in the case of Portland cement. However, great care should be taken in the use of high alumina cement, and it must not be mixed with any other type of cement, since the heat given off on setting is greater than with other cement.

12. ***Super Sulphated Cement (IS : 6909)*** : Super sulphated cement is made from well granulated blastfurnace slag (80 to 85%), calcium sulphate (10 to 15%) and Portland cement (1 to 2%) and is ground finer than the Portland cement. One of its most important properties is its low total heat of hydration. It is, therefore, very suitable for construction of dams and mass concreting work. Concrete made from super sulphated cement may expand if cured in water and may shrink if the concrete is cured in air. Another big advantage of super sulphated cement is its comparatively high resistance to chemical attack.

13. ***Natural cement*** : Natural cements are those cements which are manufactured from naturally occurring *cement rocks* which have compositions similar to the artificial mix of argillaceous and calcareous materials from which Portland cement is manufactured. However, the natural cement rocks are burned at somewhat lower temperatures than those used for the production of Portland cement clinker. The properties of such cement depend upon the composition of the natural cement rock.

14. ***Masonry Cement*** : For a long time, lime gauged with sand was used for mortar for laying brick work. However, in order to increase the strength and rapidity of gaining strength it became common to mix Portland cement with the lime. The usual proportions of cement : lime : sand may range from 1 : 1 : 6 for heavy loads to 1 : 3 : 12 for light loads. Cement sand mortars are too harsh, while lime makes the mortar easier to work. In order to avoid the necessity for mixing cement and lime, masonry cements have recently been introduced. According to Wuerpel, most successful masonry cement are composed of Portland cement clinker, lime stone, gypsum and air-entraining agent. These constituents are ground to an even greater fineness than that of high early strength Portland cement. The plasticity and workability of masonry cements are imparted by the lime stone and air-entraining agents. The ease of working masonry cements and their water retentive properties help to increase their adhesion to bricks or other building units and this is further assisted by the fact their shrinkage is fairly low.

15. ***Trief Cement*** : Trief cement is practically the same as blastfurnace cement except that the blastfurnace slag is ground wet and separately from the cement. Wet grinding results in a fine product, with a specific surface of at least 3000 cm^2/g. Due to this, the slow rate of gain of strength normally associated with blastfurnace cement is avoided and strength from early ages equal to those of ordinary Portland cement are obtained. This cement has smaller shrinkage and a smaller heat of evolution while setting than ordinary Portland cement.

16. ***Expansive Cement*** : Expansive cement expands while hardening. Ordinary concrete shrinks while hardening, resulting in shrinkage cracks. This can be avoided by mixing expansive cement with the normal cements in the concrete, which will neither shrink nor expand. Another useful application of expansive cement is in repair work where the opened up joints can be filled with this cement so that after expansion a tight joint is obtained. Expansive cements have been used in France for underpinning and for the repair of bomb damaged arch bridges.

17. ***Oil Well Cements*** : In the drilling of oil wells, cement is used to fill the space between the steel lining tube and the wall of the well, and to grout up porous strata and to prevent water or gas from gaining access to oil-bearing strata. The cement used may be subject to very high pressure, and the temperature may rise to 400°F. Cement used must be capable of being pumped for up to about 3 hours. It must also harden quickly after setting. These properties can be achieved by (*a*) adjusting the composition of the cement and (*b*) by adding retarders to ordinary Portland cement. In case (*a*), the proportion of Fe_2O_3 is adjusted so that it is above that required to combine with all the Al_2O_3 to form tetra calcium alumino-ferrite $4CaO.Al_2O_3.Fe_2O_3$. The proportion of tricalcium aluminate $3CaO.Al_2O_3$ formed is therefore very small and the setting time is accordingly increased. Setting times of upto 4 hours at a temperature of 200°F and 6 hours at a temp. of 70°F can be obtained with a Portland cement containing no tricalcium aluminate. By the use of retarders setting time of upto $6\frac{1}{2}$ hours at temperatures of upto 220°F can be obtained.

1.3. SPECIFICATIONS FOR PORTLAND CEMENT

For the quality control of Portland cement used for plain and reinforced concrete, the Indian Standard Institution has recommended the following specifications and tests : (1) Chemical composition (2) fineness (3) soundness (4) setting time (5) compressive strength, and (6) heat of hydration.

The specifications for the above requirements, recommended by the Indian Standard are given below.

1. ***Chemical Requirements***

When tested in accordance with the methods given in IS : 4032–1968 (Methods of chemical analysis of hydraulic cement), ordinary cement and rapid hardening Portland cement shall comply with the following chemical requirements ;

(*a*)	Ratio of percentage of lime to percentages of silica, alumina and iron oxide; when calculated by the formula $\frac{CaO - 0.7\ SO_3}{2.8\ SiO_2 + 1.2Al_2O_3 + 0.65\ Fe_2O_3}$	Not greater than 1.02 and not less than 0.66.
(*b*)	Ratio of percentage of alumina to that of iron oxide	Not less than 0.66.
(*c*)	Weight of insoluble residue	Not more than 2 percent
(*d*)	Weight of magnesia	Not more than 6 percent
(*e*)	Total sulphur content calculated as sulphuric anhydride (SO_3)	Not more than 2.75 percent
(*f*)	Total loss on ignition	Not more than 4 percent

When tested in accordance with methods given in IS : 4032-1968, Low Heat Portland cement shall comply with the following requirements as to its chemical composition :

The percentage of lime, after deduction of that necessary to combine with the sulphuric anhydride present, shall be :

not more than $2.4(SiO_2)+1.2(Al_2O_3)+0.65(Fe_2O_3)$

and not less than $1.9(SiO_2)+1.2(Al_2O_3)+0.65(Fe_2O_3)$.

Each symbol in brackets refers to the percentage (by weight of total cement) of the oxide, excluding any contained in the insoluble residue. In all other respects, low heat Portland cement shall comply with requirements specified in *(b), (c), (d), (e)* and *(f)* above.

2. *Fineness* : When tested for fineness in terms of specific surface, by Blaine's air permeability method as described in IS : 4031-1968 (Methods of physical analysis of hydraulic cements), the cement shall comply with the following requirements :

Type of Cement	*Specific Surface*
Ordinary	Not less than 2250
Rapid-hardening	Not less than 3250
Low heat	Not less than 3200

3. *Soundness*

When tested by the 'Le Chatelier' method described in IS : 4031–1968, unaerated ordinary rapid hardening and low heat Portland cement shall not have an expansion of more than 10 mm.

In the event of the cement failing to comply with the above requirements, a further test shall be made by the 'Le Chatelier' method, from another portion of the same sample after aeration, by being spread out to a depth of 75 mm at a relative humidity of 50 to 80 percent for a total period of 7 days, when the expansion of each of the three types of cements mentioned above shall not be more than 5 mm.

When specified by purchaser at the time of placing the order, unaerated ordinary, rapid hardening and low heat Portland cements shall not have an expansion of more than 0.3 percent when tested by the autoclave test described in IS : 4031–1968.

All cements having a magnesia content more than 2 percent shall be tested for soundness by autoclave test and shall comply with the requirements specified in the above para.

4. *Setting time*

The setting time of the cements, when tested by Vicat apparatus method shall conform to the following requirements :

	Ordinary	*Rapid hardening*	*Low heat*
(a) Initial Setting time in minutes, *not less than*	30	30	60
(*b*) Final setting time in minutes, *not more than*	600	60	600

5. *Compressive strength*

The average compressive strength of atleast three mortar cubes (area of face 50 cm^2) composed of one part of cement, three parts of standard sand (conforming to IS : 650–1966) by weight and $\frac{P}{4} + 3.0$ percent (combined weight of cement plus sand) water, and prepared, stored and tested in the manner described in IS : 4031–1968, shall be as follows :

	Ordinary (N/mm^2)	*Rapid hardening* (N/mm^2)	*Low heat* (N/mm^2)
(*a*) 24 h ± 30 min., *not less than*	—	15.7	—
(*b*) 72 h ± 1 h, *not less than*	15.7	27.0	9.8
(*c*) 168 h ± 2 h, *not less than*	21.6	—	15.7
(*d*) 672 h ± 4 h, *not less than*	—	—	34.3

Alternatively, the cement may be accepted based on the compressive strength limits indicated in the para below :

The average compressive strength of atleast three mortar cubes (area of face 50 cm^2) composed of one part of cement, three parts of sand by weight, and $\frac{P}{4} + 3.5$ percent (of combined weight of cement plus sand) water, and prepared, stored and tested in the manner described in IS : 4031–1968 shall be as follows :

	Ordinary (N/mm^2)	*Rapid hardening* (N/mm^2)	*Low heat* (N/mm^2)
(*a*) 24 h ± 30 min., *not less than*	—	11.3	—
(*b*) 72 h ± 1 h, *not less than*	11.3	20.6	6.9
(*c*) 168 h ± 2 h, *not less than*	17.2	—	11.5
(*d*) 672 h ± 4 h, *not less than*	—	—	26

Where P is the percentage of water required to produce a paste of standard consistency, to be used for the determination of the water content of mortar for the compressive strength tests and for the determination of soundness and setting time shall be obtained by the method described in IS : 4031–1968.

6. *Heat of Hydration* : The requirement shall apply only to low heat cement. When tested according to the method described in IS : 4031–1968, the heat of hydration of low heat Portland cement shall be as follows:

(a) *7 days : not more than* 65 Calories per gram, and

(b) *28 days : not more than* 75 calories per gram

Comparison of properties of various type of cements : Table 1.2 gives the *physical requirements* of various common types of cements, as specified by the Indian Standards. Similarly, Table 1.3 gives the chemical requirements for various types of cements.

TABLE 1.2. INDIAN STANDARD SPECIFICATION FOR CEMENTS : PHYSICAL REQUIREMENTS

	Type of Cement	*Ordinary Portland*	*Rapid Hardening Portland*	*Low heat Portland*	*Portland blast Furnace Slag*	*Portland Pozzolana*	*High alumina*	*Super Sulphated*
	Indian Standard	*269-1967*	*269-1967*	*269-1967*	*455-1967*	*1489-1967*	*6452-1972*	*6909-1973*
1. Fineness	Residue by Wt. on 90-micron IS sieve	$\ngtr$ 10 %	$\ngtr$ 5 %	—	$\ngtr$ 10 %	$\ngtr$ 5 %	—	—
	Min. specific surface (Blain's air permeability) cm^2/gm.	2550	3253	3200	2250	3000	2250	4000
2. Soundness	Expansion[1] (Lechatelier)	$\ngtr$ 10 mm	$\ngtr$ 10 mm	$\ngtr$ 10 mm	$\ngtr$ 10 mm	$\ngtr$ 10 mm	$\ngtr$ 5 mm	$\ngtr$ 5 mm
	Autoclave test[2]	$\ngtr$ 0.8%	$\ngtr$ 0.8%	$\ngtr$ 0.8%	$\ngtr$ 0.8%	$\ngtr$ 0.8%	—	—
3. Setting time	Initial (minutes)	$\nless$ 30	$\nless$ 30	$\nless$ 60	$\nless$ 30	$\nless$ 30	$\nless$ 30	$\nless$ 30
	Final (minutes)	$\ngtr$ 600	$\ngtr$ 600	$\ngtr$ 600	$\ngtr$ 600	$\ngtr$ 600	$\ngtr$ 600	$\ngtr$ 600
4. Min. compressive strength[4] (N/mm^2)	24 hours[3]	—	16	—	—	—	30	—
	72 hours[3]	16	27.5	10	16	—	35	15
	168 hours[3]	22	—	16	22	32	—	22
	762 hours[3]	—	—	35	—	31	—	30
				Heat of hydration to be not more than 65 cal/gm at 7 days and 75 cal/gm at 28 days		Average drying shrinkage of mortars to be not more than 0.15%		

Notes :

1. In the event of the cements failing to comply with this requirement, the test is repeated with another portion of the same sample after aeration for 7 days, when the expansion for the first five types shall not be more than 5 mm.

2. When specefied by purchaser and for all cements having a magnesia contents more than 3 per cent.

3. There should be a progressive increase in strength from the strength at 24, 72 or 168 hours.

4. Alternatively, see amendment 1 of IS : 269-1967, amendment 2 of IS : 455-1967 and amendment 2 of IS : 1489-1967 for revised compressive strength

TABLE 1.3. INDIAN STANDARD SPECIFICATIONS FOR CEMENTS : CHEMICAL REQUIREMENTS

Type of Cement	*Ordinary and rapid hardening Portland*	*Low heat Portland*	*Portland blast furnace Slag*	*Portland pozzolana*	*High alumina*	*Super sulphated*
Indian Standard	*269-1967*	*269-1967*	*455-1967*	*1489-1972*	*6452-1972*	*6909-1973*
1. Insoluble residue	2.0%	2.0%	$(x'+0.025)+2.0$	$x+2.0\,\frac{(100-x)}{100}$	—	4%
2. Megnisia (Mgo)	6%	6%	8%	6%	—	10%
3. Sulphur						
as sulphuric anhydride (SO_2)	2.75%	2.75%	3%	2.75	—	6%,
as sulphide (S)	—	—	1.5%	—	—	1.5%
4. Manganic oxide (Mn_2O_3)	—	—	2%	—	—	
5. Loss on ignition	4%	4%	4%	5%	—	—

where : x is the declared percentage of pozzalana in the given cement.

x' is the declared percentage of granulated blast furnace slage in the cement.

1.4. AGGREGATES

Aggregates is a general term applied to those inert or chemically inactive materials which, when bonded together by cement, form concrete. Most of the aggregates used are naturally occurring aggregates such as crushed rock, gravel and sand. Artificial and processed aggregates may be broken brick of crushed air-cooled blastfurnace slag. Light weight aggregates, such as pumice, furnace clinker, coke, breeze, sawdust, foamed slag, expanded slates, expanded vermiculite etc. are also used for the production of concrete of low density.

Classification : Aggregates may be divided into two groups : (*a*) coarse aggregates and (*b*) fine aggregate. Aggregates less than 4.75 mm are known as *fine aggregates* while those more than 4.75 mm in size are known as *coarse aggregate.* For large and important works it has become usual to separate the coarse aggregate also into two or more sizes, and these fractions are kept separate until the proper quantity of each has been weighed out for a batch of concrete. *All in aggregate,* that is to say, aggregate as it comes from the pit or river bed, is sometimes used for unimportant works.

Quality of aggregate : Natural aggregate used for concrete construction is required to comply with the norms laid down in IS : 383–1970 '*Specification for coarse and fine aggregates from natural sources for concrete*'.

Some of the important characteristics of aggregates are : (1) strength (2) size (3) particle shape (4) surface texture (5) grading (6) impermeability (7) cleanliness (8) chemical inertness (8) physical and chemical stability at high temperatures (10) coefficient of thermal expansion, and (11) cost.

Aggregate should be chemically inert, strong, hard, durable, of limited porosity, free from adherent coatings clay lumps, coal, and coal residues and should contain no organic or other admixture that may cause corrosion of the reinforcement, or impair the strength or durability of the concrete. The limits of the content of deleterious materials in aggregate are given in Table 1.4.

TABLE 1.4 : LIMITS OF CONTENT OF DELETERIOUS MATERIALS * (IS 383 : 1970)

Deleterious substances	*Fine Aggregate*		*Coarse Aggregates*	
	uncrushed	*crushed*	*uncrushed*	*crushed*
Coal and lignite	1.00	2.90	1.00	1.00
Clay lumps	1.00	1.00	1.00	1.00
Soft fragments	—	—	3.00	—
Materials finer than 75μ IS sieve	3.00	15.00	3.00	3.00
Shale	1.50	—	—	—
Total of percentage of all deleterious materials**	1.00	2.00	5.00	5.00

* Percentage by weight of aggregate. ** Mica is excluded

The strength of concrete depends upon the strength of aggregate. The strength of an aggregate depends upon the nature of the base rock. Granite aggregate provides greater strength than pumice or burnt clay aggregates. Strength quality of an aggregate is usually specified in terms of (*i*) crushing strength (*ii*) aggregate crushing value, or (*iii*) impact value.

Crushing strength is determined on a rock sample of some specific size. A minimum of 80 N/mm^2 as crushing strength is normally acceptable for good concrete aggregate.

The *aggregate crushing value* is given by :

$$\text{Aggregate crushing value} = 100\,\frac{W_2}{W_1}$$

where W_1 = Weight of surface dry sample of aggregate passing 12.5 mm IS sieve and retained on 10 mm IS sieve.

W_2 = Weight of the material passing through 2.36 mm IS sieve after a specified load is applied in a specified manner.

A *higher* aggregate crushing value indicates a weak aggregate. Its value should not exceed 45 for plain cement concrete, 40 for R.C.C. for buildings and 30 for prestressed concrete.

The *aggregate impact value* is given by :

$$\text{Aggregate impact value} = 100\,\frac{W_2}{W_1}$$

Where W_1 = Weight of oven dry sample of aggregate passing through 12.5 mm IS sieve but retained on 10 mm IS sieve.

W_2 = Weight of fractured aggregate finer than 2.36 mm IS sieve, formed after 15 impacts of a special hammer falling through 300 mm.

A higher aggregate impact value indicates weaker concrete. An impact value of 30 indicates good quality aggregate, though a value upto 45 may sometimes be accepted.

The size of coarse aggregate used depends upon the nature of work. The coarse aggregate must be small enough to enable it to be worked between and around all reinforcements and into all corners of the work. For R.C.C. work, the maximum size of aggregate is limited to 20 mm to 25 mm. A coarse aggregate may have three shapes : rounded, irregular and angular. For a concrete of given workability, rounded aggregates require least water cement ratio while angular aggregates require highest water cement ratio. The particle shape is thus very important, since the water cement ratio governs greatly the strength of concrete. Similarly a concrete made with aggregates of rough surface is stronger than with smooth one. Grading of aggregates greatly affects strength and imperviousness of concrete. If the coarse and fine aggregates are well graded, the percentage of voids is considerably reduced. The voids of the fine aggregates are then occupied by the cement paste while the voids of coarse aggregate are filled with the mortar consisting sand, cement and water. The imperviousness of aggregates is an essential requirement, especially when the concrete is used for water retaining structures.

This is also essential in other R.C. works of permanence, otherwise air and moisture would penetrate with the result that outer concrete would spall out. Aggregates must be clean and free from clay, silt, fine dust etc. so that proper mixing is possible. Dirt or other adherent coating would weaken the adhesion between the individual particles in a hardened concrete. Impurities, such as traces of sulphur or unburnt coal etc. may cause swelling due to chemical action, or may attack the reinforcement. The aggregate should have a thermal expansion similar to that of cement matrix. To summarise, the aggregate should be composed of inert mineral matter, should have high resistance to attrition, should be clean, free from any adhering coating, dense, durable and sufficiently strong to enable the full strength of the cement matrix to be developed.

Coarse aggregate : The material retained on 4.75 mm sieve is termed as coarse aggregate. Crushed stone and natural gravel are the common materials used as coarse aggregate for concrete. Natural gravels can be quarried from pits where they have been deposited by alluvial or glacial action, and are normally composed of flint, quartz, schist and igneous rocks. Coarse aggregates are obtained by crushing various types of granites (such as syenites, dolerites, diorites, quartzites etc.), schist, gneiss, crystalline hard lime stone and good quality sand stones. When very high strength concrete is required, a very fine-grained granite is perhaps the best aggregate. Coarse grained rocks make harsh concrete, and need high proportion of sand and high water-cement ratio to get reasonable degree of workability. Harder types of sand stones, having fine grained texture, are suitable as coarse aggregate, but softer varieties should be used with caution. Concrete made with sand stone aggregate give trouble due to cracking, because of high degree of shrinkage. Similarly hard and close-grained crystalline lime stones are very suitable for aggregate, is cheap, but should be used only in plain concrete. The bricks should be clean, hard, well burnt and free from mortar and should not contain more than half percent of soluble sulphates. It should not be used for reinforced concrete work, since it is porous and may corrode the reinforcement. Blast furnace slag, coal ashes, coke breeze etc. may also be used as aggregate to obtain light weight and insulating concrete of low strength.

Fine aggregate : The material smaller than 4.75 mm size is called fine aggregate. Natural sands are generally used as fine aggregate. Sand may be obtained from pits, river, lake or sea shore. When obtained from pits, it should be washed to free it from clay and silt. Sea shore sand may contain chlorides which may cause efflorescence, and may cause corrosion of reinforcement. Hence it should be thoroughly washed before use. Similarly, if river sand contains impurities such as mud etc. it should be washed before use. Angular grained sand produces good and strong concrete, because it has good interlocking property, while round grained particlels of sand do not afford such interlocking

Grading of aggregates : Gradation of the aggregates is almost as important as its quality is. The grading of the aggregates has marked effect on the workability, uniformity, and finishing qualities of concrete. The grading of coarse aggregate may be varied through wider limits than that of sand without appreciably affecting the workability of concrete. The following Indian Standard sieves are generally used for grading aggregates :

Fine mesh wire cloth : 4.75 mm, 2.36 mm, 1.18 mm, 600 micron, 300-micron, 150-micron, 75-micron.

Square hole perforated plate : 80-mm, 63-mm, 50-mm, 40-mm, 31.5 mm, 25-mm, 20-mm, 16-mm, 12.5 mm, 10-mm, 6.3 mm, 4.75-mm

Table 1.5 gives the grading limits for fine aggregate as specified in IS 383 : 1970. Table 1.6 and 1.7 give, respectively, the grading limits of coarse aggregate as well as for all-in-aggregate as specified in IS 383 : 1970.

TABLE 1.5 : GRADING LIMITS FOR FINE AGGREGATES (IS 383 : 1970)

IS Sieve designation	*Percentage Passing*			
	Grading Zone I	*Grading Zone II*	*Grading Zone III*	*Grading Zone IV*
10 mm	100	100	100	100
4.75 mm	90–100	90–100	90–100	95–100
2.36 mm	60–95	75–100	85–100	95–100
1.18 mm	30–70	55–90	75–100	90–100
600 micron	15–34	35–39	60–79	80–100
300 micron	5–20	8–30	12–40	15–50
150 micron	0–10	0–10	0–10	0–15

TABLE 1.6. GRADING LIMITS FOR COARSE AGGREGATES (IS 383 ; 1970)

IS sieve designa-tion	*Percentage passing, for single size aggregat of a nominal size*						*Percentage passing, for graded aggregate of nominal size*			
	63 mm	*40 mm*	*20 mm*	*16 mm*	*12.5mm*	*10 mm*	*40 mm*	*20 mm*	*16mm*	*12.5 mm*
80 mm	100	—	—	—	—	—	100	—	—	—
63 mm	85-100	—	—	—	—	—	—	—	—	—
40 mm	0-30	100	100	—	—	—	95-100	100	—	—
20 mm	0-5	85-100	85-100	100	—	—	30-70	90-100	100	100
16 mm	—	0-20	—	85-100	100	—	—	—	90-100	—
12.5 mm	—	—	—	—	85-100	100	—	—	—	90-100
10 mm	0-5	—	0-20	0-30	0-45	**85-100**	10-35	25-25	30-70	40-85
4.75 mm	—	0-5	0-5	0-10	0-10	**0-20**	0-10	0-10	0-10	0-10
2.36 mm	—	—	—	—	—	0-5	—	—	—	—

According to IS 383 : 1970, the fine aggregate (*i.e.* sand) is graded into four zones, as shown in Fig. 1.1. A fine aggregate whose grading falls outside the limits of any particular grading zones or sieves other than 600-micron I.S. Sieves is regarded as falling within that grading zone provided the total of the excess on the other four sieves is not greater than 5%. Sands of zones 1 to 3 may be used for concrete work. Zone-4 sand should not be used except in special mixes.

TABLE 1.7. GRADING LIMIT FOR ALL IN AGGREGATE (IS 383 : 1970)

IS Sieve designation	*Percentage passing, for all in aggregate of*	
	40 mm nominal size	*20 mm nominal size*
80 mm	100	–
40 mm	95–100	100
20 mm	45–57	95–100
4.75 mm	25–45	30–50
600 micron	8–30	10–25
150 micron	0–6	0–6

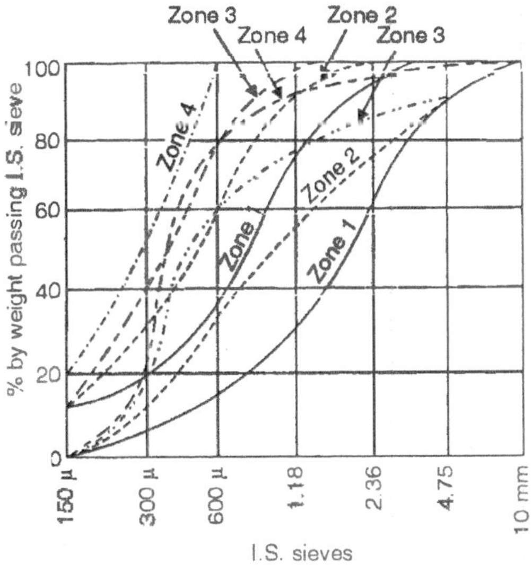

FIG 1.1
ZONE GRADING OF FINE AGGREGATES.

Fineness Modulus

The *fineness modulus* of an aggregate is an index number which is roughly proportional to the average size of the particles in the aggregate. The coarser the aggregate, the higher the fineness modulus. The fineness modulus is obtained by adding the percentage of the weight of material retained on the following I.S. sieves and dividing it by 100 :

80 mm, 40 mm, 20 mm, 10 mm, 4.75 mm, 2.36 mm, 1.18 mm, 600-micron, 300-micron and 150-micron (total 10 sieves).

TABLE 1.8. (*a*) DETERMINATION OF FINENESS MODULUS

IS Sieve	*Coarse aggregate (10 kg)*			*Fine aggregate (1 kg)*		
	Weight retainded (kg.)	*Total wt. retainded (kg.)*	*% weight retainded*	*Weight retained (kg.)*	*Total wt. retained (kg.)*	*% weight retained*
80 mm	0.0	0.0	0.0	0.0	0.0	0.0
40 mm	0.0	0.0	0.0	0.0	0.0	0.0
20 mm	3.5	3.5	35.0	0.0	0.0	0.0
10 mm	3.0	6.5	65.0	0.0	0.0	0.0
4.75 mm	2.8	9.3	93.0	0.0	0.0	0.0
2.36 mm	0.7	10.0	100.0	0.1	0.1	10.0
1.18 mm	0.0	10.0	100.0	0.25	0.35	35.0
600 micron	0.0	10.0	100.0	0.35	0.70	70.0
300 micron	0.0	10.0	100.0	0.20	0.90	90.0
150 micron	0.0	10.0	100.0	0.10	1.00	100.0
	Sum	:	693.0	Sum		305.0
Fineness modulus	$\frac{693.0}{100} = 6.93$			$\frac{305.0}{100} = 3.05$		

Table 1.8(*a*) illustrates the method of determining fineness modulus of both coarse and fine aggregates. It has been found that certain values of fineness moduli for the fine and coarse aggregates give good workability, with a minimum quantity of cement. The limits of fineness moduli are given in Table 1.8(*b*)

TABLE 1.8 (*b*) LIMITS OF FINENESS MODULI

Maximum size of aggregate		*Fineness modulus*	
		max.	*min.*
(a) Fine aggregate		2.0	3.5
(b) Coarse aggregate	(*i*) 20 mm	6.0	6.9
	(*ii*) 40 mm	6.9	7.5
	(*iii*) 75 mm	7.5	8.0
	(*iv*) 150 mm	8.0	8.5
(c) Mixed aggregate	(*i*) 20 mm	4.7	5.1
	(*ii*) 25 mm	5.0	5.5
	(*iii*) 32 mm	5.2	5.7
	(*iv*) 40 mm	5.4	5.9
	(*v*) 75 mm	5.8	6.3
	(*vi*) 150 mm	6.5	7.0

1.5. WATER

Water acts as lubricant for the fine and coarse aggregates and acts chemically with cement to form the binding paste for the aggregate and reinforcement. Water is also used for curing the concrete after it has been cast into the forms.

Water used for mixing and curing shall be clean and free from injurious amount of oils, acids, alkalis, salts, sugar, organic materials or other substances that may be deleterious to concrete or steel. Potable water is generally considered satisfactory for mixing concrete. As a guide, the following concentrations represent the maximum permissible values :

(*a*) To neutralize 100 ml sample of water, using phenolphthalein as an indicator, it should not require more than 5 ml of 0.02 normal NaOH.

(*b*) To neutralize 100 ml sample of water, using mixed indicator, it should not require more than 25 ml of 0.02 normal H_2SO_4.

(*c*) The permissible limits of solids shall be as given in Table 1.9.

TABLE 1.9 PERMISSIBLE LIMITS FOR SOLIDS (IS 456 : 2000)

S.N.	*Type of solid*	*Permissible limit (max.)*
(*i*)	Organic	200 mg/l
(*ii*)	Inorganic	300 mg/l
(*iii*)	Sulphates (as SO_3)	400 mg/l
(*iv*)	Clorides (as Cl)	2000 mg/l for concrete not containing embeded steel and 500 mg/l for reinforced concrete work
(*v*)	Suspended matter	2000 mg/l

In case of doubt regarding development of strength, the suitability of water for making concrete shall be ascertained by the compressive strength and initial setting time tests, with the following instructions :

(*i*) The sample of water taken for testing shall represent the water proposed to be used for concreting, due account being paid to seasonal variation. The sample shall not receive any treatment before testing other than that envisaged in the regular supply of water proposed for use in concrete. The sample shall be stored in a clean container previously rinsed out with similar water.

(*ii*) Average 28 days compressive strength of atleast three 150 mm concrete cubes prepared with water proposed to be used shall not be less than 90 percent of the average strength of three similar concrete cubes prepared, cured and tested in accordance with the requirements of IS 516.

(*iii*) The initial setting time of test block made with appropriate cement and the water proposed to be used shall not be less than 30 min. and shall not differ by ± 30 min. from the initial setting time of control block prepared with the same cement and distilled water.

pH value : The pH value of water shall not be less than 6.

Sea water : Mixing and curing with sea water is not recommended because of presence of harmful salts in sea water. Under unavoidable circumstances, sea water may be used for mixing or curing in plain concrete with no embedded steel after having given due consideration to possible disadvantages and precautions including use of appropriate cement system.

Water for curing : Water found satisfactory for mixing is also suitable for curing concrete. However, water used for curing should not produce any objectionable stain or unsightly deposit on the concrete surface. The presence of tannic acid or iron compounds is objectionable.

1.6. MEASUREMENT OF MATERIALS

The materials used for preparation of concrete are (*i*) cement (*ii*) fine aggregate (*iii*) coarse aggragate and (*iv*) Water. Their accurate measurement before mixing is very important so that the required quantities in the proportion of the concrete mix are obtained.

1. *Cement.*

It is preferable to measure cement in terms of its weight, and not in terms of volume. The volume of cement changes with the conditions of measurement. In our country, cement is supplied in bags, each bag weighing 50 kg. Under normal conditions, the volume of cement in the bag is considered equivalent to 34.5 Litres. However, if the same cement is shovelled, the bag may measure upto 42 litres. Before mixing, therefore, cement is measured in terms of weight.

2. *Fine aggregates.* : Fine aggregate (*i.e.* sand) may be measured by weight for accurate works, and by volume for ordinary works. However, when dry sand absorbs water from atmosphere, or when water is mixed to it artificially, its volume increases. This increase in volume due to moisture in sand is known as 'bulking of sand'. Water particles lubricate the sand particles, causing surface tension, and due to this, particles are pulled apart. Thus increase in volume results. This increase in volume depends on the gradation of sands, but may be taken to be maximum at a moisture content of about 4% by weight of dry sand. Further increase in moisture results in decrease in the percent increase of volume, as shown in Fig. 1.2. The bulking increases with fineness, and may be about 25% by volume.

Due to this, if sand is measured by volume, bulking should be properly accounted for. In order to determine unit weight of sand and percentage bulking corresponding to a given moisture content, a measuring cylinder, 15 cm in diameter and 87 cm high is used. The cylinder may first be weighed empty. It is then filled with water and weighed. The difference of the two weights would give the volume of the cylinder.

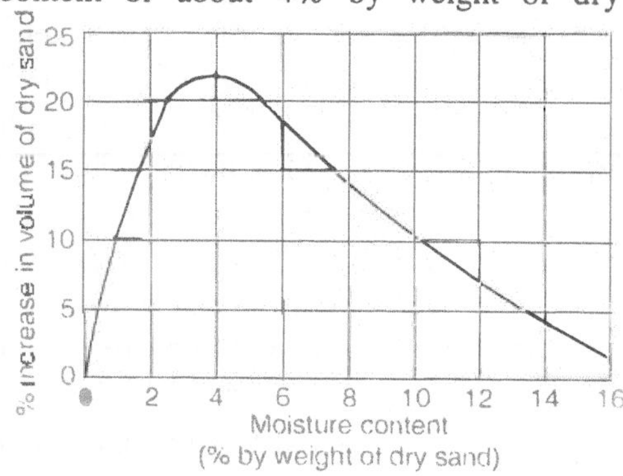

FIG 1.2. BULKING CURVE FOR SAND

To determine the unit weight of loose dry sand, fill it gently with the sand, without tamping, and level the top surface by striking off the excess sand. Weigh the cylinder. The difference between this weight and the empty weight would give the weight of loose dry sand. Dividing this with the volume of cylinder, the unit weight of loose dry sand is known. To know unit weight of dense dry sand, fill the sand in the cylinder in three equal layers, each layer being compacted 25 times with 16 mm dia. rod 60 cm long, provided at its lower end. The tamping rod should penetrate well below the layer being compacted. After filling, the level of sand is struck to the level of upper end of the cylinder, and the cylinder is weighed. Dividing the weight of compacted sand in the cylinder to the volume, unit weight of dry compacted sand is known.

To determine the percent bulking at various moisture contents, take about 6 litres of dry compacted sand, weigh it and dump it in a mixing pan. Mix it with x percent (say), by weight of dry sand, in water thoroughly. Fill the cylinder with this wet sand, without compacting, and strike off the excess sand. Weigh the cylinder, to get the weight of wet sand in the cylinder. Repeat the experiment with various values x, varying from 0 to 20 percent.

Let W_1 = Weight of one cubic metre of compacted dry sand (*i.e.* unit weight of dry compacted sand).

W_3 = Weight of one cubic metre of loose wet sand.

x = percentage of water added.

W_2 = Weight of dry sand contained in one cubic metre of wet loose sand.

Now W_3 = Weight of dry sand + weight of water

$$= W_2 + \frac{x}{100} \times W_2 = W_2\left[1 + \frac{x}{100}\right]$$

Hence

$$W_2 = \frac{W_3}{1 + \frac{x}{100}} \qquad \ldots(1.1)$$

$$\text{Therefore, percentage bulking} = \frac{\text{increase in volume}}{\text{original volume}} = \frac{W_1 - W_2}{W_2} \times 100 \quad \ldots(1.2)$$

$$\text{Bulking factor} = \frac{W_1}{W_2} \quad \ldots(1.3\ a)$$

With the help of Eqs. 1.1 and 1.2 , curve of Fig 1.2 can be plotted at various moisture contents.

In order to determine the bulking of a given sand sample in the field, fill a graduated cylinder with the sand, upto a certain mark. Now fill the cylinder to the top, with water. The sand will settle in the cylinder to another mark. Determine the decrease in the volume. To determine the bulking in the laboratory, first determine the volume V_1 of sand corresponding to a mark in the graduated cylinder. Dry the sand and determine its decreased volume V_2.

$$\text{The percentage bulking} = \frac{V_1 - V_2}{V_2} \times 100 \quad \ldots(1.3\ b)$$

Thus, knowing the percentage bulking at the site, actual volume of corresponding dry sand can be estimated by subtracting from the measured volume of sand the increase in volume due to bulking. For accurate and large scale works, sand is always measured by weight and necessary allowance is made for the hygroscopic moisture in the sand.

3. ***Coarse aggregate.*** : There is no problem of bulking in coarse aggregate. and hence it may be measured either by volume or by weight. However, the weight of a given volume of aggregate is influenced by the size of the measuring box. Hence for accurate and large scale works, measurement should be done by weight. The unit weight of coarse aggregate in loose and dry state is found exactly in the same manner as for fine aggregate, except that a bigger container is used. Since the size of container has effect on the determination, Indian Standards specify the following container for carrying out the tests :

1. Maximum size of aggregate 5 mm to 40 mm : 15 litre capacity cylinder of 25 cm diameter.
2. Maximum size of aggregate over 40 mm : 40 litre capacity cylinder of 35 cm diameter.

4. *Water.*

Water is normally measured by volume, and specified as so many litres per bag of cement. For a given quantity of water to be mixed in concrete, adjustment should be made for the amount of water present in sand and aggregate. The amount of the water present in the aggregate, due to hygroscopic action etc. should be subtracted from the total required quantity of water. However, if the aggregate is dry, and found to absorb water, extra water should be *added* to account for this. The percentage absorption should be determined first.

1.7. WATER CEMENT RATIO

Water-cement ratio is the ratio of volume of water mixed in concrete to volume of cement used. The strength and workability of concrete depend to a great extent on the amount of water used. For a given proportion of the materials, there is an amount of water which gives the greatest strength. Amount of water less than this optimum water decreases the strength and about 10 percent less may be insufficient to ensure complete setting of cement. Similarly, more than the optimum water increases the workability but decreases the strength. An increase in 10% above the optimum may decrease the strength approximately by 15%, while an increase in 50% may decrease the strength to one half. The use of an excessive amount of water not only produces low strength but increases shrinking and decreases density and durability.

According to Abram's water-cement ratio law, for any given conditions of test the strength of workable concrete mix is dependent only on the water-cement ratio. Lesser the water-cement ratio in a workable mix greater will be its strength. From Abram's law, it follows that provided the concrete is fully compacted, the strength is not affected by aggregate shape, type or surface texture, or the aggregate grading, the workability and the richness of the mix.

According to Powers, cement does not combine chemically with more than half the quantity of water in the mix. Cement requires about 1/5 to 1/4 of its weight of water to become completely hydrated. This suggests that if water cement ratio is less than 0.4 to 0.5, complete hydration will not be secured. Some practical values of water cement ratio for structural reinforced concrete are about 0.45 for 1 : 1 : 2 concrete, 0.50 for 1 : $1\frac{1}{2}$: 3 concrete and 0.55 to 0.60 for 1 : 2 : 4 concrete. However, concrete vibrated by efficient mechanical vibrators require less water cement ratio, and hence have more strength. Sometimes, plasticising agents may be mixed to increase the workability of the mix. For such concrete, therefore, water-cement ratio is reduced, resulting in an increase in the strength.

1.8. PROPERTIES AND TESTS ON CONCRETE

The important properties of concrete, which govern the design of a concrete mix are (*i*) strength (*ii*) durability (*iii*) workability and (*iv*) economy. The aim of proportioning a concrete mix will be to find the economic proportions of cement, coarse aggregate, fine aggregate and water so as to get a mix of given strength, proper workability and durability.

1.9. DURABILITY OF CONCRETE

Durability is the property of concrete by virtue of which it is capable of resisting its disintegration and decay. The concrete should be durable with proper regard to the various weathering conditions such as action of atmospheric gases, moisture changes, temperature variations. Disintegration and decay of concrete may be due to the following reasons :

(1) Use of unsound cement, which due to some delayed chemical reactions, undergo volume changes after the concrete has hardened.

(2) Use of less durable aggregate, which may either react with cement, or may be reacted upon by atmospheric gases.

(3) Entry of harmful gases and salts through excessive pores and voids present in unsound concrete, causing its disintegration.

(4) Freezing and thawing of water sucked through the cracks or crevices. by capillary action causing its disintegration.

(5) Expansion and contraction resulting from temperature changes or alternate wetting and drying.

As stated earlier, water required for chemical reaction is about 25% of the weight of cement. Hence excess water present in concrete later evaporates, leaving voids and pores. These pores or voids are later responsible for decay of concrete. Hence for durable concrete, water cement ratio should be as small as possible to get a workable mix. A well compacted concrete has less voids and pores and has more durability. The entrainment of air in concrete has been found to increase very considerably the resistance of concrete to freezing and thawing. The improvement in this respect is due to relief, occassioned by the minute dispersed air bubbles which act as expansion chambers, of stresses and pressures, caused by temperature and moisture changes and by expansion of the moisture contained in concrete on freezing. Vinsol resin is sometimes mixed with concrete to have the property of entrapping innumerable minute air bubbles in concrete. In order to prevent Vinsol resin reacting chemically with the cement, and to make it soluble in water, it is first neutralised by the addition of sodium hydroxide which converts it into a soap. The quantity of resin required for such purpose is extremely small ranging from 0.005 to 0.05 of 1 percent of the weight of cement.

RECOMMENDATIONS OF IS 456 : 2000 ON DURABILITY OF CONCRETE

1. General

A durable concrete is one that performs satisfactorily in the working environment during its anticipated exposure conditions during service. The materials and mix proportions specified and used should be such as to maintain its integrity and, if applicable, to protect embedded metal from corrosion.

1.1. One of the main characteristics influencing the durability of concrete is its permeability to the ingress of water, oxygen, carbon dioxide, chloride, sulphate and other potentially deleterious substances. Impermeability is governed by the constituents and workmanship used in making the concrete. With normal-weight aggregates a suitably low permeability is achieved by having an adequate cement content, sufficiently low free water/cement ratio, by ensuring complete compaction of the concrete, and by adequate curing.

The factors influencing durability include :

(*a*) the environment ;
(*b*) the cover to embedded steel;
(*c*) the type and quality of constituent materials;
(*d*) the cement content and water/cement ratio of the concrete;
(*e*) workmanship, to obtain full compaction and efficient curing; and
(*f*) the shape and size of the member.

The degree of exposure anticipated for the concrete during its service life together with other relevant factors relating to mix composition, workmanship, design and detailing should be considered. The concrete mix to provide adequate durability under these conditions should be chosen taking account of the accuracy of current testing regimes for control and compliance specified in the Code.

2. Requirements for Durability

2.1. Shape and Size of Member : The shape or design details of exposed structures should be such as to promote good drainage of water and to avoid standing pools and rundown of water. Care should also be taken to minimize any cracks that may collect or transmit water. Adequate curing is essential to avoid the harmful effects of early loss of moisture. Member profiles and their intersections with other members shall be designed and detailed in a way to ensure easy flow of concrete and proper compaction during concreting.

Concrete is more vulnerable to deterioration due to chemical or climatic attack when it is in thin sections, in sections under hydrostatic pressure from one side only, in partially immersed sections and at corners and edges of elements. The life of the structure can be lengthened by providing extra cover to steel, by chamfering the corners or by using circular cross-sections or by using surface coating which prevent or reduce the ingress of water, carbon dioxide or aggressive chemicals.

2.2. Exposure Conditions

2.2.1. General environment : The general environment to which the concrete will be exposed during its working life is classified into five levels of severity, that is mild, moderate, severe, very severe and extreme as described in Table 1.10.

TABLE 1.10. ENVIRONMENTAL EXPOSURE CONDITIONS

S.No. *(1)*	*Environment* *(2)*	*Exposure Conditions* *(3)*
(i)	**Mild**	Concrete surfaces protected against weather or aggressive conditions; except those situated in coastal area.
(ii)	**Moderate**	Concrete surfaces sheltered from severe rain or freezing whilst wet Concrete exposed to condensation and rain Concrete continuously under water Concrete in contact or buried under non-aggressive soil/ground water Concrete surfaces sheltered from saturated salt air in coastal area
(iii)	**Severe**	Concrete surfaces exposed to severe rain; alternate wetting and drying or occasional freezing whilst wet or severe condensation. Concrete complëtely immersed in sea water Concrete exposed to coastal environment
(iv)	**Very severe**	Concrete surfaces exposed to sea water spray; corrosive fumes or severe freezing conditions whilst wet Concrete in contact with or buried under aggressive sub-soil/ground water
(v)	**Extreme**	Surface of members in tidal zone Members in direct contact with liquid/solid aggressive chemicals

2.2.2. Abrasive : Specialist literatures may be referred to for durability requirements of concrete surfaces exposed to abrasive action for example in case of machinery and metal types.

2.2.3. Freezing and thawing : Where freezing and thawing actions under wet conditions exit, enhanced durability can be obtained by the use of suitable air entraining admixtures. When concrete lower than grade M 50 is used under these conditions, the mean total air content by volume of the fresh concrete at the time of delivery into the construction should be as per Table 1.11.

TABLE 1.11

Nominal Maximum Size Aggregate (mm)	*Entrained Air Percentage*
20	5 ± 1
40	4 ± 1

Since air entrainment reduces the strength, suitable adjustments may be made in the mix design for achieving required strength.

2.2.4. Exposure to sulphate attack : Table 1.12 gives recommendations for the type of cement, maximum free water/cement ratio and minimum cement content, which are required at different sulphate concentrations in near-neutral ground water having pH of 6 to 9.

For the very high sulphate concentrations in Class 5 conditions, some form of lining such as polyethylene or polychloroprene sheet; or surface coating based on asphalt, chlorinated rubber, epoxy; or polyurethane materials should also be used to prevent access by the sulphate solution.

TABLE 1.12 REQUIREMENTS FOR CONCRETE EXPOSED TO SULPHATE ATTACK

Sl No.	*Class*	*Concentration of Sulphates Expressed as SO_3*			*Type of Cement*	*Dense, Fully Compacted Concrete, made with 20 mm Nominal Maximum Size Aggregates Complying with IS 383*	
		In Soil		*In Ground Water*			
		Total SO_3	*SO_3 in 2:1 Water : Soil Extract*			*Minimum Cement Content kg/m^3*	*Maximum Face Water-Cement Ratio*
		Percent	*g/l*	*g/l*			
(1)	*(2)*	*(3)*	*(4)*	*(5)*	*(6)*	*(7)*	*(8)*
(i)	1	Traces (<0.2)	Less than 1.0	Less than 0.3	Ordinary Portland cement or Portland slag cement or Portland pozzolana cement	280	0.55
(ii)	2	0.2 to 0.5	1.0 to 1.9	0.3 to 1.2	Ordinary Portland cement or Portland slag cement or Portland pozzolana cement	330	0.50
					Supersulphated cement or sulphate resisting Portland cement	310	0.50
(iii)	3	0.5 to 1.0	1.9 to 3.1	1.2 to 2.5	Supersulphated cement or sulphate resisting Portland cement	330	0.50
					Portland pozzolana cement or Portland slag cement	350	0.45
(iv)	4	1.0 to 2.0	3.1 to 5.0	2.5 to 5.0	Supersulphated or sulphate resisting Portland cement	370	0.45
(v)	5	More than 2.0	More than 5.0	More than 5.0	Sulphate resisting Portland cement or supersulphated cement with protective coatings	400	0.40

NOTES :

1. Cement content given in this table is irrespective of grades of cement.

2. Supersulphated cement is generally restricted where the prevailing temperature is above 40°C.

3. Supersulphated cement givens an acceptable life provided that the concrete is dense and prepared with a water-cement ratio of 0.4 or less, in mineral acids, down to pH 3.5.

4. The cement contents given in col 6 of this table are the minimum recommended. For SO_3 contents near the upper limit of any class, cement contents above these minimum are advised.

5. For severe conditions, such as thin sections under hydrostatic pressure on one side only and sections partly immersed, considerations should be given to a further reduction of water-cement ratio.

6. Portland slag cement conforming to IS 455 with slag content more than 50 percent exhibits better sulphate resisting properties.

7. Where chloride is encountered along with sulphates in soil or ground water, ordinary Portland cement with C_3A content from 5 to 8 percent shall be desirable to be used in concrete, instead of sulphate resisting cement. Alternatively, Portland slag cement conforming to IS 455 having more than 50 percent slag or a blend of ordinary Portland cement and slag may be used provided sufficient information is available on performance of such blended cements in these conditions.

2.3. Requirement of Concrete Cover

2.3.1. The protection of the steel in concrete against corrosion depends upon an adequate thickness of good quality concrete.

2.3.2. The nominal cover to the reinforcement shall be provided as per IS 456: 2000 (see Chapter 11 and Chapter 16)

2.4. Concrete Mix Proportions

2.4.1. General : The free water-cement ratio is an important factor in governing the durability of concrete and should always be the lowest value. Appropriate values for minimum cement content and the maximum free water-cement ratio are given in Table 1.13 for different exposure conditions. The minimum cement content and maximum water-cement ratio apply to 20 mm nominal maximum size aggregate. For other sizes of aggregate they should be changed as given in Table 1.14.

TABLE 1.13. MINIMUM CEMENT CONTENT, MAXIMUM WATER-CEMENT RATIO AND MINIMUM GRADE OF CONCRETE FOR DIFFERENT EXPOSURES WITH NORMAL WEIGHT AGGREGATES OF 20 MM NOMINAL MAXIMUM SIZE

Sl. No.	*Exposure*	*Plain Concrete*			*Reinforced Concrete*		
		Minimum Cement Content kg/m³	*Maximum Free Water-Cement Ratio*	*Minimum Grade of Concrete*	*Minimum Cement Content kg/m³*	*Maximum Free Water-Cement Ratio*	*Minimum Grade of Concrete*
(1)	(2)	(3)	(4)	(5)	(6)	(7)	(8)
(i)	*Mild*	220	0.60	-	300	0.55	M 20
(ii)	*Moderate*	240	0.60	M 15	300	0.50	M 25
(iii)	*Severe*	250	0.50	M 20	320	0.45	M 30
(iv)	*Very Severe*	260	0.45	M 20	340	0.45	M 35
(v)	*Extreme*	280	0.40	M 25	360	0.40	M 40

NOTES : 1. Cement content prescribed in this table is irrespective of the grade of cement and it is inclusive of additions. The additions such as fly ash or ground granulated blast furnace slag may be taken into the concrete composition with respect to the cement content and water-cement ratio if the suitability is established and as long as the maximum amounts taken into account do not exceed the limit of pozzolana and slag specified in IS 1489 (Part 1) and IS 455 respectively. 2. Minimum grade for plain concrete under mild exposure condition is not specified.

2.4.2. *Maximum cement content* : Cement content not including fly ash and ground granulated blast furnace slag in excess of 450 kg/m^3 should not be used unless special consideration has been given in design to the increased risk of cracking due to drying shrinkage in thin sections, or to early thermal cracking and to the increased risk of damage due to alkali silica reactions.

TABLE 1.14 ADJUSTMENTS OF MINIMUM CEMENT CONTENT FOR AGGREGATES OTHER THAN 20 mm NOMINAL MAXIMUM SIZE

Sl No.	*Nominal maximum Aggregated Size (mm)*	*Adjustments to Minimum Cement Contents in Table 1.3 (kg/m^3)*
(1)	*(2)*	*(3)*
(i)	10	+ 40
(ii)	20	0
(iii)	40	− 30

TABLE 1.15 LIMITS OF CHLORIDE CONTENT OF CONCRETE

Sl No.	*Type or Use of Concrete*	*Maximum Total Acid Soluble Chloride Content Expressed as kg/m^3 of Concrete*
(1)	*(2)*	*(3)*
(i)	Concrete containing metal and steam cured at elevated temperature and pre-stressed concrete	0.4
(ii)	Reinforced concrete or plain concrete containing embedded metal	0.6
(iii)	Concrete not containing embedded metal or any material requiring protection from chloride	3.0

2.5. *Mix Constituents*

2.5.1. *General* : For concrete to be durable, careful selection of the mix and materials is necessary, so that deleterious constituents do not exceed the limits.

2.5.2. *Chlorides in concrete* : Whenever there is chloride in concrete there is an increased risk of corrosion of embedded metal. The higher the chloride content, or if subsequently exposed to warm moist conditions, the greater the risk of corrosion. All constituents may contain chlorides and concrete may be contaminated by chlorides from the external environment. To minimize the chances of deterioration of concrete from harmful chemical salts, the levels of such harmful salts in concrete coming from concrete materials, that is, cement, aggregates water and admixtures, as well as by diffusion from the environment should be limited. The total amount of chloride content (as Cl) in the concrete at the time of placing shall be as given in Table 1.15.

The total acid soluble chloride content should be calculated from the mix proportions and the measured chloride contents of each of the constituents. Wherever possible, the total chloride content of the concrete should be determined.

2.5.3. *Sulphates in concrete* : Sulphates are present in most cements and in some aggregates ; excessive amounts of water-soluble sulphate from these or other mix constituents can cause expansion and disruption of concrete. To prevent this, the total water-soluble sulphate content of the concrete mix, expressed as SO_3, should not exceed 4 percent by mass of the cement in the mix. The sulphate content should be calculated as the total from the various constituents of the mix.

The 4 percent limit does not apply to concrete made with supersulphated cement complying with IS 6909.

2.5.4. Alkali-aggregate reaction : Some aggregates containing particular varieties of silica may be susceptible to attack by alkalis (Na_2O and K_2O) originating from cement or other sources, producing an expansive reaction which can cause cracking and disruption of concrete. Damage to concrete from this reaction will normally only occur when all the following are present together.

(*a*) A high moisture level, within the concrete ;

(*b*) A cement with high alkali content, or another source of alkali:

(*c*) Aggregate containing an alkali reactive constituent.

Where the service records of particular cement/aggregate combination are well established, and do not include any instances of cracking due to alkali-aggregate reaction, no further precautions should be necessary. When the materials are unfamiliar, precautions should taken one or more of the following forms :

(*a*) Use of non-reactive aggregate from alternate sources.

(*b*) Use of low alkali ordinary Portland cement having total alkali content not more than 0.6 percent (as Na_2O equivalent).

Further advantage can be obtained by use of fly ash (Grade 1) conforming to IS 3812 or granulated blastfurnace slag conforming to IS 12089 as part replacement of ordinary Portland cement (having total alkali content as Na_2O equivalent not more than 0.6 percent), provided fly ash content is at least 20 percent or slag content is at least 50 percent.

(*c*) Measures to reduce the degree of saturation of the concrete during service such as use of impermeable membranes.

(*d*) Limiting the cement content in the concrete mix and thereby limiting total alkali content in the concrete mix. For more guidance specialist literatures may be referred.

2.6. Concrete in Aggressive Soils and Water

2.6.1. General : The destructive action of aggressive waters on concrete is progressive. The rate of deterioration decreases as the concrete is made stronger and more impermeable, and increases as the salt content of the water increases. Where structures are only partially immersed or are in contact with aggressive soils or waters on one side only, evaporation may cause serious concentrations of salts with subsequent deterioration, even where the original salt content of the soil or water is not high.

NOTE : Guidance regarding requirements for concrete exposed to sulphate attack is given in 2.2.4

2.6.2. Drainage : At sites where alkali concentrations are high or may become very high, the ground water should be lowered by drainage so that it will not come into direct contact with the concrete.

Additional protection may be obtained by the use of chemically resistant stone facing or a layer of a plaster of Paris covered with suitable fabric, such as jute thoroughly impregnated with bituminous material.

2.7. Compaction, Finishing and Curing

Adequate compaction without segregation should be ensured by providing suitable workability and by employing appropriate placing and compacting equipment and procedures. Full compaction is particularly important in the vicinity of construction and movement joints and of embedded water bars and reinforcement. Good finishing practices are essential for durable concrete. Overworking the surface and the addition of water/cement to aid in finishing should be avoided; the resulting laitance will have impaired strength and durability and will be particularly vulnerable to freezing and thawing under wet conditions. It is essential to use proper and adequate curing techniques to reduce the permeability of the concrete and enhance its durability by extending the hydration of the cement, particularly in its surface zone.

2.8. Concrete in Sea-Water

Concrete in sea-water or exposed directly along the sea-coast shall be at least M 20 Grade in the case of plain concrete and M 30 in case of reinforced concrete. The use of slag or pozzolana cement is advantageous under such conditions.

2.8.1. Special attention shall be given to the design of the mix to obtain the densest possible concrete; slag, broken brick, soft limestone, soft sandstone, or other porous or weak aggregates shall not be use.

2.8.2. As far as possible, preference shall be given to precast members unreinforced, well-cured and hardened, without sharp corners, and having trowel-smooth finished surface free from crazing, cracks or other defects; plastering should be avoided.

2.8.3. No construction joints shall be allowed within 600 mm below low water-level or within 600 mm of the upper and lower planes of wave action. Where unusually severe conditions or abrasions are anticipated, such parts of the work shall be protected by bituminous or silico-fluoride coating or stone facing bedded with bitumen.

2.8.4. In reinforced concrete structures, care shall be taken to protect the reinforcement from exposure to saline atmosphere during storage, fabrication and use. It may be achieved by treating the surface of reinforcement with cement wash or by suitable methods.

1.10. WORKABILITY OF CONCRETE

It is difficult to properly define and measure the 'workability' of concrete, despite its being the most important property. In its simplest form, the term 'workability' may be defined as the ease with which concrete may be mixed, handled, transported, placed in position and compacted. According to Indian Standard (IS : 1199), *workability of concrete is that property of concrete which determines the amount of internal work necessary to produce full compaction.* The greatest single factor affecting the workability is the amount of water in the mix. A workable concrete does not show any bleeding or segreggation. Bleeding of concrete takes place when excess of water in the mix comes up at the surface, causing small pores through the mass of concrete. Segregation is caused when coarse aggregate separate out from the finer materials, resulting in large voids, less durability and less strength.

Several tests which have been developed to measure the workability of concrete are: (1) Slump test (2) Compacting factor test (3) Vee-Bee test and (4) Vibro-workability test. These have been described in detail separately.

Slump test is probably the simplest and commonly used test, though it is not the true guide to workability. In this test, concrete is compacted in a vessel of the shape of the frustum of a cone and open at both the ends. Concrete is compacted with the help of standard tamping rod, in four equal layers. Immediately after the vessel is filled, it is raised vertically, without giving any jerks etc. The concrete in the vessel becomes free and therefore slump. The vertical settlement, measured in mm, is termed as slump. The following table gives a rough guide of workability of concrete, in terms of slump for various types of work :

Type of work	*Slump (mm)*
1. Concrete for road work	20 to 30
2. Ordinary R.C.C. work for beams and slabs etc.	50 to 100
3. Columns, retaining walls and thin vertical sections	75 to 150
4. Vibrated concrete	12 to 25
5. Mass concrete	25 to 50

The compaction factor test measures the workability of concrete in terms of internal energy required to compact the concrete fully. In this test concrete is compacted in a lower cylindrical mould by making it to fall through two vertically placed hoppers. The weight of concrete in mould is determined. The theoretical weight of materials, required to fill the mould without air voids is also calculated from the knowledge of the proportions of the mix. The compacting factor is then calculated by dividing the observed weight of concrete in the mould by the theoretical weight. A concrete of low workability is represented by a compaction factor of about 0.85, of medium workability for a compaction factor of 0.92 and of good workability for a compaction factor of 0.95.

The various factors which influence the workability of concrete are (1) water in the mix (2) maximum size of particles (3) ratio of coarse and fine aggregates (4) particles interference (5) particle interlocking and (6) admixtures. Out of these, water in the mix is greatest single factor affecting the workability. Addition of water increases workability. The larger the maximum aggregate size and coarser the grading the smaller is the amount of water required for a given workability. In general, the grading requiring the least amount of water for a given workability will be that which gives the smallest surface area for a given amount of aggregate. A smooth rounded aggregate requires less water for a given workability than the irregular shaped aggregate. For a given aggregate-cement ratio, if the quantity of coarse aggregate is increased, the total surface area is reduced and hence more water would be available for lubrication, for a constant water cement ratio, resulting in increase in workability.

RECOMMENDATIONS OF IS 456 : 2000 ON WORKABILITY OF CONCRETE

1. The concrete mix proportions chosen should be such that the concrete is of adequate

workability for the placing conditions of the concrete and can properly be compacted with the mean available. Suggested ranges of workability of concrete measured in accordance with IS 1199 are given in Table 1.16.

TABLE 1.16. WORKABILITY OF CONCRETE

Placing Conditions		Degree of Workability	Slump (mm)
(1)		(2)	(3)
(i)	Blinding concrete	Very low	See Para 1.1 below
(ii)	Shallow sections		
(iii)	Pavements using pavers,		
(iv)	Mass concrete	Low	25-75
(v)	Lightly reinforced sections in slabs; beam wall columns,		
(vi)	Floors;		
(vii)	Hand placed pavements ;		
(viii)	Canal lining;		
(ix)	Strip footings		
(x)	Heavily reinforced sections in slabs; beams; walls; columns	Medium	50-100
(xi)	Slipform work		75-100
(xii)	Pumped concrete	High	100-150
(xiii)	Trench fill,		
(xiv)	In-situ piling	Very high	See Para 1.2 below
(xv)	Tremie concrete		

NOTE : For most of the placing conditions, internal vibrators (needle vibrators) are suitable. The diameter of the needle shall be determined based on the density and spacing of reinforcement bars and thickness of section. For tremie concrete, vibrators are not required to be used

1.1 In the 'very low' category of workability where strict control is necessary, for example pavement quality concrete, measurement of workability by determination of compacting factor will be more appropriate than slump and value of compacting factor of 0.75 to 0.80 is suggested.

1.2. In the 'very high' category of workability, measurement of workability by determination of flow will be appropriate.

1.11. METHODS OF PROPORTIONING CONCRETE MIXES

1. ARBITRARY METHOD

This method is adopted only for work of small magnitude or of moderate importance. The combined aggregate should be dense and should have least voids. For this, the quantity of fine aggregates should be sufficient to fill the voids of coarse aggregate. The ratio of coarse aggregate to fine is found to lie between $1\frac{1}{2}$ to $2\frac{1}{2}$, for a dense mix of aggregates. However, a common practice is to take the quantities of fine and coarse aggregate in the proportion of 1 : 2, and hence to express the quantities of cement, sand and coarse aggregate in the proportions of $1 : n : 2n$ by volume. The ratios of 1 : 1 : 2 and 1 : 1.2 : 2.4 are considered suitable for very high strength concrete, the ratios $1 : 1\frac{1}{2} : 3$ and

1 : 2 : 4 are used for normal reinforced concrete work and ratios 1 : 3 : 6 and 1 : 4 : 8 are used for foundations and mass concrete work. The amount of water to be used in the above mixes is decided on the basis of workability of the mix. The workability depends upon the type of work and the method of compaction. In this method, there is no rigid control over the strength of the mix. However, because of simplicity in the design, the method is widely used for all works of small magnitude.

2. MINIMUM VOIDS METHOD

In this method, the voids of coarse aggregate and fine aggregate are determined separately. The quantity of sand used should be such that it completely fills the voids of the coarse aggregate. Similarly, the quantity of cement used should be such that it fills the voids of sand, so that a dense mix having minimum voids is obtained. However, in actual practice, the quantity of sand used in the mix is kept 10% more than the voids in the coarse aggregate and the quantity of cement is taken 15% more than the voids in the sand. To the mix of cement, sand and coarse aggregate so obtained, sufficient water is added to make the mix workable. However, this method does not give satisfactory result because the presence of cement, sand and water separates the constituents of the coarse aggregate, thereby increasing its voids determined previously in absence of sand and cement. Similarly, the voids of sand are increased due to the addition of cement and water. Hence we do not always get a dense concrete. At the same time, the grading of aggregates has not been done so as to require least amount of water (and hence least w/c ratio) resulting in higher strength.

3. MAXIMUM DENSITY METHOD

This method of minimum voids was later improved by Fuller, to get a grading of materials to get maximum density. Based on wide scale experiments, he gave the following expression for the grading of materials :

$$P = 100\left(\frac{d}{D}\right)^{\frac{1}{2}} \qquad \text{...(1.4)}$$

where D = maximum size of aggregate

P = percentage by weight, of material finer than diameter d.

The coarse and fine aggregates should be fully graded according to the above rule. For example, let the maximum size of coarse aggregate be 20 mm and the maximum size of fine aggregate be 4 mm, the percentage of material finer than 4 mm is given by

$$P = 100\left(\frac{4}{20}\right)^{\frac{1}{2}} = 44.7\%$$

i.e. 44.7 kg of the aggregate, including the weight of cement, are to be mixed with 55.3 kg of coarse aggregate. The quantity of various intermediate sizes should also correspond to this formula.

Let us prepare the mix having a ratio of cement to aggregates (fine + Coarse) as 1 : 6 by weight.

$\therefore$ Quantity of cement in 100 kg. of mix $= \frac{100}{7} = 14.3$ kg

$\therefore$ Quantity of sand $= 44.7 - 14.3 = 30.4$ kg.

Hence the ratio of cement, sand and coarse aggregate by weight will be 14.3 : 30.4 : 55.3. Let us assume the unit weights of cement, fine aggregate and coarse aggregate as 1440, 1750 and 1600 kg per cubic metre respectively. Then the ratio of the three constituents, by volume will be

$$\frac{14.3}{1440} : \frac{30.4}{1750} : \frac{55.3}{1600}$$

or $$1 : \left(\frac{30.4}{1750} \times \frac{1440}{14.30}\right) : \left(\frac{55.3}{1600} \times \frac{1440}{14.30}\right)$$

or $$1 : 1.75 : 3.48$$

or $$1 : 1\tfrac{3}{4} : 3\tfrac{1}{2} \text{ (nominally)}$$

After having decided the proportions of various materials, sufficient quantity of water is added to make the mix workable. Table 1.17 gives the grading of mixed aggregate for 40 mm and 20 mm maximum size of aggregate. The method is not so popular since grading cannot be accurately achieved in field, and there is no control over the strength.

TABLE 1.17. GRADING OF MIXED AGGREGATE

Max. Size of coarse aggregate	*Percentage passing the I.S size*								
	40 mm	*20 mm*	*10 mm*	*4 mm*	*2 mm*	*1 mm*	*500 micron*	*250 micron*	*125 micron*
40 mm	100	71	50	32	22	16	11	8	6
20 mm	—	100	71	44	32	22	16	11	8

4. FINENESS MODULUS AND WATER CEMENT RATIO METHOD

(a) ***Fineness modulus.***

It has been observed that strength of mix is dependent wholly on the water cement ratio while the grading of the particles is important from workability and economic point of view. The grading of particles by Fullers formula, to get maximum density, is difficult and sometimes uneconomical to achieve in practice. Fineness modulus method essentially is a substitute for Fuller maximum density method, aimed at standardisation of the grading of aggregates. The term *fineness modulus*, suggested by Abram, is a numerical index of fineness of both fine as well as coarse aggregates.

Certain values of fineness modulus for mixed aggregates are found to give the best result. Let p be the desired fineness modulus for a mix of fine and coarse aggregates. If p_1 and p_2 are the fineness moduli of fine and coarse aggregates respectively, than the proportion R of finer aggregate to the combined aggregate, by weight is given by :

$$R = \frac{p_2 - p}{p - p_1} \times 100 \qquad \ldots(1.5)$$

For example of Table 1.8 (a) if the desired fineness modulus of the combined aggregate is 5.3, we have

$$R = \frac{6.93 - 5.3}{5.3 - 3.05} \times 100 = 72.5\%$$

(*b*) *Abram's water-cement ratio law.*

Abram's water-cement ratio (w/c ratio) law states that for any given conditions of test, the strength of workable concrete mix is dependent only on the water-cement ratio. It follows from this law that provided the concrete is fully compacted, the strength is not affected by aggregate shape, type or surface texture, or the aggregate grading, the workability and the richness of mix. We know that workability of mix (defined as 'that property of the concrete which determines the amount of useful internal work necessary to produce full compaction') is dependent on the amount of water in the mix. But amount of water in the mix, corresponding to a given strength, is governed by the water-cement ratio law. Hence the only way to increase the quantity of water to increase the workability of the mix is to increase the amount of cement also. However, as the grading of aggregate does not affect the strength of the concrete directly, the object must be to choose the grading to give the best workability with lowest water content. The grading of the particles should, therefore, be such that in the fully compacted state the total surface area of particles of aggregates as well as voids in them are the least. This means that larger the maximum aggregate size and the coarser the grading, the smaller is the amount of water required for a given workability. However, beyond a certain limit, the further increase in the maximum size and coarseness of grading results in harsh and under-sanded mixtures causing honeycombing, thus requiring more cement for smoothness. On the other hand, increase in the proportion of fine aggregate (*i.e.* sand) gives smooth mix but requires more cement and

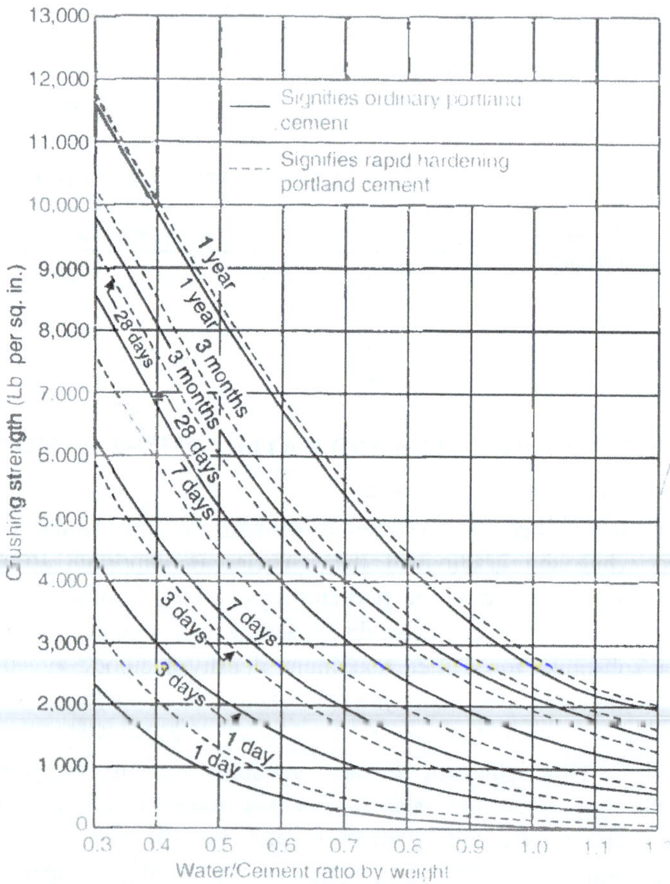

FIG 1.3. RELATION BETWEEN CRUSHING STRENGTH AND WATER-CEMENT RATIO FOR FULLY COMPACTED CONCRETE.

hence results, in uneconomical mix. Between these two limits lies the grading, which can be either determined by Fuller's maximum optimum density method or the fineness modulus method.

According to Abram's law, the strength of mix increases with the decrease in the water cement ratio. In terms of crushing strength after 7 days curing, the law can be expressed as follows :

$$p_7 = \frac{984}{7^x} \qquad \ldots(1.6\ a)$$

where p_7 = cylinder crushing strength, in kg/cm², after 28 days curing ;
x = water cement ratio by volume.

In the above expression, the constants 984 and 7 may vary slightly with the quality of aggregates and cement, method of curing and method of testing.

Expressed in terms of strength after 28 days curing, the law can be written as:

$$p_{28} = \frac{984}{4^x} \qquad \ldots(1.6\ b)$$

Fig 1.3 shows the relations between the crushing strength and water cement ratio (by weight) for various periods of curing. Fig 1.4 gives the relationship between 28 days compressive strength(cube) of concrete mixes with different water-cement ratios by weight and the 7 days compressive strength of cement (IS : 456–1964). Both Figs. 1.3 and 1.4 may be used as a guide for the selection of proper water cement ratio for a mix of given strength.

(c) ***Procedure for design of mix.***

The procedure for the design of mix can be summarised as follows:

1. For the requirements of strength of the mix, choose suitable water cement ratio from Fig 1.3 and 1.4.

2. Determine the maximum size of the aggregate available. Also determine the fineness modulus of both coarse and fine aggregates.

3. Determine the grading of the aggregates by Fuller's maximum density method. If this does not correspond to the grading of available materials, try to improve that grading to make it similar to that obtained by Fuller's method. If this is difficult to achieve, design the grading by

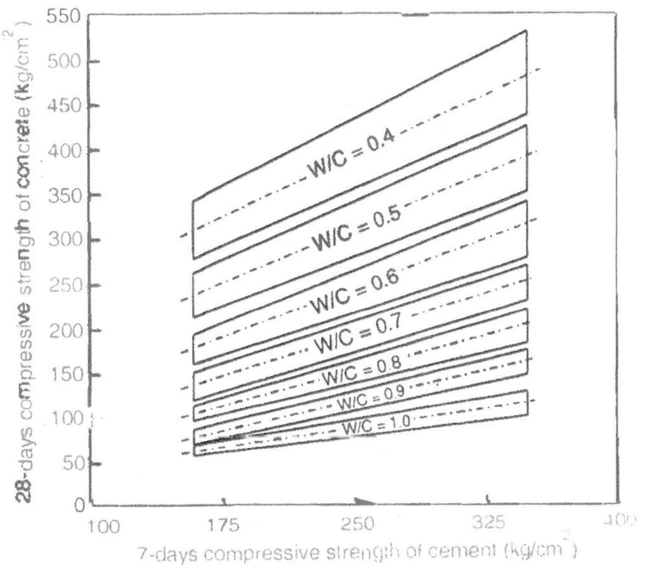

FIG 1.4 RELATIONSHIP BETWEEN COMPRESSIVE STRENGTH OF CONCRETE MIXES WITH DIFFERENT WATER-CEMENT RATIO AND THE 7 DAY COMPRESSIVE STRENGTH OF CEMENT

fineness modulus method. For that, choose suitable fineness modulus of combined aggregate and determine the ratio of fine aggregates to coarse aggregatess by Eq 1.5.

4. Determine the workability of the mix in terms of slump or compaction factor, required for the work.

5. Fix the ratio of cement to that of combined aggregate, mix quantity of water corresponding to the water cement ratio determined in step 1, and determine the workability of mix. Change the cement-aggregate ratio by trial to get the desired workability.

6. Determine the actual proportions of cement, fine aggregates, coarse aggregate and water, from the above steps, so that economical concrete of desired strength and workability is obtained.

1.12. GRADES OF CONCRETE AND CHARACTERISTIC STRENGTH (IS 456 : 2000)

Indian standard IS 456 : 2000 specifies *fifteen* grades of concrete (Table 1.18), designated as M 10, M 15, M 20, M 25, M 30, M 35, M 40, M 45, M 50, M 55, M 60, M 65, M 70, M 75 and M 80. In the designation of concrete mix, letter M refers as the mix and the number to the specified *characteristic strength* (f_{ck}) of 15 cm cube at 28 days, expressed in N/mm^2. The *characteristic strength* is defined as the strength of the material below which not more than 5% of the test results are expected to fall.

IS 456 : 2000 further recommends that minimum grade of concrete shall be not less than M 20 in reinforced concrete work.

TABLE 1.18. GRADES OF CONCRETE (IS 456 : 2000)

Group	*Grade Designation*	*Specified characteristic compressive strength of 15 mm cube at 28 days* (N/mm^2)
Ordinary concrete	M10	10
	M15	15
	M20	20
Standard concrete	M25	25
	M30	30
	M35	35
	M40	40
	M45	45
	M50	50
	M55	55
High strength concrete	M60	60
	M65	65
	M70	70
	M75	75
	M80	80

For concrete of compressive strength greater than M 55, design parameters given in the standard may not be applicable and values may be obtained from specialized literatures and experimental results.

Increase in strength with age

There is normally a gain of strength beyond 28 days. The quantum of increase depends upon the grade and type of cement, curing and environmental conditions etc. The design should be based on 28 days characteristic strength of concrete unless there is evidence to justify a higher strength for a particular structure due to age. For concrete of grade M 30 and above, the ratio of increase of compressive strength with age shall be based on actual investigations.

Tensile strength of concrete

The flexural and splitting tensile strength shall be obtained as described in IS 516 and IS 5816 respectively. When the designer wishes to use an estimate of tensile strength from the compressive strength, the following formula may be used :

Flexural strength, $f_{cr} = 0.7 \sqrt{f_{ck}}$ N/mm^2 ...(1.7)

where f_{ck} is the characteristic cube compressive strength of concrete in N/mm^2.

Elastic deformation

The modulus of elasticity is primarily influenced by the elastic properties of the aggregate and to a lesser extent by the conditions of curing and age of concrete, the mix proportions and the type of cement. The modulus of elasticity is normally related to the compressive strength of concrete.

The modulus of elasticity of concrete can be assumed as follows :

$$E_c = 5000 \sqrt{f_{ck}} \quad \text{...(1.8)}$$

where E_c is the short term modulus of elasticity in N/mm^2

Actual measured values may differ by ± 20 percent from the values obtained from the above expression.

Shrinkage

The total shrinkage of concrete depends upon the constituents of concrete, size of the member and environmental conditions. For a given humidity and temperature, the total shrinkage of concrete is most influenced by the total amount of water present in the concrete at the time of mixing and, to a lesser extent, by the cement content.

In the absence of test data, the approximate value of total shrinkage strain for design may be taken as 0.0003. For more information see IS 1343.

Creep of concrete

Creep of concrete depends in addition to the factors listed above under shrinkage, on the stress in the concrete, age at loading and the duration of loading. As long as the stress in concrete does not exceed one third of its characteristic compressive strength, creep may be assumed to be proportional to the stress.

In the absence of experimental data and detail information on the effect of the variables, the ultimate creep strain may be estimated from the following values (Table 1.19) of creep coefficient (that is ultimate creep strain divided by elastic strain at the age of the loading); for long span structure, it is advisable to determine actual creep strain, likely to take place. Note that the ultimate creep strain estimated as described above does not include the elastic strain.

TABLE 1.19. VALUES OF CREEP COEFFICIENT

Age at loading	*Creep coefficient*
7 days	2.2
28 days	1.6
1 year	1.1

Thermal coefficient

The coefficient of thermal expansion depends on nature of cement, the aggregate, the cement content, the relative humidity and the size of sections. The value of coefficient of thermal expansion for concrete with different aggregates may be taken from Table 1.20.

TABLE 1.20. VALUES OF COEFFICIENT OF THERMAL EXPANSION FOR CONCRETE

Type of aggregate	*Coefficient of thermal expansion for concrete / °C*
1. Quartzite	1.2 to 1.3×10^{-5}
2. Sandstone	0.9 to 1.2×10^{-5}
3. Granite	0.7 to 0.95×10^{-5}
4. Basalt	0.8 to 0.95×10^{-5}
5. Limestone	0.6 to 0.9×10^{-5}

1.13. CONCRETE MIX PROPORTIONING (IS 456 : 2000)

1. Mix Proportion

The mix proportions shall be selected to ensure the workability of the fresh concrete and when concrete is hardened, it shall have the required strength, durability and surface finish.

1.1. The determination of the proportions of cement, aggregates and water to attain the required strengths shall be made as follows :

(*a*) By designing the concrete mix; such concrete shall be called 'Design mix concrete', or (*b*) By adopting nominal concrete mix: such concrete shall be called 'Nominal mix concrete'.

Design mix concrete is preferred to nominal mix. If design mix concrete cannot be used for any reason on the work for grade of M 20 or lower, nominal mixes may be used with the permission of engineer-in-charge, which, however, is likely to involve a higher cement content.

1.2. ***Information Required***

In specifying a particular grade of concrete, the following information shall be included:

(*a*) Type of mix, that is, design mix concrete or nominal mix concrete :
(*b*) Grade designation ;
(*c*) Type of cement ;
(*d*) Maximum nominal size of aggregate ;
(*e*) Minimum cement content (for design mix concrete);
(*f*) Maximum water-cement ratio;
(*g*) Workability;
(*h*) Mix proportion (for nominal mix concrete);
(*j*) Exposure conditions
(*k*) Maximum temperature of concrete at the time of placing;
(*m*) Method of placing ; and
(*n*) Degree of supervision.

1.3. In appropriate circumstances, the following additional information may be specified:

(*a*) Type of aggregate,
(*b*) Maximum cement content, and
(*c*) Whether an admixture shall or shall not be used and the type of admixture and the condition of use.

2. Design Mix Concrete

2.1. As the guarantor of quality of concrete used in the construction, the constructor shall carry out the mix design and the mix so designed (not the method of design) shall be approved by the employer within the limitations of parameters and other stipulations laid down by this standard.

2.2. The mix shall be designed to produce the grade of concrete having the required workability and a characteristic strength not less than appropriate values given in Table 1.18. The target mean strength of concrete mix should be equal to the characteristic strength plus 1.65 times the standard deviation.

2.3. Mix design done earlier not prior to one year may be considered adequate for later work provided there is no change in source and the quality of the materials.

2.4. Standard Deviation

The standard deviation for each grade of concrete shall be calculated, separately.

2.4.1 Standard deviation based on test strength of sample

(*a*) *Number of test results of samples* - The total number of test strength of samples required to constitute an acceptable record for calculation of standard deviation shall be not less than 30. Attempts should be made to obtain the 30 samples, as early as possible, when a mix is used for the first time.

(*b*) *In case of significant changes in concrete* - When significant changes are made in the production of concrete batches (for example changes in the materials used, mix design, equipment or technical control), the standard deviation value shall be separately calculated for such batches of concrete.

(*c*) *Standard deviation to be brought up to data* - The calculation of the standard deviation shall be brought up to data after every change of mix design.

2.4.2. Assumed standard deviation

Where sufficient test results for a particular grade of concrete are not available, the value of standard deviation given in Table 1.21 may be assumed for design of mix in the first instance. As soon as the results of sample are available, actual calculated standard deviation shall be used and the mix designed property.

TABLE 1.21. ASSUMED STANDARD DEVIATION

Grade of concrete	*Assumed standard Deviation* N/mm^2
M 10 M 15	3.5
M 20 M 25	4.0
M 30 M 35 M 40 M 45 M 50	5.0

However, when adequate past records for a similar grade exist and justify to the designer a value of standard deviation different from that shown in Table 1.21, it shall be permissible to use that value.

NOTE : The above values correspond to the site control having proper storage of cement: weight batching of all materials; controlled addition of water ; regular checking of all materials, aggregate gradings and moisture content; and periodical checking of workability and strength. Where there is deviation from the above, the values given in the above table shall be increased by 1 N/mm^2.

3. Nominal Mix Concrete : Nominal mix concrete may be used for concrete of M 20 or lower. The proportions of materials for nominal mix concrete shall be in accordance with Table 1.22.

TABLE 1.22. PROPORTIONS FOR NOMINAL MIX CONCRETE

(Clauses 3 and 3.1)

Grade of Concrete	*Total Quantity of Dry Aggregates by Mass per 50 kg of Cement; to be Taken as the Sum of the Individual Masses of Fine and Coarse Aggregates; kg; Max*	*Proportion of Fine Aggregate to Coarse Aggregate (by Mass)*	*Quantity of Water per 50 kg of Cement; Max litres*
(1)	(2)	(3)	(4)
M 5	800	Generally 1:2 but subject to an upper limit of $1:1\frac{1}{2}$ and a lower limit of $1:2\frac{1}{2}$	60
M 7.5	625		45
M 10	480		34
M 15	330		32
M 20	250		30

NOTE : The proportion of the fine to coarse aggregates should be adjusted from upper limit to lower limit progressively as the grading of fine aggregates becomes finer and the maximum size of coarse aggregate becomes larger. Graded coarse aggregate shall be used.

Example : For an average grading of fine aggregate the proportions shall be $1:1\frac{1}{2}$, 1:2 and $1:2\frac{1}{2}$ for maximum size of aggregates 10 mm, 20 mm and 40 mm respectively.

3.1 The cement content of the mix specified in Table 1.22 for any nominal mix shall be proportionately increased if the quantity of water in a mix has to be increased to overcome the difficulties of placement and compaction, so that the water-cement ratio as specified is not exceeded.

1.14. PRODUCTION OF CONCRETE (IS 456:2000)

1. Quality Assurance Measures

1.1. In order that the properties of the completed structure be consistent with the requirements and the assumptions made during the planning and the design, adequate quality assurance measures shall be taken. The construction should result in satisfactory strength, serviceability and long term durability so as to lower the overall life-cycle cost. Quality assurance in construction activity relates to proper design, use of adequate materials and components to be supplied by the producers, proper workmanship in the execution of works by the contractor and ultimately proper care during the use of structure including timely maintenance and repair by the owner.

1.2. Quality assurance measures are both technical and organizational. Some common cases should be specified in a general Quality Assurance Plan which shall identify the key elements necessary to provide fitness of the structure and the means by which they are to be provided and measured with the overall purpose to provide confidence that the realized project will work satisfactorily in service fulfilling intended needs. The job of quality control and quality assurance would involve quality audit of both the inputs as well as the outputs. Inputs are in the form of materials for concrete; workmanship in all stages of batching, mixing, transportation, placing, compaction and curing; and the related plant, machinery and equipments; resulting in the output in the form of concrete in place. To ensure proper performance, it is necessary that each step in concreting which will be covered by the next step is inspected as the work proceeds.

1.3. Each party involved in the realization of a project should establish and implement a Quality Assurance Plan, for its participation in the project. Supplier's and subcontractor's activities shall be covered in the plan. The individual Quality Assurance Plan shall fit into the general Quality Assurance Plan. A Quality Assurance Plan shall define the tasks and responsibilities of all persons involved, adequate control and checking procedures, and the organization and maintaining adequate documentation of the building process and its results. Such documentation should generally include.

(*a*) test reports and manufacturer's certificate for materials, concrete mix design details;
(*b*) pour cards for site organization and clearance for concrete placement:
(*c*) record of site inspection of workmanship, field test;
(*d*) non-conformance reports, change orders;
(*e*) quality control charts; and
(*f*) statistical analysis.

NOTE : Quality control charts are recommended wherever the concrete is in continuous production over considerable period.

2. Batching : To avoid confusion and error in batching consideration should be given to using the smallest practical number of different concrete mixes on any site or in any one plant. In batching concrete, the quantity of both cement and aggregate shall be determined by mass; admixture, if solid, by mass; liquid admixture may however be measured in volume or mass: water shall be weighed or measured by volume in a calibrated tank (*see also* IS 4925).

Ready-mixed concrete supplied by ready-mixed concrete plant shall be preferred. For large and medium project sites the concrete shall be sourced from ready mixed concrete plants or from on site or off site batching and mixing plants (see IS 4926)

2.1. Except where it can be shown to the satisfaction of the engineer-in-charge that supply of properly graded aggregate of uniform quality can be maintained over a period of work, the grading of aggregate should be controlled by obtaining the coarse aggregate in different sizes and blending them in the right proportions when required, the different sizes being stocked in separate stock-piles. The material should be stock-piled for several hours preferably a day before use. The grading of coarse and fine aggregate should be checked as frequently as possible, the frequency for a given job being determined by the engineer-in-charge to ensure that the specified grading is maintained.

2.2. The accuracy of the measuring equipment shall be within ± 2 percent of the quantity of cement being measured and within ± 2 percent of the quantity of aggregate, admixture and water being measured.

2.3. Proportion/Type and grading of aggregates shall be made by trial in such a way so as to obtain densest possible concrete. All ingredients of the concrete should be used by mass only.

2.4. Volume batching may be allowed only where weight-batching is not practical and provided accurate bulk densities of materials to be actually used in concrete have earlier been established. Allowance for bulking shall be made in accordance with IS 2386 (Part 3). The mass volume relationship should be checked as frequently as necessary, the frequency for the given job being determined by engineer-in-charge to ensure that the specified grading is maintained.

2.5. It is important to maintain the water-cement ratio constant at its correct value. To this end, determination of moisture contents in both fine and coarse aggregates shall be made as frequently as possible, the frequency for a given job being determined by the engineer-in-charge according to weather conditions. The amount of the added water shall be adjusted to compensate for any observed variations in the moisture contents. For the determination of moisture content in the aggregates, IS 2386 (Part 3) may be referred to. To allow for the variation in mass of aggregate due to variation in their moisture content, suitable adjustments in the masses of aggregates shall also be made. In the absence of exact data, only in the case of nominal mixes, the amount of surface water may be estimated from the values given in Table 1.23.

TABLE 1.23. SURFACE WATER CARRIED BY AGGREGATE

(Clause 2.5)

Sl. No.	*Aggregate*	*Approximate Quantity of Surface Water,*	
		Percent by Mass	*l/m³*
(1)	*(2)*	*(3)*	*(4)*
(i)	Very wet sand	7.5	120
(ii)	Moderately wet sand	5.0	80
(iii)	Moist sand	2.5	40
(iv)	Moist gravel or crushed rock*	1.25-2.5	20-40

*Coarser the aggregate, less the water it will carry.

2.6. No substitutions in materials used on the work or alterations in the established proportions, except as permitted in 2.4 and 2.5 shall be made without additional test to show that the quality and strength of concrete are satisfactory.

3. Mixing

Concrete shall be mixed in a mechanical mixer. The mixer should comply with IS 1791 and IS 12119. The mixers shall be fitted with water measuring (metering) devices. The mixing shall be continued until there is a uniform distribution of the materials and the mass is uniform in colour and consistency. If there is segregation after unloading from the mixer, the concrete should be remixed.

3.1. For guidance, the mixing time shall be at least 2 min. For other types of more efficient mixers, manufacturers recommendation shall be followed; for hydrophobic cement it may be decided by the engineer-in-charge.

3.2. Workability should be checked at frequent intervals (see IS 1199).

3.3. Dosages of retarders, plasticisers and superplasticisers shall be restricted to 0.5, 1.0 and 2.0 percent respectively by weight of cementitious materials and unless a higher value is agreed upon between the manufacture and the constructor based on performance test.

1.15. FORMWORK (IS 456 : 2000)

1. General : The formwork shall be designed and constructed so as to remain sufficiently rigid during placing and compaction of concrete, and shall be such as to prevent loss of slurry from the concrete. For further details regarding design, detailing, etc. reference may be made to IS 14687. The tolerances on the shapes, lines and dimensions shown in the drawing shall be within the limits given in Table 1.24.

TABLE 1.24. TOLERANCES

(a) Deviation from specified dimensions of cross-section of columns and beams	+ 12 mm – 6
(b) Deviation from dimensions of footings	
(1) Dimension in plan	+ 50 mm – 12
(2) Eccentricity	0.02 times the width of the footing in the direction of deviation but not more than 50 mm
(3) Thickness	± 0.05 times the specified thickness

These tolerances apply to concrete dimensions only, and not to positioning of vertical reinforcing steel or dowels.

2. Cleaning and Treatment of Formwork

All rubbish, particularly, clippings, shavings and sawdust shall be removed from the interior of the forms before the concrete is placed. The face of form work in contact with the concrete shall be cleaned and treated with form release agent. Release agents should be applied so as to provide a thin uniform coating to the forms without coating the reinforcement.

3. Stripping Time

Forms shall not be released until the concrete has achieved a strength of at least twice the stress to which the concrete may be subjected at the time of removal of formwork. The strength referred to shall be that of concrete using the same cement and aggregates and admixture, if any with the same proportions and cured under conditions of temperature and moisture similar to those existing on the work.

3.1. While the above criteria of strength shall be the guiding factor for removal of formwork, in normal circumstances where ambient temperature does not fall below 15°C and where ordinary Portland cement is used and adequate curing is done, following striking period (Table 1.25) may deem to satisfy the guideline given in para 3.

TABLE 1.25. STRIPPING TIME

Type of Formwork	*Minimum Period Before Striking Formwork*
(*a*) Vertical formwork to columns; wallsbeams	16 – 24 h
(*b*) Soffit formwork to slabs (Props to be refixed immediately after removal of formwork)	3 days
(*c*) Soffit formwork to beams (Props to be refixed immediately after removal of formwork)	7 days
(*d*) Props to slabs : (1) Spanning up to 4.5 m (2) Spanning over 4.5 m	 7 days 14 days
(*e*) Props to beams and arches : (1) Spanning up to 6 m (2) Spanning over 6 m	 14 days 21 days

For other cements and lower temperature, the stripping time recommended above may be suitable modified.

3.2. The number of props left under, their sizes and disposition shall be such as to be able to safely carry the full dead load of the slab, beam or arch as the case may be together with any live load likely to occur during curing or further construction.

3.3. Where the shape of the element it such that the formwork has re-entrant angles, the formwork shall be removed as soon as possible after the concrete has set, to avoid shrinkage cracking occurring due to the restraint imposed.

1.16. TRANSPORTING, PLACING, COMPACTION AND CURING (IS 456 : 2000)

1. Transporting and Handling

After mixing, concrete shall be transported to the formwork as rapidly as possible by methods which will prevent the segregation or loss of any of the ingredients or ingress of foreign matter or water and maintaining the required workability.

1.1. During hot or cold weather, concrete shall be transported in deep containers. Other suitable methods to reduce the loss of water by evaporation in hot weather and heat loss in cold weather may also be adopted.

2. Placing

The concrete shall be deposited as nearly as practicable in its final position to avoid rehandling. The concrete shall be placed and compacted before initial setting of concrete commences and should not be subsequently disturbed. Methods of placing should be such as to preclude segregation. Care should be taken to avoid displacement of reinforcement or movement of formwork. As a general guidance, the maximum permissible free fall of concrete may be taken as 1.5 m.

3. Compaction

Concrete should be thoroughly compacted and fully worked around the reinforcement, around embedded fixtures and into corners of the formwork.

3.1. Concrete shall be compacted using mechanical vibrators complying with IS 2505, IS 2506, IS 2514 and IS 4656. Over vibration and under vibration of concrete are harmful and should be avoided. Vibration of very wet mixes should also be avoided.

Whenever vibration has to be applied externally, the design of formwork and the disposition of vibrators should receive special consideration to ensure efficient compaction and to avoid surface blemishes.

4. Construction Joints and Cold Joints

Joints are a common source of weakness and therefore, it is desirable to avoid them. If this is not possible, their number shall be minimized. Concreting shall be carried out continuously up to construction joints, the position and arrangement of which shall be indicated by the designer. Construction joints should comply with IS 11817.

Construction joints shall be placed at accessible locations to permit cleaning out of laitance, cement slurry and unsound concrete, in order to create rough/uneven surface. It is recommended to clean out laitance and cement slurry by using wire brush on the surface of joint immediately after initial setting of concrete and to clean out the same immediately thereafter. The prepared surface should be in a clean saturated surface dry condition when fresh concrete is placed, against it.

In the case of construction joints at locations where the previous pour has been cast against shuttering, the recommended method of obtaining a rough surface for the previously poured concrete is to expose the aggregate with a high pressure water jet or any other appropriate means.

Fresh concrete should be thoroughly vibrated near construction joints so that mortar from the new concrete flows between large aggregates and develop proper bond with old concrete.

Where high shear resistance is required at the construction joints, shear keys may be provided.

Sprayed curing membranes and release agents should be thoroughly removed from joints surfaces.

5. Curing

Curing is the process of preventing the loss of moisture from the concrete whilst maintaining a satisfactory temperature regime. The prevention of moisture loss from the concrete is particularly important if the water-cement ratio is low, if the cement has a high rate of strength development, if the concrete contains granulated blast furnace slag or pulverised fuel ash. The curing regime should also prevent the development of high temperature gradients within the concrete.

The rate of strength development at early ages of concrete made with supersulphated cement is significantly reduced at lower temperatures. Supersulphated cement concrete is seriously affected by inadequate curing and the surface has to be kept moist for at least seven days.

5.1. Moist Curing : Exposed surfaces of concrete shall be kept continuously in a damp or wet condition by ponding or by covering with a layer of sacking, canvas, hessian or similar materials and kept constantly wet for at least seven days from the date of placing concrete in case of ordinary Portland Cement and at least 10 days where mineral admixture or blended cement are used. The period of curing shall not be less than 10 days for concrete exposed to dry and hot weather conditions. In the case of concrete where mineral admixtures or blended cements are used, it is recommended that above minimum periods may be extended to 14 days.

5.2. Membrane Curing : Approved curing compounds may be used in lieu of moist curing with the permission of the engineer-in-charge. Such compounds shall be applied to all exposed surfaces of the concrete as soon as possible after the concrete has set. Impermeable membranes such as polyethylene sheeting covering closely the concrete surface may also be used to provide effective barrier against evaporation.

5.3. For the concrete containing Portland pozzolana cement. Portland slag cement or mineral admixture, period of curing may be increased.

6. Supervision

It is exceedingly difficult and costly to alter concrete once placed. Hence, constant and strict supervision of all the items of the construction is necessary during the progress of the work, including the proportioning and mixing of the concrete. Supervision is also of extreme importance to check the reinforcement and its placing before being covered.

6.1. Before any important operation, such as concreting or stripping of the formwork is started, adequate notice shall be given to the construction supervisor.

1.17. SAMPLING AND STRENGTH OF DESIGNED CONCRETE MIX (IS 456 : 2000)

1. General

Samples from fresh concrete shall be taken as per IS 1199 and cubes shall be made, cured and tested at 28 days in accordance with IS 516.

1.1. In order to get a relatively quicker idea of the quality of concrete, optional test on beams for modulus of rupture at 72 ± 2 h or at 7 days, or compressive strength tests at 7 days may be carried out in addition to 28 days compressive strength test. For this purpose the values should be arrived at based on actual testing. In all cases, the 28 days compressive strength specified in Table 2 shall alone be the criterion for acceptance or rejection of the concrete.

2. Frequency of Sampling

Sampling Procedure : A random sampling procedure shall be adopted to ensure that each concrete batch shall have a reasonable chance of being tested that is, the sampling should be spread over the entire period of concreting and cover all mixing units.

Frequency : The minimum frequency of sampling of concrete of each grade shall be in accordance with Table 1.26.

TABLE 1.26.

Quantity of Concrete in the Work, m^3	*Number of Samples*
1 - 5	1
6 - 15	2
16 - 30	3
31 - 50	4
51 and above	4 plus one additional sample for each additional 50 m^3 or part thereof

NOTE :- At least one sample shall be taken from each shift. Where concrete is produced at continuous production unit, such as ready-mixed concrete plant, frequency of sampling may be agreed upon mutually by suppliers and purchasers.

3. Test Specimen

Three test specimens shall be made for each sample for testing at 28 days. Additional samples may be required for various purposes such as to determine the strength of concrete at 7 days or at the time of striking the formwork, or to determine the duration of curing, or to check the testing error. Additional samples may also be required for testing samples cured by accelerated methods as described in IS 9103. The specimen shall be tested as described in IS 516.

4. Test Results of Sample

The test realists of the sample shall be the average of the strength of three specimens. The individual variation should not be more than ± 15 percent of the average. If more, the test results of the sample are invalid.

1.18. ACCEPTANCE CRITERIA (IS 456 : 2000)

1. Compressive Strength : The concrete shall be deemed to comply with the strength requirements when both the following condition are met :

(*a*) The mean strength determined from any group of four consecutive test results complies with the appropriate limits in col 2 of Table 1.27

(*b*) Any individual test result complies with the appropriate limits in col 3 of Table 1.27.

2. Flexural Strength : When both the following conditions are met, the concrete complies with the specified flexural strength.

(*a*) The mean strength determined from any group of four consecutive test results exceeds the specified characteristic strength by at least 0.3 N/mm^2.

(*b*) The strength determined from any test result is not less than the specified characteristic strength less 0.3 N/mm^2.

3. Quantity of Concrete Represented by Strength Test Results : The quantity of concrete represented by a group of four consecutive test results shall include the batches from which the first and last samples were taken together with all intervening batches.

For the individual test result requirements given in col 2 of Table 1.27 or in item (*b*) of para 2, only the particular batch from which the sample was taken shall be at risk.

Where the mean rate of sampling is not specified the maximum quantity of concrete that four consecutive test results represent shall be limited to 60 m^3.

4. If the concrete is deemed not to comply persuant to para 3, the structural adequacy of the part affected shall be investigated and any consequential action as needed shall be taken.

5. Concrete of each grade shall be assessed separately.

6. Concrete is liable to be rejected if it is porous or honey-combed, its placing has been interrupted without providing a proper construction joint, the reinforcement has been displaced beyond the tolerances specified, or construction tolerances have not been net. However, the hardened concrete may be accepted after carrying out suitable remedial measures to the satisfaction of the engineer incharge.

TABLE 1.27. CHARACTERISTIC COMPRESSIVE STRENGTH COMPLIANCE REQUIREMENT (IS 456 : 2000)

Specified grade	*Mean of the group of 4 Non-overlapping consecutive test results (N/mm^2)*	*Individual test results (N/mm^2)*
(1)	(2)	(3)
M 15	$\geq f_{ck}$ + 0.825 × established standard deviation (rounded off to the nearest 0.5 N/mm^2) or f_{ck} + 3 N/mm^2, whichever is greater	$\geq (f_{ck} - 3)$ N/mm^2
M 20 or above	$\geq f_{ck}$ + 0.825 × established standard deviation (rounded off to the nearest 0.5 N/mm^2) or f_{ck} + 4 N/mm^2, whichever is greater	$\geq (f_{ck} - 4)$ N/mm^2

Note :- In the absence of established value of standard deviation, the values given in table 1.21 may be assumed, and attempt should be made to obtain results of 30 samples as early as possible to establish the value of standard deviation.

1.19. STEEL REINFORCEMENT

Steel reinforcement used in reinforced concrete may be of the following types :

(*a*) 1. Mild steel bars conforming to IS 432 (Part I) : 1966.

2. Hot rolled mild steel deformed bars conforming to IS 1139 : 1966.

(*b*) 1. Medium tensile steel conforming to IS 432 (Part I) : 1966.

2. Hot rolled medium tensile steel deformed bars conforming to IS 1139 : 1966.

(*c*) 1. Hot rolled high yield strength deformed bars (HYSD bars) conforming to IS 1139 : 1966.

2. Cold-worked steel high strength deformed bars conforming to IS 1786 : 1979 (Grade Fe 415 and Grade Fe 500).

(*d*) 1. Hard drawn steel wire fabric comforming to IS 1566 : 1967.

2. Rolled steel made from structural steel conforming to IS 226 : 1975.

The most important characteristic of a reinforcing bar is its *stress strain curve* and the important property is the *yield stress* or 0.2% *proof stress*, as the case may be. The modulus of elasticity E for these steels may be taken as 2×10^5 N/mm^2.

Type (*a*) bars have yield strength of 250 N/mm^2. Hence it is sometimes known as Fe 250 steel, having characteristic strength of 250 N/mm^2. The stress-strain curve of mild steel bars is shown in Fig. 1.5.

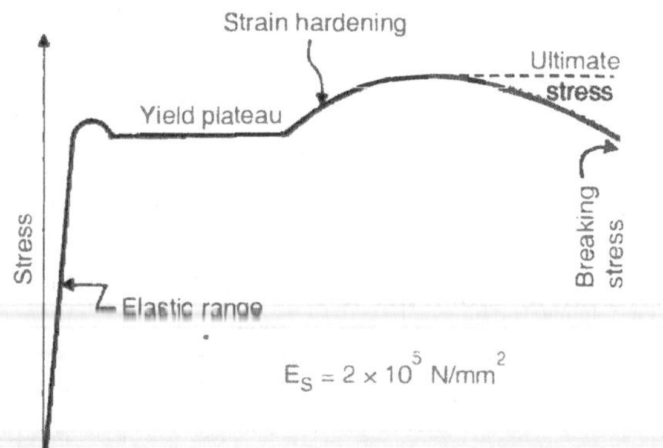

FIG 1.5 STRESS STRAIN CURVE FOR MILD STEEL.

Type (*c*) bars are high yield strength bars, also known as, HYSD bars. These may be *hot rolled* high yield strength bars or *cold worked* steel high strength deformed bars. The latter are also known as CTD bars (cold twisted deformed bars) or *Tor Steel*, and are available in two grades : (i) Fe 415 (or Tor 40) and (ii) Fe 500 (Tor 50) having 0.2% proof stress as 415 N/mm^2 and 500 N/mm^2 respectively.

The hot rolled mild steel bars, in common use for nearly six decades, has low yield stress but has significant *yield plateau* having plastic strain nearly 10 times the limiting strain at yield point. Due to such an excessive yielding, there is considerable readjustment of stress in concrete so as to maintain equilibrium of internal stresses, resulting in excessive cracking and deformations in the structural member. The idealised stress strain curve for mild steel bars, is shown in curve 1 of Fig. 1.6.

The *yield plateau* can be eliminated by straining the bar beyond the yield plateau either by twisting or *stretching* and then unloading . This process is known as cold working, and the *Tor-Steel reinforcement bars*, available in India are of this type. By this process, higher yield stress can be attained with much smaller strains. A *twisted bar* has considerable increased yield stress, about 50% to 100% more than the ordinary mild steel bars. Their use can permit higher working stress and hence considerable saving in quantity of steel can be achieved. Bond between concrete and steel can be improved by use of *deformed bars*. A deformed bar is a bar of steel provided with lugs or ribs deformation on surface of the bar to minimise the slippage of the bar in concrete, as shown in Fig. 1.7.

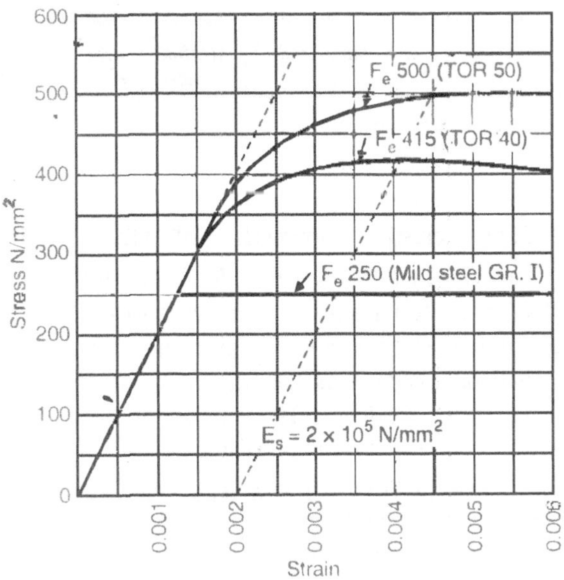

FIG. 1.6 STRESS STRAIN CURVE FOR CTD BARS

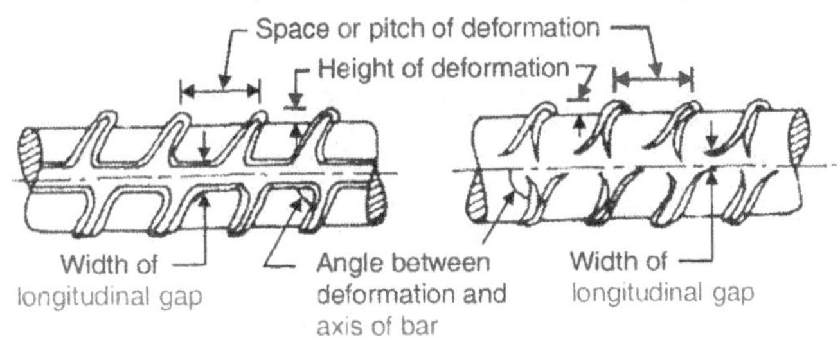

FIG. 1.7 DEFORMED BARS

The process of cold working has the disadvantage that useful ductility of mild steel is reduced. It is to be noted that if the cold worked bar is reheated, the original yield plateau is recovered. Since cold-worked bars lack a well-defined yield plateau, yield stress is specified at 0.2% *proof stress*.

In reinforced concrete, a long time trend is evident towards the use of high strength materials, both steel and concrete. In big housing projects, *High yield strength deformed bars* (HYSD bars) are in common use, having yield stress (0.2 percent proof stress) equal to 415 N/mm^2 for grade Fe 415. For grade Fe 500 HYSD bars, the yield stress is as high as 500 N/mm^2. *Modern high strength steels* of Fe 415 and Fe 500 grades require the simultaneous use of high strength concrete of M 20 grade and higher.

Percent Elongation. Minimum requirement of *% elongation* should be as follows :

TABLE 1.28.

	M.S. bars Grade I	*Medium tensile steel bars*	*Medium tensile steel bars*	*Medium steel deformed bars*	*HYSD and CTD bars*
Dia < 10mm	20	17	23	20	14.5
Dia ≥ 10 mm	23	20	23	20	14.5

Characteristic Strength of Steel Reinforcement

The *characateristic strength* (f_y) means that value of strength below which not more than 5% of the test results are expected to fall. Until the relevant Indian Standard specifications for reinforcing steel are modified to include the concept of characteristic strength, the characteristic value shall be assumed as the minimum yield/0.2 percent proof stress specified in the relevant Indian Standard Codes. The values of characteristic strength for three common types of reinforcements are given in Table 1.29.

TABLE 1.29. CHARACTERISTIC STRENGTH OF STEEL REINFORCEMENT

Type of reinforcement	*Yield stress or 0.2% proof stress (N/mm²)*	*Characteristic strength f_y (N/mm²)*
1. Mild steel bars conforming to Grade I of IS : 432 (Part I) or deformed m.s. bars of I.S. 1139	250 (average)	250
2. High yield strenth deformed bars conforming to IS 1109 or Grade Fe 415 of IS 1786 : 1979	415	415
3. High yield strength deformed bars of Grade Fe 500 of IS 1786 : 1979	500	500

PROBLEMS

1. Discuss in brief various types of cements used for construction.
2. Discuss salient specifications for Portland cement used for reinforced concrete.
3. Write a note on 'aggregates' used for cement concrete.
4. Explain the following:

(i) Determination of fineness modulus (ii) Grading aggregates (iii) Quality of water for cement concrete (iv) Durability of concrete (v) Workability of Concrete (vi) Water cement Ratio

5. Explain in brief various types of reinforcement used for R.C.C.
6. Discuss the salient properties of various grades of concrete as per IS ; 456–2000. What are the corresponding nominal mixes ?

2

Design Concepts : Limit State Design

2.1. METHODS OF DESIGN

A reinforced concrete structure should be so designed that it fulfills its intended purpose during its intended life time with :

(*i*) adequate *safety*, in terms of strength and stability.

(*ii*) adequate *serviceability* in terms of stiffness and durability, and

(*iii*) reasonable *economy*

Thus, all structures must be designed to be safe, serviceable and economical.

Reinforced concrete members can be designed by one of the following two methods:

1. The method of theoretical computations using accepted theories/procedures approved by the Codes of Practice,

2. The method of experimental investigations on models or full size structure or element.

Commonly used structures are designed by theoretical methods. The following design methods are used for the design of reinforced concrete structures/elements :

(*a*) The *Modular Ratio Method* or the *Working Stress Method* (WSM)

(*b*) The *Load Factor Method* or the *Ultimate Load Method* (ULM)

and (*c*) The *Limit State Method* (LSM)

2.2. WORKING STRESS METHOD (WSM)

This is the traditional method of design, used not only for reinforced concrete but also for structural steel and timber. Close to about hundred years old, the method is based on *linear elastic theory* or the classical *elastic theory*. This method of design was evolved around 1900 and was the first theoretical method accepted by National Codes of Practice for the design of reinforced concrete sections. This method ensures adequate safety by suitably restricting the stresses in the materials (i.e. concrete and steel) induced by the expected *working loads* on the structure. The assumption of linear elastic behaviour is considered justifiable since the specified *permissible* (or allowable) stresses are kept well below the ultimate strength of the material. The ratio of yield stress of the steel reinforcement or the cube strength of the concrete to the corresponding *permissible* or *working stress* is usually called the *factor of safety*. The WSM uses a factor of safety of about 3 with respect to the cube strength of concrete and a factor of safety of about 1.8 with respect to the yield strength of steel.

Reinforced concrete is a composite material. The WSM assumes *strain compatibility*, whereby the strain in the reinforcing steel is assumed to be equal to that in the adjoining concrete to which it is bonded. Consequently the stress in steel in *linearly* related to the stress in adjoining concrete by a constant factor, called the *modular ratio* defined as the ratio of the modulus of elasticity of steel to that of concrete. The WSM is therefore also known as the *modular ratio method*.

Demerits of WSM

Most structures designed in accordance with WSM have been generally performing satisfactorily for many years. However, the method has the following demerits :

1. The WSM does not show the real strength nor gives the *true factor of safety* of the structure under failure.

2. The modular ratio design results in larger percentage of compression steel than that given by the limit state design, thus leading to un-economic design.

3. Because of creep and non-linear stress-strain relationship, concrete does not have *definite* modulus of elasticity.

4. The WSM fails to discriminate between different types of loads that act simultaneously but have different uncertainties.

Merits of WSM

Inspite of the above defects, the WSM has the *advantage* of its *simplicity*, both in concept as well as in application. The design usually results in relatively large sections of structural members, compared to the ULM. Due to this, structures designed by WSM give better serviceability performance (i.e. less deflection, less crack width etc.) under working loads. WSM is the only method available when one has to investigate the R.C. section for service stresses and for the serviceability states of deflection and cracking. *It is essential to have a knowledge of WSM since it forms a part of limit state design (LSD) for a serviceability condition.*

The basics of WSM are explained in Appendix-A.

2.3. ULTIMATE LOAD METHOD (ULM)

The ultimate load method (ULM) was evolved in 1950 as an alternative to the WSM. The method is based on the *ultimate strength* of reinforced concrete at *ultimate load*. The ultimate load is obtained by enhancing the service load by some factor referred to as *load factor* for giving a desired margin of safety. Hence the method is also referred to as the *load factor method* or the *ultimate strength method*. The ULM was introduced as alternative to WSM in the ACI Code in 1956, the British Codes in 1957 and the Indian Code in 1964.

In the ULM method, stress condition at the state of impending *collapse* of the structure is analysed, thus using the non-linear stress-strain curves of concrete and steel. The *safety measure* in the design is obtained by the use of proper *load factor*. This makes it possible to use different load factors under combined loading conditions. It is to be carefully noted that *satisfactory strength performance at ultimate loads does not guarantee satisfactory serviceability performance at normal service loads*. Since the method utilizes a large reserve of strength in plastic region (inelastic region) and of ultimate strength of member, the resulting section is very slender or thin. This gives rise to excessive deformation and cracking. Also, the method does not take into consideration the effects of creep and shrinkage.

Merits of ULM

1. While the WSM uses only the nearly linear part of stress-strain curve, the ULM uses fully the actual stress-strain curve. In other words, the stress block parameters are defined by the actual stress-strain curve.

2. The load factor gives the exact margin of safety against collapse.

3. The method allows to use different load factors for different types of loads and the combination thereof.

4. The failure load computed by ULM matches with the experimental results.

5. The method is based on the ultimate strain as the failure criteria.

6. The method utilises the reserve of strength in the plastic region.

Demerits of ULM

1. The method does not take into consideration the serviceability criteria of deflection and cracking.

2. The use of high strength reinforcing steel and concrete results in increase of deflection and crack width.

3. The method does not take into consideration the effects of creep and shrinkage.

4. In the ULM, the distribution of stress resultants at ultimate load is taken as the distribution at service loads magnified by the load factor(s). This is erroneous since significant *redistribution* of stress resultants takes place as the loading is increased from service loads to ultimate loads.

To summarise, *the ultimate load method ensures safety at ultimate loads but disregards the serviceability at service loads.*

2.4. LIMIT STATE METHOD (LSM)

We have seen that while the WSM gives satisfactory performance of the structure at working loads, it is *unrealistic* at ultimate state of collapse. Similarly, while the ULM provides realistic assessment of safety, it does not guarantee the satisfactory serviceability requirements at service loads. *An ideal method is the one which takes into account not only the ultimate strength of the structure but also the serviceability and durability requirements.* The newly emerging 'limit state method' of design is oriented towards the simultaneous satisfaction of all these requirements. In the limit state method, a structure is designed for safety against collapse (i.e. for ultimate strength to resist ultimate load) and checked for its serviceability at working loads, thus rendering the structure fit for its intended use. Thus, the LSM includes consideration of a structure at both the working and the ultimate load levels with a view to satisfy the requirements of safety and serviceability.

The European Committee for Concrete (CEB) and the International Federation for Prestressing (FIP) were amongst the earliest to introduce the philosophy of *limit state method*, which is reliability-based in concept. 'Recommendations for an International Code of Practice for Reinforced Concrete' known as the *blue book* was published in 1963 by CEB and the complimentary report 'International Recommendations for the Design of Concrete Structures' known as *red book*, was published in 1970 by CEB along with FIP. These were subsequently revised by CEB-FIP as the 'Model Codes for Concrete Structures' as model for national Codes to follow. The LSM was introduced in the British Code in 1973 and the Indian Code in 1978. However, in United States, it was introduced in 1971 in a slightly different format of '*strength and serviceability design*'.

The acceptable limit of safety and serviceability requirements, before failure occurs, is called a limit state. A *limit state* is a state of impending failure, beyond which a structure ceases to perform its intended function satisfactorily, in terms of either safety or serviceability, i.e. it either collapses or becomes unserviceable. The aim of design is to achieve acceptable *probabilities* that the structure will not become unfit for the use for which it is intended, that is, it will not reach a limit state. As per IS 456 : 2000, all limit states (see § 2.5) shall be considered in design to ensure an adequate degree of safety and serviceability. In general, the structure shall be designed on the basis of the *most critical limit state* and shall be checked for other limit states.

For ensuring the above objectives, the design should be based on *characteristic values* for material strengths and applied loads, which take into account the variations in the material strengths and in the loads to be supported. The characteristic values should be based on *statistical data* if available; where such data are not available, they should be based on experience. The '*design values*' are derived from the characteristic values through the use of *partial safety factors*, one for material strengths and the other for loads.

2.5. TYPES OF LIMIT STATES

As defined earlier, a *limit state* is a state of impending failure beyond which a structure ceases to perform its intended function satisfactorily, in terms of safety or serviceability. The acceptable limit for the safety and serviceability requirements before failure occurs is called a *limit state*. Two categories of limit states are considered in design.

1. Limit states of collapse

(*i*) Limit state of collapse in flexure
(*ii*) Limit state of collapse in compression
(*iii*) Limit state of collapse in compression and uniaxial bending
(*iv*) Limit state of collapse in compression and biaxial bending
(*v*) Limit state of collapse in shear
(*vi*) Limit state of collapse in bond
(*vii*) Limit state of collapse in torsion
(*viii*) Limit state of collapse in tension

2. Limit states of serviceability

(*i*) Limit state of deflection
(*ii*) Limit state of cracking
(*iii*) Other limit states, such as vibration, fire resistance, durability etc.

Limit state of collapse

The limit state of collapse of the structure or part of the structure could be assessed from *rupture* of one or more critical sections and from *buckling* due to elastic or plastic *instability* (including the effects of sway where appropriate) or *overturning*. The resistance to bending, shear, torsion and axial loads (or combinations thereof) at every section shall not be less than the appropriate value at that section produced by the probable most unfavorable combination of loads on the structure using the appropriate *partial safety factors*.

The limit state of collapse in flexure, shear, bond and torsion have been discussed in detail in Chapters 3, 7, 8 and 9 respectively.

2. Limit state of serviceability

The limit state of serviceability relate to the performance or behaviour of structure at *working loads*. The limit state of serviceability in deflection and cracking has been discussed in detail in Chapter 10.

Normally, design is based on the considerations of limit states of collapse on ultimate loads and on serviceability limit states of deflection and cracking under service loads. *Durability* in taken care of by prescribing appropriate grade of concrete, nominal cover for various exposure condition, cement content etc.

2.6. CHARACTERISTIC AND DESIGN VALUES AND PARTIAL SAFETY FACTORS

1. Characteristic strength of materials

The term *characteristic strength* means that value of the strength of material below which not more than a minimum acceptable percentage of the test results are expected to fall. Most of the design Codes in the World, including IS Code, have adopted the minimum acceptable percentage as 5% for reinforced concrete structures. This implies that there in only 5% probability or chance of the actual strength being less than the characteristic strength or in other words, the characteristic strength has 95% reliability.

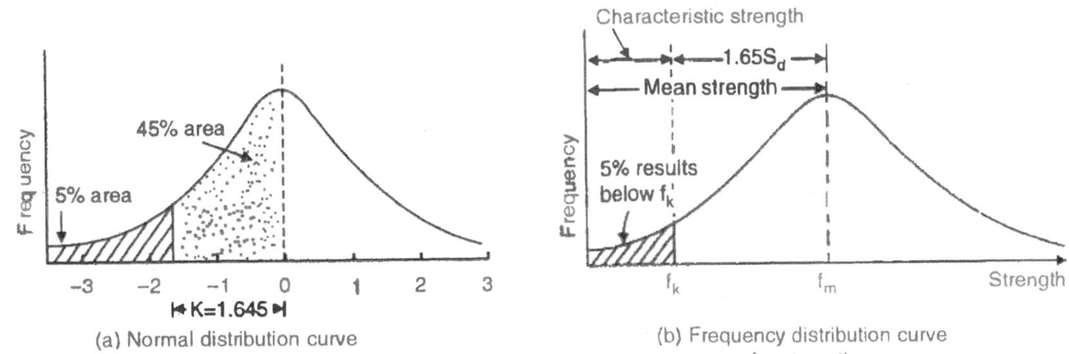

(a) Normal distribution curve

(b) Frequency distribution curve for strength

FIG. 2.1.

Fig. 2.1(*a*) shows the normal distribution curve (or normal probability curve). The value of K corresponding to 5% area of the curve is 1.645. Indian Standard has adopted the value of K as 1.65. Fig. 2.1(*b*) shows frequency distribution curve of strength of material (concrete or steel). As the *design strength* should be lower than the *mean strength* (f_m), we have

Characteristic strength = [Mean strength] − K × [Standard deviation]

or
$$f_k = f_m - K\,S_d \qquad \ldots(2.1)$$

where f_k = characteristic strength of the material.

f_m = mean strength ; K = constant = 1.645 ≏ 1.65

S_d = standard deviation for a set of test results

The value of standard deviation (S_d) is given by

$$S_d = \sqrt{\frac{\Sigma \delta^2}{n-1}} \quad \ldots(2.2)$$

where δ = Deviation of the individual test strength from the average or mean strength of n samples.

n = number of test results.

The design Codes always prescribe a minimum number of test results for arriving at the standard deviation. IS 456 : 2000 has recommended minimum value of n equal to 30. The above approach of defining the material strength shows clearly the quality of material and is therefore, conductive of *quality control*. The standard deviation is indicative of quality.

(a) Characteristic strength of concrete

The characteristic strength of concrete is denoted by f_{ck} (N/mm^2), and its value for different grades of concrete is given in Table 2.1, as per IS 456 : 2000. In the Table, the symbol M, used for the *designation* of concrete mix, refers to the mix and the number refers to the specified characteristic compressive strength of 150 mm size cube at 28 days expressed in N/mm^2.

There is normally a gain of strength beyond 28 days. The quantum of increase depends upon the grade and type of cement, curing and environmental conditions etc. The *design values should be based on 28 days characteristic strength of concrete unless there is evidence to justify a higher strength for a particular structure due to age*. For concrete of grade M 30 and above, the rate of increase of compressive strength with age shall be based on actual investigations.

TABLE 2.1. CHARACTERISTIC STRENGTH (f_{ck}) FOR VARIOUS GRADES OF CONCRETE (IS 456 : 2000)

Group	*Grade Designation*	*Specified characteristic compressive strength of 150 mm cubes at 28 days (N/mm^2)*
Ordinary Concrete	M10	10
	M15	15
	M20	20
Standard Concrete	M25	25
	M30	30
	M35	35
	M40	40
	M45	45
	M50	50
	M55	55
High Strength Concrete	M60	60
	M65	65
	M70	70
	M75	75
	M80	80

Note : For concrete of compressive strength greater than M 55, the design parameters given in the standard may not be applicable and the values may be obtained from the specified literature and experimental results.

(b) Characteristic strength of steel

Until the relevant Indian Standard Specifications for reinforcing steel are modified to include the concept of characteristic strength, the characteristic value shall be assumed as the minimum yield stress/0.2 percent proof stress specified in the relevant Indian Standard Specification. The characteristic strength of steel is designated by symbol f_y (N/mm^2).

2. Characteristic loads

A *characteristic load* is defined as that value of load which has a 95 percent probability of not being exceeded during the life of the structure.

Fig. 2.2 shows the frequency distribution curve for a load.

As the *design load* should be more than average load obtained from statistics, we have

$$F_k = F_m + K\,S_d \qquad ....(2.3)$$

where F_k = characteristic load

F_m = mean load

K = constant = 2.645 ≏ 2.65

S_d = standard deviation for the load.

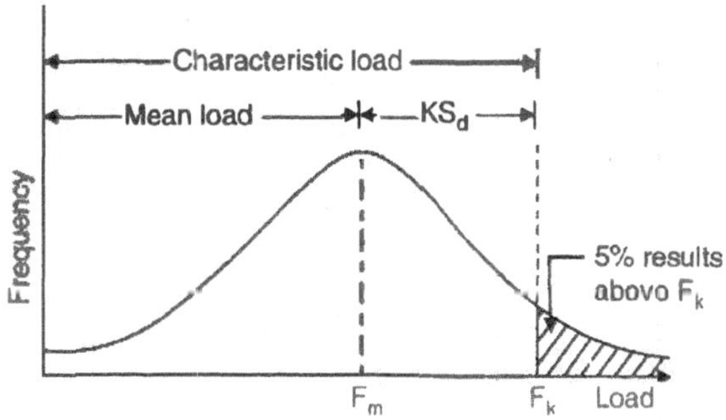

FIG. 2.2. FREQUENCY DISTRIBUTION CURVE FOR LOAD

Thus the characteristic value of a particular load can be calculated theoretically. However, research for determining actual loading on structures has not yielded adequate data to enable us to compute theoretical values of variations for arriving at the actual loading on a structure. IS 456 : 2000 states that 'since the data are not available to express loads in statistical terms, the dead loads (DL) given in IS 875 (Part 1), imposed load (IL) or live load (LL) given in IS 875 (Part 2), wind loads (WL) given in IS 875 (Part 3), snow load (SL) as given in IS 875 (Part 4) and seismic forces (EL) given in IS 1893 shall be assumed as the *characteristic loads*'.

3. Design strength of materials

The design strength of materials (f_d) is given by

$$f_d = \frac{f_k}{\gamma_m} \qquad ...(2.4)$$

where f_k = characteristic strength of material (*i.e.* f_{ck} for concrete and f_y for steel)

γ_m = partial safety factor appropriate to the material and the limit state being considered.

4. Design loads : The design load (F_d) is given by

$$F_d = F_k\,.\,\gamma_f \qquad ...(2.5)$$

where F_k = characteristic load, and

γ_f = partial safety factor appropriate to the nature of loading and the limit state being considered.

The design load given by Eq. 2.5, by multiplying the characteristic load by the partial safety factor for load, is also known as *factored load.*

5. Partial safety factor (γ_m) for materials : When assessing the strength of a structure or structural member for the limit state of collapse, the values of partial safety factor, γ_m should be taken as 1.5 for concrete and 1.15 for steel. A higher value of partial safety factor (γ_{mc} = 1.5) for concrete has been adopted because there are greater chances of variation of strength of concrete due to improper compaction, inadequate curing, improper batching and mixing and variations in properties of ingredients. The chances of variations in the strength of reinforcement are known to be small and hence a lower value (γ_{ms} = 1.15) has been adopted. It should be noted that these γ_m values are already incorporated in the equations and tables given in IS 456 : 2000 for the limit state design.

Thus, in the limit state method, the *design stress* (f_{ds}) for steel reinforcement is given by $f_{ds} = f_y / \gamma_{ms} = f_y / 1.15 \simeq 0.87 f_y$ ...(2.6)

This is reflected by the stress-strain curves shown in Figs. 2.4, 2.5 and 2.6 for mild steel, Fe 415 steel and Fe 500 steel respectively.

According to IS Code (IS 456 : 2000) for design purposes the compressive strength of concrete *in the structure* shall be assumed to be 0.67 times the characteristic strength of concrete in cube and the partial safety factor $\gamma_{mc} = 1.5$ shall be applied in addition to this. Thus, the *design stress* (f_{dc}) in concrete is given by

$$f_{dc} = 0.67 f_{ck} / \gamma_{mc} = 0.67 f_{ck} / 1.5 = 0.446 f_{ck} \simeq 0.45 f_{ck} \quad ...(2.7)$$

This is reflected in the stress-strain curve for concrete shown in Fig. 2.3.

6. **Partial safety factor (γ_f) for loads** : The partial safety factors (γ_f) for loads, as per IS 456 : 2000 are given in Table 2.2.

TABLE 2.2. PARTIAL SAFETY FACTOR γ_f FOR LOADS

Load Combinations	*Limit state of collapse*			*Limit state of serviceability*		
	DL	IL	WL	DL	IL	WL
(1)	*(2)*	*(3)*	*(4)*	*(5)*	*(6)*	*(7)*
DL+IL	1.5	1.5	1.0	1.0	1.0	—
DL+WL	1.5 or 0.9*	—	1.5	1.0	—	1.0
DL+IL+WL	1.2	1.2	1.2	1.0	0.8	0.8

Note : 1. While considering earthquake effects, substitute EL for WL. (*This value is to be considered when stability against overturning or stress reversal is critical).

2. For the limit state of serviceability, the values of γ_f given in this table are applicable for short term effects. While assessing the long term effects due to creep, the dead load and that part of the live load likely to be permanent may only be considered.

3. Imposed load (IL) is also commonly known as live load (LL).

It is to be noted that while the British Code (BS 8110) uses a factor of 1.4 for DL and 1.6 for IL, IS Code has recommended a common factor of 1.5 for both DL and IL for convenience of using the same structural analysis for both elastic (working stress) design and limit state design. Due to this simplification, the factored load, shear moment etc. in the limit design will be 1.5 times the corresponding value used for elastic design. Thus, using a suffix u for limit state design, we have the following values for dead and live loads

$$w_u = 1.5\, w \;;\quad W_u = 1.5\, W \;;\quad M_u = 1.5\, M \;;\quad V_u = 1.5\, V \;;\quad P_u = 1.5\, P \quad ...(2.8)$$

Comparison with working stress method

For the working stress method, given in Appendix A, we have

$$\gamma_f = 1.0 \text{ for all loads and } \gamma_{ms} = \frac{f_y}{\sigma_{st}} = \frac{250}{140} = 1.79 \text{ for mild steel bars}$$

and $\gamma_{ms} = 415/230 = 1.80$ for HYSD bars of grade Fe 415.

Thus average $\gamma_{ms} = 1.795$. Also, $\gamma_{mc} = \dfrac{f_{cu}}{\sigma_{cbc}} \approx 3.0$ for all grades of concrete.

Hence $\gamma_f \times \gamma_{ms} = 1.0 \times 1.795 = 1.795$ for reinforcement and $\gamma_f \times \gamma_{mc} = 1.0 \times 3.0 = 3.0$ for concrete.

For the limit state design, taking maximum compressive stress in concrete $= 0.67 f_{cu}$, and $\gamma_f = 1.5$; $\gamma_{ms} = 1.15$ and $\gamma_{mc} = 1.5$, we have, $\gamma_f \times \gamma_{ms} = 1.5 \times 1.15 = 1.73$ for reinforcement which is slightly less than 1.795 for working stress design, and $\gamma_f \times \gamma_m = 1.5 \times \frac{1.5}{0.67} = 3.36$ for concrete, which is more than 3.0 for working stress design.

2.7. IDEALISED STRESS-STRAIN CURVE FOR CONCRETE

The IS Code permits the use of any appropriate curve for relationship between the *compressive* stress and strain distribution in concrete, subject to the condition that it results in the prediction of strength in substantial agreement with test results. An acceptable stress-strain curve (Fig. 2.3) given in Fig. 21 of IS Code (IS 456 : 2000) forms the basis of design in this book. The compressive strength of *concrete in the actual structure* is assumed to be $0.67 f_{ck}$ (curve 2 of Fig. 2.3). The factor 0.67 is introduced to account for the *size effect* based on the assumption that the concrete in the structure develops a strength of only 0.67 times the strength of the cube. The partial safety factor $\gamma_m = 1.5$ shall be applied in addition to this. Thus, the design curve (curve 3 of Fig. 2.3) is obtained by using a partial safety factor $\gamma_m = 1.5$. Thus the maximum compressive stress in concrete for design purposes is equal to $0.67 f_{ck}/\gamma_m = 0.67 f_{ck}/1.5 = 0.446 f_{ck} \simeq 0.45 f_{ck}$. It is to be noted that each curve is parabolic in the initial portion upto a strain of 0.002. At a strain of 0.002 (i.e. 0.2% strain), the stress remains constant with increasing load, until a strain of 0.35% is reached, when the concrete is said to have failed.

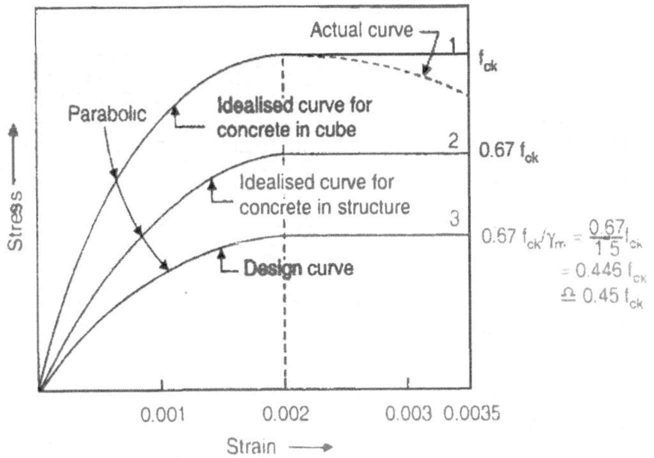

FIG. 2.3. STRESS-STRAIN CURVE FOR CONCRETE

2.8. STRESS STRAIN RELATIONSHIP FOR STEEL

The stress-strain curve for mild steel is shown in Fig. 2.4 while the stress strain curve for Fe 415 and Fe 500 steel are shown in Fig. 2.5 and Fig. 2.6 respectively. The modulus of elasticity of all types of reinforcing steel, E_s is taken as 2×10^5 N/mm^2. Also, *the stress strain relationship for steel in tension and compression is assumed to be the same.*

For mild steel, the stress is proportional to strain upto yield point and thereafter the strain

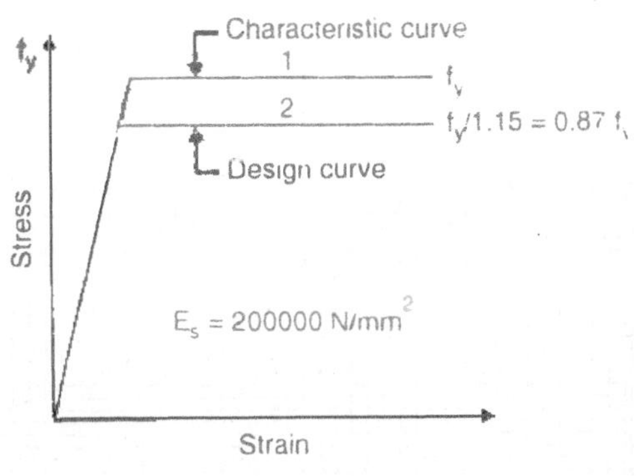

FIG. 2.4. STRESS STRAIN CURVE FOR MILD STEEL

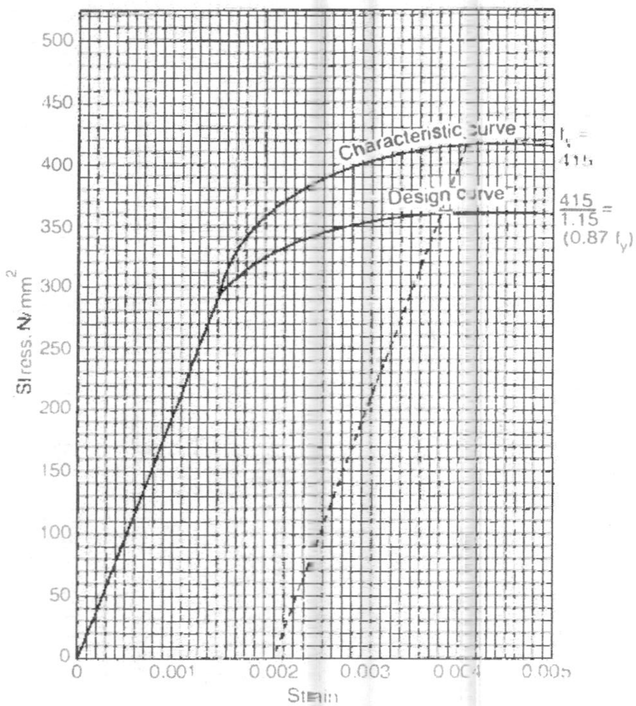
Characteristic curve
Design curve
f_y = 415
$\frac{415}{1.15}$ = (0.87 f_y)
Stress, N/mm²
Strain
500
450
400
350
300
250
200
150
100
50
0
0
0.001
0.002
0.003
0.004
0.005

FIG 2.5

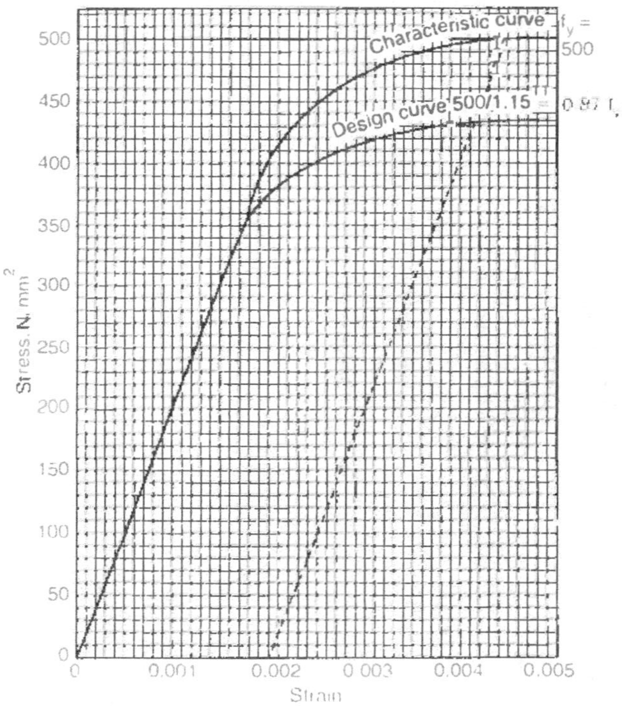
Characteristic curve
Design curve 500/1.15 = 0.87 f_y
f_y = 500
Stress, N/mm²
Strain
500
450
400
350
300
250
200
150
100
50
0
0
0.001
0.002
0.003
0.004
0.005

FIG. 2.6

increases at constant stress. The design yield stress of steel is equal to f_y/γ_m. With a value of 1.15 for γ_m, the design yield stress f_{yd} becomes 0.87 f_y. For cold worked bars (Fe 415 and Fe 500 HYSD bars), there is no definite yield point ; hence yield stress is taken as 0.2 percent proof stress. The stress-strain curves (Fig. 2.5 and Fig. 2.6) for these two types of steels are linear upto a stress of 0.8 f_y and the strains are elastic. Thereafter, the stress-strain curves are defined as given below :

Stress	*Inelastic strain*	*Stress*	*Inelastic strain*
0.80 f_y	Nil	0.95 f_y	0.0007
0.85 f_y	0.0001	0.975 f_y	0.001
0.90 f_y	0.0003	1.0 f_y	0.002

The stress-strain curve for design purposes is obtained by substituting f_{yd} for f_y in the above, where. $f_{yd} = f_y/\gamma_{ms} = f_y/1.15 \simeq 0.87 f_y$. The salient points on the design stress strain curve for two types of cold-worked steel are given in Table 2.3.

TABLE 2.3. SALIENT POINTS ON THE DESIGN STRESS-STRAIN CURVE FOR COLD WORKED BARS

(f_{yd} = Design yield stress ; f_y = Characteristic stress)

Stress level	*Fe 415 steel : f_y = 415 N/mm²*		*Fe 500 steel : f_y = 500 (N/mm²)*	
	Strain	*Stress (N/mm²)*	*Strain*	*Stress (N/mm²)*
0.8 f_{yd}	0.00144	288.7	0.00174	347.8
0.85 f_{yd}	0.00163	306.7	0.00195	369.6
0.90 f_{yd}	0.00192	324.8	0.00226	391.3
0.95 f_{yd}	0.00241	342.8	0.00277	413.0
0.975 f_{yd}	0.00276	351.8	0.00312	423.9
1.0 f_{yd}	0.00380	360.9	0.00417	434.8

Note 1. $f_{yd} = f_y/\gamma_{ms} = f_y/1.15 = 0.87 f_y$

Note 2. Linear interpolation may be made for intermediate values.

2.9. METHODS OF ANALYSIS

Application of *limit analysis* to reinforced concrete structures is not easy because of the limitation imposed on the amount of redistribution of moments by available *rotation capacity* of R.C. sections, and by the serviceability requirements. Indian Standard Code permits the analysis of structures by *linear elastic theory* to calculate the *internal actions* (i.e. B.M., S.F., torsion etc.) produced by the *design loads (i.e. factored loads). The material strength to be assumed shall be the characteristic values in the determination of elastic properties of members irrespective of limit state being considered. Redistribution* of calculated moments may be made as given below.

Redistribution of moments in continuous beams and frames

The redistribution of moments may be carried out satisfying the following *conditions:*

(*a*) Equilibrium between the internal forces and the external loads is maintained.

(*b*) The ultimate moment of resistance provided at any section of a member is not less than 70 percent of moment at that section obtained from an elastic maximum moment diagram covering all appropriate combination of loads.

(*c*) The elastic moment at any section in a member due to a particular combination of loads shall not be reduced by more than 30 percent of the numerically largest moment given any where by the elastic maximum moments diagram for the particular member, covering all appropriate combination of loads.

(*d*) At sections where the moment capacity after redistribution is less than that from the elastic maximum moment diagram, the following relationship shall be satisfied.

$$\frac{x_u}{d} + \frac{\delta M}{100} \leq 0.6 \qquad \ldots(2.9)$$

where x_u = depth of neutral axis ; d = effective depth

δM = percentage reduction in moment

(*e*) In structures in which the structural frame provides the lateral stability, the reduction in moment allowed by condition (*c*) above shall be restricted to 10 percent for structures over 4 storeys in height.

For detailed treatment on redistribution of moments, see Chapter 21.

Example 2.1. *Compression tests on 150 mm cubes of concrete gave the following results:*

28, 32, 30, 26, 31, 33, 27, 26, 34, 28, 32, 33 N/mm²

Determine the characteristic strength of concrete.

Solution : Here, $n = 12$. Arrange the test results in ascending order and find the mean value f_m. From this mean value, compute deviation $\delta\ (= f - f_m)$ and δ^2 for each value. The computation are arranged in tabular form below.

f	26	26	27	28	28	30	31	32	32	33	33	34	$\Sigma f = 360$ $f_m = 360/12$ $= 30$ N/mm²
δ	− 4	− 4	− 3	− 2	− 2	0	+ 1	+ 2	+ 2	+ 3	+ 3	+ 4	
δ^2	16	16	9	4	4	0	1	4	4	9	9	16	$\Sigma \delta^2 = 92$

$$\therefore \qquad S_d = \sqrt{\frac{\Sigma \delta^2}{n-1}} = \sqrt{\frac{92}{12-1}} = 2.89$$

$$\therefore \qquad f_{ck} = f_m - K\,S_d = 30 - 1.645 \times 2.89 \approx \mathbf{25.2\ N/mm^2}$$

Example 2.2. *The desired characteristic strength of a mix is 20 N/mm². Taking the standard deviation recommended by IS 456 : 2000, determine the average strength of 150 mm size cubes of concrete.*

Solution : Desired f_{ck} of mix is 20 N/mm². Hence it is M 20 grade concrete. From Table 8 of IS 456 : 2000 (Table 1.21 of this book), let as adopt standard deviation of 4 N/mm².

From Eq. 2.1. $\qquad f_m = f_{ck} + K\,S_d = 20 + 1.645 \times 4 = \mathbf{26.58\ N/mm^2}$

Example 2.3. *The one-way slab of a reading room is estimated to have a thickness of 125 mm along with thickness of finishings as 75 mm. The slab is to carry a live* *load of 4 kN/m². Determine the factored loads (a) for strength, and (b) for serviceability limit states.*

Solution

Let us assume unit weight of concrete as 25 kN/m³ and that of finishing as 22 kN/m³.

Dead load due to self weight of slab $= 0.125 \times 1 \times 1 \times 25 = 3.125$ kN/m²

Dead load due to finishing $= 0.075 \times 1 \times 1 \times 22 = 1.65$ kN/m²

Total DL $= 4.775$ kN/m²

Live Load, LL $= 4$ kN/m²

(*a*) ***Factored load for strength***

Hence from Table 2.2, factored load $w_u = 1.5\,(DL + LL)$

$= 1.5\,(4.775 + 4) \simeq$ **13.16 kN/m²**

(*b*) *Factored load for serviceability*

From Table 2.2, factored load, $w_s = w = 1\,(DL + LL) = 1\,(4.775 + 4) \simeq$ **8.78 kN/m²**

Example 2.4. *A 6 m high column, fixed at base and free at top, is subjected to the following loads :*

Total dead load = 60 kN

Total imposed load = 150 kN

Wind load = 3.5 kN per m height

Determine factored loads for (a) limit state of collapse, and (b) limit state of serviceability.

Solution

Total DL = 60 kN; Total LL = imposed load = 150 kN

Total WL = 3.5 × 6 = 21 kN, acting at 3 m above base.

(*a*) ***Factored load for limit state of collapse*** (i.e. for strength and stability)

(*i*) *Combination of dead and live loads*

$$P_u = 1.5\,(DL + LL) = 1.5\,(60 + 150) = \mathbf{315\ kN}$$

(*ii*) *Combination of dead and wind loads* : For stability/overturning

Vertical load : $P_u = 0.9\,DL = 0.9 \times 60 =$ **54 kN**

Horizontal load, $H_u = 1.2\,WL = 1.2 \times 21 =$ **25.2 kN**

(*iii*) *Combination of dead, live and wind loads*

Vertical load : $P_u = 1.2\,(DL + LL) = 1.2\,(60 + 150) =$ **252 kN**

Horizontal load : $H_u = 1.2\,WL = 1.2 \times 21 =$ **25.2 kN**

(*b*) *Factored load for limit state of serviceability*

(*i*) *Combination of dead and live loads*

$$P_s = 1.0\,(DL + LL) = 1.0\,(60 + 150) = \mathbf{210\ kN}$$

(*ii*) *Combination of dead and wind loads*

Vertical load $P_s = 1.0\,DL = 1.0 \times 60 =$ **60 kN**

Horizontal load $H_s = 1.0\text{ WL} = 1.0 \times 21 = \mathbf{21\ kN}$

(*iii*) *Combination of dead, live and wind loads*

Vertical load $P_s = 1.0\text{ DL} + 0.8\text{ LL} = 1.0 \times 60 + 0.8 \times 150 = 180\text{ kN}$

Horizontal load $H_s = 0.8\text{ WL} = 0.8 \times 21 = \mathbf{16.8\ kN}$

PROBLEMS

1. Define the following

(*i*) Limit state of collapse (*ii*) Factored load (*iii*) Limit state of serviceability (*iv*) Characteristic strength of steel.

2. Differentiate between working stress method and limit state method

3. What do you understand by characteristic strength of concrete ? How it is determined from a number of test results on 150 mm size cubs of concrete ?

4. Write a note on partial safety factors for material. Compare WSM and LSM with respect of overall factor of safety obtainable in the design. using HYSD bars.

6. What is the desired average strength of 150 mm cubes of M 15 concrete. for a standard deviation of 3 N/mm^2 ? (19.94 N/mm^2)

3

Limit State of Collapse in Flexure : I Singly Reinforced Rectangular Sections

3.1. INTRODUCTION

Flexure or *bending* is commonly encountered in structural elements like beams, plates, shells etc. under the action of transverse loads. Flexure is also encountered in *columns* and *walls* when these are subjected to either eccentric loading or lateral pressure along with vertical loads. Similarly, flexure also occurs in *retaining walls* subjected to lateral earth pressure and in *water tanks* subjected to hydrostatic water pressure. In most of these structural elements, *flexure* usually occurs along with transverse *shear*. Some times, other structural actions such as *torsion*, *axial compression* (or *axial tension*) etc. are encountered along with flexure and shear. However, this chapter deals only with the flexural action; other actions have been dealt with in separate chapters. Finally, the complete design of beams and slabs have been discussed in Chapter 11, taking into account the collective action of bending, shear, torsion, bond etc, in accordance with the provisions of IS 456 : 2000.

3.2. ASSUMPTIONS IN LIMIT STATE DESIGN IN FLEXURE

The ultimate moment of resistance or the strength at the limit state of flexural collapse of reinforced concrete beams (clause 38.1 of IS 456 : 2000) is based on following assumptions. These assumption are valid only for shallow beams where the span to depth ratio is greater than 2.5.

Assumptions in limit state of collapse in flexure

(*a*) Plane sections normal to the axis remain plane after bending.

(*b*) Maximum strain in concrete at the outermost compression fibre is taken as 0.0035 in bending.

(*c*) The relationship between the compressive stress distribution in concrete and the strain in concrete may be assumed to be rectangle, trapezoid, parabola or any other shape which results in the prediction of strength in substantial agreement with the results of tests. An acceptable stress-strain curve is given in Figs. 2.3 and 3.1. For design purposes, the *compressive strength of concrete in structure* shall be assumed to be 0.67 times the characteristic strength of concrete in cube. The partial safety factor γ_{mc} = 1.5 shall be applied in addition to this.

(*d*) The tensile strength of the concrete is ignored.

(*e*) The stresses in the reinforcement are derived from representative stress-strain curve for the type of steel used (Figs. 2.4, 2.5 and 2.6). For design purposes, the partial safety factor γ_{ms} = 1.15 shall be applied.

(*f*) The maximum strain in the tension reinforcement in the section at failure shall not be less than $\frac{f_y}{1.15\,E_s} + 0.002$

where f_y = characteristic strength of steel, and E_s = modulus of elasticity of steel.

Comments on the above assumptions: Assumption (*a*) implies linear variation of strain across the depth of the section (See Fig. 3.1*bii*). This has been verified by numerous tests on reinforced concrete beams. In assumption (*b*), the maximum strain in concrete is the strain at which the section reaches its maximum moment capacity. The value of such a strain is liable to large variations depending upon rate of deformation, loading time, grade of concrete, shape of cross-section, percentage of reinforcement etc. The Code adopts a value of 0.0035 which will give conservative results in most cases of pure bending. Assumption (*c*) giving acceptable stress-strain curve in concrete (Fig. 2.3) gives the shape of compressive stress distribution in concrete [See Fig. 3.1 *b* (*iii*)]. Such a stress distribution curve across the depth, popularly known as the *stress block* is a combination of a parabola and a rectangle. Assumption (*d*) is similar to the one made for the working stress method, in which the tensile strength of concrete is neglected. Assumption (*e*) is with reference to stress strain curve for steel reinforcement. For cold twisted bars which do not have definite yield point, the characteristic strength f_y is taken corresponding to 0.2% proof strain. Assumption (*f*) is intended to ensure ductile failure, that is, the tensile reinforcement at the critical section has to undergo a certain degree of inelastic deformation before the concrete fails in compression. The expresson for minimum value of maximum strain in steel, laid down by the Code, has the first term, $f_y/1.15\,E_s$, which corresponds to the yield stress f_y for the case of mild steel and other types which have well defined yield point. However, the cold twisted steel (HYSD bars) do not have a well defined yield point and the yield stress f_y is taken as the conventional value of 0.2 percent proof stress. Hence a strain of 0.002 is added so that there is sufficient yielding of steel before failure at a constant stress. It is to be noted that assumption (*b*) and (*f*) govern the maximum depth of neutral axis in flexural members.

3.3. STRESS BLOCK PARAMETERS

The various theories advanced for estimating the ultimate flexural resistance of reinforced concrete sections can be broadly classified into two categories. In the first category of theories (such as those of Suenson, Menaeh, Kempton, Dyson and Whitney) no assumptions are made regarding the distribution of strain across the section, and neutral axis is located empirically, for analysing compression failures. These theories are based on equilibrium only. The second category of theories (such as those by Stussi or Indian Standard) *include the strain compatibility*. A limiting strain in concrete is used as a criterion for the crushing of concrete in flexural compression.

The stress-strain behaviour of concrete under compression is generally obtained from cylinder or cube of concrete subjected to longitudinal compressive loading. Whereas the stress and strains are uniform for a cube (or for a cylinder), they vary across the depth of a flexural member. The ratio between the maximum flexural compression in a beam and the crushing strength of a cube is likely to be different from unity. The I.S. Code (456 - 2000) recommends the compressive strength of concrete in the structure equal to 0.67 times the characteristic strength. The corresponding stress-strain diagram is shown in Fig. 3.1(*a*), which is the same as curve 2 of Fig. 2.3.

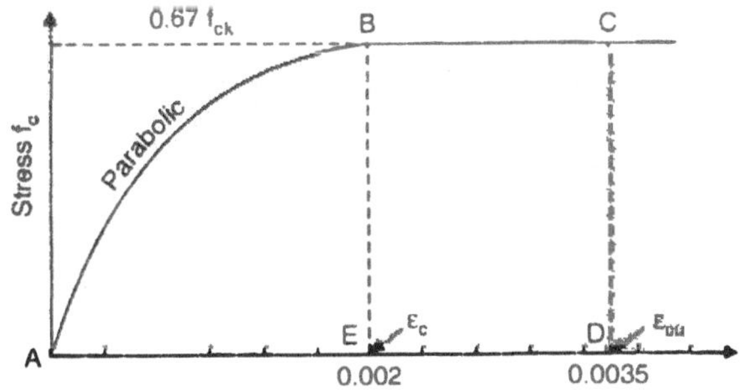

(a) Stress-strain curve for concrete

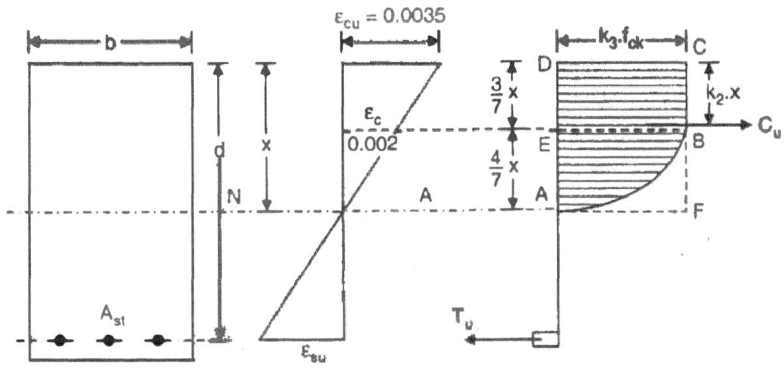

(i) Section (ii) Strain dia (iii) Stress block

(b) Strain and stress variations across section

FIG. 3.1

Fig. 3.1(*bii*) shows the strain diagram, while Fig. 3.1(*biii*) shows the variation of stresses across the depth. The stress diagram *ABCDEA* [Fig. 3.1(*biii*)] has parabolic shape from *A* to *B* (corresponding to the curve *AB* of Fig. 3.1(*a*), and then linear from *B* to *C* at a constant stress of 0.67 f_{ck} corresponding to line *BC* of Fig. 3.1(*a*). The total compressive force C_u and its location below the top fibre can be expressed in terms of stress block factors (or parameters) k_1, k_2 and k_3 as follows :

k_1 = *shape factor*, defined as the ratio between the area of stress block *ABCD* and area of rectangle *AFCD* [Fig 3.1 (*biii*)]

$$= \frac{\text{Area } ABCD}{\text{Area } AFCD}$$

or $$= \frac{\text{Area } ABCD}{\text{Area } (x \times DC)}$$

Now ultimate strain in concrete, $\varepsilon_{cu} = AD = 0.0035$ [Fig. 3.1(*a*)] while the strain $\varepsilon_c = AE$, after which concrete yields at constant stress of $0.67\ f_{ck}$. is found to be equal to 0.002 for all grades concrete.

$$\text{Ratio}\quad \frac{AE}{AD} = \frac{\varepsilon_c}{\varepsilon_{cu}} = \frac{0.002}{0.0035} = \frac{4}{7}. \qquad \text{Also,}\quad \frac{ED}{AD} = \frac{0.0035 - 0.002}{0.0035} = \frac{0.0015}{0.0035} = \frac{3}{7}$$

$$\therefore \qquad AE = \frac{4}{7}AD. \text{ and } ED = \frac{3}{7}AD.$$

Now Area $ABCD$ = Area ABE + Area $BCDE$ [Fig. 3.1 (*biii*)]

$$= \left(\frac{2}{3} \times AE \times BE\right) + (ED \times CD) = \left(\frac{2}{3} \times \frac{4}{7} AD \times CD\right) + \left(\frac{3}{7} AD \times CD\right)$$

$$= \frac{8}{21}(AD \times CD) + \frac{9}{21}(AD \times CD) = \frac{17}{21} \times \text{ Area } AFCD$$

$$\therefore \qquad k_1 = \frac{17}{21} = 0.8095 \simeq 0.81 \qquad \ldots(3.1)$$

The resultant compressive force is located at a depth $k_2\ x$ below the top fibre, where x is the depth of the netural axis.

$$\text{Now} \qquad k_2 .\ AD = \frac{(\text{Area } ABE) \times \bar{x}_1 + (\text{Area } BCDE) \times \bar{x}_2}{\text{Area } ABCD}$$

$$k_2 .\ AD = \frac{\frac{8}{21}\left(ED + \frac{3}{8}AE\right) + \frac{9}{21}\left(\frac{1}{2}DE\right)}{\frac{17}{21}} = \frac{\frac{8}{21}\left(\frac{3}{7} + \frac{3}{8} \times \frac{4}{7}\right)AD + \frac{9}{21}\left(\frac{1}{2} \times \frac{3}{7}\right)AD}{\frac{17}{21}}$$

$$\text{or} \qquad k_2 = \frac{\frac{36}{7} + \frac{27}{14}}{17} = \frac{99}{238} = 0.416 \simeq 0.42. \qquad \ldots(3.2)$$

Hence resultant compressive force is located at a depth of $k_2 x$ (= 0.416 x) below the top fibre. Let the maximum ordinate (DC) of the stress block be $k_3 f_{ck}$. This is the maximum compressive stress in flexure on the structure, in relation to characteristic strength f_{ck} . IS : 456-2000 recommends $k_3 = 0.67$. The compressive force C_u in concrete as given by

$$C_u = b \times \text{area } ABCD \text{ [Fig. 3.1 (biii)]} = k_1 \times x \times DC \times b = k_1\ k_3 f_{ck}\ .bx \qquad \ldots(3.3)$$

The moment of resistance is given by

$$M_u = C_u(d - k_2\ x) = k_1\ k_3 f_{ck}\ bd^2 \left(\frac{x}{d}\right)\left(1 - k_2 \frac{x}{d}\right) \qquad \ldots(3.4)$$

3.4. DESIGN STRESS BLOCK PARAMETERS (IS 456 : 2000)

The general expressions for C_u and M_u can be easily modified by incorporating partial safety factors γ_{mc} and γ_{ms} for concrete and steel. The stress strain curve for concrete is corresponding to curve 3 of Fig. 2.3 while the stress strain curve for steel are corresponding to curves 2 of Figs. 2.4, 2.5 and 2.6. Fig. 3.2(*b*) shows the strain diagram while Fig. 3.2(*c*) shows the stress block *for design* as recommended by IS : 456 – 2000. The maximum stress

$$DC = \frac{k_3 f_{ck}}{\gamma_{mc}} = \frac{0.67 f_{ck}}{1.5} = 0.446 f_{ck}. \simeq 0.45 f_{ck}.$$

Factor $k_2 = 0.416 \simeq 0.42$ while depth of neutral axis is x_u . The strain ε_{cu} in concrete = 0.0035.

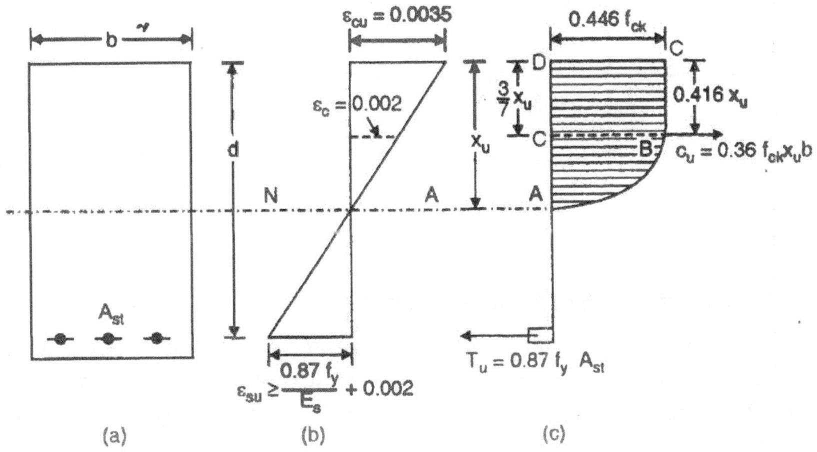

FIG. 3.2. STRAIN AND STRESS DISTRIBUTION FOR DESIGN. (IS : 456 - 2000)

The maximum stress in steel is limited to $\frac{f_y}{\gamma_{ms}} = \frac{f_y}{1.15} \simeq 0.87 f_y$

Area of design stress block (c_u) is given by

$$c_u = \text{area } ABCD = k_1 \times x_u \times DC = k_1 \times x_u \times \frac{k_3 f_{ck}}{\gamma_{mc}}$$

Substituting $k_1 = 0.8095$, $k_3 = 0.67$ and $\gamma_{mc} = 1.5$,

we get $$c_u = 0.3616 f_{ck} . x_u \simeq 0.36 f_{ck} \, x_u \qquad \text{...[3.5(a)]}$$

Total compressive force for a rectangular beam of width b is given by

$$C_u = 0.36 f_{ck} . x_u . b \qquad \text{...(3.5)}$$

The maximum strain ε_{cu} in concrete is taken equal to 0.0035.
The Code requires that at failure the strain in reinforcement (ε_{su}) should not be less than, $\frac{0.87 f_y}{E_s} + 0.002$, thus ensuring a stress of $\frac{f_y}{1.15}$ ($\simeq 0.87 f_y$). This limits the depth x_u of N.A. to its maximum of limiting value. The depth x_u of the N.A. is given by

$$x_u = \frac{\varepsilon_{cu}}{\varepsilon_{cu} + \varepsilon_{su}} d \qquad \text{...(3.6)}$$

Substituting minimum value of $\varepsilon_{su} = 0.002 + \frac{0.87 f_y}{E_s}$, we get,

$$\frac{x_{u\,(max.)}}{d} = \frac{0.0035}{0.0055 + \frac{0.87 f_y}{E_s}} = \frac{0.0035 E_s}{0.0055 E_s + 0.87 f_y} \qquad \text{...(3.7 a)}$$

Substituting $E_s = 2 \times 10^5$ N/mm^2,

we get $$\frac{x_{u\,(max)}}{d} = \frac{700}{1100 + 0.87 f_y} \qquad \text{...[3.7]}$$

The limiting values of the depth of the neutral axis, for different grades of steel, based on the above expression are as follows (Table 3.1)

TABLE 3.1 LIMITING VALUES OF DEPTH OF N.A.

f_y	250	415	500
ε_{su}	0.0031	0.0038	0.0042
$\frac{x_{u\,(max)}}{d}$	0.53 (0.531)	0.48 (0.479)	0.46 (0.456)

Note: The values of $x_{u\,max}/d$ in the bracket are the exact values.

3.5. MODES OF FAILURE : TYPES OF SECTIONS

Based on extensive test results, *maximum strain criteria* has been accepted as the failure criteria for a reinforced concrete element in flexure. A reinforced concrete member is considered to have failed when the strain ε_c in extreme compression fibre reaches its ultimate value of ε_{cu} (= 0.0035). At this stage, the actual strain (ε_s) in steel can have the following values :

(*i*) Equal to failure strain ε_{su}, corresponding to *balanced section*

(*ii*) More than failure strain ε_{su}, corresponding to *under-reinforced section*

(*iii*) Less than failure strain ε_{su}, corresponding to *over-reinforced section*

Thus, based on percentage of steel reinforcement in the tension zone of the section, we may have *balanced failure, under-reinforced failure* and *over reinforced failure*, corresponding to the balanced, under-reinforced and over-reinforced sections respectively. It should be clearly noted that the Code specifies that the maximum strain in the tension reinforcement in the section at failure shall not be less than $\left[\frac{f_y}{1.15\,E_s} + 0.002\right]$.

For a singly reinforced rectangular beam, having area of reinforcement A_{st}, we have

From Eq. 3.5, $C_u = 0.36\, f_{ck}\, x_u\, b$

Tensile force in reinforcement, $T_u = 0.87\, f_y\, A_{st}$

For equilibrium of forces, we have $C_u = T_u$

$\therefore \quad 0.36\, f_{ck}\, x_u\, .\, b = 0.87\, f_y\, A_{st}$

or
$$\frac{x_u}{d} = \frac{0.87\, f_y\, A_{st}}{0.36\, f_{ck}\, bd} = 2.417\, \frac{f_y}{f_{ck}} \cdot \frac{A_{st}}{bd} = 2.417\, p_t\, \frac{f_y}{f_{ck}} \quad \ldots(3.8)$$

where $p_t = A_{st}/bd =$ reinforcement ratio.

Thus, for given section, the actual value of x_u/d (i.e. the position of N.A.) can be determined from Eq. 3.8. Three cases may arise:

Case 1 : x_u/d equal to the limiting value, $x_{u\,,\,max}/d$: Balanced section

Case 2 : x_u/d less the limiting value : under-reinforced section

Case 3 : x_u/d more than limiting value : over-reinforced section

All the above three conditions / positions of N.A. are depicted in Fig. 3.3.

1. Balanced section (Fig. 3.3 *a*) : In the balanced section, the strain in steel and strain in concrete reach their maximum values simultaneously. Thus, in the balanced section

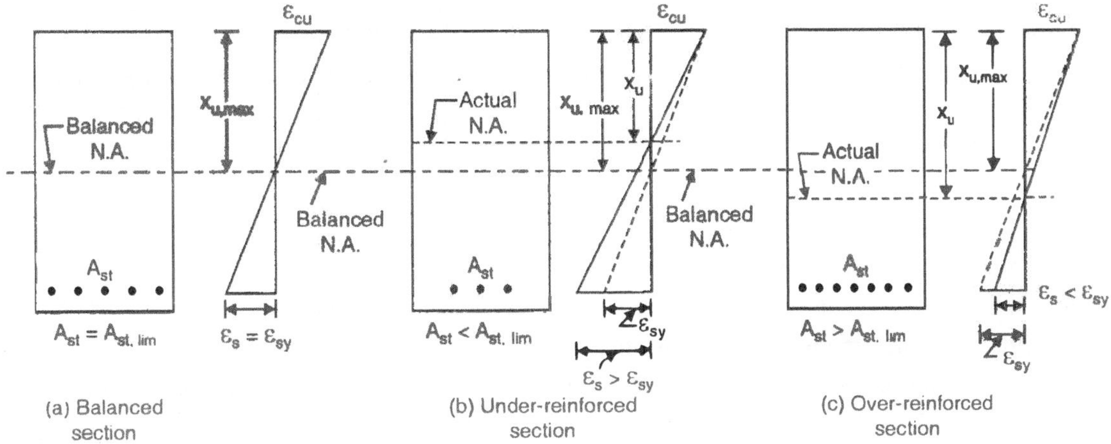

FIG. 3.3. TYPES OF SECTIONS

$\varepsilon_c = \varepsilon_{cu}$ and $\varepsilon_s = \varepsilon_{sy}$ (or ε_{su}). The percentage of steel in this section is known as *critical* or *limiting steel percentage* ($p_{t,\,lim.}$) The depth of N.A. $x_u = x_{u,\,max}$.

2. Under reinforced section (Fig. 3.3 *b*) : An under reinforced section is the one in which steel percentage (p_t) is less than critical or limiting percentage ($p_{t,\,lim}$). Due to this, the actual N.A. is above the balanced N.A. and $x_u < x_{u,\,max}$.

3. Over reinforced section (Fig. 3.3 *c*) : In an over-reinforced section, the steel percentage (p_t) is more than limiting percentage ($p_{t,\,lim}$), due to which N.A. falls below the balanced N.A. and $x_u > x_{u,\,max}$. Because of higher percentage of steel, yield does not take place in steel and failure occurs when the strain in extreme fibres in concrete reaches its ultimate value. Thus, the failure takes place due to crushing of concrete while strain in steel still remains below the yield strain ε_{su}. It should be noted that compression failure is *sudden*, and therefore not desirable. The Code recommends that if x_u/d is found to be greater than the limiting value, *the beam should be redesigned.*

3.6. BALANCED SECTION

(a) Limiting reinforcement percentage ($p_{t,\,lim.}$)

A balanced section has the tensile reinforcement ratio (p_t) equal to the limiting ratio ($p_{t,\,lim}$), the value the which can be found by equating the forces of tension and compression.

Thus $\quad p_{t,\,lim}\, bd\,(0.87\, f_y) = 0.36\, f_{ck}\, b\, x_{u,\,max}$

or
$$p_{t,\,lim} = \frac{0.36}{0.87}\,\frac{f_{ck}}{f_y}\cdot\frac{x_{u,\,max}}{d} = 0.4138\,\frac{f_{ck}}{f_y}\cdot\frac{x_{u,\,max}}{d} \simeq 0.414\,\frac{f_{ck}}{f_y}\cdot\frac{x_{u,\,max}}{d} \quad \ldots(3.9)$$

The values of tensile reinforcement percent ($100\,p_{t,\,lim}$) for various values of f_{ck} and f_y are given in Table 3.2.

TABLE 3.2. VALUES OF $p_{t,\,lim}$ (%)

Grade of concrete	f_{ck} (N/mm²)	Values of 100 $p_{t,\,lim}$ for		
		f_y = 250 N/mm²	f_y = 415 N/mm²	f_y = 500 N/mm²
M 15	15	1.319	0.716	0.566
M 20	20	1.758	0.955	0.754
M 25	25	2.198	1.194	0.943
M 30	30	2.638	1.433	1.132
M 35	35	3.077	1.672	1.320
M 40	40	3.517	1.910	1.509

Note: According to ACI and Australian Codes, normal structures should have $p_{t,\,max} \not> 0.75\, p_{t,\,lim}$

(b) Limiting moment of resistance

In a balanced section, the steel reinforcement reaches its yield stress at the same instant when the ultimate strain is reached in the concrete. The stress block is fully developed as shown in Fig. 3.2 (c). The moment of resistance M_u reaches its limiting value $M_{u,\,lim}$ which can be determined by taking the moment of compressive force about the centre of tensile reinforcement.

Now $C_u = (b \,.\, x_{u,\,max}) \times (0.36 f_{ck}) = 0.36 f_{ck}\, b \,.\, x_{u,\,max}$; Lever arm $= d - 0.416\, x_{u,\,max}$

$$\therefore \quad M_{u,\,lim} = 0.36 f_{ck}\, b \,.\, x_{u,\,max}\,(d - 0.416\, x_{u,\,max}) \qquad \ldots(3.10\ a)$$

or
$$M_{u,\,lim} = 0.36 f_{ck} \frac{x_{u,\,max}}{d}\left(1 - 0.416 \frac{x_{u,\,max}}{d}\right) bd^2 \qquad \ldots(3.10)$$

or
$$\frac{M_{u,\,lim}}{bd^2} = R_{u,\,lim} = 0.36 f_{ck} \,.\, \frac{x_{u,\,max}}{d}\left(1 - 0.416 \frac{x_{u,\,max}}{d}\right) \qquad \ldots(3.11)$$

In the above expressions, the values of $x_{u,\,max}/d$ for mild steel, Fe 415 steel and Fe 500 steel are 0.531, 0.479 and 0.456 respectively (Table 3.1). Substituting these values, we get the following simplified expressions for the three types of steel.

(*i*) ***For mild steel (Fe 250)*** : $x_{u,\,max}/d = 0.531$

$$M_{u,\,lim} = 0.1489 f_{ck}\, bd^2 = R_{u\,lim}\, bd^2 \quad (\text{where } R_{u,\,lim} = 0.1489 f_{ck}) \qquad \ldots(3.11\ a)$$

(*ii*) ***For Fe 415 Steel*** : $x_{u,\,max}/d = 0.479$

$$M_{u,\,lim} = 0.1381 f_{ck}\, bd^2 = R_{u,\,lim}\,.\, bd^2 \quad (\text{where } R_{u,\,lim} = 0.1381 f_{ck}) \qquad \ldots(3.11\ b)$$

(*iii*) ***For Fe 500 steel*** : $x_{u,\,max}/d = 0.456$

$$M_{u,\,lim} = 01330 f_{ck}\, bd^2 = R_{u,\,lim}\, bd^2 \quad \text{where } R_{u,\,lim} = 0.1330 f_{ck}) \qquad \ldots(3.11\ c)$$

Table 3.3 gives the values of $R_{u,\,lim}$ for various grades of concrete.

TABLE 3.3. VALUES OF $R_{u,\,lim}$ FOR BALANCED SECTION

Concrete Grade	Reinforcement		
	Fe 250 ($R_{u,\,lim}$ = 0.1489 f_{ck})	Fe 415 ($R_{u,\,lim}$ = 0.1381 f_{ck})	Fe 500 ($R_{u,\,lim}$ = 0.1330 f_{ck})
M 15	2.233	2.071	1.995
M 20	2.978	2.761	2.660
M 25	3.722	3.452	3.325
M 30	4.467	4.142	3.990
M 35	5.211	4.833	4.655
M 40	5.956	5.523	5.320

The readers are advised to compare these values of $R_{u\,.\,lim}$ with 1.5 R values of working stress method (see Appendix-A). It will be seen that $R_{u\,.\,lim}$ is greater than 1.5 R, resulting in smaller concrete section. *It will be seen from Table 3.2 for $p_{t\,.\,lim.}$ that the balanced design gives smallest concrete section and maximum area of reinforcement.* Since the cost of steel is quite high, an *under-reinforced section*, which gives larger concrete area and smaller value of p_t, is always desirable in the limit state design. This can be achieved by (*i*) providing b or d or both greater than those required by balanced design, or (*ii*) using lesser values of x_u/d than the limiting value, or (*iii*) assuming lesser value of $R_{u\,.\,lim}$ than the limiting value given in Table 3.3.

3.7. UNDER REINFORCED SECTION

An under-reinforced section is the one in which the steel percentage is less than the limiting percentage ($p_{t\,.\,lim.}$) given by Eq. 3.9 or Table 3.2. For such a section, x_u will be less than $x_{u\,.\,max}$, and the actual N.A. will shifted up. The strain in steel at the limit state of collapse will be more than $\left(\frac{0.87 f_y}{E_s} + 0.002\right)$, and the stress in steel will fully reach to its maximum value of 0.87 f_y. The moment of resistance for such a case is found on the basis of maximum stress in steel.

Thus, M_u = Tensile force × lever arm

or
$$M_u = 0.87 f_y \,.\, A_{st}\,(d - 0.416\, x_u) = 0.87 f_y A_{st} \,.\, d\left(1 - 0.416\,\frac{x_u}{d}\right) \qquad \text{...(3.12 } a)$$

Substituting the value of $\frac{x_u}{d}$ from Eq. 3.8, we get

$$M_u = 0.87 f_y A_{st}\, d\left[1 - 0.416\,\frac{0.87}{0.36} \times \frac{f_y}{f_{ck}}\frac{A_{st}}{bd}\right] = 0.87 f_y A_{st}\, d\left[1 - 1.005\,\frac{f_y}{f_{ck}}\frac{A_{st}}{bd}\right] \qquad \text{...(3.12 } b)$$

or
$$M_u \simeq 0.87 f_y A_{st}\, d\left[1 - \frac{f_y}{f_{ck}}\frac{A_{st}}{bd}\right] \qquad \text{...(3.12)}$$

or
$$\frac{M_u}{bd^2} = 0.87 f_y\, p_t\left[1 - \frac{f_y}{f_{ck}}\, p_t\right] \qquad \text{...(3.12 } c)$$

Note : Though the value of $0.416 \times \frac{0.87}{0.36}$ works out to be 1.005, the actual value is $0.416 \times \frac{0.87}{0.3616} = 1.0009 \simeq 1$. Hence Eq. 3.12 can be adopted with sufficient accuracy.

Thus, for a given value of p_t and for given set of values of characteristic strength of concrete (i.e. f_{ck}) and steel (i.e. f_y), the factor M_u/bd^2 can be obtained.

In actual practice, a designer is interested to know the value of p_t for the design moment M_u and for a given section dimensions b and d, i.e. for a given value of M_u/bd^2. The values of p_t can be obtained for different values of M_u/bd^2 for a particular mix of concrete and a particular steel. The values of M_u/bd^2 for three concrete mixes (M 20, M 25 and M 30) and for three types of steel (Fe 250, Fe 415 and Fe 500) are given in Tables 3.4, 3.5 and 3.6 repectively.

TABLE 3.4. FLEXURE : REINFORCEMENT PERCENTAGE, p_t FOR SINGLY REINFORCED SECTIONS
(f_{ck} = 20 N/mm²)

M_u/bd^2 N/mm²	f_y (N/mm²) 250	415	500	M_u/bd^2 N/mm²	f_y (N/mm²) 250	415	500
0.30	0.140	0.085	0.070	*2.22*	1.203	0.725	0.602
0.35	0.164	0.099	0.082	*2.24*	1.216	0.733	0.608
0.40	0.188	0.114	0.094	*2.26*	1.230	0.741	0.615
0.45	0.213	0.128	0.106	*2.28*	1.243	0.749	0.621
0.50	0.237	0.143	0.119	*2.30*	1.256	0.757	0.628
0.55	0.262	0.158	0.131	*2.32*	1.270	0.765	0.635
0.60	0.286	0.172	0.143	*2.34*	1.283	0.773	0.642
0.65	0.311	0.187	0.156	*2.36*	1.297	0.781	0.648
0.70	0.336	0.203	0.168	*2.38*	1.311	0.790	0.655
0.75	0.361	0.218	0.181	*2.40*	1.324	0.798	0.662
0.80	0.387	0.233	0.193	*2.42*	1.338	0.806	0.669
0.85	0.412	0.248	0.206	*2.44*	1.352	0.814	0.676
0.90	0.438	0.264	0.219	*2.46*	1.366	0.823	0.683
0.95	0.464	0.280	0.232	*2.48*	1.380	0.831	0.690
1.00	0.490	0.295	0.245	*2.50*	1.394	0.840	0.697
1.05	0.517	0.311	0.258	*2.52*	1.408	0.848	0.704
1.10	0.543	0.327	0.272	*2.54*	1.423	0.857	0.711
1.15	0.570	0.343	0.285	*2.56*	1.437	0.866	0.719
1.20	0.597	0.359	0.298	*2.58*	1.451	0.874	0.726
1.25	0.624	0.376	0.312	*2.60*	1.466	0.883	0.733
1.30	0.651	0.392	0.326	*2.62*	1.481	0.892	0.740
1.35	0.679	0.409	0.339	*2.64*	1.495	0.901	0.748
1.40	0.707	0.426	0.353	*2.66*	1.510	0.910	0.755
1.45	0.735	0.443	0.367	*2.68*	1.525	0.919	
1.50	0.763	0.460	0.382	*2.70*	1.540	0.928	
1.55	0.792	0.477	0.396	*2.72*	1.555	0.937	
1.60	0.821	0.494	0.410	*2.74*	1.570	0.946	
1.65	0.850	0.512	0.425	*2.76*	1.585	0.955	
1.70	0.879	0.530	0.440	*2.78*	1.601		
1.75	0.909	0.547	0.454	*2.80*	1.616		
1.80	0.939	0.565	0.469	*2.82*	1.632		
1.85	0.969	0.584	0.484	*2.84*	1.647		
1.90	1.000	0.602	0.500	*2.86*	1.663		
1.95	1.030	0.621	0.515	*2.88*	1.679		
2.00	1.062	0.640	0.531	*2.90*	1.695		
2.02	1.074	0.647	0.537	*2.92*	1.711		
2.04	1.087	0.655	0.543	*2.94*	1.727		
2.06	1.099	0.662	0.550	*2.96*	1.743		
2.08	1.112	0.670	0.556	*2.98*	1.760		
2.10	1.125	0.678	0.562				
2.12	1.138	0.685	0.569				
2.14	1.151	0.693	0.575				
2.16	1.164	0.701	0.582				
2.18	1.177	0.709	0.588				
2.20	1.190	0.717	0.595				

TABLE 3.5. FLEXURE : REINFORCEMENT PERCENTAGE, p_t FOR SINGLY REINFORCED SECTIONS
(f_{ck} = 25 N/mm²)

M_u/bd^2 N/mm²	f_y (N/mm²)			M_u/bd^2 N/mm²	f_y (N/mm²)		
	250	415	500		250	415	500
0.30	0.140	0.084	0.070	**2.55**	1.358	0.818	0,679
0.35	0.164	0.099	0.082	**2.60**	1.390	0.837	0.695
0.40	0.188	0.113	0.094	**2.65**	1.422	0.857	0.711
0.45	0.211	0.127	0.106	**2.70**	1.455	0.876	0.727
0.50	0.236	0.142	0.118	**2.75**	1.487	0.896	0.744
0.55	0.260	0.156	0.130	**2.80**	1.520	0.916	0.760
0.60	0.284	0.171	0.142	**2.85**	1.554	0.936	0.777
0.65	0.309	0.186	0.154	**2.90**	1.587	0.956	0.794
0.70	0.333	0.201	0.167	**2.95**	1.621	0.977	0.811
0.75	0.358	0.216	0.179	**3.00**	1.655	0.997	0.828
0.80	0.383	0.231	0.191	**3.05**	1.690	1.018	0.845
0.85	0.408	0.246	0.204	**3.10**	1.725	1.039	0.863
0.90	0.433	0.261	0.216	**3.15**	1.760	1.061	0.880
0.95	0.458	0.276	0.229	**3.20**	1.796	1.082	0,898
1.00	0.483	0.291	0.242	**3.25**	1.832	1.104	0.916
1.05	0.509	0.307	0.255	**3.30**	1.869	1.126	0.935
1.10	0.535	0.322	0.267	**3.32**	1.884	1.135	0.942
1.15	0.561	0.338	0.280	**3.34**	1.899	1.144	
1.20	0.587	0.353	0.293	**3.36**	1.914	1.153	
1.25	0.613	0.369	0.306	**3.38**	1.929	1.162	
1.30	0.639	0.385	0.320	**3.40**	1.944	1.171	
1.35	0.666	0.401	0.333	**3.42**	1.959	1.180	
1.40	0.692	0.417	0.346	**3.44**	1.974	1.189	
1.45	0.719	0.433	0.359	**3.46**	1.989		
1.50	0.746	0.449	0.373	**3.48**	2.005		
1.55	0.773	0.466	0.387	**3.50**	2.020		
1.60	0.800	0.482	0.400	**3.52**	2.036		
1.65	0.828	0.499	0.414	**3.54**	2.051		
1.70	0.856	0.515	0.428	**3.56**	2.067		
1.75	0.883	0.532	0.442	**3.58**	2.083		
1.80	0.911	0.549	0.456	**3.60**	2.099		
1.85	0.940	0.566	0.470	**3.62**	2.115		
1.90	0.968	0.583	0.484	**3.64**	2.131		
1.95	0.997	0.601	0.498	**3.66**	2.147		
2.00	1.026	0.618	0.513	**3.68**	2.163		
2.05	1.055	0.635	0.527	**3.70**	2.179		
2.10	1.084	0.653	0.542	**3.72**	2.196		
2.15	1.114	0.671	0.557	**3.74**			
2.20	1.143	0.689	0.572				
2.25	1.173	0.707	0.587				
2.30	1.204	0.725	0.602				
2.35	1.234	0.743	0.617				
2.40	1.265	0.762	0.632				
2.45	1.296	0.781	0.648				
2.50	1.327	0.799	0.663				

TABLE 3.6. FLEXURE : REINFORCEMENT PERCENTAGE, p_t FOR SINGLY REINFORCED SECTIONS

(f_{ck} = 30 N/mm²)

M_u/bd^2 N/mm²	f_y (N/mm²)			M_u/bd^2 N/mm²	f_y (N/mm²)		
	250	415	500		250	415	500
0.30	0.140	0.084	0.070	2.55	1.319	0.794	0.659
0.35	0.163	0.098	0.082	2.60	1.348	0.812	0.674
0.40	0.187	0.113	0.093	2.65	1.378	0.830	0.689
0.45	0.211	0.127	0.105	2.70	1.408	0.848	0.704
0.50	0.235	0.141	0.117	2.75	1.438	0.866	0.719
0.55	0.259	0.156	0.129	2.80	1.469	0.885	0.734
0.60	0.283	0.170	0.141	2.85	1.499	0.903	0.750
0.65	0.307	0.185	0.153	2.90	1.530	0.922	0.765
0.70	0.331	0.200	0.166	2.95	1.561	0.940	0.781
0.75	0.356	0.214	0.178	3.00	1.592	0.959	0.796
0.80	0.380	0.229	0.190	3.05	1.624	0.978	0.812
0.85	0.405	0.244	0.202	3.10	1.656	0.997	0.828
0.90	0.429	0.259	0.215	3.15	1.687	1.017	0.844
0.95	0.454	0.274	0.227	3.20	1.720	1.036	0.860
1.00	0.479	0.289	0.240	3.25	1.752	1.055	0.876
1.05	0.504	0.304	0.252	3.30	1.785	1.075	0.892
1.10	0.529	0.319	0.265	3.35	1.818	1.095	0.909
1.15	0.555	0.334	0.277	3.40	1.851	1.115	0.925
1.20	0.580	0.350	0.290	3.45	1.884	1.135	0.942
1.25	0.606	0.365	0.303	3.50	1.918	1.156	0.959
1.30	0.631	0.380	0.316	3.55	1.952	1.176	0.976
1.35	0.657	0.396	0.329	3.60	1.986	1.197	0.993
1.40	0.683	0.411	0.342	3.65	2.021	1.218	1.011
1.45	0.709	0.427	0.355	3.70	2.056	1.239	1.028
1.50	0.735	0.443	0.368	3.75	2.091	1.260	1.046
1.55	0.762	0.459	0.381	3.80	2.127	1.281	1.063
1.60	0.788	0.475	0.394	3.85	2.163	1.303	1.081
1.65	0.815	0.491	0.407	3.90	2.199	1.325	1.099
1.70	0.841	0.507	0.421	3.95	2.236	1.347	1.118
1.75	0.868	0.523	0.434	4.00	2.273	1.369	
1.80	0.895	0.539	0.448	4.05	2.310	1.391	
1.85	0.922	0.556	0.461	4.10	2.348	1.414	
1.90	0.950	0.572	0.475	4.15	2.386		
1.95	0.977	0.589	0.488	4.20	2.424		
2.00	1.005	0.605	0.502	4.25	2.463		
2.05	1.032	0.622	0.516	4.30	2.532		
2.10	1.060	0.639	0.530	4.35	2.542		
2.15	1.088	0.656	0.544	4.40	2.583		
2.20	1.116	0.673	0.558	4.45	2.623		
2.25	1.145	0.690	0.572				
2.30	1.173	0.707	0.587				
2.35	1.202	0.724	0.601				
2.40	1.231	0.742	0.605				
2.45	1.260	0.759	0.630				
2.50	1.289	0.777	0.645				

3.8. TYPES OF PROBLEMS

There may be three types of problems in singly reinforced concrete beams :

Type I : Determination of moment of resistance of a given section.

Type II : Determination of actual stresses in the given section subjected to a given B.M.

Type III : Design of a given section to resist a given ultimate design moment.

3.9. PROCEDURE FOR FINDING MOMENT OF RESISTANCE

The moment of resistance of a singly reinforced rectangular section is determined as follows :

(*a*) Determine the depth of N.A. from the following equation :

$$\frac{x_u}{d} = \frac{0.87 f_y A_{st}}{0.36 f_{ck}\, bd} = 2.417 \frac{f_y}{f_{ck}} \cdot \frac{A_{st}}{bd} \qquad \ldots(3.8)$$

(*b*) If the value of x_u/d is less than the limiting (or max.) value given by Eq. 3.7, calculate

$$M_u \simeq 0.87 f_y A_{st}\, d \left(1 - \frac{A_{st}}{bd} \cdot \frac{f_y}{f_{ck}}\right) \qquad \ldots(3.12)$$

(*c*) If the value of x_u/d is equal to the limiting value, the moment of resistance of the section is given by the following expression :

$$M_{u\,.\,lim.} = 0.36 f_{ck} \frac{x_{u\,,\,max}}{d}\left(1 - 0.416 \frac{x_{u\,,\,max}}{d}\right) bd^2 \qquad \ldots(3.10)$$

(*d*) If x_u/d is greater than the limiting value, the section is *over-reinforced*. The moment of resistance of such a section is limited to $M_{u\,,\,lim}$ given by Eq. 3.10. It should be noted that in case of over-reinforced section, not only that there is wastage of excess steel, the concrete reaches its ultimate capacity before steel yields resulting in *sudden* failure. *The Code recommends that such a section should be redesigned.*

Variation of moment of resistance with p_t : The variation of moment of resistance M_u with reinforcement ratio p_t is given by Eq. 3.12 (*c*).

$$\frac{M_u}{bd^2} = 0.87 f_y p_t \left[1 - \frac{f_y}{f_{ck}} p_t\right] \qquad \ldots(3.12\ c)$$

Thus, M_u increases with the increase in the value of p_t. When p_t reaches limiting value $p_{t\,,\,lim}$ given by Eq. 3.9, M_u also reaches the limiting or maximum value $M_{u\,,\,lim}$ given by Eq. 3.10, rearranged as follows :

$$\frac{M_{u\,.\,lim.}}{bd^2} = R_{u\,,\,lim.} = 0.36 f_{ck} \frac{x_{u\,,\,max}}{d}\left(1 - 0.416 \frac{x_{u\,,\,max}}{d}\right) \qquad \ldots(3.11)$$

For Fe 415 steel having $f_y = 415$ N/mm^2, we have $\frac{x_{u\,,\,max}}{d} = 0.479$.

Hence $\quad p_{t\,,\,lim.} = \frac{0.414}{f_y} \cdot \frac{x_{u\,.\,max}}{d} \cdot f_{ck} = \frac{0.414}{415} \times 0.479 f_{ck} = 4.778 \times 10^{-4} f_{ck}$

For M 20 concrete $\quad p_{t\,.\,lim} = 4.778 \times 10^{-4} \times 20 = 9.56 \times 10^{-3}$

Also, $\quad \frac{M_{u\,,\,lim.}}{bd^2} = 0.36 \times 20 \times 0.479\,(1 - 0.416 \times 0.479) = 2.761 = R_{u\,.\,lim}$

For other values of p_t, M_u/bd^2 can be computed from Eq. 3.12 (c).

Thus,
$$\frac{M_u}{bd^2} = 0.87 f_y p_t \left(1 - \frac{f_y}{f_{ck}} p_t\right)$$

$$= 0.87 \times 415\, p_t \left(1 - \frac{415}{20} p_t\right)$$

or
$$\frac{M_u}{bd^2} = 361.05\, p_t (1 - 20.75\, p_t)$$

The values of M_u/bd^2 can thus be computed for different values of p_t, ranging from $p_t = 0$ to $p_t = p_{t.lim} = 9.56 \times 10^{-3}$. Fig. 3.4 shows the variation of M_u/bd^2 with p_t for Fe 415 steel and M 20 concrete.

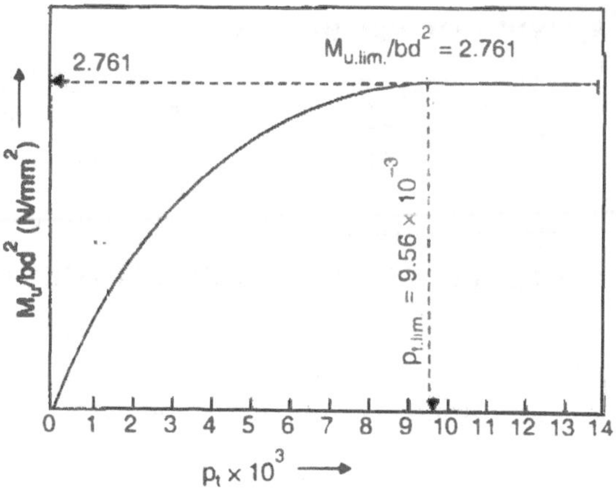

FIG. 3.4. VARIATION OF M_u/bd^2 WITH p_t (M 20 CONCRETE : Fe 415 STEEL)

3.10. DESIGN OF RECTANGULAR BEAM SECTION

The design of a section consists of determination of (*i*) cross-sectional dimensions b and d, and (*ii*) area of steel, so as to develop a given moment of resistance. Though the objective of a designer would be to design a *balanced section* so that the ultimate stresses in both the materials are developed simultaneously, such a design may not be the most economical. It has been stated in the previous para that a balanced design gives smallest concrete section and maximum area of reinforcement. Since the cost of steel is very high in comparison to that of concrete, a balanced design may not be economical. Also, for practical considerations, sometimes, it may be necessary to fix some uniform cross-sectional dimensions. In such a case, the design may result in a singly reinforced balanced section, under-reinforced section or doubly reinforced section. *However, if the section has to be singly reinforced, an under-reinforced section is always more desirable in the limit state design.*

1. Design to determine cross-sectional dimensions and reinforcement. This is the most usual case of design in which the ultimate moment of resistance (M_u) to be developed by the section is given, and it is required to determine b, d and A_{st}. The design is done in the following steps.

Step 1. Determine the limiting depth of N.A. by Eq. 3.7

$$x_{u(max)} = \frac{0.0035}{0.0055 + 0.87 f_y/E_s} d \qquad \ldots[3.7\ (a)]$$

or
$$\frac{x_{u(max)}}{d} = \frac{700}{1100 + 0.87 f_y} \qquad \ldots[3.7\ (b)]$$

Step 2. Choose some suitable *ratio* of d and b. The value of d/b in the range of 1.5 to 3 is usually taken.

Step 3. Find d from the relation (Eq. 3.11).

$$M_u = 0.36 f_{ck} \frac{x_{u\,(max)}}{d}\left(1 - 0.416 \frac{x_{u\,(max)}}{d}\right) bd^2 = R_{u,\,lim}\, bd^2 \quad ...(3.10)$$

Step 4. Knowing b and d, determine area of reinforcement from the relation

$$M_u = 0.87 f_y A_{st} (d - 0.416\, x_{u\,(max)}) \quad ...[3.12(a)]$$

Alternatively, find $p_{t,\,lim.}$ from Eq. 3.9 : $p_{t,\,lim.} = 0.414 \frac{f_{ck}}{f_y} \frac{x_{u\,(max)}}{d}$ ...(3.9)

and $A_{st} = p_{t,\,lim} \times bd$. The values of $p_{t,\,lim.}$ are given in Table 3.2.

2. Design to determine area of reinforcement if *b* and *d* are known. In this case M_u, b and d are given. To start with, we have to ascertain whether it will result in a singly reinforced section or a doubly reinforced section. The design is done in the following steps:

Step 1. Determine the limiting moment of resistance from Eq. 3.10

$$M_{u\,lim} = 0.36 f_{ck} \frac{x_{u\,(max)}}{d}\left(1 - 0.416 \frac{x_{u\,(max)}}{d}\right) bd^2 = R_{u,\,lim}\, bd^2$$

Step 2. (*a*) If $M_u = M_{u\,lim}$, it will result in a balanced section, and area of steel (or $p_{t,\,lim}$) is found either from Eq. 3.9 or from Table 3.2.

(*b*) If $M_u > M_{u\,lim}$, it will result in a doubly reinforced section (see chapter 4).

(*c*) If $M_u < M_{u\,lim}$, it will result in an *under-reinforced* section, and the design is done in the following steps :

Step 3. If $M_u < M_{u\,lim}$, determine actual x_u from relationship :

$$M_u = 0.36 f_{ck}\, b\, x_u (d - 0.416\, x_u)$$

This will result in a quadratic equation in terms of x_u the solution for which is

$$x_u = 1.2\left[1 - \sqrt{1 - \frac{4.62\, M_u}{f_{ck}\, bd^2}}\right] d \quad ...(3.13\ a)$$

Step 4. Determine area of steel from the relation

$$M_u = 0.87 f_y A_{st} (d - 0.416\, x_u)$$

or

$$A_{st} = \frac{M_u}{0.87 f_y (d - 0.416\, x_u)} \quad ...(3.12)$$

Alternatively, find p_t from Eq. 3.8 rearranged as follows :

$$p_t = 0.414 \frac{f_{ck}}{f_y} \frac{x_u}{d} \quad ...[3.8\ (a)]$$

$\therefore$ $A_{st} = p_t\, bd$

Alternatively, steps 3 and 4 can be avoided and A_{st} can be found from Eq. 3.12 applicable for under-reinforced section.

$$M_u = 0.87 f_y A_{st}\, d\left(1 - \frac{A_{st}}{bd} \frac{f_y}{f_{ck}}\right) \quad ...(3.12)$$

This gives a quadratic equation in terms of A_{st}, the solution of which works out as under :

$$A_{st} = \frac{0.5 f_{ck}}{f_y}\left[1 - \sqrt{1 - \frac{4.6\, M_u}{f_{ck}\, bd^2}}\right] bd \quad ...(3.13)$$

3.11. SLAB SPANNING IN ONE DIRECTION : ONE WAY SLAB

Fig. 3.5(*a*) shows a room of size $l \times L$, in which size L is much greater than l. The roof of the room consists of a reinforced concrete slab. When uniformly distributed load is applied to the slab, the bending generally takes place about two axes, parallel to directions of L and l. However, if the ratio L/l is greater than 1.5, 84% of the total load is carried by the shorter span (l) and 16% by the longer span (L). Hence if the ratio L/l is the greater than 1.5, it is usual to consider the bending of the slab to take place in the span l, as shown in Fig. 3.5(*b*). Such a slab is, therefore, known as a *slab spanning in one direction*. If however, L/l is less than 1.5, bending takes place in both the directions. The analysis of slabs spanning in both the directions have been given in Chapter 12.

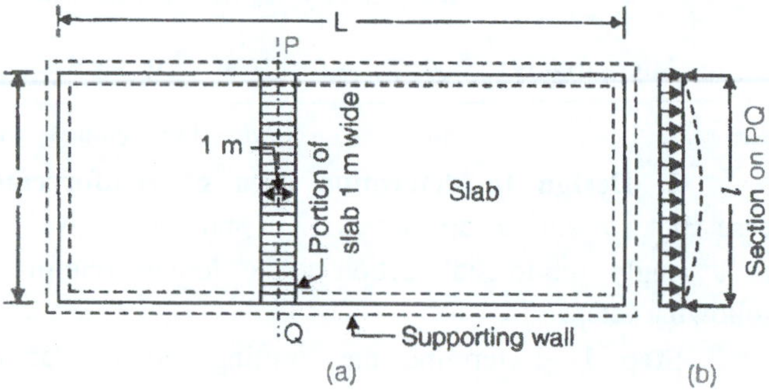

FIG. 3.5.

The analysis of the slab spanning in one direction is done by assuming it to be a beam of 1 m width, as shown in Fig. 3.5(*a*) though the slab width is continuous and is not composed of individual beams of 1 m width. The reinforcement etc. are calculated for 1 m width and the bars are distributed accordingly. The formulae for determination of N.A., moment of resistance and steel reinforcement are the same as found for singly reinforced beam, except that b is taken to be equal to 1000 mm.

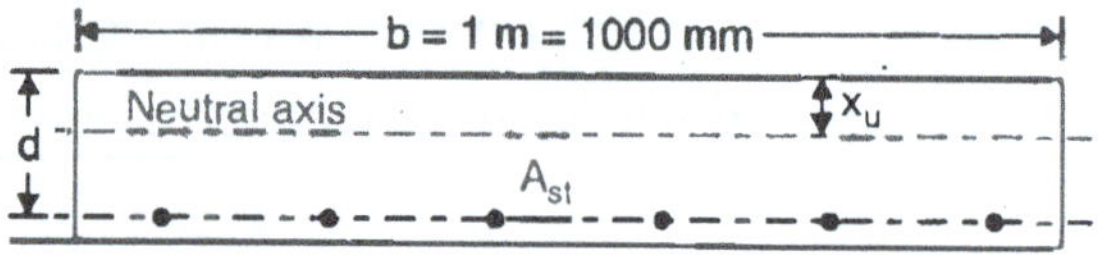

FIG. 3.6.

Fig. 3.6 shows a slab having b = 1 m = 1000 mm and an effective depth d. For a balanced design

$$d = \sqrt{\frac{M_u}{1000\, R_u}} \quad \text{where } M_u \text{ is in N-mm and } R_u = R_{u,\,lim} \qquad \ldots(3.14)$$

The reinforcement for an under-reinforced section is given by

$$A_{st} = \frac{0.5 f_{ck}}{f_y}\left[1 - \sqrt{1 - \frac{4.6\, M_u}{f_{ck}\,(1000\, d^2)}}\right] 1000\, d \qquad \ldots(3.15)$$

Thus the area of steel per 1000 mm width of the slab is known

Let φ = diameter of the bar chosen (mm)

A_φ = area of one bar (mm^2) ; s = spacing of bars (mm). c/c

Then No. bars $= \dfrac{1000}{s}$ and total area $= \dfrac{1000}{s} \cdot A_\varphi$

Equating this to the total area (A_{st}) required, $A_{st} = \dfrac{1000}{s} A_\varphi$

or
$$s = \frac{1000\, A_\varphi}{A_{st}} \qquad \ldots(3.16)$$

The complete procedure for the design of slabs spanning in one direction is given in Chapter 11. See § 11.4 and Example 11.5 for illustration.

Example 3.1. M.R. of singly reinforced section : *Find the M.R. of a singly reinforced concrete beam of 200 mm width and 400 mm effective depth, reinforced with 4 bars of 16 mm dia. of Fe 415 steel. Take M 20 concrete. Use I.S. Code method. Redesign the beam if necessary.*

Solution : $A_{st} = 4\,\frac{\pi}{4}\,(16)^2 = 804.25 \text{ mm}^2$: $p_t = \frac{A_{st}}{bd} = \frac{804.25}{200 \times 400} = 0.0101$

For Fe 415 steel bars, $f_y = 415 \text{ N/mm}^2$. For M 20 concrete, $f_{ck} = 20 \text{ N/mm}^2$.

Hence from Eq. 3.8, $\frac{x_u}{d} = 2.417\, p_t \frac{f_y}{f_{ck}} = 2.417 \times 0.0101 \times \frac{415}{20} = 0.504$

The limiting value of x_u/d is given by Eq. 3.7

$$\frac{x_{u,\,max}}{d} = \frac{700}{1100 + 0.87 \times 415} = 0.479$$

Thus, the actual N.A. depth is more than the limiting one. Such a beam is over-reinforced and hence is *undesirable*, since the failure is sudden and without warning. The Code recommends that such a beam should be *redesigned*. The limiting moment of resistance for such a beam is found from Eq. 3.10 :

$$M_{u,\,lim.} = 0.36\, f_{ck} \frac{x_{u,\,max}}{d}\left(1 - 0.416\,\frac{x_{u,\,max}}{d}\right) bd^2$$

$$= 0.36 \times 20 \times 0.479\,(1 - 0.416 \times 0.479)\,200\,(400)^2 = 88.37 \times 10^6 \text{ N-mm} = \mathbf{88.37 \text{ kN-m}}$$

Redesign : $\frac{x_{u,\,max}}{d} = 2.417\, p_t \frac{f_y}{f_{ck}}$

or $p_t = \left(\frac{x_{u,\,max}}{d}\right)\frac{f_{ck}}{f_y} \times \frac{1}{2.417} = \frac{0.479 \times 20}{415 \times 2.417} = 9.551 \times 10^{-3}$

$\therefore$ $A_{st} = p_t\, bd = 9.551 \times 10^{-3} \times 200\,(400) = \mathbf{764.1 \text{ mm}^2}$

(as against 804.25 mm^2 at present).

Example 3.2. M.R. of singly reinforced section. *For the beam of Example 3.1, find the M.R. if the reinforcement consists of 3-16 mm dia. bars of Fe 415 steel.*

Solution: $A_{st} = 3 \times \frac{\pi}{4}\,(16)^2 = 603.19 \text{ mm}^2$; $p_t = \frac{603.19}{200 \times 400} = 7.54 \times 10^{-3}$

$\therefore$ $\frac{x_u}{d} = 2.417\, p_t \frac{f_y}{f_{ck}} = 2.417 \times 7.54 \times 10^{-3} \times \frac{415}{20} = 0.378;$

$\frac{x_{u\,max}}{d} = 0.479$, as before.

Hence the beam is *under-reinforced*. The M.R. is given by Eq. 3.12 :

$$M_u = 0.87\, f_y\, A_{st}\, d\left(1 - \frac{f_y}{f_{ck}}\frac{A_{st}}{bd}\right) = 0.87 \times 415 \times 603.19 \times 400\left(1 - \frac{415}{20} \times \frac{603.19}{200 \times 400}\right)$$

$= 73.48 \times 10^6$ N-mm $=$ **73.48 kN-m**

(This is less than $M_{u,\,lim.} = 88.37$ kN-m found in the previous example)

Example 3.3. Determination of actual stresses. *For the beam section of example 3.1, determine the actual stresses when the section is subjected to the limiting moment of resistance.*

Solution : Given : $A_{st} = 804.25 \text{ mm}^2$ and $M_u = M_{u,lim.} = 88.37 \times 10^6$ N-mm

As found earlier, $\frac{x_{u,max}}{d} = 0.479$ and $\frac{x_u}{d} = 0.504$

It is very clear that if x_u/d is taken $= 0.504$, the stress in concrete will exceed $0.45 f_{ck}$. However, at $M_u = M_{u,lim}$, x_u will be equal to $x_{u,max}$ and the stress in concrete will be equal to $0.45 f_{ck} = 0.45 \times 20 = 9 \text{ N/mm}^2$. In such a circumstance, the stress f_s in steel will be less than f_y. From equilibrium of forces, $T = C$

$$\therefore \quad A_{st} \cdot f_s = 0.36 f_{ck} \cdot b \cdot x_u$$

$$\therefore \quad f_s = \frac{0.36 f_{ck} \cdot b \cdot x_u}{A_{st}}, \text{ where } x_u = x_{u,max}.$$

or
$$f_s = \frac{0.36 f_{ck} bd (x_{u,max}/d)}{A_{st}} = \frac{0.36 \times 20 \times 200 \times 400 (0.479)}{804.25}$$

$$= \mathbf{343.1 \ N/mm^2} \text{ (against a value of } f_y = 415 \text{ N/mm}^2)$$

The corresponding strain in steel is

$$\varepsilon_s = \frac{0.0035 (d - x_{u\,max})}{x_{u,max}} = \frac{0.0035 (1 - x_{u,max}/d)}{x_{u,max}/d} = \frac{0.0035 (1 - 0.479)}{0.479} = 0.0038$$

Check : Lever arm $= d - 0.416 x_{u,max} = d\left(1 - 0.416 \frac{x_{u,max}}{d}\right)$

$$= 400 (1 - 0.416 \times 0.479) = 320.29 \text{ mm}$$

$$\therefore \quad M_u = C \times \text{lever arm} = (0.36 f_{ck}\, b\, x_{u\,max}) \times 320.29$$

$$= 0.36 \times 20 \times 200 (0.479 \times 400) \times 320.29 = 88.37 \times 10^6 \text{ N-mm}$$

Also,
$$M_u = T \times \text{lever arm} = A_{st} f_s \text{ (Lever arm)}$$

$$= 804.25 \times 343.1 \times 320.29 = 88.37 \times 10^6 \text{ N-mm}$$

Each one of these is equal to $M_{u,lim.}$

Example 3.4. Design of a singly reinforced beam *Design a balanced singly reinforced concrete beam section for an applied moment of 60 kN-m. The width of the beam is limited to 175 mm. Use M 20 concrete and Fe 415 steel bars.*

Solution : $M_D = 60$ kN-m $= 60 \times 10^6$ N-mm

Since $\gamma_f = 1.5$ for both dead and live loads, $M_{uD} = 1.5 \times 60 \times 10^6 = 90 \times 10^6$ N-mm

For a balanced design, we have from Eq. 3.7

$$\frac{x_{u,max}}{d} = \frac{700}{1100 + 0.87 f_y} = \frac{700}{1100 + 0.87 \times 415} = 0.479$$

The moment of resistance of a balanced beam section is given by

$$M_{u,lim} = 0.36 f_{ck} \cdot \frac{x_{u,max}}{d}\left(1 - 0.416 \times \frac{x_{u,max}}{d}\right) bd^2$$

Equating $M_{u,lim}$ and M_{uD}, we get $90 \times 10^6 = 0.36 \times 20 \times 0.479 (1 - 0.416 \times 0.479)\, 175\, d^2$

From which, we get $d = \mathbf{432}$ **mm**

The reinforcement is given by Eq. 3.8 : $\frac{x_{u,max}}{d} = 2.417\, p_{t,lim} \cdot \frac{f_y}{f_{ck}}$

$$\therefore \quad p_{t,lim} = \frac{x_{u,max}}{d} \cdot \frac{f_{ck}}{f_y} \times \frac{1}{2.417} = \frac{0.479 \times 20}{415 \times 2.417} = 9.55 \times 10^{-3}$$

(This is the same as given in Table 3.2 for $f_{ck} = 20$ and $f_y = 415$.

$$\therefore \quad A_{st} = 9.55 \times 10^{-3} \times 175 \times 432 = \mathbf{722\ mm^2}$$

Example 3.5. Design of singly reinforced beam: *Find the reinforcement for the beam of Example 3.4, if the effective depth of the beam is kept equal to 500 mm.*

Solution : Since the depth provided by the beam is more, if will be under-reinforced beam. The limiting ultimate moment of resistance of this beam is given by Eq. 3.10.

$$M_{u,lim} = 0.36\, f_{ck} \frac{x_{u,max}}{d}\left(1 - 0.416 \frac{x_{u,max}}{d}\right) bd^2 = 0.36 \times 20 \times 0.479\,(1 - 0.416 \times 0.479)\, bd^2$$

$$= 2.761\, bd^2 \text{ (same as the value given in Table 3.3 for } f_{ck} = 20 \text{ and } f_y = 415)$$

$$= 2.761 \times 175\,(500)^2 = 120.8 \times 10^6 \text{ N-mm}$$

$M_{uD} = 90 \times 10^6$ N-mm only, and the reinforcement has to be found for this value.

From Eq. 3.12, $\quad M_u = M_{uD} = 0.87\, f_y \,.\, A_{st} \,.\, d\left(1 - \frac{f_y}{f_{ck}} \frac{A_{st}}{bd}\right)$

or $$90 \times 10^6 = 0.87 \times 415\, A_{st} \times 500\left(1 - \frac{415}{20} \times \frac{A_{st}}{175 \times 500}\right)$$

or $\quad A_{st}\,(1 - 2.371 \times 10^{-4} A_{st}) = 498.54 \quad$ or $\quad A_{st}^2 - 4217\, A_{st} + 2102302 = 0$

From which $\quad A_{st} \simeq \mathbf{578\ mm^2}$

Alternatively from Eq. 3.13,

$$A_{st} = \frac{0.5\, f_{ck}}{f_y}\left[1 - \sqrt{1 - \frac{4.6\, M_u}{f_{ck}\, bd^2}}\right] bd = \frac{0.5 \times 20}{415}\left[1 - \sqrt{1 - \frac{4.6 \times 90 \times 10^6}{20 \times 175\,(500)^2}}\right] 175 \times 500$$

$= \mathbf{578\ mm^2}$ against 722 mm^2 found in the previous example.

Example 3.6. Design of singly reinforced beam *A beam, simply supported over an effective span of 7 m carries a live load of 20 kN/m. Design the beam, using M 20 concrete and HYSD bars of grade Fe 415. Keep the width equal to half the effective depth. Assume unit weight of concrete as 25 kN/m^3.*

Solution : Self weight = 0.3 × 0.6 × 1 × 25 = 4.5 kN/m ; Live load = 20 kN/m

∴ Total load $\quad w = 4.5 + 20 = 24.5$ kN/m

∴ Ultimate design load, $\quad w_{uD} = 1.5 \times 24.5 = 36.75$ kN/m

$$\therefore \quad M_{uD} = \frac{36.75\,(7)^2}{8} = 225.09 \text{ kN-m} = 225.09 \times 10^6 \text{ N-mm}$$

For M 20 concrete and Fe 415 steel, $f_{ck} = 20$ N/mm^2 and $f_y = 415$ N/mm^2, respectively.

$$\therefore \quad \frac{x_{u,max}}{d} = \frac{700}{1100 + 0.87\, f_y} = \frac{700}{1100 + 0.87 \times 415} = 0.479$$

Let d = effective depth, so that $b = 0.5\,d$. Also, from Eq. 3.10,

$$M_{u,\,lim} = 0.36 \times 20 \times 0.479\,(1 - 0.416 \times 0.479) \times 0.5\,d \times d^2 = 1.381\,d^3$$

Equating this to M_{uD}, we get $1.381\,d^3 = 225.09 \times 10^6$

From which we get $d = 546$ mm and $b = d/2 = 273$ mm

From Eq. 3.9, $$p_{t,\,lim} = 0.414\,\frac{f_{ck}}{f_y} \cdot \frac{x_{u,\,max}}{d} = 0.414 \times \frac{20}{415} \times 0.479 = 9.55 \times 10^{-3}$$

$$\therefore \quad A_{st} = 9.55 \times 10^{-3} \times 273 \times 546 = \mathbf{1424\ mm^2}$$

Example 3.7. Design of singly reinforced beam : *A reinforced concrete beam has width equal to 300 mm and total depth equal to 700 mm, with a cover of 40 mm to the centre of the reinforcement. Design the beam if it is subjected to a total bending moment of 150 kN-m. Use M 20 concrete and HYSD bars of grade 415.*

Solution : $b = 300$ mm; $d = 700 - 40 = 660$ mm

Taking $\gamma_f = 1.5$, $M_{uD} = 150 \times 1.5 = 225$ kN-m $= 225 \times 10^6$ N-mm

$$\frac{x_{u,\,max}}{d} = \frac{700}{1100 + 0.87\,f_y} = \frac{700}{1100 + 0.87 \times 415} = 0.479$$

$$M_{u,\,lim} = 0.36 \times 20 \times 0.479\,(1 - 0.416 \times 0.479)\,bd^2$$

$$= 2.761\,bd^2 = 2.761 \times 300\,(660)^2 = 360.8 \times 10^6 > M_{uD}$$

Hence compression reinforcement is not required. Also, the beam is under-reinforced.

Now, $$M_{uD} = 0.87\,f_y\,A_{st}\,d\left(1 - \frac{f_y}{f_{ck}}\frac{A_{st}}{bd}\right)$$

$$\therefore \quad 225 \times 10^6 = 0.87 \times 415\,A_{st}\,(660)\left[1 - \frac{415}{20} \times \frac{A_{st}}{300 \times 660}\right]$$

or $A_{st}\left[1 - 1.048 \times 10^{-4}\,A_{st}\right] = 944.2$ or $A_{st}^2 - 9542\,A_{st} + 9.0097 \times 10^6 = 0$

From which $A_{st} = \mathbf{1062.5\ mm^2}$. Alternatively, from Eq. 3.13, we get

$$A_{st} = \frac{0.5\,f_{ck}}{f_y}\left[1 - \sqrt{1 - \frac{4.6\,M_u}{f_{ck}\,bd^2}}\right]bd = \frac{0.5 \times 20}{415}\left[1 - \sqrt{1 - \frac{4.6 \times 225 \times 10^6}{20 \times 300\,(660)^2}}\right]300 \times 660$$

$$\simeq \mathbf{1063\ mm^2}$$

PROBLEMS

1. A beam section 300 mm wide and 500 mm deep is reinforced with a tension reinforcement of 3000 mm^2 at an effective cover of 30 mm. Determine the ultimate moment of resistance of beam section. Use M 15 concrete and Fe 250 grade steel.

2. Solve problem 1 if the grade of concrete is M 20 and grade of steel is Fe 415.

3. Solve problem 1 for M 15 – Fe 415 combination.

4. Design a balanced singly reinforced concrete beam with a span of 5 m to carry a dead load of 25 kN/m and a working live load of 20 kN/m. Use M 20 concrete and Fe 415 grade steel.

5. Redesign the beam of problem 2 if the cross-section of the beam is limited to 300 mm width and 600 mm effective depth. Use M 20 mix and Fe 415 grade steel.

4

Limit State of Collapse in Flexure : II Doubly Reinforced Rectangular Sections

4.1. NECESSITY

We have seen in the previous chapter that a singly reinforced section has a limiting value of moment of resistance, corresponding to limiting value of steel reinforcement. However, if the applied moment M_{ua} is larger than $M_{u\,lim}$, two alternatives will be available: (*i*) to increase the depth of the section or (*ii*) to provide compression reinforcement. In many cases, the maximum value of depth of the section may be limited or restricted from architectural or other considerations. In that case, the only alternative that will be left will be to provide reinforcement in the compression zone, giving rise to a *doubly reinforced section*.

A doubly reinforced section of a beam is, therefore, provided in the following circumstances.

1. When there are architectural restrictions (such as head room requirements, appearances etc.) on the depth of otherwise singly reinforced section.

2. Restriction in the depth at the location of beam at plinth level, along with the provision of ventilator between the ground level and the bottom of plinth beam.

3. In a continuous beam floor system, where the beam acts as a T-beam in the midspan and acts as a rectangular beam at the supports where the B.M. may be much greater than at the mid span.

4. Where it is required to increase the stiffness of the beam.

6. It is found that the compression steel increases the rotation capacity and ductility. Structures with high ductility respond better to seismic forces.

4.2. STRESS BLOCK AND LOCATION OF N.A.

Fig. 4.1(*a*) shows a doubly reinforced section having compression reinforcement at a depth of d' below the outermost compression fibre. Fig. 4.1(*b*) shows the strain diagram while Fig. 4.1(*c*) the stress block. Let,

A_{st} = Total reinforcement at tension face : A_{sc} = Reinforcement in compression side.

x_u = Depth of N.A. : C_{cu} = Compressive force in concrete $= 0.36\, f_{ck}\, x_u\, b$

C_{su} = Compressive force in compression steel $= f_{sc}\, A_{sc}$

f_{sc} = *Design stress* in compression reinforcement read off from the stress strain curve corresponding to the strain ε_{sc} in compression reinforcement.

ε_{sc} = Strain in compression reinforcement $= \dfrac{0.0035\,(x_u - d')}{x_u} = 0.0035\left(1 - \dfrac{d'}{x_u}\right)$ ...(4.1)

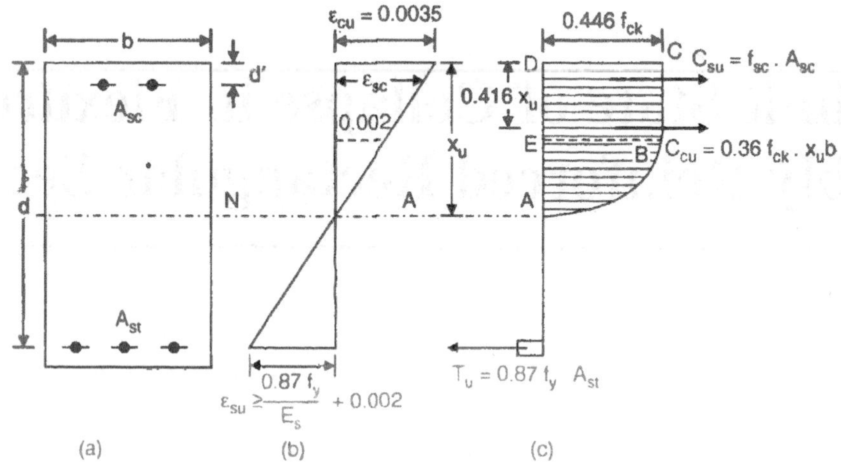

FIG. 4.1. DOUBLY REINFORCED SECTION.

Total compressive force is given by

$$C_u = C_{cu} + C_{su} = 0.36\, f_{ck}\, x_u\, b + f_{sc}\, A_{sc} \quad ...(4.2)$$

(neglecting the loss of concrete area occupied by compressive steel).

Total tensile force is given by

$$T_u = 0.87\, f_y\, A_{st} \quad ...(4.3)$$

In order to locate the N.A., equate the total compressive force to the total tensile force:

$$\therefore \quad 0.36\, f_{ck}\, x_u\, b + f_{sc}\, A_{sc} = 0.87\, f_y\, A_{st} \quad ...(4.4)$$

From the above relation, x_u can be found. However, for the solution of the above equation, an iterative procedure will have to be adopted, since f_{sc} depends upon ε_{sc}, which in turn depends upon x_u. If the loss of compressive area, occupied by the compressive steel is taken into account, Eqs. 4.2 and 4.4 are modified as under :

$$C_u = 0.36\, f_{ck}\, x_u\, b + f_{sc}\, A_{sc} - 0.446\, f_{ck}\, A_{sc} = 0.36\, f_{ck}\, x_u\, b + (f_{sc} - 0.446\, f_{ck})\, A_{sc} \quad ...[4.2(a)]$$

and

$$0.36\, f_{ck}\, x_u\, b + (f_{sc} - 0.446\, f_{ck})\, A_{sc} = 0.87\, f_y\, A_{st} \quad ...[4.4(a)]$$

Note: Normally, the term $0.446\, f_{ck}\, A_{sc}$ is very small, and can be neglected without causing any appreciable error.

Iterative procedure for computation of x_u

Step 1. Compute the depth of N.A. of a balanced section given by strain compatibility [Fig. 4.1(*b*)] :

$$x_{u,max} = \frac{0.0035}{0.0055 + 0.87\, f_y / E_s} = \frac{700}{1100 + 0.87\, f_y} \quad ...(4.3)$$

(Taking $E_s = 2 \times 10^5\ \text{N/mm}^2$)

It is to be noted that this expression is the same as Eq. 3.7 used for a singly reinforced section.

Step 2. For the given doubly reinforced section, *assume* x_u equal to $x_{u\,max}$

Step 3. Compute the value of ε_{sc} from Eq. 4.1.

Step 4 For HYSD bars, compute value of f_{sc} from the stress strain curve of steel or Table 2.3, corresponding to this value of ε_{sc} . For mild steel bars, $f_{sc} = E_s \times \varepsilon_{sc} \leq 0.87 f_y$

Step 5. Substitute the value of f_{sc} in Eq. 4.4 or 4.4(*a*) and compute the modified value of x_u .

Step 6. Repeat steps 3 to 5 till convergence for the value of x_u is achieved.

Note. 1. A slightly different form of iterative procedure has been illustrated in Example 4.1.

Note 2 : ***For mild steel bars***, $f_{sc} = E_s \times \varepsilon_{sc} \leq 0.87 f_y$.

Taking $E_s = 2 \times 10^5$ N/mm^2, we get $f_{sc} = 2 \times 10^5 \times 0.0035 \left(1 - \dfrac{d'}{x_u}\right) \leq 0.87 f_y$

or $700\left(1 - \dfrac{d'}{x_u}\right) \leq 0.87 \times 250 \leq 217.5$ or $d' \not> 0.689\, x_u$

In most of the cases, doubly reinforced sections are provided when $M_u > M_{u\,.\,lim}$ and therefore, x_u is normally equal to $x_{u\,.\,lim} = 0.53\, d$ for mild steel bars. Assuming the lowest value of $x_{u\,.\,lim} = 0.3\, d$ corresponding to maximum 30% redistribution of moments,

$$d' \not> 0.689 \times 0.3\, d \not> 0.21\, d.$$

Since in actual practice d′ hardly exceeds 0.2 d, f_{sc} in most of the cases can be taken equal to 0.87 f_y for mild steel bars.

Note 3 For ***HYSD bars***, the value of f_{sc} is obtained corresponding to ε_{sc} using Table 2.3.

Note 4. From Table 2.3, we find that a stress of $0.87 f_y$ ($= 1.0 f_{yD}$) reaches at a strain of 0.00380 for Fe 415 and 0.00417 for Fe 500 steel. Hence the stress in steel never reaches a value of $0.87 f_y$ prior to the crushing of concrete, because concrete crushes/fails at a strain of 0.0035. *In other words, the strength of compression steel is never fully utilised, specially for HYSD bars.*

4.3. TYPES OF PROBLEMS

As in the case of singly reinforced section, there may be three types of problems in doubly reinforced sections :

Type I : ***Analysis problem*** : Determination of moment of resistance of the section, given the material properties (f_{ck} and f_y) and the sectional properties (i.e. b , d , d' , A_{st} and A_{sc})

(*a*) Using mild steel bars (*b*) Using HYSD bars

Type II: ***Analysis problem*** : Determination of actual stresses in the given section, subjected to a given B.M.

Type III : Design of a given section to resist given ultimate design moment :

Given : b , d , d' and M_u

To find (*a*) Both A_{st} and A_{sc} and (*b*) A_{st} if A_{sc} is also given.

4.4. DETERMINATION OF MOMENT OF RESISTANCE

For a given section, the depth of neutral axis, x_u can be determined as illustrated in article 4.2. If this value of x_u is compared to the limiting depth ($x_{u\,max}$) of the N.A. axis, three cases may arise:

Case 1. x_u/d less than limiting value

Case 2. x_u/d equal to the limiting value.

Case 3. x_u/d greater than limiting value.

Case 1. x_u/d less than limiting value

This case corresponds to the under-reinforced section in working stress method. The moment of resistance is given by

$$M_u = 0.36\, f_{ck}\, x_u\, b\, (d - 0.416\, x_u) + A_{sc}\, f_{sc}\, (d - d') \qquad ...[4.5(a)]$$

or

$$M_u = 0.36\, f_{ck}\, \frac{x_u}{d}\left(1 - 0.416\,\frac{x_u}{d}\right) bd^2 + A_{sc}\, f_{sc}\, (d - d') \qquad ...(4.5)$$

The centroid $\bar{x}$ of the compressive forces, from the top fibre is given by

$$\bar{x} = \frac{0.36\, f_{ck}\, x_u\, b \times 0.416\, x_u + f_{sc}\, A_{sc}\, d'}{0.36\, f_{ck}\, x_u\, b + f_{sc}\, A_{sc}} \qquad ...[4.5(b)]$$

$$\therefore \quad \text{Lever arm } a = d - \bar{x} = d - \frac{0.36\, f_{ck}\, x_u\, b \times 0.416\, x_u + f_{sc}\, A_{sc}\, d'}{0.36\, f_{ck}\, x_u\, b + f_{sc}\, A_{sc}} \qquad ...(4.6)$$

Hence, with respect to tensile force,

$$M_u = 0.87\, f_y\, A_{st} \times a \qquad ...(4.7)$$

The moment of resistance found by Eqs 4.5 and 4.7 should be equal.

The above equations have been obtained by *neglecting* the loss of concrete occupied by compressive steel. The equations are modified as under if this loss is taken into account:

$$M_u = 0.36\, f_{ck}\, x_u b\, (d - 0.416\, x_u) + (A_{sc}\, f_{sc} - 0.446\, f_{ck}\, A_{sc})\, (d - d') \qquad ...(4.8)$$

where x_u is determined on the basis of Eq. 4.4(*a*)

$$\bar{x} = \frac{0.36\, f_{ck}\, x_u\, b\, (0.416\, x_u) + (f_{sc} - 0.446\, f_{ck})\, A_{sc}\, d'}{0.36\, f_{ck}\, x_u\, b + (f_{sc} - 0.446\, f_{ck})\, A_{sc}} \qquad ...(4.9)$$

$a = d - \bar{x}$ and $M_u = 0.87\, f_y\, A_{st} \times a$

Case 2. x_u/d equal to limiting value. This case corresponds to a *balanced design* in which the steel reinforcement reaches its yield stress at the same instant when ultimate strain is reached in concrete. The moment of resistance M_u reaches its limiting value $M_{u\,lim}$ and is given by

$$M_{u\,lim} = 0.36\, f_{ck}\, x_{u\,lim}\, b\, (d - 0.416\, x_{u\,lim}) + A_{sc}\, f_{sc}\, (d - d') \qquad ...(4.10)$$

where f_{sc} = stress in compressive steel, found from stress strain curve, corresponding to ε_{sc} given by Eq. 4.1 in which $x_u = x_{u.lim}$.

If, however, loss of concrete area occupied by compressive steel in taken into account.

$$M_{u\,lim} = 0.36\, f_{ck}\, x_{u\,lim}\, b\, (d - 0.416\, x_{u\,lim}) + (f_{sc} - 0.446\, f_{ck})\, A_{sc}\, (d - d') \qquad ...[4.10(a)]$$

Case 3. x_u/d greater than the limiting value This corresponds to an over-reinforced section. Failure occurs when strain in extreme fibres in concrete reaches its limiting value.

while strain in tensile steel remain below the yield strain ε_{su}. The moment of resistance of such a section is limited to $M_{u\,lim}$ given by Eq. 4.10. It should be noted that compression failure is *sudden*, and therefore not desirable. The Code recommends that if x_u/d is found to be greater than the limiting value, the *section should be redesigned.*

4.5. DESIGN OF A DOUBLY REINFORCED SECTION

We have seen earlier that a doubly reinforced section is required when the applied moment M_{ua} is larger than $M_{u\,lim}$, given by Eq. 4.10 for a singly reinforced rectangular section. The difference between the applied moment M_{ua} and the limiting moment $M_{u\,lim}$ is carried by additional tensile reinforcement A_{st2} and compression reinforcement A_{sc}. Thus, a doubly reinforced section is equivalent to a singly reinforced balanced section and a section with additional tension and compression reinforcement, as illustrated in Fig. 4.2.

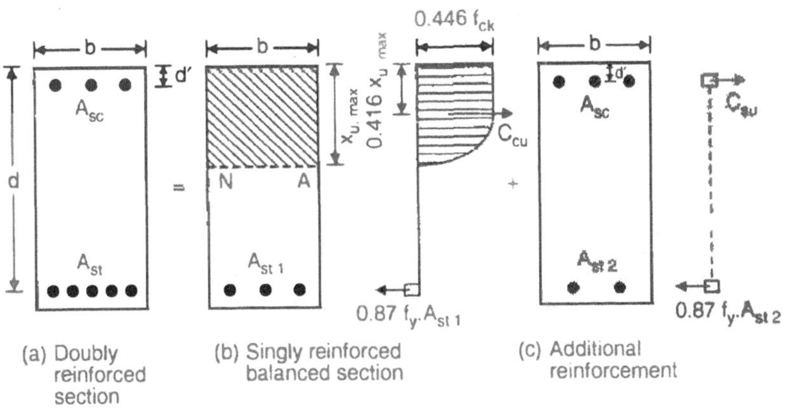

FIG. 4.2. DOUBLY REINFORCED SECTION

Let A_{st1} = area of tensile reinforcement for a singly reinforced section, for $M_{u,lim}$

$= p_{t,lim} \cdot bd.$

A_{st2} = additional tensile reinforcement . given by

$$A_{st2} = \frac{M_{ua} - M_{u.lim.}}{0.87 f_y (d - d')} = \frac{A_{sc} f_{sc}}{0.87 f_y} \qquad \ldots(4.11)$$

The compression reinforcement A_{sc} is given by

$$A_{sc} = \frac{M_{ua} - M_{u.lim.}}{f_{sc} (d - d')} \qquad \ldots(4.12)$$

If however, loss of concrete area, occupied by compressive steel is taken into account, we have

$$A_{sc} = \frac{M_{ua} - M_{u.lim.}}{(f_{sc} - 0.446 f_{ck}) (d - d')} \quad \left(\text{where } d' < \frac{3}{7} x_u\right) \qquad \ldots[4.12(a)]$$

where f_{sc} = *design stress* in compression reinforcement read off from the stress-strain curve corresponding to the strain ε_{sc} in the compression reinforcement.

$$\varepsilon_{sc} = \text{strain in compression reinforcement} = \frac{0.0035\,(x_{u\,max} - d')}{x_{u.max}} \qquad \ldots(4.13)$$

and $x_{u\,max}$ = limiting value of x_u given by Eq. 3.7 or Eq. 4.3

Design of the beam when A_{sc} is given

In many circumstances, specially in case of a continuous beam, some of the bottom bars of mid-span tensile reinforcement are continued at bottom over the intermediate support. This may result in an excess of the requirement as compression reinforcement. The section therefore, no longer remains balanced but becomes *under reinforced* and A_{st1} is no more equal to $A_{st,lim}$. for a singly reinforced section.

Procedure : Given b, d, d', M_{uD} and A_{sc}. It is required to find A_{st}.

Step 1. Assuming f_{sc}, compute $M_{u2} = f_{sc} A_{sc} (d - d')$

$$\therefore \quad M_{u1} = M_u - M_{u2}$$

Step 2. Determine x_u from relation : $M_{u1} = 0.36 f_{ck} b x_u (d - 0.416 x_u)$

This gives rise to a quadratic equation for x_u, the solution for which is

$$x_u = 1.2\left[1 - \sqrt{1 - \frac{4.62 M_{u1}}{f_{ck} bd^2}}\right] d \qquad \ldots(4.14)$$

Step 3. Determine $x_{u,max}$ for given values of f_{ck} and f_y.

Check that $\quad x_u < x_{u,max}$

Also, find $\quad M_{u,lim} = 0.36 f_{ck} (x_{u,max}) b (d - 0.416 x_{u,max})$

Check that $\quad M_1 < M_{u,lim}$.

If both the conditions are satisfied, go to steps 4(*a*) and 4(*b*). If not, go to steps 5.

Step 4 (*a*) Check for f_{sc}, assumed in step 1.

For Fe 250 steel, $\quad f_{sc} = 700\left(1 - \frac{d'}{x_u}\right) \not> 0.87 f_y$

For Fe 415 steel, $\quad \varepsilon_{sc} = 0.0035\left(1 - \frac{d'}{x_u}\right)$ and f_{sc} is obtained corresponding to ε_{sc}

Step 4(b) Compute $\quad A_{st1} = \frac{M_{u1}}{0.87 f_y (d - 0.416 x_u)}$ and $A_{st2} = \frac{f_{sc} A_{sc}}{0.87 f_y}$

Total $A_{st} = A_{st1} + A_{st2}$. See example 4.5 for illustration.

Step 5. If $M_1 > M_{u,lim}$, the original area A_{sc} falls short of the actual area A_{sc} for a doubly reinforced beam for that design moment M_{uD}. In that case, redesign the beam by the process out lined in § 4.5. The additional area A_{sc} required to be provided will be equal to the total A_{sc} required for the doubly reinforced beam minus area A_{sc} already available. See Example 4.6 for illustration.

4.6. DESIGN AIDS FOR DOUBLY REINFORCED BEAM

As discussed earlier, the total moment of resistance of doubly reinforced section consists of the moment of resistance provided by the singly reinforced balanced section plus additional moment of resistance provided by the compression steel and the remaining tensile steel. Thus, with reference to Fig. 4.2 :

$$M_u = M_{u1} + M_{u2}$$

where M_{u1} = ultimate moment of resistance offered by singly reinforced balanced section $= M_{u,lim}$ (Fig. 4.2 *b*)

$M_{u2} = M_u - M_{u1}$ = ultimate moment of resistance offered by section of Fig. 4.2 (*c*) consisting of a couple formed by compression steel, and remaining tensile steel A_{st2} at a lever arm $d - d'$.

Thus, $M_u = M_{u,\,lim.} + M_{u2}$. Dividing both sides by bd^2, we get

$$\frac{M_u}{bd^2} = \frac{M_{u\,.\,lim}}{bd^2} + \frac{M_{u2}}{bd^2} \quad \ldots(4.15)$$

or

$$R_u = R_{u\,.\,lim.} + R_{u2} \quad \ldots(4.15\ a)$$

In design problems, b, d and M_u are known, and hence R_u is known. Also, for given material properties, f_{ck} and f_y are known, from which $R_{u\,,\,lim.}$ can be calculated. Hence

$R_{u2} = R_u - R_{u\,,\,lim.}$, from which R_{u2} is known.

Now $M_{u2} = 0.87\, f_y \,.\, A_{st2}\,(d - d')$

$$\therefore \quad R_{u2} = \frac{M_{u2}}{bd^2} = 0.87\, f_y \frac{A_{st2}}{bd^2}(d - d') = 0.87\, f_y\, p_{t2}\,(1 - d'/d)$$

or

$$p_{t2} = \frac{R_{u2}}{0.87\, f_y\,(1 - d'/d)} \quad \ldots(4.16)$$

Now $A_{st1} = A_{st\,,\,lim.} = \frac{p_{t\,,\,lim}}{bd}$; $A_{st2} = \frac{p_{t2}}{bd}$ and $A_{st} = A_{st1} + A_{st2}$

Also,

$$A_{sc} = \frac{0.87\, f_y\, A_{st2}}{f_{sc} - 0.45\, f_{ck}} \quad \ldots(4.17)$$

$$p_c = A_{sc}/bd \quad \ldots(4.17\ a)$$

The above equations are useful in preparing design aids for doubly reinforced sections. Tables 4.1, 4.2, 4.3 and 4.4 give the design aids for finding p_t and p_c for following combinations ($f_{ck}' = 20$ N/mm^2 ;Fe 250 steel), ($f_{ck} = 20$ N/mm^2; Fe 415 steel), ($f_{ck} = 25$ N/mm^2 ; Fe 250 steel) and ($f_{ck} = 25$ N/mm^2 ; Fe 415 steel). For other combinations of material strength, reference may be made to SP : 16-1980.

Example 4.1. M.R. of doubly reinforced beam

Find the M.R. of an existing beam of M 15 concrete, 200 mm wide and 400 mm effective depth, reinforced with 4 bars(mild steel) of 20 mm dia. for tension and 2 bars of 20 mm dia. for compression. The effective cover to compressive reinforcement is 50 mm.

Solution

Given $f_{ck} = 15$ N/mm^2 ; $f_y = 250$ N/mm^2; $b = 200$ mm ; $d = 400$ mm ; $d' = 50$ mm ;

$$A_{st} = 4\frac{\pi}{4}(20)^2 = 1256.6 \text{ mm}^2 \quad ; \quad A_{sc} = 2\frac{\pi}{4}(20)^2 = 628.3 \text{ mm}^2$$

Let us assume that $d' < \frac{3}{7}x_u$. Then, $T_u = 0.87\, f_y\, A_{st} = 0.87 \times 250 \times 1256.6 = 273310$ N ...(*i*)

$$C_u = 0.36\, f_{ck}\, x_u\, b + f_{sc}\, A_{sc} - 0.446\, f_{ck}\, A_{sc} = 0.36 \times 15 \times 200\, x_u + f_{sc} \times 628.3 - 0.446 \times 15 \times 628.3$$

$$= 1080\, x_u + 628.3\, f_{sc} - 4203 \text{ N} \quad \ldots(ii)$$

where f_{sc} = stress in the compressive reinforcement to be read off from the stress strain curve for strain $\varepsilon_{sc} = \frac{0.0035\,(x_u - d')}{x_u}$

TABLE 4.1. FLEXURE : REINFORCEMENT PERCENTAGES FOR DOUBLY REINFORCED SECTIONS

(f_{ck} = 20 N/mm^2 : f_y = 250 N/mm^2)

M_u/bd^2 N/mm^2	d'/d = 0.05		d'/d = 0.10		d'/d = 0.15		d'/d = 0.20	
	p_t	p_c	p_t	p_c	p_t	p_c	p_t	p_c
2.99	1.765	0.005	1.765	0.005	1.765	0.006	1.766	0.006
3.00	1.769	0.010	1.770	0.011	1.771	0.011	1.771	0.012
3.05	1.794	0.035	1.796	0.037	1.798	0.039	1.80[illegible]	0.042
3.10	1.818	0.061	1.821	0.064	1.825	0.068	1.829	0.072
3.15	1.842	0.086	1.847	0.091	1.852	0.096	1.858	0.102
3.20	1.866	0.111	1.872	0.117	1.879	0.124	1.886	0.132
3.25	1.891	0.136	1.898	0.144	1.906	0.152	1.915	0.162
3.30	1.915	0.162	1.923	0.171	1.933	0.181	1.944	0.192
3.35	1.939	0.187	1.949	0.197	1.960	0.209	1.973	0.222
3.40	1.963	0.212	1.974	0.224	1.987	0.237	2.001	0.252
3.45	1.987	0.237	2.000	0.250	2.014	0.265	2.030	0.282
3.50	2.012	0.263	2.026	0.277	2.041	0.293	2.059	0.312
3.55	2.036	0.288	2.051	0.304	2.068	0.322	2.088	0.342
3.60	2.060	0.313	2.077	0.330	2.095	0.350	2.116	0.372
3.65	2.084	0.338	2.102	0.357	2.122	0.378	2.145	0.402
3.70	2.108	0.364	2.128	0.384	2.149	0.406	2.174	0.432
3.75	2.133	0.389	2.153	0.410	2.177	0.434	2.203	0.462
3.80	2.157	0.414	2.179	0.437	2.204	0.463	2.231	0.492
3.85	2.181	0.439	2.204	0.464	2.231	0.491	2.260	0.522
3.90	2.205	0.464	2.230	0.490	2.258	0.519	2.289	0.552
3.95	2.229	0.490	2.256	0.517	2.285	0.547	2.318	0.582
4.00	2.254	0.515	2.281	0.544	2.312	0.576	2.346	0.612
4.05	2.278	0.540	2.307	0.570	2.339	0.604	2.375	0.642
4.10	2.302	0.565	2.332	0.597	2.366	0.632	2.404	0.672
4.15	2.326	0.591	2.358	0.624	2.393	0.660	2.433	0.701
4.20	2.351	0.616	2.383	0.650	2.420	0.688	2.461	0.731
4.25	2.375	0.641	2.409	0.677	2.447	0.717	2.490	0.761
4.30	2.399	0.666	2.434	0.703	2.474	0.745	2.519	0.791
4.35	2.423	0.692	2.460	0.730	2.501	0.773	2.578	0.821
4.40	2.447	0.717	2.486	0.757	2.528	0.801	2.576	0.851
4.45	2.472	0.742	2.511	0.783	2.555	0.830	2.605	0.881
4.50	2.496	0.767	2.537	0.810	2.582	0.858	2.634	0.911
4.55	2.520	0.793	2.562	0.837	2.609	0.886	2.663	0.941
4.60	2.544	0.818	2.588	0.863	2.637	0.914	2.691	0.971
4.65	2.568	0.843	2.613	0.890	2.664	0.942	2.720	1.001
4.70	2.593	0.868	2.639	0.917	2.691	0.971	2.749	1.031
4.75	2.617	0.894	2.664	0.943	2.718	0.999	2.778	1.061
4.80	2.641	0.919	2.690	0.970	2.745	1.027	2.806	1.091
4.85	2.665	0.944	2.716	0.997	2.772	1.055	2.835	1.121
4.90	2.689	0.969	2.741	1.023	2.799	1.083	2.864	1.151
4.95	2.714	0.995	2.767	1.050	2.826	1.112	2.893	1.181
5.00	2.738	1.020	2.792	1.077	2.853	1.140	2.921	1.211
5.05	2.762	1.045	2.818	1.103	2.880	1.168	2.950	1.241
5.10	2.786	1.070	2.843	1.130	2.907	1.196	2.979	1.271
5.15	2.811	1.096	2.869	1.157	2.934	1.225	3.008	1.301

TABLE 4.2. FLEXURE : REINFORCEMENT PERCENTAGES FOR DOUBLY REINFORCED SECTIONS
(f_{ck} = 20 N/mm^2 ; f_y = 415 N/mm^2)

M_u/bd^2 N/mm^2	d'/d = 0.05		d'/d = 0.10		d'/d = 0.15		d'/d = 0.20	
	p_t	p_c	p_t	p_c	p_t	p_c	p_t	p_c
2.77	0.958	0.002	0.958	0.002	0.959	0.003	0.959	0.003
2.80	0.967	0.011	0.968	0.012	0.968	0.013	0.969	0.015
2.90	0.996	0.042	0.998	0.045	1.001	0.049	1.004	0.054
3.00	1.025	0.072	1.029	0.077	1.034	0.084	1.038	0.093
3.10	1.055	0.103	1.060	0.109	1.066	0.119	1.073	0.132
3.20	1.084	0.133	1.091	0.142	1.099	0.154	1.108	0.171
3.30	1.113	0.164	1.122	0.174	1.131	0.190	1.142	0.210
3.40	1.142	0.194	1.152	0.207	1.164	0.225	1.177	0.249
3.50	1.171	0.224	1.183	0.239	1.197	0.260	1.212	0.288
3.60	1.200	0.255	1.214	0.271	1.229	0.295	1.246	0.327
3.70	1.230	0.285	1.245	0.304	1.262	0.331	1.281	0.366
3.80	1.259	0.316	1.276	0.336	1.294	0.366	1.315	0.405
3.90	1.288	0.346	1.306	0.369	1.327	0.401	1.350	0.444
4.00	1.317	0.376	1.337	0.401	1.360	0.437	1.385	0.483
4.10	1.346	0.407	1.368	0.433	1.392	0.472	1.419	0.522
4.20	1.375	0.437	1.399	0.466	1.425	0.507	1.454	0.561
4.30	1.405	0.468	1.429	0.498	1.457	0.542	1.489	0.600
4.40	1.434	0.498	1.460	0.530	1.490	0.578	1.523	0.640
4.50	1.463	0.528	1.491	0.563	1.523	0.613	1.558	0.679
4.60	1.492	0.559	1.522	0.595	1.555	0.648	1.593	0.718
4.70	1.521	0.589	1.553	0.628	1.588	0.683	1.627	0.757
4.80	1.550	0.620	1.583	0.660	1.620	0.719	1.662	0.796
4.90	1.580	0.650	1.614	0.692	1.653	0.754	1.696	0.835
5.00	1.609	0.680	1.645	0.725	1.686	0.789	1.731	0.874
5.10	1.638	0.711	1.676	0.757	1.718	0.825	1.766	0.913
5.20	1.667	0.741	1.707	0.790	1.751	0.860	1.800	0.952
5.30	1.696	0.772	1.737	0.822	1.783	0.895	1.835	0.991
5.40	1.725	0.802	1.768	0.854	1.816	0.930	1.870	1.030
5.50	1.755	0.832	1.799	0.887	1.849	0.966	1.904	1.069
5.60	1.784	0.863	1.830	0.919	1.881	1.001	1.939	1.108
5.70	1.813	0.893	1.861	0.952	1.914	1.036	1.974	1.147
5.80	1.842	0.924	1.891	0.984	1.946	1.071	2.008	1.186
5.90	1.871	0.954	1.922	1.016	1.979	1.107	2.043	1.225
6.00	1.900	0.985	1.953	1.049	2.012	1.142	2.078	1.264
6.10	1.930	1.015	1.984	1.081	2.044	1.177	2.112	1.303
6.20	1.959	1.045	2.014	1.114	2.077	1.213	2.147	1.342
6.30	1.988	1.076	2.045	1.146	2.109	1.248	2.181	1.381
6.40	2.017	1.106	2.076	1.178	2.142	1.283	2.216	1.421
6.50	2.046	1.137	2.107	1.211	2.175	1.318	2.251	1.460
6.60	2.075	1.167	2.138	1.243	2.207	1.354	2.285	1.499
6.70	2.105	1.197	2.168	1.276	2.240	1.389	2.320	1.538
6.80	2.134	1.228	2.199	1.308	2.272	1.424	2.355	1.577
6.90	2.163	1.258	2.230	1.340	2.305	1.459	2.389	1.616
7.00	2.192	1.289	2.261	1.373	2.338	1.495	2.424	1.655
7.10	2.221	1.319	2.292	1.405	2.370	1.530	2.459	1.694

TABLE 4.3. FLEXURE : REINFORCEMENT PERCENTAGES FOR DOUBLY REINFORCED SECTIONS

(f_{ck} = 25 N/mm^2 ; f_y = 250 N/mm^2)

M_u/bd^2 N/mm^2	d'/d = 0.05		d'/d = 0.10		d'/d = 0.15		d'/d = 0.20	
	p_t	p_c	p_t	p_c	p_t	p_c	p_t	p_c
3.73	2.202	0.002	2.202	0.003	2.202	0.003	2.203	0.003
3.75	2.212	0.013	2.213	0.013	2.213	0.014	2.214	0.015
3.80	2.236	0.038	2.238	0.040	2.240	0.043	2.243	0.045
3.85	2.260	0.064	2.264	0.067	2.267	0.071	2.272	0.076
3.90	2.284	0.089	2.289	0.094	2.294	0.100	2.300	0.106
3.95	2.309	0.115	2.315	0.121	2.322	0.128	2.329	0.136
4.00	2.333	0.140	2.340	0.148	2.349	0.157	2.358	0.167
4.05	2.357	0.166	2.366	0.175	2.379	0.185	2.387	0.197
4.10	2.381	0.191	2.391	0.202	2.403	0.214	2.415	0.227
4.15	2.406	0.217	2.417	0.229	2.430	0.242	2.444	0.258
4.20	2.430	0.242	2.443	0.256	2.457	0.271	2.473	0.288
4.25	2.454	0.268	2.468	0.283	2.484	0.299	2.502	0.318
4.30	2.478	0.293	2.494	0.310	2.511	0.328	2.530	0.348
4.35	2.502	0.319	2.519	0.337	2.538	0.356	2.559	0.379
4.40	2.527	0.344	2.545	0.364	2.565	0.385	2.588	0.409
4.45	2.551	0.370	2.570	0.391	2.592	0.414	2.617	0.439
4.50	2.575	0.395	2.596	0.417	2.619	0.442	2.645	0.470
4.55	2.599	0.421	2.621	0.444	2.646	0.471	2.674	0.500
4.60	2.623	0.447	2.647	0.471	2.673	0.499	2.703	0.530
4.65	2.648	0.472	2.673	0.498	2.700	0.528	2.732	0.561
4.70	2.672	0.498	2.698	0.525	2.727	0.556	2.760	0.591
4.75	2.696	0.523	2.724	0.552	2.754	0.585	2.789	0.621
4.80	2.720	0.549	2.749	0.579	2.782	0.613	2.818	0.651
4.85	2.744	0.574	2.775	0.606	2.809	0.642	2.847	0.682
4.90	2.769	0.600	2.800	0.633	2.836	0.670	2.875	0.712
4.95	2.793	0.625	2.826	0.660	2.863	0.699	2.904	0.742
5.00	2.817	0.651	2.851	0.687	2.890	0.727	2.933	0.773
5.05	2.841	0.676	2.877	0.714	2.917	0.756	2.962	0.803
5.10	2.866	0.702	2.903	0.741	2.944	0.784	2.990	0.833
5.15	2.890	0.727	2.928	0.768	2.971	0.813	3.019	0.864
5.20	2.914	0.753	2.954	0.795	2.998	0.841	3.048	0.894
5.25	2.938	0.778	2.979	0.822	3.025	0.870	3.077	0.924
5.30	2.962	0.804	3.005	0.848	3.052	0.898	3.105	0.955
5.35	2.987	0.829	3.030	0.875	3.079	0.927	3.134	0.985
5.40	3.011	0.855	3.056	0.902	3.106	0.955	3.163	1.015
5.45	3.035	0.880	3.081	0.929	3.133	0.984	3.192	1.045
5.50	3.059	0.906	3.107	0.956	3.160	1.012	3.220	1.076
5.55	3.083	0.931	3.133	0.983	3.187	1.041	3.249	1.106
5.60	3.108	0.957	3.158	1.010	3.214	1.070	3.278	1.136
5.65	3.132	0.982	3.184	1.037	3.242	1.098	3.307	1.167
5.70	3.156	1.008	3.209	1.064	3.269	1.127	3.335	1.197
5.75	3.180	1.033	3.235	1.091	3.296	1.155	3.364	1.227
5.80	3.204	1.059	3.260	1.118	3.323	1.184	3.393	1.258
5.85	3.229	1.085	3.286	1.145	3.350	1.212	3.422	1.288
5.90	3.253	1.110	3.311	1.172	3.377	1.241	3.450	1.318

TABLE 4.4. FLEXURE : REINFORCEMENT PERCENTAGES FOR DOUBLY REINFORCED SECTIONS

(f_{ck} = 25 N/mm^2 ; f_y = 415 N/mm^2)

M_u/bd^2 N/mm^2	d'/d = 0.05		d'/d = 0.10		d'/d = 0.15		d'/d = 0.20	
	p_t	p_c	p_t	p_c	p_t	p_c	p_t	p_c
3.46	1.197	0.002	1.197	0.002	1.197	0.003	1.197	0.003
3.50	1.209	0.014	1.210	0.015	1.210	0.017	1.211	0.019
3.60	1.238	0.045	1.240	0.048	1.243	0.052	1.246	0.058
3.70	1.267	0.076	1.271	0.081	1.276	0.088	1.281	0.097
3.80	1.296	0.106	1.302	0.113	1.308	0.123	1.315	0.137
3.90	1.325	0.137	1.333	0.146	1.341	0.159	1.350	0.176
4.00	1.355	0.167	1.363	0.178	1.373	0.194	1.385	0.215
4.10	1.384	0.198	1.394	0.211	1.406	0.230	1.419	0.254
4.20	1.413	0.229	1.425	0.244	1.439	0.265	1.454	0.294
4.30	1.442	0.259	1.456	0.276	1.471	0.301	1.488	0.333
4.40	1.471	0.290	1.487	0.309	1.504	0.336	1.523	0.372
4.50	1.500	0.320	1.517	0.341	1.536	0.372	1.558	0.412
4.60	1.530	0.351	1.548	0.374	1.569	0.407	1.592	0.451
4.70	1.559	0.382	1.579	0.407	1.602	0.443	1.627	0.490
4.80	1.588	0.412	1.610	0.439	1.634	0.478	1.662	0.530
4.90	1.617	0.443	1.641	0.472	1.667	0.514	1.696	0.569
5.00	1.646	0.474	1.671	0.504	1.699	0.549	1.731	0.608
5.10	1.675	0.504	1.702	0.537	1.732	0.585	1.766	0.648
5.20	1.705	0.535	1.733	0.570	1.765	0.620	1.800	0.687
5.30	1.734	0.565	1.764	0.602	1.797	0.656	1.835	0.726
5.40	1.763	0.596	1.795	0.635	1.830	0.691	1.869	0.766
5.50	1.792	0.627	1.825	0.667	1.862	0.727	1.904	0.805
5.60	1.821	0.657	1.856	0.700	1.895	0.762	1.939	0.844
5.70	1.851	0.688	1.887	0.733	1.928	0.798	1.973	0.884
5.80	1.880	0.718	1.918	0.765	1.960	0.833	2.008	0.923
5.90	1.909	0.749	1.948	0.798	1.993	0.869	2.043	0.962
6.00	1.938	0.780	1.979	0.830	2.025	0.904	2.077	1.002
6.10	1.967	0.810	2.010	0.863	2.058	0.940	2.112	1.041
6.20	1.996	0.841	2.041	0.896	2.091	0.975	2.147	1.080
6.30	2.026	0.871	2.072	0.928	2.123	1.011	2.181	1.120
6.40	2.055	0.902	2.102	0.961	2.156	1.046	2.216	1.159
6.50	2.084	0.933	2.133	0.993	2.188	1.082	2.251	1.198
6.60	2.113	0.963	2.164	1.026	2.221	1.118	2.285	1.238
6.70	2.142	0.994	2.195	1.059	2.254	1.153	2.320	1.277
6.80	2.171	1.024	2.226	1.091	2.286	1.189	2.354	1.316
6.90	2.201	1.055	2.256	1.124	2.319	1.224	2.389	1.356
7.00	2.230	1.086	2.287	1.157	2.351	1.260	2.424	1.395
7.10	2.259	1.116	2.318	1.189	2.384	1.295	2.458	1.434
7.20	2.288	1.147	2.349	1.222	2.417	1.331	2.493	1.474
7.30	2.317	1.177	2.380	1.254	2.449	1.366	2.528	1.513
7.40	2.346	1.208	2.410	1.287	2.482	1.402	2.562	1.552
7.50	2.376	1.239	2.441	1.320	2.514	1.437	2.597	1.591
7.60	2.405	1.269	2.472	1.352	2.547	1.473	2.632	1.631
7.70	2.434	1.300	2.503	1.385	2.580	1.508	2.666	1.670
7.80	2.463	1.330	2.534	1.417	2.612	1.544	2.701	1.709

The depth x_u is best found by trial solution. We know that when $f_y = 250$ N/mm², the value of $\frac{x_{u,max}}{d} = 0.531$, for a singly reinforced section. However, for most of the cases the value of x_u/d is less than this, and for doubly reinforced sections, x_u/d is still much less since the N.A. will be shifted up due to the presence of compressive steel.

As a first trial, let us assume $x_u = 0.4\,d = 0.4 \times 400 = 160$ mm.

$$\therefore \qquad \varepsilon_{sc} = \frac{0.0035\,(160 - 50)}{160} = 0.0024$$

The stress-strain curve for mild steel bars is linear, with the yield point reaching at 0.2 percent strain. Hence $f_{sc} = 0.87\,f_y = 0.87 \times 250 = 217.5$ N/mm². Substituting in (*ii*), we get

$$C_u = (1080 \times 160) + (628.3 \times 217.5) - 4203$$
$$= 172800 + 136655 - 4203 = 309455 - 4203 = 305252 \text{ N}$$

(**Note** : It will be seen that $0.446\,f_{ck}\,A_{sc} = 4203$ N only in comparison to total C_u, and hence can be neglected). We find that $C_u > T_u$. In the second trial, therefore, let us reduce x_u from the 160 mm to 135 mm.

(**Note:** f_{sc} is constant $= 0.87\,f_y$ when strain ε_{sc} is greater than 0.002).

$$\varepsilon_{sc} = \frac{0.0035\,(135 - 50)}{135} = 0.0022$$

$$\therefore \qquad f_{sc} = 0.87\,f_y = 0.87 \times 250 = 217.5$$

$$\therefore \qquad C_u = 1080 \times 135 + 628.3 \times 217.5 - 4203 = 145800 + 136655 - 4203 = 278252 \text{ N}$$

This is quite near to $T_u = 273310$ N. However, adopt $x_u = 130$ mm.

$$\therefore \qquad C_u = 1080 \times 130 + 136655 - 4203 = 140400 + 132452 = (C_{cu}) + (C_{su})$$

$$\therefore \qquad M_u = C_{cu}\,(d - 0.416\,x_u) + C_{su}\,(d - d')$$
$$= 140400\,(400 - 0.416 \times 130) + 132452\,(400 - 50)$$
$$= 48.58 \times 10^6 + 46.53 \times 10^6 = 94.94 \times 10^6 \text{ N-mm} = \mathbf{94.94 \text{ kN-m}} = M_{uD}$$

$\therefore$ Design moment $M_D = 94.94/1.5 = 63.29$ kN-m

Note 1. The moment of resistance of the same beam without compression reinforcement will be found to be 71.4 kN-m only.

Note 2. The moment of resistance of the doubly reinforced beam, by the working stress method will be as under : (see Appendix A)

For M 15 concrete, with mild steel bars, $k = 0.404$; $j = 0.865$; $R = 0.874$.

For the N.A. location, $\frac{bn^2}{2} + (m_c - 1)\,A_{sc}\,(n - d') = m\,A_{st}\,(d - n)$

or $\qquad \frac{200}{2}\,n^2 + (1.5 \times 19 - 1)\,628.3\,(n - 50) = 19 \times 1256.6 \times (400 - n)$

or $\qquad 100\,n^2 + 17278\,n - 863912 = 9550160 - 23875\,n$

or $\qquad n^2 + 411.54\,n - 104141 = 0$. From which $n = 177$ mm

Critical depth $n_c = 0.404 \times 400 = 161.6$ mm. The actual N.A. is below critical ; hence the stress in concrete will reach its maximum value first.

$\therefore \quad c = \sigma_{cbc} = 5 \ \text{N/mm}^2.$

$$c' = c\,\frac{b - d'}{n} = \frac{5}{177}(177 - 50) = 3.59 \ \text{N/mm}^2$$

$$M_r = bn\,\frac{c}{2}\left(d - \frac{n}{3}\right) + (m_c - 1)\,A_{sc}\,.\,c'(d - d')$$

$$= 200 \times 177 \times \frac{5}{2}\left(400 - \frac{177}{3}\right) + (1.5 \times 19 - 1)\,628.3 \times 3.59 \times (400 - 50)$$

$$= 30.179 \times 10^6 + 21.717 \times 10^6 = 51.896 \times 10^6 \ \text{N-mm} \simeq 51.9 \ \text{kN-m}$$

This is less than $M_D = 63.29$ kN-m found from the limit analysis. The ratio of the two moments is 1.23.

Example 4.2. M.R. of doubly reinforced beam. *Determine the moment of resistance of an existing beam having the following data: $b = 350$ mm ; $d = 900$ mm ; $d' = 50$ mm ;*

Tension reinforcement : 5-20 mm HYSD bars (Fe 415) ; compression reinforcement 2-20 mm HYSD bars (Fe 415). Grade of concrete : M 15.

Solution. $A_{st} = 5\,\frac{\pi}{4}(20)^2 = 1570.8 \ \text{mm}^2$; $A_{sc} = 2\,\frac{\pi}{4}(20)^2 = 628.3 \ \text{mm}^2$.

$$T_u = 0.87\,f_y\,A_{st} = 0.87 \times 415 \times 1570.8 = 567120 \ \text{N} \quad \ldots(i)$$

$$C_u = 0.36\,f_{ck}\,x_u\,b + f_{sc}\,A_{sc} - 0.446\,f_{ck}\,A_{sc}$$

$$= 0.36 \times 15 \times 350\,x_u + 628.3\,f_{sc} - 0.446 \times 15 \times 628.3$$

$$= 1890\,x_u + 628.3\,f_{sc} - 4203 \quad \ldots(ii)$$

Let us assume $x_u = 230$ mm. Hence $\frac{3}{7}x_u = \frac{3}{7} \times 230 = 98.6 \ \text{mm} > d'$.

$$\varepsilon_{cu} = \frac{0.0035\,(x_u - d')}{x_u} = \frac{0.0035\,(230 - 50)}{230} = 0.00274.$$

Hence from stress-strain curve (Fig. 2.5 or Table 2.3), we get $f_{sc} = 351 \ \text{N/mm}^2$

$\therefore C_u = 1890 \times 230 + (628.3 \times 351) - 4203 = 434700 + 220533 - 4203 = 651030$ N

This is much more than $T_u = 567120$ N. Hence take $x_u = 190$ mm.

$$\varepsilon_{cu} = \frac{0.0035\,(190 - 50)}{190} = 0.002580. \text{ Hence, } f_{sc} \approx 347 \ \text{N/mm}^2 \ \text{(Table 2.3)}$$

$$\therefore \quad C_u = 1890 \times 190 + (628.3 \times 347) - 4203$$

$$= 359100 + 218020 - 4203 = \underset{(C_{cu})}{359100} + \underset{(C_{su})}{213817} = 572917 \simeq T_u$$

$$\therefore \quad M_u = C_{cu}\,(d - 0.416\,x_u) + C_{su}\,(d - d')$$

$$= 359100\,(900 - 0.416 \times 190) + 213817\,(900 - 50)$$

$$= 294.8 \times 10^6 + 181.7 \times 10^6 = 476.5 \times 10^6 \ \text{N-mm} = \mathbf{476.5 \ kN\text{-}m}.$$

Example 4.3. Design of beam : Fe 415 steel :

A concrete beam has 350 mm breadth and 700 mm effective depth. Design the beam if it is subjected to a super-imposed bending moment of 400 kN-m. Use HYSD bars of Fe 415 grade and concrete of M 20 grade. Take $d' = 50$ mm.

Solution : Using $\gamma_f = 1.5$, $M_{uD} = 400 \times 1.5 = 600$ kN-m $= 600 \times 10^6$ N-mm.

For Fe 415 steel, $\dfrac{x_{u,max}}{d} = \dfrac{700}{1100 + 0.87 \times 415} = 0.479$

$\therefore \quad x_{u,max} = 0.479 \times 700 = 335.3$ mm

From Eq. 3.9, $\quad p_{t,lim.} = 0.414 \times \dfrac{20}{415} \times 0.479 = 0.00956 = 0.956\,\%$

From Eq. 3.10, $M_{u,lim.} = 0.36 f_{ck} \dfrac{x_{u,max}}{d}\left(1 - 0.416 \dfrac{x_{u,max}}{d}\right) bd^2$

$$= 0.36 \times 20 \times 0.479\,(1 - 0.416 \times 0.479)\,350\,(700)^2 = 473.6 \times 10^6 \text{ N-mm}$$

$$A_{st1} = p_{t,lim.}\, bd = 0.00956 \times 350 \times 700 = 2342.2 \text{ mm}^2$$

$$A_{st2} = \frac{M_{uD} - M_{u,lim.}}{0.87 f_y (d - d')} = \frac{(600 - 473.6)\,10^6}{0.87 \times 415\,(700 - 50)} = 538.6 \text{ mm}^2$$

Total $\quad A_{st} = A_{st1} + A_{st2} = 2342.2 + 538.6 = \mathbf{2880.8\ mm^2}$

$$A_{sc} = \frac{M_{uD} - M_{u,lim}}{f_{sc}(d - d')} = \frac{(600 - 473.6)\,10^6}{f_{sc}\,(700 - 50)} = \frac{0.1945 \times 10^6}{f_{sc}}$$

$$\varepsilon_{sc} = \frac{0.0035\,(x_{u\,max} - d')}{x_{u,max}} = \frac{0.0035\,(335.3 - 50)}{335.3} = 0.00298$$

Hence from Table 2.3, for Fe 415 bars, $f_{sc} \simeq 353.7$ N/mm² for $\varepsilon_{sc} = 0.00298$

$$\therefore \quad A_{sc} = \frac{0.1945 \times 10^6}{353.7} = \mathbf{549.9\ mm^2}$$

If the loss of concrete area, occupied by compressive steel is taken into account,

$$A_{sc} = \frac{M_{uD} - M_{u,lim.}}{(f_{sc} - 0.446 f_{ck})(d - d')} = \frac{(600 - 473.6)\,10^6}{(353.7 - 0.446 \times 20)(700 - 50)} = \mathbf{564\ mm^2}$$

Values from design tables

$$\frac{M_u}{bd^2} = \frac{M_{uD}}{bd^2} = \frac{600 \times 10^6}{350\,(700)^2} = 3.4985 \simeq 3.5 \text{ N/mm}^2$$

$d'/d = 50/700 = 0.071$ Hence from Table 4.2, we get

$$p_t = 1.171 + \frac{1.183 - 1.171}{5.0} \times 2.1 = 1.176\%$$

$$p_c = 0.224 + \frac{0.239 - 0.224}{5.0} \times 2.1 = 0.2303\,\%$$

$\therefore \quad A_{st} = \dfrac{1.176}{100} \times 350 \times 700 \approx 2879$ mm² and $A_{sc} = \dfrac{0.2303}{100} \times 350 \times 700 \simeq 564$ mm²

These values are almost the same as found by analytical method.

Example 4.4. Design of beam : Mild steel bars : *Design the reinforcement for a reinforced concrete beam 300 mm wide and 400 mm deep of grade M 20 to resist an ultimate moment of 150 kN-m, using mild steel bars of grade Fe 250.*

Solution : Given $\quad b = 300$ mm ; $D = 400$ mm; $M_{uD} = 150$ kN-m;

$f_{ck} = 20$ N/mm²; $f_y = 250$ N/mm².

Let us assume cover to tensile reinforcement as 50 mm and cover to compressive reinforcement as 35 mm, each to the centre of reinforcement.

$$\therefore \quad d = 400 - 50 = 350 \text{ mm}; \quad d' = 35 \text{ mm}$$

For Fe 250 steel, $\dfrac{x_{u,max}}{d} = \dfrac{700}{1100 + 0.87 \times 250} = 0.531$

$$\therefore \quad x_{u,max} = 0.531 \times 350 \simeq 186 \text{ mm}$$

From Eq. 3.9, $p_{t,lim.} = 0.414 \times \dfrac{20}{250} \times 0.531 = 0.01759 = 1.759\%$

From Eq. 3.10, $M_{u,lim} = 0.36 f_{ck} \dfrac{x_{u,max}}{d}\left(1 - 0.416 \dfrac{x_{u,max}}{d}\right) bd^2$

$$= 0.36 \times 20 \times 0.531 (1 - 0.416 \times 0.531)\, 300\, (350)^2 = 109.4 \times 10^6 \text{ N-mm}$$

Since $M_{u,lim.} < M_{uD}$, the section will be doubly reinforced.

$$\therefore \quad M_{u2} = M_{uD} - M_{u,lim.} = (150 - 109.4)\, 10^6 = 40.6 \times 10^6 \text{ N-mm.}$$

$$A_{st1} = p_{t,lim.}\, bd = 0.01759 \times 300 \times 350 \simeq 1847 \text{ mm}^2$$

$$A_{st2} = \frac{M_{u2}}{0.87 f_y (d - d')} = \frac{40.6 \times 10^6}{0.87 \times 250\, (350 - 35)} = 592.6 \text{ mm}^2$$

Total $A_{st} = A_{st1} + A_{st2} = 1847 + 592.6 = \mathbf{2439.6\ mm^2}$

$$A_{sc} = \frac{M_{u2}}{(f_{sc} - 0.446 f_{ck})(d - d')} = \frac{40.6 \times 10^6}{(f_{sc} - 0.446 \times 20)(350 - 35)} = \frac{0.1289 \times 10^6}{f_{sc} - 8.92}$$

$$\varepsilon_{sc} = \frac{0.0035\,(x_{u,max} - d')}{x_{u\,max}} = \frac{0.0035\,(186 - 35)}{186} = 0.00284$$

Since this is more than 0.002, $f_{sc} = 0.87 f_y = 0.87 \times 250 = 217.5 \text{ N/mm}^2$

$$\therefore \quad A_{sc} = \frac{0.1289 \times 10^6}{217.5 - 8.92} = \mathbf{618\ mm^2}$$

Values from design tables

$$\frac{M_u}{bd^2} = \frac{M_{uD}}{bd^2} = \frac{150 \times 10^6}{300 \times (350)^2} = 4.08 \text{ N/mm}^2\ ;\ d'/d = 35/350 = 0.1$$

$$p_t = 2.307 + \frac{2.332 - 2.307}{5} \times 3 = 2.322\,\%$$

and

$$p_c = 0.570 + \frac{0.597 - 0.570}{5} \times 3 = 0.5862\%$$

$$\therefore \quad A_{st} = \frac{2.322}{100} \times 300 \times 350 = \mathbf{2438\ mm^2}$$

$$A_{sc} = \frac{0.5862}{100} \times 300 \times 350 = \mathbf{615.5\ mm^2}$$

Example 4.5. Design of beam : A_{sc} given: *A reinforced concrete beam of M 20 grade concrete, 300 mm wide and 500 mm deep is required to resist a super-imposed moment of 152 kN-m at an intermediate support of a continuous beam. Using mild steel bars, calculate A_{st} at top, if 4 Nos. 16 mm dia. bars are required to be continued at bottom from one span to the other. Assume effective cover to compression steel as 45* *mm and that to the tension steel as 50 mm. (b) Redesign the beam as doubly reinforced beam, if no A_{sc} is available from adjacent spans.*

Solution : Given $f_{ck} = 20 \text{ N/mm}^2$; $f_y = 250 \text{ N/mm}^2$; $b = 300$ mm; $D = 500$ mm

$d' = 45$ mm ; $d = 500 - 50 = 450$ mm; $A_{sc} = 4$ bars of 16 mm φ;

$M_{uD} = 1.5 \times 152 = 228$ kN-m ; $A_{sc} = 4 \frac{\pi}{4} (16)^2 = 804.25 \text{ mm}^2$

Step 1 : Assume $f_{sc} = 0.87 f_y = 0.87 \times 250 = 217.5 \text{ N/mm}^2$

$\therefore$ $M_{u2} = f_{sc} \cdot A_{sc} (d - d') = 217.5 \times 804.25 (450 - 45) = 70.84 \times 10^6$ N-mm

and $M_{u1} = M_{uD} - M_{u2} = (228 - 70.84) 10^6 = 157.16 \times 10^6$ N-mm

Step 2 : But $M_{u1} = 0.36 f_{ck} bx_u (d - 0.416 x_u)$, the solution of which is

$$\therefore \quad x_u = 1.2 \left[1 - \sqrt{1 - \frac{4.62 \times 157.16 \times 10^6}{20 \times 300 (450)^2}} \right] 450 = 197.4 \text{ mm}$$

Step 3 :

$$x_{u,max} = \frac{700}{1100 + 0.87 \times 250} \times 450 = 239 \text{ mm}. \quad \text{Hence } x_u < x_{u,max}$$

Also, $M_{u,lim} = 0.36 \times 20 \times 239 (450 - 0.416 \times 239) \times 300 \simeq 181 \times 10^6$ N-mm.

Hence $M_{u1} < M_{u,lim}$.

Step 4. Check for f_{sc}.

$$f_{sc} = 700 \left(1 - \frac{d'}{x_u}\right) = 700 \left(1 - \frac{45}{197.4}\right) = 540.4 \text{ N/mm}^2 \not< 0.87 f_y \not< 0.87 \times 250$$

Hence $f_{sc} = 0.87 \times 250 = 217.5 \text{ N/mm}^2$, and assumption 1 is correct.

Step 5

$$A_{st2} = \frac{f_{sc} \cdot A_{sc}}{0.87 f_y} = \frac{0.87 f_y \cdot A_{sc}}{0.87 f_y} = A_{sc} = 804.25 \text{ mm}^2$$

$$A_{st1} = \frac{M_{u1}}{0.87 f_y (d - 0.416 x_u)} = \frac{157.16 \times 10^6}{0.87 \times 250 (450 - 0.416 \times 197.4)} = 1964.15 \text{ mm}^2$$

$\therefore$ Total $A_{st} = A_{st1} + A_{st2} = 1964.15 + 804.25 =$ **2768.4 mm²**

(b) Design as a doubly reinforced beam if no A_{sc} is available from adjacent spans

$$\frac{x_{u,max}}{d} = \frac{700}{1100 + 0.87 \times 250} = 0.531$$

$$x_{u,max} = 0.531 \times 450 = 239 \text{ mm}$$

$$p_{t,lim.} = 0.414 \times \frac{20}{250} \times 0.531 = 0.01759$$

$$M_{u,lim} = 0.36 \times 20 \times 0.531 (1 - 0.416 \times 0.531) \times 300 (450)^2$$

$$= 181 \times 10^6 \text{ N-mm}$$

A_{st1} for $M_{u,lim} = p_{t,lim}\, bd = 0.01759 \times 300 \times 450 = 2374.2 \text{ mm}^2$

$$A_{st2} = \frac{(228 - 181) 10^6}{0.87 \times 250 (450 - 45)} = 533.6 \text{ mm}^2$$

Total $A_{st} = A_{st1} + A_{st2} = 2374.2 + 533.6 =$ **2907.8 mm²** and $$A_{sc} = \frac{M_{uD} - M_{u,lim}}{(f_{sc} - 0.446 f_{ck})(d - d')} = \frac{(228 - 181)\,10^6}{(f_{sc} - 0.446 \times 20)(450 - 45)} = \frac{0.116 \times 10^6}{f_{sc} - 8.92}$$

$$\varepsilon_{sc} = \frac{0.0035\,(x_{u,max} - d')}{x_{u,max}} = \frac{0.0035\,(0.531 \times 450 - 45)}{0.531 \times 450} = 0.00284$$

Since this is more than 0.002, $f_{sc} = 0.87 f_y = 0.87 \times 250 = 217.5$ N/mm²

$$\therefore \qquad A_{sc} = \frac{0.116 \times 10^6}{217.5 - 8.92} = \mathbf{556\ mm^2}$$

Example 4.6. Design of beam : A_{sc} given but small.

Redesign the beam of example 4.5 if only 2 Nos. 16 mm dia. bars are available at bottom in the form of compression reinforcement.

Solution : Given $A_{sc} = 2\,\frac{\pi}{4}(16)^2 = 402.1$ mm²

Step 1. Assume $f_{sc} = 0.87 f_y = 0.87 \times 250 = 217.5$ N/mm²

$\therefore$ $M_{u2} = f_{sc} A_{sc} (d - d') = 217.5 \times 402.1\,(450 - 45) = 35.42 \times 10^6$ N-mm

and $M_{u1} = M_{uD} - M_{u2} = (228 - 35.42)\,10^6 = 192.58 \times 10^6$ N-mm

Step 2 But $M_{u1} = 0.36 f_{ck}\, b\, x_u (d - 0.416 x_u)$, from which

$$x_u = 1.2\left[1 - \sqrt{1 - \frac{4.62 \times 192.58 \times 10^6}{20 \times 300\,(450)^2}}\right] \times 450 = 260.6 \text{ mm}$$

Step 3 $x_{u,max} = 239$ mm as before. $\therefore$ $x_u > x_{u,max}$

Also, $M_{u,lim} = 181 \times 10^6$ N/mm², as before. $\therefore$ $M_{u1} > M_{u,lim}$

Hence the beam is to be redesigned as an ordinary doubly reinforced beam, neglecting the existing A_{sc}.

Step 4. The design of beam is given in part (*b*) of example 4.5 in which we have

$$A_{st} = \mathbf{2907.8\ mm^2} \text{ and } A_{sc} = 556 \text{ mm}^2$$

Since already available $A_{sc} = 402.1$ mm²,

Additional A_{sc} required $= 556 - 402.1 =$ **153.9 mm²**

PROBLEMS

1. A rectangular beam 300 mm wide and 600 mm effective depth is reinforced with a tensile reinforcement of 9000 mm² and compressive reinforcement of 3000 mm². The compressive reinforcement has an effective cover of 50 mm. Determine its ultimate moment of resistance. Use M 15 grade concrete and Fe 250 grade steel.

2. Solve problem 1 for M 15 – Fe 415 combination.

3. Solve problem 1 for M 20 – Fe 415 combination.

4. A concrete beam has 300 mm breadth and 500 mm effective depth. Design the beam if it is subjected to a super-imposed bending moment of 200 kN-m. Use M 20 grade concrete and Fe 415 steel.

5

Limit State of Collapse in Flexure : III Non-Rectangular Sections

5.1. INTRODUCTION

Though beams of rectangular section, T-section and L-section are quite common in buildings and other structures, beams of non-rectangular sections are used in some of the following special structures :

1. Edge beams of bunkers and shell roofs
2. Beams with recess at corners to support precast slabs.
3. Beams with internal ducts
4. Beams with circular sections
5. Beams with triangular sections
6. Beams with trapezoidal sections, commonly encountered in footings.

The concrete stress block parameters, derived in Chapter 3 are applicable only when the compression zone of the beam section is rectangular in shape. However when the concrete compression zone is non-rectangular, the magnitude of the compressive force and its location from top fibre of the section can be found by proceeding from the first principles. Before laying down the general procedure, both for analysis as well as design, let us first study, in general, the properties of concrete compression zone of any given shape under the stress diagram of the Code, along with concrete stress block parameters for some basic shapes.

5.2. EQUATION OF CONCRETE STRESS DIAGRAM AND STRAIN-STRESS RELATIONSHIP

Fig. 5.1 (*a*) shows the linear strain diagram for concrete, from the top of the section, upto the N.A. and Fig. 5.1 (*b*) shows the corresponding rectangular-cum-parabolic stress diagram specified by IS 456 : 2000

The design strength at the extreme fibre in compression is given by

$$h = f_{c\,,\,\max} = 0.67\, f_{ck}/\gamma_{mc} = 0.67\, f_{ck}/1.5 \simeq 0.446\, f_{ck}$$

The stress diagram is rectangular from the top fibre upto depth $= \frac{3}{7} x_u$, and parabolic for the remaining depth of $\frac{4}{7} x_u \; (= a)$, where x_u is the depth of N.A. The equation of the parabolic stress diagram with origin E on the N.A. (Fig. 5.1 *c*) is given by

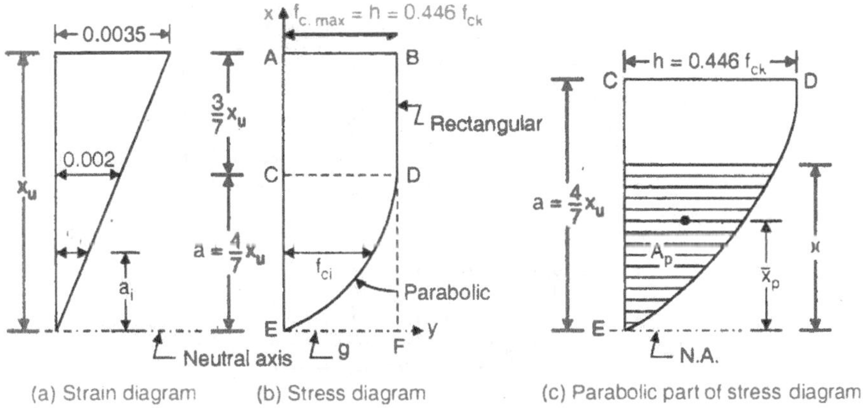

FIG. 5.1. PROPERTIES OF CONCRETE STRESS DIAGRAM

$$y = \frac{hx}{a^2}(2a - x) \quad , \text{ for the range } x = 0 \text{ to } \quad x = a = \frac{4}{7}x_u \qquad \ldots(5.1)$$

The area (A_p) of any portion of parabola (Fig. 5.1 *c*) is given by

$$A_p = \int_0^x y\,dx = \int_0^x \frac{hx}{a^2}(2a - x)\,dx = \frac{h}{a^2}\left[ax^2 - \frac{x^3}{3}\right]_0^x$$

or
$$A_p = \frac{hx^2(3a - x)}{3a^2} \qquad \ldots(5.2)$$

The distance of the centroid of the parabola, from origin E is given by

$$\bar{x}_p = \left(\int_0^x y\,x\,dx\right) \div A_p = \left[\int_0^x \frac{hx(2a - x)}{a^2}\,.\,x\,dx\right] \div A_p$$

or
$$\bar{x}_p = \frac{h}{a^2}\left[\frac{2ax^3}{3} - \frac{x^4}{4}\right] \div A_p = \frac{x(8a - 3x)}{4(3a - x)} \qquad \ldots(5.3)$$

The variation of strain (ε_i) at any distance a_i from the N.A. is given by

$$\varepsilon_i = \frac{0.0035}{x_u}a_i \qquad \ldots(5.4\ a)$$

The relation between strain ε_i and stress f_{ci} is given by following expression

For $\varepsilon_i < 0.002$, $f_{ci} = 446\,f_{ck}\,\varepsilon_i\,(1 - 250\,\varepsilon_i)$ ...(5.4 *b*)

or Alternatively,
$$f_{ci} = 0.446\,f_{ck}\left[\frac{2a_i}{a} - \left(\frac{a_i}{a}\right)^2\right] \qquad \ldots(5.4)$$

For $\varepsilon_i \geq 0.002$, $f_{ci} = 0.446\,f_{ck}$ ...(5.4 *c*)

5.3. DETERMINATION OF CONCRETE COMPRESSIVE FORCE AND ITS LOCATION

Fig. 5.2 (*a*) shows the concrete compression zone of any given shape. Let b_x be the width of the section at any distance x from the N.A. The elementary area (δA) of width b_x and thickness dx is evidently equal to b_x . dx. Fig. 5.2 (*b*) shows the stress distribution

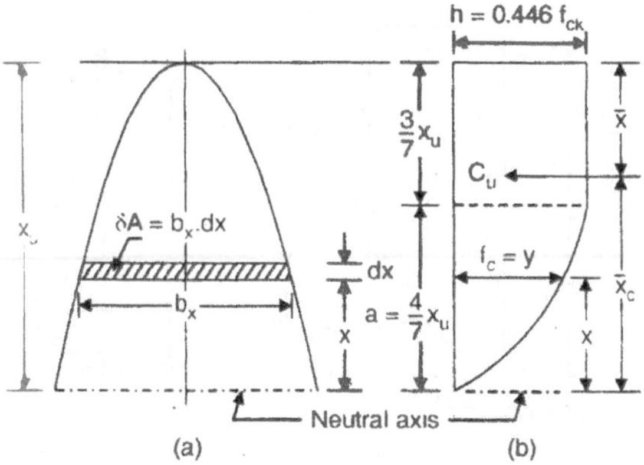

FIG. 5.2. CONCRETE COMPRESSION ZONE OF ANY GIVEN SHAPE WITH STRESS DIAGRAM OF CODE

diagram as per the Code (IS 456 : 2000), which is rectangular with $h = 0.446 f_{ck}$ upto depth $\frac{3}{7}x_u$ below the maximum compressive fibre and parabolic for the remaining depth $a = \frac{4}{7}x_u$, the stress at the N.A. being zero. If $f_c = y$ be the stress at a distance x from the N.A., the total compressive force C_u is given by

$$C_u = \int_0^a b_x\, y dx + \int_a^{x_u} b_x\, h\, dx \qquad \text{...(5.5 a)}$$

where b_x depends upon the geometry of the compression zone and y is given by Eq. 5.1.

Also, taking moments of the compressive forces about N.A., we get

$$\bar{x}_c = \frac{\int_0^a b_x\, y\, x\, dx + \int_a^{x_u} b_x\, h\, x\, dx}{\int_0^a b_x\, y\, dx + \int_a^{x_u} b_x\, h\, dx} \qquad \text{...(5.5 b)}$$

If, however, $\bar{x}$ is the distance of concrete compressive force from top face, we have $\bar{x} = x_u - \bar{x}_c$ ...(5.5 c)

5.4. STRESS BLOCK PARAMETERS FOR THREE BASIC SHAPES OF COMPRESSION ZONES OF CONCRETE

We shall now derive stress block parameters of following three basic shapes of compression zones of concrete (*i*) rectangular zone (*ii*) triangular zone, and (*iii*) inverted triangular zone. Zones of other shapes can be splitted into these basic zones and total compressive force C_u and its location can then be easily found.

1. Rectangular compression zone

Fig. 5.3 (*a*) shows rectangular compression zone of concrete, of top width b and depth x_u. Fig. 5.3 (*b*) shows the stress block recommended by the Code. The stress diagram is rectangular for an initial depth of $\frac{3}{7}x_u$ and parabolic for the remaining depth $a = \frac{4}{7}x_u$. Corresponding to this, the concrete compression zone can be splitted into two parts, as marked in Fig.

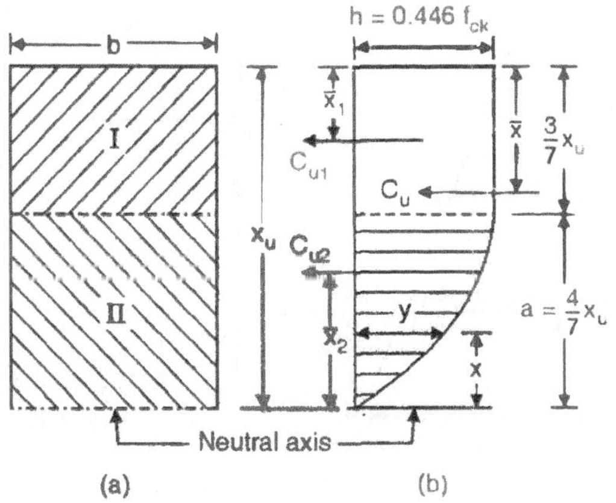

FIG. 5.3. RECTANGULAR COMPRESSION ZONE

5.3 (*a*). If C_{u1} and C_{u2} are the compressive forces, respectively for the above two parts, we have

$$C_{u1} = b \times \frac{3}{7} x_u \times h, \text{ acting at } \bar{x}_1 = \frac{1}{2} \times \frac{3}{7} x_u \text{ from the top}$$

Also, $$C_{u2} = b \times A_p = b \times \frac{ha^2 (3a - x)}{3a^2} = \frac{bh\, a^2 (3a - a)}{3a^2} = \frac{2}{3} bh\, a = \frac{8}{21} bh\, x_u$$

acting at $$\bar{x}_2 = \frac{x\,(8a - 3x)}{4\,(3a - x)} = \frac{a\,(8a - 3a)}{4\,(3a - a)} = \frac{5}{8} a = \frac{5}{14} x_u \text{ above N.A.}$$

$$\therefore \text{ Total } C_u = \frac{3}{7} h\, bx_u + \frac{8}{21} h\, b\, x_u = \frac{17}{21} \times 0.446 f_{ck}\, bx_u = 0.361 f_{ck}\, b\, x_u \quad \text{...(5.6 a)}$$

acting at $$\bar{x} = \frac{\frac{3}{7} h\, b\, x_u \times \frac{3}{14} x_u + \frac{8}{21} h\, b\, x_u \left(x_u - \frac{5}{14} x_u \right)}{\frac{3}{7} h\, b\, x_u + \frac{8}{21} h\, b\, x_u} = 0.416\, x_u \quad \text{...(5.6 b)}$$

These two expressions are the same as Eq. 3.5 and 3.2 derived in Chapter 3.

2. Triangular compression zone

Fig. 5.4 (*a*) shows the triangular compression zone of concrete with width b_u at the N.A. and height x_u. Here again, the compression zone can be divided into two Parts: Part I consisting of triangular portion *ADE* for depth $\frac{3}{7} x_u$ upto which the stress diagram is rectangular with $h = 0.446 f_{ck}$, and Part II consisting of trapezoidal portion *DECB* of depth $a = \frac{4}{7} x_u$, for which the stress diagram is parabolic. Let C_{u1} and C_{u2} be the compressive forces on there two parts.

Now $$b' = \frac{\frac{3}{7} x_u}{x_u} b_u = \frac{3}{7} b_u\,; \qquad \text{Also,} \quad b_x = \frac{x_u - x}{x_u} b_u \;; \quad h = 0.446 f_{ck}$$

$$\therefore \quad C_{u1} = \left(\frac{1}{2} \times \frac{3}{7} b_u \times \frac{3}{7} x_u \right) h = \frac{9}{2 \times 49} h\, b_u\, x_u = 0.041 f_{ck}\, b_u\, x_u$$

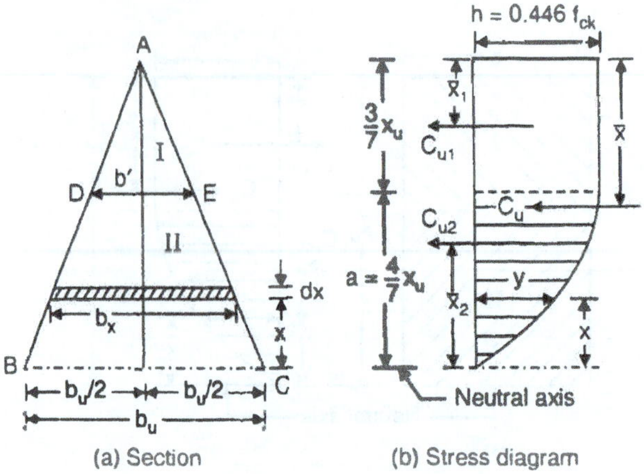

(a) Section (b) Stress diagram

FIG. 5.4. TRIANGULAR COMPRESSION ZONE.

acting at $\bar{x}_1 = \frac{2}{3} \times \frac{3}{7} x_u = \frac{2}{7} x_u$ from top.

$$C_{u2} = \int_0^a y\, b_x\, dx = \int_0^a \frac{h}{a^2} (2a\,x - x^2) \frac{(x_u - x)}{x_u} b_u\, dx$$

$$= \frac{h\, b_u}{a^2\, x_u} \int_0^a (2a\, x_u\, .\, x - x_u\, x^2 - 2a\, x^2 + x^3)\, dx$$

$$= \frac{hb_u}{a^2\, x_u} \left[2\, a\, x_u \left(\frac{x^2}{2} \right) - x_u \frac{x^3}{3} - \frac{2a\, x^3}{3} + \frac{x^4}{4} \right]_0^a = \frac{h\, b_u\, a}{x_u} \left(\frac{2}{3} x_u - \frac{5}{12} a \right)$$

$$= \frac{0.446\, f_{ck}\, b_u}{x_u} \times \frac{4}{7} x_u \left(\frac{2}{3} x_u - \frac{5}{12} \times \frac{4}{7} x_u \right) = 0.1092\, f_{ck}\, b_u\, x_u$$

and $$\bar{x}_2\, .\, C_{u2} = \int_0^a y\, b_x\, x dx = \int_0^a \frac{h}{a^2} (2ax - x^2) \frac{(x_u - x)}{x_u} b_u\, .\, x\, dx$$

$$= \frac{h\, b_u}{a^2\, x_u} \int_0^a (2a\, x_u\, x^2 - x_u\, x^3 - 2\, ax^3 + x^4)\, dx$$

$$= \frac{h\, b_u}{a^2\, x_u} \left[2a\, x_u \left(\frac{x^3}{3} \right) - x_u \left(\frac{x^4}{4} \right) - 2a \left(\frac{x^4}{4} \right) + \frac{x^5}{5} \right]_0^a = \frac{h\, b_u\, a^2}{x_u} \left(\frac{5}{12} x_u - \frac{3}{10} a \right)$$

$$= \frac{0.446\, f_{ck}\, b_u}{x_u} \left(\frac{4}{7} x_u \right)^2 \left[\frac{5}{12} x_u - \frac{3}{10} \times \frac{4}{7} x_u \right] = 0.0357\, f_{ck}\, b_u\, x_u^2$$

$\therefore$ $$\bar{x}_2 = \frac{0.0357\, f_{ck}\, b_u\, x_u^2}{0.1092\, f_{ck}\, b_u\, x_u} = 0.327\, x_u$$

$\therefore$ Total $C_u = C_{u1} + C_{u2} = 0.041\, f_{ck}\, b_u\, x_u + 0.1092\, f_{ck}\, b_u\, x_u \simeq 0.15\, f_{ck}\, b_u\, x_u$ ...(5.7 a)

Acting at $$\bar{x} = \frac{0.041\, f_{ck}\, b_u\, x_u \left(\frac{2}{7} x_u \right) + 0.1092\, f_{ck}\, b_u\, x_u\, (x_u - 0.327\, x_u)}{0.041\, f_{ck}\, b_u\, x_u + 0.1092\, f_{ck}\, b_u\, x_u}$$

or $\bar{x} = 0.567\, x_u$ ...(5.7 b)

3. Inverted triangular compression zone

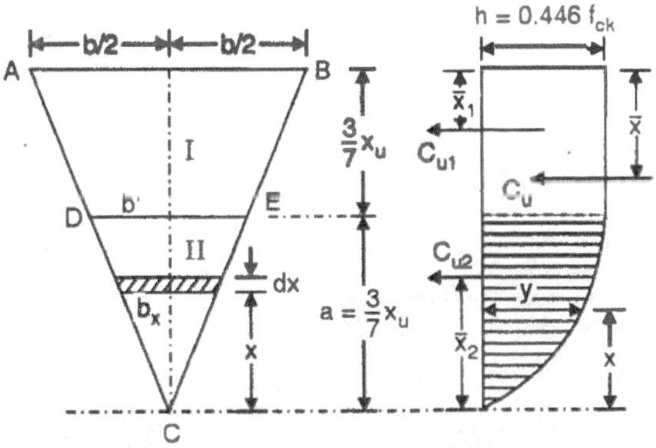

(a) Section (b) Stress diagram

FIG. 5.5. INVERTED TRIANGULAR COMPRESSION ZONE

Fig. 5.5 (*a*) shows the inverted triangular compression zone of concrete, with top width b and depth/height x_u. Here again, the section can be divided into two zones : Zone I consisting of trapezoidal portion *ABED* of depth $\frac{3}{7}x_u$ corresponding to which the stress diagram is rectangular, and zone II consisting of triangular portion *DEC*, for height $a = \frac{4}{7}x_u$ corresponding to which the stress diagram is parabolic. Let C_{u1} and C_{u2} be the compressive forces on these two parts.

Now $$b' = DE = \frac{ab}{x_u} \quad \text{and} \quad b_x = \frac{xb}{x_u}$$

$$C_{u1} = \frac{1}{2}\left(b + \frac{ab}{x_u}\right)\frac{3}{7}x_u\, h = \frac{1}{2}\frac{b\,h}{x_u}(x_u + a)\frac{3}{7}x_u = \frac{1}{2} \times 0.446\, f_{ck}\,\frac{b}{x_u}\,\frac{3}{7}x_u\left(x_u + \frac{4}{7}x_u\right)$$

or $$C_{u1} = 0.1502\, f_{ck}\, b\, x_u \qquad \text{...(1)}$$

Acting $$\bar{x}_1 = \frac{3}{7}\frac{x_u}{3}\left(\frac{b + 2\,b'}{b + b'}\right) = \frac{x_u}{7}\left[\frac{b + \frac{2ab}{x_u}}{b + \frac{ab}{x_u}}\right] = \frac{x_u}{7}\left[\frac{x_u + 2\,a}{x_u + a}\right] = 0.1948\, x_u \qquad \text{...(2)}$$

$$C_{u2} = \int_0^a y\, b_x\, dx = \int_0^a \frac{h}{a^2}(2ax - x^2)\,.\,\frac{xb}{x_u}\,dx = \frac{hb}{a^2\,x_u}\int_0^a (2a\,x^2 - x^3)\,dx$$

$$= \frac{hb}{a^2\,x_u}\left[2a\,\frac{x^3}{3} - \frac{x^4}{4}\right]_0^a = \frac{5}{12}\,\frac{h\,b\,a^2}{x_u} = \frac{5}{12}\,\frac{b}{x_u}\,(0.446\, f_{ck})\left(\frac{4}{7}x_u\right)^2$$

or $$C_{u2} = 0.0607\, f_{ck}\, b\, x_u \qquad \text{...(3)}$$

Also, $$\bar{x}_2 . C_{u2} = \int_0^a y\, b_x\, x dx = \int_0^a \frac{h}{a^2}(2ax - x^2)\frac{xb}{x_u}\,x\,.\,dx = \frac{hb}{a^2\,x_u}\int_0^a (2ax^3 - x^4)\,dx$$

$$= \frac{hb}{a^2\,x_u}\left[2a\,\frac{x^4}{4} - \frac{x^5}{5}\right]_0^a = \frac{3}{10}\,\frac{h\,b\,a^3}{x_u} = \frac{3}{10}\,\frac{b}{x_u}\,(0.446\, f_{ck})\left(\frac{4}{7}x_u\right)^3 = 0.02497\, f_{ck}\, b\, x_u^2$$

$$\therefore \qquad \bar{x}_2 = \frac{0.02497\, f_{ck}\, b\, x_u^2}{0.0607\, f_{ck}\, b\, x_u} = 0.4114\, x_u \qquad \ldots(4)$$

$$C_u = C_{u1} + C_{u2} = (0.1502 + 0.0607)\, f_{ck}\, b\, x_u \simeq 0.211\, f_{ck}\, b\, x_u \qquad \ldots(5.8\ a)$$

and
$$\bar{x} = \frac{0.1502\, f_{ck}\, b\, x_u\,(0.1948\, x_u) + 0.0607\, f_{ck}\, b\, x_u\,(x_u - 0.4114\, x_u)}{0.1502\, f_{ck}\, b x_u + 0.0607\, f_{ck}\, b\, x_u} = 0.308\, x_u$$

5.5. RECTANGULAR SECTION WITH TOP RECESS OR WITH INTERNAL DUCT

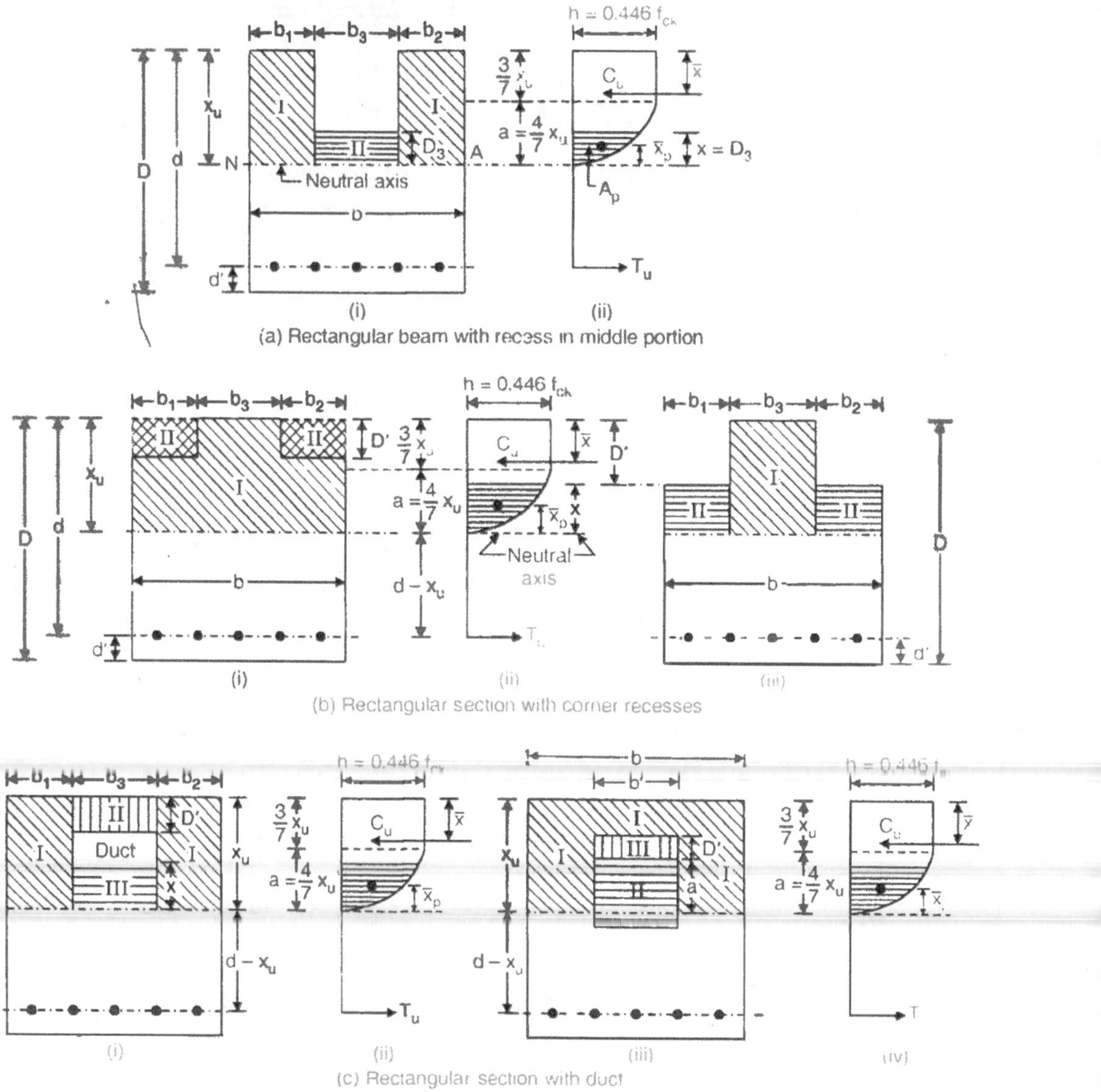

FIG. 5.6. RECTANGULAR SECTION WITH RECESS OR DUCT

Fig. 5.6 (*a*) shows a rectangular section with recess in the middle portion. The section above the N.A. is divided into two parts : part I (two numbers) of full depth x_u and part II of depth D_3 less than a, the depth of the parabola. The total compression C_u is given by

$$C_u = C_{u1} + C_{u2} \qquad \text{...(5.9 a)}$$

where $C_{u1} = 0.36\, f_{ck}\,(b_1 + b_2)\, x_u$, acting at $\bar{x}_1 = 0.416\, x_u$ from top ...(5.9 *b*)

and $$C_{u2} = b_3 \times \text{area of shaded parabola} = b_3 \times \frac{h\, x^2\,(3a - x)}{3a^2} \qquad \text{...(5.9 c)}$$

acting at $\bar{x}_p = \dfrac{x\,(8a - 3x)}{4\,(3a - x)}$ from the N.A.

where $h = 0.446\, f_{ck}$, $a = \dfrac{4}{7} x_u$ and $x =$ height of II part $= D_3$

Fig. 5.6 (*bi*) shows a rectangular section with cuts at its two corners. The section is divided into two parts : Part I consisting of the whole section of width b and depth x_u, and Part II (negative) consisting of the two cuts of width b_1 and b_2 and depth D'. The total compressive force is given by

$$C_u = C_{u1} - C_{u2} \qquad \text{...(5.10 a)}$$

where $$C_{u1} = 0.36\, f_{ck}\, b_3\, x_u \qquad \text{...(5.10 b)}$$

and $$C_{u2} = 0.446\, f_{ck}\,(b_1 + b_2)\, D' \qquad \text{...(5.10 c)}$$

Eq. 5.10 (*c*) is valid only when D' is less than $\frac{3}{7} x_u$. If, however D' is greater than $\frac{3}{7} x_u$, as shown in Fig. 5.4 (*b iii*), the section above the N.A. is divided into two parts : Part I consisting of width b_3 and depth x_u, and Part II (two numbers) of width b_1 and b_2 and depth $x\ (< a)$. The total compressive force C_u is given by

$$C_u = C_{u1} + C_{u2} \qquad \text{...(5.11 a)}$$

where $$C_{u1} = 0.36\, f_{ck}\, b_3\, x_u \qquad \text{...(5.11 b)}$$

and $$C_{u2} = (b_1 + b_2) \times \text{area of shaded parabola} = (b_1 + b_2)\frac{h\, x^2\,(3a - x)}{3a^2} \qquad \text{...(5.11 c)}$$

Fig. 5.6 (*c i*) shows a rectangular section with a duct, lying well above the assumed position of N.A. The entire section above the N.A. is divided into three parts : Part I (two numbers) of width $b_1 + b_2$ and full depth x_u, Part II (above the duct) of width b_3 and depth D' where D' is less than $\frac{3}{7} x_u$, and Part III (below the duct) of width b_3 and depth x, where x is less than $\frac{4}{7} x_u$. The total compressive force C_u is given by

$$C_u = C_{u1} + C_{u2} + C_{u3} \qquad \text{...(5.12 a)}$$

where $$C_{u1} = 0.36\, f_{ck}\,(b_1 + b_2)\, x_u \qquad \text{...(5.12 b)}$$

$$C_{u2} = 0.446\, f_{ck}\, b_3\, D' \qquad \text{...(5.12 c)}$$

and $$C_{u3} = b_3 \times \text{area of shaded parabola} = b_3 \frac{h x^2\,(3a - x)}{3a^2} \qquad \text{...(5.12 d)}$$

If, however, the duct is so positioned that it extends beyond the assumed position of the N.A., the section situated above the N.A. is divided into three parts : Part I consisting of the entire portion of width b and depth x_u, Part II (negative) of width b' and height $a\left(=\frac{4}{7}x_u\right)$, situated above N.A., and part III (negative) of width b' and remaining height D', lying entirely under the rectangular portion of stress diagram. The total compressive force C_u is given by.

$$C_u = C_{u1} - C_{u2} - C_{u3} \quad ...(5.13\ a)$$

where
$$C_{u1} = 0.36\, f_{ck}\, b\, x_u \quad ...(5.13\ b)$$

$$C_{u2} = b' \times \text{ area of full parabola } = b' \times \frac{2ha}{3} \quad ...(5.13\ c)$$

and
$$C_{u3} = 0.446\, f_{ck}\, b'\, D' \quad ...(5.13\ d)$$

5.6. BEAM OF TRIANGULAR SECTION

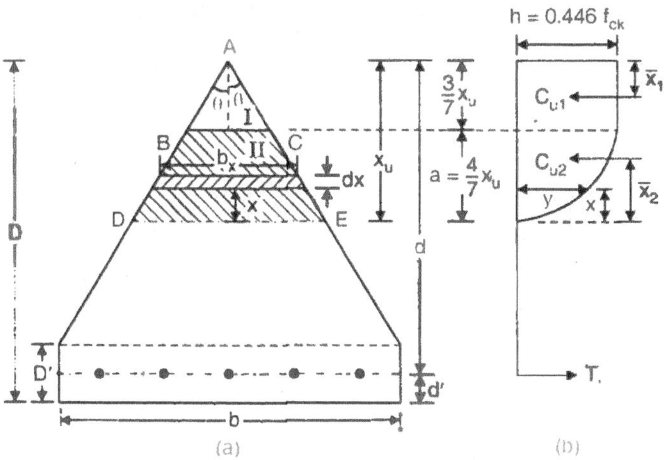

FIG. 5.7. BEAM OF TRIANGULAR SECTION

Fig. 5.7 (a) shows a beam of triangular section, of bottom width b and depth D, with its apex A upwards. Let θ let the angle made by each side with vertical. Let x_u be the depth of N.A. from apex A. Fig. 5.7 (b) shows the rectangular cum-parabolic stress diagram, which is rectangular upto depth $\frac{3}{7}x_u$, with maximum stress $h = 0.446\, f_{ck}$, and parabolic for the remaining depth $a = \frac{4}{7}x_u$. The entire section above the N.A. can be divided into two parts : Part I, consisting of the triangle ABC of depth $\frac{3}{7}x_u$, for which the stress distribution is rectangular with constant stress $h = 0.446\, f_{ck}$, and Part II consisting of trapezium $BCED$ for which stress distribution is parabolic. Let C_{u1} and C_{u2} be the compressive forces corresponding to these two parts.

(a) ***Computation of compressive force C_{u1} on triangular part ABC***

Width
$$BC = 2 \times \frac{3}{7}x_u \tan\theta = \frac{6}{7}x_u \tan\theta$$

$\therefore$ Area $\quad ABC = A_1 = \frac{1}{2} \times \frac{6}{7} x_u \tan\theta \times \frac{3}{7} x_u = \frac{9}{49} x_u^2 \tan\theta$

$\therefore$ Force $\quad C_{u1} = h \times A_1 = 0.446\, f_{ck} \times \frac{9}{49} x_u^2 \tan\theta$ ...(5.14 *a*)

Acting at $\quad \bar{x}_1 = \frac{2}{3}\left(\frac{3}{7} x_u\right) = \frac{2}{7} x_u$ from top ...(5.14 *b*)

(*b*) ***Computation of compressive force C_{u2} on trapezoidal part BCED***

Consider a strip of width b_x and thickness dx, at a distance x above the N.A. The width $b_x = 2\,(x_u - x) \tan\theta$. The stress f_c at the level is given by

$$f_c = y = \frac{hx}{a^2}(2a - x) = \frac{h}{a^2}(2ax - x^2) \quad \text{where } h = 0.446\, f_{ck} \text{ and } a = \frac{4}{7} x_u$$

$$\therefore \quad C_{u2} = \int_0^a f_c\, b_x\, dx = \int_0^a \frac{h}{a^2}(2ax - x^2)\, 2\,(x_u - x) \tan\theta\, dx$$

$$= \frac{2h \tan\theta}{a^2}\left[a x_u\, x^2 - x_u \frac{x^3}{3} - 2a\frac{x^3}{3} + \frac{x^4}{4} \right]_0^a$$

or $$C_{u2} = \frac{ah\tan\theta}{6}(8\, x_u - 5a)$$

Substituting $\quad h = 0.446\, f_{ck}$ and $a = \frac{4}{7} x_u$, we get

$$C_{u2} = 0.446\, f_{ck} \times \frac{24}{49} x_u^2 \tan\theta \qquad ...(5.14\ c)$$

Acting at $$\bar{x}_2 = \left[\int_0^a (f_c\, b_x\, dx)\, x \Big/ C_{u2} \right]$$

where $$\int_0^a (f_c\, b_x\, dx)\, x = \int_0^a \frac{h}{a^2}(2ax - x^2)\, 2\,(x_u - x)\tan\theta\, .\, x\, dx$$

$$= \frac{2\,h\tan\theta}{a^2}\int_0^a (2a\, x_u\, .\, x^2 - x_u\, x^3 - 2a\, x^3 + x^4)\, dx$$

$$= \frac{2\,h\tan\theta}{a^2}\left[2a x_u \frac{x^3}{3} - x_u\frac{x^4}{4} - 2a\frac{x^4}{4} + \frac{x^5}{5} \right]_0^a$$

$$= \frac{a^2\, h\tan\theta}{30}(25\, x_u - 18\, a)$$

$$\therefore \quad \bar{x}_2 = \frac{a^2\, h\tan\theta\,(25\, x_u - 18\, a)/30}{ah\tan\theta\,(8\, x_u - 5a)/6} = \frac{(25\, x_u - 18\, a)\, a}{5\,(8\, x_u - 5a)}$$

Substituting $\quad a = \frac{4}{7} x_u$, we get

$$\bar{x}_2 = \frac{(25\, x_u - 18 \times 4\, x_u/7)\, 4\, x_u/7}{5\,(8\, x_u - 5 \times 4\, x_u/7)} = \frac{103}{315} x_u = 0.327\, x_u \qquad ...(5.14\ d)$$

Hence total $C_u = C_{u1} + C_{u2} = 0.446\, f_{ck} \times \frac{9}{49} x_u^2 \tan\theta + 0.446\, f_{ck} \times \frac{24}{49} x_u^2 \tan\theta$

$$\therefore \quad C_u = 0.3\, f_{ck}\, x_u^2 \tan\theta \qquad \ldots(5.15)$$

Also, taking moments of C_{u1} and C_{u2} about the centre of tension steel, we get

$$M_{u1} = C_{u1}\left(d - \frac{2}{7}x_u\right) = 0.446\, f_{ck} \times \frac{9}{49} x_u^2 \tan\theta \left(d - \frac{2}{7}x_u\right)$$

and

$$M_{u2} = C_{u2}(d - x_u + \bar{x}_2) = C_{u2}(d - x_u + 0.327\, x_u) = C_{u2}(d - 0.673\, x_u)$$

$$= 0.446\, f_{ck} \times \frac{24}{49} x_u^2 \tan\theta\,(d - 0.673\, x_u)$$

$\therefore$ Total

$$M_u = C_{u1}\left(d - \frac{2}{7}x_u\right) + C_{u2}(d - 0.673\, x_u)$$

or

$$M_u = 0.446\, f_{ck} \frac{9}{49} x_u^2 \tan\theta \left(d - \frac{2}{7}x_u\right) + 0.446\, f_{ck} \times \frac{24}{49} x_u^2 \tan\theta\,(d - 0.673\, x_u)$$

or

$$M_u = 0.446\, f_{ck}\, x_u^2 \tan\theta \left[\frac{9}{49}\left(d - \frac{2}{7}x_u\right) + \frac{24}{49}(d - 0.673\, x_u)\right]$$

or

$$M_u = 0.446\, f_{ck}\, x_u^2 \tan\theta \left[\frac{9}{49}d - \frac{18}{49 \times 7}x_u + \frac{24}{49}d - \frac{24 \times 0.673}{49}x_u\right]$$

From which we get, $M_u = 0.3\, f_{ck}\, x_u^2 \tan\theta\,(d - 0.5674\, x_u)$ ...(5.16)

5.7. BEAM OF TRAPEZOIDAL SECTION

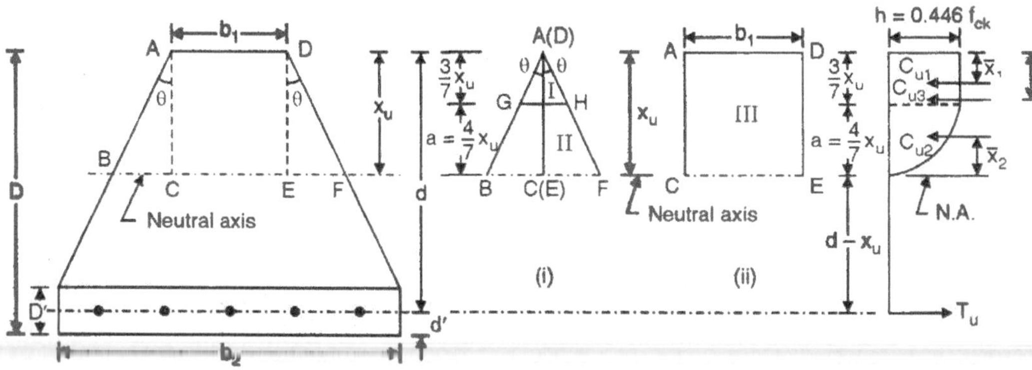

(a) Given section (b) Break-up sections (c) Stress resultants

FIG. 5.8 BEAM OF TRAPEZOIDAL SECTION

Fig. 5.8 (*a*) shows the beam of trapezoidal section, with its smaller side b_1 upwards. The portion of the section situated above the neutral axis can be divided to two parts: First part consisting of two triangles *ABC* and *DEF*, and the second part consisting of the rectangle *ADEC*. Fig. 5.8 (*bi*) shows the first part of height x_u formed by joining the two triangles *ABC* and *DEF*, both the triangles being identical, while Fig. 5.6 (*b ii*) shows the second part consisting of the rectangle *ADEC* of height x_u and width b_1. Fig. 5.8 (*c*) shows the stress diagram along with stress resultants.

The analysis of the first part, formed by joining the two outer triangles, has been done in the previous article, by splitting it further into two portions : Portion I. (triangle

AGH) and Portion II (trapezium *GHFB*). The compressive forces C_{u1} and C_{u2} on these two parts are given by Eq. 5.14 (*a*) and 5.14 (*c*) respectively, as under

$$C_{u1} = 0.446\, f_{ck} \times \frac{9}{49} x_u^2 \tan\theta \quad \text{acting at} \quad \bar{x}_1 = \frac{2}{7} x_u \text{ from top}$$

$$C_{u2} = 0.446\, f_{ck} \frac{24}{49} x_u^2 \tan\theta \quad \text{acting at} \quad \bar{x}_2 = 0.327\, x_u \text{ from N.A.}$$

Total force on triangle, $C_u = C_{u1} + C_{u2} = 0.3\, f_{ck}\, x_u^2 \tan\theta$ ...(5.17)

The moments of the these forces about the N.A. are given by

$$M_{u1} = 0.446\, f_{ck} \frac{9}{49} x_u^2 \tan\theta \left(d - \frac{2}{7} x_u \right)$$

$$M_{u2} = 0.446\, f_{ck} \frac{24}{49} x_u^2 \tan\theta\, (d - 0.673\, x_u)$$

Total $\qquad M_u = M_{u1} + M_{u2} = 0.3\, f_{ck}\, x_u^2 \tan\theta\, (d - 0.5674\, x_u)$ ...(5.18)

Now consider the rectangular part *ADEC*, shown in Fig 5.6 (*b ii*).

The force C_{u3} on this portion is given by

$$C_{u3} = 0.36\, f_{ck}\, b_1\, x_u \qquad ...(5.19)$$

acting at $\qquad \bar{x}_3 = 0.416\, x_u$ from top ....(5.19 *a*)

Hence moment of this force, about centre of tension reinforcement is given by

$$M_{u3} = C_{u3}\,(d - \bar{x}_3) = 0.36\, f_{ck}\, b_1\, x_u\, (d - 0.416\, x_u) \qquad ...(5.20)$$

Hence the total compressive force on the whole trapezoidal section is given by

$$C_u = C_{u3} + C_{u1} + C_{u2} = 0.36\, f_{ck}\, b_1\, x_u + 0.3\, f_{ck}\, x_u^2 \tan\theta \qquad ...(5.21)$$

Total moment of these forces, about the centre of tensile reinforcement is given by

$$M_u = M_{u3} + (M_{u1} + M_{u2}) = 0.36\, f_{ck}\, b_1\, x_u\, (d - 0.416\, x_u) + 0.3\, f_{ck}\, x_u^2 \tan\theta\, (d - 0.5674\, x_u) \quad ...(5.22)$$

5.8. PROCEDURE FOR ANALYSIS / DESIGN OF SECTION

In the case of concrete compression zone of non-rectangular section, it is better to proceed from the first principles. The procedures for the analysis/design of such a section are given below.

PROCEDURE FOR ANALYSIS OF THE SECTION

In most of the analysis problems, it is required to find the moment of resistance (M_u) of the given section, where all the dimensions, including the area of tension steel, along with its location are completely known. The analysis is done by trial and error in the following steps.

1. Assume some suitable value of depth x_u of N.A., starting with x_u slightly less than $x_{u\,.\,max}$

2. Divide the section above the N.A. into a number of regular parts. Each part should be such that either it lies in the rectangular portion of stress diagram or else it *touches* the N.A. so that computation of compressive force (C_u) of that part, due to parabolic stress distribution from the origin (i.e. N.A.) of the parabola becomes easy.

3. Determine the compressive force (C_{u1} , C_{u2} ......etc.) over each part. For the part over which stress distribution is rectangular, the compressive force (C_u) is found directly by taking product of constant stress $h = 0.446\ f_{ck}$ and the corresponding sectional area. The compressive force over the other parts, over which stress distribution is parabolic, is found by integration, as illustrated in Eq. 5.5. Thus, the total compressive force C_u is known.

4. Compute total tensile force $T_u = 0.87\ f_y\ A_{st}$.

5. Equate C_u to T_u. If the two forces are balanced (i.e. equal), the assumed value of x_u is correct. If $C_u > T_u$, reduce x_u while if $C_u < T_u$, increase x_u. Repeat till we obtain $C_u = T_u$.

6. For this correct value of x_u, compute the total moment of resistance (M_u) by taking the algebraic sum of moments of various compressive forces C_{u1} , C_{u2} ...... etc. about the centroid of tension steel.

PROCEDURE FOR DESIGN OF THE SECTION

In the design problem, the design moment $M_u = M_{uD}$ is known, along with dimensions of the section and it is required to find the area of tension steel (A_{st}). The design is done in the following steps.

1. Assume $x_u = x_{u\,.\,max}$

2. Divide the section above the N.A. into a number of regular parts, each part touching the N.A. Determine the compressive force (C_{u1} , C_{u2} ....etc) over each part.

3. Find the moment of each compressive part (M_{u1} , M_{u2} .....) about the centre of tensile steel, and find the algebraic sum of these moments, so that $M_{u\,.\,max}$ for the whole section is known.

4. Compare $M_{u\,.\,max}$, so obtained, with the given design moment M_{uD}. If $M_{u\,.\,max} \neq M_{uD}$, repeat the process by changing the value of x_u suitably, so that M_u equal M_{uD}. Thus correct value of x_u is obtained.

5. For this value of x_u, take the sum of compressive forces (C_{u1} , C_{u2},....). Thus, C_u, is known. Equating this to T_u, we get,

$$C_u = T_u = 0.87\ f_y\ A_{st}\ ,\ \text{from which } A_{st} \text{ is known.}$$

Example 5.1. ***Moment of resistance of rectangular beam section with central recess at top :***

Fig. 5.9 (a) shows a rectangular beam with a central recess at the top. The beam is reinforced with 3 bars of 20 mm φ, placed at a cover of 50 mm to the centre of steel. Determine the moment of the resistance of the section. Use M 20 concrete and Fe 415 steel.

Solution :

Given $D = 500$ mm ; $d = 500 - 50 = 450$ mm; $b_1 = b_2 = 100$ mm ; $b_3 = 120$ mm

$$A_{st} = 3 \times \frac{\pi}{4} (20)^2 = 942.48 \text{ mm}^2 \quad ; \text{ For Fe 415 steel,}$$

$$x_{u\,.\,max} = \frac{700}{1100 + 0.87\ f_y}\ d = \frac{700}{1100 + 0.87 \times 415}\ d = 0.479\ d = 0.479 \times 450 = 215.6 \text{ mm}$$

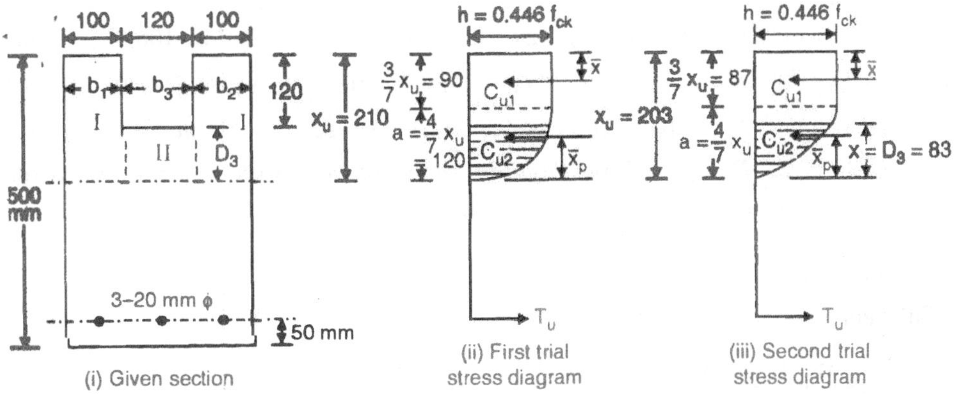

FIG. 5.9.

First trial : Let $x_u = 210$ mm

From Eq. 5.9 (*b*), $C_{u1} = 0.36\, f_{ck}\,(b_1 + b_2)\, x_u = 0.36 \times 20\,(100 + 100)\, 210 \times 10^{-3} = 302.4$ kN

Now $a = \frac{4}{7} x_u = \frac{4}{7} \times 210 = 120$ mm ; Also, $\frac{3}{7} x_u = \frac{3}{7} \times 210 = 90$ mm

$D_3 = 210 - 120 = 90$ mm $< a$. Also, $h = 0.446\, f_{ck} = 0.446 \times 20 = 8.92$ N/mm²

$$\therefore \quad C_{u2} = b_3 \times \frac{h\, x^2\,(3a - x)}{3a^2} \quad \text{where } x = D_3 = 90 \text{ mm} \not> a\ (= 120 \text{ mm})$$

$$\therefore \quad C_{u2} = \frac{120 \times 8.92\,(90)^2\,(3 \times 120 - 90)}{3\,(120)^2} \times 10^{-3} = 54.189 \text{ kN}$$

$\therefore$ Total $C_u = C_{u1} + C_{u2} = 302.4 + 54.189 = 356.589$ kN

Total tension $T_u = 0.87\, f_y\, A_{st} = 0.87 \times 415 \times 942.48 \times 10^{-3} = 340.28$ kN

Since $C_u > T_u$, decrease x_u.

Second trial : Next $x_u \simeq \frac{340.28}{356.589} \times 210 = 200.4$ mm. Let $x_u = 203$ mm

$$\therefore \quad a = \frac{4}{7} x_u = \frac{4}{7} \times 203 = 116 \text{ mm}; \quad D_3 = x = 203 - 120 = 83 \text{ mm}$$

$$\therefore \quad C_{u1} = 0.36 \times 20\,(100 + 100)\, 203 \times 10^{-3} = 292.32 \text{ kN}$$

$$C_{u2} = \frac{120 \times 8.92\,(83)^2\,(3 \times 116 - 83)}{3\,(116)^2} \times 10^{-3} = 48.407 \text{ kN}$$

$\therefore$ Total $C_u = C_{u1} + C_{u2} = 292.32 + 48.407 = 340.727$ kN $\simeq T_u\ (= 340.28$ kN)

Hence no further trial is necessary.

Now, C_{u1} acts at $\bar{x} = 0.416\, x_u = 0.416 \times 203 \simeq 84.45$ mm from top

and C_{u2} acts at $\bar{x}_p = \frac{x\,(8a - 3x)}{4\,(3a - x)} = \frac{83\,(8 \times 116 - 3 \times 83)}{4\,(3 \times 116 - 83)} = 53.17$ mm from N.A.

Taking moments about the centre of tensile steel,

$$M_{u1} = C_{u1}(d - \bar{x}) = 292.32\ (450 - 84.45) \times 10^{-3} \simeq 106.86 \text{ kN-m}$$

and $$M_{u2} = C_{u2}(d - x_u + \bar{x}_p) = 48.407\ (450 - 203 + 53.17) \times 10^{-3} \simeq 14.53 \text{ kN-m}$$

$\therefore$ Total $$M_u = M_{u1} + M_{u2} = 106.86 + 14.53 = \mathbf{121.39} \text{ kN-m}$$

Example 5.2. ***Design of rectangular beam section with central recess at top.***

For the rectangular beam section with a central recess at the top shown in Fig. 5.10 (a), design the reinforcement, for an ultimate moment of 120 kN-m. Use M 20 concrete and Fe 415 steel.

Solution

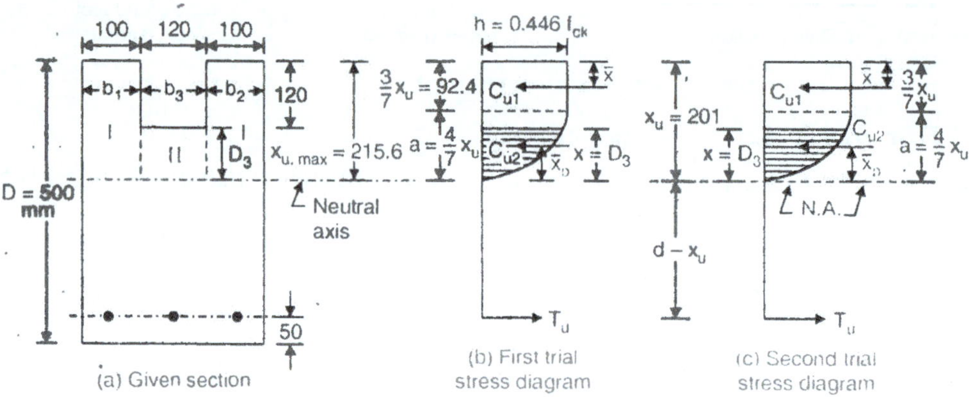

FIG. 5.10

Given : $D = 500$ mm; $d = 500 - 50 = 450$ mm; $b_1 = b_2 = 100$ mm ; $b_3 = 120$ mm

For Fe 415 steel, $x_{u,\max} = 0.479\, d = 0.479 \times 450 \simeq 215.6$ mm

First trial : Let $x_u = x_{u,\max} = 215.6$ mm.

$$\therefore \quad \frac{3}{7}x_u = \frac{3}{7} \times 215.6 = 92.4 \text{ mm} \; ; \quad a = \frac{4}{7}x_u = \frac{4}{7} \times 215.6 = 123.2 \text{ mm}$$

$$D_3 = 215.6 - 120 = 95.6 \text{ mm} < a \; ; \; h = 0.446 f_{ck} = 0.446 \times 20 = 8.92 \text{ N/mm}^2$$

$$C_{u1} = 0.36 f_{ck}(b_1 + b_2)\, x_u = 0.36 \times 20\ (100 + 100)\ 215.6 \times 10^{-3} = 310.464 \text{ kN}$$

acting at $x = 0.416\, x_u = 0.416 \times 215.6 = 89.69$ mm from top.

$$\therefore \quad M_{u1} = C_{u1}(d - \bar{x}) = 310.464\ (450 - 89.69)\ 10^{-3} = 111.863 \text{ kN-m}$$

Also, $$C_{u2} = \frac{b_3\, h\, x^2 (3a - x)}{3a^2}, \quad (\text{where } x = D_3 = 95.6 \text{ mm})$$

$$= \frac{120 \times 8.92\ (95.6)^2\ (3 \times 123.2 - 95.6)}{3\ (123.2)^2} \times 10^{-3} = 58.867 \text{ kN}$$

Acting at $$\bar{x}_p = \frac{x\,(8a - 3x)}{4\,(3a - x)} = \frac{95.6\ (8 \times 123.2 - 3 \times 95.6)}{4\ (3 \times 123.2 - 95.6)} \simeq 60.9 \text{ mm from N.A.}$$

$$\therefore \quad M_{u2} = C_{u2}(d - x_u + \bar{x}_p) = 58.867\ (450 - 215.6 + 60.9) \times 10^{-3} = 17.383 \text{ kN-m}$$

Total $$M_{u,\max} = M_{u1} + M_{u2} = 111.863 + 17.383 = 129.246 \text{ kN-m}$$

Given design moment $= M_{uD} = 120$ kN-m

Since $M_{u\,.\,max} > M_{uD}$, reduce x_u slightly. Next $x_u \triangleq \frac{120}{129.246} \times 215.6 \triangleq 200.2$ mm

(Note : If $M_{uD} > M_{u\,.\,max}$, doubly reinforced section will be required)

Second trial Let $x_u = 201$ mm so that $a = \frac{4}{7} x_u = \frac{4}{7} \times 201 = 114.86$ mm

$$D_3 = 210 - 120 = 81 \text{ mm } < a\,;\ x = D_3 = 81 \text{ mm}$$

$\therefore$ $C_{u1} = 0.36 \times 20\,(100 + 100)\,201 \times 10^{-3} = 289.44$ kN

acting at $\bar{x} = 0.416 \times 201 = 83.62$ mm from top.

$\therefore$ $M_{u1} = 289.44\,(450 - 83.62) \times 10^{-3} = 106.045$ kN-m

Also, $C_{u2} = \dfrac{120 \times 8.92\,(81)^2\,(3 \times 114.86 - 81)}{3\,(114.86)^2} = 46.77$ kN

acting at $\bar{x}_p = \dfrac{81\,(8 \times 114.86 - 3 \times 81)}{4\,(3 \times 114.86 - 81)} = 52.32$ mm from N.A.

$\therefore$ $M_{u2} = 46.77\,(450 - 201 + 52.32) \times 10^{-3} = 14.093$ kN-m

Total $M_{ur} = M_{u1} + M_{u2} = 106.045 + 14.093 = 120.138$ kN-m $\triangleq M_{uD} = 120$ kN-m

Hence adept $x_u = 201$ mm for which, we have

$$C_u = C_{u1} + C_{u2} = 289.44 + 46.77 = 336.21 \text{ kN}$$

Total tensile force, $T_u = 0.87 f_y A_{st} = 0.87 \times 415\, A_{st}$

Equating the two, we get $A_{st} = \dfrac{336.21 \times 10^3}{0.87 \times 415} = \mathbf{931.2\ mm^2}$

Example 5.3. *Moment of resistance of rectangular beam with a duct*

Fig. 5.11 (ai) shows a rectangular beam with a duct. The beam is reinforced with 3 bars of 20 mm φ placed at a cover of 50 mm to the centre of steel. Determine the moment of resistance of the section. Assume M 20 concrete and Fe 415 steel.

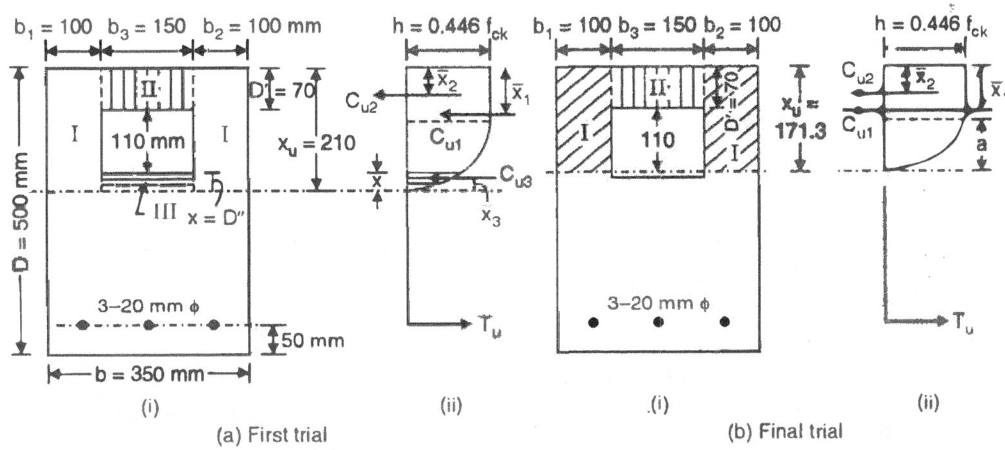

FIG. 5.11

Solution : Given $D = 500$ mm; $d = 500 - 50 = 450$ mm; $A_{st} = 3 \times \frac{\pi}{4}(20) = 942.48$ mm^2

Divide the section above N.A. into three parts, I, II and III as shown.

For Fe 415 steel, $x_{u\,.\,max} = 0.479\, d = 0.479 \times 450 \simeq 215.6$ mm

$$h = 0.446\, f_{ck} = 0.446 \times 20 = 8.92 \text{ N/mm}^2$$

Total $$C_u = C_{u1} + C_{u2} + C_{u3}$$

First trial : Let $x_u = 210$ mm. Fig. 5.11 (*aii*) shows the corresponding stress diagram.

$$\frac{3}{7}x_u = \frac{3}{7} \times 210 = 90 \text{ mm}; \quad \frac{4}{7}x_u = \frac{4}{7} \times 210 = 120 \text{ mm} = a\,;\; x = 210 - 70 - 110 = 30 \text{ mm}$$

$$C_{u1} = 0.36\, f_{ck}\,(b_1 + b_2)\, x_u = 0.36 \times 20\,(100 + 100)\, 210 \times 10^{-3} = 302.4 \text{ kN}$$

$$C_{u2} = 0.446\, f_{ck}\, b_3\, D' = 0.446 \times 20 \times 150 \times 70 \times 10^{-3} = 93.66 \text{ kN}$$

$$C_{u3} = b_3 \times \frac{h\, x^2\,(3a - x)}{3a^2} = \frac{150 \times 8.92\,(30)^2\,(3 \times 120 - 30)}{3\,(120)^2} \times 10^{-3} = 9.199 \text{ kN}$$

$\therefore$ Total $C_u = C_{u1} + C_{u2} + C_{u3} = 302.4 + 93.66 + 9.199 = 405.259$ kN

$$T_u = 0.87\, f_y\, A_{st} = 0.87 \times 415 \times 942.48 \times 10^{-3} = 340.282 \text{ kN}$$

Since $C_u > T_u$, decrease x_u.

Second trial Next $x_u \simeq \dfrac{340.282 \times 210}{405.259} \simeq 176$ mm. Let $x_u = 175$ mm

$$\therefore \quad \frac{3}{7}x_u = \frac{3}{7} \times 175 = 75 \text{ mm}; \quad a = \frac{4}{7}x_u = \frac{4}{7} \times 175 = 100 \text{ mm}; \quad x = D'' = 175 - 180 = -10 \text{ mm}$$

Hence a part of duct of 10 mm depth falls below N.A. Thus, the section is divided only in two parts as shown in Fig. 5.11 (*bi*).

$$C_{u1} = 0.36 \times 20\,(100 + 100)\, 175 \times 10^{-3} = 252 \text{ kN}$$

$$C_{u2} = 0.446\, f_{ck}\, b_3\, D' = 0.446 \times 20 \times 150 \times 70 \times 10^{-3} = 93.66 \text{ kN}$$

$\therefore$ Total $C_u = 252 + 93.66 = 345.66$ kN $> T_u = 340.282$

Final trial Let $x_u = 171.3$ mm ; $\frac{3}{7}x_u = 73.41$ mm ; $\frac{4}{7}x_u = a = 97.89$ mm

$$\therefore \quad C_{u1} = 0.36 \times 20 \times 200 \times 171.3 \times 10^{-3} = 246.672 \text{ kN},$$

acting at $\bar{x}_1 = 0.416 \times 171.3 = 71.26$ mm

and $C_{u2} = 93.66$ kN, as before,

acting at $\bar{x}_2 = 0.5\, D' = 0.5 \times 70 = 35$ mm below top.

$\therefore$ Total $C_u = 246.672 + 93.66 = 340.332$ kN $\simeq T_u$ ($= 340.282$ kN)

Hence adopt $x_u = 171.3$ mm

Taking moments about the centre of tensile steel, we get

$$M_{u1} = C_{u1}\,(d - \bar{x}_1) = 246.672\,(450 - 71.26) \times 10^{-3} = 93.425 \text{ kN-m}$$

$$M_{u2} = C_{u2}\,(d - \bar{x}_2) = 93.66\,(450 - 35) \times 10^{-3} = 38.869 \text{ kN-m}$$

$\therefore$ Total $M_u = M_{u1} + M_{u2} = 93.425 + 38.869 =$ **132.294** kN-m

Example 5.4. ***Design of rectangular beam section with corner recesses***

A rectangular beam section, 350 mm wide and 500 mm deep has recesses of size 100 mm wide and 100 mm deep at the top two corners. Design the reinforcement if it is subjected to an ultimate moment of 120 kN-m. Use M 20 concrete and Fe 415 steel. Provide 50 mm effective cover.

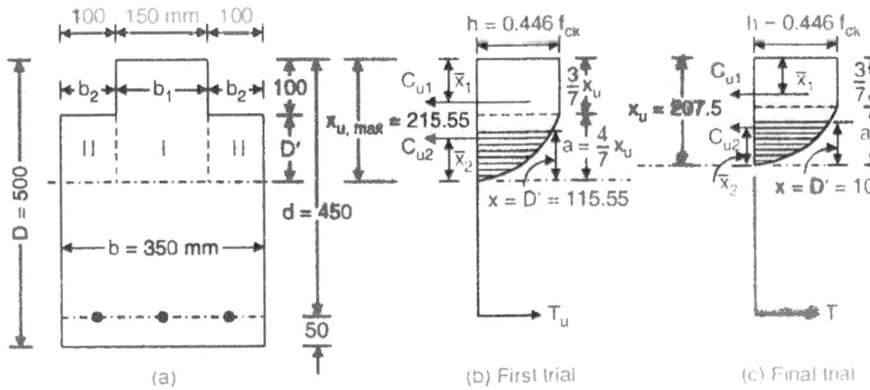

FIG. 5.12

Solution : Fig. 5.12 (*a*) shows the given section, with two corner recesses, each of 100 mm width and 100 mm depth.

For Fe 415 steel, $x_{u\,.\,max} = 0.479\,d = 0.479 \times 450 = 215.55$ mm

Divide the section above N.A. into parts I and II, as marked in Fig. 5.12 (*a*).

First trial Let $x_u = x_{u\,.\,max} = 215.55$ mm

$$\frac{3}{7}x_u = \frac{3}{7} \times 215.55 \simeq 92.38 \text{ mm}; \quad a = \frac{4}{7}x_u = \frac{4}{7} \times 215.55 = 123.17 \text{ mm}$$

$$D' = x_u - 100 = 215.55 - 100 = 115.55 \text{ mm} < a\,;\; x = D' = 115.55 \text{ mm}$$

Fig. 5.12 (*b*) shows the stress diagram where $h = 0.446 \times 20 = 8.92$ N/mm^2

$\therefore$ $$C_{u1} = 0.36\,f_{ck}\,b_1\,x_u = 0.36 \times 20 \times 150 \times 215.55 \times 10^{-3} = 232.794 \text{ kN}$$

acting at $\bar{x}_1 = 0.416\,x_u = 0.416 \times 215.55 = 89.67$ mm from top, and

$$C_{u2} = 2\,b_2 \times \frac{h\,x^2\,(3a - x)}{3a^2} = 2 \times 100 \times \frac{8.92\,(115.55)^2\,(3 \times 123.17 - 115.55)}{3\,(123.17)^2} \times 10^{-3} = 132.913 \text{ kN}$$

acting at $$\bar{x}_2 = \frac{x\,(8a - 3x)}{4\,(3a - x)} = \frac{115.55\,(8 \times 123.17 - 3 \times 115.55)}{4\,(3 \times 123.17 - 115.55)} = 72.65 \text{ mm from N.A.}$$

Taking moments about the centre of tensile steel, we get

$$M_{u1} = C_{u1}\,(d - \bar{x}_1) = 232.794\,(450 - 89.67) \times 10^{-3} = 83.883 \text{ kN-m}$$

$$M_{u2} = C_{u2}\,(d - x_u + \bar{x}_2) = 132.913\,(450 - 215.55 + 72.65) \times 10^{-3} = 40.818 \text{ kN-m}$$

$\therefore$ Total $M_{u\,.\,max} = M_{u1} + M_{u2} = 83.883 + 40.818 = 124.701$ kN-m

Required $M_{uD} = 120$ kN-m. Hence $M_{u\,.\,max} > M_{uD}$,

Hence reduce x_u slightly.

Note : If M_{uD} were more than $M_{u.max}$, doubly reinforced section will be required.

Second trial : Let next $x_u = \frac{120}{124.701} \times 215.55 \simeq 207.5$ mm. Let $x_u = 208.5$ mm

$\frac{3}{7} \times x_u = \frac{3}{7} \times 208.5 = 89.36$ mm; $a = \frac{4}{7} x_u = 119.14$ mm; $D' = 208.5 - 100 = 108.5$ mm $= x$

$$C_{u1} = 0.36 \times 20 \times 150 \times 208.5 \times 10^{-3} = 225.18 \text{ kN}$$

acting at $\bar{x}_1 = 0.416 \times 208.5 = 86.74$ mm below top.

$$\therefore \quad M_{u1} = C_{u1}(d - \bar{x}_1) = 225.18(450 - 86.74) \times 10^{-3} = 81.8 \text{ kN-m}$$

$$C_{u2} = 2 \times 100 \times \frac{8.92(108.5)^2(3 \times 119.14 - 108.5)}{3(119.14)^2} \times 10^{-3} = 122.766 \text{ kN}$$

acting at $\bar{x}_2 = \frac{108.5(8 \times 119.14 - 3 \times 108.5)}{4(3 \times 119.14 - 108.5)} = 68.39$ mm above N.A.

$$\therefore \quad M_{u2} = C_{u2}(d - x_u + \bar{x}_2) = 122.766(450 - 208.5 + 68.39) \times 10^{-3} = 38.044 \text{ kN-m}$$

$$\therefore \quad M_{ur} = M_{u1} + M_{u2} = 81.8 + 38.044 = 119.844 \text{ kN-m}$$

Since M_{ur} is approximately equal to M_{uD} (= 120 kN-m), the above value of $x_u = 208.5$ mm is acceptable.

$$\therefore \quad C_{u1} = 225.18 \text{ kN and } C_{u2} = 122.766 \text{ kN}$$

Total $\quad C_u = 225.18 + 122.766 = 347.946$ kN ...(*i*)

$T_u = 0.87 f_y A_{st}$ ...(*ii*)

Equating the two, we get

$$A_{st} = \frac{C_u}{0.87 f_y} = \frac{347.946 \times 10^3}{0.87 \times 415} = \mathbf{963.7 \text{ mm}^2}$$

Example 5.5. ***Design of trapezoidal beam section***

The trapezoidal footing for a column has a top width of 400 mm, bottom width of 1200 mm and total depth of 660 mm. Design the reinforcement for a design ultimate moment of 500 kN-m. The footing has rectangular section at the bottom, for a height of 120 mm. Use M 20 concrete and Fe 415 steel and provide an effective cover of 60 mm.

Solution

Fig. 5.13 (*a*) shows the trapezoidal section with the following dimensions :

$b_1 = 400$ mm, $b_2 = 1200$ mm; $D = 660$ mm; $d = 660 - 60 = 600$ mm ; $D' = 120$ mm

$$\tan\theta = \frac{(b_2 - b_1)/2}{(D - D')} = \frac{(1200 - 400)}{2(660 - 120)} = 0.7407$$

The section situated above the neutral axis is divided into two parts : first part consists of two triangles and the second part consists of central rectangle. Again the two triangular portions can be subdivided into two parts : part I consists of top portion of triangles, upto a depth of $\frac{3}{7} x_u$, for which stress distribution is rectangular and part II

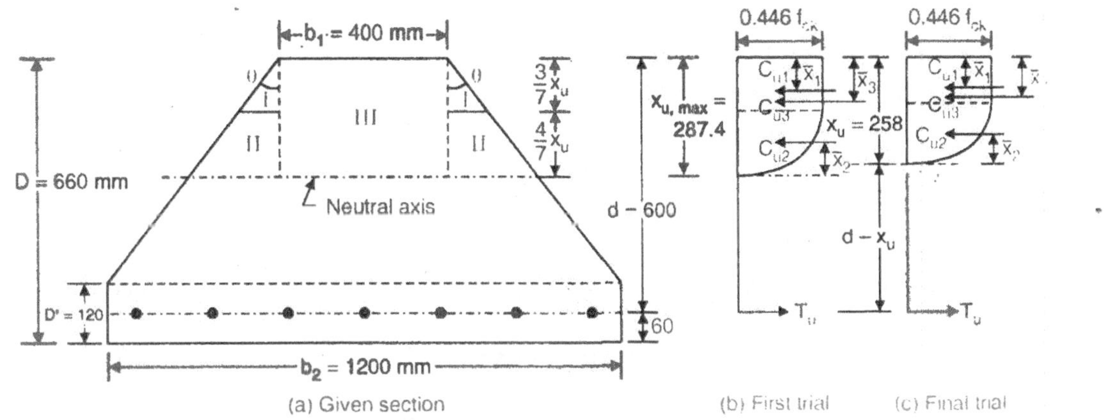

FIG. 5.13.

consists of bottom portions of the triangles, of depth $a = \frac{4}{7} x_u$, for which the stress distribution is parabolic.

Let C_{u1} = compressive force corresponding to top portion of triangles, of height $\frac{3}{7} x_u$

C_{u2} = compressive force corresponding to bottom portion of triangles, of height $\frac{4}{7} x_u$

C_{u3} = compressive force of central rectangle of width b_1 and depth x_u.

Total force on triangle, $C_u = C_{u1} + C_{u2} = 0.3\, f_{ck}\, x_u^2 \tan\theta$. ...(5.17)

and moment of this force about centre of steel is given by

$$M_u = M_{u1} + M_{u2} = 0.3\, f_{ck}\, x_u^2 \tan\theta\, (d - 0.5674\, x_u) \quad ...(5.18)$$

Also, for rectangular portion, $C_{u3} = 0.36\, f_{ck}\, b_1\, x_u$ ...(5.19)

and its moment about centre of tension steel, $M_{u3} = 0.36\, f_{ck}\, b_1\, x_u\, (d - 0.416\, x_u)$ ...(5.20)

First trial : Let $x_u = x_{u\,.\,\max} = 0.479\, d = 0.479 \times 600 = 287.4$ mm

$\therefore\ C_u = C_{u1} + C_{u2} = 0.3 \times 20\, (287.4)^2 \times 0.7407 \times 10^{-3} = 367.085$ kN

and $M_{u1} + M_{u2} = C_u\, (d - 0.5674\, x_u) = 367.085\, (600 - 0.5674 \times 287.4) \times 10^{-3} = 160.39$ kN-m

Also, $C_{u3} = 0.36\, f_{ck}\, b_1\, x_u = 0.36 \times 20 \times 400 \times 287.4 \times 10^{-3} = 827.712$ kN

$\therefore\ M_{u3} = C_{u3}\, (d - 0.416\, x_u) = 827.712\, (600 - 0.416 \times 287.4) \times 10^{-3} = 397.667$ kN-m

$\therefore\ M_{u\,.\,\max} = 160.39 + 397.667 = 558.057$ kN-m

Since $M_{u\,.\,\max} > M_{uD}$ (= 500 kN-m), reduce x_u.

Note : If M_{uD} were greater than $M_{u\,.\,\max}$, doubly reinforced section would be required.

Second trial : Let $x_u \simeq \frac{500}{558.057} \times 287.4 \simeq 257.5$ mm. Take $x_u = 258$ mm

$\therefore\ C_u = C_{u1} + C_{u2} = 0.3 \times 20\, (258)^2 \times 0.7407 \times 10^{-3} = 295.824$ kN

$M_u = M_{u1} + M_{u2} = 295.824\, (600 - 0.5674 \times 258) \times 10^{-3} = 134.189$ kN-m

$$C_{u3} = 0.36 \times 20 \times 400 \times 258 \times 10^{-3} = 743.04 \text{ kN}$$

and $$M_{u3} = 743.04\,(600 - 0.416 \times 258)\,10^{-3} = 366.075 \text{ kN-m}$$

Total $M_{ur} = 134.189 + 366.075 = 500.264$ kN-m $\simeq M_{uD}$

Hence adopt $x_u = 258$ mm.

Total $$C_u = 295.824 + 743.04 = 1038.864 \text{ kN} \quad \ldots(i)$$

Total $$T_u = 0.87\, f_y\, A_{st} = 0.87 \times 415\, A_{st} \quad \ldots(ii)$$

Equating the two, we get $A_{st} = \dfrac{C_u}{0.87 \times 415} = \dfrac{1038.864 \times 10^{-3}}{0.87 \times 415} = \mathbf{2877.4\ mm^2}$

PROBLEMS

1. Derive expressions for stress block parameters for a rectangular section.
2. Derive expressions for stress block parameters for a triangular section.
3. Derive expressions for stress block parameters for an inverted triangular section.
4. Find the moment of resistance of the beam of trapezoidal section shown in Fig. 5.14. Use M 20 concrete and Fe 415 steel.

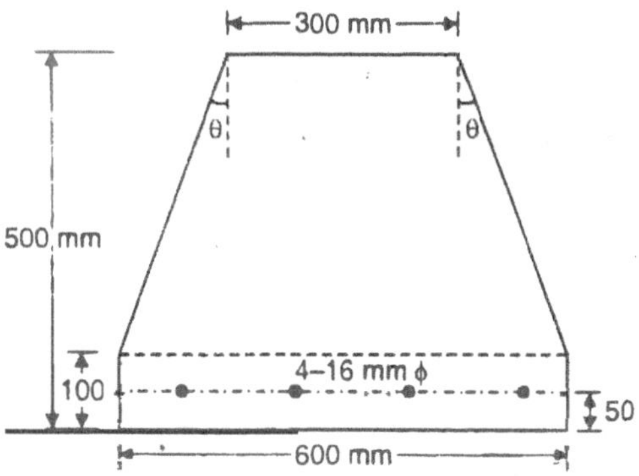

FIG. 5.14

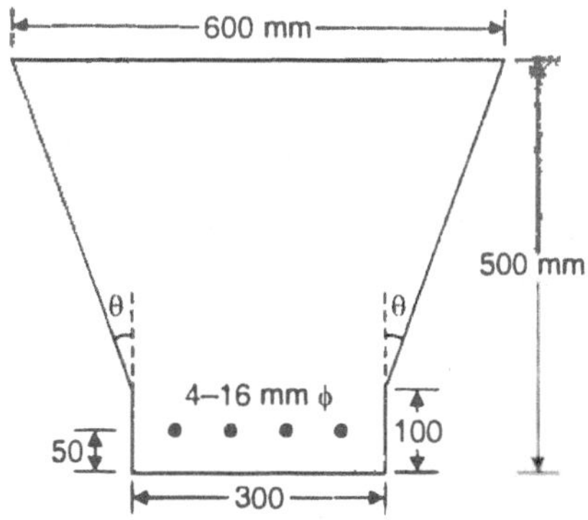

FIG. 5.15

5. Find the moment of resistance of the beam of trapezoidal section shown in Fig. 5.15. Use M 20 concrete and Fe 415 steel.

6

Limit State of Collapse in Flexure : IV Flanged Sections

6.1. INTRODUCTION

We have seen that the concrete below the N.A. does not resist any bending moment but simply serves to embed the tensile steel. Also, the portion of the concrete just above the N.A. carries only very little compressive force since the intensity of compressive stress there is of a very small magnitude. This suggests that the section of the beam should be such that it has greater width at the top (compression side) in comparison to the width below the neutral axis. Such a section is known as a T-section as shown in Fig. 6.1 (*aii*).

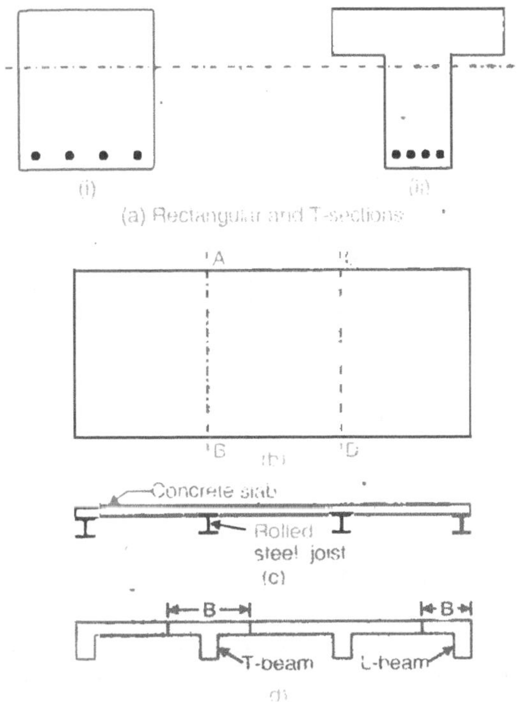

FIG. 6.1

In actual practice, a beam of T-section, known as T-beam, is more common than the rectangular section. Fig. 6.1 (*b*) shows the plan of a big size room. The roof system of the room may be of two types : (*i*) a number of beams of rolled steel section with a concrete roof slab, both acting independently, as shown in Fig 6.1 (*c*) or (*ii*) a number of R.C. beams with a roof slab acting monolithically [Fig. 6.1 (*d*)]. The second system is more common because in this system a portion of the slab of width B acts monolithically with the rectangular section of the beam, thus giving rise to T-beam. In fact, the slab forms the compression flange of the T-beam, while the comparatively narrow rectangular section forms the web of the T-beam.

In such a system, the roof slab is designed to bend over the span BD and the reinforcement for positive and negative bending moment is provided in the direction at right angles to the direction AB of the beam. The beam, on the other hand, takes uniformly distributed load from the slab, and bends over the span AB. The slab, which forms the upper part of the T-beam, is stressed laterally in compression, but this does not reduce its capacity to carry longitudinal compression as a part of the beam.

6.2. T-BEAM

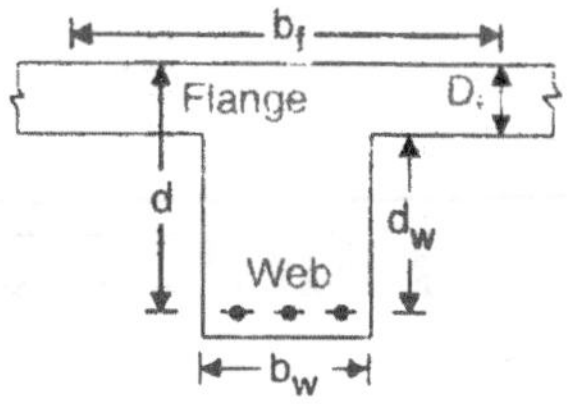

FIG. 6.2.

Fig. 6.2 shows the following four important dimensions of a T-beam section : (*i*) Breadth of the flange (b_f) (*ii*) Thickness of the flange (D_f) (*iii*) Breadth of the rib (b_w) (*iv*) Depth of the rib (d_w)

(1) Breadth of the Flange (b_f) : The breadth of the flange of the T-beam is that portion of the slab which acts monolithic with the beam and which resists the compressive stresses.

According to Indian Standard Code (IS 456 : 2000), a slab which is assumed to act as a flange of a T-beam (or of a L-beam) shall satisfy the following :

(*a*) The slab be cast integrally with the web, or the web and the slab shall be effectively bonded together in any other manner, and

(*b*) If the main reinforcement of the slab is parallel to the beam, transverse reinforcement shall be provided. Such reinforcement shall not be less than 60% of the main reinforcement at the mid-span of the slab (Fig. 6.3). The reinforcement is provided at the top of the slab.

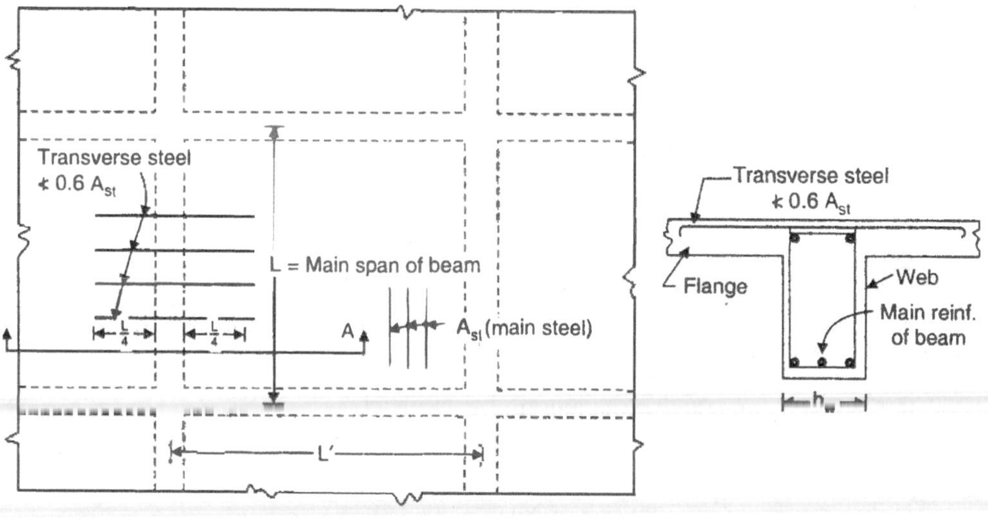

FIG. 6.3. PROVISION OF TRANSVERSE STEEL FOR FLANGED SECTION.

Note. Reinforcement in flanges of T-beams (and also of L-beams) shall satisfy the above requirement 1 (*b*). Where flanges are in tension, a part of the main tension reinforcement shall be distributed over the effective flange width or a width equal to one length of the span, whichever is smaller. If the effective flange width exceeds one length of span, nominal longitudinal reinforcement shall be provided in the outer portion of flange.

In absence of more accurate determinations, the effective width b_f of the flange may be taken as the least of the following :

(a) $b_f = \frac{l_0}{6} + 6\,D_f + b_w$ ...(6.1)

(b) $b_f = b_w + \frac{1}{2}$ (sum of clear distances to the adjacent beams on either side) ...(6.2)

(c) For isolated beams, the effective flange width shall be obtained as below but in no case greater than actual width :

$$b_f = \frac{l_0}{\left(\frac{l_0}{b}\right) + 4} + b_w \qquad ...(6.3)$$

where b_f = effective width of flange

l_0 = distance between points of zero moments in the beam

b_w = breadth of web

D_f = thickness of flange and b = actual width of flange.

Note : For continuous beams and frames, l_0 may be assumed as 0.7 times the effective span.

2. Thickness of the Flange (D_f) : The thickness of the flange is taken equal to the total thickness of the slab, including cover. The slab spans in a direction transverse to the span of the beam, and, therefore, the thickness of the slab is determined on the basis of its bending in transverse direction. As far as design of the T-beam is concerned, D_f is the fixed dimension known from the considerations of bending etc. of the slab.

3. Breadth of the Rib or Web (b_w) : The breadth of the rib is equal to width of the portion of the beam in the tensile zone. This width should be sufficient to accommodate the tensile reinforcement with proper spacing between the bars. The width of the beam should also be sufficient to provide lateral stability ; from this point of view, it should atleast be equal to one-third of the height (depth) of the rib. From architectural point of view, the breadth of the rib should be equal to the width of the column supporting the beam.

4. Depth of Rib or Web (d_w) : The depth of rib is the vertical distance between the bottom of the flange and the centre of the tensile reinforcement, and is dependent on the effective depth (d) of the beam. The effective depth (d) of the T-beam is the distance between the top of the flange and the centre of the tensile reinforcement. Thus $d = d_w + D_f$. The assumed overall depth of the T-beam may be taken as 1/12 to 1/15 of the span when it is simply supported at the ends. When it is continuous, the assumed overall depth may be taken as 1/15 to 1/20 of span for light loads, 1/12 to 1/15 of span for medium loads and 1/10 to 1/12 of span for heavy loads.

6.3. L-BEAM

In the case of T-beam, the rectangular beam has flanges on both the sides and hence load is transferred to it in vertical plane passing through the middle of its width. However, in the case of monolithic beam-slab construction, the end beams have flanges to one side only, as shown in Fig. 6.4(*a*). Such a beam having the shape of letter 'L' is known as L-beam.

Fig. 6.4 (*b*) shows the loading on the beam received from the slab. Evidently, the beam is not symmetrically loaded. Due to this, an eccentric load W is transferred to the beam. This eccentric load may be considered as central load W plus a moment M_t, as shown in Fig. 6.4(c). The central load W causes bending moment in the beam, which is jointly resisted by the rectangular portion of the beam as well as the flange (of width B), similar to that of a T-beam. The moment M_t causes *torsion* in the beam, and is known as the torsional moment which is resisted by the rectangular portion alone, and the flange does not contribute to any torsional moment of resistance. Separate torsional resistance has to be provided. According to Indian Standard Code (IS : 456–2000), the width B (or b_f) of the slab, acting monolithic with the beam, forming the flange of the L-beam is taken as the least of the following:

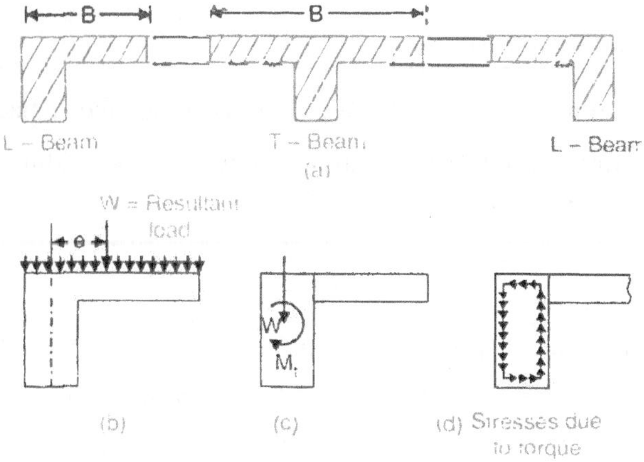

FIG. 6.4. L-BEAM

(a) $$b_f = \frac{l_0}{12} + b_w + 3\,D_f \qquad \ldots(6.4)$$

(b) For isolated beam, $$b_f = \frac{0.5\,l_0}{\dfrac{l_0}{b} + 4} + b_w. \qquad \ldots(6.5)$$

(but in no case greater than the actual width of flange) where l_0 is the distance between points of zero moments in the beam. For continuous beams and frames, l_0 may be assumed as 0.7 times the effective span.

The location of the N.A. and the calculations of the moment of resistance of L-beam are done exactly in the same way as that for T-beam, and all the equations developed for T-beam are also applicable for L-beam. The design for bending moment is also done exactly in the way as that for T-beam. However, additional reinforcement has to be provided to resist shear induced by torsional moment. The design procedure for L-beams is given Chapters 9 and 11.

6.4. STRESS BLOCK AND NEUTRAL AXIS

Consider a T-beam having flange width b_f, web width b_w and flange thickness D_f, as shown in Fig. 6.5(a). The beam is singly reinforced. The stress block, shown in Fig. 6.5(c) is similar to the one for the rectangular section, with maximum compressive stress equal to $0.446\,f_{ck}$. The strain diagram, shown in Fig. 6.5(b) is triangular, as for the rectangular section, with a maximum strain in concrete equal to 0.0035. The stress block shown in Fig. 6.5(c) has a rectangular portion $DCBE$ and parabolic portion EBA with depth $DE = \frac{3}{7} x_u$ and level EB is corresponding to a strain of 0.002 in concrete. The remaining portion $EA = \frac{4}{7} x_u$. Assuming that the N.A. falls in the flange, its position is given by equating the total compressive force to the total tensile force :

$$\therefore \quad 0.36\,f_{ck}\,b_f\,x_u = 0.87\,f_y\,A_{st}$$

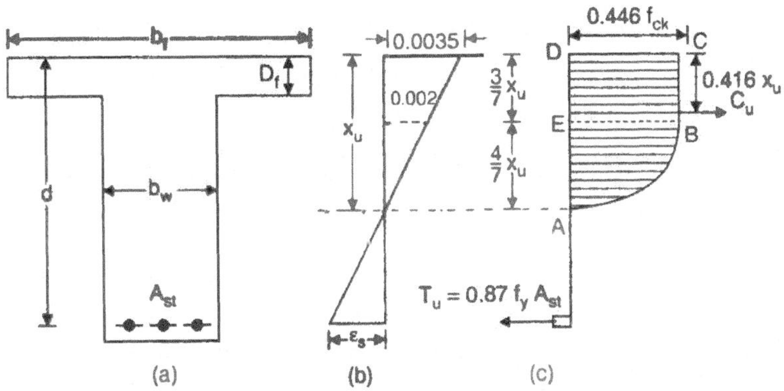

FIG. 6.5. FLANGED BEAM

or $$\frac{x_u}{d} = \frac{0.87\ f_y\ A_{st}}{0.36\ f_{ck}\ b_f\ d} = 2.417\ p_t\ \frac{f_y}{f_{ck}} \qquad \ldots(6.6)$$

If this value of x_u comes out to be less than D_f , it will be treated as a rectangular section of width b_f. The section will then be under-reinforced, balanced or over-reinforced, depending upon whether $\frac{x_u}{d}$ is less than, equal to or more than $\frac{x_{u\,max}}{d}$. The depth of critical N.A. is given Eq. 3.7 : $$\frac{x_{u\,max}}{d} = \frac{0.0035}{0.0055 + 0.87\ f_y / E_s} = \frac{700}{1100 + 0.87\ f_y}$$

If x_u found by Eq. 6.6 comes out to be more than D_f , the N.A. will fall outside the flange, and compression of the web will have to be taken into account. The method of locating the depth of N.A. for such a case has been discussed in § 6.6.

6.5. MOMENT OF RESISTANCE WHEN $x_u < D_f$

If $x_u < D_f$, the section will behave as a rectangular section with width equal to b_f. However, if x_u is greater than D_f, the analysis will depend upon whether x_u is greater than $\frac{7}{3} D_f$ or lesser than $\frac{7}{3} D_f$, *i.e.*, whether the N.A. falls outside the rectangular portion of the stress block [Fig. 6.5 (c)] or it falls inside the rectangular portion *DCBE*. Thus, in all, three cases may arise in the analysis of a T-beam.

Case 1. $x_u < D_f$

Case 2. $D_f \le \frac{3}{7} x_u \left(\text{or } \frac{D_f}{x_u} \le 0.43 \right)$

Case. 3. $D_f > \frac{3}{7} x_u \left(\text{or } \frac{D_f}{x_u} > 0.43 \right)$

We will be considering the first case in this article and the other two cases in the next article.

Case 1. $x_u < D_f$: In this case, the neutral axis is entirely within the flange. Hence the moment of resistance can be found from the equations applicable for rectangular section substituting b_f for b. Here again three cases may arise depending upon whether x_u is less than, equal to or greater than the limiting value $x_{u\,max}$. When $x_u < x_{u\,max}$, we have the case of *under-reinforced* section and the moment of resistance is given by Eq. 3.12 of rectangular section.

$$M_u = 0.87\, f_y\, A_{st}\, d \left(1 - \frac{f_y}{f_{ck}} \frac{A_{st}}{b_f\, d} \right) \qquad \ldots(6.7)$$

When x_u / d is equal to $x_{u\,max} / d$, we have the *balanced* rectangular section, and the moment of resistance M_u, which reaches its $M_{u\,lim}$ is given by Eq. 3.10.

$$M_{u\,lim} = 0.36\, f_{ck} \frac{x_{u.max}}{d} \left(1 - 0.416 \frac{x_{u.max}}{d} \right) b_f\, d^2 \qquad \ldots(6.8)$$

When $\frac{x_u}{d}$ exceeds $\frac{x_{u\,max}}{d}$, we have the *over-reinforced* section. The failure takes place due to crushing of concrete while strain ε_s in steel still remains below the yield strain ε_{su}. *Therefore, the moment of resistance of such a section is limited to* $M_{u\,lim}$ given by Eq. 6.8. However, the Code recommends that such a section should be *redesigned* because compression failure is *sudden* and therefore not desirable.

When $x_u = D_f$, the moment of resistance is given by

$$M_{u1} = 0.87\, f_y\, A_{st}(d - 0.416\, D_f) = 0.87\, f_y\, A_{st}\, d \left(1 - 0.416 \frac{D_f}{d} \right) \qquad \ldots(6.9)$$

or Alternatively, $M_{u1} = 0.36\, f_{ck}\, b_f\, D_f\,(d - 0.416\, D_f)$ ...[6.9(*a*)]

6.6. MOMENT OF RESISTANCE WHEN N.A. FALLS IN THE WEB.

When the value of x_u, found from Eq. 6.6 exceeds D_f, the N.A. falls in the web, and *a new value of x_u will have to be found by equating the total compressive force to the total tensile force.* Fig. 6.6 (*a*) shows the flanged section in which the N.A. falls outside the flange. The total compressive force is equal to C_u which can be thought to be equal to the sum of compressive force C_{uw} developed by web contribution [Fig. 6.6(*b*)] and the compressive force C_{uf} developed by contribution of remaining flange width $(b_f - b_w)$ [Fig. 6.6(*c*)].

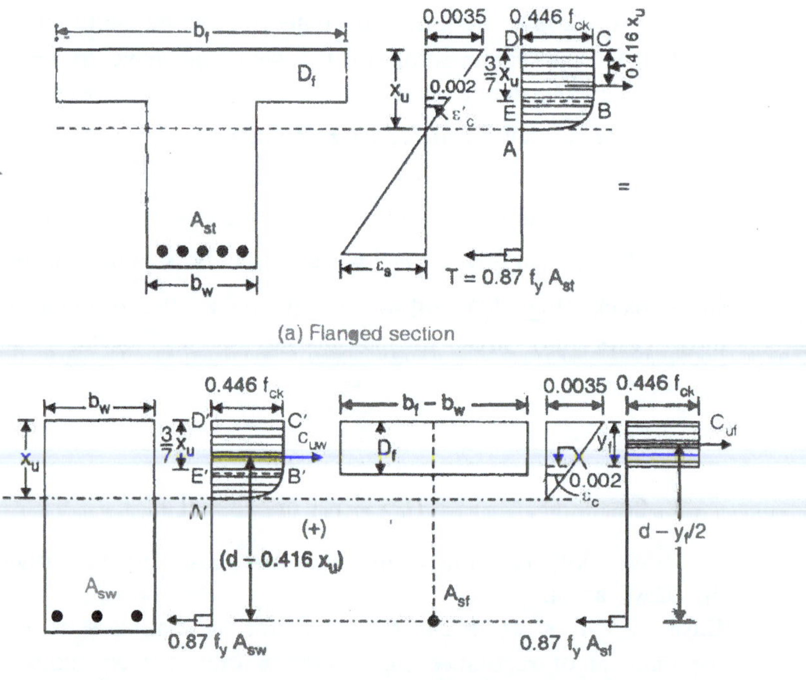

(a) Flanged section

(b) Web contribution (c) Flange contribution

FIG. 6.6. SINGLY REINFORCED FLANGED SECTION

The compressive force C_{uw} from the web contribution is given by

$$C_{uw} = 0.36 f_{ck} x_u b_w \qquad \text{...(6.10)}$$

If out of the total tensile reinforcement A_{st}, a reinforcement A_{sw} is the component required to balance the compressive force C_{uw} and the remaining reinforcement A_{sf} is required to balance the compressive force C_{uf} in the flange, we have $A_{st} = A_{sw} + A_{sf}$, and the tensile force T_{uw} to balance the compressive force C_{uw} will be equal to $0.87 f_y A_{sw}$

Hence $$0.36 f_{ck} x_u b_w = 0.87 f_y A_{sw} \qquad \text{...[6.10 (a)]}$$

The flange contribution C_{uf} is shown in Fig. 6.6 (*c*). The compressive force C_{uf} can be computed by approximating the actual stress block (rectangular-cum-parabolic) by a rectangular stress block of depth y_f, having the same peak stress value of $0.446 f_{ck}$.

The depth of equivalent rectangular stress block is given by

$$y_f = 0.15 x_u + 0.65 D_f \quad \text{(but not greater than } D_f\text{)} \qquad \text{...(6.11)}$$

The compressive force contribution of the flange is given by

$$C_{uf} = 0.446 f_{ck} (b_f - b_w) y_f \qquad \text{...(6.12)}$$

The tensile force T_{uf} to balance the compressive force C_{uf} will be equal to $0.87 f_y A_{sf}$.

The total compressive force C_u is then given by

$$C_u = 0.36 f_{ck} . x_u b_w + 0.446 f_{ck} (b_f - b_w) y_f \qquad \text{...(6.13)}$$

The total tensile force $T_u = 0.87 f_y A_{st}$ ...(6.14)

Hence equating the two, we get,

$$0.36 f_{ck} x_u b_w + 0.446 f_{ck} (b_f - b_w) y_f = 0.87 f_y A_{st} \qquad \text{...(6.15)}$$

It should be clearly noted that the value of y_f in Eq. 6.15 in terms of x_u is given by Eq. 6.11. Hence the value of x_u found from Eq. 6.15 is valid only if the value of y_f, calculated from Eq. 6.11 by using this value of x_u, is equal to or less than D_f. If however, the value of y_f, comes out to be greater than D_f, the value of x_u should be *recalculated* from Eq. 6.15, taking the value of $y_f = D_f$. The modified expression is that case will be

$$0.36 f_{ck} x_u b_w + 0.446 f_{ck} (b_f - b_w) D_f = 0.87 f_y A_{st} \qquad \text{...[6.15 (a)]}$$

Thus, the value of x_u is known. This value of x_u should now be compared with the limiting value $x_{u\,max}$. If $x_u < x_{u.max}$, we have *under reinforced section.* If $x_u = x_{u\,max}$, we have the *balanced section* while if $x_u > x_{u,\,max}$, we have *over-reinforced section.* We will take each one of these cases separately.

(*a*) UNDER-REINFORCED SECTION ($x_u < x_{u.max}$)

Here again, two cases may arise, depending upon the comparison of x_u with D_f

Case 2. $D_f \le \frac{3}{7} x_u \left(\text{or } \frac{D_f}{x_u} \le 0.43 \right)$

Case 3. $D_f > \frac{3}{7} x_u \left(\text{or } \frac{D_f}{x_u} > 0.43 \right)$

(**Note.** Case 1 was corresponding to the situation when $x_u < D_f$ as discussed in § 6.5).

Case 2. $D_f \le \frac{3}{7} x_u$ $\left(\text{or } \frac{D_f}{x_u} \le 0.43\right)$

If D_f is equal to or less than $\frac{3}{7} x_u$, the depth *DE* of the stress block will be greater than D_f. Hence the flange will lie completely within the rectangular portion of the stress block, and the strain ε_c' in concrete, at the junction of web and flange will be more than 0.002. The compressive forces C_{uw} and C_{uf} are given by

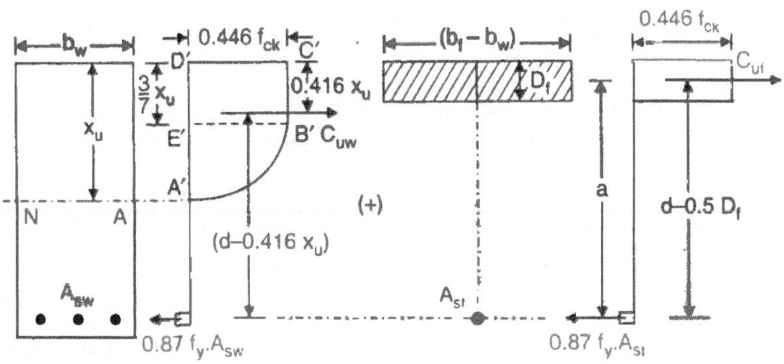

(a) Web contribution (b) Flange contribution

FIG. 6.7. M_u FOR $\frac{D_f}{x_u} \le 0.43$.

$$C_{uw} = 0.36 f_{ck}\, x_u\, b_w = 0.87 f_y A_{sw} \qquad \text{...[6.10}(a)]$$

and

$$C_{uf} = 0.446 f_{ck} (b_f - b_w) D_f = 0.87 f_y A_{sf} \qquad \text{...[6.12}(a)]$$

The moment of resistance of web is given by

$$M_{uw} = C_{uw} (d - 0.416\, x_u) = 0.36 f_{ck} \frac{x_u}{d} b_w \left(1 - 0.416 \frac{x_u}{d}\right) d^2 \qquad \text{...[6.16}(a)]$$

The moment of resistance of the flange is given by

$$M_{uf} = C_{uf} (d - 0.5\, D_f) = 0.446 f_{ck} (b_f - b_w) D_f (d - 0.5\, D_f) \qquad \text{...[6.16}(b)]$$

Hence
$$M_u = 0.36 \frac{x_u}{d} \left(1 - 0.416 \frac{x_u}{d}\right) f_{ck}\, b_w\, d^2 + 0.446 f_{ck} (b_f - b_w) D_f (d - 0.5\, D_f) \qquad \text{...(6.16)}$$

When $D_f = \frac{3}{7} x_u \left(i.e., \text{ when } x_u = \frac{7}{3} D_f\right)$ the moment of resistance is given by

$$M_{u2} = 0.36 f_{ck}\, b_w \left(\frac{7}{3} D_f\right)\left(d - 0.416 \times \frac{7}{3} D_f\right) + 0.446 f_{ck} (b_f - b_w) \times D_f (d - 0.5\, D_f) \qquad \text{...[6.16}(c)]$$

Case 3. $D_f > \frac{3}{7} x_u$ $\left(\text{or } \frac{D_f}{x_u} > 0.43\right)$. Since D_f is greater than $\frac{3}{7} x_u$ in this case, the depth *DE* of the stress block will be smaller than the depth of the flange. Due to this, the distribution of compressive stress in the flange will not be rectangular. However, the distribution of stress in web will be the same as an Case 2. The total moment of resistance will be the sum of the contribution of the web and the flange. The actual distribution is shown in Fig. 6.8(*b*). This actual stress block can be replaced by a rectangular stress block of depth y_f, having the same peak stress value of $0.446 f_{ck}$. The depth of equivalent rectangular stress block is given by

$$y_f = 0.15\, x_u + 0.65\, D_f \quad \text{(but not greater than } D_f). \qquad \text{...(6.11)}$$

The idealisation of actual stress block for the flange into a simple rectangular stress block of height y_f is based on the consideration that the compressive force and its centre of gravity of the idealised rectangular block and actual stress block compare within the acceptable limits of accuracy. The compressive forces C_{uw} and C_{uf} are given by

$$C_{uw} = 0.36 f_{ck}\, x_u\, b_w = 0.87 f_y A_{sw} \quad \text{(as before)} \qquad \text{...[6.10}(a)]$$

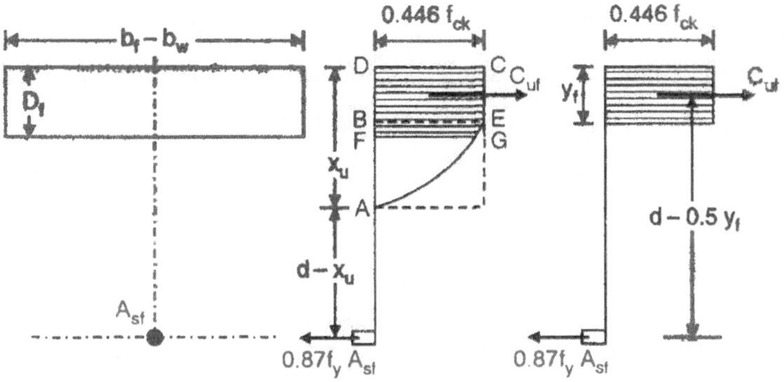

(a) (b) Actual stress block for flange (c) Equivalent stress block

FIG. 6.8. FLANGE CONTRIBUTION WHEN $D_f > \frac{3}{7}x_u$

and $$C_{uf} = 0.446\, f_{ck}\, (b_f - b_w)\, y_f = 0.87\, f_y\, A_{sf} \quad ...(6.12)$$

The moment of resistance of the web is given by

$$M_{uw} = C_{uw}\,(d - 0.416\ x_u) = 0.36\, f_{ck}\,\frac{x_u}{d}\, b_w \left(1 - 0.416\ \frac{x_u}{d}\right) d^2 \quad ...[6.17(a)]$$

The moment of resistance of the flange is given by

$$M_{uf} = C_{uf}\,(d - 0.5\ y_f) = 0.446\, f_{ck}\,(b_f - b_w)\, y_f\,(d - 0.5\ y_f) \quad ...[6.17(b)]$$

Hence, $$M_u = 0.36\,\frac{x_u}{d}\left(1 - 0.416\ \frac{x_u}{d}\right) f_{ck}\, b_w\, d^2 + 0.446\, f_{ck}\,(b_f - b_w)\, y_f\,(d - 0.5\ y_f) \quad ...(6.17)$$

(*b*) BALANCED SECTION ($x_u = x_{u,max}$).

When $x_u = x_{u\ max}$ the moment of resistance is given by Eq. 6.17, by replacing x_u by $x_{u.max}$ Thus,

$$M_{u.lim} = 0.36\,\frac{x_{u.max}}{d}\left(1 - 0.416\,\frac{x_{u.max}}{d}\right) f_{ck}\, b_w\, d^2 + 0.446\, f_{ck}\,(b_f - b_w)\, y_f\,(d - 0.5\ y_f) \quad ...(6.18)$$

where $y_f = 0.15\, x_{u.max} + 0.65\, D_f$ (subject to a maximum of D_f). ...[6.11(*b*)]

(*c*) OVER-REINFORCED SECTION ($x_u > x_{u.max}$) : When x_u is greater than $x_{u\ max}$ the section will be *over-reinforced*. The failure takes place due to crushing of concrete while strain ε_s in steel still remains below the yield strain ε_{su} The moment of resistance of such a section is limited to $M_{u\ lim}$ given by Eq. 6.18. The over-reinforced-section results in wastage of excess steel provided, and the concrete reaches its ultimate capacity before steel yields, resulting in a sudden failure. Such a section should therefore be redesigned.

6.7. IS CODE PROCEDURE FOR FINDING MOMENT OF RESISTANCE (IS : 456 - 2000)

IS Code procedure for finding moment of resistance of the flanged section depends upon the following *three cases.*

Case 1. When $x_u < D_f$

Case 2. When $(x_{u\ max} = x_u) > D_f$

Case 3. When $x_{u\ max} > x_u > D_f$

Case 1. When $x_u < D_f$: If $x_u < D_f$ the N.A. falls entirely in the flange. Hence the moment of resistance is computed, treating the section as a rectangular section, taking $b = b_f$

(*a*) The value of x_u is found by Eq. 6.6 :

$$\frac{x_u}{d} = \frac{0.87 f_y A_{st}}{0.36 f_{ck} b_f d} \qquad \text{...(6.6)}$$

(*b*) If the value of $\frac{x_u}{d}$ is less than $\frac{x_{u.max}}{d}$, we have

$$M_u = 0.87 f_y A_{st} d \left(1 - \frac{A_{st} f_y}{b_f d f_{ck}}\right) \qquad \text{...(6.19)}$$

(*c*) If the value of $\frac{x_u}{d}$ is equal to $\frac{x_{u.max}}{d}$, the moment of resistance of the section is given by :

$$M_{u\ lim} = 0.36 \frac{x_{u.max}}{d}\left(1 - 0.416 \frac{x_{u.max}}{d}\right) b_f d^2 f_{ck} \qquad \text{...(6.20)}$$

(*d*) If $\frac{x_u}{d}$ is greater than $\frac{x_{u.max}}{d}$, the section should be redesigned.

All these equations and recommendations are as per para G–2.1 of the I.S. Code.

Case 2. $(x_{u,max} = x_u) > D_f$: Balanced section

This is the case when x_u is greater than D_f but at the same time it is equal to $x_{u.max}$. It is thus the case of a *balanced section*. In such a case, the moment of resistance of the section, known as *limiting moment of resistance* will depend upon the *value* of D_f/d ratio either lesser than or greater than a certain *value* discussed below.

Taking $x_u = x_{u.max}$, and $\frac{x_{u,max}}{D_f} = \frac{7}{3}$

or $\frac{x_{u,max}/d}{D_f/d} = \frac{7}{3}$ or $\frac{D_f}{d} = \frac{3}{7}\frac{x_{u,max}}{d}$

The values of D_f/d, dividing *case* 2 into *case* 2 *(a)* and *case* 2 *(b)*, for different grades of steel are as under :

Grade of steel	$x_{u,max}/d$	$D_f/d = \frac{3}{7} x_{u,max}/d$
Fe 250	0.531	0.228
Fe 415	0.479	0.205
Fe 500	0.456	0.195

Indian Standard Code IS 456 : 2000 recommends only a single value of $D_f/d = 0.2$ for all the grades of steel. Thus, we have two dividing cases:

Case 2 (a) : $(x_{u,max} = x_u) > D_f$ *and* $D_f/d \le 0.2$

and *Case 2 (b)* : $(x_{u,max} = x_u) > D_f$ *and* $D_f/d > 0.2$

Case 2. (*a*) $(x_{u,max} = x_u) > D_f$ and $\frac{D_f}{d} \le 0.2$: The moment of resistance is given by

$$M_{u.lim.} = 0.36 \frac{x_{u.max}}{d}\left(1 - 0.416 \frac{x_{u.max}}{d}\right) f_{ck} b_w d^2 + 0.446 f_{ck} (b_f - b_w) D_f \left(d - \frac{D_f}{2}\right) \qquad \text{...(6.21)}$$

Thus, in this case, the I.S. Code has taken the depth y_f of the equivalent rectangular stress block equal to its maximum value of D_f. This equation is the same as given in para G-2.2 of the Code. Note that, this equation is same as Eq. 6.18. where y_f has been taken equal to D_f

Case 2. (*b*) $(x_{u,max} = x_u) > D_f$ **and** $\frac{D_f}{d} > 0.2$: The moment of resistance is given by

$$M_{u.lim.} = 0.36 \frac{x_{u.max}}{d}\left(1 - 0.416 \frac{x_{u.max}}{d}\right) f_{ck}\, b_w\, d^2 + 0.446 f_{ck}\,(b_f - b_w)\, y_f\left(d - \frac{y_f}{2}\right) \quad ...(6.22)$$

where $y_f = (0.15\, x_{u\,max} + 0.65\, D_f)$ (But not greater than D_f).

The above equation is the same as given in para G-2.2.1 of the Code. Also, this equation is same as Eq. 6.18.

Case 3. $x_{u.max} > x_u > D_f$: This is a case of under-reinforced section in which the N.A. falls outside the flange. Here again, the moment of resistance will depend upon whether the value of $\frac{D_f}{x_u}$ is equal to or less than 0.43(or 3/7) or the value of $\frac{D_f}{x_u}$ is greater than 0.43.

Case 3. (*a*) $x_{u,max} > x_u > D_f$ **and** $\frac{D_f}{x_u} < 0.43$: When D_f is equal to or less than $\frac{3}{7}x_u$ (or $0.43x_u$), the depth *DE* of the stress block will be greater than D_f. Thus the flange will fall entirely within the rectangular portion of the stress block. Hence the moment of resistance is given by

$$M_u = 0.36 \frac{x_u}{d}\left(1 - 0.416 \frac{x_u}{d}\right) f_{ck}\, b_w\, d^2 + 0.446 f_{ck}\,(b_f - b_w)\, D_f\left(d - \frac{D_f}{2}\right) \quad ...(6.23)$$

This is same as Eq. 6.16. Also, this equation is the same as given in para G-2.2 of the Code except that $x_{u\,max}$ has been replaced by x_u

Case 3. (*b*) $x_{u,max} > x_u > D_f$ **and** $\frac{D_f}{x_u} > 0.43$: When D_f is equal to or greater than $\frac{3}{7}x_u$ (or $0.43\, x_u$) the depth *DE* of the stress block will be smaller than the depth of the flange. Due to this, the distribution of compressive stress in the flange will not be rectangular. The moment of resistance is given by :

$$M_u = 0.36 \frac{x_u}{d}\left(1 - 0.416 \frac{x_u}{d}\right) f_{ck}\, b_w\, d^2 + 0.446 f_{ck}\,(b_f - b_w)\, y_f\left(d - \frac{y_f}{2}\right) \quad ...(6.24)$$

where $y_f = (0.15\, x_u + 0.65\, D_f)$ (But not greater than D_f)

The above equation is the same as Eq. 6.17. Also, it is the same as given in para G-2.2.1 of the Code, except that $x_{u\,max}$ has been replaced by x_u

Note 1 : When the N.A. falls outside the flange, as concluded from the operation of Eq. 6.6, a new value of x_u should be found, by taking the compressive force of the web into account, as discussed in § 6.6. Though the Code has not specifically mentioned this, it is implied that this new value of x_u should be used while comparing it with

$x_{u\,max}$ or D_f. The value of x_u for such a case can be found either by Eq. 6.15 or by Eq. 6.15 (*a*), as the case may be

Note 2. In para G-2.2.1, while computing the value of y_f, the value of x_u has been erroneously printed in the place of $x_{u,max}$. Hence the value of y_f should be taken as equal to, $(0.15\, x_{u.max} + 0.65\, D_f)$, subject to a maximum of D_f.

Note 3. The Code has not specifically dealt with the case when x_u is greater than D_f and is also greater than $x_{u.max}$. Hence, when $x_u > x_{u\,max} > D_f$, the section will be *over-reinforced.* The failure will take place due to the crushing of concrete while strain ε_s in steel still remains below the yield strain ε_{su} Therefore, the moment of resistance of such a section is limited to $M_{u\,lim}$, and is given by Eq. 6.21 $\left(\text{if } \frac{D_f}{d} \le 0.2\right)$ or by Eq. 6.22 if $\frac{D_f}{d} > 0.2$, as the case may be. Such a section should be redesigned because compression failure is *sudden* and therefore not desirable.

Use of design aids for determining maximum moment of resistance ($M_{u\,lim}$)

The maximum moment of resistance is obtained when the depth of neutral axis is $x_{u\,max}$. When the thickness of flange is small, that is less than 0.2 *d* (*i.e.*, case 2*a* discussed above), the stress in the flange will be uniform or nearly uniform, and the centroid of the compressive force in the flange can be taken at $D_f/2$ from the extreme fibre, thus giving rise to Eq. 6.21 for $M_{u\,lim}$.

$$M_{u\,lim} = 0.36\, f_{ck}\, .\, b_w\, .\, x_{u\,max}\left(d - 0.416\, x_{u,max}\right) + 0.446\, f_{ck}\,(b_f - b_w)\, D_f\left(d - \frac{D_f}{2}\right) \quad \ldots(6.21)$$

However, when the flange thickness is greater than 0.2 *d*, the above expression is not correct because the stress distribution in the flange would not be uniform. The expression given in para G–2.2.1 of the Code, expressed in the form of Eq. 6.22 below, is an approximation which makes an allowance for the variation of stress in the flange :

$$M_{u\,lim} = 0.36\, f_{ck}\, .\, b_w\, .\, x_{u.max}\left(d - 0.416\, x_{u\,max}\right) + 0.446\, f_{ck}\,(b_f - b_w)\, y_f\left(d - \frac{y_f}{2}\right) \quad \ldots(6.22)$$

where $y_f = (0.15\, x_{u.\,max.} + 0.65\, D_f)$, but not greater than D_f

Dividing both sides by $f_{ck}\, b_w\, d^2$,

$$\frac{M_{u.lim}}{f_{ck}\, b_w\, d^2} = \frac{0.36\, f_{ck}\, .\, b_w\, .\, x_{u.max}\left(d - 0.416\, x_{u\,max}\right)}{f_{ck}\, b_w\, d^2} + 0.446\left(\frac{b_f}{b_w} - 1\right)\frac{y_f}{d}\left(1 - \frac{y_f}{2d}\right) \quad \ldots(6.22a)$$

where $\frac{y_f}{d} = 0.15\,\frac{x_{u,\,max.}}{d} + 0.65\,\frac{D_f}{d}$, but $\frac{y_f}{d} < \frac{D_f}{d}$

Using the above expression, the values of moment of resistance factor $\frac{M_{u\,lim}}{f_{ck}\, b_w\, d^2}$ for different values of $\frac{b_f}{b_w}$ and $\frac{D_f}{d}$ have been worked out and given in Tables 6.1, 6.2 and 6.3 for three grades of steel having f_y = 250 N/mm^2, 415 N/mm^2 and 500 N/mm^2 respectively.

TABLE 6.1. FLEXURE : LIMITING MOMENT OF RESISTANCE FACTOR, $M_{u,\,lim}/b_w d^2 f_{ck}$ FOR SINGLY REINFORCED T-BEAMS, N/mm^2

($f_y = 250\ N/mm^2$)

D_f/d	b_f/b_w									
	1.0	*2.0*	*3.0*	*4.0*	*5.0*	*6.0*	*7.0*	*8.0*	*9.0*	*10.0*
0.06	0.149	0.175	0.201	0.227	0.253	0.279	0.305	0.331	0.357	0.383
0.07	0.149	0.179	0.209	0.239	0.270	0.300	0.330	0.360	0.390	0.420
0.08	0.149	0.183	0.218	0.252	0.286	0.320	0.355	0.389	0.423	0.457
0.09	0.149	0.187	0.226	0.264	0.302	0.341	0.379	0.417	0.456	0.494
0.10	0.149	0.191	0.234	0.276	0.318	0.361	0.403	0.446	0.488	0.530
0.11	0.149	0.195	0.242	0.288	0.334	0.381	0.427	0.474	0.520	0.566
0.12	0.149	0.199	0.250	0.300	0.350	0.401	0.451	0.501	0.551	0.602
0.13	0.149	0.203	0.257	0.312	0.366	0.420	0.474	0.528	0.583	0.637
0.14	0.149	0.207	0.265	0.323	0.381	0.439	0.497	0.555	0.614	0.672
0.15	0.149	0.211	0.273	0.335	0.397	0.458	0.520	0.582	0.644	0.706
0.16	0.149	0.215	0.280	0.346	0.412	0.477	0.543	0.609	0.674	0.740
0.17	0.149	0.218	0.288	0.357	0.427	0.496	0.565	0.635	0.704	0.773
0.18	0.149	0.222	0.295	0.368	0.441	0.514	0.587	0.660	0.733	0.806
0.19	0.149	0.226	0.302	0.379	0.456	0.532	0.609	0.686	0.763	0.839
0.20	0.149	0.229	0.310	0.390	0.470	0.550	0.631	0.711	0.791	0.872
0.21	0.149	0.233	0.317	0.400	0.484	0.568	0.652	0.736	0.820	0.903
0.22	0.149	0.236	0.324	0.411	0.498	0.586	0.673	0.760	0.848	0.935
0.23	0.149	0.240	0.330	0.421	0.511	0.602	0.692	0.783	0.873	0.964
0.24	0.149	0.242	0.334	0.427	0.520	0.613	0.705	0.798	0.891	0.984
0.25	0.149	0.244	0.339	0.434	0.529	0.624	0.719	0.814	0.909	1.003
0.26	0.149	0.246	0.343	0.440	0.538	0.635	0.732	0.829	0.926	1.023
0.27	0.149	0.248	0.348	0.447	0.546	0.645	0.745	0.844	0.943	1.043
0.28	0.149	0.250	0.352	0.453	0.555	0.656	0.758	0.859	0.961	1.062
0.29	0.149	0.253	0.356	0.460	0.563	0.667	0.770	0.874	0.978	1.081
0.30	0.149	0.255	0.360	0.466	0.572	0.677	0.783	0.889	0.995	1.100
0.31	0.149	0.257	0.365	0.472	0.580	0.688	0.796	0.903	1.011	1.119
0.32	0.149	0.259	0.369	0.479	0.588	0.698	0.808	0.918	1.028	1.138
0.33	0.149	0.261	0.373	0.485	0.597	0.709	0.820	0.932	1.044	1.156
0.34	0.149	0.263	0.377	0.491	0.605	0.719	0.833	0.947	1.061	1.175
0.35	0.149	0.265	0.381	0.497	0.613	0.729	0.845	0.961	1.077	1.193
0.36	0.149	0.267	0.385	0.503	0.621	0.739	0.857	0.975	1.093	1.211
0.37	0.149	0.269	0.389	0.509	0.629	0.749	0.869	0.989	1.109	1.229
0.38	0.149	0.271	0.393	0.515	0.637	0.759	0.880	1.002	1.124	1.246
0.39	0.149	0.273	0.397	0.521	0.644	0.768	0.892	1.016	1.140	1.264
0.40	0.149	0.275	0.401	0.526	0.652	0.778	0.904	1.029	1.155	1.281
0.41	0.149	0.277	0.404	0.532	0.660	0.787	0.915	1.043	1.170	1.298
0.42	0.149	0.279	0.408	0.538	0.667	0.797	0.926	1.056	1.186	1.315
0.43	0.149	0.280	0.412	0.543	0.675	0.806	0.938	1.069	1.200	1.332
0.44	0.149	0.282	0.416	0.549	0.682	0.815	0.949	1.082	1.215	1.349
0.45	0.149	0.284	0.419	0.554	0.689	0.825	0.960	1.095	1.230	1.365

TABLE 6.2. FLEXURE : LIMITING MOMENT OF RESISTANCE FACTOR, $M_{u,lim}/b_w d^2 f_{ck}$ FOR SINGLY REINFORCED T-BEAMS, N/mm^2

(f_y = 415 N/mm^2)

D_f/d	b_f/b_w									
	1.0	*2.0*	*3.0*	*4.0*	*5.0*	*6.0*	*7.0*	*8.0*	*9.0*	*10.0*
0.06	0.138	0.164	0.190	0.216	0.242	0.268	0.294	0.320	0.346	0.372
0.07	0.138	0.168	0.198	0.228	0.259	0.289	0.319	0.349	0.379	0.409
0.08	0.138	0.172	0.207	0.241	0.275	0.309	0.344	0.378	0.412	0.446
0.09	0.138	0.176	0.215	0.253	0.291	0.330	0.368	0.406	0.445	0.483
0.10	0.138	0.180	0.223	0.265	0.308	0.350	0.392	0.435	0.477	0.519
0.11	0.138	0.184	0.231	0.277	0.324	0.370	0.416	0.463	0.509	0.555
0.12	0.138	0.188	0.239	0.289	0.339	0.390	0.440	0.490	0.541	0.591
0.13	0.138	0.192	0.247	0.301	0.355	0.409	0.463	0.518	0.572	0.626
0.14	0.138	0.196	0.254	0.312	0.370	0.428	0.487	0.545	0.603	0.661
0.15	0.138	0.200	0.262	0.324	0.386	0.448	0.509	0.571	0.633	0.695
0.16	0.138	0.204	0.269	0.335	0.401	0.466	0.532	0.598	0.663	0.729
0.17	0.138	0.207	0.277	0.346	0.416	0.485	0.554	0.624	0.693	0.762
0.18	0.138	0.211	0.284	0.357	0.430	0.503	0.576	0.649	0.723	0.796
0.19	0.138	0.215	0.291	0.368	0.445	0.522	0.598	0.675	0.752	0.828
0.20	0.138	0.218	0.299	0.379	0.459	0.540	0.620	0.700	0.780	0.861
0.21	0.138	0.221	0.305	0.388	0.471	0.554	0.638	0.721	0.804	0.887
0.22	0.138	0.224	0.309	0.395	0.480	0.566	0.651	0.737	0.822	0.908
0.23	0.138	0.226	0.314	0.402	0.489	0.577	0.665	0.753	0.841	0.928
0.24	0.138	0.228	0.318	0.408	0.498	0.588	0.678	0.768	0.859	0.949
0.25	0.138	0.230	0.323	0.415	0.507	0.600	0.692	0.784	0.876	0.969
0.26	0.138	0.233	0.327	0.422	0.516	0.611	0.705	0.800	0.894	0.989
0.27	0.138	0.235	0.331	0.428	0.525	0.622	0.718	0.815	0.912	1.008
0.28	0.138	0.237	0.336	0.435	0.534	0.632	0.731	0.830	0.929	1.028
0.29	0.138	0.239	0.340	0.441	0.542	0.643	0.744	0.845	0.946	1.047
0.30	0.138	0.241	0.344	0.448	0.551	0.654	0.757	0.860	0.963	1.066
0.31	0.138	0.243	0.349	0.454	0.559	0.664	0.770	0.875	0.980	1.085
0.32	0.138	0.245	0.353	0.460	0.568	0.675	0.782	0.890	0.997	1.104
0.33	0.138	0.248	0.357	0.466	0.576	0.685	0.795	0.904	1.014	1.123
0.34	0.138	0.250	0.361	0.473	0.581	0.696	0.807	0.919	1.030	1.142
0.35	0.138	0.252	0.365	0.479	0.592	0.706	0.819	0.933	1.046	1.160
0.36	0.138	0.254	0.369	0.485	0.600	0.716	0.831	0.947	1.063	1.178
0.37	0.138	0.256	0.373	0.491	0.608	0.726	0.843	0.961	1.079	1.196
0.38	0.138	0.258	0.377	0.497	0.616	0.736	0.855	0.975	1.094	1.214
0.39	0.138	0.260	0.381	0.503	0.624	0.746	0.867	0.989	1.110	1.232
0.40	0.138	0.262	0.385	0.508	0.632	0.755	0.879	1.002	1.126	1.249
0.41	0.138	0.263	0.389	0.514	0.640	0.765	0.890	1.016	1.141	1.267
0.42	0.138	0.265	0.393	0.520	0.647	0.775	0.902	1.029	1.156	1.284
0.43	0.138	0.267	0.396	0.526	0.655	0.784	0.913	1.042	1.172	1.301
0.44	0.138	0.269	0.400	0.531	0.662	0.793	0.924	1.055	1.187	1.318
0.45	0.138	0.271	0.404	0.537	0.670	0.803	0.936	1.068	1.201	1.334

TABLE 6.3. FLEXURE : LIMITING MOMENT OF RESISTANCE FACTOR, $M_{u,\lim}/b_w d^2 f_{ck}$ FOR SINGLY REINFORCED T-BEAMS, N/mm^2

($f_y = 500\ N/mm^2$)

D_f/d	b_f/b_w									
	1.0	2.0	3.0	4.0	5.0	6.0	7.0	8.0	9.0	10.0
0.06	0.133	0.159	0.185	0.211	0.237	0.263	0.289	0.315	0.341	0.367
0.07	0.133	0.163	0.193	0.223	0.254	0.284	0.314	0.344	0.374	0.404
0.08	0.133	0.167	0.202	0.236	0.270	0.304	0.339	0.373	0.407	0.441
0.09	0.133	0.171	0.210	0.248	0.286	0.325	0.363	0.401	0.440	0.478
0.10	0.133	0.175	0.218	0.260	0.303	0.345	0.387	0.430	0.472	0.514
0.11	0.133	0.179	0.226	0.272	0.318	0.365	0.411	0.458	0.504	0.550
0.12	0.133	0.183	0.234	0.284	0.334	0.385	0.435	0.485	0.536	0.586
0.13	0.133	0.187	0.241	0.296	0.350	0.404	0.458	0.513	0.567	0.621
0.14	0.133	0.191	0.249	0.307	0.365	0.423	0.481	0.540	0.598	0.656
0.15	0.133	0.195	0.257	0.319	0.381	0.442	0.504	0.566	0.628	0.690
0.16	0.133	0.199	0.264	0.330	0.396	0.461	0.527	0.593	0.658	0.724
0.17	0.133	0.202	0.272	0.341	0.411	0.480	0.549	0.619	0.688	0.757
0.18	0.133	0.206	0.279	0.352	0.425	0.498	0.571	0.644	0.717	0.791
0.19	0.133	0.210	0.286	0.363	0.440	0.516	0.593	0.670	0.747	0.823
0.20	0.133	0.213	0.292	0.372	0.452	0.532	0.611	0.691	0.771	0.850
0.21	0.133	0.215	0.297	0.379	0.461	0.543	0.625	0.707	0.789	0.871
0.22	0.133	0.217	0.302	0.386	0.470	0.555	0.639	0.723	0.808	0.892
0.23	0.133	0.220	0.306	0.393	0.479	0.566	0.653	0.739	0.826	0.912
0.24	0.133	0.222	0.311	0.400	0.488	0.577	0.666	0.755	0.844	0.933
0.25	0.133	0.224	0.315	0.406	0.497	0.589	0.680	0.771	0.862	0.953
0.26	0.133	0.226	0.320	0.413	0.506	0.600	0.693	0.786	0.880	0.973
0.27	0.133	0.229	0.324	0.420	0.515	0.611	0.706	0.802	0.897	0.993
0.28	0.133	0.231	0.328	0.426	0.524	0.622	0.719	0.817	0.915	1.012
0.29	0.133	0.233	0.333	0.433	0.532	0.632	0.732	0.832	0.932	1.032
0.30	0.133	0.235	0.337	0.439	0.541	0.643	0.745	0.847	0.949	1.051
0.31	0.133	0.237	0.341	0.445	0.550	0.654	0.758	0.862	0.966	1.070
0.32	0.133	0.239	0.346	0.452	0.558	0.664	0.770	0.877	0.983	1.089
0.33	0.133	0.241	0.350	0.458	0.566	0.675	0.783	0.891	1.000	1.108
0.34	0.133	0.243	0.354	0.464	0.575	0.685	0.795	0.906	1.016	1.127
0.35	0.133	0.245	0.358	0.470	0.583	0.695	0.808	0.920	1.033	1.145
0.36	0.133	0.248	0.362	0.476	0.591	0.705	0.820	0.934	1.049	1.163
0.37	0.133	0.250	0.366	0.483	0.599	0.716	0.832	0.949	1.065	1.181
0.38	0.133	0.252	0.370	0.489	0.607	0.725	0.844	0.962	1.081	1.199
0.39	0.133	0.254	0.374	0.494	0.615	0.735	0.856	0.976	1.097	1.217
0.40	0.133	0.255	0.378	0.500	0.623	0.745	0.868	0.990	1.112	1.235
0.41	0.133	0.257	0.382	0.506	0.630	0.755	0.879	1.004	1.128	1.252
0.42	0.133	0.259	0.386	0.512	0.638	0.764	0.891	1.017	1.143	1.270
0.43	0.133	0.261	0.389	0.518	0.646	0.774	0.902	1.030	1.158	1.287
0.44	0.133	0.263	0.393	0.523	0.653	0.783	0.913	1.043	1.174	1.304
0.45	0.133	0.265	0.397	0.529	0.661	0.793	0.925	1.056	1.188	1.320

6.8. TYPES OF PROBLEMS

There may be following types of problems in flanged beams :

(a) Analysis problems

1. ***Type I problem*** : Given (*i*) Material properties (i.e. f_{ck} and f_y)
and (*ii*) Sectional properties (i.e. b_f, D_f, b_w, D, A_{st})
To find : Moment of resistance (M_u)
(See examples 6.1, 6.2, and 6.3)

2. ***Type II problem*** : Given (*i*) Material properties (i.e. f_{ck} and f_y)
(*ii*) Concrete section properties (i.e. b_f, D_f, b_w, D)
To find : (*i*) Limiting moment $M_{u\ lim}$
(*ii*) Limiting area of steel ($A_{st,\ lim.}$ or $A_{st,\ max}$)
(See example 6.4)

(b) Design problems

3. ***Type III problem*** : Given (*i*) Material properties (i.e. f_{ck} and f_y)
and (*ii*) Ultimate design moment (M_{uD})
To find : Area of tension steel A_{st}
(See example 6.5, 6.6, and 6.7)

4. ***Type IV problem*** : Given (*i*) Material properties (i.e. f_{ck} and f_y)
and (*ii*) Ultimate design moment (M_{uD})

To find : A_{st} and A_{sc} : See § 6.9 and 6.10 on doubly reinforced T-beams. Also see example 6.11 for design of doubly reinforced T-beam.

6.9. DESIGN OF T-BEAM

The design of T-beam of given dimensions (*i.e.* b_w, b_f, d known) and given design moment, is done in the following steps :

Step 1. Assuming $x_u = D_f$, compute the moment of resistance M_{u1} from Eq. 6.9(*a*)

$$M_{u1} = 0.36\, f_{ck}\, b_f D_f (d - 0.416\, D_f) \qquad \ldots[6.9(a)]$$

Step 2. If $M_{u1} > M_{uD}$, the N.A. will fall inside the flange (*i.e.* $x_u \le D_f$). In that case the reinforcement is given by Eq. 6.7 :

$$M_u = M_{uD} = 0.87\, f_y A_{st}\, d \left(1 - \frac{f_y}{f_{ck}} \frac{A_{st}}{b_f\, d}\right) \qquad \ldots(6.7)$$

The solution of the above quadratic equation gives :

$$A_{st} = \frac{0.5\, f_{ck}}{f_y} \left[1 - \sqrt{1 - \frac{4.6\, M_{uD}}{f_{ck}\, b_f\, d^2}}\right] b_f\, d \qquad \ldots(6.7\ (a))$$

Step 3. If $M_{u1} < M_{uD}$, the N.A. falls outside the flange. Assuming $x_u = \frac{7}{3} D_f$, compute M_{u2} from Eq. 6.16 (*c*) :

$$M_{u2} = 0.36\, f_{ck}\, b_w \left(\frac{7}{3} D_f\right)\left(d - 0.416 \times \frac{7}{3} D_f\right) + 0.446\, f_{ck}\, (b_f - b_w) \times D_f\, (d - 0.5\, D_f) \qquad \ldots[6.16(c)]$$

Step 4. If $M_{u2} < M_{uD}$, $x_u > \frac{7}{3} D_f$. Compute A_{su} from Eq. 6.19(*a*) as under :

$$C_{uw} = 0.36\, f_{ck}\, x_u\, b_w = 0.87\, f_y\, A_{sw}$$

or $$A_{sw} = \frac{0.36\, f_{ck}\, x_u\, b_w}{0.87\, f_y} \qquad \ldots(6.25)$$

Similarly, compute A_{sf} from Eq. 6.12

$$C_{uf} = 0.446\, f_{ck}\, (b_f - b_w)\, D_f = 0.87\, f_y\, A_{sf}$$

or $$A_{sf} = \frac{0.446\, f_{ck}\, (b_f - b_w)\, D_f}{0.87\, f_y} \qquad \ldots(6.26)$$

Total $\quad A_{st} = A_{sw} + A_{sf}$

When $M_{uD} >> M_{u2}$, it is always advisible to compute $M_{u,lim}$, since it is likely that $x_u > x_{u,max}$. If $M_{uD} > M_{u\,lim}$, the section will be doubly reinforced.

Step 5. If $M_{u2} > M_{uD}$, $x_u < \frac{7}{3} D_f$. The design procedure in this case will be that of trial and error. The following steps are recommended :

(*i*) Assume $x_u < \frac{7}{3} D_f$

(*ii*) Compute $y_f = 0.15\, x_u + 0.65\, D_f$ (subject to a max. of D_f)

(*iii*) Compute M_u from Eq. 6.17 :

$$M_u = 0.36\, \frac{x_u}{d}\left(1 - 0.416\, \frac{x_u}{d}\right) f_{ck}\, b_w\, d^2 + 0.446\, f_{ck}\, (b_f - b_w)\, y_f\, (d - 0.5\, y_f) \qquad \ldots(6.17)$$

(*iv*) Compare M_u with M_{uD}

If $M_u = M_{uD}$, assumed value of x_u is correct.

If $M_{uD} > M_u$, increase x_u in next trial.

If $M_{uD} < M_u$, decrease x_u in next trial. Repeat till $M_u = M_{uD}$

(*v*) Knowing x_u, compute ; $C_{uw} = 0.36\, f_{ck}\, x_u\, b_w \qquad \ldots[6.10(a)]$

$$C_{uf} = 0.446\, f_{ck}\, (b_f - b_w)\, y_f \qquad \ldots(6.12)$$

$$\therefore \quad C_u = C_{uw} + C_{uf}.$$

(*vi*) Compute $$A_{st} = \frac{C_u}{0.87\, f_y} \qquad \ldots(6.27)$$

Alternatively, x_u can be found by substituting the value of y_f in terms of x_u in Eq. 6.17 and then solving the quadratic equation in terms of x_u. This value of x_u will be acceptable only if the value of y_f obtained from this is less than (or equal to) D_f.

Example 6.1 : M.R. of T-beam. *Find the moment of resistance of an existing T-beam having the following data :*

b_f = 740 mm ; d = 400 mm ; b_w = 240 mm

A_{st} = 5-20 mm φ mild steel bars

D_f = 100 mm. M 15 concrete.

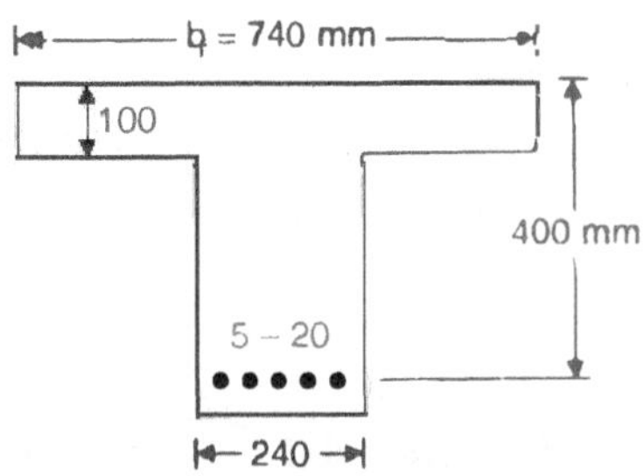

FIG. 6.9

Solution

$$A_{st} = 5\, \frac{\pi}{4}\, (20)^2 = 1570 \text{ mm}^2$$

Assuming the N.A. to fall in the flange, we get, from Eq. 6.6

$$\frac{x_u}{d} = \frac{0.87 f_y A_{st}}{0.36 f_{ck} b_f d} = \frac{0.87 \times 250 \times 1570}{0.36 \times 15 \times 740 \times 400} = 0.2136.$$

or $x_u = 0.2136 \times 400 = 85.5$ mm

$\therefore$ $x_u < D_f$. Hence the N.A. lies in the flange.

Also, $x_{u\,max} = 0.531\, d$ for Fe 250 steel $= 0.531 \times 400 = 212.4$ mm $> x_u$

Hence, the moment of resistance is given by Eq. 6.7,

$$M_u = 0.87\ f_y \cdot A_{st} \cdot d\left(1 - \frac{f_y}{f_{ck}} \cdot \frac{A_{st}}{b_f \cdot d}\right) = 0.87 \times 250 \times 1570 \times 400 \left(1 - \frac{250}{15} \times \frac{1570}{740 \times 400}\right)$$

$= 124.5 \times 10^6$ N-mm $=$ **124.5 kN-m.**

Example 6.2. M.R. of T-beam

Determine the M.R. of T-beam of example 6.1 if $D_f = 80$ mm.

Solution. Assuming the N.A. to fall in the flange, we get $x_u = 85.5$ mm, as obtained in example 6.1. Since $x_u > D_f$ (=80 mm), the N.A. falls outside the flange. Let us obtain a new value of x_u by equating the total compressive force to the total tensile force.

Total tensile force $T_u = 0.87 f_y A_{st} = 0.87 \times 250 \times 1570 = 341475$ N

Depth of equivalent stress block

$$= y_f = 0.15\, x_u + 0.65\, D_f = 0.15\, x_u + 0.65 \times 80 = 0.15\, x_u + 52$$

Hence $C_u = 0.36 f_{ck}\, x_u\, b_w + 0.446 f_{ck} (b_f - b_w)\, y_f$

$$= (0.36 \times 15\, x_u \times 240) + 0.446 \times 15\,(740 - 240)\,(0.15\, x_u + 52)$$

$$= 1296\, x_u + 501.8\, x_u + 173940 = 1797.8\, x_u + 173940$$

Equating $C_u = T_u$, we get, $1797.8\, x_u + 173940 = 341475$

From which $x_u = 93.2$ mm. This value will be acceptable only if $y_f \le D_f$

Now $y_f = 0.15\, x_u + 0.65\, D_f = 0.15 \times 93.2 + 52 \approx 66$ mm $< D_f$

Hence $x_u = 93.2$ mm is acceptable.

$x_{u\,max} = 0.531\, d$ (for Fe 250 steel) $= 0.531 \times 400 = 212.4$ mm $> x_u$

Now $\frac{3}{7} x_u = \frac{3}{7} \times 93.2 = 39.9.$

$\therefore$ $D_f > \frac{3}{7} x_u =$ or $\frac{D_f}{x_u} > 0.43$ which corresponds to case 3 (*b*) of §6.7 of IS Code procedure. Hence, the moment of resistance is given by Eq. 6.24 :

$$M_u = 0.36 \frac{x_u}{d}\left(1 - 0.416 \frac{x_u}{d}\right) f_{ck}\, b_w\, d^2 + 0.446 f_{ck} (b_f - b_w)\, y_f \left(d - \frac{y_f}{2}\right)$$

where $y_f = 0.15\, x_u + 0.65\, D_f = 66$ mm (computed above)

$$M_u = \frac{0.36 \times 93.2}{400}\left(1 - 0.416\ \frac{93.2}{400}\right) \times 15 \times 240\ (400)^2$$

$$+ 0.446 \times 15\,(740 - 240) \times 66\,(400 - 0.5 \times 66)$$

$= 43.63 \times 10^6 + 81.02 \times 10^6 = 124.65 \times 10^6$ N-mm $=$ **142.65 kN-m.**

Example 6.3 : M.R. of T-beam. *Determine the M.R. of T-beam shown in Fig. 6.10. Use M 15 concrete and Fe 415 steel.*

Solution

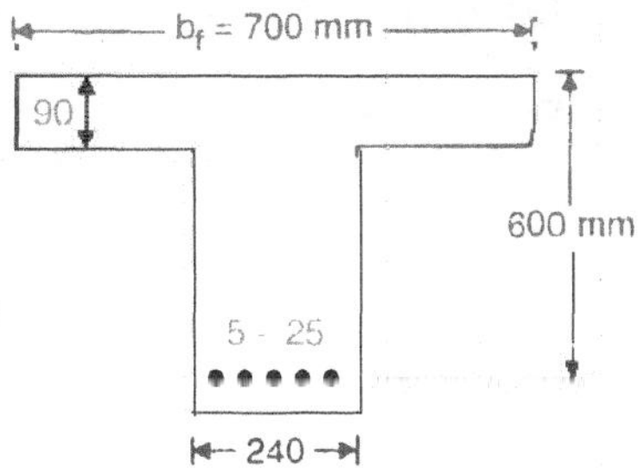

FIG. 6.10

$$A_{st} = 5 \times \frac{\pi}{4}(25)^2 = 2454.4 \text{ mm}^2$$

Assuming the N.A. to fall in the flange, we have,

$$\frac{x_u}{d} = \frac{0.87 f_y A_{st}}{0.36 f_{ck} b_f d} = \frac{0.87 \times 415 \times 2454.4}{0.36 \times 15 \times 700 \times 600} = 0.3907$$

$$\therefore \quad x_u = 0.3907 \times 600 = 234.4 \text{ mm} > D_f$$

Hence N.A. falls outside flange. To obtain a new value of x_u, equate total compressive force to the total tensile force.

$$T_u = 0.87 f_y A_{st} = 0.87 \times 415 \times 2454.4 = 886161 \quad \text{N}$$

$$y_f = 0.15 x_u + 0.65 D_f = 0.15 x_u + 0.65 \times 90 = 0.15 x_u + 58.5$$

with a maximum equal to D_f. Since approximate value of x_u found above is 234.4, $y_f \simeq 0.15 \times 234.4 + 58.5 = 93.7 > D_f$.

Hence take $\quad y_f = D_f = 90$ mm

$$C_u = 0.36 f_{ck} x_u b_w + 0.446 f_{ck}(b_f - b_w) D_f$$

$$= 0.36 \times 15 \times x_u (240) + 0.446 \times 15 (700 - 240) 90 = 1296 x_u + 276966.$$

Equating C_u and T_u, we get, $\quad 1296 x_u + 276966 = 886161$

$$\therefore \quad x_u = 470 \text{ mm.}$$

$$x_{u,max} \simeq 0.48\, d \text{ (for Fe 415 steel)} = 0.48 \times 600 = 288 \text{ mm.}$$

Hence $x_u > x_{u,max}$. Thus the beam is *over-reinforced*. Though the Code recommends that such a beam should be redesigned, its moment of resistance is limited to $M_{u,lim}$ corresponding to the balanced section.

$\therefore$ Take $\quad x_u = x_{u,max} = 288$ mm.

Again. $\dfrac{D_f}{d} = \dfrac{90}{600} = 0.15 < 0.2$. Hence moment of resistance is given by Eq. 6.21 (corresponding to para *G*-2.2 of Code) :

$$M_u = M_{u,lim} = 0.36 \frac{x_{u,max}}{d}\left(1 - 0.416 \frac{x_{u,max}}{d}\right) f_{ck} b_w d^2 + 0.446 f_{ck}(b_f - b_w) D_f \left(d - \frac{D_f}{2}\right)$$

$$= 0.36 \times \frac{288}{600}\left(1 - 0.416 \frac{288}{600}\right) 15 \times 240 (600)^2 + 0.446 \times 15 (700 - 240) \times 90 \left(600 - \frac{90}{2}\right)$$

$$= 179.2 \times 10^6 + 153.7 \times 10^6 = 332.9 \times 10^6 \text{ N-mm} = \mathbf{332.9 \text{ kN-m}}.$$

Example 6.4. Limiting M.R. and Limiting area of steel

A reinforced concrete T-beam has the following data :

Width of flange = 1600 mm; Effective depth = 350 mm

Thickness of flange = 105 mm ; Width of web = 250 mm

Grade of concrete : M 20 ; Grade of steel Fe 500

Determine (a) Limiting moment of resistance

and (b) Limiting area of tensile steel.

Solution : Given $b_f = 1600$ mm; $d = 350$ mm; $D_f = 105$ mm; $b_w = 250$ mm

For Fe 500 steel , $\frac{x_{u,\,max}}{d} = \frac{700}{1100 + 0.87 f_y} = \frac{700}{1100 + 0.87 \times 500} = 0.456$

∴ $x_{u,\,max} = 0.456\, d = 0.456 \times 350 = 159.6 \text{ mm} > D_f\,(= 105 \text{ mm})$

Also, $\frac{D_f}{d} = \frac{105}{350} = 0.3$. Since $\frac{D_f}{d} > 0.2$, it is case (*b*) of § 6.7.

$y_f = (0.15\, x_{u,\,max} + 0.65\, D_f)$, but not greater than $D_f\,(= 105 \text{ mm})$

$= 0.15 \times 159.6 + 0.65 \times 105 = 92.19$ mm, which is less than D_f.

The limiting moment of resistance is given by Eq. 6.22.

$$M_{u\,.\,lim} = 0.36 \frac{x_{u,\,max}}{d}\left(1 - 0.416 \frac{x_{u,\,max}}{d}\right) f_{ck}\, b_w\, d^2 + 0.446\, f_{ck}\,(b_f - b_w)\, y_f \left(d - \frac{y_f}{2}\right)$$

$$= 0.36 \times 0.456\,(1 - 0.416 \times 0.456) \times 20 \times 250\,(350)^2 + 0.446 \times 20\,(1600 - 250) \times 92.19\left(350 - \frac{92.19}{2}\right)$$

$$= 418.86 \times 10^6 \text{ N-mm} = \mathbf{418.86 \text{ kN-m}}$$

Also, $$A_{st\,,\,lim.} = \frac{0.36\, f_{ck}\, b_w\,.\, x_{u,\,max} + 0.446\, f_{ck}\,(b_f - b_w)\, y_f}{0.87\, f_y}$$

$$= \frac{0.36 \times 20 \times 250 \times 159.6 + 0.446 \times 20\,(1600 - 250)\, 92.19}{0.87 \times 500} = \mathbf{3212.5 \text{ mm}^2}$$

Values from design charts : Here $b_f/b_w = 1600/250 = 6.4$ and $D_f/d = 105/350 = 0.3$ Hence from Table 6.3, we get limiting moment of resistance factor = 0.6838.

∴ $M_{u\,,lim.} = 0.6838\,(b_w\, d^2\, f_{ck}) = 0.6838 \times 250\,(350)^2 \times 20 = \mathbf{418.83 \times 10^6 \text{ N-mm}}$

Example 6.5. Design of T-beam

A T-beam has the following data :

Width of flange = 750 mm ; Breadth of beam = 250 mm

Effective depth = 500 mm ; Thickness of flange = 90 mm

Applied moment = 130 kN-m

Use M 20 concrete and Fe 415 steel.

Solution : Given : $b_f = 750$ mm ; $b_w = 250$ mm; $d = 500$ mm ; $D_f = 90$ mm ; $M_D = 130$ kN-m;

Hence $M_{uD} = 1.5 \times 130 = 195 \text{ kN-m} = 195 \times 10^6$ N-mm

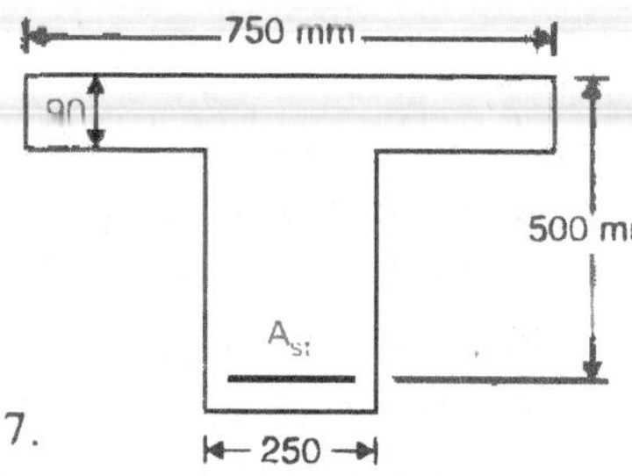

FIG. 6.11

Assuming $x_u = D_f$, we get from Eq. 6.9 (*a*)

$M_{u_1} = 0.36\, f_{ck}\, b_f\, D_f\,(d - 0.416\, D_f)$

$= 0.36 \times 20 \times 750 \times 90\,(500 - 0.416 \times 90)$

$= 224.8 \times 10^6 \text{ N-mm} > M_{uD}$

∴ $x_u < D_f$. The reinforcement is found from Eq. 6.7.

$$M_{uD} = 0.87\, f_y\, A_{st}\, d\left(1 - \frac{f_y}{f_{ck}}\,.\,\frac{A_{st}}{b_f\,.\,d}\right)$$

or $$195 \times 10^6 = 0.87 \times 415 \times 500\, A_{st}\left(1 - \frac{415}{20} \times \frac{A_{st}}{750 \times 500}\right)$$

or $A_{st}\,(1 - 5.533 \times 10^{-5} A_{st}) = 1080.2$

or $A_{st}^2 - 18072\, A_{st} + 19.52 \times 10^6$, from which $A_{st} \simeq$ **1154 mm²**

Alternatively, $$A_{st} = \frac{0.5 f_{ck}}{f_y}\left[1 - \sqrt{1 - \frac{4.6\, M_{uD}}{f_{ck}\, b_f\, d^2}}\right] b_f\, d \qquad \ldots(6.7\ a)$$

$$= \frac{0.5 \times 20}{415}\left[1 - \sqrt{1 - \frac{4.6 \times 195 \times 10^6}{20 \times 750 \times 500^2}}\right] \times 750 \times 500 = \mathbf{1154\ mm^2}$$

Example 6.6 Design of T-beam

Redesign the T-beam of example 6.5 if the applied moment is 230 kN-m.

Solution : $M_{uD} = 1.5 \times 230 = 345$ kN–m $= 345 \times 10^6$ N-mm

Since M_{uD} is more than M_{u1} ($= 224.8 \times 10^6$) determined in the previous example, $x_u > D_f$.

Let us calculate compressive force for $x_u = \frac{7}{3} D_f = \frac{7}{3} \times 90 = 210$ mm

$\therefore$ $$C_{uw} = 0.36\, f_{ck}\, x_u\, b_w = 0.36 \times 20 \times 210 \times 250 = 378000 \text{ N}$$

$$C_{uf} = 0.446\, f_{ck}\, (b_f - b_w)\, D_f = 0.446 \times 20\,(750 - 250)\, 90 = 401400 \text{ N}$$

$$M_{uw} = 378000\,(500 - 0.416 \times 210) = 155.98 \times 10^6 \text{ N-mm}$$

$$M_{uf} = 401400\,(500 - 0.5 \times 90) = 182.64 \times 10^6 \text{ N-mm}$$

$\therefore$ $$M_{u2} = M_{uw} + M_{uf} = (155.98 + 182.64)\,10^6 = 338.62 \times 10^6 \text{ N-mm}$$

Since $$M_{u2} < M_{uD}\,,\quad x_u > \frac{7}{3} D_f$$

Now, M_{uf} does not depend on x_u, and thus remains constant.

$\therefore$ $$M_{uw} = M_{uD} - M_{uf} = (345 - 182.64)\,10^6 = 162.36 \times 10^6 \text{ N-mm.}$$

Now $$M_{uw} = 0.87\, f_y\, A_{sw}\, d\left(1 - \frac{f_y\, A_{sw}}{f_{ck}\, b_w\, d}\right) \qquad \ldots(6.28)$$

$\therefore$ $$162.36 \times 10^6 = 0.87 \times 415 \times A_{sw} \times 500\left(1 - \frac{415}{20} \times \frac{A_{sw}}{250 \times 500}\right)$$

or $A_{sw}\,(1 - 1.66 \times 10^{-4} A_{sw}) = 899$

or $A_{sw}^2 - 6024\, A_{sw} + 5.418 \times 10^6 = 0$; From which $A_{sw} \simeq 1100.5$ mm²

Alternatively, $$A_{sw} = \frac{0.5 f_{ck}}{f_y}\left[1 - \sqrt{1 - \frac{4.6\, M_{uw}}{f_{ck}\, b_w d^2}}\right] b_w\, .\, d \qquad \ldots(6.28\ a)$$

$$= \frac{0.5 \times 20}{415}\left[1 - \sqrt{1 - \frac{4.6 \times 162.36 \times 10^6}{20 \times 250\,(500)^2}}\right] 250 \times 500 \simeq 1100.5 \text{ mm}^2$$

Also, $$A_{sf} = \frac{C_{uf}}{0.87\, f_y} = \frac{401400}{0.87 \times 415} \simeq 1111.8 \text{ mm}^2$$

$\therefore$ Total $$A_{st} = A_{sw} + A_{sf} = 1100.5 + 1111.8 = 2212.3 \text{ mm}^2$$

It is now necessary to check that actual x_u is less than the critical value.

$$x_u = \frac{0.87\, f_y\, A_{sw}}{0.36\, f_{ck}\, b_w} = \frac{0.87 \times 415 \times 1100.5}{0.36 \times 20 \times 250} = 220.7 \text{ mm}$$

$$x_{u,\,max} \simeq 0.48 \times 500 = 240 \text{ mm} > x_u \quad \text{Hence OK.}$$

Example 6.7. Design of T-beam

Redesign the T-beam of Example 6.5 if the applied moment is 196 kN-m.

Solution $\quad M_{uD} = 1.5 \times 196 = 294$ kN-m

In example 6.5, $M_{u1} = 224.8$ kN–m ; In example 6.6, $M_{u2} = 338.62$ kN-m

$\therefore \quad M_{u1} < M_{uD} < M_{u2}$. Hence for this case $x_u < \frac{7}{3} D_f < 210$ mm

The design procedure will be essentially that of trial and error.

Trial 1: Let us assume $x_u = 150$ mm. Using I.S. Code method.

$$y_f = 0.15\, x_u + 0.65\, D_f = 0.15 \times 150 + 0.65 \times 90 = 81 \text{ mm} \le D_f. \quad \text{Hence OK.}$$

$$M_u = 0.36\, \frac{x_u}{d}\left(1 - 0.416\, \frac{x_u}{d}\right) f_{ck}\, b_w\, d^2 + 0.446\, f_{ck}\, (b_f - b_w)\, y_f \times (d - 0.5\, y_f) \quad \ldots(6.17)$$

$$= 0.36 \times \frac{150}{500}\left(1 - 0.416 \times \frac{150}{500}\right) \times 20 \times 250\, (500)^2 + 0.446 \times 20\, (750 - 250)\, 81\, (500 - 0.5 \times 81)$$

$$= 118.15 \times 10^6 + 165.99 \times 10^6 = 284.14 \times 10^6 \text{ N-mm} < M_{uD}$$

Trial 2. Let $x_u = 160$ mm

$$y_f = 0.15 \times 160 + 0.65 \times 90 = 82.5 \text{ mm} \le D_f$$

$$M_u = 0.36 \times \frac{160}{500}\left(1 - 0.416 \times \frac{160}{500}\right) \times 20 \times 250\, (500)^2 + 0.446 \times 20 \times (750 - 250)\, 82.5\, (500 - 0.5 \times 82.5)$$

$$= 124.83 \times 10^6 + 168.80 \times 10^6 = 293.63 \simeq M_{uD}$$

Hence adopt $x_u = 160$ mm

Now $\quad C_{uw} = 0.36\, f_{ck}\, x_u\, b_w = 0.36 \times 20 \times 160 \times 250 = 288000$ N

$\quad C_{uf} = 0.446\, f_{ck}\, (b_f - b_w)\, y_f = 0.446 \times 20\, (750 - 250)\, 82.5 = 367950$ N

$\therefore$ Total compression $C_u = 288000 + 367950 = 655950$ N

$$\therefore \quad A_{st} = \frac{C_u}{0.87\, f_y} = \frac{655950}{0.87 \times 415} = \mathbf{1816.8 \text{ mm}^2}$$

Note : x_u can be alternatively obtained by substituting the value of y_f in terms of x_u in Eq. 6.17 and then solving the quadratic equation for x_u.

Thus, $\quad y_f = 0.15\, x_u + 0.65 \times 90 = 0.15\, x_u + 58.5$

Hence from Eq. 6.17, putting $M_u = M_{uD} = 294 \times 10^6$ N-mm, we get

$$294 \times 10^6 = 0.36\, \frac{x_u}{500}\left(1 - \frac{0.416\, x_u}{500}\right) 20 \times 250\, (500)^2 + 0.446 \times 20\, (750 - 250) \times (0.15\, x_u + 58.5) \times [500 - 0.5\, (0.15\, x_u + 58.5)]$$

Simplifying and arranging, we get $x_u^2 - 1496\,x_u + 214295 = 0$

From which $x_u = 160.4$ mm. This value of x_u will be acceptable only if the value of y_f obtained from this is less than D_f.

Thus, $y_f = 0.15 \times 160.4 + 0.65 \times 90 = 82.56 \text{ mm} < D_f$

Hence the exact value of $x_u = 160.4$ mm

$$C_{uw} = 0.3 \times 20 \times 160.4 \times 250 = 288720 \text{ N}$$

$$C_{uf} = 0.446 \times 20\,(750 - 250)\,82.56 = 368218 \text{ N}$$

Total $C_u = 288720 + 368218 = 656938$ N

$$\therefore \quad A_{st} = \frac{656938}{0.87 \times 415} = \mathbf{1819.5\ mm^2}$$

Example 6.8. Design of T-beam

Design the reinforcement for a T-beam for the following data :

Effective span	:	*8 m ; Ends simply supported.*
Spacing of beams	:	*3.3 m centre to centre.*
Thickness of slab	:	*130 mm*
Width of web	:	*300 mm*
Total Depth	:	*450 mm*
Live load on the floor	:	*10 kN/m²*
Floor finish load	:	*0.75 kN/m².*

The beam also supports a partition wall which transmits a load of 12 kN/m run. Use M 20 concrete and Fe 415 steel

Solution : Weight of slab $= 25 \times 0.13 = 3.25 \text{ kN/m}^2$

Weight of finishing $= 0.75 \text{ kN/m}^2$

Live load $= 10.0 \text{ kN/m}^2$

Total 14.0 kN/m^2.

Load on beam, transferred from slab $= 14 \times 3.3 = 46.2$ kN/m

Load transferred from wall = 12 kN/m

Weight of rib $= 0.3\,(0.45 - 0.13) \times 25 = 2.4$ kN/m

Total working load $w = 60.6$ kN/m.

Load factor = 1.5.

$\therefore \quad w_u = 1.5 \times 60.6 = 90.9$ kN/m

Ultimate design moment

$$M_{ub} = \frac{w_u L^2}{8} = \frac{90.9\,(8)^2}{8}$$

$= 727.2$ kN-m. $= 727.2 \times 10^6$ N-mm.

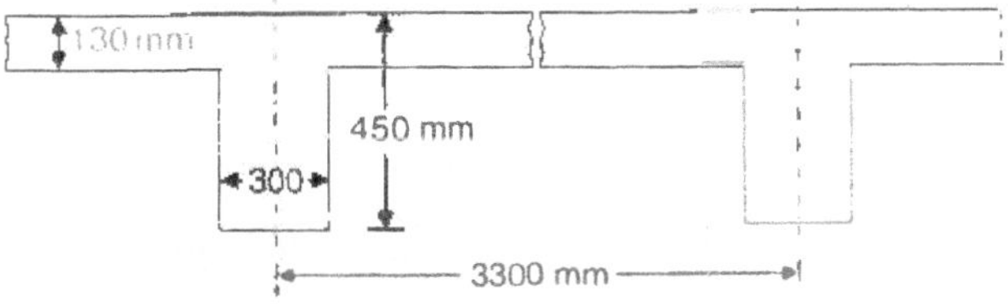

FIG. 6.12

The flange width b_f given by,

$$b_f = \frac{l_0}{6} + b_w + 6\,D_f = \frac{8}{6} + 0.3 + 6\,(0.13) = 2.413 \text{ m} = 2413 \text{ mm}$$

$d = D - \text{cover} = 450 - 80 \text{ (say)} = 330$ mm

Step 1. Assuming $x_u = D_f$, we get from Eq. 6.9(*a*)

$$M_{u1} = 0.36\, f_{ck}\, b_f\, D_f\, (d - 0.416\, D_f)$$

$$= 0.36 \times 20 \times 2413 \times 130\, (330 - 0.416 \times 130) = 623.18 \times 10^{-6} \text{ N-m.}$$

Step 2. Since $M_{uD} > M_{u1}$, N.A. falls outside the flange

Assuming $x_u = \frac{7}{3} D_f \approx 303.3$ mm, we get from Eqs. 6.10 and 6.12.

$$C_{uw} = 0.36\, f_{ck}\, x_u\, b_w = 0.36 \times 20 \times 303.3 \times 300 = 655128 \text{ N.}$$

$$y_f = 0.15 \times 303.3 + 0.65 \times 130 = 45.5 + 84.5 = 130 \text{ mm, but } \leq D_f.$$

$$C_{uf} = 0.446\, f_{ck}\, (b_f - b_w)\, y_f = 0.446 \times 20\, (2413 - 300)\, 130 = 2450235 \text{ N}$$

Hence $M_{uw} = C_{uw}\, (d - 0.416\, x_u) = 655128\, (330 - 0.416 \times 303.3) = 133.53 \times 10^6$ N-mm

and $M_{uf} = C_{uf}\, (d - 0.5\, D_f) = 2450235\, (330 - 0.5 \times 130) = 649.31 \times 10^6$ N-mm

$$M_{u2} = M_{uw} + M_{uf} = (133.53 + 649.31)\, 10^6 = 782.84 \times 10^6 \text{ N-mm}$$

Since $M_{uD} < M_{u2}$, $x_u < \frac{7}{3} D_f$. Hence the design procedure is done by trial and error.

Step 3. *Let us follow I.S. Code method :*

(*i*) Assume $x_u < \frac{7}{3} D_f < 303.3$ mm. Let us assume $x_u = 230$ mm.

Then $y_f = 0.15\, x_u + 0.65\, D_f = 0.15 \times 230 + 0.65 \times 130 = 119$ mm

and $M_u = 0.36 \frac{x_u}{d}\left(1 - 0.416 \frac{x_u}{d}\right) f_{ck}\, b_w\, d^2 + 0.446\, f_{ck}\, (b_f - b_w) \times y_f\, (d - 0.5\, y_f)$

$$= 0.36 \times \frac{230}{330}\left(1 - 0.416\, \frac{230}{330}\right) \times 20 \times 300\, (330)^2$$

$$+ 0.446 \times 20\, (2413 - 300)\, 119\, (330 - 0.5 \times 119)$$

$$= 116.41 \times 10^6 + 606.71 \times 10^6 = 723.12 \times 10^6 \text{ N-mm}$$

This is quite close to $M_{uD} = 727.2 \times 10^6$ N-mm. However, in the second trial, let $x_u = 235$ mm.

$$y_f = (0\ 15 \times 235) + 0.65 \times 130 = 119.75$$

$$M_u = 0.36 \times \frac{235}{330}\left(1 - 0.416 \times \frac{235}{330}\right) \times 20 \times 300\, (330)^2$$

$$+ 0.446 \times 20\, (2413 - 300)\, 119.75\, (330 - 0.5 \times 119.75)$$

$$= 117.89 \times 10^6 + 609.68 \times 10^2 = 727.57 \times 10^6 \simeq M_{uD} \quad \text{Hence } x_u = 235 \text{ mm}$$

$\therefore$ $C_{uw} = 0.36\, f_{ck}\, x_u\, b_w = 0.36 \times 20 \times 235 \times 300 = 507600$ N

$C_{uf} = 0.446\, f_{ck}\, (b_f - b_w)\, y_f = 0.446 \times 20\, (2413 - 300)\, 119.75 = 2257043$ N

$\therefore$ $C_u = C_{uw} + C_{uf} = 507600 + 2257043 = 2764643$ N

$\therefore$ $A_{st} = \dfrac{C_u}{0.87 f_y} = \dfrac{2764643}{0.87 \times 415} = \mathbf{7657\ mm^2}$

6.10. ANALYSIS OF DOUBLY REINFORCED T-BEAMS

We have seen in the previous articles that a singly reinforced-section has limiting value of moment of resistance. However, if the applied moment is more than $M_{u\,lim}$ two alternatives will be available : (*i*) to increase the depth of the section, or (*ii*) to provide compressive reinforcement. In many cases the maximum value of depth of section may be limited or restricted from architectural or other considerations. In that case, the only alternative that will be left will be to provide reinforcement in the compression zone, giving rise to *doubly reinforced-section*.

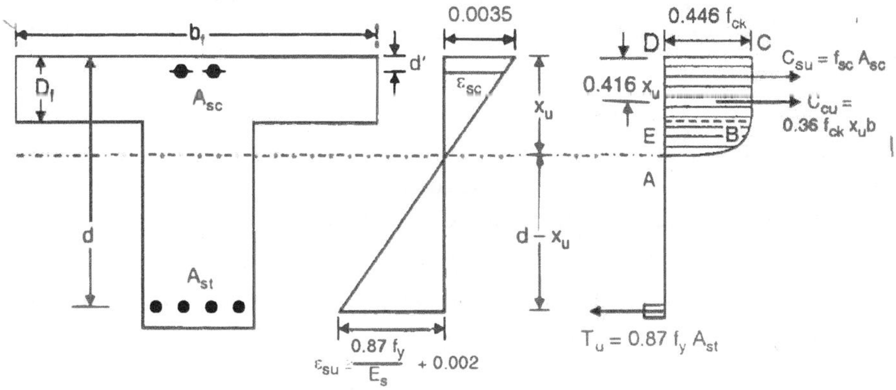

FIG. 6.13. DOUBLY REINFORCED T-BEAM

Fig. 6.13(*a*) shows a doubly reinforced T-beam with compression reinforcement A_{sc} provided at a cover of d'. The strain in the compressive reinforcement is ε_{sc} as shown in Fig. 6.13(*b*). The compressive force contribution of A_{sc} will be $C_{su} = f_{sc} A_{sc}$ [Fig. 6.13(*c*)] where f_{sc} is the compressive stress in steel at a strain of ε_{sc} and is found from the stress strain curve, where :

$$\varepsilon_{sc} = \frac{0.0035\,(x_u - d')}{x_u} \qquad \text{...(6.29)}$$

DETERMINATION OF x_u : For a given section, x_u can be determined by equating the total tensile force to the total compressive force. Assuming that the N.A. lies in the flange.

$$0.36 f_{ck} b_f x_u + f_{sc} A_{sc} = 0.87 f_y A_{st} \qquad \text{...(6.30)}$$

If the loss of concrete area occupied by compressive steel is taken into account, we have

$$0.36 f_{ck} b_f x_u + (f_{sc} - 0.446 f_{ck}) A_{sc} = 0.87 f_y A_{st} \qquad \text{...[6.30(a)]}$$

(This is subject to the condition that $d' < \frac{3}{7} x_u$ so that stress in concrete at the level of compressive steel is equal to $0.446 f_{ck}$.)

Stress f_{sc} is found from stress-strain curve, corresponding to the value of ε_{sc} given by Eq. 6.29. Thus, f_{sc} indirectly depends upon x_u. Hence the following ***iterative procedure*** is adopted:

Step 1. Find $x_{u.max}$ for a balanced section from strain compatibility :

$$x_{u.max} = \frac{0.0035}{0.0055 + 0.87 f_y / E_s} = \frac{700}{1100 + 0.87 f_y}$$

Step 2. Assume $x_u = x_{u\,max}$ or D_f (whichever is less) and compute ε_{sc} from Eq. 6.29.

Step 3. Find f_{sc} from stress-strain curve, corresponding to the above value of ε_{sc}

(Alternatively, in the first trial, take f_{sc} slightly less than 0.87 f_y).

Step 4. Substitute this value of f_{sc} in Eq. 6.30 (or Eq. 6.30 *a*) and compute the modified value of x_u.

Step 5. Repeat step 2 to 4, till convergence of the value of x_u is achieved.

If $x_u < D_f$, the N.A. lies in the flange and hence Eq. 6.30 [or 6.30(*a*)] is valid. If however, x_u comes out to be greater than D_f, the N.A. will be outside the flange, *and a new value of* x_u *will have to be found* by equating the total compressive force to the total tensile force. The determination of x_u will be done exactly in the same manner as done for the singly reinforced T-beam discussed in § 6.6. The total compressive force C_u will consists of (*i*) web contribution, C_{uw} and (*ii*) flange contribution C_{uf}, where

$$C_{uw} = 0.36 f_{ck} x_u b_w + (f_{sc} - 0.446 f_{ck}) A_{sc} = 0.87 f_y A_{sw} \quad \ldots(6.31)$$

The compressive force C_{uf} can be computed by approximating the actual stress block into an equivalent rectangular block of height y_f given by

$$y_f = 0.15 x_u + 0.65 D_f \text{ (subject to a max. of } D_f \text{)}$$

$$\therefore \quad C_{uf} = 0.446 f_{ck} (b_f - b_w) y_f = 0.87 f_y A_{sf} \quad \ldots(6.32)$$

Equating the total compressive force to the total tensile force, we get

$$\therefore \quad 0.36 f_{ck} x_u b_w + (f_{sc} - 0.446 f_{ck}) A_{sc} + 0.446 f_{ck} (b_f - b_w) y_f = 0.87 f_y A_{st} \quad \ldots(6.33)$$

Since f_{sc} depends upon x_u, an iterative procedure will have to be adopted. Also, y_f depends upon x_u, and hence the final value of x_u is valid only if the value of y_f calculated by using this value of x_u, is less than or equal to D_f. If this value of y_f comes out to be greater than D_f, the value of x_u should be *recalculated* from Eq. 6.33 by taking $y_f = D_f$. Thus, the value of x_u is known. This value of x_u should be compared with $x_{u\,max}$ to find whether the section is under reinforced, balanced or over-reinforced.

1. MOMENT OF RESISTANCE WHEN $x_u < D_f$: If x_u comes out to be less than D_f, from Eq. 6.30 or 6.30(*a*), the moment of resistance can be found by treating the beam to be of a doubly reinforced rectangular section of width b_f. Again, if $x_u < x_{u\,max}$, the section is under-reinforced, and the moment of resistance is given by

$$M_u = 0.36 f_{ck} x_u b_f (d - 0.416 x_u) + (f_{sc} - 0.446 f_{ck}) A_{sc} (d - d') \quad \ldots(6.34)$$

When $x_u = D_f$, we get

$$M_{u1} = 0.36 f_{ck} b_f D_f (d - 0.416 D_f) + (f_{sc} - 0.446 f_{ck}) A_{sc} (d - d') \quad \ldots[6.34(a)]$$

If $x_u = x_{u.max}$, we have the balanced section, for which,

$$M_{u.lim} = 0.36 f_{ck} x_{u.max} b_f (d - 0.416 x_{u.max}) + (f_{sc} - 0.446 f_{ck}) A_{sc} (d - d') \quad \ldots(6.35)$$

If $x_u > x_{u\,max}$, the section is *over-reinforced*, and its moment resistance M_u will be limited to $M_{u\,lim}$, given by Eq. 6.35. However, the Code recommends that such section should be redesigned.

2. MOMENT OF RESISTANCE WHEN $x_u > D_f$

The value of x_u obtained from Eq. 6.33 should be compared with $x_{u\,max}$ to find whether the section is under-reinforced, balanced or over-reinforced.

The procedure for finding M_u will be the same as that discussed for singly reinforced section except that an additional moment of resistance contribution by compressive steel $= (f_{sc} - 0.446 f_{ck}) A_{sc}(d - d')$ will have to be taken into account.

(a) Thus, for Under-reinforced Section $(x_u < x_{u\,max})$,

If $D_f / x_u \le 0.43$ or 3/7, we have Case 2, for which

$$M_{uw} = 0.36 f_{ck} \frac{x_u}{d} b_w \left(1 - 0.416 \frac{x_u}{d}\right) d^2 + (f_{sc} - 0.446 f_{ck}) A_{sc}(d - d') \quad \ldots[6.36(a)]$$

$$M_{uf} = 0.446 f_{ck} (b_f - b_w) D_f (d - 0.5 D_f) \quad \ldots[6.36(b)]$$

and
$$M_u = 0.36 f_{ck} \frac{x_u}{d} b_w \left(1 - 0.416 \frac{x_u}{d}\right) d^2 + (f_{sc} - 0.446 f_{ck}) A_{sc}(d - d') + 0.446 f_{ck} (b_f - b_w) D_f (d - 0.5 D_f) \quad \ldots(6.36)$$

When $D_f = \frac{3}{7} x_u$ or $x_u = \frac{7}{3} D_f$, we have

$$M_{u2} = 0.36 f_{ck} b_w \left(\frac{7}{3} D_f\right)\left(d - 0.416 \times \frac{7}{3} D_f\right) + (f_{sc} - 0.446 f_{ck}) A_{sc} (d - d') + 0.446 f_{ck} (b_f - b_w) D_f (d - 0.5 D_f) \quad \ldots[6.36(c)]$$

(b) Similarly, When $x_u < x_{u\,max}$ **and** $D_f / x_u > 0.43$ or 3/7 we have Case 3, for which

$$M_{uw} = 0.36 f_{ck} \frac{x_u}{d} b_w \left(1 - 0.416 \frac{x_u}{d}\right) d^2 + (f_{sc} - 0.446 f_{ck}) A_{sc} (d - d') \quad \ldots[6.37(a)]$$

and
$$M_{uf} = 0.446 f_{ck} (b_f - b_w) y_f (d - 0.5 y_f) \quad \ldots[6.37(b)]$$

and
$$M_u = 0.36 \frac{x_u}{d}\left(1 - 0.416 \frac{x_u}{d}\right) f_{ck} b_w d^2 + (f_{sc} - 0.446 f_{ck}) A_{sc} (d - d') + 0.446 f_{ck}(b_f - b_w) y_f (d - 0.5 y_f) \quad \ldots(6.37)$$

3. MOMENT OF RESISTANCE WHEN $(x_u = x_{u\,max}) > D_f$

The limiting moment of resistance can be computed by modifying the equations (Eq. 6.21 and 6.22) suggested by IS Code, by including the moment of resistance term of the compression steel. Thus, we have the following modified equations.

(a) When $\frac{D_f}{d} \le 0.2$.

$$M_{u\,lim} = 0.36 \frac{x_{u\,max}}{d}\left(1 - 0.416 \frac{x_{u\,max}}{d}\right) f_{ck} b_w d^2 + (f_{sc} - 0.446 f_{ck}) A_{sc} (d - d') + 0.446 f_{ck} (b_f - b_w) D_f \left(d - \frac{D_f}{2}\right) \quad \ldots(6.38)$$

where f_{sc} is found from stress-strain curve corresponding to ε_{sc} given by Eq. 6.29 in which $x_u = x_{u\,max}$.

(b) When $\frac{D_f}{d} > 0.2$.

$$M_{u\,lim} = 0.36 \frac{x_{u\,max}}{d}\left(1 - 0.416 \frac{x_{u\,max}}{d}\right) f_{ck} b_w d^2 + (f_{sc} - 0.446 f_{ck}) A_{sc} (d - d') + 0.446 f_{ck} (b_f - b_w) y_f \left(d - \frac{y_f}{2}\right)$$

where $y_f = (0.15\, x_{u\,max} + 0.65\, D_f) \le D_f$ ...(6.39)

4. MOMENT OF RESISTANCE WHEN $(x_u > x_{u.max}) > D_f$

If $x_u > x_{u.max}$ the section is over-reinforced, and its moment of resistance M_u will be limited to $M_{u,lim}$ given by Eq. 6.38 (when $D_f/d \le 0.2$) or by Eq. 6.39 (when $D_f/d > 0.2$). However, the Code recommends that such a section should be redesigned.

6.11. DESIGN OF DOUBLY REINFORCED T-BEAM

A doubly reinforced section will be required when $M_{uD} > M_{u.lim}$ For a T-beam of given cross-sectional dimensions, (*i.e.* given b_w, b_f, D_f and d), the design is done in the following steps.

Step 1. Assuming $x_u = D_f$, and assuming singly reinforced section, compute M_{u1} from Eq. 6.9 (*a*).

Step 2. If $M_{u1} < M_{uD}$. the N.A. will fall outside the flange.

Taking $x_u = \frac{7}{3} D_f$ and taking singly reinforced section, compute M_{u2} from Eq. 6.16(*c*)

Step 3. If $M_{u2} < M_{uD}$, $x_u > \frac{7}{3} D_f$. Compute $M_{u.lim}$ by taking $x_u = x_{u,max}$, either from Eq. 6.21 (if $D_f/d \le 0.2$) or from Eq. 6.22 (if $D_f/d \ge 0.2$).

If $M_{u.lim} < M_{uD}$, a doubly reinforced section will be required.

Step 4. (*a*) Compute A_{st1} for $M_{u.lim.}$ as under :

$$A_{sw} = \frac{0.36 f_{ck}\, x_{u\,max}\, b_w}{0.87 f_y} \quad \text{and} \quad A_{sf} = \frac{0.446 f_{ck}\,(b_f - b_w)\, D_f}{0.87 f_y}$$

$$\therefore \quad A_{st1} = [0.36 f_{ck}\, x_{u,max}\, b_w + 0.446 f_{ck}\,(b_f - b_w)\, D_f] \div (0.87 f_y) \qquad \ldots(6.40)$$

(*b*) Compute the additional tensile area A_{st2} from the relation

$$A_{st2} = \frac{M_{uD} - M_{u.lim.}}{0.87 f_y\,(d - d')} \qquad \ldots(6.41)$$

(*c*) Compute the compressive area A_{sc} corresponding to A_{st2}, by the relation.

$$A_{sc} = \frac{M_u - M_{u.lim}}{(f_{sc} - 0.446 f_{ck})\,(d - d')} \qquad \ldots(6.42)$$

Example 6.9. M.R. of doubly reinforced T-beam *Find the moment of resistance of T-beam of example 6.3, if the beam also carries a compression reinforcement in the form of 3 bars of 25 mm dia. provided with a cover of 40 mm from the centre of bars.*

Solution

$A_{st} = 2454.4\ \text{mm}^2$ (as before) ;

$A_{sc} = 3 \times \frac{\pi}{4} (25)^2 = 1472.6\ \text{mm}^2$.

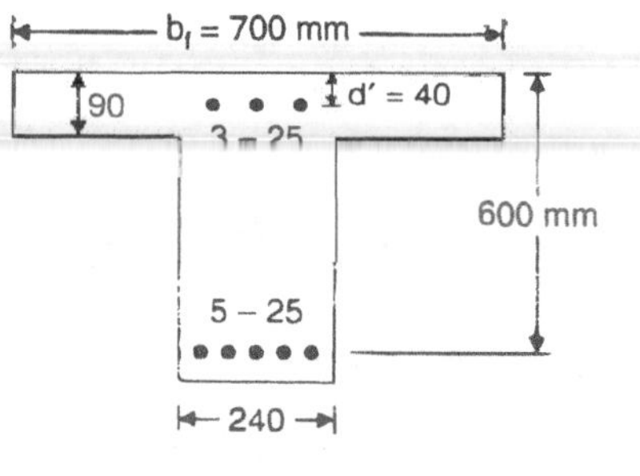

FIG. 6.14

Assuming the N.A. to fall in the flange, we get, from Eq. 6.30(*a*).

$$0.36 f_{ck}\, b_f\, x_u + (f_{sc} - 0.446 f_{ck})\, A_{sc} = 0.87 f_y\, A_{st} \qquad \ldots[6.30(a)]$$

Take $f_{sc} < 0.87 f_y < 0.87 \times 415 < 361$, in the first trial.

Hence take $f_{sc} = 350\ \text{N/mm}^2$ (say)

$\therefore \quad 0.36 \times 15 \times 700\, x_u + (350 - 0.446 \times 15) \times 1472.6 = 0.87 \times 415 \times 2454.4$

From which $x_u = 100.7$ mm $> D_f$

(Note that d' is less than $\frac{3}{7} x_u$ so that stress in concrete at A_{sc} level is $0.446 f_{ck}$)

Hence the N.A. lies outside the flange. To obtain a new value of x_u, equate total compressive force to the total tensile force.

$$T_u = 0.87 f_y A_{st} = 0.87 \times 415 \times 2454.4 = 886161 \text{ N}$$

$$y_f = 0.15\, x_u + 0.65\, D_f = 0.15\, x_u + 0.65 \times 90 = 0.15\, x_u + 58.5$$

with a maximum equal to D_f. Since the approximate value of x_u found above is 100.7 mm,

$$y_f = 0.15 \times 100.7 + 58.5 = 73.6 \text{ mm} < D_f.$$

$$\therefore \quad C_u = 0.36 f_{ck} x_u b_w + (f_{sc} - 0.446 f_{ck}) A_{sc} + 0.446 f_{ck} (b_f - b_w) y_f$$

or $$C_u = 0.36 \times 15\, x_u \times 240 + (f_{sc} - 0.446 \times 15)\, 1472.6 + 0.446 \times 15\, (700 - 240) \times (0.15\, x_u + 58.5)$$

or $$C_u = 1296\, x_u + 1472.6 f_{sc} - 9852 + 461.6\, x_u + 180028 = 1757.6\, x_u + 170176 + 1472.6 f_{sc}$$

Equating C_u and T_u, we get $1757.6\, x_u + 170176 + 1472.6 f_{sc} = 886161$

or $$x_u = 407.4 - 0.8378 f_{sc} \qquad \ldots(i)$$

In the above equation, f_{sc} can be obtained from stress-strain curve, corresponding to ε_{sc} which itself depends upon x_u, as is evident from Eq. 6.29 :

$$\varepsilon_{sc} = \frac{0.0035\,(x_u - d')}{x_u} \qquad \ldots(ii)$$

In the first approximation, taking $x_u = 100.7$ mm,

$$\varepsilon_{sc} = \frac{0.0035\,(100.7 - 40)}{100.7} = 0.0021$$

From stress-strain curve for Fe 415 (Fig. 2.5) or from Table 2.3, we get $f_{sc} \approx 328$ N/mm^2.

Hence from (*i*), $x_u = 407.4 - 0.8378 \times 328 = 132.6$ mm

For this new value of x_u, $y_f = 0.15 \times 132.6 + 58.5 = 78.4 \text{ mm} < D_f$. Hence O.K.

In the next trial, recalculate ε_{sc} by taking $x_u = 132.6$ mm, from (*ii*)

$$\therefore \quad \varepsilon_{sc} = \frac{0.0035\,(132.6 - 40)}{132.6} = 0.0024$$

From Table 2.3, we get $f_{sc} = 342$ N/mm^2 for $\varepsilon_{sc} = 0.0024$

Hence from (*i*), $x_u = 407.4 - 0.8378 \times 342 = 120.8$ mm.

In the next trial, $$\varepsilon_{sc} = \frac{0.0035\,(120.8 - 40)}{120.8} = 0.00234$$

From Table 2.3, $f_{sc} = 340$ N/mm^2 for $\varepsilon_{sc} = 0.00234$

Hence from (*i*), $x_u = 407.4 - 0.8378 \times 340 = 122.5$ mm

In the next trial, the solution will converge to a value of $x_u = 122$ mm, for which $\varepsilon_{sc} = 0.00235$ and $f_{sc} \approx 340.5$ N/mm^2

$$y_f = 0.15\, x_u + 58.5 = 0.15 \times 122 + 58.5 = 76.8 \text{ mm}$$

$$x_{u,max} \simeq 0.48\, d \text{ (for Fe 415 steel)} \simeq 0.48 \times 600 = 288 \text{ mm}$$

$\therefore \quad x_u < x_{u,max}$ and the section is under-reinforced.

Again, $\frac{3}{7} x_u = \frac{3}{7} \times 122 = 52.3$ mm $< D_f$. Hence $D_f > \frac{3}{7} x_u$

or $\quad \frac{D_f}{x_u} > 0.43$. Hence M_u is given by Eq. 6.37

$$M_u = 0.36 \frac{x_u}{d}\left(1 - 0.416 \frac{x_u}{d}\right) f_{ck}\, b_w\, d^2 + (f_{sc} - 0.446 f_{ck})\, A_{sc}(d - d') + 0.446 f_{ck}\,(b_f - b_w)\, y_f\,(d - 0.5\, y_f) \qquad \ldots(6.37)$$

$$M_u = 0.36 \frac{122}{600}\left(1 - 0.416 \frac{122}{600}\right) 15 \times 240\,(600)^2 + (340.5 - 0.446 \times 15) \times 1472.6 \times (600 - 40) + 0.446 \times 15\,(700 - 240) \times 76.8\,(600 - 0.5 \times 76.8)$$

$$= 86.84 \times 10^6 + 275 \times 10^6 + 132.73 \times 10^6 = 494.57 \times 10^6 \text{ N-mm} = \mathbf{494.57 \text{ kN-m.}}$$

Example 6.10. M.R. of doubly reinforced T-beam

Find the moment of resistance of the T-beam of Example 6.3 if the beam carries a compression reinforcement in the form of 2 bars of 20 mm dia. with d' = 40 mm and tensile reinforcement consists of 6 bars of 25 mm dia.

Solution.

$$A_{st} = 6 \times \frac{\pi}{4}(25)^2 = 2945.2 \text{ mm}^2$$

$$A_{sc} = 2 \frac{\pi}{4}(20)^2 = 628.3 \text{ mm}^2$$

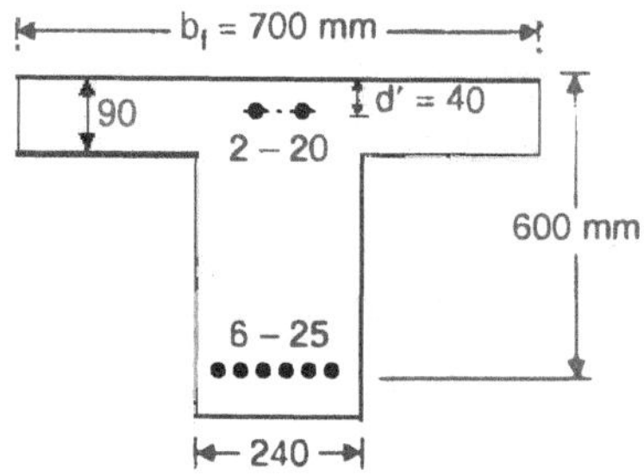

FIG. 6.15

Assuming the N.A. to fall in flange, we have, from Eq. 6.30(*a*).

$$0.36 f_{ck}\, b_f\, x_u + (f_{sc} - 0.446 f_{ck})\, A_{sc} = 0.87 f_y\, A_{st}$$

Taking $f_{sc} < 0.87 f_y < 0.87 \times 415 < 361$ or equal to 350 N/mm^2 in the first trial, we get

$$0.36 \times 15 \times 700\, x_u + (350 - 0.446 \times 15)\, 628.3 = 0.87 \times 415 \times 2945.2$$

From which $x_u = 224.2$ mm $> D_f$

(Note that d' is less than $\frac{3}{7} x_u$ so that stress in concrete at the level of compressive steel is equal to $0.446 f_{ck}$).

Hence the N.A. lies outside the flange. To obtain a new value of x_u, equate T_u and C_u .

$$T_u = 0.87 f_y A_{st} = 0.87 \times 415 \times 2945.2 = 1063365 \text{ N}$$

$$y_f = 0.15\, x_u + 0.65\, D_f = 0.15\, x_u + 0.65 \times 90 = 0.15\, x_u + 58.5$$

with a maximum equal to D_f. Since the approximate value of x_u found above is 224.2 mm,

$$y_f = 224.2 \times 0.15 + 58.5 = 92.1 > D_f \text{ . Hence take } y_f = D_f$$

$\therefore \quad C_u = 0.36 f_{ck}\, x_u\, b_w + (f_{sc} - 0.446 f_{ck})\, A_{sc} + 0.446 f_{ck}(b_f - b_w) D_f$

or $\quad C_u = 0.36 \times 15\, x_u \times 240 + (f_{sc} - 0.446 \times 15)\, 628.3 + 0.446 \times 15\,(700 - 240)\, 90$

or $\quad C_u = 1296\, x_u + 628.3 f_{sc} - 4203 + 276966 = 1296\, x_u + 272763 + 628.3 f_{sc}$

Equating C_u and T_u, $1296\, x_u + 272763 + 628.3\, f_{sc} = 1063365$

or $$x_u = 610 - 0.4848\, f_{sc} \quad \ldots(i)$$

In the above equation, f_{sc} can be obtained from stress-strain curve, corresponding to ε_{sc} which itself depends upon x_u, as is evident from Eq. 6.29.

$$\varepsilon_{sc} = \frac{0.0035\,(x_u - d')}{x_u} \quad \ldots(ii)$$

In the first approximation, taking $x_u \approx 224.2$ mm

$$\varepsilon_{sc} = \frac{0.0035\,(224.2 - 40)}{224.2} = 0.00282$$

From stress-strain curve for Fe 415 (Fig. 2.5) or from Table 2.3, we get :

$$f_{sc} \approx 353 \text{ N/mm}^2.$$

Hence from (*i*), $x_u = 610 - 0.4848 \times 353 = 439$ mm

Substituting this value of x_u in (*ii*), we get $\varepsilon_{sc} = \dfrac{0.0035\,(439 - 40)}{439} = 0.0032$

$\therefore$ $f_{sc} = 356$ N/mm^2 from Table 2.3.

$\therefore$ Next value of $x_u = 610 - 0.4848 \times 356 \approx 437$ mm

Since the value has converged, take $x_u = 437$ mm for which $f_{sc} = 356$ N/mm^2

Now, $x_{u,max} = 0.48\, d = 0.48 \times 600 = 288$ mm. Hence $x_u > x_{u,max}$.

Thus, the beam is over-reinforced. Though the Code recommends that such a beam should be *redesigned*, its moment of resistance is limited to $M_{u,lim}$ corresponding to the balanced section. $\therefore$ Take $x_u = x_{u,max} = 288$ mm.

For this value of x_u, $\varepsilon_{sc} = \dfrac{0.0035\,(288 - 40)}{288} = 0.003.$

$\therefore$ $$f_{sc} \approx 354 \text{ N/mm}^2 .$$

Again $\dfrac{D_f}{d} = \dfrac{90}{600} = 0.15 < 0.2$. Hence moment of resistance is given by Eq. 6.38 (corresponding to para G-2.2 of Code).

$$M_u = M_{u\,lim} = 0.36\, \frac{x_{u,max}}{d}\left(1 - 0.416\, \frac{x_{u,max}}{d}\right) f_{ck}\, b_w\, d^2$$
$$+ (f_{sc} - 0.446\, f_{ck})\, A_{sc}\,(d - d') + 0.446\, f_{ck}\,(b_f - b_w)\, D_f\left(d - \frac{D_f}{2}\right)$$

$$= 0.36 \times \frac{288}{600}\left(1 - 0.416 \times \frac{288}{600}\right) 15 \times 240\,(600)^2 + (354 - 0.446 \times 15) \times 628.3\,(600 - 40)$$
$$+ 0.446 \times 15\,(700 - 240) \times 90 \times (600 - 0.5 \times 90)$$

$$= 179.2 \times 10^6 + 122.2 \times 10^6 + 153.7 \times 10^6 = 455.1 \times 10^6 \text{ N-mm} = \mathbf{455.1\ kN\text{-}m}.$$

Example 6.11. Design of doubly reinforced T-beam

Redesign the T-beam of example 6.5 if the applied moment is 300 kN-m.

Solution : $M_{uD} = 1.5 \times 300 = 450$ kN-m

Step 1 : Assuming $x_u = D_f$, and *assuming* singly reinforced section, compute M_{u1}.

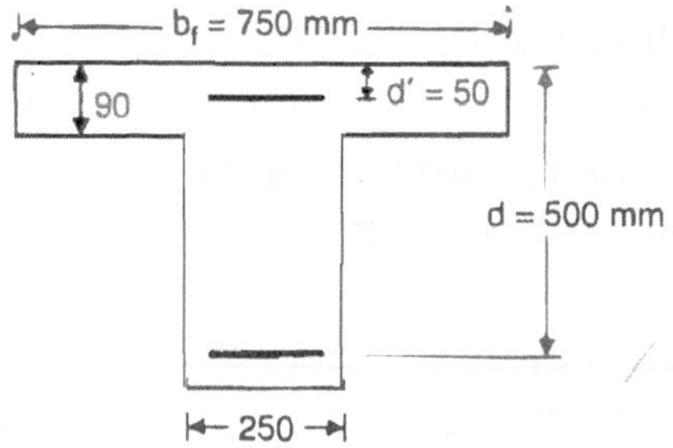

FIG. 6.16

$$M_{u1} = 0.36\, f_{ck}\, b_f\, D_f\, (d - 0.416\, D_f)$$
$$= 0.36 \times 20 \times 750 \times 90\, (500 - 0.416 \times 90)$$
$$= 224.80 \times 10^6 \text{ N-mm} < M_{uD}$$

Step 2 : Since $\mathbf{M_{uD} > M_{u1}}$, N.A. will fall out side the flange.

Taking $x_u = \frac{7}{3} D_f = \frac{7}{3} \times 90 = 210$ mm, compute M_{u2} from Eq. 6.16 (*c*), treating the section as singly reinforced.

$$M_{u2} = 0.36\, f_{ck}\, b_w \left(\frac{7}{3} D_f\right)\left(d - 0.416 \times \frac{7}{3} D_f\right) + 0.446\, f_{ck}\, (b_f - b_w)\, D_f\, (d - 0.5\, D_f)$$
$$= 0.36 \times 20 \times 250\, (210)\, (500 - 0.416 \times 210) + 0.446 \times 20\, (750 - 250)\, 90\, (500 - 0.5 \times 90)$$
$$= 155.98 \times 10^6 + 182.64 = 338.62 \times 10^6 \text{ N-mm} < M_{uD}$$

Step 3 : Since $M_{uD} > M_{u2}$, $x_u > \frac{7}{3} D_f > 210$ mm. Compute $M_{u,\,lim.}$ by taking $x_u = x_{u,\,max}$

$$\frac{x_{u,\,max}}{d} = 0.479 \text{ (for Fe 415)}; \quad x_{u,\,max} = 0.479 \times 500 = 239.5 \text{ mm.}$$

Also, $$\frac{D_f}{d} = \frac{90}{500} = 0.18 < 0.2.$$

Hence compute $M_{u,\,lim.}$ from Eq. 6.21, treating the section as singly reinforced.

$$M_{u,\,lim} = 0.36 \frac{x_{u,\,max}}{d}\left(1 - 0.416 \frac{x_{u,\,max}}{d}\right) f_{ck}\, b_w\, d^2 + 0.446\, f_{ck}\, (b_f - b_w)\, D_f \left(d - \frac{D_f}{2}\right)$$
$$= 0.36 \times 0.479\, (1 - 0.416 \times 0.479)\, 20 \times 250\, (500)^2 + 0.446 \times 20\, (750 - 250)\, 90 \left(500 - \frac{90}{2}\right)$$
$$= 172.60 \times 10^6 + 182.64 \times 10^6 = 355.24 \times 10^6 < M_{uD}$$

Since $M_{uD} > M_{u,\,lim.}$ *a doubly reinforced section will be required.*

Step 4. Design of doubly reinforced section

(*a*) *Computation of A_{st1} for $M_{u,\,lim.}$* :

$$A_{sw} = \frac{0.36\, f_{ck}\, x_{u,\,max}\, b_w}{0.87\, f_y} = \frac{0.36 \times 20 \times 239.5 \times 250}{0.87 \times 415} = 1194 \text{ mm}^2$$

$$A_{sf} = \frac{0.446\, f_{ck}\, (b_f - b_w)\, D_f}{0.87\, f_y} = \frac{0.446 \times 20\, (750 - 250)\, 90}{0.87 \times 415} \simeq 1112 \text{ mm}^2$$

$$\therefore \quad A_{st1} = 1194 + 1112 = 2306 \text{ mm}^2$$

(*b*) *Computation of additional tensile area A_{st_2}* : Taking $d' = 50$ mm (say)

$$A_{st2} = \frac{M_{uD} - M_{u,\,lim.}}{0.87\, f_y\, (d - d')} = \frac{(450 - 355.24)\, 10^6}{0.87 \times 415\, (500 - 50)} \simeq 583 \text{ mm}^2$$

(*c*) *Computation of compressive steel area A_{sc} corresponding to A_{st2}*

$$A_{sc} = \frac{M_u - M_{u,\,lim.}}{(f_{sc} - 0.446\, f_{ck})\, (d - d')} = \frac{(450 - 355.24)\, 10^6}{(f_{sc} - 0.446 \times 20)\, (500 - 50)} = \frac{210577}{f_{sc} - 8.92}$$

$$\varepsilon_{sc} = \frac{0.0035\,(x_{u,\,max.} - d\,')}{x_{u,\,max}} = \frac{0.0035\,(239.5 - 50)}{239.5} = 0.00276$$

Hence from Table 2.3 for Fe 415 bars, $f_{sc} = 351.8\ N/mm^2$ for $\varepsilon_{sc} = 0.00276$

$$\therefore \qquad A_{sc} = \frac{210577}{351.8 - 8.92} \simeq \mathbf{614\ N/mm^2}$$

(*d*) *Total tensile area*

$$A_{st} = A_{st1} + A_{st2} = 2306 + 583 = \mathbf{2889\ mm^2}$$

PROBLEMS

1. Find the moment of resistance of a T-beam having the following data
   Width of flange : 800 mm ; Thickness of slab : 120 mm
   Width of rib : 200 mm ; Effective depth : 400 mm
   Tensile steel area : 3500 mm^2. Use M 15 concrete and Fe 250 steel.
2. Solve problem 1 with M 15 – Fe 415 combination.
3. Find the moment of resistance of a T-beam section having $b_w = 300$ mm, $b_f = 1600$ mm, $D_f = 100$ mm and $d = 510$ mm. The reinforcement consists of 4 bars of 25 mm dia. Use M 15 concrete and Fe 415 Steel.
4. Solve problem 3 if the reinforcement consists of 6 bars of 25 mm dia.
5. Solve problem 3 if the reinforcement consists of 8 bars of 25 mm dia.
6. Solve problem 3 if the beam also carries compressive reinforcement in the form of 2 bars of 16 mm dia.
7. Solve problem 4, if the beam also carries compressive reinforcement in the form of 4 bars of 16 mm dia.
8. For the T-beam of the dimensions of problem 1, find the reinforcement for an applied B.M. of 100 kN-m.

7

Limit State of Collapse in Shear : Design for Shear

7.1. SHEAR STRESS IN R.C. BEAMS

When a beam is loaded with transverse loads, the bending moment varies from section to section. Shearing stresses in the beam are caused by this variation of bending moment along the span. Due to the variation of B.M. at two sections distant dx apart, there are unequal bending stresses at the same fibre. This inequality of bending stresses produces a tendency in each horizontal fibre to slide over adjacent horizontal fibre causing *horizontal shear stress*, which is accompanied by *complimentary shear stress* in vertical direction. This chapter deals with the limit state of collapse in *flexural shear*, i.e. shear associated with a varying bending moment. Commonly, flexural shear is simply referred to as *'shear'*. Failure of a R.C. beam in flexural shear may not lead to an immediate *collapse* by itself. However, flexural shear can significantly reduce the *flexural-strength* (i.e. moment bearing capacity) as well as *ductility*. Hence the state of impending shear failure is treated by the Code as the *limit state of collapse in shear*, for design purposes.

Shear stresses in homogeneous beams (Fig. 7.1 *a*)

In the case of a homogeneous beam (i.e. a concrete beam without reinforcement), the shear stress q at any point, as given by the elastic theory, is

$$q = \frac{V}{Iz}(A\bar{y})$$

where V = shear force at the section

I = moment of inertia of the beam section

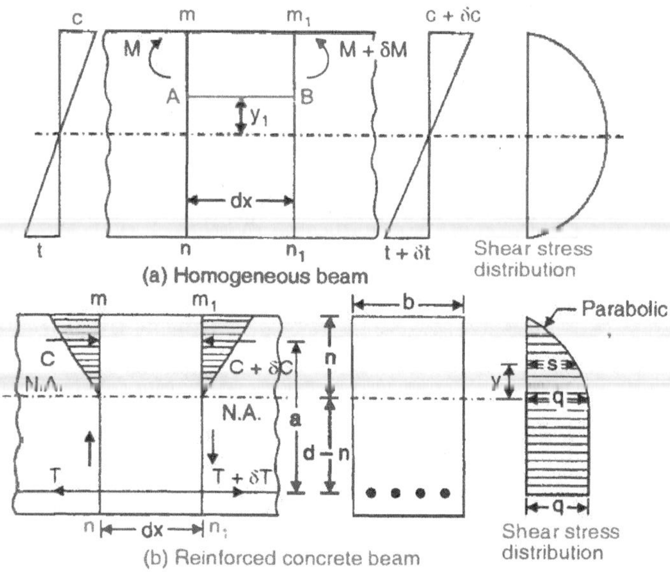

FIG. 7.1. SHEAR STRESS IN HOMOGENEOUS BEAM BY ELASTIC THEORY

z = width of the section

$A\bar{y}$ = moment of area above the section, about N.A.

Shear stresses in R.C. beams

Fig. 7.1(*b*) shows two sections *mn* and m_1n_1 distant *dx* apart, of a R.C. beam. Let the bending moment at *mn* be M and that at $m_1\,n_1$ be $M + \delta M$. Due to this variation of bending moment in a length δx, the compressive forces in concrete at these two sections will be C and $C + \delta C$, and the tensile forces in steel will be T and $T + \delta T$ as shown. Let us assume that the shear force V remains constant in the length *dx*. Let us assume that concrete does not take any tension. Consider any plane between N.A. and centre of steel reinforcement. Let q be the intensity of shear at the plane. Hence total horizontal shear $= q.b.\delta x$ at that layer. Also, the total horizontal force that tends to slide this layer past the adjacent one is equal to $(T + \delta T) - (T) = \delta T$. Hence equating the two, we get

$$q \,.\, b \,.\, \delta x = \delta T \qquad \text{or} \qquad q = \frac{\delta T}{b \,.\, \delta x} \qquad \ldots(i)$$

In order to get the value of q, it is first essential to determine the value of δT. For equilibrium $\Sigma M = 0$,

$$\therefore \qquad (T + \delta T)\, jd = Tjd + V \,.\, \delta x \qquad \therefore \qquad \delta T = \frac{V \,.\, \delta x}{jd} \qquad \ldots(ii)$$

Substituting in (*i*), we get $q = \dfrac{V}{bjd}$ where $jd = a$ = lever arm ...(7.1)

This value of q is evidently the same for any layer between the N.A. and the centre of steel. Thus the shear stress distribution below the N.A. is rectangular. The shear stress distribution above the N.A. is parabolic, as in the case of homogeneous beams. The intensity of shear stress s at any layer distant y above the N.A. is given by

$$s = \frac{V}{Ib} \int_{y}^{n} b \,.\, y \,.\, dy = \frac{V}{2I}\,(n^2 - y^2) \qquad \ldots(7.2)$$

where I is the moment of inertia of the R.C. beam section about the N.A. and can be found the following expression:

$$I = \frac{1}{3} b \,.\, n^3 + m\, A_{st}\,(d - n)^2 \qquad \ldots(7.3)$$

The complete shear stress distribution diagram is shown in Fig. 7.1(*b*). The intensity of maximum shear (q) can also be found be equating the external shear force to the shearing resistance of the beam, as under :

$$V = (\text{Area of shear stress diagram}) \times b$$

$$= \left\{\left(\frac{2}{3} qn\right) + q\,(d - n)\right\} b = \left(\frac{2}{3} qn + qd - qn\right) b = qb\left(d - \frac{n}{3}\right) = q \,.\, bjd$$

Hence $$q = \frac{V}{b \,.\, jd} \qquad \ldots(7.1)$$

Eq. 7.1, obtained from the stress based approach, gives the value of maximum shear stress q at N.A. and for all layers below N.A. in a cracked reinforced beam. Though in pure flexure, we neglect the concrete area below N.A., the same concrete between the cracks is needed for shear transfer between steel reinforcement and the compression zone. Also the flexural cracks (assumed vertical in pure flexure) are not vertical in the shear-flexural zone under consideration. We have also neglected the shear deformations, which are really responsible for the mobilization of aggregate interlock at the cracks and dowel forces in the reinforcement. Due to these reasons, the *stress-based approach* followed above in the derivation of Eq. 7.1 may not represent the real behaviour of the beam in shear and flexure. Hence Eq. 7.1 , representing conventional shear stress formula $q = V/bjd$, has now been abandoned. Indian Standard Code (IS 456 : 2000) recommends to use *nominal shear stress* given by the expression.

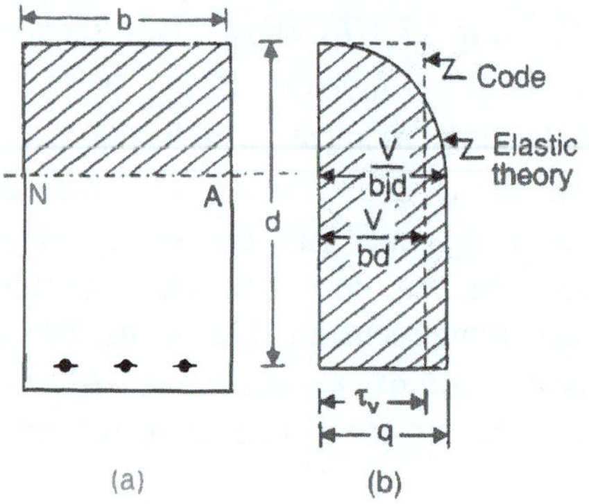

FIG. 7.2.

$$\tau_v = \frac{V}{bd} \quad ..(7.4\ a)$$

where τ_v = nominal shear stress

V = shear force at the section, due to design loads.

Thus in Fig. 7.2(*b*), the maximum shearstr ess q ($= V/bjd$) obtained from elastic theory, is greater than the *nominal shear stress* (or average shear stress) τ_v suggested by the Code.

Nominal shear stress in Limit State (IS 456 : 2000) : In the case of limit state of shear, Eq. 7.4 (*a*) can be written as

$$\tau_v = \frac{V_u}{bd} \quad ...(7.4)$$

where V_u = ultimate shear force (factored shear force)

τ_v = nominal shear stress at the limit state

Beams of varying depth : In the case of beam of varying depth, Eq. 7.4 is modified as

$$\tau_v = \frac{V_u \pm \dfrac{M_u}{d}\tan\beta}{bd} \quad ...(7.4\ a)$$

where M_u = bending moment at the section, and

β = angle between the top and the bottom edges of the beam.

The negative sign in the formula applies when the bending moment M_u increases numerically in the same direction as the effective depth d increases, and the positive sign when the moment decreases numerically in this direction.

7.2. EFFECTS OF SHEAR : DIAGONAL TENSION

We know that the bending stress and the shearing stress vary across the cross section of R.C. beam. Below the N.A., the bending stress (f) is tensile while the shear stress q is constant. Fig. 7.3(*b*) shows a small element, taken from the portion below the N.A. This element is subjected to a longitudinal tensile stress f and horizontal shear stress q alongwith the vertical complementary shearing stress q.

$$p = \frac{f}{2} \pm \sqrt{\left(\frac{f}{2}\right)^2 + q^2} \quad ...(7.5)$$

and the inclination of the principal plane with horizontal is given by

$$\tan 2\theta = -\frac{2q}{f} \quad ...[7.5(a)]$$

The major principal stress p_1 is tensile and is given by

$$p_1 = \frac{f}{2} \pm \sqrt{\left(\frac{f}{2}\right)^2 + q^2} \quad ...[7.5(b)]$$

At the supports, where bending stress f is practically zero, the value of the principal (tensile) stress is equal to the shear stress q, and it is inclined at 45° to the horizontal, *i.e.* it acts diagonally. Hence it is known as *diagonal tension*. At the centre of the beam, where shearing stresses are practically negligible, the principal tensile stress is equal to the longitudinal tensile stress f, its direction will be horizontal and the cracks will be vertical.

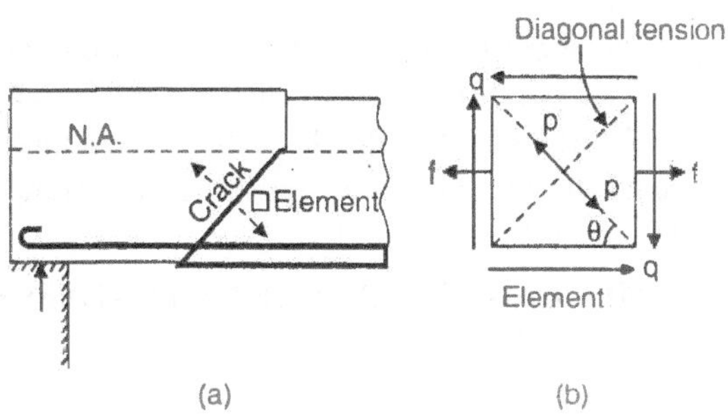

FIG. 7.3. DIAGONAL TENSION.

Concrete is quite strong in shear, but the diagonal tension, which is caused by the combined action of longitudinal tension and the transverse shearing stresses, is to be resisted by the provision of *shear reinforcement or diagonal (or inclined tension) reinforcement or web reinforcement.*

7.3. REGIONS OF CRACKS IN BEAMS

As stated earlier, at any section in a beam, there exists both bending moment (M) as well as shear (V). Depending upon the ratio of bending moment (M) to shear force (V) at different sections, there may be three regions of creacks in the beam (Fig. 7.4):

(*a*) Region I : Region of *flexure cracks*
(*b*) Region II : Region of *flexure-shear cracks*
(*c*) Region III : Region of *web shear cracks* or *diagonal tension cracks*

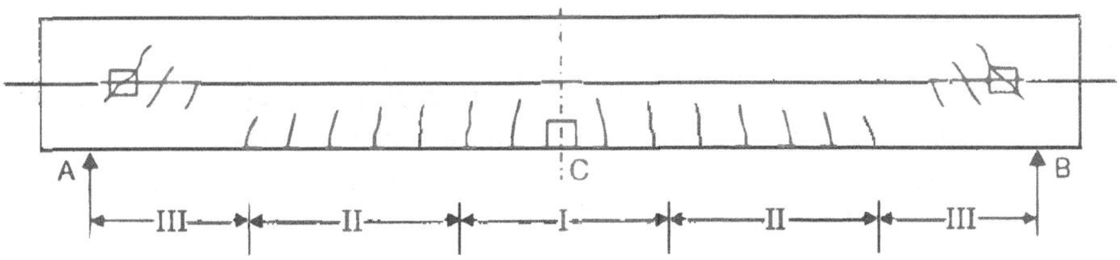

FIG. 7.4. DIFFERENT REGIONS OF CRACKS IN BEAM

(a) Region I : Region of flexural cracks : This region normally occurs adjacent to mid-span where B.M. is *large* and shear force is either zero or very small. The principal planes are perpendicular to the beam axis. when the principal tensile stress reaches the *tensile strength* of concrete (which is quite low) *tensile cracks develop vertically*. These cracks are known as *flexural cracks* resulting primarily due to flexure.

(b) Region II : Regions of flexure-shear cracks : These regions are near the quarter span, to both the sides, where B.M. is considerable and at the same time shear force is significant. The cracks in this region are initiated at the tension face. travel vertically (due to flexure) and gradually tend to develop in the inclined direction towards the N.A., as the shear stress goes on increasing towards the N.A. Since the cracks develop under the combined action of bending moment and shear, these cracks are known as *flexure-shear cracks*.

(c) Region III : Regions of diagonal tension cracks : These regions are adjacent to each support of a beam, where shear force is predominant. Since shear stress is maximum at the N.A., inclined cracks start developing at the N.A. along the diagonal of an element subjected to the action of pure shear. Hence these cracks are known as *diagonal tension cracks* or *web-shear cracks.*

7.4. MECHANISMS OF SHEAR TRANSFER IN R.C. BEAM WITHOUT SHEAR REINFORCEMENT

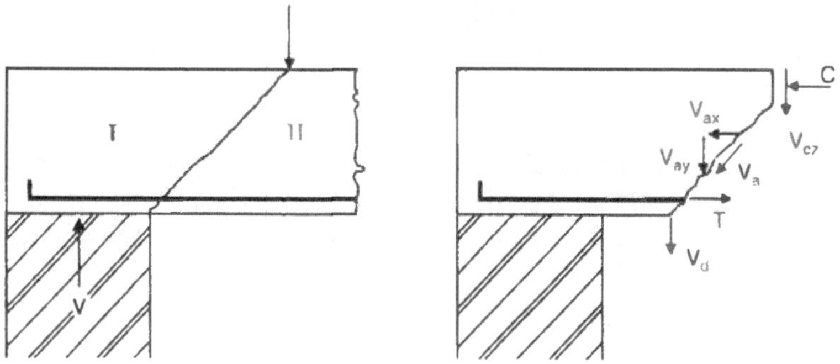

(a) Diagonal tension crack (b) Flexural shear crack

FIG. 7.5. MECHANISMS OF SHEAR TRANSFER

Shear is transferred between two adjacent planes in a R.C. beam by the following mechanisms :

(*a*) Shear resistance V_{cz} of the *uncracked* portion of concrete

(*b*) Vertical component V_{ay} of the *interface shear* or *aggregate interlock* force V_a

and (*c*) Dowel force V_d in the tension reinforcement, due to *dowel action*

Thus, $$V = V_{cz} + V_{ay} + V_d \qquad ...(7.6)$$

The relative contribution of each of the above three mechanisms depend upon the stage of loading and extent of cracking. In the initial stage, before the flexural cracking starts, the entire shear is resisted by the shear resistance of the concrete (i.e. $V = V_{cz}$). As the flexural cracking starts, interface shear comes into action, resulting in the redistribution of stresses. Further extension of flexural crack results in sharing the shear by the dowel force V_d of the tension reinforcement. Thus, at the final stage of collapse, the shear is borne by all the three mechanisms, expressed by Eq. 7.6.

For beams of rectangular section, without shear reinforcement, the proportion of shear transferred by the above three mechanisms are in the range of 20 to 40% for V_{cz}, 30 to 50% for V_{ay} and 15 to 25% by V_d.

7.5. SHEAR SPAN

At any section, magnitude and direction of maximum principal tensile stress, and hence the development and growth of inclined cracks depend upon the relative magnitudes of flexural stress f and the shear stress τ. Since f and τ at any section are *proportional* to M/bd^2 and V/bd respectively, we have

$$\frac{f}{\tau} = \frac{A_1\, M/bd^2}{A_2\, V/bd} = A_3 \frac{M}{V\,.\,d} \quad \ldots(7.7)$$

where A_1, A_2, A_3 are the constants of proportionality.

For a beam subjected to concentrated load, the ratio M/V at the critical section subjected to *maximum* V is expressed by a distance a_v called *shear span* between the support and the load (Fig. 7.6). The distance a_v ($= M/V$) represents the dominance of shear over flexure. *Many tests on shear have established that diagonal tension is a function of M/V and hence is function of shear span* a_v. It also depends on the effective depth d as clear from the following expression

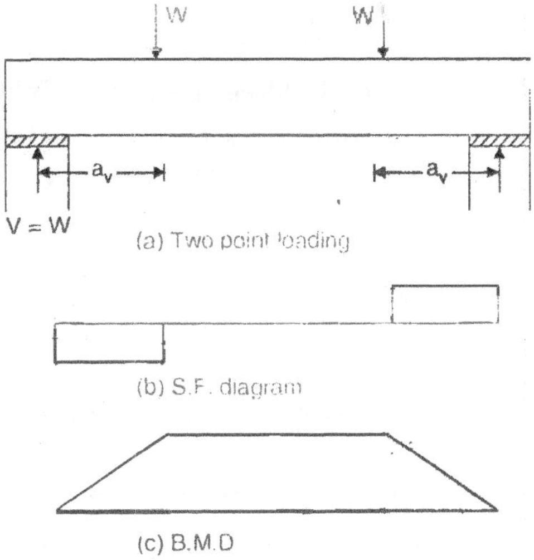

FIG. 7.6. SHEAR SPAN

$$\frac{f}{\tau} \propto \frac{a_v}{d} \quad \ldots(7.8)$$

Thus, the diagonal tension failure (i.e. shear failure) is a function of ratio a_v/d. It is also interesting to note that the dimensionless ratio a_v/d (or M/Vd) provides a measure of the relative magnitudes of the flexural stress and shear stress, and hence enables the prediction of the mode of failure in the beam in flexural shear.

7.6. MODES OF SHEAR FAILURE

The shear failure of a R.C. beam, without shear reinforcement is governed by a_v/d ratio. Depending upon the value of a_v/d, ratio, a beam may experience following types of shear failure.

1. Case I : $a_v/d < 1$: Splitting or compression failure (deep beams)
2. Case II : $1 < a_v/d < 2.8$: Shear compression or shear tension failure
3. Case III : $2.8 < a_v/d < 6$: Diagonal tension failure.

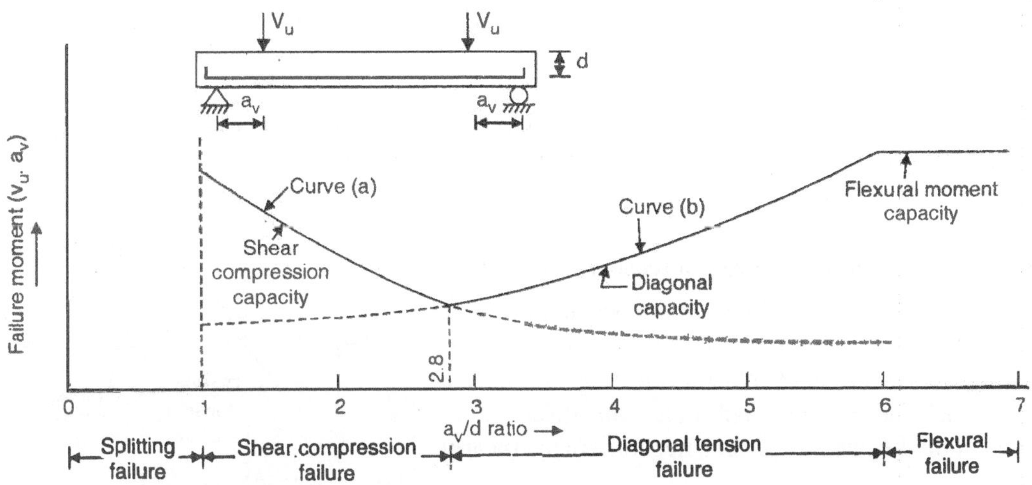

FIG. 7.7 EFFECT OF a_v/d ON SHEAR STRENGTH OF R.C. BEAM

4. Case IV : $a_v/d > 6$: Flexure failure

Fig. 7.7 shows the plot of failure moment ($V_u \cdot a_v$) versus a_v/d ratio.

CASE I : $a_v/d < 1$ (deep beams) : Splitting or compression failure

This case corresponds to a deep beam without shear reinforcements where the inclined cracking transforms the beam into a *tied arch* (Fig. 7.8 *a*). The load is carried by (*i*) direct compression in the concrete between the load and reaction point by crushing of concrete and by (*ii*) tension in the longitudinal steel by yielding or fracture or anchorage failure or bearing failure.

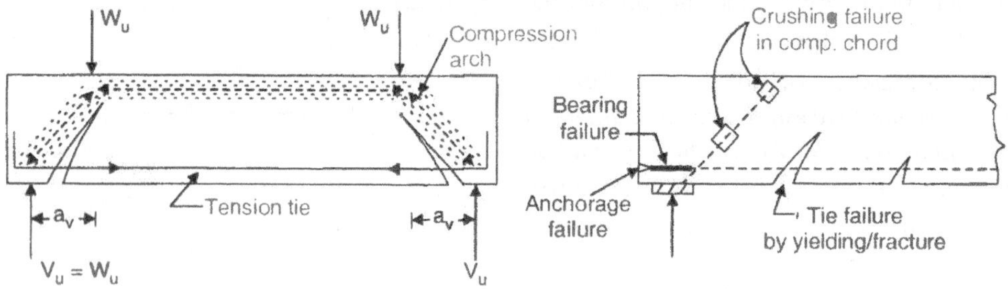

(a) Tied Arch Action (b) Failure modes

FIG. 7.8. CASE I : $a_v/d < 1$ (DEEP BEAMS)

CASE II : $1 < a_v/d < 2.8$: Shear compression/tension failure

This case is common in short beams with a_v/d ratio between 1 to 2.8, where failure is initiated by an inclined crack – more commonly a *flexural shear crack*. Fig. 7.9 (*a*) shows the *shear compression failure* due to vertical compressive stresses developed in the vicinity of the load. Similarly, the vertical compressive stress over the reaction limits the bond splitting and diagonal cracking along the steel. The crack extends towards the tension reinforcement and then propagates along the reinforcement (Fig. 7.9 *b*) resulting in the failure of the beam by *anchorage failure* of tension reinforcement called *shear tension failure*.

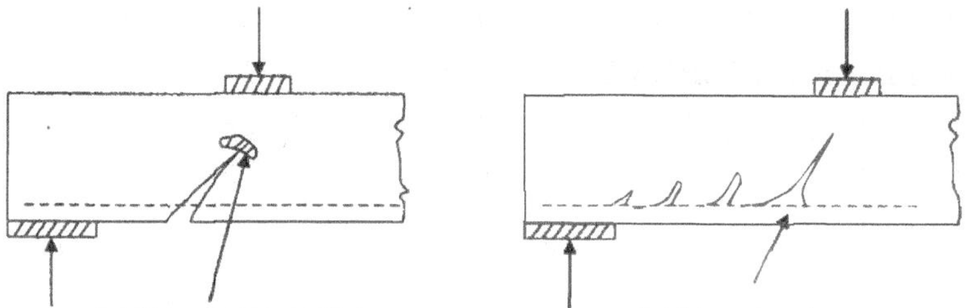

(a) Shear compression failure (b) Shear tension failure

FIG. 7.9. CASE II : $1 < a_v/d < 2.8$

Case III $2.8 < a_v/d < 6$: Diagonal tension failure

Diagonal tension failure occurs when the shear span to the effective depth ratio is in the range of 2.8 to 6. *Normal beams* have a_v/d ratio in excess of 2.8. Such beams may fail either in shear or in flexure. In the beginning, several flexural cracks (nearly

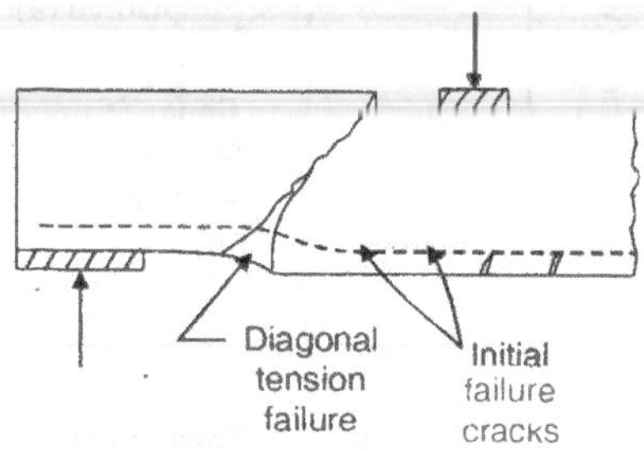

FIG. 7.10. CASE III : $2.8 < a_v/d < 6$

vertical) are formed and then the diagonal flexural crack starts from the last flexural crack.

The crack then extends in the compression zone in the inclined direction, first steeply and then at a flat slope. Shortly before reaching the critical failure, the bottom inclined crack gets widened. If shear (web) reinforcement is not provided, the cracks extend rapidly to the top of the beam and failure occurs *rapidly*. Addition of shear reinforcement enhances the shear strength considerably.

Case IV : $a_v/d > 6$: Flexural failure

Flexural failure is encountered when a_v/d ratio is greater than 6. Two cases may be encountered ; (*i*) under-reinforced beam and (*ii*) over reinforced beam. In the case of under-reinforced beam, tension reinforcement is less than the limiting one due to which failure is initiated by yielding of tension reinforcement, leading to the ultimate failure due to crushing of concrete in compression zone. Such *a ductile failure* is known as *flexural tension failure* which is quite slow giving enough warning. In the over-reinforced sections, failure occurs due to crushing of concrete in compression zone before yielding of tension reinforcement. Such a failure, known as *flexural compression failure*, is quite sudden.

7.7. FACTORS AFFECTING SHEAR RESISTANCE OF A R.C. MEMBER

The *shear resistance* of rectangular beams, *without shear reinforcement* depends upon the following factors.

1. Grade of concrete : Higher grade of concrete has higher characteristic strength which in turn results in (*i*) higher tensile strength (*ii*) greater dowel shear resistance, (*iii*) greater aggregate interlock capacity, and (*iv*) greater concrete strength in compression zone. Hence *shear resistance increases with the increase in the grade of concrete.*

2. Percentage and grade of longitudinal tensile reinforcement : The increase in percentage (p_t) of longitudinal tensile reinforcement results in the increase in dowel shear (V_d). Due to this reason, the design Codes make the shear strength (τ_c) of concrete a function of p_t and grade of concrete (see Table 7.1). However, higher grade of steel results in lesser shear resistance of R.C. beam because the percentage of steel (p_t) corresponding to a higher grade steel is less than that required for a low grade steel, say mild steel.

3. Ratio of shear span to effective depth (i.e. a_v/d ratio) : As discussed in the previous article, for a_v/d ratio between 6 and 2.8, the shear capacity, being governed by inclined crack resistance, decrease with decrease in a_v/d ratio (curve *b* of Fig. 7.7). However, for a value of a_v/d less than 2.8, the shear capacity, being dependent on *shear-compression* or *shear-bond capacity*, increases rapidly. The minimum shear capacity is at a_v/d ratio around 2.8.

4. Compressive force : Presence of axial compressive force result in increase of shear capacity. The effect of axial compression on the design shear strength has been taken into account by I.S. Code by increasing the design shear strength by a modification factor δ. (See Eq. 7.11).

5. Compressive reinforcement : The shear resistance is found to increase with the increase in the percentage of compressive steel (p_c).

6. Axial tensile force : Axial tensile force reduces marginally the shear resistance of concrete, as is evident from Eq. 7.12.

7. Shear reinforcement : The shear resistance of a R.C. beam increases with the increase in *shear reinforcement ratio*. This is due to two reasons (*i*) concrete gets confined between stirrup spacing, and (*iii*) the shear/web reinforcement itself provides shear resistance, V_{us}.

7.8. DESIGN SHEAR STRENGTH OF CONCRETE WITHOUT SHEAR REINFORCEMENT (IS 456 : 2000)

The magnitude of design shear strength (τ_c) depends basically on the grade of concrete (f_{ck}) and the percentage (p_t) of tension steel. As per IS 456 : 2000, the design shear strength of concrete in beams *without shear reinforcement* shall be as given in Table 7.1.

TABLE 7.1. DESIGN SHEAR STRENGTH (τ_c) OF CONCRETE, (N/mm²)

$100\frac{A_{st}}{bd}$	Grade of concrete					
	M 15	*M 20*	*M 25*	*M 30*	*M 35*	*M 40 and above*
≤0.15	0.28	0.28	0.29	0.29	0.29	0.30
0.25	0.35	0.36	0.36	0.37	0.37	0.38
0.50	0.46	0.48	0.49	0.50	0.50	0.51
0.75	0.54	0.56	0.57	0.59	0.59	0.60
1.00	0.60	0.62	0.64	0.66	0.67	0.68
1.25	0.64	0.67	0.70	0.71	0.73	0.74
1.50	0.68	0.72	0.74	0.76	0.78	0.79
1.75	0.71	0.75	0.78	0.80	0.82	0.84
2.00	0.71	0.79	0.82	0.84	0.86	0.88
2.25	0.71	0.81	0.85	0.88	0.90	0.92
2.50	0.71	0.82	0.88	0.91	0.93	0.95
2.75	0.71	0.82	0.90	0.94	0.96	0.98
3.0 and above	0.71	0.82	0.92	0.96	0.99	1.01

Note : The term A_{st} is the area of longitudinal tension reinforcement which continues at least one effective depth beyond the section being considered except at support where the full area of tension reinforcement may be used provided it is properly anchored.

Analytical expression for design shear strength

The values of τ_c given in the above Table by the Code are based on the following semi empirical expression (SP 24, 1983) :

$$\tau_c = \frac{0.85\sqrt{0.8 f_{ck}}\,(\sqrt{1 + 5\beta} - 1)}{6\beta} \quad \ldots(7.9)$$

where $\beta = \frac{0.8 f_{ck}}{6.89\, p_t}$, but not less than 1 ...(7.9 *a*)

$p_t = \frac{100\, A_{st}}{bd}$ (percentage steel in rib width only)

$0.8 f_{ck}$ = cylinder strength in terms of cube strength

0.85 = reduction factor similar to $1/\gamma_m$

The formula in BS 8110 for design shear strength of concrete is slightly different, and is given by the expression

$$\tau_c = 0.79\,(p_t)^{1/3}\left(\frac{400}{d}\right)^{1/4}\left(\frac{1}{\gamma_m}\right)\left(\frac{f_{ck}}{25}\right)^{1/3} \quad \ldots(7.10)$$

where $\left(\frac{400}{d}\right)$ = the correction factor for depth and should not be less than 1

$\left(\frac{f_{ck}}{25}\right)$ = the correction factor for the strength of concrete and should not be greater than 40

$\gamma_m = 1.25$

p_t = percentage steel, the value of which should not exceed 3

Design shear strength for solid slabs

For solid slabs, the design shear strength for concrete shall be $\tau_c \cdot k$, where k has the values given in Table 7.2

TABLE 7.2 VALUES OF k (IS 456 : 2000)

Overall depth of slab (mm)	300 or more	275	250	225	200	175	150 or less
k	1.00	1.05	1.10	1.15	1.20	1.25	1.30

Note : The above provision shall not apply to flat slabs.

Shear strength of members under axial compression (IS 456 : 2000)

For members subjected to axial compression P_{uc}, the design shear strength of concrete, given in Table 7.1, shall be multiplied by the following factor :

$$\delta = 1 + \frac{3\,P_{uc}}{A_g \cdot f_{ck}}, \text{ but not exceeding } 1.5 \quad \ldots(7.11)$$

where P_{uc} = factored axial compressive force in Newtons

A_g = gross area of concrete section in mm^2, and

f_{ck} = characteristic compressive strength of concrete, in N/mm^2

Shear strength of members under axial tension (ACI Code, 1989):

Though it is evident that there is some reduction in design shear strength of a member under axial tension, IS Code (IS 456 : 2000) does not explicitly mention this case. However, the following simplified expression for δ, based on ACI Code (1989) may be used :

$$\delta = 1 - \frac{P_{ut}}{3.45\,A_g} \quad \ldots(7.12)$$

where P_{ut} = factored axial tensile force in Newtons.

Maximum shear stress in concrete with shear reinforcement (IS 456 : 2000)

(*a*) ***Maximum shear stress in beams*** :

Under no circumstances, even with shear reinforcement, shall the nominal shear stress (τ_v) in beams exceed $\tau_{c,max}$ given is Table 7.3.

TABLE 7.3. MAXIMUM SHEAR STRESS. $\tau_{c,\,max}$, (N/mm^2)

Grade of concrete	M 15	M 20	M 25	M 30	M 35	M 40 and above
$\tau_{c,\,max}$ N/mm^2	2.5	2.8	3.1	3.5	3.7	4.0

(b) Maximum shear stress in solids slabs

For solid slabs, the nominal shear stress shall not exceed half the appropriate values given in Table 7.3

7.9. WEB REINFORCEMENT FOR DIAGONAL TENSION

As stated earlier, proper reinforcement must be provided to resist the diagonal tension. The shear resisted by shear reinforcement can be worked out by considering the equilibrium of forces across a potential diagonal crack, which is assumed to be inclined at an angle of 45° with axis of the beam. Fig. 7.11 shows a diagonal crack *AB*. Let the web reinforcement be inclined at angle α with the axis of the beam, and be spaced at distance s_v apart. Let the diagonal crack *AB* intersect *n* number of *web reinforcing bars.*

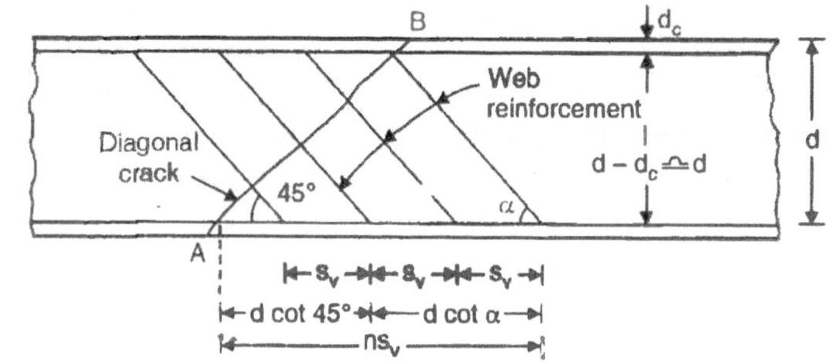

FIG. 7.11 SHEAR RESISTED BY WEB STEEL

Let V_{us} = Ultimate shear carried by shear (or web) reinforcement

f_{yd} = design yield stress in web steel $= 0.87 f_y$

n = number of bars/links crossing the crack

α = inclination of web steel

A_{sv} = total cross-section area of each set of bar or link.

The web reinforcement is anchored to the main tensile steel at the bottom, and to the holding bars (at a cover d_c) at the top. Hence the vertical component of length of inclined bar $= (d - d_c)$. Since d_c in normally quite small in comparison to d, we can take $(d - d_c \simeq d)$, as marked in Fig. 7.11. Now, for equilibrium

Shear carried by shear reinforcement = Sum of vertical components of tensile forces developed in shear reinforcement

$$\therefore \quad V_{us} = n\, A_{sv}\, f_{yd} \sin\alpha \qquad \ldots(7.13\ a)$$

In order to get the value of n, we have from geometry,

$$n\, s_v = d \cot 45° + d \cot\alpha \quad \text{or} \quad n = \frac{d \cot 45° + d \cot\alpha}{s_v} = \frac{d\,(1 + \cot\alpha)}{s_v} \qquad \ldots(7.13\ b)$$

Substituting the value of n and $f_{yd}\,(= 0.87 f_y)$ in Eq. 7.13 (a) we get

$$V_{us} = \frac{d\,(1 + \cot\alpha)}{s_v} \cdot A_{sv}\,(0.87 f_y) \sin\alpha = \frac{0.87 f_y\, A_{sv}\, d}{s_v} (\sin\alpha + \cos\alpha) \qquad \ldots(7.14)$$

Rearranging the above, we get

$$s_v = \frac{0.87 f_y\, A_{sv}\, d}{V_{us}} (\sin\alpha + \cos\alpha) \qquad \ldots(7.14\ a)$$

The above equation gives the spacing of bars inclined at α with horizontal. Here A_{sv} = (area of cross-section of bar) × (No. of legs) $= A_{\varphi} \times$ number of legs.

Special cases

(*i*) ***Bars inclined at 45°*** (i.e. $\alpha = 45°$)

If $\alpha = 45°$,
$$V_{us} = \frac{0.87\, f_y\, A_{sv}\, d}{s_v} (\sqrt{2}) \quad ...(7.15)$$

or
$$s_v = \frac{0.87\, f_y\, A_{sv}\, d}{V_{us}} \sqrt{2} \quad ...(7.15\ a)$$

(*ii*) ***Bars inclined at 90° (i.e. vertical stirrups)***

$$V_{us} = \frac{0.87\, f_y\, A_{sv}\, d}{s_v} \quad ...(7.16)$$

or
$$s_v = \frac{0.87\, f_y\, A_{sv}\, d}{V_{us}} \quad ...(7.16\ a)$$

(*iii*) ***Single bar or single group of bars***

For a single bar, or single group of bars, all bent up at the same cross-section, we get from Eq. 7.13 (*a*) taking $n = 1$

$$V_{us} = 0.87\, f_y \,.\, A_{sv} \sin \alpha \quad ...(7.13)$$

7.10. TYPES OF SHEAR REINFORCEMENT

Shear reinforcement is necessary if the nominal shear stress (τ_v) exceeds the design shear stress τ_c. In general, shear reinforcement is provided in any one of the following three forms :

(*a*) Vertical stirrups
(*b*) Bent up bars along with the stirrups
(*c*) Inclined stirrups

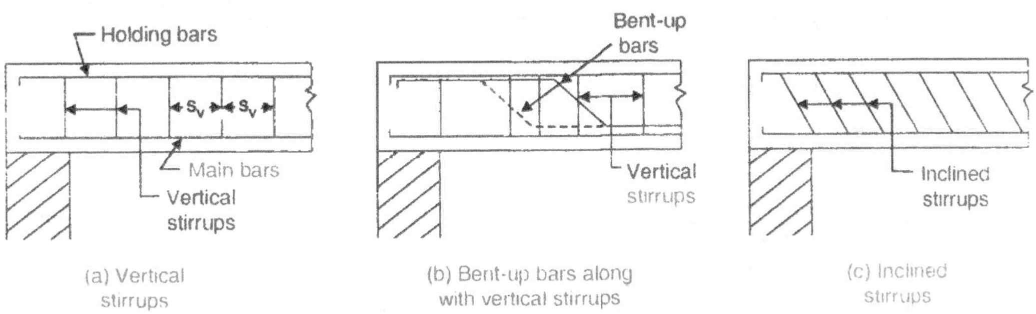

FIG. 7.12. TYPES OF SHEAR REINFORCEMENT

Where bent-up bars are provided, their contribution towards shear resistance shall not be more than half that of total shear reinforcement.

The total external shear V_u is jointly resisted by concrete as well as shear reinforcement and is represented by the expression.

$$V_u = V_{uc} + V_{us}$$

or $$V_{us} = V_u - V_{uc}$$

where V_{uc} = shear strength of concrete $= \tau_c \cdot bd$

7.11. ACTION OF R.C. BEAM WITH SHEAR REINFORCEMENT : TRUSS ANALOGY

As stated earlier, when the nominal shear stress (τ_v) in a R.C. beam exceeds the design shear stress τ_c, shear reinforcement, also known as web reinforcement is provided to resist balance of shear force V_{us} ($= V_u - V_{uc}$). The behaviour of a R.C. beam with shear reinforcement is reprepresented by *truss analogy*, also known as *lattice girder effect*. Fig. 7.13 (*a*) shows a steel truss in which the compression members (marked by *C*) are shown by *thick lines* and the tension members (marked by *T*) are shown by *thin lines*. Similar to a simple truss of Fig. 7.13 (*a*), the action of a reinforced concrete beam with *vertical stirrups* is shown in Fig. 7.13 (*b*) wherein the bottom longitudinal tensile steel form the bottom chord tension member, the vertical stirrups form vertical tension members and the

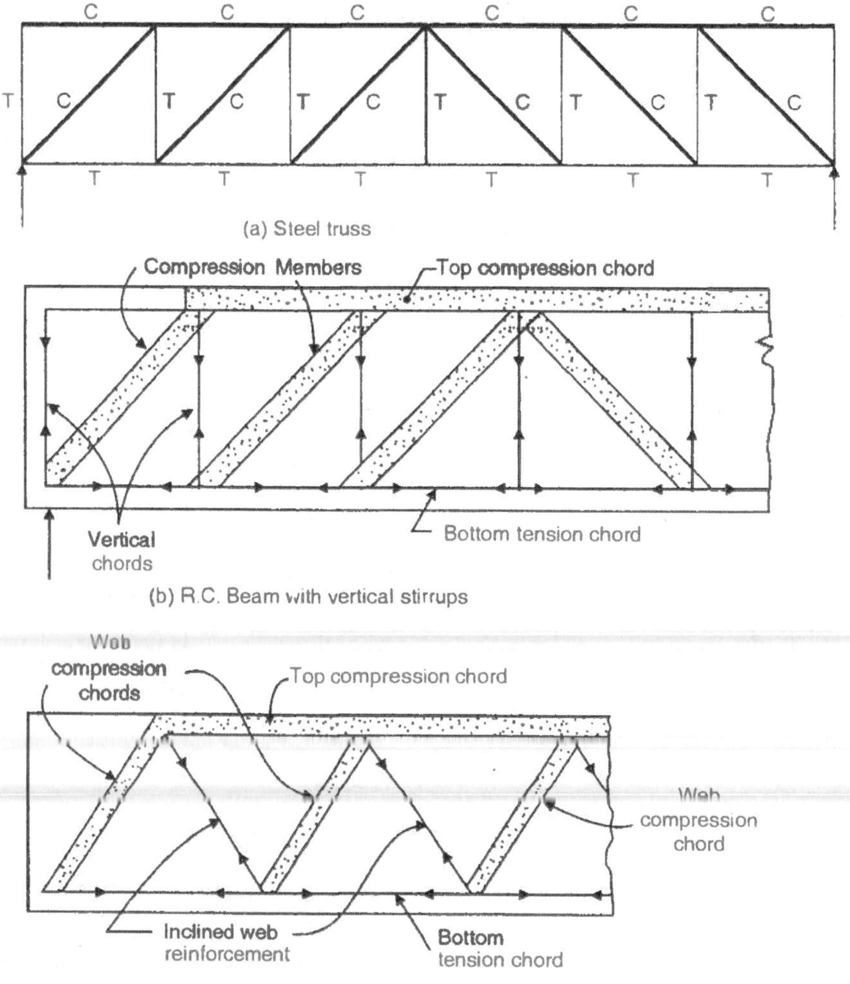

(a) Steel truss

(b) R.C. Beam with vertical stirrups

(c) R.C. beam with inclined shear reinforcement

FIG. 7.13. TRUSS ANALOGY IN R.C. BEAM WITH SHEAR REINFORCEMENT

concrete acts as the diagonal and top chord compression members. Similarly, Fig. 7.13 (c) shows an *analogous truss* to depict the action of R.C. beam with inclined shear reinforcement, either in the form of inclined stirrups or in the form of bent up bars.

Systems of lattice girder effect : The resistance of shear at any section of a beam reinforced with inclined bars may be calculated on the assumption that the inclined bars form the tension members of a system of the lattice girder in which the concrete forms the compression members. There are two systems of lattice girders : (*i*) single system (*ii*) double system

(*i*) ***Single System*** (Fig. 7.14) : In this system, the bars are bent at 45° and the imaginary compression members (known as *struts*) *AB, CD* etc. are assumed to be at $67\frac{1}{2}°$ with horizontal. Member *AC* is an imaginary horizontal portion of the inclined bar acting as *tension chord* and is assumed to have the same cross-sectional area as of inclined bars. The member *BD* is an imaginary *compression chord*. It is assumed that the distance between the top compression chord and bottom tension chord is equal to lever arm. In the triangle *ABC, BC* and *AC* are tension members while *AB* is compression member. From geometry of triangle *ABC*,

$$AC = BC = a\sqrt{2} = 1.414\ a$$

Thus in order that the inclined bars are to be effective, the bars nearest to the support are bent at 45° at a distance of 1.414*a* from the support. Since the triangle *ABC* is isosceles, the total force in the inclined bar *BC* is equal to the total force in the horizontal bar. Thus, the force in the inclined bar is $A_{sv}\,f_{yd}$ where A_{sv} is the area of cross-section of the bent up bars. The vertical component of this force to resist the shear force is equal to :

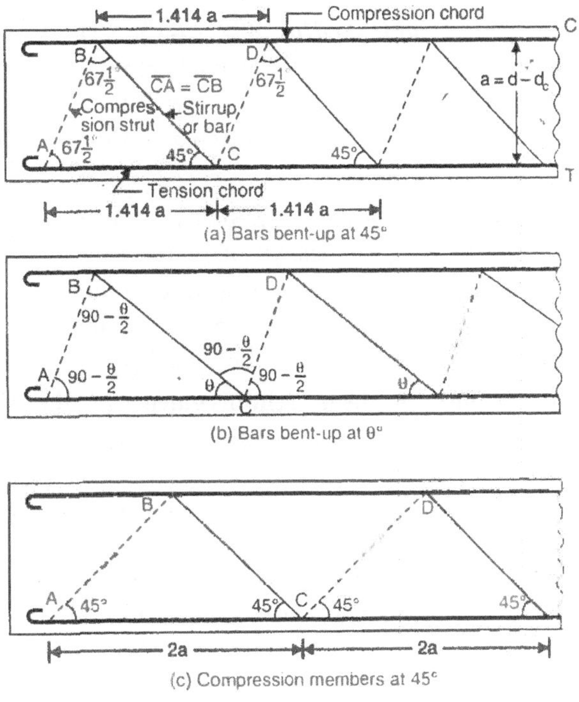

FIG. 7.14. SINGLE SYSTEM OF LATTICE GIRDER

$$V_{us} = A_{sv} \,.\, f_{yd} \,.\, \sin 45° = 0.707\ A_{sv} \,.\, f_{yd} \quad (\text{ where } \quad f_{yd} = 0.87\, f_y\ ) \qquad \ldots(7.17\ a)$$

The inclined bar is effective for a length equal to lever arm on either side measured along the N.A.

If, however, the bars are bent up at an angle θ, the compression member is assumed to be inclined at 90° − θ/2 with the horizontal as shown in Fig. 7.14(*b*).

Sometimes, the imaginary compression members are assumed to be inclined at 45° as shown in Fig. 7.14 (*c*). In this case, the lengths *AB* and *BC* will be equal, and the spacing will be equal to 2*a*. The total force in the bar *BC* can be calculated from the force triangle *ABC* as under :

$$\frac{\text{Force in } BC}{\text{Force in } AC} = \frac{\sin 45°}{\sin 90°} \qquad \therefore \qquad \text{Force in } BC = \frac{A_{sv} \cdot f_{yd} \cdot \sin 45°}{\sin 90°} = A_{sv} \frac{f_{yd}}{\sqrt{2}}$$

This shows that the permissible stresses in BC is reduced to $\frac{f_{yd}}{\sqrt{2}}$.

This ensures that the stress in the horizontal portion of the inclined bars is equal to the permissible value of f_{yd}

Hence the vertical S.F. resisted by $BC = V_{us} = A_{sv} \cdot \frac{f_{yd}}{\sqrt{2}} \cdot \sin 45°$

or
$$V_{us} = 0.707 \cdot A_{sv} \cdot \frac{f_{yd}}{\sqrt{2}} \qquad \text{...(7.17 } b\text{)}$$

(*ii*) ***Double system*** **(Fig. 7.15)** : In this system, the bars are bent up at spacing equal to half the spacing of the single system. The compression members are assumed to be inclined at $67\frac{1}{2}°$, but the spacing between them is assumed to be equal to 1.414 a, while the spacing between the inclined bars is kept equal to 0.707a, as shown in Fig. 7.15.

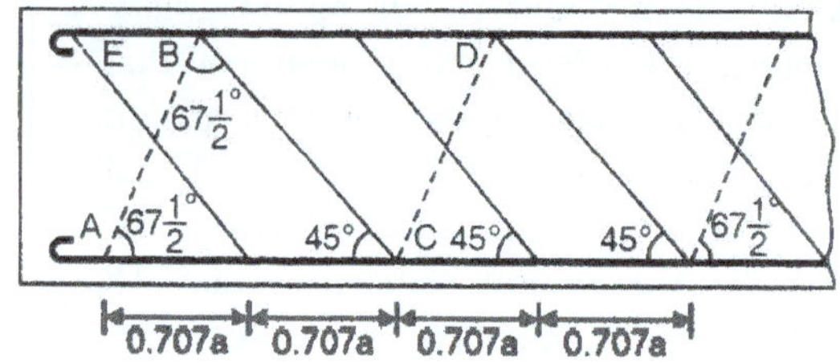

FIG 7.15. DOUBLE SYSTEM.

The shear resistance of such a system will evidently be equal to twice that of the single system *i.e.*, it will be equal to : $2 \times 0.707\ A_{sv} \cdot f_{yd}$. (where $f_{yd} = 0.87\, f_y$)

Contribution of shear reinforcement to the shear strength of a R.C. beam

Provision of shear or web reinforcement enhances the shear strength of a R.C. beam in the following ways :

1. Bearing a part of shear by the *truss action*
2. Increasing the aggregate inter-lock force (V_a) by limiting the opening of the diagonal tension crack.
3. Increasing the dowel action force (V_d) by providing support to the longitudinal bars crossing the shear crack
4. Increasing the strength of concrete due to *confinement* of concrete by closely spaced shear reinforcement, specially in the form or vertical or inclined stirrups.

7.12. VERTICAL STIRRUPS

Shear reinforcement in the form of vertical *stirrups* consists of 5 mm to 15 mm diameter steel bars bent round the tensile reinforcement where it is anchored to 6 to 12 mm dia. anchor bars or holding bars. Depending upon the magnitude of the shear stress to be resisted, a stirrup may be one legged, two legged, four legged or multi-legged, as shown in Fig. 7.16.

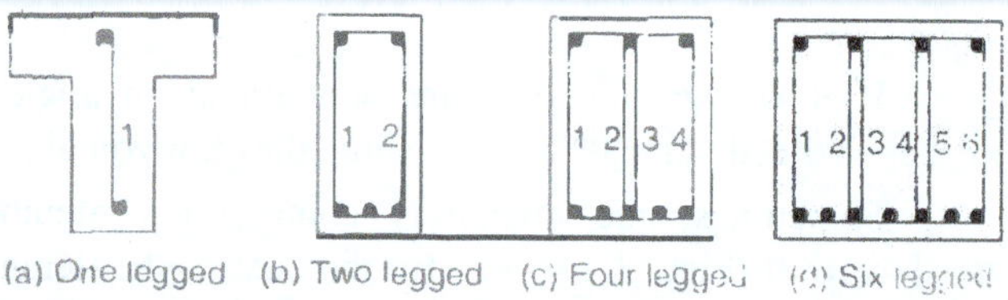

FIG. 7.16. FORMS OF VERTICAL STIRRUPS

The strength of shear reinforcement in the form of vertical stirrups is given by

$$V_{us} = \frac{0.87 f_y A_{sv} d}{s_v} \quad ...(7.16)$$

or Spacing
$$s_v = \frac{0.87 f_y A_{sv} d}{V_{us}} \quad ...(7.16\ a)$$

Alternatively,
$$s_v = \frac{0.87 f_y \cdot A_{sv}}{(\tau_v - \tau_c) b} \quad ...(7.16\ b)$$

Here, $A_{sv} = m A_{\phi}$ where m = No. of legs in the stirrup and A_{ϕ} = area of cross-section of the bar of the stirrup

Eq. 7.16(a) can also be *alternatively* obtained from Fig 7.17. Let us assume that in absence of shear reinforcement, the beam fails in diagonal tension, the inclination of the tension crack being at 45° to the axis of the beam and extended up to a horizontal distance equal to $(d - d_c) \simeq d$

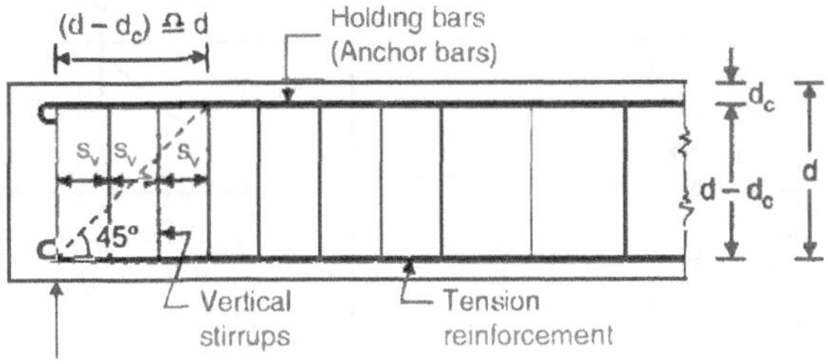

FIG. 7.17. SPACING OF VERTICAL STIRRUPS

∴ No. of stirrups resisting shear force $= \frac{d}{s_v}$

$$\therefore \left[\begin{array}{l}\text{Shear} \\ \text{resisted by stirrups}\end{array}\right] = \left[\begin{array}{l}\text{shear resisted} \\ \text{by each stirrup}\end{array}\right] \times \left[\begin{array}{l}\text{No. of} \\ \text{stirrups}\end{array}\right]$$

or
$$V_{us} = 0.87 f_y \cdot A_{sv} \times \frac{d}{s_v}$$

or
$$s_v = \frac{0.87 f_y \cdot A_{sv} \cdot d}{V_{us}} = \frac{0.87 f_y \cdot A_{sv} \cdot d}{V_u - V_{uc}} \quad ...(7.16\ a)$$

Spacing diagram for vertical stirrups. The spacing of the stirrups at any section along the length of the beam can be determined from Eq. 7.16, since the net shear V_{us} at that section can be easily determined. This will give variable spacings all along the length of the beam. These variable spacings can be *averaged* out by the construction of *spacing diagram*.

Procedure. (1) Determine net shear force at various sections distant x_1, x_2, x_3 etc. from the support and calculate the spacings at these sections from Eq. 7.16 (a).

(2) Plot the spacing diagram with the calculated spacing as ordinate (Fig. 7.18).

(3) Calculate the spacing of the first stirrup, on the basis of net S.F. at the support. Set the first stirrup at half the distance of this spacing, from the support.

(4) Where this ordinate at (1) intersects the spacing diagram, set a line at 45°, meeting the span line in (2). This will give the position of the second stirrup. Repeat this procedure to get the position of other stirrups.

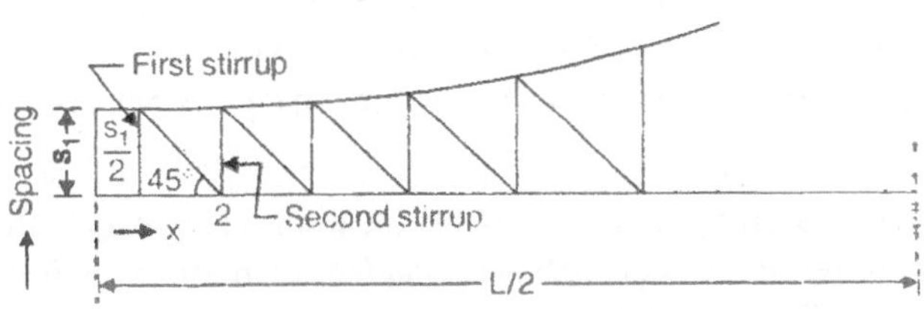

FIG. 7.18 SPACING DIAGRAM.

7.13. CRITICAL SECTION FOR DESIGN SHEAR : IS 456 : 2000

A support to a beam may offer either a *compressive reaction or a tensile reaction*. Fig. 7.19 (*a*) shows some typical situations, such as in the case of a beam resting on wall (Fig. 7.19 *ai*) or on column (Fig. 7.19 *aii* and 7.19 *aiii*), where the support offers a compressive reaction. The critical section for such a support is at a distance *d* from the face of the support because the diagonal crack nearest to the support starts from the face of the support and runs upwards at 45° (approx.) towards the load.

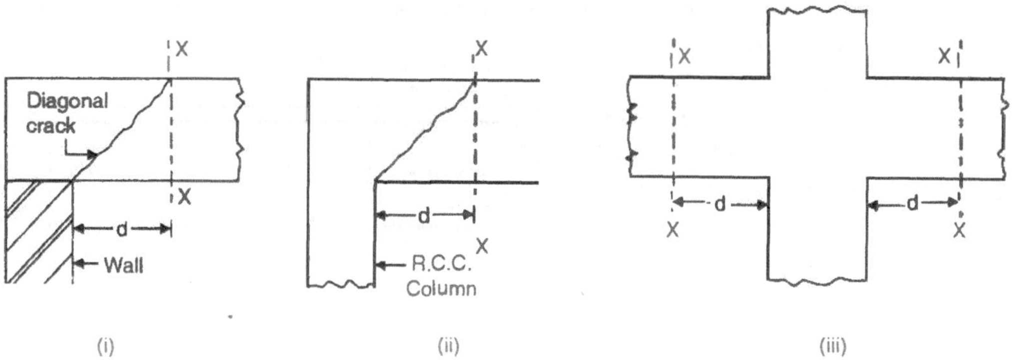

(a) Critical section X-X at d from the face of the support

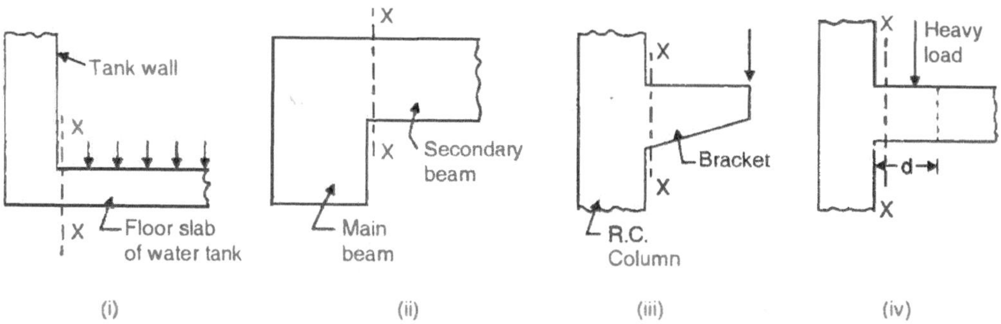

(b) Critical section X-X at the face of the support

FIG. 7.19. CRITICAL SECTION FOR SHEAR

Similarly Figs. 7.19 *b(i), (ii), (iii)* show situations where the support offers a tensile reaction; the critical section for such a situation is at the face of the support. Also, in a case where there is a concentrated load between the face of the support and the distance *d* (Fig. 7.14 *b iv*), the critical section is at the face of the support.

Note : The above provisions are applicable for beams generally carrying uniformly distributed load or where the principal load is located father than 2*d* from the face of the support.

7.14. MINIMUM SHEAR REINFORCEMENT (IS 456 : 2000)

The shear reinforcement in the form of stirrups remain *unstressed* till the diagonal crack occurs at the critical location. However, the instant a diagonal crack occurs, the web reinforcement receives *sudden* increase in stress. If web reinforcement is not provided, shear failure may occur without giving any warning. The Code, therefore, specifies that all the beams should be provided with atleast some minimum reinforcement called *nominal shear reinforcement* even if nominal shear stress (τ_v) is less than the design shear stress (τ_c) of concrete.

Reasons for providing minimum shear reinforcement

1. It prevents *sudden* shear failure with the formation of diagonal tension crack, and imparts *ductility* to provide sufficient warning of impending failure. Thus *brittle shear failure* is prevented.

2. It guards against any sudden failure of a beam if concrete cover bursts and bond to tension steel is lost.

3. It holds the main reinforcements in place while pouring the concrete. Thus minimum requirement of cover and clear distance between longitudinal bars are maintained.

4. It acts as necessary ties for the compression steel (if any) and makes it effective.

5. It prevents pressing down of the longitudinal reinforcement, thereby maintaining the *dowel capacity*.

6. It confines the concrete, thereby increasing its strength and rotation capacity.

7. It prevents failure that can be caused by tension due to shrinkage and thermal stresses and internal cracking in the beam.

As per IS 456 : 2000, minimum shear reinforcement in the form of stirrups shall be provided such that

$$\frac{A_{sv}}{b\, s_v} \geq \frac{0.4}{0.87\, f_y} \qquad \text{...(7.18)}$$

where A_{sv} = total cross-sectional area of stirrup legs effective in shear.
s_v = stirrup spacing along the length of the member
b = breadth of beam or breadth of the web of flanged beam.
f_y = characteristic strength of stirrup reinforcement in N/mm², which *shall not be taken greater than 415 N/mm²*.

Hence spacing based on minimum shear reinforcement is given by

$$s_v \leq \frac{0.87\, f_y \,.\, A_{sv}}{0.4\, b} \leq \frac{2.175\, f_y\, A_{sv}}{b} \qquad \text{...(7.18 } a)$$

However, where the maximum shear stress calculated is less than half the permissible value, and in members of minor structural importance such as lintels, this provision need not be complied with.

Shear resistance of minimum shear reinforcement

The shear resistance of minimum reinforcement envisaged in Eq. 7.18 is found by substituting the value of $\frac{0.87\, f_y \,.\, A_{sv}}{s_v} = 0.4\, b$ (Eq. 7.18 *a*) into Eq. 7.16(*a*) for vertical stirrups.

Thus, $$V_{us,\,min} = \left(0.87\, \frac{f_y\, A_{sv}}{s_v}\right) d = (0.4\, b)\, d = 0.4\, bd \qquad \text{...(7.19)}$$

Thus, shear carried by concrete and that carried by minimum stirrups is given by

$$V_{u,\,min} = \tau_c \,.\, bd + 0.4\, bd \qquad \text{...(7.20)}$$

7.15. MAXIMUM SPACING OF SHEAR REINFORCEMENT (IS 456 : 2000)

To avoid shear failure, the diagonal crack must be intercepted by at least one stirrup. Substituting $n = 1$ and $\alpha = 90°$ in Eq. 7.13 (*a*), we get $s_v = d$. However, actual tests on R.C. beam with shear reinforcement have shown that the links which intersect the diagonal crack near the top are relatively ineffective. Therefore, the maximum spacing of vertical stirrups is limited to 0.75 *d*. Indian Standard Code (IS 456 : 2000) gives the following recommendation for maximum spacing of shear reinforcement :

The maximum spacing of shear reinforcement measured along the axis of the member shall not exceed 0.75 *d* for vertical stirrups and *d* for inclined stirrups at 45° where *d* is the effective depth of the section under consideration. *In no case shall the spacing exceed 300 mm.*

7.16. ENHANCED SHEAR STRENGTH OF SECTIONS CLOSE TO SUPPORTS (IS 456 : 2000)

1. General

Shear failure at section of beams and cantilevers *without* shear reinforcement will normally occur on plane inclined at an angle 30° to the horizontal, such as the failure plane *AC* shown dotted in Fig. 7.20 (*a*). If, however, the failure plane is *forced* to be inclined more steeply than this because the section *X-X* considered in Fig. 7.20 (*a*) is close to the support, or for other reasons, the *shear force required to produce failure is increased.* This occurs when the ratio a_v/d is less than 2. As is evidenced by extensive test results on R.C. beams without shear reinforcement, when a_v/d less than 2, the shear capacity increases parabolically while for a_v/d ratio more than 2, the shear capacity almost remains constant (Fig. 7.20 *b*).

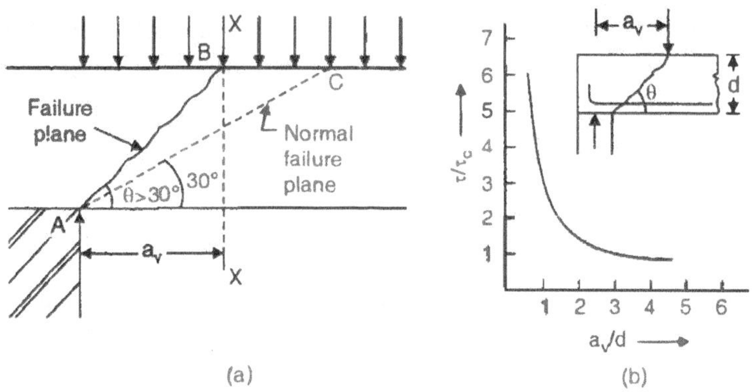

FIG. 7.20. ENHANCEMENT OF SHEAR STRENGTH

2. Enhancement of shear strength

As per IS 456 : 2000, the *enhancement* of shear strength may be taken into account in the design of sections near a support by increasing design shear strength (τ_c) of concrete to $2\,d\,\tau_c/a_v$ provided that the design stress at the face of support remains less than the value $\tau_{c,\,max}$ given in Table 7.3.

Thus, enhanced design shear strength $\tau_{ce} = \tau_c \cdot \dfrac{2d}{a_v}$ ...(7.21)

(But not greater than $\tau_{c,\,max}$)

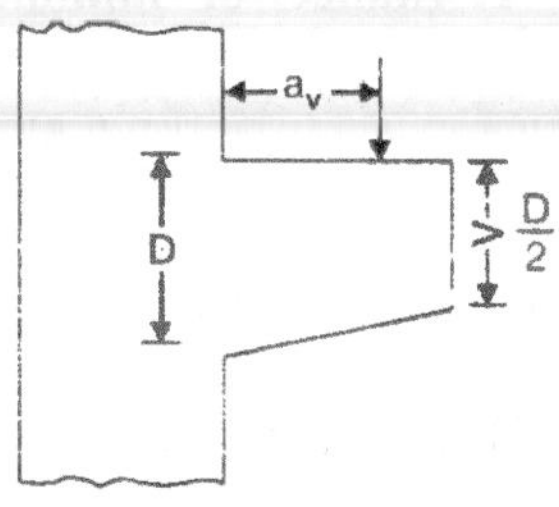

FIG. 7.21.
BRACKET HAVING $a_v/d < 1$

Account may be taken of enhancement in any situation where the section considered in closer to the face of the support or concentrated load than twice the effective depth, d. Such an enhancement may be particularly useful for the design of corbels (or brackets), or pile caps where a_v/d ratio may be less than unily. To be effective, tension reinforcement should extend on each side of the point where it is intersected by a possible failure plane for a distance at least equal to the effective depth, or be provided with an equivalent anchorage.

3. Shear reinforcement for sections close to supports

If shear reinforcement is required, the total area of this is given by

$$\Sigma A_{sv} = \frac{a_v b\left(\tau_v - \dfrac{2 d \tau_c}{a_v}\right)}{0.87 f_y} \geq \frac{0.4\, a_v b}{0.87 f_y} \quad \ldots(7.22)$$

This reinforcement should be provided within the middle three quarters of a_v. Where a_v is less than d, horizontal shear reinforcement will be effective than vertical.

4. Simplified approach for enhanced shear strength near supports

The approach given in para, 1, 2 and 3 above may be used for all beams. However, for beams carrying generally uniform load or where the principal load is located farther than $2\,d$ from the face of the support, the shear stress may be calculated at a section at a distance d from the face of the support. The value of τ_c is calculated in accordance with Table 7.1 and appropriate shear reinforcement is provided at section closer to the support, no further check for shear at such sections is required.

In the above treatment, a_v is either the *shear span* or the distance of the concentrated load from the face of the support.

Example 7.1. Design of shear stirrups of Fe 250 grade steel. *A reinforced concrete beam 250 mm wide and 400 mm effective depth is subjected to ultimate design shear force of 150 kN at the critical section near supports. The tensile reinforcement at the section near supports is 0.5 percent. Design the shear stirrups near the supports. Also, design the minimum shear reinforcement at the mid span. Assume concrete of grade M 20 and mild steel bars of Fe 250 grade.*

Solution : Given : $b = 250$ mm ; $d = 400$ mm; $A_{st}/bd = 0.5\% = 0.005$

$$\tau_v = \frac{V_u}{bd} = \frac{150 \times 10^3}{250 \times 400} = 1.5 \text{ N/mm}^2$$

From Table 7.1, $\tau_c = 0.48$ N/mm² for M 20 concrete and $100\, A_{st}/bd = 0.5$

Also, from Table 7.3, $\tau_{c,\,max} = 2.8$ N/mm² for M 20 concrete.

Thus, τ_v is less than $\tau_{c,\,max}$, but greater than τ_c. Hence shear reinforcement is necessary.

$$V_{uc} = \tau_c\, bd = 0.48 \times 250 \times 400 = 48000 \text{ N}$$

Hence $\qquad V_{us} = V_u - V_{uc} = 150000 - 48000 = 102000$ N

The shear resistance of nominal stirrups is given by

$$V_{us,\,min} = 0.4\, bd = 0.4 \times 250 \times 400 = 40000 \text{ N} < V_{us}$$

Hence nominal stirrups are *not* sufficient at the section near supports.

From Eq. 7.16 (*a*), $$s_v = \frac{0.87\, f_y\, A_{sv}}{V_{us}} \cdot d$$

Using two legged stirrups of 10 mm dia. bars, $A_{sv} = 2\,\frac{\pi}{4}\,(10)^2 = 157.08\ \text{mm}^2$

$$\therefore \qquad s_v = \frac{0.87 \times 250 \times 157.08}{102000} \times 400 \approx 134 \quad \text{mm}$$

Alternatively, from 7.16 (*b*), $$s_v = \frac{0.87\, f_y\, A_{sv}}{(\tau_v - \tau_c)\, b} = \frac{0.87 \times 250 \times 157.08}{(1.5 - 0.48)\, 250} = 134 \quad \text{mm}$$

Maximum spacing = 0.75 d or 300 mm, which ever is less.

Hence provide 10 mm dia. two legged stirrups @ 130 mm c/c at the section near supports.

At mid-span, the spacing of minimum shear reinforcement for 10 mm φ – 2 lgd stirrups is given by (Eq. 7.18 *a*) :

$$s_v = 0.87\,\frac{f_y\, A_{sv}}{0.4\, b} = \frac{0.87 \times 250 \times 157.08}{0.4 \times 250} = 341.6 \quad \text{mm}$$

However, maximum spacing is limited to 0.75 d or 300 mm which ever is less. Hence $s_v = 300$ mm.

Hence provide 10 mm dia. two legged stirrups @ 300 mm c/c at the mid-span.

Example 7.2. Design of shear stirrups of Fe 415 grade steel.
Redesign shear stirrups for example 7.1, using Fe 415 steel for shear stirrups.

Solution : As found in example 7.1, $\tau_v = 1.5\ \text{N/mm}^2$; $\tau_c = 0.48\ \text{N/mm}^2$; $V_{us} = 102000$ N.

Using two legged stirrups of 8 mm dia. bar, $A_{sv} = 2\,\frac{\pi}{4}\,(8)^2 = 100.53\ \text{mm}^2$

$$\therefore \qquad s_v = \frac{0.87 \times 415 \times 100.53}{102000} \times 400 = 142.3 \quad \text{mm}$$

Maximum spacing = 300 mm.

Hence provide 8 mm φ 2 lgd stirrups @ 140 mm c/c near supports.

At mid-span, using minimum shear reinforcement,

$$s_v = \frac{0.87 \times 415 \times 100.53}{0.4 \times 250} \approx 363 \quad \text{mm}$$

However, maximum spacing is limited to 0.75 d or 300 mm.

Hence provide 8 mm dia. two legged stirrups @ 300 mm c/c at the mid-span.

Example 7.3. Design of shear reinforcement. Bent up bars with vertical stirrups
A simply supported beam, 300 mm wide and 600 mm effective depth carries a uniformly distributed load of 74 kN/m including its own weight over an effective span of 6 m. The reinforcement consists of 5 bars of 25 mm diameter. Out of these, two bars can be safely bent up at 1 m distance from the support. Design the shear reinforcement for the beam.

Given : Grade of concrete : M 20 ; Grade of steel : Fe 415

Assume width of supports = 400 mm.

Solution : $w_u = 1.5\, w = 1.5 \times 74 = 111$ kN/m

$\therefore$ Max. shear $\quad V_{u,\,max} = \dfrac{w_u L}{2} = \dfrac{111 \times 6}{2} = 333$ kN

According to IS Code, the critical section for shear occurs at a distance of d (= 600 mm) from the face of the support, i.e. at a distance of 400/2 + 600 = 800 mm from the centre of support.

Design shear $\quad V_{uD} = 333 - 111 \times 0.8 = 244.2$ kN

$$\therefore \quad \tau_v = \frac{244.2 \times 1000}{300 \times 600} \approx 1.36 \text{ N/mm}^2$$

$$A_{st} \text{ at support} = (5-2)\frac{\pi}{4}(25)^2 = 1472.6 \text{ mm}^2$$

$$\therefore \quad \frac{100\,A_{st}}{bd} = \frac{100 \times 1472.6}{300 \times 600} = 0.818\,\%$$

Hence from Table 7.1 $\quad \tau_c = 0.56 + \dfrac{0.62 - 0.56}{0.25}(0.818 - 0.75) = 0.576$ N/mm^2

Also, from Table 7.3. $\quad \tau_{c,\,max} = 2.8$ N/mm$^2 > \tau_v$

Since $\tau_v > \tau_c$, shear reinforcement is necessary.

Shear resistance of concrete $\quad V_{uc} = 0.576 \times 300 \times 600 = 103680$ N = 103.68 kN

$$\therefore \quad V_{us} = V_{uD} - V_{uc} = 244.2 - 103.68 = 140.52 \text{ kN}$$

Area of cross-section of two bent-up bars $= A_{sv} = 2\,\dfrac{\pi}{4}(25)^2 = 981.7$ mm^2

Shear resistance of two bent-up bars $= 0.87\,f_y \,.\, A_{sv} \sin\alpha$

$= 0.87 \times 415 \times 981.7 \sin 45° = 250630$ N = 250.63 kN

However, the maximum shear resistance, to be assigned to the bent up bars should not exceed half the resistance to be provided by the shear reinforcement, i.e. it should not exceed $\frac{1}{2} \times 140.52 = 70.26$ kN. Hence the remaining resistance to be provided by the vertical stirrups = 140.52 − 70.26 = 70.26 kN.

$\therefore \quad V_{us} = 70.26$ kN for vertical stirrups

From Eq. 7.16 (*a*), $\quad s_v = \dfrac{0.87\,f_y\,A_{sv}\,.\,d}{V_{us}}$

Using 8 mm dia 2-lgd stirrups, $A_{sv} = 2 \times \dfrac{\pi}{4}(8)^2 = 100.54$ mm^2

$$\therefore \quad s_v = \frac{0.87 \times 415 \times 100.54 \times 600}{70260} \approx 310 \text{ mm}$$

Also, spacing corresponding to minimum steel is given by

$$s_v = \frac{0.87\,f_y\,A_{sv}}{0.4\,b} = \frac{0.87 \times 415 \times 100.54}{0.4 \times 300} = 302.5 \text{ mm}$$

However, max. $\quad s_v = 0.75\,d$ or 300 mm (whichever is less) = 300 mm

Hence provide 8 mm dia. 2-lgd stirrups @ 300 mm c/c from the face of the support to the point at which the bars are bent. These two bars are bent up at a distance of $1.414 \times a \simeq 1.414 \times 0.9 \times 600 \simeq 760$ mm from the face of the support (*i.e.* at a distance of

760 + 200 = 960 mm from the centre of the support). Beyond this point, bent up bars are not available.

At 0.96 m away from the support, $V_u = 333 - 111 \times 0.96 = 226.44$ kN

$$A_{st} = 5\frac{\pi}{4}(25)^2 = 2454.4 \text{ mm}^2 \; ; \qquad \frac{100\,A_{st}}{bd} = \frac{100 \times 2454.4}{300 \times 600} = 1.36\,\%$$

Hence from Table 7.1, $$\tau_c = 0.67 + \frac{0.72 - 0.67}{0.25}(1.36 - 1.25) = 0.692 \text{ N/mm}^2$$

$$V_{uc} = 0.692 \times 300 \times 600 = 124560 \text{ N} = 124.56 \text{ kN}$$

$$\therefore \quad V_{us} = V_u - V_{uc} = 226.44 - 124.56 = 101.88 \text{ kN}$$

The spacing of 8 mm dia. 2-lgd stirrups is given by

$$s_v = \frac{0.87 \times 415 \times 100.54 \times 600}{101880} = 213.8 \text{ mm}$$

Hence provide these @ 210 mm c/c.

To locate the section where maximum spacing of 300 mm, corresponding to minimum shear reinforcement is provided, proceed as follows :

Let the distance of the section be x m from the mid-span.

$$V_u = \frac{333}{3}x = 111\,x \text{ kN} = 111000\,x \text{ N}$$

$$V_{uc} = 124560 \text{ N (as before);} \quad \therefore \quad V_{us} = 111000\,x - 124560$$

Now $$s_v = \frac{0.87 f_y \cdot A_{sv} \cdot d}{V_{us}} \quad \text{or} \quad 300 = \frac{0.87 \times 415 \times 100.54 \times 600}{111000\,x - 124560}$$

This gives $x \simeq 1.776$ m from mid-span (or 1.224 m from the centre of the support).

Summary

	Distance from centre of support	*Stirrups*	*Spacing*
1	0 to 0.76 m	8 φ 2 lgd	300 mm c/c
2.	0.76 to 1.224 m	8 φ 2 lgd	210 mm c/c
3.	1.224 m to 3 m	8 φ 2 lgd	300 mm c/c

Example 7.4. Design of shear reinforcement : Vertical Stirrups

A simply supported beam, 300 mm wide and 500 mm effective depth carries a uniformly distributed load of 50 kN/m, including its own weight over an effective span of 6 m. Design the shear reinforcement in the form of vertical stirrups. Assume that the beam contains 0.75% reinforcement throughout the length. The concrete is of M 20 grade and steel for stirrups is of Fe 250 grade. Take width of supports as 400 mm.

Solution : $w_u = 1.5 \times 50 = 75$ kN/m

$$V_{u,max} = \frac{w_u L}{2} = \frac{75 \times 6}{2} = 225 \text{ kN}$$

The critical section lies at a distance of $d = 500$ mm from the face of the support or at a distance of 500 + 400/2 = 700 mm from the centre of the support.

$$\therefore \quad V_{uD} = 225 - 75 \times 0.7 = 172.5 \text{ kN}$$

and $$\tau_v = \frac{172.5 \times 10^3}{300 \times 500} = 1.15 \text{ N/mm}^2$$

From Table 7.1, for $100 A_s/bd = 0.75\%$, we get $\tau_c = 0.56$ N/mm² for M 20 concrete.

$$\therefore \quad V_{uc} = 0.56 \times 300 \times 500 = 84000 \text{ N} = 84 \text{ kN}$$

Also, $\tau_{v,max} = 2.8$ N/mm² for M 20 concrete. Since $\tau_v < \tau_{v,max}$ it is OK.

However, $\tau_v > \tau_c$; hence shear reinforcement is necessary.

$$V_{us} = V_{uD} - V_{uc} = 172500 - 84000 = 88500 \text{ N}$$

Using 10 mm φ 2-lgd vertical stirrups, $A_{sv} = 2\frac{\pi}{4}(10)^2 = 157.1$ mm²

$$\therefore \text{ Spacing } s_v = \frac{0.87 f_y \cdot A_{sv} \cdot d}{V_{us}} = \frac{0.87 \times 250 \times 157.1 \times 500}{88500} = 193 \text{ mm} \simeq 190 \text{ mm (say)}$$

Spacing corresponding to minimum shear reinforcements is

$$s_v = \frac{0.87 f_y A_{sv}}{0.4 b} = \frac{0.87 \times 250 \times 157.1}{0.4 \times 300} = 284.7 \text{ m} \simeq 280 \text{ mm (say)}$$

However in no case should the spacing exceed $0.75 d = 0.75 \times 500 = 375$ mm, or 300 mm whichever is less. Hence the spacing is to vary from 190 mm at the end section to 280 mm at a section distant x m (say) from the mid-span. Let us locate this section where the S.F. is V_{ux}.

$$\therefore \quad V_{ux} = \frac{V_{u,max}}{3} x = \frac{225000}{3} x = 75000 x$$

$$V_{us} = V_{ux} - V_{uc} = 75000 x - 84000$$

$$s_v = 280 = \frac{0.87 \times 250 \times 157.1 \times 500}{75000 x - 84000}$$

from which, we get $x = 1.93$ m from mid-span or 1.07 m from supports. Hence provide 8 mm φ 2 lgd stirrups at a spacing of 190 mm c/c from supports to a section distant 1.07 m from the centre of either supports. For the remaining length, provide the stirrups @ 280 mm c/c.

Example 7.5. *Shear reinforcement for a cantilever*

Design the shear reinforcement for a R.C. cantilever beam shown in Fig. 7.22. Carrying U.D.L. of 80 kN/m inclusive of its own weight. Consider concrete of grade M 20 and steel of grade Fe 415. Assume a cover of 50 mm to the centre of reinforcement.

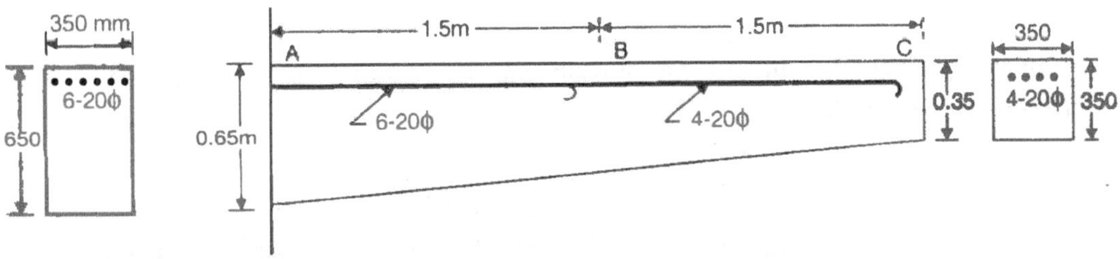

FIG. 7.22

Solution : $w_u = 1.5 \times 80 = 120$ kN/m

(a) Design at the support

For a cantilever beam, the critical section is at the face of the support, i.e. at a section at A. Effective depth $d = 650 - 50 = 600$ mm.

$$V_u \text{ at } A = w_u \cdot L = 120 \times 3 = 360 \text{ kN}$$

Design shear, $V_{ud} = V_u \pm \dfrac{M_u}{d} \tan \beta$ ....(7.4 *a*)

$$M_u = w_u L^2/2 = 120\,(3)^2/2 = 540 \text{ kN-m} = 540 \times 10^3 \text{ kN-mm}$$

$\tan \beta$ = tangent of slope of the bottom edge of the beam.

$= (650 - 350)/3000 = 0.1$

Since the B.M. increases in the same direction as d increases, we will use minus sign in Eq. 7.4 (*a*). Hence $V_{ud} = 360 - \dfrac{540 \times 10^3}{600} \times 0.1 = 270$ kN

$$\therefore \quad \tau_v = \frac{270 \times 10^3}{350 \times 600} = 1.286 \text{ N/mm}^2$$

$$\frac{100\,A_{st}}{bd} = \frac{100 \times 6 \times \frac{\pi}{4}(20)^2}{350 \times 600} = 0.898$$

Hence from Table 7.1, $\tau_c = 0.56 + \dfrac{0.62 - 0.56}{0.25}(0.898 - 0.75) = 0.5955$ N/mm²

Since $\tau_v > \tau_c$, shear reinforcement is necessary.

Also, $\tau_{c,\,max} = 2.8$ N/mm². Since $\tau_v < \tau_{c,\,max}$, it is OK.

$$V_{uc} = \tau_c\, bd = 0.5955 \times 350 \times 600 = 125055 \text{ N}$$

$$\therefore \quad V_{us} = V_{uD} - V_{uc} = 270000 - 125055 = 144945 \text{ N}$$

Using 8 mm φ 2 lgd stirrups, $A_{sv} = 2 \times \dfrac{\pi}{4}(8)^2 = 100.53$ mm²

$$\therefore \quad s_v = \frac{0.87 f_y A_{sv}}{V_{us}} d = \frac{0.87 \times 415 \times 100.53}{144945} \times 600 = 150.25 \text{ mm}$$

Spacing corresponding to minimum steel is given by

$$s_v = \frac{0.87 f_y A_{sv}}{0.4\,b} = \frac{0.87 \times 415 \times 100.53}{0.4 \times 350} = 259.2 \text{ mm} \simeq 250 \text{ mm}$$

However, $s_{v,\,max} = 0.75\,d$ or 300 mm (which ever is less)

Hence provide 8 mm φ 2 lgd stirrups @ 150 mm c/c at the supports.

(b) Design at point B

Effective depth $d = \dfrac{650 + 350}{2} - 50 = 450$ mm

$$V_u = w_u \cdot \frac{L}{2} = 120 \times \frac{3}{2} = 180 \text{ kN}; \quad M_u = \frac{120\,(1.5)^2}{2} = 135 \text{ kN-m}$$

$$V_{uD} = V_u - \frac{M_u}{d} \tan\beta = 180 - \frac{135 \times 10^3}{450} \times 0.1 = 150 \ \text{kN}$$

$$\tau_v = \frac{V_{us}}{bd} = \frac{150 \times 10^3}{350 \times 450} = 0.9524 \ \text{N/mm}^2$$

$$\frac{100\, A_{st}}{bd} = \frac{100 \times 4 \times \frac{\pi}{4}(20)^2}{350 \times 450} = 0.798$$

$$\therefore \quad \tau_c = 0.56 + \frac{0.62 - 0.56}{0.25}(0.798 - 0.75) = 0.5715 \ \text{N/mm}^2$$

Since $\tau_v > \tau_c$, shear reinforcement is necessary, Also $\tau_v < \tau_{c\,,\,max}$.

$$V_{uc} = \tau_c\, bd = 0.5715 \times 350 \times 450 = 90011 \ \text{N}$$

$$\therefore \quad V_{us} = V_{uD} - V_{uc} = 150000 - 90011 = 59989 \ \text{N}$$

Using 8 mm φ 2 lgd stirrups, $A_\varphi = 100.53$ mm^2, as before

$$\therefore \quad s_v = \frac{0.87 \times 415 \times 100.53 \times 450}{59989} = 272.3 \ \text{mm}$$

Spacing corresponding to minimum steel = 259.2 mm ≏ 250 mm as before.

Also, $s_{v\,,\,max} = 0.75\, d = 0.75 \times 450 = 337.5$ or 300 whichever is less.

Hence provide 8 mm φ 2 lgd stirrups @ 250 mm c/c from B to C.

PROBLEMS

1. A R.C. beam supports a total uniformly distributed load of 80 kN inclusive of its own weight and has 16 cm width and 32 cm effective depth. Determine (*i*) maximum shear stress (*ii*) spacing of 10 mm diameter two legged stirrups near the support. Use M 20 concrete and Fe 415 steel.

2. A beam has an effective depth of 40 cm. At a particular section, it has 8 mm φ 2 lgd stirrups provided at 20 cm centre to centre. Calculate the sheare resistance of the stirrups. Use M 20 concrete and Fe 415 steel.

3. A R.C. beam supports a total uniformly distributed load of 160 kN. The beam has 20 cm width and 36 cm effective depth. Determine the spacing of 10 mm diameter two legged vertical stirrups, in addition to two main reinforcing bars of 16 mm bent up at 45° to resist stear. Use M 20 concrete and Fe 415 steel.

8

Limit State of Collapse in Bond

8.1. INTRODUCTION : DEFINITION

One of the most important assumption in the behaviour of reinforced concrete structures is that there is proper 'bond' between concrete and reinforcing bars. When a R.C. element is loaded, the load is first borne by concrete and it is then transferred to steel reinforcement. This transfer of force from concrete to steel (and vice versa) can be effected only when there is no relative movement or slip or sliding between them when any one of these two is strained. *The force which prevents this slippage between the two constituent materials is known as bond.* In fact, bond is responsible for providing '*strain compatibility*' and '*composite action*' of concrete and steel. It is through the action of *bond resistance* that the axial stress (tensile or compressive) in a reinforcing bar can undergo variation from point to point along its length. This is required to accommodate the variation in bending moment along the length of the flexural member.

When steel bars are embedded in concrete, the concrete, after setting, adheres to the surface of the bar and thus resists any force that tends to pull or push this rod. The intensity of this adhesive force is called *bond stress*. The bond stresses are the *longitudinal shearing stresses* acting on the surface between the steel and concrete, along its length. Hence bond stress is also known as the *interfacial shear*. *Hence bond stress is the shear stress acting parallel to the reinforcing bar on the interface between the bar and the concrete.*

8.2. MECHANISMS OF BOND RESISTANCE

The bond between steel and concrete comprise of the following three resistances :

(*i*) Pure adhesive resistance (or chemical adhesion)

(*ii*) Frictional resistance

and (*iii*) Mechanical resistance

(*i*) Pure adhesive resistance : It is provided by the relatively weak adhesion of the *chemical gum* produced by concrete during setting.

(*ii*) Frictional resistance : Pure adhesive resistance is provided by the relatively weak adhesion of the 'chemical gum' produced by concrete during setting. Frictional resistance is provided by the shrinkage of the concrete thereby gripping the steel bars. When loaded, the adhesive resistance is broken first. But even after the adhesion is broken, friction between the two materials continues to provide a considerable bond resistance. The hooks and other means of *anchoring* steel bar do not prevent the initial slipping ; but they delay the collapse of structure as the member acts as a tied arch with the uncracked concrete representing the arch and the anchored bars as the tie rods. The frictional resistance is low for a smooth bar surface, such as that of cold-rolled steel, and is higher for rough surfaces. Hence it is essential that the bars must be clean and free from dirt, grease, scale and loose rust so that pure adhesive resistance and frictional resistance can be developed fully. Anything that destroys the ability of the concrete to grip the steel may prove to be serious because it will prevent the stress in the latter from being fully developed.

(*iii*) **Mechanical resistance** : The *mechanical resistance* is provided by *deformed bars* only, and not by the plain bars. The deformed bars have lugs, or corrugations, and give higher bond resistance by providing an interlock between steel and concrete. In the case of deformed bars, adhesion and friction become minor elements, and the bond strength is primarily dependent on the bearing (compressive) strength of the concrete against the lugs or corrugations. The plain bars slip easily and cause localised serious cracking while the deformed bars tend to distribute the deformation in the concrete in a series of closely spaced hair cracks.

8.3. TYPES OF BOND

Bond stress along the length of a reinforcing bar may be induced under two loading situations, and accordingly bond stresses are of two types :

1. *Flexural bond* or *Local bond*

and 2. *Anchorage bond* or *development bond*

Flexural bond (τ_{bf}) is one which arises from the change in tensile force carried by the bar, along its length, due to *change in bending moment* along the length of the member. Evidently, flexural bond is critical at points where the shear ($V = dM/dx$) is significant. Since this occurs at a particular section, flexural bond stress is known as *local bond stress* (Fig. 8.1. *b*)

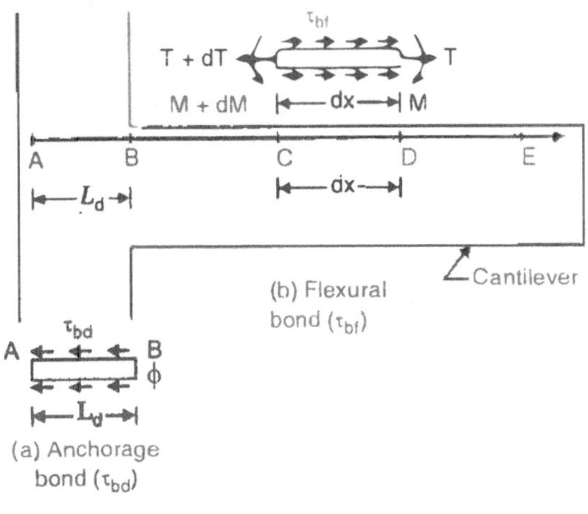

FIG. 8.1. TWO TYPES OF BOND

'Anchorage bond' (τ_{bd}) is that which arises over the *length of anchorage* provided for a bar. It also arises near the end or cutoff point of a reinforcing bar. The anchorage bond resists the *'pulling out'* of the bar if it is in tension or 'pushing in' of the bar if it is in compression. Fig. 8.1. (*a*) shows the situation of anchorage bond over a length AB (= L_d). Since bond stresses are *developed* over a specified length L_d, anchorage bond stress is also known as ***development bond*** *stress*.

8.4. FLEXURAL BOND STRESS

Fig. 8.2 shows two sections *ab* and *cd*, spaced *dx* apart of a reinforced concrete beam. Let M be the bending moment at section *ab* and $M + dM$ be the bending moment at section *cd*. Similarly let T and $T + dT$ be the tensile forces developed in steel reinforcement at *ab* and *cd* respectively.

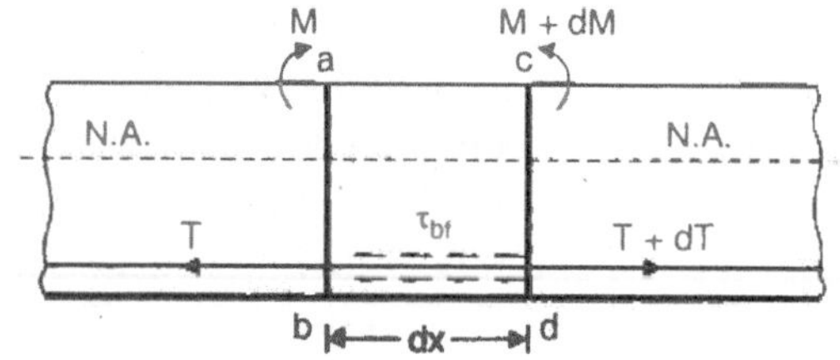

FIG. 8.2.

Now $M = T \cdot a$

and $M + dM = (T + dT)\, a$

Hence $dM = dT \cdot a$

or $dT = \dfrac{dM}{a}$ (where a = lever arm)...(*i*)

The difference in the tensile force (dT) in steel bars between sections is resisted by bond stress developed in the length dx.

Let τ_{bf} = bond stress developed in concrete around the steel reinforcement.

$\Sigma 0$ = sum of the perimeters of the steel bars resisting tension.

Total surface area of bars = $dx\, \Sigma 0$. Hence, $\tau_{bf}\, [dx\, \Sigma 0] = dT$ ...(*ii*)

Substituting the value of dT in (*i*), we get, $\tau_{bf}\, (dx\, \Sigma 0) = dM / a$

or $$\tau_{bf} = \frac{dM}{dx} \cdot \frac{1}{a\, \Sigma 0} \qquad \text{But} \quad \frac{dM}{dx} = \text{shear force} = V$$

Hence $$\tau_{bf} = \frac{V}{a\, \Sigma 0} \qquad ...(8.1)$$

The stress τ_{bf} at a particular section is called *Flexural* or *local bond stress.*

General expression for bond stress : Eq 8.1 is applicable only when all the reinforcing bars placed are equidistant from the neutral axis. However, there are cases, such as the case of a beam of circular section, in which it is not possible to place the bars equidistant from the N.A. Also, Eq. 8.1 is not applicable for finding bond stress in the reinforcement provided in the compression zone of the beam. In such cases, the expression derived below can be used to find bond stress in the individual bars. Let a reinforcing bar of diameter φ be placed at a distance y from the N.A. Let dM be the change in bending moment in a length dx. Hence change in the stress in the bar due to change in the bending moment dM is equal to $m. \dfrac{dM}{I} . y$, where I the moment of inertia of the equivalent section and m is the modular ratio. ($= E_s / E_c$)

$\therefore$ Force causing slip = area $\times$ stress $= \dfrac{\pi}{4} \varphi^2 . m \dfrac{dM}{I} . y$ ...(*i*)

Bond force resisting the slip $= \tau_{bf} . \pi \varphi . dx$. ...(*ii*)

Equating the two, we get, $\tau_{bf} . \pi \varphi . dx = \dfrac{\pi}{4} \varphi^2 . m \dfrac{dM}{I} y$

or $$\tau_{bf} = \frac{dM}{dx} \frac{my}{I} \cdot \frac{\varphi}{4} \qquad \text{But} \quad \frac{dM}{dx} = \text{shear force} = V$$

$$\therefore \quad \tau_{bf} = \frac{V}{4I}\, my \,.\, \varphi \qquad ...(8.2)$$

8.5. ANCHORAGE BOND STRESS

Eq. 8.1 and 8.2 give the *localised flexural bond stress* at a *particular section*. However, bond stresses are *developed* over a specified length and any attempt to visualise bond on the basis of stresses at a particular cross-section is bound to be erroneous. In the case of bond, it has been recently concluded that the concept of *anchorage bond* is more meaningful than concept of flexural bond stresses of localised nature. *Current design methods disregard such local bond stresses, even though they may result in localized slip between steel and concrete adjacent to the cracks.* Thus, the concepts of *development length* (in the case of flexure) and *anchorage length* (in case of tension or compression) have gained prominence.

Fig. 8.3 shows a steel bar embedded in concrete, and subjected to a tensile force T. Due to this force, there will be a tendency of the bar to slip out and this tendency is resisted by the bond stress developed over the perimeter of the bar, along its length of embedment. Let us assume that average, uniform bond stress is developed along the length. The required length, necesssary to develop full resisting force, is called *anchorage length* in case of axial tension(or compression) and *development length* in the case of flexural tension, and is designated by symbol L_d.

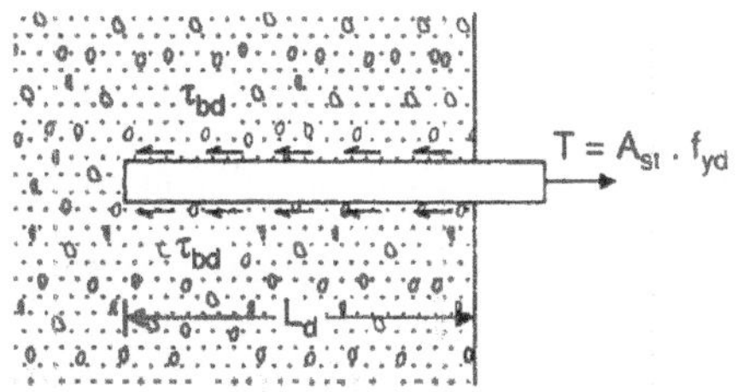

FIG. 8.3

Hence if φ is the nominal diameter of the bar, we have

$$\frac{\pi}{4}\varphi^2 \cdot f_{yd} = \tau_{bd} \cdot \pi\varphi \cdot L_d \quad (\text{ where } f_{yd} = \text{design stress in steel} = 0.87 f_y)$$

or
$$L_d = \frac{\varphi \cdot f_{yd}}{4\,\tau_{bd}} = \frac{0.87 f_y}{4\,\tau_{bd}} \cdot \varphi \qquad \ldots(8.3)$$

The design bond stress τ_{bd} is specified in the Code (Table 8.1) for various grades of concrete.

Eq. 8.3 shows that a bar must extend a length L_d beyond any section at which it is required to develop its full strength so that sufficient bond resistance is mobilized. For M 20 concrete, $\tau_{bd} = 1.2$ N/mm^2 while for mild steel bars, $f_y = 250$ N/mm^2. Hence from Eq. 8.3, $L_d = \frac{0.87 \times 250}{4 \times 1.2} = 45.3\,\varphi$.

8.6. DESIGN BOND STRESS (IS 456 - 2000)

The design bond stress in limit state method for *plain bars in tension* shall be as given below (Table 8.1) :

TABLE 8.1. DESIGN BOND STRESS FOR PLAIN BARS IN TENSION (IS 456 : 2000)

Grade of concrete	M 20	M 25	M 30	M 35	M 40 and above
Design bond stress τ_{bd} *(N/mm²)*	1.2	1.4	1.5	1.7	1.9

Design bond stresses for deformed bars in tension : For deformed bars conforming to IS 1786, these values shall be increased by 60%.

Design bond stress for bars in compression : For bars in compression, the values of bond stress for bars in tension shall be increased by 25%.

8.7. DEVELOPMENT LEGTH OF BARS (IS 456 : 2000)

The development length is defined as the length of the bar required on either side of the section under consideration, to develop the required stress in steel at that section through bond. The development length L_d is given by

$$L_d = \frac{\varphi\ \sigma_s}{4\ \tau_{bd}} = k_d\ \varphi \qquad \ldots(8.4)$$

where φ = nominal diameter of the bar

σ_s = stress in bar at the section considered at design load

k_d = development length factor $= \dfrac{\sigma_s}{4\ \tau_{bd}}$

Note : The development length includes the anchorage values of hooks in tension reinforcement.

Taking $\sigma_s = 0.87 f_y$ at the collapse stage, $k_d = \dfrac{0.87 f_y}{4\ \tau_{bd}}$ ...(8.5)

For bars in compression, the value of τ_{bd} given in Table 8.1 are to be increased by 25%. Hence development length (L_{dc}) for bar in compression is given by

$$L_{dc} = \frac{\varphi\ \sigma_{sc}}{5\ \tau_{bd}} \qquad \ldots(8.4\ a)$$

Hence the values of k_d for bars in compression will be $= \dfrac{0.87 f_y}{5\ \tau_{bd}}$

Table 8.2. gives the values of *development length factor* (k_d) for various grades of concrete and various grades of steel, both in tension as well as compression. The values have been *rounded-off to the higher side.*

TABLE 8.2. VALUES OF DEVELOPMENT LENGTH FACTOR (k_d)

1. Grade of concrete	M 20			M 25			M 30		
2. Grade of steel	Fe 250	Fe 415	Fe 500	Fe 250	Fe 415	Fe 500	Fe 250	Fe 415	Fe 500
3. Bars in tension	46	47	57	39	41	49	37	38	46
4. Bars in comp.	37	38	46	31	33	39	29	31	37
1. Grade of concrete	M 35			M 40 and above					
2. Grade of steel	Fe 250	Fe 415	Fe 500	Fe 250	Fe 415	Fe 500			
3. Bars in tension	32	34	40	29	30	36			
4. Bars in comp.	26	27	32	23	24	29			

Note : When the actual reinforcement provided is more than that theoretically required, so that the actual stress (σ_s) in steel is less than the full design stress $(0.87 f_y)$, the development length required may be *reduced* by the following relation :

Reduced development length $L_{dr} = L_d \left(\dfrac{A_{st}\ required}{A_{st}\ provided}\right)$ ...(8.4 *b*)

This principle is used in the design of footing and other short bending members where bond is critical. *By providing more steel, the bond requirements are satisfied.*

Bars bundled in contact : The development length of each bar of bundled bars shall be that for the individual bar, increased by 10% for two bars in contact, 20% for three bars in contact and 33% for four bars in contact.

8.8. STANDARD HOOKS AND BENDS FOR END ANCHORAGE : ANCHORAGE LENGTH

The development length required at the end of a bar is known as *anchorage length*. However, in the case of development length, the force in the bar is developed by transfer of force from concrete to steel, while in the case of anchorage length, there is dissipation of force from steel to concrete.

Quite often, space available at the end of beam is limited to accommodate the full development length L_d. In that case, hooks or bends are provided. The anchorage value (L_a) of hooks or bend is accounted as contribution to the development length L_d.

Fig. 8.4 (*ai*) shows a semi-circular hook, fully dimensioned, with respect to a factor K. The value of K is taken as 2 in the case of mild steel conforming to IS : 432–1966, (specifications for Mild-Steel and Medium Tensile Steel bars and Hard-Drawn steel wires for concrete reinforcement) or IS : 1139–1959, (specifications for 'Hot rolled mild steel and medium tensile steel deformed bars for concrete reinforcement'). The hook with K = 2 is shown in Fig. 8.4(*aii*) with equivalent horizontal length of the hook. For the case of Medium Tensile Steel conforming to IS : 432–1966 or IS : 1139–1959, K is taken as 3. In the case of cold worked steel conforming to IS : 1986–1961, (specifications for cold twisted steel bars for concrete reinforcement), K is taken as 4. In the case of bars above 25 mm, however, it is desirable to increase the value of K to 3, 4 and 6 respectively.

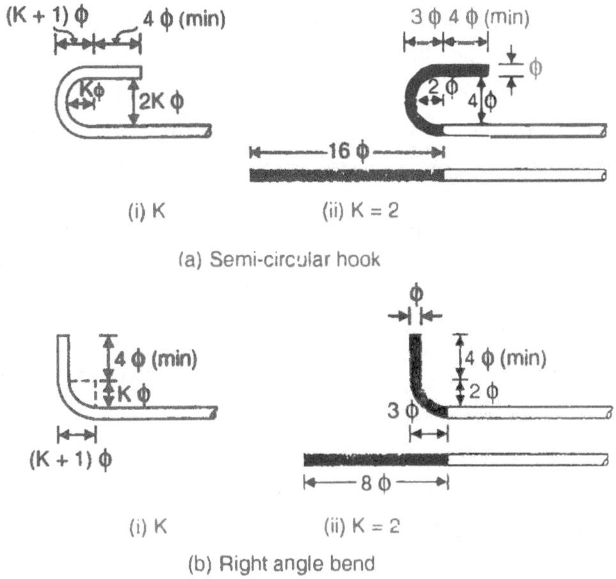

FIG. 8.4

Fig. 8.4 (*b*) shows a right angled bend, with dimensions in terms of K, the value of which may be taken as 2 for ordinary mild steel for diameters below 25 mm and 3 for diameters above 25 mm.

In the case of deformed bars, the value of bond stress for various grades of concrete is greater by 60% than the plain bars. Hence deformed bars may be used without hooks, provided anchorage requirements are adequately met with. Where hooks are formed in deformed bars, the internal radius of the bend should be at least three times the diameter of the bar. The length of straight bar beyond the end of the curve should be atleast four times the diameter of the bars.

8.9. CODE REQUIREMENTS FOR ANCHORING REINFORCING BARS (IS 456 : 2000)

(*i*) **Anchoring Bars in Tension.** Deformed bars may be used without end anchorages provided development length requirement is satisfied. Hooks should normally be provided for plain bars in tension. *The anchorage value of bend shall be taken as 4 times the diameter of the bar for each 45° bend subject to a maximum of 16 times the diameter* *of the bar.* The anchorage value of a standard U-type hook shall be equal to 16 times the diameter of the bar.

***(ii)* Anchoring Bars in Compression.** The anchorage length of straight bar in compression shall be equal to the development length of bars in compression. The projected length of hooks, bends and straight lengths beyond bends if provided for a bar in compression, shall be considered for development length.

***(iii)* Anchoring Shear Reinforcement :**

Inclined bars. The development length shall be as for bars in tension ; this length shall be measured as under : (1) in tension zone from the end of the sloping or inclined portion of the bar, and (2) in the compression zone, from mid depth of the beam.

Stirrups. Notwithstanding any of the provisions of this standard, in case of secondary reinforcement, such as stirrups and transverse ties, complete development lengths and anchorage shall be deemed to have been provided when the bar is bent through an angle of atleast 90° round a bar of atleast its own diameter and is continued beyond the end of the curve for a length of atleast eight diameters, or when the bar is bent through an angle of 135° and is continued beyond the end of curve for a length of atleast six bar diameters or when the bar is bent through an angle of 180° and is continued beyond the end of the curve for a length atleast four bar diameters.

8.10. CHECKING DEVELOPMENT LENGTH OF TENSION BARS

As stated earlier, the computed stress (σ_s) in a reinforcing bar, at every section must be *developed* on both the sides of the section. This is done by providing development length L_d (computed from Eq. 8.4 corresponding to σ_s) to both the sides of the section. Such a development length is usually available at mid-span location where positive (or sagging) B.M. is maximum for simply supported beams. Similarly, such a development length is usually available at the intermediate support of a continuous beam where negative (or hogging) B.M. is maximum. Hence no special checking may be necessary in such locations. *However special checking for development length is essential* at the following locations :

1. At simple supports
2. At cantilever supports
3. In flexural members that have relatively short spans
4. At points of contraflexure
5. At lap splices
6. At points of bar cutoff
7. For stirrups and transverse ties.

We shall consider some of the above cases in the forthcoming sections.

8.11. DEVELOPMENT LENGTH REQUIREMENTS AT SIMPLE SUPPORTS : DIAMETER OF REINFORCING BARS

The Code stipulates that at the simple supports (and at the point of inflection), the positive moment tension reinforcement shall be *limited* to a diameter such that

$$L_d \leq \frac{M_1}{V} + L_0 \qquad \ldots(8.6)$$

where L_d = development length computed for design stress f_{yd} (= 0.87 f_y) from Eq. 8.4

M_1 = Moment of resistance of the section assuming all reinforcement at the section to be stressed to f_{yd} (= $0.87 f_y$)

V = Shear force at the section due to design loads

L_0 = sum of anchorage beyond the centre of support and the equivalent anchorage value of any hook or mechanical anchorage at the simple support (At the point of inflexion, L_0 is limited to d or 12ϕ which ever is greater).

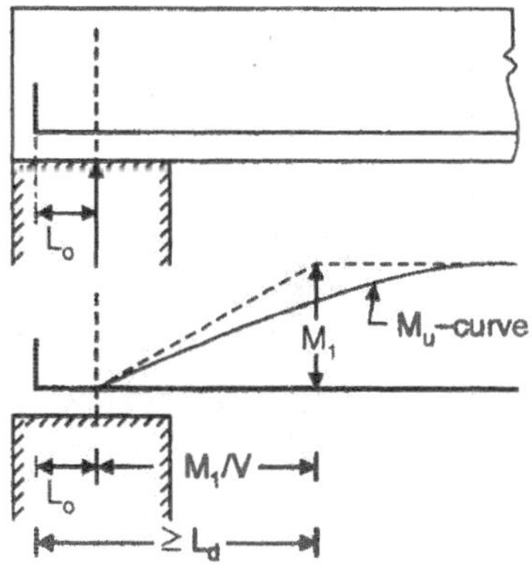

FIG. 8.5.
DEFINITION SKETCH FOR EQ. 8.6.

The Code further recommends that the value of M_1/V in Eq. 8.6 may be increased by 30% when the ends of the reinforcement are confined by a compressive reaction. This condition of confinement of reinforcing bars may not be available at all the types of simple supports.

Four types of simple supports are shown in Fig. 8.6. In Fig. 8.6 (*a*), the beam is simply supported on a wall which offers a compressive reaction which confines the ends of reinforcement. Hence a factor of 1.3 will be applicable with the term M_1/V. Similarly in Fig. 8.6 (*b*), the slab is simply supported on a wall and hence factor 1.3 will be applicable. However, in Fig. 8.6 (*c*) and (*d*), though a simple support is available, the reaction does not confine the ends of the reinforcement and hence the factor 1.3 will not be applicable with M_1/V term. Similarly, for the case of slab connected to a beam [Fig. 8.6 (*e*)] or for the case of a secondary beam connected to a main beam [8.6 (*f*)] tensile reaction is induced and hence factor 1.3 will not be available.

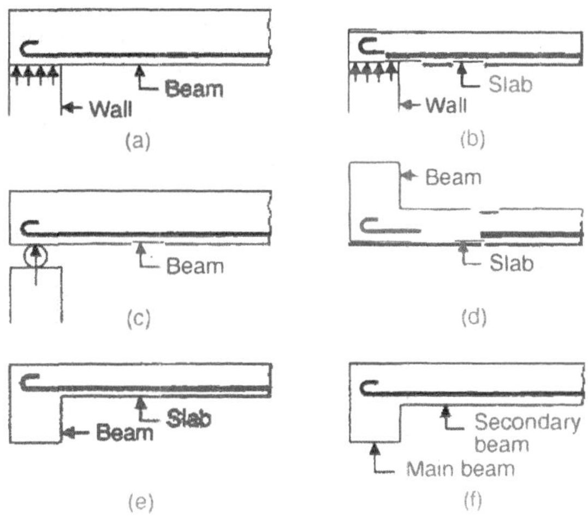

FIG. 8.6

Thus, at simple supports where the compressive reaction confines the ends of reinforcing bars, we have

$$L_d \leq 1.3 \frac{M_1}{V} + L_0 \quad \quad \ldots(8.6\ a)$$

Computation of the moment of resistance M_1 of bars available at supports :

In Eq. 8.6, M_1 = moment of resistance of the section corresponding to the area of steel (A_{st1}) continued into the support and stressed to design stress equal to $0.87 f_y$.

$$\therefore \quad M_1 = 0.87 f_y \cdot A_{st1} (d - 0.416\, x_u) \quad \quad \ldots(8.7)$$

where

$$x_u = \frac{0.87 f_y A_{st1}}{0.36 f_{ck}\, b} \quad \quad \ldots(8.7\ a)$$

Computation of length L_0 :

For the computation of L_0, the support width should be known. Fig. 8.7 (*a*) and (*b*) show a beam with end support, with a standard hook and a 90° bend respectively. Let x' be the side cover to the hook (or bend), and x_0 be the distance of the *beginning* of the hook (or bend) from the centre line of the support.

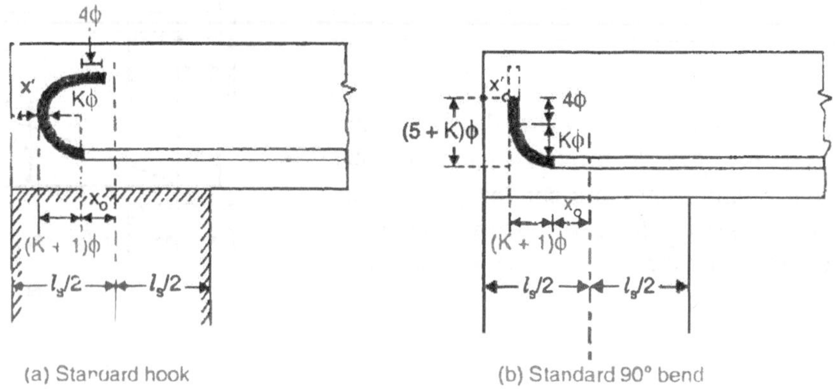

FIG. 8.7. COMPUTATION OF L_0

(*a*) Case 1 : Standard hook at the end (Fig. 8.7 *a*) : The dark portion shows the hook which has an anchorage value of 16 φ (IS 456 : 2000), for all types of steel. The distance of the beginning of the hook from its apex of the semi-circle is equal to $(K + 1)\,\varphi$. For mild steel bars, $K = 2$ and for HYSD bars, $K = 4$. Hence this distance is equal to 3 φ for mild steel bars and 5 φ for HYSD bars. Let l_s be the width of the support.

Then $$L_0 = x_0 + 16\,\varphi \quad \text{where} \quad x_0 = \frac{l_s}{2} - x' - (K+1)\,\varphi$$

$$\therefore \quad L_0 = \left(\frac{l_s}{2} - x' - (K+1)\,\varphi\right) + 16\,\varphi = \frac{l_s}{2} - x' + (15 - K)\,\varphi \qquad \text{...(8.8)}$$

Taking $K = 2$ for mild steel bars, $$L_0 = \frac{l_s}{2} - x' + 13\,\varphi \qquad \text{...(8.8 } a)$$

Taking $K = 4$ for HYSD bars, $$L_0 = \frac{l_s}{2} - x' + 11\,\varphi \qquad \text{...(8.8 } b)$$

(*b*) Case 2 : 90° standard bend (Fig. 8.7 *b*) : The dark portion shows the 90° bend which has an anchorage value of 8 φ (IS 456 : 2000) for all types of steel. Here also, the distance of beginning of the hook from its apex of the semi-circle is equal to $(K + 1)\,\varphi$.

Then $$L_0 = x_0 + 8\,\varphi \quad \text{where} \quad x_0 = \frac{l_s}{2} - x' - (K+1)\,\varphi$$

$$\therefore \quad L_0 = \left(\frac{l_s}{2} - x' - (K+1)\,\varphi\right) + 8\,\varphi = \frac{l_s}{2} - x' + (7 - K)\,\varphi \qquad \text{...(8.9)}$$

Taking $K = 2$ for mild steel bars, $$L_0 = \frac{l_s}{2} - x' + 5\,\varphi \qquad \text{...(8.9 } a)$$

Taking $K = 4$ for HYSD bars $$L_0 = \frac{l_s}{2} - x' + 3\,\varphi \qquad \text{...(8.9 } b)$$

Remedies to get development length : If the check for the satisfaction of Eq. 8.6 is not obtained, following remedial measures may be adopted to satisfy the check :

1. Reduce the diameter φ of the bar, thereby reducing the value of L_d, keeping the area of steel at the section unchanged. This is the standard procedure envisaged in the Code, i.e. reducing the value of L_d by *limiting* the diameter of the bar to such a value that Eq. 8.6 is satisfied.

2. Increasing the value of L_0 by providing *extra length* of the bend over and above the standard value $(5 + K)\varphi$ shown by dotted lines in Fig. 8.6 (*b*).

3. By increasing the number of bars (there by increasing A_{st1}) to be taken into the support. This method is uneconomical.

4. By providing adequate mechanical anchorage.

We shall discuss the first remedial method in the following section.

Limiting bar diameter to satisfy Eq. 8.6 or 8.6 (*a*) : IS 456 : 2000 stipulates that at simple supports having compression reaction reinforcement shall be limited to a diameter such that

$$L_d \le \frac{1.3\, M_1}{V} + L_0 \qquad \text{...(8.6 } a\text{)}$$

Substituting $L_d = \dfrac{f_{yd}\,\varphi}{4\,\tau_{bd}}$, in the above condition, we get $\dfrac{f_{yd}\cdot\varphi}{4\,\tau_{bd}} \le 1.3\dfrac{M_1}{V} + L_0$

or

$$\varphi \le \frac{4\,\tau_{bd}}{f_{yd}}\left(1.3\frac{M_1}{V} + L_0\right) \le \frac{4\,\tau_{bd}}{0.87\,f_y}\left(1.3\frac{M_1}{V} + L_0\right) \qquad \text{...(8.10)}$$

Similarly for simple supports not offering compressive reaction,

$$\varphi \le \frac{4\,\tau_{bd}}{f_{yd}}\left(\frac{M_1}{V} + L_0\right) \le \frac{4\,\tau_{bd}}{0.87\,f_y}\left(\frac{M_1}{V} + L_0\right) \qquad \text{...(8.10 } a\text{)}$$

8.12. DEVELOPMENT LENGTH AT POINT OF INFLEXION

Fig. 8.8 shows the conditions at a point of inflection (P.I.) As already indicated in § 8.11, the Code states that the following condition be satisfied

$$\left(\frac{M_1}{V} + L_0\right) \ge L_d \qquad \text{...(8.11)}$$

where L_0 should not be greater than d or 12 φ whichever is greater, and V is the shear force at the point of inflexion.

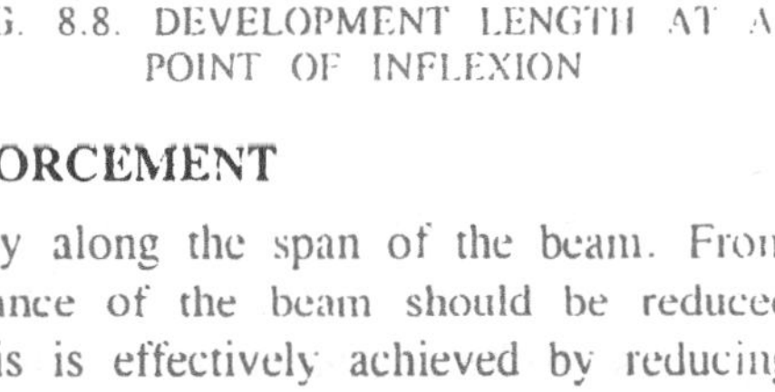

FIG. 8.8. DEVELOPMENT LENGTH AT A POINT OF INFLEXION

8.13. CONDITIONS FOR CURTAILMENT OF REINFORCEMENT

In most of the cases, the B.M. varies appreciably along the span of the beam. From the point of view of economy, the moment of resistance of the beam should be reduced along the span according to the variation of B.M. This is effectively achieved by reducing the area of reinforcement, i.e. by curtailing the reinforcement provided for maximum B.M. In general, all steel, whether in tension or in compression, should extend d or 12 φ (which ever is greater) beyond the theoretical point of cut off (TPC).

1. Rules for continuation of positive moment reinforcement in the span region of supports (IS : 456 : 2000)

(*a*) At least one third the positive moment reinforcement in *simple members* and one fourth the positive moment reinforcement in *continuous members* shall extend along the same face of the member into the support to a length equal to L_d 3.

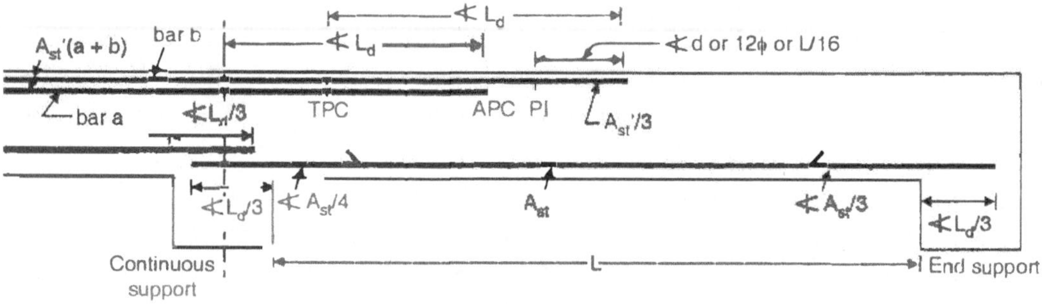

FIG. 8,9. RULES FOR CONTINUATION OF BARS INTO AND FROM SUPPORTS

(*b*) When a flexural member is part of the primary lateral load resisting system, the positive reinforcement required to be extended into the support as described in (*a*) shall be anchored to develop its design stress in tension at the face of the support.

(*c*) At simple supports Eq. 8.6 (or 8.6 *a*) should be satisfied.

2. Rules for continuation of negative moment reinforcement beyond the point of inflexion (P.I.) : At least one third of the total reinforcement provided for negative moment at the support (i.e. $A_{st}'/3$) shall extend beyond the point of inflexion (P.I.) for a distance not less than the effective depth (*d*) or 12 φ or one sixteenth of the clear span which ever is greater.

3. Rules for continuation of a curtailed bar beyond theoretical point of cutoff

As per IS Code, every bar that is to be curtailed shall continue for a distance equal to the effective depth of the member or 12 φ, whichever is greater, beyond the theoretical point of cutoff, except at the simple support or end of cantilever. Thus the distance between the *theoretical point of cut-off* (TPC) and the *actual point of cutoff* (APC) should be greater than *d* or 12 φ. Also, the distance of the actual point of cutoff (APC) from the point of max. B.M. should be at least L_d.

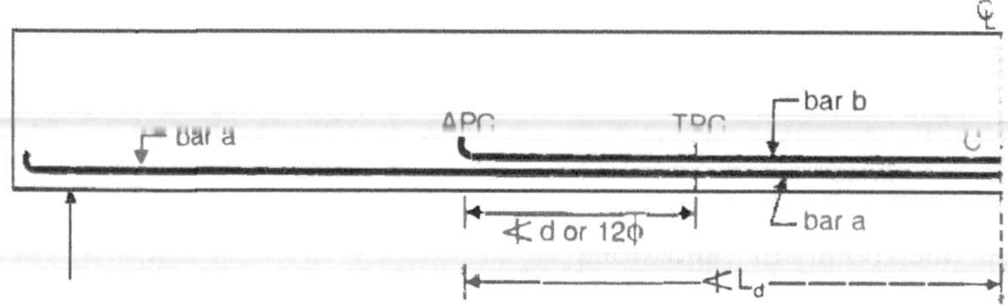

FIG. 8.10

4. Rules for maintaining shear capacity at region of cutoff of flexural reinforcement in tension zone : Curtailment of flexural tension reinforcement results in the loss of shear strength in the region of cutoff, and hence it is necessary to make provision to guard against such loss. Flexural reinforcement shall not be terminated in a tension zone unless any one of the following conditions is satisfied :

(*a*) The shear at the cutoff point does not exceed two thirds that permitted, including the shear strength of web reinforcement. In other words, the total *shear capacity* shall be atleast 1.5 times the applied shear at the point of curtailment, thus

$$V_u \ngtr \frac{2}{3}(V_{uc} + V_{us}) \quad \text{or} \quad V_{uc} + V_{us} \geq 1.5\, V_u \qquad \ldots(8.12)$$

Where V_{uc} = shear capacity of concrete, based on continuing reinforcement only.
V_{us} = shear capacity of shear reinforcement
V_u = applied shear at the point of curtailment.

(*b*) Stirrup area in excess of that required for shear and torsion is provided along each terminated bar over a distance from cutoff point equal to three fourth the effective depth of the member. Excess area of shear reinforcement is given by :

$$\text{Excess} \quad A_{sv} \geq \frac{0.4\, b\, s_v}{f_y} \qquad \ldots(8.13)$$

where

$$s_v \nless \frac{d}{8\,\beta_b} \ngtr \frac{0.87 f_y A_{sv}}{0.4\, b} \qquad \ldots(8.14)$$

$$\beta_b = \frac{\text{area of bars cutoff at the section}}{\text{total area of bars at the section}} \qquad \ldots(8.14\ a)$$

(*c*) For 36 mm or smaller bars, the continuing bars provide double the area required for flexure at the cutoff point and the shear does not exceed three fourth that permitted.

Thus, $$M_{ur} \geq 2\, M_u \qquad \ldots(8.15\ a)$$

and $$V_{uc} + V_{us} \geq 1.33\, V_u \qquad \ldots(8.15\ b)$$

where M_{ur} = moment of resistance of remaining (or continued) bars
M_u = B.M. at cutoff point ; V_u = S.F. at cutoff point

5. Scheme of curtailment of tension reinforcement in simply supported beams

Some bars of tensile reinforcement can be curtailed if not required from flexure point of view. The distance x, measured from the centre of the span, at which N_x number of bars (out of that N_c bars) can be curtailed, is given by Eq. 8.16.

$$x = \frac{L}{2}\sqrt{\frac{N_x}{N_c}} \qquad \ldots(8.16)$$

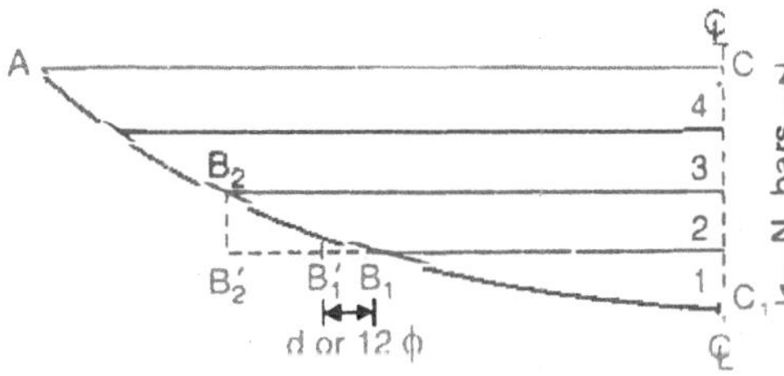

FIG. 8.11. CURTAILMENT OF BARS

Alternatively, a simple graphical method can be used for such curtailment. Fig. 8.11 shows the bending moment diagram for half the span of the beam, in which N_c (=4, say) number of bars are required at the mid span. Each bar has area equal to A_{st}/N_c, where A_{st} is the total area of tensile reinforcement. The stress in the bars is close to the design value f_{sd} (= 0.87 f_y) at the midspan, but decreases with the B.M. towards the point B_1 at which one of the bars can be curtailed and the remaining bars are stressed to the design value. However, one of the bars can be terminated at point B_1 only if the length $C_1 B_1$ is equal to or greater than the development length L_d for that bar. Similarly, the second bar can be curtailed at point B_2 if $B_1 B_2'$ is equal to or greater than L_d for the second bar. The Code specifies that the reinforcement shall extend the point at which it is no longer required to resist flexure, for a distance equal to effective depth d or 12 φ whichever is greater. Thus, for the first bar, if $B_1 C_1 > L_d$, it should extend, beyond B_1, to a point B_1' such that $B_1 B_1'$ is equal to d or 12 φ whichever is more.

8.14. REINFORCEMENT SPLICING

Splicing of a reinforcing bar becomes essential in the field due to either requirement of construction or due to non-availability of bars of desired length. The purpose of *splicing* is to transfer effectively the axial force from the terminating bar to the connecting (continuing) bar with the same line of action at the junction (Fig. 8.12 *a*). The following are the recommendations of IS Code (IS 456 : 2000)

(*a*) Lap splices shall not be used for bars larger than 36 mm; for larger diameters, bars may be welded. In cases where welding is not practicable, lapping of bars larger than 36 mm may be permitted, in which case additional spirals should be provided around the lapped bars (Fig. 8.12 *d*)

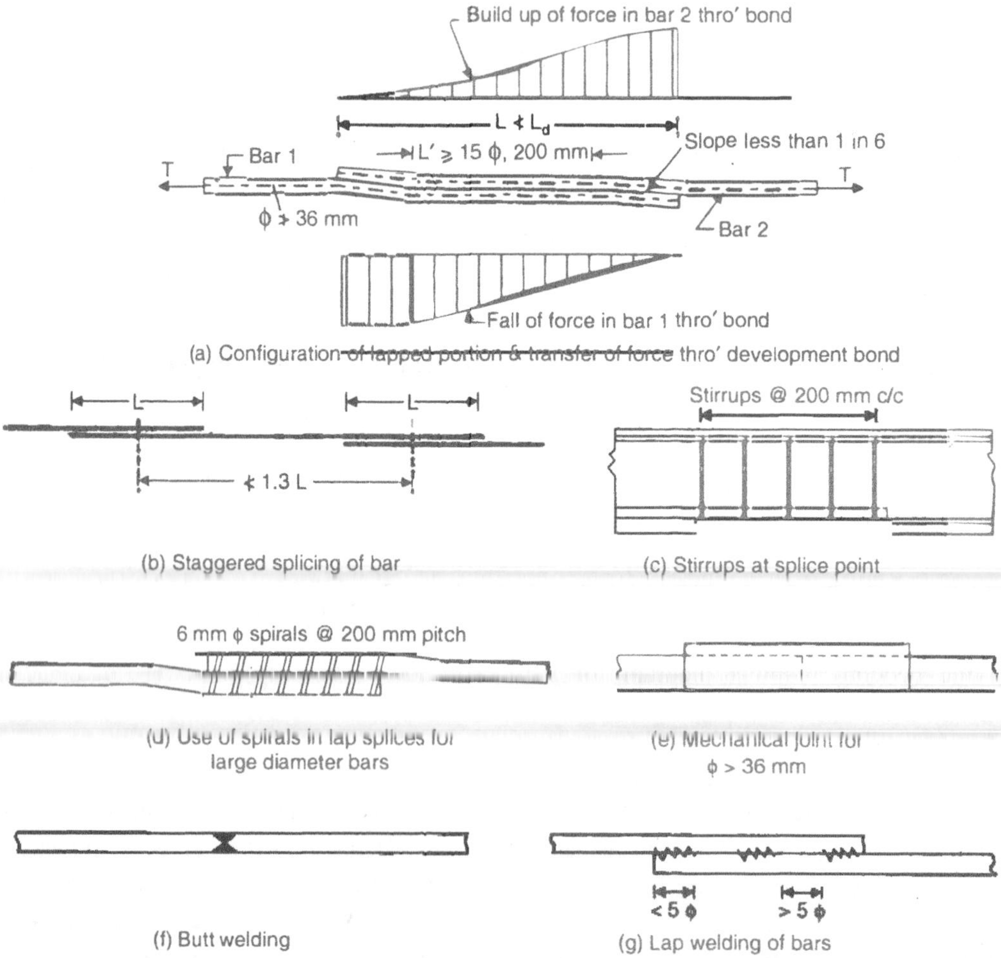

FIG. 8.12. REINFORCEMENT SPLICING

(*b*) Lap splices shall be considered as staggered if the centre to centre distance of the splices is not less than 1.3 times the lap length calculated as described in (*c*).

(*c*) The lap length including anchorage value of hooks for bars in *flexural tension shall be* L_d *or* 30 φ whichever is greater and for *direct tension* shall be $2\,L_d$ or 30 φ whichever is greater. The *straight length* (L') of the lap shall not be less than 15 φ or 200 mm (Fig. 8.12 *a*). The following provisions shall also apply :

(1) Top of a section as cast and the minimum cover is less than twice the diameter of the lapper bar, the lapped length shall be increased by a factor of 1.4.

(2) Corner of a section and minimum cover to either face is less than twice the diameter of the lapped bar or where the clear distance between adjacent laps is less than 75 mm or 6 times the diameter of lapper bar, whichever is greater, the lap length should be increased by a factor of 1.4.

Where both conditions (1) and (2) apply , the lap length should be increased by a factor of 2.0.

Note : Splices in tension members shall be enclosed in spirals made of bars not less than 6 mm diameter with pitch not more than 100 mm.

(*d*) The lap length in compression shall be equal to the development length in compression, but not less than 24 φ.

(*e*) When bars of two different diameter are to be spliced, the lap length shall be calculated on the basis of diameter of the smaller bar.

(*f*) When splicing of welded wire fabric is to carried out, lap splices of wires shall be made so that overlap measured between the extreme cross wires shall be not less than spacing of cross wires plus 100 mm.

(*g*) In case of bundled bars, lapped splices of bundled bars shall be made by splicing one bar at a time : such individual splices within a bundle shall be staggered.

Strength of welds : The following values may be used where the strength of the weld has been proved by tests to be at least as great as that of the parent bars.

(*a*) ***Splices in compression***: For welded splices and mechanical connection, 100 percent of the design strength of joined bars.

(*b*) ***Splices in tension***

(1) 80 percent of the design strength of welded bars (100 percent if welding is strictly supervised and if at any cross-section of the member not more than 20 percent of the tensile reinforcement is welded).

(2) 100 percent of design strength of mechanical connection.

End-bearing splices : End-bearing splices shall be used only for bars in compression. The end of the bars shall be square cut and concentric bearing ensured by suitable devices.

Illustrative Examples :

For illustrative examples to explain the application of Eqs. 8.6, 8.7 and 8.8, in actual designs, see various Examples of Chapter 11.

9

Limit State of Collapse in Torsion

9.1. INTRODUCTION

Axial loads, flexure, shear and torsion are the basic loading situations for which independent theories have been developed, and the more complicated interactive loading situations have been analysed as combinations of these basic effects. A common example of torsional loading is that of *peripheral beams* in each floor of a multistoreyed building, in which the slab is cast monolithic with the beam giving rise to L-beam configuration (see chapter 6). Another example of torsional loading is that of a ring beam provided at the bottom of an elevated circular water tank. Such a ring beam is subjected to bending moment, shear force and torsional moment. The beams supporting cantilevered canopy slabs are also subjected to significant torsional loading. Other prominent examples of torsional loadings are edge beams of concrete shell roofs, and helicoidal staircases.

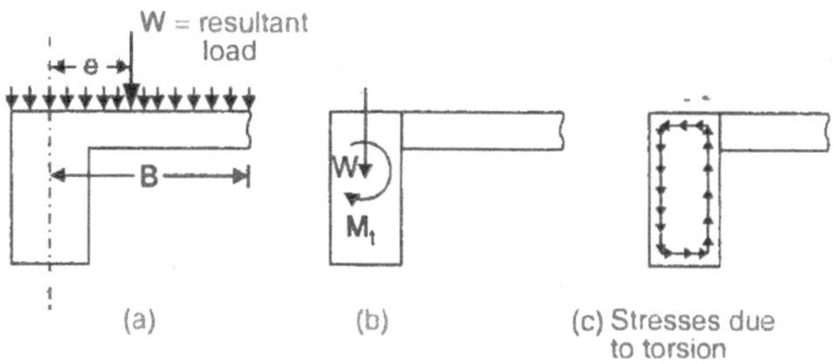

FIG. 9.1. TORSIONAL LOADING IN L BEAMS.

Fig. 9.1(*a*) shows the loading on the peripheral beam received from the slab. Due to monolithic beam-slab construction, the end beam has a flange to one side, giving rise to L-beam configuration. A portion of the slab, of length *B*, acts as flange of the L-beam. Evidently, such an L-beam is not symmetrically loaded. Due to load on the flange, an eccentric load *W* is transferred to the beam. This eccentric load may be considered as a central load *W* plus a moment M_t as shown in Fig. 9.1(*b*). The central load *W* causes bending moment in the beam, which is jointly resisted by the rectangular portion of the beam as well as the flange (of width *B*). The torsional moment M_t causes torsion in the beam which is resisted by the rectangular portion alone and the flange does not contribute to any torsional moment of resistance.

9.2. TORSIONAL RESISTANCE : ELASTIC BEHAVIOUR

When a shaft of circular section is subjected to torsional moment (or torque) M_t, the maximum shearing stress induced in the material is given by the classical formula:

$$\frac{M_t}{J}=\frac{N\theta}{l}=\frac{f_s}{R} \qquad ...(9.1)$$

where M_t = applied torque ; J = polar moment of inertia.
N = shear modulus ; l = length of the shaft.
f_s = maximum shear stress at the extreme fibres.
R = radius of circular shaft.

However, circular sections are very rarely used in reinforced concrete construction. When a beam of rectangular and other non-circular section is subjected to torsion, considerable warping of the cross-section takes place, and plane sections do not remain plane.

For a beam of rectangular section subjected to torsional moment M_t, the maximum torsional shear stress occurs at the middle of the *long side*, and its magnitude is given by

$$\tau_t = \tau_{max} = \frac{M_t}{\alpha\, b^2 D} = \frac{M_t}{Z_t} \qquad ...(9.2)$$

The torsional shear stress at the middle of the *short side* of the rectangular section is given by : $\tau_s = \eta\, \tau_{max}$ . ...(9.3)

where D = depth or long side of the section.
b = short side of the section.
Z_t = polar or torsional section modulus
α , η = coefficient depending upon D/b ratio (Table 9.1)

TABLE 9.1

D/b	1.0	1.5	2.0	3.0	6.0	10.0	∞
α	0.208	0.231	0.246	0.267	0.299	0.313	0.333
η	1.00	0.859	0.795	0.753	0.743	0.742	0.742

For the design purposes, the *torsional section modulus* for rectangular sections may be taken as :

$$Z_t = \alpha\, b^2 D \qquad ...[9.4(a)]$$

$$Z_t \approx \frac{b^2 D}{3 + \frac{2b}{D}} \qquad ...[9.4(b)]$$

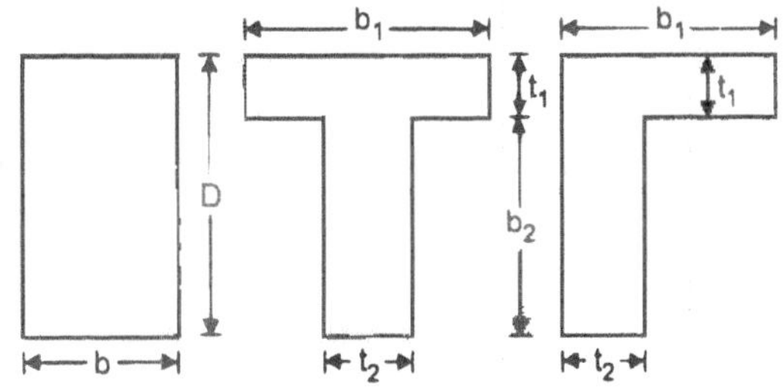

(a) Rectangle (b) T-section (c) L-section

FIG. 9.2.

For the T and L-sections, Cowan gives the following expressions for maximum torsional shear stress :

$$\tau_{max} = \frac{3\, M_t \, . \, t_i}{\Sigma\, b_i\, t_i^3} \qquad ...(9.5)$$

where t_i is t_1 or t_2 .

9.3. INDIAN STANDARD RECOMMENDATION ON DESIGN FOR TORSION (IS 456 : 2000)

Torsional reinforcement is not calculated separately from that required for bending and shear. Instead, the total longitudinal reinforcement is determined for a fictitious bending moment which is a function of actual bending moment and torsion. Similarly, shear reinforcement (web reinforcement) is determined for a fictitious shear which is a function of actual shear and torsion.

1. ***Shear and Torsion*** : Equivalent shear V_e is calculated from the expression

$$V_e = V_u + 1.6\frac{T_u}{b} \quad \ldots(9.6)$$

where V_e = equivalent shear ; V_u = shear

T_u = torsional moment ; b = breadth of the beam.

Equivalent nominal shear stress to be given by

$$\tau_{ve} = \frac{V_e}{bd} \quad \ldots(9.7)$$

The values of τ_{ve} shall not exceed the values of $\tau_{c.max}$ given in Table 7.3.

2. ***Reinforcement in Members Subjected to Torsion*** : If the equivalent nominal shear stress τ_{ve} does not exceed τ_c given in Table 7.1, minimum shear reinforcement shall be provided as per Eq. 7.18. If τ_{ve} exceeds τ_c, reinforcement is required. Reinforcement consists of (*a*) longitudinal reinforcement and (*b*) transverse reinforcement :

(*a*) Longitudinal Reinforcement : The longitudinal reinforcement shall be designed to resist an equivalent bending moment, M_{e1} given by

$$M_{e1} = M_u + M_t \quad \ldots(9.8)$$

where M_u = bending moment at the cross-section, and

$$M_t = T_u\left(\frac{1 + D/b}{1.7}\right) \quad \ldots(9.9)$$

T_u = torsional moment ; D = overall depth of beam.

If the numerical value of M_t given by Eq. 9.9 exceeds the numerical value of M_u, longitudinal reinforcement shall be provided on the flexural compression face, such that beam can also withstand an equivalent B.M. M_{e2} given by

$$M_{e2} = M_t - M_u \quad \ldots(9.10)$$

In which M_{e2} is taken as acting in the opposite sense to the moment M_u.

(*b*) Transverse Reinforcement : Two-legged closed hoops enclosing the corner longitudinal bars shall have an area of cross-section A_{sv} given by

$$A_{sv} = \frac{T_u\, s_v}{b_1\, d_1\,(0.87 f_y)} + \frac{V_u\, s_v}{2.5\, d_1\,(0.87 f_y)} \quad \ldots(9.11)$$

However, the total reinforcement shall not be less than $\dfrac{(\tau_{ve} - \tau_c)\, b\, s_v}{0.87 f_y}$.

where s_v = spacing of stirrups.

b_1 = centre to centre distance between corner bars in the direction of width.

d_1 = centre to centre distance between corner bars in the direction of depth.

Note 1. Distribution of torsional reinforcement

When a member is designed for torsion, torsion reinforcement shall be provided as below.

(*a*) The transverse reinforcement for torsion shall be rectangular closed stirrups placed perpendicular to the axis of the member. The spacing of the stirrups shall not exceed the least of x_1, $\dfrac{x_1 + y_1}{4}$ or 300 mm, where x_1, and y_1, are respectively the short and long dimension of the stirrup.

(*a*) Longitudinal reinforcement shall be placed as close as practicable to the corners of the cross-section and in all cases, there shall be at least one longitudinal bar in each corner of the ties. When the cross-sectional dimension of the member exceeds 450 mm, additional longitudinal bars shall be provided to satisfy the requirements of minimum reinforcement and spacing given in note 2.

Note 2 : Side face reinforcement

Where the depth of web of a beam exceeds 450 mm, the side face reinforcement shall be provided along the two faces. The total area of such reinforcement shall not be less than 0.1 percent of the web area and shall be distributed equally on two faces at a spacing not exceeding 300 mm or web thickness which ever is less.

Example 9.1 : Resistance in pure torsion

A rectangular beam, 300 mm wide and 400 mm deep is reinforced with 2 Nos 12 mm dia. bars at top and 2 Nos. 16 mm dia. bars at the bottom, each provided at an effective cover of 40 mm. Assuming concrete of M 20 grade and steel of Fe 415 grade, determine the resistance of the beam in pure torsion.

Solution : Given : $b = 300$ mm, $D = 400$ mm, top steel = 2 – 12 mm φ bars; bottom steel = 2 – 16 mm φ bars; $f_{ck} = 20$ N/mm²; $f_y = 415$ N/mm². It is required to find torsional moment of resistance, T_u.

Since the beam section is under pure torsion, the torsional moment of resistance (T_u) will by governed by M_{e1} or M_{e2} whichever is lesser. However, since the top steel has lesser area, M_{e2} will govern the torsional moment of resistance.

For Top steel (for M_{e2}*),* $A_{st} = 2 \times \frac{\pi}{4}(12)^2 = 226.19 \text{ mm}^2$

$$d = 400 - 40 - 360 \text{ mm} \quad ; \quad b = 300 \text{ mm}$$

$$\therefore \quad p_t = \frac{226.19 \times 100}{300 \times 360} = 0.209$$

For M 20 concrete and Fe 415 steel,

$$p_{t\,.\,lim.} = 0.414\,\frac{f_{ck}}{f_y}\,\frac{x_{u\,.\,max}}{d} \times 100 = 0.414 \times \frac{20}{415} \times 0.479 \times 100 = 0.955\% \quad \text{(see Table 3.2)}$$

Since actual $p_t < p_{t\,.\,lim.}$, the section is *under-reinforced*, for which

$$M_{e2} = M_{ur} = 0.87\,f_y\,A_{st}\,d\left(1 - \frac{f_y}{f_{ck}}\,\frac{A_{st}}{bd}\right) \qquad \text{...(3.12 } b\text{)}$$

$$= 0.87 \times 415 \times 226.19 \times 360\left(1 - \frac{415}{20} \times \frac{226.19}{300 \times 360}\right) = 28.122 \times 10^6 \text{ N-mm}$$

But $\quad M_{e2} = M_t - M_u$. Since $M_u = 0$, we have

$$M_t = M_{e2} = 28.122 \times 10^6 \text{ N-mm} = 28.122 \text{ kN-m}$$

$$\therefore \quad T_u = \frac{1.7\,M_t}{1 + D/b} = \frac{1.7 \times 28.122}{1 + 400/300} = \mathbf{20.489 \text{ kN-m}}$$

Example 9.2. Torsional resistance of beam subjected to B.M. and S.F.

A reinforced concrete rectangular beam 300 mm wide and 425 mm deep is reinforced with 4 bars of 20 mm dia. at bottom and 4 bars of 12 mm dia. at top. Determine the strength of the beam section in torsion, if it is also subjected to an ultimate bending moment of 60 kN-m and ultimate shear of 15 kN. Assume M 20 grade concrete, Fe 415 grade steel and an effective cover of 40 mm from each face.

Solution : Given : $b = 300$ mm; $D = 425$ mm ; $M_u = 60$ kN-m ; $V_u = 15$ kN; $f_{ck} = 20$ N/mm^2; $f_y = 415$ N/mm^2.

1. Computation of M_{e1} with bottom steel as tension steel

The section is doubly reinforced, with $A_{st} = 4 \times \frac{\pi}{4}(20)^2 = 1256.63$ mm^2 and $A_{sc} = 4 \times \frac{\pi}{4}(12)^2 = 452.39$ mm^2. Also, $d = 425 - 40 = 385$ mm. For the determination of x_u, we have to follow the iterative procedure outlined in § 4.2 and Example 4.2.

$$x_{u\,.\,max} = 0.479\,d = 0.479\,(385) = 184.4 \text{ mm.}$$

Now, Total tensile force $T_u = 0.87\,f_y\,A_{st} = 0.87 \times 415 \times 1256.63 = 453706$ N ...(1)

Total $C_u = 0.36\,f_{ck}\,x_u\,b + f_{sc}\,A_{sc} - 0.446\,f_{ck}\,A_{sc}$

$$= 0.36 \times 20 \times 300\,x_u + 452.39\,f_{sc} - 0.446 \times 20 \times 452.39$$

$$= 2160\,x_u + 452.39\,f_{sc} - 4035 \quad \text{...(2)}$$

Let us assume $x_u = 150$ mm. Hence $\frac{3}{7}x_u = \frac{3}{7} \times 150 = 64.3$ mm $> d'$

$$\varepsilon_{cu} = \frac{0.0035\,(x_u - d')}{x_u} = \frac{0.0035\,(150 - 40)}{150} = 0.00257$$

Hence from Table 2.3, we get $f_{sc} \simeq 347.5$ N/mm^2

Hence $C_u = 2160 \times 150 + 452.39 \times 347.5 - 4035 = 477170$ N

Since this is more than T_u (= 453706) reduce x_u. Let $x_u = 140$ mm.

$$\therefore \quad \varepsilon_{cu} = \frac{0.0035\,(140 - 40)}{140} = 0.00250$$

Hence from Table 2.3, we get $f_{sc} \simeq 345$ N/mm^2

$\therefore$ $C_u = 2160 \times 140 + 452.39 \times 345 - 4035 = 454440$ N

Since this is only slightly more than T_u, adopt final value of $x_u = 139$ mm

$$\therefore \quad \varepsilon_{cu} = \frac{0.0035\,(139 - 40)}{139} = 0.00249$$

Hence from Table 2.3, we have $f_{sc} = 344.9$ N/mm^2

Now $C_{cu} = 0.36\,f_{ck}\,x_u\,b = 0.36 \times 20 \times 139 \times 300 = 300240$ N

$C_{su} = (f_{sc} - 0.446\,f_{ck})\,A_{sc} = (344.9 - 0.446 \times 20)\,452.39 = 151994$ N

$\therefore$ $M_{ur} = M_{e1} = C_{cu}\,(d - 0.416\,x_u) + C_{su}\,(d - d')$

$$= 300240\,(385 - 0.416 \times 139) + 151994\,(385 - 40) \simeq 150.67 \times 10^6 \text{ N-mm}$$

$$= 150.67 \text{ kN-m.}$$

2. Computation of M_{e2} with top steel as tension steel

$\therefore$ $A_{st} = 452.39$ mm^2 and $A_{sc} = 1256.63$ mm^2

Since the section is heavily reinforced towards compression side, x_u will be very much less than $x_{u.max}$.

Now $$T_u = 0.87 f_y \; A_{st} = 0.87 \times 415 \times 452.39 = 163335 \text{ N} \quad ...(iii)$$

$$C_u = 0.36 f_{ck} x_u b + f_{sc} A_{sc} - 0.446 f_{ck} A_{sc}$$
$$= 0.36 \times 20 \times 300 x_u + 1256.63 f_{sc} - 0.446 \times 20 \times 1256.63$$
$$= 2160 x_u + 1256.63 f_{sc} - 11209 \quad ...(iv)$$

In the first trial, let $x_u = 70$ mm; $\frac{3}{4} x_u = \frac{3}{7} \times 70 = 30$ mm $< d'$ (= 40 mm)

Since $\frac{3}{4} x_u$ is less than d', the stress in concrete around compressive steel will be less than $0.446 f_{sc}$ ($= 0.446 \times 20 = 8.92$ N/mm^2). Let us assume stress in concrete around compressive steel as 7 N/mm^2 (approximately). Since this quantity is quite small, the error involved in this assumption is also negligible.

$$\therefore \quad \varepsilon_{cu} = \frac{0.0035\,(x_u - d')}{x_u} = \frac{0.0035\,(70 - 40)}{70} = 0.0015$$

Hence from Table 2.3, we get $f_{sc} \simeq 297.5$ N/mm^2

$$\therefore \quad C_u = 2160 \times 70 + 1256.63 \times 297.5 - 7 \times 1256.63 = 516251 \text{ N}$$

Since $C_u > T_u$, reduce x_u. Let $x_u = 43$ mm

$$\varepsilon_{cu} = \frac{0.0035\,(43 - 40)}{40} = 0.0002625$$

Since this strain is quite low, the stress-strain curve for steel will be linear for this range. Hence

$$f_{sc} = \varepsilon_{cu} \,.\, E_s = 0.0002625 \times 2 \times 10^5 = 52.5 \text{ N/mm}^2$$

Neglecting the loss of concrete area occupied by compression steel,

$$C_u = C_{cu} + C_{su} = 0.36 f_{ck} x_u b + f_{sc} A_{sc}$$
$$= 0.36 \times 20 \times 43 \times 300 + 52.5 \times 1256.63 = 92880 + 65973 = 158853 \text{ N}$$

Since $C_u < T_u$ (= 163335), let the final value of $x_u = 43.5$ mm.

$$\therefore \quad \varepsilon_{cu} = \frac{0.0035\,(43.5 - 40)}{43.5} = 0.002816$$

$$\therefore \quad f_{sc} = \varepsilon_{cu} E_s = 0.002816 \times 2 \times 10^5 = 56.32 \text{ N/mm}^2$$

$$\therefore \quad C_u = C_{cu} + C_{su} = 0.36 \times 20 \times 43.5 \times 300 + 56.32 \times 1256.63$$
$$= 93960 + 70775 = 164735 \text{ N}$$

$$\therefore \quad M_{ur} = M_{e2} = C_{cu}\,(d - 0.416\, x_u) + C_{su}\,(d - d')$$
$$= 93960\,(385 - 0.416 \times 43.5) + 70775\,(385 - 40)$$
$$= 58.89 \times 10^6 \text{ N-mm} = 58.89 \text{ kN-m}$$

3. *Computation of M_t on the basis of M_{e1}*

$$M_{e1} = M_u + M_t$$

$$\therefore \quad M_t = M_{e1} - M_u = 150.67 - 60 = 90.67 \text{ kN-m}$$

4. *Computation of M_t on the basis of M_{e2}*

$$M_{e2} = M_t - M_u$$

$$\therefore \quad M_t = M_{e2} + M_u = 58.89 + 60 = 118.89 \text{ kN-m}$$

5. *Computation of torsional moment T_u*

Allowable M_t will be lesser of the two values obtained in (3) and (4) above.

∴ Allowable $M_t = 90.67$ kN-m.

Hence $$T_u = \frac{1.7\,M_t}{1 + D/b} = \frac{1.7 \times 90.67}{1 + 425/300} = \mathbf{63.78 \text{ kN-m}}$$

Example 9.3. Design for Torsion.

Determine the reinforcement required for a rectangular beam section with the following data: Width of section : 300 mm ; Depth of section : 500 mm

Factored B.M : 65 kN-m ; Factored torsional moment : 40 kN-m

Factored shear force : 70 kN. Use M 15 grade concrete and Fe 415 grade steel.

Solution.

1. ***Equivalent Shear*** : Total depth $D = 500$ mm. Assume 20 mm dia. bars and 25 mm nominal cover and 10 mm dia. striups.

$$\therefore \quad d = 500 - 25 - 10 - 20/2 = 455 \text{ mm.}$$

$$V_e = V_u + 1.6\,\frac{T_u}{b} = 70 + 1.6\frac{40}{0.3} = 283.333 \text{ kN.}$$

$$\therefore \quad \tau_{ve} = \frac{283.333 \times 1000}{300 \times 455} = 2.07 \text{ N/mm}^2$$

From Table 7.3, $\tau_{c.max} = 2.5$ N/mm². Hence the section does not require revision.

2. ***Longitudinal Reinforcement***. Equivalent B.M. is given by

$$M_{e1} = M_u + M_t$$

where $$M_t = T_u\left(\frac{1 + D/b}{1.7}\right) = 40\left[\frac{1 + 0.5/0.3}{1.7}\right] = 62.745 \text{ kN-m}$$

$$\therefore \quad M_{e1} = 65 + 62.745 = 127.745 \text{ kN-m}$$

The moment of resistance of a balanced section is given by

$$M_{u\,lim.} = 0.36\,f_{ck}\,\frac{x_{u.max}}{d}\left(1 - 0.416\,\frac{x_{u.max}}{d}\right) bd^2$$

where $$\frac{x_{u.max}}{d} = 0.479 \text{ for Fe 415 steel}$$

$$\therefore \quad M_{u\,lim} = 0.36\,f_{ck}\,(0.479)\,(1 - 0.416 \times 0.479)\,bd^2 = 0.138\,f_{ck}\,bd^2$$

or $$M_{u\,lim} = 0.138 \times 15\,(300)\,(455)^2 = 128.56 \times 10^6 \text{ N-mm.}$$

Since $M_{e1} < M_{u.lim}$, the section is under-reinforced. Hence

$$M_u = M_{e1} = 0.87\,f_y\,A_{st}\,d\left(1 - \frac{f_y}{f_{ck}}\frac{A_{st}}{bd}\right)$$

or $$127.745 \times 10^6 = 0.87 \times 415\,A_{st} \times 455\left(1 - \frac{415}{15} \times \frac{A_{st}}{300 \times 455}\right)$$

or $\quad A_{st}(1 - 202.7 \times 10^{-6} A_{st}) = 777.62$

or $\quad A_{st}^2 - 4934 A_{st} + 3.837 \times 10^6 = 0$. From which $A_{st} = 967.9 \text{ mm}^2$

No. of 20 mm dia. bars = 967.9/314 = 3.08

Hence provide 4 bars of 20 mm dia., having total $A_{st} = 4 \times 314 = 1256 \text{ mm}^2$.

Again, as M_t is less than M_u, we need not consider M_{u2} Therefore provide only two bars of 12 mm dia. on the compression face to act as holding bars for transverse reinforcement.

3. *Transverse Reinforcement*

$$\frac{100 A_s}{bd} = \frac{100 \times 1256}{300 \times 455} \approx 0.92 \%$$

Hence from Table 7.1, $\tau_c = 0.54 + \dfrac{0.60 - 0.54}{0.25} \times 0.17 = 0.58 \text{ N/mm}^2$

Since $\tau_{ve} > \tau_c$, transverse shear reinforcement is essential. The area of two legs of the stirrups should satisfy the following :

$$A_{sv} = \frac{T_u s_v}{b_1 d_1 (0.87 f_y)} + \frac{V_u s_v}{2.5 d_1 (0.87 f_y)}$$

Assuming dia. of stirrups as 10 mm,

$$d_1 = 500 - \left(25 + 10 + \frac{20}{2}\right) - \left(25 + 10 + \frac{12}{2}\right) = 414 \text{ mm}$$

$$b_1 = 300 - 2\left(25 + 10 + \frac{20}{2}\right) = 210 \text{ mm.}$$

$$\therefore \quad \frac{A_{sv}(0.87 f_y)}{s_v} = \frac{40 \times 10^6}{210 \times 414} + \frac{70 \times 1000}{2.5 \times 414} = 527.7 \text{ N/mm.}$$

Also, area of all the legs of the stirrups should satisfy the following condition

$$A_{sv} = \frac{(\tau_{ve} - \tau_c)\, b\, s_v}{0.87 f_y}$$

or $$\frac{A_{sv}(0.87 f_y)}{s_v} = (\tau_{ve} - \tau_c)\, b = (2.07 - 0.58)\, 300 = 447 \text{ N/mm.} \quad \ldots(ii)$$

Choosing the greater one of the two, we have

$$\frac{A_{sv}}{s_v} = \frac{527.7}{0.87 f_y} = \frac{527.7}{0.87 \times 415} = 1.462.$$

Hence spacing of 10 mm dia. 2-legged stirrups is

$$s_v = \frac{A_{sv}}{1.462} = \frac{2 \times 78.54}{1.462} = 107 \text{ mm.}$$

Hence provide the stirrups @ 100 mm c/c.

As per Code requirements, this spacing should not exceed x_1, $(x_1 + y_1)/4$ and 300 mm.

where x_1 = short dimension of stirrup = 300 − 2 (25 + 5) = 240 mm

y_1 = long dimension of stirrup = 500 − 2 (25 + 5) = 440 mm

$$\frac{x_1 + y_1}{4} = \frac{240 + 440}{4} = 170 \text{ mm.}$$

Hence 10 mm dia. two legged stirrups @ 100 mm c/c satisfy the above Code requirements.

*4. **Side face reinforcement*** : Since the depth of web of the beam (= 500 mm) is more than 450 mm, side face reinforcement is necessary.

Area of side face reinforcement = $0.001\ bD = 0.001 \times 300 \times 500 = 150\ \text{mm}^2$. Hence area of steel on each face = $75\ \text{mm}^2$. Hence provide 1 – 10 mm φ bar on each face, at mid-depth, providing area of steel on each face = $78.5\ \text{mm}^2$.

10

Limit State of Serviceability in Deflection and Cracking

10.1. INTRODUCTION

In addition to safety, economy and durability, the serviceability of the structure is one of the most important criteria of structural design. *Serviceability is the fitness of the structure to serve the desired function satisfactorily under service loads.* Various considerations that come under the purview of serviceability are : (*i*) deflection, (*ii*) cracking, (*iii*) vibrations (*iv*) slenderness (*v*) impermeability (*vi*) acoustic, and (*vii*) thermal insulation. However, we shall consider only two important limit states of serviceability in this chapter : (1) deflection, and (2) cracking. It should be clearly borne in mind that a design which adequately satisfies the 'safety' requirements need not automatically satisfy the serviceability requirements. Two of the most important serviceability conditions are as follows :

1. The member should not undergo excessive deformation (i.e. limit state of deflection)

2. The crack width at the surface of the R.C. member should not be more that specified by the Code (i.e. limit state of cracking).

As per IS 456 : 2000, the combinations of loads for serviceability conditions should be the largest of the following.

(*i*) 1.0 DL + 1.0 LL

(*ii*) 1.0 DL + 1.0 WL

(*iii*) 1.0 DL + 0.8 LL + 0.8 WL (or EL)

IS : 456 - 2000 gives two *approaches* for safety against excessive deflection of bending members : (*i*) empirical method of restricting the span to effective depth ratio and (*ii*) computation of deflection satisfying the specified allowable limits of deflection. Cracking of concrete in tension zone can not be avoided; however, it can be limited by adhering to the Codal requirements of minimum steel area and maximum spacing of bars in the member. Further, proper and prolonged curing of concrete and good quality of mix. should be ensured for the control of cracking. In addition to this, the actual width of cracks should be calculated by theoretical method and checked whether they satisfy the requirements of the Codes for the given environmental conditions.

10.2. LIMIT STATE OF DEFLECTION

Limits of deflection are prescribed on the basis of several factors such as (*i*) aesthetic/ psychological discomfort, (*ii*) effect on geometry and shape of the structure, (*ii*) deformation of doors, windows and other fixures (*iv*) cracks in attached walls, floor and ceiling finishes, (*v*) ponding of roof slabs and (*vi*) limiting crack widths, which are otherwise increased with increased deflections. Design based on ultimate load method/limit state method result in thinner or slender sections. Such sections deflect more than those designed by the working stress method. The problem of deflection is further amplified due to the increasing use of HYSD bars. Deflection of a R.C. structure can be divided in the following categories.

(*i*) *short term*, instantaneous or immediate deflection.

(*ii*) *long term* deflection.

(*iii*) *total deflection.*

Short term deflection or instantaneous deflection is due to initial elastic deformation of the R.C. member due to the load and permanent imposed load under service conditions. *Long term deflection* is caused due to (*i*) creep and shrinkage under sustained loading, and (*ii*) additional elastic deflection due to temporary live loads. *Total deflection* is the sum of short term and long term deflections. It is this quantity that requires overall control. If a_T is the total deflection, we have

$$a_T = a_i + a_{lt} = a_i + (a_{cs} + a_{cc\,(perm)} + a_{tp}) \quad \ldots(10.1)$$

where a_T = total deflection

a_i = immediate or short term deflection

a_{lt} = long time deflection

a_{cs} = deflection due to shrinkage

$a_{cc\,(perm)}$ = deflection due to creep, under permanent loads

a_{tp} = deflection due to temperature

Where as excessive short term deflections can be effectively countered by *cambering*, this is generally not done in case of long terms deflections.

10.3. SHORT TERM DEFLECTION

Short term deflection of a structure is an instantaneous elastic deflection (a_i) due to service loads. The factors affecting the short term deflections are as follows :

(*i*) magnitude of live load and its mode of distribution (i.e. concentrated, uniformly distributed or other variations)

(*ii*) span of the structure

(*iii*) type of end restraints

(*iv*) cross-sectional properties, including percentages of tensile and compressive reinforcement (if any)

(*v*) grade of concrete and its characteristic strength

(*vi*) type of reinforcement and magnitude of stress in steel reinforcement.

and (*vii*) amount and extent of flexural cracking.

Computations of short term deflection. due to applied *service loads* are generally based on the assumption of linear elastic behaviour, and for this purpose reinforced concrete is treated as a homogeneous material. Various formulae for deflection usually take the following form :

$$a_i = K_w \frac{Wl^3}{E_c I_{eff}} = K_m \frac{M_c l^2}{E_c I_{eff}} \quad \ldots(10.2)$$

where W = total load on the structural element

M_c = maximum moment or critical moment

K_w or K_m = constants, depending upon load distribution, condition of end restraints, and variation in flexural rigidity ($E_c I_{eff}$)

a_i = mid-span defection for simply supported beam (which is usually very close to the maximum deflection) or end deflection of a cantilever beam.

I_{eff} = effective moment of inertia

E_c = short term modulus of elasticity of concrete $= 5000\sqrt{f_{ck}}$ N/mm² ...(10.3)

Table 10.1 gives the value of K_m and M_c for some common cases of beams.

TABLE 10.1. FORMULAE FOR DEFLECTION AT CRITICAL SECTION FOR PRISMATIC BEAMS

$$a_i = K_m \frac{M_c l^2}{E_c I_{eff}}$$

S.N.	Beam	K_m	M_c	S.N.	Beam	K_m	M_c
1.	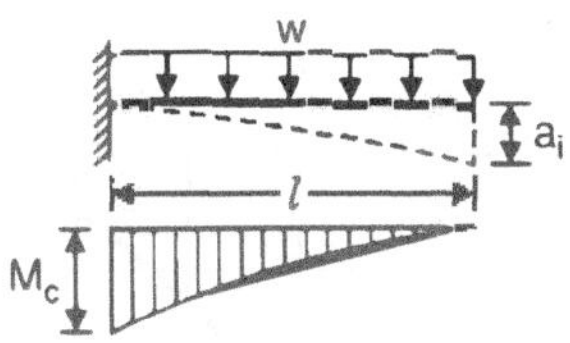	$\frac{1}{4}$	$\frac{wl^2}{2}$	5.	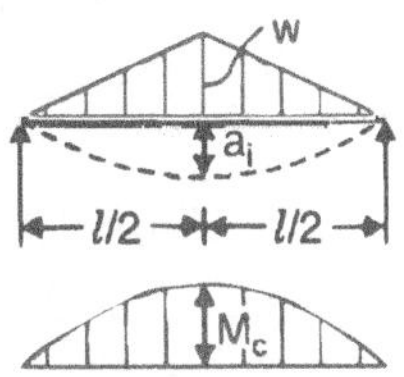	$\frac{1}{10}$	$\frac{wl^2}{12}$
2.	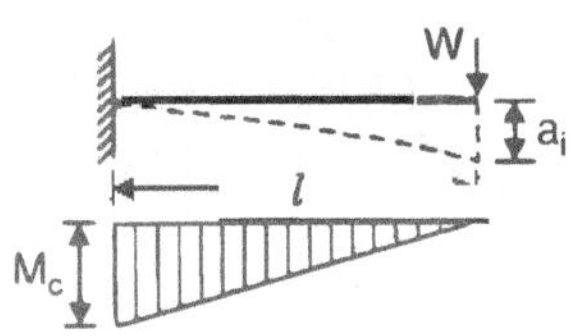	$\frac{1}{3}$	Wl	6.	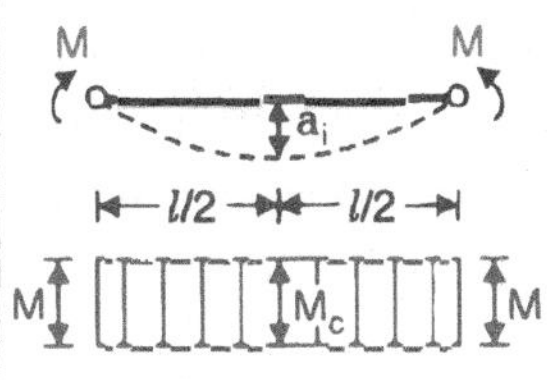	$\frac{1}{8}$	M
3.	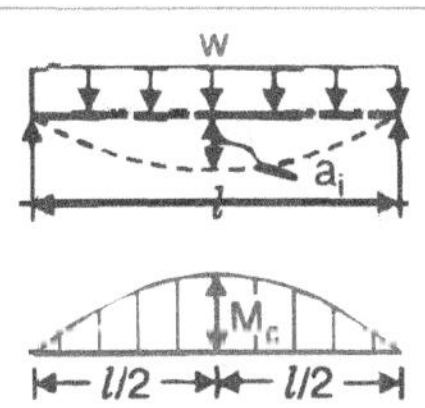	$\frac{5}{48}$	$\frac{wl^2}{8}$	7.	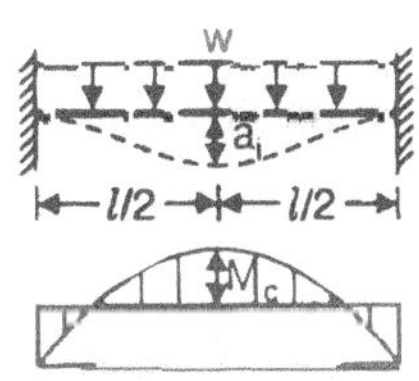	$\frac{1}{16}$	$\frac{wl^2}{24}$
4.	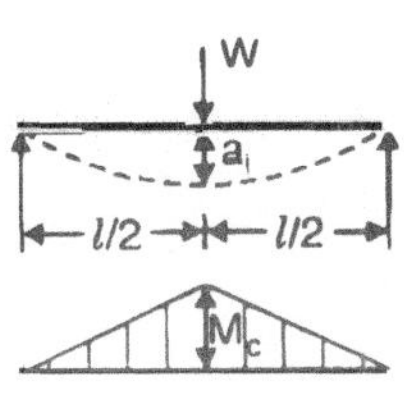	$\frac{1}{12}$	$\frac{Wl}{4}$	8.	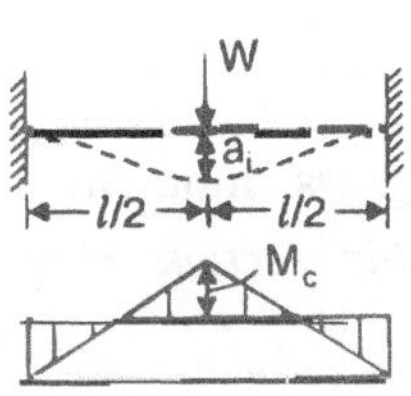	$\frac{1}{24}$	$\frac{Wl}{8}$

In Eq. 10.2., I_{eff} is the effective moment of inertia, given by

$$I_{eff} = \frac{I_r}{1.2 - \frac{M_r}{M_c} \cdot \frac{z}{d}\left(1 - \frac{x}{d}\right)\frac{b_w}{b}} \quad \ldots(10.4)$$

But $I_r \leq I_{eff} \leq I_{gr}$ ...(10.4 *a*)

where I_r = moment of inertia of cracked section.

M_r = cracking moment, equal to $\frac{f_{cr} I_{gr}}{y_t}$, where f_{cr} is the modulus of rupture of concrete, I_{gr} is the moment of inertia of the gross section about the centroidal axis, neglecting the reinforcement, and y_t is the distance from centroidal axis of gross section, neglecting the reinforcement, to extreme fibre in tension.

M_c = maximum moment (critical moment) under service loads

f_{cr} = modulus of rupture of concrete $= 0.7\sqrt{f_{ck}}$ N/mm² ...(10.5)

z = lever arm $= d - x/3$ (by elastic theory)

x = depth of N.A. ; d = effective depth

b_w = breadth of web, and b = breadth of compression face

For continuous beams, deflection shall be calculated using the values of I_r, I_{gr} and M_r modified by the following equation.

$$X_e = k_1\left[\frac{X_1 + X_2}{2}\right] + (1 - k_1) X_0 \quad \ldots(10.6)$$

where X_e = modified value of X

X_1, X_2 = values of X at supports.

X_0 = value of X at mid-span

k_1 = coefficient given in Table10.2 (or Fig. 10.1). Values dependent on k_2

$$k_2 = \frac{M_1 + M_2}{M_{F1} + M_{F2}}$$

M_1, M_2 = actual support moments

M_{F1}, M_{F2} = fixed end moments

X = value of I_r, I_{gr} or M_r, as appropriate

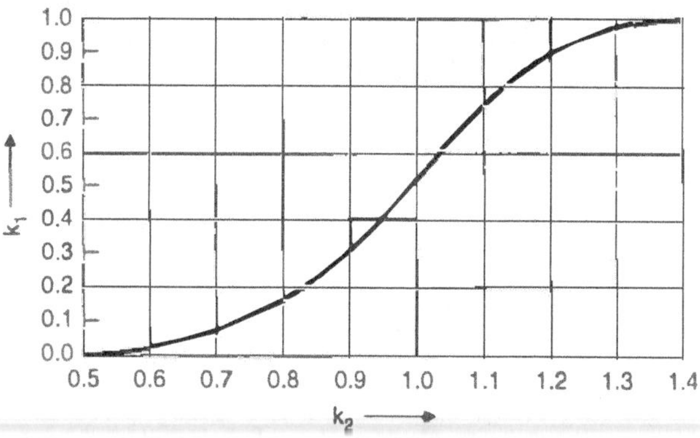

FIG. 10.1. VALUE OF k_1

TABLE 10.2 VALUES OF COEFFICIENT k_1

$k_2 \rightarrow$	*0.5 or less*	*0.6*	*0.7*	*0.8*	*0.9*	*1.0*	*1.1*	*1.2*	*1.3*	*1.4*
$k_1 \rightarrow$	0	0.03	0.08	0.16	0.30	0.50	0.73	0.91	0.97	1.0

Determination of gross moment of inertia :

(*a*) ***Rectangular section*** (Fig. 10.2 *a*)

Ignoring reinforcement, $I_{gr} = \frac{bD^3}{12}$ ...(10.7) (*b*) ***Flanged section***

$$x = \frac{b_w D^2/2 + (b_f - b_w) D_f^2/2}{b_w D + (b_f - b_w) D_f} \quad ...(10.8\ a)$$

$$I_{gr} = \frac{b_f x^3}{3} - \frac{(b_f - b_w)(x - D_f)^3}{3} + \frac{b_w (D - x)^3}{3} \quad ...(10.8)$$

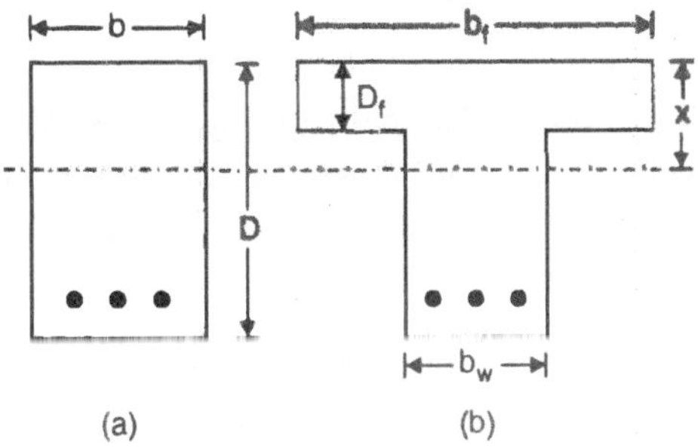

FIG. 10.2. UNCRACKED SECTION

Determination of moment of inertia of cracked section (I_r)

(a) Singly reinforced rectangular section (Fig. 10.3 a)

The neutral axis is located by taking moments of *effective areas* about N.A. Thus,

$$\frac{b x^2}{2} = m A_{st} (d - x) \quad \text{where } m = \text{modular ratio} = E_s/E_c \quad ...(10.9\ a)$$

Also,
$$I_r = \frac{bx^3}{3} + m A_{st} (d - x)^2 \quad ...(10.9)$$

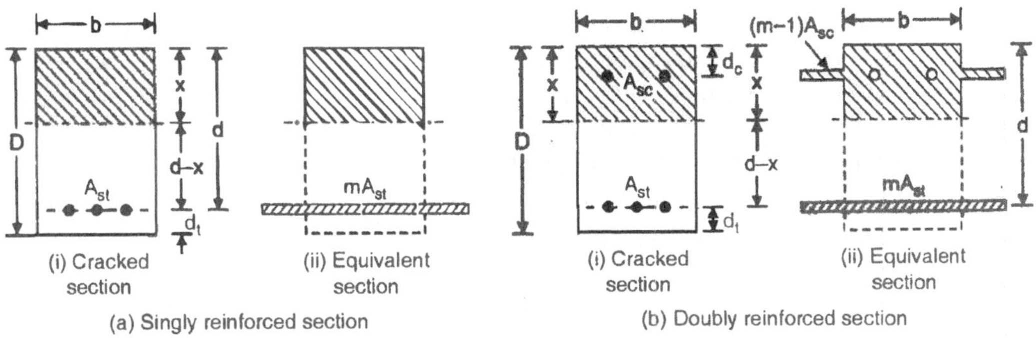

FIG. 10.3. RECTANGULAR SECTION

(*b*) ***Doubly reinforced rectangular section (Fig 10.3 b)***

Taking moments of effective areas about N.A. we have

$$\frac{bx^2}{2} + (m - 1) A_{sc} (x - d_c) = mA_{st} (d - x) \quad ...(10.10\ a)$$

Also,
$$I_r = \frac{bx^3}{3} + (m - 1) A_{sc} (x - d_c)^2 + m A_{st} (d - x)^2 \quad ...(10.10)$$

(*c*) ***Singly reinforced flanged section***

Assuming N.A. lying in the web,

$$\bar{x} = \frac{b_f D_f^2/2 + m A_{st} d}{b_f D_f + m A_{st}} \quad ...(10.11\ a)$$

and
$$I_r = \frac{b_f x^3}{3} - \frac{(b_f - b_w)(x - D_f)^3}{3} + mA_{st} (d - x)^2 \quad ...(10.11)$$

(Ignoring the area of web between the flange and the web)

When $x < D_f$, I_r is the same as that for singly reinforced rectangular section, but replacing b by b_f.

Note : In the working stress method (see Appendix), equivalent area of compression steel (A_{sc}) is multiplied by the factor ($1.5\,m - 1$) to take into account the long term effects of creep and shrinkage. However, in the above expression (Eqs. 10.10), we have used a multiplying factor of ($m - 1$) only, since we are computing short term deflection only.

10.4. LONG TERM DEFLECTION

Long term deflections are the additional deflections of a R.C. member under the effect of sustained loads. The long term deflections may mainly be due to the following :

(*i*) Deflection due to differential shrinkage (a_{cs})

(*ii*) Deflection due to creep under sustained loading (a_{cr})

(*iii*) Deflection due to temperature effects (a_{tp})

(*iv*) Deflection due to additional live loads (a_{adl})

The phenomenon due to shrinkage and creep causes formation of new cracks and widening of existing cracks. The combined long term deflection due to shrinkage, creep and temperature effects may be as large as two to three times the short term or immediate deflection due to dead and live loads. The computation of total deflection due to combined effects of load, creep and shrinkage are quite laborious and complicated. Hence empirical methods are usually employed considering the effects of load, shrinkage and creep independently. The resultant deflection is then obtained by adding all the three.

Deflection due to shrinkage (IS 456 - 2000)

Shrinkage of an isolated plain concrete member is its *volumetric free contraction*, causing shortening of the member. However, reinforcing bars, embedded into it to resist tension, resists this shortening, resulting in the production of curvature, commonly known as *shrinkage curvature*. Due to this, compressive stresses are developed in steel and tensile stresses are developed in concrete and the member deflects. The shrinkage deflection enhances the deflection due to loads. However, the presence of steel in compression side obviously prevents shrinkage to a greater extent.

The deflection due to shrinkage a_{cs} may be computed from the following equation

$$a_{cs} = k_3\, \psi_{cs}\, l^2 = k_3\, k_4\, \varepsilon_{cs}\, (l^2/D) \quad ...(10.12)$$

where k_3 = constant depending upon the support additions.

= 0.5 for cantilevers

= 0.125 for simply supported members

= 0.086 for members continuous at one end, and

= 0.063 for fully continuous members.

$$\psi_{cs} = \text{shrinkage curvature} = k_4 \frac{\varepsilon_{cs}}{D} \quad ...(10.13)$$

ε_{cs} = ultimate shrinkage strain of concrete. In absence of test data, the absolute value of total shrinkage strain for design may be taken as 0.0003.

$$k_4 = 0.72 \times \frac{p_t - p_c}{\sqrt{p_t}} \leq 1.0 \quad (\text{for } 0.25 \leq p_t - p_c < 1.0) \quad ...(10.14)$$

and $k_4 = 0.65 \dfrac{p_t - p_c}{\sqrt{p_t}} \le 1.0$ (for $p_t - p_c > 1.0$) ...(10.15)

$$p_t = \frac{100\,A_{st}}{bd} \;;\quad p_c = 100\,\frac{A_{sc}}{bd}$$

D = total depth of the section ; l = length of the span.

Deflection due to creep (IS : 456 - 2000)

Creep is the phenomenon owing to which compressive strain in concrete keep on increasing under sustained loading. The strain in tension steel also increases depending upon the reduction in the lever arm. In flexural members, the distribution of creep strains across the depth is non-uniform, with a practically linear variation, resulting in *creep curvature* (ψ_{cp}), which is additive to initial elastic curvature. This results in the increase in deflection, and depends mainly on the magnitude of permanent load on the structure. Instead of determining the creep deflection ($a_{cc\,(perm)}$) directly, in terms of the initial elastic deflection, $a_{i\,(perm)}$ (due to permanent load), IS Code procedure involves the explicit calculations of deflection $a_{i+cc\,(perm)}$ due to the permanent load plus creep, using an effective modulus of elasticity $E_{ce} = E_c/(1+\theta)$.

Thus, as per IS 456 : 2000, the creep deflection due to permanent loads, $a_{cc\,(perm.)}$ may be obtained from the following equation :

$$a_{cc\,(perm)} = a_{i+cc\,(perm.)} - a_{i\,(perm.)} \qquad \text{...(10.16)}$$

where $a_{i+cc\,(perm.)}$ = initial plus creep deflection due to permanent loads obtained using an elastic analysis with an effective modulus of elasticity E_{ce}

$$E_{ce} = \frac{E_c}{1+\theta} \quad \text{where} \quad \theta = \text{creep coefficent} \qquad \text{...(10.17)}$$

$a_{i\,(perm.)}$ = short term deflection due to permanent load using E_c.

Creep coefficient (θ)

In the absence of experimental data and detailed information on the effect of variables, the ultimate creep strain may be estimated from the following values of creep coefficient (that is ultimate creep strain/elastic strain at the age of loading. For long span structure, it is advisable to determine actual creep strain likely to take place

TABLE 10.3 VALUES OF CREEP COEFFICIENT (θ)

Age of loading	*Creep coefficient* (θ)
7 days	2.2
28 days	1.6
1 year	1.1

Note : The ultimate creep strain, estimated as described above include the elastic strain.

Deflection due to temperature

A bending member may deflect if there is difference in temperatures between the top and bottom faces of the member. In roof members, there may be substantial temperature differential between the top face which is exposed to the sun and the bottom face which may be under air-conditioned atmosphere. Such a deflection is usually of a *hogging* nature and is subtractive from deflection due to gravity loads, shrinkage and creep. However, for cantilevered structures, this deflection may be additive. Similarly, in very cold weather, the deflection due to temperature may be additive to the deflections due to gravity loads, shrinkage and creep.

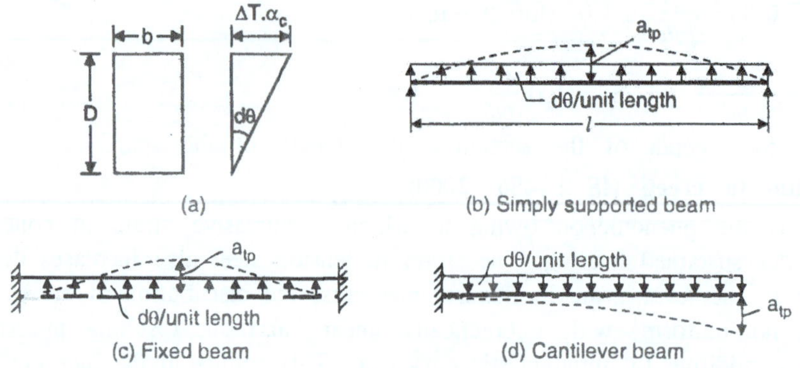

FIG. 10.4. DEFLECTION DUE TO TEMPERATURE

Refer Fig. 10.4 (*a*). If ΔT is the temperature differential, the temperature differential $d\theta$, which acts as the *loading* over the bending member, is given by

$$d\theta = \Delta T \,.\, \alpha_c / D, \text{ per unit length} \quad \ldots(10.18)$$

For a simply supported beam,

$$a_{tp} = d\theta \,.\, \frac{l^2}{8} = \frac{\Delta T \,.\, \alpha_c}{D} \,.\, \frac{l^2}{8} \quad \ldots(10.19)$$

For a fixed beam,

$$a_{tp} = d\theta \,.\, \frac{l^2}{24} = \frac{\Delta T \,.\, \alpha_c}{D} \,.\, \frac{l^2}{24} \quad \ldots(10.20)$$

For a cantilever,

$$a_{tp} = d\theta \,.\, \frac{l^2}{2} = \frac{\Delta T \,.\, \alpha_c}{D} \,.\, \frac{l^2}{2} \quad \ldots(10.21)$$

where a_{tp} = deflection due to temperature

$d\theta$ = temperature differential ; α_c = coefficient of thermal expansion for concrete.

Total deflection : Total deflection a_T is given by

$$a_T = a_i + a_{lt} = a_i + (a_{cs} + a_{cc\,(perm)} + a_{tp}) \quad \ldots(10.1)$$

10.5. CONTROL OF DEFLECTION

IS : 456 - 2000 has given two *approaches* for safety against deflection of bending members (beam and slabs) :

(*i*) specifying allowable limits of deflection in terms of span-deflection ratio.

and (*ii*) specifying span to effective depth ratio for control of deflection.

1. First approach : Deflection limits by span-deflection ratio

IS : 456 2000 has specified two *allowable limits* of deflection for safety of members against excessive deflection :

(*a*) The final deflection due to all loads including the effects of temperature, creep and shrinkage, and measured from the as-cast level of the supports of floors, roofs and all other horizontal members, should not normally exceed span/250.

(*b*) The deflection including the effects of temperature, creep and shrinkage occurring after erection of partitions and application of finishes should not normally exceed span/350 or 20 mm, whichever is less.

2. Second approach : Span to effective depth ratios

The first approach is based on allowable span/deflection ratio. The second approach is based on span-effective depth ratio, which is indirectly related to the span-deflection ratio, as shown below :

For a simply supported beam with uniformly distributed load w, the maximum deflection a is given by

$$a = \frac{5}{384}\frac{wl^4}{EI} = \left(\frac{5}{48}\right)\left(\frac{wl^2}{8}\right)\left(\frac{l^2}{EI}\right) = \frac{5}{48}\left(\frac{M_{max}}{I}\right)\frac{l^2}{E}$$

Substituting $M/I = f/y$ and $y = d/2$, we get

$$\frac{a}{l} = \frac{5}{48}\left(\frac{2f}{d}\right)\left(\frac{l}{E}\right) = \frac{5}{24}\left(\frac{f}{E}\right)\left(\frac{l}{d}\right)$$

Since f and E are constant, the deflection-span ratio is a function of span-effective depth ratio. Taking allowable values of $f = 5\ N/mm^2$, $E = 10 \times 10^3\ N/mm^2$, and $a/l = 1/350$, we get

$$\frac{l}{d} = \frac{1}{350} \times \frac{24}{5} \times \frac{10 \times 10^3}{5} \simeq 27.$$

Basic 'span to effective depth ratios'

As per IS 456 : 2000, the vertical deflection limits may generally be assumed to be satisfied provided that the span to effective depth ratio are not greater than the following.

Cantilever	:	7
Simply supported	:	20
Continuous	:	26

The above recommendations are valid only when the spans are upto 10 m. For spans above 10 m, above values may be multiplied by 10/span in metres, except for cantilevers in which case deflection calculations should be made.

Modification factors for basic span - depth ratios.

For a given case, basic span-depth ratios are modified to take into account the following:

(*i*) Effect of span greater than 10 m (modification factor F_l)

(*ii*) Effect of percentage of tension steel and service stress in steel (modification factor F_t)

(*iii*) Effect of percentage of compression steel (modification factor F_c)

(*iv*) Effect of beam section i.e. rectangular or flanged (modification factor F_b)

$$\therefore \quad \left(\frac{l}{d}\right)_{final} = \left(\frac{l}{d}\right)_{basic} \times F_l \times F_t \times F_c \times F_b \qquad \text{...(10.22)}$$

(*i*) ***Span factor*** F_l

$$F_l = \frac{10}{\text{span in meters}} \not> 1 \qquad \text{...(10.23)}$$

(*ii*) ***Factor*** F_t : This factor depends on the type of steel used (corresponding to service stress in steel) and the percentage of tension reinforcement at the point of maximum deflection. Its value is obtained from Fig. 10.5 (*a*).

where $$f_s = 0.58 f_y \times \frac{\text{Area of cross section of steel required}}{\text{Area of cross section of steel provided}}$$

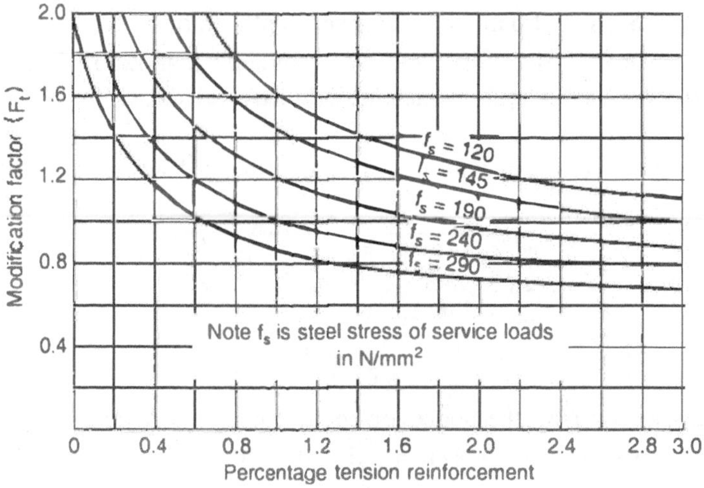

$$f_s = 0.58\, f_y \frac{\text{Area of cross-section of steel required}}{\text{Area of cross-section of steel provided}}$$

(a) Factor F_t

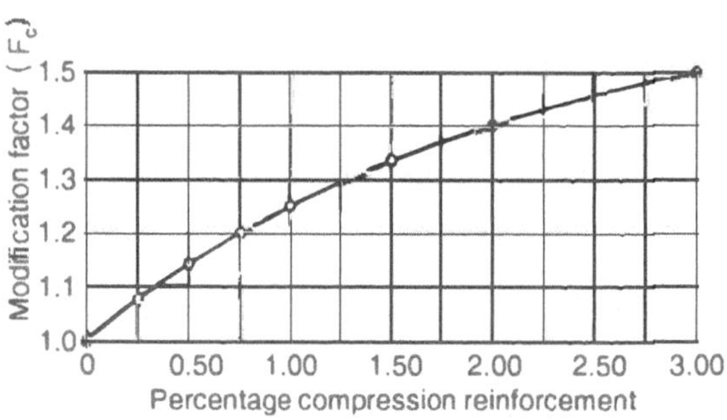

(b) Factor F_c

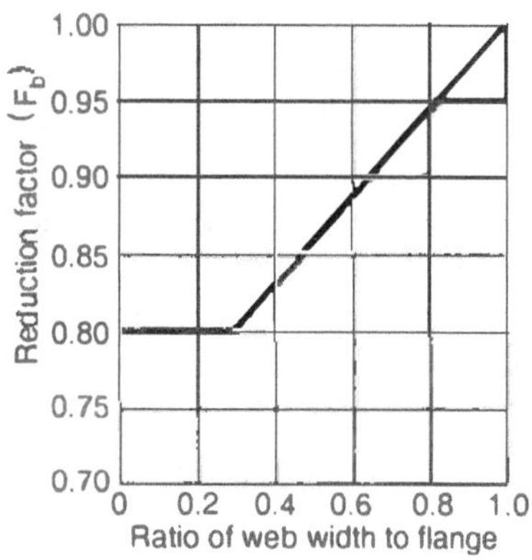

(c) Factor F_b

FIG. 10.5. VARIOUS MODIFICATION FACTORS

Alternatively, the value of F_t may be computed from the following equation given in SP 24:

$$F_t = \frac{1}{0.225 + 0.00322\, f_s + 0.625 \log_{10} p_t} \qquad \ldots(10.24)$$

where $p_t = 100\, A_{st}/bd$.

(iii) Factor F_c : This factor takes into account the effect of compression steel provided at the point of maximum deflection. Its value of may taken either from Fig. 10.5 (*b*) or computed from the following formula given in SP 24 :

$$F_c = \frac{1.6\, p_c}{p_c + 0.275} \le 1.5 \qquad \ldots(10.25\ a)$$

where $p_c = 100\, A_{sc}/bd$. The corresponding expression in BS 8110 (1985) is

$$F_c = 1 + \frac{p_c}{3 + p_c} \leq 1.5 \qquad \text{...(10.25)}$$

(*iv*) ***Factor*** F_b : This is applicable for flanged beam. As the factors f_t and F_c for flanged beams are calculated for the effective flange width b_f, a reduction factor F_b is used to allow for the reduced area in the tension zone. Its value can either be taken from Fig. 10.5 (*c*) or computed from the following equation by both IS and BS.

$$F_b = 0.8 + \frac{2}{7}\left(\frac{b_w}{b_f} - 0.3\right) \nless 0.8 \qquad \text{...(10.26)}$$

Span-depth ratios for two way slabs : For two way slabs of shorter spans (upto 3.5 m) with mild steel reinforcement, the span to *overall depth* ratios (i.e. *l/D* ratio) given below may generally be assumed to satisfy vertical deflection limits for loading class upto 3 kN/m^2.

Simply supported slabs	35
Continuous slabs	40

For high strength deformed bars of grade Fe 415, the values given above should be multiplied by 0.8.

10.6. MEASURES FOR REDUCING DEFLECTION

If the computed deflection for a designed member/structure comes out to be greater than the allowable deflection, the following measures may be taken to reduce the deflection.

1. *Increase* the depth of section.
2. *Increase the camber* : This should not be greater than $l/250$
3. *Use of richer concrete mix* : This may reduce deflection by about 10%.
4. *Use of Fe 250 steel* in place of HYSD bars : This may reduce deflection by about 25%.
5. *Increase the width* of section.
6. *Delay* in erection of partitions and finishes ; This will reduce creep deflection by about 30%.
7. *Use of T-section.*

Example 10.1. *A flanged beam of T-section is simply supported over an effective span of 8 m. The beam has effective flange width of 1400 mm, thickness of flange as 150 mm, breadth of web as 300 mm and effective depth of 450 mm. It is reinforced with 4 bars of 25 mm dia. in tension and 3 bars of 16 mm dia in compression, of Fe 415 grade. Check the beam for deflection, if the concrete is of M 20 grade.*

Solution : $A_{st} = 4 \times \frac{\pi}{4}(25)^2 = 1963.5 \text{ mm}^2$; $A_{sc} = 3 \times \frac{\pi}{4}(16)^2 = 603.2 \text{ mm}^2$

$$p_t = \frac{100\,A_{st}}{b_f\,d} = \frac{100 \times 1963.5}{1400 \times 450} = 0.31\%; \qquad p_c = \frac{100\,A_{sc}}{b_f\,d} = \frac{100 \times 603.2}{1400 \times 450} = 0.096\%$$

Basic $\frac{l}{d} = 20$ for simply supported beam.

$F_l = 1$, since span is less than 10 m.

$F_t \simeq 1.48$ for $p_t = 0.31\%$ and for $f_s = 0.58\,f_y = 0.58 \times 415 \simeq 240 \text{ N/mm}^2$

Alternatively from Eq. 10.24,

$$F_l = \frac{1}{0.225 + 0.00322\, f_s + 0.625 \log_{10} p_t} = \frac{1}{0.225 + 0.00322 \times 240 + 0.625 \log_{10} 0.31} = 1.47$$

$F_c = 1.03$ for $p_c = 0.096$

Alternatively, from Eq. 10.25,

$$F_c = \left[1 + \left(\frac{p_c}{3 + p_c}\right)\right] \le 1.5 = 1 + \frac{0.096}{3 + 0.096} = 1.03$$

Now $\quad \frac{b_w}{b_f} = \frac{300}{1400} = 0.214$. Hence $F_b = 0.8$

$$\therefore \quad \left(\frac{l}{d}\right)_{allowable} = \left(\frac{l}{d}\right)_{basic} \times F_l \times F_l \times F_c \times F_b = 20 \times 1 \times 1.47 \times 1.03 \times 0.8 = 24.2$$

Actual $\quad \frac{l}{d} = \frac{8000}{450} = 17.8 < 24.2 \quad$ Hence safe.

Example 10.2. *A cantilever of 3.5 m span is 300 mm wide and 600 mm deep. It is subjected to a maximum B.M. of 125 kN-m, due to uniformly distributed service loads out of which 50% moment is due to permanent loads. The beam is reinforced with 4 bars of 20 mm diameter at an effective cover of 50 mm, in the tension zone. Check the beam for deflection, using M 20 concrete and Fe 415 steel.*

Solution : Given : $l = 3.5$ m; $b = 300$ mm; $D = 600$ mm; $d = 550$ mm.

$$A_{st} = 4\, \frac{\pi}{4} (20)^2 = 1256.64 \text{ mm}^2; \quad M_c = 125 \text{ kN-m} = 125 \times 10^6 \text{ N-mm}$$

$b_w / b = 1$ for rectangular beam

The total deflection (a_T) is given by $\quad a_T = a_i + a_{cs} + a_{cc\,(perm.)}$

(*i*) ***Immediate or elastic deflection*** (a_i)

For cantilever, $\quad a_i = \frac{wl^4}{8\, E_c\, I_{eff}} = \frac{Ml^2}{4\, E_c\, I_{eff}}$

Here, $\quad E_c = 5000\sqrt{f_{ck}} = 5000\sqrt{20} = 22360 \text{ N/mm}^2$

$$I_{gr} = \frac{bD^3}{12} = \frac{300\,(600)^3}{12} = 54 \times 10^8 \text{ mm}^4$$

$$f_{cr} = 0.7\sqrt{f_{ck}} = 0.7\sqrt{20} = 3.13 \text{ N/mm}^2$$

$$M_r = \frac{f_{cr} \cdot I_{gr}}{y_t} = \frac{3.13 \times 54 \times 10^8}{600/2} = 56.34 \times 10^6 \text{ N-mm}$$

To find the depth of N.A., take moment of effective area about the N.A.

Thus $\quad bx\left(\frac{x}{2}\right) = m \times A_{st}\,(d - x), \quad$ where $\quad m = \frac{E_s}{E_c} = \frac{2 \times 10^5}{22360} = 8.94$

$$\therefore \quad \frac{300\, x^2}{2} = 8.94 \times 1256.64\,(550 - x) \quad \text{or} \quad x^2 + 74.896\, x - 41193 = 0$$

From which $\quad x = 168.9$ mm and $z = d - x/3 = 550 - 168.9/3 \simeq 493.7$ mm

$$I_r = \frac{bx^3}{3} + m\, A_{st}\,(d - x)^2 = \frac{300\,(168.9)^3}{3} + 8.94 \times 1256.64\,(550 - 168.9)^2$$

$$= 21.135 \times 10^8 \text{ mm}^4$$

$$\therefore \quad I_{eff} = \frac{I_r}{1.2 - \frac{M_r}{M_c}\frac{z}{d}\left(1 - \frac{x}{d}\right)\frac{b_w}{b}} = \frac{21.135 \times 10^8}{1.2 - \frac{56.34 \times 10^6}{125 \times 10^6}\left(\frac{493.7}{550}\right)\left(1 - \frac{168.9}{550}\right) 1}$$

$$= 22.981 \times 10^8 \text{ mm}^4 < I_{gr}$$

Hence $\quad I_{eff} = 22.981 \times 10^8$ N-mm

$$\therefore \quad a_i = \frac{M_c\, l^2}{4\, E_c\, I_{eff}} = \frac{125 \times 10^6 (3500)^2}{4 \times 22360 \times 22.981 \times 10^8} = 7.45 \text{ mm}$$

(ii) Shrinkage deflection (a_{cs})

$$a_{cs} = k_3\, \psi_{cs}\, l^2 \qquad \ldots(10.12)$$

where $\quad k_3 = 0.5$ (for cantilever); $l = 3.5$ m $= 3500$ mm

$$p_t = \frac{100\, A_{st}}{bd} = \frac{100 \times 1256.64}{300 \times 550} = 0.762\,\%; \quad p_c = 0$$

$$\therefore \quad k_4 = 0.72\, \frac{p_t - p_c}{\sqrt{p_t}} \le 1.0 \quad (\text{for } p_t - p_c < 1.0) = \frac{0.72 \times 0.762}{\sqrt{0.762}} = 0.629 < 1$$

$$\psi_{cs} = k_4\, \frac{\varepsilon_{cs}}{D} = \frac{0.629 \times 0.0003}{600} = 3.14 \times 10^{-7}$$

Hence $\quad a_{cs} = 0.5 \times 3.14 \times 10^{-7} (3500)^2 = 1.93$ mm

(iii) Creep deflection $(a_{cc\,,\,per})$

$$a_{cc\,(perm)} = a_{i\,+\,cc\,(perm)} - a_{i\,(perm.)} \qquad \ldots(10.16)$$

Assuming age of concrete as 28 days, creep coefficient $\theta = 1.6$.

$$\therefore \quad E_{ce} = \frac{E_c}{1 + \theta} = \frac{22360}{1 + 1.6} = 8600 \text{ N/mm}^2$$

Modular ratio $\quad m = \frac{E_s}{E_{CE}} = \frac{2 \times 10^5}{8600} = 23.26$

Let us now calculate x, z, I_r and I_{eff}, for this value of m.

We have $\quad \frac{bx^2}{2} = m\, A_{st}\, (d - x) \quad$ or $\quad \frac{300\, x^2}{2} = 23.26 \times 1256.64\, (550 - x)$

From which $\quad x = 244.14$ mm; $z = 550 - 244.14/3 = 468.62$ mm

$$\therefore \quad I_r = \frac{bx^3}{3} + m\, A_{st}\, (d - x)^2 = \frac{300\,(244.14)^3}{3} + 23.26 \times 1256.64\, (550 - 244.14)^2$$

$$= 41.896 \times 10^8 \text{ mm}^4$$

$$I_{eff}' = \frac{I_r}{1.2 - \frac{M_r}{M_c'}\frac{z}{d}\left(1 - \frac{x}{d}\right)\frac{b_w}{b}}$$, where M_c' is due to permanent loads.

$$= \frac{41.896 \times 10^8}{1.2 - \frac{56.34 \times 10^6}{0.5 \times 125 \times 10^6}\left(\frac{468.62}{550}\right)\left(1 - \frac{244.14}{550}\right) \times 1} = 54.21 \times 10^8 \not> I_{gr}$$

Hence $\quad I_{eff}' = I_{gr} = 54.00 \times 10^8$ mm^4

$$\therefore \quad a_{i+cc\,(perm)} = \frac{M_c'\,l^2}{4\,E_{ce}\,I_{eff}'}, \text{ where } M_c' = \text{ is due to permanent loads}$$

$$= \frac{0.5 \times 125 \times 10^6\,(3500)^2}{4 \times 8600 \times 54.0 \times 10^8} = 4.12 \text{ mm}$$

$$a_{i\,(perm.)} = \frac{M_c'\,l^2}{4\,E_c\,I_{eff}} = \frac{0.5 \times 125 \times 10^6\,(3500)^2}{4 \times 22360 \times 22.981 \times 10^8} = 3.72 \text{ mm}$$

$$\therefore \quad a_{cc\,(perm.)} = a_{i+cc\,(perm.)} - a_{i\,(perm.)} = 4.12 - 3.72 = 0.40 \text{ mm}$$

(iv) Total deflection

$$a_T = a_i + a_{cs} + a_{cc\,(perm.)} = 7.15 + 1.93 + 0.40 = \mathbf{9.48} \text{ mm}$$

Total permissible settlement $= \frac{l}{250} = \frac{3500}{250} = 14$ mm $> a_T$. Hence safe.

Example 10.3. *A simply supported one way slab of 4 m span carries a live load of 3 N/m² and the load of floor finish as 1.25 kN/m². The slab, having a total depth of 150 mm is reinforced with 8 mm φ bars @ 100 mm c/c at a nominal cover of 20 mm. Assuming a permanent load equal to dead load plus 20% of live load, compute the total maximum deflection and check it as per Code requirements. Use M 20 concrete and Fe 415 steel.*

Solution : Given : $D = 150$ mm ; $d = 150 - 20 - 8/2 = 126$ mm

$$A_{st} = \frac{1000 \times A_\varphi}{s} = \frac{1000 \times 50.26}{100} = 502.6 \text{ mm}^2 \; ; \; p_t = \frac{100 \times 502.6}{1000 \times 126} \simeq 0.4\%$$

Dead load $= 0.15 \times 25 + 1.25 = 5$ kN/m²

Permanent load = Dead load $+ 0.2$ LL $= 5 + 0.2 \times 3 = 5.6$ kN/m²

Temporary load $= 0.8$ LL $= 0.8 \times 3 = 2.4$ kN/m²

Total load = Dead load + LL $= 5 + 3 = 8$ kN/m²

$$f_{cr} = 0.7\sqrt{f_{ck}} = 0.7\sqrt{20} = 3.13 \text{ N/mm}^2$$

$$I_{gr} = \frac{1}{12}bD^3 = \frac{1}{12} \times 1000\,(150)^3 = 281.25 \times 10^6 \text{ mm}^4 \; ; \; y_t = D/2 = 75 \text{ mm}$$

$$\therefore \quad M_r = \frac{f_{cr}\,.\,I_{gr}}{y_t} = \frac{3.13 \times 281.25 \times 10^6}{75} = 11.74 \times 10^6 \text{ N-mm}$$

$$E_c = 5000\sqrt{f_{ck}} = 5000\sqrt{20} = 22360 \text{ N/mm}^2$$

$$m = E_s/E_c = 2 \times 10^5/22360 = 8.94$$

1. Short term deflections

The depth of N.A. is given by

$$\frac{bx^2}{2} = m\,A_{st}\,(d - x) \quad \text{or} \quad \frac{1000\,x^2}{2} = 8.94 \times 502.6\,(126 - x)$$

or $\quad x^2 + 8.986 - 1132.3 = 0$ from which $x = 29.46$ mm

$\therefore$ Lever arm $z = d - x/3 = 126 - 29.46/3 = 116.18$ mm

$$I_r = \frac{bx^3}{3} + m\,A_{st}\,(d-x)^2 = \frac{1000\,(29.46)^3}{3} + 8.94 \times 502.6\,(126 - 29.46)^2$$

$$= 50.4 \times 10^6 \text{ mm}^4$$

Deflection due to total load

Total load = 8 kN/m^2 ; $M_{c\,.\,total} = \frac{8\,(4)^2}{8} = 16$ kN-m = 16×10^6 N-mm

$$I_{eff\,(t)} = \frac{50.4 \times 10^6}{1.2 - \frac{11.74 \times 10^6}{16 \times 10^6} \times \frac{116.18}{126}\left(1 - \frac{29.46}{126}\right) \times 1} = 73.94 \times 10^6 \leq I_{gr}$$

$$a_{i\,(total)} = K_m \frac{M_{cc\,(total)}\,l^2}{E_c\,I_{eff(t)}} = \frac{5}{48} \times \frac{16 \times 10^6 \times (4000)^2}{22360 \times 73.94 \times 10^6} = 16.13 \text{ mm}$$

Deflection due to permanent load

Permanent load = 5.6 kN/m^2; $M_{c\,(perm.)} = \frac{5.6\,(4)^2}{8} = 11.2$ kN-m = 11.2×10^6 N-mm.

$$I_{eff\,(p)} = \frac{50.4 \times 10^6}{1.2 - \frac{11.74 \times 10^6}{11.2 \times 10^6} \times \frac{116.18}{126}\left(1 - \frac{29.46}{126}\right) \times 1} = 109.69 \times 10^6 \leq I_{gr}$$

$$\therefore \quad a_{i\,(perm.)} = K_m \frac{M_{c\,(perm)}\,l^2}{E_c\,I_{eff\,(p)}} = \frac{5}{48} \times \frac{11.2 \times 10^6\,(4000)^2}{22360 \times 109.69 \times 10^6} = 7.61 \text{ mm}$$

2. Deflection due to shrinkage (a_{cs})

$$a_{cs} = k_3\,\psi_{cs}\,l^2$$

where $k_3 = 0.125$ for simply supported beam.

$p_t = 0.4\%$; $p_c = 0$

$$k_4 = \frac{0.72\,p_t - p_c}{\sqrt{p_t}} = \frac{0.72 \times 0.4}{\sqrt{0.4}} = 0.4554 < 1$$

$$\varepsilon_{cs} = 0.0003$$

$$\psi_{cs} = k_4 \frac{\varepsilon_{cs}}{D} = 0.4554 \times \frac{0.0003}{150} = 0.91 \times 10^{-6}$$

Hence $a_{cs} = 0.125 \times 0.91 \times 10^{-6}\,(4000)^2 = 1.82$ mm

3. Deflection due to creep ($a_{cc\,,\,perm}$)

$$a_{cc\,(perm)} = a_{i\,+\,cc\,(perm)} - a_{i\,(perm.)}$$

Assuming age of concrete as 28 days, $\theta = 1.6$

$$\therefore \quad E_{ce} = \frac{E_c}{1+\theta} = \frac{22360}{1+1.6} = 8600$$

$$\therefore \quad m = E_s/E_{ce} = 2 \times 10^5/8600 = 23.26$$

Let us now calculate x , z , I_r and I_{eff} for this value of m.

$$\frac{bx^2}{2} = m\,A_{st}\,(d-x) \quad \text{or} \quad \frac{1000}{2}x^2 = 23.26 \times 502.6\,(126 - x)$$

or $x^2 + 23.38\,x - 2945 = 0$, from which $x = 43.83$ mm

$\therefore$ $z = d - x/3 = 126 - 43.83/3 = 111.39$ mm

$$I_r = \frac{bx^3}{3} + m\,A_{st}\,(d-x)^2 = \frac{1000\,(43.83)^3}{3} + 23.26 \times 502.6\,(126 - 43.83)^2 = 107 \times 10^6 \text{ mm}^4$$

$$M_{c\,(perm.)} = 11.2 \times 10^6 \text{ N-mm}$$

$$I_{eff\,(p)} = \frac{107 \times 10^6}{1.2 - \dfrac{11.74 \times 10^6}{11.2 \times 10^6} \times \dfrac{111.39}{126}\left(1 - \dfrac{43.83}{126}\right) \times 1} = 179.63 \times 10^6 \not> I_{gr}$$

$$\therefore \quad a_{i+cc(perm.)} = \frac{K_m\,M_{c\,(perm)}\,l^2}{E_{ce}\,I_{eff\,(p)}} = \frac{5}{48} \times \frac{11.2 \times 10^6\,(4000)^2}{8600 \times 179.63 \times 10^6} = 12.08 \text{ mm}$$

$$a_{i\,(perm.)} = 7.61 \text{ mm (determined earlier)}$$

$$\therefore \quad a_{cc\,(perm.)} = a_{i\,+\,cc\,(perm.)} - a_{i\,(perm)} = 12.08 - 7.61 = 4.47 \text{ mm}$$

4. Total deflection

$$a_T = a_{i\,(total)} + a_{cs} + a_{cc\,(perm)} = 16.13 + 1.82 + 4.47 = \mathbf{22.42} \text{ mm}$$

Allowable deflection $= l/250 = 4000/250 = 16$ mm. Hence the slab is *unsafe*.

10.7. LIMIT STATE OF CRACKING

Cracks appear in reinforced concrete members due to the following reasons :

1. *Structural cracking* due to bending action under applied loads :

   (*i*) Flexural cracks due to flexural tensile stress.

   (*ii*) Shear or diagonal tension cracks

   (*iii*) Splitting cracks along reinforcement due to bond and anchorage failure

2. *Cracking due to volumetric changes* initiated by shrinkage, creep, thermal and chemical effects.

3. *Cracking due to additional curvature* because of continuity effects, settlement of supports etc.

4. *Cracking due to stress concentration* caused by curtailment and splicing of bars.

5. *Cracking due to corrosion* of reinforcement.

In this chapter, we shall discuss only the cracking of concrete and determination of *crack width* due to structural causes, i.e. those caused due to direct or flexural stresses due to bending and shear. Crack width due to other causes can not be easily predicted. However, cracks caused due to these causes can be controlled by ensuring good quality concrete, prolonged curing, proper detailing of shrinkage and distribution reinforcements and adhering to the requirements of minimum steel area and maximum spacing of steel bars.

Adverse effects of cracking:

Cracking of concrete has the following adverse effects :

(*i*) Bad appearance of exposed surface of R.C. members.

(*ii*) Feeling of lack of safety of the member.

(*iii*) Reduction in imperviousness of exposed surface, leading to leakage problems, specially in roofs and walls of water tanks.

(*iv*) Spalling of concrete cover and consequent corrosion of steel reinforcement.

(*v*) Reduction in the stiffness of the members and consequent increase in deflection.

Limits on cracking : IS Code recommendations

Control of cracking is required not only from aesthetic reasons but more importantly for *durability* and for protection of reinforcement. The primary objective of crack control is to limit the width of individual cracks. Although flexural cracking is inevitable, it is desirable to aim for a large number of well distributed fine hair-line cracks, rather than a few but wide cracks.

As per IS 456 : 2000, cracking of concrete should not adversely affect the appearance or durability of the structure; the acceptable limits of cracking would vary with the type of structure and environment. Where specific attention is required to limit the designed crack width of a particular value, crack width calculation may be done, using formulae given below.

The practical objective of calculating crack width is merely to give guidance to the design in making appropriate structural arrangements and in avoiding gross errors in design, which might result in concentration and excessive width of flexural crack.

The surface width of the crack should not, in general, exceed 0.3 mm in members where cracking is not harmful and does not have any serious adverse effects upon the preservation of reinforcing steel, nor upon the durability of the structure. In members where cracking in the tensile zone is harmful, either because they are exposed to moisture or in contact soil or ground water, an upper limit of 0.2 mm is suggested for maximum width of cracks. For particularly aggressive environment, such as the 'severe' category, the assessed surface width of cracks should not in general, exceed 0.1 mm.

Empirical method of crack control

The empirical method of crack control is based on the premise that the cracks will be within the allowable limits, if the detailing is done according to the rules laid down by the Code. Following are the various items of detailing steel reinforcement.

1. Rules for maximum and minimum spacing of reinforcement
2. Rules for maximum and minimum areas of steel.
3. Rules for curtailment of reinforcing bars.
4. Rules for anchorage of reinforcing bars.
5. Rules for cover to reinforcement
7. Rules for maximum and minimum dia. of steel bars.
8. Limiting the tensile stress in reinforcing steel.

Calculation of crack width (IS : 456 - 2000)

Provided that the strain in the tension reinforcement is limited to $0.8\, f_y/E_s$, the design surface crack width may be calculated from the following equation :

$$W_{cr} = \frac{3\, a_{cr}\, \varepsilon_m}{1 + \dfrac{2\,(a_{cr} - C_{\min})}{D - x}} \qquad \ldots(10.27)$$

where W_{cr} = design surface crack width

a_{cr} = distance from the point considered to the surface of the nearest longitudinal bar

ε_m = average steel strain at the level considered, allowing for stiffening effect of concrete in tension zone. Its value is obtained from Eq. 10.28.

C_{min} = minimum cover to the longitudinal bar.

D = overall depth of the member.

x = depth of the neutral axis.

The average steel strain ε_m may be calculated on the basis of the following assumptions, illustrated in Fig. 10.6.

(*i*) The concrete and the steel are both considered to be fully elastic in tension and in compression.

(*ii*) The elastic modulus of the steel may be taken as 200 kN/mm^2.

(*iii*) The elastic modulus of the concrete is derived from the following equation both in compression and in tension

$$E_c = 5000\sqrt{f_{ck}}$$

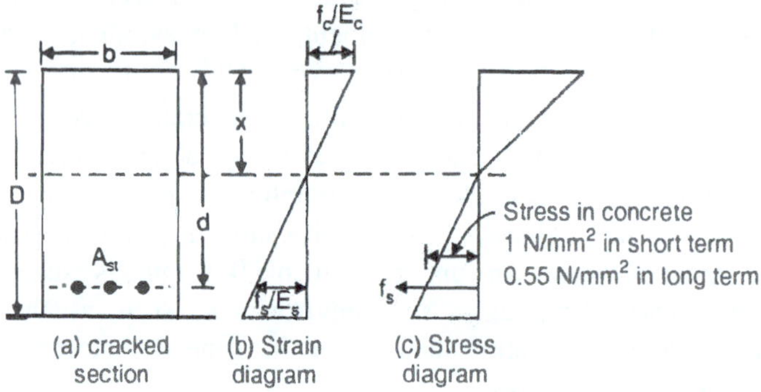

FIG. 10.6. ASSUMPTIONS

Alternatively, as an approximation, it will normally be satisfactory to calculate the steel stress on the basis of a cracked section and then reduce this by an amount equal to the tensile force generated by the triangular distribution, having a value of zero at the N.A. and a value at the centroid of the tension steel of 1 N/mm^2 instantaneously, reducing to 0.55 N/mm^2 in long term, acting over the tension zone divided by the steel area. For a rectangular tension zone, this gives

$$\varepsilon_m = \varepsilon_1 - \frac{b\,(D-x)\,(a-x)}{3\,E_s\,A_{st}\,(d-x)} \qquad \ldots(10.28)$$

where f_c = maximum compressive stress in the concrete

f_t = tensile stress in the reinforcement

E_s = modulus of elasticity of the reinforcement

A_{st} = area of tension reinforcement.

b = width of the section at the centroid of the tension steel.

ε_1 = strain at the level considered, calculated ignoring the stiffening of the concrete in the tension zone.

a = distance from compression face to the point at which the crack width is being calculated, and

d = effective depth

A negative value of ε_m indicates that the section is uncracked. Fig. 10.7 shows the definition sketch, depicting various parameters used in Eqs. 10.27 and 10.28.

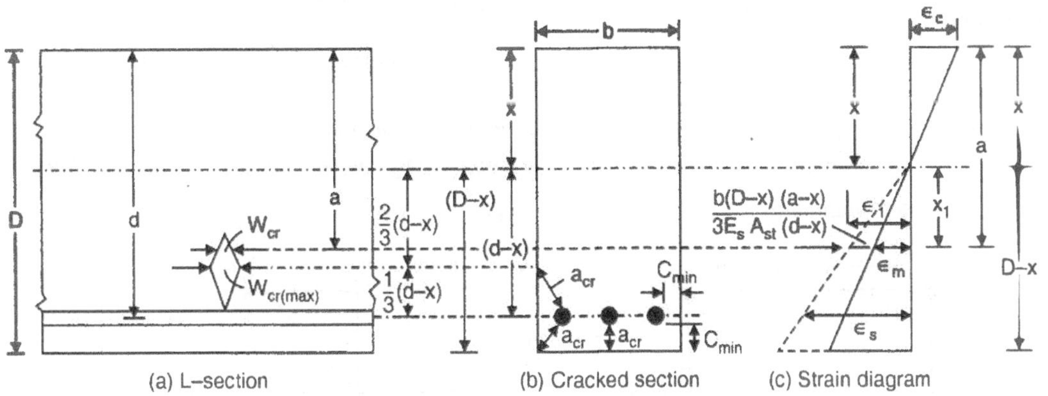

FIG. 10.7. PARAMETERS OF CRACK WIDTH COMPUTATIONS

The strain ε_1 at the level under consideration can be computed by the use of usual flexure formula, taking modulus of concrete equal to half the short term modulus of concrete (i.e. $E_{ce} = \frac{1}{2} E_c$) Thus,

$$\varepsilon_1 = \frac{M_s}{E_{ce}\, I_c}\, x_1 = \frac{M_s}{\left(\frac{E_c}{2}\right) I_c}\, x_1 \qquad \text{...(10.29)}$$

where M_s = B.M. at service load

I_c = moment of inertia of cracked section

x_1 = distance, measured from elastic N.A. to the point at which the strain ε_1 is required.

Magnitude of crack widths at various locations

1. Crack width immediately under the reinforcing bars

Immediately under the bar on tension face, $a_{cr} = C_{min}$. Hence from Eq. 10.27,

$$W_{cr} = 3\, C_{\min}\, \varepsilon_m \qquad \text{...(10.30)}$$

2. Crack width away from the bar

When the point under consideration is very much away from the bar, a_{cr} becomes quite large, and Eq. 10.28 approaches the following limit.

$$W_{cr} = 1.5\,(D - x)\, \varepsilon_m \qquad \text{...(10.31)}$$

3. Maximum value of crack width

Maximum crack width occurs at a distance of $\frac{1}{3}(d - x)$ from the centre of bar.

4. Crack width at N.A.

At the N.A., $a = x$ and $\varepsilon_m = 0$. Hence $W_{cr} = 0$

Example 10.4. *A rectangular beam, 300 mm × 500 mm, having an effective span of 6 m, is subjected to uniformly distributed load of 16 kN/m, inclusive of its own weight. The beam is reinforced width 3 bars of 20 mm dia, at an effective cover of 50 mm. Assuming M 20 concrete and Fe 415 steel, calculate the surface crack width at the following locations :*

(*a*) *at a point A directly under a bar on tension face*

(*b*) *at the bottom corner* (*B*) of the beam

(*c*) *at a point C on tension face, mid-way between two bars*

(*d*) *at a point D distant* $\frac{2}{3}(d-x)$ *from N.A., where crack width is likely to be maximum.*

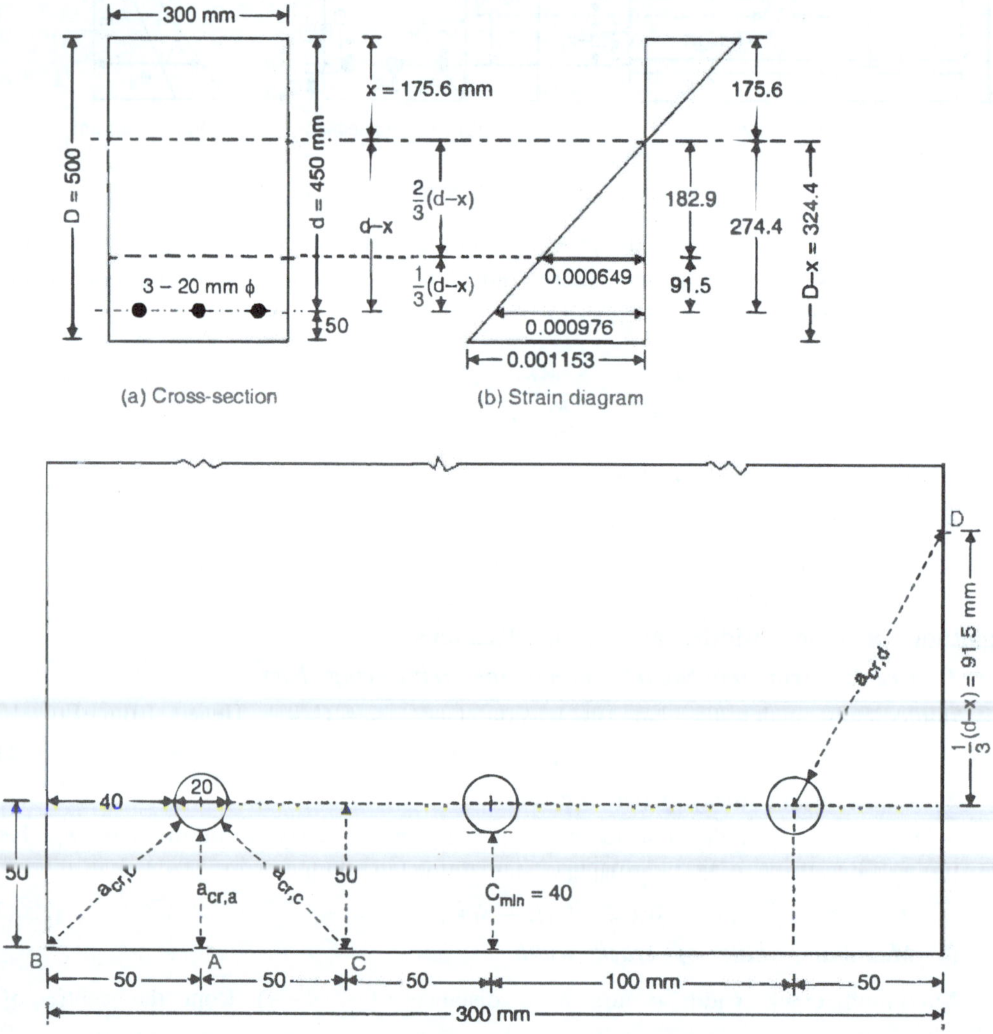

FIG. 10.8.

Solution : $A_{st} = 3 \times \frac{\pi}{4} (20)^2 = 942.48 \text{ mm}^2$.

Service load, $w_s = 16$ kN/m; $\therefore$ B.M. at service load, $M_s = \frac{16\,(6)^2}{8} = 72$ kN-m

$$E_c = 5000 \sqrt{f_{ck}} = 5000 \sqrt{20} = 22360 \text{ N/mm}^2; \quad E_{ce} = \frac{1}{2} \times 22360 = 11180 \text{ N/mm}^2$$

$$m = E_s / E_{ce} = 2 \times 10^5 / 11180 = 17.89$$

1. Determination of N.A. at working loads

We have $\frac{bx^2}{2} = m A_{st} (d - x)$ or $\frac{300\, x^2}{2} = 17.89 \times 942.48\,(450 - x)$

or $x^2 + 112.4\, x - 50580 = 0$, which gives $x = 175.6$ mm

$$\therefore \quad d - x = 450 - 175.6 = 274.4 \text{ mm} \; ; \; D - x = 500 - 175.6 = 324.4 \text{ mm}$$

$$\frac{2}{3}(d - x) = \frac{2}{3}(450 - 175.6) = 182.9 \text{ mm} = a - x$$

$$\frac{1}{3}(d - x) = \frac{1}{3}(450 - 175.6) = 91.5 \text{ mm}$$

$$I_c = \frac{1}{3} bx^3 + m A_{st} (d - x)^2 = \frac{1}{3} \times 300\,(175.6)^3 + 17.89 \times 942.48\,(450 - 175.6)^2$$

$$= 1811 \times 10^6 \text{ mm}^4$$

Fig. 10.8 (*b*) shows the strain diagram. Strain at any distance x_1 below N.A. is given by

$$\varepsilon_1 = \frac{M_s\,(x_1)}{E_{ce}\, I_c}$$

$\therefore$ Strain ε_1 at $x_1 = 324.4$ mm is : $\varepsilon_{1\,(324.4)} = \frac{72 \times 10^6 \times 324.4}{11180 \times 1811 \times 10^6} = 0.001153$

Strain ε_1 at $x_1 = 274.4$ mm is : $\varepsilon_{1\,(274.4)} = \frac{72 \times 10^6 \times 274.4}{11180 \times 1811 \times 10^6} = 0.000976$

Strain ε_1 at $x_1 = 182.9$ mm is : $\varepsilon_{1\,(182.9)} = \frac{72 \times 10^6 \times 182.9}{11180 \times 1811 \times 10^6} = 0.000649$

Fig. 10.8 (*c*) shows the location of the four point *A, B, C, D,* where crack width is required.

2. Determination of crack width at point A directly under a bar on tension face

$a_{cr.\,a} = 50 - 10 = 40$ mm; $\varepsilon_1 = \varepsilon_{1\,(324.4)} = 0.001153$; $a = D = 500$ mm

From Eq. 10.28, $\varepsilon_m = \varepsilon_1 - \frac{b\,(D - x)\,(a - x)}{3\, E_s\, A_{st}\,(d - x)} = 0.001153 - \frac{300\,(500 - 175.6)\,(500 - 175.6)}{3 \times 2 \times 10^5 \times 942.48(450 - 175.6)}$

$$= 0.000949$$

$\therefore$ Crack width $\quad W_{cr} = 3\, C_{min}\, \varepsilon_m = 3 \times 40 \times 0.000949 =$ **0.114 mm**

3. Determination of crack width at point B at bottom corner of the beam

$a_{cr.\,b} = \sqrt{(50)^2 + (50)^2} - 20/2 = 60.71$ mm; $\varepsilon_1 = \varepsilon_{1\,(324.4)} = 0.001153$; $a = D = 500$ mm

$\therefore \quad \varepsilon_m = 0.000949$, as for point A ; $C_m = 40$ mm

$$\therefore \quad W_{cr} = \frac{3\, a_{cr}\, \varepsilon_m}{1 + \dfrac{2\,(a_{cr} - C_m)}{(D - x)}} = \frac{3 \times 60.71 \times 0.000949}{1 + \dfrac{2\,(60.71 - 40)}{324.4}} = \mathbf{0.153\ mm}$$

4. Determination of crack width at point C on tension face, midway between two bars

$a_{cr,\,c} = \sqrt{50^2 + 50^2} - 20/2 = 60.71$ mm ; $\varepsilon_1 = \varepsilon_{1\,(324.4)} = 0.001153$; $a = D = 500$ mm

$\therefore \quad \varepsilon_m = 0.000949$, as for point A; $C_{\min} = 40$ mm

$$\therefore \quad W_{cr} = \frac{3\, a_{cr}\, \varepsilon_m}{1 + \dfrac{2\,(a_{cr} - C_{\min})}{D - x}} = \frac{3 \times 60.71 \times 0.000949}{1 + \dfrac{2\,(60.71 - 40)}{324.4}} = \mathbf{0.153\ mm}$$

5. Determination of crack width at point D distant $\frac{2}{3}(d - x)$ from N.A.

$a_{cr,\,d} = \sqrt{(91.5)^2 + (50)^2} - 20/2 = 94.27$ mm ; $\varepsilon_1 = \varepsilon_{1\,(182.9)} = 0.000649$

$D - x = 324.4$ mm ; $a - x = 182.9$ mm; $d - x = 274.4$ mm

$$\therefore \quad \varepsilon_m = 0.000649 - \frac{300 \times 324.4 \times 182.9}{3 \times 2 \times 10^5 \times 942.48\,(274.4)} = 0.000534$$

$$\therefore \quad W_{cr} = \frac{3\, a_{cr}\, \varepsilon_m}{1 + \dfrac{2\,(a_{cr} - C_{\min})}{D - x}} = \frac{3 \times 94.27 \times 0.000534}{1 + \dfrac{2\,(94.27 - 40)}{324.4}} = \mathbf{0.113\ mm}$$

Hence maximum crack width ≈ 0.153 mm < 0.3 mm. Hence safe.

PROBLEMS

1. A rectangular cantilever beam of 3 m span is 300 mm wide and 600 mm deep, and is reinforced width 3 bars of 25 mm φ at tension side, placed at an effective cover of 50 mm. It subjected to a max. B.M. of 120 kN-m at the fixed end, out of which 50% is due to permanent loads. Check the beam for deflection. Assume M 20 concrete and Fe 415 steel.

2. A one way slab of 3.5 m effective span is subjected to a live load of 3 kN/m^2 and a surface finish of load of 1 kN/m^2. The slab is 150 mm thick and is reinforced with 8 mm φ bars @ 125 mm c/c. Check the slab for deflection if one third of the live load is permanent load. Assume M 20 concrete and Fe 415 steel.

3. Explain how you determine the deflection due to shrinkage and creep

4. Explain how can you control cracking of reinforced concrete structural element.

11

Design of Beams and Slabs

11.1. DESIGN OF SIMPLY SUPPORTED SINGLY REINFORCED BEAMS

(A) BASIC RULES FOR DESIGN (IS : 456–2000)

We have already discussed the theory of singly reinforced beams. While designing the beams, following important rules must be kept in mind :

1. **Effective Span**. Unless otherwise specified, the effective span of a member shall be as follows :

(*a*) ***Simply supported beam or slab***. The effective span of a member that is not built integrally with its supports shall be taken as clear span plus the effective depth of slab or beam or centre to centre of supports, whichever is less.

(*b*) ***Continuous beam or slab***. In case of continuous beam or slab, if the width of the support is less than 1/12 of the clear span, the effective span shall be as in (*a*) above. If the supports are wider than 1/12 of the clear span or 600 mm whichever is less, the effective span shall be taken as under :

(*i*) For the end span with one end fixed and the other continuous or for intermediate spans, the effective span shall be clear span between supports, and

(*ii*) For the end span with one end free and the other continuous, the effective span shall be equal to the clear span plus half the effective depth of the beam or slab or the clear span plus half the width of the discontinuous support, whichever is less.

(*iii*) In the case of spans with roller or rocker bearings, the effective span shall always be the distance between the centres of bearings.

(*c*) ***Cantilever*** : The effective length of cantilever shall be taken as its length to the face of the support plus half the effective depth except where it forms the end of a continuous beam where the length to the centre of the support shall be taken.

(*d*) ***Frames***. In the analysis of a continuous frame, centre to centre distance shall be used.

2. **Control of Deflection**. The deflection of a structure or part there of shall not adversely affect the appearance or efficiency of the structure or finishes or partitions. For beams and slabs, the vertical deflection limits may generally be assumed to be satisfied provided that the span to depth ratios are not greater than the values obtained as below:

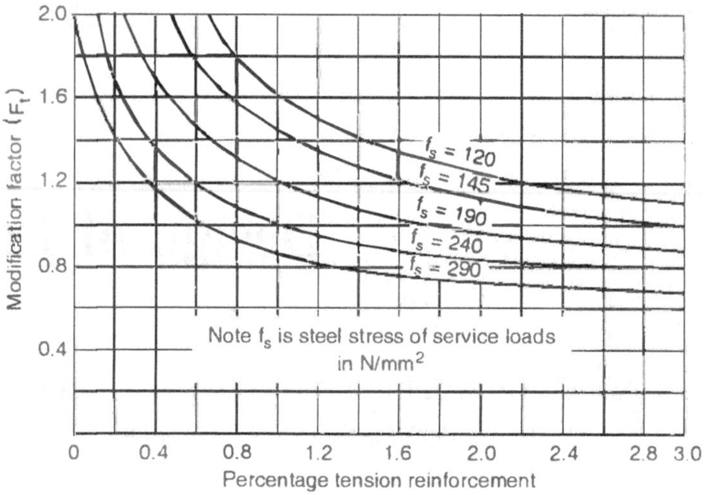

$$f_s = 0.58\, f_y \frac{\text{Area of cross-section of steel required}}{\text{Area of cross-section of steel provided}}$$

FIG. 11.1 MODIFICATION FACTOR (F_t) FOR TENSION REINFORCEMENT.

(*a*) Basic values of span to effective depth ratios for span upto 10 m :

Cantilever	7
Simply supported	20
Continuous	26

(*b*) For spans above 10 m, the values in (*a*) may be multiplied by factor F_l = 10/span (⨏) in metres, except for cantilever in which case deflection calculations should be made.

(c) Depending on the area and the type of steel for tension reinforcement, the values in (*a*) or (*b*) shall be multiplied by modification factor F_t as per Fig.11.1. In Fig. 11.1.

f_s = the steel stress of service loads

$$= 0.58\, f_y \frac{\text{Area of cross-section of steel required}}{\text{Area of cross-section of steel provided}}$$

To start with, f_s may be taken equal to $0.58\, f_y$

Thus, for Fe 500 steel,	$f_s = 0.58 \times 500 = 290 \text{ N/mm}^2$
For Fe 415 steel,	$f_s = 0.58 \times 415 \approx 240 \text{ N/mm}^2$
For Fe 250 steel,	$f_s = 0.58 \times 250 \approx 145 \text{ N/mm}^2$

Since the factor of 0.58 has been arrived from the fact that $f_s = \dfrac{f_y}{1.5 \times 1.15} = 0.58\, f_y$, graphs given in Fig. 11.1 for $f_s = 290 \text{ N/mm}^2$, $f_s = 240 \text{ N/mm}^2$ and $f_s = 145 \text{ N/mm}^2$ can be directly used for steel of grade Fe 500, Fe 415 and Fe 250 respectively *corresponding to the percentage of steel required.* Initially, p_t% is not known. Hence assume an initial values of p_t, and find the modification factor F_t . Since it is desirable to use *under-reinforced* section, the value of p_t should be assumed to be equal to 40 to 50% of $p_{t\,.\,lim}$ for mild steel bars and equal to about 30% of $p_{lim.}$ for HYSD bars.

Finally, when the beam has been designed, the required p_t (%) is known. Also, A_{st} required and A_{st} actually provided are also known. Hence one can compute f_s as :

$$f_s = 0.58\, f_y \frac{\text{Area of cross-section required}}{\text{area of cross-section provided}} = 0.58\, f_y \,.\, r_a$$

Since the ratio r_a is always less than 1, the value of f_s, to be taken to determine F_t , will be less than $0.58\, f_y$.

For example, let $f_y = 415$ and let $r_a = 0.9$

Then $f_s = 0.58\, f_y \,.\, r_a = 0.58 \times 415 \times 0.9 \approx 217\ \text{N/mm}^2$

Hence the final value of modification factor has to be found corresponding to an *imaginary curve* of $f_s = 217\ \text{N/mm}^2$ value, which falls between $f_s = 240$ and $f_s = 190$ values.

Alternatively, the value of F_t may be computed from Eq. 10.24.

(*d*) Depending on the area of compression reinforcement, the value of span to depth ratio be further modified as per modification factor F_c given in Fig. 11.2. Alternatively, the value of F_c may be computed from Eq. 10.25.

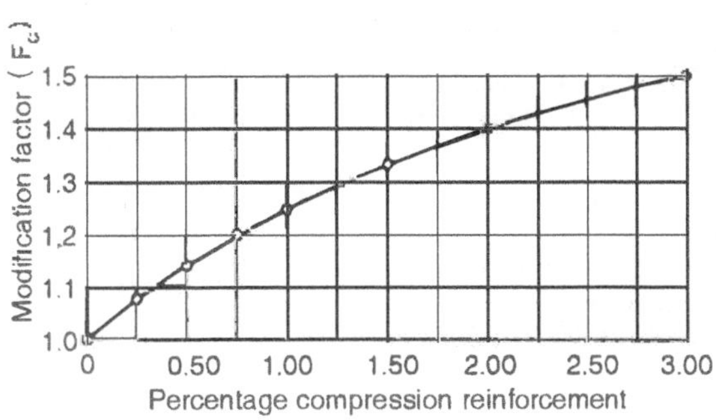

FIG. 11.2. MODIFICATION FACTOR F_c FOR COMPRESSION REINFORCEMENT.

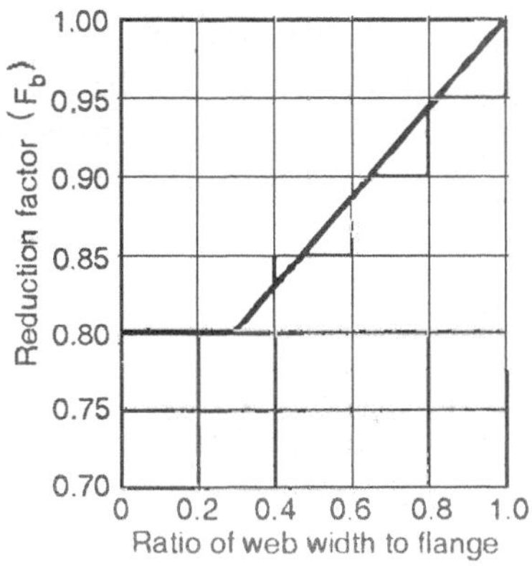

FIG. 11.3 REDUCTION FACTOR F_b FOR FLANGED BEAMS.

(*e*) For flanged beams, the values of (*a*) or (*b*) be modified by factor F_b as per Fig. 11.3 and the reinforcement percentage for use in Fig 11.1 and 11.2 should be based on area of section of equal to $b_f\, d$. Alternatively, the value of F_b may be computed from Eq. 10.26.

Thus, finally,
$$\left(\frac{l}{d}\right)_{final} = \left(\frac{l}{d}\right)_{basic} \times F_l \times F_t \times F_c \times F_b \qquad \ldots(11.1)$$

Note : Where deflections are required to be calculated, the method given in Appendix C of the Code (IS : 456-2000) may be followed.

Solid Slabs : The above provisions for beams also apply for slabs.

Note 1. For slabs spanning in two directions, the shorter of the two spans should be used for calculating the span to effective depth ratios.

Note 2. For two-way slabs of small spans (up to 3.5 m), with mild steel reinforcement, the span to overall depth ratio given below may generally be assumed to satisfy vertical deflection limits for loading class upto 3 kN/m^2.

Simply supported slabs : 35 Continuous slabs : 40

For high strength deformed bars of grade Fe 415, the values given above should be multiplied by 0.8.

3. Slenderness Limits for Beams to Ensure Lateral Stability. : A simply supported or continuous beam shall be so proportioned that the clear distance between the lateral restraints does not exceed 60 b or 250 b^2/d whichever is less, where d is the effective depth of the beam and b the breadth of the compression face midway between the lateral restraints. For a cantilever, the clear distance from the free end of the cantilever to the lateral restraint shall not exceed 25 b or 100 b^2/d whichever is less.

4. Critical Sections for Moments and Shear. For monolithic construction, the moments computed at the face of the supports shall be used in the design of the member at those sections. For non-monolithic construction the design of the member shall be done keeping in view para 1.

Critical section for shear. The shear computed at the face of the support shall be used in the design of the member at that section except as in case mentioned below.

When the reaction in the direction of the applied shear introduces compression into the end region of the member, sections located at the distance less than d from the face of the support may be designed for the same shear as that computed at distance d (see Fig. 11.4). Also, see § 7.12 and Fig. 7.19.

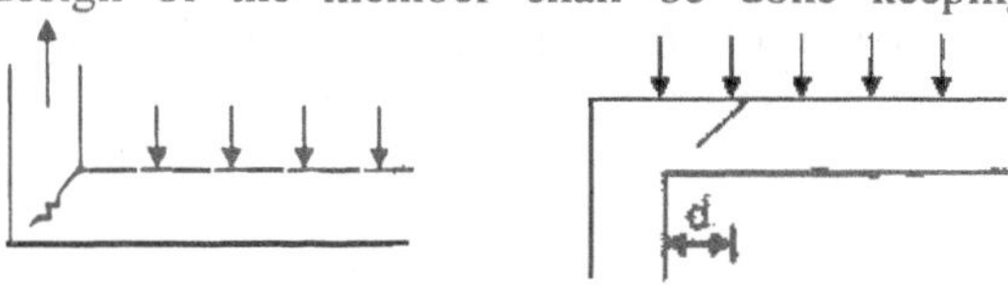

(a) Critical section at face of the support

(b) Critical section at distance d from the face of the support

FIG. 11.4. CRITICAL SECTION FOR SHEAR.

5. Reinforcement in Beams

(a) Tension reinforcement

(*i*) *Minimum reinforcement*. The minimum area of the tension reinforcement shall not be less than that given by the following :

$$\frac{A_{st}}{bd} = \frac{0.85}{f_y}$$ where f_y is the characteristic strength of reinforcement in N/mm^2.

(*ii*) *Maximum reinforcement*. The maximum area of tension reinforcement shall not exceed 0.04 bD.

(*b*) ***Compression reinforcement***. The maximum area of compression reinforcement shall not exceed 0.04 bD. Compression reinforcement in beams shall be enclosed by stirrups for effective lateral restraint.

(*c*) ***Side face reinforcement***. Where the depth of the web in a beam exceeds 750 mm, side face reinforcement shall be provided along the two faces. The total area of such reinforcement shall be not less than 0.1 percent of the web area and shall be distributed equally on two faces at a spacing not exceeding 300 mm or web thickness whichever is less.

(*d*) ***Transverse reinforcement in beams for shear and torsion***. *The transverse reinforcement in beams shall be taken around the outer-most tension and compression bars. In T-beams and L-beams, such reinforcement shall pass around longitudinal bars located close to the outer face of the flange.*

(*e*) ***Maximum spacing of shear reinforcement***. The maximum spacing of shear reinforcement measured along the axis of the member shall not exceed 0.75 d for vertical stirrups and d for inclined stirrups at 45°, where d is the effective depth of the sections under consideration. In no case shall the spacing exceed 300 mm.

(*f*) ***Minimum shear reinforcement***. Minimum shear reinforcement in the form of stirrups shall be provided such that

$$\frac{A'_{sv}}{b \cdot s_v} \geq \frac{0.4}{0.87 f_y}$$

where A_{sv} = total cross-sectional area of stirrup legs effective in shear

s_v = stirrup spacing along the length of member

b = breadth of the beam or breadth of the web of flanged beam, and

f_y = characteristic strength of the stirrup reinforcement in N/mm^2 which shall not be taken greater than 415 N/mm^2.

Based on the above $$s_v \leq \frac{0.87 f_y A_{sv}}{0.4 b} \leq \frac{2.175 f_y A_{sv}}{b}$$

(*h*) ***Maximum diameter***. The diameter of the reinforcing bars shall not exceed one eighth of the total thickness of the slab.

6. Curtailment of Tension Reinforcement in Flexural Member : See also § 8.13.

(***a***) For curtailment, reinforcement shall extend beyond the point at which it is no longer required to resist flexure for a distance equal to the effective depth of the member or 12 times the bar diameter, whichever is greater except at simple support or end of cantilever. In addition, clauses (*b*) to (*f*) given below shall be satisfied.

Note. A point at which reinforcement is no longer required to resist flexure is where the resistance moment of the section, considering only the continuing bars, is equal the design moment.

(***b***) Flexural reinforcement shall not be terminated in a tension zone unless any one of the following conditions is satisfied :

(*i*) The shear at the cut-off point does not exceed two thirds that permitted, including the shear strength of web reinforcement provided.

(*ii*) Stirrup area in excess of that required for shear and torsion is provided along each terminated bar over a distance from the cut-off point equal to three fourths the effective depth of the member. The excess stirrup area shall be not less than $0.4\ bs/f_y$, where b is the width of the beam, s is the spacing and f_y is the characteristic strength of reinforcement in N/mm^2. The resulting spacing shall not exceed $d/8\ \beta_0$ where β_0 is the ratio of the area of bars cut off to the total area of bars at the section, and d is the effective depth.

(*iii*) For 36 mm and smaller bars, the continuing bars provide double the area required for flexure at the cut-off point and the shear does not exceed three-fourth that permitted.

(*c*) ***Positive moment reinforcement.*** (*i*) At least one-third the positive moment reinforcement in simple members and one-fourth the positive moment reinforcement in continuous members shall extend along the same face of the member into the support, to a length equal to $L_d/3$.

(*ii*) When a flexural member is part of the primary lateral load resisting system, the positive reinforcement required to be extended into the support as described in (*i*) above shall be anchored to develop its design stress in tension at the face of the support.

7. Development length requirements at supports : At simple supports and at point of inflection, positive moment tension reinforcement shall be limited to a diameter such that L_d computed for f_{yd} (by the relation $L_d = \varphi f_{yd}/4\,\tau_{bd}$) does not exceed $\left(\frac{M_1}{V} + L_0\right)$, where

M_1 = moment of resistance of the section assuming all reinforcement at the section to be stressed to f_{yd}.

V = S.F. at the section due to design loads

L_0 = Sum of the anchorage beyond the centre of the support and the equivalent anchorage value of any hook or mechanical anchorage at simple support ; and at a point of inflection, L_0 is limited to the effective depth of the member or 12 φ whichever is greater, and

φ = diameter of bars.

The value of M_1/V in the above expression may be increased by 30 percent when the ends of the reinforcement are confined by a compression reaction. (See Fig. 8.6)

Computation of L_0. For the computation of L_0, the support width should be known. Fig 11.5 shows a beam with end support, in which x' is the side cover and x_0 is the distance of the *beginning* of the hook or bend from the centre line of the support.

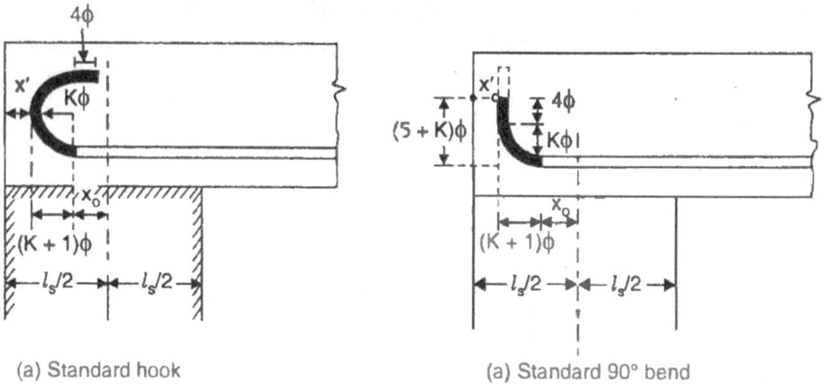

FIG. 11.5. COMPUTATION OF L_0

If no hook is provided, as in case of deformed bars, $L_0 = \frac{l_s}{2} - x'$. However, if a hook or bend is provided, L_0 is computed as under.

(*a*) Case 1 : Provision of standard hook at the end (Fig. 11.5 *a*)

The dark portion shows the hook which has an anchorage value of 16 φ (IS 456: 2000), for all types of steel. The distance of the beginning of the hook from its apex of the semi-circle is equal to $(K+1)\varphi$. For mild steel bars, $K = 2$ and for HYSD bars, $K = 4$. Hence this distance is equal to 3φ for mild steel bars and 5φ for HYSD bars. Let l_s be the width of the support.

Then $L_0 = x_0 + 16\,\varphi$ where $x_0 = \frac{l_s}{2} - x' - (K+1)\,\varphi$

$$\therefore \quad L_0 = \left(\frac{l_s}{2} - x' - (K+1)\,\varphi\right) + 16\,\varphi = \frac{l_s}{2} - x' + (15 - K)\,\varphi \qquad \text{...(8.8)}$$

Taking $K = 2$ for mild steel bars, $L_0 = \frac{l_s}{2} - x' + 13\,\varphi$ ...(8.8 *a*)

Taking $K = 4$ for HYSD bars, $L_0 = \frac{l_s}{2} - x' + 11\,\varphi$ ...(8.8 *b*)

(*b*) Case 2 : Provision of 90° standard bend (Fig. 11.5 *b*) : The dark portion shows the 90° bend which has an anchorage value of 8φ (IS : 456- 2000), for all types of steel.

Here also, the distance of beginning of the hook from its apex of the semi-circle is equal to $(K + 1)\,\varphi$.

Then $L_0 = x_0 + 8\,\varphi$ where $x_0 = \frac{l_s}{2} - x' - (K+1)\,\varphi$

$$\therefore \quad L_0 = \left(\frac{l_s}{2} - x' - (K+1)\,\varphi\right) + 8\,\varphi = \frac{l_s}{2} - x' + (7-K)\,\varphi \qquad \text{...(8.9)}$$

Taking $K = 2$ for mild steel bars, $L_0 = \frac{l_s}{2} - x' + 5\,\varphi$ ...(8.9 *a*)

Taking $K = 4$ for HYSD bars, $L_0 = \frac{l_s}{2} - x' + 3\,\varphi$ ...(8.9 *b*)

Remedies to get development length

If the check for the satisfaction of Eq. 8.6 or 8.6 *a* is not obtained, following remedial measures may be adopted to satisfy the check :

1. Reduce the diameter φ of the bar, thereby reducing the value of L_d, keeping the area of steel at the section unchanged. This is the standard procedure envisaged in the Code, i.e. reducing the value of L_d by *limiting* the diameter of the bar to such a value that Eq. 8.6 or 8.6 *a* is satisfied.

2. Increasing the value of L_0 by providing *extra length* of the bend over and above the standard value $(5 + K)\,\varphi$ shown by dotted lines in Fig. 11.5 (*b*).

3. By increasing the number of bars (thereby increasing A_{st1}) to be taken into the support. This method is uneconomical.

4. By providing adequate mechanical anchorage.

We shall discuss the first remedial method in the following section.

Limiting bar diameter to satisfy Eq. 8.6 or 8.6 (*a*)

IS 456 : 2000 stipulates that at simple supports having compressive reaction, the positive moment reinforcement shall be limited to a diameter such that

$$L_d \le \frac{1.3\,M_1}{V} + L_0 \qquad \text{...(8.6 } a\text{)}$$

Substituting $L_d = \frac{f_{yd}\,\varphi}{4\,\tau_{bd}}$, in the above condition, we get

$$\frac{f_{yd}\cdot\varphi}{4\,\tau_{bd}} \le 1.3\,\frac{M_1}{V} + L_0 \quad \text{or} \quad \varphi \le \frac{4\,\tau_{bd}}{f_{yd}}\left(1.3\,\frac{M_1}{V} + L_0\right) \le \frac{4\,\tau_{bd}}{0.87\,f_y}\left(1.3\,\frac{M_1}{V} + L_0\right) \qquad \text{...(8.10)}$$

Similarly for simple supports not offering compressive reaction,

$$\varphi \le \frac{4\,\tau_{bd}}{f_{yd}}\left(\frac{M_1}{V} + L_0\right) \le \frac{4\,\tau_{bd}}{0.87\,f_y}\left(\frac{M_1}{V} + L_0\right) \qquad \text{...(8.10 } a\text{)}$$

7. **Spacing of Reinforcement**. For the purpose of the clause, the diameter of a round bar shall be its nominal diameter and in the case of bars which are not round, or in case of deformed bars or crimped bars, the diameter shall be taken as the diameter of a circle giving an equivalent effective area. Where spacing limitations and minimum concrete cover (see clause 9) are based on bar diameter, a group of bars bundled in contact shall be treated as a single bar of diameter derived from the total equivalent area.

(*a*) ***Minimum distance between individual bars*** : The following shall apply for spacing of bars :

(*i*) The horizontal distance between two parallel main reinforcing bars shall usually be not less than the greatest of the following :

(1) The diameter of the bar if the diameters are equal (2) the diameter of the larger bar if the diameters are unequal, and (3) 5 mm more than the nominal maximum size of coarse aggregate.

Note. This does not preclude the use of larger size of aggregates beyond the congested reinforcement in the same member ; the size of aggregates may be reduced around congested reinforcement to comply with this provision.

(*ii*) Greater horizontal distance than the minimum specified in (*i*) should be provided wherever possible. However when needle vibrators are used, the horizontal distance between bars of a group may be reduced to two-thirds the nominal maximum size of the coarse aggregate, provided that sufficient space is left between groups of bars to enable the vibrator to be immersed.

(*iii*) Where there are two or more rows of bars, the bars shall be vertically in line and the minimum vertical distance between the bars shall be 15 mm, two-thirds the nominal maximum size of aggregate or the maximum size of bar whichever is the greatest.

(*b*) ***Maximum distance between bars in tension*** : Unless the calculation of crack widths shows that a greater spacing is acceptable, the following rules shall be applied to flexural members in normal internal or external conditions of exposure. The horizontal distance between parallel reinforcement bars, or groups, near the tension face of a beam shall not be greater than the value given in Table 11.1 depending on the amount of redistribution carried out in analysis and the characteristic strength of the reinforcement.

TABLE 11.1. CLEAR DISTANCE BETWEEN BARS

f_y	*Percentage redistribution to or from section considered*				
	– 30	*– 15*	*0*	*+15*	*+30*
	Clear distance between bars				
N/mm²	*mm*	*mm*	*mm*	*mm*	*mm*
250	215	260	360	300	300
415	125	155	180	210	235
500	105	130	150	175	135

Note. The spacings given in the Table 11.1 are not applicable to members subjected to particularly aggressive environments unless in the calculation of the moment of resistance, f_y has been limited to 300 N/mm^2 in limit state design.

8. Nominal Cover to Reinforcement

1. ***Nominal cover*** : Nominal cover is the design depth of concrete cover to all steel reinforcements, *including links*. It is the dimension used in design and indicated in the drawings. *It shall not be less than the diameter of the bar.*

2. ***Nominal cover to meet durability requirement*** : Minimum values for the nominal cover of normal weight aggregate concrete which should be provided to all reinforcement, including links depending on the condition of exposure shall be as given in Table 11.2.

TABLE 11.2 NOMINAL COVER TO MEET DURABILITY REQUIREMENTS

Exposure	*Nominal concrete cover in mm (Not less than)*
1. Mild	20
2. Moderate	30
3. Severe	45
4. Very severe	50
5. Extreme	75

Note : 1 For main reinforcement upto 12 mm dia bar for mild exposure, the nominal cover may be reduced by 5 mm.

2. For exposure conditions 'severe' and 'very severe', reduction of 5 mm may be made, where concrete grade is M 35 and above.

3. Nominal cover to meet specified period of fire resistance : Minimum values of nominal cover of normal weight aggregate concrete to be provided to all reinforcement including links to meet specified period of fire resistance shall be as given in Table 11.3.

TABLE 11.3 NOMINAL COVER TO MEET SPECIFIED PERIOD OF FIRE RESISTANCE

Fire resistance (h)	*Nominal cover (mm)*						
	Beams		*Slabs*		*Ribs*		*Columns*
	Simply supported	*Continuous*	*Simply supported*	*Continuous*	*Simply supported*	*Continues*	
0.5	20	20	20	20	20	20	40
1	20	20	20	20	20	20	40
1.5	20	20	25	20	35	20	40
2	40	30	35	25	45	35	40
3	60	40	45	35	55	45	40
4	70	50	55	45	65	55	40

Note : Cases that lie below the bold lines require attention to the additional measures necessary to reduce the risks of spalling.

4. Tolerance for cover : Unless specified otherwise, actual concrete cover should not deviate from the required nominal cover by $\begin{matrix}+10\\0\end{matrix}$ mm.

9. Anchoring Reinforcing Bars

(*a*) *Anchoring bars in tension*

(*i*) Deformed bars may be used without end anchorages provided development length requirement is satisfied. Hooks should normally be provided for plain bars in tension.

(*ii*) *Bends and hooks*. The anchorage value of bend shall be taken as 4 times the diameter of the bar for each 45° bend subject to a maximum of 16 times the diameter of the bar. The anchorage value of a standard U-type hook shall be equal to 16 times the diameter of the bar.

(*b*) ***Anchoring bars in compression*** : The anchoring length of straight bar in compression shall be equal to the development length of bar in compression. The projected length of hooks, bends, straight length beyond bends if provided for a bar in compression shall be considered for development length.

10. Moment and Shear Coefficients for Continuous Beams and Slabs

(*a*) Unless more exact estimates are made, for beams and slabs of uniform cross-section which support substantially uniformly distributed loads over three or more spans which do not differ by more than 15 percent of the longest, the bending moments and shear forces used in design may be obtained by using coefficients given in Table 11.4 and Table 11.5 respectively. For moment at supports where two unequal spans meet or in case where the spans are not equally loaded, the average of the two values for the negative moment at the support may be taken for design.

(*b*) Where a member is built into a masonry wall which develops only partial restraint, the member shall be designed to resist a negative moment at the face of the support of $Wl/24$ where W is the total design load and l is the effective span, or such other restraining moment as may be shown to be applicable. For such a condition, shear coefficient given in Table 11.5 at the end support may be increased by 0.05.

TABLE 11.4. BENDING MOMENT COEFFICIENTS

Type of load	Span moment		Support moment	
	Near middle of end span	*At middle of interior span*	*At support next to the end support*	*At other interior supports*
1. Dead load and imposed load (fixed)	$+\frac{1}{12}$	$+\frac{1}{16}$	$-\frac{1}{10}$	$-\frac{1}{12}$
2. Imposed load (not fixed)	$+\frac{1}{10}$	$+\frac{1}{12}$	$-\frac{1}{9}$	$-\frac{1}{9}$

Note. For obtaining the B.M. , the coefficient shall be multiplied by the total design load and effective span.

TABLE 11.5. SHEAR FORCE COEFFICIENTS

Type of load	At end support	At support next to the end support		At all other Interior supports
		Outer side	*Inner side*	
1. Dead load and imposed load (fixed)	0.4	0.6	0.55	0.5
2. Imposed load (not fixed)	0.45	0.6	0.6	0.6

Note. For obtaining the S.F., the coefficient shall be multiplied by the total design load.

11. Unit Weight of Concrete

Indian Standard Code (IS : 456 – 2000) specifies that *unless more accurate calculations are warranted*, the unit weights of plain concrete and reinforced concrete made with standard gravel or crushed natural stone aggregate may be taken as 24000 N/m^3 (24 kN/m^3) and 25000 N/m^3 (25 kN/m^3) respectively.

(B) DESIGN PROCEDURE

The design of simply supported singly reinforced beam is done in the following steps:

1. Determine the limiting depth of N.A. by Eq. 3.7 (*a*) or 3.7.

$$\frac{x_{u,\,max}}{d}=\frac{0.0035}{0.0055+0.87\,f_y/E_s}=\frac{700}{1100+0.87\,f_y}$$

2. Choose some suitable ratio of d and b, which may be in the range of 1.5 to 3.

From limit state of serviceability (deflection), after applying modification factors, compute d. With this value of d estimate D, based on nominal cover, assumed value of main steel dia (φ) and stirrup steel dia. Thus, b and D are known—these values being used to compute dead weight of beam per unit length. Compute B.M. ($M_{ur} = M_{u,\,lim.}$).

3. Determine d from Eq. 3.11

$$\frac{M_{u,\,lim}}{bd^2} = R_u = 0.36\, f_{ck}\,\frac{x_{u,\,max}}{d}\left(1 - 0.416\,\frac{x_{u,\,max}}{d}\right)$$

4. Knowing b, estimate D on the basis of main tension steel dia., shear stirrups and nominal cover requirements.

Round off the value of D to a practical value, and redetermine d on the basis of nominal cover requirements and assumed diameters of main steel bars and stirrups. Also, round off the value of b, keeping in mind nominal side cover and distance between the main tension bars.

Thus b and d are completely known. Note that this value of d will be slightly more than the one for balanced design, because of the above mentioned rounding off process.

5. Find A_{st} from Eq. 3.13 applicable for under-reinforced section :

$$A_{st} = \frac{0.5\, f_{ck}}{f_y}\left[1 - \sqrt{1 - \frac{4.6\, M_u}{f_{ck}\, bd^2}}\right] bd \qquad \ldots(3.13)$$

The reinforcement so computed should be equal to or more than the minimum reinforcement given by

$$\frac{A_{st}}{bd} = \frac{0.85}{f_y}$$

6. Curtailment of reinforcement : Curtail the reinforcement at other locations where it is no longer required from the bending moment point of view.

Let M = bending moment at the centre of the span.

M_x = bending moment at other section distant x from the centre

N = numbers of bars at the centre of the span

N_x = number of bars curtailed at the section distant x from the centre

$N - N_x$ = number of bars at the given section distant x from the centre

Since A_{st} is proportional to the bending moment, we have

$$\therefore \quad \frac{N - N_x}{N} = \frac{M_x}{M} \quad \text{or} \quad 1 - \frac{N_x}{N} = \frac{M_x}{M}$$

or

$$\frac{N_x}{N} = 1 - \frac{M_x}{M} = \frac{M - M_x}{M} \qquad \ldots(i)$$

If the beam carries U.D.L., $M = \dfrac{wL^2}{8}$ and $M_x = \left(\dfrac{wL^2}{8} - \dfrac{wx^2}{2}\right)$

$$\therefore \quad M - M_x = \frac{wx^2}{2} \quad \text{or} \quad \frac{M - M_x}{M} = \frac{wx^2}{2} \Big/ \frac{wL^2}{8} = \frac{4\,x^2}{L^2} \qquad \ldots(ii)$$

Equating (*i*) and (*ii*),

$$\frac{N_x}{N} = \frac{4\,x^2}{L^2}$$

or $$x = \frac{L}{2}\sqrt{\frac{N_x}{N}} \quad \text{...(11.2)}$$

Thus, the above expression gives the distance x from the centre where N_x bars can be curtailed out of total bars N. However, the bars have to develop full tension at the centre of span, and hence the distance x should be equal to or greater than the development length L_d of the bar which is to be curtailed. The Code also specifies that the reinforcement shall extend the point at which it is no longer required to resist flexure, for distance equal to effective depth d or 12 φ whichever is greater. Hence the bar should be curtailed at a distance $x + d$ (or $x + 12\varphi$) from the centre ; and this distance should be greater than L_d. See § 8.13 and Fig. 8.11 also for graphical method.

Another important recommendation of the Code is that in simple beams, one third of the positive reinforcement should be continued to the supports, and the diameter should be so chosen that Eq. 8.6 or 8.6(*a*) is satisfied.

However, if the bar is to be bent up, it may be safely bent up at a distance x given by Eq. 11.2.

7. Check for shear and design of shear reinforcement

Find Max. $V_u = \frac{w_u\, l}{2}$ and $\tau_v = \frac{V_u}{bd}$.

For the known value of $100\, A_{st}/bd$ at the support, find τ_c from Table 7.1.

Find $V_{uc} = \tau_c \cdot bd$ and $V_{us} = V_u - V_{uc}$

Find the spacing of stirrups from Eq. 7.16 *a*

$$s_v = \frac{0.87\, f_y\, A_{sv}\, d}{V_{us}} \quad \text{...}(i)$$

Also find spacing corresponding to minimum shear reinforcement, from Eq. 7.18(*a*)

$$s_v \le \frac{0.87\, f_y\, A_{sv}}{0.4\, b} \le \frac{2.175\, f_y\, A_{sv}}{b} \quad \text{...}(ii)$$

Also, maximum $s_v =$ Least of (0.75 d or 300 mm) ...(*iii*)

Thus, the spacing of stirrups at supports will be minimum of (*i*), (*ii*), (*iii*)

Also, determine spacing corresponding to minimum shear reinforcement at mid-span (where shear in minimum) from (*ii*) and (*iii*) above.

8. Check for development length at supports

The Code stipulates that at simple supports, where reinforcement is confined by a compressive reaction, the diameter of the reinforcement be such that

$$1.3\,\frac{M_1}{V} + L_0 \ge L_d \quad \text{...(8.6 } a)$$

where $M_1 = 0.87\, f_y\, A_{st1}\, (d - 0.416\, x_u)$ and $x_u = \frac{0.87\, f_y\, A_{st1}}{0.36\, f_{ck}\, b}$

A_{st1} = Remaining area of main reinforcement at supports

Compute L_0 from Eqs. 8.8 (*a*), (*b*), or 8.9 (*a*), (*b*), as the case be, as explained in §8.11 (Fig. 8.7). Compute L_d from Eq. 8.4. Check that Eq. 8.6 (*a*) is satisfied.

Design Example 11.1. Design of singly reinforced beam

A reinforced concrete beam is supported on two walls 750 mm thick, spaced at a clear distance of 6 m. The beam carries a super-imposed load of 9.8 kN/m. Design the beam using M 20 concrete mix and HYSD bars of Fe 415 grade.

Solution

Step 1. ***Limiting depth of N.A.***

For Fe 415 steel, $f_y = 415\ \text{N/mm}^2$. For M 20 concrete, $f_{ck} = 20\ \text{N/mm}^2$

$\frac{x_{u\,max}}{d}$ depends only on the grade of steel and is independent on grade of concrete.

For Fe 415 steel, $\frac{x_{u\,max}}{d} = \frac{700}{1100 + 0.87 \times 415} = 0.479.$

Step 2. ***Computation of design B.M.*** : The minimum depth of beam is based on limit state of serviceability requirements. For simply supported beam, $L/d = 20$. For a balanced section, $p_{t\,.\,\text{lim}} \bumpeq 0.96\%$ (Table 3.2). Also for Fe 415 steel, $f_s = 0.58 \times 415 \bumpeq 240\ \text{N/mm}^2$. Hence modification factor $F_t \bumpeq 1$.

$$\therefore \qquad d = L / 20 \approx 6000 / 20 = 300 \text{ mm.}$$

This is the minimum value of *d*. Actual value of *d*, based on bending may be more than this. In working stress design method, *d* is kept equal to $l/10$ for preliminary computations. However, the limit state design method results in smaller depth. Hence take

$$d = \frac{l}{15} = \frac{6000}{15} = 400 \text{ mm. ;} \qquad D = 400 + 30 = 430 \text{ mm (say); Keep } b = 300 \text{ mm.}$$

$$\therefore \qquad \text{Self weight of beam} = 0.43 \times 0.3 \times 1 \times 25000 = 3225 \text{ N/m}$$

$$\text{External load} = 9800 \text{ N/m.}$$

$$\therefore \qquad \text{Total } w = 9800 + 3225 = 13025 \text{ N/m}$$

$$w_u = 1.5\,w = 1.5 \times 13025 = 19538 \text{ N/m}$$

Effective span $L = l + d = 6 + 0.4 = 6.4$ m, which is smaller than centre to centre distance of 6.75 m between the supports.

$$M_u = \frac{w_u \,.\, L^2}{8} = \frac{19538(6.4)^2}{8} = 100035 \text{ N-m} \bumpeq 100.04 \times 10^6 \text{ N-mm}$$

Step 3. ***Computation of effective depth d***

$$M_u = 0.36\, f_{ck}\, \frac{x_{u\,.\,max}}{d}\left(1 - 0.416\, \frac{x_{u\,.\,max}}{d}\right) bd^2 = R_u \,.\, bd^2$$

where $R_u = 0.36 \times 20 \times 0.479\,(1 - 0.416 \times 0.479) = 2.761$

(Compare this with value given in Table 3.3)

$$\therefore \qquad d = \sqrt{\frac{M_u}{R_u.\, b}} = \sqrt{\frac{100.04 \times 10^6}{2.761 \times 300}} = 347 \text{ mm.}$$

Providing 25 mm nominal cover and using 20 mm bars with 8 mm dia strrups

$$D = 347 + 25 + 8 + 20/2 = 390 \text{ mm.}$$

However, keep D = 430 mm. Since this value is the one assumed earlier, no revised computations for self load are recommended.

$d = 430 - 25 - 8 - 20/2 = 387$ mm. Since available d (=387 mm) is more than the one required for balanced design, we will have an under-reinforced section.

Step 4. Steel reinforcement : The reinforcement, for an under-reinforced section, is given by Eq. 3.13 :

$$A_{st} = \frac{0.5 f_{ck}}{f_y}\left[1 - \sqrt{1 - \frac{4.6 M_u}{f_{ck}\, bd^2}}\right] bd = \frac{0.5 \times 20}{415}\left[1 - \sqrt{1 - \frac{4.6 \times 100.04 \times 10^6}{20 \times 300 \times (387)^2}}\right] 300 \times 387$$

$= 843.5 \quad mm^2$

No. of 20 mm dia. bars = 843.5 / 314.15 = 2.68. Provide 3-20 mm dia. bars at mid span.

Actual $A_{st} = 3 \times 314.15 = 942.5$ mm^2.

Hence actual moment of resistance available at mid span is

$$M_{ur} = 0.87 f_y \,.\, A_{st} \,.\, d\left(1 - \frac{f_y}{f_{ck}} \,.\, \frac{A_{st}}{bd}\right) \quad \ldots(3.12)$$

$$= 0.87 \times 415 \times 942.5 \times 387\left(1 - \frac{415}{20} \times \frac{942.5}{300 \times 387}\right) \simeq 110 \times 10^6 \text{ N/mm}$$

$M_{u\_lim} = 2.761 \; bd^2 = 2.761 \times 300(387)^2 = 124.1 \times 10^6$ N/mm

Since $M_{ur} < M_{u\,.\,lim}$, the design is OK.

The Code recommends that at least one third the positive moment reinforcement in simple members shall extend along the same face of the member into the support. However bend one bar upwards, at a distance x_1 from the centre of support.

B.M. at $x_1 = \frac{w_u \,.\, L}{2} x_1 - \frac{w_u x_1^2}{2}$. This should be 2/3 of the maximum B.M.

$\therefore \quad \frac{w_u \,.\, L}{2} x_1 - \frac{w_u \,.\, x_1^2}{2} = \frac{2}{3}\,\frac{w_u L^2}{8}$, which gives $x_1 = 0.211 L = 0.211 \times 6.4 = 1.35$ m

However, the bars are to be taken further by a distance d or 12 φ. Hence distance of point from the centre of support = 1.35 − 0.387 = 0.963 m , or at 0.963 − 0.2 $\simeq$ 0.75 m from the face of the support. Hence bend one bar up, at a distance of 0.75 m from the face of the support, and continue the remaining two bars into the support.

Step 5. Shear reinforcement : The critical section for shear is at a distance of d (=0.387 m) from the face of the support. Note that distance of theoretical centre of support from the face = 0.387/2 $\simeq$ 0.2 m only.

$$\therefore \quad V_{uD} = \frac{w_u \; L}{2} - w_u\left(0.2 + 0.387\right) = \frac{19538 \times 6.4}{2} - 19538\,[0.2 + 0.387] = 51053 \text{ N}$$

$$\therefore \quad \tau_v = \frac{V_{uD}}{bd} = \frac{51053}{300 \times 387} \approx 0.44 \text{ N/mm}^2$$

$$\frac{100 A_{st}}{bd} \text{ (at support)} = \frac{100}{300 \times 387}(314.15 \times 2) = 0.54\%.$$

Hence from Table 7.1, $\quad \tau_c \approx 0.5$ N/mm^2.

Since $\tau_v < \tau_c$, no shear reinforcement is necessary. However, as per Code recommendations, provide nominal shear reinforcement, given by the expression

$$\frac{A_{sv}}{b \,.\, s_v} > \frac{0.4}{0.87 f_y} \quad \text{or} \quad s_v = \frac{2.175\ A_{sv} \,.\, f_y}{b}.$$

Choosing 8 mm φ bars, $A_{sv} = 2 \times \frac{\pi}{4}(8)^2 = 100.5\ \text{mm}^2$

$$\therefore \qquad s_v = \frac{2.175 \times 100.5 \times 415}{300} = 302\ \text{mm}.$$

Max. s_v = least of (0.75 d or 300 mm) = 0.75 d = 0.75 × 387 = 290 mm.

Hence provide 8 mm φ 2-lgd stirrups @ 290 mm c/c throughout the length of the beam. Provide 2-10 mm φ holding bars at the top.

Step 6. Check for development length at the ends. : Since the beam is supported on wall, the compressive reaction will confine the reinforcement. Hence

$$L_d \le 1.3\ \frac{M_1}{V} + L_0$$

Only two bars are available at the supports.

$$x_u = \frac{0.87 f_y \,.\, A_{st1}}{0.36\ f_{ck} \,.\, b} = \frac{0.87 \times 415\ (2 \times 314.15)}{0.36 \times 20 \times 300} = 105\ \text{mm}$$

$$\therefore \quad M_1 = 0.87\ f_y\, A_{st1}(d - 0.416\ x_u) = 0.87 \times 415(2 \times 314.15)(387 - 0.416 \times 105)$$

$$\simeq 77.88 \times 10^6\ \text{N-mm}$$

$$V_u = \frac{w_u \,.\, L}{2}\ \text{N} = \frac{19538 \times 6.4}{2} = 62522$$

$$L_d = \frac{0.87 f_y}{4\tau_{bd}} \times \varphi = \frac{0.87 \times 415}{4 \times 1.2 \times 1.6} \times \varphi = 47\varphi$$

$= 47 \times 20 = 940$ mm

L_0 = sum of anchorage beyond centre line of support and anchorage value of hook.

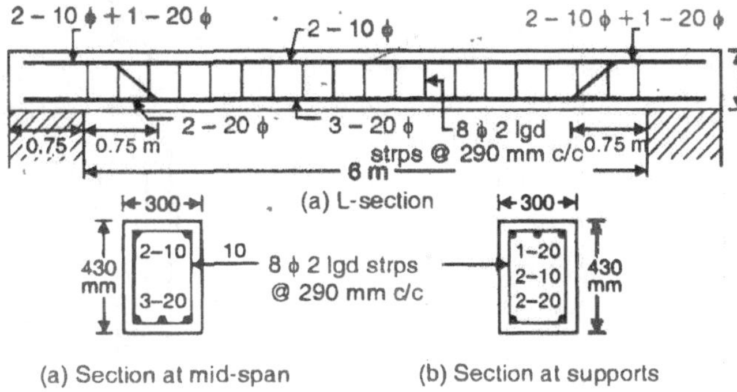

FIG 11.6

If no hook is provided, $L_0 = \frac{L_s}{2} - x' = \frac{750}{2} - 40 = 335$ mm,

where x' = side cover assumed to be 40 mm. The Code further recommends that each bar should be extended by a distance of $L_d/3 = 940/3 = 313$ mm beyond the face of the support. Available distance, beyond the face of support $= L_0 + \frac{L_s}{2} = 335 + \frac{750}{2} = 710$ mm, which is more than $L_d / 3$. Hence no hook is required. Hence $L_0 = 335$ mm.

$$\therefore\ 1.3\ \frac{M_1}{V_u} + L_0 = \left(1.3 \times \frac{77.88 \times 10^6}{62522}\right) + 335 = 1954\ \text{mm} > L_d.\ \text{Hence O.K.}$$

Step 7. Details of Reinforcement etc. The details of reinforcement etc. are shown in Fig. 11.6.

11.2. DESIGN OF LINTEL BEAMS

Lintels are provided over the openings of doors, windows, almirahs, etc. Generally, they support the load of the wall over it, and sometimes also the live load are transferred by the sub-roof of the room. Following five cases may arise from point of view of distribution of load over the lintels :

(1) When the length of wall on each side is more than half the effective span of the lintel.

(2) When the length of wall to one side is less than half the effective span.

(3) When the length of wall on both sides is less than half the effective span.

(4) When there are openings over the lintel.

(5) When there is load carrying slab over the lintel.

Case I. ***When the length of wall on each side is more than half the effective span of the lintel.***

This is most general case. Because of arch-action in the masonry, all the load of the wall above the lintel is not transferred to the lintel. It is assumed that the load transferred is in the form of equilateral triangle, and the load on the lintel is equal to the weight of the masonry in the triangular portion, as shown in Fig 11.7(*a*). If, however, the height of the wall above the lintel is insufficient (*i.e.* if the apex of the triangle falls above the top of the wall), whole of the rectangular load above the lintel is taken to act on the lintel, as shown in Fig. 11.7(*b*). Let :

H = height of wall above the lintel.

l = actual opening.

L = effective length of the opening.

h = effective height of masonry, above the lintel.

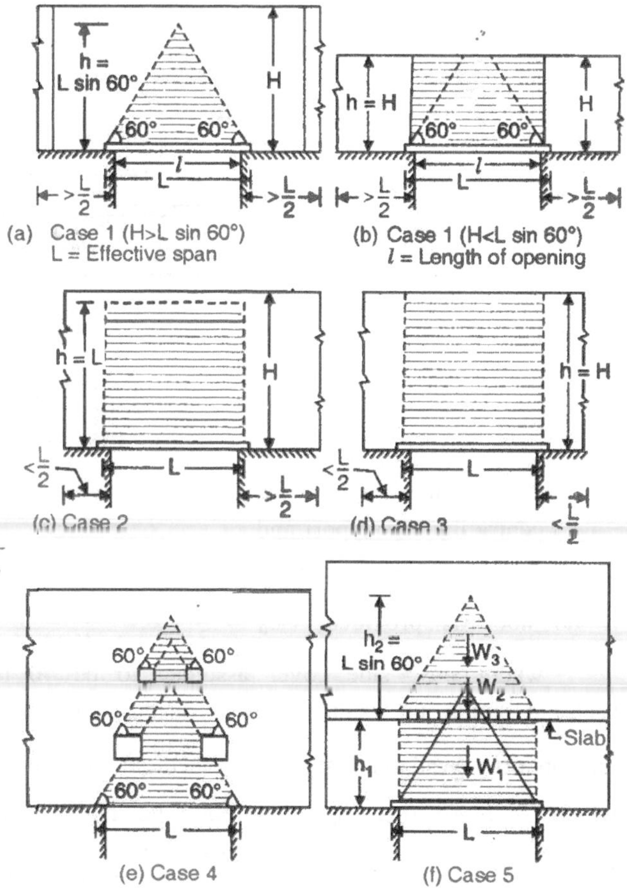

FIG. 11.7. VARIOUS CASES OF LOADING ON LINTELS.

Thus, $h = L \sin 60° = L \cdot \frac{\sqrt{3}}{2} = \frac{L}{2}\sqrt{3}.$

$\therefore$ Area of triangle

$$= \frac{1}{2} \times L \times \frac{L}{2}\sqrt{3} = \frac{L^2}{4}\sqrt{3}.$$

Case 2. ***When the length of the wall to one side is less than half the effective span.***

Fig 11.7 (*c*) shows the situation where the length of wall to one side is less than half the effective span, but the length to the other side is more than half the effective span. In that case, the load transferred to the lintel will be equal to the weight of masonry contained in the rectangle of height h equal to the effective span.

Case 3. *When the length of the walls to each side is less than half the effective span.*

This is shown in Fig. 11.7(*d*). The load acting on the lintel will be equal to the weight of the masonry contained in the rectangle of height h equal to the full height H of the wall.

Case 4. *When there are openings on the lintel.*

If there are openings, due to the provision of ventilators etc. and if these openings are intersected by the 60° lines, the loading will be calculated by allowing dispersion lines at 60° from the top edges of the openings, as shown in Fig. 11.7(*e*). The total load on the lintel will be equal to the weight of the masonry contained in the shaded area.

Case 5. *When there is load carrying slab falling within the dispersion triangle [Fig. 11.7(f)].*

If the roof slab is provided at a level well above the apex of the dispersion triangle, the uniformly distributed load carried by the slab is *not* transferred to the lintel. If, however, the slab intersects the dispersion triangle, three types of loads are transferred to the lintel:

(*i*) Load W_1 due the weight of the masonry contained in the rectangle of height h_1 equal to the height of the slab above the lintel.

(*ii*) Load W_2 carried by the slab, in a length L.

(*iii*) Load W_3 due to the weight of masonry contained in the equilateral triangle above the slab.

The design of the lintel is similar to that of R.C. beam discussed in § 11.1. The width of lintel is normally kept equal to width of the wall.

The max. bending moment M_1 due to the triangular load is taken as $M_1 = W L/6$ where W = total weight of masonry contained in the equilateral triangle. The max. bending moment M_2 due the uniformly distributed self-weight w of the lintel is : $M_2 = w L^2/8$

$$\therefore \qquad \text{Total } M = \frac{W L}{6} + \frac{wL^2}{8}$$

The maximum shear will act at the edge of the opening and is given by

$$F = \frac{W}{2} + \frac{wl}{2}$$

Design Example 11.2. Design of lintel beam

Design a R.C. lintel beam to be provided over an opening of 2 m wide. The height of masonry above the lintel is 3 m. The opening is centrally located in a long wall, 300 mm thick. The unit weight of masonry may be taken as 19 kN/m³. Use M 20 concrete and Fe 415 steel.

Solution

Step 1. Limiting depth of N.A.

For Fe 415 steel, $f_y = 415\ \text{N/mm}^2$. For M 20 concrete, $f_{ck} = 20\ \text{N/mm}^2$.

$$\frac{x_{u,\,max}}{d} = \frac{700}{1100 + 0.87 \times 415} = 0.479$$

Step 2. Computation of design B.M.

From deflection point of view, neglecting modification factor, $L/d = 20$

$\therefore\ d = L/20 \simeq 2200/20 = 110$ mm. Actual d from the flexure point of view will be much less. Hence assume $D = 150$ mm for computation of dead weight.

Width of lintel = width of wall = 300 mm

Self weight of lintel, per m length $= 0.15 \times 0.3 \times 1 \times 25000 = 1125\ \text{N} = w$

Effective span will be the minimum of the following.

(*i*) Centre to centre of bearing : Providing a bearing of 200 mm,

$$L = 2 + 0.2 = 2.2 \text{ m}$$

(*ii*) Clear span + Effective depth : Assuming $d = 120$ mm,

$$L = 2 + 0.12 = 2.12 \text{ m}$$

Hence adopt $L = 2.12$ m

Height of equilateral triangle, assuming 60° dispersion

$$= L \sin 60° = \frac{L}{2}\sqrt{3} = \frac{2.12}{2}\sqrt{3} = 1.84 \text{ m}$$

This is less than the height of the wall above the lintel. Hence load on the lintel will be equal to the weight of the masonry contained in the triangular portion.

$$\therefore \qquad W = \left(\frac{1}{2} \times 2.12 \times 1.84\right)(0.3 \times 19000) = 11118 \text{ N}$$

$$\therefore \qquad \text{B.M.} = \frac{WL}{6} + \frac{wL^2}{8} = \frac{11118 \times 2.12}{6} + \frac{1125\,(2.12)^2}{8} = 4560.4 \text{ N-m}$$

$$\therefore \qquad M_u = 1.5\,M = 1.5 \times 4560.4 \simeq 6841 \text{ N-m} = 6841 \times 10^3 \text{ N-mm}$$

Step 3. Computation of effective depth

$$M_u = 0.36\, f_{ck} \frac{x_{u,\,max}}{d}\left(1 - 0.416\,\frac{x_{u,\,max}}{d}\right) = R_u\, bd^2$$

where $\qquad R_u = 0.36 \times 20 \times 0.479\,(1 - 0.416 \times 0.479) = 2.761$

$$\therefore \qquad d = \sqrt{\frac{M_u}{R_u\, b}} = \sqrt{\frac{6841 \times 10^3}{2.761 \times 300}} = 90.9 \text{ mm}$$

From deflection point of view $d = 110$ mm, as found earlier.

Hence keep $D = 150$ mm. Using 10 mm φ main bars, 8 mm φ links and 25 nominal cover, available $d = 150 - 25 - 8 - 10/2 = 112$ mm.

Step 4. Determination of steel reinforcement

Since actual d is more than 90.9 mm, we will have under-reinforced section for which

$$A_{st} = \frac{0.5\, f_{ck}}{f_y}\left[1 - \sqrt{1 - \frac{4.6\, M_u}{f_{ck}\, bd^2}}\right] bd = \frac{0.5 \times 20}{415}\left[1 - \sqrt{1 - \frac{4.6 \times 6841 \times 10^3}{25 \times 300 \times 112^2}}\right] \times 300 \times 112$$

$\approx 150\ \text{mm}^2$

Minimum tension steel is given by

$$(A_{st})_{min} = \frac{0.85}{f_y} \times bd = \frac{0.85}{415} \times 300 \times 112 = 68.9 \text{ mm}^2$$

$$\text{No. of 10 mm } \varphi \text{ bars} = \frac{150}{\frac{\pi}{4}(10)^2} = 1.91.$$

Provide 2 bars. Actual A_{st} provided $= 2 \times \frac{\pi}{4}(10)^2 = 157.1 \text{ mm}^2$

Since there are only 2 bars, no curtailment or bending the bar upwards is recommended.

Step 5. Check for shear and design of shear reinforcement

$$V = \frac{W}{2} + \frac{wl}{2} = \frac{11118}{2} + \frac{1125 \times 2}{2} = 6684 \text{ N}$$

$$\tau_v = \frac{V_u}{bd} = \frac{6684 \times 1.5}{300 \times 112} = 0.298 \text{ N/mm}^2$$

At supports $\quad \dfrac{100\, A_s}{bd} = \dfrac{100 \times 157.1}{300 \times 112} = 0.468\%$

Hence from Table 7.1, $\tau_c = 0.36 + \dfrac{0.48 - 0.36}{0.25}(0.468 - 0.25) = 0.465 \text{ N/mm}^2$

Since $\tau_c > \tau_v$, no shear reinforcement is necessary. However provide nominal shear reinforcement, given by expression

$$s_v \le \frac{2.175\, A_{sv} f_y}{b}$$

Choosing 6 mm φ mild steel bars, $A_{sv} = 2 \times \frac{\pi}{4}(6)^2 = 56.55 \text{ mm}^2$

$$\therefore \quad s_v \le \frac{2.175 \times 56.55 \times 250}{300} \le 102.4 \text{ mm}$$

subject to a maximum of least of $0.75\, d$ ($= 0.75 \times 112 = 84$ mm) or 300 mm.

However, since the structure is of minor importance, provide 6 mm φ 2-lgd stirrups of mild steel @100 mm c/c. Provide 2–8 mm φ holding bars at top.

Step 6. Check for development length at the ends : Since the lintel beam is supported on wall, the compressive reaction will confine the reinforcement.

Hence $\quad L_d \le 1.3 \dfrac{M_{1u}}{V_u} + L_0$

At supports $\quad x_u = \dfrac{0.87 f_y A_{st1}}{0.36 f_{ck}\, b} = \dfrac{0.87 \times 415 \times 157.1}{0.36 \times 20 \times 300} = 26.3 \text{ mm}$

$$\therefore \quad M_{1u} = 0.87 f_y A_{st1}(d - 0.416\, x_u)$$

$$= 0.87 \times 415 \times 157.1\,(112 - 0.416 \times 26.3) = 5.732 \times 10^6 \text{ N-mm}$$

$$V_u = 1.5 \times 6684 \text{ N} = 10026 \text{ N}$$

$$L_d = 47\, \varphi = 47 \times 10 = 470 \text{ mm}$$

Taking the bars straight into supports, without any hook or bend, and providing side cover $x' = 25$ mm, we have

$$L_0 = \frac{L_s}{2} - x' = \frac{200}{2} - 25 = 75 \text{ mm}$$

$$\therefore \quad 1.3\frac{M_{1u}}{V_u} + L_0 = 1.3 \times \frac{5.732 \times 10^6}{10026} + 75 \simeq 818 > L_d \quad . \text{ Hence OK.}$$

Step 7. Details of reinforcement etc

The details of reinforcement etc. are shown in Fig. 11.8.

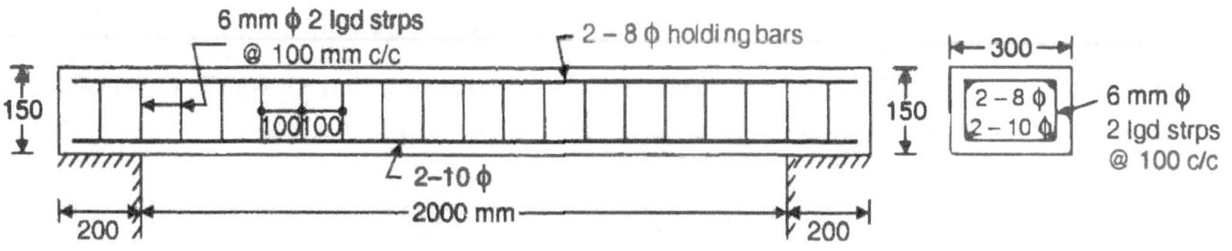

FIG. 11.8

Design Example 11.3 : Design of lintel beam : *Fig 11.9 (a) shows the cross-section of a wall of room, 5 m wide and 18 m long from inside. Design the lintel for the window having a opening of 2 m. The reinforced concrete slab 16 cm thick, may be assumed to transfer half the load to the wall shown. Use the following data for the design.*

(i) Weight of lime concrete terracing $= 19\ kN/m^3$

(ii) Weight of masonry $= 20\ kN/m^3$

(iii) Weight of reinforced concrete $= 25\ kN/m^3$

(iv) Super-imposed load on the slab $= 1.6\ kN/m^2$

The window may be assumed to be centrally located along the length of the room. Use M 20 concrete and Fe 415 steel.

Step 1. Loading and bending moment : Let us assume the overall depth of the lintel to be 300 mm, and a bearing of 300 mm on either side. Hence effective span of lintel will be equal to $2 + 0.3 = 2.3$ m. Assuming the dispersion to be along 60° lines, the height of the load triangle, measured above the top of the lintel $= 2.3 \sin 60° = 2.3\sqrt{3}/2 \simeq 2$ m, which is more than the height (= 1.5 m) of the slab above the lintel. Hence the load of the slab will also to transferred to the lintel.

Fig 11.9 (*b*) shows the elevation, showing all the heights. If we construct equilateral triangle above the top of the slab, its apex will fall very much above the top of the parapet wall. Hence the weight of the whole wall above the lintel will be transferred to the lintel. Thus, the

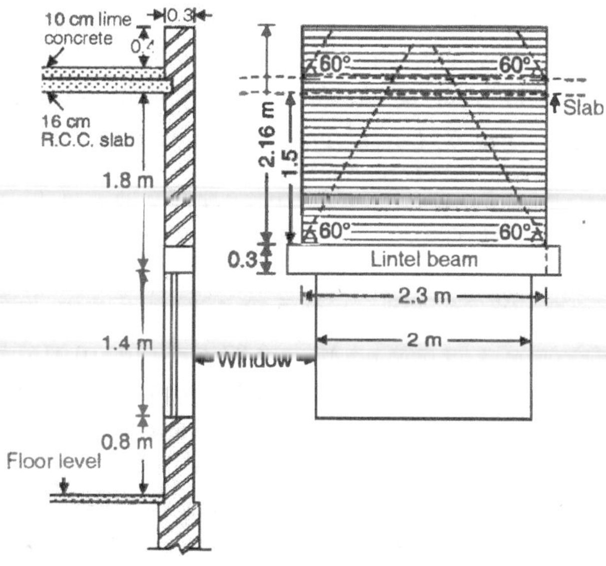

(a) Section (b) Elevation

FIG. 11.9

loads per metre length of the lintel will consist of the following.

(*i*) *Weight of the wall*

Weight of wall per metre run $= 0.3 \times 2.16 \times 20000 = 12960$ N

(*ii*) *Load transferred by the slab*

Consider 1 m strip of the slab of 5 m span

Live load on the strip $= \frac{1}{2}(5 \times 1) \times 1600 \qquad = 4000$ N

Dead weight of 16 cm thick slab $= \frac{1}{2}(5 \times 1 \times 0.16)\, 25000 \qquad = 10000$ N

Dead weight of lime concrete $= \frac{1}{2}(5 \times 1 \times 0.1)\, 19000 \qquad = 4750$ N

Total load per metre run of the lintel $= 18750$ N

(*iii*) *Self weight of lintel*

Weight per metre run $= (0.3 \times 0.3 \times 1)\, 25000 = 2250$ N

Total $\quad w = 12960 + 18750 + 2250 = 33960$ N/m

$\therefore \quad w_u = 1.5\, w = 1.5 \times 33960 = 50940$ N/m

$$M_u = \frac{w_u L^2}{8} = \frac{50940\,(2.3)^2}{8} = 33684 \text{ N-m} = 33.684 \times 10^6 \text{ N-mm}$$

Step 2. Limiting depth of N.A.

For Fe 415 steel, $\dfrac{x_{u,\,max}}{d} = \dfrac{700}{1100 + 0.87 \times 415} = 0.479$

Step 3. Computation of effective depth

$$M_u = 0.36\, f_{ck} \frac{x_{u,\,max}}{d}\left(1 - 0.416 \frac{x_{u,\,max}}{d}\right) bd^2 = R_u\, bd^2$$

where $\quad R_u = 0.36 \times 20 \times 0.479\,(1 - 0.416 \times 0.479) = 2.761$

$$\therefore \quad d = \sqrt{\frac{M_u}{R_u\, b}} = \sqrt{\frac{33.684 \times 10^6}{2.761 \times 300}} \simeq 202 \text{ mm}$$

Keep $D = 300$ mm. Providing 25 mm nominal cover, 12 mm φ main bars and 8 mm dia. stirrups, available $d = 300 - 25 - 8 - 12/2 = 261$ mm.

Step 4. Computation of reinforcement

Since actual d is more than 202 mm, we will have under-reinforced section for which

$$A_{st} = \frac{0.5\, f_{ck}}{f_y}\left[1 - \sqrt{1 - \frac{4.6\, M_u}{f_{ck}\, bd^2}}\right] bd = \frac{0.5 \times 20}{415}\left[1 - \sqrt{1 - \frac{4.6 \times 33.684 \times 10^6}{20 \times 300\,(261)^2}}\right] \times 300 \times 261$$

$$= 400 \text{ mm}^2$$

$$\therefore \text{ No. of 12 mm φ bars} = \frac{400}{\frac{\pi}{4}(12)^2} = 3.54$$

However, provide 4-12 mm φ bars. Actual $A_{st} = 4 \times \frac{\pi}{4}(12)^2 = 452.4 \text{ mm}^2$

Hence actual moment of resistance available at mid span is

$$M_{ur} = 0.87\, f_y\, A_{st}\, d\left(1 - \frac{f_y}{f_{ck}} \frac{A_{st}}{bd}\right) = 0.87 \times 415 \times 452.4 \times 261 \left(1 - \frac{415}{20} \times \frac{452.4}{300 \times 261}\right)$$

$$= 37.52 \times 10^6 \text{ N-mm}$$

$$M_{u,\,lim} = 2.761\, bd^2 = 2.761 \times 300\,(261)^2 = 56.42 \times 10^6 \text{ N-mm}$$

Since $M_{ur} < M_{u,\,lim}$, the design is OK.

Since the Code recommends to provide atleast $\frac{1}{3} A_{st}$ at the upper face of the beam, at supports, bend two bars upwards, at a distance x_1, from the centre of the support.

B.M. at $x_1 = \frac{w_u L}{2} x_1 - \frac{w_u x_1^2}{2}$. This should be half of max. B.M.

$$\therefore \quad \frac{w_u L}{2} x_1 - \frac{w_u x_1^2}{2} = \frac{1}{2} \frac{w_u L^2}{8}, \text{ which gives } x_1 = 0.1464\, L = 0.1464 \times 2.3 = 0.34 \text{ m}$$

However, the bars are to be taken further by a distance of d (= 261 mm) or 12 φ (= 144 mm). Hence distance of point of bending from the centre of support = 0.34 − 0.261 ≃ 0.18 m or = 0.18 − 0.3/2 = 0.03 m from the face of the support. Continue remaining two bars into the support.

Step 5. Shear reinforcement : The critical section for shear is at a distance d from the face of the support. However, since the span is very small, we can take the critical section at the face of the support itself, without much error.

$$\therefore \quad V_u = \frac{w_u\, l}{2} = \frac{50940 \times 2}{2} = 50940 \text{ N}$$

$$\tau_v = \frac{V_u}{bd} = \frac{50940}{300 \times 261} = 0.65 \text{ N/mm}^2, \text{ which is less than } \tau_{c,\,max} \,(= 2.8)$$

At supports, $\frac{100\, A_s}{bd} = \frac{100\,(\frac{1}{2} \times 452.4)}{300 \times 261} = 0.29\,\%$

Hence from Table 7.1, $\tau_c = 0.36 + \frac{0.48 - 0.36}{0.5 - 0.25}(0.29 - 0.25) \simeq 0.38 \text{ N/mm}^2$

$$\therefore \quad V_{uc} = \tau_c\, bd = 0.38 \times 300 \times 261 = 29754 \text{ N}$$

$V_{us} = V_u - V_{uc} = 50940 - 29754 = 21186$ N; Using 8 mm φ stirrups,

Spacing of stirrup is $s_v = \frac{0.87\, f_y \,.\, A_{sv}\, d}{V_{us}}$ where $A_{sv} = 2\, \frac{\pi}{4} (8)^2 = 100.53 \text{ mm}^2$.

$$s_v = \frac{0.87 \times 415 \times 100.53 \times 261}{21186} = 447 \text{ mm}$$

For nominal stirrups, $s_v = \frac{2.175\, A_{sv}\, f_y}{b} = \frac{2.175 \times 100.53 \times 415}{300} = 302 \text{ mm}$

subject to maximum of least of 0.75 d (= 0.75 × 261 = 196) or 300 mm.

However, provide 8 mm φ 2-lgd stirrups @ 200 mm c/c throughout the length of the lintel. Provide 2 − 8 φ holding bars at the top.

Step 6. Check for development length at the ends

Since only half the bars are available at supports,

$$x_u = \frac{0.87\, f_y\, A_{st1}}{0.36\, f_{ck}\, b} = \frac{0.87 \times 415 \times (0.5 \times 452.4)}{0.36 \times 20 \times 300} = 37.8 \quad \text{mm}$$

$$M_{1u} = 0.87\, f_y\, A_{st1}(d - 0.416\, x_u) = 0.87 \times 415\,(0.5 \times 452.4)\,(261 - 0.416 \times 37.8)$$

$$= 20.03 \times 10^6 \text{ N-mm}$$

$$V_u = \frac{w_u \,.\, L}{2} = \frac{50940 \times 2.3}{2} = 58581 \quad \text{N}$$

Taking the bars straight into supports without provision of any hook or bend, and keeping 25 mm side cover,

$$L_0 = \frac{L_s}{2} - x' = \frac{300}{2} - 25 = 125 \quad \text{mm}$$

$$L_d = 47\, \varphi = 47 \times 12 = 564 \quad \text{mm}$$

$$1.3\, \frac{M_{1u}}{V_u} + L_0 = 1.3 \times \frac{20.03 \times 10^6}{58581} + 125 = 444 + 125 = 569 \text{ mm} \simeq L_d$$

Hence the Code requirement is just satisfied. However, provide 90° bend at the ends, thus increasing L_0.

Step 7. Details of reinforcement etc

The details of reinforcement etc are shown in Fig. 11.10

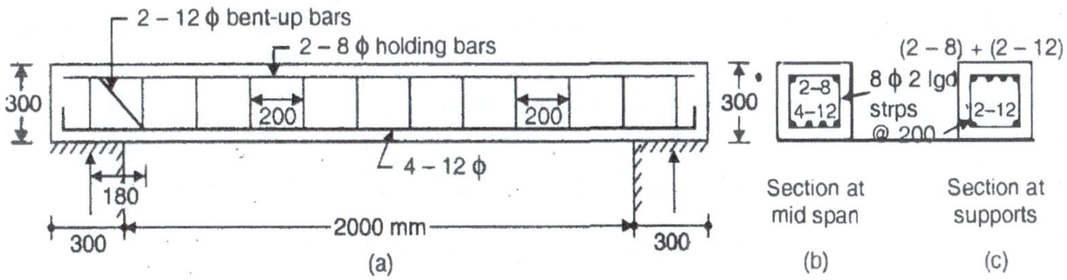

FIG. 11.10

11.3. DESIGN OF CANTILEVER BEAM

When a cantilever beam or slab is subjected to downward loads, it bends downwards with its convexity upwards. Due to this, the upper fibres of the section are subjected to tensile stress and the lower fibres are subjected to compressive stress. Hence the main reinforcement is provided at the *upper face,* and not at the lower face. The B.M. varies from a zero value at the free end to maximum value at the fixed end (support). Hence the depth of the section and the steel reinforcement is varied as per requirements of bending moment. In order that the whole cantilever does not overturn, it should be properly fixed in the wall. The reinforcement should be embedded in the support for a distance of L_d from the edge of the support.

When curtailment of bar is done, it should be kept in mind that the embedment of the bar is not less than L_d from the point it is needed the most, nor less than 12 dia. (or *d*) from the point it can be spared from bending moment point of view.

Effective length : The effective length of a cantilever shall be taken as its length to the face of the support plus half the effective depth except where it forms the end of a continuous beam where the length to the centre of the support shall be taken.

Design Example 11.4. Design of a cantilever beam.

A cantilever beam projects 2.5 m beyond the fixed end, and carries a super-imposed load of 12 kN per metre run. Design the cantilever, using M 20 concrete and Fe 415 steel. The width of support is 350 mm.

Solution. ***Step 1. Design constants and limiting depth of N.A.***

For Fe 415 steel, $f_y = 415 \text{ N/mm}^2$. For M 20 concrete, $f_{ck} = 20 \text{ N/mm}^2$.

$$\frac{x_{u,max}}{d} \text{ (for Fe 415 steel)} = \frac{700}{1100 + 0.87 \times 415} = 0.479$$

$$R_u = 0.36 f_{ck} \cdot \frac{x_{u,max}}{d}\left(1 - 0.416 \frac{x_{u,max}}{d}\right) = 0.36 \times 20 \times 0.479 (1 - 0.416 \times 0.479) = 2.761$$

Step 2. ***Computation of design B.M. and S.F.***

The minimum depth of beam is based on limit state of serviceability requirement.

For a cantilever, basic $L/d = 7$. Assuming $p_t = 0.7$ %, we get modification factor $F_t \simeq 1.1$ from Fig. 11.1. Also, assuming effective depth $d \simeq 350$ mm, effective span $L = 2.5 + 0.35/2 = 2.675$ m. Hence, $d = L/7 \times 1.1 \simeq 2675/7 \times 1.1 \simeq 348$ mm

However, keep D = 400 mm at fixed end and 200 mm at free end. Keep width b = 250 mm.

Self weight of beam $= \frac{1}{2}(0.4 + 0.2) \times 0.25 \times 2.5 \times 25000 = 4688$ N acting at $\frac{0.4 + 2 \times 0.2}{0.4 + 0.2} \times \frac{2.5}{3} = 1.11$ m from fixed end.

$$\therefore \quad M = (4688 \times 1.11) + \frac{12000(2.5)^2}{2} = 42704 \text{ N-m}$$

$$\therefore \quad M_u = 1.5M = 1.5 \times 42704 = 64056 \text{ N-m} \simeq 64.06 \times 10^6 \text{ N-mm}$$

S.F. V at the edge of support $= 4688 + (12000 \times 2.5) = 34688$ N

$\therefore \quad V_u = 1.5\, V = 1.5 \times 34688 = 52032$ N

Step 3. Computation of effective depth: For a balanced section, $M_u = R_u bd^2 = 2.761\, bd^2$

$$\therefore \quad d = \sqrt{\frac{M_u}{2.761 b}} = \sqrt{\frac{64.06 \times 10^6}{2.761 \times 250}} = 305 \text{ mm.}$$

However, keep D = 400 mm, so that using 16 mm dia. bars, and 8 mm dia stirrups with 25 mm nominal cover, $d = 400 - 25 - 8 - 8 = 359$ mm. This is more than the one required for the balanced section and also more than that required from deflection point of view. Hence, we will have an under reinforced section. Keep D = 200 mm at free end.

Step 4. Steel Reinforcement

The reinforcement for an under-reinforced section is given by :

$$A_{st} = \frac{0.5 f_{ck}}{f_y}\left[1 - \sqrt{1 - \frac{4.6 M_u}{f_{ck} bd^2}}\right] bd = \frac{0.5 \times 20}{415}\left[1 - \sqrt{1 - \frac{4.6 \times 64.06 \times 10^6}{20 \times 250(359)^2}}\right] \times 250 \times 359$$

$= 569 \text{ mm}^2$

∴ No. of 16 mm dia. bars = 569 / 201.06 = 2.83. Hence provide 3 bars of 16 mm dia.

Actual A_{st} = 3 × 201.06 = 603.2 mm^2. Hence actual moment of resistance available is

$$M_{ur} = 0.87 f_y \cdot A_{st}\, d\left(1 - \frac{f_y}{f_{ck}} \cdot \frac{A_{st}}{bd}\right) = 0.87 \times 415 \times 603.2 \times 359\left(1 - \frac{415}{20} \times \frac{603.2}{250 \times 359}\right)$$

$$= 67.28 \times 10^6 \text{ N-mm} > M_u.$$

Also, $M_{u,\,lim} = 2.761\, bd^2 = 2.761 \times 250\,(359)^2 = 88.96 \times 10^6$ N-mm.

Since $M_{ur} < M_{u,\,lim}$, the design is OK.

Curtailment of reinforcement : Since the B.M. decreases to zero at the end, let us curtail one bar at a distance x mm from free end.

Effective depth at free end ≈ 200 − 25 − 8 − 8 = 159 mm

Effective depth d_x from x free end $= 159 + \left(\frac{359 - 159}{2500}\right) x = (159 + 0.08\,x)$ mm.

Assuming B.M.D. to be parabolic, the B.M. at this section may be *approximately* taken to be equal to $\left(\frac{x}{2500}\right)^2 \times 64.06 \times 10^6 \simeq 10.25\, x^2$ N-mm.

Area of remaining 2 bars = 2 × 201.06 ≈ 402.1 mm^2

$$M_{ur} = 0.87 f_y \cdot A_{st}\left[d_x - \frac{f_y}{f_{ck}} \cdot \frac{A_{st}}{b}\right] = 0.87 \times 415 \times 402.1\left[(159 + 0.08\,x) - \frac{415}{20} \times \frac{402.1}{250}\right]$$

$$= 145178\,[125.6 + 0.08\,x] = 18.234 \times 10^6 + 11614\,x$$

$$\therefore \quad 10.25\,x^2 = 18.234 \times 10^6 + 11614\,x$$

or $\quad x^2 - 1134\,x - 1.780 \times 10^6 \approx 0.$ From which $x \approx 2016$ mm

$$d_x = 159 + 0.08 \times 2016 \simeq 320 \text{ mm.}$$

Hence one bar can be curtailed at a distance = 2500 − 2016 + 320 ≈ 804 mm away from the face of fixed end. However, minimum distance at which any bar can be curtailed from the fixed end is L_d so that all the bars can develop their full tensile strength at the fixed end.

Now $\quad L_d = \frac{0.87 f_y}{4\tau_{bd}} \cdot \varphi = \frac{0.87 \times 415}{4 \times 1.6 \times 1.2}\varphi = 47\,\varphi = 47 \times 16 = 752$ mm < 804 mm.

Hence one bar can be curtailed at a distance of 804 mm from the fixed end. However, curtail one bar at 810 mm from the fixed end. Continue the remaining two bars upto the end. At the bottom, provide 2 − 10 φ bar as holding bars at the bottom face.

Step 5. Shear reinforcement : V_u = 52032 N

Nominal shear stress $\tau_v = \dfrac{V_u - \dfrac{M_u}{d}\tan\beta}{bd}$, where $\tan\beta = \dfrac{400 - 200}{2500} = 0.08$. Hence,

$$\tau_v = \frac{52032 - \dfrac{64.06 \times 10^6}{359} \times 0.08}{250 \times 359} = 0.42 \text{ N/mm}^2. \qquad p_t = \frac{100 A_s}{bd} = \frac{100 \times 603.2}{250 \times 359} = 0.67\%$$

$\tau_c \simeq 0.5$ N/mm^2 > τ_v.

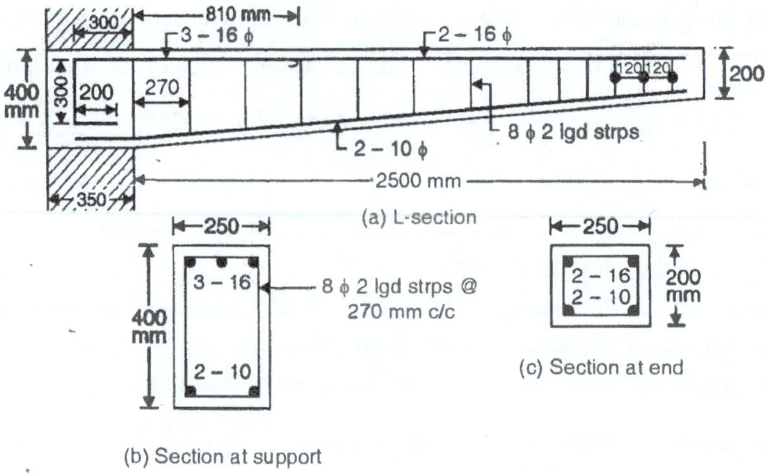

FIG. 11.11

Hence provide only nominal shear reinforcement, given by Eq. 7.18 *a*.

$$s_v = \frac{2.175\, A_{sv} \,.\, f_y}{b}$$

Using 8 mm φ 2 lgd. stirrups, $A_{sv} = 2 \times 50.3 = 100.6$.

Hence, $$s_v = \frac{2.175 \times 100.6 \times 415}{250} = 363 \text{ mm.}$$

However, max. $s_v = 0.75\, d = 0.75 \times 359 \simeq 270$ mm. Hence provide 8 mm φ, 2-lgd stirrups @ 270 mm c/c, at the supports. Reduce this spacing gradually, to a value $= 0.75 \times 159 \simeq 120$ mm at the ends.

Step 6. Embedment of reinforcement in the support.

In order to develop full tensile strength at the face of the support, each of the three bars must be embedded into the support by a length equal to $L_d = 752$ mm. This could be best achieved by providing two bends of 90°, as shown in Fig. 11.11, where the anchorage value of each 90° bend $= 8\varphi = 8 \times 16 = 128$ mm.

Step 7. Details of Reinforcement : Shown in Fig. 11.11.

11.4. DESIGN OF ONE WAY SLAB

(*a*) LONGITUDINAL AND TRANSVERSE REINFORCEMENTS

The analysis of slab spanning in one direction is done by assuming it to be a beam of 1 m width. The reinforcement etc. are calculated for 1 m width and bars are distributed accordingly. Thus, $d = \sqrt{\dfrac{M_u}{1000\, R_u}}$, where M_u is in N-mm ...(3.14)

Reinforcement for an under-reinforced section is given by

$$A_{st} = \frac{0.5 f_{ck}}{f_y}\left[1 - \sqrt{1 - \frac{4.6\, M_u}{f_{ck}\,(1000\, d^2)}}\right] 1000\, d \qquad \text{...(3.15)}$$

The spacing of the main reinforcement is given by $s = \dfrac{1000\, A_\phi}{A_{st}}$ mm ...(3.16)

In addition to the main reinforcement (called longitudinal reinforcement), *transverse reinforcement* (also called distribution reinforcement) is also provided in a direction at right angles to the span of the slab. The transverse reinforcement is provided to serve the following purposes:

(*i*) It distributes the effects of point load on the slab more evenly and uniformly.
(*ii*) It distributes the shrinkage and temperature cracks more evenly.
(*iii*) It keeps the main reinforcement in position.

(*b*) BASIC RULES FOR DESIGN (IS : 456 – 2000)

1. ***Effective span***. The basic rules for the effective span of a slab are the *same* as given by beams (clause A-1 of § 11.1). Thus, in the case of freely supported slab, the effective span is taken equal to the distance between centre to centre of supports, or the clear distance between the supports plus the effective depth of the slab whichever is less.

2. ***Control of deflection***. The same rules apply, as given for beams, in clause A-2 of § 11.1. Thus, the *basic* values of span to effective depth ratios, for span upto 10 m are as follows:

Cantilever	7
Simply supported	20
Continuous	26

3. ***Minimum reinforcement***. The mild steel reinforcement in either direction in slabs shall not be less than 0.15% of the total cross-sectional area. However, this value can be reduced to 0.12% when high strength deformed bars or welded wire fabric are used.

4. ***Maximum diameter***. The diameter of the reinforcing bars shall not exceed one eighth of the total thickness of the slab.

5. ***Spacing of bars***. The horizontal distance between parallel main reinforcement bars shall not be more than three times the effective depth of a solid slab or 300 mm, whichever is smaller. The horizontal distance between parallel reinforcement bars provided against shrinkage and temperature shall not be more than five times the effective depth of the slab or 450 mm whichever is smaller. The horizontal distance between two parallel main reinforcing bars shall usually be not less than the greatest of the following :

(*i*) The diameter of the bar if the diameters are equal.
(*ii*) The diameter of the larger bar if the diameters are unequal, and
(*iii*) 5 mm more than the nominal maximum size of coarse aggregate.

6. ***Cover to reinforcement :*** The same rules apply, as given in clause A-8 of §11.1

7. ***Curtailment of tension reinforcement***. The same rules, as given for beams in clause A-6 of § 11.1 apply for slabs also. Thus, for curtailment, reinforcement shall extend beyond the point at which it is no longer required to resist flexure, for a distance equal to effective depth or 12 times the bar diameter whichever is greater. except at supports or end of cantilever.

At least one-third the positive moment reinforcement in simple members and one-fourth of the positive moment reinforcement in continuous members shall extend along the same face of the member into the support, to a length equal to $L_d/3$.

At simple supports, and at point of inflexion, positive moment tension reinforcement shall be limited to such a diameter that $\left(\frac{M_1}{V} + L_0\right) \geq L_d$.. The value of M_1/V in the expression may be increased by 30% when the ends of the reinforcement are confined by a compressive reaction.

At least one-third of the total reinforcement provided for negative moment at the support shall extend beyond the point of inflexion, for a distance not less than the effective depth of the member or 12 φ or one-sixteenth of the clear span whichever is greater.

(c) BAR BENDING SCHEME

As per IS 456 : 2000, atleast one third of the maximum positive reinforcement should extend along the same face of the slab into the support, to a length equal to $L_d/3$. Some positive reinforcement should also be bent up, near support, to take up negative B.M. which may develop due to partial fixidity. Unfortunately, IS 456 : 2000 does not give any clear cut guidance about the bar bending and cutoff details, as given in CP 110. Due to this, different practices about bar bending is followed. One *practice* is to bend alternate bars up at a distance of 0.15 L (or $L/7$) from the centre of the supports, so that bars are available at the top face, for atleast a length equal to 0.1 L (or $L/10$) from the centre of the support. The other *popular practice* is to bend alternate bars up at a distance of 0.15 l (*or* $l/7$) from the *face of the support*, where B.M. reduces to less than half its maximum values; this will ensure in majority of cases, that the bars available at the upper face is more than $L/10$ from the centre of support.

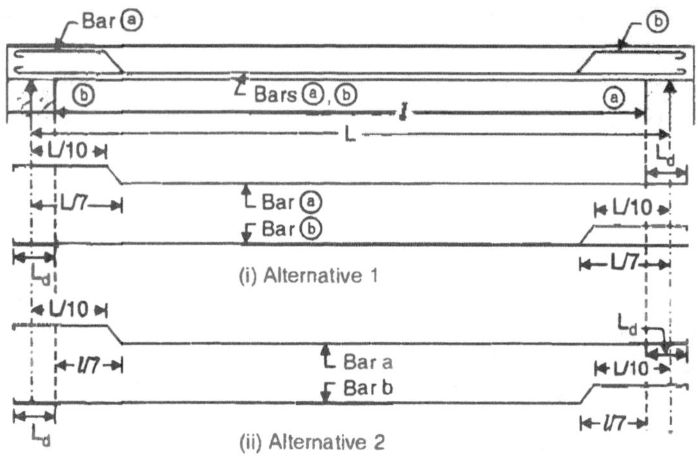

FIG. 11.12. BAR BENDING SCHEME IN ONE WAY SLAB

Design Example 11.5 : Design of one way slab

Design a R.C. slab for a room having inside dimensions 3 m × 7 m. The thickness of supporting wall is 300 mm. The slab carries 75 mm thick lime concrete at its top, the unit weight of which may be taken as 20 kN/m³. The live load on the slab may be taken as 2 kN/m². Assume the slab to be simply supported at the ends. Use M 20 concrete and Fe 415 steel.

Solution

Step 1. Design constants and limiting depth of N.A. : For Fe 415 steel, $f_y = 415$ N/mm².

For M 20 concrete, $f_{ck} = 20$ N/mm². $\frac{x_{u,max}}{d}$ (for Fe 415 steel) $= \frac{700}{1100 + 0.87 \times 415} = 0.479$

$$R_u = 0.36 f_{ck} \frac{x_{u,max}}{d}\left(1 - 0.416 \frac{x_{u,max}}{d}\right) = 0.36 \times 20 \times 0.479\ (1 - 0.416 \times 0.479) = 2.761$$

Step 2. Computation of design B.M. and S.F. The minimum depth of slab is based on serviceability requirement. For a simply supported slab, basic $L/d = 20$.

In limit state design, $p_{t,\,lim} = 0.955\%$ for M 20 – Fe 415 combination. However, assuming $p_t = 0.35\%$ for an *under-reinforced* section, we get modification factor $\simeq 1.5$ from Fig. 11.1.

Taking $L \simeq 3.2$ m, we have $d = \dfrac{3200}{20 \times 1.5} \simeq 107$ mm

Keep D = 125 mm. Using 15 mm nominal cover and 8 mm dia. bars, $d = 125 - 15 - 4 = 106$ mm

Effective span $L = (l + d) = 3 + 0.106 \simeq 3.1$ m (or equal to c/c distance between supports, whichever is less).

Self weight of slab	$= 0.125 \times 25000$	$= 3125\ \text{N/m}^2$
Live load	$= 2\ \text{kN/m}^2$	$= 2000\ \text{N/m}^2$
Weight of lime terracing	$= 0.075 \times 20000$	$= 1500\ \text{N/m}^2$
Total		$w = 6625\ \text{N/m}^2$

$$\therefore \quad w_u = 1.5w = 1.5 \times 6625 = 9938\ \text{N/m}^2$$

$$M_u = \frac{w_u\, l^2}{8} = \frac{9938\ (3.1)^2}{8} = 11.94\ \text{N-m} = 11.94 \times 10^6\ \text{N-mm}$$

$$V_u = \frac{w_u\,.\,L}{2} = 9938\ \frac{(3.1)}{2} = 15404\ \text{N}$$

Step 3. Computation of effective depth d : For a balanced section,

$$\therefore \quad d = \sqrt{\frac{M_u}{R_u\,.\,b}} = \sqrt{\frac{11.94 \times 10^6}{2.761 \times 1000}} = 66\ \text{mm}$$

Actual d available = 106 mm (based on the limit state of serviceability requirements). Hence the section will be *under-reinforced*.

Step 4. Steel Reinforcement. For an under-reinforced section,

$$A_{st} = \frac{0.5 f_{ck}}{f_y}\left[1 - \sqrt{1 - \frac{4.6 M_u}{f_{ck}\, bd^2}}\right] bd = \frac{0.5 \times 20}{415}\left[1 - \sqrt{1 - \frac{4.6 \times 11.94 \times 10^6}{20 \times 1000\ (106)^2}}\right] 1000 \times 106$$

$$= 334\ \text{mm}^2$$

$$(A_{st})_{min} = \frac{0.12}{100} \times 1000 \times 125 = 150\ \text{mm}^2 < A_{st}$$

Hence spacing of 8 φ mm bars as main reinforcement is

$$s = \frac{1000\, A_{\varphi}}{A_{st}} = \frac{1000 \times 50.3}{334} = 150.6\ \text{mm c/c}$$

However, provide 8 mm φ bars @ 150 mm c/c. This spacing is less than lesser of $3d$ or 300 mm.

Actual $\quad A_{st} = \dfrac{1000 \times 50.3}{150} = 335.3\ \text{mm}^2$ and $p_t = \dfrac{335.3}{1000 \times 106} \times 100 = 0.32\%$.

Spacing of 8 mm φ bars as distribution reinforcement is

$$s = \frac{1000\, A_{\varphi}}{A_{sd}} = \frac{1000 \times 50.3}{150} = 335 \text{ mm} \simeq 300 \text{ mm}$$

(which is less than max. spacing $5d = 5 \times 106$ mm = 530 or 450 mm whichever is smaller).

Bend alternate bars of main reinforcement up, at a distance of $l/7 = 3000/7 = 430$ mm from the face of the support. Continue the remaining half bars into the support.

Step 5. Check for limit state of serviceability for deflection : Basic $L/d = 20$. Actual $p_t = 0.32\%$. Hence for Fe 415 steel, $f_s = 0.58 f_y \times \dfrac{A_{st}\text{ required}}{A_{st}\text{ provided}} = 0.58 \times 415 \times \dfrac{334}{335.3} = 239.77$ N/mm². Hence from Fig. 11.1, for $p_t = 0.32\%$ and $f_s = 239.77$ N/mm², we get $F_t \simeq 1.6$. Hence required $d = 3100/(20 \times 1.6) \simeq 97$ mm, against available $d = 106$ mm. Hence safe.

Step 6. Check for shear : The critical section for shear is at a distance of d (= 106 mm) from the face of the support or at a distance of $d + \dfrac{d}{2} = 106 + \dfrac{106}{2} \approx 159$ mm from the centre of support. $V_{uD} = V_u - w_u(0.159) = 15404 - 9938\ (0.159) = 13824$ N.

$$\tau_v = \frac{V_{uD}}{bd} = \frac{13824}{1000 \times 106} = 0.13 \text{ N/mm}^2 \; ; \quad \frac{100\, A_s}{bd} \text{ (at supports)} = \frac{1}{2} \times 0.32 = 0.16\%$$

Hence $\tau_c \approx 0.285$ N/mm² from Table 7.1 Also k = 1.30 (from Table 7.2)

Since $\tau_v \ll k\tau_c$, the section is safe in shear.

Note : Slabs are normally found safe in shear. Hence it will be reasonably correct if we take face of the support as the critical section for shear. Thus, at the face of support, $V_u = \frac{1}{2} \times 9938 \times 3 = 14907$ N and $\tau_v = 14907/(1000 \times 106) = 0.14$ N/mm², against 0.13 N/mm² found earlier.

Step 7. Check for development length and anchorage at ends

$$L_d = \frac{0.87\, f_y}{4\tau_{bd}}\,\varphi = \frac{0.87 \times 415}{4 \times 1.2 \times 1.6}\,\varphi = 47\,\varphi = 47 \times 8 = 376 \text{ mm}.$$

The Code requires that the reinforcement should extend by a length equal to $L_d / 3 = 376 / 3 \approx 125$ mm beyond the face of support. This suggests that the width of support should not be less than 125 + 25 = 150 mm. In the present case, the support width is 300 mm and hence the anchorage condition is satisfied. Also, at the simple support like this, where compressive reaction confines the ends of bars,

$$1.3\frac{M_{1u}}{V_u} + L_0 >> L_d$$

Here $V_u = 15404$ N, at the centre of support.

M_{1u} = moment of resistance of the remaining bars which is computed as under:

$$x_u = \frac{0.87\, f_y\, A_{st1}}{0.36\, f_{ck}\, .\, b} = \frac{0.87 \times 415 \times (335.3/2)}{0.36 \times 20 \times 1000} = 8.4 \text{ mm}$$

$$\therefore \quad M_{1u} = 0.87\, f_y\, A_{st1}(d - 0.416\, x_u) = 0.87 \times 415 \times \frac{335.3}{2}(106 - 0.416 \times 8.4)$$

$$= 6.2 \times 10^6 \text{ N-mm}$$

Providing 90° bend and a clear cover = 25 mm at the side end,

$$L_0 = \frac{l_s}{2} - x' + 3\varphi = \frac{300}{2} - 25 + 3 \times 8 = 149 \text{ mm.}$$

Hence, $$1.3\frac{M_{1u}}{V_u} + L_0 = 1.3 \times \frac{6.2 \times 10^6}{15404} + 149 = 672 \text{ mm} > L_d$$

Hence the Code requirements are satisfied.

Note : In limit state design, effective depth required from the point of view of bending will be very much less than the one required from deflection point of view. This will result in an *under-reinforced* section. Hence while taking the modification factor from Fig. 11.1, the value of p_t should be assumed equal to 40 to 50% of $p_{t,lim}$ for mild steel bars, and equal to about 30% of $p_{t,lim}$ for HYSD bars.

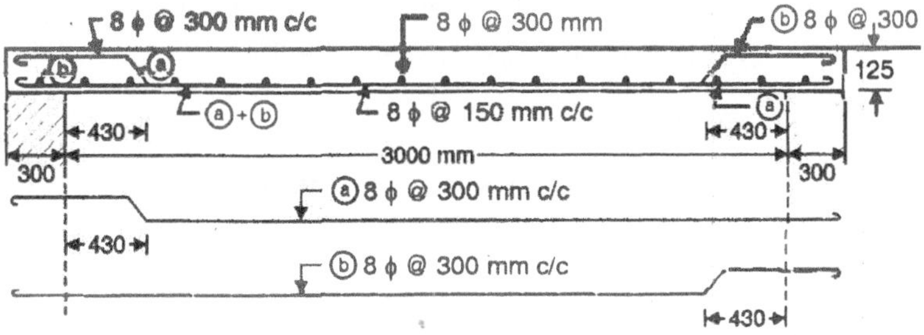

FIG. 11.13.

Step. 8. Details of reinforcement. Shown in Fig. 11.13.

11.5. DESIGN OF CANTILEVER SLAB / CHHAJAH

A cantilever slab bends downwards, with the result that tension is developed at the upper face. Hence reinforcement is also provided at the upper face. The span of the slab is taken equal to the actual length of overhang plus half the effective depth. If the width of the cantilever is long, 1 metre width of the cantilever is taken for the design purposes. However, if the width of the cantilever is short, whole of the width may be taken as the width of the slab for the design purposes.

Design Example 11.6 Design of a cantilever slab

Design a cantilever slab having an overhang of 1.25 m. Take live load intensity of 1000 N/m² on the cantilever, and the weight of finishing on the top of slab as 800 N/m². Use M 20 concrete and Fe 415 steel.

Solution

Step 1. Design constants and limiting depth of N.A.

For Fe 415 steel, $f_y = 415 \text{ N/mm}^2$. For M 20 concrete, $f_{ck} = 20 \text{ N/mm}^2$

$$\frac{x_{u,max}}{d} \text{ (for Fe 415 steel)} = \frac{700}{1100 + 0.87 \times 415} = 0.479$$

$$\therefore \quad R_u = 0.36 f_{ck}\frac{x_{u,max}}{d}\left(1 - 0.416\frac{x_{u,max}}{d}\right) = 0.36 \times 20 \times 0.479\,(1 - 0.416 \times 0.479) = 2.761$$

Step 2. Computation of design B.M. and S.F.

The minimum depth of slab is based on serviceability requirements. For a cantilever slab, $L/d = 7$. For M 20 - Fe 415 combination, $p_{t,lim} \simeq 0.96\%$, but for under-reinforced section let us assume $p_t = 0.2\%$. Hence from Fig. 11.1, we get modification factor $\simeq 1.68$. Assuming $d \simeq 120$ mm, effective length (L) of cantilever $= 1.25 + 0.12/2 = 1.31$ m

$$\therefore \quad d = 1310/(7 \times 1.68) \simeq 111 \text{ mm.}$$

Provide $D = 130$ mm at the fixed end, and reduce it to 100 mm at the free end. Hence average total thickness $= 115$ mm.

$\therefore$	Dead weight per $m^2 = 0.115 \times 1 \times 1 \times 25000$	$= 2875$ N
	Dead weight of finishing $= 800 \times 1 \times 1$	$= 800$ N
	Live load per $m^2 = 1000 \times 1 \times 1$	$= 1000$ N
	Total	$= 4675$ N

$$\therefore \quad w_u = 1.5\, w = 1.5 \times 4675 \simeq 7013 \text{ N}$$

$$M_u = w_u l^2/2 = 7013\,(1.25)^2/2 = 5479 \text{ N-m} = 5479 \times 10^6 \text{ N-mm}$$

$$V_u = w_u \, . \, l = 7013 \times 1.25 = 8766 \text{ N}$$

Step 3. Computation of effective depth

$$d = \sqrt{\frac{M_u}{R_u\, b}} = \sqrt{\frac{5479 \times 10^6}{2.761 \times 1000}} \simeq 46 \text{ mm}$$

However, from the requirements of deflection, provide $D = 130$ mm. Providing nominal cover of 15 mm and using 8 mm dia. bars, available $d = 130 - 15 - 8/2 = 111$ mm.

Hence the section will be under-reinforced.

Reduce D to 100 mm at the free end.

Step 4. Computation of steel reinforcement : For an under-reinforced section

$$A_{st} = \frac{0.5\, f_{ck}}{f_y}\left[1 - \sqrt{1 - \frac{4.6\, M_u}{f_{ck}\, bd^2}}\right] bd = \frac{0.5 \times 20}{415}\left[1 - \sqrt{1 - \frac{4.6 \times 5.479 \times 10^6}{20 \times 1000\,(111)^2}}\right] \times 1000 \times 111$$

$$\simeq 141 \text{ mm}^2$$

$$(A_{st})_{min} = \frac{0.12 \times 1000 \times 130}{100} = 156 \text{ mm}^2 = A_{sd}.$$

Hence spacing of 8 mm φ main reinforcement $s = \dfrac{1000 \times 50.3}{156} = 322$ mm.

Maximum spacing $= 3\,d = 3 \times 111 = 333$ mm or 300 mm whichever is smaller.

Hence provide 8 mm φ main bars @ 300 mm c/c.

Actual $\quad A_{st} = \dfrac{1000 \times 50.3}{300} \simeq 168 \text{ mm}^2$; $\quad p_t = \dfrac{168 \times 100}{1000 \times 111} = 0.152\%$

Also, provide distribution reinforcement in the form of 8 mm φ bars @ 300 mm c/c.

Step 5. Check for shear

Slabs are normally safe in shear.

$$\tau_v = \frac{V_u}{bd} = \frac{8766}{1000 \times 111} \simeq 0.08 \text{ N/mm}^2 \text{ (Neglecting the taper)}$$

$$\frac{100\,A_s}{bd} = \frac{100 \times 168}{1000 \times 111} = 0.151. \quad \text{Hence } \tau_c \simeq 0.28 \text{ N/mm}^2$$

Also, $k = 1.3$ Hence $\tau_v << k\,\tau_c$

Step 6. Check for development length and anchorage at ends

$$L_d \simeq 47\,\varphi = 47 \times 8 = 376 \text{ mm}$$

The Code requires that the reinforcement should extend by a length equal to $L_d/3$ $= 376/3 \simeq 125$ mm beyond the face of the support. This suggests that the width of support should not be less than $125 + 25 = 150$ mm. In the present case, width of support $= 300$ mm and hence the anchorage condition is satisfied. Also, at the simple support like this where compressive reaction confines the ends of bars,

$$1.3\,\frac{M_{1u}}{V_u} + L_0 \geq L_d$$

Here $V_u = 8766$

$$x_u = \frac{0.87\,f_y\,A_{st1}}{0.36\,f_{ck}\,b} = \frac{0.87 \times 415 \times 168}{0.36 \times 20 \times 1000} \simeq 8.4 \text{ mm}$$

$$M_{1u} = 0.87\,f_y\,A_{st1}\,(d - 0.416\,x_u) = 0.87 \times 415 \times 168\,(111 - 0.416 \times 8.4) = 6.52 \times 10^6 \text{ N-mm}$$

Providing 90° bend and a clear cover of 25 mm at the side end,

$$L_0 = \frac{l_s}{2} - x' + 3\,\varphi = \frac{300}{2} - 25 + 3 \times 8 = 149 \text{ mm}$$

$$\therefore \quad 1.3\,\frac{M_{1u}}{V_u} + L_0 = \frac{1.3 \times 6.52 \times 10^6}{8766} + 149 \simeq 1115 \text{ mm} >> L_d$$

Hence Code requirements are satisfied.

Step 7. Details of reinforcement : Shown in Fig. 11.14.

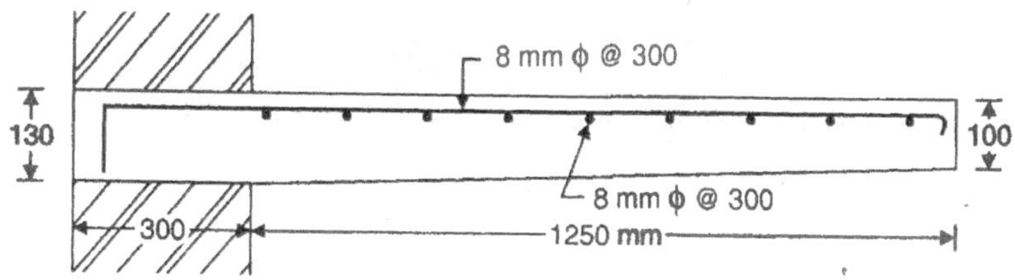

FIG. 11.14

11.6. DESIGN OF LINTEL WITH SUNSHADE

Design Example 11.7 ***Design of lintel beam with sunshade***

A lintel, covering a verandah opening of 3 m, carries a sunshade of 1 m span, monolithic with it. The bottom of the lintel and the sunshade are at the same level, as shown in Fig. 11.15. The verandah roof of 120 mm thickness carries a live load of 1500 N/m². Design the lintel and the sunshade. Use M 20 concrete and Fe 415 steel. Take thickness of water proofing as 100 mm with unit weight of 20 kN/m³. Also, take the unit weight of masonry as 20 kN/m³.

Step 1 Design Constants and Limiting Depth of N.A.

$f_y = 415\ \text{N/mm}^2;\ f_{ck} = 20\ \text{N/mm}^2$

For Fe 415 steel,

$$\frac{x_{u,max}}{d} = \frac{700}{1100 + 0.87 \times 415} = 0.479$$

$$R_u = 0.36 f_{ck} \frac{x_{u,max}}{d}\left(1 - 0.416 \frac{x_{u,max}}{d}\right)$$

$$= 0.36 \times 20 \times 0.479\,(1 - 0.416 \times 0.479)$$

$$= 2.761$$

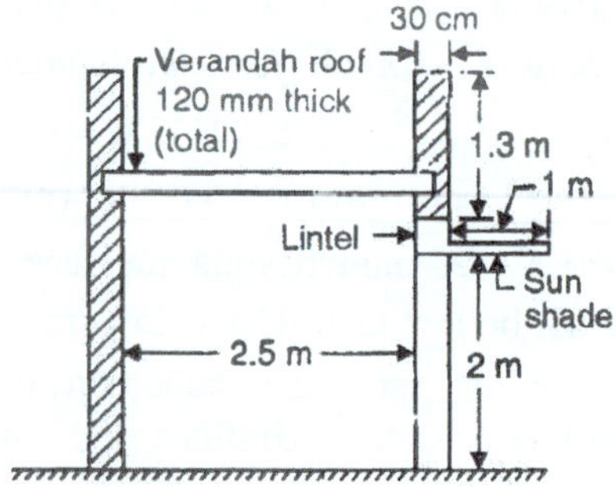

FIG. 11.15

Step 2. Design of Sun Shade

Let us design the sunshade for a live load of $1000\ \text{N/m}^2$. Let the total thickness (average) be 100 mm. Also, let the thickness of water proofing be 25 mm. Consider one metre length of the sunshade.

Live load per m run		= 1000 N.
Dead load per metre run	$= 1 \times 1 \times 0.1 \times 25000$	= 2500 N
Weight of water proofing/m run	$= 1 \times 1 \times 0.025 \times 20000$	= 500 N
	Total	w = 4000 N/m

$$\therefore \quad w_u = 1.5\, w = 1.5 \times 4000 = 6000\ \text{N/m}$$

$$M_u = \frac{w_u l^2}{2} = \frac{6000\,(1)^2}{2} = 3000\ \text{N-m} = 3 \times 10^6\ \text{N-mm}$$

$$V_u = w_u\, l = 6000 \times 1 = 6000\ \text{N}$$

$$d = \sqrt{\frac{M_u}{R_u b}} = \sqrt{\frac{3 \times 10^6}{2.761 \times 1000}} \simeq 33\ \text{mm}$$

Thickness from serviceability requirement : For a cantilever, $L/d = 7$. For M 20 – Fe 415 combination, $p_{t,\,lim.} \simeq 0.96\%$, but for under-reinforced section let us assume $p_t = 0.2\%$. Hence from Fig. 11.1, we get modification factor $\simeq 1.68$. Assuming $d \simeq 100$ mm, effective length (L) of cantilever $= 1 + 0.1/2 = 1.05$ m.

$$\text{Hence} \qquad d = \frac{1050}{7 \times 1.68} \simeq 90\ \text{mm}$$

Hence provide $D = 110$ mm. Taking nominal cover as 15 mm and using 8 mm bars, available $d = 110 - 25 - 8/2 = 91$ mm. Reduce the total thickness to 90 mm at the ends.

Computation of steel reinforcement

Since the available $d\ (= 91\ \text{mm})$ is much more than the one required from B.M. point of view, we have an *under-reinforced* section. Hence

$$A_{st} = \frac{0.5 f_{ck}}{f_y}\left[1 - \sqrt{1 - \frac{4.6\, M_u}{f_{ck}\, b d^2}}\right] bd = \frac{0.5 \times 20}{415}\left[1 - \sqrt{1 - \frac{4.6 \times 3 \times 10^6}{20 \times 1000\,(91)^2}}\right] \times 1000 \times 91$$

$$= 93.4\ \text{mm}^2$$

$$(A_{st})_{min.} = \frac{0.12\, bD}{100} = \frac{0.12 \times 1000 \times 110}{100} = 132 \text{ mm}^2 = A_{sd}.$$

$\therefore$ Spacing of 8 mm φ main reinforcement $= \dfrac{1000 \times 50.3}{132} = 381$ mm

Max. spacing $= 3d = 3 \times 91 = 273$ mm, or 300 mm, which ever is less.

Hence provide 8 mm φ bars @ 270 mm c/c.

$$\text{Actual } A_{st} = \frac{1000 \times 50.3}{270} = 186.3 \text{ mm}^2 \quad ; \qquad p_t = \frac{186.3 \times 100}{1000 \times 91} = 0.20\,\%.$$

Also, provide distribution reinforcement, 8 mm φ @ 300 mm c/c.

Check for shear : $\tau_v = \dfrac{V_u}{bd} = \dfrac{6000}{1000 \times 91} = 0.07 \text{ N/mm}^2$

For $p_t = 0.2\,\%$, $\tau_c \simeq 0.32 \text{ N/mm}^2$. Also, $k = 1.3$ Hence $k\,\tau_c >> \tau_v$

Check for development length

$$L_d \simeq 47\,\varphi = 47 \times 8 = 376 \text{ mm.} \; ; \qquad V_u = 6000 \text{ N}$$

$$x_u = \frac{0.87\, f_y\, A_{st1}}{0.36\, f_{ck}\, b} = \frac{0.87 \times 415 \times 186.3}{0.36 \times 20 \times 1000} = 9.3 \text{ mm}$$

$$M_{1u} = 0.87\, f_y\, A_{st1}\,(d - 0.416\, x_u) = 0.87 \times 415 \times 186.3\,(91 - 0.416 \times 9.3) = 5.86 \times 10^6 \text{ N-mm}$$

Providing 90° bend and a clear cover of 35 mm at the side end,

$$L_0 = \frac{l_s}{2} - x' + 3\,\varphi = \frac{300}{2} - 35 + 3 \times 8 = 139 \text{ mm}$$

$$\therefore \quad 1.3\,\frac{M_{1u}}{V_u} + L_0 = \frac{1.3 \times 6.86 \times 10^6}{6000} + 139 \simeq 1408 \text{ mm} >> L_d$$

Hence Code requirements are satisfied.

Step 3. Design of Lintel Beam : Width of lintel beam = width of wall = 300 mm

For computation of self weight, let total depth of lintel beam = 1/10 span = 300 mm. Consider 1 m length of the beam.

(*i*)	Dead load of verandah roof $= \frac{1}{2}(2.5 \times 1 \times 0.12 \times 25000)$	= 3750 N
(*ii*)	Dead load of water proofing on roof $= \frac{1}{2}(2.5 \times 1 \times 0.1 \times 20000)$	= 2500 N
(*iii*)	Live load from verandah roof $= \frac{1}{2}(2.5 \times 1 \times 1500)$	= 1875 N
(*iv*)	Dead weight of masonry over lintel $= 0.3 \times 1.3 \times 1 \times 20000$	= 7800 N
(*v*)	Self weight of lintel beam $= 0.3 \times 0.3 \times 1 \times 25000$	= 2250 N
(*vi*)	Dead load of sunshade $= 1 \times 1 \times 0.1 \times 25000$	= 2500 N
(*vii*)	Dead load of finishing on sunshade $= 1 \times 1 \times 0.025 \times 20000$	= 500 N

Total $= w = 21175$ N $\simeq 21200$ N/m

$$\therefore \qquad w_u = 21200 \times 1.5 = 31800 \text{ N/m}$$

Note : The live load on the sunshade is not likely to be there simultaneously with live load on verandah roof and hence has not been included. Similarly, torsional moment due to load on sunshade will be extremely small and has been neglected.

Assuming a bearing of 300 mm on either end, $L = 3 + 0.3 = 3.3$ m

$$M_u = \frac{w_u L^2}{8} = \frac{31800\,(3.3)^2}{8} = 43288 \text{ N-m} = 43.288 \times 10^6 \text{ N-mm}$$

$$d = \sqrt{\frac{M_u}{R_u\, b}} = \sqrt{\frac{43.288 \times 10^6}{2.761 \times 300}} \simeq 229 \text{ mm}$$

Provide $D = 300$ mm. Keeping nominal cover of 25 mm, 12 mm φ main bars and 8 mm φ rings, available $d = 300 - 25 - 8 - 12/2 = 261$ mm.

Computation of main reinforcement

Since available d (= 261 mm) is more than that required for balanced section, we have an under-reinforced section, for which

$$A_{st} = \frac{0.5 f_{ck}}{f_y}\left[1 - \sqrt{1 - \frac{4.6\, M_u}{f_{ck}\, bd^2}}\right] bd = \frac{0.5 \times 20}{415}\left[1 - \sqrt{1 - \frac{4.6 \times 43.288 \times 10^6}{20 \times 300\,(261)^2}}\right] 300 \times 261$$

$$\simeq 536 \text{ mm}^2$$

Hence No. of 12 mm φ bars $= \dfrac{536}{113.1} = 4.74$. Hence provide 5-12 mm φ bars. Out of these, bend 2 bars up at a distance of $L/7 = 3.3/7 = 0.47$ m from the edge of support. Remaining $A_{st1} = 3 \times \dfrac{\pi}{4}(12)^2 = 339.3 \text{ mm}^2$.

Actual area provided at mid-span $= 5 \times 113.1 = 566 \text{ mm}^2$

Check for deflection : For simply-supported beam, $L/d = 20$.

For $\dfrac{100\, A_s}{bd} = \dfrac{100 \times 566}{300 \times 261} \simeq 0.73\%$, modification factor $\simeq 1.05$.

$$\therefore \qquad d = \frac{L}{20 \times 1.05} = \frac{3300}{20 \times 1.05} = 157 << \text{available } 261 \text{ mm. Hence OK.}$$

Check for shear : The critical section for shear is at a distance of d (= 0.261 m) from the face of the support.

$$\therefore \qquad V_{uD} = \frac{w_u L}{2} - w_u\left(\frac{0.30}{2} + 0.261\right) = \frac{31800 \times 3.3}{2} - 31800 \times 0.411 = 39400 \text{ N}$$

$$\tau_v = \frac{V_{uD}}{bd} = \frac{39400}{300 \times 261} = 0.503 \text{ N/mm}^2$$

At supports, $\dfrac{100\, A_s}{bd} = \dfrac{100 \times 339.3}{300 \times 261} = 0.43\%$. Hence from Table 7.1, $\tau_c \simeq 0.45 \text{ N/mm}^2$

Since τ_v and τ_c are nearly equal, provide nominal shear reinforcement. Providing 8 mm φ 2 lgd stirrups,

$$s_v = \frac{2.175\, A_{sv}\, f_y}{b} = \frac{2.175 \times 100.5 \times 415}{300} \simeq 300 \text{ mm}$$

Max. s_v = lesser of $0.75\, d$ (= 0.75 × 261 = 195 mm) or 300 mm.

Hence provide 8 φ 2 lgd stirrups @ 190 mm c/c throughout the length of the beam. Provide 2-10 mm φ holding bars at the top.

Check for development length at the ends : Since the beam is supported on wall, the compressive reaction will confine the reinforcement. Hence

$$L_d \leq 1.3 \frac{M_{1u}}{V_u} + L_0$$

$$x_u = \frac{0.87 f_y A_{st1}}{0.36 f_{ck} b} = \frac{0.87 \times 415 \times 339.3}{0.36 \times 20 \times 300} = 56.7 \text{ mm}$$

$$M_{1u} = 0.87 f_y A_{st1} (d - 0.416 x_u) = 0.87 \times 415 \times 339.3 (261 - 0.416 \times 56.7) = 29.08 \times 10^6 \text{ N-mm}$$

$$V_u = \frac{w_u L}{2} = \frac{31800 \times 3.3}{2} = 52470 \text{ N}$$

$$L_d = \frac{0.87 f_y}{4 \tau_{bd}} \varphi = \frac{0.87 \times 415}{4 \times 1.2 \times 1.6} \varphi = 47 \varphi = 47 \times 12 = 564 \text{ mm}$$

If no hook or bend is provided, $L_0 = \frac{L_s}{2} - x' = \frac{300}{2} - 30 = 120$ mm

where x' = side cover assumed to be 30 mm. The Code further recommends that each bar should extend by a distance of $L_d/3 = 564/3 = 188$ mm beyond the face of the support. Available distance beyond the face of the support $= L_0 + \frac{L_s}{2} = 120 + \frac{300}{2} = 270$ mm, which is more than $L_d/3$. Hence no hook is required, and the value of L_0 (= 120 mm) computed above is correct.

$$\therefore \quad 1.3 \frac{M_{1u}}{V_u} + L_0 = \frac{1.3 \times 29.08 \times 10^6}{52470} + 120 = 720 + 120 = 840 \text{ mm} >> L_d. \text{ Hence OK.}$$

Step 4 : Details of reinforcement

The details of reinforcement etc are shown in Fig. 11.16.

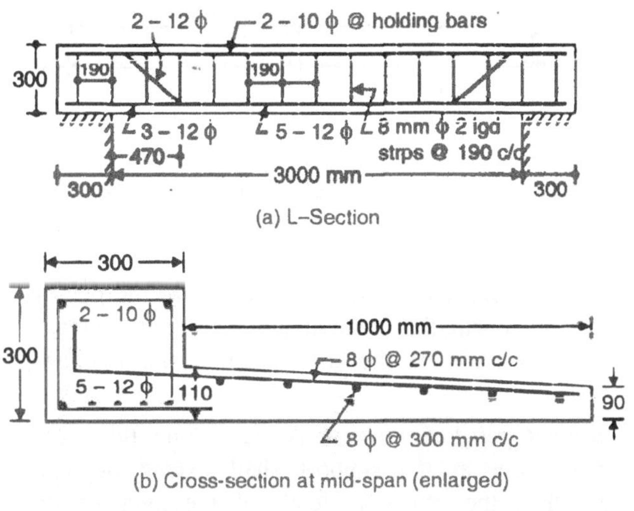

FIG. 11.16

11.7. DESIGN OF ONE WAY CONTINUOUS SLAB

When a slab is continuous over several spans, negative (*i.e.* hogging) bending moment is induced over the intermediate supports and hence reinforcement has to be provided at the top of the slab portion over the intermediate supports. These intermediate supports may either be monolithic with the slab, or may be of some different material such as rolled steel joist etc. Following are some of the *basic* rules for design, as per IS : 456–2000.

1. ***Effective span.*** In the case of continuous slab, if the width of the support is less than 1/12 of the clear span, the effective span shall be the distance between centres of supports, or the clear distance between supports plus the effective depth of the slab, whichever is smaller. If the supports are wider than 1/12 the clear span or 600 mm, whichever is less, the effective span shall be taken as under :

(*a*) For end span with one end fixed and the other continuous, or for intermediate spans, the effective span shall be the clear span between supports ; and

(*b*) For end span with one end free and the other continuous, the effective span shall be equal to the clear span plus half the effective depth of the slab or the clear span plus half the width of the continuous support, whichever is less.

Note : In the case of spans with roller or rocker bearings, the effective span shall always be the distance between the centres of bearings.

2. ***Limiting stiffness.*** The ratio of span to effective depth should not exceed 26. Modification factor for tension reinforcement will be extra as per Fig. 11.1.

3. ***Moment and shear coefficients.*** Same rules as for continuous beams, given in clause *A*-10 of § 11.1. Thus, the B.M. values are indicated on Fig. 11.17.

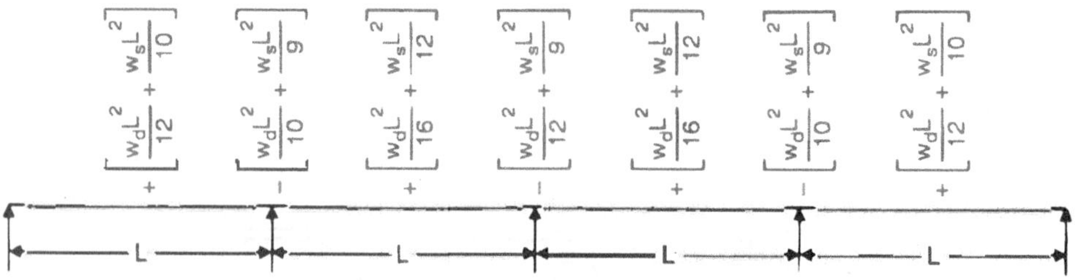

FIG. 11.17

4. ***Reinforcement and its arrangement*** : The rules for main reinforcement and transverse/distribution reinforcement are the same as discussed for simply supported one way slab. However, the Code (IS 456 : 2000) makes the following recommendations for continuation/curtailment of reinforcement.

(*a*) ***Positive moment reinforcement*** : At least *one third* the positive moment reinforcement in simple members and one fourth the positive moment reinforcement in continuous members shall extend along the same face of the member into the member, to a length equal to $L_d/3$.

(*b*) ***Negative moment reinforcement*** : At least one third of the total reinforcement provided for negative moment at the support shall extend beyond the point of inflexion for a distance not less than the effective depth of the member or 12 φ or one sixteenth of the clear span whichever is greater.

Except for the above two recommendations, IS 456 : 2000 is silent in regard to simplified rules for curtailment of reinforcement is one way continuous slab. In absence of any such recommendation for IS Code, various practices are followed, and one such practice is given in Fig. 11.18.

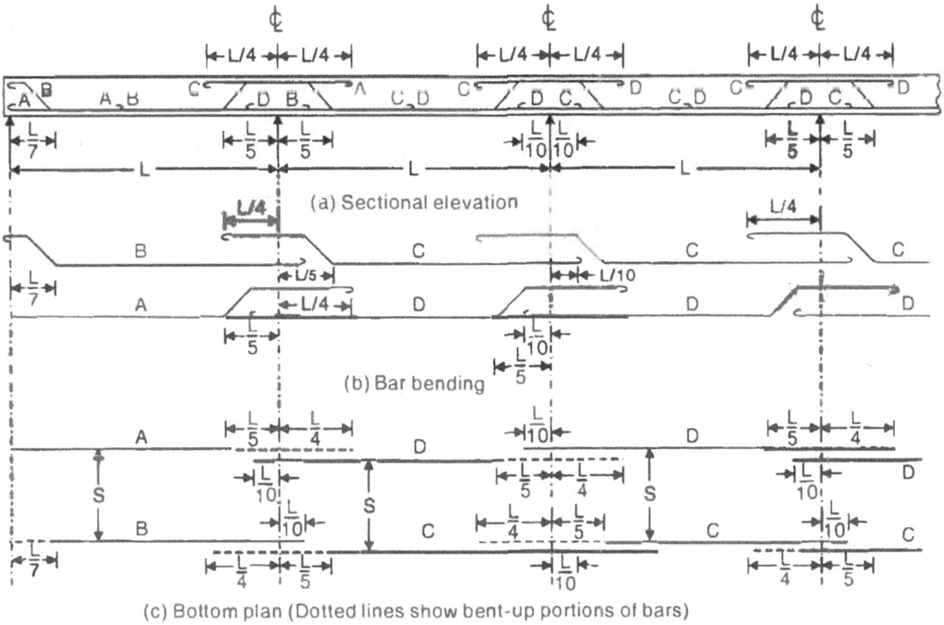

FIG. 11.18

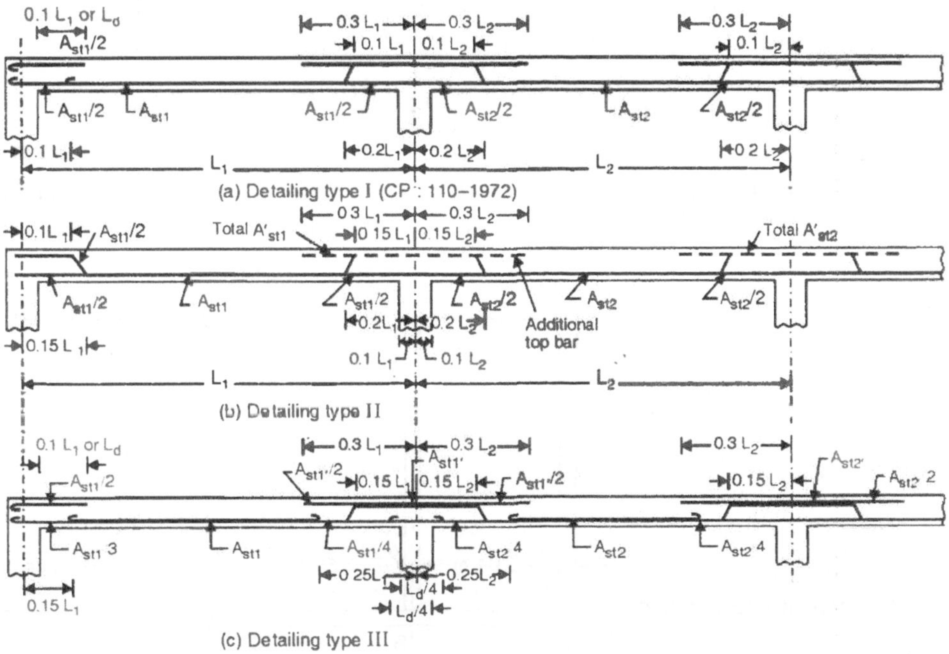

FIG. 11.19. ALTERNATIVE FORMS OF REINFORCEMENT DETAILING

However, CP 110-1972 has given some simplified rules, as given in Fig. 11.19 (*a*). Two more forms of reinforcement detailing are shown in Fig. 11.19 (*b*) and (*c*) respectively. It may be noted that detailing type II shown in Fig. 11.19 (*b*) is similar to the one shown in Fig. 11.18, except that the bend up point is at a distance of 0.25 *L* from the centre of support (in place of *L/5*) and the negative reinforcement extends to a distance of 0.3 *L* to the either side of the support (in place of *L/4*). In detailing type III, no bar is bent up, and the positive reinforcement and the negative reinforcement are provided separately at appropriate faces. The curtailment of positive reinforcement in type III detailing is as per IS Code recommendations.

Design Example 11.8. Design of one way continuous slab

Design a continuous R.C. slab for a hall 6.5 m wide and 13.5 m long. The slab is supported on R.C.C. beams, each 240 mm wide which are monolithic. The ends of the slab are supported on walls, 300 mm wide. Design the slab for a live load of 2 kN/m². Assume the weight of roof finishing equal to 1.5 kN/m². Use M 20 concrete and Fe 415 steel.

Solution.

Step 1. Arrangement of spans : Let us arrange the spans as shown in Fig. 11.20. Assume d = 100 mm. The width of intermediate supports is less than 1/12 of the clear span. Hence the effective span will be as under :

(*i*) *For end spans* : L = centre to centre of supports $= 3.04 + \frac{0.3 + 0.24}{2} = 3.31$ m

or L = clear span $+ d = 3.04 + 0.10 = 3.14$ m, whichever is less. Hence $L = 3.14$ m

(*ii*) *For intermediate spans* :

L = centre to centre of supports $= 3.35 + 0.24 = 3.59$ m

or L = clear span $+ d$ $= 3.35 + 0.10 = 3.45$ m whichever is less. Hence $L = 3.45$ m

Thus, in this case, the lengths of effective spans are not equal. However, the difference is very much less than 0.15% of the longer spans ; hence they may be treated as being equal for the purposes of using the B.M. coefficients given in the I.S. Code.

Step 2. Fixation of d and D : From the limit state of serviceability, the effective depth is given by : $L/d = 26$ for continuous spans and $L/d = 20$ for simply supported spans

Hence for end span,

$L/d \approx \frac{1}{2}(26 + 20) = 23.$

Assume a modification factor (corresponding to $p_t \sim 0.25\%$) equal to 1.6 for Fe 415 steel. Hence, on the basis of end span,

$$d = \frac{L}{23 \times 1.6} = \frac{3140}{23 \times 1.6} \approx 86 \text{ mm}$$

On the basis of intermediate spans,

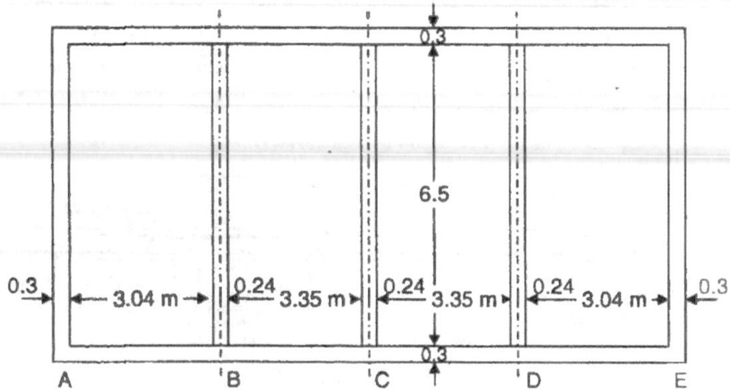

FIG. 11.20. ARRANGEMENT OF SPANS.

$d = \frac{L}{26 \times 1.6} = \frac{3450}{26 \times 1.6} = 83$ mm. Keep $D = 110$ mm, so that $d = 110 - 15 - 4 = 91$ mm.

Step 3. Design constants and limiting depth of N.A.

For Fe 415 steel, $f_y = 415$ N/mm^2. For M 20 concrete, $f_{ck} = 20$ N/mm^2.

$$\frac{x_{u,max}}{d} \text{ (For Fe 415 steel) } = \frac{700}{1100 + 0.87 \times 415} = 0.479$$

$$\therefore \quad R_u = 0.36 f_{ck} \frac{x_{u,max}}{d}\left(1 - 0.416 \frac{x_{u,max}}{d}\right) = 0.36 \times 20 \times 0.479(1 - 0.416 \times 0.479) = 2.761$$

Step 4. Computation of design B.M. and effective depth.

Self weight of slab	$= 0.11 \times 25000$	$= 2750$ N/m^2
Weight of roof finishing	$= 1.5$ kN/m^2	$= 1500$ N/m^2
	Total w_d	$= 4250$ N/m^2

Live load $w_s = 2$ kN/m^2 = 2000 N/m^2

For end spans

$$M \text{ (near the centre)} = \frac{w_d L^2}{12} + \frac{w_s L^2}{10} = \frac{4250\,(3.14)^2}{12} + \frac{2000(3.14)^2}{10} = 5464 \text{ N-m}$$

$$= 5.464 \times 10^6 \text{ N-mm}$$

$$M \text{ (at support next to end support)} = -\left[\frac{w_d L^2}{10} + \frac{w_s L^2}{9}\right] = -\left[\frac{4250(3.14)^2}{10} + \frac{2000(3.14)^2}{9}\right]$$

$$= -6382 \text{ N-m} = -6.382 \times 10^6 \text{ N-mm}$$

For Intermediate spans

$$M \text{ (at the middle)} = +\frac{w_d L^2}{16} + \frac{w_s L^2}{12} = \frac{4250\,(3.45)^2}{16} + \frac{2000(3.45)^2}{12} = 5145 \text{ N-m}$$

$$= 5.145 \times 10^6 \text{ N-mm}$$

$$M \text{ (at interior supports)} = -\left[\frac{w_d L^2}{12} + \frac{w_s L^2}{9}\right] = -\left[\frac{4250\,(3.45)^2}{12} + \frac{2000\,(3.45)^2}{9}\right]$$

$$= -6861 \text{ N-m} = -6.861 \times 10^6 \text{ N-mm}$$

Out of the four values of moments, the effective depth will be determined for max. value.

$\therefore$ Design moment $= 6.861 \times 10^6$ N-mm.

Hence $M_{uD} = 1.5 \times 6.861 \times 10^6 = 10.291 \times 10^6$ N-mm

$$\therefore \quad d = \sqrt{\frac{M_{uD}}{R_u . b}} = \sqrt{\frac{10.291 \times 10^6}{2.761 \times 1000}} = 62 \text{ mm.}$$

Available d, determined from limit state of serviceability, is equal to 91 mm. Hence the section will be *under-reinforced.*

Step 5. Determination of reinforcement : It is proposed to provide type II detailing shown in Fig. 11.19, wherein there is a provision for different reinforcements at various critical locations.

(*a*) *At Middle of end span*

$$M_{uD1} = 1.5 \times 5.464 \times 10^6 = 8.196 \times 10^6 \text{ N-mm}$$

$$A_{st1} = \frac{0.5 f_{ck}}{f_y}\left[1 - \sqrt{1 - \frac{4.6 M_u}{f_{ck} bd^2}}\right] bd$$

$$= \frac{0.5 \times 20}{415}\left[1 - \sqrt{1 - \frac{4.6 \times 8.196 \times 10^6}{20 \times 1000(91)^2}}\right] 1000 \times 91 = 265.7 \text{ mm}^2$$

$\therefore$ s-8 mm φ bars $= \dfrac{1000 \times 50.3}{265.7} = 189.3$ mm.

Max. spacing $= 3d = 3 \times 91 = 273$ mm. Hence provide 8 mm φ bars @ 180 mm c/c.

Actual A_{st1} provided $= \dfrac{1000 \times 50.3}{180} = 279.4$ mm^2

$$p_t = \frac{100 \times 279.4}{1000 \times 91} = 0.31\% \quad \left(A_{st.min} = \frac{0.12}{100} \times 1000 \times 110 = 132 \text{ mm}^2\right)$$

Bend half bars up at a distance of $0.15\,L = 0.15 \times 3140 = 471$ mm from the centre of support, or from $471 - d/2 = 471 - 101/2 \approx 420$ mm from the face of the support. Continue remaining half bars into the support for a distance $L_d/3$. At the other end, bend half bars up at a distance of $0.25\,L = 0.25 \times 3140 = 785$ from the centre of the support, and continue the remaining half into the support, for a distance or $L/10 \approx 320$ mm from the face of the support.

(*b*) *At middle of intermediate span*

$$M_{uD2} = 1.5 \times 5.145 \times 10^6 = 7.718 \times 10^6 \text{ N-mm}$$

$$\therefore \quad A_{st2} = \frac{0.5 \times 20}{415}\left[1 - \sqrt{1 - \frac{4.6 \times 7.718 \times 10^6}{20 \times 1000\,(91)^2}}\right] 1000 \times 91$$

$$= 249.2 \text{ mm}^2 \quad (A_{st.min} = 132 \text{ mm}^2).$$

However keep $A_{st2} = A_{st1}$ because of practical considerations, and also because of the reason that the section is already very much under-reinforced.

(*c*) *At support next to end support*

$$M_{uD1}' = 1.5 \times 6.382 \times 10^6 = 9.573 \times 10^6 \text{ mm}$$

$$\therefore \quad A_{st1}' = \frac{0.5 \times 20}{415}\left[1 - \sqrt{1 - \frac{4.6 \times 9.573 \times 10^6}{20 \times 1000\,(91)^2}}\right] 1000 \times 91 = 314 \text{ mm}^2.$$

Area of steel available in the form of bent up bars from the adjacent spans

$$= \tfrac{1}{2}(A_{st1} + A_{st2}) = 279.4 \text{ mm}^2.$$

Additional steel required $= 314 - 279.4 = 34.6$ mm^2

and s-8 mm φ bars $= \dfrac{1000 \times 50.3}{34.6} = 1450$ mm.

However, this has to be in multiple of earlier spacing of 180 mm. Hence provide extra 8 mm φ bars @ 900 mm c/c for length equal to $0.3L = 0.3 \times 3140 \approx 950$ mm to one side and $0.3 \times 3450 = 1050$ mm 1050 mm to the other side of the support.

(*d*) *At interior supports*

$$M_{uD2}' = 1.5 \times 6861 \times 10^6 = 10.292 \times 10^6 \text{ N-mm.}$$

$$\therefore \quad A_{st2}' = \frac{0.5 \times 20}{415}\left[1 - \sqrt{1 - \frac{4.6 \times 10.292 \times 10^6}{20 \times 1000\,(91)^2}}\right] \times 1000 \times 91 = 340 \text{ mm}^2.$$

Steel available in the form of bent-up bars from adjacent spans = 279.4 mm^2.

$\therefore$ Additional steel required = 340 − 279.4 = 60.6 mm^2

$$\therefore \quad s\text{ - 8 mm } \varphi = \frac{1000 \times 50.3}{60.6} = 830 \text{ mm.}$$

However, this has to be in multiple of earlier spacing of 180 mm c/c. Hence provide 8 φ extra bars @ 720 mm c/c, for a length $= 0.3L = 0.3 \times 3450 \approx 1050$ mm to the either side of the support.

Distribution reinforcement

$$A_{sd} = \frac{0.12}{100} \times 1000 \times 110 = 132 \text{ mm}^2. \quad \therefore \quad s\text{ - 8 mm } \varphi = \frac{1000 \times 50.3}{132} \approx 381 \text{ mm}$$

However provide 8 mm φ bars @ 375 mm c/c.

Step 5. Check for anchorage and development length at end support.

The end support is a free support : $L_d = 47\,\varphi = 47 \times 8 \approx 376$ mm

$$L_d/3 = 376/3 \simeq 126 \text{ mm.}$$

Width of support available = 300 mm. Length by which half the bars can be taken inside the support = 300 − 25 = 275 mm > 126 mm. Hence safe.

Again, at the end support, the following relation should satisfied

$$1.3\,\frac{M_{1u}}{V_u} + L_0 \geq L_d$$

where $V_u = 1.5[0.4\ w_d L + 0.45 w_s \,.\, L]$

$$= 1.5[0.4 \times 4250 \times 3.14 + 0.45 \times 2000 \times 3.14] = 12246 \text{ N.}$$

Area of steel available at supports $= A_{st1}/2 = 279.4/2 \approx 139.7$ mm^2

$$x_u = \frac{0.87 f_y \,.\, A_{st}}{0.36 f_{ck} \,.\, b} = \frac{0.87 \times 415 \times 139.7}{0.36 \times 20 \times 1000} = 7 \text{ mm}$$

$$\therefore \quad M_{1u} = 0.87\ f_y A_{st}(d - 0.416\,x_u) = 0.87 \times 415(139.7)(91 - 0.416 \times 7)$$

$$= 4.443 \times 10^6 \text{ N-mm}$$

$$\therefore \quad L_0 = \frac{b_s}{2} - x' = \frac{300}{2} - 25 = 125 \text{ mm.}$$

$$1.3\,\frac{M_{1u}}{V_u} + L_0 = 1.3 \times \frac{4.443 \times 10^6}{12246} + 125 = 596 \text{ mm.}$$

$$L_d = 376 \text{ mm. Hence safe.}$$

Step 6. Check for shear. Normally, slabs are found to be safe in shear. However, let us check for shear at the support next to the end support, where S.F. is maximum.

$V_u = 1.5\ [0.6\ w_d L + 0.6\ w_s L] = 1.5 \times 0.6 \times 3.14\ [4250 + 2000] = 17662$ N.

Though the critical section will be at a distance of d from the support, where S.F. will be slightly less, we will adopt the same value of V_u as that the centre of the support.

$$\tau_v = \frac{17662}{1000 \times 91} = 0.194 \text{ N/mm}^2; \quad \frac{100 A_s}{bd} = 0.31\% \quad \text{Hence } \tau_c \simeq 0.24 \text{ N/mm}^2$$

$\therefore$ $\tau_v < 1.3\ \tau_c$. Hence safe.

Step 7. Check for deflection

For the centre of end span, $p_t = 0.31\%$. $\therefore$ Modification factor ≈ 1.5

$$\therefore \quad d = \frac{L}{\frac{1}{2}(20 + 26) \times 1.5} = \frac{3140}{23 \times 1.5} = 91 \text{ mm}$$

Available d = 91 mm. Hence safe.

Step 8 Details of reinforcement etc. : Shown in Fig. 11.21.

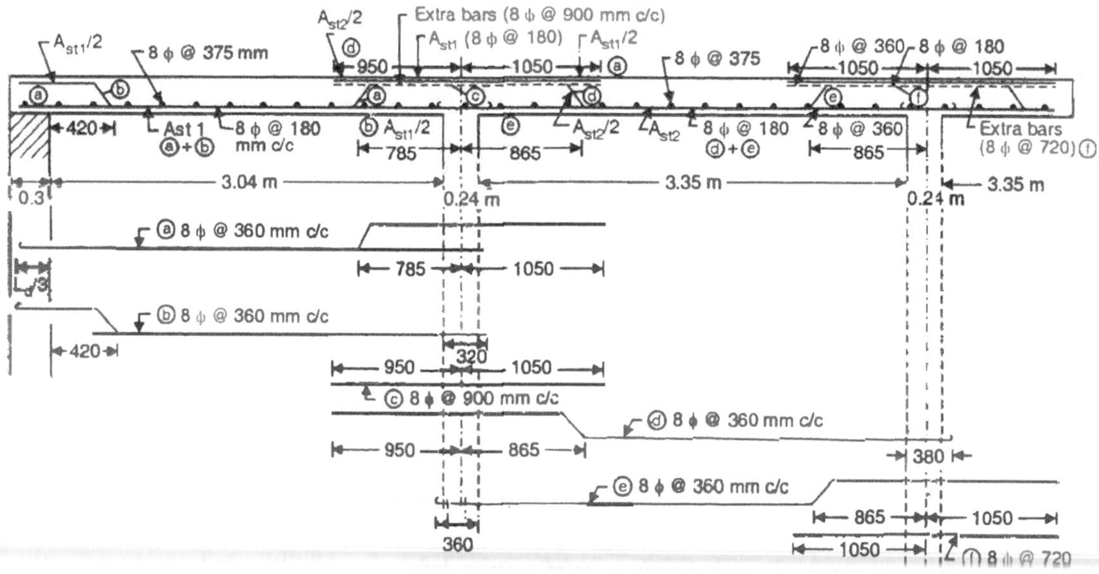

FIG. 11.21.

11.8. DESIGN OF DOUBLY REINFORCED BEAM

We have seen earlier that a doubly reinforced section is required when the applied moment M_{ua} is larger than $M_{u,lim}$ for a singly reinforced rectangular section. The difference between the applied moment M_{ua} and the limiting moment $M_{u,lim}$ is carried by additional tensile reinforcement A_{st2} and compression reinforcement A_{sc}. Thus a doubly reinforced section is equivalent to a singly reinforced balanced section and a section with additional tension and compression reinforcement as illustrated in Fig. 4.2.

The theory of doubly reinforced beam has already been discussed in Chapter 4. In addition to the basic rules for the design of singly reinforced beams given in § 11.1, the following two important points are note worthy (IS 456 : 2000).

1. Percentage of compression reinforcement

The maximum area of compression reinforcement shall not exceed 0.04 *BD*. Compression reinforcement in beams shall be enclosed by stirrups for effective lateral restraint.

2. Anchoring bars in compression

The anchoring length of a straight bar in compression shall be equal to the development length of the bar in compression. The projected length of hooks, bends and straight length beyond bend if provided for a bar in compression shall be considered for development length.

3. Modification factor for compression reinforcement

To control the deflection, the basic *L/d* ratios are given in § 11.1, along with the modification factor (F_t) corresponding to the percentage tension reinforcement (Fig 11.1). However, depending on the area of compression reinforcement, the value of span to depth ratio shall be further modified by modification factor (F_c) given in Fig. 11.2 corresponding to the percentage of compression reinforcement.

Design Procedure

The design of a simply supported doubly reinforced beam is done in the following steps :

1. Determine the design constants and limiting depth on N.A. by the expression

$$\frac{x_{u,\,max}}{d} = \frac{700}{1100 + 0.87\,f_y}$$

Since b and d are already known, determine (*i*) $x_{u,\,max}$ (*ii*) R_u and (*iii*) $M_{u,\,lim.}$

2. Compute design B.M. If $M_{uD} > M_{u,\,lim.}$, a doubly reinforced section will be required.

3. Determine tensile reinforcement by the following expressions

$$p_{lim.} = 0.414\,\frac{f_{ck}}{f_y}\cdot\frac{x_{u,\,max}}{d}$$

$$A_{st1} = p_{lim.} \times bd.$$

$$A_{st2} = \frac{M_{uD} - M_{u,\,lim}}{0.87\,f_y\,(d - d')}$$

$$\text{Total } A_{st} = A_{st1} + A_{st2}$$

4. Determine compression reinforcement

$$A_{sc} = \frac{M_{uD} - M_{u,\,lim}}{(f_{sc} - 0.446\,f_{ck})\,(d - d')},$$ where f_{sc} is corresponding to ε_{sc}

$$\varepsilon_{sc} = \frac{0.0035\,(x_{u,\,max} - d')}{x_{u,\,max}}$$

5. Curtail tensile and compressive reinforcement.

Find $$(L_d)_t = \frac{0.87\,f_y}{4\,\tau_{bd}}\,\varphi \quad \text{and} \quad (L_d)_c = \frac{0.87\,f_y}{4\,(1.25\,\tau_{bd})}\,\varphi$$

6. Design shear reinforcement

7. Check for development length at support.

Design Example 11.9. Design of a doubly reinforced beam.

Design a reinforced concrete beam supported on two walls 500 mm thick, spaced at a clear distance of 6 m. The beam carries a super-imposed load of 30 kN/m. The size of the beam is restricted to 300 mm × 500 mm. Use M 20 concrete and Fe 415 steel.

Solution

Step 1. Computation of design constants and limiting depth of N.A.

For Fe 415 steel, $f_y = 415$ N/mm² and $\frac{x_{u.max}}{d} = \frac{700}{1100 + 0.87 \times 415} = 0.479$

For M 20 concrete, $f_{ck} = 20$ N/mm².

$$R_u = 0.36 f_{ck} \cdot \frac{x_{u.max}}{d}\left(1 - 0.416\frac{x_{u.max}}{d}\right) = 0.36 \times 20 \times 0.479\,(1 - 0.416 \times 0.479) = 2.761$$

Here $d \approx 500 - 25 - 10 - 8 = 457$ mm, using 20 mm φ bars and 8 mm φ rings

$\therefore$ $M_{u.lim} = R_u bd^2 = 2.761 \times 300\,(457)^2 = 173.05 \times 10^6$ N-mm,

$x_{u.max} = 0.479\, d = 0.479 \times 457 = 218.9$ mm.

Step 2. Computation of design B.M. : $L \approx 6000 + 457 = 6457$ mm < 6500 mm

$\therefore$ Available $L/d = 6457/457 = 14.13$

Max. permissible value of L/d, even without modification factor = 20.

Hence deflection requirements are satisfied.

Dead load of beam = 0.3 × 0.5 × 25000 = 3750 N/m

Total $w = 3750 + 30{,}000 = 33750$ N/m.

$w_u = 1.5 \times 33750 = 50625$ N

$$M_{uD} \text{ (at midspan)} = \frac{w_u L^2}{8} = \frac{50625\,(6.457)^2}{8} = 263838 \text{ N-m} \approx 263.84 \times 10^6 \text{ N-mm}.$$

Since $M_{uD} > M_{u.lim.}$ a doubly reinforced section will be required.

Step 3. Determination of tensile reinforcement.

$$p_{t,\,lim} = 0.414\,\frac{f_{ck}}{f_y}\,\frac{x_{u.max}}{d}$$

$$= 0.414 \times \frac{20}{415} \times 0.479 = 0.0095569$$

$\therefore$ $A_{st1} = p_{t,\,lim} \times bd = 0.0095569 \times 300 \times 457 = 1310$ mm²

The cover to the centre of compressive reinforcement $(d') = 25 + 10 + 8 = 43$ mm

$$\therefore \quad A_{st2} = \frac{M_{uD} - M_{u.lim}}{0.87 f_y (d - d')} = \frac{(263.84 - 173.05)10^6}{0.87 \times 415\,(457 - 43)} = 607.4 \text{ mm}^2$$

$\therefore$ Total $A_{st} = A_{st1} + A_{st2} = 1310 + 607.4 = 1917.4$ mm²

(Max. reinforcement $= 0.04\, bD = 0.04 \times 300 \times 500 = 6000$ mm² $> A_{st}$). Hence O.K.

N – 20 mm φ bars $= \frac{1917.4}{314.2} = 6.1 \approx 7$ and N – 25 mm φ bars $= \frac{1917.4}{490.9} = 3.91 \approx 4$.

Hence provide 4 bars of 25 mm dia.

Step 4. Determination of compressive reinforcement

$$A_{sc} = \frac{M_{uD} - M_{u.lim}}{(f_{sc} - 0.446\, f_{ck})\,(d - d')} = \frac{(263.84 - 173.05)10^6}{(f_{sc} - 0.446 \times 20)(457 - 43)} = \frac{219300}{(f_{sc} - 8.92)}$$

$$\varepsilon_{sc} = \frac{0.0035\,(x_{u,max} - d')}{x_{u.max}} = \frac{0.0035(218.9 - 43)}{218.9} = 0.00281$$

Hence from Table 2.3, for Fe 415 steel, $f_{sc} \approx 352$ N/mm² for $\varepsilon_{sc} = 0.00281$

$\therefore$ $A_{sc} = \dfrac{219300}{352 - 8.92} = 639.2$ mm². Provide 2 – 16 mm φ and 1 – 20 mm φ bars having total $A_{sc} = 2 \times 201.06 + 1 \times 314.2 = 716.32$ mm².

Step 5. Curtailment of tensile and compressive reinforcement

Development length (L_d) of 25 dia. bars in tension.

$$= \frac{0.87\, f_y}{4\,\tau_{bd}}\,\varphi = \frac{0.87 \times 415}{4 \times 1.2 \times 1.6}\,\varphi = 47\,\varphi = 47 \times 25 = 1175 \text{ mm}$$

Development length (L_d) of 20 mm dia. bars in compression.

$$= \frac{0.87\, f_y}{4(1.25\,\tau_{bd})}\,\varphi = \frac{0.87 \times 415}{5 \times 1.92}\,\varphi = 37.6\,\varphi \approx 750 \text{ mm}$$

Hence the tension steel cannot be curtailed at less than 1.18 m from the centre of span, while the compression steel cannot be curtailed at less than 0.75 m from the centre. Let us curtail 1 compression bar and two tension bars at a theoretical point of cut off distant 1.20 m from the centre of span.

$\therefore$ Actual point of cut-off from the centre of the span will be larger of the two values: (*i*) $1.2 + d = 1.2 + 0.457 = 1.657$ m. (*ii*) $1.2 + 12\,\varphi = 1.2 + 12 \times 0.025 = 1.50$ m

Keep the *actual* point of cut-off at 1.70 m from centre of span. Distance of theoretical point of cut off from centre of span = 1.70 – 0.457 = 1.243 m. Hence distance of theoretical point of cut-off from centre of support = 6.465 / 2 – 1.243 ≈ 2.2 m. It can be shown that the moment of resistance of the section at this point will be larger than the actual B.M. at this location.

Step 6. Shear reinforcement : The critical section for shear is at a distance of d (= 457 mm) from the face of the support, or at a distance of $457 + \dfrac{457}{2} = 685.5$ mm ≈ 0.69 m from the centre of support.

$$V_u = \left(50625 \times \frac{6.457}{2}\right) - (50625 \times 0.69) = 128511 \text{ N}$$

$$\tau_v = \frac{V_u}{bd} = \frac{128511}{300 \times 457} \approx 0.94 \text{ N/mm}^2.; \qquad \frac{100\,A_s}{bd} = \frac{100\,(2 \times 490.9)}{300 \times 457} = 0.72\%$$

Hence from Table 7.1. $\tau_c = 0.34$ N/mm². Hence shear reinforcement is necessary. Let us provide 8 mm φ 2 lgd stirrups, having $A_{sv} = 100.5$ mm²

$$\therefore \qquad s_v = \frac{0.87\, f_y\, .\, A_{sv}}{(\tau_v - \tau_c)b} = \frac{0.87 \times 415 \times 100.5}{(0.94 - 0.34) \times 300} = 201 \text{ mm}$$

Spacing based on minimum shear reinforcement :

$$s_v = \frac{2.175\ A_{sv} f_y}{b} = \frac{2.175 \times 100.5 \times 415}{300} = 302 \text{ mm}$$

Spacing based on effective depth of beam

$$s_v = 0.75\ d = 0.75 \times 457 = 342 \text{ mm or } 300 \text{ mm which ever is less.}$$

Hence provide 8 mm φ 2 lgd strip @ 200 mm c/c near support and increase this gradually to 300 mm c/c towards the centre.

Step 7. Check for development length at support : At the support, the tensile reinforcement should satisfy the following relation : $1.3 \dfrac{M_{1u}}{V_u} + L_0 \geq L_d$

There are two tensile bars available at supports, having

$$A_{st1} = 2 \times 490.9 = 981.7 \text{ mm}^2$$

$$x_u = \frac{0.87\ f_y\ A_{st1}}{0.36\ f_{ck}\ .\ b} = \frac{0.87 \times 415 \times 981.7}{0.36 \times 20 \times 300} = 164 \text{ mm}$$

$$\therefore \quad M_{1u} = 0.87\ f_y\ A_{st1}(d - 0.416\ x_u) = 0.87 \times 415 \times 981.7\ (457 - 0.416 \times 164)$$

$$= 137.8 \times 10^6 \text{ N-mm}$$

$$V_u = 50625 \times \frac{6.457}{2} = 163443 \text{ N}$$

$L_d = 47\ \varphi = 47 \times 25 = 1175$ mm for tension and $L_{dc} = 38\ \varphi = 38 \times 16 = 608$ mm for comp.

Minimum embedment required from the face of the support $= L_d / 3 = 1175 / 3 = 392$ mm. Width of support available $(b_0) = 500$ mm. Providing end cover x' of 40 mm, and providing 90° bend to the tensile bars,

$$L_0 = (500 - 40 - 228.6 + 3\ \varphi) = 231.4 + 3 \times 25 = 306.4 \text{ mm}$$

$$1.3 \frac{M_{1u}}{V_u} + L_0 = 1.3 \times \frac{137.8 \times 10^6}{163443} + 306.4 = 1402 \text{ mm} > L_d. \text{ Hence O.K.}$$

The compression reinforcement at the top is embedded into the support by a length $= b_0 -$ end cover $= 500 - 40 = 460$ mm, which is more than $\dfrac{L_{dc}}{3}\ (= \dfrac{608}{3} \simeq 203$ mm).

Step 8. Details of reinforcement etc. As shown in Fig. 11.22.

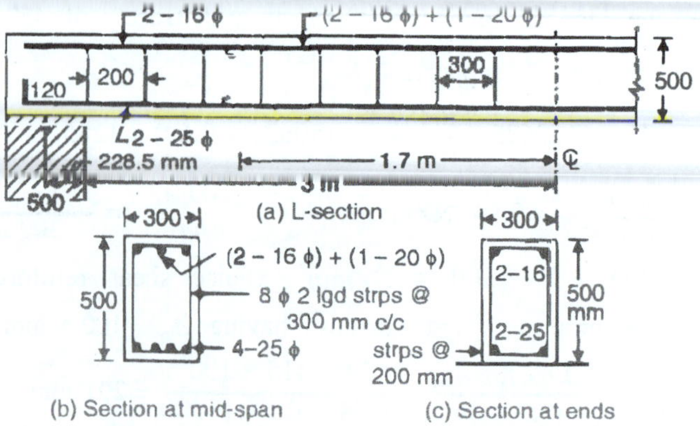

FIG. 11.22. DOUBLY REINFORCED BEAM.

11.9. DESIGN OF T-BEAM

The theory of T-beam has been discussed in detail in Chapter 6. The rules for selection of width of flange and breadth of rib have been given in Chapter 6. The general notes for design, given in § 11.1 for rectangular beams also apply for T and L beams. However, for flanged beams, the basic span to effective depth ratio shall be further modified as per Fig. 11.3 and the reinforcement percentage for use in Fig. 11.1 and 11.2 should be based on area of section equal to $b_f\, d$.

Design Procedure

The design of a T-beam is done in the following steps :

1. Assume some suitable value of width of web based on criteria discussed in § 6.2.

2. Determine the effective width b_f of the flange as discussed in § 6.2..

3. For computation of self weight, assume total depth (D) of the beam, equal to 1/13 to 1/15 of the span when it is simply supported at the ends. When it is continuous, the assumed over all depth may be taken as 1/15 to 1/20 of the span for light loads, 1/12 to 1/15 for medium loads and 1/10 to 1/12 of span for heavy loads. Compute load on the beam.

4. Assume effective span of the beam and compute ultimate design B.M. (M_{uD}) and ultimate S.F. (V_u)

5. Compute effective depth of beam by taking 2/3 of the depth required for maximum sagging B.M., treating it as a balanced section. Thus

$$d \simeq \frac{2}{3}\sqrt{\frac{M_u}{R_u \,.\, b}}$$

Estimate total depth (D) of the beam as under :

$$D = d + \text{nominal cover} + \text{dia. } (\varphi_s) \text{ of stirrups } + \frac{\varphi}{2} \text{ of estimated main bar}$$

Round off this value of D, and recalculate d as under

$$d = D - \text{nominal cover} - \varphi_s - \varphi/2$$

Thus, effective depth d of the T-beam has been fixed.

6. At this stage, b_w , b_f , D_f and d are known. Assuming $x_u = D_f$, compute moment of resistance M_{u1} from Eq. 6.9 (a)

$$M_{u1} = 0.36\, f_{ck}\, b_f\, D_f\, (d - 0.416\, D_f) \qquad \ldots(6.9\ a)$$

7. If $M_{u1} > M_{uD}$, the N.A. will fall inside the flange (i.e. $x_u \le D_f$). In that case, the reinforcement is given by Eq. 6.7

$$M_u = M_{uD} = 0.87\, f_y\, A_{st}\, d\left[1 - \frac{f_y}{f_{ck}}\frac{A_{st}}{b_f \,.\, d}\right] \qquad \ldots(6.7)$$

The solution of the above quadratic equation gives

$$A_{st} = \frac{0.5\, f_{ck}}{f_y}\left[1 - \sqrt{1 - \frac{4.6\, M_{uD}}{f_{ck}\, b_f\, d^2}}\right] b_f \,.\, d \qquad \ldots(6.7\ a)$$

8. If $M_{u1} < M_{uD}$, the N.A. falls outside the flange. Assuming $x_u = \frac{7}{3} D_f$, compute M_{u2} from Eq. 6.16 (*c*).

$$M_{u2} = 0.36 f_{ck} \cdot b_w \left(\frac{7}{3} D_f\right)\left(d - 0.416 \times \frac{7}{3} D_f\right) + 0.446 f_{ck}.(b_f - b_w) D_f \left(d - 0.5 D_f\right) \quad \text{...(6.16 } c)$$

If $M_{u2} < M_{uD}$, $x_u > \frac{7}{3} D_f$, compute A_{sw} and A_{sf} as under

$$A_{sw} = \frac{0.36 f_{ck} x_u b_w}{0.87 f_y} \quad \text{and} \quad A_{sf} = \frac{0.446 f_{ck} (b_f - b_w) D_f}{0.87 f_y}$$

Total $A_{st} = A_{sw} + A_{sf}$

When $M_{uD} >> M_{u2}$ it is always advisable to compute $M_{u,lim}$ since it is likely that $x_u > x_{u,max}$. If $M_{uD} > M_{u,lim}$, the section will be doubly reinforced.

9. If $M_{u2} > M_{uD}$, $x_u < \frac{7}{3} D_f$. The design procedure is this case will be that of trial and error, with the following steps :

(*i*) Assume $x_u < \frac{7}{3} D_f$

(*ii*) Compute $y_f = 0.15 x_u + 0.65 D_f$ (subject to a max. of D_f)

(*iii*) Compute M_u from Eq. 6.17:

$$M_u = 0.36 \frac{x_u}{d}\left(1 - 0.416 \frac{x_u}{d}\right) f_{ck} b_w d^2 + 0.446 f_{ck} (b_f - b_w) y_f (d - 0.5 y_f)$$

(*iv*) Compare M_u with M_{uD}. If $M_u = M_{uD}$, assumed value of x_u is correct

If $M_{uD} > M_u$, increase x_u in next trial

If $M_{uD} < M_u$, decrease x_u in next trial

Repeat till $M_u = M_{uD}$

(*v*) Knowing x_u, compute $C_u = 0.36 f_{ck} x_u b_w + 0.446 f_{ck} (b_f - b_w) y_f$

and $$A_{st} = \frac{C_u}{0.87 f_y}$$

Alternatively, x_u can be found by substituting the value of y_f in terms of x_u in Eq. 6.17 and then solving the quadratic equation in terms of x_u. This value of x_u will be acceptable only of the value of y_f obtained from this is less than (or equal to) D_f.

10. Check for shear and design the shear reinforcement, exactly in the same way as done of rectangular beam.

11. Check for anchorage and development length at the supports.

Design Example 11.10 : Design of T-beam

The floor of a hall measures 16 m × 6 m to the faces of the supporting walls. The floor consists of three beams spaced at 4 m centre to centre, and the slab thickness is 120 mm. The floor carries a uniformly distributed load of 5 kN/m², inclusive of the floor finishes. Design the intermediate beam. Use M 20 concrete and Fe 415 steel. The support width may be assumed equal to 500 mm.

Solution

Step 1 : Determination of b_w

The width of the web should be sufficient to accommodate the tensile reinforcement. Assume that the reinforcement consists of 5 bars of 25 mm diameter, and the stirrups consists of 8 mm φ wire. Keeping a nominal cover of 25 mm at the sides and 25 mm clear distance between the bars, total width $b_w = (5 \times 25) + (4 \times 25) + (2 \times 25) + (2 \times 8) = 291$ mm. Keep $b_w = 300$ mm.

Step 2. Computation of width of flange (b_f)

The width of flange is taken the least of the following :

(*i*) $$b_f = \frac{l_0}{6} + b_w + 6\,D_f \quad (\text{where } l_0 = 6 + 0.5 = 6.5 = 6500 \text{ mm})$$

$$= \frac{6500}{6} + 300 + 6 \times 120 \simeq 2100 \text{ mm}$$

or (*ii*) $$b_f = b_w + \frac{1}{2} \text{ (sum of clear distances to adjacent beams on either side)}$$

$$= 300 + \frac{1}{2}[2\,(4000 - 300)] = 4000 \text{ mm. Hence adopt } b_f = 2100 \text{ mm.}$$

Step 3

Effective span = 6 + 0.5 = 6.5 m = 6500 mm

$$\text{Assume } D = \frac{1}{13} \text{ of span} = \frac{6500}{13} = 500 \text{ mm (say)}$$

Dead load of slab/m² = $0.12 \times 1 \times 1 \times 25000$ = 3000 N/m²

Super-imposed load/m² = 5000 N/m²

Total = 8000 N/m²

Load per metre run of beam = 8000×4 = 32000 N/m

Self weight of web/m run = $0.3 \times 0.38 \times 1 \times 25000$ = 2850 N/m

Total load $w = 34850 \simeq 35000$ N/m (say)

$$w_u = 1.5\,w = 1.5 \times 35000 = 52500 \text{ N/m}$$

Step 4. Computation of M_u and V_u

$$\therefore \quad M_u = \frac{w_u L^2}{8} = \frac{52500\,(6.5)^2}{8} \simeq 277265 \text{ N-m} = 277.265 \times 10^6 \text{ N-mm}$$

$$V_u = \frac{w_u\,l}{2} = 52500 \times \frac{6}{2} = 157500 \text{ N}$$

Step 5. Fixation of effective depth and total depth

$$d = \frac{2}{3}\sqrt{\frac{M_u}{R_u\,.\,b}} = \frac{2}{3}\sqrt{\frac{277.265 \times 10^6}{2.761 \times 300}} = 386$$

Also, from stiffness point of view, $L/d = 20$.

$$p_{lim.} = 0.955\,\% \simeq 1\%, \text{ for which, modification factor } F_t = 1$$

Also, $b_w/b_f = \frac{300}{2100} \simeq 0.14$ for which reduction factor $F_b = 0.8$ (Fig. 11.3)

Hence $d = \frac{L}{20 \times 1 \times 0.8} = \frac{6500}{20 \times 1 \times 0.8} = 406$ mm

Adopting $d = 406$ mm (greater of the two), using 25 mm φ main bars, 8 mm φ stirrups and providing 25 mm nominal cover, $D = 406 + 25 + 8 + 12.5 = 451.5$ mm.

However, keep $D = 450$ mm, so that $d = 450 - 25 - 8 - 25/2 = 404.5$ mm.

Thus, for the T-beam, we have $b_w = 300$ mm, $b_f = 2100$ mm; $D = 450$ mm and $d = 404.5$ mm.

Step 6. Assume $x_u = D_f = 120$ mm

$$M_{u1} = 0.36 f_{ck}\, b_f\, D_f\,(d - 0.416\, D_f) = 0.36 \times 20 \times 2100 \times 120\,(404.5 - 0.416 \times 120)$$
$$= 643.35 \times 10^6 \text{ N-mm}$$

Step 7. Since $M_{u1} > M_{uD}$, N.A. falls inside the flange, i.e. $x_u \leq D_f$

Hence $M_u = M_{uD} = 0.87 f_y A_{st}\, d\left[1 - \frac{f_y}{f_{ck}}\frac{A_{st}}{b_f\, d}\right]$ or

$$A_{st} = \frac{0.5 f_{ck}}{f_y}\left[1 - \sqrt{1 - \frac{4.6 M_{uD}}{f_{ck}\, b_f\, d^2}}\right] b_f\, d = \frac{0.5 \times 20}{415}\left[1 - \sqrt{1 - \frac{4.6 \times 277.265 \times 10^6}{20 \times 2100\,(404.5)^2}}\right] \times 2100 \times 404.5$$
$$= 1997 \text{ mm}^2$$

Choosing 25 mm φ bars, number of bars $= 1997/490.9 = 4.07$

Provide 5–25 mm φ bars. Actual $A_{st} = 5 \times 490.9 = 2454.5 \text{ mm}^2$

Bend 2 bars up at 45° at a distance of 1.414 (0.9 d) $= 1.414 \times 0.9 \times 404.5 \simeq 500$ mm from the edge of the support.

Step 8 $$\tau_v = \frac{V_u}{b_w\, d} = \frac{157500}{300 \times 404.5} \simeq 1.3 \text{ N/mm}^2$$

$$\frac{100\, A_s}{b_w\, d} = \frac{100 \times 3 \times 490.9}{300 \times 404.5} \simeq 1\%$$

For 1% reinforcement, $\tau_c = 0.62 \text{ N/mm}^2$; Also $\tau_{c,\,max} = 2.8 \text{ N/mm}^2$. Hence safe.

Since $\tau_v > \tau_c$, shear reinforcement is necessary

$$V_{uc} = \tau_c\, b_w\, .\, d = 0.62 \times 300 \times 404.5 = 75236 \text{ N}$$

$\therefore$ $$V_{us} = 157500 - 75236 = 82264 \text{ N}$$

Shear resistance of 2 bent up bars, is given by

$$V_{us1} = 0.87 f_y A_{sv} \sin 45° = 0.707 \times 0.87 \times 415\,(2 \times 490.9) = 250616 \text{ N}$$

However, maximum resistance assigned to the bent-up bars $= \frac{1}{2} V_{us} \simeq 41132$ N

Hence provide vertical stirrups for the remaining 41132 N shear $= V_{us2}$

Using 8 mm φ 2-lgd stirrups, $A_{sv} = 2\,\frac{\pi}{4}(8)^2 = 100.5 \text{ mm}^2$

$\therefore$ $$s_v = \frac{0.87 f_y A_{sv}\, d}{V_{us2}} = \frac{0.87 \times 415 \times 100.5 \times 404.5}{41132} = 356.8 \text{ mm}$$

Maximum spacing corresponding to nominal stirrups is given by

$$s_v = \frac{2.175\, A_{sv} f_y}{b_w} = \frac{2.175 \times 100.5 \times 415}{300} = 302 \text{ mm}$$

Also, the spacing should be less than lesser of 0.75 d (= 0.75 × 404.5 = 303) or 300 mm.

Hence provide 8 mm φ 2 lgd stirrups @ 300 mm c/c throughout. Provide 2-12 mm φ bars top as anchor bars.

Step 9. Check for development length at supports : The Code envisages that the diameter of bars should be so selected that following relation is satisfied

$$1.3\, \frac{M_{1u}}{V_u} + L_0 \geq L_d.$$

$$A_{st1} = \text{area of remaining bars at supports} = 3 \times 490.9 = 1472.7 \text{ mm}^2$$

$$x_u = \frac{0.87\, f_y\, A_{st1}}{0.36\, f_{ck}\, b_w} = \frac{0.87 \times 415 \times 1472.7}{0.36 \times 20 \times 300} = 246.2 \text{ mm}$$

$$M_{1u} = 0.87\, f_y\, A_{st1}\, (d - 0.416\, x_u) = 0.87 \times 415 \times 1472.7\, (404.5 - 0.416 \times 246.2)$$

$$= 160.62 \times 10^6 \text{ N-mm}$$

$$V_u = \frac{w_u\, L}{2} = 52500 \times \frac{6.5}{2} = 170625 \text{ N}$$

$$L_d = 47\, \varphi = 47 \times 25 = 1175 \text{ mm}$$

Taking the bars straight into supports without any hook or bend, and providing side cover $x' = 30$ mm we have

$$L_0 = \frac{L_s}{2} - x' = \frac{500}{2} - 30 = 220 \text{ mm.}$$

$$\therefore \quad 1.3\, \frac{M_{1u}}{V_u} + L_0 = 1.3\, \frac{160.62 \times 10^6}{170625} + 220 = 1223 + 220 = 1443 \text{ mm} > L_d$$

Hence Code requirements are satisfied.

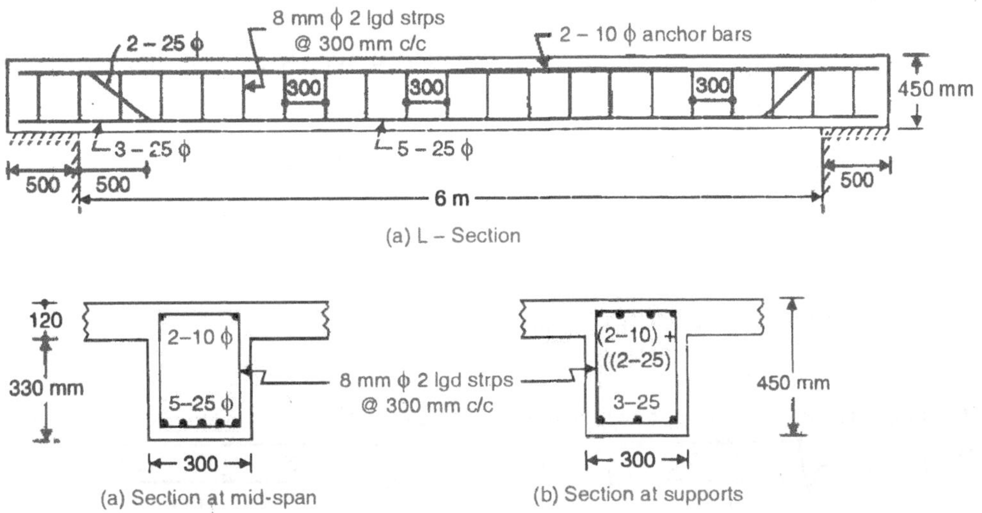

FIG. 11.23

Step 10. Details of reinforcement etc

The details of reinforcement etc. are shown in Fig. 11.23

11.10. DESIGN OF CYCLE STAND SHADE

Design Example 11.11. *A cycle stand shade consists of a R.C. slab which cantilevers 3 metres on each side of a central R.C. beam and is monolithic with it. The R.C. beam is simply supported on columns 400 mm wide, at the ends, over a clear span of 6 metres. Design the shade for a super-imposed load of 2000 N/m². Use M 20 concrete and Fe 415 steel.*

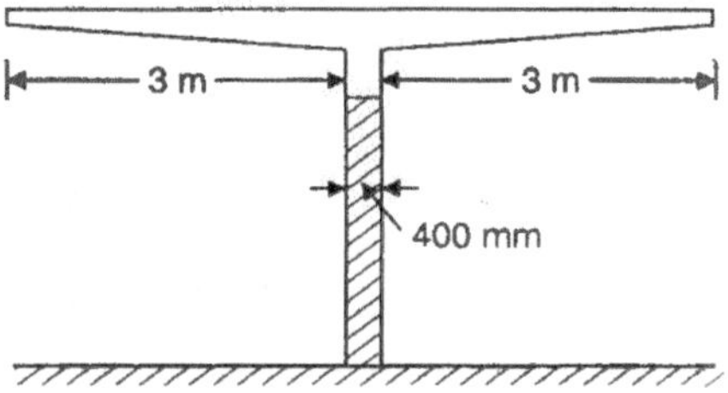

FIG. 11.24

Solution : For M 20 concrete, $f_{ck} = 20\ \text{N/mm}^2$, For Fe 415 steel, $f_y = 415\ \text{N/mm}^2$

(A) Design of cantilevering slab

For stiffness, the ratio of span to depth of cantilever = 7. For a balanced design, the percentage reinforcement for M 20 concrete = 0.955% (Table 3.2). However, for an under-reinforced slab, let us assume $p_t = 0.2\%$, corresponding to which we get modification factor $F_t \simeq 1.7$ for Fe 415 steel. Assuming $d \simeq 270$ mm, effective $L = 3000 + 270/2 = 3135$ mm. Hence $d = \dfrac{L}{7 \times 1.7} = \dfrac{3135}{11.9} = 264$ mm.

Let us keep a total depth of 290 mm and reduce it to 100 mm at the ends. Keeping 15 mm nominal cover and using 10 mm φ bars, available $d = 290 - 15 - 10/2 = 270$ mm at the supports and $d = 100 - 15 - 10/2 = 80$ mm at free end.

Mean thickness of slab $= \frac{1}{2}(290 + 100) = 195$ mm

Dead load of slab per sq. meter $= 0.195 \times 1 \times \times 25000 = 4875$ N

Superimposed load = 2000 N/m². Hence total load = 6875 N/m².

Hence $w_u = 6875 \times 1.5 = 10312.5\ \text{N/m}^2$

$$M_u = w_u\, l^2/2 = 10312.5\,(3)^2/2 \simeq 46406\ \text{N-m} = 46.406 \times 10^6\ \text{N-mm}$$

and $$V_u = w_u\, l = 10312.5 \times 3 \simeq 30938\ \text{N}$$

For Fe 415 steel, $$\frac{x_{u,\,max}}{d} = \frac{700}{1100 + 0.87 \times 415} = 0.479$$

$$\therefore \quad R_u = 0.36 f_{ck} \frac{x_{u,\,max}}{d}\left(1 - 0.416 \frac{x_{u,\,max}}{d}\right) = 0.36 \times 20 \times 0.479\,(1 - 0.416 \times 0.479) = 2.761$$

$$\therefore \quad d = \sqrt{\frac{M_u}{R_u\, d}} = \sqrt{\frac{46.406 \times 10^6}{2.761 \times 1000}} = 129.6\ \text{mm}$$

However, from deflection point of view, we want $D = 290$ mm so that available $d = 270$ mm. Hence we will have *under-reinforced section* for which A_{st} is given by

$$A_{st} = \frac{0.5 f_{ck}}{f_y}\left[1 - \sqrt{1 - \frac{4.6 M_u}{f_{ck}\, bd^2}}\right] bd = \frac{0.5 \times 20}{415}\left[1 - \sqrt{1 - \frac{4.6 \times 46.406 \times 10^6}{20 \times 1000\,(270)^2}}\right] \times 1000 \times 270$$

$$= 495\ \text{mm}^2$$

Using 10 mm φ bars having $A_\varphi = 78.54\ \text{mm}^2$, $s = \dfrac{1000 \times 78.54}{495} = 158.6$ mm

Hence provide 10 mm φ bars @150 mm c/c.

Actual $\quad A_{st} = \dfrac{1000 \times 78.54}{150} = 523.6\ \text{mm}^2$; $p_t\,(\%) = \dfrac{523.6 \times 100}{1000 \times 270} = 0.19\%$

Let us curtail alternate bars at half the cantilever length.

B.M. at the point $= \dfrac{w_u x^2}{2} = \dfrac{10312.5(1.5)^2}{2} \times 1000 = 11.601 \times 10^6$ N-mm

Effective depth of the slab $= \dfrac{270 + 80}{2} = 175$ mm

$$\therefore \quad A_{st} \text{ required} = \frac{0.5 \times 20}{415}\left[1 - \sqrt{1 - \frac{4.6 \times 11.601 \times 10^6}{20 \times 1000\,(175)^2}}\right] 1000 \times 175 = 187.9\ \text{mm}^2$$

Actual A_{st} available $= \frac{1}{2} \times 523.6 = 261.8\ \text{mm}^2$ Hence OK

$$A_{sd} = \frac{0.12\, b\,.\,D}{100} = \frac{1.2 \times 1000 \times 290}{100} = 348\ \text{mm}^2 \text{ at fixed end}$$

Using 8 mm φ bars, $s = \dfrac{1000 \times 50.3}{348} = 144$ mm. Hence provide these @140 mm c/c.

At half the cantilever length, $D = \frac{1}{2}(290 + 100) = 195$ mm

$$\therefore \quad A_{sd} = \frac{0.12 \times 1000 \times 195}{100} = 234\ \text{mm}^2$$

$$\therefore \quad s = \frac{1000 \times 50.3}{234} = 210\ \text{mm.}$$

Hence provide these @ 200 mm c/c.

Check for shear : $\quad \tau_v = \dfrac{V_u}{bd} = \dfrac{30938}{1000 \times 270} = 0.114\ \text{N/mm}^2 < \tau_c$

(B) Design of beam

The beam will be designed as T-beam with tapered flanges. However, we can approximately assume the tapered flanges to be equivalent to flanges of uniform average thickness $= \frac{1}{2}(290 + 100) = 195$ mm $= D_f$.

Width of web $b_w =$ width of column $= 400$ mm

Total width of flange $= b = 6.4$ m $= 6400$ mm.

Let the effective span of beam $= l_0 \simeq 6 + 0.4 = 6.4$ m

For an isolated T-beam, the width of flange is

$$b_f = \frac{l_0}{\left(\dfrac{l_0}{b}\right) + 4} + b_w = \frac{6400}{\dfrac{6400}{6400} + 4} + 400 = 1680\ \text{mm}$$

Assume the overall depth $D = l/15 = 6000/15 = 400$ mm for the purpose of computation of dead load.

(*i*) Dead load of slab $= 0.195 \times 6 \times 25000 = 29250$ N

(*ii*) Dead load of beam $= 0.4 \times 0.4 \times 25000 = 4000$ N

(*iii*) Superimposed load $= 2000\,(3 + 3 + 0.4) \times 1 = 12800$ N

Total $w \simeq 46000$ N/m

$$\therefore \qquad w_u = 1.5 \times 46000 = 69000 \text{ N and } M_u = \frac{69000\,(6.4)^2}{8} = 353280 \text{ N-m}$$

Fixation of effective depth and total depth

$$\text{Let} \qquad d \simeq \frac{2}{3}\sqrt{\frac{M_u}{R_u\, b_w}} = \frac{2}{3}\sqrt{\frac{353.28 \times 10^6}{2.761 \times 400}} \simeq 377 \text{ mm}$$

Also, from stiffness point of view, $L/d = 20$. Taking $p_t = p_{t,\,lim} = 0.955\%$, we get modification factor $F_t = 1$ (Fig. 11.1). Also $b_w/b_f = 400/1680 = 0.24$, for which reduction factor $F_b = 0.8$ (Fig. 11.3). Hence $d = \dfrac{L}{20 \times 1 \times 0.8} = \dfrac{6500}{20 \times 1 \times 0.8} = 406$ mm.

Hence provide $D = 460$ mm, so that using main bars of 30 mm φ, stirrups of 8 mm φ and nominal cover of 30 mm, available $d = 460 - 30 - 8 - 30/2 = 407$ mm. Due to this increased depth, revised dead load of beam $= 0.4 \times 0.46 \times 25000 = 4600$ N/m and total $w \simeq 46600$ N/m,

$$\therefore \qquad w_u = 1.5 \times 46600 = 69900 \text{ N/m}$$

$$\text{and} \qquad M_u = 69900\,(6.4)^2/8 = 357888 \text{ N-m} \simeq 357.9 \times 10^6 \text{ N-mm}$$

Determination of reinforcement : Assume $x_u = D_f = 195$ mm

$$M_{u1} = 0.36\, f_{ck}\, b_f\, D_f\,(d - 0.416\, D_f) = 0.36 \times 20 \times 1680 \times 195\,(407 - 0.416 \times 195)$$

$$= 768.7 \times 10^6 \text{ N-mm}$$

Since $M_{u1} > M_u$, N.A. falls inside the flange, i.e. $x_u \le D_f$

Hence $M_u = M_{uD} = 0.87\, f_y\, A_{st}\, d\left[1 - \dfrac{f_y}{f_{ck}}\dfrac{A_{st}}{b_f\, d}\right]$, which gives

$$A_{st} = \frac{0.5\, f_{ck}}{f_y}\left[1 - \sqrt{1 - \frac{4.6\, M_{uD}}{f_{ck}\, b_f\, d^2}}\right] b_f\, d = \frac{0.5 \times 20}{415}\left[1 - \sqrt{1 - \frac{4.6 \times 357.9 \times 10^6}{20 \times 1680\,(407)^2}}\right] \times 1680 \times 407$$

$$= 2650 \text{ mm}^2$$

Using 30 mm φ bars having $A_\varphi = 706.86$ mm², number of bars $= 2650/706.86 = 3.75$

Using 25 mm φ bars having $A_\varphi = 490.87$ mm², number of bars $= 2650/490.87 = 5.4$

Hence provide 6 bars of 25 mm φ. Actual $A_{st} = 490.87 \times 6 = 2945$ mm² and actual $p_t = A_{st}/(b_f \,.\, d) = 2945 \times 100/(1680 \times 407) = 0.43\%$.

Check for shear and design of shear reinforcement : Bend 2 bars up at 45° at a distance of $1.414\,(0.9\,d) = 1.414 \times 0.9 \times 407 \simeq 500$ mm from the face of the support.

The critical section for shear is at a distance of d (= 407 mm) from the face of the support, or at a distance of $\dfrac{500}{2} + 407 \simeq 650$ mm from the centre of the support, where S.F. is given by

$$V_u = \frac{w_u L}{2} - w_u x = 69900\left(\frac{6.4}{2} - 0.65\right) = 178245 \text{ N}$$

$$\tau_v = \frac{V_u}{b_w d} = \frac{178245}{400 \times 407} \simeq 1.1 \text{ N/mm}^2$$

$$\frac{100 A_s}{b_w d} = \frac{100 (4 \times 490.87)}{400 \times 407} = 1.2\% \text{. Hence } \tau_c = 0.66 \text{ N/mm}^2$$

Since $\tau_v > \tau_c$, shear reinforcement is necessary.

$$V_{uc} = \tau_c b_w d = 0.66 \times 400 \times 407 = 107448 \text{ N}$$

$$\therefore \quad V_{us} = V_u - V_{uc} = 178245 - 107448 \simeq 70800 \text{ N}$$

Shear resistance of 2 bent-up bars is given by

$V_{us1} = 0.87 f_y A_{sv} \sin 45° = 0.87 \times 415 (2 \times 490.9) \times 0.707 = 250616$ N

However, maximum resistance assigned to bent up bars $= \frac{1}{2} V_{us} = 35400$ N

Hence provide vertical stirrups for the remaining 35400 N shear $= V_{us2}$.

Shear resistance of nominal stirrups, $V_{us,\,min} = 0.4\, b_w \,.\, d = 0.4 \times 400 \times 407 = 65120$ N

Hence nominal stirrups will be sufficient, the spacing of which, using 8 mm φ 2-lgd bars, is given by

$$s_v \le \frac{2.175 A_{sv} f_y}{b_w} \le \frac{2.175 \times 100.5 \times 415}{400} = 226 \text{ mm}$$

Also, the spacing should be less than lesser of 0.75 d or 300 mm.

Hence provide 8 mm φ 2 lgd stirrups @ 220 mm c/c throughout. Provide 2-12 mm φ bars at top as anchor bars.

Check for development length at supports

$$1.3 \frac{M_{1u}}{V_u} + L_0 \ge L_d$$

$$A_{st1} = 4 \times 490.87 = 1963.5 \text{ mm}^2$$

$$x_u = \frac{0.87 f_y A_{st1}}{0.36 f_{ck} b_w} = \frac{0.87 \times 415 \times 1963.5}{0.36 \times 20 \times 400} = 246 \text{ mm}$$

$$M_{1u} = 0.87 f_y A_{st1} (d - 0.416 x_u) = 0.87 \times 415 \times 1963.5 (407 - 0.416 \times 246) = 215.8 \times 10^6$$

$$V_u = \frac{w_u L}{2} = \frac{69900 \times 6.5}{2} = 227175 \text{ N}$$

Taking the bars straight into supports, without any hook or bend and providing side cover $x' = 30$ mm, we have

$$L_0 = \frac{L_s}{2} - x' = \frac{500}{2} - 30 = 220 \text{ mm}$$

$$\therefore \quad 1.3 \frac{M_{1u}}{V_u} + L_0 = 1.3 \frac{215.8 \times 10^6}{227175} + 220 \simeq 1455 \text{ mm}$$

$$L_d = 47\,\varphi = 47 \times 25 = 1175 \text{ mm}$$

Hence Codes requirements are satisfied.

Details of reinforcement etc : Given in Fig. 11.25

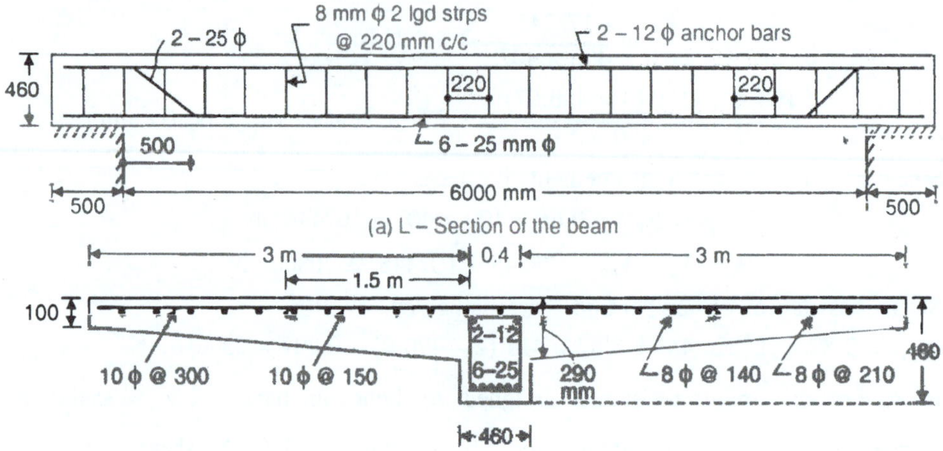

(a) L – Section of the beam

(b) Cross – section at mid-span

FIG. 11.25

11.11. DESIGN OF T-BEAM ROOF

Design Example 11.12 *A hall measures 10 m × 6 m from inside, and has walls 400 mm thick. Design a suitable R.C. T-beam roof to carry a super-imposed load of 2000 N/m². Use M 20 concrete and Fe 415 steel.*

Solution

Let us keep end spans of slab slightly less than the middle span so that approximately equal max. B.M. is induced in all the three spans. Assuming the width of web (b_w) of T-beam as 300 mm, the following spans of the slab are selected:

(*i*) End span = 3 m (clear)

(*ii*) Intermediate span = 3.4 m (clear).

The clear span of each T-beam will be 6 m while the R.C. slab monolithic with the beam will be continuous over the three spans.

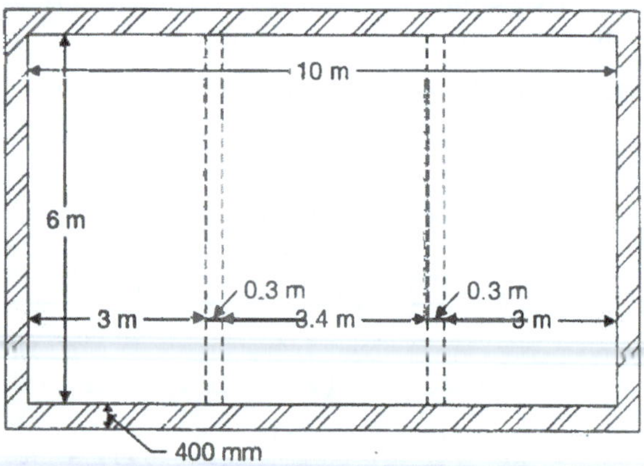

FIG. 11.26

For M 20 - Fe 415 combination, we have $f_{ck} = 20$ N/mm²; $f_y = 415$ N/mm²

$$\frac{x_{u,max}}{d} = \frac{700}{1100 + 0.87 \times 415} = 0.479$$

$$R_u = 0.36\, f_{ck} \frac{x_{u,max}}{d}\left(1 - 0.416 \frac{x_{u,max}}{d}\right) = 0.36 \times 20 \times 0.479\,(1 - 0.416 \times 0.479)$$

$$= 2.761$$

(A) Design of continuous slab

1. Fixation of D and d : For stiffness, $L/d = 26$ for continuous slab. Using an under-reinforced section and $p_t = 0.25\%$, we have modification factor $\simeq 1.6$.

Also, assuming total thickness of slab = 100 mm and effective depth as 80 mm, effective end span $= 3 + \frac{1}{2} \times 0.08 = 3.04$ m

Hence $$d = \frac{\text{span}}{26 \times 1.6} = \frac{3040}{26 \times 1.6} = 73 \text{ mm for end spans}$$

and $$d = \frac{3400}{26 \times 1.6} = 82 \text{ mm for intermediate span.}$$

Let us keep $D = 100$ mm. Using 8 mm φ bars and a nominal cover of 15 mm, available $d = 100 - 15 - 4 = 81$ mm.

$$\text{Dead load / m}^2 = 0.1 \times 1 \times 1 \times 25000 = 2500 \text{ N/m}^2 = w_d$$

$$\text{Superimposed load / m}^2 = 2000 \text{ N/m}^2 = w_s$$

2. Computation of B.M. and effective depth from flexure

(*i*) *For end span* : Near middle of end span

$$M \text{ due to dead load} = \frac{w_d L^2}{12} = \frac{2500\,(3.04)^2}{12} \times 1000 = 1.925 \times 10^6 \text{ N-mm}$$

$$M \text{ due to superimposed load} = \frac{w_s L^2}{10} = \frac{2000\,(3.04)^2}{10} \times 1000 = 1.848 \times 10^6 \text{ N-mm}$$

$\therefore$ Total $M = 3.773 \times 10^6$ N-mm; $M_u = 1.5 \times 3.773 \times 10^6 \simeq 5.66 \times 10^6$ N-mm

(*ii*) *For intermediate span (at the middle)*

$$M_d = \frac{w_d L^2}{16} = \frac{2500\,(3.4)^2}{16} \times 1000 = 1.806 \times 10^6 \text{ N-mm}$$

$$M_s = \frac{w_s L^2}{12} = \frac{2000\,(3.4)^2}{12} \times 1000 = 1.927 \times 10^6 \text{ N-mm}$$

$\therefore$ Total $M = 3.733 \times 10^6$ N-mm ; $M_u = 1.5 \times 3.733 \times 10^6 \simeq 5.6 \times 10^6$ N-mm

(*iii*) *At support next to end support*

$$M_d = \frac{w_d L^2}{10} = \frac{2500\,(3.04)^2}{10} \times 1000 = 2.310 \times 10^6 \text{ N-mm}$$

$$M_s = \frac{w_s L^2}{9} = \frac{2000\,(3.04)^2}{9} \times 1000 = 2.054 \times 10^6 \text{ N-mm}$$

$\therefore$ Total $M = 4.364 \times 10^6$ N-mm ; $M_u = 1.5 \times 4.364 \times 10^6 = 6.546 \times 10^6$ N-mm

Hence $(M_u)_{max} = 6.546 \times 10^6$ N-mm $= M_{uD}$

$$d = \sqrt{\frac{M_{uD}}{R_u\, b}} = \sqrt{\frac{6.546 \times 10^6}{2.761 \times 1000}} = 48.7 \text{ mm}$$

However, from deflection point of view, we want $D = 100$ mm so that $d = 81$ mm. Hence we will have an under-reinforced section.

3. Computation of reinforcement

$$A_{st} = \frac{0.5 f_{ck}}{f_y}\left[1 - \sqrt{1 - \frac{4.6 M_{uD}}{f_{ck}\, bd^2}}\right] b.d = \frac{0.5 \times 20}{415}\left[1 - \sqrt{1 - \frac{4.6 \times 6.546 \times 10^6}{20 \times 1000\,(81)^2}}\right] \times 1000 \times 81$$

$$= 238.5 \text{ mm}^2$$

Spacing of 8 mm φ bars, $s = \frac{1000\, A_{\varphi}}{A_{st}} = \frac{1000 \times 50.3}{238.5} = 210.9$ mm

However provide 8 mm φ bars @ 200 mm c/c at the middle of each span. Bend up alternative bars from the bottom at a distance of $L/5 = 3040/5 = 600$ mm from the centre of intermediate support and at $L/7 = 450$ mm from the end support. Thus reinforcement available at each intermediate support will be 8 mm φ bars @ 200 mm c/c. Actual A_{st} provided $= 1000 \times 50.3/200 = 251.5 \text{ mm}^2$ and $p_t\,(\%) = 251.5 \times 100/(1000 \times 100) \simeq 0.25\%$.

Distribution reinforcement : $A_{sd} = \frac{0.12 \times 1000}{100} D = 1.2\, D = 1.2 \times 100 = 120 \text{ mm}^2$

Using 8 mm φ, $s = 1000 \times 50.3/120 = 419$ mm. Hence provide these @ 400 mm c/c.

4. Check for shear

$$V = 0.4\, w_d\, L + 0.45\, w_s\, L \text{ at the outer support.}$$
$$= (0.4 \times 2500 + 0.45 \times 2000)\, 3.04 = 5776 \text{ N}$$
$$V_u = 1.5 \times 5776 = 8664 \text{ N}$$
$$\therefore \quad \tau_v = \frac{V_u}{bd} = \frac{8664}{1000 \times 81} = 0.11 \text{ N/mm}^2$$

This is much less than the permissible value of 1.3×0.28 N/mm² even for a minimum reinforcement of 0.15 %, (see Table 7.1). Hence safe.

5. Check for development length at the end support

At the end supports alternate bars are bent up. Hence available $A_{st1} = \frac{1}{2} \times 251.5 \simeq 125.7 \text{ mm}^2$.

$$x_u = \frac{0.87 f_y A_{st1}}{0.36 f_{ck}\, b} = \frac{0.87 \times 415 \times 125.7}{0.36 \times 20 \times 1000} = 6.3 \text{ mm}$$

$$M_{1u} = 0.87 f_y A_{st1}\,(d - 0.416\, x_u) = 0.87 \times 415 \times 125.7\,(81 - 0.416 \times 6.3)$$
$$= 3.56 \times 10^6 \text{ N-mm}$$

$$L_d = 47\,\varphi = 47 \times 8 = 376 \text{ mm}$$

Taking the bars straight into supports, without any bend or hook,

$$L_0 = \frac{l_s}{2} - x' = \frac{400}{2} - 20 \text{ (say)} = 180 \text{ mm}$$

$$\therefore \quad 1.3\,\frac{M_{1u}}{V_u} + L_0 = 1.3 \times \frac{3.56 \times 10^6}{8664} + 180 = 534 + 180 = 714 \text{ mm} > L_d$$

(B) Design of T-beams

1. Determination of b_w and D

Let the width of web $= b_w = 300$ mm. Effective span $(l_0) = 6.4$ m

$$b_f = \frac{l_0}{6} + b_w + 6\ D_f = \frac{6400}{6} + 300 + 6 \times 100 = 1966 \text{ mm}$$

Assume over all depth $D = l/15 = 6000/15 = 400$ mm for the purpose of computation of dead load. Hence depth of web $= 400 - 100 = 300$ mm.

2. *Computation of load and B.M.*

$$\text{Load from slab} = (w_d + w_s)\left[\frac{3+3.4}{2} + 0.3\right] - (2500 + 2000)\ 3.5 = 15750 \text{ N/m}$$

Weight of web $= 0.3 \times 0.3 \times 25000 = 2250$ N/m

$\therefore$ Total $w = 15750 + 2250 = 18000$ N/m; $w_u = 18000 \times 1.5 = 27000$ N/m

$$\therefore \quad M_u = \frac{w_u\,(6.4)^2}{8} = \frac{27000\,(6.4)^2}{8} \times 1000 = 138.24 \times 10^6 \text{ N-mm}$$

3. *Fixation of effective depth and total depth*

$$\text{Let} \quad d \simeq \frac{2}{3}\sqrt{\frac{M_u}{R_u\,b_w}} = \frac{2}{3}\sqrt{\frac{138.24 \times 10^6}{2.761 \times 300}} = 272 \text{ mm}$$

Also, from stiffness point of view, $L/d = 20$. Taking $p_t = p_{t,\,lim} \simeq 1\%$, we get modification factor $F_t = 1$ (Fig. 11.1). Also, $b_w/b_f = 300/1966 = 0.153$, for which reduction factor $F_b = 0.8$ (Fig. 11.3). Hence $d = 6400/(20 \times 1 \times 0.8) = 400$ mm.

Hence provide $D = 450$ mm, so that using main bars of 20 mm φ, stirrups of 8 mm φ and nominal cover of 25 mm, available $d = 450 - 25 - 8 - 20/2 = 407$ mm.

Due to this increased depth, weight of web $= 0.3 \times 0.35 \times 25000 = 2625$ N/m, and total $w = 15750 + 2625 = 18365$ N/m.

Hence $w_u = 1.5 \times 18375 \simeq 27560$ N/m and $M_u = 27560\,(6.4)^2/8 = 141.1 \times 10^5$ N-m $= M_{uD}$.

4. *Determination of reinforcement*

Assume $x_u = D_f = 100$ mm

$$M_{u1} = 0.36\,f_{ck}\,b_f\,D_f\,(d - 0.416\,D_f) = 0.36 \times 20 \times 1966 \times 100\,(407 - 0.416 \times 100) = 517.2 \times 10^6$$

Since $M_{u1} > M_{uD}$, N.A. falls inside the flange, i.e. $x_u \le D_f$

$$\text{Hence} \quad M_u = M_{uD} = 0.87\,f_y\,A_{st}\,d\left[1 - \frac{f_y}{f_{ck}}\frac{A_{st}}{b_f\,d}\right], \text{ which gives}$$

$$A_{st} = \frac{0.5\,f_{ck}}{f_y}\left[1 - \sqrt{1 - \frac{4.6\,M_{uD}}{f_{ck}\,b_f\,d^2}}\right] b_f\,d = \frac{0.5 \times 20}{415}\left[1 - \sqrt{1 - \frac{4.6 \times 141.1 \times 10^6}{20 \times 1966\,(407)^2}}\right] \times 1966 \times 407$$

$$= 986 \text{ mm}^2$$

Using 20 mm φ bars, number of bars $= 986/314.16 = 3.14$. However, provide 4 bars of 20 mm φ giving actual $A_{st} = 314.16 \times 4 = 1256.6$ mm^2 and

$$\text{actual} \quad p_t = \frac{1256.6 \times 100}{b_f\,.\,d} = \frac{1256.6 \times 100}{1966 \times 407} \simeq 0.16\% \text{ only.}$$

5. *Check for shear and design of shear reinforcement* : Bend 2 bars up at 45° at a distance of 1.414 (0.9 d) $= 1.414 \times 0.9 \times 407 \simeq 500$ mm from the face of the support.

The critical section for shear is at a distance of d (= 407 mm) from the face of the support or at a distance of $\frac{400}{2} + 407 \simeq 600$ mm from the centre of the support, where S.F. is given by

$$V_u = \frac{w_u L}{2} - w_u x = \frac{27560 \times 6.4}{2} - 27560 \times 0.6 = 71656 \text{ N}$$

$$\tau_v = \frac{V_u}{b_w d} = \frac{71656}{300 \times 407} \simeq 0.59 \text{ N/mm}^2$$

$$\frac{100 A_s}{b_w d} = \frac{100 (2 \times 314.16)}{300 \times 407} = 0.51 \text{ N/mm}^2.$$

Hence $\tau_c \simeq 0.48 \text{ N/mm}^2$.

Since $\tau_v > \tau_c$, shear reinforcement is necessary

$$V_{uc} = \tau_c b_w d = 0.48 \times 300 \times 407 = 58608 \text{ N}$$

$$\therefore \quad V_{us} = 71656 - 58608 = 13048 \text{ N}.$$

Apart from shear resistance of bent up bars, the shear resistance of nominal stirrups $= 0.4\, b_w d = 0.4 \times 300 \times 407 = 48840$ N. Hence provide only nominal shear stirrups consisting of 8 mm φ 2 lgd stirrups at spacing

$$s_v \le \frac{2.175 A_{sv} f_y}{b_w} = \frac{2.175 \times 100.5 \times 415}{300} = 302.5 \text{ mm}$$

Hence provide 8 mm φ 2 lgd stirrups @ 300 mm c/c throughout. Provide 2-12 mm φ bars at top as anchor bars.

6. Check for development length at supports

At supports, $1.3 \frac{M_{1u}}{V_u} + L_0 \ge L_d$.

Here $A_{st1} = 2 \times 314.16 = 628.32 \text{ mm}^2$

$$x_u = \frac{0.87 f_y A_{st1}}{0.36 f_{ck} b_w} = \frac{0.87 \times 415 \times 628.32}{0.36 \times 20 \times 300} = 105 \text{ mm}$$

$$M_{1u} = 0.87 f_y A_{st1} (d - 0.416 x_u) = 0.87 \times 415 \times 628.32 (407 - 0.416 \times 105)$$

$$= 82.4 \times 10^6 \text{ N-mm}$$

$$V_u = \frac{w_u L}{2} = \frac{27560 \times 6.4}{2} = 88192 \text{ N}$$

Taking the bars straight into the support, without any hook or bend,

$$L_0 = \frac{L_s}{2} - x' = \frac{400}{2} - 30 \text{ (say)} = 170 \text{ mm}$$

$$L_d = 47 \varphi = 47 \times 20 = 940 \text{ mm}$$

$$1.3 \frac{M_{1u}}{V_u} + L_0 = \frac{1.3 \times 82.4 \times 10^6}{88192} + 170 = 1214 + 170 = 1384 \text{ mm} > L_d$$

7. Details of reinforcement etc. : Fig. 11.27

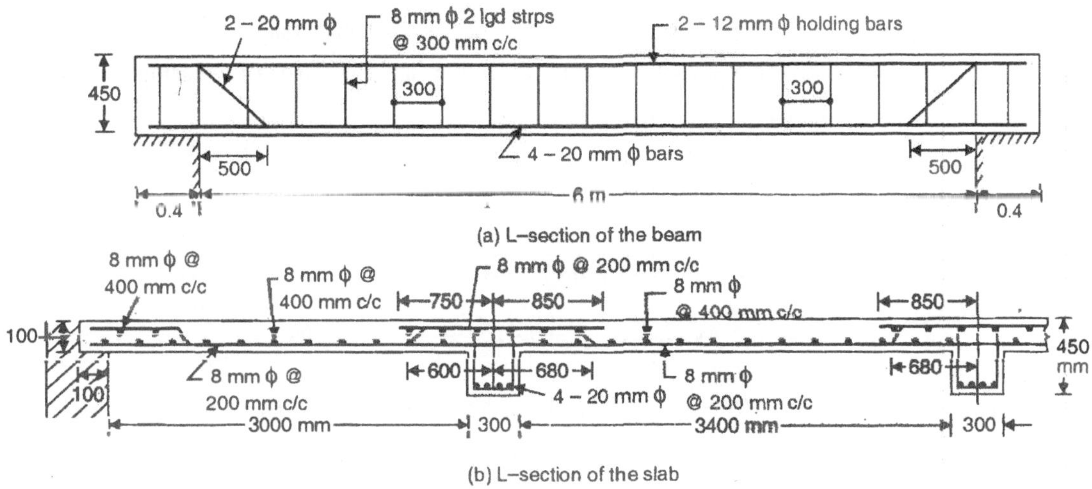

FIG. 11.27

11.12. DESIGN OF INVERTED T-BEAM ROOF

We have seen that if the size of a room is large, we may use monolithic beam-slab construction forming the T-beam roof. However, in some cases, the provision of the beam (i.e. rib) below the slab may be undesirable from architectural or other point of view. In that case, the beam is provided above the slab, forming what is known as the *inverted* T-beam system. In such a case, the ceiling, composed of the roof slab, is plane and the beam projects above the top of the slab. The slab is thus provided at the 'tension side' of the beam and is therefore not helpful in resisting the tensile stresses with the beam. The beam, therefore, acts as a simple rectangular beam, which may either be singly reinforced or doubly reinforced. The depth of the beam will be equal to the depth measured between the top of the beam and the bottom of the slab. Though the slab is cast monolithic with the beam it is tied to the beam by providing suitable vertical ties or stirrups. The design of the slab, however, is done in the same manner as for an ordinary T-beam roof.

Design example 11.13. *A hall measures 10 m × 6 m from inside and has walls 400 mm thick. Design a suitable inverted T-beam roof to carry a super-imposed load of 200 N/m². Use M 20 concrete and Fe 415 steel.*

Solution : This example is the same as example 11.12 except that the system is inverted. However, since the slab is continuous over the two beams, its design will be the same as done in example 11.12.

Design of rectangular beam

(*i*) ***Loading per metre run and bending moment***

Total thickness of slab = 100 mm. Let the width of the beam = 300 mm

$$\therefore \quad \text{Load from slab} = (2500 + 2000)\left(\frac{3}{2} + 0.3 + \frac{3.4}{2}\right) = 15750 \text{ N}$$

Since the system is inverted, the beam will act as a rectangular beam and not as a T-beam. Let us assume total depth of beam as $L/12 = 6400/12 \simeq 530$ mm.

$\therefore$ Weight of rib = 0.3 (0.53 − 0.1) × 25000 = 3225 N

$\therefore$ Total load/m run = 15750 + 3225 = 18975 N ≃ 19000 N/m (say)

$\therefore$ $w_u = 19000 \times 1.5 = 28500$ N/m; Effective span = c/c of bearing = 6.4 m

$$M_u = \frac{w_u\,(6.4)^2}{8} = 28500\,\frac{(6.4)^2}{8} = 145920 \text{ N-m} = 145.92 \times 10^6 \text{ N-mm}$$

(ii) Computation of effective depth and total depth

For Fe 415 steel, $\dfrac{x_{u,\,max}}{d} = \dfrac{700}{1100 + 0.87 \times 415} = 0.479$

$$M_u = 0.36\, f_{ck}\,\frac{x_{u,\,max}}{d}\left(1 - 0.416\,\frac{x_{u,\,max}}{d}\right) = R_u\, bd^2$$

where $R_u = 0.36 \times 20 \times 0.479\,(1 - 0.416 \times 0.479) = 2.761$

$$\therefore \quad d = \sqrt{\frac{M_u}{R_u\, b}} = \sqrt{\frac{145.92 \times 10^6}{2.761 \times 300}} \simeq 420 \text{ mm}$$

From deflection point of view, neglecting modification factor,

$$d = L/20 = 6400/20 = 320 \text{ mm only.}$$

However keep $D = 500$ mm. Providing 25 mm nominal cover and using 20 mm bars (say) and 8 mm dia. rings, $d = 500 - 25 - 8 - 20/2 = 457$ mm.

(iii) Determination of reinforcement

Since the actual effective depth provided is more than the balanced one, we have an under-reinforced section, for which

$$A_{st} = \frac{0.5\, f_{ck}}{f_y}\left[1 - \sqrt{1 - \frac{4.6\, M_u}{f_{ck}\, bd^2}}\right] bd = \frac{0.5 \times 20}{415}\left[1 - \sqrt{1 - \frac{4.6 \times 145.92 \times 10^6}{20 \times 300\,(457)^2}}\right] \times 300 \times 457$$

$$= 1053 \text{ mm}^2$$

$\therefore$ No. of 20 mm φ bars $= \dfrac{1053}{314.16} = 3.35$. However, provide 4 bars of 20 mm φ.

Actual area provided = 4 × 314.16 = 1256.6 mm². Bend 2 bars up at a distance of $\sqrt{2}\, a = 1.414 \times 0.9 \times 457 = 581.6$ mm ≃ 580 mm from the edge of the support.

(iv) Check for shear and design of shear reinforcement

$$V_u = \frac{w_u\, L}{2} = \frac{28500 \times 6.4}{2} = 91200 \text{ N}$$

The critical section for shear is at a distance of d (= 0.457 m) from the face of the support, where the S.F. is given by

$$V_{uD} = V_u - w_u\,(0.2 + 0.457) = 91200 - 28500 \times 0.657 = 72476 \text{ N}$$

$$\tau_v = \frac{V_{uD}}{bd} = \frac{72476}{300 \times 457} = 0.53 \text{ N/mm}^2$$

At supports, $\dfrac{100\, A_s}{bd} = \dfrac{100\,(0.5 \times 1256.6)}{300 \times 457} = 0.458\,\%$

Hence from Table 7.1, $\tau_c = 0.36 + \dfrac{0.48 - 0.36}{0.25}(0.458 - 0.25) = 0.46\ \text{N/mm}^2$

We observe that τ_v is only marginally more than τ_c. The bent-up bars will take care of this difference. However, provide nominal shear reinforcement given by

$$s_v \le \frac{2.175\, A_{sv} f_y}{b}$$

Choosing 8 mm φ HYSD bars, for two-legged stirrups, $A_{sv} = 2 \times \frac{\pi}{4}(8)^2 = 100.53\ \text{mm}^2$

$$\therefore \qquad s_v \le \frac{2.175 \times 100.53 \times 415}{300} = 302 \quad \text{mm}$$

Hence provide 8 mm φ 2-lgd stirrups @ 300 mm c/c throughout the length of the beam.

(v) *Check for development length at the supports*

At supports, $\quad 1.3\dfrac{M_{1u}}{V_u} + L_0 \ge L_d$

At supports, $\quad x_u = \dfrac{0.87 f_y A_{st1}}{0.36 f_{ck}\, b} = \dfrac{0.87 \times 415 \times 0.5 \times 1256.6}{0.36 \times 20 \times 300} = 105 \quad \text{mm}$

$$\therefore \qquad M_{1u} = 0.87 f_y A_{st1}(d - 0.416 \times x_u) = 0.87 \times 415 \times 0.5 \times 1256.6\,(457 - 0.416 \times 105)$$

$$= 93.76 \times 10^6 \ \text{N-mm}$$

$$V_u = 91200 \ \text{N}$$

$$L_d = 47\varphi = 47 \times 20 = 940 \quad \text{mm}$$

Taking the bars straight into supports, without any hook or bend, and providing side cover $x' = 30$ mm, we have

$$L_0 = \frac{L_s}{2} - x' = \frac{400}{2} - 30 = 170 \quad \text{mm}$$

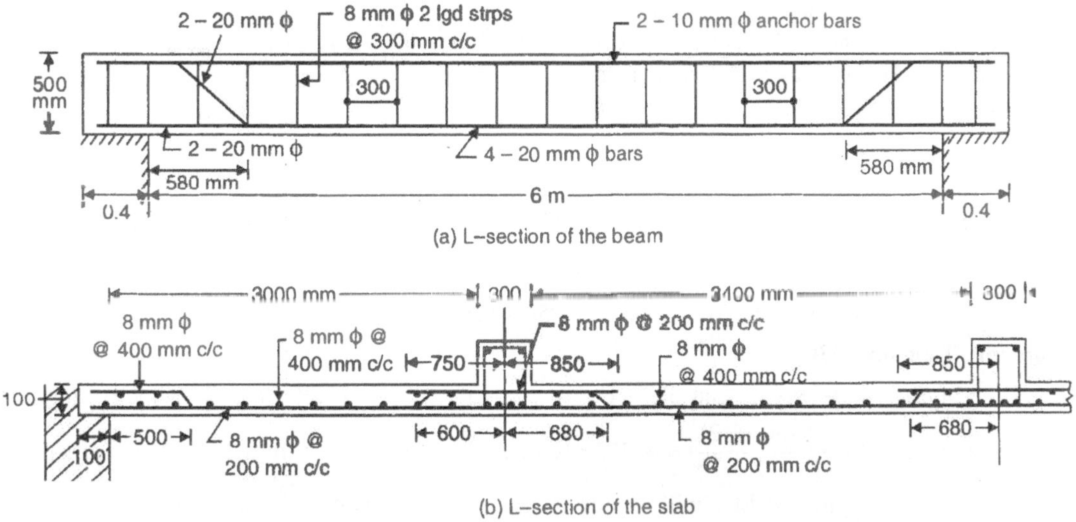

FIG. 11.28

$$\therefore \quad 1.3\,\frac{M_{1u}}{V_u} + L_0 = 1.3 \times \frac{93.76 \times 10^6}{91200} + 170 = 1506 > L_d. \text{ Hence OK.}$$

The details of reinforcement etc. are shown in Fig. 11.28.

11.13. DESIGN OF OVERHANGING T-BEAM ROOF

Design Example 11.14 : *A hall measures 10 m × 6 m from inside and has walls 400 mm thick. The beams of the T-beam roof project out by 2 m on one side of the room. The slab is monolithic with beam, and extends to the full length of the beam. Design a suitable roof system to carry a super-imposed load of 2 kN/m². Use M 20 concrete and Fe 415 steel.*

Solution : The problem is similar to example 11.12 except that the beams overhang to one side. Let us arrange the beams as shown in Fig. 11.29. Since the spans of the slab are the same as in example 11.12, its design will be the same as done in example 11.12. We will use the same set of stresses. Hence the thickness of the slab = 100 mm.

Design of beam

The walls are 400 mm thick. Hence Fig. 11.30, showing the spans etc may be drawn. The major portion of the beam *AB* will act as a T-beam, while the portion *BC* of the overhanging beam will act as a rectangular beam.

For the T-beam portion, $D \simeq \frac{1}{14}$ span

$= \frac{1}{14} \times 6400 \simeq 460$ mm.

However, for the overhanging portion, where the beam acts as a rectangular beam, the depth required from the B.M. point of view may be more. Hence take $D = 500$ mm throughout, for the purpose of computation of dead load.

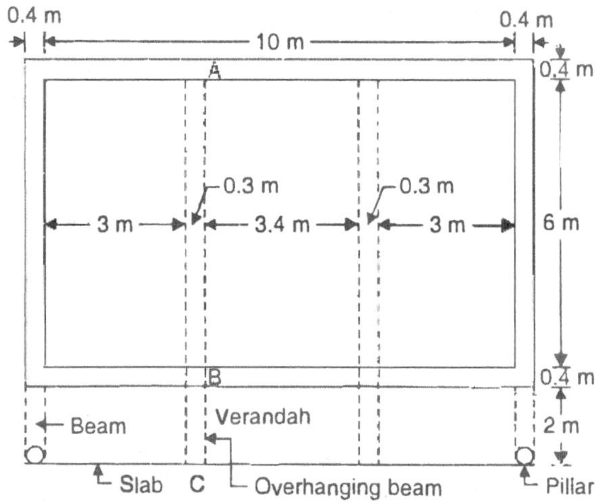

FIG. 11.29

$$\text{Super-imposed load from the slab} = (2500 + 2000)\left(\frac{3}{2} + 0.3 + \frac{3.4}{2}\right) = 15750 \text{ N}$$

Weight of rib = 0.3 × 0.4 × 25000 = 3000 N/m

∴ Total w = 15750 + 3000 = 18750 N/m

Design of T-beam *AB*

1. Bending moments : The B.M.D. is shown in Fig. 11.30 (*b*). The max. hogging B.M. at support $B = \frac{18750\,(2.2)^2}{2} = 45375$ N-m $= 45.375 \times 10^6$ N-mm

The max. sagging B.M. will occur some where between *A* and *B*.

$$R_B = \frac{18750\,(8.6)^2}{2} \times \frac{1}{6.4} = 108340 \text{ N}$$

$$\therefore \quad R_A = \frac{18750 \times 8.6\,(6.4 - 4.3)}{6.4}$$

$$= 52910$$

Measuring x from A, we have

$$M_x = 52910\,x - 18750\,\frac{x^2}{2}$$

For maxima,

$$\frac{dM_x}{dx} = 52910 - 18750\,x = 0,$$

from which $x = 2.822$ m. Hence,

$$M_{max} = 52910\,(2.822) - \frac{18750}{2}\,(2.822)^2$$

$= 74645$ N-mm $= 74.645 \times 10^6$ N-mm

The distance at which B.M. is zero is given by

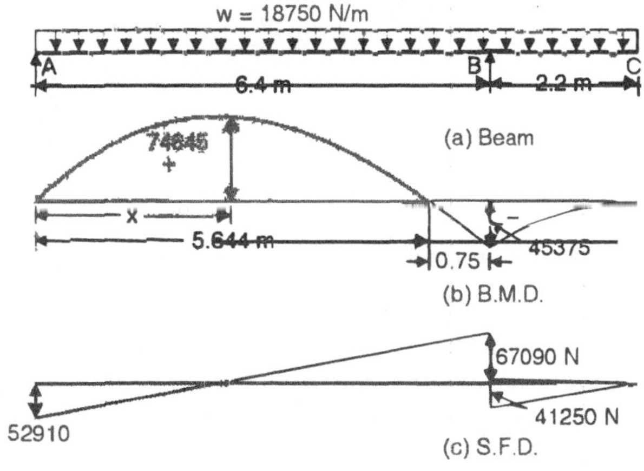

FIG. 11.30

$$52910\,x - 18750\,\frac{x^2}{2} = 0, \quad \text{from which } x = 5.644 \text{ m}$$

The B.M.D. and S.F.D. are shown in Fig. 11.30 (*b*) and (*c*) respectively.

Hence design B.M. $= M_{uD} = 1.5\,M_{max} = 1.5 \times 74.645 \times 10^6 \simeq 112 \times 10^6$ N-mm

Design S.F. $= V_u = 1.5 \times 67090 = 100635$ N.

2. *Determination of b_w and b_f*

Let the width of web, $b_w = 300$ mm. Span $(l_0) = 5.644$ m

$$b_f = \frac{l_0}{6} + b_w + 6\,D_f = \frac{5644}{6} + 300 + 6 \times 100 \simeq 1840 \text{ mm}$$

3. *Fixation of effective depth and total depth*

Let
$$d \simeq \frac{2}{3}\sqrt{\frac{M_{uD}}{R_u\,b_w}} = \frac{2}{3}\sqrt{\frac{112 \times 10^6}{2.761 \times 300}} \simeq 245 \text{ mm}$$

Also, from stiffness point of view, $L/d = 20$. Taking $p_t = p_{lim.} \simeq 1\%$, we get modification factor $F_t = 1$ (Fig. 11.1). Also, $b_w/b_f = 300/1840 = 0.163$ for which reduction factor $F_b = 0.8$ (Fig.11.3). Hence $d = 6400/(20 \times 1 \times 0.8) = 400$ mm. However keep $D = 450$ mm, so that using main bars of 20 mm φ, stirrups of 8 mm φ and nominal cover of 25 mm, available $d = 450 - 25 - 8 - 20/2 = 407$ mm.

4. *Determination of reinforcement*

Assume $x_u = D_f = 100$ mm

$$M_{u1} = 0.36\,f_{ck}\,b_f\,D_f\,(d - 0.416\,D_f) = 0.36 \times 20 \times 1840 \times 100\,(407 - 0.416 \times 100)$$

$$= 484 \times 10^6 \text{ N-mm}$$

Since $M_{u1} > M_{uD}$, N.A. falls inside the flange, i.e. $x_u \le D_f$

Hence
$$M_u = M_{uD} = 0.87\,f_y\,A_{st}\,d\left[1 - \frac{f_y}{f_{ck}} \cdot \frac{A_{st}}{b_f\,d}\right], \text{ from which}$$

$$A_{st} = \frac{0.5 f_{ck}}{f_y}\left[1 - \sqrt{1 - \frac{4.6\, M_{uD}}{f_{ck}\, b_f\, d^2}}\right] b_f\, d = \frac{0.5 \times 20}{415}\left[1 - \sqrt{1 - \frac{4.6 \times 112 \times 10^6}{20 \times 1840\,(407)^2}}\right] \times 1840 \times 407$$

$$= 780 \text{ mm}^2$$

Note that this is more than min. $A_{st} = \frac{0.85}{f_y} . b_w . d = \frac{0.85}{415} \times 300 \times 407 = 250 \text{ mm}^2$

Using 16 mm φ bars, No. of bars $= 780/201.06 = 3.88 \simeq 4$

Hence provide 4-16 mm φ bars giving actual $A_{st} = 4 \times 201.06 \simeq 804 \text{ mm}^2$

At the simply support ends, bend two bars up, at a distance of $1.414\, a = 1.414 \times 0.9 \times 407 = 520$ mm from the face of the support. Similarly, bend two bars up near support B, at a distance of 520 mm from the face of the support. These bent up bars may be taken right up to end C, to resist part of tension developed in beam BC. The remaining 2 bars at the bottom may be taken straight into the bottom of beam BC, to act as holding bars, and also to act as compression reinforcement, in BC, if required.

5. Check for shear and design of shear reinforcement at B

The critical section for shear lies at a distance of d (= 0.407 m) from the face of the support B, in span BA, where the design S.F. is given by

$$V_{uD} = 1.5\,[67090 - 18750(0.2 + 0.407)] \simeq 83563 \text{ N}$$

$$\therefore \quad \tau_v = 83563/(300 \times 407) = 0.68 \text{ N/mm}^2$$

At support B, $\frac{100\, A_s}{b_w\, d} = \frac{100 \times 0.5 \times 804}{300 \times 407} = 0.33\%$

Hence for 0.33% reinforcement, $\tau_c \simeq 0.40 \text{ N/mm}^2$

$V_{uc} = \tau_c\, bd = 0.40 \times 300 \times 407 = 48840$ N $\quad \therefore \quad V_{us} = 82157 - 48840 = 33317$ N

Shear resistance of 2 bent up bars is given by

$$V_{us1} = 0.87 f_y A_{sv} \sin 45° = 0.707 \times 0.87 \times 415\,(402) = 102615 \text{ N}$$

However, maximum resistance assigned to bent up bars $= \frac{1}{2} V_{us} \simeq 16658$ N

Hence provide vertical stirrups for the remaining 16659 N $= V_{us2}$

Using 8 mm φ 2 lgd stirrups, $A_{sv} = 2\,\frac{\pi}{4}\,(8)^2 = 100.5 \text{ mm}^2$

$$\therefore \quad s_v = \frac{0.87 f_y A_{sv}\, d}{V_{us2}} = \frac{0.87 \times 415 \times 100.5 \times 407}{16659} = 886 \text{ mm}$$

Maximum spacing corresponding to nominal stirrups is given by

$$s_v \leq \frac{2.175\, A_{sv} f_y}{b_w} = \frac{2.175 \times 100.45 \times 415}{300} = 302 \text{ mm}$$

Also, the spacing should be less than of $0.75\, d$ ($0.75 \times 407 = 305$) or 300 mm. Hence provide 8 mm φ 2 lgd stirrups @ 300 mm c/c throughout. Provide 2-12 mm φ bars at top as anchor bars.

6. Check for development length at support A

At support A, $A_{st1} =$ area of remaining steel $= 804/2 = 402 \text{ mm}^2$

$$\therefore \quad x_u = \frac{0.87 f_y A_{st1}}{0.36 f_{ck} b_w} = \frac{0.87 \times 415 \times 402}{0.36 \times 20 \times 300} = 67.2 \text{ mm}$$

$$M_{1u} = 0.87 f_y A_{st1} (d - 0.416 x_u) = 0.87 \times 415 \times 402 (407 - 0.416 \times 67.2)$$

$$= 55.02 \times 10^6 \text{ N-mm}$$

$$V_u = 52910 \times 1.5 = 79365 \text{ N}$$

$$L_d = 47 \varphi = 47 \times 16 = 752 \text{ mm}$$

Taking the bars straight into the support A, without any hook or bend and providing side cover $x' = 30$ mm, we have

$$L_0 = L_s/2 - x' = 400/2 - 30 = 170 \text{ mm}$$

$$\therefore \quad 1.3 \frac{M_{1u}}{V_u} + L_0 = 1.3 \times \frac{55.02 \times 10^6}{79365} + 170 \simeq 1071 \text{ mm} > L_d. \text{ Hence OK.}$$

Design of Beam BC

The beam BC will be designed as an ordinary rectangular beam with its tension side upwards. However, keep the depth and width of the beam the same as that of the T-beam.

Thus, $b = 300$ mm; $D = 450$ mm.

Also, $w_u = 1.5 \times 18750 = 28125$ N; $M_u = 1.5 \times 45375 = 68063$ N-m $= 68.063 \times 10^5$ N-mm

and $V_u = 1.5 \times 41250 = 61875$ N

(i) Computation of balanced effective depth

For Fe 415 steel, $\frac{x_{u,max}}{d} = \frac{700}{1100 + 0.87 \times 415} = 0.479$

$$R_u = 0.36 f_{ck} \frac{x_{u,max}}{d}\left(1 - 0.416 \frac{x_{u,max}}{d}\right) = 0.36 \times 20 \times 0.479 (1 - 0.416 \times 0.479) = 2.761$$

$$\therefore \quad d = \sqrt{\frac{M_u}{R_u \cdot b}} = \sqrt{\frac{68.063 \times 10^6}{2.761 \times 300}} = 286.6 \text{ mm}$$

However, keep $D = 450$ mm, so that available $d = 407$ mm, as earlier.

(ii) Determination of reinforcement

Since the actual effective depth is more than the balanced one, we have an over-reinforced section for which

$$A_{st} = \frac{0.5 f_{ck}}{f_y}\left[1 - \sqrt{1 - \frac{4.6 M_u}{f_{ck} bd^2}}\right] bd = \frac{0.5 \times 20}{415}\left[1 - \sqrt{1 - \frac{4.6 \times 68.063 \times 10^6}{20 \times 300 (407)^2}}\right] \times 300 \times 407$$

$$= 507 \text{ mm}^2$$

A_{st} already available in the form of 2-16 mm φ bars $= 804/2 = 402$ mm^2. Also A_{st} available from 2-12 mm φ anchor bars from span $AB = 2 \times 113 = 226$ mm^2.

Hence no additional reinforcement is necessary.

(iii) Check for shear and design of shear reinforcement

The critical section for shear is the face of the support, where S.F. is given by

$$V_{uD} = 1.5 \times 18750 \times 2 = 56250 \quad \text{N}$$

$$\tau_v = 56250/(300 \times 407) = 0.46 \text{ N/mm}^2$$

$$\frac{100\,A_s}{bd} = \frac{100\,(3 \times 201)}{300 \times 407} = 0.49\%; \quad \text{Hence} \quad \tau_c \simeq 0.48 \text{ N/mm}^2$$

Since $\tau_v < \tau_c$, no shear reinforcement is necessary. However, provide nominal stirrups in the form of 8 mm φ 2 lgd stirrups @ 300 mm c/c.

The details of reinforcement etc. are shown in Fig. 11.31.

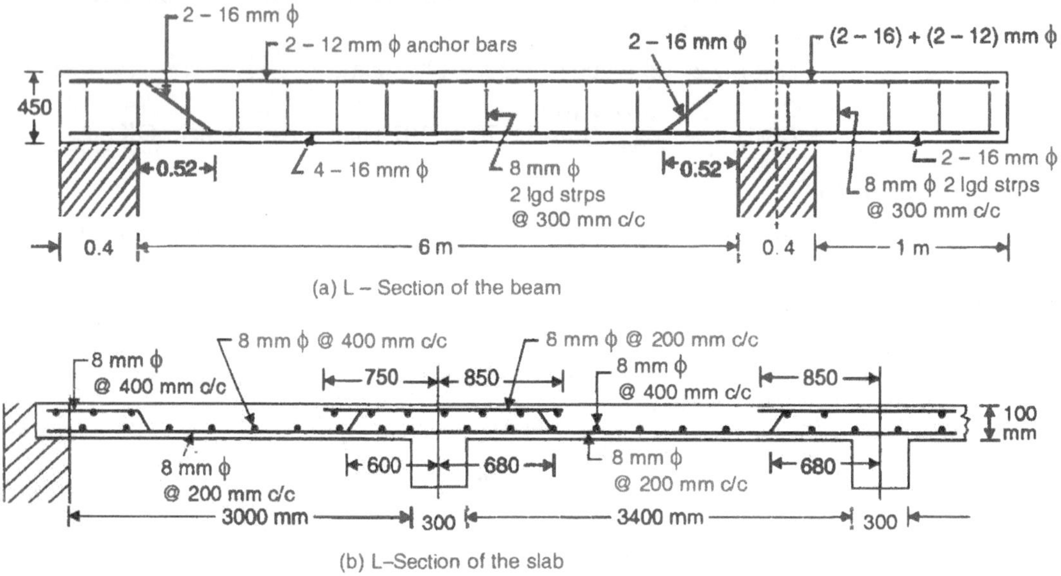

FIG. 11.31

11.14. DESIGN OF CANTILEVER CANOPY

Design Example 11.15. *Design a R.C. cantilever canopy having 2.5 m effective over-hang and 6 m wide as shown in Fig. 11.32. The canopy is supported on two beams cantilevered out from walls and spaced at 3.5 m centre to centre, and over hangs each beam by 1.25 m from each side. The beams and the slab are cast monolithic to give a flat soffit. The super-imposed load on the slab may be taken as 1500 N/m². If the height of wall over the anchoring length of the beam is 5 m, calculate the anchoring length of the beam required for a factor of safety of 1.5. The width of the supporting walls is 600 mm. Use M 20 concrete and Fe 415 steel.*

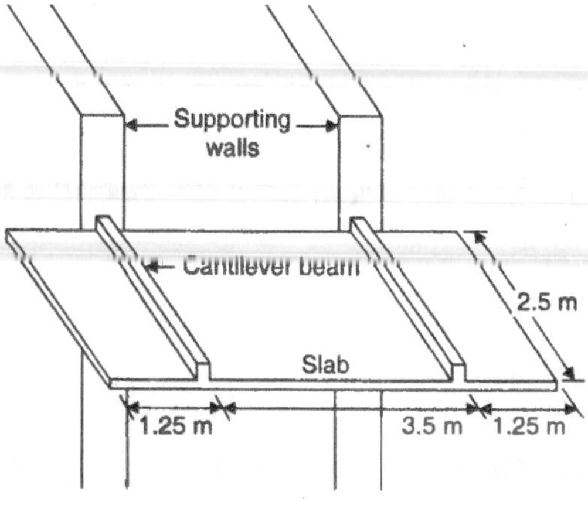

FIG. 11.32

Solution : (a) Design of slab

(*i*) ***Fixation of d and D*** : For stiffness, $L/d = 26$ for a continuous slab. Using under-reinforced section and $p_t = 0.25\,\%$, we have modification factor $F_t = 1.6$. Also, assuming $D = 100$ mm and $d = 80$ mm, effective span = clear span = 3.2 m.

$$\therefore \qquad d = \frac{\text{span}}{26 \times 1.6} = \frac{3200}{26 \times 1.6} = 77 \text{ mm}$$

Hence, let us keep $D = 100$ mm. Using 8 mm φ bars and a nominal cover of 15 mm, available $d = 100 - 15 - 4 = 81$ mm.

(*ii*) *Computation of B.M. and S.F.*

Dead load / m^2 = $0.1 \times 1 \times 1 \times 25000$	= 2500 N/m^2
Super-imposed load	= 1500 N/m^2
Total w	= 4000 N/m^2

The B.M. and S.F. diagrams are shown in Fig. 11.33.

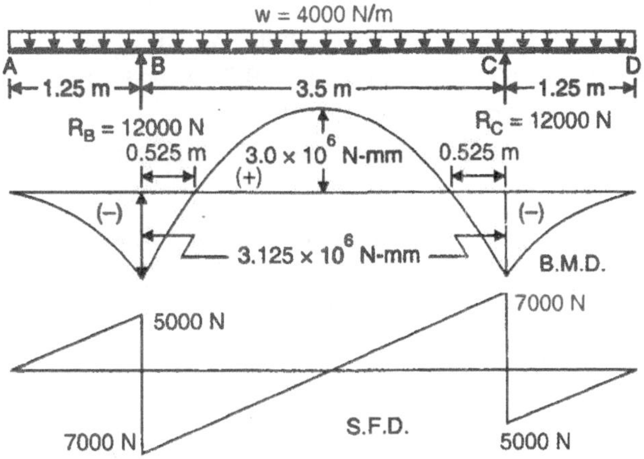

FIG. 11.33

$$\text{Max. hogging B.M.} = \frac{4000\,(1.25)^2}{2} \times 1000 = 3.125 \times 10^6 \text{ N-mm}$$

$$\text{Max. sagging B.M.} = \left[\,12000 \times 1.75 - \frac{4000\,(3)^2}{2}\,\right] 1000 = 3 \times 10^6 \text{ N-mm}$$

Let the point of contraflexure occur at x from support B. Its value is given by

$$12000\,x - \frac{4000\,(1.25 + x)^2}{2} = 0$$

or $$x^2 - 3.5\,x + 1.5625 = 0 \text{ from which } x = 0.525 \text{ m}$$

$\therefore$ Design B.M., $M = 3.125 \times 10^6$ N-mm; $M_u = 1.5 \times 3.125 \times 10^6 \simeq 4.69 \times 10^6$ N-mm.

(*iii*) *Computation of effective depth from flexure*

For Fe 415 steel, $\dfrac{x_{u,\,max}}{d} = \dfrac{700}{1100 + 0.87 \times 415} = 0.479$

$\therefore \quad R_u = 0.36\, f_{ck} \frac{x_{u,max}}{d}\left(1 - 0.416 \frac{x_{u,max}}{d}\right) = 0.36 \times 20 \times 0.479\,(1 - 0.416 \times 0.479) = 2.761$

$\therefore \quad d = \sqrt{\frac{M_u}{R_u\, b}} = \sqrt{\frac{4.69 \times 10^6}{2.761 \times 1000}} = 41.2 \text{ mm}$

However, from deflection point of view, we want $D = 100$ mm so that $d = 81$ mm.

(*iv*) ***Computation of reinforcement*** : For an under-reinforced section,

$$A_{st} = \frac{0.5\, f_{ck}}{f_y}\left[1 - \sqrt{1 - \frac{4.6\, M_u}{f_{ck}\, bd^2}}\right] bd = \frac{0.5 \times 20}{415}\left[1 - \sqrt{1 - \frac{4.6 \times 4.69 \times 10^6}{20 \times 1000\,(81)^2}}\right] \times 1000 \times 81$$

$$= 167.6 \text{ mm}^2$$

Minimum reinforcement $= 0.12\, D = 0.12 \times 100 = 120 \text{ mm}^2$

$\therefore$ Spacing of 8 mm φ bars, $s = \frac{1000\, A_\varphi}{A_{st}} = \frac{1000 \times 50.3}{167.6} \simeq 300$ mm

Hence provide 8 mm φ bars @ 300 mm c/c in all the three spans. For *AB* and *CD*, provide reinforcement at the top and take the bars insides the span *BC* for a distance equal to 525 mm or $L_d = 47\,\varphi = 47 \times 8 = 376$ mm, which ever is more. However, take these bars inside span *BC* by 600 mm. For span *BC*, provide reinforcement at the bottom and take it upto supports *B* and *C*.

This provision will meet the requirements of development length.

(*v*) ***Distribution reinforcement***

$$A_{sd} = 0.12\, D = 0.12 \times 100 = 120 \text{ mm}^2$$

Using 8 mm φ bars, $s = \frac{1000 \times 50.27}{120} = 418.9$ mm.

Hence provide these @ 400 mm c/c.

(vi) Check for shear

$$V_u = 1.5 \times 7000 = 10500 \text{ N};\ \tau_v = 10500/(1000 \times 81) = 0.13 \text{ N/mm}^2$$

This is much less than the permissible value of $1.3 \times 0.28 = 0.364$ N/mm^2 even for a minimum reinforcement of 0.15 %. Hence safe.

(b) Design of beams : The beam and slab are cast monolithic. Since the compression is in bottom fibres, flange action will be effective and the beam will act as a T-beam.

(*i*) ***Determination of*** b_w, b_f **and** D

Let $b_w = 300$ mm; Span $l_0 \simeq 2.5$ m, $D_f = 100$ mm

$$b_f = \frac{l_0}{6} + b_w + 6\, D_f = \frac{2500}{6} + 300 + 6 \times 100 \simeq 1316 \text{ mm}$$

and b_f = cantilever projection of slab $+ \frac{1}{2}$ spacing of beam $= 1250 + 1750 = 3000$ mm

Hence adopt $b_f = 1316$ mm

From stiftness point of view, ignoring modification factors, $L/d = 7$

$\therefore \quad d = L/7 = 2500/7 = 357$ mm.

However take $D = 400$ mm for the purpose of computation of dead load.

(*ii*) ***Computation of dead load and B.M.***

$$\text{Load from slab} = \frac{6}{2} \times 1 \times 4000 = 12000 \ \text{N/m}$$

$$\text{Self weight of web} = (0.4 - 0.1)\, 0.3 \times 1 \times 25000 = 2250 \ \text{N/m}$$

$$\text{Total} \qquad w = 12000 + 2250 = 14250 \ \text{N/m}.$$

$$\therefore \qquad M = \frac{wl^2}{2} = \frac{14250\,(2.5)^2}{2} \times 1000 = 44.53 \times 10^6 \ \text{N-mm};$$

$$M_u = 1.5 \times 44.53 \times 10^6 = 66.8 \times 10^6 \ \text{N-mm}$$

$$V = wl = 14250 \times 2.5 = 35625 \ \text{N}; \quad V_u = 1.5 \times 35625 = 53438 \ \text{N}$$

(*iii*) ***Fixation of effective depth and total depth***

$$\text{Let} \qquad d \simeq \frac{2}{3}\sqrt{\frac{M_u}{R_u\, b_w}} = \frac{2}{3}\sqrt{\frac{66.8 \times 10^6}{2.761 \times 300}} \simeq 190 \ \text{mm}$$

However, from stiffness point of view, $L/d = 7$. Taking $p_t = 0.5\%$, we get modification factor $F_t = 1.15$ (Fig. 11.1). Also, $b_w/b_f = 300/1316 = 0.228$ for which reduction factor $F_b = 0.8$ (Fig. 11.3). Taking $d \simeq 450$ mm, effective span $L = 2500 + 450/2 = 2725$ mm. Hence $d = 2725/(7 \times 1 \times 0.8) = 424$ mm.

However provide $D = 500$ mm, so that using main bars of 20 mm φ, stirrups of 8 mm φ and nominal cover of 25 mm, available $= 500 - 25 - 8 - 20/2 = 457$ mm.

Due to this increased depth, weight of web $= 0.3 \times 0.4 \times 1 \times 25000 = 3000$ N/m and total $w = 12000 + 3000 = 15000$ N/m.

$$\text{Hence} \qquad M_{uD} = \frac{1.5 \times 15000\,(2.5)^2}{2} \times 1000 = 70.312 \times 2 \times 10^6 \ \text{N-mm}$$

$$\text{and} \qquad V_{uD} = 1.5 \times 15000 \times 2.5 = 56250 \ \text{N}.$$

(*iv*) ***Determination of reinforcement*** : Assume $x_u = D_f = 100$ mm.

$$M_{u1} = 0.36\, f_{ck},\, b_f\, D_f\,(d - 0.416\, D_f) = 0.36 \times 20 \times 1316 \times 100\,(457 - 0.416 \times 100)$$

$$= 393 \times 10^6 \ \text{N-mm}$$

Since $M_{u1} > M_{uD}$, N.A. fall inside the flange, i.e. $x_u \le D_f$

$$\text{Hence} \qquad M_u = M_{uD} = 0.87\, f_y\, A_{st}\, d\left[1 - \frac{f_y}{f_{ck}}\frac{A_{st}}{b_f\, d}\right], \text{ from which}$$

$$A_{st} = \frac{0.5\, f_{ck}}{f_y}\left[1 - \sqrt{1 - \frac{4.6\, M_{uD}}{f_{ck}\, b_f\, d^2}}\right] b_f\, d$$

$$= \frac{0.5 \times 20}{415}\left[1 - \sqrt{1 - \frac{4.6 \times 70.312 \times 10^6}{20 \times 1316\,(457)^2}}\right] 1316 \times 457 = 432.8 \ \text{mm}^2$$

Using 16 mm φ bars, No. of bars $= 432.8/201 = 2.15$

Hence provide 3-16 mm φ bars, providing $A_{st} = 201 \times 3 = 603 \ \text{mm}^2$

(*v*) ***Check for shear and design of shear reinforcement***

$$\tau_v = \frac{V_{uD}}{b_w\, d} = \frac{56250}{300 \times 457} = 0.41 \ \text{N/mm}^2$$

$$\frac{100\,A_s}{b_w\,d} = \frac{100 \times 603}{300 \times 457} \simeq 0.44\,\%. \quad \text{Hence} \quad \tau_c \simeq 0.45 \text{ N/mm}^2$$

Since $\tau_c > \tau_v$, shear reinforcement is not required. However, provide nominal shear stirrups consisting of 8 mm φ 2 lgd stirrups at spacing

$$s_v \leq \frac{2.175\,A_{sv}\,f_y}{b_w} = \frac{2.175 \times 100.5 \times 415}{300} = 302.5 \text{ mm}$$

Hence provide 8 mm φ 2 lgd stirrups @ 300 mm c/c throughout the length. Provide 2-10 mm φ holding bars at the bottom.

(*vi*) ***Length of embedment of the beam*** : Let the length of embedment of the beam be x m in the wall. Weight of wall on the embedded length $= 5 \times 0.3\,x\,(22500)$ N

Its moment about the edge of the support $= 5 \times 0.3 \times 22500 \times \dfrac{x^2}{2}$ N-m $= 16875\,x^2$ N-m

Factor of safety = 1.5.

External moment causing rotation $= \dfrac{15000\,(2.5)^2}{2} = 46.875 \times 10^3$ N-m.

Hence, for stability, $16875\,x^2 = 1.5 \times 46.875 \times 10^3$

This gives $x = 2.04$ m. However, embed the beam by 2.1 m. If this is considered excessive, increase the width of beam to 600 mm in the embedded portion.

This will give $x = \sqrt{\dfrac{2 \times 1.5 \times 46.875 \times 10^3}{5 \times 0.6 \times 22500}} = 1.44 \text{ m} \simeq 1.5 \text{ m}$

The details of reinforcements etc are shown in Fig. 11.34.

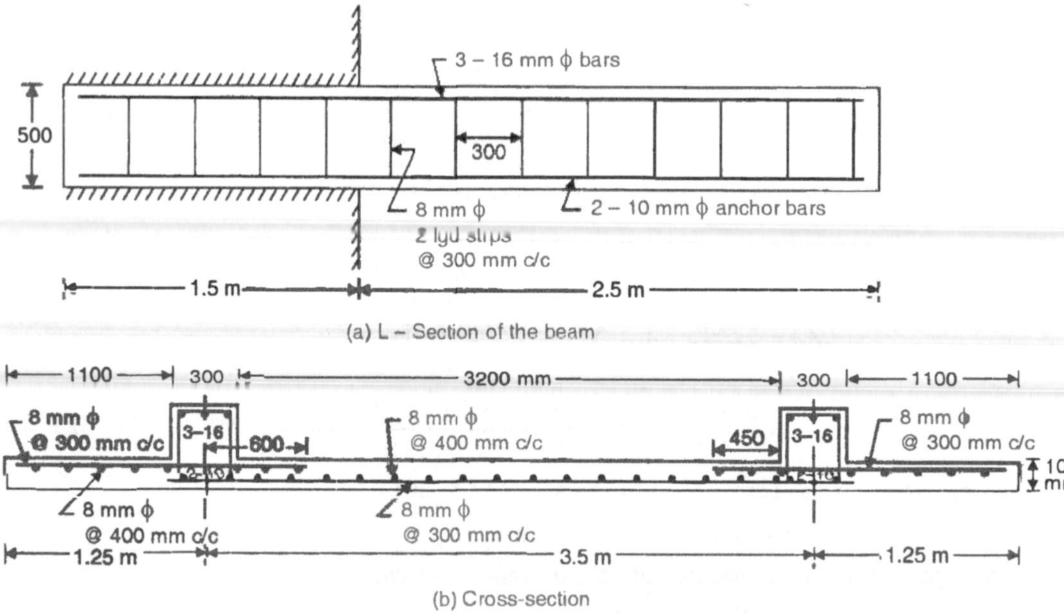

FIG. 11.34

11.15. DESIGN OF L-BEAM : DESIGN FOR TORSION

The design principles for L-beam are similar to those for T-beam, except that additional torsional reinforcement is provided to resist torsional moment. The width b_f of the slab, acting monolithic with the beam, forming the flange of the L-beam is taken as the least of the following :

(*i*) $$b_f = \frac{l_0}{12} + b_w + 3\,D_f$$

(*ii*) For isolated L-beam, $$b_f = \frac{0.5\,l_0}{\frac{l_0}{b} + 4} + b_w$$

but in no case greater than the actual width of the flange, where l_0 is the distance between points of zero moments in the beam. For continuous beams and frames, l_0 may be assumed to be 0.7 times the effective span.

The location of N.A. and calculations of the moment of resistance of L-beam are done exactly in the same way as that for a T-beam and all the equations developed for T-beam are also application for L-beam.

Design for shear and torsion : The design for shear and torsion is done in the following steps :

1. Calculate shear force V_u and torsional moment T_u
2. Compute the equivalent shear stress V_e and τ_{ve} from the expressions

$$V_e = V_u + 1.6\,\frac{T_u}{b} \qquad \text{and} \qquad \tau_{ve} = \frac{V_e}{b_w \,.\, d}$$

The value of τ_{ve} should not exceed $\tau_{c\,.\,max}$ given in Table 7.3. If it exceeds, the section should be redesigned by increasing the concrete area.

3. If τ_{ve} does not exceeds τ_c given in Table 7.1, provide nominal shear stirrups given by the equation $s_v \leq \dfrac{2.175\,A_{sv}\,f_y}{b_w}$.

4. If τ_{ve} exceeds τ_c, provide shear and torsion reinforcement in the form of longitudinal as well as transverse reinforcement discussed in steps 5 and 6 below.

5. ***Longitudinal reinforcement*** : It is designed to resist an equivalent B.M. M_{e_1} given by $M_{e_1} = M_u + M_T$

where $M_u =$ B.M. at the cross-section and $M_T = T_u\left(\dfrac{1 + D/b_w}{1.7}\right)$

If M_T exceeds M_u, provide longitudinal reinforcement on the compression face also for a moment M_{e2} given by $M_{e2} = M_T - M_u$, the moment M_{e2} being taken as acting in the opposite sence to the moment M_u.

6. ***Transverse reinforcement*** : Find the area of transverse reinforcement A_{sv} given by

$$A_{sv} = \frac{T_u\,s_v}{b_1\,d_1\,(0.87\,f_y)} + \frac{V_u\,s_v}{2.5\,d_1\,(0.87\,f_y)}, \text{ subject to a minimum of } \frac{(\tau_{ve} - \tau_c)\,b_w\,s_v}{0.87\,f_y}$$

where b_1 = centre to centre distance between corner bars in the direction of the width.

d_1 = centre to centre between corner bars, in the direction of depth.

The spacing s_v should not exceed the least of x_1, $(x_1 + y_1)/4$ and 300 mm, where x_1 and y_1 are respectively the short and long dimensions of the stirrup.

7. ***Side face reinforcement*** : Where the depth of the web in a beam exceeds 450 mm, side face reinforcement shall be provided along the two faces. The total area of such reinforcement shall be not less than 0.1 percent of the web area and shall be distributed equally on the two faces at spacing not exceeding 300 mm or web thickness which ever is less.

Design Example 11.16. *Fig. 11.35 shows the plan and section of a canopy to be provided over an entrance opening of a building. Design the beam and the slab. Use M 20 concrete and Fe 415 steel. the super-imposed load on the slab may be taken as 1200 N/m².*

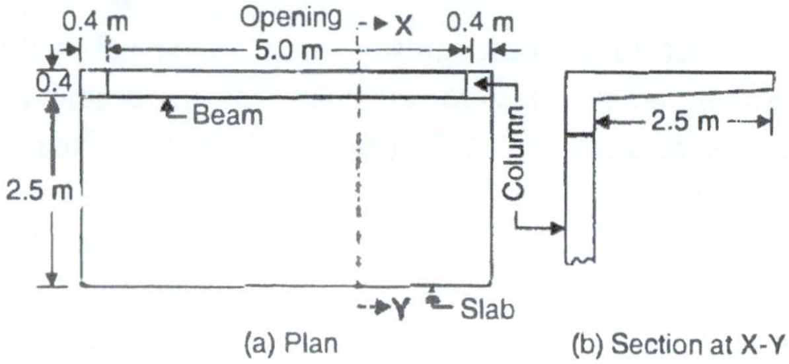

FIG. 11.35

Solution : Since the beam and the slab are monolithic, the beam will act as L-beam and part of the slab will resist the compression of the beam. The breadth of the beam will be kept equal to the width of the column, i.e. 400 mm.

(a) Design of the canopy slab

(*i*) ***Computation of loading, B.M. and S.F.***

Let the average depth the slab be 120 mm, for the purpose of computing dead weight. Dead load due to self weight $= 0.12 \times 1 \times 1 \times 25000 = 3000$ N/m²

Super-imposed load = 1200 N/m²

Total $\quad w = 3000 + 1200 = 4200$ N/m²

$$\therefore \quad M = \frac{4200\,(2.5)^2}{2} \times 1000 = 13.125 \times 10^6 \text{ N-mm;}$$

$$M_u = 1.5 \times 13.125 \times 10^6 \simeq 19.69 \times 10^6 \text{ N-mm}$$

$$V = 4200 \times 2.5 = 10500 \text{ N}; \quad V_u = 1.5 \times 10500 = 15750 \text{ N}$$

(*ii*) ***Computation of effective depth and total depth***

For Fe 415 steel, $\dfrac{x_{u,\,max}}{d} = \dfrac{700}{1100 + 0.87 \times 415} = 0.479$

$$R_u = 0.36\, f_{ck} \frac{x_{u,\,max}}{d}\left(1 - 0.416\frac{x_{u,\,max}}{d}\right) = 0.36 \times 20 \times 0.479\,(1 - 0.416 \times 0.479) = 2.761$$

$$d = \sqrt{\frac{M_u}{R_u\, b}} = \sqrt{\frac{19.69 \times 10^6}{2.761 \times 1000}} = 84.4 \text{ mm}$$

However, for stiffness, the ratio of span to depth of cantilever = 7. For an under-reinforced slab, assuming $p_t = 0.2\%$, we get a modification factor of 1.7 for Fe 415 steel. Also, assuming $d \simeq 200$ mm, effective span $L = 2.5 + 0.2/2 = 2.6$ m. Hence $d = \dfrac{2600}{7 \times 1.7} = 219$ mm.

Hence keep $D = 250$ mm at the supports and reduce it to 100 mm at the free end. Using 15 mm nominal cover and 8 mm φ bars, available $d = 250 - 15 - 4 = 231$ mm at the fixed end and $d = 100 - 15 - 4 = 81$ mm at the free end.

Average total thickness $= (250 + 100)/2 = 175$ mm against assumed value of 120 mm. Hence dead load due to revised dimension $= 0.175 \times 25000 = 4375$ N.

Thus, total $w = 4375 + 1200 = 5575 \text{ N/m}^2$.

Revised $$M = \frac{5575\,(2.5)^2}{2} \times 1000 = 17.422 \times 10^6 \text{ N-mm}$$

and $$M_u = 1.5 \times 17.422 \times 10^6 = 26.132 \times 10^6 \text{ N-mm}$$

Revised $V = 5575 \times 2.5 = 13938$ N and $V_u = 1.5 \times 13938 = 20906$ N.

(iii) Computation of steel reinforcement : For an under-reinforced section,

$$A_{st} = \frac{0.5 f_{ck}}{f_y}\left[1 - \sqrt{1 - \frac{4.6 M_u}{f_{ck}\, bd^2}}\right] bd$$

$$= \frac{0.5 \times 20}{415}\left[1 - \sqrt{1 - \frac{4.6 \times 26.132 \times 10^6}{20 \times 1000\,(231)^2}}\right] \times 1000 \times 231 = 322.8 \text{ mm}^2$$

Spacing of 8 mm φ bars, is $s = \dfrac{1000 \times 50.26}{322.8} = 155.7$ mm

Hence provide 8 mm φ @ 150 mm c/c.

Available $A_{st} = \dfrac{1000 \times 50.26}{150} = 335 \text{ mm}^2$; $p_t = \dfrac{335 \times 100}{1000 \times 231} = 0.145\,\%$

(iv) *Computation of distribution reinforcement*

$$A_{sd} = \frac{0.12}{100} \times \text{breadth} \times \text{average depth} = \frac{0.12}{100} \times 1000 \times 175 = 210 \text{ mm}^2$$

Spacing of 8 mm φ bars $= 1000 \times \dfrac{50.27}{210} = 218$ mm.

However provide 8 mm φ bars @ 210 mm c/c.

(v) ***Check for development length at the support*** : At the support, $A_{st1} = 335 \text{ mm}^2$

$$x_u = \frac{0.87 f_y A_{st1}}{0.36 f_{ck}\, b} = \frac{0.87 \times 415 \times 335}{0.36 \times 20 \times 1000} = 16.8 \text{ mm}$$

$$M_{1u} = 0.87 f_y A_{st1}(d - 0.416\, x_u) = 0.87 \times 415 \times 335\,(231 - 0.416 \times 16.8) = 27.09 \times 10^6 \text{ N-mm}$$

$V_u = 15750$ N ; $L_d = 47\,\varphi = 47 \times 8 = 376$ mm

Taking the bars straight into supports, without any hook or bend and providing side cover $x' = 20$ mm, we have $L_0 = 100/2 - 20 = 30$ mm

$$\therefore \quad 1.3\,\frac{M_{1u}}{V_u} + L_0 = 1.3 \times \frac{27.09 \times 10^6}{15750} + 30 = 2266 \text{ mm} > L_d \text{ . Hence OK.}$$

(b) Design of L-beam

(i) Fixation of b_w, b_f and D

Let $b_w = 400$ mm; Effective span $l_0 \simeq 5 + 0.4 = 5.4$ m

For an isolated beam, $b = 400 + 2500 = 2900$ mm

$$b_f = \frac{0.5\, l_0}{\left(\frac{l_0}{b}\right) + 4} + b_w = \frac{0.5 \times 5400}{\frac{5400}{2900} + 4} + 400 \simeq 860 \text{ mm.}$$

Thickness of slab at $(0.86 - 0.4) = 0.46$ m from the face of the beam

$$= 100 + \frac{250 - 100}{2.5}(2.5 - 0.46) = 222.4 \text{ mm}$$

$$\therefore \text{ Average } \quad D_f = \frac{250 + 222.4}{2} \simeq 236 \text{ mm}$$

Assume over all depth $= l/13 = 6000/13 \simeq 460$ mm for the purpose of computation of dead load.

(ii) *Computation of load and B.M.*

Load from the slab $= 2.5 \times 1 \times 5575 \simeq 13938$ N/m

Dead load of beam / m $= 0.46 \times 0.4 \times 1 \times 25000 = 4600$ N/m

Total $w = 13938 + 4600 \simeq 18540$ N/m.

Assuming partial fixidity, $M = \frac{wL^2}{10} = \frac{18540\,(5.4)^2}{10} = 54060$ N-m

$\therefore \quad M_u = 1.5 \times 54060 \times 10^3 = 81.09 \times 10^6$ N-mm

Also, $\quad V = wL = 18540 \times 2.5 = 46350; \quad V_u = 1.5 \times 46350 = 69525$ N

(iii) *Fixation of effective depth and total depth*

Let $\quad d \simeq \frac{2}{3}\sqrt{\frac{M_u}{R_u\, b_w}} = \frac{2}{3}\sqrt{\frac{81.09 \times 10^6}{2.761 \times 400}} = 271$ mm

Also, from stiffness point of view $L/d = 20$. Taking $p_t = p_{t,\,lim.} \simeq 1\%$, we get modification factor $F_t = 1$ (Fig. 11.1).

Also $b_w/b_f = 400/860 = 0.465$, for which reduction factor $F_b \simeq 0.85$ (Fig. 11.3). Hence $d = 5400/(20 \times 1 \times 0.85) = 318$ mm.

Depth from shear point of view : The total thickness should be such that the beam is safe in shear due to both bending as well as torsion.

$$T = \left[5575 \times 2.5\left(\frac{2.5}{2} + \frac{0.4}{2}\right)\right]\frac{5.4}{2} \text{ N-m} = 54565 \text{ N-m} = 54.565 \times 10^6 \text{ N-mm}$$

$$V = \frac{wl}{2} = \frac{18540\,(5)}{2} = 46350 \text{ N}$$

$\therefore \quad T_u = 1.5 \times 54.565 \times 10^6 = 81.848 \times 10^6$ N-mm and $V_u = 1.5 \times 46350 = 69525$ N

$$V_e = V_u + 1.6\,\frac{T_u}{b_w} = 69525 + 1.6 \times \frac{81.848 \times 10^6}{400} = 396915 \text{ N}$$

$\tau_{ve} = \frac{V_e}{b_w\, d}$. Limiting this to $\tau_{c,\,max} = 2.8$ N/mm^2 for M 20 concrete, we get

$$d = \frac{V_e}{b_w\, \tau_{c\,max}} = \frac{396915}{400 \times 2.8} = 354 \text{ mm.}$$

However, keep $D = 450$ mm, so that providing 25 mm nominal cover, and using 16 mm φ main bars and 8 mm dia. stirrups, available $d = 450 - 25 = 8 - 16/2 = 409$ mm.

(iv) Determination of Longitudinal reinforcement

$$M_{e_1} = M_u + M_T, \text{ where } M_u = 81.09 \times 10^6 \text{ N-mm}$$

$$M_T = T_u \left(\frac{1 + D/b_w}{1.7} \right) = 81.848 \times 10^6 \left[\frac{1 + 450/400}{1.7} \right] = 102.31 \times 10^6 \text{ N-mm}$$

$$\therefore \quad M_{e_1} = (81.09 + 102.31)\, 10^6 = 183.4 \times 10^6 \text{ N-mm}$$

Let us find the longitudinal reinforcement for this *equivalent* B.M.

For Fe 415 - M 20 combination, $R_u = 2.761$

$$\therefore \quad M_{u\,,\,lim.} = R_u\, b_w\, d^2 + 0.446\, f_{ck}\, (b_f - b_w)\, y_f\, (d - 0.5\, y_f) \qquad \ldots(6.18)$$

where $y_f = 0.15\, x_{u.\,max.} + 0.65\, D_f = 0.15 \times 0.479 \times 409 + 0.65 \times 236 = 182.8 \text{ mm} < D_f$

$$\therefore \quad M_{u\,,\,lim.} = 2.761 \times 400\, (409)^2 + 0.446 \times 20\, (860 - 400) \times 182.8\, (409 - 0.5 \times 182.8)$$

$$= 423 \times 10^6 \text{ N-mm}$$

Since $M_{e_1} < M_{u\,,\,lim.}$, the section is under-reinforced, for which

$$A_{st1} = \frac{0.5 f_{ck}}{f_y} \left[1 - \sqrt{1 - \frac{4.6\, M_{e_1}}{f_{ck}\, b_f\, d^2}} \right] b_f\, d = \frac{0.5 \times 20}{415} \left[1 - \sqrt{1 - \frac{4.6 \times 183.4 \times 10^6}{20 \times 860\, (409)^2}} \right] \times 860 \times 409$$

$$= 1350 \text{ mm}^2$$

Using 16 mm φ bars, No. of bars $= 1350/201 \simeq 7$.

Hence provide 7 bars of 16 mm φ at the bottom of the beam, throughout its length.

Also, $M_{e_2} = M_T - M_u = 102.31 \times 10^6 - 81.09 \times 10^6 = 21.22 \times 10^6$ N-mm.

Since the section is under-reinforced for M_{e_2}, the area of steel is given by

$$A_{st2} = \frac{0.5 f_{ck}}{f_y} \left[1 - \sqrt{1 - \frac{4.6\, M_{e_2}}{f_{ck}\, b_w\, d^2}} \right] b_w\, d = \frac{0.5 \times 20}{415} \left[1 - \sqrt{1 - \frac{4.6 \times 21.22 \times 10^6}{20 \times 400\, (409)^2}} \right] \times 400 \times 409$$

$$= 146 \text{ mm}^2$$

This is quite small. However, provide 3 bars of 16 mm φ at top which will also work as holding bars for shear stirrups.

(v) Determination of transverse reinforcement

Transverse reinforcement will be provided in the form of vertical stirrups. Let the top 16 mm φ corner bars be provided at a nominal cover of 25 mm. Let us use 4-lgd stirrups of 10 mm φ bar, having $A_{sv} = 4 \times \frac{\pi}{4} (10)^2 = 314 \text{ mm}^2$.

b_1 = centre to centre distance between corner bars in the direction of width
$= 400 - 2\,(25 + 10 + 8) = 314$ mm

d_1 = centre to centre distance between corner bars in the direction of depth
$= 450 - 2\,(25 + 10 + 8) = 364$ mm.

Now, $$A_{sv} = \frac{T_u\, s_v}{b_1\, d_1\, (0.87\, f_y)} + \frac{V_u\, .\, s_v}{2.5\, d_1\, (0.87\, f_y)}$$

$$\therefore \quad 314 = \left(\frac{81.848 \times 10^6}{314 \times 364\ (0.87 \times 415)} + \frac{69525}{2.5 \times 364 \times 0.87 \times 415} \right) s_v. \text{ From which } s_v = 143 \text{ mm}$$

As per Code requirements, this spacing should not exceed x_1, $(x_1 + y_1)/4$ and 300 mm

where x_1 = short dimension of stirrup = 400 − 2 (16 + 5) = 358 mm

y_1 = long dimension of stirrup = 450 − 2 (16 + 5) = 408 mm

$\therefore \quad (x_1 + y_1)/4 = (358 + 408)/4 = 191.5$ mm

Hence provide 10 mm φ 4-lgd stirrups @ 140 mm c/c at supports.

However, at the mid-span of the beam, where both V_u and T_u are zero, provide 8 mm φ 4 lgd stirrups @ 300 mm c/c for the middle 2.5 m length of the beam.

The details of reinforcement etc. are shown in Fig. 11.36.

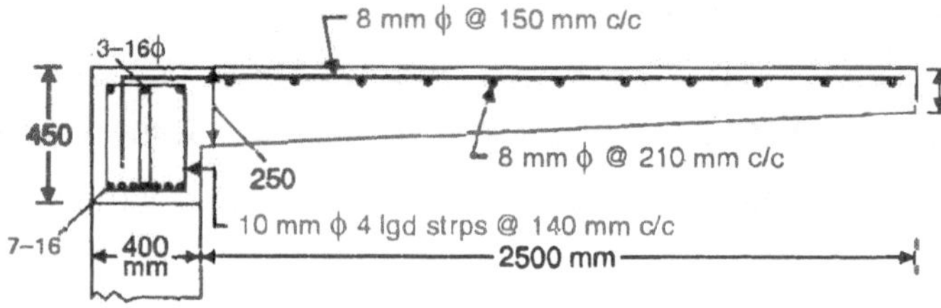

FIG. 11.36

Design Example 11.17 *Solve example 11.16 if the canopy slab is flush with the bottom of the beam so as to have a plane soffit.*

Solution : Since the slab is attached to the bottom of the beam, it will be in tension zone, and hence flange action will not be available. Thus, the beam will act as a simple rectangular beam. The design of slab will remain the same as in the previous example.

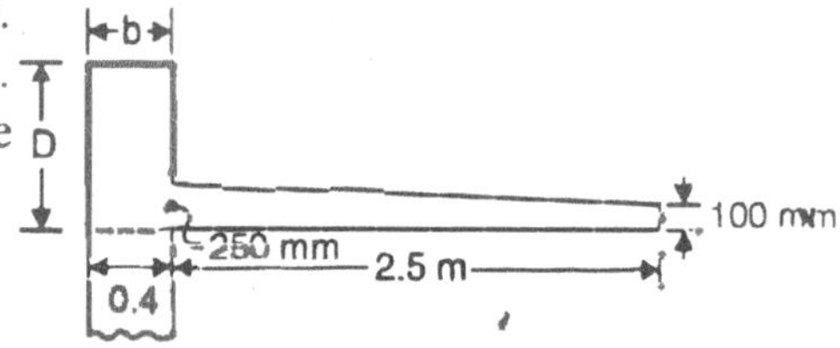

FIG. 11.37

Design of slab

(i) *Loading and bending moment*

Width of beam = 400 mm

Let us assume the total depth of beam $= L/12 = 5400/12 = 450$ mm

$\therefore$ Dead load of beam/m $= 0.4 \times 0.45 \times 1 \times 25000 = 4500$ N/m.

Load from the slab $= 2.5 \times 1 \times 5575 = 13938$ N/m

$\therefore$ Total $w = 13938 + 4500 = 18438$ N/m. Effective span $= 5 + 0.4 = 5.4$ m

Assuming partial fixidity, $M = \dfrac{w\,l^2}{10} = \dfrac{18438\,(5.4)^2}{10} \times 1000 = 53.765 \times 10^6$ N-mm

$$\therefore \quad M_u = 1.5 \times 53.765 \times 10^6 = 80.65 \times 10^6 \text{ N-mm.}$$

$$V = \frac{wL}{2} = \frac{18438 \times 5.4}{2} = 49783 \quad ; \quad V_u = 1.5 \times 49783 = 74674 \text{ N}$$

(ii) *Computation of effective depth and total depth*

For Fe 415 steel, $\dfrac{x_{u,\,max}}{d} = \dfrac{700}{1100 + 0.87 \times 415} = 0.479$

$$R_u = 0.36 f_{ck} \frac{x_{u,max}}{d}\left(1 - 0.416 \frac{x_{u,max}}{d}\right) = 0.36 \times 20 \times 0.479 \,(1 - 0.416 \times 0.479) = 2.761$$

$$\therefore \quad d = \sqrt{\frac{M_u}{R_u\, b}} = \sqrt{\frac{80.65 \times 10^6}{2.761 \times 400}} = 270 \text{ mm.}$$

Depth from shear point of view : The total thickness should be such that beam is safe in shear due to both bending as well as torsion.

$$T = \left[5575 \times 2.5\left(\frac{2.5}{2} + \frac{0.4}{2}\right)\right]\frac{5.4}{2} = 54565 \text{ N-m} = 54.565 \times 10^6 \text{ N-mm}$$

$$T_u = 1.5 \times 54.565 \times 10^6 = 81.85 \times 10^6 \text{ N-mm} \quad ; \quad V_u = 74674 \text{ N}$$

$$V_e = V_u + 1.6 \frac{T_u}{b_w} = 74674 + 1.6 \times \frac{81.85 \times 10^6}{400} = 402064 \text{ N}$$

$\tau_{ve} = \frac{V_e}{b_w\, d}$. Limiting this to $\tau_{c,max} = 2.8 \text{ N/mm}^2$ for M 20 concrete, we get

$$d = \frac{V_e}{b_w\, \tau_{c,max}} = \frac{402064}{400 \times 2.8} = 359 \text{ mm.}$$

However, since M_{e_1} will be much more than M_u, provide $D = 450$ mm, so that keeping nominal cover of 25 mm, and using 16 mm φ main bars and 10 mm φ rings, available $d = 450 - 25 - 10 - 16/2 = 407$ mm.

(*iii*) *Determination of longitudinal reinforcement*

$$M_{e_1} = M_u + M_T \quad \text{where} \quad M_u = 80.65 \times 10^6 \text{ N-mm}$$

$$M_T = T_u\left(\frac{1 + D/b}{1.7}\right) = 80.65 \times 10^6\left(\frac{1 + 450/400}{1.7}\right) = 100.81 \times 10^6 \text{ N-mm}$$

$$\therefore \quad M_{e_1} = (80.65 + 100.81)\,10^6 = 181.46 \times 10^6 \text{ N-mm.}$$

Let us find the longitudinal reinforcement for this *equivalent* B.M.

$$M_{u,lim.} = R_u\, bd^2 = 2.761 \times 400\,(407)^2 = 183 \times 10^6 \text{ N-mm.}$$

Since $M_{e_1} < M_{u,lim.}$, the section is under reinforced, for which

$$A_{st1} = \frac{0.5 f_{ck}}{f_y}\left[1 - \sqrt{1 - \frac{4.6 M_{e_1}}{f_{ck}\, bd^2}}\right] bd = \frac{0.5 \times 20}{415}\left[1 - \sqrt{1 - \frac{4.6 \times 181.46 \times 10^6}{20 \times 400\,(407)^2}}\right] \times 400 \times 407$$

$$= 1536 \text{ mm}^2$$

Using 16 mm φ bars, No. of bars $= 1536/201 \simeq 8$

Hence provide 8 Nos. 16 mm φ bars in two rows at the bottom of the beam, throughout the length of the beam.

Also, $M_{e_2} = M_T - M_u = (100.81 - 80.65)\,10^6 = 20.16 \times 10^6$

$$A_{st2} = \frac{0.5 f_{ck}}{f_y}\left[1 - \sqrt{1 - \frac{4.6 M_{e_2}}{f_{ck}\, bd^2}}\right] bd = \frac{0.5 \times 20}{415}\left[1 - \sqrt{1 - \frac{4.6 \times 20.16 \times 10^6}{20 \times 400\,(407)^2}}\right] \times 4000 \times 407$$

$$= 140 \text{ mm}^2$$

This is quite small. However, provide 3 bars of 16 mm φ at the top which will also work as holding bars for shear stirrups.

(iv) Determination of transverse reinforcement

Transverse reinforcement will be provided in the form of vertical stirrups. Let the top 16 mm φ corner bars be provided at a nominal cover of 25 mm. Let us use 4-lgd stirrups of 10 mm φ bars, having $A_{sv} = 4 \times \frac{\pi}{4}(10)^2 = 314\ \text{mm}^2$.

b_1 = centre to centre distance between corner bars in the direction width
= 400 − 2 (25 + 10 + 8) = 314 mm

d_1 = centre to centre distance between corner bars in the direction of depth
= 450 − 2 (25 + 10 + 8) = 364 mm

Now $$A_{sv} = \frac{T_u\, s_v}{b_1\, d_1\,(0.87 f_y)} + \frac{V_u\, .\, s_v}{2.5\, d_1\,(0.87 f_y)}$$

or $$314 = \left[\frac{81.85 \times 10^6}{314 \times 364\,(0.87 \times 415)} + \frac{74674}{2.5 \times 364\,(0.87 \times 415)}\right] s_v$$

From which $s_v = 142$ mm

As per Code requirements, this spacing should not exceed x_1, $(x_1 + y_1)/4$ and 300 mm

where x_1 = short dimension of stirrup = 400 − 2 (16 + 5) = 358 mm

y_1 = long dimension of stirrup = 450 − 2 (16 + 5) = 408 mm

∴ $(x_1 + y_1)/4 = (358 + 408)/4 = 191.5$ mm

Hence provide 10 mm φ 4 lgd stirrups @140 mm c/c at supports. However, at the mid span of the beam, where both V_u and T_u are zero, provide 8 mm φ 4 lgd stirrups @ 300 mm c/c for the middle 2.5 m length of the beam.

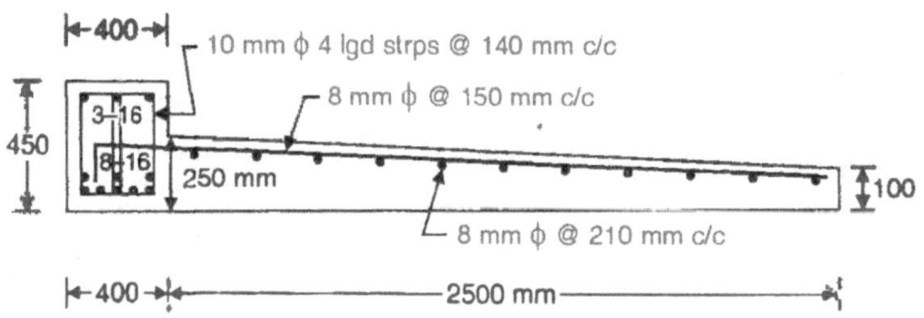

FIG. 11.38

The details of reinforcement etc. are shown in Fig. 11.38.

PROBLEMS

1. Design a reinforced concrete beam, supported on two walls 60 cm thick, over a clear opening of 5 m. The beam carries a superimposed load of 15 kN/m. Use M 20 mix and HYSD bars.

2. Design a reinforced concrete lintel over the openings of two windows, each 2 m wide separated by a wall of 50 cm. The height of wall over the lintel is 3 m, and thickness of wall is 40 cm. The lintel is to be provided with a sunshade projecting out by 80 cm and cast monolithically with the lintel. Use M 20 mix and HYSD reinforcement. The live load on the sun shade may be taken as 1500 N/m^2.

3. Design a simply supported reinforced concrete beam, carrying a superimposed load of 20 kN per metre run over a clear opening of 6 m. In addition to this the beam carries two concentrated loads of 30 kN each at a distance of 1.5 m from the face for each support. The width of supporting wall is 60 cm. Use M 20 concrete and Fe 415 steel.

4. Design a R.C. cantilever projecting out 3 m beyond the fixed end and carrying a superimposed load of 15 kN. The cantilever also carries a concentrated load of 5 kN at the free end. Use M 20 mix. Take $\sigma_{st} = 230$ N/mm^2.

5. Design a R.C.C. floor slab for a room having inside dimensions 4 m × 10 m and supported on all sides by a 40 cm thick brick wall. The superimposed load may be taken as 3 kN/m^2. Use M 20 mix and HYSD bars.

6. Fig. 11.39 shows the cross-section of a roof slab, the supports being 4 m part. Design the slab for a live load of 2000 N/m^2. The slab is to be finished with 30 mm thick granolithic flooring. Use M 20 concrete and HYSD bars.

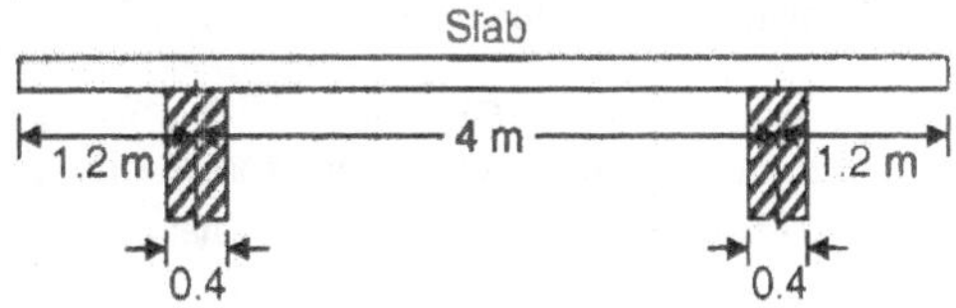

FIG. 11.39

7. Design the roof for a hall, 6 m wide and 11 m long. The slab is supported on R.C.C. beams, 30 cm wide and arranged as shown in Fig. 11.40. The live load on the roof may be taken at 3500 N/m^2 and floor finishing load as 700 N/m^2 Use M 20 mix. and HYSD bars.

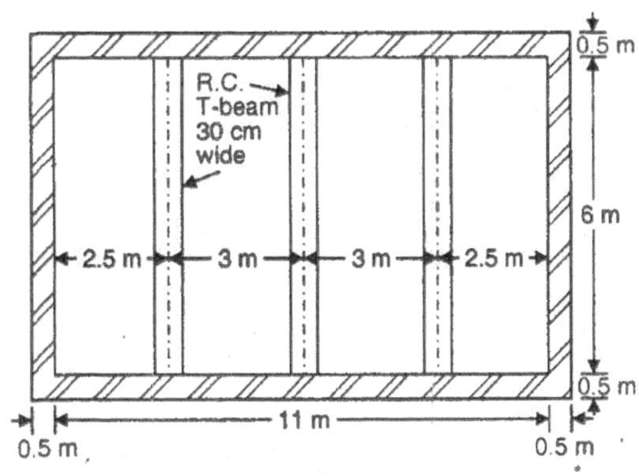

FIG. 11.40

8. Design a doubly reinforced concrete beam, supporting a uniformly distributed superimposed load of 40 kN/m over a clear span of 6 m. The beam carries two concentrated loads of 20 kN each at a distance of 2 m from the face of the support. The overall depth and width of the beam are restricted to 90 cm and 40 cm respectively. Use M 20 mix. and Fe 415 steel.

9. Design completely a R.C.C., roof of a railway platform, consisting of a R.C. slab which cantilevers 2.5 metres on each side of a central R.C. beam (Fig. 11.41). The slab is monolithic with the beam. The beam is supported on columns 5 metres apart, and 30 cm wide. The superimposed load may be taken as 2000 N/m^2. Use M 20 concrete and Fe 415 steel.

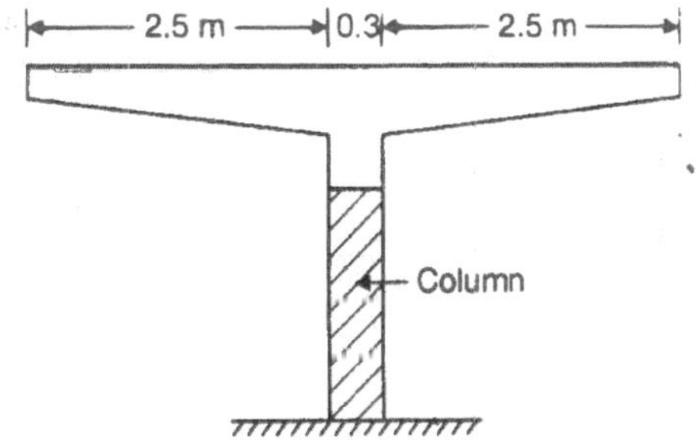

FIG. 11.41

10. The beam of a T-beam roof, over a room 6 m × 18 m projects out by 2 m on side of the room. The roof slab which is 12 cm thick extends to the full length of the beams and its monolithic with them. The beams are spaced 3 metres centre to centre and the web which projects below

the slab is 30 cm wide. Design the T-beams completely, if the superimposed load on roof is 4000 N/m². The walls of the room are 40 cm thick. Use M 20 mix. and HYSD reinforcement.

11. Design a R.C.C., T-beam and a R.C.C. balcony for a lecture theatre having the following data:
(*i*) Size of the lecture theatre....15 m × 9 m × 6 m high
(*ii*) Size of the balcony.....1.5 m clear with 80 cm high and 1.5 cm thick R.C.C. railing.
(*iii*) Height of balcony.....3.5 m from floor.
(*iv*) Thickness of brick wall.....40 cm.
(*v*) Live load for roof.....1500 N/m².
(*vi*) Live load for balcony ....4000 N/m². Use M 20 mix. and Fe 415 reinforcement.

12. Design as inverted T-beam roof for data of problem 7.

13. A R.C. cantilevered canopy 3.5 metre effective overhang and 5.6 metre wide is supported on two beams cantilevered out from wall and spaced at 3.2 metres centre to centre so that the slab overhangs each beam by 1.2 metre from the centre of each beam. The ribs of the beams are integrated with the slab so as to have flat soffit. Design the canopy completely, for a superimposed load of 1500 N/m². Use M 20 mix. and Fe 415 steel.

14. Design the porch slab and the beam shown in Fig. 11.42 for a live load of 1500 N/m², inclusive of the floor finishes, but exclusive of the self-weight of the slab. Use M 20 mix and Fe 415 steel.

15. A R.C.C. bent for a city bus stand is shown in Fig. 11.43. The total length of the bus stand is 11 m. The bents are spaced at 3 m centres so that the slab overhangs by 1 m on either side beyond the centres of bent ribs. Design the slab as well as the bents, for a live load of 2000 N/m² inclusive of the finishes on the slab. The ribs are to be provided above the roof. Use M 20 mix. and HYSD bars.

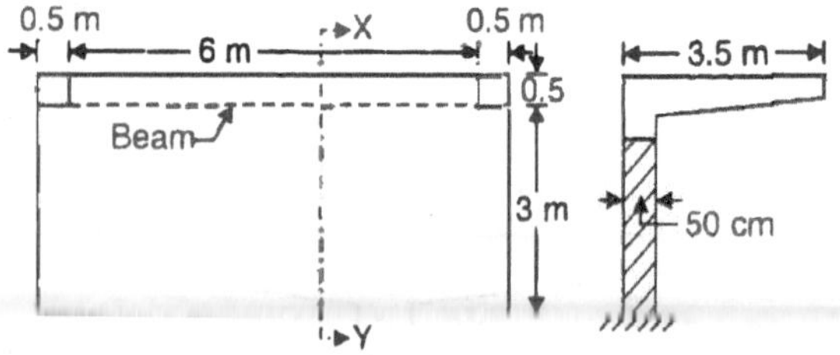

(a) Plan (b) Section at X-Y

FIG. 11.42

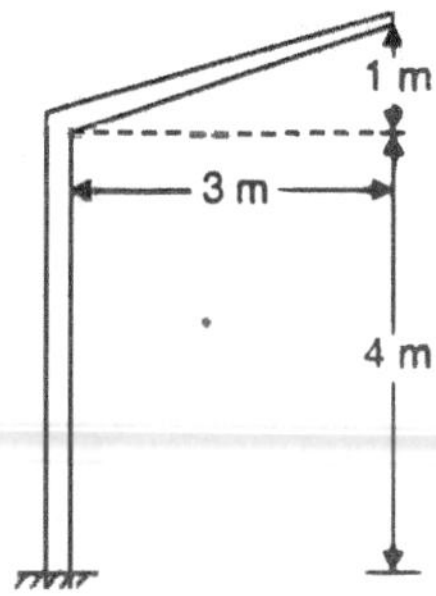

FIG. 11.43

12

Two Way Slabs

12.1. INTRODUCTION

In Chapters 3 and 11, we have discussed the theory and design of slabs spanning in one direction, *i.e.* slabs supported on two parallel edges. In such a slab, bending takes place along the short span, and the B.M. carried along the longer span is usually a very small fraction of the B.M. carried along the short span. When the slab is supported on all the four edges, and when the ratio of long span to short span is small (say less than 2), bending takes place along both the spans. Such a slab is known as a two-way slab or a slab spanning in two directions. The maximum bending moment and deflection for such a slab is much smaller than that of a one-way slab and hence a thinner slab is required. However, reinforcement has to be provided in both the directions. When such a slab is loaded, the corners get lifted up. If the corners are held down, by fixidity at the wall support etc., the bending moment and deflection are further reduced, thus requiring still thinner slab. In that case, special torsional reinforcement at the corners has to be provided to check the cracking of corners. We will divide the two-way slabs in the following three heads :

1. Slabs simply supported on the four edges, with corners not held down, and carrying uniformly distributed load.
2. Slabs simply supported on the four edges, with corners held down and carrying uniformly distributed load.
3. Slabs with edges fixed or continuous and carrying uniformly distributed load.

12.2. SLAB SIMPLY SUPPORTED ON THE FOUR EDGES, WITH CORNERS NOT HELD DOWN AND CARRYING U.D.L.

Such slabs are commonly used in single storey buildings, where the supporting walls do not cause any fixidity. When such a slab is loaded to failure, a crack pattern at the bottom of slab, shown in Fig. 12.1 is observed. As evident from the crack pattern, the bending takes place in the direction *cd* for the portion *EF* of the slab, while near the corners(say *A*) bending takes place along *ab*, perpendicular to the crack direction *AE*. However, the analysis of such a slab done on the basis of directions of cracks is tedious. Grashoff-Rankine gave the following simplified method.

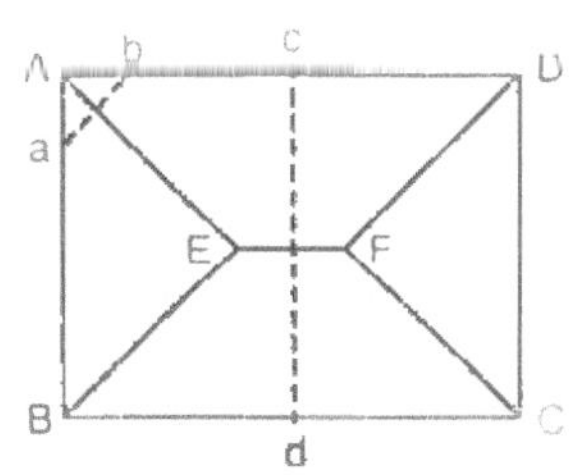

FIG. 12.1

GRASHOFF-RANKINE METHOD

Let L = Long span of a slab ; B = Short span of slab

w = Uniformly distributed load per unit area.

In this method, the bending moments carried in the long and short directions are determined on the basis of maximum deflection at the centre of the slab. Consider two middle strips EF and GH of unit width, parallel to long and short spans respectively.

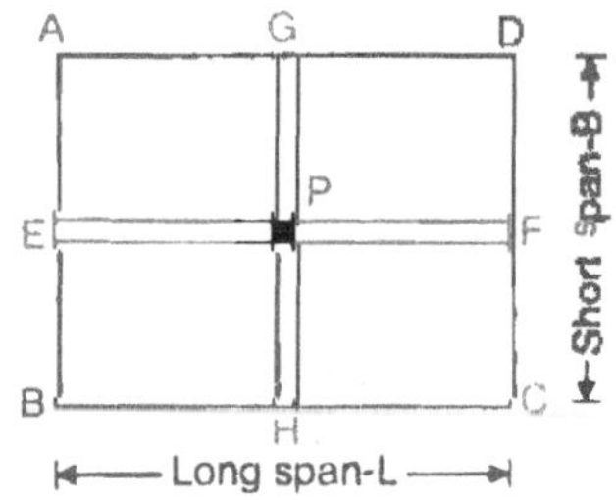

FIG. 12.2.

Let w_L = Load carried by the slab in long direction L

w_B = Load carried by the slab in short direction B.

$$\therefore \qquad w = w_L + w_B \qquad \ldots(1)$$

Consider the deflection of point P, at the intersection of both the strips.

For strip EF, $\delta_P = \frac{5}{384}\frac{w_L L^4}{EI}$; For strip GH, $\delta_P = \frac{5}{384}\frac{w_B B^4}{EI}$

Equating the two, we get, $\frac{5}{384}\frac{w_L L^4}{EI} = \frac{5}{384}\frac{w_B B^4}{EI}$

$$\text{or} \qquad w_L = w_B \left(B/L \right)^4 \qquad \ldots(2)$$

Substituting in (1), we get, $w = w_B \left(\frac{B}{L}\right)^4 + w_B$

$$\text{From which,} \qquad w_B = \frac{w}{1+\left(\frac{B}{L}\right)^4} \qquad \ldots[12.1(a)]$$

$$\text{Hence} \qquad w_L = \frac{w}{1+\left(\frac{B}{L}\right)^4}\left(\frac{B}{L}\right)^4 = \frac{w}{1+\left(\frac{L}{B}\right)^4} \qquad \ldots[12.2(b)]$$

Since $L/B > 1$, we note that $w_B > w_L$. Putting $L/B = r$, we have

$$w_L = \frac{w}{1+r^4} = r_L\, w \qquad \ldots(12.1) \qquad \text{and} \qquad w_B = \frac{w\, r^4}{1+r^4} = r_B\, w \qquad \ldots(12.2)$$

where $r_L = \frac{1}{1+r^4}$ and $r_B = \frac{r^4}{1+r^4}$, and depend upon the value of L/B ratio. Table 12.1 gives the values of r_L and r_B for various values of r. From the Table it will be seen that when $L/B = r$ exceeds 2, the load carried in the longer direction is only 6% while that carried in the shorter direction is 94%. Hence for the slabs in which $L/B > 2$, one may design it as a slab supported on two edges, without any serious error. It should be clearly noted that the above relations hold good only at the middle of the slab, where conditions of maximum B.M. and maximum deflection exist. However, at other points, this proportion of w_B and w_L does not exist, and the design is done by approximately assuming the same ratio as given above. The maximum bending moments in the two directions are:

$$M_L = w_L \frac{L^2}{8} = \frac{w}{1+r^4}\frac{L^2}{8} = w\, r_L \frac{L^2}{8} \qquad \ldots(12.3)$$

$$M_B = w_B \frac{B^2}{8} = \frac{w\, r^4}{1+r^4}\frac{B^2}{8} = w\, r_B \frac{B^2}{8} \qquad \ldots(12.4)$$

TABLE 12.1 VALUES OF r_B AND r_L

$r = L/B$	r_B	r_L	$r = L/B$	r_B	r_L
1.00	0.50	0.50	*1.40*	0.79	0.21
1.05	0.55	0.45	*1.50*	0.84	0.16
1.10	0.59	0.41	*1.60*	0.87	0.13
1.15	0.64	0.36	*1.75*	0.90	0.10
1.20	0.68	0.32	*2.00*	0.94	0.06
1.25	0.71	0.29	*2.50*	0.97	0.03
1.30	0.74	0.26	*3.00*	0.98	0.02

Normally, the bending moment in the short span will be more than the bending moment on the long span, and hence the slab thickness will be determined on the basis of M_B. However, reinforcement in the short span is placed *below* the reinforcement in the long span. As per IS 456 : 2000 minimum total thickness of slab should satisfy the stiffness requirements.

Alternate bars can be bent up at 1/7th span from the centre of the supports, in each direction. If both the spans are equal (for square slab), the effective depth should be taken up to the centre of the top layer of reinforcement.

Shear force : The relations for w_L and w_B developed above give reasonably good results from bending moment point of view only. The reactions at the supports obtained by the above values of w_L and w_B differ very much from those observed experimentally. This is because near the corners, the slab spans in a direction *ab* at right angles at the 45° lines *AE*, *EB* etc. It has been observed that almost the entire load of the triangle *AEB* is transferred to the short support *AB*, while the load in trapezium *AEFD* is transferred to the long support *AD*.

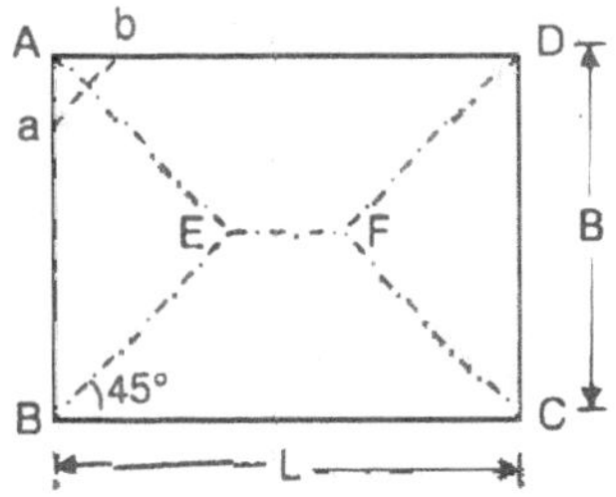

FIG. 12.3

Thus total reaction on the edge *AB* will be equal to $\frac{1}{2} w\, B\,.\, B/2 = w\, B^2/4$. The *average* reaction per unit width, along *AB*, will therefore be $w\, B/4$. However, the *maximum* reactions, per unit width along *AB* will occur near the centre of *AB*, and its value may be taken as $w\, B/3$ for all practical purposes.

Similarly, the S.F. along the long span *AD* or *BC*, per unit width may be taken as $w.B.\dfrac{r}{2+r}$, subject to a maximum of 0.5 $w.B.$ when r exceeds 2.

Design Example 12.1. *Design a R.C. slab for room measuring 4 m × 5 m from inside. The slab carries a live load of 2000 N/m² and is finished with 20 mm thick granolithic topping. Use M 20 concrete and Fe 415 steel. The slab is simply supported at all the four edges, with corners free to lift.*

Solution

1. Design constants and limiting depth of N.A.

Given : $f_{ck} = 20$ N/mm² for M 20 concrete and $f_y = 415$ N/mm² for Fe 415 steel.

For Fe 415, $$\frac{x_{u\,.\,max}}{d} = \frac{700}{1100 + 0.87 \times 415} = 0.479$$

$$R_u = 0.36\, f_{ck} \frac{x_{u\,.\,max}}{d}\left(1 - 0.416 \frac{x_{u\,.\,max}}{d}\right) = 0.36 \times 20 \times 0.479\,(1 - 0.416 \times 0.479) = 2.761$$

2. Computation of loading and bending moment

From deflection point of view, $L/d = 20$, for simply supported slab. Let us assume $p_t = 0.2\%$ for an under-reinforced section. Hence from Fig 11.1, we get modification factor $\simeq 1.68$. Hence $L/d = 20 \times 1.68 = 33.6$ and $d = L/33.6 \simeq 4120/33.6 \simeq 123$ mm. Keeping 20 mm nominal cover and using 8 mm φ bars, $D = 123 + 20 + 4 = 147$ mm. However, take $D = 150$ mm for the purposes of calculating self weight. The load per square metre of the slab will be as under :

1.	Self weight $= 0.15 \times 1 \times 1 \times 25000$	=	3750 N
2.	Wt. of flooring $= 0.02 \times 1 \times 1 \times 24000$	=	480 N
3.	Live load @ 2000 N/m^2	=	2000 N
	Total	=	6230 N

Taking effective depth $\simeq$ 120 mm, we have $L = 5 + 0.12 = 5.12$ m and $B = 4 + 0.12 = 4.12$ m

$$r = L/B = 5.12/4.12 = 1.2427;\quad r^4 = 2.385$$

Hence $$w_B = w \frac{r^4}{1 + r^4} = 6230 \frac{2.385}{1 + 2.385} \simeq 4390 \text{ N}$$

and $$w_L = w \frac{1}{1 + r^4} = 6230 \frac{1}{1 + 2.385} \simeq 1840\text{N}$$

$$\text{Total} = w_B + w_L = 4390 + 1840 = 6230 = w$$

$\therefore$ $$w_{uB} = 1.5 \times 4390 = 6585 \text{ N and } w_{uL} = 1.5 \times 1840 = 2760$$

$$M_{uB} = w_{uB} \frac{B^2}{8} = \frac{6585 \times (4.12)^2}{8} = 13972 \text{ N-m} = 13.972 \times 10^6 \text{ N-mm}$$

$$M_{uL} = w_{uL} \frac{L^2}{8} = \frac{2760\,(5.12)^2}{8} = 9044 \text{ N-mm} = 9.044 \times 10^6 \text{ N-mm}$$

3. Computation of effective depth and total depth

$$d = \sqrt{\frac{M_{u\,,\,max}}{R_u\,.\,b}} = \sqrt{\frac{13.972 \times 10^6}{2.761 \times 1000}} = 71.1 \text{ mm}$$

However, from deflection point of view, provide $D = 150$ mm so that keeping nominal cover = 20 mm and providing 8 mm dia. bars, available $d = 150 - 20 - 4 = 126$ mm for shorter span and $d = 126 - 8 = 118$ mm for longer span.

4. Computation of steel reinforcement : Since actual d is more than d required from B.M., we have an under-reinforced section for which

$$A_{st.B} = \frac{0.5 f_{ck}}{f_y}\left[1 - \sqrt{1 - \frac{4.6 M_{uB}}{f_{ck} bd^2}}\right] bd$$

$$= \frac{0.5 \times 20}{415}\left[1 - \sqrt{1 - \frac{4.6 \times 13.972 \times 10^6}{20 \times 1000 (126)^2}}\right] 1000 \times 126 = 324.6 \text{ mm}^2$$

$\therefore$ Spacing of 8 mm dia. bars, $s_B = \frac{1000 A_\varphi}{A_{st.B}} = \frac{1000 \times 50.3}{324.6} = 154.9$ mm

Hence provide 8 mm φ bars @150 mm c/c along the short span. Bend-up alternate bars at $B/7 = 4.12/7 = 0.59$ m from the centre of each support.

For long span, $A_{st.L} = \frac{0.5 \times 20}{415}\left[1 - \sqrt{1 - \frac{4.6 \times 9.044 \times 10^6}{20 \times 1000 (118)^2}}\right] 1000 \times 118 = 221 \text{ mm}^2$

Spacing $s_L = \frac{1000 A_\varphi}{A_{st.L}} = \frac{1000 \times 50.3}{221} = 227.6$ mm

However, provide 8 mm φ bars @ 220 mm c/c. Bend-up alternate bars at $L/7 = 5.12/7 \simeq 0.73$ m from the centre of each support.

5. Check for shear : S.F. on the short edge per unit length is given by

$$V_{uB} = \frac{1}{3} w_u B = \frac{1}{3} \times 1.5 \times 6230 \times 4.12 \simeq 12834 \text{ N}$$

Similarly, $V_{uL} = w_u B \frac{r}{2 + r} = 1.5 \times 6230 \times 4.12 \times \frac{1.2427}{2 + 1.2427} = 14755$ N

$\therefore \tau_{vB} = \frac{V_{uB}}{bd_L} = \frac{12834}{1000 \times 118} = 0.109 \text{ N/mm}^2$ and $\tau_{vL} = \frac{V_{uL}}{bd_B} = \frac{14755}{1000 \times 126} = 0.117 \text{ N/mm}^2$

Note carefully that the effective depth of slab at short edge is 118 mm while that at long edge is 126 mm. Permissible shear stress τ_c (Table 7.1 and 7.2) is much more than these values. Hence safe.

5. Check for development length

Let the slab have a bearing of 200 mm at the ends. Let us check for development length at the end of short span. Available $A_{st1} = \frac{1000 \times 50.3}{300} = 167.7 \text{ mm}^2$

$$x_u = \frac{0.87 f_y A_{st1}}{0.36 f_{ck} b} = \frac{0.87 \times 415 (167.7)}{0.36 \times 20 \times 1000} = 8.41 \text{ mm}$$

$$M_{1u} = 0.87 f_y A_{st1} (d - 0.416 x_u) = 0.87 \times 415 \times 167.7 (126 - 0.416 \times 8.41)$$

$$= 7.417 \times 10^6 \text{ N-mm}$$

$$V = V_{uL} = 14755 \text{ N}$$

$$L_0 = \frac{l_s}{2} - x' = \frac{200}{2} - 20 = 80 \text{ mm (with no end hooks)}$$

$$L_d = 47 \varphi = 47 \times 8 = 376 \text{ mm}$$

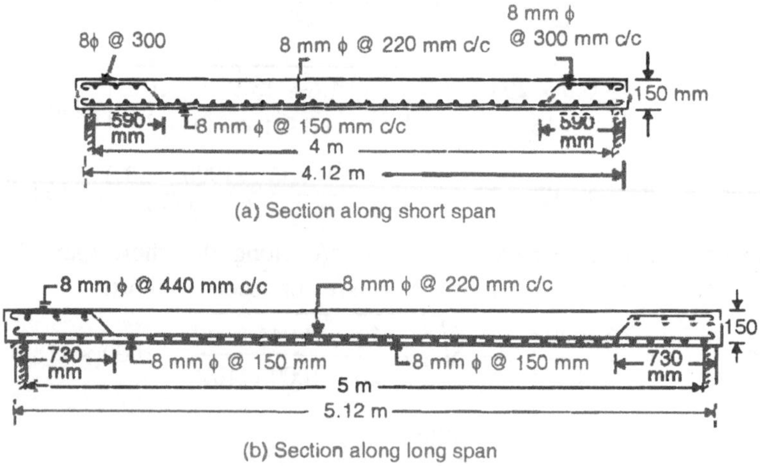

FIG 12.4

$$1.3\frac{M_{1u}}{V_u}+L_0 = 1.3\times\frac{7.417\times10^6}{14755}+80 \simeq 733 \quad \text{mm} \ > L_d$$

Hence the Code requirements are satisfied. Similar check can be applied at the end of the long span. The sections of the slab along the short and long spans are shown in Fig. 12.4.

12.3. SLAB SIMPLY SUPPORTED ON THE FOUR EDGES WITH CORNERS HELD DOWN AND CARRYING U.D.L.

Such a case arises when the slab is built into brick walls or when they are cast monolithically into thin beam, provided the slab is not continuous over its edges. The exact analysis of such a slab, based on theory of elasticity, is difficult and complicated. We shall discuss in brief the following three methods of analysis of such slabs :

1. Pigeaud's method. 2. Marcus's method. 3. I.S. Code method.(see § 12.6)

1. PIGEAUD'S METHOD

When corners are held down, the maximum bending moment and maximum deflection are further reduced. However, holding down of corners super-imposes a twisting moment in the slab, for which special reinforcement is needed at top and bottom faces of the slab at each corner. The behaviour of such a slab is similar to that of a uniformly loaded thin elastic plate, and has been analysed by Pigeaud. According to this method, the bending moments M_B and M_L along the short and long spans, per unit width are given by the following expressions:

$$M_B = r_B'\frac{w B^2}{8} = \frac{w_B B^2}{8} \quad \ldots(12.5) \qquad \text{where} \quad w_B = r_B' w \quad \ldots[12.5(a)]$$

$$M_L = r_L'\frac{w L^2}{8} = \frac{w_L L^2}{8} \quad \ldots(12.6) \qquad \text{where} \quad w_L = r_L' w \quad \ldots[12.6(a)]$$

The coefficients r_B' and r_L' depend on L/B ratio and are given in Table 12.2

TABLE 12.2. VALUES OF r_B' AND r_L'

$r = L/B$	r'_B	r_L'	$r = L/B$	r'_B	r_L'
1	0.3	0.3	*1.4*	0.53	0.13
1.05	0.33	0.27	*1.5*	0.58	0.11
1.1	0.36	0.24	*1.6*	0.63	0.09
1.15	0.39	0.22	*1.75*	0.68	0.07
1.2	0.42	0.19	*2.0*	0.76	0.05
1.25	0.45	0.17	*2.5*	0.87	0.03
1.3	0.48	0.15	*3.0*	0.94	0.02

2. MARCUS'S METHOD

Dr. Marcus evolved a simplified but approximate method by which maximum bending moment in a slab with corners held down can be determined. According to this method, the bending moments M_B and M_L calculated by Rankine-Grashoff method are multiplied by a reduction factor C, the value of which is always less than unity since the positive bending moments in the slab are reduced due to corner restraint. The value of C depends upon L/B ratio, and is given by

$$C = 1 - \frac{5}{6}\frac{r^2}{1+r^4} \qquad \text{...(12.7)}$$

The value of C for various values of L/B ratio are tabulated in Table 12.3.

TABLE 12.3 VALUE OF FACTOR C

$r = L/B$	C	$r = L/B$	C
1	0.861	*1.4*	0.888
1.05	0.862	*1.5*	0.898
1.1	0.864	*1.6*	0.907
1.15	0.866	*1.75*	0.919
1.2	0.871	*2.0*	0.935
1.25	0.874	*2.5*	0.957
1.3	0.879	*3.0*	0.970

Thus, the midspan bending moment per unit width are given by

$$M_B = \left(1 - \frac{5}{6}\frac{r^2}{1+r^4}\right)\frac{r^4}{1+r^4}\frac{w B^2}{8} \qquad \text{...[12.8(a)]}$$

or $$M_B = C\frac{r^4}{1+r^4}\frac{w B^2}{8} = C\, r_B \frac{w B^2}{8} \qquad \text{...(12.8)}$$

$$M_L = \left(1 - \frac{5}{6}\frac{r^2}{1+r^4}\right)\frac{1}{1+r^4}\frac{w L^2}{8} \qquad \text{...(12.9(a)]}$$

or $$M_L = C\frac{1}{1+r^4} \cdot \frac{w L^2}{8} = C \,.\, r_L \,.\, \frac{w L^2}{8} \qquad \text{...(12.9)}$$

The resulting bending moments calculated by Marcus's method and by the exact theory, with Poisson's ratio equal to zero, are almost identical. If, however Poisson's ratio is taken equal to 0.15, the resulting bending moments are given by the following expressions:

$$M_B = C\, r_B \left(1 + \frac{0.15}{r^2}\right)\frac{w B^2}{8} \qquad \text{...[12.10(a)]} \qquad \text{and} \qquad M_L = C\, r_L \left(0.15 + \frac{1}{r^2}\right)\frac{w L^2}{8} \qquad \text{...[12.10(b)]}$$

Torsional reinforcement at corners. To resist torsional moments induced at the held-down corners, it is necessary to provide a mesh reinforcement at both the faces of the slab, at each corner. The quantity and direction of such mesh reinforcement required is not definitely known either by experimental evidence or by theoretical analysis. According to common practice both top and bottom reinforcements should consist of two layers of bars parallel to the sides of the slab and extending in these directions for a distance of one-fifth of the shorter span. The area of the bars in each of the four layers, per unit width of the slab should be three quarters of the area required for the maximum positive moment in the slab. To provide such mesh reinforcement effectively, it is advisable to use the same diameters of bars, as used at the mid-span, and then bend these bars up through 180° in the corner strips.

Shear Force. The shear force, per unit width, along the long and short span is calculated by the same method explained in § 12.2.

Thus, $$F_B = \frac{1}{3} w B \qquad \text{...[12.11(a)]}$$

$$F_L = w B \frac{r}{2+r} \quad \text{(subject to maximum of } 0.5\, w\, B) \qquad \text{...[12.11(b)]}$$

Design Example 12.2. *Solve Example 12.1, assuming that the corners are hold down. Use Marcus's method.*

1. Design constants and limiting depth of N.A.

As in the previous example, $\frac{x_{u,\max}}{d} = 0.479$ and $R = 2.761$

2. Computation of loading and B.M.

As in the previous example, we have

$B = 4 + 0.12 = 4.12$ m; $L = 5 + 0.12 = 5.12$; $r = 1.243$; $r^2 = 1.544$; $r^4 = 2.385$

$w = 6230$; $w_u = 1.5 \times 6230 = 9345$

$$M_{uB} = \left[1 - \frac{5}{6}\frac{r^2}{1+r^4}\right]\frac{r^4}{1+r^4}\frac{w_u B^2}{8} = \left[1 - \frac{5}{6}\frac{1.544}{1+2.385}\right] \times \frac{2.385}{1+2.385} \times \frac{9345\,(4.12)^2}{8} \times 1000$$

$$= 0.62 \times 0.7046 \times 19828 \times 1000 = 8.66 \times 10^6 \text{ N-mm}$$

$$M_{uL} = \left[1 - \frac{5}{6}\frac{r^2}{1+r^4}\right]\frac{1}{1+r^4} \cdot \frac{w_u L^2}{8} = \left[1 - \frac{5}{6}\frac{1.544}{1+2.385}\right] \times \frac{1}{1+2.385} \times \frac{9345\,(5.12)^2}{8} \times 1000$$

$$= 0.62 \times 0.2954 \times 30622 \times 1000 = 5.609 \times 10^6 \text{ N-mm}$$

3. Computation of effective depth and total depth

$$d = \sqrt{\frac{M_{u,\max}}{R_u \cdot b}} = \sqrt{\frac{8.66 \times 10^6}{2.761 \times 1000}} = 56 \text{ mm.}$$

However, from deflection point of view, provide $D = 150$ mm so that keeping a nominal cover of 20 mm and providing 8 mm dia. bars, available $d = 150 - 20 - 4 = 126$ mm for shorter span and $d = 126 - 8 = 118$ mm for longer span.

4. Computation of steel reinforcement

Since actually provided d is more than that required from bending point of view, we have an under-reinforced section, for which

$$A_{stB} = 0.5\frac{f_{ck}}{f_y}\left[1 - \sqrt{1 - \frac{4.6\, M_{uB}}{f_{ck}\, bd^2}}\right] bd = \frac{0.5 \times 20}{415}\left[1 - \sqrt{1 - \frac{4.6 \times 8.66 \times 10^6}{20 \times 1000\,(126)^2}}\right] \times 1000 \times 126$$

$$= 196.8 \text{ mm}^2$$

Spacing of 8 mm φ bars, $s_B = \frac{1000 \times A_\varphi}{A_{stB}} = \frac{1000 \times 50.3}{196.8} = 255.5$ mm

Hence provide 8 mm φ bars @ 250 mm c/c along the short span. Bend-up alternate bars up at $B/7 = 4.12/7 = 0.59$ m from the centre of each support.

For long span, $$A_{st\,L} = \frac{0.5 \times 20}{415}\left[1 - \sqrt{1 - \frac{4.6 \times 5.609 \times 10^6}{20 \times 1000\,(118)^2}}\right] \times 1000 \times 118 = 134.9 \text{ mm}^2$$

Spacing $$s_L = \frac{1000\,A_\varphi}{A_{st\,L}} = \frac{1000 \times 50.3}{134.9} = 372.8 \text{ mm}$$

Maximum spacing = 300 mm.

Hence provide 8 mm φ bars @ 300 mm c/c. Bend-up alternate bars at $L/7 = 5.12/7 \simeq 0.73$ m from the centre of each support.

5. ***Check for shear*** : As found in Example 12.1, $V_{uB} = 12834$ N and $V_{uL} = 14755$ N

$$\therefore \quad \tau_{vB} = \frac{V_{uB}}{bd_L} = \frac{12834}{1000 \times 118} = 0.11 \text{ N/mm}^2 \quad \text{and} \quad \tau_{vL} = \frac{V_{uL}}{bd_B} = \frac{14755}{1000 \times 126} = 0.12 \text{ N/mm}^2$$

Permissible shear stress τ_c (Tables 7.1 and 7.2) is much more than these values. Hence safe.

6. ***Check for development length at supports*** : At simple supports, $1.3\,\frac{M_{1u}}{V_u} + L_0 \geq L_d$

Let the slab have a bearing of 200 mm at each end.

For the ends of the short span :

$$V_u = V_{uL} = 14755 \; N; \quad L_0 = \frac{l_s}{2} - x' = \frac{200}{2} - 20 = 80 \text{ mm (with no hooks)}$$

$$A_{st1B} = \frac{1000 \times 50.3}{2 \times 250} = 100.6 \text{ mm}^2; \quad L_d = 47 \times 8 = 376 \text{ mm}$$

$$x_u = \frac{0.87\,f_y\,A_{st1B}}{0.36\,f_{ck}\,b} = \frac{0.87 \times 415 \times 100.6}{0.36 \times 20 \times 1000} = 5.04 \text{ mm}$$

$$M_{1u} = 0.87\,f_y\,A_{st1B}\,(d_B - 0.416\,x_u) = 0.87 \times 415 \times 100.6\,(126 - 0.416 \times 5.04) = 4.5 \times 10^6 \text{ N-mm}$$

$$\therefore \quad 1.3\,\frac{M_{1u}}{V_u} + L_0 = 1.3 \times \frac{4.5 \times 10^6}{14755} + 80 = 476 \text{ mm} > L_d$$

Hence the Code requirements are satisfied.

For the ends of long span

$$V = V_{uB} = 12834 \text{ N}; \quad L_0 = \frac{l_s}{2} - x' = \frac{200}{2} - 20 = 80 \text{ mm (with no hooks)}$$

$$A_{st1L} = \frac{1000 \times 50.3}{2 \times 300} = 83.8 \text{ mm}^2; \quad L_d = 47 \times 8 = 376 \text{ mm}$$

$$x_u = \frac{0.87\,f_y\,A_{st1L}}{0.36\,f_{ck}\,b} = \frac{0.87 \times 415 \times 83.8}{0.36 \times 20 \times 1000} = 4.2 \text{ mm}$$

$$M_{1u} = 0.87\,f_y\,A_{st1L}\,(d_L - 0.416\,x_u) = 0.87 \times 415 \times 83.8\,(118 - 0.416 \times 4.2)$$
$$= 3.517 \times 10^6 \text{ N-mm}$$

$$\therefore \quad 1.3\,\frac{M_{1u}}{V_u} + L_0 = 1.3 \times \frac{3.517 \times 10^6}{12834} + 80 = 436 \text{ mm} > L_d.$$

Hence the Code requirements are satisfied.

Though the Code requirements are satisfied, it is *preferable* to carry all the bars at the bottom of slab right upto the edge and then bend them up through 180° so that all the bars are available at the top face of the slab, near the edges. Distance of the outer edge of the slab from the centre of the support = 200 – 120/2 = 140 mm. Hence extend the bent-up bars to a distance of $L/7 + 140 = 5120/7 + 140 \simeq 870$ mm from the outer edge of slab, for the long span and to a distance of $B/7 + 140 = 4120/7 + 140 \simeq 730$ mm from the outer edge of the slab. For the end strip of 1000 mm, the bent-up bars will be extended to a distance of 1000 mm from the outer edge of the slab, so that these bars may serve as torsional reinforcement (see next step).

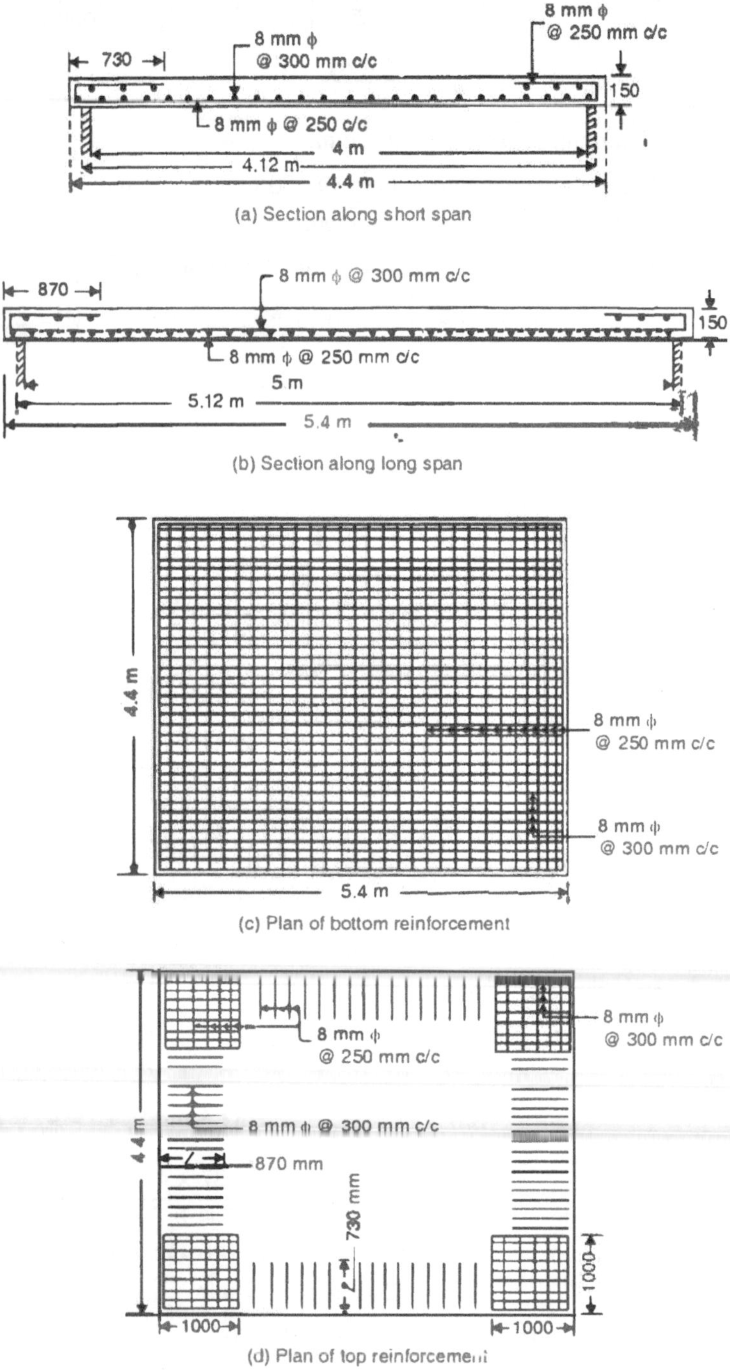

FIG. 12.5

7. *Torsional reinforcement at corners*

Area of torsional reinforcement $= 3/4 \times 196.8 = 147.6 \text{ mm}^2$.

Size of mesh-reinforcement $= \frac{1}{5} \times 4120 = 824$ mm from the face of the support or $824 + 200 \simeq 1000$ mm from the outer edge of the slab. It is preferable to provide the same diameter of the bar as provided for the main reinforcement and at spacing which is a convenient fraction or multiple of the spacing provided for the main reinforcement for the positive bending moments. Using 8 mm φ bars, spacing

= 1000 × 50.3 / 147.6 = 340 mm. The main reinforcement has been provided @ 250 mm c/c in the short span and @ 300 mm c/c in the long span. Thus, at the bottom face of the corner mesh reinforcement, no additional reinforcement is necessary. Similarly, at the top face of the slab, bent-up bars are available in the respective spacing of 250 and 300 mm in both the directions. *Hence no additional reinforcement is required.* However, these bent-up bars are extended to a distance of 100 mm both ways. The details of reinforcement etc. are shown in Fig. 12.5.

12.4. SLAB WITH EDGES FIXED OR CONTINUOUS AND CARRYING U.D.L.

The following three methods may be used for the analysis of slabs with fixed edges, or continuous slabs :

1. Pigeaud's coefficients method. 2. Marcus's method. 3. I.S. Code method (see § 12.6).

1. PIGEAUD'S COEFFICIENTS METHOD

In this method, the load w_B and w_L causing bending moments are calculated by Eqs. 12.5(*a*) and 12.6(*a*), using coefficients r_B' and r_L' from Table 12.2 for the value of r based on the basis of reduced values of long and short span lengths depending on the end conditions of the slab. Table 12.4 gives the span reduction factors f_L and f_B by which r should be multiplied for use in Table 12.2. The maximum positive and negative bending moment in the slab are also given in Table 12.4.

TABLE 12.4. PIGEAUD'S VALUES FOR SLAB WITH FIXED OR FREE EDGES

Case	*Condition of support*	*Reduction factors for r*		*Positive moments*		*Negative moments*	
		f_L	f_B	M_L	M_B	M_L	M_B
1.	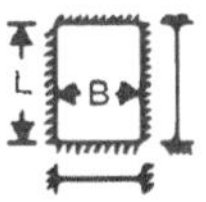	$\frac{2}{3}$	$\frac{2}{3}$	$\frac{1}{10} w_L.L^2$	$\frac{1}{10} w_B.B^2$	$\frac{1}{8} w_L.L^2$	$\frac{1}{8} w_B.B^2$
2.	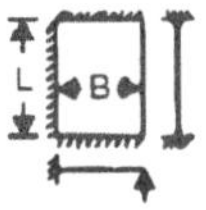	$\frac{2}{3}$	$\frac{4}{5}$	$\frac{1}{10} w_L.L^2$	$\frac{1}{9} w_B.B^2$	$\frac{1}{8} w_L.L^2$	$\frac{1}{8} w_B.B^2$
3.	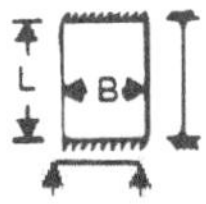	$\frac{2}{3}$	1	$\frac{1}{10} w_L.L^2$	$\frac{1}{8} w_B.B^2$	$\frac{1}{8} w_L.L^2$	Zero, but design moment is taken as $\frac{1}{24} w_B.B^2$
4.	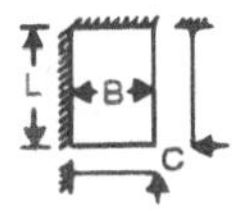 Corner C held down	$\frac{4}{5}$	$\frac{4}{5}$	$\frac{1}{9} w_L.L^2$	$\frac{1}{9} w_B.B^2$	$\frac{1}{8} w_L.L^2$	$\frac{1}{8} w_B.B^2$
5.	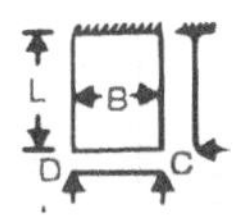 Corners C,D held down	$\frac{4}{5}$	1	$\frac{1}{9} w_L.L^2$	$\frac{1}{8} w_B.B^2$	$\frac{1}{8} w_L.L^2$	Zero, but design moment is taken as $\frac{1}{24} w_B.B^2$

Table 12.4 describes 5 cases of end conditions: (1) all the four edges fixed (or continuous) (2) three edges fixed or continuous while one edge along long span is freely supported

(3) two edges along short span are fixed while the other two edges along long span are free (4) two adjacent edges are fixed while the other two are free, with the common corner held down, and (5) one edge along short span is fixed while the three edges are free, with the opposite corners held down. For cases (1), (2) and (3), corners are always held down. For cases (4) and (5), if the free corners are not held down, the value of r may be calculated using reduction factors f_B and f_L, but the values of w_B and w_L are determined using Grashoff-Rankine formulae (or Table 12.1). For case (4), the positive moments at the centre of span and negative moment at the fixed edges will be equal to $\frac{1}{10} w_B B^2$ for short span and $\frac{1}{10} w_L L^2$ for long span, when the free corner is not held down. Similarly, in case (5), if the free corners are not held down, the positive and negative moments each will be equal to $\frac{1}{10} w_L L^2$ for long span, while the positive moment for span will be equal to $\frac{1}{8} w_B B^2$.

2. MARCUS'S METHOD

In this method, the loads w_B and w_L along the two spans are calculated exactly in the same manner as done in Grashoff-Rankine method, except that proper expressions for central deflection are used. After having determined w_B and w_L, span moments M_B and M_L are calculated as per support conditions. However, due to torsion of corners, the span moments are further reduced by multiplying factors given by Marcus (Table 12.5).

TABLE 12.5 MARCUS'S VALUES FOR SLAB WITH FIXED OR FREE EDGES

Case	Condition of support	Load coefficient		Span moments (+)		Multiplying factors for span moment		Support moments	
		w_B/w	w_L/w	M_B	M_L	For M_B	For M_L	M_B'	M_L'
1.	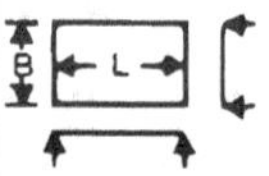	$\frac{r^4}{1+r^4}$	$\frac{1}{1+r^4}$	$\frac{1}{8} w_B B^2$	$\frac{1}{8} w_L L^2$	$1-\frac{5}{6}\cdot\frac{r^2}{1+r^4}$	$1-\frac{5}{6}\cdot\frac{r^2}{1+r^4}$	zero	zero
2.	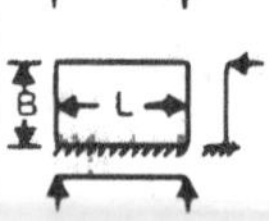	$\frac{5r^4}{2+5r^4}$	$\frac{2}{2+5r^4}$	$\frac{9}{128} w_B B^2$	$\frac{1}{8} w_L L^2$	$1-\frac{75}{32}\cdot\frac{r^2}{2+5r^4}$	$1-\frac{5}{3}\cdot\frac{r^2}{2+5r^4}$	$\frac{1}{8} w_B B^2$ and zero	zero
3.	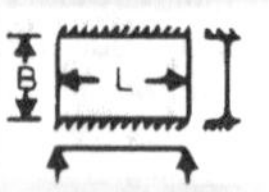	$\frac{5r^4}{1+5r^4}$	$\frac{1}{1+5r^4}$	$\frac{1}{24} w_B B^2$	$\frac{1}{8} w_L L^2$	$1-\frac{25}{18}\cdot\frac{r^2}{1+5r^4}$	$1-\frac{5}{6}\cdot\frac{r^2}{1+5r^4}$	$\frac{1}{12} w_B B^2$ each	zero
4.	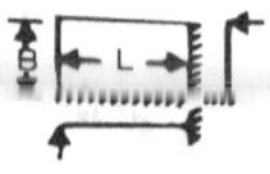	$\frac{r^4}{1+r^4}$	$\frac{1}{1+r^4}$	$\frac{9}{128} w_B B^2$	$\frac{9}{128} w_L L^2$	$1-\frac{15}{32}\cdot\frac{r^2}{1+r^4}$	$1-\frac{15}{32}\cdot\frac{r^2}{1+r^4}$	$\frac{1}{8} w_B B^2$ and zero	$\frac{1}{8} w_L L^2$ and zero
5	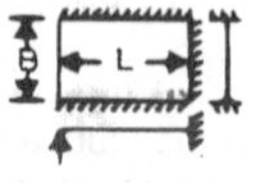	$\frac{2r^4}{1+2r^4}$	$\frac{1}{1+2r^4}$	$\frac{1}{24} w_B B^2$	$\frac{9}{128} w_L L^2$	$1-\frac{5}{9}\cdot\frac{r^2}{1+2r^4}$	$1-\frac{15}{32}\cdot\frac{r^2}{1+2r^4}$	$\frac{1}{12} w_B B^2$ each	$\frac{1}{8} w_L L^2$ and zero
6.	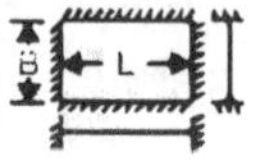	$\frac{r^4}{1+r^4}$	$\frac{1}{1+r^4}$	$\frac{1}{24} w_B B^2$	$\frac{1}{24} w_L L^2$	$1-\frac{5}{18}\cdot\frac{r^2}{1+r^4}$	$1-\frac{5}{18}\cdot\frac{r^2}{1+r^4}$	$\frac{1}{12} w_B B^2$ each	$\frac{1}{12} w_L L^2$ each

For example, let us take the case of slab which has edges of short span fixed, while the long span is freely supported. Then the deflection at the middle of long span will be $\frac{5}{384}\frac{w_L L^4}{EI}$ while that at the middle of short span will be $\frac{1}{384}\frac{w_B B^4}{EI}$ Equating the two, we get

$$\frac{5}{384} w_L \frac{L^4}{EI} = \frac{1}{384}\frac{w_B \cdot B^4}{EI}$$

or

$$w_B = 5w_L(L/B)^4 = 5w_L r^4 \qquad \ldots(1)$$

Also

$$w_B + w_L = w \qquad \ldots(2)$$

Hence from (1) and (2), $w_L = \frac{1}{1+5r^4} w$ ...(12.12) and $w_B = \frac{5r^4}{1+5r^4} w$ ...(12.13)

Similarly, values of w_L and w_B for other end conditions may be calculated. Table 12.5 gives the values of w_L/w and w_B/w for various end conditions, in terms of factor r, the span ratio. The values of span moments (positive moments) along with the multiplying factors for the span moments are also given. However, the support moments (fixed end moments) can be calculated. When a support is free, support moment will be $\frac{1}{12}w_B B^2$ or $\frac{1}{12}w_L L^2$ as the case may be. If one end of a span is fixed while the other end is free, the fixed end moment will be $\frac{1}{8} w_B B^2$ or $\frac{1}{8} w_L L^2$, as the case may be.

12.5. OTHER CASES OF SLABS

1. Non-rectangular slabs. Slabs with non-rectangular panels may be approximately analysed by treating the slab as an equivalent two way slab or a circular slab, depending upon the shape of the panel. If the shape of the panel is trapezoidal, the slab may be designed as an equivalent rectangular slab having one span (L) as the centre to centre distance between the parallel sides, and the other span (B) as perpendicular distance between the parallel sides, as marked in Fig. 12.6(a). If a panel is triangular or regular polygon in shape, it may be treated as an equivalent circular slab having a diameter equal to the mean of circumscribed and inscribed circles, as shown in Fig. 12.6 (b) and (c).

(a) Equivalent rectangular slab B × L for trapezoidal panel (b) Equivalent circular slab for triangular panel (c) Equivalent circular slab for regular polygon panel

FIG. 12.6. EQUIVALENT SLABS FOR NON-RECTANGULAR PANELS.

2. Loads other than U.D.L. If the slab does not carry uniformly distributed load, but carries concentrated load, spread on a small specified areas or locations of slab, it may be analysed by using curves based on Pigeaud's theory.

Example 12.3 *Fig. 12.7 shows the plan of the roof of a building. The super-imposed load of the slab may be taken as 3000 N/m². Design all the slab panels. Use Marcus's method for design. Take M 20 concrete and Fe 415 steel.*

Solution : The four corner slabs *ABFE, CDHG, KLPO* and *IJNM* have the same edge conditions, i.e. each slab panel is free at two adjacent edges, with the corner held down, and fixed at other two adjacent edges. Hence design of only one slab out of these four need be done.

Similarly, the four slabs *EFJI, BFGC, GHLK* and *JKON* have the same edge conditions, i.e. free at one edge and fixed (or continuous) at the other edges. Hence we will design only one slab out of the four. Then remains the slab *FGKJ*, which is fixed (or continuous) on all the four edges.

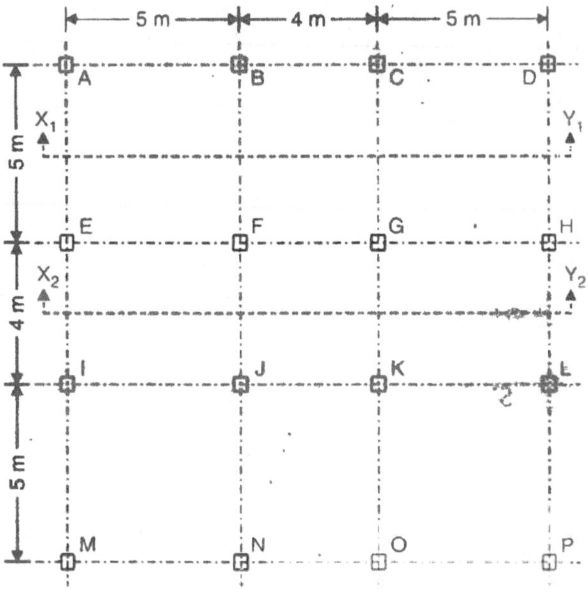

FIG. 12.7.

Step 1. Design constants and limiting depth of N.A.

Given : f_{ck} = 20 N/mm² for M 20 concrete and f_y = 415 N/mm² for Fe 415 steel.

For Fe 415 steel,

$$\frac{x_{u,max}}{d} = \frac{700}{1100 + 0.87 \times 415} = 0.479$$

$$R_u = 0.36 f_{ck} \frac{x_{u,max}}{d}\left(1 - 0.416 \frac{x_{u,max}}{d}\right) = 0.36 \times 20 \times 0.479\,(1 - 0.416 \times 0.479) = 2.761$$

Step 2. Design of slab *ABFE* : This is a square panel of effective size 5 m × 5 m. Let *AB* = *L* = 5 m and *AE* = *B* = 5 m. Edges *BF* and *EF* are continuous while *AB* and *AE* are freely supported with corner a held down to column at *A* (Fig. 12.8)

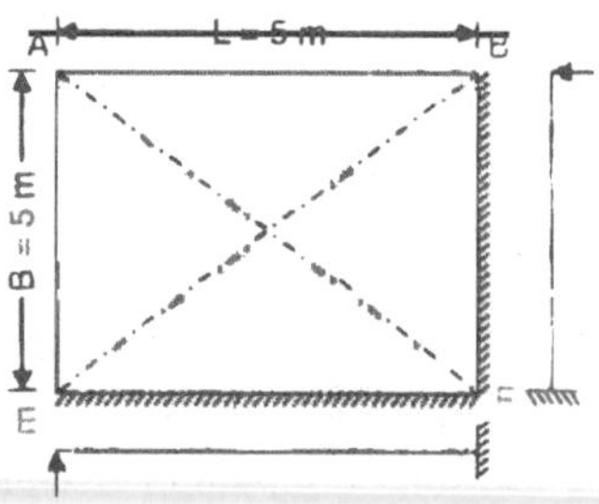

FIG. 12.8.

(*i*) ***Computation of loading and B.M.*** : From the deflection point of view, *L/d* = 20 for simply supported slab and *L/d* = 26 for continuous slab. Since two edges are free and other two edges are continuous, take $L/d = \frac{1}{2}(20 + 26) = 23$. Assuming $p_t = 0.2\%$ for under reinforced section, we get modification factor = 1.68 from Fig. 11.1. Hence *L/d* = 23 × 1.68 = 38.64 and *d* = *L*/38.64 = 5000/38.64 = 129.4 mm. However, since this panel is a two way slab (and that too a square one), a slightly lower value of *d* may be assumed. Hence take *D* = 150 mm for the purpose of calculating self weight. The load per square metre of the slab panel will be as under:

Self weight = 0.15 × 25000 = 3750 N/m²

Super-imposed load = 3000 N/m²

Total w = 6750 N/m² and w_u = 1.5 × 6750 = 10125 N/m²

$$r = L/B = 5/5 = 1;\ r^2 = 1 \text{ and } r^4 = 1$$

From Table 12.5 (case 4), we have

$$w_{uB} = \frac{r^4}{1+r^4} \times 10125 = \frac{1}{1+1} \times 10125 = 5062.5 \quad \text{N/m}$$

and

$$w_{uL} = \frac{1}{1+r^4} \times 10125 = \frac{1}{1+1} \times 10125 = 5062.5 \quad \text{N/m}$$

$$\text{Multiplying factor for span moment} = 1 - \frac{15}{32}\frac{r^2}{1+r^4} = 1 - \frac{15}{32} \times \frac{1}{1+1} = 0.7656$$

$$\therefore \quad M_{uB} = M_{uL} = \frac{9}{128} w_{uB} B^2 \times 0.7656 = \frac{9}{128} \times 5062.5\,(5)^2 \times 0.7656 = 6813 \quad \text{N-m}$$

$$\simeq 6.813 \times 10^6 \quad \text{N-mm}$$

$$\text{Support moment are : } M_{uB}' = M_{uL}' = \frac{w_{uB} B^2}{8} = \frac{5062.5\,(5)^2}{8} \times 1000 = 15.82 \times 10^6 \quad \text{N-mm}$$

(ii) Computation of effective depth and total depth

$$d = \sqrt{\frac{M_{uB}'}{R_u\, b}} = \sqrt{\frac{15.82 \times 10^6}{2.761 \times 1000}} = 75.7 \quad \text{mm}$$

However, d required from deflection point of view is much more, and thus governs the design. Hence keep $D = 150$ mm, so that keeping a nominal cover of 20 mm and providing 8 mm dia bars is each direction, available $d = 150 - 20 - 4 = 126$ mm in one span and $d = 126 - 8 = 118$ mm in the other span, both the spans being equal.

(iii) ***Computation of steel reinforcement*** : Since actually provided d is more than that required from bending point of view, we have an under-reinforced section, for which reinforcement at the centre of each span is given by

$$A_{stB} = A_{stL} = \frac{0.5 f_{ck}}{f_y}\left[1 - \sqrt{1 - \frac{4.6\, M_u}{f_{ck}\, bd^2}}\right] bd$$

$$= \frac{0.5 \times 20}{415}\left[1 - \sqrt{1 - \frac{4.6 \times 6.813 \times 10^6}{20 \times 1000\,(118)^2}}\right] \times 1000 \times 118 = 165 \text{ mm}^2$$

$$\therefore \quad s - 8\,\varphi \text{ at the centre of span} = \frac{1000 \times 50.3}{165} = 305 \quad \text{mm}$$

Hence provide 8 φ bars @ 300 mm c/c in the centre of the span. Bend alternate bars up at a distance of $L/5 = 1$ m from the fixed end and at a distance of $L/7 = 5000/7 \simeq 700$ mm from the free edge.

Reinforcement at the fixed end is given by

$$A_{stB}' = A_{stL}' = \frac{0.5 f_{ck}}{f_y}\left[1 - \sqrt{1 - \frac{4.6\, M_u'}{f_{ck}\, bd^2}}\right] bd$$

$$= \frac{0.5 \times 20}{415}\left[1 - \sqrt{1 - \frac{4.6 \times 15.82 \times 10^6}{20 \times 1000\,(118)^2}}\right] \times 1000 \times 118 = 400 \text{ mm}^2$$

Area of steel available at the fixed end from half bent up bars of one span

$$= \frac{1000 \times 50.3}{600} = 83.8 \text{ mm}^2$$

Area of steel available from half bent-up bars of other adjacent span $\simeq$ 83.8 mm^2

$\therefore$ Remaining area to be provided at fixed edges, to take up negative B.M.

$$= 400 - 2 \times 83.8 = 232 \text{ mm}^2.$$

$\therefore$ Spacing of 8 mm φ *extra* bars at top of each fixed edge $= \frac{1000 \times 50.3}{232} = 216$ mm

However, provide these extra bars at top of each fixed edge @ 150 mm c/c. These bars are to be provided for a length $= L/4 = 5/4 = 1.25$ m to each side of the fixed support.

Torsional reinforcement has to be provided at corner A, since it is held down.

$$A_{st} = \frac{3}{4} \times 165 = 123.8 \text{ mm}^2$$

This has to be provided for a length of $B/5 = 1$ m in both directions and at both the faces. Using 8 mm φ bars, spacing $s = 1000 \times 50.3/123.8 = 406$ mm c/c. However it is preferable to keep a spacing equal to 300 mm c/c, i.e. same as that provided at the centre of the span. It should be noted that the bent up bars are already available at 600 mm c/c. The remaining bars need be provided @ 600 mm c/c, so that over all spacing will be 300 mm c/c.

Torsional reinforcement will also be provided at corners E and B, the area being half of that provided at A. However, as per Code requirements, maximum spacing should not exceed 300 mm. Hence provide 8 mm φ bars @ 300 mm c/c at these corners, for a mesh size of 1 m $\times$ 1 m. No torsional reinforcement is needed at F, because of symmetrical fixidity conditions there.

(*iv*) ***Check for shear*** : Since it is a square panel, the distribution of S.F. on all the four parals will be triangular as shown in Fig. 12.8.

$$\text{Total reaction on fixed edge} = \frac{5}{8} w_{uB} \,.\, B \,.\, L = \frac{5}{8} \times 5062.5\,(5)^2 = 79102 \text{ N}$$

$$\text{Total reaction on free edge} = \frac{3}{8} w_{uB}\, BL = \frac{3}{8} \times 5062.5\,(5)^2 = 47461 \text{ N}$$

The distribution of each of these is triangular.

$$\text{Now maximum } F = \text{Total reaction} \times \frac{2}{B}$$

$$\therefore \quad \text{Max. S.F. at fixed edge} = 79102 \times \frac{2}{5} = 31641 \text{ N}$$

$$\text{and} \quad \text{Max. S.F. at free edge} = 47461 \times \frac{2}{5} = 18984 \text{ N}$$

$$\text{Nominal shear stress at fixed edge} = \frac{31641}{1000 \times 118} = 0.268 \text{ N/mm}^2$$

$$\text{Nominal shear stress at free edge} = \frac{18984}{1000 \times 118} = 0.161 \text{ N/mm}^2$$

Min. reinforcement ratio at fixed edge $= \dfrac{100 \times 400}{1000 \times 118} = 0.339\%$.

Permissible shear stress, even at minimum reinforcement is 0.28 N/mm^2. Hence safe.

(*v*) ***Check for development length at free edge*** : The Code envisages that the following equation should be satisfied at the free-edge :

$$1.3 \frac{M_{1u}}{V_u} + L_0 \geq L_d$$

where $V_u = 18984\ N$; $l_s = 200$ mm (say); $L_d = 47\ \varphi = 47 \times 8 = 376$ mm

$$L_0 = \frac{l_s}{2} + x' = \frac{200}{2} - 20 = 80 \text{ mm (with no hooks);}$$

$$A_{st1} = \frac{1000 \times 50.3}{2 \times 300} = 83.83 \text{ mm}^2$$

$$x_u = \frac{0.87\, f_y\, A_{st1}}{0.36\, f_{ck}\, b} = \frac{0.87 \times 415 \times 83.83}{0.36 \times 20 \times 1000} = 4.2 \text{ mm}$$

$$M_{1u} = 0.87\, f_y\, A_{st1}\, (d - 0.416\, x_u) = 0.87 \times 415 \times 83.83\, (118 - 0.416 \times 4.2)$$

$$= 3.519 \times 10^6 \text{ N-mm}$$

$$\therefore \quad 1.3 \frac{M_{1u}}{V_u} + L_0 = 1.3 \times \frac{3.519 \times 10^6}{18984} + 80 = 321 \text{ mm} < L_d$$

Hence the Code requirements are not satisfied at the free end. In order to satisfy the Code requirements, carry all the bars at the bottom of the slab right upto the edge and then bend them up through 180° so that all the bars are available at the top face of the slab, near the free edge. In that case, $A_{st1} = 2 \times 83.83 = 167.66$ mm^2

$$x_u = \frac{0.87 \times 415 \times 167.66}{0.36 \times 20 \times 1000} = 8.41 \text{ mm}$$

$$M_{1u} = 0.87 \times 415 \times 167.66\, (118 - 0.416 \times 8.41) = 6.931 \times 10^6 \text{ N-mm}$$

and $$1.3 \frac{M_{1u}}{V_u} + L_0 = 1.3 \times \frac{6.931 \times 10^6}{18984} + 80 = 555 \text{ mm} > L_d$$

Step 3. Design of slab *EFJI* : The slab is freely supported on the short edge *EI* and is continuous on the remaining three edges (Fig. 12.9)

$L = 5$ m and $B = 4$ m ; $r = L/B = 5/4 = 1.25$;

$r^2 = 1.5625$ and $r^4 = 2.4414$

(*i*) ***Computation of loading and B.M.***

We take the same loading, i.e. $w_u = 10125\ N/m^2$.

From Table 12.5, case 5,

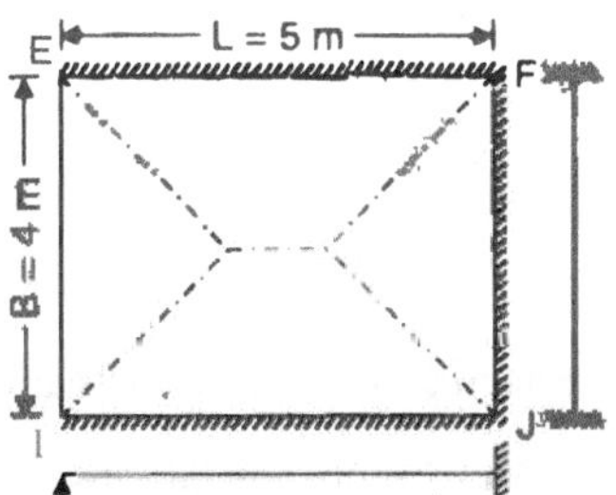

FIG. 12.9.

$$w_{uB} = \frac{2r^4}{1 + 2\, r^4}\, w_u = \frac{2 \times 2.4414}{1 + 2 \times 2.4414} \times 10125 = 8404\ N/m^2$$

$$w_{uL} = \frac{1}{1+2\,r^4}\, w_u = \frac{1}{1+2\times 2.4414} \times 10125 = 1722 \text{ N/m}^2$$

Multiplying factor for $M_{uB} = 1 - \frac{5}{9}\frac{r^2}{1+2\,r^4} = 1 - \frac{5}{9} \times \frac{1.5625}{1+2\times 2.4414} = 0.8524$

Multiplying factor for $M_{uL} = 1 - \frac{15}{32}\frac{r^2}{1+2\,r^4} = 1 - \frac{15}{32} \times \frac{1.5625}{1+2\times 2.4414} = 0.8755$

$\therefore\ M_{uB} = \frac{1}{24}\, w_{uB}\, B^2 \times$ multiplying factor $= \frac{1}{24} \times 8404\ (4^2)\ (0.8524) \times 1000 = 4.776 \times 10^6$ N-mm

$M_{uL} = \frac{9}{128}\, w_{uL}\, L^2 \times$ multiplying factor $= \frac{9}{128} \times 1722\ (5)^2\ (0.8755) \times 1000 = 2.65 \times 10^6$ N-mm

Fixed end moment at each end (*EF* or *IJ*) of short span is

$$M_{uB}' = \frac{1}{12}\, w_{uB}\, .\, B^2 = \frac{1}{12} \times 8404\ (4)^2 \times 1000 = 11.205 \times 10^6 \text{ N-mm}$$

Fixed end moment at and *FJ* of long span is

$$M_{uL}' = \frac{1}{8}\, w_{uL}\, .\, L^2 = \frac{1}{8} \times 1722\ (5)^2 \times 1000 = 5.381 \times 10^6 \text{ N-mm}$$

(*ii*) ***Computation of effective and total thickness***

The thickness will be designed for the maximum of the above moments, i.e. for 11.205×10^6 N-mm

$$\therefore \qquad d = \sqrt{\frac{11.205 \times 10^6}{1000 \times 2.761}} = 63.7 \text{ mm}$$

However, provide an overall depth $D = 150$ mm, from deflection point of view, as provided is the panel *ABFE*. Keeping a nominal cover of 20 mm and providing 8 mm dia. bars in each direction, available $d = 150 - 20 - 4 = 126$ mm in short span and $d = 126 - 8 = 118$ mm in the long span.

(*iii*) ***Computation of steel reinforcement***

Since d actually provided is more than that required from B.M. point of view, we have an under-reinforced section, for which reinforcement at the centre of each span is given by

$$A_{stB} = \frac{0.5 \times 20}{415}\left[1 - \sqrt{1 - \frac{4.6 \times 4.776 \times 10^6}{20 \times 1000\ (126)^2}}\right] \times 1000 \times 126 = 106.9 \text{ mm}^2$$

Minimum reinforcement $= 1.2\, D = 1.2 \times 150 = 180$ mm^2

$\therefore$ Spacing of 8 mm φ bars, $s_B = \frac{1000 \times 50.3}{180} = 279$ mm

Maximum permissible spacing = 300 mm. However, provide 8 mm φ bars @ 270 mm c/c. Bend half bars up near the supports at a distance $= B/5 = 4/5 = 0.8$ m from each fixed end. Reinforcement for negative moment for short span is

$$A_{stB}' = \frac{0.5 \times 20}{415}\left[1 - \sqrt{1 - \frac{4.6 \times 11.205 \times 106}{20 \times 1000\ (126)^2}}\right] 1000 \times 126 = 257.4 \text{ mm}^2$$

Out of this, area available from half bent-up bars from each adjacent span

$$= \frac{1000 \times 50.3}{270} = 186.3 \text{ mm}^2$$

$\therefore$ Remaining area = 257.4 − 186.3 = 71.1 mm²

$$\text{Spacing of 8 mm } \varphi \text{ bars} = \frac{1000 \times 50.3}{71.1} = 707 \text{ mm.}$$

However, provide these @ 540 mm c/c for a distance of $B/4$ = 1 m from each fixed edge.

For long span, $A_{stL} = \frac{0.5 \times 20}{415}\left[1 - \sqrt{1 - \frac{4.6 \times 2.65 \times 10^6}{20 \times 1000\,(118)^2}}\right] 1000 \times 118 = 62.9 \text{ mm}^2$

Minmum reinforcement = 1.2 D = 1.2 × 150 = 180 mm², i.e. 8 mm bars @ 270 mm c/c. Bend half the bars up near the support at a distance of $L/7 = 5/7 \simeq 0.7$ m from free edge and at $L/5$ m = 1 m from fixed edge.

Also, $A_{stL}' = \frac{0.5 \times 20}{415}\left[1 - \sqrt{1 - \frac{4.6 \times 5\,381 \times 10^6}{20 \times 1000\,(118)^2}}\right] \times 1000 \times 118 = 129.3 \text{ mm}^2$

Minimum reinforcement = 1.2 D = 1.2 × 150 = 180 mm². Area available from the bent up bars from both adjacent spans will be more than this.

Torsional reinforcement : Both ends E and I are continuous. Hence provide area of reinforcement of half of that provided at A. However, provide 8 mm φ bars @ 270 mm c/c, so that some of the bent-up bars may be utilized for torsional reinforcement. This is to be provided as a mesh of dimension 1 m × 1 m in each direction.

(iv) Check for shear : Reaction R at each edge = $\frac{1}{2} w_{uB}\, B\, L = \frac{1}{2} \times 8404 \times 4 \times 5 = 84040$ N. The distribution of these is trapezoidal. Let the maximum shear force per unit width be V.

$$\therefore \quad \frac{VB}{2} + V(L - B) = R$$

or
$$V = \frac{2R}{2L - B} = \frac{2R}{L + (L - B)} = \frac{2 \times 84040}{5 + (5 - 4)} = 28013 \text{ N} \quad \ldots(i)$$

Reaction on the fixed short edge $= \frac{5}{8} w_{uL}\, BL = \frac{5}{8} \times 1722 \times 4 \times 5 = 21525$ N

Assuming this to be distributed triangularly, $V = 2 \times \frac{21525}{4} = 10762.5$ ...(*ii*)

Reaction at the free short edge $= \frac{3}{8} w_{uL}\, BL = \frac{3}{8} \times 1722 \times 4 \times 5 = 12915$ N

$$\therefore \quad V = \frac{1}{4}(12915 \times 2) = 6457.5 \text{ N}$$

Hence maximum $V = 28013$ N and $\tau_v = \frac{28013}{1000 \times 126} = 0.222 \text{ N/mm}^2$

From Tables 7.1 and 7.2, it will be seen that the permissible shear stress τ_c is much greater than this. Hence safe.

(*v*) ***Check for development length at free edge***

Along the free short edge (i.e. free support of long span),

$V = V_u = 6457.5$ N and A_{st1} (i.e. 8 mm φ bars @ 540 mm c/c) $= \dfrac{1000 \times 50.3}{540} = 93.1 \text{ mm}^2$

$$l_s = 200 \text{ mm (say)}; \quad L_d = 47\,\varphi = 47 \times 8 = 376 \text{ mm.}$$

$$L_0 = \frac{l_s}{2} - x' = \frac{200}{2} - 20 = 80 \text{ mm (with no hooks)}$$

$$x_u = \frac{0.87 f_y A_{st1}}{0.36 f_{ck} b} = \frac{0.87 \times 415 \times 93.1}{0.36 \times 20 \times 1000} = 4.67 \text{ mm}$$

$$M_{1u} = 0.87 f_y A_{st1} (d - 0.416 x_u) = 0.87 \times 415 \times 93.1 (118 - 0.416 \times 4.67)$$

$$= 3.90 \times 10^6 \text{ N-mm}$$

$$1.3 \frac{M_{1u}}{V_u} + L_0 = 1.3 \times \frac{3.90 \times 10^6}{6457.5} + 80 \simeq 865 \text{ mm} > L_d$$

Hence Code requirements are satisfied.

Step 4. Design of slab *FGKJ* (Fig. 12.10)

This slab is continuous on all the four edges.

$$L = B = 4 \text{ m}; \quad r = 1$$

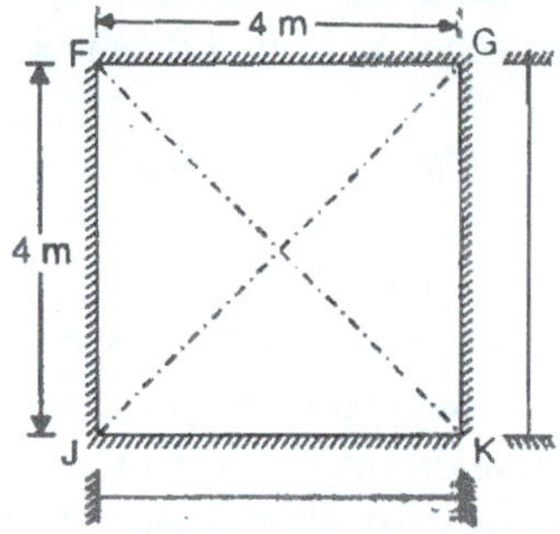

FIG. 12.10

(*i*) ***Computation of loading and B.M.***

We take the same loading, i.e.

$$w_u = 10125 \text{ N/m}^2$$

From Table 12.5, Case 6

$$w_{uB} = w_{uL} = \frac{1}{2} w_u = 5062.5 \text{ N/m}^2$$

Multiplying factor for each span moment $= 1 - \dfrac{5}{18} \dfrac{r^2}{1 + r^4} = 1 - \dfrac{5}{18} \times \dfrac{1}{1 + 1} = 0.8611$

Span moments $\quad M_{uB} = M_{uL} = \dfrac{1}{24} w_{uB} B^2 \times$ multiplying factor

$$= \frac{1}{24} \times 5062.5 (4)^2 \times 0.8611 \times 1000 = 2.906 \times 10^6 \text{ N-mm}$$

Support moments $\quad M_{uB}' = M_{uL}' = \dfrac{1}{12} w_{uB} B^2 = \dfrac{1}{12} \times 5062.5\,(4)^2 \times 1000 = 6.75 \times 10^6$ N-mm

(*ii*) ***Computation of effective depth and total thickness***

Keep overall depth = 150 mm as before. Using 8 mm bars and providing a nominal cover of 20 mm, the minimum available $d = 150 - 20 - 4 - 8 = 118$ mm.

(*iii*) ***Computation of steel reinforcement***

Since d actually provided is more than that required for bending, we have an under-reinforced section.

Hence $\quad A_{stB} = A_{stL} = \dfrac{0.5 \times 20}{415} \left[1 - \sqrt{1 - \dfrac{4.6 \times 2.906 \times 10^6}{20 \times 1000\,(118)^2}} \right] 1000 \times 118 = 69.1 \text{ mm}^2$

However, minimum reinforcement $= 1.2 \times 150 = 180 \text{ mm}^2$.

Hence provide 8 mm φ bars @ 270 mm c/c. Bend half the bars up near the support at a distance $= B/5 = 0.8$ m from the edge.

For fixing moments,

$$A_{stB}' = A_{stL}' = \frac{0.5 \times 20}{415}\left[1 - \sqrt{1 - \frac{4.6 \times 6.75 \times 10^6}{20 \times 1000\,(118)^2}}\right] \times 1000 \times 118 = 163.2 \text{ mm}^2$$

However, minimum reinforcement $= 1.2 \times 150 = 180 \text{ mm}^2$

Area available from bent up bars from both the adjacent spans

$$= \frac{1000 \times 50.3}{270} = 186 \text{ mm}^2. \text{ Hence OK.}$$

No torsional reinforcement is required for this panel.

(*iv*) ***Check for shear***

Total reaction on any edge $= \frac{1}{4} w_u B \cdot L = \frac{1}{4} \times 10125 \times 4 \times 4 = 40500$ N

$$V_{max} = \frac{40500 \times 2}{4} = 20250 \text{ N and } \tau_v = \frac{20250}{1000 \times 118} = 0.172 \text{ N/mm}^2$$

This is less than permissible shear stress even at the minimum reinforcement. Since it is an interior panel, it is not necessary to check the development length. The section of the panels along $X_1 - Y_1$ and $X_2 - Y_2$ are shown in Fig. 12.11.

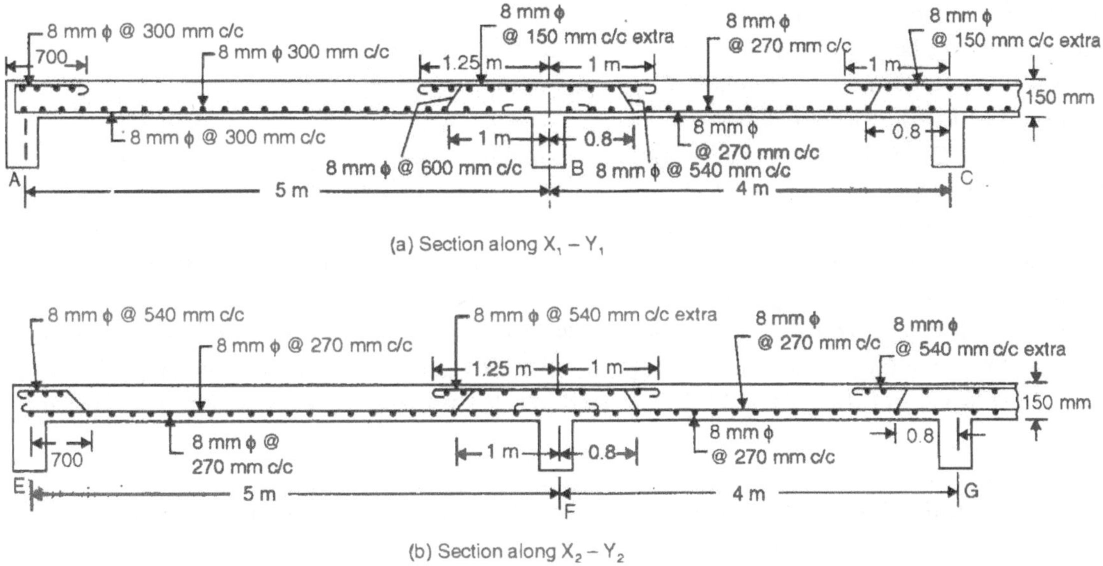

FIG. 12.11.

12.6. INDIAN STANDARD CODE METHOD (IS : 456 - 2000)

Indian Standard Code (IS : 456 - 2000) lays down design rules for two cases of slabs spanning in two directions : (*a*) restrained slabs (*b*) simply supported slabs.

Note : The most commonly used *elastic methods* are based on Pigeaud's or Westergaard's theory and the most commonly used *limit state of collapse method* is based on Johansen's yield line theory.

(*a*) RESTRAINED SLABS.

When the corners of a slab are prevented from lifting, the slab may be designed as specified in para *a*–1 to *a*–11 below.

(a–1). The maximum bending moments pei unit width in a slab are given by the following equations : $M_x = \alpha_x w l_x^2$ and $M_y = \alpha_y . w . l_x^2$

where α_x and α_y are coefficients given in Table 12.6.

w = total load per unit area

M_x , M_y = moments on strips of unit width spanning l_x and l_y respectively, and

l_x and l_y = lengths of shorter span and longer span respectively.

(a–2). Slabs are considered as divided in each direction into middle strips and edge strips as shown Fig. 12.12, the middle strips being three quarters of the width and each edge strip one-eighth of the width.

(a–3). The maximum moments calculated as in *a*–1 apply on to the middle strips and no redistribution shall be made.

(a–4). Tension reinforcement provided at mid-span in the middle strip shall extend in the lower part of the slab to within 0.25 l of a continuous edge, or 0.15 l of a discontinuous edge.

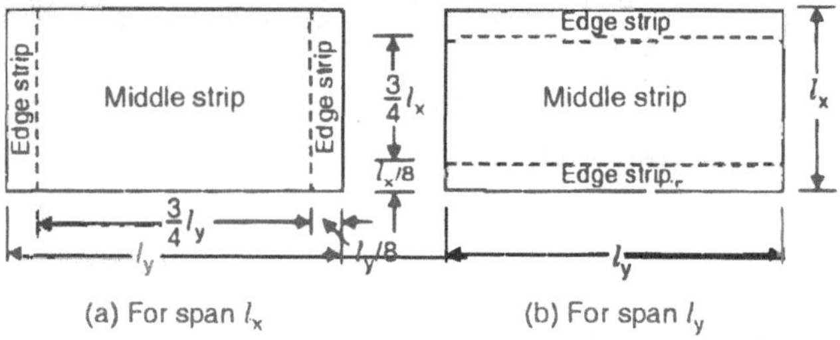

(a) For span l_x (b) For span l_y

FIG. 12.12. DIVISION OF SLAB INTO MIDDLE AND EDGE STRIPS

(a–5). Over the continuous edge of the middle strip, the tension reinforcement shall extend in the upper part of the slab a distance of 0.15 l from the supports, and at least 50 percent shall extend a distance of 0.3 l .

(a–6). At a discontinuous edge, negative moments may arise. They depend on the degree of fixidity at the edge of the slab but, in general, tension reinforcement equal to 50 percent of that provided at mid span extending 0.1 l into the span will be sufficient.

(a–7). Reinforcement in edge strip, parallel to that edge shall comply with minimum given in Chapter 11 (*i.e.* @ 0.15% for mild steel bars and @ 0.12% for HYSD bars) and the requirements for torsion given in *a* – 8 , *a* – 9 and *a* – 10.

(a–8). Torsion reinforcement shall be provided at any corner when the slab is simply supported on both edges meeting at that corner. It shall consist of top and bottom reinforcement, each with layers of bars placed parallel to the sides of the slab and extending from the edges to a minimum distance of one fifth of the shorter span. The area of reinforcement in each of these four layers shall be three quarters of the area required for the maximum mid-span moment in the slab.

(a–9). Torsion reinforcement equal to half that described in *a*–8 shall be provided at a corner contained by edges over only one which the slab is continuous.

(a–10). Torsion reinforcement need not be provided at any corner contained by edges over both of which slab is continuous.

(a–11). Where l_y/l_x is greater than 2, the slabs shall be designed as spanning one way.

TABLE 12.6 BENDING MOMENT COEFFICIENTS FOR RECTANGULAR PANELS SUPPORTED ON FOUR SIDES WITH PROVISION FOR TORSION AT CORNERS (IS : 456 - 2000)

Case No.	*Type of panel and moment*	*Short span coefficient α_x for values of l_y/l_x*								*Long span coefficient α_y for all values of l_y/l_x*
		1.0	*1.1*	*1.2*	*1.3*	*1.4*	*1.5*	*1.75*	*2.0*	
1.	***Interior Panels***									
	Negative moment at continuous edge	0.032	0.037	0.043	0.047	0.051	0.053	0.060	0.065	0.032
	Positive moment at mid-span	0.024	0.028	0.032	0.036	0.039	0.041	0.045	0.049	0.024
2.	***One short edge discontinuous***									
	Negative moment at continuous edge	0.037	0.043	0.048	0.051	0.055	0.057	0.064	0.068	0.037
	Positive moment at mid-span	0.028	0.032	0.036	0.039	0.041	0.044	0.048	0.052	0.028
3.	***One long edge discontinuous***									
	Negative moment at continuous edge	0.037	0.044	0.052	0.057	0.063	0.067	0.077	0.085	0.037
	Positive moment at mid-span	0.028	0.033	0.039	0.044	0.047	0.051	0.059	0.065	0.028
4.	***Two adjacent edges discontinuous***									
	Negative moment at continuous edge	0.047	0.053	0.060	0.065	0.071	0.075	0.084	0.091	0.047
	Positive moment at mid-span	0.035	0.040	0.045	0.049	0.053	0.056	0.063	0.069	0.035
5.	***Two short edges discontinuous***									
	Negative moment at continuous edge	0.045	0.049	0.052	0.056	0.059	0.060	0.065	0.069	—
	Positive moment at mid-span	0.035	0.037	0.040	0.043	0.044	0.045	0.049	0.052	0.035
6.	***Two long edges continuous***									
	Negative moment at continuous edge	—	—	—	—	—	—	—	—	0.045
	Positive moment at mid-span	0.035	0.043	0.051	0.057	0.063	0.068	0.080	0.088	0.035
7.	***Three edges discontinuous*** (one long edge continuous)									
	Negative moment at continuous edge	0.057	0.064	0.071	0.076	0.080	0.084	0.091	0.097	—
	Positive moment at mid-span	0.043	0.048	0.053	0.057	0.060	0.064	0.069	0.073	0.043
8.	***Three edges discontinuous*** (one short edge continuous)									
	Negative moment at continuous edge	—	—	—	—	—	—	—	—	0.057
	Positive moment at mid-span	0.043	0.051	0.059	0.065	0.071	0.076	0.087	0.096	0.043
9.	***Four edges discontinuous***									
	Positite moment at mid span	0.056	0.064	0.072	0.079	0.085	0.089	0.100	0.107	0.056

Restrained Slab with Unequal Conditions at Adjacent Panels

In some cases, the support moments calculated from Table 12.6 for adjacent panels may differ significantly. The following procedure may be adopted to adjust them :

(a) Calculate the sum of the moments at midspan and supports (neglecting signs).

(b) Treat the values from Table 12.6 as fixed end moments.

(c) According to relative stiffness of adjacent spans, distribute the fixed end moments across the supports, giving new support moments.

(d) Adjust midspan moment such that, when added to the support moments from (*c*), neglecting signs, the total should be equal to that from (*a*).

If the resulting support moments are significantly greater than the value from Table 12.6, the tension steel over the supports will need to be extended further. The procedure should be as follows :

(1) Take the span moment as parabolic between the supports : its naximum value is as found from (*d*).

(2) Determine the points of contraflexure of the new support moments [from (*c*)] with the span moment [from (1)].

(3) Extend half the support tension steel at each end to atleast an effective depth or 12 bar diameters beyond the nearest point of contraflexure.

(4) Extend the full area of the support tension steel at each end to half the distance from (3).

(*b*) SIMPLY SUPPORTED SLABS.

(b–1). When simply supported slabs do not have adequate provision to resist torsion at corners and to prevent the corners from lifting, the maximum moments for unit width are given by the following equations :

$$M_x = \alpha_x w l_x^2 \quad \text{and} \quad M_y = \alpha_y w l_x^2$$

where, M_x, M_y, l_x, l_y are the same as those in *a*–1, and α_x and α_y are moment coefficients given in Table 12.7.

TABLE 12.7 BENDING MOMENT COEFFICIENTS FOR SLABS SPANNING IN TWO DIRECTIONS AT RIGHT ANGLES, SIMPLY SUPPORTED ON FOUR SIDES (IS : 456 – 2000)

l_y/l_x	1.0	1.1	1.2	1.3	1.4	1.5	1.75	2.0	2.5	3.0
α_x	0.062	0.074	0.084	0.093	0.099	0.104	0.113	0.118	0.122	0.124
α_y	0.062	0.061	0.059	0.055	0.051	0.046	0.037	0.029	0.020	0.014

(b–2). At least 50 percent of the tension reinforcement provided at mid-span should extend to the supports. The remaining 50 percent should extend to within $0.1\ l_x$ or $0.1\ l_y$ of the support, as appropriate.

Details of Reinforcement. Fig. 12.13 shows the details of reinforcement. Half the positive moment reinforcement $(A_{st})_x$ is bent upwards, at a distance $a_x \leq 0.15 l_x$ from the centre of the support while the remaining half is continued upto the end. The bent up bars or the negative reinforcement (top reinforcement) $(A_{st})_x'$ should be available for distance $a_x' \geq 0.1\ l_x$ from the centre of the support.

Similarly, in the span l_y, half the positive moment reinforcement $(A_{st})_y$ is bent up, at a distance $a_y \leq 0.15\ l_y$ from the centre of the support. The remaining half is continued upto the end. The bent bars or the negative reinforcement $(A_{st})_y'$ is available at the top of the slab for a distance $a_y' \geq 0.1\ l_y$ from the centre of the slab.

The corner reinforcement, at top and bottom of each corner is provided in the form of square mesh of size $0.2\ l_x \times 0.2\ l_x$.

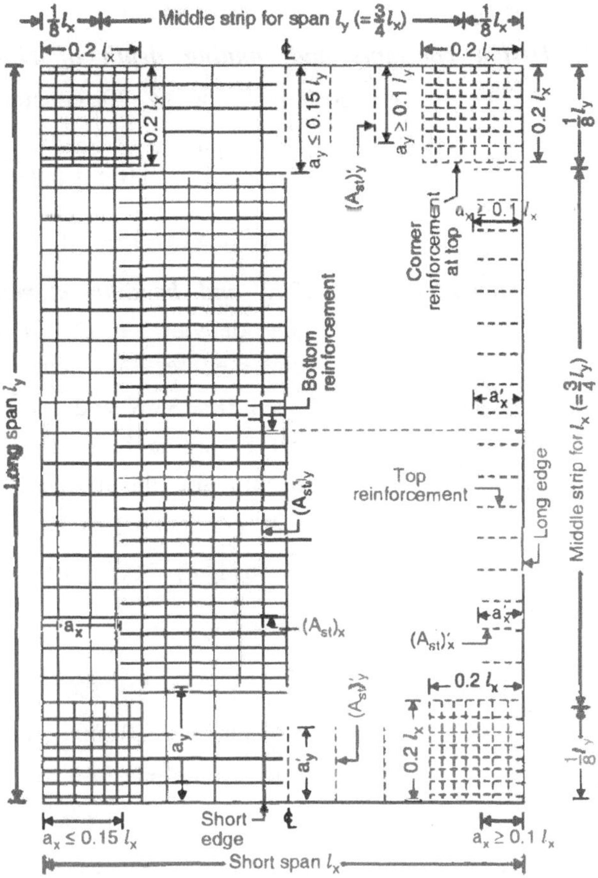

FIG. 12.13.

Load on supporting beams.

The loads on the beams supporting solid slabs spanning in two directions at right angles and supporting uniformly distributed loads, may be assumed as marked in Fig. 12.14.

Thus total reaction on short edge AB(or CD) will be equal to $\frac{1}{2}\ w\ l_x\ \frac{l_x}{2} = (w\ l_x^2)/4$. The *average reaction* per unit width along short edge will therefore be $w\ l_x/4$. However the *maximum reaction*(hence max. S.F.) per unit width along AB will occur near the centre of AB and its value may be taken as $w\ l_x/3$ for all practical purposes.

Similarly, S.F. along the edge AD or AC, (*i.e.* ends of short span of slab) per unit width may be taken as $w\ l_x \frac{r}{2+r}$, subject to a maximum of $0.5\ w\ l_x$ when r exceeds 2.

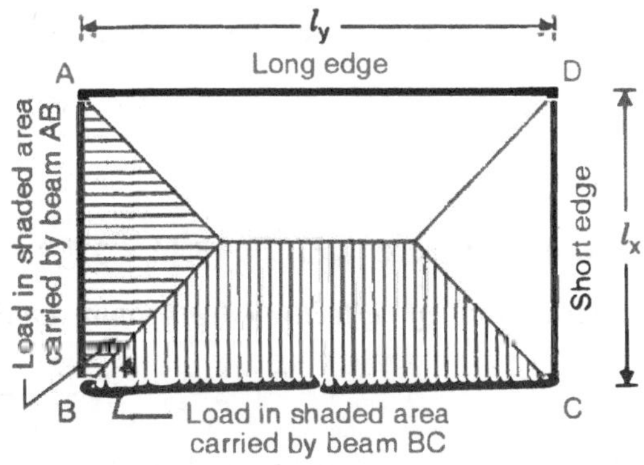

FIG. 12.14. LOADS CARRIED BY SUPPORTED BEAM.

Example 12.4. *Design a R.C. slab for a room measuring 5 m × 6 m size. The slab is simply supported on all the four edges, with corners held down and carries a superimposed load of 3000 N/m², inclusive of floor finishes etc. Use M 20 mix, Fe 415 steel and IS Code method.*

Solution

1. Design constants and limiting depth of N.A.

Give : $f_{ck} = 20 \text{ N/mm}^2$ for M 20 concrete and $f_y = 415 \text{ N/mm}^2$ for Fe 415 steel.

For Fe 415, $$\frac{x_{u,\max}}{d} = \frac{700}{1100 + 0.87 \times 415} = 0.479$$

$$R_u = 0.36 f_{ck} \frac{x_{u,\max}}{d}\left(1 - 0.416 \frac{x_{u,\max}}{d}\right) = 0.36 \times 20 \times 0.479\,(1 - 0.416 \times 0.479) = 2.761$$

2. Computation of loading and bending moment

From deflection point of view, $L/d = 20$ for simply supported slab. Let us assume $p_t = 0.2\%$ for an under-reinforced section. Hence from Fig. 11.1, we get modification factor $\simeq 1.68$. Hence $L/d = 20 \times 1.68 = 33.6$ and $d = L/33.6 = 5000/33.6 = 148.8$ mm. Providing 20 mm nominal cover and 8 mm φ bars, $D = 148.8 + 20 + 4 = 172.8$ mm. Hence assume an overall depth of 180 mm for the purposes of computing dead weight.

(*i*)	Weight of slab per $\text{m}^2 = 0.18 \times 1 \times 1 \times 25000$	$= 4500$
(*ii*)	Super-imposed load @ 3000 N/m^2	$= 3000$
	Total w	$= 7500 \text{ N/m}^2$

Hence $\quad w_u = 1.5\, w = 1.5 \times 7500 = 11250 \text{ N/m}^2$

Taking an effective depth of 150 mm, we have effective $l_y = 6 + 0.15 = 6.15$ m and effective $l_x = 5 + 0.15 = 5.15$ m. Hence $r = l_y / l_x = 6.15/5.15 \simeq 1.2$.

This is case 9 of Table 12.6, from which $\alpha_x = 0.072$ and $\alpha_y = 0.056$

$\therefore$ $$M_{ux} = \alpha_x\, w_u\, l_x^2 = 0.072 \times 11250\,(5.15)^2 = 21483 \text{ N-m} = 21.483 \times 10^6 \text{ N-mm}$$

$$M_{uy} = \alpha_y\, w_u\, l_x^2 = 0.056 \times 11250\,(5.15)^2 = 16709 \text{ N-m} = 16.709 \times 10^6 \text{ N-mm}.$$

For short span width of middle strip $= \frac{3}{4} l_y = \frac{3}{4} \times 6.15 \simeq 4.61$ m

Width of edge strip $= \frac{1}{2}(6.15 - 4.61) = 0.77$ m.

For long span, width of middle strip $= \frac{3}{4} l_x = \frac{3}{4} \times 5.15 \simeq 3.87$

Width of edge strip $= \frac{1}{2}(5.15 - 3.87) = 0.64$ m.

3. Computation of effective depth and total depth

$$d = \sqrt{\frac{M_{ux}}{R_u\, b}} = \sqrt{\frac{21.483 \times 10^6}{2.761 \times 1000}} = 88.2 \text{ mm}.$$

However, from the requirements of stiffness (deflection), keep $D = 180$ mm. Keeping nominal cover of 20 mm and providing 8 mm φ bars, available d from short span $= 180 - 20 - 4 = 156$ mm and that for long span, $d = 156 - 8 = 148$ mm.

4. Computation of steel reinforcement for short span

Since actually provided d is more than that required from bending, we have an under reinforced section. Hence

$$A_{stx} = \frac{0.5 f_{ck}}{f_y}\left[1 - \sqrt{1 - \frac{4.6 M_{ux}}{f_{ck}\, bd^2}}\right] bd$$

$$= \frac{0.5 \times 20}{415}\left[1 - \sqrt{1 - \frac{4.6 \times 21.483 \times 10^6}{20 \times 1000\,(156)^2}}\right] 1000 \times 156 = 403.2 \text{ mm}^2$$

Spacing of 8 mm φ bars, $s_x = \dfrac{1000 \times 50.3}{403.2} = 124.7$ mm.

However, use 8 mm φ bars @ 120 mm c/c for the middle strip of width 4.61 m. Bend half the bars up at a distance $= 0.15\, l_x = 0.15 \times 5150 \simeq 770$ mm from the centre of the support, or at a distance of 770 + 80 = 850 mm from the edge of the slab (assuming a bearing of 160 mm on the wall). Available length of bars at the top = 770 − (160 − 20) = 630 mm from the centre of the support, assuming bending of bars at 45°. The length is more than $0.1\, l_x$ (= 0.1 × 5150 = 515 mm) required by the Code. Hence length of top bars from the edge of slab = 630 + 80 = 710 mm. Edge strip is of length 0.77 m. The reinforcement in the edge strips is $= 1.2\, D = 1.2 \times 180 = 216 \text{ mm}^2$.

Hence spacing of 8 mm φ bars, $s = \dfrac{1000 \times 50.3}{216} = 232$ mm c/c.

However, it is more convenient to keep the spacing as the simple fraction of the edge strip or simple multiple of spacing for the middle strip. Hence keep $s = 770/4 = 192.5$ mm c/c.

5. Computation of reinforcement for long span

$$A_{sty} = \frac{0.5 \times 20}{415}\left[1 - \sqrt{1 - \frac{4.6 \times 16.709 \times 10^6}{20 \times 1000\,(148)^2}}\right] 1000 \times 148 = 327.9 \text{ mm}^2$$

Using 8 mm φ bars, $s = \dfrac{1000 \times 50.3}{327.9} = 153.4$ mm.

Hence provide 8 mm φ bars @ 150 mm c/c in the middle strip of width 3.87 m. Bend half the bars of the middle strip up at a distance of $0.15\, l_y = 0.15 \times 6150 \simeq 920$ mm from the centre of the support or 920 + 80 = 1000 mm from the edge of the slab. Available length of bars of top = 920 − (150 − 20) = 790 mm from the centre of the support or 790 + 80 = 870 mm from the edge of the slab.

For the edge strip of width 0.64 m, $A_{st} = 0.12\, D = 0.12 \times 180 = 216 \text{ mm}^2$ giving rise to a spacing $s = 1000 \times 50.3/216 = 232$ mm c/c. Hence provide 8 mm φ bar @ 640/3 $\simeq$ 213 mm c/c.

6. Check for shear and development length in short span.

S.F. at long edges $= w_u\, l_x \dfrac{r}{2 + r} = 11250 \times 5.15 \times \dfrac{1.2}{2 + 1.2} = 21727$ N

∴ Nominal shear stress at long edges $= \dfrac{21727}{1000 \times 156} = 0.139 \text{ N/mm}^2$

At the long edges, the diameter of bars should be so restricted that the following requirement is satisfied : $1.3\,\dfrac{M_{u1}}{V_u} + L_0 \geq L_d$

$$A_{st1} \text{ at supports of short span } = \frac{1000 \times 50.3}{240} = 209.6 \text{ mm}^2$$

$$x_u = \frac{0.87\, f_y\, A_{st1}}{0.36\, f_{ck}\, b} = \frac{0.87 \times 415 \times 209.6}{0.36 \times 20 \times 1000} = 10.5 \text{ mm}$$

$$M_{1u} = 0.87\, f_y\, A_{st1}\,(d - 0.416\, x_u) = 0.87 \times 415 \times 209.6\,(156 - 0.416 \times 10.5) = 11.475 \times 10^6 \text{ N-mm}$$

$V_u = 21727$ N ; $L_d = 47\,\varphi = 47 \times 8 = 376$ mm

Assume that support width $l_s = 160$ mm and a side cover of 20 mm.

Providing no hooks, $L_0 = \dfrac{l_s}{2} - x' = \dfrac{160}{2} - 20 = 60$ mm. Hence

$$1.3\,\frac{M_{1u}}{V_u} + L_0 = 1.3 \times \frac{11.475 \times 10^6}{21727} + 60 = 747 \text{ mm} > L_d.$$

Hence Code requirements are satisfied.

Also the Code requires that the positive reinforcement should extend into the support atleast by $L_d/3$. Hence minimum support width $= \dfrac{L_d}{3} + x' = \dfrac{360}{3} + 20 = 140$ mm. In our case, the support width is 160 mm.

7. Check for shear and development length in long span

S.F. at short edges $= \frac{1}{3}\, w_u\, l_x = \dfrac{1}{3} \times 11250 \times 5.15 = 19313$ N/m

$$\therefore \text{ Nominal shear stress} = \frac{19313}{1000 \times 148} = 0.13 \text{ N/mm}^2 \text{ (safe)}$$

$$A_{st} = \frac{1000 \times 50.3}{300} = 167.7 \text{ mm}^2 \quad ; \quad x_u = \frac{0.87 \times 415 \times 167.7}{0.36 \times 20 \times 1000} = 8.4 \text{ mm}$$

$$M_{1u} = 0.87 \times 415 \times 167.7\,(148 - 0.416 \times 8.4) = 8.749 \times 10^6 \text{ N-mm}$$

$$V_u = 19313 \text{ N} \;;\quad L_0 = \frac{160}{2} - 20 = 60 \text{ mm}$$

$$L_d = 47 \times 8 = 376 \text{ mm}$$

$$\therefore \quad 1.3\,\frac{M_{1u}}{V_u} + L_0 = 1.3\,\frac{8.749 \times 10^6}{19313} + 60 = 649 \text{ mm} > L_d$$

Hence Code requirements are satisfied.

8. Torsional reinforcement at corners : Size of torsional mesh $= \dfrac{l_x}{5} = \dfrac{5.15}{5} \simeq 1.03$ m from the centre of support or $1.03 + 0.08 \simeq 1.10$ from the edge of the slab. Area of torsional reinforcement $= \frac{3}{4} A_{stx} = \frac{3}{4} \times 403.2 = 302.4$ mm². Spacing of 8 mm φ bar is

$$s = \frac{1000 \times 50.3}{302.4} = 166 \text{ mm}.$$

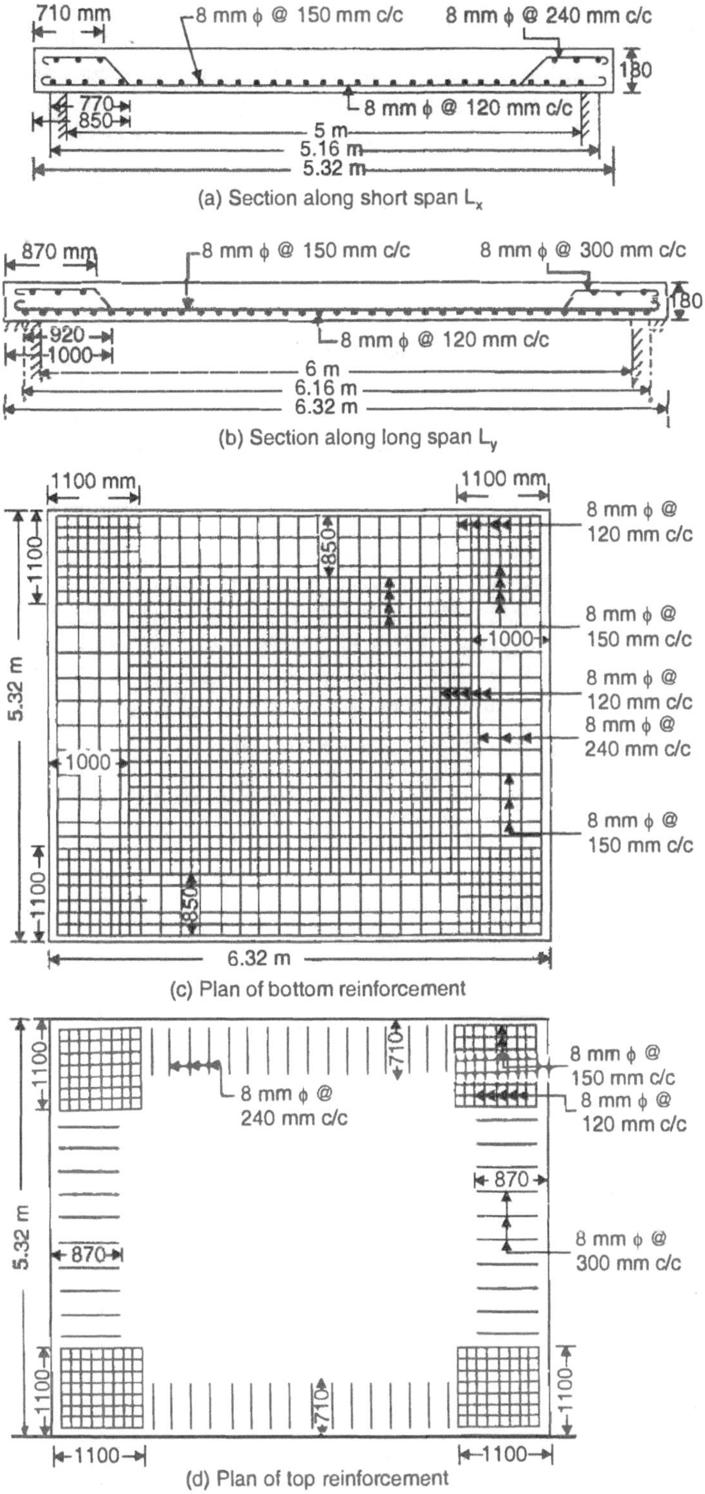

(a) Section along short span L_x

(b) Section along long span L_y

(c) Plan of bottom reinforcement

(d) Plan of top reinforcement

FIG. 12.15

However, it is preferable to use the same spacing as provided for main reinforcement. In the short span, main reinforcement in the middle strip has been provided @ 120 mm c/c while for the edge strip, it has been provided @ 192.5 mm c/c. Hence provide 8 mm φ bars @ 120 mm c/c in the short span direction. In the long span, main reinforcement is @ 150 mm c/c. Hence provide torsional reinforcement @ 150 mm c/c. This reinforcement is to be provided at both faces of the slab, at each corner.

The details of reinforcement etc. are shown in Fig. 12.15.

PROBLEMS

1. Design a R.C. slab for a room 4 m × 4 m measuring from inside. The thickness of wall is 400 mm. The superimposed load, exclusive of the self weight of the slab, is 2 kN/m^2. The slab may be assumed to be simply supported at all the four edges, with corners free to lift. Use M 20 mix. and Fe 415 steel.
2. Redesign problem 1 assuming that the corners are held down. Use Marcus's method.
3. Design a simply supported R.C. slab over a room 4.5 m × 6 m from inside, assuming that the corners are not free to lift. The thickness of all the four walls is 400 mm. The live load on the floor is 2 kN/m^2. The floor carries a floor finish which weighs 8.5 kN/m^2. Use M 20 mix and Fe 415 steel. Use Marcus's method.
4. Redesign Problem 3 using I.S. Code method.
5. Redesign problem 1, assuming that the slab is fixed on all the four edges.
6. A R.C. floor is supported on R.C. columns 250 mm square at the corners of a rectangle 6 m × 7.5 m with R.C. beams connecting the columns along the perimeter of the rectangle. Design the floor as a two-way reinforced slab. Also design one of the 6 m span beams as L-beam. Live load on floor = 8 kN/m^2. Use M 20 mix and Fe 415 steel.
7. An underground room of a dimensions 10 m × 12 m between the existing walls 500 mm thick is to be provided with a R.C.C. roof using one column at the centre of the room and four beams, so that the slab is divided in four panels of 5 m × 6 m. The roof is to carry a live load of 5 kN/m^2. Design the slab and beams. Use M 20 mix and Fe 415 steel.

13

Circular Slabs

13.1. INTRODUCTION

Circular slabs are used for the following purposes : (*i*) roof of a room or hall circular in plan (*ii*) floor of circular water tank or tower (*iii*) roof of pump house constructed above tube well (*iv*) roof of a traffic control post at the intersection of roads, etc. The bending of such a slab is essentially different from a rectangular slab where bending takes place in distinctly two perpendicular directions along the two spans. When a circular slab simply supported at the edge is loaded with uniformly distributed load, it bends in the form of a saucer, due to which stresses are developed both in the radial as well as the in circumferential directions. The tensile radial and circumferential stresses develop towards the *convex* side of the saucer, and hence reinforcement need be provided at the convex face of the slab. Theoretically, reinforcement should be provided both in the radial as well as circumferential directions, but this arrangement would cause congestion and anchoring problem at the centre of the slab. Hence an alternative method of providing reinforcement is adopted : reinforcement is provided in the form of a mesh of bars having equal area of cross-section in both the directions, the area being equal to that required for the bigger of the radial and circumferential moments. However, if the stresses near the edge are not negligible, or if the edge is fixed, radial and circumferential reinforcements near the edge become essential.

The exact analysis of slab, based on theory of elasticity and assuming Poisson's ratio equal to zero, is beyond the scope of the present book ; only final equations are given here. Sometimes, empirical formulae are used for bending moments and shear force etc. We shall discuss in brief the following cases :

1. Slab freely supported at edges and carrying U.D.L. 2. Slab fixed at edges and carrying U.D.L. 3. Slab simply supported at the edges, with load W uniformly distributed along the circumference of a concentric circle. 4. Slab simply supported at edges, with U.D.L. inside a concentric circle. 5. Slab simply supported at the edges with a central hole and carrying U.D.L. 6. Slab simply supported at the edges, with a central hole and carrying W distributed along the circumference of a concentric circle.

13.2. SLAB FREELY SUPPORTED AT EDGES AND CARRYING U.D.L.

Let w = uniformly distributed load.

a = radius of slab.

M_r = radial bending moment at any point at radius r from the centre of the slab.
$(M_r)_c$ &$(M_r)_e$ = radial moments at centre and edge respectively.

M_θ = circumferential bending moment at any point at radius r from the centre of the slab.
$(M_\theta)_c$&$(M_\theta)_e$ = circumferential moments at centre and edge respectively.

The circumferential moment (M_θ) distribution diagram along any diameter is shown in Fig. 13.1(b). The moment varies parabolically with a maximum value of $\frac{3}{16}wa^2$ at the middle to a minimum value of $\frac{2}{16}wa^2$ at the edges. Similarly, the radial moment (M_r) distribution along any diameter is shown in Fig. 13.1 (c). The moment varies from $\frac{3}{16}wa^2$ at the middle to zero at the edges. The values of various moments, per unit width are given below :

$$(M_\theta)_c = +\frac{3}{16}wa^2 \quad \text{...(13.1)}$$

$$(M_\theta)_e = +\frac{2}{16}wa^2 \quad \text{...(13.2)}$$

$$M_\theta = +\frac{w}{16}(3a^2 - r^2), \text{ at any radius } r \quad \text{...(13.3)}$$

$$(M_r)_c = +\frac{3}{16}wa^2 \quad \text{...(13.4)}$$

$$(M_r)_e = 0 \quad \text{...(13.5)}$$

$$M_r = \frac{3}{16}w(a^2 - r^2) \quad \text{...(13.6)}$$

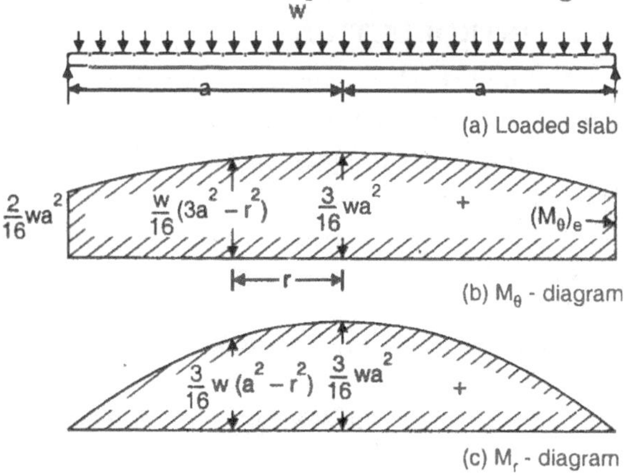

FIG. 13.1

The radial shear force F_r at any radius r is given by

$$F_r = \frac{1}{2}w \, . \, r \text{ (per unit width)} \quad \text{...(13.7)}$$

The circumferential shear force is zero every where. C.E. Reynolds assumes that there is redistribution of moments in the slab due to plasticity of materials, during failure, and therefore recommends maximum values of M_θ and M_r as $\frac{1}{9}wa^2$ instead of $\frac{3}{16}wa^2$ given above.

13.3. SLABS FIXED AT EDGES AND CARRYING U.D.L.

Fig. 13.2 (b) shows the circumferential moment distribution diagram, where, M_θ varies parabolically from a maximum value at the centre to zero at the edge. The moment is *positive* throughout. Fig. 13.2(c) shows the radial moment distribution diagram. M_r varies from a maximum positive value at the centre to zero at a distance $(a/\sqrt{3}) = 0.577\,a$ from the centre, and then becomes negative at the edges. The various values of moments and shear per unit width are as under:

$$(M_\theta)_c = +\frac{1}{16}wa^2 \quad \text{...(13.8)}$$

$$M_\theta = +\frac{1}{16}w(a^2 - r^2) \quad \text{...(13.9)}$$

$$(M_\theta)_e = 0 \quad \text{...(13.10)}$$

$$(M_r)_c = +\frac{1}{16}\,wa^2 \qquad \ldots(13.11)$$

$$M_r = +\frac{1}{16}\,w\,(a^2 - 3r^2) \qquad \ldots(13.12)$$

$$(M_r)_e = -\frac{2}{16}\,wa^2 \qquad \ldots(13.13)$$

$$F_r = \frac{1}{2}\,w\,.\,r \text{ (per unit width)} \qquad \ldots(13.14)$$

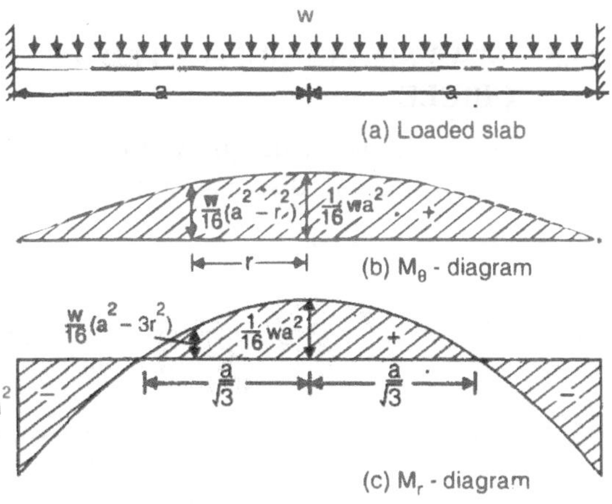

FIG. 13.2.

Slabs partially fixed at the edges. This case is intermediate between the case of a freely supported slab and a fixed slab. Hence the moments may be taken as the average of the corresponding moment of the two cases. The values of various moments per unit width are as follows. For radial moment the point of contraflexure occurs at a radius $r = a\sqrt{2/3}$. Thus

$$(M_r)_c = (M_\theta)_c = +\frac{2}{16}\,wa^2 \qquad \ldots(13.15)$$

$$(M_r)_e = -\frac{1}{16}\,wa^2 \qquad \ldots(13.16)$$

$$(M_\theta)_e = +\frac{1}{16}\,wa^2 \qquad \ldots(13.17)$$

13.4. SLAB SIMPLY SUPPORTED AT THE EDGES WITH LOAD W UNIFORMLY DISTRIBUTED ALONG THE CIRCUMFERENCE OF A CONCENTRIC CIRCLE

Both M_r as well as M_θ are constant from $r = 0$ to $r = b$. When r is greater than b (*i.e.* outside the load circle), both M_θ and M_r decrease parabolically, to values $(M_\theta)_e$ and zero respectively at the edges. Various values of moments etc. per unit width are as follows:

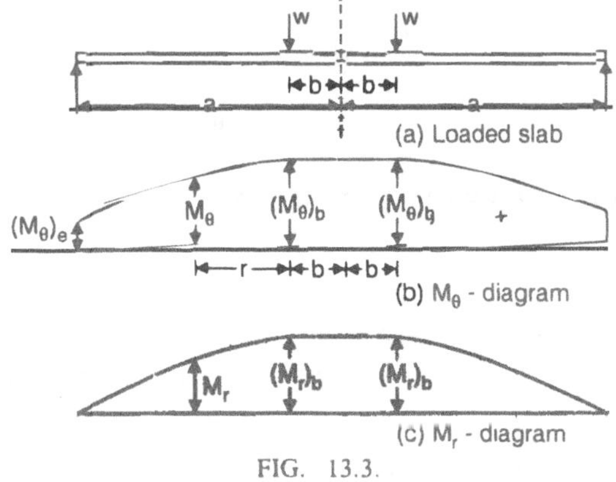

FIG. 13.3.

For $r < b$

$$M_r = (M_r)_b = M_\theta = (M_\theta)_b$$

$$= \frac{W}{8\pi}\left[2\log_e\left(\frac{a}{b}\right) + 1 - \left(\frac{b}{a}\right)^2\right] \qquad \ldots(13.18)$$

$$F_r = 0 \qquad (13.19)$$

For $r > b$

$$M_\theta = \frac{W}{8\pi}\left[2\log_e\left(\frac{a}{r}\right) - \left(\frac{b}{r}\right)^2 + 2 - \left(\frac{b}{a}\right)^2\right] \qquad \ldots(13.20)$$

$$M_r = \frac{W}{8\pi}\left[2\log_e\left(\frac{a}{r}\right) - \left(\frac{b}{a}\right)^2 + \left(\frac{b}{r}\right)^2\right] \qquad \ldots(13.21)$$

$$F_r = \frac{W}{2\pi r} \qquad \ldots(13.22)$$

13.5. SLAB SIMPLY SUPPORTED AT EDGES, WITH U.D.L. INSIDE A CONCENTRIC CIRCLE

Let b be the radius of concentric circle carrying U.D.L. per unit area. The bending moments M_θ and M_r diagrams are shown in Fig. 13.4(b) and (c) respectively. The various values of moments per unit width are as follows :

For $r < b$,

$$M_r = -\frac{3}{16} wr^2 + \frac{1}{4} wb^2 \left[1 - \log_e\left(\frac{b}{a}\right) - \frac{b^2}{4a^2}\right] \quad \ldots(13.23)$$

$$M_\theta = -\frac{1}{16} wr^2 + \frac{1}{4} wb^2 \left[1 - \log_e\left(\frac{b}{a}\right) - \frac{b^2}{4a^2}\right] \quad \ldots(13.24)$$

$$(M_r)_c = +\frac{1}{4} wb^2 \left[1 - \log_e\left(\frac{b}{a}\right) - \frac{b^2}{4a^2}\right] \quad \ldots[13.23(a)]$$

$$(M_r)_b = (M_r)_c = \frac{3}{16} wb^2 \quad \ldots[13.23(b)]$$

$$(M_\theta)_c = +\frac{1}{4} wb^2 \left[1 - \log_e\left(\frac{b}{a}\right) - \frac{b^2}{4a^2}\right] \quad \ldots[13.24(a)]$$

$$(M_\theta)_b = (M_\theta)_c = \frac{1}{16} wb^2 \quad \ldots[13.24(b)]$$

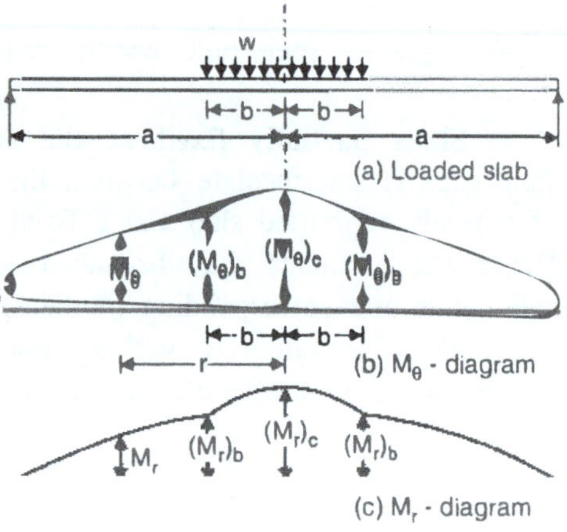

FIG. 13.4

$$F_r = \frac{wr}{2} \quad \ldots(13.25)$$

For $r > b$

$$M_r = -wb^2 \left[\frac{1}{4} \log_e\left(\frac{r}{a}\right) + \frac{b^2}{16}\left(\frac{1}{a^2} - \frac{1}{r^2}\right)\right] \quad \ldots(13.26)$$

$$M_\theta = -wb^2 \left[\frac{1}{4} \log_e\left(\frac{r}{a}\right) - \frac{1}{4} + \frac{b^2}{16}\left(\frac{1}{a^2} + \frac{1}{r^2}\right)\right] \quad \ldots(13.27)$$

$$\therefore \quad (M_r)_e = 0 \quad \ldots(13.28)$$

$$(M_\theta)_e = -wb^2 \left(-\frac{1}{4} + \frac{b^2}{8a^2}\right) \quad \ldots(13.29) \qquad F_r = \frac{wb^2}{2r} \quad \ldots(13.30)$$

13.6. SLAB SIMPLY SUPPORTED AT EDGES, WITH A CENTRAL HOLE AND CARRYING U.D.L.

Let b be the radius of the hole. The B.M. diagrams are shown in Fig. 13.5. The various values of the moments per unit width are as under :

$$M_\theta = -\frac{1}{16} wr^2 + \frac{wb^2}{4} \left[\log_e\left(\frac{r}{a}\right) + \frac{3}{4}\left(-\frac{1}{3} + \frac{a^2}{b^2} + \frac{a^2}{r^2}\right) - \frac{a^2 + r^2}{r^2} \frac{b^2}{a^2 - b^2} \log_e \frac{a}{b}\right] \quad \ldots(13.31)$$

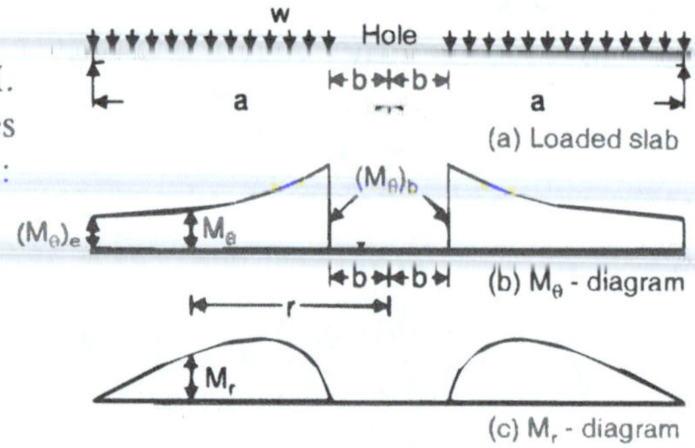

FIG. 13.5

$$M_r = -\frac{3}{16} wr^2 + \frac{wb^2}{4} \left[\log_e \frac{r}{a} + \frac{3}{4}\left(1 + \frac{a^2}{b^2} - \frac{a^2}{r^2}\right) + \frac{a^2 - r^2}{r^2} \frac{b^2}{a^2 - b^2} \log_e \frac{a}{b}\right] \quad \ldots(13.32)$$

$$F_r = \frac{wr}{2} - \frac{w\,b^2}{2\,r} \qquad ...(13.33)$$

13.7. SLAB SIMPLY SUPPORTED AT THE EDGES WITH A CENTRAL HOLE AND CARRYING W DISTRIBUTED ALONG THE CIRCUMFERENCE OF A CONCENTRIC CIRCLE

Let b = radius of hole and c = radius of load circle. The bending moment M_θ and M_r diagrams are shown in Fig. 13.6 (*b*) and (*c*) respectively. Various values per unit width are given below :

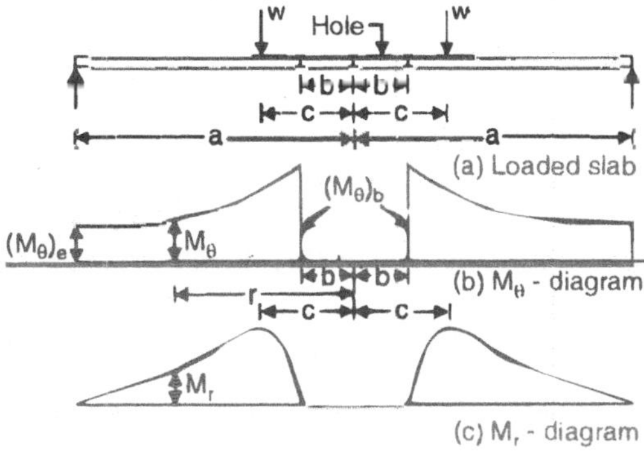

FIG. 13.6

For r < c

$$M_\theta = \frac{W}{4\pi} \cdot \frac{a^2}{a^2 - b^2}\left(1 + \frac{b^2}{r^2}\right)\left[\log_e \frac{a}{c} + \frac{1}{2} - \frac{c^2}{2a^2}\right] \qquad ...(13.34)$$

$$M_r = \frac{W}{4\pi} \cdot \frac{a^2}{a^2 - b^2}\left(1 - \frac{b^2}{r^2}\right)\left[\log_e \frac{a}{c} + \frac{1}{2} - \frac{c^2}{2a^2}\right] \qquad ...(13.35)$$

$$F_r = 0 \qquad ...(13.36)$$

For r > c

$$M_\theta = \frac{W}{4\,\pi}\left[\log_e \frac{c}{r} + \frac{1}{2} + \frac{a^2}{a^2 - b^2} \cdot \frac{r^2 + b^2}{r^2}\left(\log_e \frac{a}{c} + \frac{1}{2} - \frac{c^2}{2\,a^2}\right) - \frac{c^2}{2\,r^2}\right] \qquad ...(13.37)$$

$$M_r = \frac{W}{4\pi}\left[\log_e \frac{c}{r} - \frac{1}{2} + \frac{a^2}{a^2 - b^2} \cdot \frac{r^2 - b^2}{r^2}\left(\log_e \frac{a}{c} + \frac{1}{2} - \frac{c^2}{2a^2}\right) + \frac{c^2}{2\,r^2}\right] \qquad ...(13.38)$$

$$F = \frac{W}{2\,\pi r} \qquad ...(13.39)$$

Example 13.1. *A circular room has 5 m diameter from inside. Design a circular roof slab for room, to carry a superimposed load of 3750 N/m². Assume that the slab is simply supported at the edges. Use M 20 concrete and HYSD bars of Fe 415 grade.*

Solution

1. Design constants and limiting depth of N.A.

Given : $f_{ck} = 20\ N/mm^2$ for M 20 concrete and $f_y = 415\ N/mm^2$ for Fe 415 steel.

$$\text{For Fe 415,}\quad \frac{x_{u\,.\,max}}{d} = \frac{700}{1100 + 0.87 \times 415} = 0.479$$

$$R_u = 0.36\, f_{ck}\, \frac{x_{u\,.\,max}}{d}\left(1 - 0.416\,\frac{x_{u\,.\,max}}{d}\right) = 0.36 \times 20 \times 0.479\,(1 - 0.416 \times 0.479) = 2.761$$

2. Computation of loading and bending moment

From deflection point of view, $L/d = 20$ for simply supported one way slab. Unfortunately, Code has not given any recommendation for circular slab. Let us arbitrarily fix L/d ratio for circular slab equal to $1\frac{1}{3}$ times that for a one way slab. Assuming under-reinforced section with $p_t = 0.2\%$, we get a modification factor $\simeq 1.68$. Hence $L/d = (20 \times 4\ /3) \times 1.68$

≏ 45. Hence $d = L/45 = 5000/45 = 111$ mm. Providing 20 mm nominal cover and using 8 mm φ bars, we get $D = 111 + 20 + 4 = 135$ mm. Hence assume $D = 135$ mm for computation of dead load.

∴ Self weight $= 0.135 \times 1 \times 1 \times 25000 = 3375 \text{ N/m}^2$

Super-imposed load $= 3750 \text{ N/m}^2$

Total $w = 7125 \text{ N/m}^2$

$$\therefore \quad w_u = 1.5\, w = 1.5 \times 7125 \simeq 10688 \text{ N/m}^2$$

$$\therefore \quad (M_{ur})_c = (M_{u\theta})_c = +\frac{3}{16} w_u a^2 = +\frac{3}{16} \times 10688\,(2.5)^2 \times 1000 = 12.525 \times 10^6 \text{ N-mm.}$$

$$(M_{u\theta})_e = \frac{2}{16} w_u a^2 = \frac{2}{16} \times 10688\,(2.5)^2 \times 1000 = 8.35 \times 10^6 \text{ N-mm}$$

3. Computation of effective depth and total depth

$$d = \sqrt{\frac{M_{u\,,\,\max}}{R_u \,.\, b}} = \sqrt{\frac{12.525 \times 10^6}{2.761 \times 1000}} = 67.4 \text{ mm}$$

However from deflection point of view, provide $D = 135$ mm, so that keeping nominal cover = 20 mm and providing 8 mm dia. bars, available $d = 135 - 20 - 4 = 111$ mm for one layer and $111 - 8 = 103$ mm for the other layer.

4. Computation of steel reinforcement

Since actual d is more than that required from B.M. , we have an under-reinforced section for which radial and circumferential reinforcement required at the centre is

$$(A_{st})_c = \frac{0.5 f_{ck}}{f_y}\left[1 - \sqrt{1 - \frac{4.6\, M_{uc}}{f_{ck}\, bd^2}}\right] bd = \frac{0.5 \times 20}{415}\left[1 - \sqrt{1 - \frac{4.6 \times 12.525 \times 10^6}{20 \times 1000\,(103)^2}}\right] \times 1000\,(103)$$

$= 363.3 \text{ mm}^2$

$$\text{Spacing of 8 mm φ bars} = \frac{1000 \times 50.3}{363.3} = 138 \text{ mm.}$$

Hence provide both the reinforcements in the form of a mesh consisting of 8 mm φ bars spaced at 130 c/c in each of the two layers at right angles to each other.

Near the edges, the bars do not have proper anchorage since they are free. There will be slipping tendency and hence the bars will not be capable of taking any tension. At the edges, radial tensile stress is zero, but circumferential tensile stresses exist because $(M_{u\theta})_e = 8.35 \times 10^6$ N-mm. The tendency of slipping can be avoided by providing extra circumferential reinforcement in the form of rings placed in a width equal to the development length of the mesh bars. Available d for rings $= 103 - 8 = 95$ mm.

$$\therefore \quad (A_{st\theta})_e = \frac{0.5 \times 20}{415}\left[1 - \sqrt{1 - \frac{4.6 \times 8.35 \times 10^6}{20 \times 1000\,(95)^2}}\right] 1000 \times 95 = 258 \text{ mm}^2$$

$$\text{and spacing } s \text{ of 8 mm dia. bars} = \frac{1000 \times 50.3}{258} = 194.9 \text{ mm} \simeq 190 \text{ mm}$$

$(M_\theta)_e$ is $\frac{2}{3}$ of the maximum moment at the centre. The circumferential steel is to be provided for a length $= \frac{2}{3} L_d$ $= \frac{2}{3} \times 47 \times 8 = 251$ mm. (i.e. the length required to develop a tensile stress of $\frac{2}{3}\sigma_s = \frac{2}{3}(0.87 \times 415)$ $= 240.7$ N/mm^2 by bond). Hence provide rings of 8 mm φ bars @ 190 m c/c, total number of rings being $= \frac{251}{190} + 1 \simeq 3$.

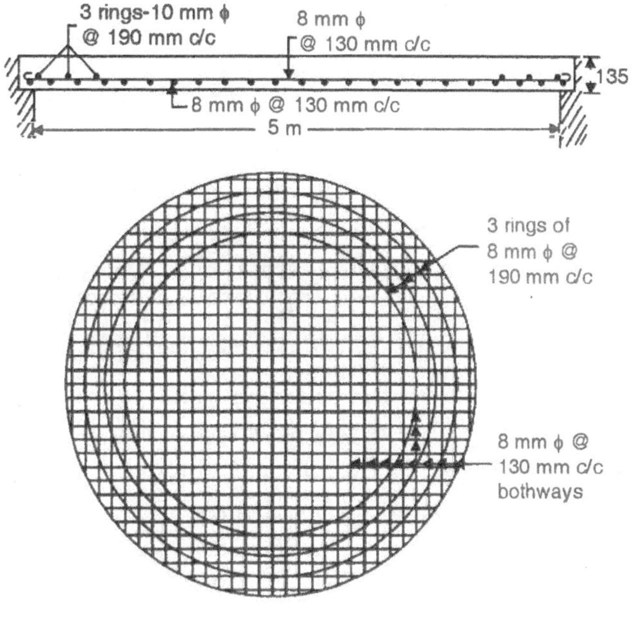

FIG. 13.7

Since the slab is quite thin, no temperature or distribution reinforcement at the top face of the slab is necessary. *However*, for a thicker slab (say greater than 200 mm), such reinforcement @ 0.12% of cross-sectional area of concrete may be provided at the top face of the slab.

The details of reinforcement etc. are shown in Fig. 13.7.

Design Example 13.2. *Solve example 13.1 assuming the slab to be partially restrained at the edges.*

Solution

1. Design constants and limiting depth of N.A.

As in Example 13.1, $\frac{x_{u,\max}}{d} = 0.479$ and $R_u = 2.761$

2. Computation of loading and bending moment

For a partially fixed rectangular slab, $L/d = \frac{1}{2}(20 + 26) = 23$. Let us arbitrarily fix L/d ratio for circular slab equal to $1\frac{1}{3}$ times that for one way slab. Taking modification factor as 1.68 (as in example 13.1), we have $L/d = 23 \times 4/3 \times 1.68 \simeq 52$, Hence $d = L/52$ $= 5000/52 = 96$ mm. Providing 20 mm nominal cover and using 8 mm φ bars, we get $D = 96 + 20 + 4 = 120$ mm. Hence assume $D = 120$ mm for computation of dead load.

Self weight $= 0.12 \times 1 \times 1 \times 25000$ $= 3000$ N/m^2

Super-imposed load $= 3750$ N/m^2

Total $w = 6750$ N/m^2

$\therefore \quad w_u = 1.5\,w = 1.5 \times 6750 = 10125$ N/m^2

Hence $(M_r)_c = (M_\theta)_c = +\frac{2}{16} w_u a^2 = +\frac{2}{16} \times 10125\,(2.5)^2 \times 1000 = 7.91 \times 10^6$ N-mm

$$(M_\theta)_e = +\frac{1}{16} w_u a^2 = +3.955 \times 10^6 \text{ N-mm}$$

$$(M_r)_e = -\frac{1}{16} w_u a^2 = -3.955 \times 10^6 \text{ N-mm.}$$

The radial moment changes sign, the point of inflection being at a radius $a\sqrt{2/3} = 2.5\sqrt{2/3} = 2.04$ m $= 2040$ mm or at a distance of $2500 - 2040 = 460$ mm from the edge of the support.

3. Computation of effective depth and total depth

$$d = \sqrt{\frac{M_{u,\max}}{R_u \cdot b}} = \sqrt{\frac{7.91 \times 10^6}{2.761 \times 1000}} = 53.5 \text{ mm}$$

However, from the deflection point of view, provide $D = 120$ mm, so that keeping nominal cover $= 20$ mm and providing 8 mm dia bars, available $d = 120 - 20 - 4 = 96$ mm for one layer and $96 - 8 = 88$ mm for the other layer.

4. Steel reinforcement at the centre

Since actual d is more than that required from B.M., we have an under-reinforced section for which radial and circumferential reinforcement at the centre are given by

$$(A_{st})_c = \frac{0.5 f_{ck}}{f_y}\left[1 - \sqrt{1 - \frac{4.6 M_{uc}}{f_{ck} bd^2}}\right] bd = \frac{0.5 \times 20}{415}\left[1 - \sqrt{1 - \frac{4.6 \times 7.91 \times 10^6}{20 \times 1000\,(88)^2}}\right] \times 1000 \times 88$$

$$= 265.8 \text{ mm}^2$$

Hence spacing of 8 mm φ bars, $s = \frac{1000 \times 50.3}{265.8} = 189$ mm. Hence provide a mesh of reinforcement consisting of 8 mm φ bars placed @ 180 mm c/c in two mutually perpendicular directions.

5. Circumferential reinforcement at the edges : At the edges, $(M_\theta)_e = +3.955 \times 10^6$ N-mm and area of steel required $= \frac{1}{2} \times 265.8 = 132.9$ mm². Since the mesh bars will not be in a position to resist this, additional ring reinforcement will have to be provided near the edges. Using 8 mm φ bars for the rings, $s = \frac{1000 \times 50.3}{132.9} = 378$ mm.

However, maximum permissible spacing = 300 mm.

Since the circumferential moment at the edges is half of that at the centre, the required development length $= \frac{1}{2} \times 47\,\varphi = \frac{1}{2} \times 47 \times 8 = 188$ mm, *i.e.* the length required to develop a tensile steel $= \frac{1}{2} \times 0.87 \times 415 = 180$ N/mm² by bond. Hence provide rings of 8 mm φ bars @ 300 mm c/c, total number of rings being $= \frac{188}{300} + 1 \simeq 2$. Hence provide 2 rings at the edges, spaced @ 300 mm c/c.

6. Radial reinforcement at the edges : Since there is negative radial moment of magnitude 3.955×10^6 N-mm at the edges, radial reinforcement has to be provided at the top face of the slab, at the edges. The point of inflection is at a distance of 460 mm from the edge of the support. Hence the radial distance, measured from the face of the support, upto which the radial reinforcement is to be provided will be the greater of the following:

(*i*) $460 + d = 460 + 96 = 556$ mm

(*ii*) $460 + 12\,\varphi = 460 + 12 \times 8 = 556$ mm

(*iii*) $L_d = 47\,\varphi = 47 \times 8 = 376$ mm

Hence provide these upto radial distance of 600 mm from the face of the support. The Code specifies that atleast one third the total negative moment reinforcement at the support shall extend beyond the point of inflexion for a distance not less than one sixteenth of clear span $\left(= \frac{1}{16} \times 5000 = 312.5 \text{ mm}\right)$.

Hence provide alternate radial bars for a distance of $460 + 312.5 = 772.5 \simeq 800$ mm from the face of support. Available $d = 120 - 20 - 4 = 96$ mm.

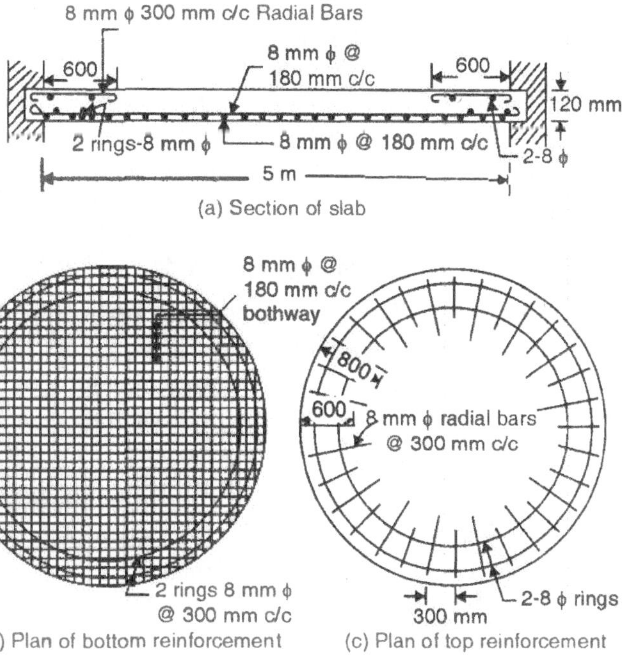

FIG. 13.8

$$A_{st} = \frac{0.5 \times 20}{415}\left[1 - \sqrt{1 - \frac{4.6 \times 3.955 \times 10^6}{20 \times 1000\,(96)^2}}\right] \times 1000 \times 96 = 117.1 \text{ mm}^2$$

Spacing of 8 mm φ bars $= \frac{1000 \times 50.3}{117.1} = 429$ mm

However, maximum permissible spacing = 300 mm.

Provide 2 rings of 8 mm φ bars to hold the above radial bars.

7. Check for shear

$$F_r = \frac{1}{2} w_u \,.\, a = \frac{1}{2} \times 10125 \times 2.5 = 12656 \text{ N.}$$

$$\therefore \qquad \tau_v = \frac{12656}{1000 \times 96} = 0.132 \text{ N/mm}^2$$

This is less than permissible shear stress, even at the minimum reinforcement. The details of reinforcement etc. are shown in Fig. 13.8.

Design Example 13.3. *A traffic control post, 2 m in diameter is supported centrally by a reinforced concrete column 300 mm in diameter. Design the circular slab for a super-imposed load of 1500 N/m². Use M 20 concrete and Fe 415 steel.*

Solution

1. Design constants and limiting depth of N.A.

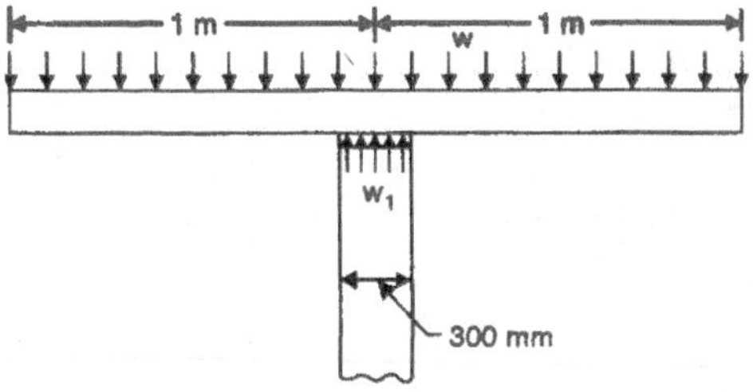

FIG. 13.9

Given : $f_{ck} = 20 \text{ N/mm}^2$ for M 20 concrete and $f_y = 415 \text{ N/mm}^2$ for Fe 415 steel,

For Fe 415, $\dfrac{x_{u\,,\,\max}}{d} = \dfrac{700}{1100 + 0.87 \times 415} = 0.479$

$$R_u = 0.36\, f_{ck}\, \frac{x_{u\,,\,\max}}{d}\left(1 - 0.416\, \frac{x_{u\,,\,\max}}{d}\right)$$

$= 0.36 \times 20 \times 0.479\,(1 - 0.416 \times 0.479) = 2.761$

2. Computation of loading and B.M.

Fig. 13.9 shows the slab supported centrally on the column and carrying U.D.L. w N/m². Let the column reaction be w_1 N/m². Hence the slab may be considered to be simply supported at the edge and subjected to (*i*) a downward U.D.L. w and (*ii*) an upward U.D.L. w_1 over a concentric circle of radius $b = 150$ mm. Hence we can use § 13.2 and § 13.5 for determining net moments on the slab.

In order to compute w and w_1, let us assume thickness of the slab, based on deflection criterion. For a cantilever, $L/d = 7$. Also, apply a factor of 4/3 for the cantilever being circular. Assuming an under-reinforced section with $p_t \simeq 0.2\%$, we get modification factor $\simeq 1.68$ from Fig. 11.1. Hence $L/d = 4/3 \times 7 \times 1.68 = 15.68$ and $d = L/15.68 = 1000/15.68 \simeq 64$ mm. Keeping a nominal cover of 20 mm and using 8 mm dia. bars, $D = 64 + 20 + 4 = 88$ mm. Hence assume $D = 90$ mm for the purpose of computing self weight.

$\therefore$ $$w = 0.09 \times 25000 + 1500 = 3750 \text{ N/m}^2$$

Hence $$w_1 = -\frac{3750\,\pi\,(1)^2}{\pi\,(0.15)^2} = -\,166667 \text{ N/m}^2$$

Minus sign has been used since w_1 acts in an upward direction.

Bending moments due to downward load w (Refer § 13.2)

$$(M_r)_e = 0$$

$$(M_\theta)_e = +\frac{2}{16}\,wa^2 = +\frac{2}{16} \times 3750\,(1)^2 \times 1000 = +\,4.688 \times 10^5 \text{ N-mm}$$

$\therefore$ $$(M_{u\theta})_e = +\,1.5 \times 4.688 \times 10^5 = +\,7.031 \times 10^5 \text{ N-mm}$$

Since the slab is cast monolithic with the column, the critical section will be at the edge of the column, where $r = 0.15$ m. Hence from Eq. 13.3 :

$$M_\theta = +\frac{w}{16}\,(3a^2 - r^2) = +\frac{3750}{16}\left[\,3 \times 1^2 - (0.15)^2\,\right] \times 1000 = +\,6.979 \times 10^5 \text{ N-mm}$$

From Eq.13.6, $M_r = +\dfrac{3}{16}\,w\,(a^2 - r^2) = +\dfrac{3 \times 3750}{16}\left[\,1^2 - (0.15)^2\,\right] \times 1000 = +\,6.873 \times 10^5$ N-mm

$\therefore$ $$M_{u\theta} = +\,1.5 \times 6.979 \times 10^5 = +\,10.468 \times 10^5 \text{ N-mm}$$

and $$M_{ur} = +\,1.5 \times 6.873 \times 10^5 = +\,10.31 \times 10^5 \text{ N-mm}$$

*(ii) **Bending moments due to downward load*** w_1

Refer § 13.5. From Eq. 13.23 at $r = b = 0.15$ m,

$$M_r = -\frac{3}{16}wr^2 + \frac{1}{4}wb^2\left[1 - \log_e\left(\frac{b}{a}\right) - \frac{b^2}{4a^2}\right]$$

Putting $w = w_1 = -166667$ and $r = b = 0.15$ m

$$M_r = +\frac{3}{16} \times 166667\ (0.15)^2 - \frac{1}{4}(166667)\ (0.15)^2\left[1 - \log_e\left(\frac{0.15}{1}\right) - \left(\frac{0.15}{2 \times 1}\right)^2\right]$$

$= 703.1 - 2710.9 = -2007.8$ N-m $= -20.078 \times 10^5$ N-mm

$\therefore \quad M_{ur} = 1.5\ M_r = -1.5 \times 20.078 \times 10^5 = -30.117 \times 10^5$ N-mm

Also, From Eq. 13.24, $M_\theta = -\frac{1}{16}wr^2 + \frac{1}{4}w\,b^2\left[1 - \log_e\left(\frac{b}{a}\right) - \frac{b^2}{4a^2}\right]$

Putting $w = w_1 = -166667$ and $r = b = 0.15$ m

$$\therefore \quad M_\theta = \frac{1}{16} \times 166667\ (0.15)^2 - \frac{1}{4} \times 166667\ (0.15)^2\left[1 - \log_e\left(\frac{0.15}{1}\right) - \left(\frac{0.15}{2 \times 1}\right)^2\right]$$

$= +234.4 - 2710.8 - 2476.4 = -2476.4$ N-m $= -24.764 \times 10^5$ N-mm

$\therefore \quad M_{u\theta} = 1.5\ M_\theta = -1.5 \times 24.764 \times 10^5 = -37.146 \times 10^5$ N-mm

The radial moment at the edge is zero, i.e. $(M_r)_e = 0$. Hence $(M_{ur})_e = 0$.

The circumferential moment at the edge is given by Eq. 13.29, where $w = w_1 = -166667$

$$(M_\theta)_e = -wb^2\left[-\frac{1}{4} + \frac{b^2}{8a^2}\right] = 166667\ (0.15)^2\left[-\frac{1}{4} + \frac{1}{8}\left(\frac{0.15}{1}\right)^2\right] = -927 \text{ N-m}$$

$= -9.27 \times 10^5$ N-mm

$\therefore \quad (M_{u\theta})_e = 1.5\ (M_\theta)_e = -1.5 \times 9.27 \times 10^5 = -13.904 \times 10^5$ N-mm

(iii) Net moments : The net moments are are tabulated below

Case	*Radial moments (N-mm)*		*Circumferential moments (N-mm)*	
	$r = 0.15$ m	$r = 1$ m	$r = 0.15$ m	$r = 1$ m
(i)	$+10.31 \times 10^5$	0	$+10.468 \times 10^5$	$+7.031 \times 10^5$
(ii)	-30.117×10^5	0	-37.146×10^5	-13.904×10^5
Net	-19.807×10^5	0	-26.678×10^5	-6.873×10^5

Thus the moments are negative throughout, as expected, since the slab will bend having convexity upwards. Hence both radial and circumferential reinforcements will be placed at the top face of the slab.

3. Computation of effective depth and total depth

The thickness will be designed for a bending moment of 26.678×10^5 N-mm.

$$d = \sqrt{\frac{M_{u\,.\,\max}}{R_u\,.\,b}} = \sqrt{\frac{26.678 \times 10^5}{2.761 \times 1000}} = 31.1 \text{ mm}$$

However, keep $D = 90$ mm, required from deflection point of view. Using 8 mm φ bars and providing 20 mm nominal cover, available $d = 90 - 20 - 4 = 66$ mm for one layer and $66 - 8 = 58$ mm for the second layer of reinforcement.

4. Computation of steel reinforcement : Let us design the reinforcement for the greater of the two moments. Since actual *d* is more than that required for bending, we have an under reinforced section for which

$$A_{st} = \frac{0.5 f_{ck}}{f_y}\left[1 - \sqrt{1 - \frac{4.6 M_{u.max}}{f_{ck} bd^2}}\right] bd = \frac{0.5 \times 20}{415}\left[1 - \sqrt{1 - \frac{4.6 \times 26.678 \times 10^5}{20 \times 1000 (58)^2}}\right] 1000 \times 58$$

$$= 133.9 \text{ mm}^2$$

Minimum reinforcement = $1.2 D = 1.2 \times 90 = 108$ mm²

Using 8 mm φ bars, spacing $s = \dfrac{1000 \times 50.3}{133.9} = 375$ mm.

Maximum permissible spacing = 300 mm.

Hence provide both the reinforcements in the form of mesh consisting of 8 mm φ bars spaced at 300 mm c/c in two directions at right angles to each other. This mesh will not be in a position to resist circumferential tensile stress at the edge. The effective depth available for rings reinforcement = 58 − 8 = 50 mm. Hence

$$A_{st} = \frac{0.5 \times 20}{415}\left[1 - \sqrt{1 - \frac{4.6 \times 6.873 \times 10^5}{20 \times 1000 (50)^2}}\right] 1000 \times 50 = 38.7 \text{ mm}^2 \text{ for } 1000 \text{ mm width.}$$

Minimum reinforcement = $1.2 D = 1.2 \times 90 = 108$ mm² for 1000 mm width. Since the circumferential moment at the edge is about $\frac{1}{4}$ of the moment at the centre, the circumferential moment is to be provided for a width $= \frac{1}{4} L_d = \frac{1}{4} \times 47 \times 8 = 94$ mm.

Hence area of steel required for this width $= \dfrac{108 \times 94}{1000} = 10.2$ mm². Using 8 mm φ bars, area $A_\varphi = 50.3$ mm². Hence one ring is *sufficient*.

5. Check for shear : Shear force at the edge of the column support is given by Eqs 13.30 and 13.7

$$F = \frac{w_1 b^2}{2r} - \frac{1}{2} wr = \frac{w_1 b^2}{2b} - \frac{1}{2} wb$$

$$= \frac{w_1 b}{2} - \frac{1}{2} wb = \frac{b}{2}(w_1 - w)$$

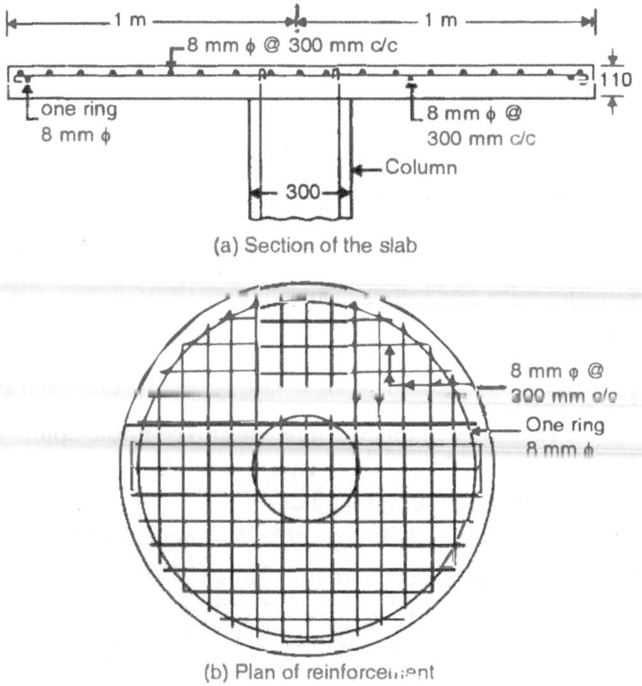

(a) Section of the slab

(b) Plan of reinforcement

FIG. 13.10

$$= \frac{0.15}{2}(166667 - 3750) = 12219 \text{ N}$$

$$\therefore \quad F_u = 1.5\,F = 1.5 \times 12219 = 18328 \text{ N}$$

$$\therefore \quad \tau_v = \frac{18328}{1000 \times 66} = 0.278 \text{ N/mm}^2$$

The permissible shear stress even at minimum reinforcement $= 1.3 \times 0.28 = 0.364$ N/mm^2. The details of reinforcement etc are shown in Fig. 13.10.

Design Example 13.4. *The base slab of an overhead circular tank is 6 m in diameter and supported centrally over a ring beam of 2 m diameter. The slab has a central hole of 1.2 m diameter. At the periphery of the central hole, the inner walls of the tank transmit a total load of 30 kN. At the outer periphery of the slab, the outer walls of the tanks transmit a total load of 150 kN. In addition to this, the slab supports a uniformly distributed water load of intensity 18350 N/m^2, over the annular area. Design the base slab, using M 20 concrete and Fe 415 steel.*

1. Design constants and limiting depth of N.A.

Given : $f_{ck} = 20$ N/mm^2 for M 20 concrete and $f_y = 415$ N/mm^2 for Fe 415 steel.

$$\text{For Fe 415,} \quad \frac{x_{u,\max}}{d} = \frac{700}{1100 + 0.87 \times 415} = 0.479$$

$$R_u = 0.36\,f_{ck}\frac{x_{u,\max}}{d}\left(1 - 0.416\frac{x_{u,\max}}{d}\right) = 0.36 \times 20 \times 0.479\,(1 - 0.416 \times 0.479) = 2.761$$

2. Computation of loading B.M. and S.F.

For a rectangular cantilever, $L/d = 7$. Let us apply a factor of 4/3 for the cantilever being circular. Because of heavy loading, let us assume, $p_t = 0.5\%$, corresponding to which we get modification factor $\simeq 1.25$. Hence $L/d = 7 \times 4/3 \times 1.25 = 11.67$ and $d = 2000/11.67 \simeq 172$ mm.

Providing 20 mm nominal cover and 16 mm φ bars, we get $D = 172 + 20 + 8 = 200$ mm. However assume $D = 300$ mm for computation of dead weight.

Hence weight

$= 0.3 \times 1 \times 1 \times 25000$

$= 7500$ N/m^2.

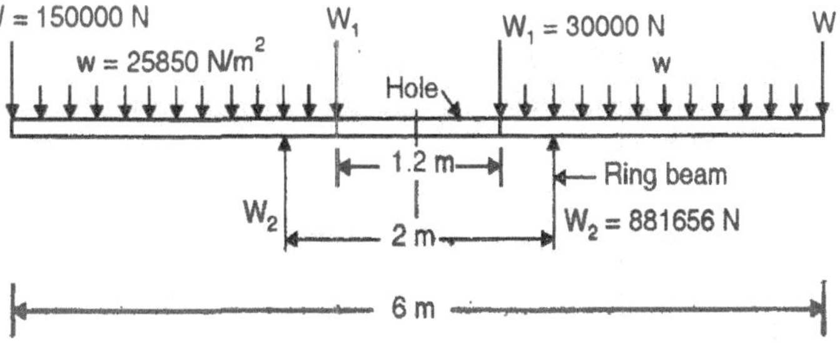

FIG. 13.11

$\therefore$ Total $w = 7500 + 18350 = 25850$ N/m^2.

The loading on the base slab is shown in Fig. 13.11. The total load (reaction) on the support is

$$W_2 = 30000 + 150000 + \frac{\pi}{4}[(6)^2 - (1.2)^2] \times 25850 = 881656 \text{ N}$$

The given slab under various loadings may be looked upon a slab (with central hole) simply supported at the periphery and loaded as follows.

1. A downward load $W_1 = 30000$ N at the periphery of the hole

2. A uniformly distributed load $w = 25850\ \text{N/m}^2$ over the annular area

3. An upward load $W_2 = 881656$ N at the outer periphery of a circle of radius 1 m.

4. A downward load $W = 150000$ N at the outer periphery of the slab. Since the imaginary supports are at the outer periphery, there will be no B.M. or S.F. due to this load W. Hence this case has not been considered.

(i) Bending moments due to W_1 (Case 1 of loading)

Refer § 13.7 the radial and circumferential moments are given by Eqs. 13.38 and 13.37 respectively. Substituting $W = W_1 = 30000$ N, $a = 3$ m, $b = c = 0.6$ m, these equations reduce as follows.

$$M_{u\theta} = \frac{1.5\,W}{4\,\pi}\left[\log_e \frac{c}{r} + \frac{1}{2} + \frac{a^2}{a^2 - b^2} \cdot \frac{r^2 + b^2}{r^2} \times \left(\log_e \frac{a}{c} + \frac{1}{2} - \frac{c^2}{2a^2}\right) - \frac{c^2}{2r^2}\right]$$

$$= \frac{1.5 \times 30000}{4\,\pi}\left[\log_e \frac{0.6}{r} + \frac{1}{2} + \frac{3^2}{3^2 - (0.6)^2} \cdot \frac{r^2 + (0.6)^2}{r^2} \times \left(\log_e \frac{3}{0.6} + \frac{1}{2} - \frac{(0.6)^2}{2\,(3)^2}\right) - \frac{1}{2}\left(\frac{0.6}{r}\right)^2\right]$$

$$= 3581\left[\log_e \frac{0.6}{r} + \frac{1}{2} + 2.176\left(\frac{r^2 + 0.36}{r^2}\right) - \frac{0.18}{r^2}\right] \qquad \ldots(i)$$

$$M_{ur} = \frac{1.5\,W}{4\,\pi}\left[\log_e \frac{c}{r} - \frac{1}{2} + \frac{a^2}{a^2 - b^2} \cdot \frac{r^2 - b^2}{r^2} \times \left(\log_e \frac{a}{c} + \frac{1}{2} - \frac{c^2}{2a^2}\right) + \frac{c^2}{2r^2}\right]$$

$$= \frac{1.5 \times 30000}{4\,\pi}\left[\log_e \frac{0.6}{r} - \frac{1}{2} + \frac{3^2}{3^2 - (0.6)^2} \cdot \frac{r^2 - (0.6)^2}{r^2} \times \left(\log_e \frac{3}{0.6} + \frac{1}{2} - \frac{(0.6)^2}{2\,(3)^2}\right) + \frac{1}{2}\left(\frac{0.6}{r}\right)^2\right]$$

$$= 3581\left[\log_e \left(\frac{0.6}{r}\right) - \frac{1}{2} + 2.176\left(\frac{r^2 - 0.36}{r^2}\right) + \frac{0.18}{r^2}\right] \qquad \ldots(ii)$$

The values of $M_{u\theta}$ and M_{ur} at the salient locations are tabulated below :

r (m)	0.6	1	1.5	2	2.5	3
$M_{u\theta}$ (N m)	+ 15585	+ 9914	+ 7262	+ 5811	+ 4818	+ 4165
M_{ur} (N-m)	0	+ 2012	+ 1760	+ 1150	+ 546	0

(ii) Bending moments due to U.D.L. w (Case 2 of loading)

The circumferential and radial moments are given by Eqs. 13.31 and 13.32 respectively

$$M_{u\theta} = -\frac{1}{16}(1.5\,wr^2) + \frac{1.5\,wb^2}{4}\left[\log_e\left(\frac{r}{a}\right) + \frac{3}{4}\left(-\frac{1}{3} + \frac{a^2}{b^2} + \frac{a^2}{r^2}\right) - \frac{a^2 + r^2}{r^2} \cdot \frac{b^2}{a^2 - b^2}\log_e \frac{a}{b}\right]$$

$$= -\frac{1.5}{16} \times 25850\,r^2 + \frac{1.5 \times 25850}{4}(0.6)^2\left[\log_e\left(\frac{r}{3}\right) + \frac{3}{4}\left(-\frac{1}{3} + \left(\frac{3}{0.6}\right)^2 + \left(\frac{3}{r}\right)^2\right)\right.$$

$$\left. - \frac{3^2 + r^2}{r^2} \cdot \frac{(0.6)^2}{3^2 - (0.6)^2}\log_e \frac{3}{0.6}\right]$$

$$= -2423\, r^2 + 3489.8 \left[\log_e \left(\frac{r}{3} \right) + \frac{3}{4} \left(24.67 + \frac{9}{r^2} \right) - \frac{9 + r^2}{r^2} \times 0.067 \right] \quad \ldots(iii)$$

$$M_{ur} = -\frac{3}{16} (1.5\, w\, r^2) + \frac{1.5\, w\, b^2}{4} \left[\log_e \frac{r}{a} + \frac{3}{4} \left(1 + \frac{a^2}{b^2} - \frac{a^2}{r^2} \right) + \frac{a^2 - r^2}{r^2} \cdot \frac{b^2}{a^2 - b^2} \log_e \frac{a}{b} \right]$$

$$= -\frac{3}{16} \times 1.5 \times 25850\, r^2 + \frac{1.5 \times 25850}{4} (0.6)^2 \left[\log_e \frac{r}{3} + \frac{3}{4} \left(1 + \left(\frac{3}{0.6} \right)^2 - \frac{9}{r^2} \right) + \frac{9 - r^2}{r^2} \frac{(0.6)^2}{9 - (0.6)^2} \log_e \left(\frac{3}{0.6} \right) \right]$$

$$= -7270.3\, r^2 + 3489.8 \left[\log_e \frac{r}{3} + \frac{3}{4} \left(26 - \frac{9}{r^2} \right) + \frac{9 - r^2}{r^2} \times 0.067 \right] \quad \ldots(iv)$$

The values of $M_{u\theta}$ and M_{ur} for salient values of r are tabulated below :

r (m)	0.6	1	1.5	2	2.5	3
$M_{u\theta}$ (N-m)	+ 117436	+ 79531	+ 66000	+ 58592	+ 51988	+ 44913
M_{ur} (N-m)	0	+ 35261	+ 39506	+ 31958	+ 18309	0

(iii) *Bending moments due to upward load* $W_2 = 881656$ ***N***

The moments are given by Eqs. 13.34, 13.35, 13.37 and 13.38, where $W = W_2 = -881656$ N, $b = 0.6$ m, $c = 1$ m and $a = 3$ m.

For $r < c$

$$M_{u\theta} = \frac{1.5\, W_2}{4\pi} \times \frac{a^2}{a^2 - b^2} \left(1 + \frac{b^2}{r^2} \right) \left[\log_e \frac{a}{c} + \frac{1}{2} - \frac{c^2}{2a^2} \right]$$

$$= \frac{1.5 \times 881656}{4\pi} \times \frac{9}{9 - 0.36} \left(1 + \frac{0.36}{r^2} \right) \left[\log_e \frac{3}{1} + \frac{1}{2} - \frac{1}{2 \times 9} \right]$$

$$= -169157 \left(1 + \frac{0.36}{r^2} \right) \quad \ldots(v)$$

$$M_{ur} = \frac{1.5\, W_2}{4\pi} \times \frac{a^2}{a^2 - b^2} \left(1 - \frac{b^2}{r^2} \right) \left[\log_e \frac{a}{c} + \frac{1}{2} - \frac{c^2}{2a^2} \right] = -169157 \left(1 - \frac{0.36}{r^2} \right) \quad \ldots(vi)$$

For $r > c$

$$M_{u\theta} = \frac{1.5\, W_2}{4\pi} \left[\log_e \frac{c}{r} + \frac{1}{2} + \frac{a^2}{a^2 - b^2} \cdot \frac{r^2 + b^2}{r^2} \left(\log_e \frac{a}{c} + \frac{1}{2} - \frac{c^2}{2a^2} \right) - \frac{c^2}{2r^2} \right]$$

$$= -\frac{1.5 \times 881656}{4\pi} \left[\log_e \frac{1}{r} + \frac{1}{2} + \frac{9}{9 - 0.36} \cdot \frac{r^2 + 0.36}{r^2} \left(\log_e 3 + \frac{1}{2} - \frac{1}{2 \times 9} \right) - \frac{1}{2\, r^2} \right]$$

$$= -105240 \left[\log_e \frac{1}{r} + \frac{1}{2} + 1.607 \left(\frac{r^2 + 0.36}{r^2} \right) - \frac{1}{2\, r^2} \right] \quad \ldots(vii)$$

and

$$M_{ur} = \frac{1.5\, W_2}{4\pi} \left[\log_e \frac{c}{r} - \frac{1}{2} + \frac{a^2}{a^2 - b^2} \cdot \frac{r^2 - b^2}{r^2} \left(\log_e \frac{a}{c} + \frac{1}{2} - \frac{c^2}{2a^2} \right) - \frac{c^2}{2r^2} \right]$$

$$= -105240\left[\log_e \frac{1}{r} - \frac{1}{2} + 1.607\left(\frac{r^2 - 0.36}{r^2}\right) + \frac{1}{2r^2}\right] \qquad \ldots(viii)$$

The values of $M_{u\theta}$ and M_{ur} for salient values of r are tabulated below :

r (m)	0.6	1	1.5	2	2.5	3
$M_{u\theta}$ (N-m)	− 338314	− 230054	− 182742	− 150860	− 126632	− 107041
M_{ur} (N-m)	0	− 108260	− 70157	− 41488	− 18748	0

(*iv*) ***Net moments*** : The combined moments for the above three cases are tabulated below:

Moment	Loading case	r (m)					
		0.6	1	1.5	2	2.5	3
$M_{u\theta}$	1	+ 15585	+ 9914	+ 7262	+ 5811	+ 4818	+ 4165
	2	+ 117436	+ 79531	+ 66000	+ 58592	+ 51988	+ 44913
	3	− 338314	− 230054	− 182742	− 150860	− 126632	− 107041
	Net	− 205293	− 140609	− 109480	− 86457	− 69826	− 57963
M_{ur}	1	0	+ 2012	+ 1760	+ 1150	+ 546	0
	2	0	+ 35261	+ 39506	+ 31958	+ 18309	0
	3	0	− 108260	− 70157	− 41488	− 18748	0
	Net	0	− 70987	− 28891	− 8380	+ 107	0

Shear force : Maximum shear force will occur at the outer edge of the ring beam, its magnitude per metre width is given by Eqs. 13.39 and 13.33 :

$$F = \frac{W_1}{2\pi r} + \left(\frac{wr}{2} - \frac{wb^2}{2r}\right) - \frac{W_2}{2\pi r} = \frac{30000}{2\pi \times 1} + \frac{25850 \times 1}{2}\left\{1 - \left(\frac{0.6}{1}\right)^2\right\} - \frac{881656}{2\pi \times 1}$$

$$= -127273 \text{ N per metre width}$$

$$\therefore \qquad F_u = 1.5\,F = -1.5 \times 127273 \simeq -190910 \text{ N per metre width}$$

3. *Computation of thickness*

$$\text{From B.M.} \quad d = \sqrt{\frac{M_{u,\,max}}{R_u\,b}} = \sqrt{\frac{205293 \times 1000}{2.761 \times 1000}} = 272.7 \text{ mm}$$

From S.F. point of view, putting $\tau_v = \tau_c = 0.48$ N/mm^2 corresponding to about 0.5% reinforcement, we get

$$d = \frac{F_u}{b\,.\,\tau_u} = \frac{190910}{1000 \times 0.48} = 398 \text{ mm}$$

Let us keep total thickness of 430 mm. Providing 20 mm nominal cover and using 16 mm φ bars for circumferential reinforcement, available $d = 430 - 20 - 8 = 402$ mm. Using 8 mm φ radial bars, available d for radial reinforcement $= 402 - 8 - 4 = 390$ mm.

4. *Design of circumferential reinforcement* : The circumferential reinforcement is provided in the form of rings near the inner edge of the slab. Since d provided is more than that required from B.M. point of view, we have an under-reinforced section, for which

$$(A_{st})_\theta = \frac{0.5\, f_{ck}}{f_y}\left[1-\sqrt{1-\frac{4.6\, M_{u\theta}}{f_{ck}\, bd^2}}\right] bd = \frac{0.5 \times 20}{415}\left[1-\sqrt{1-\frac{4.6 \times 205293 \times 10^3}{20 \times 1000\,(402)^2}}\right] \times 1000 \times 402$$

$$= 1537 \text{ mm}^2 \text{ per metre width}$$

Using 16 mm φ bars, spacing $s = \dfrac{1000 \times 201.06}{1537} = 130.8$ mm

Hence provide the rings @ 125 mm c/c near the inner edge.

At the outer edge

$$(A_{st})_\theta = \frac{0.5 \times 20}{415}\left[1-\sqrt{1-\frac{4.6 \times 57963 \times 10^3}{20 \times 1000\,(402)^2}}\right] 1000 \times 402 = 408.2 \text{ mm}^2$$

$\therefore$ Spacing $\quad s = \dfrac{1000 \times 201.06}{408.2} = 492$ mm

But maximum permissible spacing is 300 mm as per IS Code.

Hence the spacing of the rings can be gradually increased from 125 mm c/c at the inner edge to 300 mm c/c at the outer edge.

5. Design of radial reinforcement : Available $d = 390$ mm

$$(A_{st})_r = \frac{0.5 \times 20}{415}\left[1-\sqrt{1-\frac{4.6 \times 70987 \times 1000}{20 \times 1000\,(390)^2}}\right] 1000 \times 390 = 518.7 \text{ mm}^2$$

Using 8 mm φ bars, spacing $s = \dfrac{1000 \times 50.3}{518.7} \simeq 97$ mm

Hence provide radial bars @ 90 mm c/c at the ring beam support. These radial bars may be bent downwards near the inner edge.

6. Check for development length

The radial reinforcement should be checked for development length at the ring beam support where spacing is @ 90 mm c/c. Hence

$$(A_{st})_r = \frac{1000 \times 50.3}{90} = 558.9 \text{ mm}^2$$

$$x_u = \frac{0.87\, f_y\, A_{str}}{0.36\, f_{ck}\, b} = \frac{0.87 \times 415 \times 558.9}{0.36 \times 20 \times 1000}$$

$$= 28 \text{ mm}$$

$$M_{ur} = 0.87\, f_y\, A_{str}\,(d - 0.416\, x_u)$$

$$= 0.87 \times 415 \times 558.9\,(390 - 0.416 \times 28)$$

$$= 76.346 \times 10^6 \text{ N-mm}$$

$$V_{u1} = 190910 \text{ N}$$

$$L_0 = \left(\frac{2000 - 1200}{2} - 20\right) + 3\,\varphi$$

$$= 380 + 3 \times 8 = 404 \text{ mm}$$

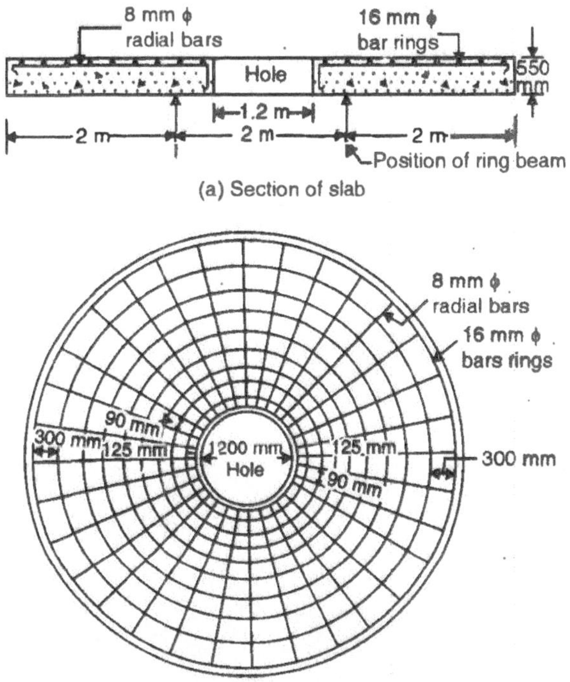

(a) Section of slab

(b) Plan of reinforcement

FIG. 13.12

$$L_d = 47\ \varphi = 47 \times 8 = 376 \quad \text{mm}$$

$$\therefore \quad 1.3\,\frac{M_{1u}}{V_u} + L_0 = 1.3 \times \frac{76.346 \times 10^6}{190910} + 404 \simeq 924 > L_d$$

Hence Code requirements are satisfied. However, bend the radial bars downwards by 90°, at the perimeter of the central hole. The details of reinforcement etc are shown in Fig. 13.12.

PROBLEMS

1. Design the roof slab for a circular room 6 metre in diameter from inside and carrying a super imposed load of 5 kN/m^2. Assume that the slab is simply supported at the edges. Use M 20 mix and Fe 415 steel. The thickness of wall is 400 mm.
2. Redesign problem 1 assuming that the slab is restrained at the edge.
3. A circular room 8 metre in diameter from inside carrries a 500 mm dia. column at its centre. Design a circular slab, simply supported at the outer edge and supported on the column at its middle. The slab carries a total uniformly distributed load of 3 kN/m^2. including its own weight. Use M 20 mix and Fe 415 steel.
4. Design the slab of a traffic control post, 1.5 m diameter and supported centrally on a column of 200 mm diameter. The total superimposed load, inclusive of its own weight may be taken as 1.8 kN/m^2. Use M 20 mix and Fe 415 steel reinforcement.

14

Flat Slabs

14.1. INTRODUCTION

A flat slab is a typical type of construction in which a reinforced slab is built monolithically with the supporting columns and is reinforced in two or more directions, without any provision of beams. The flat slab thus transfers the load directly to the supporting columns suitably spaced below the slab. Because of exclusion of beam-system in this type of construction, a plain ceiling is obtained, thus giving attractive appearance from architectural point of view. The plain ceiling diffuses the light better and is considered less vulnerable in the case of fire than the usual beam slab construction. The flat slab is easier to construct and requires cheaper form-work. Concrete is more logically used in this type of construction, and hence in the case of large spans and heavy load, the total cost is considerably less.

14.2. COMPONENTS OF FLAT SLAB CONSTRUCTION

Some of the components of flat slab construction, shown in Fig. 14.1 and 14.2 are defined below.

1. ***Drop of flat slab.*** The slab in a flat slab construction may be either with drop or without drop. *Drop* is that part of the slab around the column, which is of greater thickness than the rest of the slab. The drop may have 25% to 50% more thickness, than the rest of the slab, as shown in Fig. 14.2 *(b)*. Flat slab without drop is shown in Fig. 14.2 (*a*) and (*c*).

2. ***Capital or column head.*** Sometimes, the diameter of a supporting column is increased below the slab. This part of column with increased diameter is called 'column head'. The column head increases rigidity

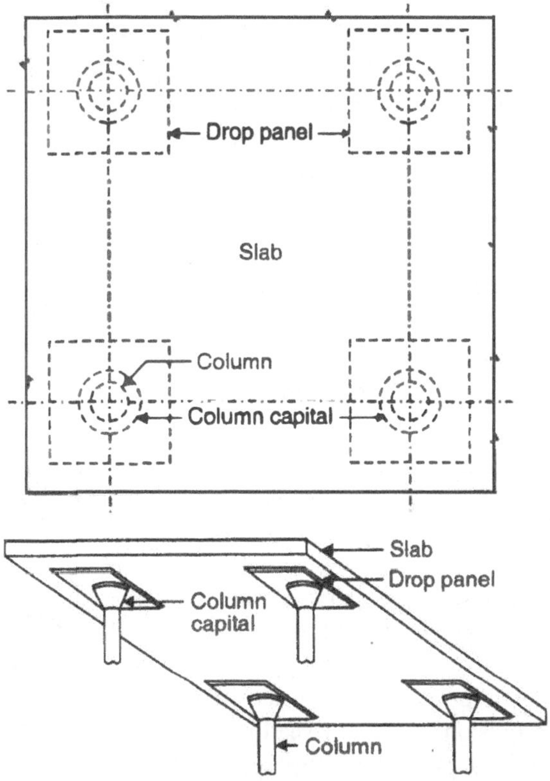

FIG. 14.1.

of the slab and 'column connection' increases resistance of the slab to shear and decreases the effective span of the slab. Fig. 14.2*(c)* shows a flat slab without drop but with column head.

3. ***Panel.*** A panel of a flat slab construction is the area enclosed between the centre lines connecting adjacent columns in two directions and the outline of the column heads as shown in Fig. 14.3.

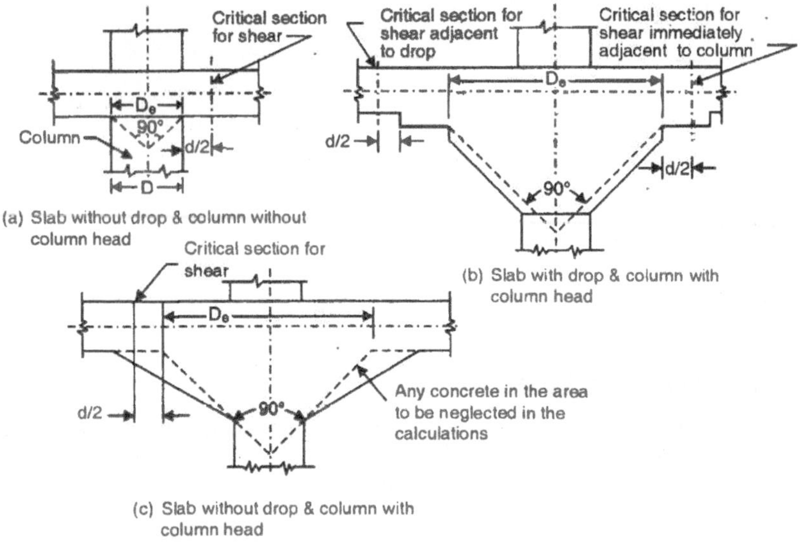

FIG. 14.2

14.3. INDIAN CODE RECOMMENDATIONS (IS : 456 - 2000)

(*a*) **General.** The following definitions shall apply :

1. ***Column strip.*** Column strip means a design strip having a width of 0.25 l_2, but not greater than 0.25 l_1 on each side of the column centre line, where l_1 is the span in the direction moments are being determined, measured centre to centre of supports and l_2 is the span transverse to l_1, measured centre to centre of supports.

2. ***Middle strip.*** Middle strip means a design strip bounded on each of its opposite sides by the column strip.

3. ***Panel.*** Panel means that part of the slab bounded on each of its four sides by the centre-line of a column or centre lines of adjacent spans.

(*b*) **Proportioning**

1. ***Thickness of Flat slab.*** The thickness of flat slab shall be generally controlled by considerations of span to effective depth ratios given in § 11.1. For slabs with drops conforming to clause *(b-2)*, span to effective depth ratios given in § 11.1 shall be applied directly; otherwise the span to effective depth ratios obtained in accordance with provisions of § 11.1 shall be multiplied by 0.9. For this purpose, the longer span shall be considered. The minimum thickness of slab shall be 125 mm.

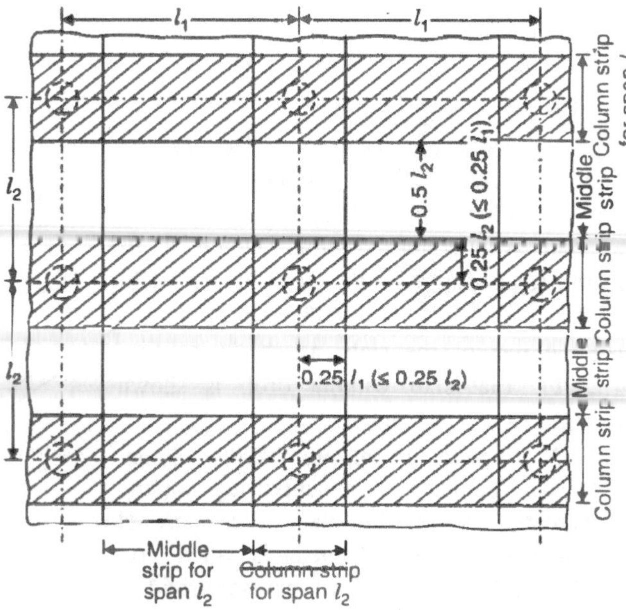

FIG. 14.3.
DIVISION OF FLAT SLAB PANEL INTO STRIPS.

2. ***Drops***. The drops when provided shall be rectangular in plan, and have a length in each direction not less than one third of the panel length in the direction. For exterior panels, the width of drops at right angles to the non-continuous edge and measured from centre line of the columns shall be equal to one half the width of drop for interior panels.

3. ***Column heads.*** When column heads are provided, that portion of a column head which lies within the largest circular cone or pyramid that has a vertex angle of 90° and can be included entirely within the outlines of the column and the column head, shall be considered for design purposes (see Fig. 14.2).

(*c*) Determination of bending moment

1. ***Methods of analysis and design***. It shall be permissible to design the slab system by one of the following design methods.

(*i*) The direct design method. (*ii*) The equivalent frame method.

2. ***Bending moments in panels with marginal beams or walls.*** Where the slab is supported by a marginal beam with depth greater than 1.5 times the thickness of the slab, or by a wall, then :

(*i*) The total load to be carried by the beam or wall shall comprise those loads directly on the wall or beam plus a uniformly distributed load equal to one-quarter of the total load on the slab.

(*ii*) The bending moments on the half column strip adjacent to the beam or wall shall be one-quarter of the bending moments for the first interior column strip.

3. ***Transfer of bending moments to columns***. When unbalanced gravity load, wind, earthquake, or other lateral loads cause transfer of bending moment between slab and column, the flexural stresses shall be investigated using a fraction, α of the moment given by

$$\alpha = \frac{1}{1 + \frac{2}{3}\sqrt{\frac{a_1}{a_2}}}$$

where a_1 = overall dimension of the critical section for shear in the direction in which moment acts, and

a_2 = overall dimension of the critical section for shear transverse to the direction in which moment acts.

A slab width between lines that are one and one-half slab or drop panel thickness, 1.5 D, on each side of the column or capital may be considered effective, D being the size of the column. Concentration of reinforcement over column head by closer spacing or additional reinforcement may be used to resist the moment on this section.

14.4. DIRECT DESIGN METHOD

1. **Limitations.** Slab systems designed by the direct design method shall fulfil the following conditions :

(*i*) There shall be minimum of three continuous spans in each direction.

(*ii*) The panel shall be rectangular, and the ratio of the longer span to the shorter span within a panel shall not be greater than 2.0.

(*iii*) It shall be permissible to offset columns to a maximum of 10 percent of the span in the direction of the offset notwithstanding the provision in (*ii*).

(*iv*) The successive span lengths in each direction shall not differ by more than one third of the longer span. The end spans may be shorter but not longer than the interior spans, and

(*v*) The design live load shall not exceed three times the design dead load.

2. Total design moment for a span

(*i*) In the direct design method, the total design moment for a span shall be determined for a strip bounded laterally by the centre line of the panel on each side of the centre line of the supports.

(*ii*) The absolute sum of the positive and average negative bending moments in each direction shall be taken as :

$$M_0 = \frac{W\,l_n}{8}, \text{ where}$$

M_0 = total moment ; W = design load on area $l_2.l_n$

l_n = clear span extending from face to face of columns, capitals, brackets or walls, but not less than 0.65 l_1

l_1 = length of span in the direction of M_0 ; and

l_2 = length of span transverse to l_1.

(*iii*) Circular supports shall be treated as square supports having the same area.

(*iv*) When the transverse span of the panels on either side of the centre line of support varies, l_2 shall be taken as the average of the transverse spans.

(*v*) When the span adjacent and parallel to an edge is being considered, the distance from the edge to the centre line of the panel shall be substituted for l_2 in (*ii*) above.

3. Negative and positive design moments

(*i*) The negative design moment shall be treated at the face of rectangular supports, circular supports being treated as square supports having the same area.

(*ii*) In an interior span, the total design moment M_0 shall be distributed in the following proportions : Negative design moment : 0.65. Positive design moment : 0.35

(*iii*) In an end span, the total design moment M_0 shall be distributed in the following proportions :

Interior negative design moment : $0.75 - \dfrac{0.10}{1 + \dfrac{1}{\alpha_c}}$

Positive design moment : $0.63 - \dfrac{0.28}{1 + \dfrac{1}{\alpha_c}}$

Exterior negative design moment : $\dfrac{0.65}{1 + \dfrac{1}{\alpha_c}}$

α_c is the ratio of flexural stiffness of the exterior columns to the flexural stiffness of the slab at a joint taken in the direction moments are being determined and is given by

$$\alpha_c = \frac{\Sigma K_c}{K_s}$$

where ΣK_c = Sum of the flexural stiffness of the columns meeting at the joint, and
K_s = flexural stiffness of the slab, expressed as moment per unit rotation.

(*iv*) It shall be permissible to modify these design moments by upto 10 percent, so long as the total design moment M_0 for the panel in the direction considered is not less than that required by 2 *(ii)* above.

(*v*) The negative moment section shall be designed to resist the larger of the two interior negative design moments determined for the spans forming into a common support unless an analysis is made to distribute the unbalanced moment in accordance with the stiffness of the adjoining parts.

4. **Distribution of bending moments across the panel width** : Bending moments at critical cross-section shall be distributed to the column strips and middle strips as stated below :

(i) Column strip : Negative moment at an interior support. At an interior support, the column strip shall be designed to resist 75 percent of the total negative moment in the panel at that support.

(ii) Column strip : Negative moment at an exterior support.

(*a*) At an exterior support, the column strip shall be designed to resist the total negative moment in the panel at the support.

(*b*) Where the exterior support consists of a column or a wall extending for a distance equal to or greater than three quarters of the value of l_2, the length of span transverse to the direction moments are being determined, the exterior negative moments shall be considered to be uniformly distributed across the length l_2.

(iii) Column strip : Positive moment for each span: For each span, the column strip shall be designed to resist 60 percent of the total positive moment in the panel.

(iv) Moment in the middle strip : The middle strip shall be designed on the following basis:

(*a*) The portion of the design moment not resisted by the column strip shall be assigned to the adjacent middle strips.

(*b*) Each middle strip shall be proportioned to resist the sum of the moment assigned to its two half middle strips.

(*c*) The middle strip adjacent and parallel to an edge supported by a wall shall be proportioned to resist twice the moment assigned to half the middle strip corresponding to the first row of interior columns.

5. **Moments in columns**

(*i*) Columns built integrally with the slab system shall be designed to resist moments arising from loads on the slab systems.

(*ii*) At an interior support, the supporting members above and below the slab shall be designed to resist the moment M given by the following equation, in direct proportion to their stiffness unless general analysis is made :

$$M = 0.08 \frac{(w_d + 0.5\, w_l)\, l_2\, l_n^2 - w_d'\, l_2'\, l_n^{2\,\prime}}{1 + \frac{1}{\alpha_c}} \quad ; \text{ where}$$

w_d, w_l = design dead and live loads respectively, per unit area

l_2 = length of span transverse to the direction of M.

l_n = length of the clear span in the direction of M, measured face to face of supports.

$\alpha_c = \frac{\Sigma K_c}{\Sigma K_s}$ and w_d', l_2' and l_n' refer to the shorter span.

TABLE 14..1 MINIMUM PERMISSIBLE VALUES OF α_c (IS : 456-2000)

Live load/Dead load	*Ratio l_2/l_1*	*Value of $\alpha_{c\,(min.)}$*
0.5	0.5 to 2.0	0
1.0	0.5	0.6
1.0	0.8	0.7
1.0	1.0	0.7
1.0	1.25	0.8
1.0	2.0	1.2
2.0	0.5	1.3
2.0	0.8	1.5
2.0	1.0	1.6
2.0	1.25	1.9
2.0	2.0	4.9
3.0	0.5	1.8
3.0	0.8	2.0
3.0	1.0	2.3
3.0	1.25	2.8
3.0	2.00	13.0

6. Effect of pattern of loading. In the direct design method, when the ratio of the live load to dead load exceed 0.5 :

(*i*) the sum of the flexural stiffness of the columns above and below the slab, ΣK_c shall be such that α_c is not less than the appropriate minimum value $\alpha_{c\,.\,min}$ specified in Table 14.1, or

(*ii*) if the sum of flexural stiffnesses of the column ΣK_c does not satisfy (*i*) above, the positive design moments for the panel shall be multiplied by the coefficient β_s given by the following equation :

$$\beta_s = 1 + \left(\frac{2 - \frac{w_d}{w_l}}{4 + \frac{w_d}{w_l}}\right)\left(1 - \frac{\alpha_c}{\alpha_{c.min}}\right)$$

α_c is the ratio of flexural stiffness of the columns above and below the slab to the flexural stiffness of the slabs at the joint taken in the direction moments are being determined and the given by :

$$\alpha_c = \frac{\Sigma K_c}{\Sigma K_s}$$

where K_c and K_s are flexural stiffness of column and slab respectively. The method of computing flexural stiffnesses of various members has been illustrated in the design example.

14.5. EQUIVALENT FRAME METHOD

1. **Assumptions.** The bending moments and shear forces may be determined by an analysis of the structure as a continuous frame and the following assumptions may be made:

(*i*) The structure shall be considered to be made up of equivalent frames on column lines taken longitudinally and transversely through the building. Each frame consists of a row of equivalent columns or supports, bounded laterally by the centre line of the panel on each side of the centre line of the columns or supports. Frames adjacent and parallel to an edge shall be bounded by the edge and the centre line of the adjacent panel.

(*ii*) Each such frame may be analysed in its entirety, or for vertical loading — each floor thereof and the roof may be analysed separately with columns being assumed fixed at their remote ends. Where slabs are thus analysed separately, it may be assumed in determining the bending moment at a given support that the slab is fixed at any support two panels distant there from, provided the slab is continuous beyond that point.

(*iii*) For the purpose of determining relative stiffness of members, the moment of inertia of any slab or column may be assumed to be that of the gross cross-section of the concrete alone.

(*iv*) Variations of moment of inertia along the axis of the slab on account of provision of drops shall be taken into account. In case of recessed or coffered slab which is made solid in the region of the columns, the stiffening effect may be ignored provided the solid part of the slab does not extend more than 0.15 l_e into the span measured from the centre line of the columns. The stiffening effect of flared column heads may be ignored.

2. Loading Pattern. (*i*) When the loading pattern is known, the structure shall be analysed for the load concerned.

(*ii*) When the live load is variable but does not exceed three quarters of the dead load, or the nature of the live load is that all panels will be loaded simultaneously, the maximum moments may be assumed to occur at all sections when full design live load is on entire slab system.

(*iii*) For other conditions of live load/dead load ratio and when all panels are not loaded simultaneously :

(*a*) maximum positive moment near mid-span of a panel may be assumed to occur when three-quarters of the full design live load is on the panel and on alternate panels; and (*b*) maximum negative moment in the slab at a support may be assumed to occur when three-quarters of the full design live load is on the adjacent panels only.

(*iv*) In no case shall design moments be taken to be less than those occuring with full design live load on all panels.

3. Negative Design Moment

(*i*) At interior supports, the critical section for negative moment in both the column strip and middle strip, shall be taken at the face of rectilinear supports but in no case at a distance greater than 0.175 l_1 from the centre of the column where l_1 is the length of the span in the direction moments are being determined, measured centre-to-centre of supports.

(*ii*) At exterior supports provided with brackets or capitals, the critical section for negative moment in the direction perpendicular to the edge shall be taken at a distance from the face of the supporting element not greater than one-half the projection of the bracket or capital beyond the face of the supporting element.

(*iii*) Circular or regular polygon shaped supports shall be treated as square supports having the same area.

4. Modification of Maximum Moment. Moments determined by means of the equivalent frame method, for slabs which fulfills the limitations of *direct design method* may be reduced in such proportion that the numerical sum of the positive and average negative moments is not less than the value of total design moment M_0 specified in para 2 of § 14.4.

5. **Distribution of Bending Moment across the Panel width**

(*i*) ***Column strip : Negative moment at an interior support***. At an interior support, the column strip shall be designed to resist 75 percent of the total negative moment in the panel at that support.

(*ii*) ***Column strip : Negative moment at an exterior support***

(*a*) At an exterior support, the column strip shall be designed to resist the total negative moment in the panel at the support.

(*b*) Where the exterior support consists of a column or a wall extending for a distance equal to or greater than three-quarters of the value of l_2, the length of span transvere to the direction moments are being determined, the exterior negative moment shall be considered to be uniformaly distributed across the length l_2.

(*iii*) ***Column strip : Positive moment for each span***: For each span, the column strip shall be designed to resist 60 percent of the total positive moment. in the panel.

(*iv*) ***Moments in the middle strip:*** The middle strip shall be designed on the following bases:

(*a*) That portion of the design moment not resisted by the column strip shall be assigned to the adjacent middle strips.

(*b*) Each middle strip shall be proportioned to resist the sum of the moments assigned to its two half middle strips.

(*c*) The middle strip adjacent and parallel to an edge supported by a wall shall be proportioned to resist twice the moment assigned to half the middle strip corresponding to the first row of interior columns.

14.6. SHEAR IN FLAT SLAB

1. **The critical section :** The critical section shall be at a distance $d/2$ from the periphery of the column /capital/drop panel, perpendicular to the plane of the slab where d is the effective depth of the section (See Fig. 14.2). The shape in plan is geometrically similar to the support immediately below the slab. See Fig. 14.4*(a)* and 14.4 *(b)*.

Note : For column sections with re-entrant angles the critical section shall be taken as indicated in Fig. 14.4*(c)* and 14.4*(d)*.

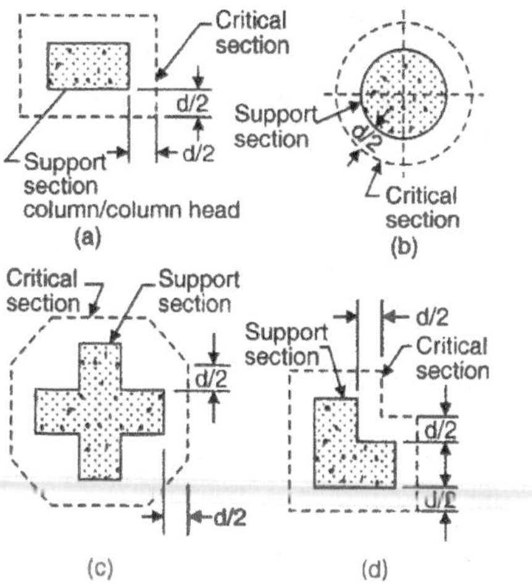

FIG 14.4 CRITICAL SECTIONS IN PLAN FOR SHEAR.

(*a*) In the case of columns near the free edge of a slab, the critical section shall be taken as shown in Fig. 14.5.

(*b*) When the openings in flat slabs are located at a distance less than ten times the thickness of the slab from a concentrated reaction or when the openings are located within the column strips, the critical sections specified in *(a)* shall be modified so that the part of

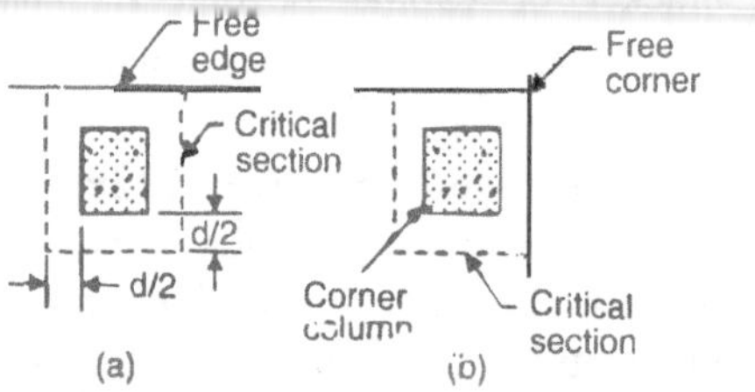

FIG. 14.5. EFFECT OF FREE EDGES ON CRITICAL SECTIONS FOR SHEAR

the periphery of the critical section which is enclosed by radial projection of the openings to the centroid of the reaction area shall be considered ineffective (See Fig. 14.6), and openings shall not encroach upon column head.

2. Calculation of shear stress. The shear stress τ_v shall be the sum of the values calculated according to (*i*) and (*ii*) below :

(*i*) The nominal shear stress in flat slabs shall be taken as $V/b_0 d$ where V is the shear force due to design load, b_0 is the periphery of the critical section and d is the effective depth.

(*ii*) When unbalanced gravity load, wind, earthquake or other forces cause transfer of bending moment between slab and column, a fraction $(1 - \alpha)$ of the moment shall be considered transferred by eccentricity of the shear about the centroid of the critical section. Shear stress shall be taken as varying linearly about the centroid of the critical section. The value of α shall be obtained from the equation given in clause c-3.

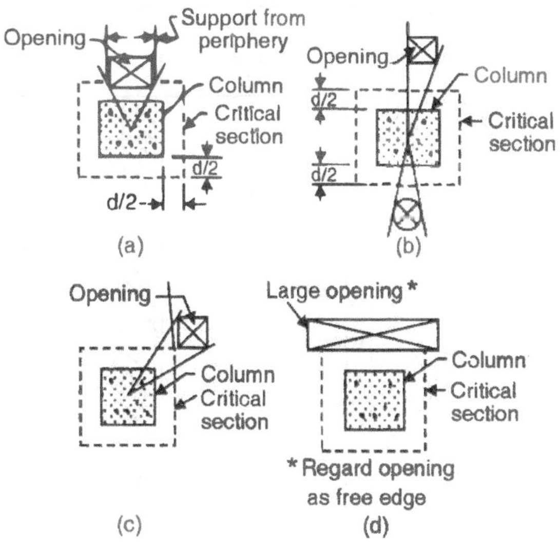

FIG. 14.6. EFFECTS OF OPENINGS ON CRITICAL SECTION FOR THE SHEAR.

3. **Permissible shear stress.**

(*i*) When shear reinforcement is not provided, the calculated shear stress at the critical section shall not exceed $k_s . \tau_c$, where

$k_s = (0.5 + \beta_c)$ but not greater than 1, β_c being the ratio of short side to long side of the column/capital ; and

$\tau_c = 0.25 \sqrt{f_{ck}}$ in limit state method of design.

(*ii*) When the shear stress at the critical section exceeds the value given in (*i*) above, but less than 1.5 τ_c, shear reinforcement shall be redesigned. Shear stresses shall be investigated at successive sections more distant from the support and shear reinforcement shall be provided up to a section where the shear stress does not exceed $0.5\ \tau_c$. While designing the shear reinforcement, the shear stress carried by the concrete shall be assumed to be $0.5\tau_c$ and reinforcement shall carry the remaining shear.

14.7. SLAB REINFORCEMENT

1. **Spacing.** The spacing of bars in a flat slab, shall not exceed 2 times the slab thickness except where a slab is of cellular or ribbed construction.

2. **Area of reinforcement.** When drop panels are used, the thickness of drop panel for determination of area of reinforcement shall be the lesser of the following:

(*i*) Thickness of drop, and

(*ii*) Thickness of slab plus one quarter the distance between edge of drop and edge of capital.

The reinforcement in flat slabs should have minimum lengths specified in Fig. 14.7.

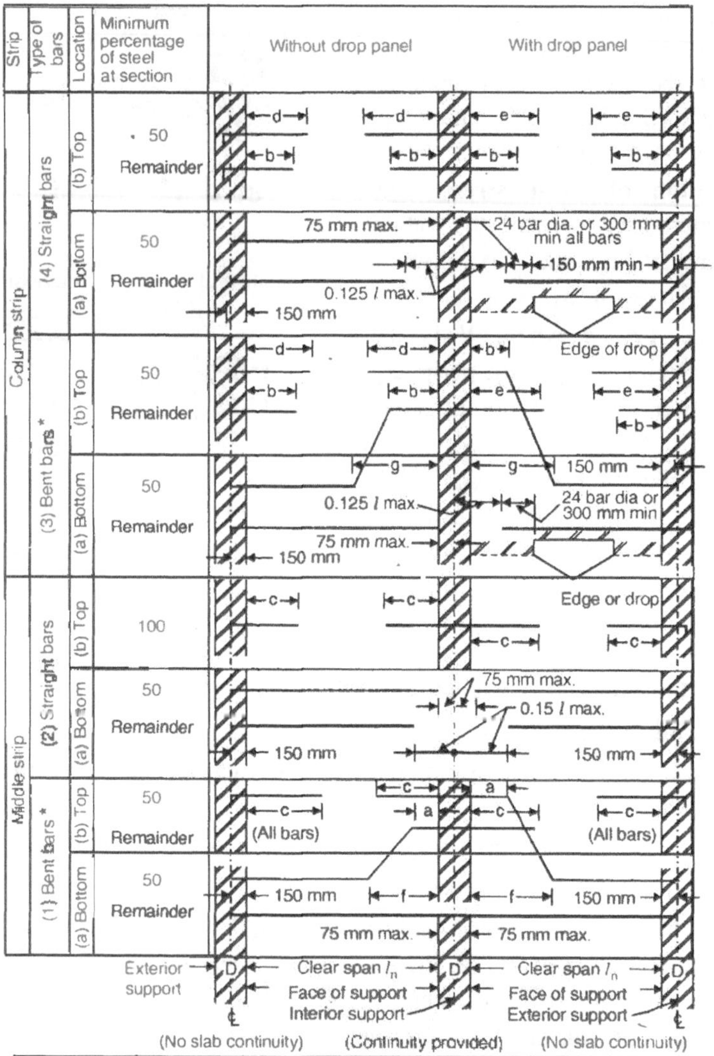

	Bar length from face of support						
	Minimum length					Maximum length	
Mark	a	b	c	d	e	f	g
Length	$0.14\ l_n$	$0.20\ l_n$	$0.22\ l_n$	$0.00\ l_n$	$0.00\ l_n$	$0.00\ l_n$	$0.01\ l_n$

FIG. 14.7. THE REINFORCEMENT DETAILS

14.8. OPENINGS IN FLAT SLAB

Openings of any size may be provided in the flat slab if it is shown by analysis that the requirement of strength and serviceability are met. However, for openings conforming to the following, no special analysis is required :

(1) Openings of any size may be placed within the middle half of the span in each direction provided the total amount of reinforcement required for the panel without the opening is maintained.

(2) In the area common to two column strips, not more than one-eighth of the width of strip in either span shall be interrupted by the openings. The equivalent of reinforcement interrupted shall be added on all sides of the openings.

(3) In the area common to one column strip and one middle strip not more than one-quarter of the reinforcement in either strip shall be interrupted by the openings. The equivalent of reinforcement interrupted shall be added on all sides of the openings.

(4) Shear requirements of § 14.6 shall be satisfied.

Note. D is the dia. of column and the dimension of rectangular column in the direction under consideration.

Design Example: *Design the interior panel of a flat slab 5.6 m × 6.6 m in size, for a super-imposed load of 7 kN/m². Provide two way reinforcement. Use M 20 concrete and Fe 415 steel.*

Solution.

1. Design constants and limiting depth of N.A.

Given : f_{ck} = 20 N/mm² for M 20 concrete and f_y = 415 N/mm² for Fe 415 steel.

For Fe 415, $\frac{x_{u,\max}}{d} = \frac{700}{1100 + 0.87 \times 415} = 0.479.$

$$R_u = 0.36\, f_{ck} \frac{x_{u\,.\max}}{d}\left(1 - 0.416\frac{x_{u\,.\max}}{d}\right) = 0.36 \times 20 \times 0.479\,(1 - 0.416 \times 0.479) = 2.761$$

2. Panel dimensions :

Length of panel, $L = 6.6$ m; Width of panel, $B = 5.6$ m

Widths of strips will be as follows :

(*i*) *Along length L* : $l_1 = L = 6.6$ m and $l_2 = B = 5.6$ m

Width of column strip $= 0.25\, l_2 = 0.25 \times 5.6 = 1.4$ m, with an upper limit of $0.25\, l_1 = 0.25 \times 6.6 = 1.65$ m. Hence keep width of column strip = 1.4 m on each side of column centre line.

Width of middle strip = 5.6 − 2.8 = 2.8 m

(*ii*) *Along width B* : $l_1 = B = 5.6$ m and $l_2 = L = 6.6$ m

Width of column strip $= 0.25\, l_2 = 0.25 \times 6.6 = 1.65$ m, with an upper limit of $0.25\, l_1 = 0.25 \times 5.6 = 1.4$ m. Hence keep width of column strip = 1.4 m on either side of column centre line.

Width of middle strip = 6.6 − 2.8 = 3.8 m

Let us provide drops also. The drops should be rectangular in plan, having a length in each direction not less than one third the panel length in that direction. Thus,

Along length L : Minimum length of drop $= \frac{1}{3} L = \frac{1}{3} \times 6.6 = 2.2$ m. However, keep it equal to total width of column strip along $B = 2.8$ m.

Along width B :

Minimum length of drop

$= \frac{1}{3} B = \frac{1}{3} \times 5.6 = 1.87$ m

However, keep it equal to the total width of column strip along $L = 2.8$ m. Let the column have a head of diameter one fifth of average span.

$$l = \frac{1}{2}(L + B)$$

$$= \frac{1}{2}(6.6 + 5.6)$$

$= 6.1$ m. Hence

$$D = l/5 = 1.2 \text{ m}$$

The panel details are shown in Fig. 14.8.

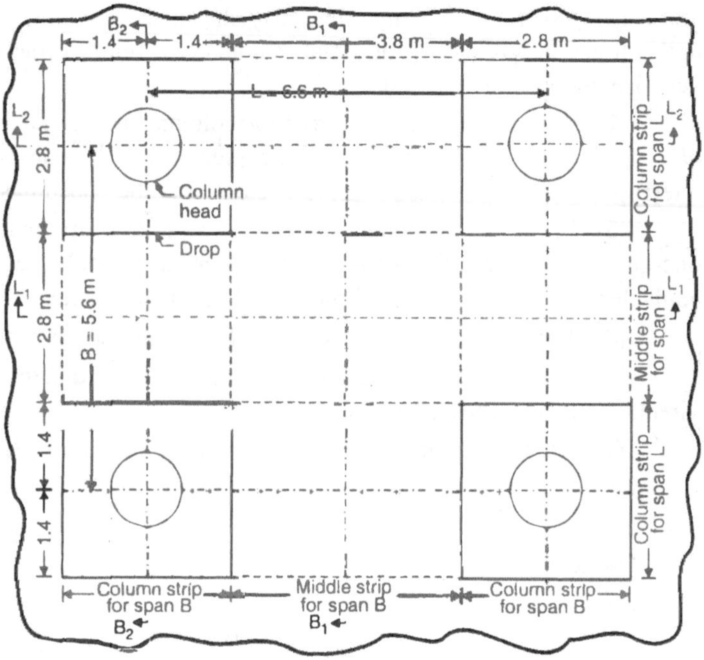

FIG. 14.8.

2. *Computation of loading*

The thickness of flat slab is generally controlled by considerations of span to effective depth ratio. For a continuous panel,

$$\frac{\text{longer span}}{d} \not> 26$$

Assuming an under-reinforced section with $p_t = 0.6\%$, we get modification factor for Fe 415 steel $\simeq 1.15$.

$$\therefore \qquad d = \frac{\text{longer span}}{26 \times 1.15} = \frac{6.6 \times 1000}{26 \times 1.15} \simeq 220 \text{ mm}$$

Assuming a nominal cover of 20 mm and using 10 mm φ bars, total thickness = 220 + 20 + 5 + 10 = 255 mm. However, let us provide a total thickness of 250 mm. The thickness of drops is normally kept 25% to 50% more than the thickness of the slab. Let us assume an average thickness of 280 mm for the calculation of dead load.

Weight of slab/m² = 0.28 × 1 × 1 × 25000 = 7000 N/m²

Self imposed load @ 7000 N/m² = 7000 N/m²

Total w = 14000 N/m²

$$\therefore \qquad w_u = 1.5\, w = 1.5 \times 14000 = 21000 \text{ N/m}^2$$

3. *Computation of moments along longer span*

For the longer span ($L = 6.6$ m), $l_1 = L = 6.6$ m; $l_2 = B = 5.6$ m.

The column head is circular of diameter 1.2 m

Hence size of equivalent square supports $= \sqrt{\frac{\pi}{4}(1.2)^2} = 1.06$ m

$\therefore \qquad l_{nL} = l_1 - 1.06 = 6.6 - 1.06 = 5.54$ m

(subject to a minimum of $0.65\, l_1 = 0.65 \times 6.6 = 4.29$ m)

$W_{uL} =$ total design load on the area $l_2\, l_{nL} = (5.6 \times 5.54)\, 21000 = 651504$ N

$$\therefore \qquad M_{uoL} = \frac{W_{uL} \cdot l_{nL}}{8} = \frac{651504 \times 5.54}{8} = 451167 \text{ N-m}$$

Apportionment of moments

Total negative moment $= 0.65\, M_{uoL} = 0.65 \times 451167 = 293258$ N-m

Total positive moment $= 0.35\, M_{uoL} = 0.35 \times 451167 = 157908$ N-m

Effect of pattern of loading

In the direct design method, when the ratio of live load to dead load exceeds 0.5, the factor $\alpha_c = \dfrac{\Sigma K_c}{\Sigma K_s}$ should be greater than $\alpha_{c\,.\,\min}$ given in Table 14.1. If not, the positive moments for the panel should be multiplied by coefficient β_0. In order to compute α_c, the flexural stiffness K_c for the columns and k_s for the slab are needed. The flexural stiffness K for any member of *uniform section* is given by

$$K = \frac{4\, E_c\, I}{l}$$

where I is the moment of inertia of concrete section, neglecting the reinforcement and l is the appropriate length of the member. For slab of uniform thickness h, span l_1, and transverse span l_2,

$$K_s = \frac{4\, E_c\, l_2\, h^3}{12\, l_1}$$

Similarly, for uniform circular column of side $D \times D$ and storey height l,

$$K_c = \frac{4\, E_c\, D^4}{12\, l}$$

If, however, the member (i.e. slab or column) has variable moment of inertia, the flexural stiffness of the member is computed from the expression

$K = \dfrac{k\, E_c\, I}{l}$, where k = stiffness factor for the section, which is more than 4.

The value of stiffness factors k for slab without drop, slab with drops and columns are given in Table 14.2, 14.3 and 14.4 respectively, for some idealised cases.

TABLE 14.2 STIFFNESS COEFFICIENTS FOR SLAB WITH VARIABLE MOMENT OF INERTIA (WITHOUT DROP)

Ratio D_1/l_1	*Stiffness factor k*	*Ratio* D_1/l_1	*Stiffness factor k*
0.0	4.00	*0.15*	4.40
0.05	4.05	*0.20*	4.72
0.10	4.18	*0.25*	5.14

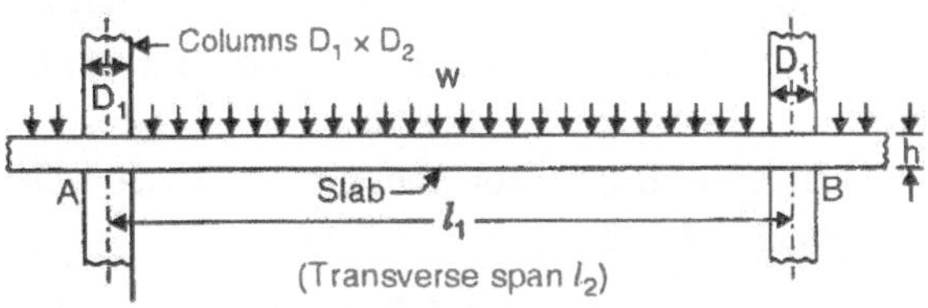

FIG. 14.9.

Note : The above table is valid for case $D_1 / l_1 = D_2 / l_2$. For other cases the values are only approximate.

TABLE 14.3. STIFFNESS COEFFICIENTS FOR SLAB WITH VARIABLE MOMENT OF INERTIA (WITH DROP)

Ratio D_1/l_1	*Stiffness factor k*	*Ratio* D_1/l_1	*Stiffness factor k*
0.00	4.78	*0.15*	5.22
0.05	4.84	*0.20*	5.55
0.10	4.98	*0.25*	5.98

Note. The above table is valid for case $D_1 / l_1 = D_2 / l_2$. For other cases, the values are only approximate.

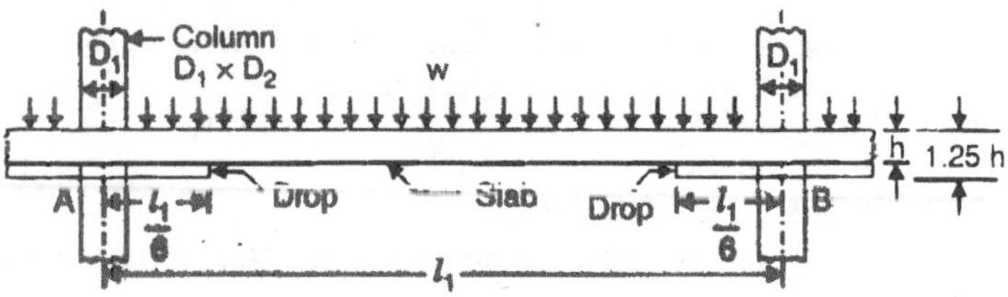

(Transverse span l_2)

FIG. 14.10.

TABLE 14.4. STIFFNESS COEFFICIENTS FOR COLUMNS WITH VARIABLE MOMENT OF INERTIA

Ratio H/l	k_{AB}	k_{BA}
0.00	4.00	4.00
0.05	4.91	4.21
0.10	6.09	4.44
0.15	7.64	4.71
0.20	9.69	5.00
0.25	12.44	5.33

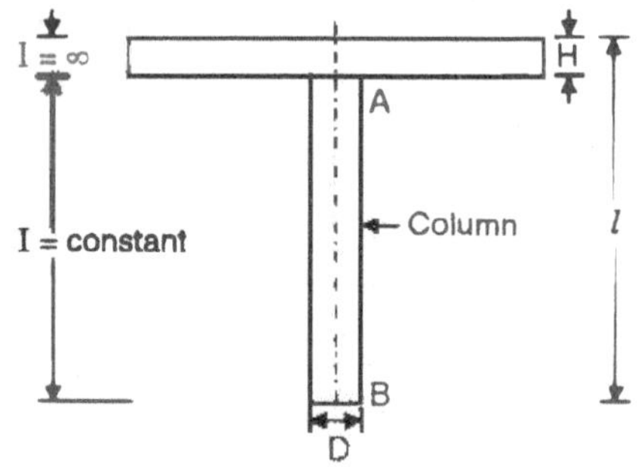

FIG. 14.11.

For the present case, thickness of slab, $h = 250$ mm $= 0.25$ m, $l_1 = 6.6$ m and $l_2 = 5.6$ m. Let the thickness of the drops be 320 mm $= 0.32$ m. Also, assume that column diameter $= 400$ mm, height of column head $= 500$ mm and diameter of column head $= 1200$ mm.

Hence $D_1 = D_2 = D =$ size of equivalent square column $= \sqrt{\frac{\pi}{4}(0.4)^2} = 0.35$ m

$$D_1/l_1 = D_2/l_2 = 0.35/6.6 = 0.053$$

Table 14.2 is applicable for slab without drops. For most of the practical cases, D/l_1 will be between 0.05 to 0.1, and hence k will vary between 4.05 to 4.18. *In the absence of the table, therefore, the value of k may be taken as 4.1*

Table 14.3 is applicable for slab with drops of width $l_1/6$ on either side. For $D/l_1 = 0.05$, $k \simeq 4.84$, against $k = 4.05$ for $D/l_1 = 0.05$ when the slab does not have drops. However, in absence of any table, one can adopt the value of $k = 5$ for all common cases for slabs with drops. For the present case, when the drops are more than $l_1/6$ on either side, assume $k = 5$.

Now $$K_s = \frac{k\, E_c\, I}{l} = \frac{k\, E_c\, l_2\, h^3}{12\, l_1} = \frac{5 \times 5.6\,(0.25)^3}{12 \times 6.6} E_c = 5.52 \times 10^{-3} E_c$$

where E_c is the modulus of elasticity of concrete.

For computing the stiffness of columns meeting at the slab, Table 14.4 may be used where H may be taken approximately as the thickness of the drop (or the slab) plus half the height of the pedestal.

$$\therefore \qquad H = 0.32 + 0.25 = 0.57 \text{ m}$$

The moment of inertia for this height H is taken to be infinite. Let us assume that the height of storey is 4 m from top of one slab to the top of other slab.

Hence $l = 4$ m. For $H/l = 0.57/4 = 0.1425$, we have $k_{AB} \simeq 7.41$ and $k_{BA} \simeq 4.67$.

For column, $\quad I = \frac{1}{12} D^4 = \frac{1}{12}(0.35)^4 = 1.25 \times 10^{-3}$

Hence for lower column, $\quad K_c = \frac{k_{AB} E_c I}{l} = \frac{7.41 \times 1.25 \times 10^{-3}}{4} E_c = 2.32 \times 10^{-3} E_c$

and for lower column, $\quad K_c = \frac{k_{BA} E_c I}{l} = \frac{4.67 \times 1.25 \times 10^{-3}}{4} E_c = 1.46 \times 10^{-3} E_c$

$$\therefore \qquad \alpha_c = \frac{\Sigma K_c}{\Sigma K_s} = \frac{(2.32 + 1.46)\, 10^{-3}}{2 \times 5.52 \times 10^{-3}} = 0.342$$

Now ratio $\quad \frac{w_l}{w_d} = \frac{7000}{7000} = 1$. Also, ratio $\frac{l_2}{l_1} = \frac{5.6}{6.6} = 0.85$

For these ratios, the values of $\alpha_{c,\,\min}$ comes out to be 0.7 from Table 14.1. Thus $\alpha_c < \alpha_{c,\,min}$. Hence positive moment has to be multiplied by coefficient β_0 given by

$$\beta_0 = 1 + \frac{\left(2 - \frac{w_d}{w_l}\right)}{\left(4 + \frac{w_d}{w_l}\right)}\left(1 - \frac{\alpha_c}{\alpha_{c,\,min.}}\right) = 1 + \left(\frac{2-1}{4+1}\right)\left(1 - \frac{0.342}{0.7}\right) = 1.1023$$

Hence positive design moment $= M_{uPL} = 1.1023 \times 157908 \simeq 174062$ N-m.

$\therefore$ Negative design moment, $M_{uNL} = 451167 - 174062 = 277105$ N-m

These moments will be apportioned to the column strips and middle strip as under:

For column strips :

– ve moment $\quad M_{u1L} = 0.75\, M_{uNL} = 0.75 \times 277105 = 207829$ N-m

+ ve moment $\quad M_{u2L} = 0.60\, M_{uPL} = 0.6 \times 174062 = 104437$ N-m

For middle strip

– ve moment $\quad M_{u3L} = M_{uNL} - M_{u1L} = 277105 - 207829 = 69276$ N-m

+ ve moment $\quad M_{u4L} = M_{uPL} - M_{u2L} = 174062 - 104437 = 69625$ N-m

5. *Computation of moments along shorter span*

For shorter span ($B = 5.6$ m), $l_1 = B = 5.6$ m; $l_2 = L = 6.6$ m; $l_{nB} = 5.6 - 1.06 = 4.54$ m

(subject to a minimum of $0.65\, l_1 = 0.65 \times 5.6 = 3.65$ m)

$W_{uB} =$ total design load on area $l_2 \,.\, l_{nB} = (6.6 \times 4.54)\, 21000 = 629244$

$$\therefore \qquad M_{uoB} = \frac{W_B \,.\, l_{nB}}{8} = \frac{629244 \times 4.54}{8} = 357096 \text{ N-m}$$

Apportionment of moments

Total negative moment $= 0.65\, M_{uoB} = 0.65 \times 357096 = 232112$ N-m

Total positive moment = $0.35\, M_{uoB} = 0.35 \times 357096 = 124984$ N-m

Effect of pattern of loading

$$D/l_1 = 0.35/5.6 = 0.0625$$

Hence from Table 14.3, $k \simeq 4.9$, Take $k = 5$ since width of drop is more than $l_1/6$ on each side of column centre.

$$K_s = \frac{k\, E_c\, l_2\, h^3}{12\, l_1} = \frac{5 \times 6.6\, (0.25)^3}{12 \times 5.6} E_c = 7.67 \times 10^{-3} E_c$$

$$\Sigma K_c = (2.32 + 1.46)\, 10^{-3} E_c, \text{ as before.}$$

$$\therefore \qquad \alpha_c = \frac{\Sigma K_c}{\Sigma K_s} = \frac{(2.32 + 1.46)\, 10^{-3}}{2 \times 7.67 \times 10^{-3}} = 0.246$$

Ratio $\dfrac{w_l}{w_d} = \dfrac{7000}{7000} = 1$ and ratio $l_2/l_1 = 6.6/5.6 = 1.179$

From Table 14.1, the value of $\alpha_{c,\min}$ comes out to be 0.772. $\therefore\ \alpha_c < \alpha_{c.\,min.}$

Hence $$\beta_0 = 1 + \frac{(2 - w_d/w_l)}{(4 + w_d/w_l)}\left(1 - \frac{\alpha_c}{\alpha_{c,\,min.}}\right) = 1 + \frac{2-1}{4+1}\left(1 - \frac{0.246}{0.772}\right) = 1.1363$$

$\therefore$ Positive design moment, $M_{uPB} = 1.1363 \times 124984 = 142015$ N-m

and Negative design moment, $M_{uNB} = 357096 - 142015 = 215081$ N-m

These moments will be apportioned to the column strips and middle strips as follows:

Column strips :

– ve moment $M_{u1B} = 0.75\, M_{uNB} = 0.75 \times 215081 = 161311$ N-m

+ ve moment $M_{u2B} = 0.6\, M_{uPB} = 0.6 \times 142015 = 85209$ N-m

Middle strip

– ve moment $M_{u3B} = M_{uNB} - M_{u1B} = 215081 - 161311 = 53770$

+ ve moment $M_{u4B} = M_{uPB} - M_{u2B} = 142015 - 85209 = 56806$

6. Determination of thickness of slab and drops

The thickness of the slab will be designed on the basis of maximum positive moment M_{u2L} in the column strip of longer span. Width of column strip of span $L = 2.8$ m.

$$\therefore \qquad d = \sqrt{\frac{M_{u2L}}{R_u\, b}} = \sqrt{\frac{104437 \times 1000}{2.761 \times 2800}} = 116.2 \text{ mm}$$

However, provide total thickness of 250 mm, as fixed from deflection criterion. Using 10 mm φ bars and a nominal cover of 20 mm, available d for longer span $= 250 - 20 - 5 = 225$ mm and for shorter span (upper layer) $= 225 - 10 = 215$ mm.

The thickness of drops d_d will be designed on the basis to maximum negative bending moment $M_{u1L} = 207829$ N-m.

$$\therefore \qquad d_d = \sqrt{\frac{M_{u1L}}{R_u\, b}} = \sqrt{\frac{207829 \times 1000}{2.761 \times 2800}} = 164 \text{ mm}$$

However, keep total thickness of drop equal to 1.25 times the thickness of slab $= 1.25 \times 250 \simeq 320$ mm. Using 10 mm φ bars and 20 mm nominal cover, available $d = 320 - 20 - 5 = 295$ mm for longer span and $295 - 10 = 285$ mm for the shorter span.

7. *Check for shear*

The critical section for the slab will be at a distance of $d/2$ from the face of the drop. Fig. 14.12 (*a*) shows the plan for the periphery of critical section for slab.

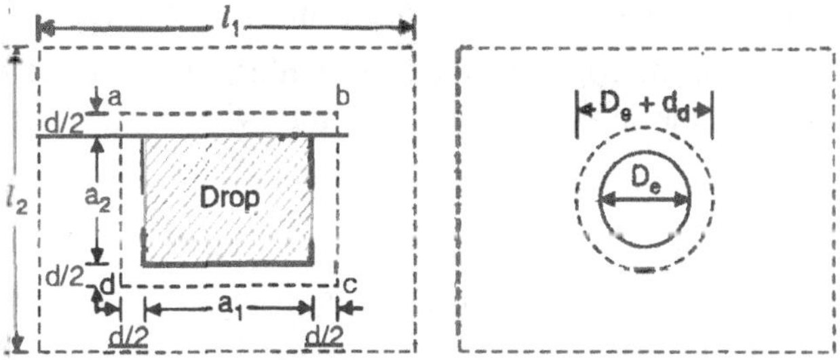

(a) Shear in slab (b) Shear in drops

FIG. 14.12 PERIPHERY OF CRITICAL SECTION.

Shear in slab is given by

$$V_s = (w_d + w_l)\left[\, l_1 l_2 - (a_1 + d)\,(a_2 + d)\,\right]$$

b_0 = periphery of critical section in slab $= 2\,(a_1 + a_2 + 2d)$

For the present case

$a_1 = a_2 = 2.8$ m $= 2800$ mm ; $d =$ effective depth of slab $= 225$ mm

$$V_s = (7000 + 7000)\,[6.6 \times 5.6 - (2.8 + 0.225)\,(2.8 + 0.225)] = 389331 \text{ N}$$

$$V_{us} = 1.5\, V_s = 1.5 \times 389331 = 583997 \text{ N}$$

$$b_0 = 2\,(2800 + 2800 + 2 \times 225) = 12100 \text{ mm}$$

$$\tau_v = \frac{V_{us}}{b_0\, d} = \frac{583997}{12100 \times 225} = 0.215 \text{ N/mm}^2$$

Again, for shear in the drops, the critical section is at a distance of $d_d/2$ from the periphery of the column head, as shown in Fig. 14.12 (*b*), where $D_e = 1.2$ m and $d_d = 0.295$ m. Shear in the drops is given by

$$V_d = (w_d + w_l)\left[\, l_1 l_2 - \frac{\pi}{4}(D_e + d_d)^2 \right] = (7000 + 7000)\left[\, 6.6 \times 5.6 - \frac{\pi}{4}(1.2 + 0.295)^2 \right]$$

$$= 492865 \text{N} \;;\; V_{ud} = 1.5 \times 492865 \simeq 739298 \text{ N}$$

b_0 = periphery of critical section in slab $= \pi\,(D_e + d_d)$

$$= \pi\,(1.2 + 0.295) = 4.697 \text{ m} = 4697 \text{ mm}$$

$$\therefore \quad \tau_v = \frac{V_{ud}}{b_0\, d_d} = \frac{739298}{4697 \times 295} = 0.534 \text{ N/mm}^2$$

Thus, shear stress is more critical in drops than in slab.

Permissible shear stress $= k_s \,.\, \tau_c$ where $k_s = (0.5 + \beta_c)$ but $\not> 1$.

β_c = ratio of short side to long side of column capital = 1 for the present case.

$\therefore \quad k_s = (0.5 + 1) = 1.5$ but $\not> 1$. Hence $k_s = 1$

$$\tau_c = 0.25\sqrt{f_{ck}} = 0.25\sqrt{20} = 1.119 \text{ N/mm}^2$$

$\therefore$ Permissible shear stress $= k_s \,.\, \tau_c = 1 \times 1.119 = 1.119 \text{ N/mm}^2$

Since permissible shear stress is more than actual shear stress in the drops as well as in the slab, both are safe and no shear reinforcement is required.

8. Reinforcement along span L : $l_1 = L = 6.6$ m. The spacing of the reinforcement is limited to twice the thickness of the slab (i.e. $2 \times 225 = 450$ mm) or 300 mm whichever is less.

For determining area of reinforcement in drops, the effective thickness of drop panel will be the maximum of the following : (*i*) effective thickness of drop = 295 mm, or (*ii*) the effective thickness of slab $+\frac{1}{4}$ the distance between the edge of drop and edge of drop and edge of capital $= 225 + \frac{1}{4}(1400 - 600) = 425$ mm. Hence adopt effective thickness of drops = 295 mm. The reinforcements are to be provided for the following moments.

$M_{u1L} = -207829$ N-m and $M_{u2L} = +104437$ N-m for column strip of width 2800 mm

$M_{u3L} = -69276$ N-m and $M_{u4L} = +69625$ N-m for middle strip of width 2800 mm

(*i*) ***Middle strip*** : For middle strip, the effective thickness of slab = 225 mm. Since the depth provided is more than that required for bending, we have an under-reinforced section, for which

$$A_{st4L} = \frac{0.5 f_{ck}}{f_y}\left[1 - \sqrt{1 - \frac{4.6 M_{u4L}}{f_{ck}\, bd^2}}\right] bd = \frac{0.5 \times 20}{415}\left[1 - \sqrt{1 - \frac{4.6 \times 69625 \times 10^3}{20 \times 2800\,(225)^2}}\right] \times 2800 \times 225$$

$= 883$ mm^2 (+ ve reinforcement)

∴ Number of 10 mm φ bars required $= 883/78.4 \simeq 12$

∴ Average spacing $= 2800/12 \simeq 233$ mm c/c

$$A_{st3L} = \frac{0.5 f_{ck}}{f_y}\left[1 - \sqrt{1 - \frac{4.6 M_{u3L}}{f_{ck}\, bd^2}}\right] bd = \frac{0.5 \times 20}{415}\left[1 - \sqrt{1 - \frac{4.6 \times 69276 \times 10^3}{20 \times 2800\,(225)^2}}\right] \times 2800 \times 225$$

$= 878.6$ mm^2 (– ve reinforcement)

∴ Number of 10 mm φ bars required $= 878.6/78.54 \simeq 12$

∴ Average spacing $= 2800/12 \simeq 233$ mm c/c.

Half of the positive reinforcement is continued upto 75 mm away from the centre of the support while the remaining half may be bent up at a distance of $f = 0.20\, l_{nL}$ $= 0.2 \times 5.54 = 1.108$ (max) $\simeq 1100$ mm from the face of the column. The bent-up bars should continue upto a distance $c = 0.22\, l_n = 0.22 \times 5.54 = 1.22$ m (min.) $\simeq 1250$ mm to the other face of the column. Since bent up bars will be available from both the adjacent spans, reinforcement for negative moment M_{u3L} will be available as 10 mm φ bars @ 233 mm c/c. It is to be ensured that such bending of bars provide a minimum length $a = 0.14\, l_n$ $= 0.14 \times 5.54 \simeq 0.78$ m is available from the first face of the column. The above reinforcement has been provided according to item (1) of Fig. 14.7.

(*ii*) ***Column strips*** : The column strips of span L have also width equal to 2800 mm. The effective thickness for +ve reinforcement is 225 mm (slab) while the effective thickness for – ve reinforcement is 295 mm (drops).

$$A_{st2L} = \frac{0.5 f_{ck}}{f_y}\left[1 - \sqrt{1 - \frac{4.6 M_{u2L}}{f_{ck}\, bd^2}}\right] bd = \frac{0.5 \times 20}{415}\left[1 - \sqrt{1 - \frac{4.6 \times 104437 \times 10^3}{20 \times 2800\,(225)^2}}\right] 2800 \times 225$$

$= 1345.9$ mm^2 (+ ve reinforcement)

$$A_{st1L} = \frac{0.5 f_{ck}}{f_y}\left[1 - \sqrt{1 - \frac{4.6\, M_{u1L}}{f_{ck}\, bd^2}}\right] bd$$

$$= \frac{0.5 \times 20}{415}\left[1 - \sqrt{1 - \frac{4.6 \times 207829 \times 1000}{20 \times 2800\,(295)^2}}\right] \times 2800 \times 295$$

$$= 2058.7 \text{ mm}^2 \quad (-\text{ve reinforcement})$$

Number of 10 mm φ bars for +ve reinforcement $= 1345.9/78.54 \simeq 18$

∴ Average spacing $= 2800/18 \simeq 155$ mm c/c.

Continue half the positive reinforcement to a distance of $0.125\,L = 0.125 \times 6.6 = 0.825$ m (max) $\simeq 800$ mm from the column centre. Available length of bar from the edge of drop $= 1400 - 800 = 600$ mm which is more than 24 φ (or 300 mm). The remaining positive reinforcement bars will be bent up to a distance $g = 0.24\, l_n = 0.24 \times 5.54 = 1.33$ m (max) $\simeq 1300$ mm from the face of the column. The bent-up bars should continue upto a distance $d = 0.3\, l_n = 0.3 \times 5.54 \simeq 1.662$ m (min.) $\simeq 1700$ mm beyond the outer face of the column (see item 3 of Fig. 14.7).

Thus, available negative reinforcement is 10 mm φ @ 155 mm c/c, giving a total reinforcement of $\frac{2800}{155} \times 78.54 \simeq 1418.7 \text{ mm}^2$ against a requirement of $A_{st1L} = 2058.7 \text{ mm}^2$. The balance of negative reinforcement $= 2058.7 - 1418.7 = 640 \text{ mm}^2$ is to be provided in the form of *straight bars* at the top.

Number of bars required $= 640/78.54 \simeq 9$.

However provide 10 bars to have an even spacing $= 2800/10 = 280$ mm c/c.

These bars should continue to a distance $e = 0.33\, l_n = 0.33 \times 5.54 = 1.828$ (min.) $\simeq 1900$ mm on either side of column face, as required in item 4(*b*) of Fig. 14.7.

9. Reinforcement along span B : $l_1 = B = 5.6$ m

(*i*) ***Middle strip*** : Width of the middle strip $= 3.8$ m $= 3800$ mm. The middle strip carries a positive moment of $M_{u4B} = 56806$ N-m at its middle and negative moment of $M_{u3B} = 53770$ N-m at its ends. Available $d = 215$ mm.

$$A_{st4B} = \frac{0.5 f_{ck}}{f_y}\left[1 - \sqrt{1 - \frac{4.6\, M_{u4B}}{f_{ck}\, bd^2}}\right] bd = \frac{0.5 \times 20}{415}\left[1 - \sqrt{1 - \frac{4.6 \times 56806 \times 10^3}{20 \times 3800\,(215)^2}}\right] \times 3800 \times 215$$

$$= 746.3 \text{ mm}^2 \quad (+\text{ve reinforcement})$$

Number of 10 mm φ bars required $= 746.3/78.54 \simeq 10$

∴ Spacing $= 3800/10 = 380$ mm c/c; Max. spacing $= 2d$ or 300 mm.

Hence, provide 10 mm φ bars @ 300 mm c/c.

$$A_{st3B} = \frac{0.5 f_{ck}}{f_y}\left[1 - \sqrt{1 - \frac{4.6\, M_{u3B}}{f_{ck}\, bd^2}}\right] bd = \frac{0.5 \times 20}{415}\left[1 - \sqrt{1 - \frac{4.6 \times 53770 \times 10^3}{20 \times 3800\,(215)^2}}\right] 3800 \times 215$$

$$= 705.7 \text{ mm}^2 \quad (-\text{ve reinforcement})$$

Number of 10 mm φ bars required $= 705.7/78.54 \simeq 9$

∴ Spacing $= 3800/9 = 422$ mm c/c. Max spacing $= 2d$ or 300 mm.

However, provide 10 mm φ bars @ 300 mm c/c, as for +ve reinforcement, so that bent up bars serve the purpose.

l_n for span $B = l_{nB} = 4.54$ m. Half the positive reinforcement is continued upto 75 mm away from the centre of the column, while the remaining half may be bent up at a distance of $f = 0.2\, l_n = 0.2 \times 4.54 = 0.908$ m (max.) ≏ 900 mm from the face of the column. The bent up bars should continue upto a distance $c = 0.22\, l_n = 0.22 \times 4.54 = 0.999$ m (min.) ≏ 1000 mm to the other face of the column.

(*ii*) ***Column strips*** : The width of column strip = 2800 mm. The available effective depth = 215 mm (slab) for positive reinforcement and = 285 mm (drops) for negative reinforcement. The strip carries a positive moment of $M_{u2B} = 85209$ N-m at its middle and negative moment of $M_{u1B} = 161311$ N-m at the ends.

$$A_{st2B} = \frac{0.5 f_{ck}}{f_y}\left[1 - \sqrt{1 - \frac{4.6\, M_{u2B}}{f_{ck}\, bd^2}}\right] bd = \frac{0.5 \times 20}{415}\left[1 - \sqrt{1 - \frac{4.6 \times 85209 \times 10^3}{20 \times 2800\,(215)^2}}\right] 2800 \times 215$$

$= 1143.3$ mm^2 (+ ve reinforcement)

$$A_{st1B} = \frac{0.5 f_{ck}}{f_y}\left[1 - \sqrt{1 - \frac{4.6\, M_{u1B}}{f_{ck}\, bd^2}}\right] bd = \frac{0.5 \times 20}{415}\left[1 - \sqrt{1 - \frac{4.6 \times 161311 \times 10^3}{20 \times 2800\,(285)^2}}\right] 2800 \times 285$$

$= 1638.2$ mm^2 (– ve reinforcement)

Number of 10 mm φ bars required for +ve reinforcement $= 1143.3/78.54$ ≏ 15

∴ Average spacing $= 2800/15 = 186.7$ mm c/c. Continue half the positive reinforcement to a distance of $0.125\, B = 0.125 \times 5.6 = 0.7$ m (max.) = 700 mm from the column centre. Available length of bar from the edge of drop $= 1400 - 700 = 700$ mm which is more than 24 φ (or 300 mm). The remaining +ve reinforcement bars will be bent up at a distance $g = 0.24\, l_n = 0.24 \times 4.54 = 1.09$ m (max.) ≏ 1000 mm from the face of the column. The bent up bars should continue upto a distance $d = 0.3\, l_n = 0.3 \times 5.54 = 1.362$ m (min.) ≏ 1400 mm beyond the other face of the column (see item 3 of Fig. 14.7).

Thus, available negative reinforcement is 10 mm φ @ 186.7 mm c/c giving a total reinforcement of $\frac{2800}{186.7} \times 78.54 = 1177.9$ mm^2 against a requirement of $A_{st1B} = 1638.2$ mm^2. The balance $= 1638.2 - 1177.9 = 460.3$ mm^2 is to be provided in the form of *straight bars* at the top.

No. of straight bars required $= 460.3/78.54$ ≏ 6. Hence spacing of straight bars $= 2800/6$ ≏ 467 mm. However, keep spacing $= 2 \times 186.7 = 373.4$ mm c/c for facility in construction. These bars should continue to a distance $e = 0.33\, l_n = 0.33 \times 4.54 = 1.5$ m (min.) = 1500 mm on either side of the column face, as required in item 4 (*b*) of Fig. 14.7.

10. Details of reinforcement

The details of reinforcement are shown in Fig. 14.13 and Fig. 14.14.

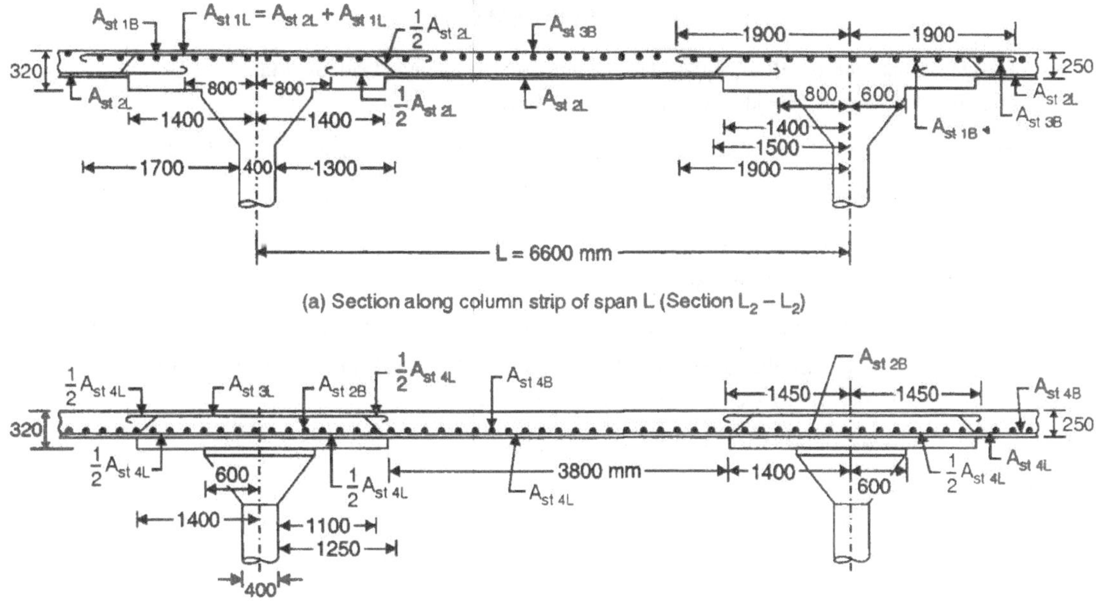

(a) Section along column strip of span L (Section $L_2 - L_2$)

(b) Section along middle strip of span L (Section $L_1 - L_1$)

FIG. 14.13

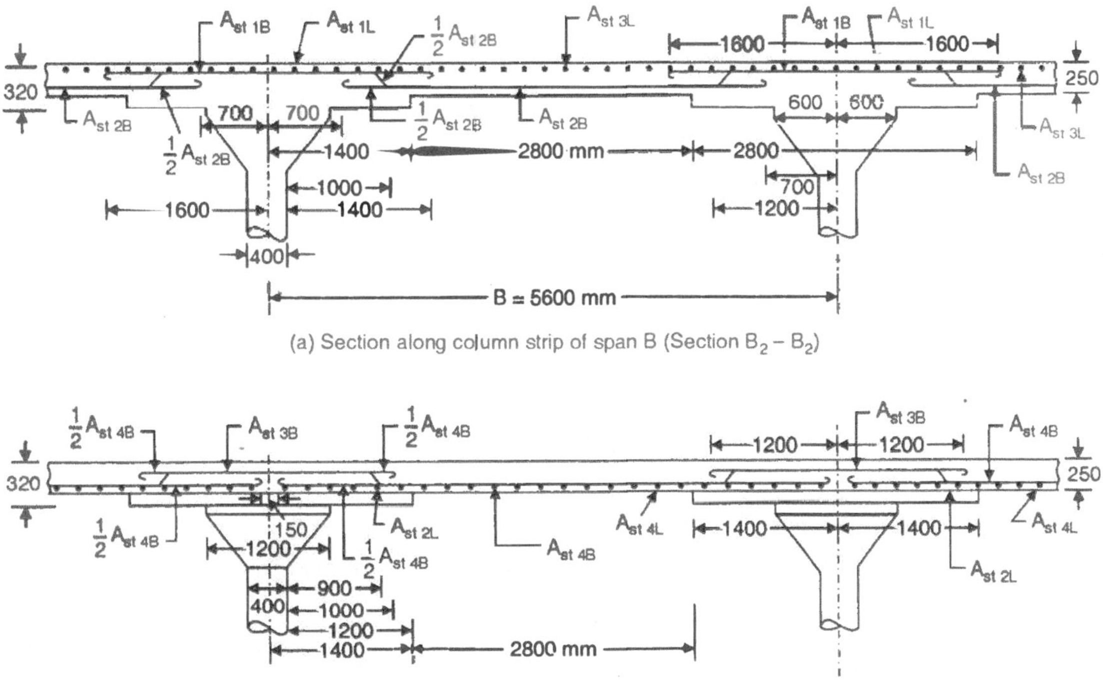

(a) Section along column strip of span B (Section $B_2 - B_2$)

(b) Section along middle strip of span B (Section $B_1 - B_1$)

FIG. 14.14.

PROBLEMS

1. Design an interior panel of a flat slab, 6 m × 6 m, for a live load of 7 kN/m^2. Use M 20 concrete and Fe 415 steel. Provide two-way reinforcement.
2. Redesign problem 1 by providing four way reinforcement as per I.S. Code of practice.
3. Redesign problem 1 if the slab has an opening of 1 m × 1 m at its middle.
4. Design an interior panel of a flat slab. 4 m × 4 m for a live load of 6 kN/m^2. Use M 20 concrete and Fe 415 steel. Provide two-way reinforcement.
5. The roof of a hall measures 26 m × 34 m, and consists of flat slab construction dividing it into 16 panels. Design all the panels of the roof for a live load of 6.5 kN/m^2. Use M 20 concrete and Fe 415 steel. Provide two way reinforcement.

15

Design of Stair Cases

15.1. INTRODUCTION

Staircases are used in almost all buildings. A staircase consists of a number of steps arranged in a series, with landings at appropriate locations, for the purpose of giving access to different floors of a building. The width of a staircase may depend on the purpose for which it is provided, and may generally vary between 1 m for residential buildings to 2 m for public buildings. A *flight* is the length of the staircase situated between two landings. The number of steps in a flight may vary between 3 to 12. The *rise* of a step and the *tread* should be so proportioned that it gives comfortable access. Generally,

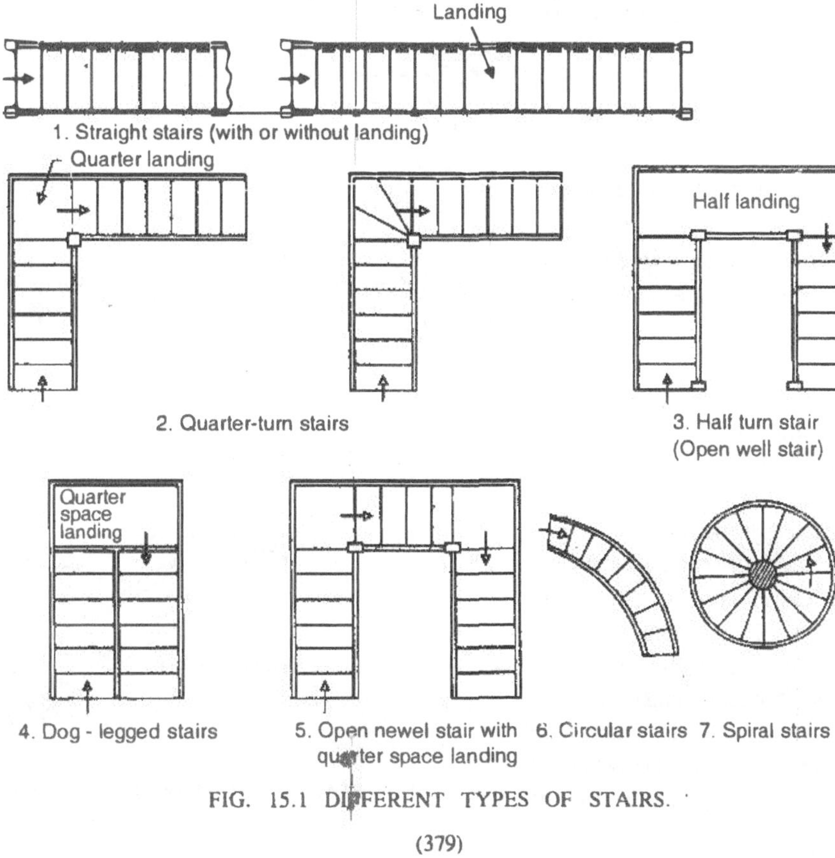

FIG. 15.1 DIFFERENT TYPES OF STAIRS.

the sum of tread plus twice the rise is kept about 500 mm and the product of tread and rise is kept about 40000 to 42000. In residential buildings, the rise may vary between 150 mm to 180 mm and tread between 200 to 250 mm. In public buildings, rise is kept between 120 to 150 mm and tread between 200 to 300 mm.

Classification of stairs. Staircases may be broadly classified in the following heads (Fig. 15.1)

(1) Straight stair.
(2) Quarter turn stair.
(3) Half turn stair (open newel type or open well stair).
(4) Dog-legged stair.
(5) Open newel stair with quarter space landing.
(6) Geometrical stairs such as circular stair, spiral stair etc .

15.2. GENERAL NOTES ON DESIGN OF STAIRS

1. **Live load on stairs.** I.S. 875 (*Code of Practice for Structural Safety of Buildings*) gives the loads for staircases. For stairs in residential buildings, office buildings, hospital wards, hostels, etc., where there is no possibility of overcrowding, the live load may be taken as 3000 N/m^2, subject to a minimum of 1300 N concentrated load at the unsupported end of each step for stairs constructed out of structurally independent cantilever step. For other public buildings liable to be overcrowded, the live load may be taken to be 5000 N/m^2.

2. **Effective span of stairs**. Stair slab may be divided into two categories, depending upon the direction in which the stair slab spans:

(*i*) Stair slab spanning horizontally. (*ii*) Stair slab spanning longitudinally.

(*i*) *Stair slab spanning horizontally*. In this category, the slab is supported on each side by side wall or stringer beam on one side and beam on the other side. Sometimes, as in the case of straight stair, the slab may also be supported on both the sides by the two side walls. The slab may also be supported horizontally by side wall on one side of each flight and the common newel on the other side between the backward and forward flights. In such a case the effective span L is the horizontal distance between centre-to-centre of supports. Each step is designed as spanning horizontally with a bending moment equal to

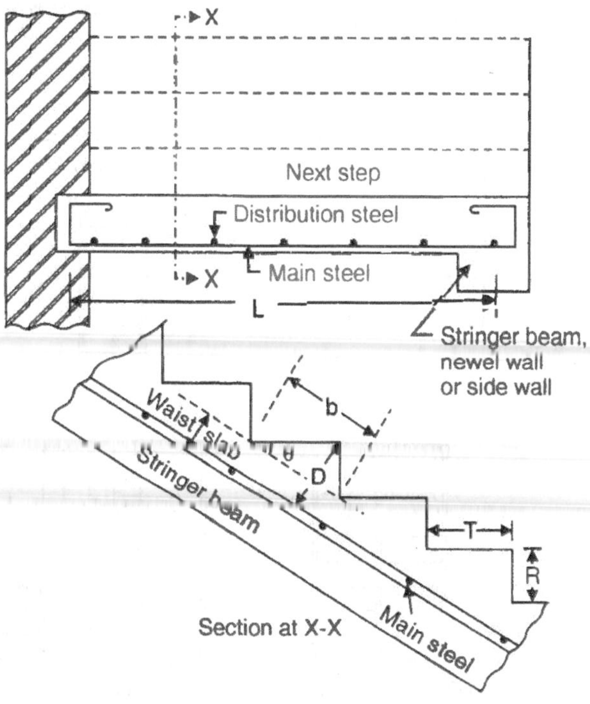

FIG. 15.2. STAIR SLAB SPANNING HORIZONTALLY.

$(w L^2)/8$. Each step is considered equivalent to a rectangular beam of width b (measured parallel to the slope of the stair) and an effective depth equal to $D/2$ as shown in Fig 15.2. Main reinforcement is provided in the direction of L, while distribution reinforcement is provided parallel to the flight direction. A waist of about 8 cm is provided.

(ii) Stair slab spanning longitudinally. In this category, the slab is supported at bottom and top of the flight and remain unsupported on the sides. Each flight of stairs is continuous, supported on beams at top and bottom or on landings. The effective span of such stairs, without stringer beams, should be taken as the following horizontal distances :

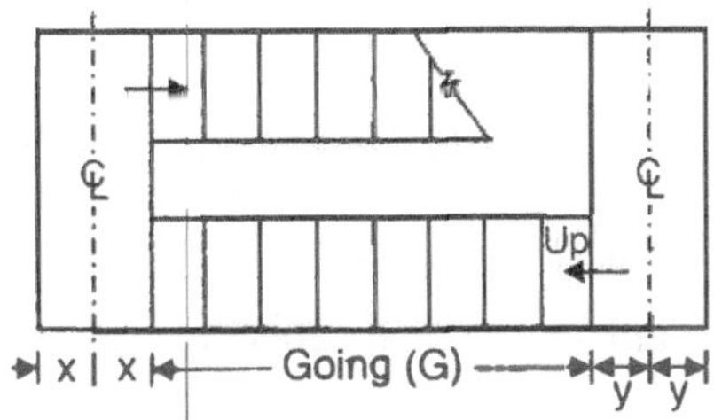

x (m)	y (m)	Span L (m)
<1	<1	G + x + y
<1	≥1	G + x + 1
≥1	<1	G + y + 1
≥1	≥1	G + 1 + 1

FIG. 15.3 EFFECTIVE SPAN FOR STAIRS SUPPORTED AT EACH END BY LANDINGS SPANNING PARALLEL WITH THE RISERS.

(*a*) where unsupported at top and bottom risers by beams spanning parallel with the risers, the distance centre-to-centre of beam ;

(*b*) where spanning on the edge of a landing slab which spans parallel to the risers (Fig.15.3), a distance equal to the 'going' of the stairs plus at each end either half the width of the landing or one metre, whichever is smaller ; and

(*c*) where the landing slab spans in the same direction as the stairs, they should be considered as acting together to form a single slab and the span determined at the distance centre-to-centre of the supporting beams or walls, the going being measured horizontally.

3. Distribution of loading on stairs

In case of stairs with open wells, where spans partly crossing at right angles occur, the load on areas common to any two such spans may be taken as one-half in each direction as shown in Fig. 15.4 (*a*). Where flights or landings are built into walls at a distance of not less than 110 mm and are designed to span in the direction of the flight, a 150 mm strip may be deducted from the loaded area and the effective breadth of the section increased by 75 mm for the purposes of design [Fig. 15.4(*b*)].

4. Estimation of dead weight

The dead weight of stair consists of (*i*) dead weight of waist slab and (*ii*) dead weight of steps.

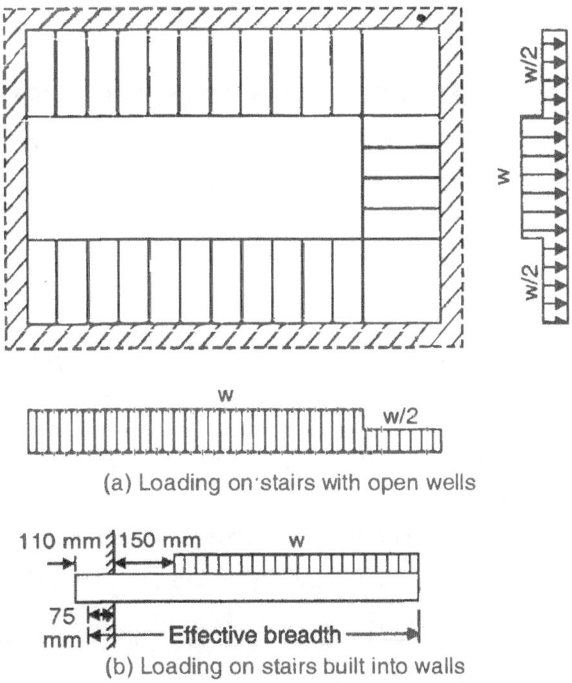

FIG. 15.4. DISTRIBUTION OF LOADING ON STAIRS.

*(i) **Dead weight of waist slab***

The dead weight w', per unit area, is first calculated at right angles to the slope. The corresponding load per unit horizontal area is then obtained by increasing w' by the ratio $\sqrt{R^2+T^2}/T$ where $R =$ rise and $T =$ tread.

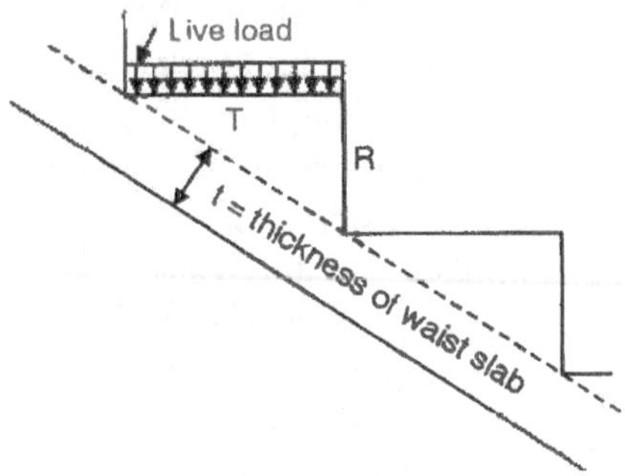

FIG. 15.3

Thus, if $t =$ thickness of waist in mm, then

$w' = \dfrac{t \times 1 \times 1}{1000} \times 25000 = 25\,t$ N/m^2 of inclined area.

Hence dead weight w_1 per unit horizontal area is given by

$$w_1 = w' \times \frac{\sqrt{R^2+T^2}}{T} = 25\,t\,\frac{\sqrt{R^2+T^2}}{T} = 25\,t\sqrt{1+\left(R/T\right)^2}$$

For example, if $R = 150$ mm, $T = 300$ mm and $t = 80$ mm. Then $w_1 = 25\,t\sqrt{1+\left(\frac{150}{300}\right)^2}$

$$= 27.95\,t = 2236 \text{ N/m}^2 \text{ of horizontal area.}$$

*(ii) **Dead weight of steps.*** The dead weight of the steps is calculated by treating the step to be equivalent *horizontal* slab of thickness equal to half the rise $\left(R/2\right)$. Thus, if w_2 is the weight of step per unit horizontal area, we have

$$w_2 = \frac{R}{2 \times 1000} \times 1 \times 1 \times 25000 = 12.5\,R \text{ N/m}^2 \text{ where } R \text{ is rise in mm.}$$

$$\text{Total } w = w_1 + w_2 \quad \text{per unit horizontal area.}$$

15.3. DESIGN OF STAIRS SPANNING HORIZONTALLY

In this type of stairs, the effective span is the horizontal distance between centre to centre of supports. Each step is designed as spanning horizontally. Each step is considered equivalent to a horizontal beam of width $b = \sqrt{R^2+T^2}$ measured parallel to the slope of the stair, and an effective depth $D/2$, as shown in Fig. 15.2. Main reinforcement is provided in the direction of span. The distribution reinforcement, provided in the form of 6 mm diameter bars at 30 cm c/c is normally adequate.

Design Example 15.1. *A straight stair in a residential building supported on wall on one side and stringer beam on the other side. The risers are 150 mm and treads are 250 mm and the horizontal span of the stairs may be taken as 1.2 metres. Design the steps. Use M 20 concrete and Fe 415 steel bars.*

Solution

*1. **Computation of design constants***

For Fe 415 steel, $f_y = 415$ N/mm^2, For M 20 concrete, $f_{ck} = 20$ N/mm^2

$$\frac{x_{u,\max}}{d} \text{ (For Fe 415 steel)} = \frac{700}{1100 + 0.87 \times 415} = 0.479$$

$\therefore \quad R_u = 0.36\, f_{ck} \frac{x_{u\,.\,max}}{d}\left(1 - 0.416 \frac{x_{u\,.\,max}}{d}\right) = 0.36 \times 20 \times 0.479\,(1 - 0.416 \times 0.479) = 2.761$

2. *Computation of step dimension* (Fig. 15.2)

$R = 150$ mm; $T = 250$ mm ; $b = \sqrt{R^2 + T^2} = \sqrt{150^2 + 250^2} = 292$ mm

Let us keep waist thickness = 80 mm

$$D = 80 + \frac{RT}{b} = 80 + \frac{150 \times 250}{292} = 208 \text{ mm}$$

Hence the effective depth of equivalent beam $= D/2 = 104$ mm

Width $b = 292$ mm ; Span $L = 1.25$ m

3. *Computation of loading and B.M.* Each step spans horizontally.

Dead load of each step per mete $= \frac{1}{2} \times \frac{150}{1000} \times \frac{250}{1000} \times 25000 \quad \simeq 469$ N/m

Dead load of waist $= \frac{80 \times 292}{10^6} \times 25000 \quad = 584$ N/m

Load of finishing $\quad = 70$ N/m (say)

Total $= 1123$ N/m

Live load @ 3000 N/m² $= (250/1000) \times 3000 \times 1 \quad = 750$ N/m

Total $w = 1873$ N/m

$\therefore \quad w_u = 1.5\, w = 1.5 \times 1873 \simeq 2810$ N/m

$$M_u = \frac{w_u L^2}{8} = \frac{2810\,(1.25)^2}{8} = 548.8 \text{ N-m} = 54.88 \times 10^4 \text{ N-mm}$$

4. *Computation of effective depth*

$$d = \sqrt{\frac{M_u}{R_u\, b}} = \sqrt{\frac{54.88 \times 10^4}{2.761 \times 292}} = 26.1 \text{ mm. But available } d = 104 \text{ mm}$$

5. *Computation of steel reinforcement* : Since available d is more than that required from B.M., we have an under-reinforced section, for which

$$A_{st} = \frac{0.5\, f_{ck}}{f_y}\left[1 - \sqrt{1 - \frac{4.6\, M_u}{f_{ck}\, bd^2}}\right] bd = \frac{0.5 \times 20}{415}\left[1 - \sqrt{1 - \frac{4.6 \times 54.88 \times 10^4}{20 \times 292\,(104)^2}}\right] 292 \times 104$$

$= 14.8 \text{ mm}^2$

However, provide minimum steel in the form of one bar of 8 mm diameter per step giving $A_{st} = 50.3$ mm². Provide distribution reinforcement in the form of 8 mm φ bars @ 450 mm c/c. The reinforcement is arranged as shown in Fig. 15.2.

15.4. DESIGN OF DOG-LEGGED STAIR

Design Example 15.2. *Design a dog-legged stair for a building in which the vertical distance between floors is 3.6 m. The stair hall measures 2.5 m × 5 m. The live load may be taken as 2500 N/m². Use M 20 concrete and Fe 415 steel bars.*

Solution

1. ***General arrangement of stair*** : Fig. 15.6 shows the plan of stair hall. Let the rise be 150 mm and tread be 250 mm. Let us keep width of each flight = 1.2 m.

Height of each flight = 3.6/2 = 1.8 m

No. of risers required = 1.8/0.15 = 12 in each flight.

No. of treads in each flight = 12 − 1 = 11

Space occupied by treads = 11 × 25 = 275 cm.

Keep width of landing equal to 1.25 m. Hence space left for passage = 5.0 − 1.25 − 2.75 = 1 m

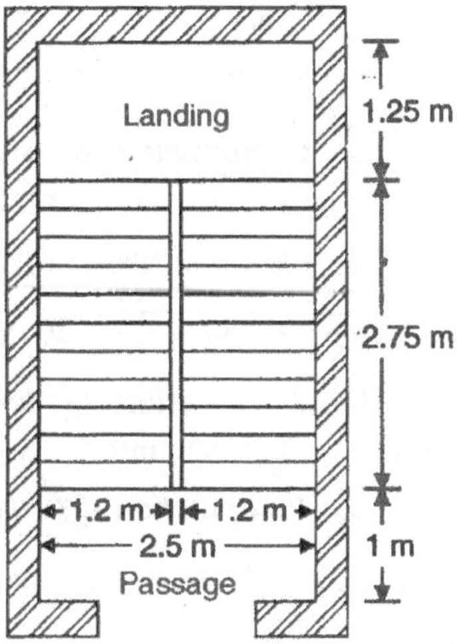

FIG. 15.6.

2. ***Computation of design constants***

For Fe 415 steel, $f_y = 415 \text{ N/mm}^2$. For M 20 concrete, $f_{ck} = 20 \text{ N/mm}^2$.

$$\frac{x_{u,\max}}{d} \text{ (for Fe 415 steel)} = \frac{700}{1100 + 0.87 \times 415} = 0.479$$

$$\therefore \quad R_u = 0.36 f_{ck} \frac{x_{u,\max}}{d}\left(1 - 0.416 \frac{x_{u,\max}}{d}\right) = 0.36 \times 20 \times 0.479\,(1 - 0.416 \times 0.479) = 2.761$$

3. ***Computation of loading and B.M.*** : The landing slab is assumed to span in the same direction as the stairs, and is considered as acting together to form a single slab. Let the bearing of the landing slab in the wall be 160 mm. The effective span $= 2.75 + 1.25 + 0.160/2 = 4.08 \simeq 4.1$ m.

Let the thickness of waist slab be equal to 200 mm (assumed at the rate of 40 to 50 mm per metre span).

$$\therefore \text{ Weight of slab } w' \text{ on slope } = \frac{20}{1000} \times 1 \times 1 \times 25000 = 5000 \text{ N/m}^2$$

$$\text{Dead weight on horizontal area, } w_1 = w' \frac{\sqrt{R^2 + T^2}}{T} = 5000 \times \frac{\sqrt{(150)^2 + (250)^2}}{250} = 5830 \text{ N/m}^2$$

Dead weight of steps is given by

$$w_2 = \frac{R}{2 \times 1000} \times 1 \times 1 \times 25000 = \frac{150}{2000} \times 25000 \qquad = 1875 \text{ N/m}$$

∴ Total dead weight per m run = 5830 + 1875	= 7705 N
Weight of finishing etc.	= 100 N (assumed)
Live load	= 2500 N
Total w	= 10305 N/m

$$\therefore \qquad w_u = 1.5 \times 10305 \simeq 15460 \text{ N/m}.$$

Note : The w on the landing portion will be 10305 − 1875 = 8805 N, since weight of steps will not come on it. However, a uniform value of w has been adopted here.

$$M_u = \frac{w_u L^2}{8} = \frac{15460\,(4.1)^2}{8} = 32485 \quad \text{N-m} \;= 32.485 \times 10^6 \quad \text{N-mm}$$

4. Design of waist slab

$$d = \sqrt{\frac{M_u}{R_u\, b}} = \sqrt{\frac{32.485 \times 10^6}{2.761 \times 1000}} = 108.5 \quad \text{mm}$$

Adopt 150 mm overall depth. Using 20 mm nominal cover and 10 mm φ bars, effective depth = 150 − 20 − 5 = 125 mm.

5. Computation of reinforcement

Since actual d provided is more than that required for B.M., we have an under-reinforced section for which

$$A_{st} = \frac{0.5 f_{ck}}{f_y}\left[1 - \sqrt{1 - \frac{4.6 M_u}{f_{ck}\, bd^2}}\right] bd = \frac{0.5 \times 20}{415}\left[1 - \sqrt{1 - \frac{4.6 \times 32.485 \times 10^6}{20 \times 1000\,(125)^2}}\right] 1000 \times 125$$

= 836.3 mm². Using 10 mm φ bars having A_φ = 78.54 mm², No. of bars required in 1.2 m width $= \dfrac{1.2 \times 836.3}{78.54} = 12.8 \simeq 13$ (say)

∴ Spacing of bars = 1200/13 = 92.3 mm

Distribution reinforcement, $A_{sd} = \dfrac{0.12 \times 150 \times 1000}{100} = 180 \text{ mm}^2$

Hence spacing of 8 mm φ bars $= \dfrac{1000 \times 50.3}{180} = 279$ mm

Hence provide 8 mm φ bars @ 250 mm c/c, i.e. one bar per step.

The main reinforcement should be bent to follow the bottom profile of the stair. However, if this pattern is followed near the landing, an angle will be formed in the bar. When the bar is stressed, it will try to throw off the concrete cover. Hence near the landing, the bars are taken straight up and then bent in the compression zone of landing. In order to take tensile stresses in the landing portion, it is desirable to use separate set of bars as shown in Fig. 15.7. However, since the B.M. is very much reduced near the landing, only half the number of bars may be provided, i.e. provides 7 No. of 10 mm φ bars. The length of each type of bar on either side of crossing should be at least L_d = 47 φ = 47 × 10 = 470 mm. The details of reinforcement etc. are shown in Fig. 15.7.

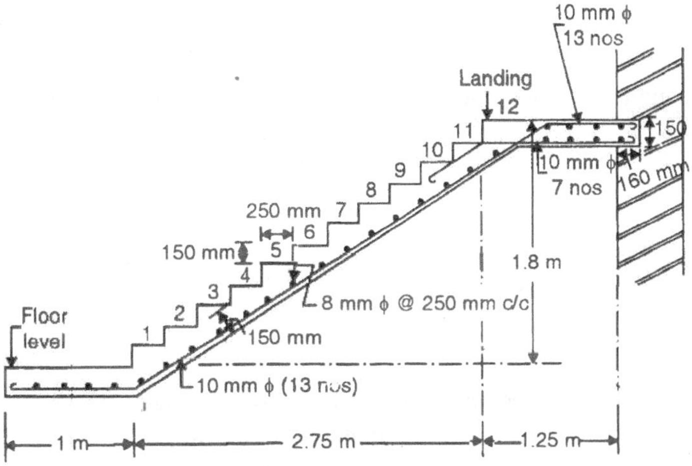

FIG. 15.7

15.5. DESIGN OF STAIR WITH QUARTER SPACE LANDING

Design Example 15.3 : *Fig. 15.8 shows the general arrangement of stair case for an office building. The tread is 300 mm and rise is 150 mm. The stairs is built-in the side wall along the flights for a distance of 120 mm. Design the stair case for a live load of 3000 N/mm², taking the span in the direction of the flight. Use M 20 concrete and Fe 415 steel.*

Solution

*1. **Design constants*** : For Fe 415 - M 20 combination,

$\frac{x_{u\,.\,\max}}{d} = 0.479$ and $R_u = 2.761$ as in example 15.1.

*2. **Effective span**:* Assume 200 mm bearing of the landing in the wall.

Effective span of flight $AB = 3 + 1.4 + 0.1$

$= 4.5$ m

Effective span of flight BC

$= 0.1 + 1.4 + 1.5 + 1.4 + 0.1 = 4.5$ m

Thus, effective span of both the flights is equal. Hence any one flight (say flight BC) may be designed and the same design may be adopted for the other flight.

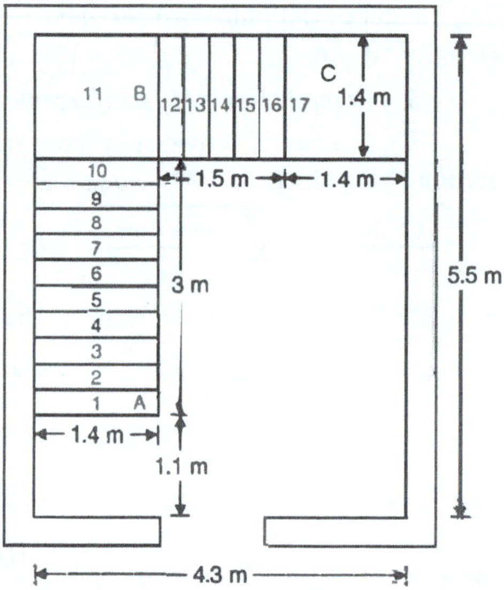

FIG. 15.8.

*3. **Loading on each flight**:*

Let the thickness of the waist slab be 200 mm.

∴ Weight w' on slope $= 0.2 \times 1 \times 1.4 \times 25000$ N/m run

Weight w_1 per horizontal metre run $= 0.2 \times 1 \times 1.4 \times 25000 \frac{\sqrt{(150)^2 + (300)^2}}{300} = 7825$ N/m

Weight of each step $= 1.4 \times \frac{0.15}{2} \times 0.3 \times 25000 = 788$ N

∴ Weight w_2 of steps per horizontal metre run $= 788\,(1000/300) = 2625$ N

Alternatively, $w_2 = \frac{0.15}{2} \times 1 \times 1.4 \times 25000 = 2625$ N

∴ Total dead weight / m run $= 7825 + 2625 = 10450$ N

Weight of finishing etc. $= 150$ N

Total $w = 10600$ N/m

For the computation of live load, consider Fig. 15.9. Since the flight is built into the side wall by a distance 120 mm (> 110 mm), the loading width $= 1.4 - 0.15 = 1.25$ m.

∴ Live load/m $= 3000 \times 1.25 = 3750$ N

Effective breadth $b = 1.4 + 0.075$

$= 1.475$ m $= 1475$ mm

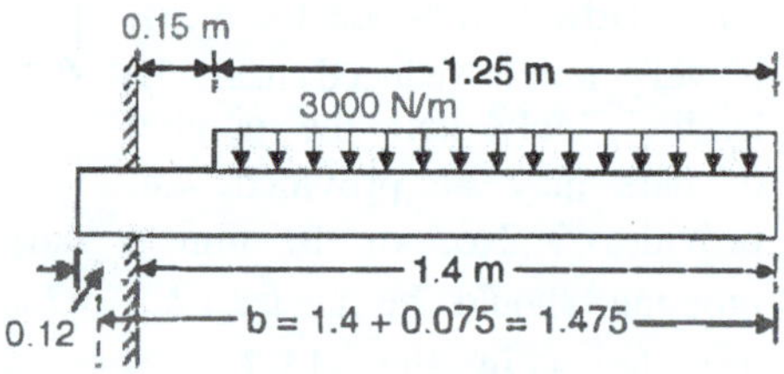

FIG. 15.9.

Hence total w per horizontal metre run $= 10600 + 3750 = 14350$ N/m

4. Design of flight B.C.

Landing is common to both the flights. Hence w for landing B $= \frac{1}{2} \times 14350$ $= 7175$ N/m, while w for landing C will be taken as 14350 N/m. The loading, B.M.D. and S.F.D. are shown in Fig. 15.10.

Reaction R_C

$$= \frac{1}{4.5}\left(\frac{7175 \times 1.5 \times 1.5}{2} + 14350 \times 3 \times 3\right)$$

$= 30494$ N.

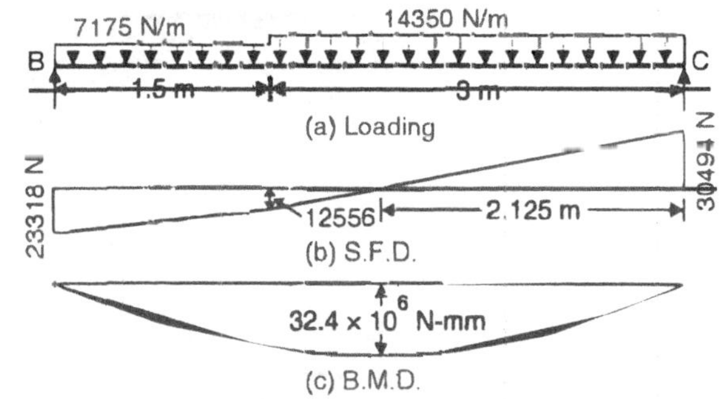

FIG. 15.10.

$R_B = 7175 \times 1.5 + 14350 \times 3 - 30494$
$= 23318$ N

S.F. is zero at a distance $= 30494/14350 = 2.125$ m from C

B.M. will be maximum where S.F. in zero

$$\therefore \quad M_{max} = \left[30494 \times 2.125 - \frac{14350\,(2.125)^2}{2}\right]1000 = 32.4 \times 10^6 \text{ N-mm}$$

$$\therefore \quad M_{u,\,max} = 1.5 \times 32.4 \times 10^6 = 48.6 \times 10^6 \text{ N-mm}$$

The breadth b of slab for design $= 1475$ mm

$$\therefore \quad d = \sqrt{\frac{M_u}{R_u\, b}} = \sqrt{\frac{48.6 \times 10^6}{2.761 \times 1475}} = 109.3 \text{ mm.}$$

However, keep total depth = 150 mm. Using 10 mm φ bars and a nominal cover of 20 mm, $d = 150 - 20 - 5 = 125$ mm. Since d actually provided is more than that required from bending, we have an under-reinforced section for which

$$A_{st} = \frac{0.5\, f_{ck}}{f_y}\left[1 - \sqrt{1 - \frac{4.6\, M_u}{f_{ck}\, bd^2}}\right] bd$$

$$= \frac{0.5 \times 20}{415}\left[1 - \sqrt{1 - \frac{4.6 \times 48.6 \times 10^6}{20 \times 1475\,(125)^2}}\right] 1475 \times 125 = 1254.6 \text{ mm}^2$$

$\therefore$ Number of 10 mm φ bars required in a width of 1475 mm $= 1254.6/78.54 = 15.97$. Hence provide 16 bars of 10 mm φ. Spacing $s = 1475/16 \simeq 92$ mm.

Distribution reinforcement $A_{sd} = 1.2 \times 150 = 180$ mm^2.

Hence spacing of 8 mm φ bars $= \dfrac{1000 \times 50.3}{180} = 279$ mm.

Hence provide 8 mm φ bars @ 250 mm c/c. The same reinforcement may be provided for both the flights. At the landing, provide reinforcement both at top as well as at bottom. The details of reinforcement etc. are shown in Fig. 15.11.

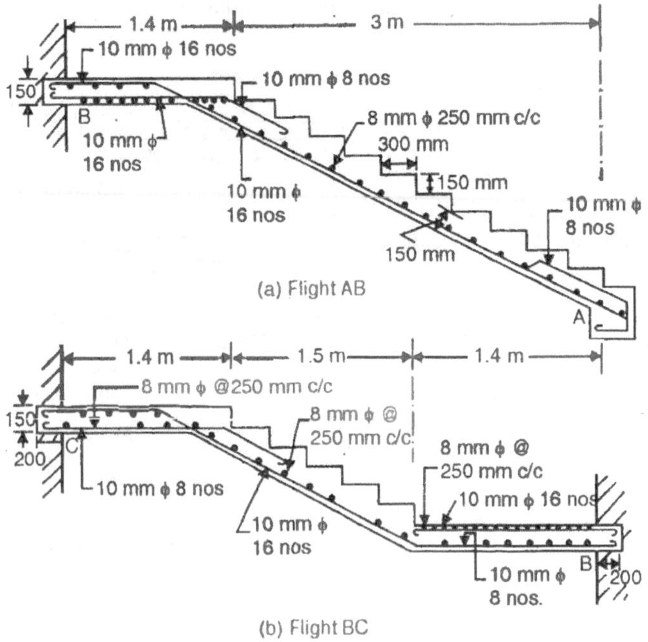

FIG. 15.11

15.6. DESIGN OF OPEN NEWEL STAIR WITH QUARTER SPACE LANDING

Design Example 15.4. *Fig. 15.12 shows the general arrangement of a staircase of a building. The risers are 150 mm and the treads are 250 mm. Design the stair case for a live load of 3000 N/m². The width of stair is 1.5 m and the width of wall is 400 mm. Use M 20 concrete and Fe 415 steel.*

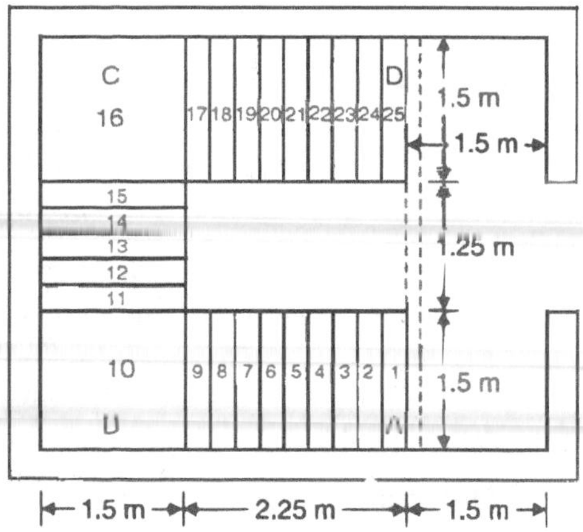

FIG. 15.12.

Solution

1. Design constants

For Fe 415 steel, $f_y = 415\ \text{N/mm}^2$

For M 20 concrete, $f_{ck} = 20\ \text{N/mm}^2$

$$\frac{x_{u,\max}}{d} = \frac{700}{1100 + 0.87 \times 415} = 0.479.$$

$$R_u = 0.36 f_{ck} \frac{x_{u,\max}}{d}\left(1 - 0.416 \frac{x_{u,\max}}{d}\right)$$

$$= 0.36 \times 20 \times 0.479\,(1 - 0.416 \times 0.479)$$

$$= 2.761$$

2. Effective span: Assume a bearing of 200 mm in walls. For flights *AB* and *CD*,

$$L = 0.1 + 2.25 + 1.5 + 0.1 = 3.95 \text{ m}.$$

For flight BC, $L = 0.1 + 1.5 + 1.25 + 1.5 + 0.1 = 4.45$ m.

3. *Computation of loading*

(*a*) *For the flight portion* : L t the waist slab be 200 mm thick. Also, let the weight of slab w' on slope $= 0.2 \times \quad \times 25000 = 5000\ \text{N/m}^2$.

Weight on horizontal area $= 50\ 0 \dfrac{\sqrt{150^2 + 250^2}}{250} \simeq 5830\ \text{N/m}^2$

Dead weight of steps is give by $w_2 = \dfrac{150}{2000} \times 25000 \simeq 1870\ \text{N}$

Weight of finishing = 100 N (ssumed). Live load $= 3000\ \text{N/m}^2$

Hence total load = 5830 + 1870 + 100 + 3000 = 10800 N/m

(*a*) *For the landing portion*

Dead load $= 0.2 \times 1 \times 1 \times 25000$	$= 5000\ \text{N/m}^2$
Weight of finishing	$= 100\ \text{N/m}^2$
Live load	$= 3000\ \text{N/m}^2$
Total	$= 8100\ \text{N/m}^2$

However, since each quarter space landing is common to both the flights, only half of the above loading, *i.e.* 4050 N/m will be taken.

4. *Design of flight AB* (Fig. 15.13)

$$R_A = \frac{1}{3.95}[4050 \times 1.6 \times 0.8 + 10800 \times 2.35 \times 2.775] \simeq 19142\ \text{N}$$

$$R_B = \frac{1}{3.95}\left[\frac{10800\,(2.35)^2}{2} + 4050 \times 1.6 \times 3.15\right] = 12718\ \text{N}$$

S.F. is zero at 19142/10800 = 1.772 m from A.

$$M_{\max} = \left[(19142 \times 1.722) - \frac{10800\,(1.772)^2}{2}\right] = 16.006 \times 10^6\ \text{N-mm.}$$

$$M_{u\,.\,\max} = 1.5 \times 16.006 \times 10^6 = 24.01 \times 10^6\ \text{N-mm}$$

$$\text{Depth} \quad d = \sqrt{\frac{M_{u\,.\,\max}}{R_c\,.\,b}} = \sqrt{\frac{24.01 \times 10^6}{2.761 \times 1000}} = 93.4\ \text{mm}$$

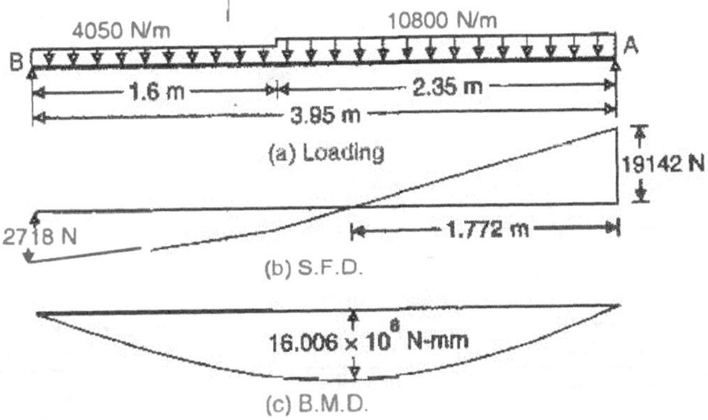

FIG. 15.13

Provide overall depth of 150 mm so that available $d = 150 - 20 - 5 = 125$ mm. with 20 mm nominal cover and 10 mm φ main bars. Since d actually provided is more than that required for bending, we have an under-reinforced section for which

$$A_{st} = \frac{0.5\, f_{ck}}{f_y}\left[1 - \sqrt{1 - \frac{4.6\, M_u}{f_{ck}\, bd^2}}\right] bd = \frac{0.5 \times 20}{415}\left[1 - \sqrt{1 - \frac{4.6 \times 24.01 \times 10^6}{20 \times 1000\,(125)^2}}\right] 1000 \times 125$$

$= 590 \text{ mm}^2$

$$\text{Spacing of 10 mm φ bars} = \frac{1000 \times 78.54}{590} = 133 \text{ mm}$$

∴ Number of bars in 1.5 m with $= 1500/133 \simeq 12$

Actual $s = 1500/12 = 125$ mm

$A_{sd} = 1.2\, D = 1.2 \times 150 = 180 \text{ mm}^2$

$$\therefore \text{ Spacing of 8 mm φ bars} = \frac{1000 \times 50.3}{180} = 279 \text{ mm.}$$

However provide these @ 250 mm c/c. $\tau_v = \dfrac{V_u}{bd} = \dfrac{1.5 \times 19142}{1000 \times 125} = 0.23 \text{ N/mm}^2$ which is less than the permissible shear stress of 0.28 N/mm² even at minimum steel.

5. *Design of flight BC (Fig. 15.14)*

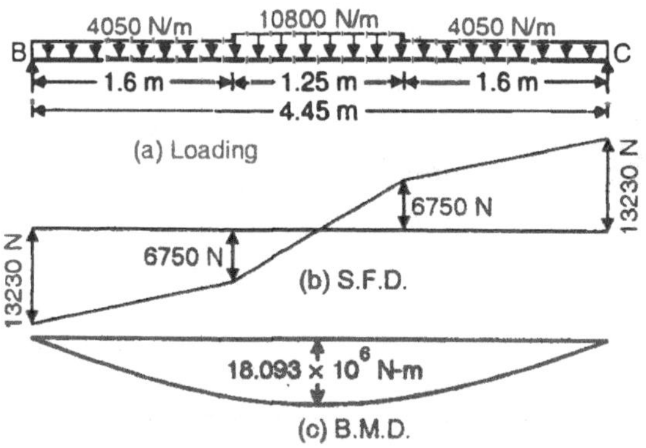

FIG. 15.14.

$$R_B = R_C = \frac{1}{2}\,[4050 \times 1.6 + 10800 \times 1.25 + 4050 \times 1.6] = 13230 \text{N}$$

$$M_{max} = \left[(13230 \times 2.225) - (4050 \times 1.6 \times 1.425) - \frac{10800\,(0.625)^2}{2}\right] \times 1000 = 18.093 \times 10^6 \text{ N-mm}$$

$\therefore \quad M_{u,max} = 1.5 \times 18.093 \times 10^6 \simeq 27.14 \times 10^6$ N-mm

$$d = \sqrt{\frac{M_{u.max}}{R_u \,.\, b}} = \sqrt{\frac{27.14 \times 10^6}{2.761 \times 1000}} = 99.2 \text{ mm}$$

Keep $D = 150$ mm so that $d = 125$ mm

$$A_{st} = \frac{0.5\, f_{ck}}{f_y}\left[1 - \sqrt{1 - \frac{4.6\, M_{u.max}}{f_{ck}\, bd^2}}\right] bd = \frac{0.5 \times 20}{415}\left[1 - \sqrt{1 - \frac{4.6 \times 27.14 \times 10^6}{20 \times 1000\,(125)^2}}\right] 1000 \times 125$$

$= 678 \text{ mm}^2$

Spacing of 10 mm φ bars is

$$s = \frac{1000 \times 78.54}{678}$$

$$= 115.8 \text{ mm}$$

Keep $s = 100$ mm

∴ No of bars

$$= 1500/100 = 15$$

Distribution reinforcement:

Provide 8 mm φ bars @ 250 mm c/c.

The details of reinforcement for both the flights are shown in Fig. 15.15.

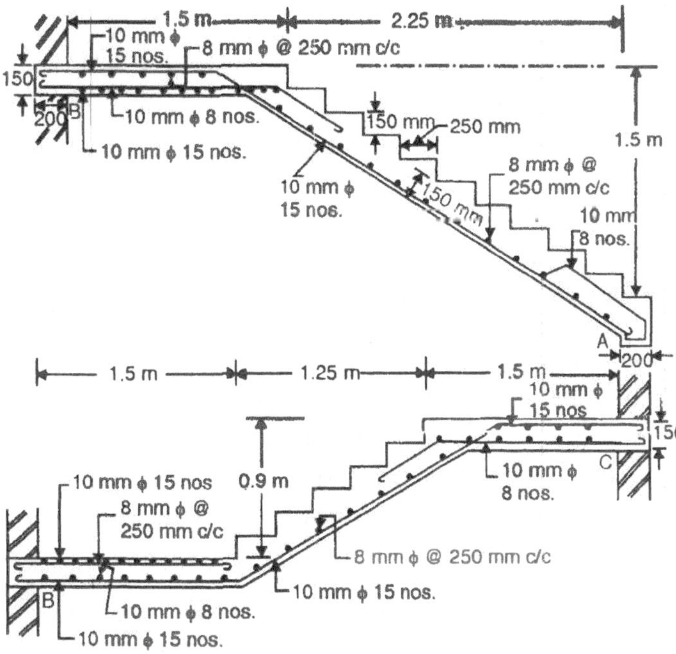

FIG. 15.15.

15.7. DESIGN OF STAIRCASE WITH CENTRAL STRINGER BEAM

Design Example 15.5 : *Fig. 15.16 shows the general layout of a staircase with central stringer beam supported on columns at B, C, and D. The rise and tread of the stairs are 100 mm and 250 mm respectively. The width of the steps is 1.40 m. Design the staircase for a live load of 4000 N/m². Use M 20 concrete and Fe 415 steel.*

Solution

1. Design constants :

For Fe 415 - M 20 combination, $\frac{x_{u,\max}}{d} = 0.479$ and $R_u = 2.761$ as in example 15.1.

2. Loading on waist slab

Let the waist slab be 80 mm thick. The weight of the waist slab on slope should be multiplied by the factor $\sqrt{R^2 + T^2}/T = \sqrt{160^2 + 250^2}/250 = 1.185$ to get the equivalent weight on horizontal plane. Consider 1 m width of slab load per metre horizontal run will be as follows:

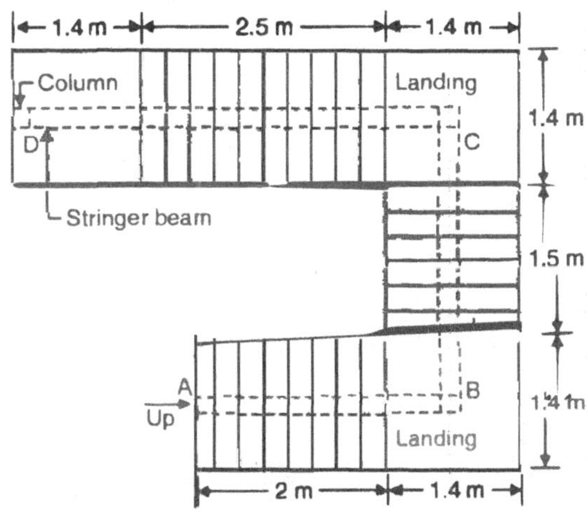

FIG. 15.16.

(*a*) Self weight $= (0.8 \times 1 \times 1 \times 25000)\ 1.185 = 2370$ N

(*b*) Weight of steps $= \frac{1}{2} \times 0.16 \times 1 \times 25000 = 2000$ N

(*c*) Finishing etc. $= 100$ N

(*d*) Live load $= 4000$ N

Total $w = 8470$ N

The loading on the landing will be lesser; however, for simplicity, we will take the same loading throughout.

3. ***Design of waist slab*** : The waist slab is supported on the central stringer beam of 200 mm width. Hence the worst condition may be when we consider concentrated live load of 1300 N to act to one side only.

$$\text{Dead weight} = 1.5\,(2370 + 2000 + 100) = 6705 \text{ N}$$

$$\text{Uniformly distributed live load} = 1.5 \times 4000 = 6000 \text{ N}$$

$$\text{Concentrated live load} = 1.5 \times 1300 = 1950 \text{ N}$$

$$\text{B.M. due to dead load} = M_{ud} = \frac{1}{2} \times 6705\,(0.6)^2 \simeq 1207 \text{ N-m}$$

$$\text{B.M. due to U.D.L.} = M_{ul1} = \frac{1}{2} \times 6000\,(0.6)^2 = 1080 \text{ N-m}$$

$$\text{B.M. due to concentrated L.L.} = M_{ul2} = 1950 \times 0.6 = 1170 \text{ N-m}$$

$$\therefore \quad M_{u\,.\,max} = M_{ud} + M_{ul2} = 1207 + 1170 = 2377 \text{ N-m} = 2.377 \times 10^6 \text{ N-mm}$$

Hence
$$d = \sqrt{\frac{M_{u\,.\,max}}{R_u\, b}} = \sqrt{\frac{2.377 \times 10^6}{2.761 \times 1000}} = 29.4 \text{ mm}$$

However, keep a minimum total depth $= 80$ mm. Keeping 20 mm nominal cover and using 10 mm φ bars, $d = 80 - 20 - 5 = 55$ mm. Since d actually provided is more than that required from bending, we have an under-reinforced section, for which

$$A_{st} = \frac{0.5 f_{ck}}{f_y}\left[1 - \sqrt{1 - \frac{4.6\, M_u}{f_{ck}\, bd^2}}\right] bd = \frac{0.5 \times 20}{415}\left[1 - \sqrt{1 - \frac{4.6 \times 2.377 \times 10^6}{20 \times 1000\,(55)^2}}\right] 1000 \times 55$$

$$= 125.7 \text{ mm}^2$$

$\therefore$ Spacing of 10 mm φ bars $= 1000 \times 78.5/125.7 = 624$ mm

However, keep spacing $= 250$ mm, i.e. one bar per step.

Distribution reinforcement, $A_{sd} = 1.2 \times 80 = 96 \text{ mm}^2$

$\therefore$ Spacing of 8 mm φ bars $= 1000 \times 50.3/96 \simeq 524$ mm, subject to a maximum of $3d$ $(= 3 \times 55 = 275$ mm) or 450 mm. Hence provide 8 φ @ 275 mm c/c.

5. ***Design of stringer beam*** : The stringer beam will act as a T-beam. Flight CD is the longest, and hence we will design the stringer beam CD.

$$\text{Effective span} = (1.4 - 0.2/2) + 2.5 + 1.4/2 = 4.5 \text{ m}$$

Now
$$b_f = \frac{l_0}{(l_0/b) + 4} + b_w \quad \text{where } l_0 = L = 4.5 \text{ m}$$

$$= \frac{4.5}{(4.5/1.5) + 4} + 0.2 = 0.824 \text{ m} = 824 \text{ mm}$$

Computation of loading, B.M., S.F. and torsional moment

The loading on the stringer beam will be as follows, assuming the web to be 200 mm wide and 200 mm deep.

(*a*) Weight of rib/m run $= (0.2 \times 0.2 \times 25000)\ 1.185 \qquad = 1185$ N

(*b*) Load from waist slab $= 8470 \times 1 \times 1.4 \qquad = 11858$ N

Total $w = 13043$ N

$$\therefore \quad w_u = 1.5 \times 13043 \simeq 19565 \text{ N}$$

Assuming partial fixidity at the ends,

$$M_{uD} = \frac{w_u L^2}{10} = \frac{19565\,(4.5)^2}{10} \times 1000 = 39.618 \times 10^6 \text{ N-mm}$$

$$V_u = \frac{w_u L}{2} = \frac{19565\,(4.5)}{2} = 44021 \text{ N}$$

The beam will also be subjected to torsional moment induced due to live load acting only to one side of step.

$$\therefore \quad T = \frac{4000\,(0.6)^2}{2} \times \frac{4.5}{2} \times 1000 = 1.62 \times 10^6 \text{ N-mm}$$

$$\therefore \quad T_u = 1.5\,T = 1.5 \times 1.62 \times 10^6 = 2.43 \times 10^6 \text{ N-mm}$$

$$\therefore \quad V_{eu} = V_u + 1.6\,\frac{T_u}{b_w} = 44021 + 1.6 \times \frac{2.43 \times 10^6}{200} = 63461 \text{ N}$$

Computation of effective depth on the basis of bending alone

$$d = \frac{2}{3}\sqrt{\frac{M_{uD}}{R_u\, b}} = \frac{2}{3}\sqrt{\frac{39.618 \times 10^6}{2.761 \times 200}} \simeq 179 \text{ mm.}$$

Computation of effective depth on the basis of shear stress

$\tau_{ve} = \dfrac{V_{eu}}{b_w\, d}$. Limiting this to $\tau_{c,\,max} = 2.8$ N/mm^2 for M 20 concrete, we get

$$d = \frac{V_{eu}}{b_w \,.\, \tau_{c,\,max}} = \frac{63461}{200 \times 2.8} \simeq 114 \text{ mm}$$

Computation of d on the basis of stiffness

Also, from stiffness point of view, $L/d = 23$ for partially fixed beam.

$$p_{lim} = 0.955\% \simeq 1\%, \text{ for which modification factor} = 1.$$

Also, $\quad b_w/b_f = 200/824 = 0.24$, for which reduction factor $= 0.8$ (Fig. 11.3)

Hence $\quad d = \dfrac{L}{23 \times 1 \times 0.8} = \dfrac{4500}{18.4} = 245$ mm.

Final Fixation of D and d

However let $D = 300$ mm. Providing 25 mm nominal cover and using 16 mm φ main bars and 8 m φ rings, available $d = 300 - 25 - 16/2 - 8 = 259$ mm. Thus, for the T-beam, we have $b_w = 200$ mm, $b_f = 824$ mm, $D = 300$ mm and $d = 259$ mm.

Determination of longitudinal reinforcement

$$M_{e1} = M_u + M_T, \quad \text{where} \quad M_u = 39.618 \times 10^6$$

and
$$M_T = T_u\left(\frac{1 + D/b_w}{1.7}\right) = 2.43 \times 10^6\left(\frac{1 + 300/200}{1.7}\right) = 3.574 \times 10^6 \text{ N-mm}$$

$\therefore$
$$M_{e1} = (39.618 + 3.574)\,10^6 = 43.192 \times 10^6 \text{ N-mm}$$

$$M_{u\,.\,\text{lim}} = R_u\, b_w\, d^2 + 0.446\, f_{ck}\,(b_f - b_w)\, y_f\,(d - 0.5\, y_f)$$

where
$$y_f = 0.15\, x_{u\,.\,\text{max}} + 0.65\, D_f = 0.15\,(0.479 \times 259) + 0.65 \times 80 \quad \text{but} \not> D_f$$
$$= 70.6 \text{ mm}$$

$\therefore$
$$M_{u\,.\,lim.} = 2.761 \times 200\,(259)^2 + 0.446 \times 20\,(824 - 200)\,70.6\,(259 - 0.5 \times 70.6)$$
$$= 37.042 \times 10^6 + 87.906 \times 10^6 = 124.948 \times 10^6 \text{ N-mm}$$

Since $M_{e1} < M_{u\,.\,\text{lim}}$, the section is under reinforced, for which

$$A_{st1} = \frac{0.5\, f_{ck}}{f_y}\left[1 - \sqrt{1 - \frac{4.6\, M_{e1}}{f_{ck}\, b_f\, d^2}}\right] b_f\, d = \frac{0.5 \times 20}{415}\left[1 - \sqrt{1 - \frac{4.6 \times 43.192 \times 10^6}{20 \times 824\,(259)^2}}\right] 824 \times 259$$
$$= 485 \text{ mm}^2$$

Number of 16 mm φ bars = 485/201.06 = 2.41 ≏ 3.

Near the column D, take these bars straight up. Provide 2-16 mm φ bars at the lower face under the landing as shown in Fig. 15.17.

Determination of tranverse reinforcement : Transverse reinforcement will be provided in the form of vertical stirrups. Let us provide 25 mm clear cover all round.

b_1 = Centre to centre distance between corner bars in the direction of width
$= 200 - 2 \times 25 - 8 = 142$ mm

d_1 = Centre to centre distance between corner bars in the direction of depth
$= 300 - 2 \times 25 - 8 = 242$ mm

Using 8 mm φ 2-lgd stirrups, $A_{sv} = 2\,\frac{\pi}{4}\,(8)^2 = 100.53$

Now
$$A_{sv} = \frac{T_u\, s_v}{b_1\, d_1\,(0.87\, f_y)} + \frac{V_u\,.\, s_v}{2.5\, d_1\,(0.87\, f_y)}$$

$\therefore$
$$100.53 = \left[\frac{2.43 \times 10^6}{142 \times 242\,(0.87 \times 415)} + \frac{44021}{2.5 \times 242\,(0.87 \times 415)}\right] s_v$$

From which $\quad s_v$ ≏ 253 mm

However, the spacing should not exceed the least of x_1, $\frac{x_1 + y_1}{4}$ and 300 mm, where

$$x_1 = \text{short dimension of stirrups} = 142 + 16 + 8 = 166 \text{ mm}$$
$$y_1 = 242 + 16 + 8 = 266 \text{ mm}$$

$$\frac{x_1 + y_1}{4} = \frac{166 + 266}{4} = 108 \text{ mm}$$

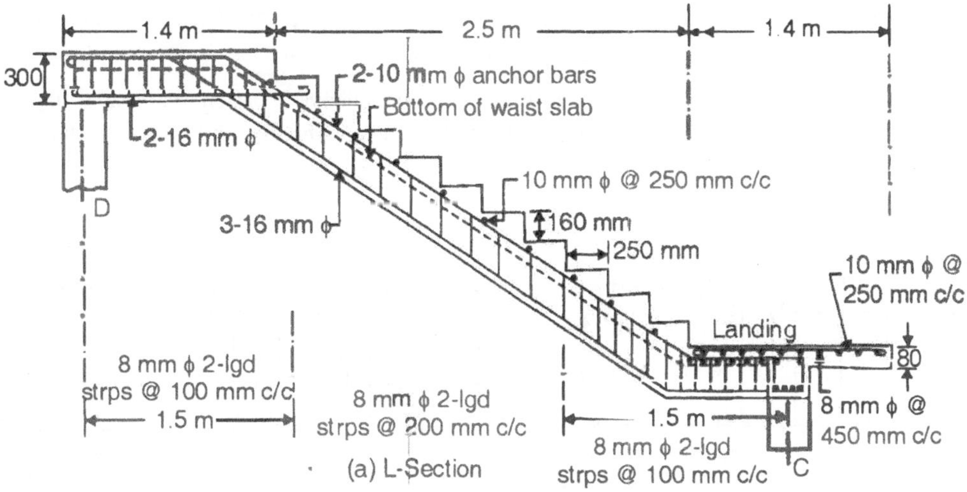

(a) L-Section

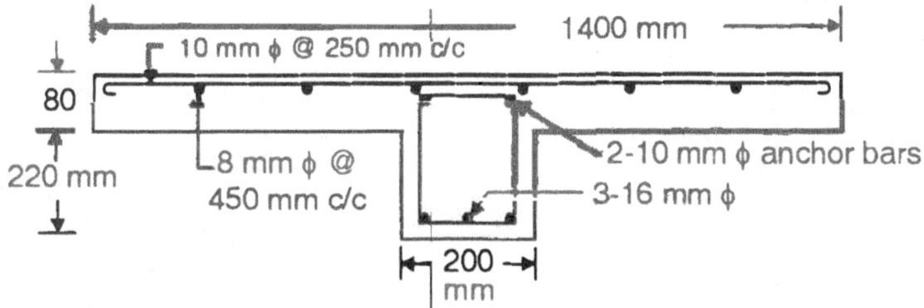

(b) Cross-section at mid-span

FIG. 15.17

Hence s_v = 253 mm is not permissible. Keep s_v = 100 mm. Increase the spacing to 200 mm c/c in the mid-span where both transverse shear as well as torsional shear are minimum. Provide 2-10 mm φ holding bars. Keep the same section for other flights. The details of reinforcement etc. are shown in Fig. 15.17.

PROBLEMS

1. Design the stairs for a public building, supported on wall on one side and stringer beam on the other side. The horizontal span of stairs if 1.4 m. The risers are 120 mm and tread are 300 mm. Use M 20 mix. and Fe 415 steel.

2. Design a suitable dog-legged stair in a public building, to be located in a staircase 6 metre long, 3.2 m wide and 3.7 m high, with a door of 1.1 m wide in each of the longitudinal walls. The doors face each other and are located with their centres at a distance of 0.9 metres from the respective corners of the staircase. Use M 20 mix. and Fe 415 steel.

3. A two storeyed building is to have a R.C. staircase from ground floor to first floor roof. The size of the staircase is 4.3 m × 4.3 m and there is one door opening in one wall and a window opening on the opposite wall. Design the staircase, giving the details of formation, R.C. slab arrangement of building, risers and treads with their top finishing with suitable sketch. The width of stair is 1.2 m and height of each storey is 3.4 m. Use M 20 mix and Fe 415 steel.

4. Fig 15.18 shows the arrangement of a staircase for an office building. The rise of steps is 150 mm. Design completely the stair alongwith the beams, landing etc. The live load may be taken as 4000 N/m². Use M 20 mix. and Fe 415 steel.

5. Fig. 15.19 shows the plan of stair hall for a public building. Design staircase with the central stringer beam. The tread is 300 mm and rise 150 mm. The live load may be taken as 4000 N/m². Use M 20 mix. and Fe 415 steel.

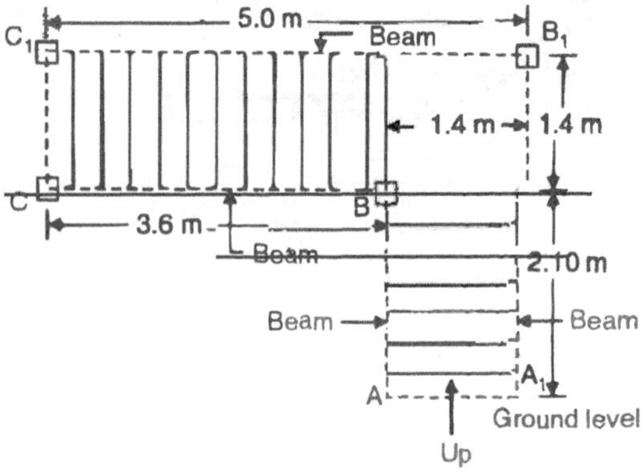

FIG. 15.18.

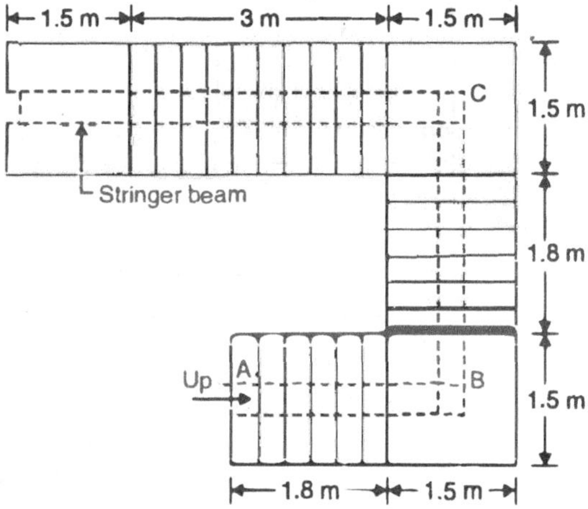

FIG. 15.19.

16

Limit State of Collapse in Compression : I Axially Loaded Columns

16.1. INTRODUCTION

A column is an important component of R.C. structures. A column, in general, may be defined as a member carrying direct axial load which causes compressive stresses of such magnitude that these stresses largely control its design. A column or strut is a compression member, the effective length of which exceeds three times the least lateral dimension. When a member carrying mainly axial load is vertical, it is termed as 'column' while if it is inclined or horizontal, it is termed as a 'strut'. Depending upon structural or architectural requirements, columns may be of various shapes, *i.e.* circular, rectangular, square, hexagonal, etc.

Concrete is strong in compression. However, longitudinal steel rods are always provided to assist in carrying the direct loads. A minimum area of longitudinal steel is provided in the column, whether it is required from load point of view or not. This is done to resist tensile stresses caused by some eccentricity of the vertical loads. There is also an upper limit of amount of reinforcement in R.C. columns, because higher percentage of steel may cause difficulties in placing and compacting the concrete. Longitudinal reinforcing bars are 'tied' laterally by 'ties' or 'stirrups' at suitable interval so that the bars do not buckle.

Functions of longitudinal and transverse reinforcements in a column

Longitudinal and transverse reinforcements are provided in a column to serve the following functions :

(a) Longitudinal reinforcement

1. To share the vertical compressive load, thereby reducing the overall size of the column.
2. To resist tensile stresses caused in the column due to (*i*) eccentric load (*ii*) moment, or (*iii*) transverse load
3. To prevent sudden brittle failure of the column
4. To impart certain ductility to the column
5. To reduce the effects of creep and shrinkage due to sustained loading.

(b) Transverse reinforcement

1. To prevent longitudinal buckling of longitudinal reinforcement
2. To resist diagonal tension caused due to transverse shear due to moment / transverse load.
3. To hold the longitudinal reinforcement in position at the time of concreting.
4. To confine the concrete, thereby preventing its longitudinal splitting.
5. To impart ductility to the column.
6. To prevent sudden brittle failure of the columns.

16.2. TYPES OF COLUMNS

Columns can mainly be divided into the following three types :

1. Column with longitudinal steel and lateral ties
2. Column with longitudinal steel and spirals.
3. Composite column.

Fig. 16.1(*a*) shows a concrete column reinforced with longitudinal bars, but with no laterals or ties. When load is applied on such a column, the concrete bulges out laterally, as shown. The bars themselves act as long slender columns and therefore tend to buckle away from the column's axis. Due to this, tension is caused in the outside shell of the concrete which opens out. The failure usually takes place suddenly. In order to check this tendency, the longitudinal reinforcement is tied transversely, at suitable intervals, with the help of ties, as shown in Fig 16.1(*b*) and (*c*). These ties check the bars from buckling and also restrain the concrete from bulging action. When the number of bars are more than four, it is preferable to use two kinds of ties [(Fig.16.1(*b*)], placing them alternatively, so that one (set *a*) holds the corner rods while the other (set *b*) holds the intermediate rods.

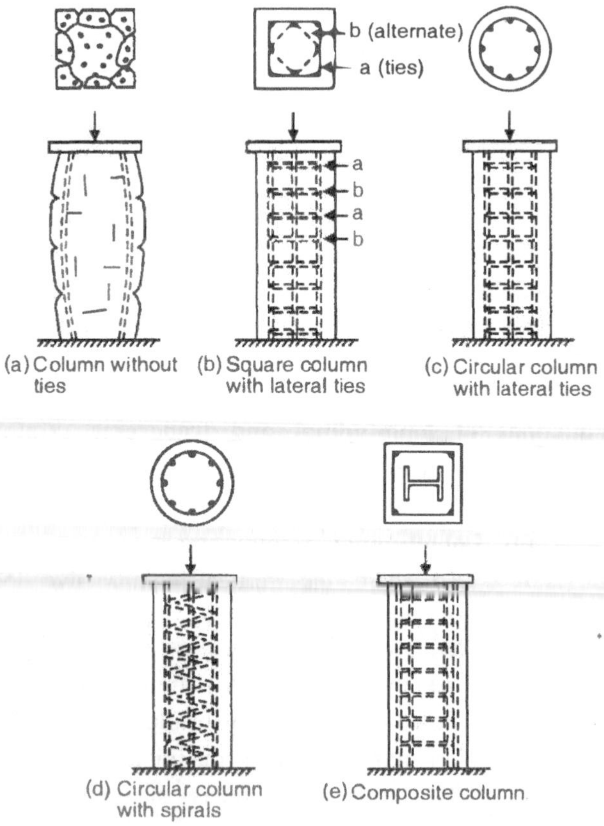

(a) Column without ties (b) Square column with lateral ties (c) Circular column with lateral ties

(d) Circular column with spirals (e) Composite column

FIG. 16.1.

Each tie has to be spliced by lapping or by bending its ends around the main rod, which is quite troublesome. In order to overcome this difficulty, the longitudinal bars are tied continuously together with the help of 'spirals' shown in Fig. 16.1(*d*). The spirals so provided serve an additional purpose of laterally supporting the concrete inside and thus has confining effect on it.

Fig. 16.1(*e*) shows a typical composite column, reinforced with a centrally placed joist and with four or more longitudinal bars. Other steel sections may also be used. However, composite columns are used only for heavy loads.

Braced and unbraced columns

In most of the cases, columns are also subjected to horizontal loads like wind, earthquake etc. If lateral supports are provided at the ends of the column, the lateral loads are borne entirely by the lateral supports. Such columns are known as *braced columns*. Other columns, where the lateral loads have to be resisted by them, in addition to axial loads and end moments, are considered as *unbraced columns*. Bracings can be in one direction or in more than one direction, depending on the likely hood of the direction of external lateral loads. Thus, a column can be braced either in x-direction, or in y-direction or in both x and y directions. A braced column does not have relative lateral movements of its two ends. Thus, a braced column is not subject to *side sway*. A pin-jointed column is a simple example of a braced column. Similarly, an unbraced column is subject to *side sway* or lateral drift, *i.e.* there is significant lateral displacement between top and bottom ends of the column.

16.3. ASSUMPTIONS IN LIMIT STATE OF COLLAPSE IN AXIAL COMPRESSION

The limit state method of design of a compression member subjected to axial compression is based on the following assumptions :

1. The plane section normal to the axis of column before deformation remain plane after deformation. This means that the strain at any point is proportional to its distance from the N.A.

2. The relationship between compressive stress distribution in concrete and strain in concrete is represented by stress-strain curve as given in Fig. 2.3. For design purposes, the compressive strength of concrete in structure is assumed to be 0.67 times the characteristic strength. The partial safety factor $\gamma_{mc} = 1.5$ is applied in addition to this. Thus, the design strength of concrete is taken as $0.67\ f_{ck}/1.5 = 0.446\ f_{ck}$.

3. The stress in reinforcement are derived from representative stress-strain curve for the type of steel used. Typical design curves are shown in Figs. 2.4, 2.5 and 2.6. For design purposes, partial safety factor $\gamma_{ms} = 1.15$ shall be applied.

4. The maximum compressive strain in concrete in axial compression in taken as 0.002, and is uniform along the section (Fig. 16.2 *b*). Hence Maximum compressive stress in concrete, assumed to be uniform across the section is taken equal to $0.446\ f_{ck}$ according to assumption 2 (Fig. 16.2*c*)

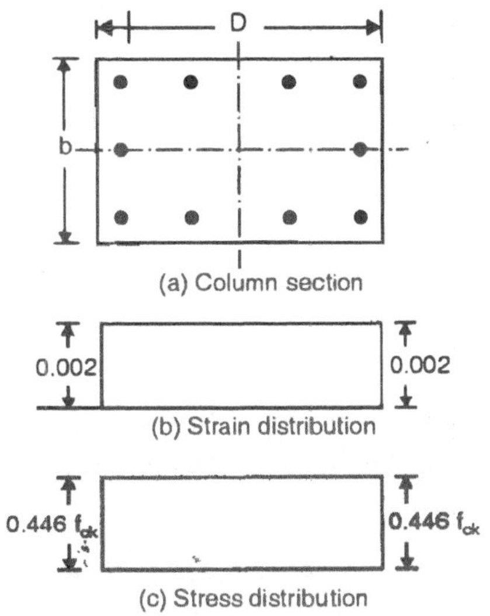

FIG. 16.2. STRAIN AND STRESS DISTRIBUTION ACROSS THE SECTION OF COLUMN SUBJECTED TO AXIAL COMPRESSION

16.4. SHORT COLUMN

A compression member may be considered as short when both the slenderness ratios l_{ex}/D and l_{ey}/D are less than 12, where

l_{ex} = effective length in bending with respect to major axis (i.e. x-axis)

l_{ex} = effective length in bending with respect to minor axis (i.e. y-axis)

D = depth of the section in respect of major axis.

b = width of the section in respect of minor axis.

16.5. SHORT AXIALLY LOADED MEMBER IN AXIAL COMPRESSION

Experiments on columns show that load carrying capacity (P_u) of an axially loaded R.C. member at collapse is made up of ultimate strength of concrete member (P_{uc}) at collapse plus the ultimate strength of steel (P_{us}) in compression. Thus

$$P_u = P_{uc} + P_{us} = \alpha_c f_{ck} A_c + \alpha_s f_y A_s \qquad ...(1)$$

where $\alpha_c f_{ck} = f_c =$ stress in concrete at failure, at uniform strain of 0.002

$\alpha_s f_y = f_s =$ stress in steel at failure at strain of 0.002

A_c = area of concrete ; A_s = area of steel reinforcement

When a short column is axially loaded, the strain distribution across the section will be rectangular. At failure, the strain in concrete will be uniform at a value of 0.002, as stated in assumption No. 4 of § 16.3. The stress in concrete will be $= 0.67 f_{ck}/\gamma_{mc}$ $= 0.67 f_{ck}/1.5 \simeq 0.446 f_{ck}$ (assumption 2). Thus α_c is taken as equal to 0.446.

When the concrete attains a limiting strain of 0.002, the stress in mild steel reinforcement may develop full design stress ($f_{yd} = 0.87 f_y$). However with cold twisted reinforcements or HYSD bars ($f_y = 415$ N/mm^2 or $f_y = 500$ N/mm^2), full design stress will not develop at a strain of 0.002. In general therefore, stress f_s in steel reinforcement at strain of 0.002, can be taken equal to $\alpha_s f_y$, where the value of α_s will depend upon the type of reinforcement, as given below :

Type of reinforcement	*Value of* α_s	*Stress in steel* (f_s)
(*i*) Mild steel	0.87	$0.87 f_y$
(*ii*) Fe 415	0.79	$0.79 f_y$
(*iii*) Fe 500	0.75	$0.75 f_y$

Hence the load carrying capacity of a member, subjected to axial load only, is given by

$$P_u = 0.446 f_{ck} A_c + \alpha_s f_y A_s \qquad ...(2)$$

For *simplicity*, I.S. Code (IS : 456 - 2000) adopts only the lowest value of α_s (= 0.75), which is for steel of Fe 500 grade. Also the Code has redesignated P_u as P_{uz} in section 39.6 of the Code, and has given the following expression in *design aids* (SP : 16 - 1980).

$$P_{uz} = 0.446 f_{ck} A_c + 0.75 f_y A_s = 0.446 f_{ck} A_g + (0.75 f_y - 0.446 f_{ck}) A_s \qquad ...(16.1)$$

where A_g = gross area of concrete

A_c = net area of concrete $= A_g - A_s$

16.6. SHORT AXIALLY LOADED COLUMN WITH MINIMUM ECCENTRICITY

The *ideal* condition of axial loading, indicated in the previous article hardly ever exists. There is always certain inherent minimum eccentricity in the columns. According to IS : 456-2000, all compression members are to be designed for a *minimum eccentricity* of the load in two principal directions. Clause 25.4 of the Code specifies the following minimum eccentricity e_{min} for the design of columns :

$$e_{min} = \frac{l}{500} + \frac{D}{30}, \text{ subject to a minimum of 20 mm}$$

where l = unsupported length of the column in the direction under consideration

D = lateral dimension of the column in the direction under consideration

If x-axis is the major axis and y-axis is the minor axis of bending, we have

$$e_{x,\,min} = \frac{l_x}{500} + \frac{D}{30} \text{ and } e_{y,\,min} = \frac{l_y}{500} + \frac{b}{30}, \text{ each not less than 20 mm}$$

where $e_{x,\,min}$ and. $e_{y,\,min}$ are minimum eccentricities for bending about x and y axes respectively and l_x and l_y = unsupported length of the column for bending in the two directions respectively.

If the value of minimum eccentricity is less than or equal to 0.05 D, clause 39.3 of the Code permits the design of short axially loaded compression member by the following equation :

$$P_u = 0.4\, f_{ck}\, A_c + 0.67\, f_y\, A_s = 0.4\, f_{ck}\, A_g + (0.67\, f_y - 0.4\, f_{ck})\, A_s \quad \text{...(16.2)}$$

It should be noted that Eq. 16.2 is obtained by reducing the capacity of column, given by Eq. 16.1, by approximately 10% thereby allowing for the minimum eccentricity of 0.05 D.

If, however, the minimum eccentricity is greater than 0.05 D, the section is designed for combined axial load and bendings, as discussed in Chapter 17.

Note : The minimum eccentricity to be adopted in the design, as specified by the Code, is 20 mm.

$$\therefore \quad 0.05\, D \geq 20 \text{ mm which gives } D \geq 20/0.05 \geq 400 \text{ mm} \quad \text{...(16.3 a)}$$

Also,

$$e_{min} = \frac{l}{500} + \frac{D}{30} \not> 0.05\, D \not> \frac{D}{20}$$

or

$$\frac{l}{500} \not> \left(\frac{D}{20} - \frac{D}{30}\right) \not> \frac{D}{60}$$

Hence

$$D \not< 0.12\, l \quad \text{...(16.3 b)}$$

Hence in all columns, which have D equal to or less than 400 mm or 0.12 l, the minimum eccentricity will be greater than 0.05 D, and hence Eq. 16.2 will not be applicable. Such sections are therefore, to be designed for the combined axial load and bending, as discussed in Chapters 17 and 18.

16.7. DESIGN CHARTS

Eq.16.2 can be rearranged as follows :

$$P_u = 0.4\, f_{ck}\left(A_g - \frac{p A_g}{100}\right) + 0.67\, f_y \frac{p A_g}{100}$$

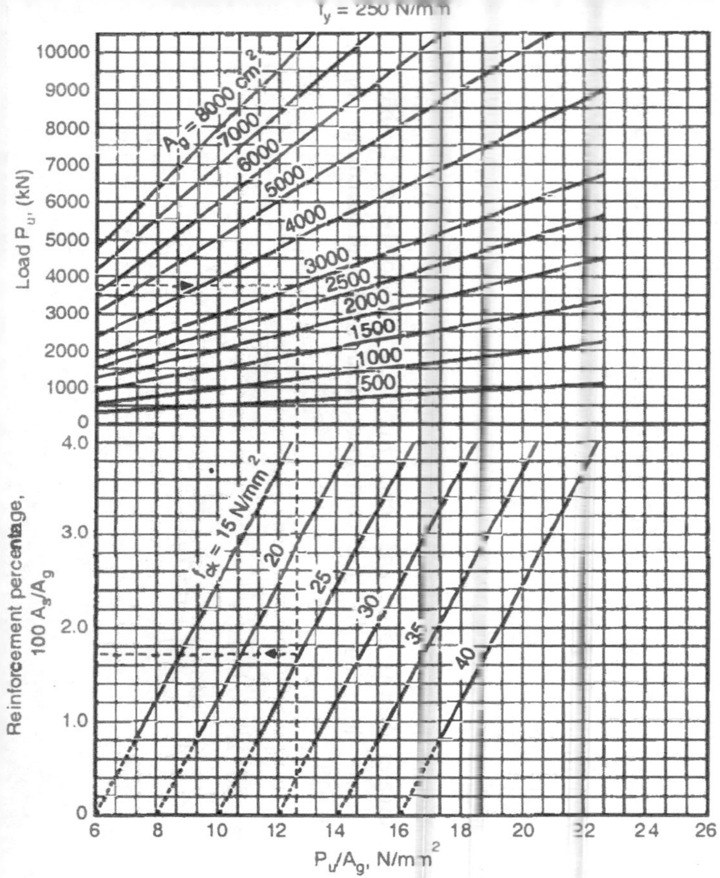

FIG. 16.3. DESIGN CHARTS FOR AXIAL COMPRESSION (MILD STEEL BARS)

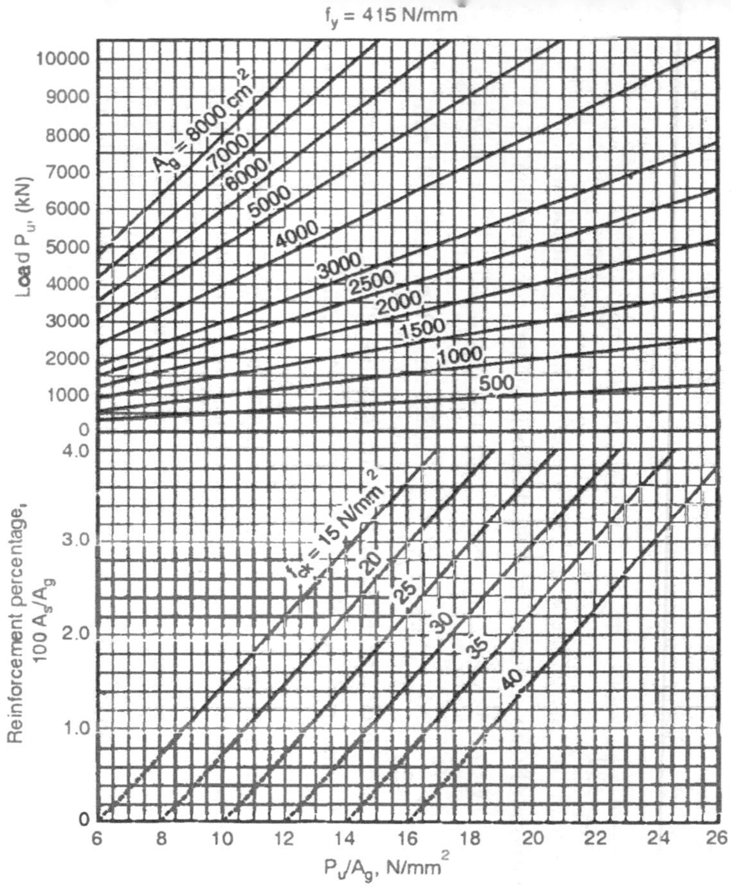

FIG. 16.4. DESIGN CHARTS FOR AXIAL COMPRESSION (HYSD BARS : Fe 415)

where, A_g = Gross area of cross section $= b \times D$ for rectangular section and

$= \frac{\pi}{4} D^2$ for a circular section.

p = percentage of reinforcement $= \frac{A_s}{A_g} \times 100$. Dividing both sides by A_g.

$$\frac{P_u}{A_g} = 0.4 f_{ck}\left(1 - \frac{p}{100}\right) + 0.67 f_y \frac{p}{100} - 0.4 f_{ck} + \frac{p}{100}(0.67 f_y - 0.4 f_{ck}) \quad \ldots(16.4)$$

Eq. 16.4 suggests that design charts can be prepared with $\frac{P_u}{A_g}$, f_{ck} and p as variables, for different types of steels (i.e. for different values of f_y). Such design charts are given in Design Aids for Reinforced Concrete to IS : 456 prepared by ISI under its publication No. SP : 16-1980. Two such charts are reproduced here in Fig. 16.3 and 16.4 for $f_y = 250$ N/mm^2 (mild steel bars) and $f_y = 415$ N/mm^2 (Fe 415 bars) respectively. The working of these charts are explained on the chart itself.

16.8. COMPRESSION MEMBERS WITH HELICAL REINFORCEMENT

The Code permits larger load in short compression members with helical reinforcement because columns with helical reinforcement have greater ductility or toughness when they are loaded concentrically or with small eccentricity. As per the Code, the strength of short compression members with helical reinforcement satisfying the requirement given below shall be taken as 1.05 times the strength of similar members with lateral ties.

Requirement. The ratio of volume of helical reinforcement (V_{hs}) to the volume of core (V_k) shall not be less than $0.36\left(\frac{A_g}{A_k} - 1\right)\frac{f_{ck}}{f_{yh}}$

where A_g = gross area of the section

A_k = area of core of the helically reinforced column measured to the outside diameter of the helix $= \frac{\pi}{4} D_k^2 - A_s$

D_k = diameter of concrete core, measured from outside of helix = $D - 2 \times$ clear cover.

f_{yh} = characteristic strength of helical reinforcement but not exceeding 415 N/mm^2.

16.9. DESIGN SPECIFICATIONS (IS : 456 - 2000)

(1) Longitudinal reinforcement. (*a*) The cross-sectional area of longitudinal reinforcement shall be not less than 0.8 percent nor more than 6 percent of the gross cross-sectional area of the column.

Note. The use of 6 percent reinforcement may involve practical difficulties in placing and compaction of concrete ; hence lower percentage is recommended. Where bars from the columns below have to be lapped with those in the column under consideration, the percentage of steel shall usually not exceed 4%.

(*b*) In any column that has a larger cross-sectional area than that required to support the load, the minimum percentage of steel shall be based upon the area of concrete required to resist the direct stress and not upon the actual area.

(*c*) The minimum number of longitudinal bars provided in a column shall be four in rectangular columns and six in circular columns.

(*d*) The bars shall not be less than 12 mm in diameter.

(*e*) A reinforced concrete column having helical reinforcement shall have at least six bars of longitudinal reinforcement.

(*f*) In a helically reinforced column, the longitudinal bars shall be in contact with the helical reinforcement and equidistant around its inner circumference.

(*g*) Spacing of longitudinal bars measured along the periphery of the column shall not exceed 300 mm.

(*h*) In case of padestals in which the longitudinal reinforcement is not taken into account in strength calculations, nominal longitudinal reinforcement not less than 0.15% of the cross-sectional area shall be provided.

Note : Padestal is a compression member, the effective length of which does not exceed three times the least lateral dimension.

(2) Transverse reinforcement

(*a*) *General*. A reinforced concrete compression member shall have transverse or helical reinforcement so disposed that every longitudinal bar nearest to the compression face has effective lateral support against buckling subject to provisions in 2(*b*). The effective lateral support is given by transverse reinforcement either in the form of circular rings capable of taking up circumferential tension or by polygonal links (lateral ties) with internal angles not exceeding 135°. The ends of the transverse reinforcement shall be properly anchored.

(*b*) *Arrangement of transverse reinforcement*

(*i*) If the longitudinal bars are not spaced more than 75 mm on either side, transverse reinforcement need only to go round corner and alternate bars for the purpose of providing effective lateral support [Fig. 16.5(*a*)].

(*ii*) If the longitudinal bars spaced at a distance of not exceeding 48 times the diameter of the tie are effectively tied in two directions, additional longitudinal bars in between these bars need to be tied in one direction by open ties. [Fig. 16.5(*b*)].

(*iii*) Where the longitudinal reinforcing bars in a compression member are placed in more than one row, effective lateral support to the longitudinal bars in the inner rows may be assumed to have been provided if (*a*) transverse reinforcement is provided for outermost row in accordance with (2) above, and (*b*) no bar of the corner row is closer to the nearest compression face than three times the diameter of the largest bar in the inner row [Fig. 16.5(*c*)].

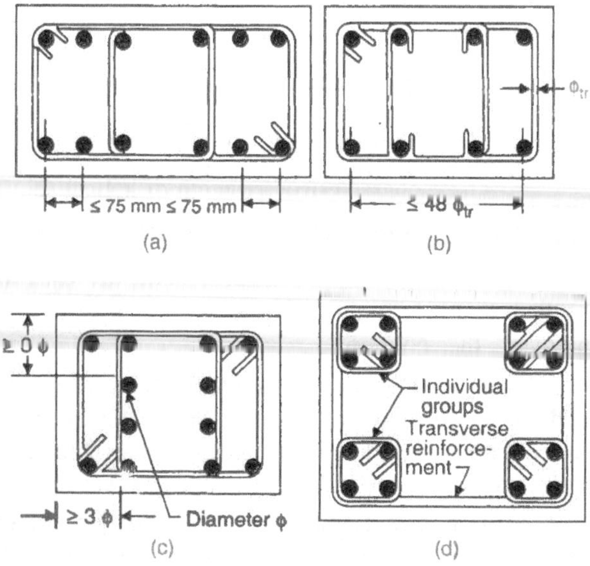

FIG. 16.5.

(*iv*) Where the longitudinal bars in a compression member are grouped (not in contact) and each group adequately tied with transverse reinforcement in accordance with clause (2), the transverse reinforcement for the compression member as a whole may be provided on the assumption that each group is a single longitudinal bar for purpose of determining the pitch and diameter of the transverse reinforcement in accordance with clause (2). The diameter of such transverse reinforcement need not, however, exceed 20 mm [see Fig. 16.5(*d*)].

(*c*) *Pitch and diameter of lateral ties*

(*i*) *Pitch.* The pitch of the transverse reinforcement shall be not more than the least of the following distances: (1) the least lateral dimension of the compression member ; (2) sixteen times the smallest diameter of the longitudinal reinforcement bar to be tied; and (3) 300 mm.

(*ii*) *Diameter.* The diameter of the polygonal links or lateral ties shall not be less than one-fourth of the diameter of the largest longitudinal bar, and in no case less than 5 mm.

(*d*) *Helical reinforcement*

(*i*) *Pitch.* Helical reinforcement shall be of regular formation with the turns of the helix spaced evenly and its ends shall be anchored properly by providing one and a half extra turns of the spiral bar. Where an increased load on the column on the strength of the helical reinforcement is allowed for, the pitch of the helical turns shall be not more than 75 mm, nor more than one sixth of the core diameter of the column, nor less than 25 mm, nor less than three times the diameter of the steel bar forming the helix.

(*ii*) *Diameter.* The diameter of the helix reinforcement shall be in accordance with clause 2 (*c ii*) given above.

(3) Cover to reinforcement. For a longitudinal reinforcing bar in a column, the nominal cover shall not be less than 40 mm, nor less than the diameter of such bar. In the case of columns of minimum dimension of 200 mm or under, whose reinforcing bars does not exceed 12 mm, a cover of 25 mm may be used.

4.1. Unsupported length of column : The unsupported length l of a compression member shall be taken as the clear distance between end restraints, except that :

(*a*) in flat slab construction, it shall be clear distance between the floor and the lower extremity of the capital, the drop panel or slab whichever is the least.

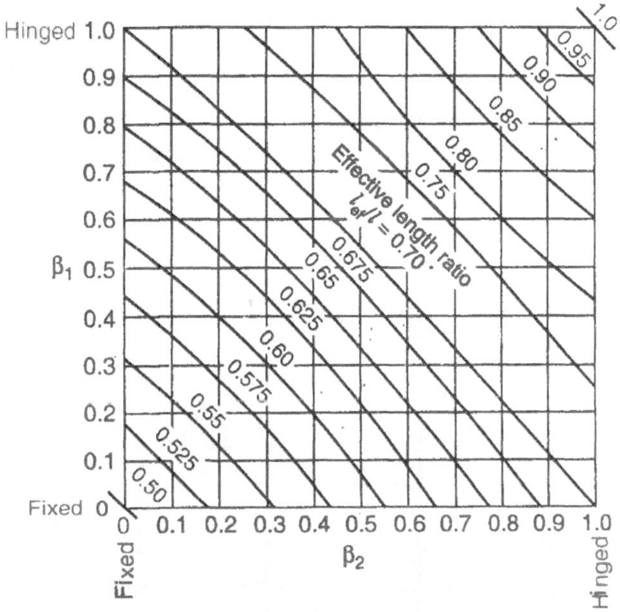

FIG. 16.6. EFFECTIVE LENGTH RATIOS FOR A COLUMN IN A FRAME WITHOUT RESTRAINT AGAINST SWAY

(*b*) in beam and slab construction, it shall be the clear distance between the floor and the underside of the shallower beam framing into the floor and the underside of the shallower beam framing into the columns in each direction at the next higher floor level.

(*c*) in columns restrained laterally by struts, it shall be the clear distance between consecutive struts in each vertical plane, provided that to be an adequate support, two such struts shall meet the columns at approximately the same level, and the angle between the vertical planes through the struts shall not vary more than 30° from a right angle. Such struts shall be of adequate dimensions and shall have sufficient anchorage to restrain the member against lateral deflection.

(*d*) in columns restrained laterally by struts or beams with brackets used at the junction, it shall be the clear distance between the floor and the lower edge of the bracket, provided that the bracket width equals that of the beam strut and is atleast half that of the column.

4.2. Effective length of columns. (*a*) In the absence of more exact analysis, the effective length of columns in framed structures may be obtained from the ratio of effective length to unsupported length l_{ef}/l given in Fig. 16.6 when relative displacement of the ends of the column is prevented and in Fig. 16.7 when relative lateral displacement is not prevented. In the latter case, it is recommended that the effective length ratio l_{ef}/l may not be taken less than 1.2.

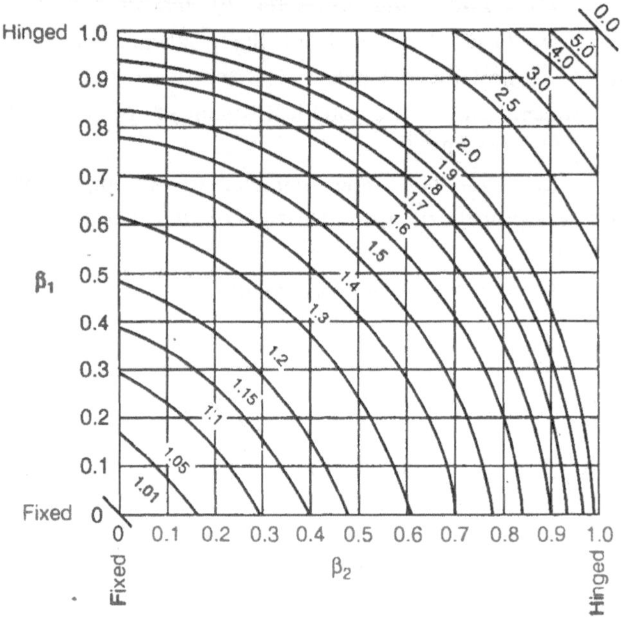

FIG. 16.7. EFFECTIVE LENGTH RATIOS FOR A COLUMN IN A FRAME WITH NO SWAY.

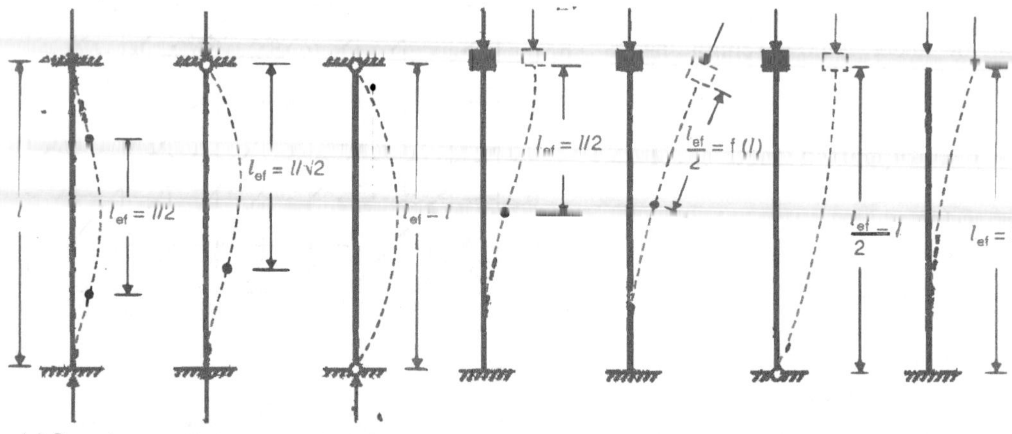

FIG. 16.8. END CONDITIONS FOR COMPRESSION MEMBERS

Note. In Figures 16.6 and 16.7, β_1 and β_2 are equal to $\dfrac{\Sigma k_c}{\Sigma k_c + \Sigma k_b}$ where the summation is done for the members framing into a joint at top and bottom respectively ; k_c and k_b being the flexural stiffness for column and beam respectively.

(*b*) For a normal usage, assuming idealised conditions, the effective length l_{ef} in a given plane may be assessed on the basis of Table 16.1 for various cases of end conditions (Fig. 16.8). In the table, l is the unsupported length of the column.

TABLE 16.1. EFFECTIVE LENGTH OF COMPRESSION MEMBERS (IS : 456 - 2000)

Case No.	*Degree of end restraint of compression member (Fig. 16.8)* (1)	*Theoretical value of effective length* (2)	*Recommended value of effective length* (3)
1.	Effectively held in position and restrained against rotation at both ends.	$0.5l$	$0.65l$
2.	Effectively held in position at both ends, restrained regainst rotation at one end,	$0.7l$	$0.80l$
3.	Effectively held in position at both ends, but not restrained against rotation.	$1.00l$	$1.00l$
4.	Effectively held in position and restrained against rotation at one end, and at the other restrained against rotation but not held in position.	$1.00\ l$	$1.20l$
5.	Effectively held in position and restrained against rotation at one end, and at the other partially restrained against rotation but not held in position.	–	$1.50l$
6.	Effectively held in position at one end but not restrained against rotation, and at the other end restrained against rotation but not held in position.	$2.00l$	$2.00l$
7.	Effectively held in position and restrained against rotation at one end but not held in position nor restrained against rotation at the other end.	$2.00l$	$2.00l$

Example 16.1. ***Load capacity of a short column :***

A concrete column is reinforced with 4 bars of 20 mm dia. Determine the ultimate load capacity of the column, using M 20 grade concrete and Fe 415 grade steel, if the size of the column is (a) 450 mm × 450 mm (b) 300 mm × 300 mm. What will be the allowable service load in each case ?

Solution. Adopting a min. eccentricity of 20 mm, the minimum value of D, for which Eq. 16.2 will be applicable, comes out to be 400 mm. Hence the column of 300 mm × 300 mm of case (*b*) will have to be solved for a case of combined axial load and bending (see Chapter 17).

For column of size 450 mm × 450 mm, $D = 450$ mm.

Hence $0.05\,D = 0.05 \times 450$ mm $= 22.5$ mm. Thus, the given minimum eccentricity of 20 mm is less than $0.05\,D$ (= 22.5 mm), and Eq. 16.2 will be applicable.

$$A_g = 450 \times 450 = 202500 \text{ mm}^2 \ ; \ A_s = 4 \times \frac{\pi}{4}(20)^2 = 1256.64 \text{ mm}^2$$

$$A_c = A_g - A_s = 202500 - 1256.64 = 201243.36 \text{ mm}^2$$

For M 20 concrete, $f_{ck} = 20$ N/mm^2 ; For Fe 415 steel, $f_y = 415$ N/mm^2

$$\therefore \quad P_u = 0.4\, f_{ck}\, A_c + 0.67\, f_y\, A_s = 0.4 \times 20 \times 201243.36 + 0.67 \times 415 \times 1256.64$$

$= 1609947 + 349408 = 1959355 \text{ N} = 1959.355 \text{ kN}$

Allowable service load $= \dfrac{P_u}{1.5} = \dfrac{1959.355}{1.5} \approx$ **1306.2 kN**

Example 16.2. ***Design of short column***

Design a short axially loaded square column, 500 mm × 500 mm for a service load of 2000 kN. Use M 20 concrete and Fe 415 grade steel.

Solution : Service load = 2000 kN, Load factor = 1.5

$P_u = 2000 \times 1.5 = 3000 \text{ kN} = 3000 \times 10^3 \text{ N}$

Min. eccentricity when the effective length of the column is not given, is equal to 20 mm. But $0.05\, D = 0.05 \times 500 = 25$ mm. Hence min. eccentricity is less than $0.05\, D$, and therefore Eq. 16.2 is applicable.

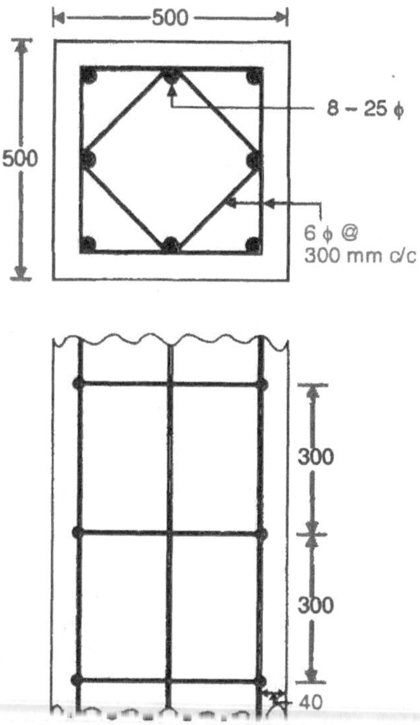

FIG. 16.9

$\therefore \quad P_u = 0.4\, f_{ck}\, A_c + 0.67\, f_y\, A_s$

$3000 \times 10^3 = 0.4 \times 20\,(500 \times 500 - A_s) + 0.67 \times 415\, A_s$

or $\quad 278.05\, A_s - 8\, A_s = 1000000$

From which $A_s = 3703.02 \text{ mm}^2$

Min. area of longitudinal reinforcement = 0.8 %

$= \dfrac{0.8}{100} \times 500 \times 500 = 2000 \text{ mm}^2$

Hence $\quad A_s = 3703.02 \text{ mm}^2$

No. of 20 mm dia. bars = 3703.02 / 314.16 = 11.79

This number is large.

No. of 25 mm dia. bars = 3703.02 / 490.87 = 7.54

Hence provide 8 Nos. of 25 mm dia. bars.

Actual $\quad A_s = 8 \times 490.87 \approx 3927 \text{ mm}^2$

Using 6 mm dia. ties, spacing is the least of the following:

(*i*) Least lateral dimension = 500 mm

(*ii*) 16 × 25 = 400 mm

(*iii*) 300 mm

Hence provide 6 mm dia. ties @ 300 mm c/c, as shown in Fig. 16.9.

Example 16.3. ***Design of short column*** : *Design a circular column to carry an axial load of 1000 kN. Use M 20 concrete and Fe 415 steel.*

Solution: $P_u = 1000 \times 1.5 = 1500 \text{ kN} = 15 \times 10^3 \text{ N}$

For M 20 concrete $f_{ck} = 20 \text{ N/mm}^2$. For Fe 415 steel, $f_y = 415 \text{ N/mm}^2$

Assuming that $e_{min} \le 0.05\, D$,

$P_u = 0.4\, f_{ck}\, A_c + 0.67\, f_y\, A_s$. Assuming 1% steel :

$1500 \times 10^3 = 0.4 \times 20\,(A_g - 0.01\, A_g) + 0.67 \times 415 \times 0.01\, A_g$

From which $A_g = 140180 = \frac{\pi}{4} D^2$

$\therefore$ $D = 422.5$ mm. Adopt $D = 425$ mm.

Area of longitudinal steel $= 0.01 \times 140180 = 1401.8 \text{ mm}^2$

No. of 16 mm dia. bars $= 1401.8/201.06 = 6.97$

Hence provide 7 bars of 16 mm dia.

Using 6 mm dia. lateral ties, spacing is the least of the following :

(*i*) Least lateral dimension = 425 mm

(*ii*) 16 × 16 = 256 mm (*iii*) 300 mm

Hence provide 6 mm dia. lateral ties @ 250 mm c/c.

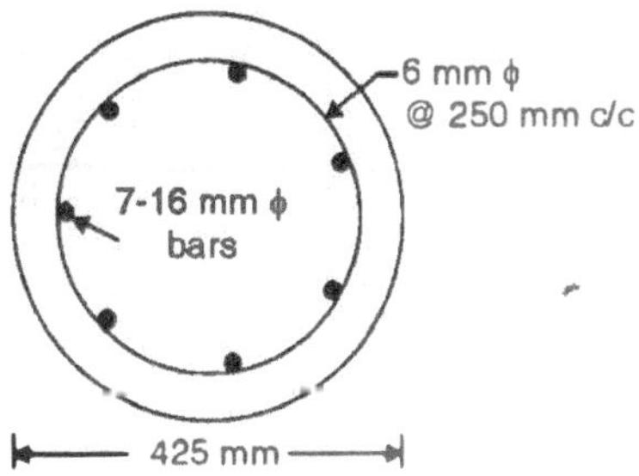

FIG. 16.10

Example 16.4. *Design of short column using helical reinforcement*

Redesign the column of Example 16.3 using helical reinforcement, M 20 grade concrete and Fe 415 grade steel.

Solution

The strength of a column with helical reinforcement is 1.05 times the strength of similar member with lateral ties.

$$P_u = \frac{1000 \times 1.5}{1.05} = 1428.57 \text{ kN}$$

Assuming that $e_{min} \le 0.05\,D$, and providing 1% steel.

$$1428.57 \times 10^3 = 0.4 \times 20\,(A_g - 0.01\,A_g) + 0.67 \times 415 \times 0.01\,A_g$$

From which $A_g = 133505 \text{ mm}^2 = \frac{\pi}{4} D^2$ and $D = 412.3$ mm

Provide overall dia. $D = 420$ mm.

$$0.05\,D = 0.05 \times 420 = 21 \text{ mm} \quad \therefore \; e_{min} < 0.05\,D$$

$A_s = 0.01 \times 133505 = 1335.1 \text{ mm}^2$. No. of 16 mm dia. bars $= 1335.1/201.06 = 6.64$

Hence provide 7 bars of 16 mm dia. giving total area $= 1407.4 \text{ mm}^2$, at a nominal cover of 40 mm. Let us use 8 mm dia. HYSD bars for helical reinforcement.

Outside dia. of helix $= D_k = 420 - 2 \times 40 = 340$ mm

A_k = area of core of helically reinforced column measured to the outside of the helix $= \frac{\pi}{4}(340)^2 - 1407.4 \simeq 89385 \text{ mm}^2$

A_g = gross area of section $= \frac{\pi}{4}(420)^2 = 138544 \text{ mm}^2$

$$\text{Factor } 0.36\left(\frac{A_g}{A_k} - 1\right)\frac{f_{ck}}{f_{yh}} = 0.36\left(\frac{138544}{89385} - 1\right)\frac{20}{415} = 0.00954 \qquad \ldots(i)$$

Dia. d of core upto centre of helix $= 420 - (2 \times 40) - 8 = 332$ mm

Let the pitch of the spiral be s mm. Volume of spiral V_{hs} per 1 mm length of column

$$= \frac{\pi\,d}{s}\left(\frac{\pi}{4}\varphi_s^2\right) = \frac{\pi \times 332}{s}\left(\frac{\pi}{4}\right)(8)^2 = \frac{52427.3}{s}$$

Volume of core per mm length $= V_k = A_k \times 1 = 89385 \times 1 = 89385$

Ratio $\frac{V_{hs}}{V_k} = \frac{52427.3}{s \times 89385} = \frac{0.5865}{s}$ ...(*ii*)

Equating (*i*) and (*ii*), we get $\frac{0.5865}{s} = 0.00954$ or $s \approx 61.5$ mm

However, the pitch should not be more than 75 mm, nor more than 1/6 core dia. (= 1/6 × 340 = 56.7 mm). Also, it should not be less 3 φ_s (= 3 × 8 = 24 mm). Hence keep the pitch equal to 55 mm.

It should be noted that since the helically reinforced columns are very ductile as compared with columns with lateral ties, they are more desirable in highly seismic zones.

Example 16.5 ***Design of circular column*** : *A circular column, 4.6 m high is effectively held in position at both the ends and restrained against rotation at one end. Design the column, to carry an axial load of 1200 kN, if its dia. is restricted to 450 mm. Use M 20 mix and Fe 415 steel.*

Solution. For a column with the above end conditions, we get effective length = 0.80 × 4600 = 3680 mm. Using 450 mm dia. column, slenderness ratio = 3680/450 = 8.18<12. Hence it is a short column.

Also, $e_{min} = \frac{l}{500} + \frac{D}{30} = \frac{3680}{500} + \frac{450}{30} = 22.36$ mm

Also, $0.05\ D = 0.05 \times 450 = 22.5$ mm

∴ $e_{min} < 0.05\ D$. Hence strength of the column is given by Eq. 16.2

$$P_u = 0.4\, f_{ck}\, A_c + 0.67\, f_y\, A_s$$

where $P_u = 1.5 \times 1200 = 1800$ kN $= 1800 \times 10^3$ N

$$A_c = \frac{\pi}{4}(450)^2 - A_s = (159043 - A_s)\ \text{mm}^2$$

∴ $1800 \times 10^3 = 0.4 \times 20\,(159043 - A_s) + 0.67 \times 415\, A_s$.

From which, $A_s = 1954$ mm²

Min. steel @ 0.8% $= \frac{0.8}{100} \times \frac{\pi}{4}(450)^2 = 1272$ mm². Hence $A_s = 1954$ mm²

∴ No. of 20 mm φ bars = 1954/314.56 = 6.22. Hence provide 7 No. 20 mm φ longitudinal bars. Using 6 mm dia. ties its spacing should not exceed the least of the following :

(*i*) Least lateral dimension = 450 mm
(*ii*) 16 × dia. of main bar = 16 × 20 = 320 mm
(*iii*) 300 mm.

Hence provide 6 mm dia. mild steel ties @ 300 mm c/c.

Example 16.6. ***Design of rectangular column*** : *Design a rectangular column of 4.5 m unsupported length, restrained in position and direction at both the ends, to carry an axial load of 1200 kN. Use M 20 concrete and Fe 415 steel.*

Solution. For a column with the above end conditions, effective length = 0.65 times the actual length = 0.65 × 4500 = 2925 mm.

Let us use 1% steel. Also let width $b = \frac{1}{2} D$.

Assuming that $e_{min} \leq 0.05\, D$ in appropriate direction, Eq. 16.2 will be applicable.

$\therefore \quad P_u = 0.4\, f_{ck}\, A_c + 0.67\, f_y\, A_s$. Hence,

$$1.5 \times 1200 \times 10^3 = 0.4 \times 20\, [A_g - 0.01\, A_g] + 0.67 \times 415 \times 0.01\, A_g .$$

From which $A_g = 168216 \text{ mm}^2$

$\therefore \quad b.\,D = \frac{1}{2} D.D = 168216$, from which $D = 580$ mm and $b = \frac{1}{2} D = 290$ mm

However, provide 300 mm × 600 mm column, from practical considerations.

$$A_s = 0.01 \times 168216 = 1682.2 \text{ mm}^2$$

For column of rectangular shape, let us provide 8 bars.

Area of each bar = 1682.2/8 = 210.27 mm². ∴ Dia. of bar = 16.36 mm

Since the actual area of concrete provided is more than required, let us provide 8 bars of 16 mm dia. Actual $A_s = 8 \times 201.06 = 1608.5 \text{ mm}^2$.

$$\% \text{ steel} = \frac{1608.5}{300 \times 600} \times 100 = 0.89\% > 0.8\% .$$

Before checking for the load capacity of the column, let us check for the slenderness and eccentricity requirements.

$$l_e = 2925 \text{ mm} \;;\; l_e/b = 2925/300 = 9.75 < 12 \;;\; l_e/D = 2925/600 = 4.875 < 12$$

Hence the column is a short column, in both the directions.

In one direction (i.e. x-direction)

$$e_{min.} = \frac{l}{500} + \frac{D}{30} = \frac{4500}{500} + \frac{600}{30} = 29 \text{ mm}$$

$$0.05\, D = 0.05 \times 600 = 30 \text{ mm}.$$

$\therefore \quad e_{min} < 0.05\, D.$

In the other direction (i.e. y-direction)

$$e_{min} = \frac{l}{500} + \frac{b}{30} = \frac{4500}{500} + \frac{300}{30} = 19 \text{ mm (subject to a min. of 20 mm)}$$

$\therefore \quad e_{min} = 20$ mm ; $0.05\, b = 0.05 \times 300 = 15$ mm ; $\therefore \quad e_{min} > 0.05\, b$

Hence Eq. 16.2 is not strictly applicable, and the column is to be designed as an eccentrically loaded column. Alternatively, let us keep $b = 400$ mm, so that $0.05\, b = 0.05 \times 400 = 20$ mm $= e_{min.}$, and redesign the column.

Redesign : $b = 400$ mm.

$\therefore \quad 400 \times D = 168216$

$\therefore \quad D = 168216/400 = 420.5$ mm.

Hence adopt a column of size 400 × 450 mm, and provide 8 bars of 16 mm dia, arranged as shown in Fig. 16.11.

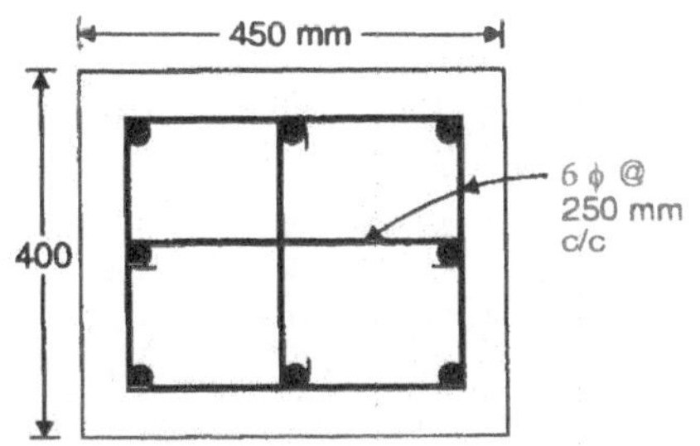

FIG. 16.11

Using 6 mm dia. ties, spacing will we the least of the following (*i*) Least lateral dimension = 400 mm

(*ii*) $16 \times \varphi = 16 \times 16 = 256$ mm

(*iii*) 300 mm. Hence provide 6 mm dia. ties @ 250 mm c/c, including two open ties as shown in Fig. 16.11.

PROBLEMS

1. Discuss various assumptions used in the limit state method of design of compression members.
2. A reinforced concrete column, 500 mm × 500 mm is reinforced with 8 bars of 20 mm dia, symmetrically arranged. Determine the service load which the column can carry, assuming it to be a short one. Use (*a*) M 15 grade concrete and Fe 250 steel (*b*) M 15 grade concrete and Fe 415 steel, and (*c*) M 20 concrete and Fe 415 steel.
3. A short R.C. column has a diameter of 450 mm, and is reinforced with 8 bars of 16 mm dia. placed with a clear cover of 40 mm. Determine load capacity of the column if the transverse reinforcement consists of (*a*) 8 mm dia. mild steel ties @ 250 mm c/c (*b*) 8 mm spirals of Fe 415 steel, arranged at a pitch of 50 mm. Use M 20 concrete and Fe 415 steel for longitudinal bars.
4. Design a short square column to carry a service load of 1600 kN. Use M 20 concrete Fe 415 steel combination.
5. Redesign a circular column for problem 4.
6. Redesign a rectangular column for problem 4, taking width of column equal to 400 mm.

17

Limit State of Collapse in Compression: II Columns with Axial load and Uniaxial Bending

17.1. INTRODUCTION

As discussed in Chapter 16, load on a column is rarely axial. Due to non-homogeneity of material, imperceptible crookedness of the member, inaccuracies in loading, inaccuracies in construction, etc, there is always some eccentricity inherent in a compression member. Due to this, IS : 456-2000 recommends that every compression member should be designed for a certain minimum eccentricity e_{min}. If this e_{min} is equal to or less than 0.05 D, the column may be designed as an axially loaded column, using Eq. 16.2. If, however, e_{min} is greater than 0.05 D, the axially loaded column has to be designed for axial load P_u and moment $M_u = P_u \cdot e_{min}$. Besides this, there are many occasions when a column is subjected to (*i*) eccentric load, or (*ii*) end moments, on account of monolithic connection of beams and columns. Such a column may be subjected to either uni-axial moment or bi-axial moments, in addition to the axial load.

17.2. BASIC ASSUMPTIONS

The limit state method of design of compression member is based on the following assumptions.

1. The plane sections normal to the axis of column before deformation remain plane after deformation. This means that the strain at any point is proportional to its distance from the N.A.

2. The relationship between compressive stress distribution in concrete and strain in concrete may be assumed to be a rectangle, trapezoid, parabola or any other shape which results in prediction of strength in substantial agreement with the results of tests. An acceptable stress-strain curve for concrete is given in Fig. 2.3. For design purposes, the *compressive strength of concrete in structure* is assumed to be 0.67 times the characteristic strength. The partial safety factor $\gamma_{mc} = 1.5$ is applied in addition to this. Thus the design strength of concrete is taken as $0.67 f_{ck}/1.5 \simeq 0.446 f_{ck}$.

3. The tensile strength of concrete is ignored.

4. The stresses in the reinforcement are derived from representative stress-strain curve for the type of steel used. Typical curves are shown in Fig. 2.4, 2.5 and 2.6. For design purposes, partial safety factor $\gamma_{ms} = 1.15$ shall be applied.

5. The maximum compressive strain in concrete in axial compression is taken as 0.002.

6. The maximum strain in concrete, at the outer most compression fibre is taken as 0.0035 in bending when N.A. lies within the section.

7. The maximum compressive strain at the highly compressed extreme fibre in concrete subjected to axial compression and bending, and where there is no tension on the section shall be 0.0035 minus 0.75 times the strain at the least compressed extreme fibre.

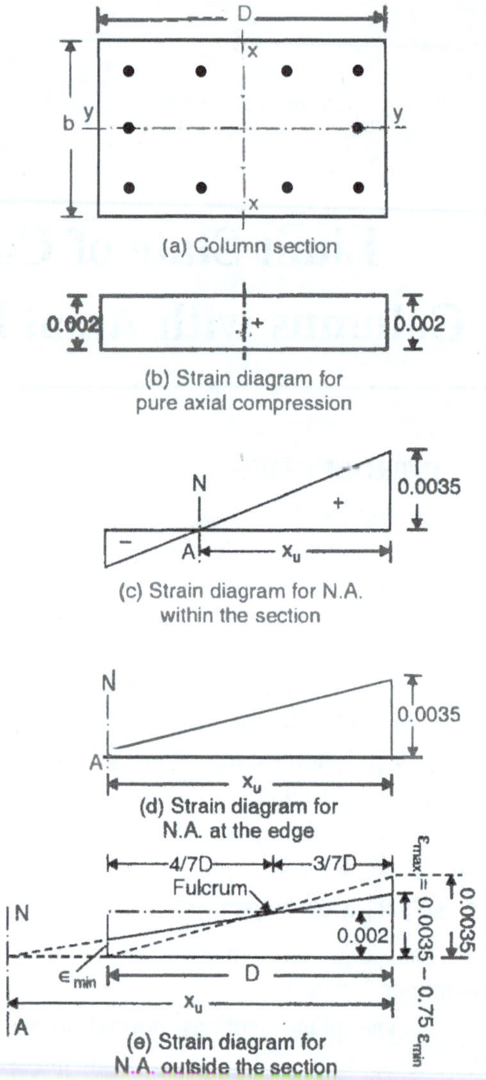

FIG. 17.1. STRAIN DISTRIBUTION ACROSS THE SECTION

According to assumption 5, for purely axial compression, the strain is assumed to be uniform, equal to 0.002, across the section, as shown in Fig. 17.1 (*b*). Also, for the case of axial compression along with bending, the maximum strain (ε_{max}) at the outer most fibre is 0.0035 and this assumption is applicable when the N.A. lies within the section (Fig. 17.1 *c*), and in the limiting case when the N.A. lies at the edge of the section. In the later case (Fig. 17.1 *d*), the strain varies from 0.0035 at the highly compressed edge to zero at the opposite edge. The strain distribution for these two cases (of assumptions 5 and 6) intersect at each other at a depth of $\frac{3}{7}D$ from the highly compressed edge. This point is assumed as *fulcrum* for the strain distribution line when the N.A. lies outside the section (Fig. 17.1 *e*). This leads to assumption No. 7 that the maximum strain at the highly compressed edge is 0.0035 minus 0.75 times the strain at the least compressed edge.

Justification of assumption No. 7

The justification inherent in assumption No. 7 can be proved easily with reference to Fig. 17.2 (*b*) where N.A. lies outside the section.

$$\varepsilon_{max} = AB = AB_2 - BB_2 = 0.0035 - BB_2 \qquad \ldots(i)$$

Now from similar triangles,

$$\frac{BB_2}{DD_2} = \frac{3D/7}{4D/7} = \frac{3}{4}$$

$$\therefore \quad BB_2 = \frac{3}{4} DD_2 = \frac{3}{4} \varepsilon_{min} \qquad \ldots(ii)$$

Substituting in (*i*)

$$\therefore \quad \varepsilon_{max} = 0.0035 - 0.75\, \varepsilon_{min} \qquad \ldots(17.1)$$

Also, $\dfrac{AB}{CO} = \dfrac{x_u}{x_u - 3D/7}$

or $AB = CO \times \dfrac{x_u}{x_u - 3D/7}$

Hence

$$\varepsilon_{max} = 0.002 \left(\frac{x_u}{x_u - 3D/7} \right)$$

$$= 0.002 \left(1 + \frac{3D/7}{x_u - 3D/7} \right) \qquad \ldots(17.1\,a)$$

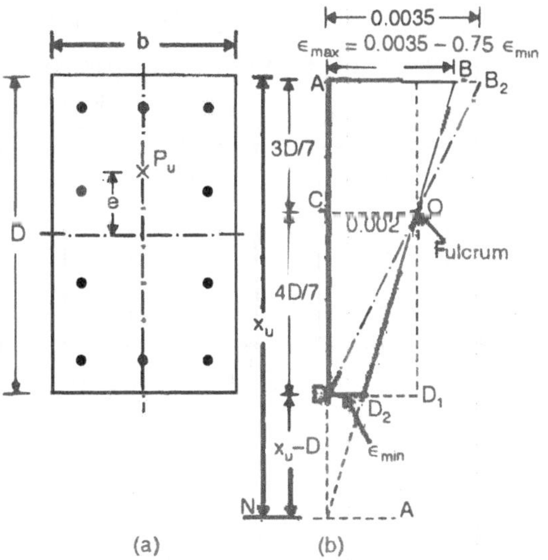

FIG. 17.2. STRAIN DISTRIBUTION FOR COLUMN UNDER AXIAL COMPRESSION AND BENDING.

17.3. COLUMN UNDER AXIAL COMPRESSION AND UNI-AXIAL BENDING

Let us now take a case of a column which is subjected to combined action of axial load (P_u) and uniaxial bending moment (M_u). This case of loading can be reduced to a single resultant load P_u acting at an eccentricity e such that $e = M_u / P_u$. The behaviour of such a column depends upon the relative magnitudes of P_u and M_u, or indirectly on the value of eccentricity e. For a column subjected to load P_u at an eccentricity e, the location of neutral axis (N.A.) will depend upon the value of eccentricity e. Depending upon the value of eccentricity and the resulting position (x_u) of N.A., we will consider the following cases :

Case I : Concentric loading : zero eccentricity or nominal eccentricity : $x_u = \infty$

Case II : Moderate eccentricity : $x_u > D$

Case III : Moderate eccentricity : $x_u = D$

Case IV : Large eccentricity : $x_u < D$

Case I shown in Fig. 17.3 (*a*) is the case of concentric loading, when eccentricity is equal to zero. This means that the column is subjected to zero bending moment with the result that the entire column section is under uniform strain (= 0.002) and hence subjected to uniform compressive stress (= 0.446 f_{ck}). The N.A. depth x_u will be equal to infinity. We shall also consider the case of nominal eccentricity ($e \le e_{min}$) under this case.

Case II shown in Fig. 17.3 (*b*) is the more common case of column subjected to P_u and M_u, with relative magnitudes such that e is moderate and the N.A. falls outside the section. This results in $x_u > D$ and consequently, the entire column will be under compression. The strain diagram will be trapezoidal with $\varepsilon = 0.002$ at a depth of $\frac{3}{7} D$ (i.e. at the fulcrum) and $\varepsilon_{max} = 0.0035 - 0.75\ \varepsilon_{min}$. The stress diagram will be rectangular upto a depth of $\frac{3}{7}D$, with uniform value of $0.446\ f_{ck}$, and parabolic for the remaining portion.

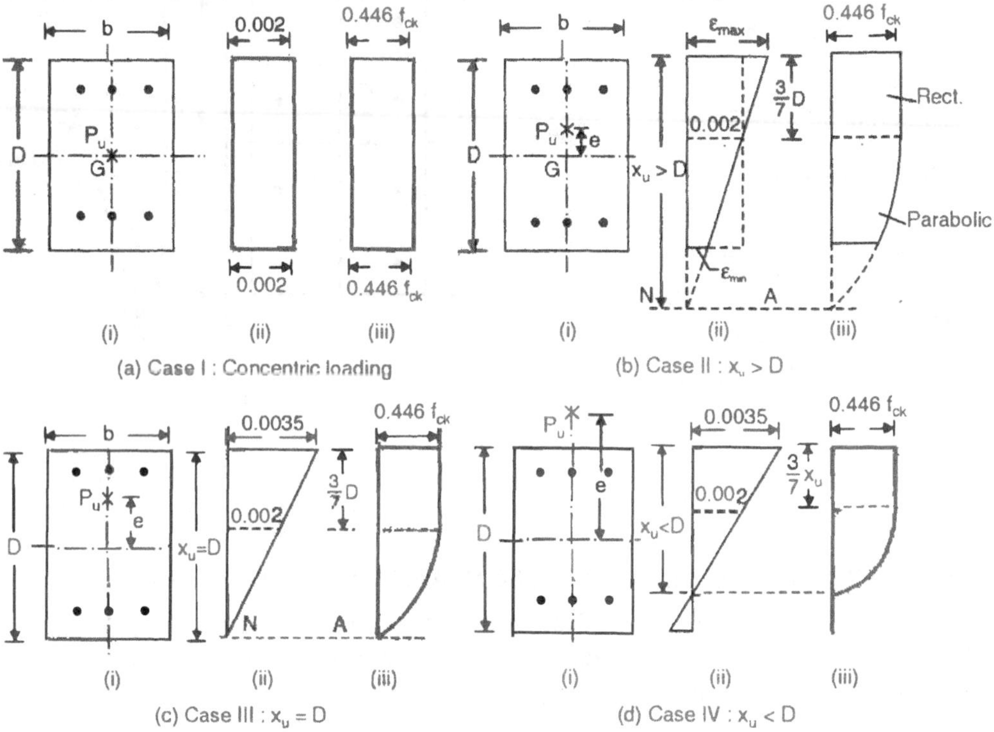

FIG. 17.3. VARIOUS CASES OF AXIAL LOADING

Case III (Fig. 17.3 *c*) is a special case of Case II, when the N.A. lies just at the outer edge, i.e. $x_u = D$. This will result in a triangular distribution of strain, with a maximum value $\varepsilon_{max} = 0.0035$ and minimum value $\varepsilon_{min} = 0$. The resulting stress diagram will be rectangular upto a depth of $\frac{3}{7}D$, with a uniform value of $0.446\ f_{ck}$ and parabolic for the remaining portion with zero value at the N.A.

Case IV (Fig. 17.3 *d*) is the case of large eccentricity, with the result that some part of column will be under tension, resulting in $x_u < D$. The maximum strain at the inner edge will be equal to 0.0035 (compressive), minimum strain at the outer edge will be tensile, and strain at the depth of $\frac{3}{7}D$ will be equal to 0.002. The resulting stress at the inner edge will be equal to $0.446\ f_{ck}$ with uniform value upto a depth of $\frac{3}{7}D$. For the remaining depth up to N.A., the stress distribution will be parabolic as in the case of beams. We shall now consider the above four cases separately in greater detail.

17.4. CASE I : $e = 0$ and $e \leq e_{min}$

Fig. 17.4 (*a*) shows a symmetrically reinforced column section axially loaded with zero eccentricity. The column is subjected to uniform strain of 0.002. Fig 17.4 (*c*) shows the stress diagram with uniform stress of $0.446\ f_{ck}$.

(*a*) **Zero eccentricity** ($e = 0$) : Fig. 17.4 Balancing the axial forces on the section,

$$P_u = P_{uc} + P_{us}$$

or $$P_u = 0.446\, f_{ck}\, A_c + f_s\, A_s$$

or $$P_u = 0.446\, f_{ck}\, (bD - A_s) + f_s\, A_s$$

or $$P_u = 0.446\, f_{ck}\, bD + A_s\, (f_s - 0.446\, f_{ck})$$

Hence

$$\frac{P_u}{f_{ck}\, bD} = 0.446 + \frac{A_s}{f_{ck}\, bD}(f_s - 0.446\, f_{ck})$$

or $$\frac{P_u}{f_{ck}\, bD} = 0.446 + \frac{p}{100\, f_{ck}}(f_s - 0.446\, f_{ck}) \quad \text{...(17.2 a)}$$

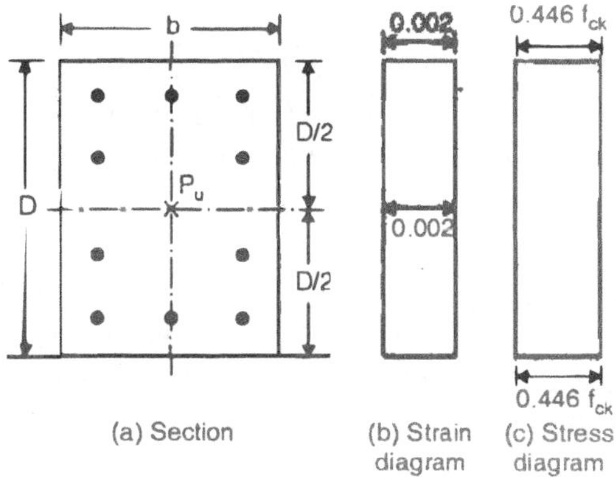

FIG. 17.4. PURE AXIAL LOADING

where p = percent of *total steel* in the section $= (A_s/bD)\, 100$

f_s = stress in steel corresponding to a strain of 0.002

$= 0.87\, f_y$ for Fe 250 (mild steel)

$= 0.79\, f_y$ for Fe 415 steel.

$= 0.75\, f_y$ for Fe 500 steel.

In general, the capacity of a member subjected to pure axial load is given by

$$P_u = 0.446\, f_{ck}\, A_c + \alpha_s\, f_y\, A_s$$

For simplicity, I.S. Code adopts only the lowest value of α_s (= 0.75) which is for steel of grade Fe 500. Also the Code has redesignated P_u as P_{uz} in section 39.6 and has given the following equation in the design aids (SP : 16-1980).

$$P_{uz} = 0.45\, f_{ck}\, A_c + 0.75\, f_y\, A_s \quad \text{...(i)} \quad \text{...(16.1)}$$

(*b*) **Nominal eccentricity** ($e \le e_{min}$) : Again, as per clause 39.3 of IS 456 - 2000, a short column, axially loaded, should be designed for minimum eccentricity e_{min} given by

$$e_{min} = \frac{l}{500} + \frac{D}{30}, \text{ subject to a minimum of 20 mm}$$

where l = unsupported length of the column

and D = lateral dimension of the column in the direction under consideration.

If the value of minimum eccentricity calculated as above is less than or equal to 0.05 D, clause 39.3 of the Code permits the design of short axially loaded compression member by following equation :

$$P_u = 0.4\, f_{ck}\, A_c + 0.67\, f_y\, A_s \quad \text{...(ii)} \quad \text{...(16.2)}$$

It should be noted that Eq. (*ii*) given above is obtained by reducing the capacity of the column given by Eq. (*i*) by approximately 10% thereby allowing for the minimum eccentricity e_{min} not exceeding 0.05 D. If however, the minimum eccentricity is greater than 0.05 D, the section should be designed for combined axial load and bending, discussed in the subsequent articles.

Eq. (*ii*) given above can be reduced to the following form :

$$\frac{P_u}{f_{ck}\, bD} = 0.4 + \frac{p}{100\, f_{ck}} (0.67\, f_y - 0.4\, f_{ck}) \qquad \ldots(17.2\; b)$$

Thus, we have two equations for P_u : Eq. 17.2 (*a*) and Eq. 17.2 (*b*) and the ISI design hand book (SP : 16-1980) uses both these equations. When the minimum eccentricity ($e_{min.}$) does not exceed 0.05 D, Eq. 17.2 (*b*) has been used, while for eccentricities less than $e_{min.}$, Eq. 17.2 (*a*) has been used. At this stage, it is essential to know the difference between Eqs. 17.2 (*a*) and 17.2 (*b*). In Eq. 17.2 (*a*), a factor 0.446 has been used for concrete and f_s corresponds to the actual stress in steel corresponding to a strain of 0.002. In Eq. 17.2 (*b*), a factor of 0.4 has been used for concrete and f_s has been taken equal to $0.67\, f_y$ for all types of steel.

In both the equations, the second term within the parenthesis represent the deduction of concrete replaced by reinforcement bars. This term is usually neglected for convenience. However, as a better approximation, a constant value corresponding to M 20 grade concrete has been used in the design charts (interaction curves) in the ISI hand book, so that the error is negligibly small over the range of concrete mixes used. An accurate consideration of this term will necessitate the preparation of separate charts for each grade of concrete which is not considered worth while.

17.5. CASE II : NEUTRAL AXIS OUTSIDE THE SECTION

This case corresponds to the situation of a column, subjected to both the axial load and moment, when the entire column section is in compression, and the strain varies across the section as shown is Fig. 17.5 (*b*). Obviously, the N.A. will lie outside the section. For such a case, the points for plotting the charts (interaction curves) are obtained by *assuming* different positions of N.A. across the section and the stress block parameters are determined, as explained below. The stresses in the reinforcement are also calculated from the known strains. Therefore, the resultant axial force and the moment about the centroid of the section are calculated and the equilibrium equations are applied, so that (*i*) axial forces on the section balance, and (*ii*) moment on the section also balance.

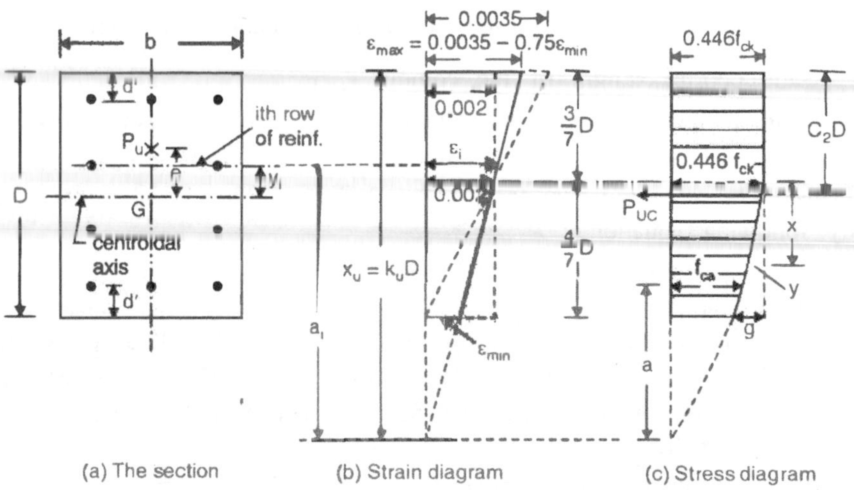

FIG. 17.5. CASE II : N.A. OUTSIDE THE SECTION ($x_u > D$)

Fig. 17.5 (*a*) shows a symmetrically reinforced section, with i^{th} row of the reinforcement bar situated at a distance y_i from the centroidal axis of the section, taken positive towards the highly compressed edge and negative towards the least compressed edge. Fig. 17.5 (*b*) shows the strain diagram, with a strain of 0.002 at the fulcrum point $\frac{3}{7}D$ from the most compressed edge. Fig. 17.5 (*c*) shows the stress diagram which is partly rectangular (upto depth $\frac{3}{7}D$) and partly parabolic (for the remaining depth), with a maximum value of 0.446 f_{ck} at the highly compressed edge. Before applying the equilibrium equations for forces and moments, let us first find the stress block parameters.

Stress block parameters

Let $x_u = k_u D =$ depth of N.A. from highly compressed edge.

C_1 = coefficient for area of the stress block.

g = difference between the stress at the highly compressed edge and the stress at the least compressed edge.

From the equation of the parabola, $y = x^2$, we get

$$g = \left(\frac{4}{7}D\right)^2 \qquad \ldots(i)$$

Also,

$$0.446\, f_{ck} = \left(k_u D - \frac{3}{7}D\right)^2 \qquad \ldots(ii)$$

Dividing Eq. (*i*) by Eq. (*ii*), we get

$$g = 0.446\, f_{ck}\left[\frac{\frac{4}{7}D}{k_u D - \frac{3}{7}D}\right]^2 = 0.446\, f_{ck}\left[\frac{4}{7\,k_u - 3}\right]^2 \qquad \ldots(17.3\ a)$$

Area of the stress block is $A = 0.446\, f_{ck}\, D - \frac{g}{3}\left(\frac{4}{7}D\right) = 0.446\, f_{ck}\, D - \frac{4}{21}\, g\, D$

or

$$A = 0.446\, f_{ck}\, D\left[1 - \frac{4}{21}\left(\frac{4}{7\,k_u - 3}\right)^2\right] = C_1 f_{ck}\, D \qquad \ldots(17.3\ b)$$

where

$$C_1 = 0.446\left[1 - \frac{4}{21}\left(\frac{4}{7\,k_u - 3}\right)^2\right] \qquad \ldots(17.36)$$

Putting

$$C_0 = \frac{8}{7}\left[\frac{4}{7\,k_u - 3}\right]^2 \qquad \ldots(17.3)$$

We get

$$C_1 = 0.446\left(1 - \frac{C_0}{6}\right) \qquad \ldots(17.4\ a)$$

The centroid of the stress block is found by taking moments about the highly compressed edge. If C_2D is the distance of the centroid of the stress block from the highly compressed edge, we have

$$A\,(C_2D) = 0.446\, f_{ck}\, D\left(\frac{D}{2}\right) - \frac{4}{21}\, gD\left[\frac{3}{7}D + \frac{3}{4}\left(\frac{4}{7}D\right)\right]$$

or

$$A\,(C_2D) = 0.446\, f_{ck}\,\frac{D^2}{2} - \frac{8}{49}\, gD^2$$

Substituting the value of g from Eq. 17.3 (a), we get

$$A \cdot C_2 D = 0.446 f_{ck} D^2 \left[0.5 - \frac{8}{49} \left(\frac{4}{7k_u - 3} \right)^2 \right] = 0.446 f_{ck} D^2 \left[0.5 - \frac{C_0}{7} \right]$$

$$\therefore \qquad C_2 = \frac{0.446 f_{ck} D^2 [0.5 - C_0/7]}{C_1 f_{ck} D^2} = \frac{0.446 f_{ck} D^2 (0.5 - C_0/7)}{0.446 (1 - C_0/6) f_{ck} D^2}$$

Hence
$$C_2 = \frac{0.5 - C_0/7}{1 - C_0/6} \qquad \ldots(17.4\ b)$$

Thus, the values of both the parameters C_1 and C_2 are known in terms of C_0, where C_0 is given by Eq. 17.3.

The values of C_0, C_1 and C_2 for different values of k_u are tabulated in Table 17.1. In this Table, the value of stress block parameters have been tabulated for values of k_u upto 10, for information only. However, for the construction of interaction diagrams (i.e. design charts), it is generally adequate to consider the values of k_u upto about 1.2.

TABLE 17.1. STRESS BLOCK PARAMETERS WHEN N.A. FALLS OUTSIDE THE SECTION ($x_u > D$)

k_u	C_0	C_1	C_2	k_u	C_0	C_1	C_2
1.00	1.143	0.361	0.416	1.21	0.611	0.401	0.460
1.01	1.104	0.364	0.419	1.22	0.596	0.402	0.461
1.02	1.067	0.367	0.423	1.23	0.581	0.403	0.462
1.03	1.032	0.369	0.426	1.24	0.567	0.404	0.463
1.04	0.998	0.372	0.429	1.25	0.553	0.405	0.464
1.05	0.966	0.374	0.431	1.26	0.540	0.406	0.465
1.06	0.936	0.376	0.434	1.27	0.527	0.407	0.466
1.07	0.907	0.379	0.436	1.28	0.515	0.408	0.467
1.08	0.879	0.381	0.439	1.29	0.503	0.409	0.467
1.09	0.853	0.383	0.441	1.30	0.419	0.409	0.468
1.10	0.828	0.384	0.443	1.40	0.395	0.417	0.475
1.11	0.804	0.386	0.445	1.50	0.325	0.422	0.480
1.12	0.781	0.388	0.447	1.60	0.272	0.426	0.483
1.13	0.758	0.390	0.448	1.70	0.231	0.429	0.486
1.14	0.737	0.391	0.450	1.80	0.198	0.431	0.488
1.15	0.717	0.393	0.452	1.90	0.172	0.433	0.489
1.16	0.698	0.394	0.453	2.0	0.151	0.435	0.491
1.17	0.679	0.396	0.454	3.0	0.056	0.442	0.497
1.18	0.661	0.397	0.456	4.0	0.029	0.444	0.498
1.19	0.644	0.398	0.457	10.0	0.004	0.446	0.500
1.20	0.627	0.399	0.458	∞	0.000	0.446	0.500

Equations of equilibrium : Balancing the axial forces on the section, we have

$$P_u = P_{uc} + P_{us} = C_1 f_{ck} bD + \sum_{i=1}^{n} \frac{p_i bD}{100} (f_{si} - f_{ci}) \qquad \ldots(17.5\ a)$$

where $p_i = \frac{100\, A_{si}}{bD}$, where A_{si} is the area of reinforcement in i^{th} row

f_{si} = stress in the ith row of reinforcement, compression being positive and tension being negative.

f_{ci} = stress in concrete at the level of ith row of reinforcement

n = number of rows of reinforcement

Rewriting Eq. 17.5 (*a*), $$\frac{P_u}{f_{ck}\, bD} = C_1 + \sum_{i=1}^{n} \frac{p_i}{100\, f_{ck}} (f_{si} - f_{ci}) \qquad \ldots(17.5)$$

Similarly, taking moments of forces about the centroid of the section, we get

$$M_u = M_{uc} + M_{us} = M_{uc} + \sum_{i=1}^{n} M_{usi}$$

or $$M_u = C_1 f_{ck}\, bD \left(\frac{D}{2} - C_2 D \right) + \sum_{i=1}^{n} \frac{p_i\, b\, D}{100} (f_{si} - f_{ci})\, y_i \qquad \ldots(17.6\ a)$$

where y_i = distance of the bar in the ith row, from the centroid of the section, taken positive towards the highly compressed edge and negative towards least compressed edge.

Dividing both the sides by $f_{ck}\, bD^2$, we get

$$\frac{M_u}{f_{ck}\, bD^2} = C_1\, (0.5 - C_2) + \sum_{i=1}^{n} \frac{p_i}{100\, f_{ck}} (f_{si} - f_{ci}) \left(\frac{y_i}{D} \right) \qquad \ldots(17.6)$$

Variation of strain across the section : The values of stress in steel• (f_{si}) and that in concrete (f_{ci}) are obtained corresponding to ε_i at the ith row of reinforcement, from the respective stress-strain curves of steel and concrete.

The value ε_i, is obtained from the following linear expression, based on the premise that the reference strain is taken equal to 0.002 at a depth of $3D/7$ from the highly compressed edge, as indicated in Fig. 17.5 (*b*). Thus

$$\varepsilon_i = \frac{0.002\, a_i}{x_u - 3D/7} = \frac{0.002\, (x_u - 0.5\, D + y_i)}{x_u - 3\, D/7} \qquad \ldots(17.7)$$

where a_i = distance of ith row of reinforcing bars from the N.A. $= x_u - D/2 + y_i$

y_i = distance of ith row of reinforcing bars from the centroidal axis, with proper algebraic sign.

Determination of value of f_{si}

(a) For mild steel bars of Fe 250 grade

(*i*) For $\varepsilon_i > \frac{0.87\, f_y}{E_s} = \frac{0.87 \times 250}{2 \times 10^5} > 0.0010875$, $f_{si} = 0.87\, f_y = 0.87 \times 250 = 217.5\ \text{N/mm}^2$

(In other words, $\varepsilon_{su} = 0.0010875$ for Fe 250)

(*ii*) For $\varepsilon_i < \frac{0.87\, f_y}{E_s} < 0.0010875$, $f_{si} = \varepsilon_i \,.\, E_s$

(*b*) ***For Fe 415 bars***

(*i*) For $\varepsilon_i > \frac{0.8 f_{yd}}{E_s} > \frac{0.8 \times 0.87 f_y}{E_s} > \frac{0.8 \times 0.87 \times 415}{2 \times 10^5} > 0.00144$

f_{si} is obtained either from stress-strain curve or from Table 2.3.

(*ii*) For $\varepsilon_i < \frac{0.8 f_{yd}}{E_s} < 0.00144$, $f_{si} = \varepsilon_i E_s$

(*c*) ***For Fe 500 bars***

(*i*) For $\varepsilon_i > \frac{0.8 f_{yd}}{E_s} > \frac{0.8 \times 0.87 f_y}{E_s} > \frac{0.8 \times 0.87 \times 500}{2 \times 10^5} > 0.00174$

f_{si} is obtained either from stress-strain curve or from Table 2.3.

(*ii*) For $\varepsilon_i < 0.8 f_{yd}/E_s < 0.00174$, $f_{si} = \varepsilon_i E_s$

Determination of value of f_{ci}

The value of f_{ci} corresponding to ε_i can be found from the equation of the parabolic part of idealised stress-strain curve for concrete, expressed as under.

$$f_{ci} = 0.446 f_{ck} \left[\frac{2 \varepsilon_i}{0.002} - \left(\frac{\varepsilon_i}{0.002} \right)^2 \right] \quad \text{for } \varepsilon_i < 0.002 \qquad \ldots(17.8\ a)$$

or

$$f_{ci} = 446 f_{ck}\, \varepsilon_i (1 - 250\, \varepsilon_i) \quad \text{for } \varepsilon_i < 0.002 \qquad \ldots(17.8)$$

However, for $\varepsilon_i \geq 0.002$, $f_{ci} = 0.446 f_{ck}$

Alternatively, f_{ci} can be directly found from the following expression obtained by substituting the value of ε_i (given by Eq. 17.7) in Eq. 17.8.

$$f_{ci} = 0.446 f_{ck} \left[\frac{2 a_i}{(x_u - \frac{3}{7} D)} - \frac{a_i^2}{(x_u - \frac{3}{7} D)^2} \right] \qquad \ldots(17.8\ b)$$

where a_i = distance of ith row from the N.A.

17.6. CASE III : NEUTRAL AXIS ALONG THE EDGE ($x_u = D$)

This is the special version of case II, where the N.A. falls along the outer edge, giving $x_u = D$. Consequently, the strain ε_{min} at the outer edge (i.e. at the N.A.) will be equal to zero and the strain at the inner edge or highly stressed edge will therefore be equal to 0.0035 according to assumption No. 7 as shown in Fig. 17.6 (*b*). The column

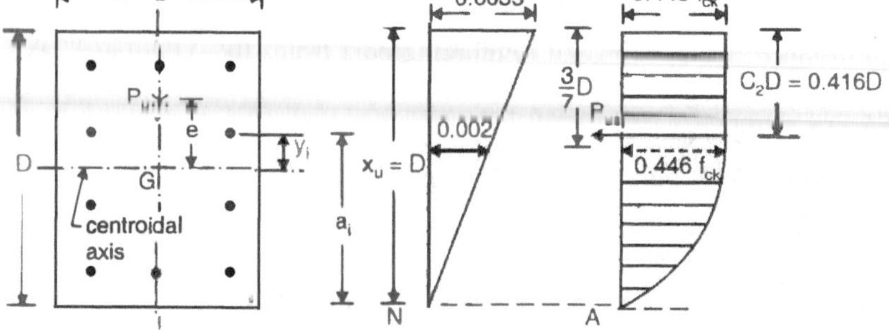

(a) The section (b) Strain diagram (c) Stress diagram

FIG. 17.6. CASE III ; N.A. AT THE OUTER EDGE ($x_u = D$)

is said to be on the *verge of cracking* for this position of N.A. Fig. 17.6 (*c*) shows the stress diagram for concrete. The equilibrium condition of forces gives

$$P_u = P_{uc} + \sum_{i=1}^{n} P_{usi} = 0.36 f_{ck}\, bD + \sum_{i=1}^{n} (f_{si} - f_{ci})\, A_{si}$$

or

$$P_u = 0.36 f_{ck}\, bD + \sum_{i=1}^{n} \frac{p_i\, bD}{100} (f_{si} - f_{ci})$$

or

$$\frac{P_u}{f_{ck}\, bD} = C_1 + \sum_{i=1}^{n} \frac{p_i}{100 f_{ck}} (f_{si} - f_{ci}) \quad ...(17.9)$$

where $C_1 = 0.36$ (see Table 17.1 also)

Also, balancing the moments of forces about the centroidal axis

$$M_u = M_{uc} + \sum_{i=1}^{n} M_{usi} = P_{uc}\left(\frac{D}{2} - 0.416\, D\right) + \sum_{i=1}^{n} A_{si}\, (f_{si} - f_{ci})\, y_i$$

or

$$M_u = 0.36 f_{ck}\, bD\left(\frac{D}{2} - 0.416\, D\right) + \sum_{i=1}^{n} \frac{p_i\, bD}{100} (f_{si} - f_{ci})\, y_i$$

or

$$\frac{M_u}{f_{ck}\, bD^2} = C_1\,(0.5 - C_2) + \sum_{i=1}^{n} \frac{p_i}{100 f_{ck}} (f_{si} - f_{ci}) \left(\frac{y_i}{D}\right) \quad ...(17.10)$$

where $C_1 = 0.36$ and $C_2 = 0.416$ (see Table 17.1 also)

Variation of strain across the section : The variation of strain across the section is triangular, with $\varepsilon_{max} = 0.0035$ at the inner edge and $\varepsilon_{min} = 0$ at the outer edge (i.e. at the N.A.). Hence at the ith row, distant a_i from the N.A. or y_i from the centroidal axis, ε_i is given by

$$\varepsilon_i = \frac{0.0035}{D} \cdot a_i = \frac{0.0035}{D}\left(\frac{D}{2} + y_i\right) \quad ...(17.11)$$

Determination of values of f_{si} : The values of f_{si} corresponding to ε_i at the ith row of reinforcement is determined from the stress-strain curve for the type of steel reinforcement. Alternatively, f_{si} can be determined from Table 2.3 corresponding to ε_i, by linear interpolation.

Determination of f_{ci} : The value of f_{ci} can be determined from the following equations:

For $\varepsilon_i < 0.002$, $f_{ci} = 446 f_{ck}\, \varepsilon_i\, (1 - 250\, \varepsilon_i)$ ...(17.8)

For $\varepsilon_i \geq 0.002$, $f_{ci} = 0.446 f_{ck}$

Alternatively, for $\varepsilon_i < 0.002$, $f_{ci} = 0.446 f_{ck}\left[\frac{2a_i}{\frac{4}{7}D} - \left(\frac{a_i}{\frac{4}{7}D}\right)^2\right]$ ...(17.8 *c*)

17.7. CASE IV : NEUTRAL AXIS LYING WITHIN THE SECTION ($x_u < D$)

This case arises when the eccentricity is *moderate* or *large*, due to which some part of the section is under tension. The stress block is the same as that for a beam, except that in this case the entire depth is effective, and $x_u = D$ in the limiting case. Hence x_u

is related to the total depth D and not to the effective depth (d) as taken in beams. Fig. 17.7 (a) shows the section, in which N.A. is within the section. Fig. 17.7 (b) shows the strain diagram, in which the maximum strain is 0.0035 at the highly compressed edge, and there is tension below the N.A. Fig. 17.7 (c) shows the stress diagram, which is similar to that for a beam.

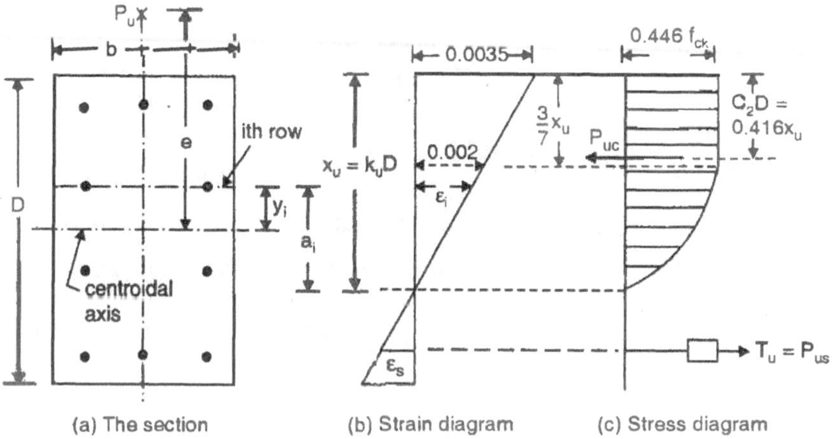

FIG. 17.7. CASE IV : N.A. WITHIN THE SECTION ($x_u < D$)

Evidently, the stress block parameters are as under :

(i) Area of the stress block $= 0.36\, f_{ck}\, k_u\, D = C_1\, f_{ck}\, D$ where $C_1 = 0.36\, k_u$

(ii) Distance of the c.g. of compressive force from highly compressed edge $= 0.416\, x_u = 0.416\, k_u\, D = C_2\, D$ where $C_2 = 0.416\, k_u$

It should be clearly noted that unlike flexural members, there is no limitation on the depth ($x_u = k_u D$) of the neutral axis in compression members, and in extreme case, k_u can be equal to one. Balancing the axial forces on the section, we have

$$P_u = P_{uc} + \sum_{i=1}^{n} P_{usi} = 0.36\, f_{ck}\, k_u\, b\, D + \sum_{i=1}^{n} (f_{si} - f_{ci})\, A_{si}$$

or $$\frac{P_u}{f_{ck}\, bD} = 0.36\, k_u + \sum_{i=1}^{n} \frac{p_i}{100\, f_{ck}} (f_{si} - f_{ci}) = C_1 + \sum_{i=1}^{n} \frac{p_i}{100\, f_{ck}} (f_{si} - f_{ci}) \quad \text{...(17.12)}$$

Similarly, taking moments of forces, about the centroid of the section.

$$M_u = M_{uc} + \sum_{i=1}^{n} M_{usi} = 0.36\, f_{ck}\, k_u\, bD\, (0.5\, D - 0.416\, k_u\, D) + \sum_{i=1}^{n} (f_{si} - f_{ci})\, A_{si}\, y_i \quad \text{...(17.13}a)$$

or $$\frac{M_u}{f_{ck}\, bD^2} = 0.36\, k_u\, (0.5 - 0.416\, k_u) + \sum_{i=1}^{n} \frac{p_i}{100\, f_{ck}} (f_{si} - f_{ci}) \frac{y_i}{D} \quad \text{...(17.13)}$$

or $$\frac{M_u}{f_{ck}\, bD^2} = C_1\, (0.5 - C_2) + \sum_{i=1}^{n} \frac{p_i}{100\, f_{ck}} (f_{si} - f_{ci}) \frac{y_i}{D} \quad \text{...(17.13 }b)$$

where, $C_1 = 0.36\, k_u$, $C_2 = 0.416\, k_u$ and $p_i = \frac{100\, A_{si}}{b\, D}$

In all these equations, f_{ci} representing the deduction for concrete replaced by steel bars can be neglected. However, as a better approximation, a constant value of f_{ci} corresponding to M 20 grade concrete is normally used.

Determination of strain ε_i at i^{th} row of reinforcement : From Fig. 17.7 (*a*) the strain ε_i is given by

$$\varepsilon_i = 0.0035 \frac{a_i}{x_u} = \frac{0.0035}{x_u}\left(x_u - \frac{D}{2} + y_i\right) \qquad \text{...(17.14)}$$

where a_i = distance of ith row from N.A., taken +ve for values above N.A. and negative for values below N.A.

y_i = distance of ith row from the centroidal axis, taken +ve for rows above the centroidal axis and −ve for rows below the centroidal axis.

Determination of f_{si} : The values of f_{si} corresponding to ε_i at the ith row of reinforcement is determined from the stress-strain curve for the type of steel reinforcement. Alternatively, f_{si} can be determined from Table 2.3 corresponding to ε_i, by linear interpolation.

Determination of f_{ci}: The value of f_{ci} can be determined from the following equations.

For $\varepsilon_i < 0.002$, $f_{ci} = 446 f_{ck}\, \varepsilon_i (1 - 250\, \varepsilon_i)$ ...(17.8)

or Alternatively,

$$f_{ci} = 0.446 f_{ck}\left[\frac{2\, a_i}{(x_u - \frac{3}{7} D)} - \left\{\frac{a_i}{(x_u - \frac{3}{7} D)}\right\}^2\right] \qquad \text{...(17.8 a)}$$

and for $\varepsilon_i \geq 0.002$, $f_{ci} = 0.446 f_{ck}$.

17.8. MODES OF FAILURE IN ECCENTRIC COMPRESSION

Though Indian Standard IS 456 : 2000, in its clause 39, uses the term *'limit state of collapse in compression'* to describe the 'limit state' of eccentrically loaded compression member, the actual failure need not necessarily occur in compression. The mode of failure depends upon the relative magnitudes of P_u and M_u and hence on the magnitude of eccentricity e. If the eccentricity $e = M_u/P_u$ is relatively small, the axial *compression behaviour* predominates, and the consequent failure is termed '*compression failure*'. On the other hand, if the eccentricity is *large*, the *flexural behaviour* predominates, and the consequent failure is termed '*tension failure*'. In between these two, there exists a '*balanced failure*' in which the yielding of the outermost row of longitudinal steel on the tension side and the attainment of maximum compressive strain in concrete, ($\varepsilon_c = 0.0035$) at the highly compressed edge of the column occur *simultaneously*. Thus, in terms of yield strain, we have the following three types of failures :

1. Compression failure

(*a*) Failure by crushing of concrete without tension on the other face

$$(\varepsilon_{c,\,max} = 0.0035 - 0.75\, \varepsilon_{c,\,min})$$

(*b*) Failure by crushing of concrete with some tension on the other face

$$(\varepsilon_{c,\,max} = 0.0035)$$

2. Tension failure :

(*c*) The tensile stress in the reinforcement reaches the yield limit before the strain in the compression concrete reaches the crushing strain.

$$(\varepsilon_{st} = \varepsilon_{sy}\ ;\ \varepsilon_c < 0.0035)$$

3. *Balanced failure*

(*d*) Yielding of outermost row of steel on the tension side and attainment of maximum compressive strain in concrete at the highly compressed edge of the column occur simultaneously.

$$(\varepsilon_{st} = \varepsilon_{sy} \ ; \ \varepsilon_c = 0.0035)$$

As indicated above, there are two types (type *a* and type *b*) of *compression failures*. Type *a* compression failure occurs when P_u is *large* and *e* is *small*, resulting in compression prevailing over the entire section, with N.A. falling out side the section. Failure occurs by *crushing* of concrete when the strain ($\varepsilon_{c\,.\,max}$) at the highly compressed edge reaches its maximum value given by Eq. 17.1. It should be carefully noted that type (*a*) compression failure is *sudden* and collapse takes place without giving any warning. *Type (b) compression failure* occurs under the condition of large eccentricity of the load (i.e. under large moment), with the result that the section is subjected to flexural action causing N.A. to lie inside the section, similar to the case of a beam. Hence, similar to the failure of a beam, the column section may have three modes of failure : (*i*) *compression failure* (or over-reinforced failure), (*ii*) *tension failure* and (*iii*) *balanced failure*. This *compression failure* is the type (*b*) compression failure which occurs by crushing of concrete ($\varepsilon_{c\,.\,max} = 0.0035$) prior to the yield of tension side steel. On the other hand, *tension failure* (i.e. type *c* failure), also called the *under-reinforced failure*, is initiated by yielding of tension steel prior to crushing of concrete. The outermost longitudinal bars in the tension side of the N.A. first under go yielding and successive inner rows (if provided), on the tension side of the N.A. may also yield in tension with increasing strain. *Eventually, collapse occurs when the concrete at the highly compressed edge gets crushed*. Lastly, the *balanced failure* occurs when the yielding of outermost row of steel ($\varepsilon_{st} = \varepsilon_{sy}$) on the tension side and the attainment of maximum compressive strain (= 0.0035) in concrete at the highly compressed edge of the column occur simultaneously. This condition occurs at a specific value of eccentricity (known as $e = e_b$) giving specific value of $x_u = x_{ub}$.

Thus, when $e < e_b$, compression failure occurs. $(x_u > x_{ub})$

when $e > e_b$, tension failure occurs $(x_u < x_{ub})$

and when $e = e_b$, balanced failure occurs $(x_u = x_{ub})$

17.9. DESIGN STRENGTH : $P_u - M_u$ INTERACTION CURVE

The design strength of an uniaxially loaded short column depends upon the eccentricity of loading. The design strength of such a column has two components :

(*i*) Axial compression component P_u

and (*ii*) Uniaxial moment component M_u ($= P_u \,.\, e$)

We have seen earlier that there exists a unique strain profile and consequent N.A. location at the ultimate limit state, corresponding to a given eccentricity (*e*) of the loading. Corresponding to this distribution of strain, one can determine the two components of strength: P_u ($= P_{uc} + P_{us}$) and M_u ($= M_{uc} + M_{us}$). This eccentricity *e* can be varied in the range $0 \le e < \infty$, to get the pairs of strength components (P_u , M_u) for each case and an *interaction curve* can be obtained. An *interaction curve* is a diagram which shows the variation of strength component P_u with different values of strength component M_u, for a column section of given proportions. Each point on the curve corresponds to the design strength values of P_u and M_u associated with a specific eccentricity (e) of loading on the given column section. If the point, given by the co-ordinates (P_u , M_u), falls within the design interaction curve, the column is *safe*, otherwise it is not. *In other words, the design interaction curve serves as a failure envelope.* A given reinforced concrete section can resist, in presence of moment M_u, an axial load P_u which is less than the pure axial load P_{uz} it can carry when $M_u = 0$. Similarly, the section will resist a moment M_u less than M_{ub} in presence of $P_u > P_{ub}$, where suffix b represents balanced failure condition. *Thus the effect of increase in one strength parameter is to reduce the other strength parameter.* Thus for a given section, there is always one particular value of design P_u accompanying a given design M_u.

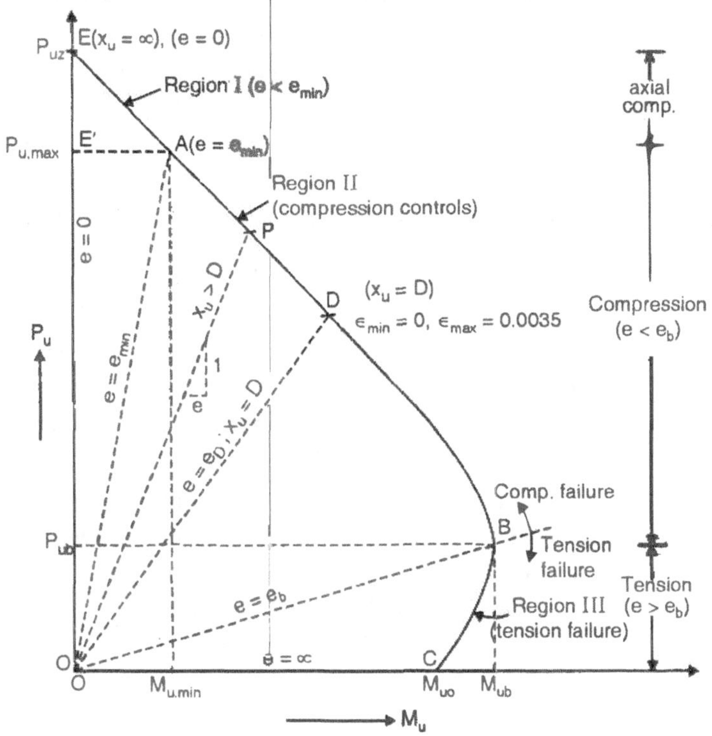

FIG. 17.8. $P_u - M_u$ INTERACTION DIAGRAM

Fig. 17.8 shows a typical $P_u - M_u$ interaction curve /diagram for a column, section of given proportions. The curve represents various pairs of values of design $P_u - M_u$ values for different eccentricities. Such a curve (*EADBC*) represents various modes of possible failure along its length. The entire curve has three prominent regions :

(*i*) Region I of minimum eccentricity

(*ii*) Region II : Compression control region

(*iii*) Region III : Tension control region.

When the relative magnitudes of P_u and M_u are such that the entire column is under compression, we have the *compression control region*, in which case the N.A. lies *outside* the section. When the relative values of P_u and M_u are such that a part of the section is in tension such that the strain in tension steel is greater than the yield point strain when the compressive strain in the concrete reaches 0.0035, we have the *tension control region*, in which case the N.A. lies inside the section.

The *compression control region*, when the entire column section is under compression, can further be subdivided in two regions: (*i*) *region I* in which the eccentricity e is less than e_{min} specified by the Code and (*ii*) *region II*, in which eccentricity e exceeds e_{min}. The first region is indicated by the line EA of the interaction curve. Point E indicates the failure load P_{uz} when the column section is subjected to a perfectly axial load, with zero eccentricity. Point A on the curve indicates the failure load of the column subjected to axial load with *nominal eccentricity* equal to e_{min}. The moment corresponding to this eccentricity is equal to $P_u \cdot e_{min}$.

The region II where compression controls, indicated by curve AB, is controlled by crushing strain in concrete, indicated by increasing moment on the column. Point B on the interaction curve is called the *balanced failure point* (P_{ub}, M_{ub}) wherein the column will fail by simultaneous occurrence of limiting strains in concrete and tension side reinforcement.

With further increase in eccentricity, we enter into the region III (curve BC), where tension controls the failure. The axial load carrying capacity of the section decreases rapidly in the zone BC as the eccentricity increases. This is because of higher tensile force of the reinforcement. When the compressive axial load is zero, corresponding to point C, the column section behaves as a doubly reinforced beam with its moment capacity (or pure moment capacity) equal to M_{u0}. A section reinforced with nearly equal steel on both faces is invariably an under-reinforced section in which the moment resisting capacity M_u is governed by yielding of tension steel (i.e. a tension failure).

Characteristics of salient points on the $P_u - M_u$ interaction curve : Table 17.2 shows the characteristics of salient points E, D, B, C on the interaction curve (Fig 17.8) corresponding to failure strain profiles marked as E, D, B, C in Fig. 17.9. Point P represents general point on the curve, with N.A. outside the section.

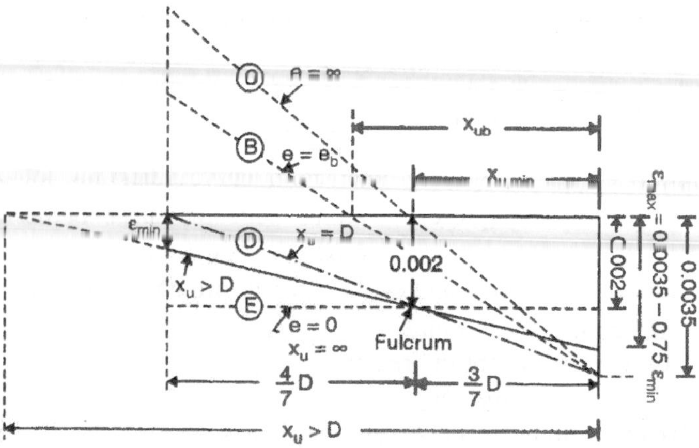

FIG. 17.9. STRAIN PROFILES UNDER LIMIT STATE IN ECCENTRIC COMPRESSION

TABLE 17.2. CHARACTERISTICS OF SALIENT POINTS ON INTERACTION CURVE

Point	P_u	M_u	e	x_u	$f_{s,min}$	$\varepsilon_{c\ max}$	$\varepsilon_{c\ min}$
E	P_{uz}	0	0	∞	f_{sc}	0.002	0.002
A	$P_{u\ max}$	$M_{u\ min}$	0.05 D	D to 1.1 D	f_{sc}	< 0.0035	< 0.002
D	P_{uD}	M_{uD}	e_D	D	0	0.0035	0
B	P_{ub}	M_{ub}	e_b	x_{ub}	$-0.87\ f_y$	0.0035	$-\varepsilon_{sy}$
C	0	M_{u0}	∞	$x_{u,min}$	$0.87\ f_y$	< 0.0035	$-\varepsilon_{sy}$

Note : 1. Negative sign denotes tension

2. $x_{u,min}$ denotes minimum depth of N.A. corresponding to a doubly reinforced section.

Design interaction curves : The method of drawing an interaction curve for a column section of known dimensions (i.e. b, d, A_s, cover etc) and known material properties (i.e. f_{ck}, f_y) has been illustrated in Examples 17.13 and 17.19.

17.10. METHOD OF ANALYSIS OF ECCENTRICALLY LOADED SHORT COLUMN

There are two types of problems in eccentrically loaded short columns: (*i*) Analysis, and (*ii*) design. The analysis consists of determination of M_u for a given value of P_u (or vice versa) or determination of e for a given value of P_u or determination of M_u, for a column of given configuration. The design consists of determination of column configuration (i.e. b, D, A_s etc) for a given set of values of P_u , M_u. The methods of design of eccentrically loaded short column have been discussed in the next article.

The analysis of eccentrically loaded short column can be done by the following methods

1. Analysis using equations 2. Analysis using interaction curves/Charts.

1. Analysis using equations : As observed in §17.5, 17.6 and 17.7, the values of P_u and M_u depend on x_u (or k_u), i.e. on the location or position of the N.A. which is not known in the beginning. The equations for (P_u, M_u) for the case of $x_u > D$ (or $k_u > 1$) are different from those for the case of $x_u < D$ ($k_u < 1$), and hence in the beginning one does not know from which equation the analysis should be started with. The solution of the problem consists of a trial and error procedure. The following procedure is suggested.

Procedure

Step 1 : To start with, assume that the N.A. lies on the outer edge, so that $x_u = D$ or $k_u = 1$. Thus for this location of N.A., we get $C_1 = 0.361$ and $C_2 = 0.416$ from Table 17.1.

Step 2 : Compute the values of P_{u1} from Eq. 17.9 by first determining the values of f_{si} and f_{ci} corresponding to the values of strains ε_i for this location of N.A.

Step 3 : Compare this value of P_{u1} with the given values of P_u

If given $P_u < P_{u1}$, actual $x_u < D$

If given $P_u > P_{u1}$ actual $x_u > D$

Step 4 : In the second trial, select new or revised value of x_u, based on the deviation or difference between P_u and P_{u1}. Obtain new values of C_1 and C_2 from Table 17.1 and compute new value P_{u2} using appropriate equation.

Step 5 : Compare new value of P_{u2} with the given value of P_u. If P_{u2} tallies with given P_u, the trial gives correct position of N.A. If not, make next trial with revised value of x_u.

Step 6 : Repeat the process till the computed value of P_u tallies with the given value of P_u. When this happens, correct value of x_u is obtained. Compute the value of M_u for this location of N.A. Examples 17.14 and 17.15 illustrate the procedure

2. Analysis using interaction curves/Charts : As seen from above, the method of analysis using basic equations is laborious and time consuming. Hence the analysis can be done with the help of interaction curves/Charts. Even clause 39.5 of IS 456 : 2000 gives the following recommendations : "The design of member subject to combined axial load and uniaxial bending will involve lengthy calculation by trial and error. In order to overcome these difficulties, interaction diagrams may be used. These have been prepared and published by BIS in SP : 16. Design aids for reinforced concrete to IS 456".

Although the above statement refers to *design*, it is equally valid in the case of *analysis*. The Charts consists of interaction diagrams for P_u and M_u in *non-dimensional* form. The non-dimensional parameters used in the interaction curves prepared by BIS in SP : 16 are $P_u/f_{ck}\,bD$, $M_u/f_{ck}\,bD^2$ and p/f_{ck} for rectangular sections and $P_u/f_{ck}\,D^2$, $M_u/f_{ck}\,D^3$ and p/f_{ck} for circular sections. Separate Charts, have been prepared for the following parameters:

(*a*) *Shape of section* (*i*) Rectangular and (*ii*) circular shapes
(*b*) *Placement of steel reinforcement:* (*i*) symmetrical placement on opposite sides, and (*ii*) uniformly spaced placement all around the periphery of the section.
(*c*) *Grade of steel* : (*i*) Fe 250, (*ii*) Fe 415, (*iii*) Fe 500
(*d*) *Ratio of cover to total depth* (*i*) $d'/D = 0.05$, (*ii*) $d'/D = 0.1$
(*iii*) $d'/D = 0.15$, and (*iv*) $d'/D = 0.2$

Thus, a large number of Charts have been presented, covering various cases. If however, a given or assumed section of a given shape does not fit in any particular Chart, two nearest Charts, for that shape may be used and the desired value may be obtained by linear interpolation. The following Charts, corresponding to commonly used Fe 415 grade reinforcement have been *reproduced* in this book for ready reference.

Shape of section	*Steel grade*	*Steel placement*	*d'/D ratio*	*Chart No. of SP 16*	*Fig. No. of the book*
(*a*) ***Rectangular***	Fe 415	Equal steel on opposite faces	0.05	31	17.10
			0.1	32	17.11
			0.15	33	17.12
			0.20	34	17.13
		Uniform steel all round	0.05	43	17.14
			0.1	44	17.15
			0.15	45	17.16
			0.2	46	17.17
(*b*) ***Circular***	Fe 415	Uniform steel all round	0.05	55	17.18
			0.1	56	17.19
			0.15	57	17.20
			0.2	58	17.21

Examples 17.1, 17.2, 17.3 and 17.4. illustrate the method of using these Charts for various types of analysis problems.

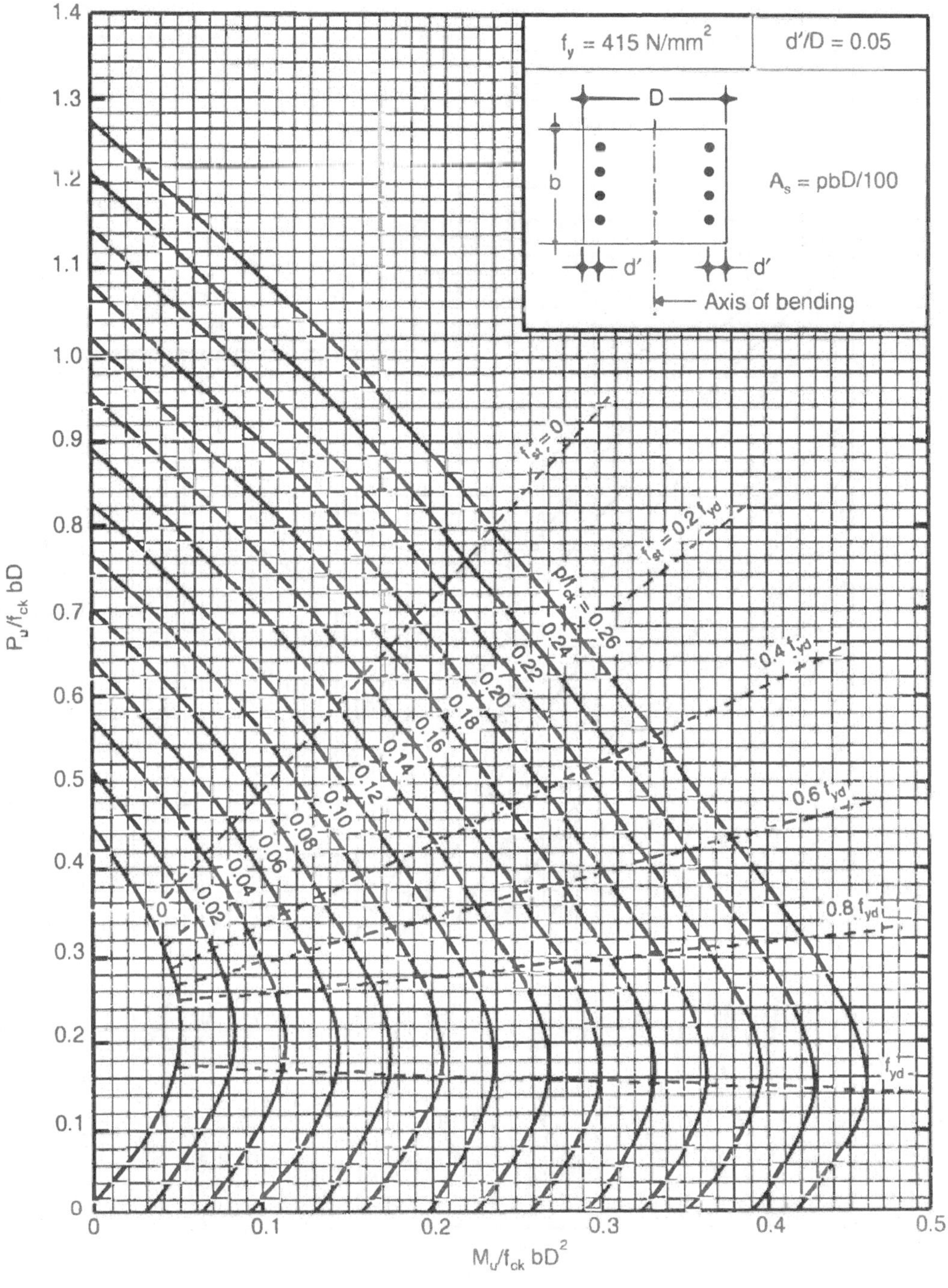

f_y = 415 N/mm^2
d'/D = 0.05
D
b
d'
d'
Axis of bending
A_s = pbD/100
P_u/f_{ck} bD
M_u/f_{ck} bD^2
f_{st} = 0
f_{st} = 0.2 f_{yd}
0.4 f_{yd}
0.6 f_{yd}
0.8 f_{yd}
f_{yd}
p/f_{ck} = 0.26
0.24
0.22
0.20
0.18
0.16
0.14
0.12
0.10
0.08
0.06
0.04
0.02
0
1.4
1.3
1.2
1.1
1.0
0.9
0.8
0.7
0.6
0.5
0.4
0.3
0.2
0.1
0
0.1
0.2
0.3
0.4
0.5

FIG. 17.10

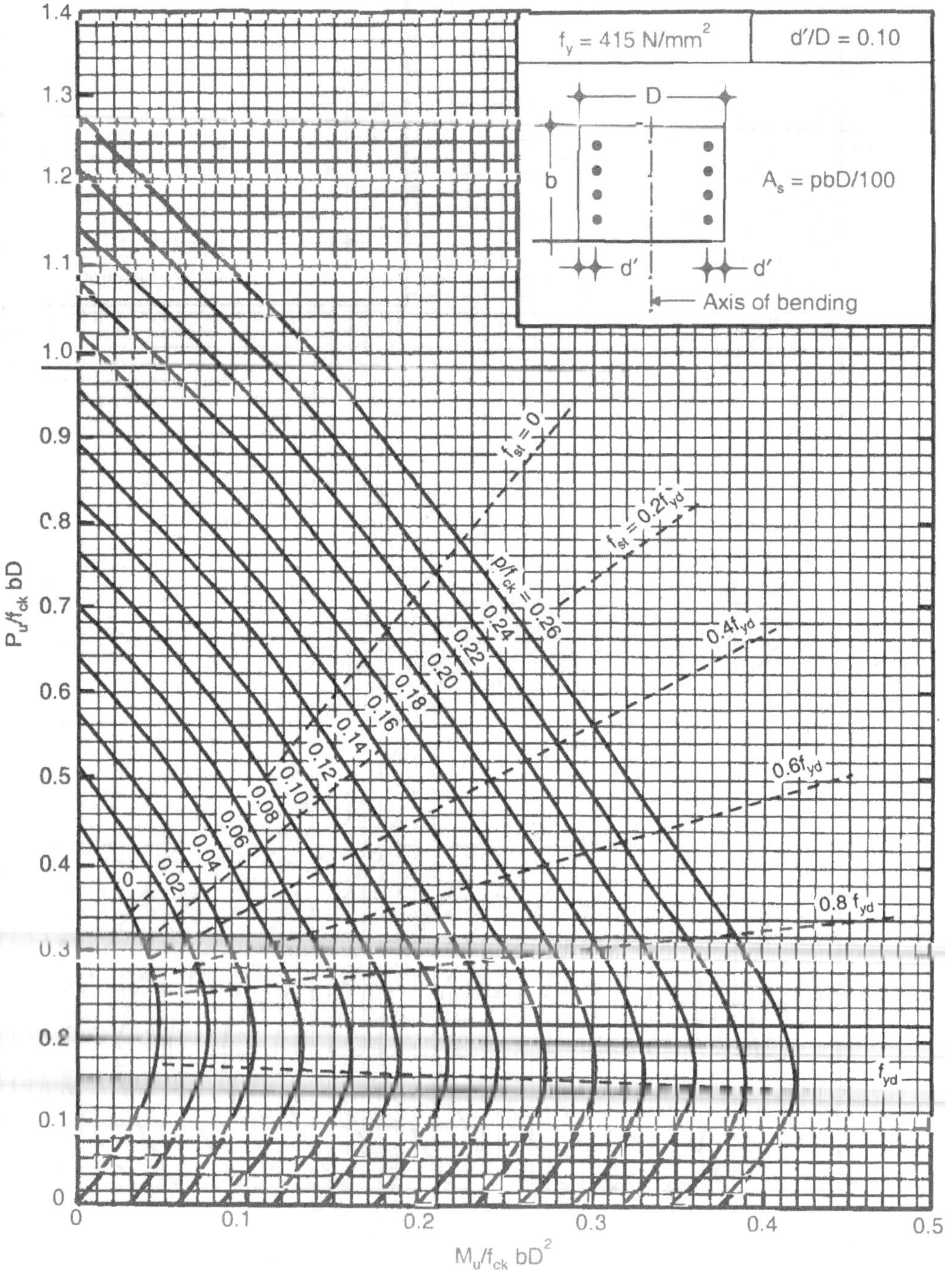

f_y = 415 N/mm^2
d′/D = 0.10
D
b
A_s = pbD/100
d′
d′
Axis of bending
P_u/f_ck bD
M_u/f_ck bD^2
f_st = 0
f_st = 0.2f_yd
0.4f_yd
0.6f_yd
0.8 f_yd
f_yd
p/f_ck = 0.26
0.24
0.22
0.20
0.18
0.16
0.14
0.12
0.10
0.08
0.06
0.04
0.02
0

FIG. 17.11

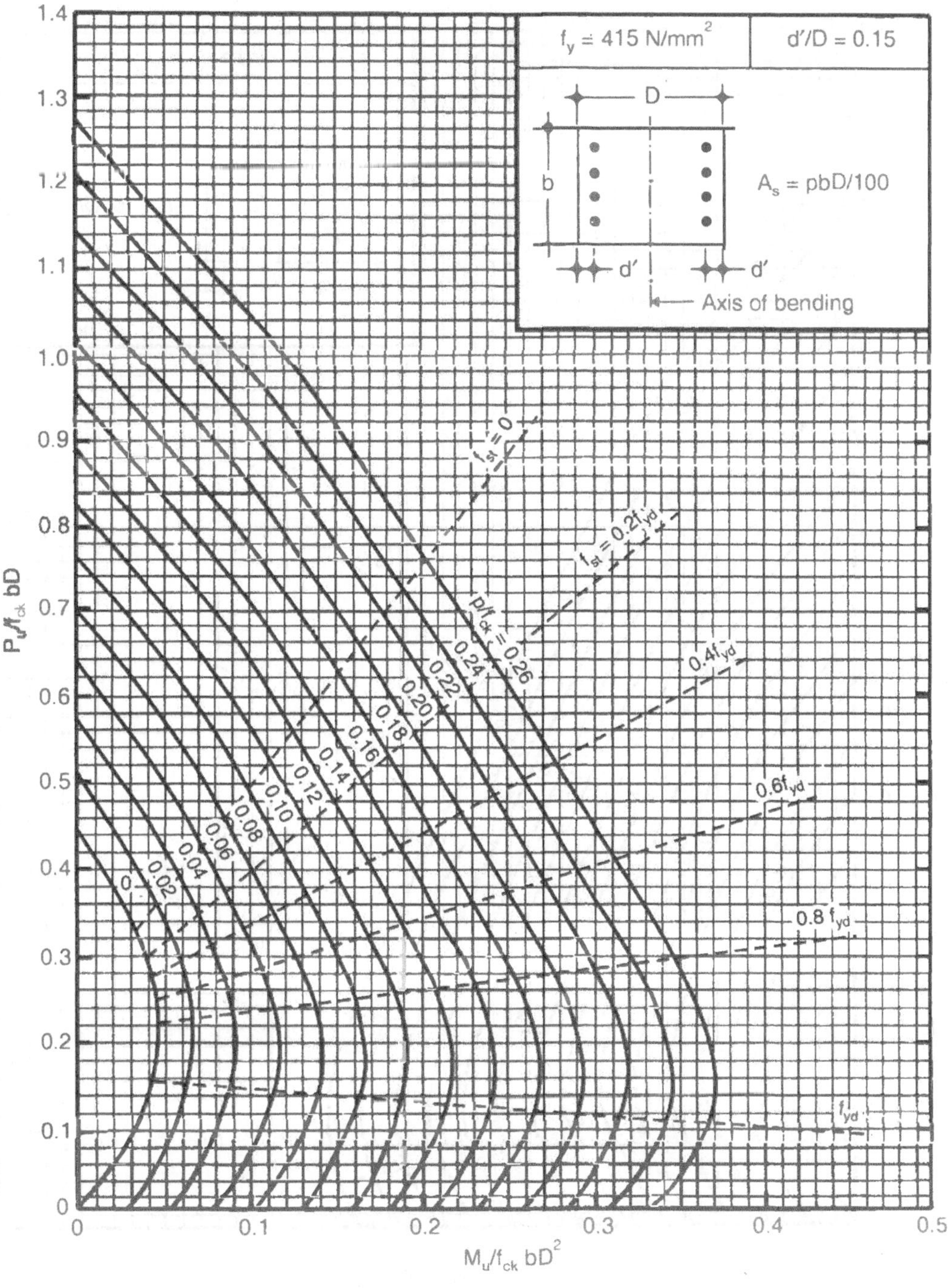

f_y = 415 N/mm^2
d′/D = 0.15
D
b
d′
d′
A_s = pbD/100
Axis of bending
$f_{st} = 0$
$f_{st} = 0.2f_{yd}$
$0.4f_{yd}$
$0.6f_{yd}$
$0.8\ f_{yd}$
f_{yd}
p/f_{ck} = 0.26
0.24
0.22
0.20
0.18
0.16
0.14
0.12
0.10
0.08
0.06
0.04
0.02
0
$P_u/f_{ck}\ bD$
$M_u/f_{ck}\ bD^2$

FIG. 17.12

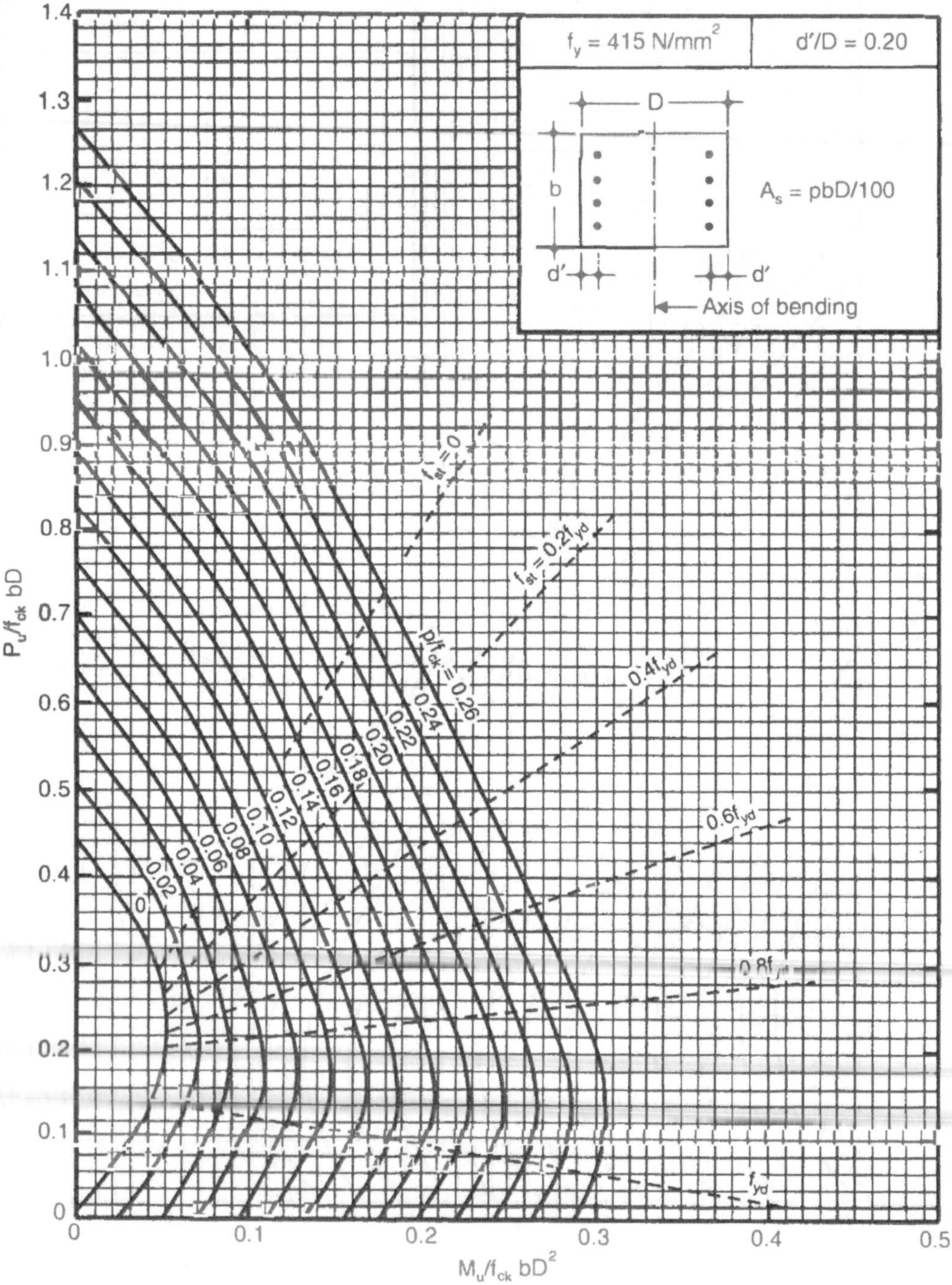

f_y = 415 N/mm^2
d'/D = 0.20
D
b
d'
d'
Axis of bending
A_s = pbD/100
f_st = 0
f_st = 0.2f_yd
0.4f_yd
0.6f_yd
0.8f_yd
f_yd
p/f_ck = 0.26
0.24
0.22
0.20
0.18
0.16
0.14
0.12
0.10
0.08
0.06
0.04
0.02
0
P_u/f_ck bD
M_u/f_ck bD^2

FIG. 17.13

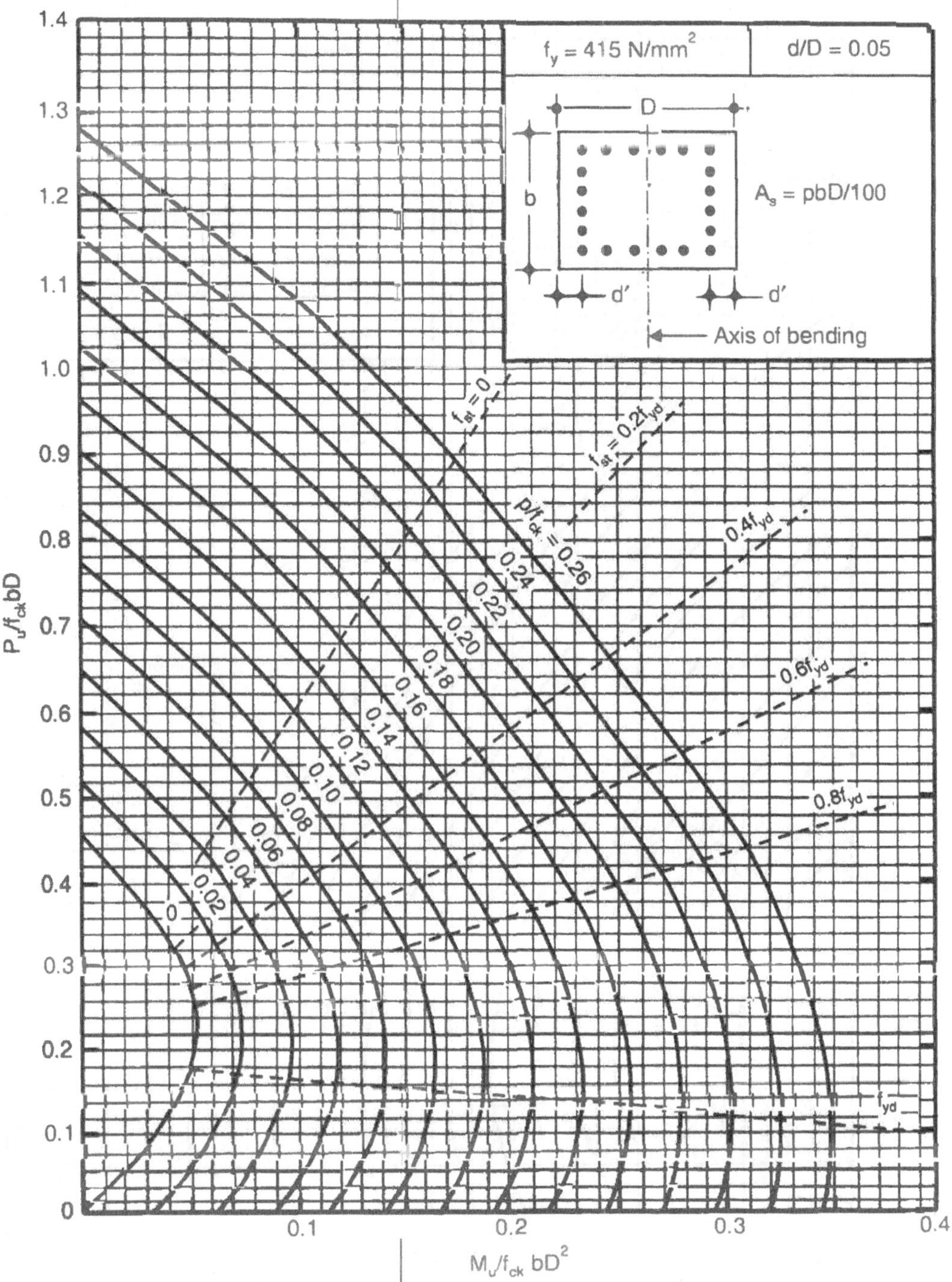
f_y = 415 N/mm²
d/D = 0.05
D
b
A_s = pbD/100
d'
d'
Axis of bending
f_{st} = 0
f_{st} = 0.2f_{yd}
0.4f_{yd}
0.6f_{yd}
0.8f_{yd}
f_{yd}
p/f_{ck} = 0.26
0.24
0.22
0.20
0.18
0.16
0.14
0.12
0.10
0.08
0.06
0.04
0.02
0
$P_u/f_{ck}bD$
$M_u/f_{ck}bD^2$
1.4
1.3
1.2
1.1
1.0
0.9
0.8
0.7
0.6
0.5
0.4
0.3
0.2
0.1
0
0.1
0.2
0.3
0.4

FIG. 17.14

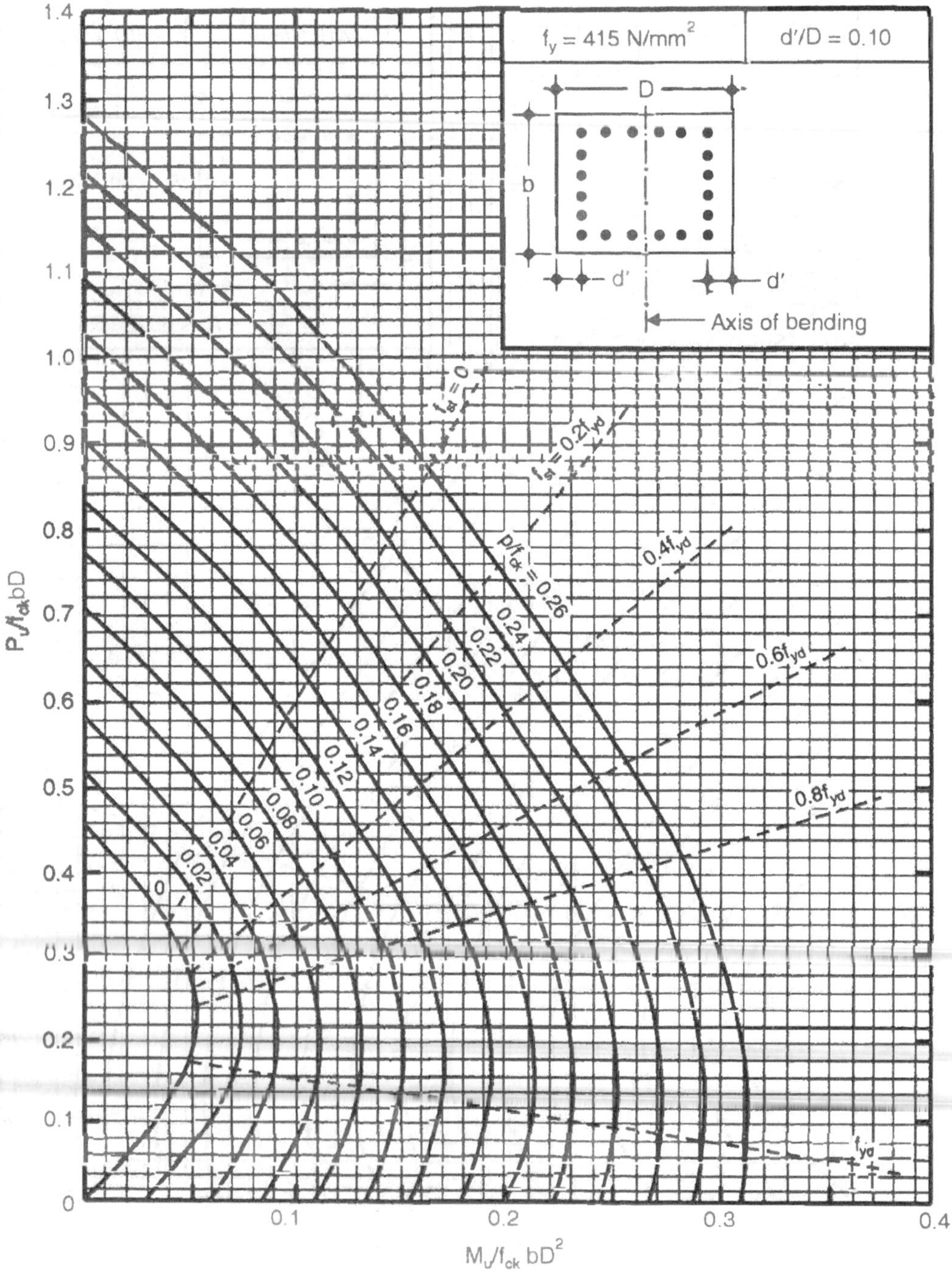
f_y = 415 N/mm²
d'/D = 0.10
D
b
d'
d'
Axis of bending
f_{st} = 0
f_{st} = 0.2f_{yd}
0.4f_{yd}
0.6f_{yd}
0.8f_{yd}
f_{yd}
p/f_{ck} = 0.26
0.24
0.22
0.20
0.18
0.16
0.14
0.12
0.10
0.08
0.06
0.04
0.02
0
$P_u/f_{ck}bD$
$M_u/f_{ck}bD^2$
0
0.1
0.2
0.3
0.4
0.5
0.6
0.7
0.8
0.9
1.0
1.1
1.2
1.3
1.4

FIG. 17.15

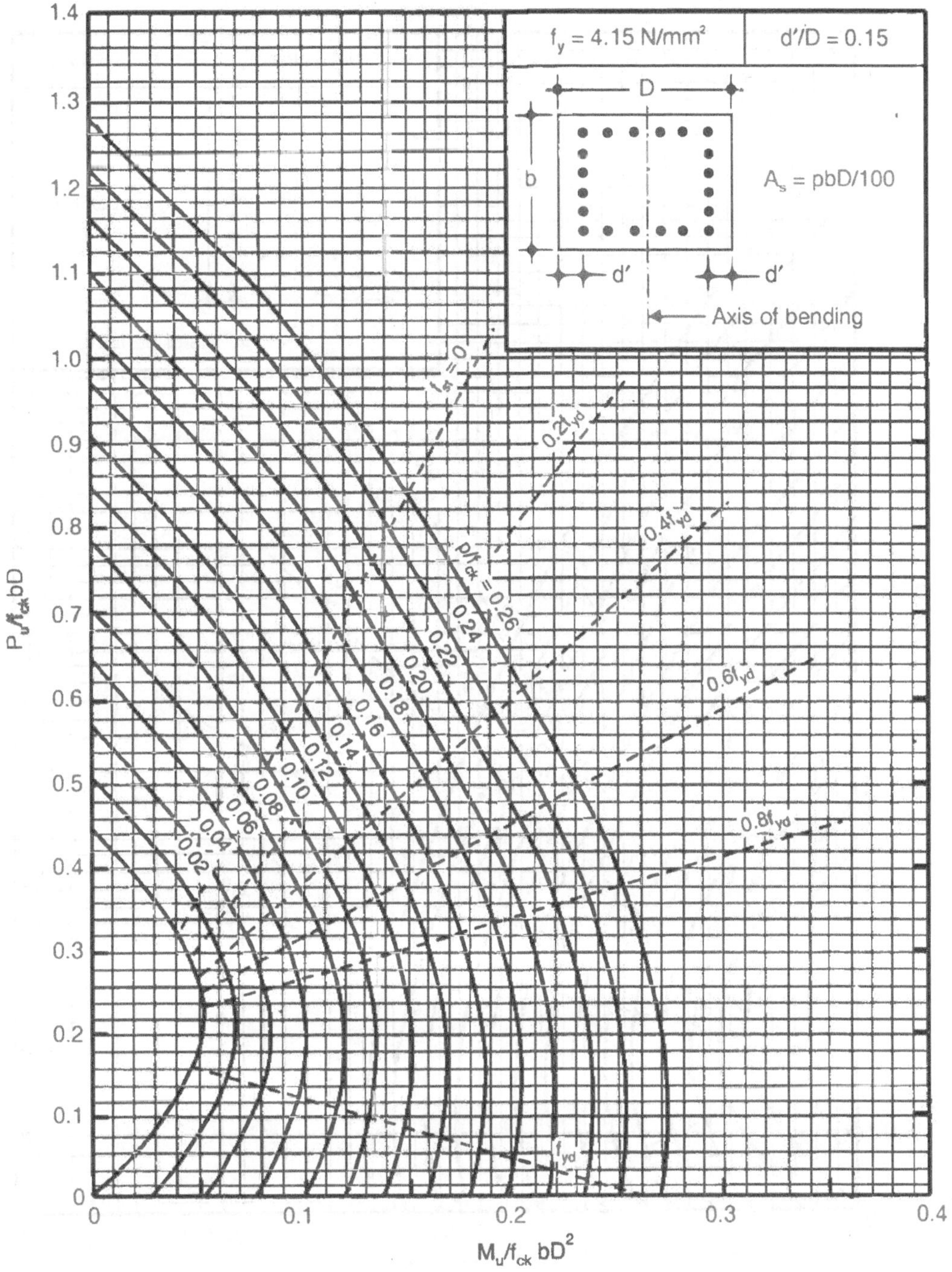

f_y = 4.15 N/mm²
d′/D = 0.15
D
b
d′
d′
A_s = pbD/100
Axis of bending
f_{st} = 0
$0.2f_{yd}$
$0.4f_{yd}$
$0.6f_{yd}$
$0.8f_{yd}$
f_{yd}
p/f_{ck} = 0.26
0.24
0.22
0.20
0.18
0.16
0.14
0.12
0.10
0.08
0.06
0.04
0.02
$P_u/f_{ck}bD$
$M_u/f_{ck}bD^2$

FIG. 17.16

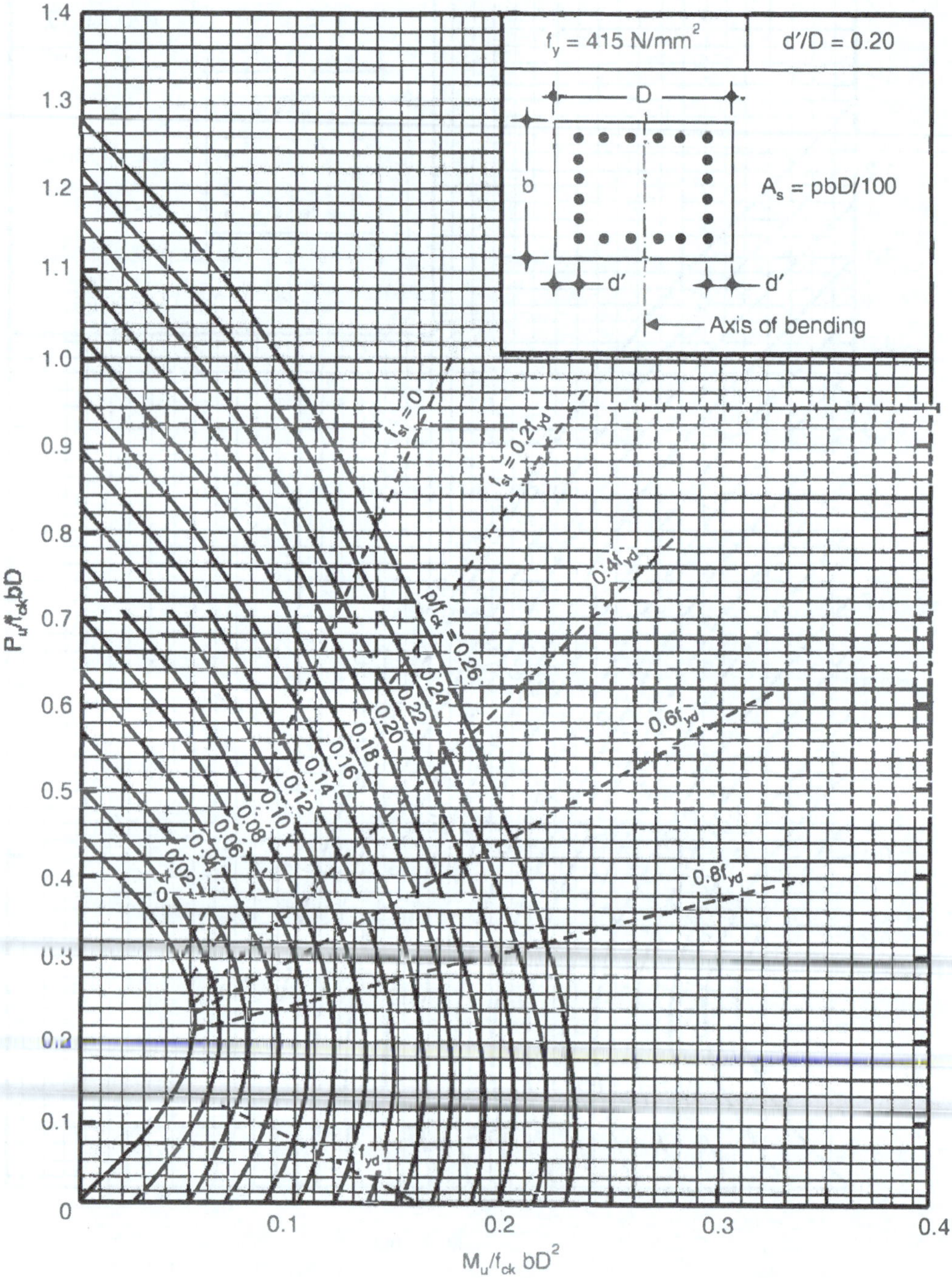

fy = 415 N/mm2
d'/D = 0.20
D
b
As = pbD/100
d'
d'
Axis of bending
fst = 0
fst = 0.2fyd
0.4fyd
0.6fyd
0.8fyd
fyd
p/fck = 0.26
0.24
0.22
0.20
0.18
0.16
0.14
0.12
0.10
0.08
0.06
0.04
0.02
0
Pu/fck bD
Mu/fck bD2
0
0.1
0.2
0.3
0.4
0.5
0.6
0.7
0.8
0.9
1.0
1.1
1.2
1.3
1.4

FIG. 17.17

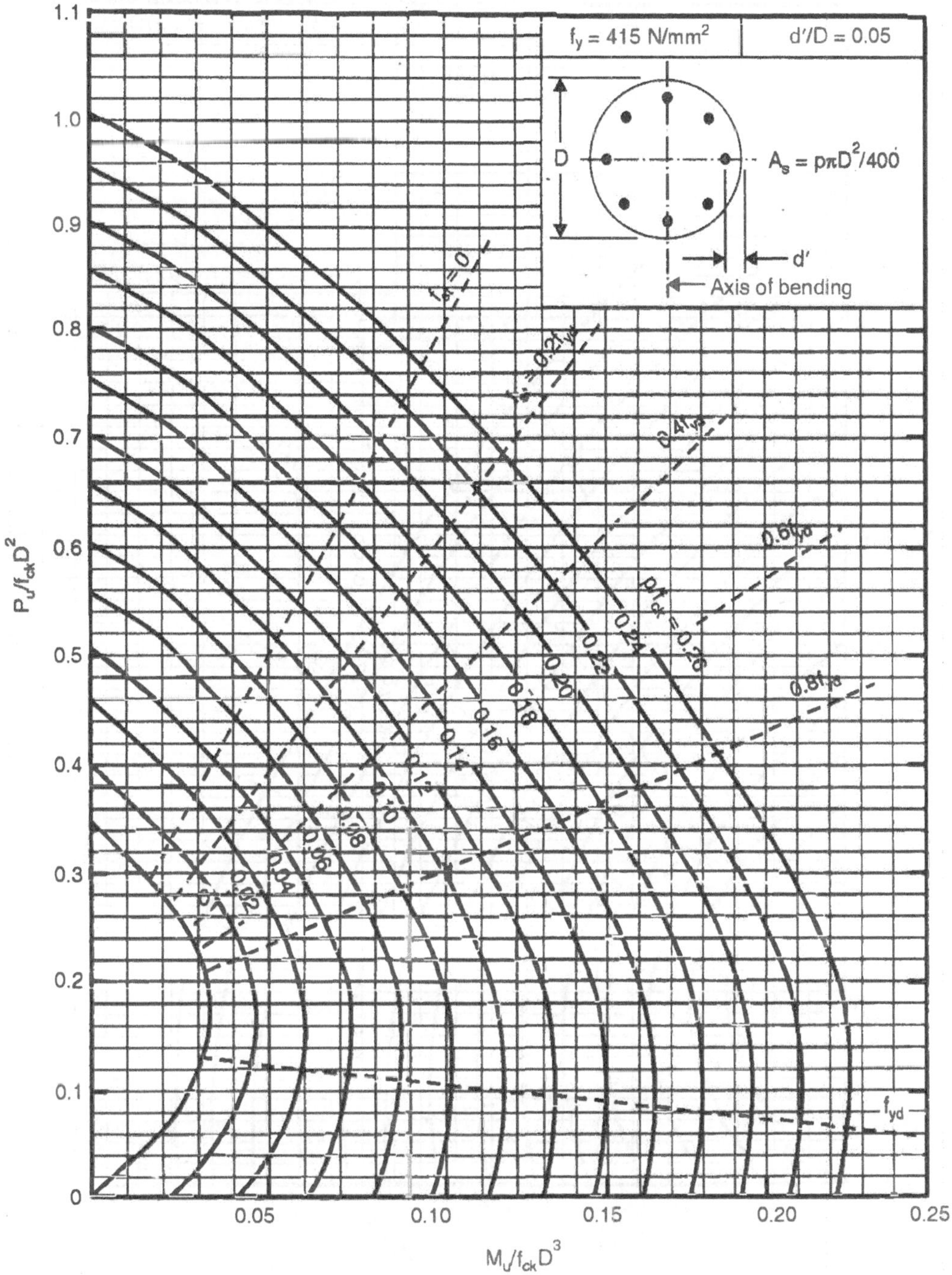
f_y = 415 N/mm²
d'/D = 0.05
D
$A_s = p\pi D^2/400$
d'
Axis of bending
$f_{st} = 0$
$f_{st} = 0.2f_{yd}$
$0.4f_{yd}$
$0.6f_{yd}$
$0.8f_{yd}$
f_{yd}
$p/f_{ck} = 0.26$
0.24
0.22
0.20
0.18
0.16
0.14
0.12
0.10
0.08
0.06
0.04
0.02
0
$P_u/f_{ck}D^2$
$M_u/f_{ck}D^3$
0
0.1
0.2
0.3
0.4
0.5
0.6
0.7
0.8
0.9
1.0
1.1
0.05
0.10
0.15
0.20
0.25

FIG. 17.18

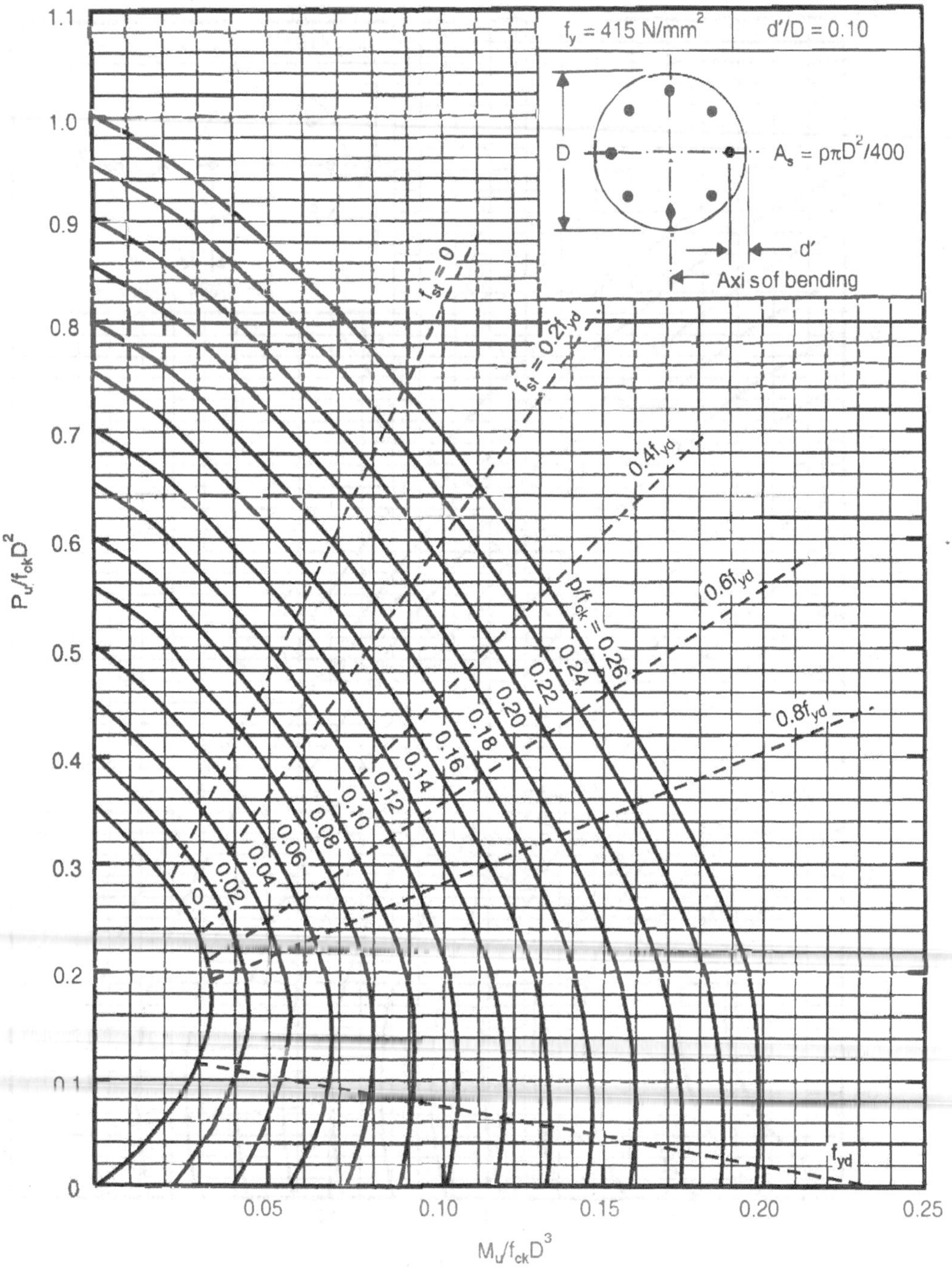

f_y = 415 N/mm^2
d'/D = 0.10
D
$A_s = \rho\pi D^2/400$
d'
Axi s of bending
$P_u/f_{ck}D^2$
$M_u/f_{ck}D^3$
$f_{st} = 0$
$f_{st} = 0.2f_{yd}$
$0.4f_{yd}$
$0.6f_{yd}$
$0.8f_{yd}$
f_{yd}
p/f_{ck} = 0.26
0.24
0.22
0.20
0.18
0.16
0.14
0.12
0.10
0.08
0.06
0.04
0.02
0

FIG. 17.19

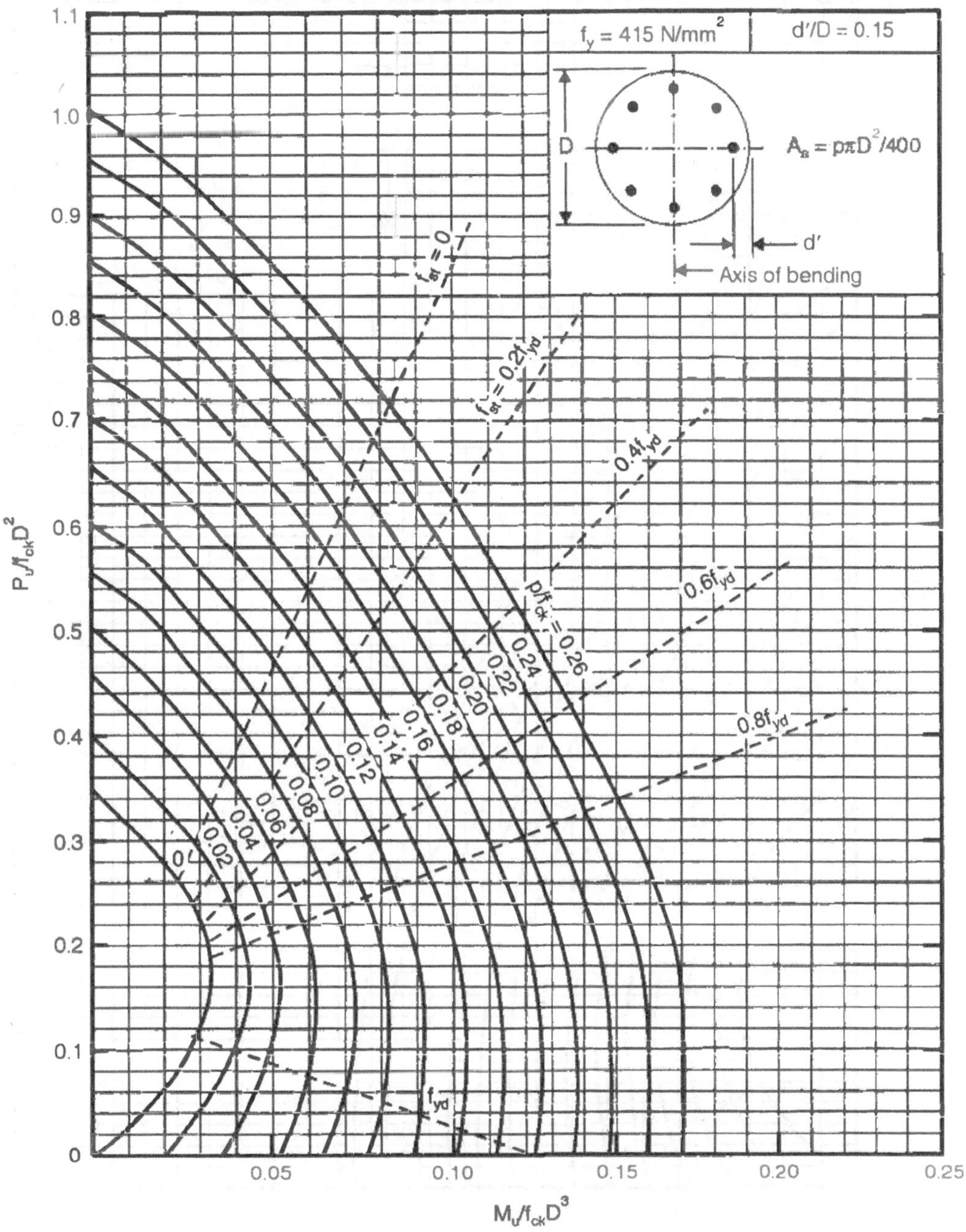

f_y = 415 N/mm²
d′/D = 0.15
D
$A_s = p\pi D^2/400$
d′
Axis of bending
$f_{st} = 0$
$f_{st} = 0.2f_{yd}$
$0.4f_{yd}$
$0.6f_{yd}$
$0.8f_{yd}$
f_{yd}
$p/f_{ck} = 0.26$
0.24
0.22
0.20
0.18
0.16
0.14
0.12
0.10
0.08
0.06
0.04
0.02
0
$P_u/f_{ck}D^2$
$M_u/f_{ck}D^3$
0
0.1
0.2
0.3
0.4
0.5
0.6
0.7
0.8
0.9
1.0
1.1
0.05
0.10
0.15
0.20
0.25

FIG. 17.20

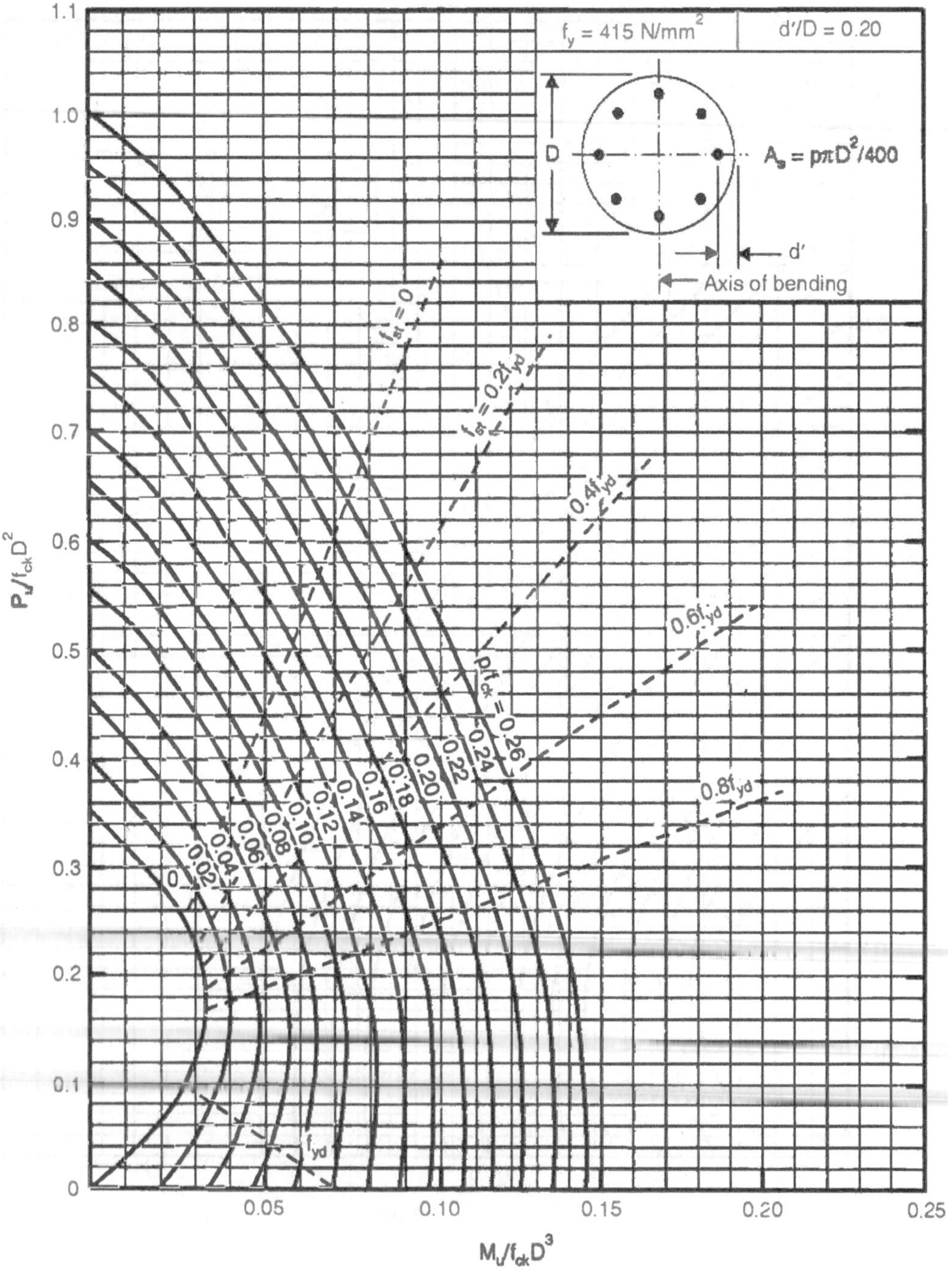

f_y = 415 N/mm²
d'/D = 0.20
D
$A_s = p\pi D^2/400$
d'
Axis of bending
$f_{st} = 0$
$f_{st} = 0.2f_{yd}$
$0.4f_{yd}$
$0.6f_{yd}$
$0.8f_{yd}$
f_{yd}
p/f_{ck} = 0.26
0.24
0.22
0.20
0.18
0.16
0.14
0.12
0.10
0.08
0.06
0.04
0.02
0
$P_u/f_{ck}D^2$
$M_u/f_{ck}D^3$
1.1
1.0
0.9
0.8
0.7
0.6
0.5
0.4
0.3
0.2
0.1
0
0.05
0.10
0.15
0.20
0.25

FIG. 17.21

17.11. METHODS OF DESIGN OF ECCENTRICALLY LOADED SHORT COLUMN

The design of eccentrically loaded short column can be done by two methods.

1. Design of column, using equations
2. Design of column, using interaction Charts.

The design consists of determination of dimensions b, D, and reinforcement ratio (p) for a predetermined cover ratio (d'/D), for given values of P_u, M_u, corrcsponding to given material properties f_{ck} and f_y.

Design of column using equations:

The design is done in the following steps :

Step 1. Fix the dimensions b and D, from the consideration of given effective length so that the column is short. Thus, keep $b > l_y/12$ and $D > l_x/12$. Also, from practical consideration, b should not be less than the width of the supported beam.

Step 2 : Assume suitable steel percentage ranging from $p = 1$ to 3 depending upon relative magnitudes of P_u and M_u.

Step 3 : Fix nominal cover (d') on the basis of exposure conditions.

Thus effective cover = nominal cover + dia. of link + $\frac{1}{2}$ dia. of bar.

Thus d'/D is known.

Steep 4 : Decide about the arrangement of steel.

(*i*) For large eccentricity, bending action will be predominant and hence provide equal steel on opposite faces, parallel to the axis of bending, This will result in the maximum contribution of moment of resistance by steel reinforcement.

(*ii*) For small eccentricity, compressive action will be more dominant and hence provide steel bars uniformly spaced all round.

Step 5 : After having fixed the configuration, assume that the N.A. is located at the outer edge, so that $x_u = D$ or $k_u = 1$. For this position, compute P_{u1}.

Step 6 : Compare P_{u1} with the given value of P_u.

(*i*) If $P_u > P_{u1}$, the compressive resistance needs to be increased by increasing x_u, or by making $k_u > 1$. Take next value of $k_u = P_u/P_{u1}$ and compute P_{u2}.

(*ii*) If $P_u < P_{u1}$, decrease the depth of N.A. by making $k_u < 1$. Take next value of $k_u = P_u/P_{u1}$ and compute P_{u2}.

Step 7. Repeat till the computed value of P_u equals the given value of P_u. For this position of N.A., compute $(M_{uc} + \Sigma M_{usi})$.

Step 8. If the given value of M_u is less than computed value of $(M_{uc} + \Sigma M_{usi})$ the section is safe. If not change the section and/or the reinforcement till both equilibrium conditions are satisfied simultaneously.

Design of column using interaction Charts :

The method of use of equations, described above is really very laborious and time consuming. The Code, therefore, recommends the use of Charts, thereby making the design extremely easy. The design is done in the following steps.

Step 1 : Decide column dimensions ($b \times D$), cover ratio (d'/D and d'/b), and mode of placement of reinforcing bars, as described in steps 1, 2 and 3 above.

Step 2 Compute $P_u/f_{ck}\,bD$ and $M_u/f_{ck}\,bD^2$.

Step 3. On the appropriate Chart, obtain the point of intersection of $P_u/f_{ck}\,bD$ and $M_u/f_{ck}\,bD^2$.

Step 4. The point of intersection will fall in between any two curves of p/f_{ck}. Interpolate the *value* of p/f_{ck} corresponding to this point of intersection.

Step 5. Compute total $A_s = f_{ck}\dfrac{bD}{100} \times$ value obtained from the Chart.

The use of Charts for design has been illustrated in Examples 17.5, 17.6 and 17.7.

Example 17.1. ***Determination of M_u for given P_u.*** *A short R.C. column 300 mm wide and 500 mm deep is reinforced with 6 bars of 20 mm diameter, arranged as shown in Fig. 17.22. Determine the bending moment M_u about an axis bisecting the depth, when it is also subjected to $P_u = 800$ kN. Assume M 20 grade concrete and Fe 415 grade steel.*

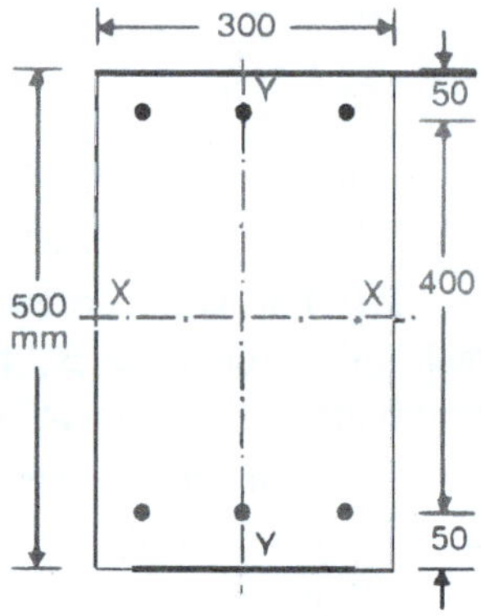

FIG. 17.22

Solution. Here $d' = 50$ mm ; $D = 500$ mm ;

$$d'/D = 50/500 = 0.1$$

Also, $f_y = 415\ \text{N/mm}^2$. Hence use Chart 32 of ISI Hand book. (or Fig. 17.11 of this book).

$$f_{ck} = 20\ \text{N/mm}^2$$

$$\frac{P_u}{f_{ck}\,bD} = \frac{800 \times 1000}{20 \times 300 \times 500} = 0.267$$

$$\therefore \quad A_s = 6 \times \frac{\pi}{4}(20)^2 \simeq 1885\ \text{mm}^2$$

$$p = \frac{A_s}{bD} \times 100 = \frac{1885}{300 \times 500} \times 100 \simeq 1.257$$

$$\frac{p}{f_{ck}} = \frac{1.257}{20} = 0.063.$$

Hence from Chart 32 of ISI Hand book,

for $\dfrac{P_u}{f_{ck}\,bD} = 0.267$ and $\dfrac{p}{f_{ck}} = 0.063$, we get $\dfrac{M_u}{f_{ck}\,bD^2} \approx 0.13$.

$$M_u = 0.13\, f_{ck}\,bD^2 = 0.13 \times 20 \times 300\,(500)^2 = 195 \times 10^6\ \text{N-mm} = \mathbf{195\ kN\text{-}m}$$

Example 17.2. ***Determination of P_u for given M_u.***
For the data of example 17.1, determine the value of P_u if $M_u = 100$ kN-m

Solution

$$\frac{M_u}{f_{ck}\,bD^2} = \frac{100 \times 10^6}{20 \times 300\,(500)^2} = 0.067$$

Hence from Chart 32 of ISI Design Hand Book, for $M_u/f_{ck}\,bD^2 = 0.067$ and $p/f_{ck} = 0.063$,

we get $P_u/f_{ck}\,bD = 0.49$.

$$\therefore \quad P_u = 0.49 \times 20 \times 300\,(500) = 1470 \times 10^3\ \text{N} = \mathbf{1470\ kN}$$

Example 17.3. ***Determination of P_u for given eccentricity.***

For the data of example 17.1, determine the value of P_u, if it is applied at an eccentricity of (a) 150 mm (b) 80 mm.

Solution (*a*) Given : $e = 150$ mm. $\therefore$ $e/D = 150/500 = 0.3$

In Chart 32, imagine a line, by holding a straight edge, joining the origin and a point having $P_u/f_{ck}\,bD = 1.0$ and $M_u/f_{ck}\,bD^2 = (P_u/f_{ck}\,bD)\,(e/D) = 0.3$. The point of intersection of this line with visually interpolated curve corresponding to $p/f_{ck} = 0.063$ gives the required point. The value of $P_u/f_{ck}\,bD$ corresponding to this point come out to be 0.36.

$$P_u = 0.36 \times 20 \times 300 \times 500 = 1080 \times 10^3 \text{ N } = \mathbf{1080\ kN}$$

(*b*) Given : $e = 80$ mm $\quad\therefore\quad$ $e/D = 80/500 = 0.16$

In Chart 32 imagine a line, by holding a straight edge, joining the origin and a point having $P_u/f_{ck}\,bD = 1.0$ and $M_u/f_{ck}\,bD^2 = (P_u/f_{ck}\,bD)\,(e/D) = 0.16$. The point of intersection of this line with visually interpolated curve corresponding to $p/f_{ck} = 0.063$ gives the required point. The value of $P_u/f_{ck}\,bD$ corresponding to this point comes out to be 0.47.

$$P_u = 0.47 \times 20 \times 300 \times 500 = 1410 \times 10^3 \text{ N } = \mathbf{1410\ kN}$$

Example 17.4. ***Determination of P_u at minimum eccentricity for given length of column.*** *For the data of example 17.1, determine the value of P_u that can be permitted at the minimum eccentricity specified by the Code, if the effective length of the column is 5 m.*

Solution Let us first find the minimum eccentricities in the two directions.

(*i*) *In the direction of larger dimension.*

$$e_{\min} = \frac{l}{500} + \frac{D}{30} = \frac{5000}{500} + \frac{500}{30} = 26.67 \text{ mm.}$$

$$\frac{e_{\min}}{D} = \frac{26.67}{500} = 0.0533 > 0.05$$

(*ii*) *In the direction of shorter dimension*

$$e_{\min} = \frac{l}{500} + \frac{b}{30} = \frac{5000}{500} + \frac{300}{30} = 20 \text{ mm}$$

$$\frac{e_{\min}}{b} = \frac{20}{300} = 0.067 > 0.05$$

Thus minimum eccentricity ratio is more than 0.05 in both the directions, and hence Eq. 16.2 cannot be applied to determine P_u.

For bending about x-axis

$$e/D = 0.0533$$

In Chart 32, imagine a line by holding a straight edge joining the origin and a point having $P_u/f_{ck}\,bD = 1.0$ and $M_u/f_{ck}\,bD^2 = (P_u/f_{ck}\,bD)\,(e/D) = 0.0533$. The point of intersection of this line, with visually interpolated curve corresponding to $p/f_{ck} = 0.063$ gives the required point. The value of $P_u/f_{ck}\,bD$ corresponding to this point comes out to be 0.58.

$$\therefore \quad P_u = 0.58 \times 20 \times 300 \times 500 = 1740 \times 10^3 \text{ N} = 1740 \text{ kN}$$

For bending about y-axis : For the purpose of using the Chart, $b = 500$ mm and $D = 300$ mm. Hence $e/D = 20/300 = 0.067$. In Chart 32, imagine a line having $P_u/f_{ck}\,bD = 1.0$ and $M_u/f_{ck}\,b\,D^2 = (P_u\,f_{ck}\,bD)\,(e/D) = 0.067$. The point of intersection of this line with visually interpolated curve corresponding to $p/f_{ck} = 0.067$ gives the required point. The value of $P_u/f_{ck}\,b\,D$ corresponding to this point comes but to be 0.565.

$$P_u = 0.565 \times 20 \times 500 \times 300 = 1695 \times 10^3 \text{ N} = 1695 \text{ kN}$$

Permissible P_u will be the lesser of the two values. Hence $P_u =$ **1695 kN**

Example 17.5. ***Design of square column with eccentric load***

Design a reinforced concrete column, 400 mm square, to carry an ultimate load of 1000 kN at an eccentricity of 160 mm. Use M 20 grade concrete and Fe 250 grade steel.

Here $b = D = 400$ mm. $d'/D = 40/400 = 0.1$

$$P_u = 1000 \text{ kN} = 1000 \times 10^3 \text{ N} \;; \qquad M_u = e\,.\,P_u = 1000 \times 10^3 \times 160$$

The non-dimensional parameters are

$$\frac{P_u}{f_{ck}\,.\,bD} = \frac{1000 \times 10^3}{20 \times 400 \times 400} = 0.3125$$

$$\frac{M_u}{f_{ck}\,.\,bD^2} = \frac{1000 \times 10^3 \times 160}{20 \times 400\,(400)^2} = 0.125$$

Referring to the interaction diagram for $f_y = 250$ N/mm^2 and $d'/D = 0.1$ i.e. Chart 28 of ISI hand book, the corresponding value of p/f_{ck} is obtained as 0.105.

$$\therefore \qquad p = 0.105 \times f_{ck} = 0.105 \times 20 = 2.1 \quad \text{or} \quad p = \frac{100\,A_s}{bD} = 2.1$$

$$A_s = \frac{2.1\,bD}{100} = \frac{2.1 \times 400 \times 400}{100} = 3360 \text{ mm}^2$$

Provide 2-25 mm dia. bars and 1-30 mm dia. bar on each face.

$$\text{Total} \qquad A_s = 4 \times \frac{\pi}{4}(25)^2 + 2 \times \frac{\pi}{4}(30)^2 = 1963.5 + 1413.7 = 3377.2 \text{ mm}^2$$

Use 6 mm dia. lateral ties, at a spacing of the least of the following :
(*i*) Least lateral dimension = 400 mm, (*ii*) 16 × 25 = 400 mm (*iii*) 300 mm.
Hence provide 6 mm dia. lateral ties at a spacing of 300 mm/cc.

Example 17.6. ***Design of circular column : Hoop reinforcement***

Design a short circular column of 500 mm dia. with the following data :
Factored load : 800 kN. Factored moment : 162.5 kN-m
Provide hoop reinforcement. Take M 20 concrete mix and use Fe 415 steel.

Solution. Let us provide 20 mm dia. bars at a clear cover of 40 mm.

$$d' = 40 + 10 = 50 \text{ mm and } d'/D = 50/500 = 0.1.$$

Hence use Chart 56 of ISI hand book (Fig. 17.19 of book). Let us assume that moment due to minimum eccentricity will be less than the given moment.

$$\frac{P_u}{f_{ck}\,D^2} = \frac{800 \times 10^3}{20\,(500)^2} = 0.16 \;; \qquad \frac{M_u}{f_{ck}\,.\,D^3} = \frac{162.5 \times 10^6}{20\,(500)^3} = 0.065$$

Referring to Chart 56 (or Fig. 17.19), we get $p/f_{ck} \approx 0.05$, for the above pair of two values.

$$\therefore \quad p = 0.05\, f_{ck} = 0.05 \times 20 = 1.0\%$$

$$\frac{100\, A_s}{\frac{\pi}{4}(500)^2} = p = 1.0 \quad \text{or} \quad A_s = \frac{1.0 \times \frac{\pi}{4}(500)^2}{100} = 1963.5 \text{ mm}^2$$

No. of 20 mm dia. bars = 1963.5/314.16 = 6.25. Hence provide 7 bars of 20 mm dia. Using 6 mm dia lateral ties, spacing will be the least of the following :

(*i*) Least lateral dimension = 500 mm (*ii*) 16 φ = 16 × 20 = 320 mm and (*iii*) 300 mm. Hence provide 6 mm dia. lateral ties @ 300 mm c/c.

Example 17.7. ***Design of circular column : Helical reinforcement***

Redesign the column of example 17.6, using helical reinforcement

Solution. According to Clause 39.4 of the Code, the strength of a compression member with helical reinforcement is 1.05 times the strength of a similar member with lateral ties. Therefore, the given load and moment should be divided by 1.05 before using the Chart.

$$\therefore \quad \frac{P_u}{f_{ck}\, D^2} = \frac{0.16}{1.05} = 0.152 \quad \text{and} \quad \frac{M_u}{f_{ck}\, D^3} = \frac{0.065}{1.05} = 0.062$$

Hence from Chart 56 (or Fig. 17.19), we get $p/f_{ck} = 0.045$

$$\therefore \quad p = 0.045 \times 20 = 0.9\%$$

$$A_s = \frac{p\, \pi\, D^2}{400} = \frac{0.9 \times \pi\, (500)^2}{400} = 1767.1 \text{ mm}^2$$

∴ No. of 20 mm dia. bars = 1767.1/314.16 = 5.62

Hence provide 6 nos. of 20 mm dia. bars.

Now $\quad A_g = \frac{\pi}{4}(500)^2 = 196349.5 \text{ mm}^2$

Let us use 8 mm dia. mild steel bars for helical reinforcement.

Core diameter = D_c = 500 − 2 × 40 + 2 × 8 = 436 mm

$$\text{Area of core} = A_c = \frac{\pi}{4}(436)^2 = 149301 \text{ mm}^2$$

$$\frac{A_g}{A_c} = \frac{196349.5}{149301} = 1.315$$

$$\text{Factor} \quad 0.36\left(\frac{A_g}{A_c} - 1\right)\frac{f_{ck}}{f_{yh}} = 0.36\,(1.315 - 1) \times \frac{20}{250} = 0.00907 \qquad \ldots(i)$$

Dia. of core upto centre of helix = 500 − (2 × 40) + 8 = 428 mm

Let the pitch of the spiral be s mm

$$V_h = \frac{\pi d}{s}\left(\frac{\pi}{4}\varphi_s^2\right) = \frac{\pi \times 428}{s} \times \frac{\pi}{4}(8)^2 = \frac{67587}{s}$$

$$V_c = A_c \times 1 = 149301$$

$$\frac{V_h}{V_c} = \frac{67587}{s \times 149301} = \frac{0.4527}{s} \qquad \ldots(ii)$$

Equating (*i*) and (*ii*), we get

$$\frac{0.4527}{s} = 0.00907 \quad \text{or} \quad s \simeq 50 \text{ mm}$$

However, the pitch should not be more than 75 mm, nor more than 1/6 core dia. $\left(= \frac{1}{6} \times 436 = 72.6 \text{ mm} \right)$. Also, it should not be less than $3\ \varphi_s$ ($= 3 \times 8 = 24$ mm). Hence keep the pitch equal to 50 mm.

Example 17.8. ***Determination of P_u and M_u for a given value of x_u***

For the column section shown in Fig. 17.23 (a) determine the design strength components corresponding to the condition of 'balanced failure'. Take M 20 concrete and Fe 415 steel.

Solution : Given $b = 300$ mm; $D = 500$ mm, $f_{ck} = 20$ N/mm^2; $f_y = 415$ N/mm^2. The details of steel reinforcement in various rows ($i = 1$ to $i = 3$) are as follows:

$$A_{s1} = A_{s2} = A_{s3} = 2 \times \frac{\pi}{4} (25)^2 = 981.75 \text{ mm}^2$$

$y_1 = -190$ mm ; $y_2 = 0$ and $y_3 = +190$ mm

Step 1: Determination of x_{ub}

Let us first locate the N.A for balanced failure condition wherein $\varepsilon_{c.max} = \varepsilon_{sy}$.

Here, $\varepsilon_{c.max} = 0.0035$

$$\varepsilon_{sy} \text{ (for Fe 415)} = 0.002 + \frac{0.87 f_y}{E_s}$$

$$= 0.002 + \frac{0.87 \times 415}{2 \times 10^5} = 0.00380$$

From strain diagram (Fig. 17.23 *b*), considering similar triangles, we get

$$x_{ub} = \frac{500 - 60}{0.0035 + 0.00380} \times 0.0035 = 210.96 \text{ mm},$$

which is less than 0.5 *D*.

(Alternatively for a balanced section, $x_{u,\max} = 0.479\, d = 0.479 \times 440 \simeq 210.8$ mm, for Fe 415 steel.

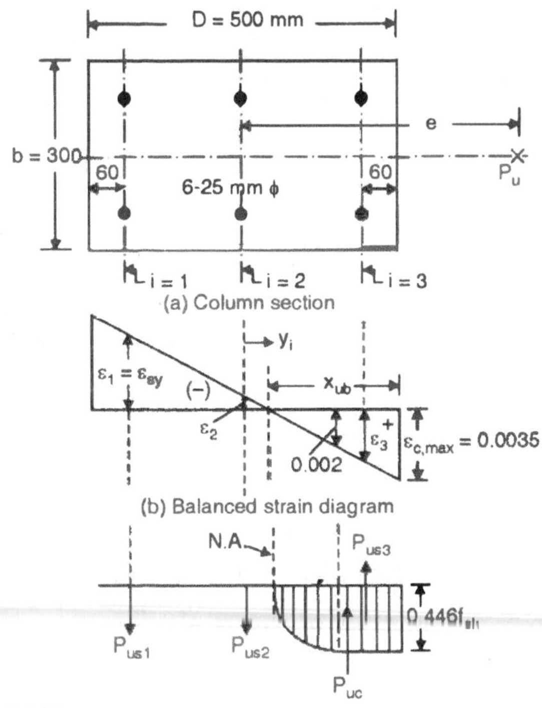

FIG. 17.23

Step 2. Determination of strains at various rows of steel bars

$$\varepsilon_1 = \varepsilon_{sy} = -0.00380 \text{ (i.e. tensile)}$$

$$\varepsilon_2 = -\frac{0.0035}{210.96} \times (250 - 210.96) = 0.000648 \text{ (i.e. tensile)}$$

$$\varepsilon_3 = +\frac{0.0035}{210.96} [210.96 - 60] = 0.002504 > 0.002$$

Step 3. Determination of stress in steel bars at various rows

$$f_{s1} = -\frac{f_y}{1.15} = -\frac{415}{1.15} = -360.9 \text{ N/mm}^2 \text{ (i.e. tensile)}$$

f_{s2} (for $\varepsilon_2 = 0.000648$) $= - E_s \,.\, \varepsilon_2 = - 2 \times 10^5 \times 0.000648 = - 129.6\ \text{N/mm}^2$

f_{s3} (for $\varepsilon_3 = 0.002504$) $= 342.8 + \dfrac{351.8 - 342.8}{276 - 241}(250.4 - 241) = + 345.2\ \text{N/mm}^2$

(By linear interpolation from Table 2.3)

Step 4. Determination of stresses in concrete at various rows of reinforcement

f_{c1} and f_{c2} are evidently tensile and hence neglected.

$f_{c3} = 0.446\, f_{ck} = 0.446 \times 20 \simeq 8.9\ \text{N/mm}^2$ (comp.)

Step 5. Determination of strength component P_u in axial compression

$$P_u = P_{uc} + \sum_{i=1}^{n} P_{usi}$$

Hence $P_{uc} = 0.36\, f_{ck}\, k_u\, bD = 0.36\, f_{ck}\, x_{ub}\,.\, b = 0.36 \times 20 \times 210.96 \times 300 = 455674\ \text{N}$

$$\sum_{i=1}^{3} P_{usi} = \sum_{i=1}^{3} (f_{si} - f_{ci})\, A_{si} = [(-360.9) + (-129.6) + (345.2 - 8.9)]\, 981.75$$

$$= -151386\ \text{N (i.e. tensile)}$$

$$\therefore \quad P_u = 455674 - 151386 = 304288\ \text{N} = \mathbf{304.288\ kN}$$

Step 6. Determination of strength component M_u in flexure

$$M_u = M_{uc} + \sum_{i=1}^{n} M_{usi}$$

where $M_{uc} = P_{us}\,(0.5\,D - 0.416\,k_u\,D) = P_{uc}\,(0.5\,D - 0.416\,x_{ub})$ from Eq. 17.13 (*a*)

$= 455674\,(0.5 \times 500 - 0.416 \times 210.96) \simeq 73.929 \times 10^6\ \text{N-mm}$

$$\sum_{i=1}^{3} M_{usi} = \sum_{i=1}^{3} P_{usi}\, y_i = [(-360.9)(-190) + (-129.6)(0) + (345.2 - 8.9)(+190)]\, 981.75$$

$= 130.05 \times 10^6\ \text{N-mm}$

Hence $M_u = (73.929 + 130.05)\, 10^6 = 203.979 \times 10^6\ \text{N-mm} = \mathbf{203.979\ kN\text{-}m}$

Note : Eccentricity $\quad e_b = \dfrac{M_u}{P_u} = \dfrac{203.979 \times 10^3}{304.288} = 670.3\ \text{mm}$

$\dfrac{e_b}{D} = \dfrac{670.3}{500} = 1.341 \quad ; \quad k_{ub} = \dfrac{x_{ub}}{D} = \dfrac{210.96}{500} = 0.422$

Hence balanced failure occurs when $e/D = 1.341$ or $k_u = 0.422$

Tension failure occurs only when $e/D > 1.341$ or $k_u < 0.422$

Compression failure occurs only when $e/D < 1.341$ or $k_u > 0.422$

Example 17.9. ***Determination of P_u and M_u for $x_u = D$***

For the column section of example 17.8, determine the strength components P_u and M_u for the limiting condition of no tension in the section.

Solution : Fig. 17.24 shows the section along with strain protile and stress resultants.

$$A_{s1} = A_{s2} = A_{s3} = 2\,\frac{\pi}{4}\,(25)^2 = 981.75 \text{ mm}^2$$

$y_1 = -190$ mm; $y_2 = 0$; $y_3 = +190$ mm

Step 1. Location of N.A.

For no tension to develop, the N.A. will pass through the outer edge of the section. Hence $x_u = D = 500$ mm.

Step 2. Determination of strains at various rows of steel bars

$$\varepsilon_{c,\min} = 0\,; \quad \varepsilon_{c,\max} = 0.0035$$

$$\varepsilon_1 = \frac{0.0035}{500} \times 60 = 0.00042$$

$$\varepsilon_2 = \frac{0.0035}{500} \times 250 = 0.00175$$

$$\varepsilon_3 = \frac{0.0035}{500} \times 440 = 0.00308$$

Step 3: Determination of stresses in steel bars at various rows

$\varepsilon_1 = 0.00042$.

Hence $f_{s1} = +E_s \,.\, e_{s1} = +2 \times 10^5 \times 0.00042$

$= +84$ N/mm^2

$\varepsilon_2 = 0.00175$.

Hence $$f_{s2} = 306.7 + \frac{342.8 - 306.7}{192 - 163}(175 - 163)$$

$= 314.2$ N/mm^2

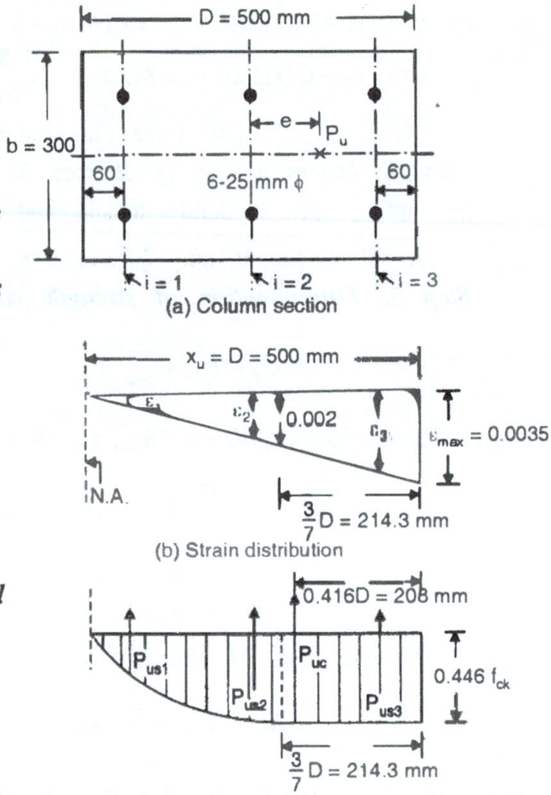

FIG. 17.24.

$\varepsilon_3 = 0.00308$. Hence $$f_{s3} = 351.8 + \frac{360.9 - 351.8}{380 - 276}(308 - 276) = 354.6 \text{ N/mm}^2$$

Step 4. Determination of stresses f_{ci} at various rows of bars

For $\varepsilon_i > 0.002$, $f_{ci} = 0.446 f_{ck}$;

For $\varepsilon_i < 0.002$, f_{ci} is obtained from Eq. 17.8.

$$f_{c1} = 446 f_{ck}\,\varepsilon_1 (1 - 250\,\varepsilon_1) = 446 \times 20 \times 0.00042\,(1 - 250 \times 0.00042) \simeq 3.4 \text{ N/mm}^2$$

$$f_{c2} = 446 \times f_{ck} \times 0.00175\,(1 - 250 \times 0.00175) = 0.439 f_{ck} = 0.439 \times 20 \simeq 8.8 \text{ N/mm}^2$$

$$f_{c3} = 0.446 f_{ck} = 0.446 \times 20 \simeq 8.9 \text{ N/mm}^2$$

Step 5 : Determination of strength component P_u in axial compression

$$P_u = P_{uc} + \sum_{i=1}^{3} P_{sui}$$

Here $$P_{uc} = 0.36 f_{ck}\,k_u\,bD = 0.36 f_{ck}\,x_u\,b = 0.36 \times 20 \times 500 \times 300 = 1080000 \text{ N}$$

$$\sum_{i=1}^{3} P_{usi} = [(84 - 3.4) + (314.2 - 8.8) + (354.6 - 8.9)] \times 981.75 = 718346 \text{ N}$$

$\therefore \qquad P_u = 1080000 + 718346 = 1798346 \text{ N} \simeq \mathbf{1798.346 \text{ kN}}$

Step 6. Determination of strength component M_u in flexure

$$M_u = M_{uc} + \sum_{i=1}^{3} M_{usi}$$

$$M_{uc} = P_{uc}\,(0.5\,D - 0.416\,k_u\,D) = P_{uc}\,(0.5\,D - 0.416\,x_u)$$
$$= 1080000\,(0.5 - 0.416)\,500 = 45.36 \times 10^6 \text{ N-mm}$$

$$M_{us} = \sum_{i=1}^{3} P_{usi}\,y_i = [(84 - 3.4)(-190) + (314.2 - 8.8)(0) + (354.6 - 8.9)(+190)]\,981.75$$
$$= 49.45 \times 10^6 \text{ N-mm}$$

$$M_u = (45.36 + 49.45)\,10^6 = 94.81 \times 10^6 \text{ N-mm} = 94.81 \text{ kN-m}$$

Note : Eccentricity e corresponding to no tension is given by

$$e = \frac{M_u}{P_u} = \frac{94.81}{1798.346} \times 10^3 = 52.74 \text{ mm} \quad ; \qquad \frac{e}{D} = \frac{52.74}{500} = 0.1054$$

Hence the entire section will be under compression if $e/D < 0.1054$

Example 17.10. ***Determination of P_u and M_u for $x_u > D$***

For the column section of Example 17.8, determine the strength components P_u and M_u corresponding to N.A. location at 1.10 D.

Solution Fig. 17.25 shows the section along with strain and stress distributions.

$$A_{s1} = A_{s2} = A_{s3} = 2\,\frac{\pi}{4}\,(25)^2 = 981.75 \text{ mm}^2$$

$y_1 = -190$ mm; $y_2 = 0$; $y_3 = +190$ mm

Step 1. Determination of stress block parameters

Given $\quad x_u = 1.10\,D$; Hence $k_u = 1.10$

From Eq. 17.3, $C_0 = \dfrac{8}{7}\left[\dfrac{4}{7 \times 1.1 - 3}\right]^2 = 0.8278$

From Eq. 17.4, $C_1 = 0.446\left(1 - \dfrac{0.8278}{6}\right) = 0.3845$

From Eq. 17.5 $\;C_2 = \dfrac{0.5 - 0.8278/7}{1 - 0.8278/6} = 0.4428$

Alternatively, from Table 17.1,

$C_1 = 0.384$ and $C_2 = 0.443$

Step 2 : Determination of strains at various locations (Fig. 17.25 b)

From Eq. 17.1 (*a*),

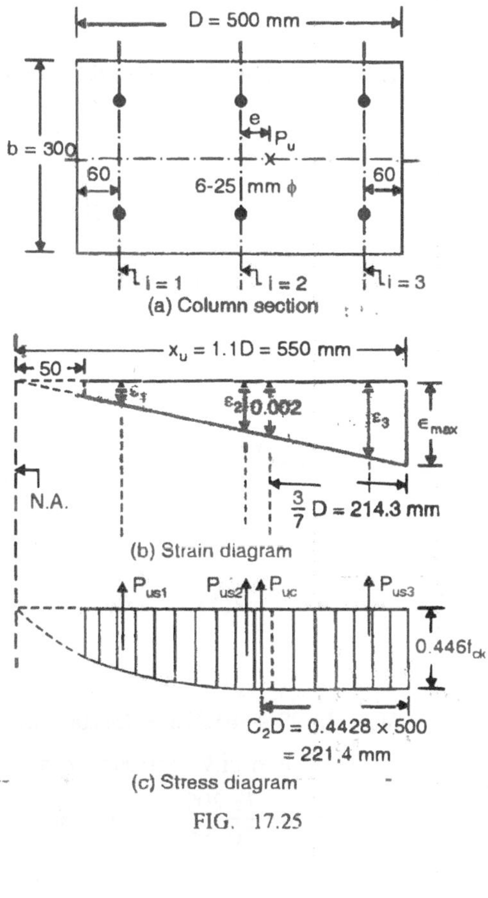

FIG. 17.25

$$\varepsilon_{max} = 0.002\left(\frac{x_u}{x_u - 3\,D/7}\right) = 0.002\left(\frac{1.1\,D}{1.1\,D - 3\,D/7}\right) = 0.003277$$

$\therefore$ $$\varepsilon_1 = \frac{0.003277}{550}(50 + 60) = 0.0006553$$

$$\varepsilon_2 = \frac{0.003277}{550}(550 - 250) = 0.001787$$

$$\varepsilon_3 = \frac{0.003277}{550}(550 - 60) = 0.002919$$

Step 3. Determination of stresses in steel bars at various locations

$$\varepsilon_1 = 0.0006553 \quad \therefore\ f_{s1} = \varepsilon_1 \,.\, E_s = 0.0006553 \times 2 \times 10^5 = 131.1 \text{ N/mm}^2$$

$$\varepsilon_2 = 0.001787 \quad \therefore\ f_{s2} = 306.7 + \frac{324.8 - 306.7}{192 - 163}(178.7 - 163) = 316.5 \text{ N/mm}^2$$

$$\varepsilon_3 = 0.002919 \quad \therefore\ f_{s3} = 351.8 + \frac{360.9 - 351.8}{380 - 276}(291.9 - 276) = 353.2 \text{ N/mm}^2$$

Step 4. Determination of stresses f_{ci} at various rows of bars .

For $\varepsilon_i > 0.002$, $f_{ci} = 0.446\, f_{ck}$; For $\varepsilon_i < 0.002$, f_{ci} is obtained from Eq. 17.8

$\therefore$ $$f_{c1} = 446\, f_{ck}\, \varepsilon_1\,(1 - 250\,\varepsilon_1) = 446 \times 0.0006553\ (1 - 250 \times 0.0006553) \times 20 = 2.4 \text{ N/mm}^2$$

$$f_{c2} = 446 \times 0.001787\ (1 - 250 \times 0.001787) \times 20 = 8.8 \text{ N/mm}^2$$

$$f_{c3} = 0.446\, f_{ck} = 0.446 \times 20 = 8.9 \text{ N/mm}^2$$

Step 5. Determination of strength component P_u in axial compression

$$P_u = P_{uc} + \sum_{i=1}^{3} P_{usi}$$

$$P_{uc} = C_1\, f_{ck}\, bD = 0.3845 \times 20 \times 300 \times 500 = 1153500 \text{ N} = 1153.5 \text{ kN}$$

$$\sum_{i=1}^{3} P_{usi} = [(131.1 - 2.4) + (316.5 - 8.8) + (353.2 - 8.9)]\ 981.75 = 766452 \text{ N} = 766.452 \text{ kN}$$

$\therefore$ $$P_u = 1153500 + 766452 = 1919952 \text{ N} = \mathbf{1919.952\ kN}$$

Step 6. Determination of strength component M_u in flexure

$$M_u = M_{uc} + \sum_{i=1}^{3} M_{usi}$$

$$M_{uc} = P_{uc}\,(0.5\,D - C_2\,D) = 1153500\ (0.5 - 0.4428)\ 500 = 32.99 \times 10^6 \text{ N-mm}$$

$$\sum_{i=1}^{3} M_{usi} = \sum_{i=1}^{3} P_{usi} \,.\, y_i = [(131.1 - 2.4)(-190) + (316.5 - 8.8)(0) + (353.2 - 8.9)(+190)]\ 981.75$$

$$= 40.216 \times 10^6 \text{ N-mm} = 40.216 \text{ kN-m}$$

$\therefore$ $$M_u = 32.99 + 40.216 = \mathbf{73.206\ kN\text{-}m}$$

Note : Eccentricity corresponding to this location is given by

$$e = \frac{M_u}{P_u} = \frac{73.206}{1919.952} \times 10^3 = 38.13 \text{ mm} \ ; \quad \therefore \quad \frac{e}{D} = \frac{38.13}{500} = 0.076$$

Example 17.11. ***Determination of M_u for pure flexure (i.e. $P_u = 0$)***

For the column section of Example 17.8, determine strength components for the case of pure flexure.

Solution : This case is the same as the case of a doubly reinforced section. For this case, the position of the N.A. will be corresponding to $k_{u,min} = x_{u,min}/D$. Fig. 17.26 shows the section along with strain and stress distributions.

As before : $A_{s1} = A_{s2} = A_{s3} = 981.75$ mm^2;

$y_1 = -190$ mm; $y_2 = 0$ and $y_3 = +190$ mm

Let the N.A. be located at $x_{u,min}$ from highly compressed edge

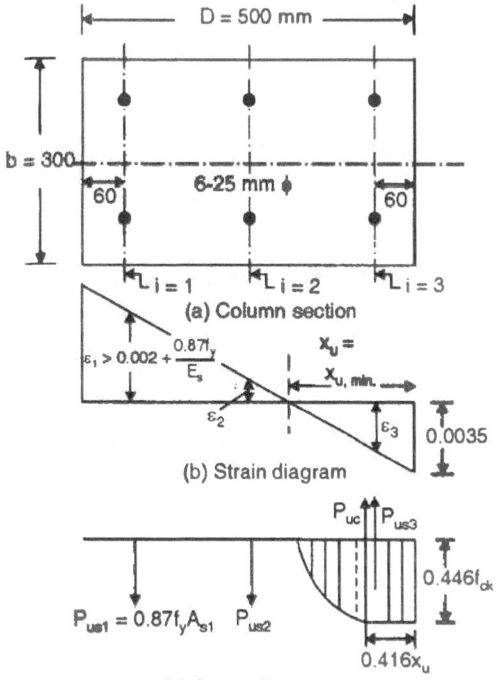

FIG. 17.26

Step 1. Determination of ε_{si}, f_{si} and f_{ci} at various locations

$$\varepsilon_1 = \varepsilon_{s1} > 0.002 + 0.87\frac{f_y}{E_s} > 0.002 + \frac{0.87 \times 415}{2 \times 10^5}$$

$$> 0.0038$$

$$\therefore \quad f_{s1} = 0.87 f_y = f_y/1.15 = 415/1.15$$

$$= 360.9 \text{ N/mm}^2 \text{ (tensile)}$$

$$\varepsilon_2 = \varepsilon_{s2} = \frac{0.0035\,(250 - x_{u,min})}{x_{u,min}} = 0.0035\left(\frac{250}{x_{u,min}} - 1\right)$$

Assuming $\frac{3}{7} x_{u,min} \geq 60$ mm or $x_{u,min} = 140$ mm, for the purpose of computing approximate value of f_{s2}, we get $\varepsilon_2 = 0.0035\left(\frac{250}{140} - 1\right) = 0.00275$.

Hence $\quad f_{s2} \simeq 351.7$ N/mm^2 from Table 2.3

Also $\quad \varepsilon_3 = 0.0035 \times \frac{x_{u,min} - 60}{x_{u,min}} = 0.0035\left(1 - \frac{60}{x_{u,min}}\right)$.

Assuming this to be smaller than $0.8 \times 0.87 f_y/E_s$, we get

$$f_{s3} = E_s\,E_s = 2 \times 10^5 \times 0.0035\left(1 - \frac{60}{x_{u,min}}\right) = 700 - \frac{42000}{x_{u,min}}$$

Now $\quad f_{c1} = 0$ (being tensile) and $f_{c2} = 0$. (being tensile)

$\quad f_{c3} = 0.446 f_{ck} = 446 \times 20 \simeq 8.9$ N/mm^2. (Assuming $x_{u,min} \geq 140$ mm).

Step 2. Determination of value of $x_{u,min}$

Total compressive force $= 0.36 f_{ck}\, b\, x_{u,min} + (f_{s3} - f_{c3})\, A_{s3}$

$$= 0.36 \times 20 \times 300\, x_{u,min} + \left(700 - \frac{42000}{x_{u,min}} - 8.9\right) 981.75$$

$$= 2160\, x_{u\,.\,\min} + 981.75\left(691.1 - \frac{42000}{x_{u\,,\,\min}}\right) \qquad \ldots(i)$$

Total tensile force $= (f_{s1} + f_{s2})\, A_s = (360.9 + 351.7)\, 981.75 = 699595$ N ...(*ii*)

Equating the two, and rearranging, we get

$$x_{u\,.\,\min}^2 - 9.772\, x_{u,\min} - 19090 = 0, \text{ from which } x_{u\,,\,\min} \simeq 143 \text{ mm}$$

The above value is based on computation of ε_2 and f_{s2} assuming $x_u = 140$ mm.

Step 3. Next trial for $x_{u\,,\,\min}$: Let $x_{u\,.\,\min} = 143$ mm

$\varepsilon_1 = 0.0038$, $f_{s1} = 360.9$ N/mm^2 (tensile) and $f_{c1} = 0$, as before

$$\varepsilon_2 = 0.0035\left(\frac{250}{143} - 1\right) = 0.00262$$

Hence $$f_{s2} = 342.8 + \frac{351.8 - 342.8}{276 - 241}(262 - 241) = 348.2 \text{ N/mm}^2 \text{ (tensile)}$$

$$\varepsilon_3 = 0.0035\left(1 - \frac{60}{143}\right) = 0.00203$$

$\therefore$ $$f_{s3} = 324.8 + \frac{342.8 - 324.8}{241 - 192}(203 - 192) = 328.8 \text{ N/mm}^2 \text{ (comp.)}$$

and $$f_{c3} = 0.446\, f_{ck} = 0.446 \times 20 = 8.9 \text{ N/mm}^2$$

Total compressive force $= 0.36 \times 20 \times 300\, x_{u\,,\,\min} + (328.8 - 8.9)\, 981.75$

Total tensile force $= (360.9 + 348.2)\, 981.75$

Equating the two, we get $x_{u\,,\,\min} \simeq 177$ mm

Next Trial : In the next trial, let us adopt an intermediate value of $x_{u\,.\,\min} = 162$ mm (say), for the purposes of computation of ε_2 , f_{s2} and ε_3 , f_{s3}.

$$\varepsilon_2 = 0.0035\left(\frac{250}{162} - 1\right) = 0.001901 \text{ mm} < \varepsilon_{sy}$$

$\therefore$ $$f_{s2} = 306.7 + \frac{324.8 - 306.7}{192 - 163}(190.1 - 163) = 323.6 \text{ N/mm}^2$$

$$\varepsilon_3 = 0.0035\left(1 - \frac{60}{162}\right) = 0.002204$$

$\therefore$ $$f_{s3} = 324.8 + \frac{342.8 - 324.8}{241 - 192}(220.4 - 192) = 335.2 \text{ N/mm}^2$$

$$f_{c3} = 0.446\, f_{ck} = 8.9 \text{ N/mm}^2$$

Hence balancing the forces, we get

$$0.36 \times 20 \times 300\, x_{u\,.\,\min} + (335.2 - 8.9)\, 981.75 = (360.9 + 323.6)\, 981.75$$

which gives $x_{u\,.\,\min} = 162.8$ mm

This is quite near to the assumed value of 162 mm. Hence, finally, take $x_{u\,.\,\min} = 162.4$ mm. Also, $k_{u\,.\,\min} = x_{u\,.\,\min}/D = 162.4/500 = 0.3248$.

Step. 5: Determination of final values of P_u ***and*** M_u

For this value of $x_{u\,.\,\min} = 162.4$ mm, we get

$$\varepsilon_2 = 0.0035\left(\frac{250}{162.4} - 1\right) = 0.001888 \text{ mm} < \varepsilon_{sy}$$

$$\therefore \quad f_{s2} = 306.7 + \frac{324.8 - 306.7}{192 - 163}(188.8 - 163) = 322.8 \text{ N/mm}^2$$

$$\varepsilon_3 = 0.0035\left(1 - \frac{60}{162.4}\right) = 0.002207$$

$$\therefore \quad f_{s3} = 324.8 + \frac{342.8 - 324.8}{241 - 192}(220.7 - 192) = 335.3 \text{ N/mm}^2$$

$$f_{c3} = 0.446\, f_{ck} = 0.446 \times 20 = 8.9 \text{ N/mm}^2$$

Hence $P_u = 0.36 \times 20 \times 300 \times 162.4 + (335.3 - 8.9)\,981.75 - 360.9 \times 981.75 - 322.8 \times 981.75$

$= 350784 + 320443 - 354313 - 316909 = 5 \text{ N} \simeq 0$

Also, taking moment of forces about A_{s1}, we get

$M_u = M_{u0} = 350784\,(440 - 0.416 \times 162.4) + 320443\,(440 - 60) - 316909\,(440 - 250)$

$= 130.647 \times 10^6 + 121.768 \times 10^6 - 60.213 \times 10^6 = 192.202 \times 10^6 \text{ N-mm} = 192.202 \text{ kN-m}$

Alternatively, taking moment of forces about the centroidal axis,

$M_{u0} = 350784\,(250 - 0.416 \times 162.4) + 320443\,(190) + (-354313)(-190)$

$= 63.998 \times 10^6 + 60.884 \times 10^6 + 67.319 \times 10^6 = 192.201 \times 10^6 \text{ N-mm} = 192.201 \text{ kN-m}$

Hence adopt $M_{u0} \simeq$ **192.2 kN-m**

Example 17.12. ***Determination of P_u and M_u for $x_u < D$.***

For the column section of Example 17.8, determine the strength components corresponding to the condition of maximum strain of 0.0035 in concrete and tensile strain of 0.002 in the outer most layer of steel.

Solution : Fig. 17.27 shows the section along with strain and stress diagrams.

As before, $A_{s1} = A_{s2} = A_{s3} = 2\,\frac{\pi}{4}(25)^2$

$= 981.75 \text{ mm}^2$

Also, $y_1 = -190$ mm ; $y_2 = 0$; $y_3 = +190$ mm.

Step 1. Location of N.A.

From Fig. 17.27 (*b*)

$$\frac{x_u}{0.0035} = \frac{440 - x_u}{0.002}$$

From which $x_u = 280$ mm

$\therefore \quad k_u = x_u/D = 280/500 = 0.56$

Hence $x_u > 0.5\,D$

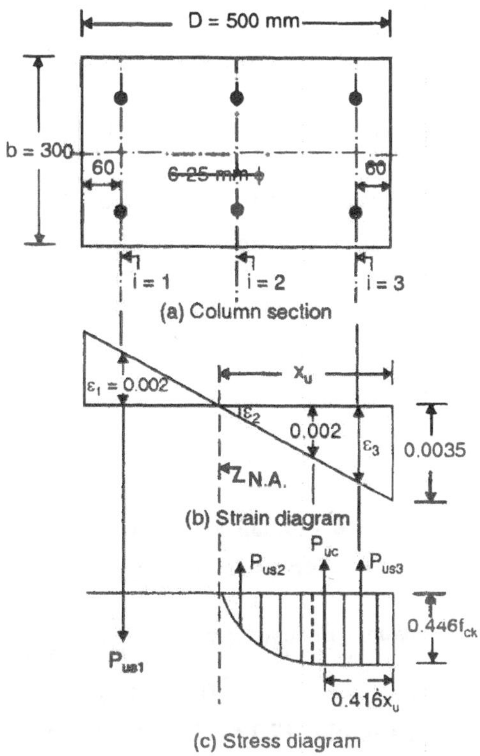

FIG. 17.27

Step 2. Determination of strains at various rows of steel bars

$$\varepsilon_1 = -0.002 \quad \text{(given)},$$

$$\varepsilon_2 = +\frac{0.0035}{280}(280 - 250) = 0.000375$$

$$\varepsilon_3 = +\frac{0.0035}{280}(280 - 60) = 0.00275$$

Step 3. Determination of f_{si} at various rows

$$f_{s1} = 324.8 + \frac{342.8 - 324.8}{241 - 192}(200 - 192) = 327.7 \text{ N/mm}^2 \quad \text{(tensile)}$$

$$f_{s2} = E_s \cdot \varepsilon_2 = 2 \times 10^5 \times 0.000375 = 75 \text{ N/mm}^2 \quad \text{(compressive)}$$

$$f_{s3} = 342.8 + \frac{351.8 - 342.8}{276 - 241}(275 - 241) = 351.5 \text{ N/mm}^2 \quad \text{(compressive)}$$

Step 4. Determination of f_{ci} at various rows

$f_{c1} = 0$ (being tensile)

$f_{c2} = 446 f_{ck} \times 0.000375 (1 - 250 \times 0.000375) = 0.1516 f_{ck} = 0.1516 \times 20 \simeq 3 \text{ N/mm}^2$

$f_{c3} = 0.446 f_{ck} = 0.446 \times 20 = 8.9 \text{ N/mm}^2$

Step 5. Determination of strength component P_u in axial compression

$$P_u = P_{uc} + \sum_{i=1}^{3} P_{usi}$$

Here $P_{uc} = 0.36 f_{ck} k_u b \cdot D = 0.36 f_{ck} x_u \cdot b = 0.36 \times 20 \times 280 \times 300 = 604800$ N

$$\sum_{i=1}^{3} P_{usi} = \sum_{i=1}^{3} (f_{si} - f_{ci}) A_{si} = [(-327.7) + (75 - 3) + (351.5 - 8.9)]\, 981.75 = 85314 \text{ N}$$

$\therefore$ $P_u = 604800 + 85314 = 690114$ N $=$ **690.114 kN**

Step 6. Determination of strength component M_u in flexure

$$M_u = M_{uc} + \sum_{i=1}^{3} M_{usi}$$

where $M_{uc} = P_{uc}(0.5\, D - 0.416\, x_u) = 604800 (0.5 \times 500 - 0.416 \times 280) = 80.753 \times 10^6$ N-mm

$$\sum_{i=1}^{3} M_{usi} = \sum_{i=1}^{3} P_{usi} \cdot y_i = [(-327.7)(-190) + (75 - 3)(0) + (351.5 - 8.9)(+190)]\, 981.75$$

$= 125.033 \times 10^6$ N-mm

Hence $M_u = (80.753 + 125.033)\, 10^6 = 205.786 \times 10^6$ N-mm $=$ **205.786 kN-m**

Note: Eccentricity $e = \dfrac{M_u}{P_u} = \dfrac{205.786}{690.114} \times 10^3 = 298.2$ mm

$\therefore$ $e/D = 298.2/500 = 0.596$ Also $k_u = x_u/D = 280/500 = 0.56$

Example 17.13. *Interaction curve $P_u - M_u$.*

For the column section of Example 17.8, draw the $P_u - M_u$ interaction curve.

Solution: Fig. 17.23 (*a*) shows the column section, where $A_{s1} = A_{s2} = A_{s3} = 981.75 \text{ mm}^2$ and $y_1 = -190$ mm; $y_2 = 0$ and $y_3 = +190$ mm. For drawing the interaction curve, let us obtain the coordinates of pairs of values (P_u, M_u) for the following cases:

(*a*) Axial load with zero eccentricity $(e = 0 \,;\, x_u = \infty)$

(*b*) Axial load with minimum eccentricity $(e = e_{min} \,;\, \infty < x_u < D)$

(*c*) Neutral axis outside the section : $x_u = 1.2\, D \,;\, k_u = 1.2$

(*d*) Neutral axis lying along outer edge : $x_u = D \,;\, k_u = 1.00$

(*e*) Neutral axis within the section : condition of maximum strain of 0.0035 in concrete and tensile strain of 0.002 in outer most layer of steel : $x_u = 0.56\, D \,;\, k_u = 0.56$

(*f*) Neutral axis location corresponding to balanced failure : $x_u = x_{ub}$

(*g*) Neutral axis location corresponding to the case of pure flexure.

(*a*) ***Axial load*** (P_{uz}) ***with zero eccentricity*** $(e = 0 \,;\, x_u = \infty)$:

From Eq. 16.1, $\quad P_{uz} = 0.446\, f_{ck}\, A_g + (0.75\, f_y - 0.446\, f_{ck})\, A_{sc}$

$$\therefore \quad P_{uz} = 0.446 \times 20 \times 300 \times 500 + (0.75 \times 415 - 0.446 \times 20)\,(3 \times 981.75)$$

$$= 1338000 + 890437 = 2228437 \text{ N} = 2228.437 \text{ kN}$$

(*b*) ***Axial load with minimum eccentricity*** $(e = e_{min})$ **:** $P_{u\,,\,max}$

From Eq. 16.2, $\quad P_{u\,.\,max} = 0.4\, f_{ck}\, bD + (0.67\, f_y - 0.4\, f_{ck})\, A_s$

$$= 0.4 \times 20 \times 300 \times 500 + (0.67 \times 415 - 0.4 \times 20)\,(3 \times 981.75)$$

$$= 1200000 + 795364 = 1995364 \text{ N} = 1995.364 \text{ kN}$$

Corresponding $M_u = P_{u\,.\,max} \cdot e_{min} = 1995.364\,(500 \times 10^{-3}/20) = 49.884$ kN-m (For $e = 0.05\, D$)

(*c*) ***Neutral axis outside the section*** : $x_u = 1.1\, D \,;\, k_u = 1.1$

This case has been dealt with in Example 17.10, from which we have

$P_u = 1919.952$ kN; $M_u = 73.206$; $e/D = 0.076$

(*d*) ***Neutral axis at outer edge***: $x_u = D \,;\, k_u = 1$

This case has been solved in Example 17.9, from which we have

$P_u = 1798.346$ kN ; $M_u = 94.81$ kN-m; $e/D = 0.1054$

(*e*) ***N.A. within the section*** : $x_u = 0.56\, D \,;\, k_u = 0.56$

This case has been dealt with in Example 17.12, from which we have

$P_u = 690.114$ kN; $M_u = 205.786$ kN-m; $e/D = 0.596$

(*f*) ***N.A. location corresponding to balanced failure*** : $x_u = x_{ub} \,;\, k_u = k_{ub} = 0.422$

From Example 17.8, we have $P_u = 304.288$ kN; $M_u = 203.979$ kN-m; $e/D = 1.341$

(*g*) ***N.A. location corresponding to pure flexure***

This case has been discussed in Example 17.11, from which we get

$P_{u0} = 0$; $M_{u0} = 192.2$ kN-m; $x_{u\,.\,min} = 162.4$ mm; $k_{u\,.\,min} = 0.3248$

The results obtained for the above cases are summarised below :

e/D	***0***	***0.05***	***0.073***	***0.1054***	***0.596***	∞
$k_u = x_u/D$	∞	∞	1.10	1.0	0.56	0.3248
P_u (kN-m)	2228.437	1995.364	1919.952	1798.346	690.114	0
M_u (kN-m)	0	49.884	73.206	94.81	205.786	192.22

Fig. 17.28 shows the interaction diagram

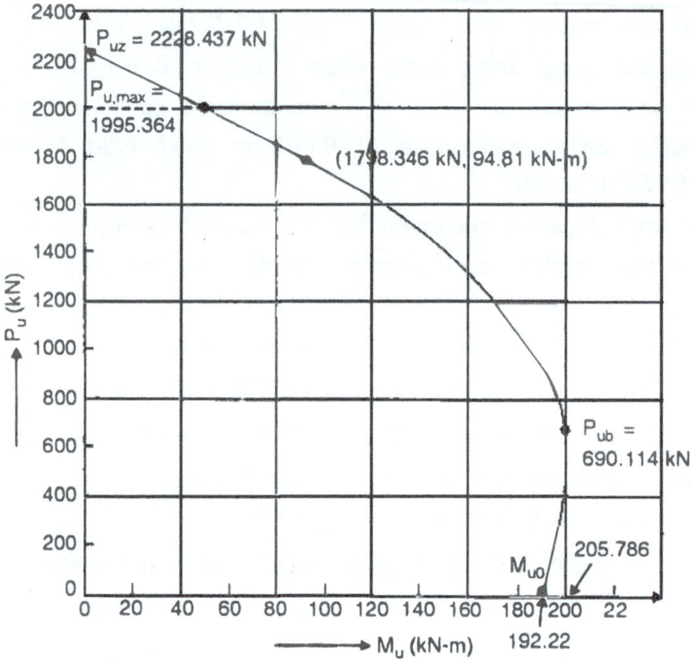

FIG. 17.28.

Example 17.14. *Determination of P_u by Analytical method*

Compute the maximum allowable ultimate axial load on a short R.C.C. column, of 250 × 375 mm size reinforced with 4 bars of 20 mm dia. of Fe 415 grade at the four corners if (a) unsupported length is not known and (b) unsupported length is 5.84 m. Assume suitable cover and M 20 grade concrete.

Solution :

Both b and D are less than 400 mm, due to which minimum eccentricity of 20 mm comes out to be more than 0.05 D. Hence the problem will be analysed on the basis of theory of eccentrically loaded column. Here $A_{s1} = A_{s2} = 2\frac{\pi}{4}(20)^2 = 628.32$ mm². Providing 40 mm nominal cover, effective cover $d' = 40 + 6 + 10 = 56$ mm, using 6 mm φ rings.

(*a*) **Case *a* : unsupported length (*l*) not given** : Since l is not known, $e_{min} = 20$ mm. Also, since the column in subjected to only eccentric loading, and since there is no *external* B.M. acting on the column, the column will buckle about minor axis (*y-y* axis), perpendicular to the shorter dimension, and hence minimum eccentricity will be from *y*-axis. Due to this, $D = 250$ m and $b = 375$ mm (Fig. 17.29).

$A_{s1} = A_{s2} = 628.32 \text{ mm}^2$; $y_1 = -(125 - 56) = -69$ mm ; $y_2 = +(125 - 56) = +69$ mm

Let us now obtain the position of the N.A. by trial and error.

*1. **First trial*** : Let $x_u = D = 250$ mm; $k_u = 1$

The corresponding strain and stress diagrams are shown in Fig. 17.29 (*b*) and (*c*) respectively.

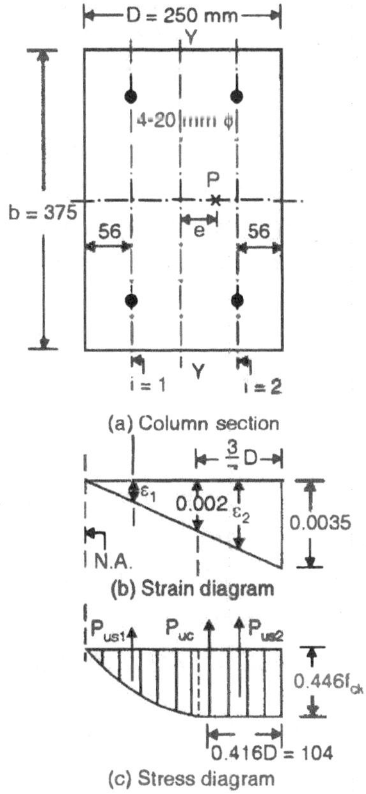

FIG. 17.29

Now $P_u = P_{uc} + \sum_{i=1}^{2} P_{usi}$ and $M_u = M_{uc} + \sum_{i=1}^{2} M_{usi}$

$$P_{uc} = 0.36\, f_{ck}\, bx_u = 0.36 \times 20 \times 375 \times 250$$

$$= 675000 \text{ N} = 675 \text{ kN}$$

$$M_{uc} = P_u\,(0.5\,D - 0.416\,x_u)$$

$$= 675000\,(0.5 \times 250 - 0.416 \times 250)$$

$$= 14175000 \text{ N-mm} = 14.175 \text{ kN-m}$$

$$\varepsilon_1 = 0.0035\,(56)/250 = 0.000784$$

$$\varepsilon_2 = 0.0035\,(250 - 56)/250 = 0.002716$$

$$f_{s1} = 0.000784 \times 2 \times 10^5 = 156.8 \text{ N/mm}^2$$

$$f_{s2} = 342.8 + \frac{351.8 - 342.8}{276 - 241}(271.6 - 241)$$

$$= 350.7 \text{ N/mm}^2, \text{ from Table 2.3}$$

$$f_{c1} = 446\,\varepsilon_1\,(1 - 250\,\varepsilon_1)\,f_{ck}$$

$$= 446 \times 0.000784\,(1 - 250 \times 0.000784)\,20$$

$$= 5.6 \text{ N/mm}^2$$

$$f_{c2} = 0.446\,f_{ck} = 0.446 \times 20 \simeq 8.9 \text{ N/mm}^2$$

$$\sum_{i=1}^{2} P_{usi} = [(156.8 - 5.6) + (350.7 - 8.9)]\,628.32 = 309762 \text{ N}$$

and $\sum_{i=1}^{2} M_{usi} = [(156.8 - 5.6)(-69) + (350.7 - 8.9)(+69)]\,628.32 = 8263288$ N-mm

∴ $P_u = 675000 + 309762 = 984762 \text{ N} = 984.762 \text{ kN}$

and $M_u = 14175000 + 8263288 = 22438288$ N-mm

∴ $e = M_u/P_u = 22438288/984762 = 22.8 \text{ mm} > 20 \text{ mm}$

*2. **Second trial*** : Since *e* obtained above is more than 20 mm, P_u, must increase. This is obtained by increasing the value of x_u. In the second trial let $x_u = 260$ mm, so that $k_u = 260/250 = 1.04$ Fig. 17.30 shows the corresponding position along with strain and stress diagrams.

$$C_0 = \frac{8}{7}\left[\frac{4}{7\,k_u - 3}\right]^2 = \frac{8}{7}\left[\frac{4}{7 \times 1.04 - 3}\right]^2 = 0.998$$

$$C_1 = 0.446\left(1 - \frac{0.998}{6}\right) = 0.3718$$

$$C_2 = \frac{0.5 - 0.998/7}{1 - 0.998/6} = 0.4287$$

$\therefore$ $$P_{uc} = C_1 f_{ck}\, bD = 0.3718 \times 20 \times 375 \times 250$$
$$= 697125 \text{ N}$$

$$M_{uc} = P_{uc}\,(0.5 - C_2)\, D = 697125\,(0.5 - 0.4287)\, 250$$
$$= 12426253 \text{ N-mm}$$

Taking reference strain of 0.002 at $(3/7)\,D = (3/7)\,250 = 107.1$ mm from the inner edge, we get :

$$\varepsilon_1 = 0.002 \times \frac{10 + 56}{260 - 107.1} = 0.000863$$

$$\varepsilon_2 = 0.002 \times \frac{260 - 56}{260 - 107.1} = 0.002668$$

$$f_{s1} = 0.000863 \times 2 \times 10^5 = 172.6 \text{ N/mm}^2$$

$$f_{s2} = 342.8 + \frac{351.8 - 342.8}{276 - 241}(266.8 - 241) = 349.4 \text{ N/mm}^2$$

$$f_{c1} = 446 \times 0.000863(1 - 250 \times 0.000863)\, 20 = 6 \text{ N/mm}^2$$

$$f_{c2} = 0.446 f_{ck} = 0.446 \times 20 \simeq 8.9 \text{ N/mm}^2$$

$\therefore$ $$\sum_{i=1}^{2} P_{usi} = [(172.6 - 6) + (349.4 - 8.9)]\, 628.32$$
$$= 318621 \text{ N}$$

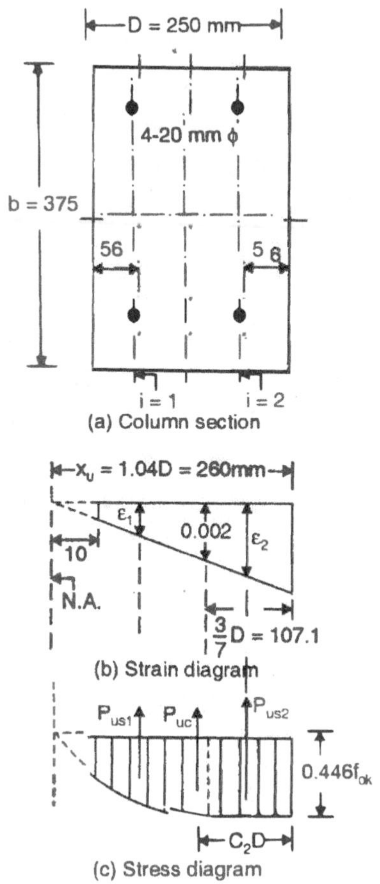

FIG. 17.30

$$\sum_{i=1}^{2} M_{usi} = [(172.6 - 6)(-69) + (349.4 - 8.9)(+69)]\, 628.32 = 7539274 \text{ N-mm}$$

$\therefore$ $$P_u = 697125 + 318621 = 1015746 = 1015.746 \text{ kN}$$

$$M_u = 12426253 + 7539274 = 19965527 \text{ N-mm}$$

$\therefore$ $$e = \frac{19965527}{1015746} = 19.66 \text{ mm}$$

This is quite close to the desired eccentricity of 20 mm. In order to get the value of P_u, corresponding to $e = 20$ mm, we can use the method of interpolation, without resorting to the third trial, on the basis of two sets of values.

Trial	e (mm)	P_u (kN)	x_u (mm)
1	22.8	984.762	250
2	19.66	1015.746	260

Hence for $e = 20$ mm $P_u = 984.762 + \dfrac{1015.746 - 984.762}{22.8 - 19.66}(22.8 \quad 20.0) \simeq$ **1012.39 kN** (*b*) **Case *b* : unsupported length $l = 5.84$ m**

$$e_{min,x} = \frac{l}{500} + \frac{D}{30} = \frac{5840}{500} + \frac{375}{30} = 24.18 \text{ mm, for bending about } x\text{-axis}$$

$$e_{min,y} = \frac{l}{500} + \frac{b}{30} = \frac{5840}{500} + \frac{250}{30} = 20.01 \text{ mm, for bending about } y\text{-axis.}$$

Out of these, we have already computed P_u for $e_{min} = 20$ mm for bending about y-axis in part (a). Let us now calculate P_u for $e = 24.18$ mm about x-axis. For this case, $b = 250$ mm and $D = 375$ mm.

Here again $A_{s1} = A_{s2} = 2 \frac{\pi}{4} (20)^2 = 628.32 \text{ mm}^2$

Also $y_1 = -(187.5 - 56) = -131.5$ mm and $y_2 = (187.5 - 56) = +131.5$ mm

1. ***First trial*** : Let $x_u = D = 375$ mm; $k_u = 1$

The corresponding strain and stress diagrams are shown in Fig. 17.31 (b) and (c) respectively.

$$P_{uc} = 0.36 f_{ck}\, b\, x_u = 0.36 \times 20 \times 250 \times 375 = 675000 \text{ N}$$

$$M_{uc} = P_u (0.5 D - 0.416 x_u) = 675000 (0.5 - 0.416) \times 375 = 21262500 \text{ N-mm}$$

$$\varepsilon_1 = 0.0035 (56)/375 = 0.0005227$$

$$\varepsilon_2 = 0.0035 (375 - 56)/375 = 0.002977$$

$$f_{s1} = 0.0005227 \times 2 \times 10^5 = 104.5 \text{ N/mm}^2$$

$$f_{s2} = 351.8 + \frac{360.9 - 351.8}{380 - 276} (297.7 - 276) = 353.7 \text{ N/mm}^2$$

$$f_{c1} = 446 \times 0.0005227 (1 - 250 \times 0.0005227)\, 20 \simeq 4.1 \text{ N/mm}^2$$

$$f_{c2} = 0.446 f_{ck} = 0.446 \times 20 = 8.9 \text{ N/mm}^2$$

$$\therefore \sum_{i=1}^{2} P_{usi} = [(104.5 - 4.1) + (353.7 - 8.9)]\, 628.32 = 279728 \text{ N-mm}$$

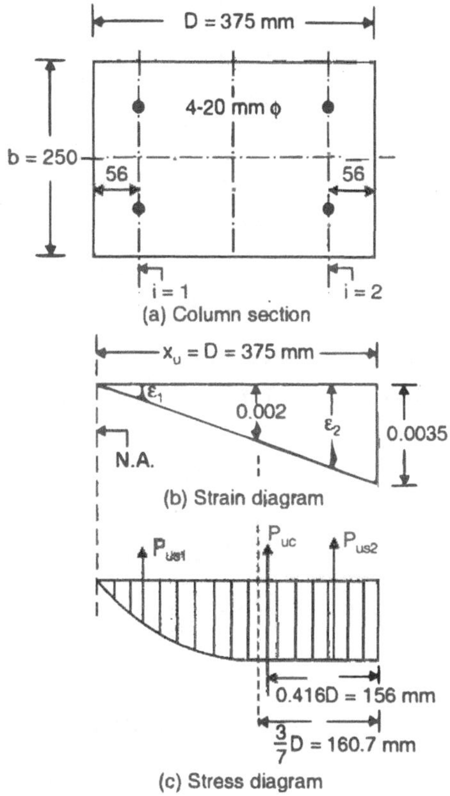

FIG. 17.31

$$\sum_{i=1}^{2} M_{usi} = [(104.5 - 4.1)(-131.5) + (353.7 - 8.9)(+131.5)]\, 628.32 = 20193325 \text{ N-mm}$$

$$\therefore \quad P_u = P_{uc} + \sum_{i=1}^{2} P_{usi} = 675000 + 279728 = 954728$$

$$M_u = M_{uc} + \sum_{i=1}^{2} M_{usi} = 21262500 + 20193325 = 41455825 \text{ N-mm}$$

$$e = \frac{M_u}{P_u} = \frac{41455825}{954728} = 43.42 \text{ mm.} \gg 24.18 \text{ mm}$$

Second trial : In the first trial, e comes out to be very much greater than the actual desired value of 24.18 mm. Hence P_u must increase, which is obtained by increasing the value of x_u. So, in the second trial, let $x_u = 440$ mm, so that $k_u = 440/375 = 1.173$. Fig. 17.32 (*b*) and (*c*) show the strain and stress diagrams respectively.

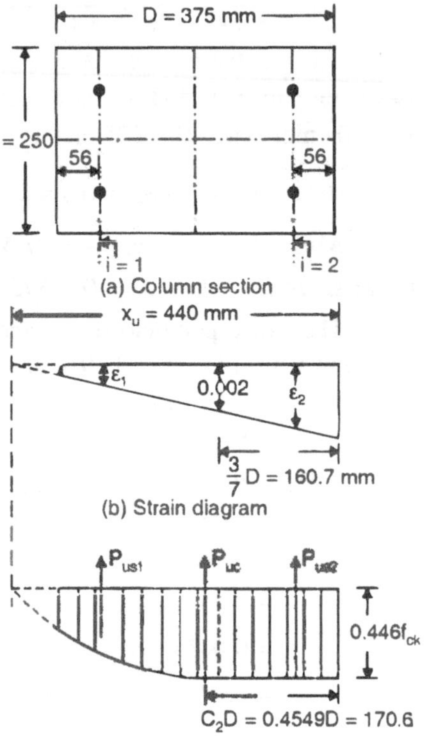

FIG. 17.32

$$C_0 = \frac{8}{7}\left[\frac{4}{7 \times 1.173 - 3}\right] = 0.6728$$

$$C_1 = 0.446\left(1 - \frac{0.6728}{6}\right) = 0.3960$$

$$C_2 = \frac{0.5 - 0.6728/7}{1 - 0.6728/6} = 0.4549$$

$$P_{uc} = C_1 f_{ck} bD = 0.396 \times 20 \times 250 \times 375$$

$$= 742500 \text{ N}$$

$$M_{uc} = P_{uc}(0.5 - C_2)D = 742500\,(0.5 - 0.4549) \times 375$$

$$= 12557531 \text{ N-mm}$$

Taking reference strain of 0.002 at $\frac{3}{7}D = \frac{3}{7} \times 375 = 160.7$ mm from the inner edge, we get

$$\varepsilon_1 = 0.002\,\frac{65 + 56}{440 - 160.7} = 0.000866.$$

$$\varepsilon_2 = 0.002\,\frac{440 - 56}{440 - 160.7} = 0.00275$$

$$f_{s1} = 0.000866 \times 2 \times 10^5 = 173.2 \text{ N/mm}^2$$

$$f_{s2} = 342.8 + \frac{351.8 - 342.8}{276 - 241}(275 - 241) = 351.5 \text{ N/mm}^2$$

$$f_{c1} = 446 \times 0.000866\,(1 - 250 \times 0.000866)20 = 6.1 \text{ N/mm}^2$$

$$f_{c2} = 0.446\, f_{ck} = 0.446 \times 20 = 8.9 \text{ N/mm}^2$$

$$\therefore \sum_{i=1}^{2} P_{usi} = [(173.2 - 6.1) + (351.5 - 8.9)]\,628.32 = 320254 \text{ N}$$

$$\sum_{i=1}^{2} M_{usi} = [(173.2 - 6.1)(-131.5) + (351.5 - 8.9)(131.5)]\,628.32 = 14500526 \text{ N-mm}$$

$$\therefore \quad P_u = P_{uc} + \sum_{i=1}^{2} P_{usi} = 742500 + 320254 = 1062754 \text{ N}$$

$$M_u = M_{uc} + \sum_{i=1}^{2} M_{usi} = 12557531 + 14500526 = 27058057 \quad \text{N-mm}$$

$$\therefore \quad e = \frac{M_u}{P_u} = \frac{27058057}{1062754} = 25.46 \quad \text{mm}$$

This is close to the desired value of 24.18 mm. However, the actual value of x_u corresponding to $e = 24.18$ mm may be obtained by linear interpolation as under, without doing any third trial.

Trial No.	e (mm)	P_u (kN)	x_u (mm)
1	43.42	954.729	375
2	25.46	1062.754	440

For $e = 24.18$ mm, we get $P_u = 954.729 + \frac{1062.754 - 954.729}{43.42 - 25.46}(43.42 - 24.18) \triangleq$ **1070.45 kN**

Allowable value of P_u

For $e_{\min,y} = 20$ mm, we get $P_u = 1012.39$ kN from case (a)

Also, for $e_{\min,x} = 24.18$ mm, we get $P_u = 1070.45$ kN from case (b)

$\therefore$ Allowable value of P_u = **1012.39 kN**

Example 17.15. ***Determination of P_u by analytical method.***

A short R.C. column of 400 mm × 500 mm section, reinforced with 4 bars of 20 mm dia. at the four corners and 2 bars of 16 mm dia. at the middle of longer sides. Determine the maximum allowable ultimate axial load, if the unsupported lengths about x-axis and y-axis are equal to 6.5 m and 3 m respectively. Take grade of concrete as M 20 and grade of steel as Fe 415. Assume cover to the centre of steel as 60 mm.

Solution : Given $l_x = 6.5$ m $= 6500$ mm; $l_y = 3$ m $= 3000$ mm

$$\therefore \quad e_{\min,x} = \frac{l_x}{500} + \frac{D}{30} = \frac{6500}{500} + \frac{500}{30} = 29.67 \quad \text{mm}$$

$$e_{\min,y} = \frac{l_y}{500} + \frac{b}{30} = \frac{3000}{500} + \frac{400}{30} = 19.33 < 20 \text{ mm}; \quad \text{Hence } e_{\min,y} = 20 \text{ mm}$$

$$\frac{e_{\min,x}}{D} = \frac{29.67}{500} = 0.0593 \quad ; \quad \frac{e_{\min,y}}{b} = \frac{20}{400} = 0.05$$

Since $\frac{e_{\min,x}}{D} > \frac{e_{\min,y}}{b}$, bending about x-axis governs the load.

Hence for bending about x-axis, $b = 400$ mm; $D = 500$ mm.

$$A_{s1} = A_{s3} = 2 \times \frac{\pi}{4}(20)^2 = 628.32 \text{ mm}^2;$$

$$A_{s2} = 2 \times \frac{\pi}{4}(16)^2 = 402.12 \text{ mm}^2$$

$$y_1 = -(250 - 60) = -190 \text{ mm}; \quad y_2 = 0 \quad ;$$

$$y_3 = +(250 - 60) = +190 \text{ mm}$$

The position of the N.A. will be located by trial and error.

First trial : Let $x_u = D = 500$ mm; $k_u = x_u/D = 1$

Fig. 17.33 (*b*) shows the strain distribution while Fig. 17.33 (*c*) shows the stress distribution.

$$P_{uc} = 0.36\, f_{ck}\, b\, x_u = 0.36 \times 20 \times 400 \times 500$$
$$= 1440000 \quad \text{N}$$
$$M_{uc} = P_{uc}\,(0.5\, D - 0.416\, x_u)$$
$$= 1440000\,(0.5 - 0.416)\,500$$
$$= 60480000 \quad \text{N-mm}$$
$$\varepsilon_1 = 0.0035\,(60)/500 = 0.00042$$
$$\varepsilon_2 = 0.0035\,(250)/500 = 0.00175$$
$$\varepsilon_3 = 0.0035\,(500 - 60)/500 = 0.00308$$
$$f_{s1} = 0.00042 \times 2 \times 10^5 = 84 \text{ N/mm}^2$$
$$f_{s2} = 306.7 + \frac{324.8 - 306.7}{192 - 163}\,(175 - 163)$$
$$= 314.2 \quad \text{N/mm}^2$$
$$f_{s3} = 351.8 + \frac{360.9 - 351.8}{380 - 276}\,(308 - 276)$$
$$= 354.6 \quad \text{N/mm}^2$$
$$f_{c1} = 446 \times 0.00042\,(1 - 250 \times 0.00042)\,20$$
$$= 3.4 \quad \text{N/mm}^2$$
$$f_{c2} = 446 \times 0.00175\,(1 - 250 \times 0.00175)\,20$$
$$= 8.8 \text{ N/mm}^2$$
$$f_{c3} = 0.446\, f_{ck} = 0.446 \times 20 = 8.9 \text{ N/mm}^2$$

$$\sum_{i=1}^{3} P_{sui} = (84 - 3.4)\,628.32 + (314.2 - 8.8)\,402.12 + (354.6 - 8.9)\,628.32 = 393802 \quad \text{N}$$

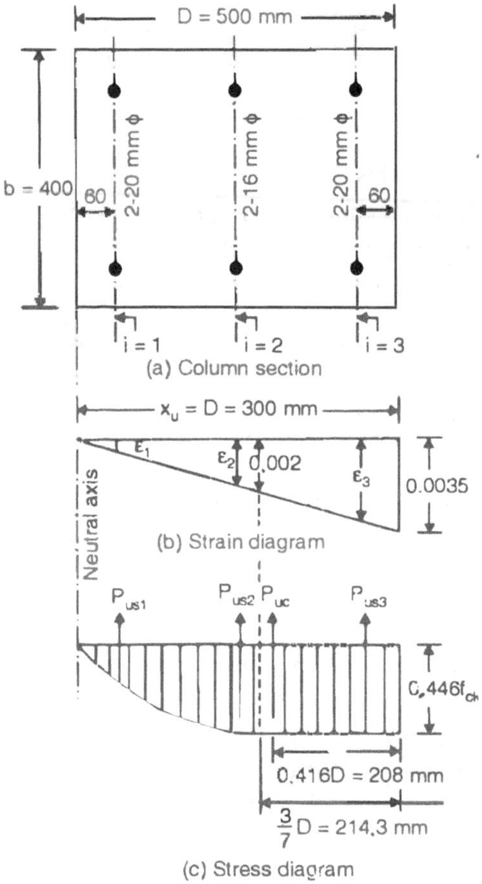

FIG. 17.33

$$\sum_{i=1}^{3} M_{usi} = (84 - 3.4)\,628.32\,(-190) + (314.2 - 8.8)\,402.12\,(0) + (354.6 - 8.9)\,628.32\,(190)$$
$$= 31647850$$

$$\therefore \quad P_u = P_{uc} + \sum_{i=1}^{3} P_{usi} = 1440000 + 393802 = 1833802 \quad \text{N} = 1833.802 \quad \text{kN}$$

$$M_u = M_{uc} + \sum_{i=1}^{3} M_{usi} = 60480000 + 31647850 = 92127850 \quad \text{N-mm}$$

$$\therefore \quad e = \frac{M_u}{P_u} = \frac{92127850}{1833802} = 50.23 \quad \text{mm}$$

This is very much larger than the desired value of 29.67 mm. Hence P_u has to be increased, which can be done by increasing the value of x_u.

Second trial. Take the new value of $x_u = 1.16\,D = 1.16 \times 500 = 580$ mm; $k_u = 1.16$. Figs. 17.34 (*b*) and (*c*) show the strain and stress diagrams respectively.

$$C_0 = \frac{8}{7}\left[\frac{4}{7 \times 1.16 - 3}\right]^2 = 0.6975$$

$$C_1 = 0.446\left(1 - \frac{0.6975}{6}\right) = 0.3941$$

$$C_2 = \frac{0.5 - 0.6975/7}{1 - 0.6975/6} = 0.4530$$

$$P_{uc} = C_1 f_{ck} bD = 0.3941 \times 20 \times 400 \times 500$$
$$= 1576400 \quad \text{N}$$

$$M_{uc} = P_{uc}(0.5 - C_2) D = 1576400(0.5 - 0.453)500$$
$$= 37045400 \quad \text{N-mm}$$

Taking reference strain of 0.002 at $\frac{3}{7}D = \frac{3}{7} \times 500 = 214.3$ mm from the inner edge, we get

$$\varepsilon_1 = 0.002\,\frac{80 + 60}{580 - 214.3} = 0.0007657$$

$$\varepsilon_2 = 0.002\,\frac{80 + 250}{580 - 214.3} = 0.001805$$

$$\varepsilon_3 = 0.002\,\frac{580 - 60}{580 - 214.3} = 0.002844$$

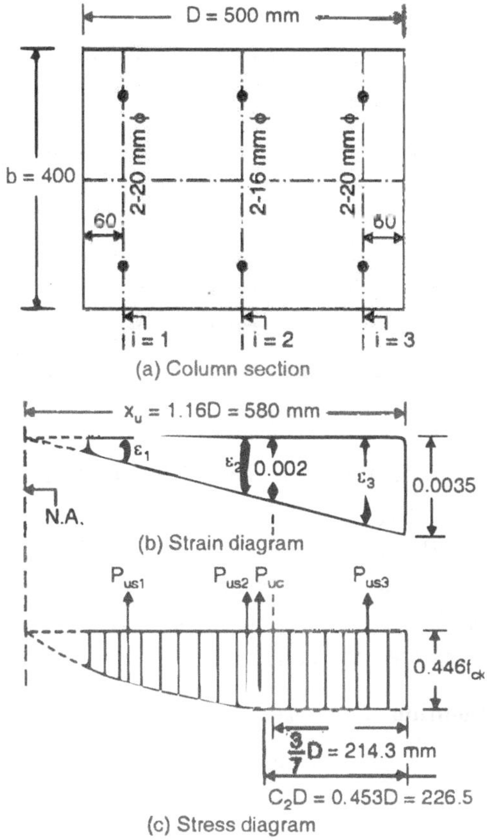

FIG. 17.34

$\therefore \quad f_{s1} = 0.0007657 \times 2 \times 10^5 = 153.1 \text{ N/mm}^2$

$$f_{s2} = 306.7 + \frac{324.8 - 306.7}{192 - 163}(180.5 - 163)$$
$$= 317.6 \text{ N/mm}^2$$

$$f_{s3} = 351.8 + \frac{360.9 - 351.8}{380 - 276}(284.4 - 276) = 352.5 \text{ N/mm}^2$$

$$f_{c1} = 446 \times 0.0007657\ (1 - 250 \times 0.0007657)\ 20 = 5.5 \text{ N/mm}^2$$

$$f_{c2} = 446 \times 0.001805\ (1 - 250 \times 0.001805)\ 20 = 8.8 \text{ N/mm}^2$$

$$f_{c3} = 0.446 f_{ck} = 0.446 \times 20 = 8.9 \text{ N/mm}^2$$

$$\therefore \quad \sum_{i=1}^{3} P_{usi} = (153.1 - 5.5)\,628.32 + (317.6 - 8.8)\,402.12 + (352.5 - 8.9)\,628.32$$
$$= 92740 + 124175 + 215891 = 432806 \quad \text{N}$$

$$\sum_{i=1}^{3} M_{usi} = 92740\,(-190) + 124175\,(0) + 215891\,(+190) = 23398690 \quad \text{N-mm}$$

$$\therefore \quad P_u = P_{uc} + \sum_{i=1}^{3} P_{usi} = 1576400 + 432806 = 2009206 \quad \text{N} = 2009.206 \quad \text{kN}$$

$$M_u = M_{uc} + \sum_{i=1}^{3} M_{usi} = 37045400 + 23398690 = 60444090 \quad \text{N-mm}$$

$$\therefore \qquad e = \frac{M_u}{P_u} = \frac{60444090}{2009206} = 30.08 \text{ mm}$$

This is quite close to the desired value of $e = 29.67$ mm. However, we will find the value of P_u, corresponding to $e = 29.67$ mm by linear interpolation from the two sets of values of (P_u, e) obtained above :

Trial	*e(mm)*	*P_u (kN)*	*x_u(mm)*
1	50.23	1833.802	500
2	30.08	2009.206	580

Hence for $\qquad e = 29.67$ mm, $P_u = 2009.206 - \dfrac{2009.206 - 1833.802}{50.23 - 30.08}(30.08 - 29.67)$

$$\simeq 2006 \text{ kN}$$

Example 17.16. ***Determination of P_u and M_u for given x_u for H-section column***

Fig. 17.35 (a) shows H-shaped column for M 20 concrete. Determine strength components P_u and M_u corresponding to a neutral axis located at 0.76 D from the highly compressed edge. The reinforcement consists of 6 bars of 25 mm φ.

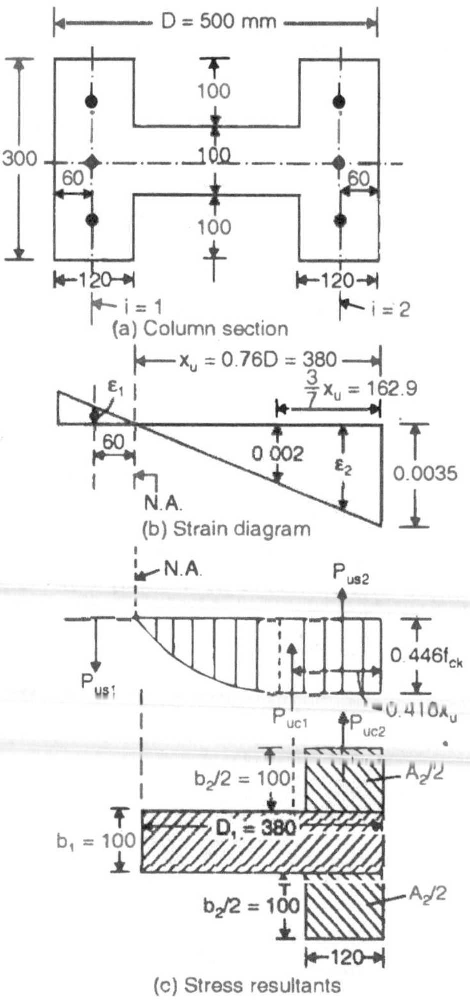

FIG. 17.35

Solution : $\qquad A_{s1} = A_{s2} = 3\,\dfrac{\pi}{4}(25)^2 = 1472.62 \text{ mm}^2$

$y_1 = -(250 - 60) = -190$ mm;

$y_2 = +(250 - 60) = +190$ mm

$x_u = 0.76\,D = 0.76 \times 500 = 380$ mm

Fig. 17.35 (*b*) shows strain diagram with $\varepsilon_{c,\,max} = 0.0035$ and strain $\varepsilon = 0.002$ at $\frac{3}{7}x_u = \frac{3}{7} \times 380 = 162.9$ mm.

$$\varepsilon_1 = -\frac{0.0035}{380} \times 60 = -0.0005526 \ ;$$

$$\varepsilon_2 = +\frac{0.0035}{380}(380 - 60) = +0.002947$$

$$f_{s1} = -0.0005526 \times 2 \times 10^5$$

$$= -110.5 \text{ N/mm}^2$$

$$f_{s2} = 351.8 + \frac{360.9 - 351.8}{380 - 276}(294.7 - 276)$$

$$= 353.4 \text{ N/mm}^2$$

$f_{c1} = 0$ (being tensile)

$f_{c2} = 0.446\,f_{ck} = 0.446 \times 20 = 8.9 \text{ mm}^2$

$P_{uc} = P_{uc1} + P_{uc2}$, for areas A_1 and A_2

$P_{uc1} = 0.36\,f_{ck}\,b_1\,x_u = 0.36 \times 20 \times 100 \times 380$

$= 273600$ N

$$P_{uc2} = 0.446\, f_{ck}\, b_2\, D_2 = 0.446 \times 20 \times 200 \times 120 = 214080 \text{ N}$$

$$\sum_{i=1}^{2} P_{usi} = \sum_{i=1}^{3} (f_{si} - f_{ci})\, A_{si} = [-110.5 + (353.4 - 8.9)]\, 1472.62 = -162724 + 507317 = 344593 \text{ N}$$

$$\therefore \quad P_u = P_{uc1} + P_{uc2} + \sum_{i=1}^{3} P_{usi} = 273600 + 214080 + 344593 = 832273 \text{ N} = \mathbf{832.273 \text{ kN}}$$

$M_{uc} = M_{uc1} + M_{uc2}$, corresponding to areas A_1 and A_2 respectively.

$$M_{uc1} = P_{uc1}\,(0.5\,D - 0.416\,x_u) = 273600\,(0.5 \times 500 - 0.416 \times 380) = 25149312 \text{ N-mm}$$

$$M_{uc2} = P_{uc2}\,(0.5\,D - 60) = 214080\,(0.5 \times 500 - 60) = 40675200 \text{ N-mm}$$

$$\sum_{i=1}^{2} M_{usi} = P_{us1}\,(y_1) + P_{us2}\,(y_2) = -162724\,(-190) + 507317\,(190) = 127307790 \text{ N-m}$$

$$\therefore \quad M_u = M_{uc1} + M_{uc2} + \sum_{i=1}^{2} M_{usi} = 25149312 + 40675200 + 127307790 = 193132302 \text{ N-mm}$$

$\simeq$ **193.132 kN-m**

Note : $e = \dfrac{M_u}{P_u} = \dfrac{193.132}{832.273} \times 10^3 = 232.05$ mm and $\dfrac{e}{D} = \dfrac{232.05}{500} = 0.464$

Example 17.17. ***Design of column using analytical method***

Design a R.C. column of 300 mm width to carry an axial ultimate load of 1500 kN. The unsupported length as well as effective depth of the column in both the directions is 3.0 m. Use M 20 concrete and Fe 415 steel. Provide an effective cover of 60 mm.

Solution Let the depth of column be 450 mm.

$l_e/b = 3000/300 = 10 < 12.$

Hence the column is short.

$$\therefore \quad e_{\min,x} = \frac{l_x}{500} + \frac{D}{30} = \frac{3000}{500} + \frac{450}{30} = 21 \text{ mm} > 20 \text{ mm}.$$

Hence $e_{\min,x} = 21$ mm.

$$e_{\min,y} = \frac{l_y}{500} + \frac{b}{30} = \frac{3000}{500} + \frac{300}{30} = 16 \text{ mm} < 20 \text{ mm}$$

$\therefore\ e_{\min,y} = 20$ mm

$$\frac{e_{\min,x}}{D} = \frac{21}{450} = 0.0467 \;; \quad \frac{e_{\min,y}}{b} = \frac{20}{300} = 0.067$$

Since $\dfrac{e_{\min,y}}{b} > \dfrac{e_{\min,x}}{D}$, $e_{\min,y}/b$ will govern the design.

$$\therefore \quad M_u = M_{u,y} = P_u \times e_{\min,y} = 1500 \times \frac{20}{1000} = 30 \text{ kN-m}$$

Let us provide 3 bars of 25 mm φ on each longer side.

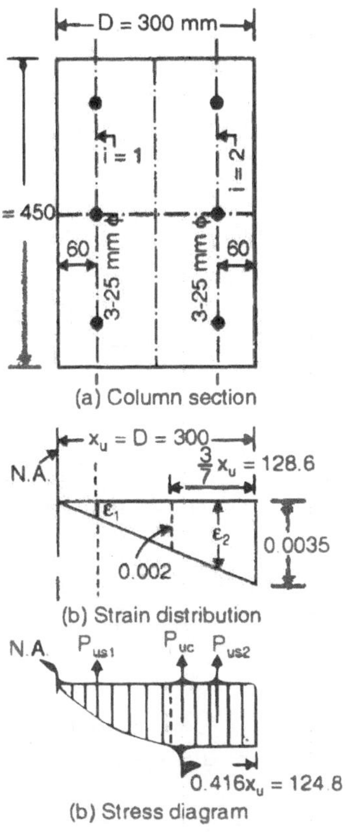

FIG. 17.36

$$\therefore \qquad A_{s1} = A_{s2} = 3 \times \frac{\pi}{4}(25)^2 = 1472.62 \text{ mm}^2$$

Also, for bending about y-axis, $b = 450$ mm; $D = 300$ mm

$$\therefore \qquad y_1 = -(150 - 60) = -90 \text{ mm} \; ; \qquad y_2 = +(150 - 60) = +90 \text{ mm}$$

Let us locate the N.A. by trial.

Trial-1 Let $k_u = 1$, so that $x_u = D = 300$ mm. Fig. 17.36 (b) shows the strain diagram.

$$\varepsilon_1 = 0.0035\,(60)/300 = 0.0007 \; ; \qquad \varepsilon_2 = 0.0035\,(300 - 60)/300 = 0.0028$$

$$f_{s1} = 0.0007 \times 2 \times 10^5 = 140 \text{ N/mm}^2$$

$$f_{s2} = 351.8 + \frac{360.9 - 351.8}{380 - 276}(280 - 276) = 352.1 \text{ N/mm}^2$$

$$f_{c1} = 446 \times 0.0007\,(1 - 250 \times 0.0007)\,20 = 5.2 \text{ N/mm}^2$$

$$f_{c2} = 0.446 f_{ck} = 0.446 \times 20 = 8.9 \text{ N/mm}^2$$

$$P_{uc} = 0.36 \times 20 \times 450 \times 300 \times 10^{-3} = 972 \text{ kN}$$

$$\sum_{i=1}^{2} P_{usi} = [140 - 5.2) + (352.1 - 8.9)] \times 1472.62 \times 10^{-3} = 703.912 \text{ kN}$$

$$P_u = 972 + 703.912 = 1675.912 \text{ kN} > 1500 \text{ kN}$$

Since computed P_u is greater than given value of P_u, $k_u < 1$

$\therefore$ Approximate revised value of $k_u = \frac{1500}{1675.912} \times 1$

$= 0.895 \simeq 0.9$

Trial 2 : Take $k_u = 0.9$.

Hence $x_u = 0.9 \times 300 = 270$ mm

Fig. 17.37 (b) shows the strain diagram.

$$\varepsilon_1 = -\frac{0.0035 \times 30}{270} = -0.000389$$

$$\varepsilon_2 = \frac{0.0035\,(270 - 60)}{270} = 0.002722$$

$$f_{s1} = 0.000389 \times 2 \times 10^5 = 77.8 \text{ N/mm}^2$$

$$f_{s2} = 342.8 + \frac{351.8 - 342.8}{376 - 241}(272.2 - 241)$$

$$= 350.8 \text{ N/mm}^2$$

$$f_{c1} = 446 \times 0.000389\,(1 - 250 \times 0.000389)\,20$$

$$= 3.1 \text{ N/mm}^2$$

$$f_{c2} = 0.446 f_{ck} = 0.446 \times 20 = 8.9 \text{ N/mm}^2$$

$$P_{uc} = 0.36 \times 20 \times 450 \times 270 \times 10^{-3} = 874.8 \text{ kN}$$

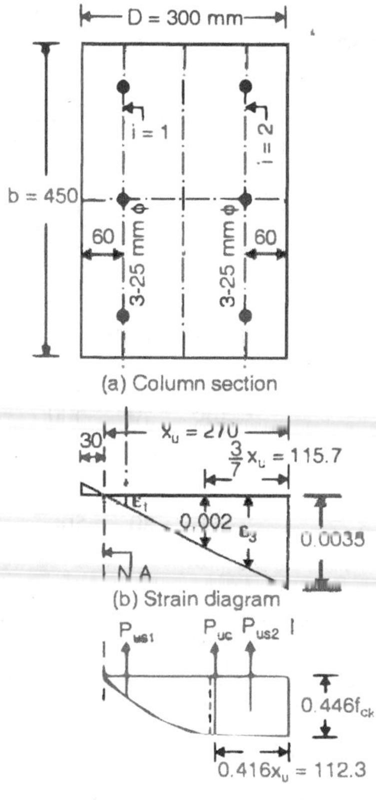

FIG. 17.37

$$\sum_{i=1}^{2} P_{usi} = [(77.8 - 3.1) + (350.8 - 8.9)]\ 1472.62 \times 10^{-3} = 613.493 \text{ kN}$$

$$\therefore \quad P_u = P_{uc} + \sum_{i=1}^{2} P_{usi} = 874.8 + 613.493 = 1488.29 \text{ kN} \simeq 1500 \text{ kN}$$

$$M_{uc} = P_{uc}\,(0.5\,D - 0.416\,x_u) = 874.8\,(150 - 0.416 \times 270) \times 10^{-3} = 32.962 \text{ kN-m}$$

$$\sum_{i=1}^{2} M_{usi} = [(77.8 - 3.1)(-90) + (350.8 - 8.9)(90)]\ 1472.62 \times 10^{-6} = 35.414 \text{ kN-m}$$

$$\therefore \quad M_u = M_{uc} + \sum_{i=1}^{2} M_{usi} = 32.962 + 35.414 = 68.376 \text{ kN-m} > 30 \text{ kN-m}$$

Hence the design is OK

Example 17.18. Design of column using analytical method

Design a short column of R.C.C. of M 20 grade, to carry an ultimate load of 1500 kN and an ultimate moment of 70 kN-m. Take width of concrete as 300 mm and effective cover as 60 mm. Use Fe 415 steel.

Solution : Let us provide a depth $D = 450$ mm and reinforcement consisting of 3-25 mm φ on each short side, as shown in Fig. 17.38 (*a*).

$$A_{s1} = A_{s2} = 3\,\frac{\pi}{4}\,(25)^2 = 1472.62 \text{ mm}^2$$

$y_1 = -(225 - 60) = -165$ mm;

$y_2 = +165$ mm

Trial 1 : Let us take $k_u = 1$ so that $x_u = D = 450$ mm. Fig. 17.38 (*b*) and (*c*) show strain and stress distributions respectively.

$$P_{uc} = 0.36\,f_{ck}\,b\,x_u$$

$$= 0.36 \times 20 \times 300 \times 450 \times 10^{-3}$$

$$= 972 \text{ kN}$$

$$\varepsilon_1 = 0.0035\,(60)/450 = 0.0004667$$

$$\varepsilon_2 = 0.0035\,(450 - 60)/450 = 0.003033$$

$$f_{s1} = 0.0004667 \times 2 \times 10^5 = 93.3 \text{ N/mm}^2$$

$$f_{s2} = 351.8 + \frac{360.9 - 351.8}{380 - 276}\,(303.9 - 276)$$

$$= 354.2 \text{ N/mm}^2$$

$$f_{c1} = 446 \times 0.0004667\,(1 - 250 \times 0.0004667)\,20$$

$$= 3.7 \text{ N/mm}^2$$

$$f_{ck} = 0.446\,f_{ck} = 0.446 \times 20 = 8.9 \text{ N/mm}^2$$

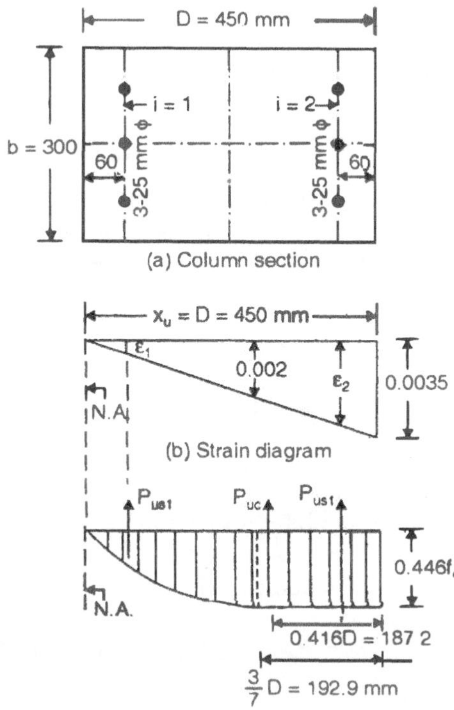

FIG. 17.38

$$\sum_{i=1}^{2} P_{usi} = [(93.3 - 3.7) + (354.2 - 8.9)]\ 1472.62 \times 10^{-3} = 640.44 \text{ kN}$$

$$\therefore \quad P_u = P_{uc} + \sum_{i=1}^{2} P_{sui} = 972 + 640.44 = 1612.44 \text{ kN} < 1700 \text{ kN}$$

Hence $k_u > 1$

***Trial* 2** : Let new $k_u = \dfrac{1700}{1612.44} \times 1 = 1.054$

Take $k_u = 1.06$ so that $x_u = 1.06 \times 450 = 477$ mm

Fig. 17.39 (*b*), (*c*) show strain and stress distributions respectively.

$$C_0 = \frac{8}{7}\left[\frac{4}{7 \times 1.06 - 3}\right]^2 = 0.9360; \quad C_1 = 0.446\left(1 - \frac{0.936}{6}\right) = 0.3764$$

$$C_2 = \frac{0.5 - 0.936/7}{1 - 0.936/6} = 0.434$$

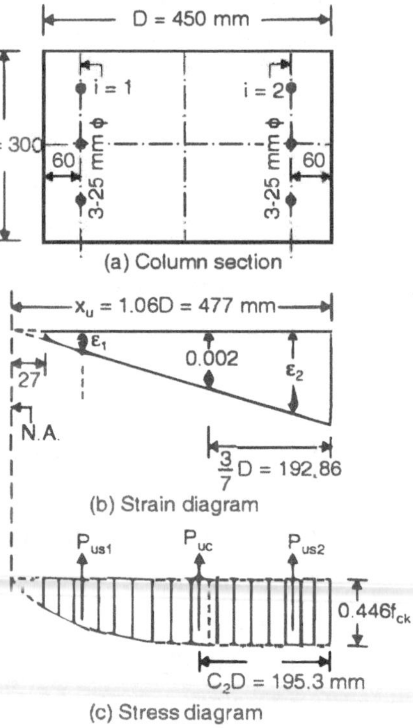

FIG. 17.39

$$P_{uc} = C_1 f_{ck}\, b\, D$$

$$= 0.3764 \times 20 \times 300 \times 450 \times 10^{-3} = 1016.28 \text{ kN}$$

Taking the reference strain of 0.002 at $\frac{3}{7}D$ $= \frac{3}{7} \times 450 = 192.86$ mm from the inner edge, we get

$$\varepsilon_1 = 0.002 \times \frac{27 + 60}{477 - 192.86} = 0.000612$$

$$\varepsilon_2 = 0.002 \times \frac{477 - 60}{477 - 192.86} = 0.002935$$

$$f_{s1} = 0.000612 \times 2 \times 10^5 = 122.4 \text{ N/mm}^2$$

$$f_{s2} = 351.8 + \frac{360.9 - 351.8}{380 - 276}(293.5 - 276)$$

$$= 353.3 \text{ N/m}^2$$

$$f_{c1} = 446 \times 0.000612\ (1 - 250 \times 0.000612) \times 20$$

$$= 4.6 \text{ N/mm}^2$$

$$f_{c2} = 0.446\, f_{ck} = 0.446 \times 20 = 8.9 \text{ N/mm}^2$$

$$\sum_{i=1}^{2} P_{usi} = [(122.4 - 4.6) + (353.3 - 8.9)]\ 1472.62 \times 10^{-3}$$

$$= 680.64 \text{ kN}$$

$$\therefore \quad P_u = P_{uc} + \sum_{i=1}^{2} P_{usi} = 1016.28 + 680.64 = 1696.92 \text{ kN} \simeq 1700 \text{ kN}$$

For this position of N.A.

$$M_{uc} = P_{uc}\,(0.5 - C_2)\,D = 1016.28\,(0.5 - 0.434) \times 450 \times 10^{-3} = 30.184 \text{ kN-m}$$

$$\sum_{i=1}^{2} M_{usi} = [(122.4 - 4.6)(-165) + (353.3 - 8.9)(165)]\ 1472.62 \times 10^{-6} = 55.06 \text{ kN-m}$$

$$\therefore \quad M_u = M_{uc} + \sum_{i=1}^{2} M_{usi} = 30.184 + 55.06 = 85.244 \text{ kN-m} > 80 \text{ kN-m}.$$

Hence the design is OK. The final section is 300 mm × 450 mm reinforced with 3 bars of 25 mm φ on each short side.

$$p = \frac{2 \times 1472.62}{300 \times 450} \times 100 = 2.18\%$$

Design of lateral ties: Diameter of lateral ties > 25/4 mm. Provide 8 mm φ ties.

Pitch : Pitch lesser than (*i*) $b = 300$ mm (*ii*) 16 × dia. of longitudinal bars = 16 × 25 = 400 mm, and (*iii*) 300 mm. Hence provide 8 mm φ lateral ties @ 300 mm c/c.

Example 17.19. ***Non-dimensional interaction curve***

Obtain an interaction curve between axial compression P_u and bending moment M_u at ultimate state in a non-dimensional form for a short R.C. column of rectangular section, reinforced with 4% steel of grade Fe 250, placed equally near opposite short faces with an effective cover of 0.1 D. Take concrete of M 20 grade.

Solution : Given : concrete grade : M 20, steel grade Fe 250 for which $f_y = 250$ N/mm^2 and $\varepsilon_{sy} = 0.0010875$.

Also, $p/f_{ck} = 4/20 = 0.2$

We will take different cases of position of N.A.

Case I (a) $k_u = \infty$: zero eccentricity

From Eq. 17.2 (*a*),

$$\frac{P_u}{f_{ck}\, bD} = 0.446 + \frac{p}{100 f_{ck}} (f_s - 0.446 f_{ck})$$

where $f_s = 0.87 f_y$ ...(17.2 *a*)

Hence,

$$\frac{P_u}{f_{ck}\, bD} = 0.446 + \frac{4}{100 \times 20} (0.87 \times 250 - 0.446 \times 20)$$

$$= 0.8632$$

Since $M_u = P_u \,.\, e = P_u \times 0 = 0, \quad \dfrac{M_u}{f_{ck}\, bD^2} = 0$

Case I (*b*) Eccentricity less than e_{min}

From Eq. 17.2 (*b*),

$$\frac{P_u}{f_{ck}\, bD} = 0.4 + \frac{p}{100 f_{ck}} (0.67 f_y - 0.4 f_{ck}),$$

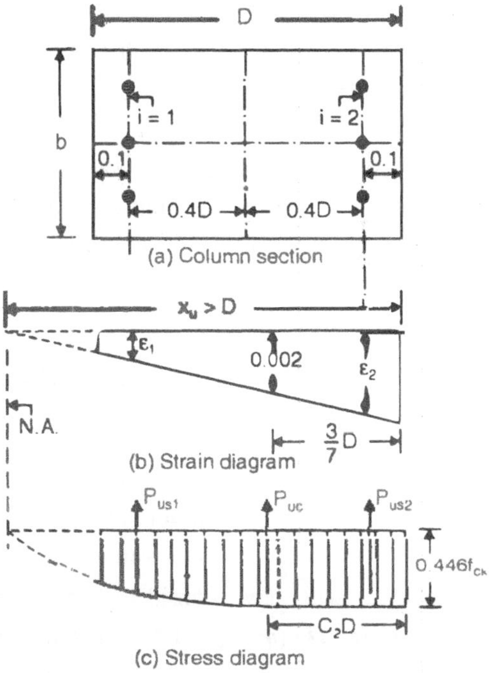

FIG. 17.40

$$\therefore \quad \frac{P_u}{f_{ck}\, bD} = 0.4 + \frac{4}{100 \times 20}(0.67 \times 250 - 0.4 \times 20) = 0.719$$

Case II : $k_u > 1$ (Fig. 17.40). From Eq. 17.5 (a),

$$P_u = C_1 f_{ck}\, bD + \sum_{i=1}^{2} \frac{p_i\, bD}{100}(f_{si} - f_{ci})$$

or

$$P_u = C_1 f_{ck}\, bD + \frac{p\, bD}{200}(f_{s1} - f_{c1}) + \frac{p\, bD}{200}(f_{s2} - f_{c2})$$

$$\therefore \quad \frac{P_u}{f_{ck}\, bD} = C_1 + \frac{p}{200\, f_{ck}}\left[(f_{s1} - f_{c1}) + (f_{s2} - f_{c2})\right]$$

where $p = 4\%$ and $f_{ck} = 20$ N/mm^2

$$\therefore \quad \frac{P_u}{f_{ck}\, bD} = C_1 + 0.001\left[(f_{s1} - f_{c1}) + (f_{s2} - f_{c2})\right] \qquad \ldots(1)$$

Also, from Eq. 17.6 (*a*),

$$M_u = C_1 f_{ck}\, bD\,(0.5 - C_2)\, D + \sum_{i=1}^{2} \frac{p_i\, bD}{100}(f_{si} - f_{ci})\, y_i$$

or

$$M_u = C_1 f_{ck}\, bD^2 (0.5 - C_2) + \frac{p\, bD}{200}(f_{s1} - f_{c1})\, y_1 + \frac{p\, bD}{200}(f_{s2} - f_{c2})\, y_2$$

$$\frac{M_u}{f_{ck}\, bD^2} = C_1 (0.5 - C_2) + \frac{p}{200\, f_{ck}}\left[(f_{s1} - f_{c1})\frac{y_1}{D} + (f_{s2} - f_{c2})\frac{y_2}{D}\right]$$

or

$$\frac{M_u}{f_{ck}\, bD^2} = C_1 (0.5 - C_2) + 0.001\left[(f_{s1} - f_{c1})\frac{y_1}{D} + (f_{s1} - f_{c2})\frac{y_2}{D}\right]$$

Here $y_1 = -0.4\, D$ and $y_2 = 0.4\, D$.

Hence $y_1/D = -0.4$ and $y_2/D = +0.4$

$$\therefore \quad \frac{M_u}{f_{ck}\, bD^2} = C_1 (0.5 - C_2) + 0.0004\left[(f_{s2} - f_{c2}) - (f_{s1} - f_{c1})\right] \qquad \ldots(2)$$

In the above equations, values of C_1 and C_2 can be obtained either from Table 17.1 or from Eqs. 17.4 and 17.5.

Also
$$\varepsilon_1 = 0.002 \times \frac{k_u D - 0.9\, D}{k_u D - 3\, D/7} = 0.002\, \frac{k_u - 0.9}{k_u - 3/7} \qquad \ldots(3)$$

$$\varepsilon_2 = 0.002 \times \frac{k_u D - 0.1\, D}{k_u D - 3\, D/7} = 0.002\, \frac{k_u - 0.1}{k_u - 3/7} \qquad \ldots(4)$$

$$f_{si} = \varepsilon_i E_s = \varepsilon_i \times 2 \times 10^5 \quad \text{for} \quad \varepsilon_i < 0.0010875$$

and
$$f_{si} = 0.87 f_y = 0.87 \times 250 = 217.5 \text{ N/mm}^2, \quad \text{for} \quad \varepsilon_i > 0.0010875$$

$$f_{ci} = 0.446 f_{ck} = 0.446 \times 20 \simeq 8.9 \text{ N/mm}^2 \quad \text{for} \quad \varepsilon_i > 0.002$$

and
$$f_{ci} = \text{value from Eq. 17.8 for } \varepsilon_i < 0.002$$

We will now consider various locations of NA. outside the section.

Case II (a) $k_u = 1.1$ From Table 17.1, $C_1 = 0.384$; $C_2 = 0.443$

$$\varepsilon_1 = 0.002 \times \frac{1.1 - 0.9}{1.1 - 3/7} = 0.000596$$

$$\varepsilon_2 = 0.002 \times \frac{1.1 - 0.1}{1.1 - 3/7} = 0.002979$$

$$f_{s1} = 0.000596 \times 2 \times 10^5 = 119.1 \text{ N/mm}^2 \quad ; \quad f_{s2} = 0.87 f_y = 217.5 \text{ N/mm}^2$$

$$f_{c1} = 446 \times 0.000596 \,(1 - 250 \times 0.000596)\, 20 = 4.5 \text{ N/mm}^2$$

$$f_{c2} = 0.446 f_{ck} = 0.446 \times 20 = 8.9 \text{ N/mm}^2$$

$$\therefore \quad \frac{P_u}{f_{ck}\, b\, D} = 0.384 + 0.001\, [(119.1 - 4.5) + (217.5 - 8.9)] = 0.707$$

and

$$\frac{M_u}{f_{ck}\, bD^2} = 0.384\,(0.5 - 0.443) + 0.0004\, [(217.5 - 8.9) - (119.1 - 4.5)] = 0.0595$$

Case II (b) $k_u = 1.5$ From Table 17.1, $C_1 = 0.422$; $C_2 = 0.480$

$$\therefore \quad \varepsilon_1 = 0.002 \times \frac{1.5 - 0.9}{1.5 - 3/7} = 0.00112 \quad \text{and} \quad \varepsilon_2 = 0.002 \times \frac{1.5 - 0.1}{1.5 - 3/7} = 0.002613$$

$$f_{s1} = 0.87 f_y = 217.5 \text{ N/mm}^2 \qquad f_{s2} = 0.87 f_y = 217.5 \text{ N/mm}^2$$

$$f_{c1} = 446 \times 0.00112 \,(1 - 250 \times 0.00112)\, 20 = 7.2 \text{ N/mm}^2$$

$$f_{c2} = 0.446 f_{ck} = 0.446 \times 20 = 8.9 \text{ N/mm}^2$$

$$\therefore \quad \frac{P_u}{f_{ck}\, b\, D} = 0.422 + 0.001\, [(217.5 - 7.2) + (2.75 - 8.9)] = 0.841$$

and

$$\frac{M_u}{f_{ck}\, bD^2} = 0.422\,(0.5 - 0.480) + 0.0004\, [(217.5 - 8.9) - (217.5 - 7.2)]$$

$$= 0.00776 \simeq \text{zero.}$$

Case III (c) $k_u = 1.0$: This is a special case when the N.A. passes through the outer edge. From Table 17.1, $C_1 = 0.361$ and $C_2 = 0.416$ for $k_u = 1.0$

$$\varepsilon_1 = 0.002 \times \frac{1 - 0.9}{1 - 3/7} = 0.00035 \quad ; \quad \varepsilon_2 = 0.002 \times \frac{1 - 0.1}{1 - 3/7} = 0.00315$$

$$f_{s1} = 0.00035 \times 2 \times 10^5 = 70 \text{ N/mm}^2 \quad ; \quad f_{s2} = 0.87 f_y = 217.5 \text{ N/mm}^2$$

$$f_{c1} = 446 \times 0.00035 \,(1 - 250 \times 0.00035)\, 20 = 2.8 \text{ N/mm}^2$$

$$f_{c2} = 0.446 f_{ck} = 0.446 \times 20 = 8.9 \text{ N/mm}^2$$

$$\therefore \quad \frac{P_u}{f_{ck}\, bD} = 0.361 + 0.001\, [(70 - 2.8) + (217.5 - 8.9)] = 0.637$$

and

$$\frac{M_u}{f_{ck}\, bD^2} = 0.361\,(0.5 - 0.416) + 0.0004\, [(217.5 - 8.9) - (70 - 2.8)] = 0.0869$$

Case III $k_u < 1$ (Fig. 17.41)

Here $C_1 = 0.36\, k_u$ and $C_2 = 0.416\, k_u$. From Eq. 17.12,

$$\frac{P_u}{f_{ck}\, bD} = C_1 + \sum_{i=1}^{2} \frac{p_i}{100 f_{ck}} (f_{si} - f_{ci})$$

or

$$\frac{P_u}{f_{ck}\, bD} = C_1 + \frac{p}{200 f_{ck}} [(f_{s1} - f_{c1}) + \frac{p}{200 f_{ck}} (f_{s2} - f_{c2})]$$

or $$\frac{P_u}{f_{ck}\, bD} = C_1 + 0.001\ [(f_{s1} - f_{c1}) + (f_{s2} - f_{c2})] \qquad ...(5)$$

From Eq. 17.13,

$$\frac{M_u}{f_{ck}\, bD^2} = C_1\ (0.5 - C_2) + \sum_{i=1}^{2} \frac{p_i}{100 f_{ck}} (f_{si} - f_{ci}) \frac{y_i}{D}$$

Here, $p = 4\%$; $f_{ck} = 20$; $y_1/D = -0.4$; $y_2/D = +0.4$

$$\frac{M_u}{f_{ck}\, bD^2} = C_1\ (0.5 - C_2) + 0.0004\ [(f_{s2} - f_{c2}) - (f_{s1} - f_{c1})] \qquad ...(6)$$

Also, $\varepsilon_1 = 0.0035\ (k_u D - 0.9\ D)/k_u D$

$= 0.0035\ (k_u - 0.9)/k_u \qquad ...(7)$

and $\varepsilon_2 = 0.0035\ (k_u D - 0.1\ D)/k_u D$

$= 0.0035\ (k_u - 0.1)/k_u \qquad ...(8)$

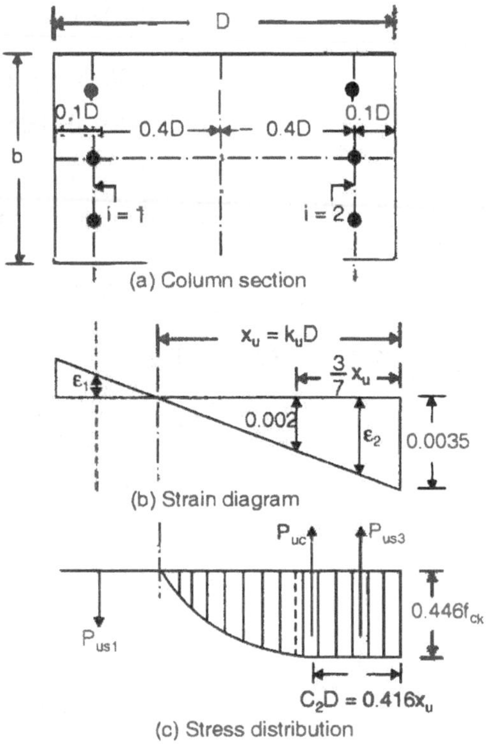

FIG. 17.41.

Case III (a) : N.A. passes through outer steel : $x_u = 0.9\ D$; $k_u = 0.9$

$C_1 = 0.36\ k_u = 0.36 \times 0.9 = 0.324$;

$C_2 = 0.416\ k_u = 0.3744$

$\varepsilon_1 = 0.0035\ (0.9 - 0.9)/0.9 = 0$;

$\varepsilon_2 = 0.0035\ (0.9 - 0.1)/0.9 = 0.003111$

$f_{s1} = \varepsilon_1 E =$ zero ; $f_{s2} = 0.87\ f_y = 0.87 \times 250 = 217.5\ \text{N/mm}^2$

$f_{c1} = 0$; $f_{c2} = 0.446\ f_{ck} = 0.446 \times 20 = 8.9\ \text{N/mm}^2$

$$\therefore \quad \frac{P_u}{f_{ck}\, bD} = 0.324 + 0.001\ [(0 - 0) + (217.5 - 8.9)] = 0.5326$$

$$\frac{M_u}{f_{ck}\, bD^2} = 0.324\ (0.5 - 0.3744) + 0.0004\ [(217.5 - 8.9) - 10 - 0)] = 0.1241$$

Case III (b) : $k_u = 0.75$

$C_1 = 0.36\ k_u = 0.36 \times 0.75 = 0.27$; $C_2 = 0.416\ k_u = 0.416 \times 0.75 = 0.312$

$\varepsilon_1 = 0.0035\ (0.75 - 0.9)/0.75 = -0.0007$

$\varepsilon_2 = 0.0035\ (0.75 - 0.1)/0.75 = 0.003033$

$f_{s1} = \varepsilon_1 E = (-0.0007) \times 2 \times 10^5 = -140\ \text{N/mm}^2$

$f_{s2} = 0.87\ f_y = 0.87 \times 250 = 217.5\ \text{N/mm}^2$

$f_{c1} = 0$ (being tensile) ;

$f_{c2} = 0.446\ f_{ck} = 0.446 \times 20 = 8.9\ \text{N/mm}^2$

$$\therefore \quad \frac{P_u}{f_{ck}\, bD} = 0.27 + 0.001\ [(-140 - 0) + (217.5 - 8.9)] = 0.3386$$

$$\frac{M_u}{f_{ck}\, bD^2} = 0.27\ (0.5 - 0.312) + 0.0004\ [(217.5 - 8.9) - (-140)] = 0.1902$$

Case III (c)* : *k_u corresponding to first yield point

At the first yield point, $f_{s1} = f_{yd} = 0.87\, f_y = 0.87 \times 250 = 217.5\ \text{N/mm}^2$

$$\therefore \qquad \varepsilon_1 = \varepsilon_{s1} = \frac{f_{s1}}{E_s} = \frac{217.5}{2 \times 10^5} = 0.0010875$$

Hence from the strain diagram,

$$\frac{k_u\, D}{0.0035} = \frac{0.9\, D - k_u\, D}{0.0010875}, \text{ from which } k_u = 0.6867$$

$\therefore$ $C_1 = 0.36\, k_u = 0.36 \times 0.6867 = 0.2472$; $C_2 = 0.416\, k_u = 0.416 \times 0.6867 = 0.2866$

$\varepsilon_2 = 0.0035\ (0.6867 - 0.1)/0.6867 = 0.00299$

$\therefore$ $f_{s2} = 0.87\, f_y = 0.87 \times 250 = 217.5\ \text{N/mm}^2$; $f_{c2} = 0.446\, f_{ck} = 8.9\ \text{N/mm}^2$

$$\therefore \qquad \frac{P_u}{f_{ck}\, bD} = 0.2472 + 0.001\ [(-217.5) + (217.5 - 8.9)] = 0.2383$$

$$\frac{M_u}{f_{ck}\, bD^2} = 0.2472\ (0.5 - 0.2866) + 0.0004\ [(217.5 - 8.9) - (-217.5)] = 0.2232$$

Case III (d) k_u corresponding to final yield point

At the final yield point, $f_{s1} = f_{yd} = 0.87\, f_y = 0.87 \times 250 = 217.5\ \text{N/mm}^2$

Maximum strain, $\varepsilon_1 = \varepsilon_{s1} = 0.002 + \dfrac{0.87\, f_y}{E_s} = 0.002 + \dfrac{0.87 \times 250}{2 \times 10^5} = 0.003088$

Hence from the strain diagram,

$$\frac{k_u\, D}{0.0035} = \frac{0.9\, D - k_u\, D}{0.003088}, \text{ from which } k_u = 0.4782$$

$\therefore$ $C_1 = 0.36\, k_u = 0.36 \times 0.4782 = 0.1721$

$C_2 = 0.416\, k_u = 0.416 \times 0.4782 = 0.1989$

$\varepsilon_2 = 0.0035\ (0.4782 - 0.1)/0.4782 = 0.002768$

$\therefore$ $f_{s2} = 0.87\, f_y = 0.87 \times 250 = 217.5\ \text{N/mm}^2$ and $f_{c2} = 0.446\, f_{ck} = 8.9\ \text{N/mm}^2$

$$\therefore \qquad \frac{P_u}{f_{ck}\, b\, D} = 0.1721 + 0.001\ [-217.5 + (217.5 - 8.9)] = 0.1632$$

$$\frac{M_u}{f_{ck}\, bD^2} = 0.1721\ (0.5 - 0.1989) + 0.0004\ [(217.5 - 8.9) - (-217.5)] = 0.2223$$

***Case III (e)* : $k_u = 0.25$**

$\therefore$ $C_1 = 0.36\, k_u = 0.36 \times 0.25 = 0.09$; $C_2 = 0.416\, k_u = 0.416 \times 0.25 = 0.104$

$\varepsilon_1 = 0.0035\ (0.25 - 0.9)/0.25 = -0.0091 > \varepsilon_{sy}$

$\varepsilon_2 = 0.0035\ (0.25 - 0.1)/0.25 = 0.0021 > \varepsilon_{sy}$

$\therefore$ $f_{s1} = -0.87\, f_y = -217.5\ \text{N/mm}^2$; $f_{s2} = +0.87\, f_y = +217.5\ \text{N/mm}^2$

$f_{c1} = 0$ (being tensile) ; $f_{c2} = 0.447\, f_{ck} = 0.447 \times 20 = 8.9\ \text{N/mm}^2$

$$\therefore \qquad \frac{P_u}{f_{ck}\, bD} = 0.09 + 0.001\ [-217.5 + (217.5 - 8.9)] = 0.0811$$

$$\frac{M_u}{f_{ck}\, bD^2} = 0.09\ (0.5 - 0.104) + 0.0004\ [(217.5 - 8.9) - (-217.5)] = 0.2061$$

Case III (f) : Pure flexure : Doubly reinforced beam action

For this case discussed in example 17.11 (Fig. 17.26), $\varepsilon_{c\,.\,max} = 0.0035$ and $\varepsilon_{s1} = \varepsilon_1 > 0.002 + 0.87 f_y/E_s > 0.003088$. $\varepsilon_2 = 0.0035\ (k_u - 0.1)/k_u$

$$f_{s2} = \varepsilon_2\ .\ E_s = 2 \times 10^5 \times 0.0035\ (k_u - 0.1)/k_u = 700 - \frac{70}{k_u}$$

$$f_{s1} = 0.87 f_y = 217.5\ \text{N/mm}^2\ \text{(tensile)}\quad f_{c1} = 0\ \text{(being tensile)}$$

$$f_{c2} = 0.446 f_{ck} = 0.446 \times 20 = 8.92\ \text{N/mm}^2,\quad C_1 = 0.36\ k_u\ \text{ and }\ C_2 = 0.416\ x_u$$

For this case of pure flexure, $P_u = 0$.

$$\therefore \qquad \frac{P_u}{f_{ck}\, bD} = 0 = C_1 + 0.001\ [(f_{s1} - f_{c1}) + (f_{s2} - f_{c2})] = 0$$

$$\therefore \qquad 0.36\ k_u + 0.001\ [(-217.5) + 700 - \frac{70}{k_u} - 8.92] = 0$$

which reduces to $k_u^2 + 1.3155\ k_u - 0.1944 = 0$, from which $k_u = 0.1341$

Hence $$f_{s2} = 700 - \frac{70}{k_u} = 700 - \frac{70}{0.1341} = 178\ \text{N/mm}^2$$

$$\therefore \qquad C_1 = 0.36\ k_u = 0.36 \times 0.1341 = 0.0483;\quad C_2 = 0.416\ k_u = 0.416 \times 0.1341 = 0.05579$$

$$\therefore \qquad \frac{M_u}{f_{ck}\, bD^2} = 0.0483\ (0.5 - 0.05579) + 0.0004\ [(178 - 8.92) - (-217.5)] = 0.1761$$

The moment corresponding to this is designated as M_{u0}.

Case III (g) Steel beam action : Pure flexure

For steel beam action, concrete has no contribution in bending action. Also, for pure flexure, $P_u = 0$. Hence

$$\frac{M_u}{f_{ck}\, bD^2} = 0.0004\ [f_{s2} - f_{s1}] = 0.0004\ [217.5 - (-217.5)] = 0.174$$

Note : It is not justified to neglect the contribution of concrete in resisting moment.

Summary of the results :

The results obtained from various cases are summarised below:

k_u	0.1341	0.25	0.4782	0.6867	0.75	0.9	1.0	1.1	1.5	∞ ($e_{min.}$)	∞ ($e = 0$)
$\frac{P_u}{f_{ck}\, bD}$	0	0.0811	0.1632	0.2383	0.3386	0.5326	0.637	0.707	0.841	0.719	0.8632
$\frac{M_u}{f_{ck}\, bD^2}$	0.1761	0.2061	0.2223	0.2232	0.1902	0.1241	0.0869	0.0595	0.0078	–	–

The values obtained above match well with the values of Chart 28 of SP 16-1980. The non-dimensional interaction curve is shown in Fig. 17.42.

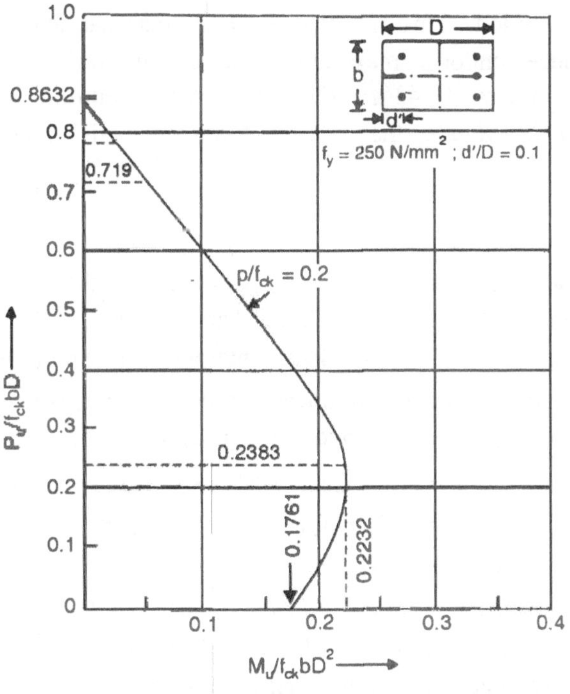

FIG. 17.42. NON-DIMENSIONAL INTERACTION CURVE.

17.12. SHORT ECCENTRICALLY LOADED COLUMN WITH UNSYMMETRICAL REINFORCEMENT

Normally, a column section subjected to permanent axial load is symmetrically reinforced. This is so because the B.M. caused due to temporary loads such as wind etc, may change sign. However, where the column is subjected to permanent bending moment along with permanent axial load (such as the column of a canopy or outer columns in a building frame), it is more useful to provide more reinforcement on tension side than that on the compression side.

Fig. 17.43(*a*) shows such a column section, having A_{s1} on tension side greater than A_{s2} provided on the compression side. Let d_1' and d_2' be the corresponding effective covers. The basic equations for P_u and M_u, developed earlier remain the same, except that moments of the forces are taken about '*plastic centroidal axis*' rather than the '*geometrical centroidal axis*'. A *plastic centroid* (G_p) is the centroid of a concrete section under uniform plastic condition with concrete reaching a uniform compressive strain

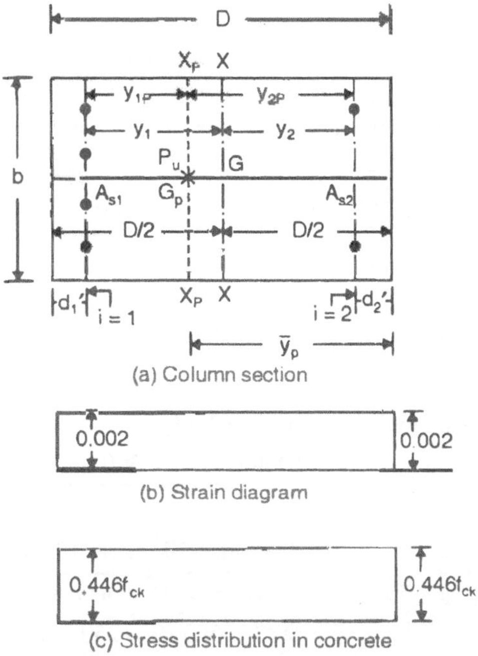

FIG. 17.43. PLASTIC CENTROIDAL AXIS

of 0.002 all over the section. This means that if an external load (P_u) acts at the plastic centroid, it will produce uniform maximum compressive strain of 0.002 over the entire concrete section. Thus in Fig. 17.43(a), G_p is the plastic centroid while C is the geometric centroid ; $X_p - X_p$ and $X - X$ are the 'plastic centroidal axis' and 'geometrical centroidal axis' respectively.

In order to determine the position of the plastic centroidal axis, uniform strain = 0.002 is taken throughout the section as shown in Fig. 17.43 (b). The stress $f_{s1} = f_{s2} = f_s$ in both the reinforcements is taken corresponding to $\varepsilon_s = \varepsilon = 0.002$, either from the stress-strain curve for the steel, or from Table 2.3, for HYSD bars. Thus, for Fe 415 steel, we get $f_s \simeq 327.7\ \text{N/mm}^2$, from Table 2.3, by linear interpolation. Similarly, for Fe 500 steel, we get $f_s = 373.1\ \text{N/mm}^2$ from Table 2.3 by linear interpolation corresponding to $\varepsilon_s = 0.002$. However, for mild steel, $f_s = 0.87 f_y = 0.87 \times 250 = 217.5\ \text{N/mm}^2$, for $\varepsilon_s = 0.002$.

Let $\bar{y}_p$ be the depth of plastic centroidal axis from the highly compressed edge. Taking moments of forces about this edge, we get

$$\bar{y}_p = \frac{(0.446\ f_{ck}\ bD)\ D/2 + A_{s1} \times f_s\ (D - d_1') + A_{s2}\ f_s\ d_2'}{0.446\ f_{ck}\ bD + (A_{s1} + A_{s2})\ f_s} \qquad \ldots(17.15)$$

Thus, the plastic centroidal axis is located. For this location, the distances y_{1p} and y_{2p} for the two rows of steel are found easily. The analysis is now done with the help of basic equations developed earlier, except for the difference that moments are taken about the plastic centroidal axis. The procedure is illustrated through Example 17.20.

Example 17.20. ***Determination of M_u for unsymmetrically reinforced R.C. column***

A short rectangular column of size 300 mm × 500 mm, reinforced with four bars of 20 mm diameter on the tension side and 2 bars of 20 mm φ on the compression side, carries an ultimate axial load of 315 kN. Determine the maximum ultimate moment it can carry about the major axis. Take the following : Grade of concrete : M 20; grade of steel : Fe 415, and effective cover = 60 mm to both the reinforcements.

Solution : Given $b = 300$ mm $D = 500$ mm; $d_1' = d_2' = 60$ mm

$$A_{s1} = 4\ \frac{\pi}{4}\ (20)^2 = 1256.64\ \text{mm}^2;\ \ A_{s2} = 2 \times \frac{\pi}{4}\ (20)^2 = 628.32\ \text{mm}^2$$

For $\varepsilon_s = 0.002$, we have $f_{s1} = f_{s2} = f_s = 324.8 + \dfrac{342.8 - 324.8}{241 - 192}\ (200 - 192) \simeq 327.7\ \text{N/mm}^2$

Location of plastic centrodal axis : Taking moments about highly compressed edge, we get from Eq. 17.15,

$$\bar{y}_p = \frac{(0.446 \times 20 \times 300 \times 500)\ 250 + 1256.64 \times 327.7\ (500 - 60) + 628.32 \times 327.7 \times 60}{0.446 \times 20 \times 300 \times 500 + 1256.64 \times 327.7 + 628.32 \times 327.7}$$

$$= \frac{334500 \times 10^3 + 181192.4 \times 10^3 + 12354 \times 10^3}{1338 \times 10^3 + 411.8 \times 10^3 + 205.9 \times 10^3} = 270\ \text{mm}$$

$\bar{y} = D/2 = 500/2 = 250$ mm

Thus, plastic centroidal axis is 20 mm below the geometric centroidal axis.

$\therefore \quad y_{1p} = -(500 - 270 - 60) = -170$ mm

$y_{2p} = +270 - 60 = +210$ mm

Let us now determine the position of N.A. corresponding to $P_u = 250$ kN, by trial and error, starting with the assumption that N.A. falls within the section.

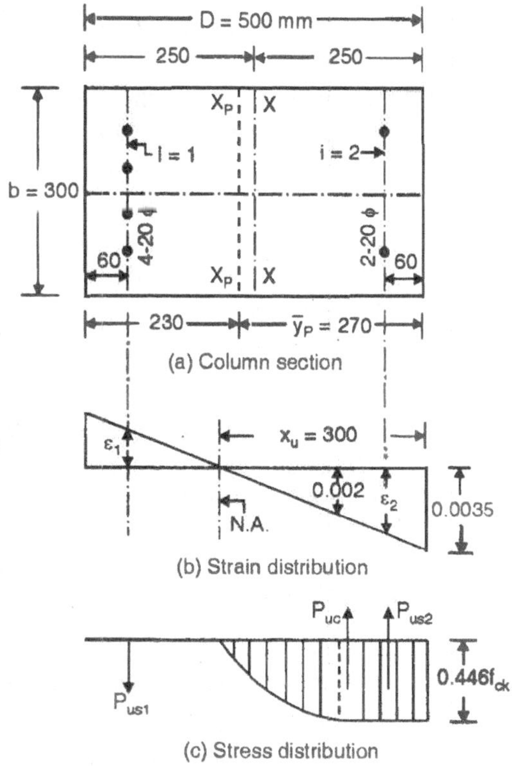

FIG. 17.44

Trial 1 : Let $x_u = 0.6\,D = 0.6 \times 500 = 300$ mm

$$\varepsilon_1 = \frac{0.0035\,(300 - 440)}{300} = -0.001633$$

$$\varepsilon_2 = \frac{0.0035\,(300 - 60)}{300} = 0.0028$$

$$f_{s1} = 306.7 + \frac{324.8 - 306.7}{192 - 163}\,(163.3 - 163)$$

$$= 306.9 \text{ N/mm}^2 \quad \text{(tensile)}$$

$$f_{s2} = 351.8 + \frac{360.9 - 351.8}{380 - 276}\,(280 - 276)$$

$$= 352.1 \text{ N/mm}^2$$

$f_{c1} = 0$ (being tensile)

$f_{c2} = 0.446\,f_{ck} = 8.9$ N/mm^2

$P_{uc} = 0.36\,f_{ck}\,b\,x_u$

$= 0.36 \times 20 \times 300 \times 300 \times 10^{-3}$

$= 648$ kN

$$\sum_{i=1}^{2} P_{usi} = [1256.64\,(-306.9) + 628.32\,(352.1 - 8.9)]\,10^{-3} \simeq -170 \text{ kN}$$

$$\therefore \quad P_u = P_{uc} + \sum_{i=1}^{2} P_{usi} = 648 - 170 = 478 \text{ kN}$$

This is much greater than the given value of 315 kN. Hence in the second trial, reduce the value of x_u

Trial 2 : Let $x_u = 0.5\,D = 250$ mm

$$\varepsilon_1 = \frac{0.0035\,(250 - 440)}{250} = -0.00266; \quad \varepsilon_2 = \frac{0.0035\,(250 - 60)}{250} = +0.00266$$

$$f_{s1} = 342.8 + \frac{351.8 - 342.8}{276 - 241}\,(266 - 241) = 349.2 \text{ N/mm}^2 \quad \text{(tensile)}$$

$f_{s2} = 349.2$ N/mm^2 (comp.) ; $f_{c1} = 0$ (being tensile) ; $f_{c2} = 0.446\,f_{ck} = 8.9$ N/mm^2

$P_{uc} = 0.36 \times 20 \times 300 \times 250 \times 10^{-3} = 540$ kN

$$\sum_{i=1}^{2} P_{usi} = [1256.44\,(-349.2) + 628.32\,(349.2 - 8.9)]\,10^{-3} = -438.75 + 213.82 = -224.93 \text{ kN}$$

$$\therefore \qquad P_u = P_{uc} + \sum_{i=1}^{2} P_{usi} = 540 - 224.93 = 315.07 \text{ kN}$$

This is quite close to the desired value of 315 kN and hence no further trial is required. For this position of N.A., the moment M_u about the plastic centroidal axis is given by

$$M_u = M_{uc} + \sum_{i=1}^{2} M_{usi} = P_{uc} (\bar{y}_p - 0.416 \, x_u) + P_{us1} (y_{1p}) + P_{us2} (y_{2p})$$

$$= [540 (270 - 0.416 \times 250) + (438.75) (-170) + 213.82 (210)] \, 10^{-3}$$

$$= 89.64 + 74.589 + 44.902 = \mathbf{209.131 \text{ kN-m}}$$

PROBLEMS

1. A short R.C. column 240 mm wide and 400 mm deep is reinforced with 6 bars of 16 mm dia., arranged equally along the two short faces at an effective cover of 50 mm. Determine the B.M. M_u about an axis bisecting the depth, when it is also subjected to $P_u = 600$ kN. Assume M 20 grade concrete and Fe 415 grade steel.

2. For the data of problem 1, determine the value of P_u if $M_u = 60$ kN-m.

3. For the data of problem 1, determine the value of P_u if it is applied at an eccentricity of (*a*) 120 mm (*b*) 60 mm.

4. For the data of problem 1, determine the value of P_u that can be permitted at the minimum eccentricity specified by the Code, if the effective length of the column is 4 m.

5. Design A R.C. column 300 mm square to carry an ultimate load of 600 kN at an eccentricity of 120 mm. Use M 20 grade concrete and mild steel bars.

6. Design a circular column of 300 mm dia. for the data of problem 5 using (*i*) hoop reinforcement (*ii*) helical reinforcement.

7. A R.C. column of M 20 concrete, 300 mm × 400 mm is reinforced with 8 bars of 20 mm φ of Fe 415 grades, uniformly distributed along all the four sides, with an effective cover of 50 mm. Determine the strength components P_u and M_u for the balanced 'failure' condition.

8. For the column section of problem 7, determine strength components P_u and M_u for the 'limiting condition' of no tension in the section.

9. For the column section of problem 7, determine the strength components P_u and M_u corresponding to N.A. location at 1.1 D.

10. For the column section of problem 8, determine the strength components corresponding to the condition of maximum strain of 0.0035 in concrete and tensile strain of 0.002 in the outer most layer of steel.

11. A short R.C. column of 300 mm × 400 mm section is reinforced with 6 bars of 16 mm dia., uniformly distributed along the four sides, at an effective cover of 50 mm. Determine the maximum allowable ultimate load if the unsupported lengths about x-axis and y-axis are equal to 5 m and 3 m respectively. Take M 25 grade concrete and Fe 415 grade steel bars.

12. Design a short column of M 20 grade concrete, to carry an ultimate load of 1200 kN and an ultimate moment of 60 kN-m about the major axis. Take width of column as 250 mm and depth as 350 mm. Use Fe 415 steel bars, at an effective cover of 50 mm.

13. A short rectangular column of 300 mm × 400 mm, reinforced width three bars of 20 mm dia on tension side and two bars of 20 mm φ on the compression side carries an axial load of 250 kN. Determine the maximum ultimate moment it can carry about major axis.
Take M 20 concrete, Fe 415 grade steel, and 50 mm effective cover.

18

Limit State of Collapse in Compression: III Columns with axial load and biaxial bending

18.1. INTRODUCTION

A column with axial load and biaxial bending is commonly found in structures, because of two major reasons : (*i*) axial load may have natural eccentricities, though small, with respect to both the axes, and (*ii*) corner columns of a building may be subjected to bending moments in both the directions along with axial load. Similarly, external facade columns under combined vertical and horizontal load are other examples of biaxial bending. Also, beams supporting helical or free-standing stairs or oscillating and rotary machinery are subjected to biaxial bending with or without axial load of either compressive or tensile nature.

18.2. BIAXIAL ECCENTRICITIES

Strictly speaking, every column should be treated as being subjected to axial compression along with biaxial bending by considering possible eccentricities of the axial load with respect to both the major axis (i.e. *x*-*x* axis) as well the minor axis (i.e. *y*-*y* axis). These eccentricities, designated as e_x and e_y with respect of *x* and *y* axes, may be atleast $e_{min.}$ (i.e. $e_{x\ ,\ \min}$ and $e_{y\ .\ \min}$), though in majority of cases of biaxial bending, these may be much more than e_{min} In fact *uniaxial bending* is an idealised approximation which is made when the *e*/*D* ratio with respect to one or the other principal axes is considered to be negligible. *Axial loading*

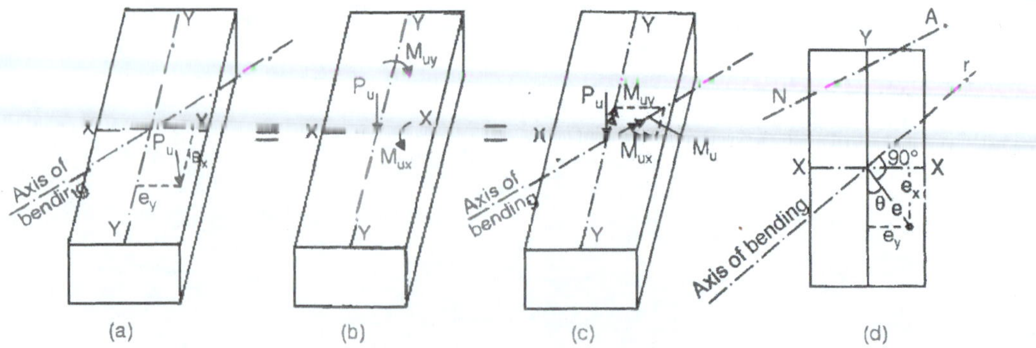

FIG. 18.1.

is a further idealised approximation which is made when *e/D* ratios with respect to both the principal axes are negligibly small.

Fig. 18.1 (*a*) shows a eccentrically loaded short column, which is subjected to an axial load P_u which has eccentricities e_x and e_y with respect to *x* and *y*-axis respectively. This eccentric loading is equivalent to an axial load P_u, along with moments M_{ux} and M_{uy} as shown in Fig. 18.1 (*b*). These two moments M_{ux} and M_{uy} about the two axes can be resolved into a single resultant moment M_u about an axis *r - r* inclined to the two axes (Fig. 18.1. *c*), as under :

$$M_u = \sqrt{M_{ux}^2 + M_{uy}^2} \quad \ldots(18.1)$$

Alternatively, the two eccentricities e_x ($= M_{ux}/P_u$) and e_y ($= M_{uy}/P_u$) can be resolved into a resultant eccentricity e ($= M_u/P_u$), given by (Fig. 18.1 *d*)

$$e = \sqrt{e_x^2 + e_y^2} \quad \ldots(18.2)$$

The resultant eccentricity (*e*) is inclined at an angle θ with the *y*-axis, given by

$$\theta = \tan^{-1} \frac{e_y}{e_x} \quad \ldots(18.3)$$

The resultant axis of bending is perpendicular to the line joining the point of loading (P_u) to the centroid of the section, as shown in Fig. 18.1 (*d*). The neutral axis (Fig. 18.1 *d*) will generally not be parallel to the axis of bending. The method of locating the position of N.A., by trial and error, is very tedious and laborious.

18.3. METHODS OF DESIGN

IS : 456 - 2000 gives two methods of design of R.C. columns subjected to axial load and biaxial bending.

(*i*) Method based on conditions of equilibrium

(*ii*) Breslar's load contour method.

Method based on conditions of equilibrium

We have seen in Fig. 18.1 (*d*) that a column subjected to axial load and biaxial bending has a neutral axis which is inclined to both *x* and *y* axes, and which is generally

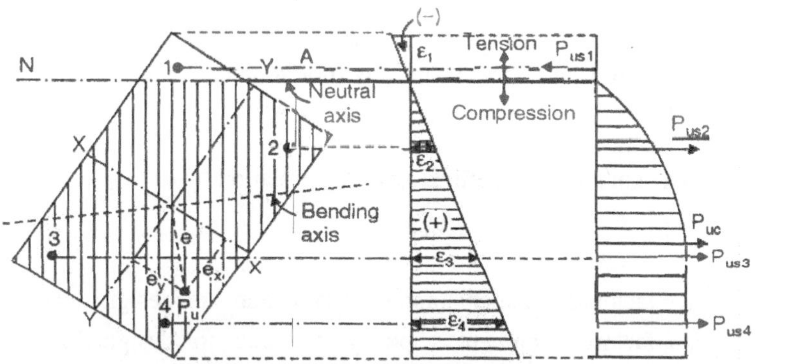

(a) Column section (b) Strain distribution (c) Stress distribution

FIG. 18.2. STRAIN AND STRESS DISTRIBUTIONS FOR A CHOSEN LOCATION OF N.A.

not parallel to the axis of bending. The first method is based on applying the conditions of equilibrium, to a given section, on the basis of assumptions of the method of limit state of collapse in combined axial load and uniaxial bending (see Chapter 17), with a suitably chosen *inclined neutral axis*. Fig. 18.2 (*b*), (*c*) show the strain and stress distributions corresponding to an *assumed* or chosen location of neutral axis. The method of analysis, explained in the previous Chapter can now be applied to this section with the assumed position of N.A. However, we have already noticed in the previous Chapter, the cumbersome nature of trial and error approach, even for the problem of axial load with uniaxial bending. The problem is further complicated due to the irregular shape of the portion of concrete in compression (shown hatched in Fig. 18.2 *a*) and computation of total compression P_u due to rectangular-cum-parabolic stress distribution over this irregular surface. Some times, approximations are made by either assuming an equivalent rectangular block of width equal to 0.45 f_{ck} and depth equal to 0.85 x_u or an equivalent stress block of width 0.36 f_{ck} over full depth of x_u with total compression acting at 0.42 x_u from the highly compressed edge. However, inspite of these approximations, the problem still remains complex, making manual computations almost impossible. This gives rise to the fact that the determination of location of exact neutral axis is an extremely laborious process of trial and error. This explains the paucity of published material based on this method.

18.4. BRESLAR'S LOAD CONTOUR METHOD

This method is based on the approximation of an *interaction surface* which may be visualised in a three-dimensional plot of $P_u - M_{ux} - M_{uy}$. Such an interaction surface is obtained by obtaining $P_u - M_u$ interaction curves for different axes of bending (Fig. 18.3 *a*). The required three dimensional interaction surface shown in Fig. 18.3 (*b*) is generated as the envelope of a number of such interaction curves for different axes of bending of a column. The interaction surface so obtained can be considered to be a *failure surface*. Any point (P_u , M_{ux} , M_{uy}) on the *failure surface*, corresponds to a *failure conditions* in a column about a neutral axis and any point *inside* the surface can be considered *safe*. Also, any point (P_u , M_{ux} , M_{uy}) that lies *outside* the surface is considered *unsafe*. The *interaction surface* (or the failure surface) has three typical traces : (*i*) trace on *x-z* plane (*ii*) trace on *y-z* plane and (*iii*) trace on any horizontal surface parallel to the *x-y* plane. The trace on *x-z* plane correspond to the interaction curve for uniaxial eccentricity with respect to *x*-axis, while the trace on *y-z* plane correspond to the interaction curve for uniaxial eccentricity with respect to *y*-axis. However, trace on any horizontal plane, parallel to *x-y* plane, at any load level (P_u) is also an interaction curve between M_{ux} and M_{uy} and is known as a *load contour* corresponding to a particular P_u/P_{uz} value. The general non-dimensional equation for the load contour is defined by the relation :

$$\left(\frac{M_{ux}}{M_{ux1}}\right)^{\alpha n} + \left(\frac{M_{uy}}{M_{uy1}}\right)^{\alpha n} = 1 \qquad \text{...(18.4)}$$

where $M_{ux} = P_u \cdot e_x$ = factored moment about *x*-axis due to design load

$M_{uy} = P_u \cdot e_y$ = factored moment about *y*-axis due to design load

M_{ux1} = Maximum uniaxial moment capacity for bending about *x*-axis with axial load P_u, when $M_{uy} = 0$

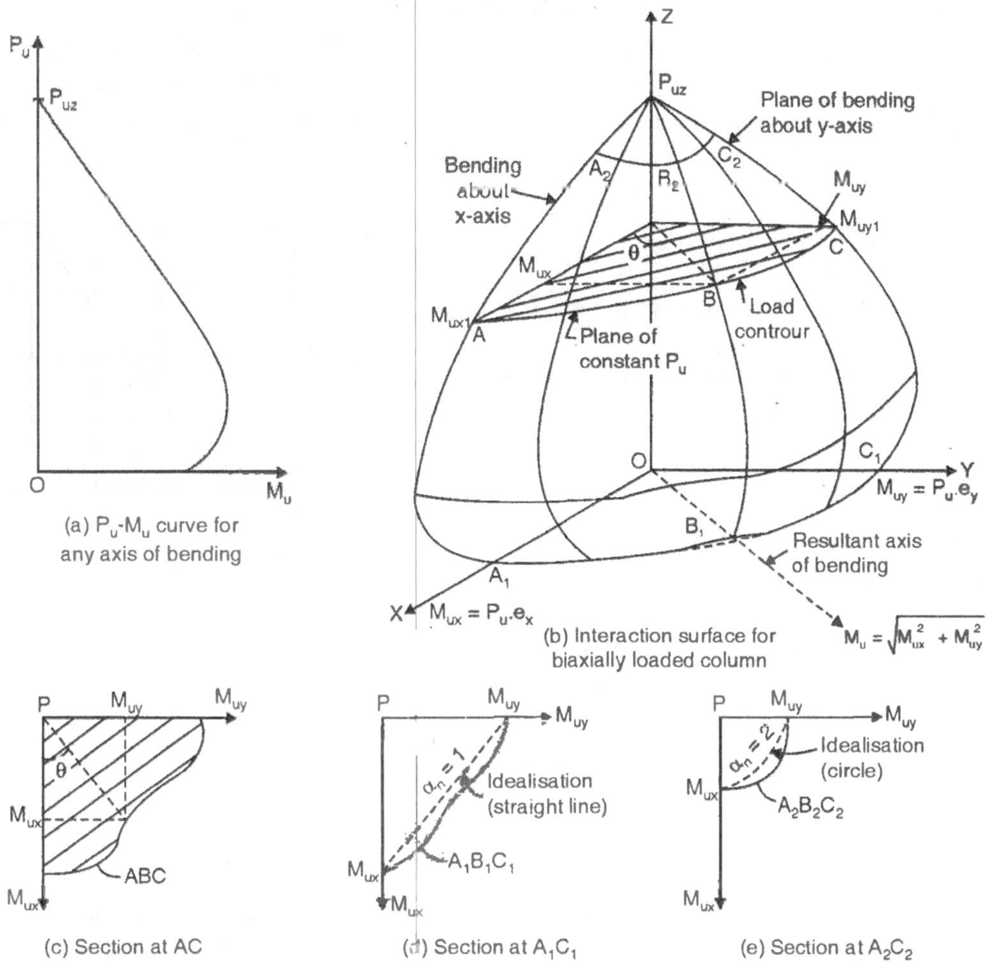

FIG. 18.3. BRESLAR'S LOAD CONTOUR METHOD

M_{uy1} = maximum uniaxial moment capacity for bending about y-axis with axial load P_u, when $M_{ux} = 0$

α_n = an exponent whose value depends on the ratio P_u/P_{uz}

P_u = factored axial load

P_{uz} = axial load capacity of the section under pure axial load, given by

$$P_{uz} = 0.45\, f_{ck}\, A_c + 0.75\, f_y\, A_s \qquad \ldots(16.1)$$

In fact the above expression approximates the equation of load contour at constant P_u. It would be a circle when $\alpha_n = 2$ (Fig. 18.3 *e*) and a straight line when $\alpha_n = 1$ (Fig. 18.3 *d*). Breslar has also suggested that in practical design, it is satisfactory to take value of α_n between 1.15 to 1.55 for rectangular section and 1.5 to 2 for square section.

18.5. METHOD SUGGESTED BY IS CODE (IS : 456 - 2000)

The method set out in clause 39.6 of the Code (IS : 456 - 2000) is based on an assumed failure surface that extends the axial load-moment diagram ($P_u - M_u$) for single axis bending in three dimensions. Such an approach is also known as *Breslar's load contour method*. According to the Code, the left hand side of Eq. 18.4 shall not exceed 1. Thus we have

$$\left(\frac{M_{ux}}{M_{ux1}}\right)^{\alpha_n} + \left(\frac{M_{uy}}{M_{uy1}}\right)^{\alpha_n} \leq 1 \qquad \ldots(18.5)$$

The Code further relates α_n to the ratio of P_u/P_{uz} as under :

P_u/P_{uz}	α_n
≤ 0.2	1.0
≥ 0.8	2.0

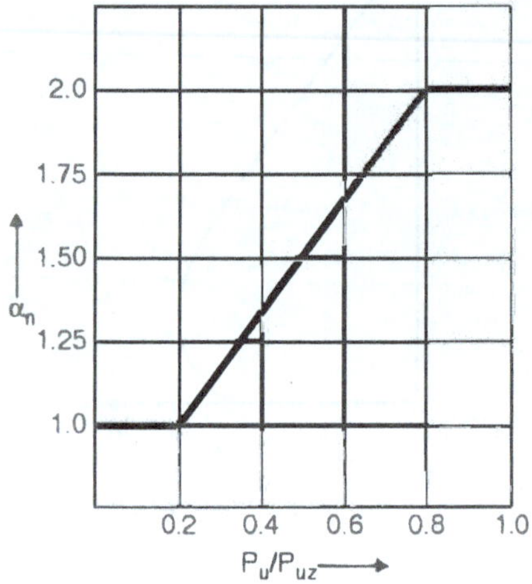

FIG. 18.4

For intermediate values of P_u/P_{uz}, linear interpolation may be done as shown in Fig. 18.4. The value of α_n can also be determined by the following equation applicable for P_u/P_{uz} between 0.2 and 0.8.

$$\alpha_n = \frac{2}{3}\left[1 + \frac{5}{2}\frac{P_u}{P_{uz}}\right] \qquad \ldots(18.6)$$

which should be within 1.0 and 2.0 as above.

As seen in Chapter 16, load P_{uz} is given by

$$P_{uz} = 0.45\, f_{ck}\, A_c + 0.75\, f_y\, A_s = 0.45\, f_{ck}\, A_g + (0.75\, f_y - 0.45\, f_{ck})\, A_s \qquad \ldots(18.7)$$

Load P_{uz} may be evaluated from Chart 63 of ISI hand book (SP 16-2000) reproduced here in Fig. 18.5. Fig. 18.6 shows graphical representation of Eq. 18.5 for appropriate value of α_n corresponding to different values of P_u/P_{uz}.

18.6. DESIGN OF COLUMN

In most of the cases of building design, the shape and size of a column are fixed in advance, for the purpose of frame analysis where the stiffnesses of various members are required. Hence the only design element which remains to be decided is the reinforcement: both its percentage as well as placement. For a column subjected to axial load along with biaxial bending, reinforcement is assumed to be distributed equally on all the sides of a rectangular section.

To start with, the reinforcement is selected for the section, for uniaxial moment M_u, given by the following equation.

$$M_u = a\sqrt{M_{ux}^2 + M_{uy}^2} \qquad \ldots(18.8)$$

The value of a may vary between 1.10 to 1.20 — lower value of a for *higher axial loading* level (P_u/P_{uz}).

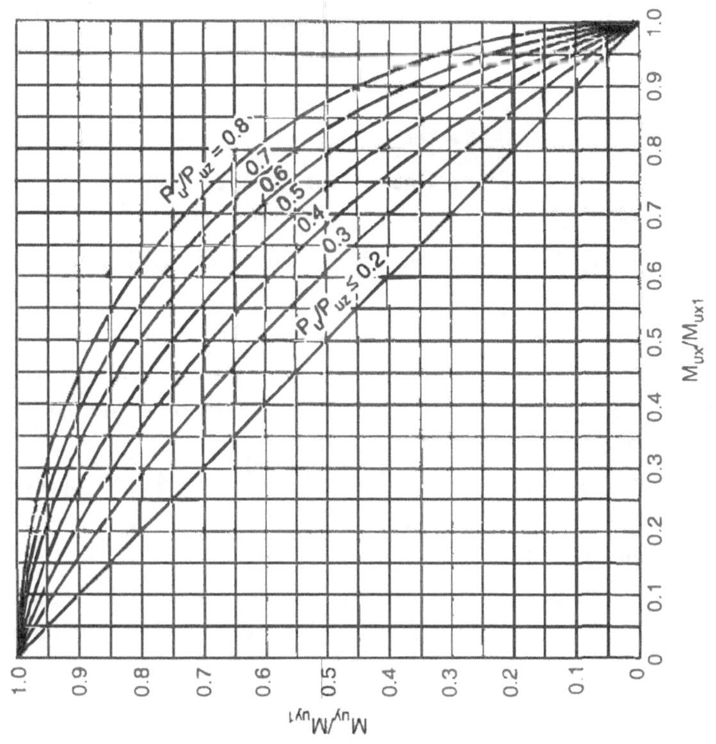

FIG. 18.6. INTERACTION CURVES FOR BIAXIAL BENDING IN COMPRESSION IN MEMBERS.

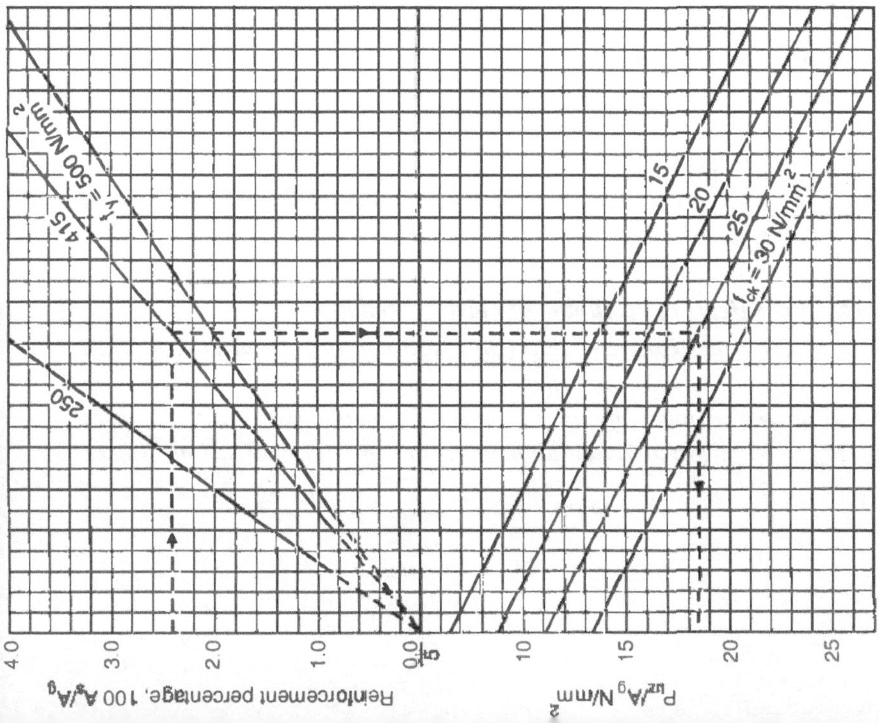

FIG. 18.5. VALUES OF P_{uz} FOR COMPRESSION MEMBERS

The column is designed for this value of M_u and given value of P_u, for bending about x-axis if $M_{ux} > M_{uy}$ and about y-axis if $M_{uy} > M_{ux}$.

Design procedure. The design is performed in the following steps :

Step 1: Assume the cross-section of the column, and the area of reinforcement along with its distribution, based on moment M_u given by Eq. 18.8.

Step 2 : Compute P_{uz} either from Eq. 18.7 or from Fig. 18.5.

Find ratio P_u/P_{uz}.

Step 3: Determine uniaxial moment capacities M_{ux1} and M_{uy1} combined with axial load P_u, using appropriate interaction curves (design Charts) for case of column subjected to axial load (P_u) and uniaxial moment.

Step 4: Compute the values of M_{ux}/M_{ux1} and M_{uy}/M_{uy1}. From Fig. 18.6, find the permissible value of M_{ux}/M_{ux1} corresponding to the above values of M_{uy}/M_{uy1} and P_u/P_{uz}. If the *actual value* of M_{ux}/M_{ux1} is less than the above value found from Fig. 18.6, the assumed section is safe. If the actual value of M_{ux}/M_{ux1} is *more* than the above value found from Fig. 18.6, the assumed section is *unsafe* and needs revision. Even if the assumed value is *over safe*, it needs revision for the sake of economy.

Alternatively, rather than using Fig. 18.6, Eq. 18.4 can be applied for checking the adequacy of the section selected in step 1.

Alternative simplified method as per BS 8110

As per BS 8110, a column subjected to P_u , M_{ux} and M_{uy} can be designed for uniaxial bending corresponding to (P_u , M_{ux}') or (P_u , M_{uy}') depending upon the following conditions.

Condition 1. When $M_{ux}/d_x > M_{uy}/d_y$, M_{ux}' controls the design, where

$$M_{ux}' = M_{ux} + \alpha \frac{d_x}{d_y} M_{uy} \qquad \text{...(18.9 } a)$$

Condition 2. When $M_{ux}/d_x < M_{uy}/d_y$, M_{uy}' controls the design, where

$$M_{uy}' = M_{uy} + \alpha \frac{d_y}{d_x} M_{ux} \qquad \text{...(18.9 } b)$$

where d_x = effective depth with respect to major axis and total depth D.

d_y = effective depth with respect to minor axis, and total depth b.

$$\alpha = \text{coefficient} = 1 - \frac{7}{6}\frac{P_u}{f_{ck}\, bD} \qquad \text{...(18.9 } c)$$

The rest of the procedure, explained in steps 2, 3 and 4 above, remains the same.

Example 18.1. *Check for safety of column of given section*

A short R.C. column of M 20 grade concrete, 300 mm × 480 mm in section is reinforced with 10 bars of 16 mm φ of Fe 415 with an effective cover of 60 mm, as shown in Fig. 18.7. Check the safety of the column if it is subjected to ultimate axial load of 1000 kN, ultimate B.M. of M_{ux} *= 80 kN-m and an ultimate B.M. of* M_{uy} *= 50 kN-m. Check the safety of the column.*

Solution

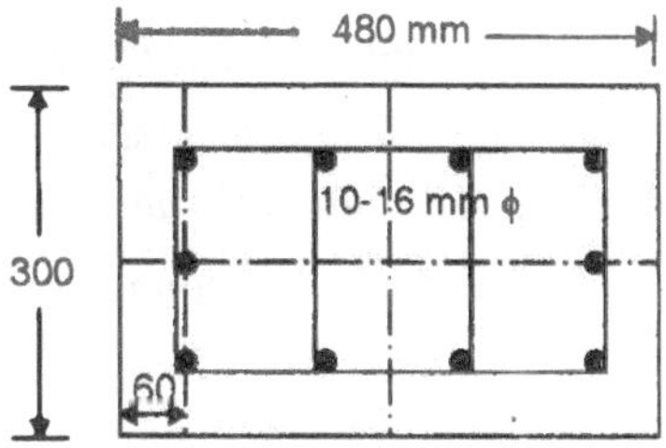

FIG. 18.7.

$$A_s = 10\,\frac{\pi}{4}\,(16)^2 = 2010.62 \text{ mm}^2$$

$$p = \frac{100\,A_s}{bD} = \frac{100 \times 2010.62}{300 \times 480} = 1.396\%$$

$d' = 60$ mm; $p/f_{ck} = 1.396/20 = 0.0698$

$$P_{uz} = 0.45\,f_{ck}\,A_g + (0.75\,f_y - 0.45\,f_{ck})\,A_s$$

$$= [0.45 \times 20 \times 300 \times 480 + (0.75 \times 415 - 0.45 \times 20)\,2010.62]\,10^{-3} = 1903.71 \text{ kN}$$

$$\therefore \quad \frac{P_u}{P_{uz}} = \frac{1000}{1903.71} \simeq 0.525\ ;\ \alpha_n = \frac{2}{3}\left[1 + \frac{5}{2} \times 0.525\right] = 1.542\ ;\ \frac{P_u}{f_{ck}\,bD} = \frac{1000 \times 10^3}{20 \times 300 \times 480} = 0.347$$

(*a*) ***For bending about x-axis*** : $D = 480$ mm; $b = 300$ mm; $d'/D = 60/480 = 0.125$. Hence we will use Chart 44 for $d'/D = 0.1$ and Chart 45 for $d'/D = 0.15$.

From Chart 44, for $P_u/f_{ck}\,bD = 0.347$ and $p/f_{ck} = 0.0698$, we get $M_u/f_{ck}\,bD^2 = 0.104$.

From Chart 45, for $P_u/f_{ck}\,bD = 0.347$ and $p/f_{ck} = 0.0698$, we get $M_u/f_{ck}\,bD^2 = 0.094$.

$\therefore$ Average value of $M_u/f_{ck}\,bD^2$ (for $d'/D = 0.125$) $= \frac{1}{2}(0.104 + 0.094) = 0.099$

$$\therefore \quad M_{ux1} = 0.099 \times 20 \times 300\,(480)^2 \times 10^{-6} = 136.868 \text{ kN-m}$$

(*b*) ***For bending about y-axis*** : $D = 300$ mm; $b = 450$ mm ; $d'/D = 60/300 = 0.2$

$\therefore$ From Chart 46, for $P_u/f_{ck}\,bD = 0.347$ and $p/f_{ck} = 0.0698$, we get $M_u/f_{ck}\,bD^2 = 0.085$

$$\therefore \quad M_{uy1} = 0.085 \times 20 \times 480\,(300)^2 \times 10^{-6} = 73.44 \text{ kN-m}$$

Substituting the values in the L.H.S. of Eq. 18.4,

$$\left(\frac{M_{ux}}{M_{ux1}}\right)^{\alpha_n} + \left(\frac{M_{uy}}{M_{uy1}}\right)^{\alpha_n} = \left(\frac{80}{136.868}\right)^{1.542} + \left(\frac{50}{73.44}\right)^{1.542} = 0.437 + 0.553 = 0.99 < 1$$

Hence column is **safe.**

Example 18.2. ***Determination of M_{uy} for given values of M_{ux} and P_u***

For the column section of Fig. 18.7, determine the value of M_{uy} if $M_{ux} = 100$ kN-m and P_u remains at the same value of 1000 kN.

Solution : As found in the previous example,

$A_s = 2010.62$ mm^2; $p = 1.396\%$; $p/f_{ck} = 0.0698$; $P_{uz} = 1903.71$ kN; $P_u/P_{uz} = 0.525$; $\alpha_n = 1.542$ and $P_u/f_{ck}\,bD = 0.347$

Also, $M_{ux1} = 136.868$ kN-m and $M_{uy1} = 73.44$ kN-m

Substituting the values in Eq. 18.4, we get

$$\left(\frac{M_{ux}}{M_{ux1}}\right)^{\alpha_n} + \left(\frac{M_{uy}}{M_{uy1}}\right)^{\alpha_n} = \left(\frac{100}{136.868}\right)^{1.542} + \left(\frac{M_{uy}}{73.44}\right)^{1.542} = 1$$

$$\therefore \quad \left(\frac{M_{uy}}{73.44}\right) = (1 - 0.6163)^{1/1.542} = 0.5373, \text{ from which}$$

$$M_{uy} = 0.5373 \times 73.44 \simeq \mathbf{39.46 \text{ kN-m}}$$

Example 18.3. ***Determination of M_{ux} for given values of M_{uy} and P_u***

For the column section of Fig. 18.7, determine the value of M_{ux} if $M_{uy} = 65$ kN-m and P_u remains at the same value of 1000 kN.

Solution As found in Example 18.1,

$A_s = 2010.62$ mm^2; $p = 1.396\,\%$; $p/f_{ck} = 0.0698$; $P_{uz} = 1903.71$ kN; $P_u/P_{uz} = 0.525$; $\alpha_n = 1.542$ and $P_u/f_{ck}\,bD = 0.347$.

Also, $M_{ux1} = 136.868$ kN-m and $M_{uy1} = 73.44$ kN-m

Substituting the values in Eq. 18.4 we get

$$\left(\frac{M_{ux}}{M_{ux1}}\right)^{\alpha_n} + \left(\frac{M_{uy}}{M_{uy1}}\right)^{\alpha_n} = \left(\frac{M_{ux}}{136.868}\right)^{1.542} + \left(\frac{65}{73.44}\right)^{1.542} = 1$$

$$\therefore \quad \frac{M_{ux}}{136.818} = (1 - 0.8284)^{1/1.542} = 0.3188$$

$$\therefore \quad M_{ux} = 0.3188 \times 136.818 = \mathbf{43.62 \text{ kN-m}}$$

Example 18.4. ***Determination of P_u for given values of M_{ux} and M_{uy}***

A short R.C. column of M 20 concrete, 300 mm × 400 mm in section is reinforced with 8 bars of 16 mm dia. of Fe 415 grade steel, with an effective cover of 60 mm, as shown in Fig. 18.8. Find the ultimate axial load which the column can carry along with biaxial moments of 50 kN-m and 40 kN-m about major and minor axes respectively.

Solution : Given $M_{ux} = 50$ kN-m ; $M_{uy} = 40$ kN-m ; $d' = 60$ mm

$$A_s = 8\,\frac{\pi}{4}(16)^2 = 1608.5 \text{ mm}^2 \;;$$

$$p = \frac{100\,A_s}{bD} = \frac{100 \times 1608.5}{300 \times 400} = 1.34\%$$

$$p/f_{ck} = 1.34/20 = 0.067$$

$$P_{uz} = [0.45 \times 20 \times 300 \times 400 + (0.75 \times 415 - 0.45 \times 20)\,1608.5]\,10^{-3} = 1566.17 \text{ kN}$$

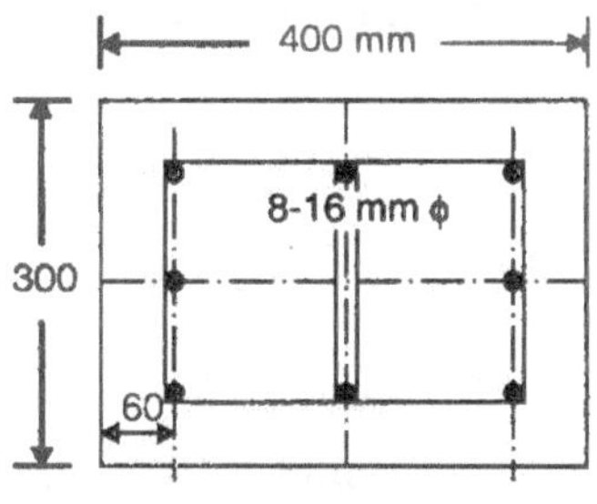

FIG. 18.8

The procedure for finding P_u will be of trial and error because M_{ux1} , M_{uy1} and α_n depend on P_u itself.

First trial for P_u

Let resultant $M_u = \sqrt{M_{ux}^2 + M_{uy}^2} = \sqrt{(50)^2 + (40)^2} = 64.03$ kN-m

Since $M_{ux} > M_{uy}$, the governing axis of bending for M_u will be x-axis.

Hence $b = 300$ mm; $D = 400$ mm; $d'/D = 60/400 = 0.15$

$$M_u/f_{ck}\,bD^2 = 64.03 \times 10^6/(20 \times 300 \times 400^2) = 0.0667$$

From Chart 45, for $d'/D = 0.15$, $p/f_{ck} = 0.067$ and $M_u/f_{ck}\,bD^2 = 0.0667$

we get $\quad P_u/f_{ck}\,bD \simeq 0.46$

$$\therefore \quad P_u = 0.46 \times 20 \times 300 \times 400 \times 10^{-3} = 1104 \text{ kN}$$

Hence let us take $P_u = 1100$ kN in the first trial.

$\therefore \quad P_u/P_{uz} = \dfrac{1100}{1566.17} = 0.7024 \quad ; \quad \alpha_n = \dfrac{2}{3}\left[1 + \dfrac{5}{2} \times 0.7024\right] = 1.837$

$P_u/f_{ck}\,bD = 1100 \times 10^3/(20 \times 300 \times 400) = 0.458$

For bending about x-axis, $D = 400$ mm ; $b = 300$ mm; $d'/D = 60/400 = 0.15$

Hence from Chart 45, for $P_u/f_{ck}\,bD = 0.458$ and $p/f_{ck} = 0.067$, we get $M_u/f_{ck}\,bD^2 = 0.073$

$\therefore \quad M_{ux1} = 0.073 \times 20 \times 300\,(400)^2 \times 10^{-6} = 70.08$ kN-m

For bending about y-axis, $D = 300$ mm; $b = 400$ mm; $d'/D = 60/300 = 0.2$

Hence from Chart 46, for $P_u/f_{ck}\,bD = 0.458$ and $p/f_{ck} = 0.067$, we get $M_u/f_{ck}\,bD^2 = 0.063$

$\therefore \quad M_{uy1} = 0.063 \times 20 \times 400\,(300)^2 \times 10^{-6} = 45.36$ kN-m.

Hence from Eq. 18.4, $\left(\dfrac{M_{ux}}{M_{ux1}}\right)^{\alpha_n} + \left(\dfrac{M_{uy}}{M_{uy1}}\right)^{\alpha_n} = m$ (say), where $m = 1$ (desired)

$$\therefore \quad \left(\frac{50}{70.08}\right)^{1.837} + \left(\frac{40}{45.36}\right)^{1.837} = 0.5378 + 0.7937 = 1.3315 >> 1$$

Hence the section is unsafe for $P_u = 1100$ kN. To make it safe, we have to reduce P_u so that m comes out to be equal to 1.

Second trial : Let $P_u = 950$ kN

$P_u/P_{uz} = 950/1566.17 = 0.6066 \quad ; \quad \alpha_n = \dfrac{2}{3}\left[1 + \dfrac{5}{2} \times 0.6066\right] = 1.678$

$P_u/f_{ck}\,bD = 950 \times 10^3/(20 \times 300 \times 400) = 0.396$

For bending about x-axis : $D = 400$ mm; $d'/D = 60/400 = 0.15$

$\therefore$ From Chart 45, for $P_u/f_{ck}\,bD = 0.396$ and $p/f_{ck} = 0.067$, $M_u/f_{ck}\,bD^2 = 0.083$

$\therefore \quad M_{ux1} = 0.083 \times 20 \times 300\,(400)^2 \times 10^{-6} = 79.68$ kN-m

For bending about y-axis ; $D = 300$ mm ; $d'/D = 60/300 = 0.2$

$\therefore$ From Chart 46, for $P_u/f_{ck}\,bD = 0.396$ and $p/f_{ck} = 0.067$, $M_u/f_{ck}\,bD^2 = 0.075$

$\therefore \quad M_{uy1} = 0.075 \times 20 \times 400\,(300)^2 \times 10^{-6} = 54$ kN-m

$$\therefore \quad \left(\frac{50}{79.68}\right)^{1.678} + \left(\frac{40}{54}\right)^{1.678} = 0.458 + 0.604 = 1.062 > 1$$

Third trial : Let $P_u = 900$ kN

$P_u/P_{uz} = 900/1566.17 = 0.575 \quad ; \quad \alpha_n = \dfrac{2}{3}\left[1 + \dfrac{5}{2} \times 0.575\right] = 1.624$

$P_u/f_{ck}\,bD = 900 \times 10^3/(20 \times 300 \times 400) = 0.375$

For bending about x-axis, $D = 400$ mm ; $d'/D = 60/400 = 0.15$

From Chart 45, for $P_u/f_{ck}\,bD = 0.375$ and $p/f_{ck} = 0.067$, $M_u/f_{ck}\,bD^2 \simeq 0.09$

$\therefore \quad M_{ux1} = 0.09 \times 20 \times 300\,(400)^2 \times 10^{-6} = 86.4$ kN-m

For bending about y-axis ; $D = 300$ mm ; $d'/D = 60/300 = 0.2$

From Chart 46, for $P_u/f_{ck}\,bD = 0.375$ and $p/f_{ck} = 0.067$, $M_u/f_{ck}\,bD^2 = 0.08$

$M_{uy1} = 0.08 \times 20 \times 400\,(300)^2 \times 10^{-6} = 57.6$ kN-m

$$\therefore \left(\frac{50}{86.4}\right)^{1.624} + \left(\frac{40}{57.6}\right)^{1.624} = 0.411 + 0.553$$

$$= 0.964 < 1$$

Final selection of P_u

Fig. 18.9 shows a plot between P_u and m. From the plot, we get $P_u = 920$ kN for $m = 1$.

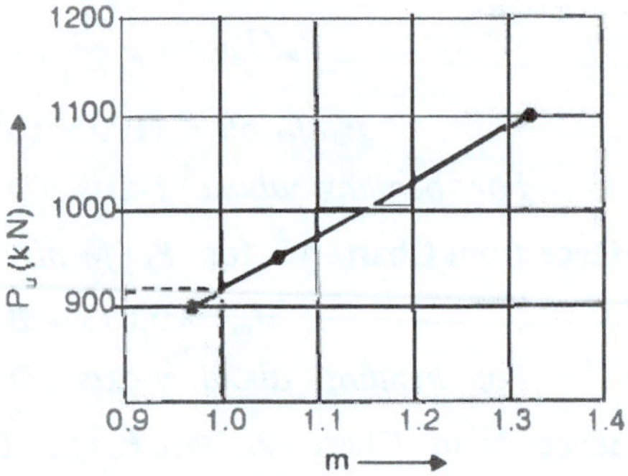

FIG. 18.9.

Example 18.5. Design of a short column subjected to biaxial bending

Determine the reinforcement for a column of a braced frame for the following data:

Size of column : 400 mm × 600 mm

P_u = 2000 kN ; M_{ux} = 160 kN-m ; M_{uy} = 120 kN-m.

Unsupported length of column = 4 m. Effective cover = 60 mm

Use M 20 concrete and Fe 415 steel.

Solution : Given : $b = 400$ mm ; $D = 600$ mm ; $l = 4$ m = 4000 mm

$f_{ck} = 20$ N/mm^2 ; $f_y = 415$ N/mm^2; $P_u = 2000$ kN ; $M_{ux} = 160$ kN-m; $M_{uy} = 120$ kN-m

$\therefore$ $l_{ex} = l_{ey} = 0.85\,l$ (for a braced column) $= 0.85 \times 4000 = 3400$ mm

$\therefore$ $l_{ey}/b = 3400/400 = 8.5 < 12$. Hence the column is short.

Let us take effective cover = 60 mm.

Step 1. Check for minimum eccentricities

$e_x = M_{ux}/P_u = 160 \times 1000/2000 = 80$ mm ; $e_y = 120 \times 1000/2000 = 60$ mm

$$e_{x\,,\,min} = \frac{l_x}{500} + \frac{D}{30} = \frac{4000}{500} + \frac{600}{30} = 28 \text{ mm} > 20 \text{ mm}$$

$$e_{y\,,\,min} = \frac{l_y}{500} + \frac{b}{30} = \frac{4000}{500} + \frac{400}{30} = 21.3 \text{ mm} > 20 \text{ mm}$$

Since the applied eccentricities are more than the minimum eccentricities, no modifications in the applied moments M_{ux} , M_{uy} are required.

Step 2. Selection of trial reinforcement : Taking $a = 1.15$,

$$M_u = a\sqrt{M_{ux}^2 + M_{uy}^2} = 1.15\sqrt{(160)^2 + (120)^2} = 230 \text{ kN-m}$$

Since $M_{ux} > M_{uy}$, design the column for uniaxial bending with $M_u = 230$ kN-m for bending about x-axis, along with axial load P_u. Hence $b = 400$ mm and $D = 600$ mm for the purpose of using Charts.

$d'/D = 60/600 = 0.1$ Hence Chart 44 will be used.

$$\frac{P_u}{f_{ck}\,bD} = \frac{2000 \times 10^3}{20 \times 400 \times 600} = 0.417 \text{ and } \frac{M_u}{f_{ck}\,bD^2} = \frac{230 \times 10^6}{20 \times 400\,(600)^2} \simeq 0.08$$

Hence from Chart 44, we get $p/f_{ck} \simeq 0.0625$ corresponding to the above two values.

$\therefore \quad p = 0.0625 f_{ck} = 0.0625 \times 20 = 1.25\% = 100 \frac{A_s}{bD}$

$\therefore \quad A_s = \frac{p\, bD}{100} = \frac{1.25 \times 400 \times 600}{100} = 3000 \text{ mm}^2$

Let us provide 10 bars, uniformly distributed along the four sides. Hence diameter of bars is given by

$$\varphi = \sqrt{\frac{3000 \times 4}{10\,\pi}} = 19.5 \text{ mm}$$

Hence provide 10 bars of 20 mm φ. Actual $A_s = 10 \times \frac{\pi}{4} (20)^2 = 3141.6 \text{ mm}^2$

$\therefore \quad p = 100 \frac{A_s}{bD} = \frac{100 \times 3141.6}{400 \times 600} \simeq 1.31\%$ and $p/f_{ck} = 1.31/20 = 0.0654$

Step 3. Uniaxial moment capacity of the above section about x-x axis

$$\frac{d'}{D} = 0.1; \quad \frac{P_u}{f_{ck}\, bD} = 0.417, \text{ as found above.}$$

Hence from Chart 44 of ISI hand book, for $P_u/f_{ck} bD = 0.417$ and $p/f_{ck} = 0.0654$, we get $M_u/f_{ck}\, bD^2 = 0.085$

$\therefore \quad M_{ux1} = 0.085 \times 20 \times 400\,(600)^2 \times 10^{-6} \simeq 244.8 \text{ kN-m}$

Step 4. Uniaxial moment capacity of the section about y-y axis

For bending about y-axis, $D = 400$ mm ; $b = 600$ mm. Hence $d'/D = 60/400 = 0.15$

Hence from Chart 45, for $P_u/f_{ck}\, bD = 0.417$ and $p/f_{ck} = 0.0654$, we get $M_u/f_{ck}\, bD^2 = 0.08$. Hence $M_{uy1} = 0.08 \times 20 \times 600\,(400)^2 \times 10^{-6} = 153.6$ kN-m.

Step 4. Computation of P_{uz}

From Chart 63, corresponding to $p = 1.31$, $f_y = 415 \text{ N/mm}^2$ and $f_{ck} = 20 \text{ N/mm}^2$, we get $P_{uz}/A_g = 13 \text{ N/mm}^2$.

$\therefore \quad P_{uz} = 13\, A_g = 13 \times 400 \times 600 \times 10^{-3} = 3120$ kN

$\therefore \quad \frac{P_u}{P_{uz}} = \frac{2000}{3120} = 0.641 \; ; \quad \frac{M_{ux}}{M_{ux1}} = \frac{160}{244.8} = 0.655 \; ; \quad \frac{M_{uy}}{M_{uy1}} = \frac{120}{153.6} = 0.781$

Hence from Chart 64, for $P_u/P_{uz} = 0.641$ and $M_{uy}/M_{uy1} = 0.781$, we get $M_{ux}/M_{ux1} = 0.53$. This value is *lower* than the actual value of 0.655. Hence the reinforcement has to be increased.

Second trial : Revised $p = 1.31 \times \frac{0.655}{0.53} = 1.619$

$\therefore \quad A_s = \frac{p\, bD}{100} = \frac{1.619 \times 400 \times 600}{100} = 3886 \text{ mm}^2$

$\therefore$ No. of 20 mm φ bars $= 3886/314.10 = 12.37$

However, provide 12 bars of 20 mm φ, uniformly distributed as shown in Fig. 18.10.

Actual $$p = \frac{12 \times 314.16}{400 \times 600} \times 100 = 1.571$$

$\therefore \quad p/f_{ck} = 1.571/20 = 0.0785$

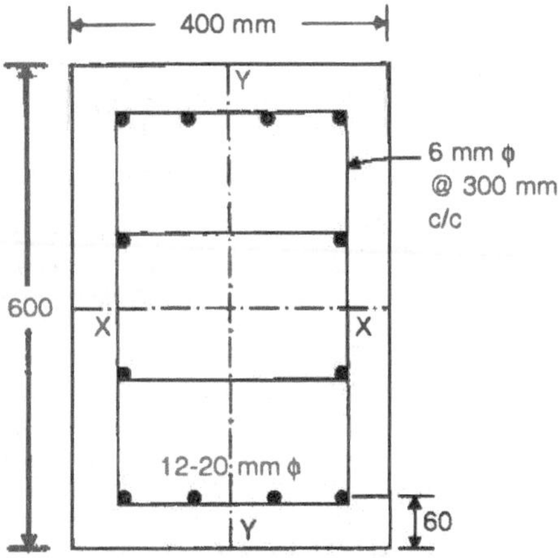

FIG. 18.10.

Check for the second trial section

We will do the checking of the second trial section partly with the help of Charts and partly with the help of equations.

For bending about x-axis : $D = 600$ mm, $b = 400$ mm, $d'/D = 60/600 = 0.1$

Hence from Chart 44, for $p/f_{ck} = 0.0785$ and $P_u/f_{ck}\,bD = 0.417$, we get $M_u/f_{ck}\,bD^2 = 0.095$

$\therefore\ M_{ux1} = 0.095 \times 20 \times 400\,(600)^2 \times 10^{-6} = 273.6$ kN-m

For bending about y-axis, $D = 400$ mm, $b = 600$ mm and $d'/D = 60/400 = 0.15$

Hence from Chart 45, for $p/f_{ck} = 0.0785$ and $P_u/f_{ck}\,bD = 0.417$,

we get $M_u/f_{ck}\,bD^2 \simeq 0.088$

$$\therefore \quad M_{uy1} = 0.088 \times 20 \times 600\,(400)^2 \times 10^{-6} \simeq 169 \text{ kN-m}$$

Now $\quad P_{uz} = 0.45\,f_{ck}\,A_g + (0.75\,f_y - 0.45\,f_{ck})\,A_s$

$$= [0.45 \times 20 \times 400 \times 600 + (0.75 \times 415 - 0.45 \times 20)\,12 \times 314.16] \times 10^{-3} = 3299.5 \text{ kN}$$

$$\therefore \quad \frac{P_u}{P_{uz}} = \frac{2000}{3295.5} = 0.606$$

Hence from Eq. 18.6, $\alpha_n = \frac{2}{3}\left[1 + \frac{5}{2} \times 0.606\right] = 1.677$

Substituting the values in the L.H.S. of Eq. 18.4,

$$\left(\frac{160}{273.6}\right)^{1.677} + \left(\frac{120}{169}\right)^{1.677} = 0.4067 + 0.5631 = 0.9698 < 1. \text{ Hence safe.}$$

Design of transverse reinforcement

Diameter of lateral ties $\nless \varphi/4 \nless 20/4$. Provide 6 mm φ ties.

Pitch $\ngtr b$ (= 400 mm) or $16\,\varphi$ (= 16 × 20 = 320 mm) or 300 mm whichever is less.

Hence provide 6 mm φ lateral ties @ 300 mm c/c.

(b) Alternative design based on BS 8110

For the present case, $b = 400$ mm ; $D = 600$ mm

$$\therefore \quad d_x = 600 - 60 = 540 \text{ mm}, \quad d_y = 400 - 60 = 340 \text{ mm}$$

$$\therefore \quad M_{ux}/d_x = 160 \times 10^6/540 = 0.2962 \times 10^6 \ ; \ M_{uy}/d_y = 120 \times 10^6/340 = 0.3529$$

Since $\quad M_{ux}/d_x < M_{uy}/d_y$, M_{uy}' governs the design

$$\alpha = 1 - \frac{7}{6}\frac{P_u}{f_{ck}\,bD} = 1 - \frac{7}{6} \times \frac{2000 \times 10^3}{20 \times 400 \times 600} = 0.5139$$

$$\therefore \quad M_{uy}' = M_{uy} + \alpha \frac{d_y}{d_x} M_{ux} = 120 + 0.5139 \times \frac{340}{540} \times 160 = 171.77 \text{ kN-m}$$

Hence design the column for $P_u = 2000$ kN and $M_{uy}' = 171.77$ kN-m

$$P_u/f_{ck}\, bD = 2000 \times 10^3/(20 \times 400 \times 600) = 0.417 \;;$$

For bending about y-axis, $D = 400$; $\therefore$ $d/D = 60/400 = 0.15$

Hence from Chart 45, for $\dfrac{P_u}{f_{ck}\, bD} = 0.417$ and $\dfrac{M_u}{f_{ck}\, bD^2} = \dfrac{171.77 \times 10^6}{20 \times 600\,(400)^2} = 0.0895$

we get $p/f_{ck} \simeq 0.078$. Hence $p = 0.078 \times 20 = 1.56\%$.

$$\therefore \quad A_s = \frac{p\, bD}{100} = \frac{1.56 \times 400 \times 600}{100} = 3744 \text{ mm}^2$$

$\therefore$ Number of 20 mm φ bars = 3744/314.16 = 11.9

Hence provide 12 bars of 20 mm diameter.

Actual area provided 12 × 314.16 = 3770 mm^2

The column can now be checked for biaxial bending, for this reinforcement, as explained in part (*a*) above.

PROBLEMS

1. A R.C. rectangular column of size 240 mm × 300 mm is reinforced with 4 bars of 20 mm dia., provided one at each corner with an effective cover of 60 mm. Check the safety of the column if it is subjected to $P_u = 350$ kN, $M_{ux} = 32$ kN-m and $M_{uy} = 20$ kN. Assume M 20 concrete and Fe 415 grade steel reinforcement.

2. For the column section of problem 1, determine the permissible value of M_{ux} if it is subjected to $M_{uy} = 12$ kN-m, with the same value of $P_u = 350$ kN.

3. For the column section of problem 1, determine the permissible value of M_{uy} if it is subjected to $M_{ux} = 25$ kN-m, with the same value of $P_u = 350$ kN.

4. Find the reinforcement of a R.C.C. rectangular short column of size 240 mm × 300 mm if it is subjected to $P_u = 500$ kN, $M_{ux} = 50$ kN-m and $M_{uy} = 30$ kN-m. Assume M 20 concrete and Fe 415 grade steel. Provide an effective cover of 60 mm.

19

Limit State of Collapse in Compression : IV Slender Columns

19.1. INTRODUCTION

As seen in Chapter 16, a compression member may be considered as slender or long when the slenderness ratios l_{ex}/D or and l_{ey}/b are more than 12. Thus, if $l_{ex}/D > 12$, the column is considered to be slender for bending about x-x axis, while if $l_{ey}/b > 12$, the column is considered to be slender for bending about y-y axis. The difference between a short column and a slender column should be clearly understood. When a short column is loaded even with an axial load, the lateral deflection is either zero or very small. In contrast to this, when a slender column is loaded even with an axial load, the *lateral deflection* Δ, measured from the original centre line along its length, becomes appreciable. This makes the axial load eccentric at the central section of the column by a value Δ, subjecting the section to a B.M. equal to $P_u \cdot \Delta$ in addition to an axial load. Hence separate consideration of this *additional B.M.* has to be made. The axial compression due to axial

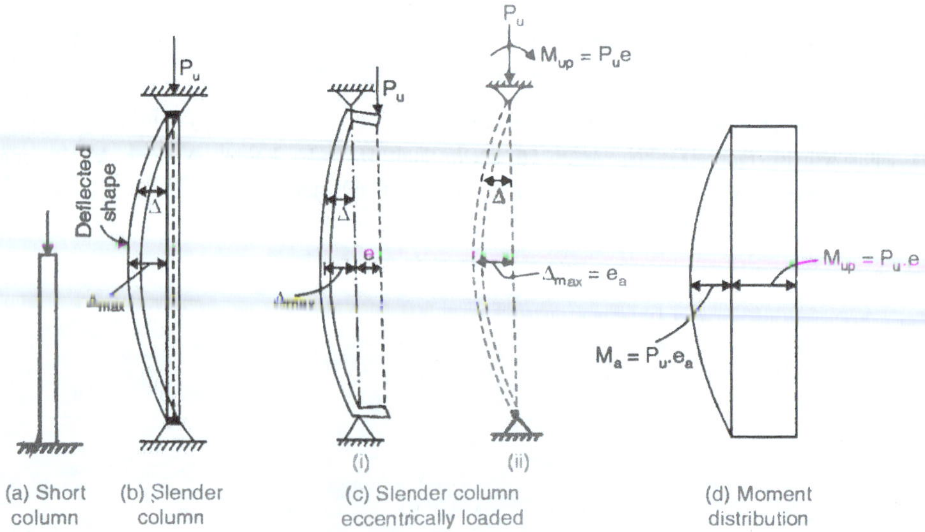

FIG. 19.1. EFFECTS OF SLENDERNESS

load is called the *primary effect* while the bending caused due to lateral deflection of slender column is called the *secondary effect*.

Similar effect takes place on an eccentrically loaded slender column. The moment due to the primary effect is $M_{up} = P_u \times e$ where e is the original eccentricity of loading. Since the column deflects appreciably, the column is subjected to additional eccentricity $e_a = \Delta_{max}$ giving rise to additional moment $M_a = P_u \cdot e_a$ due to this secondary effect. This secondary moment is also known as $P - \Delta$ moment, which has a variation along the height of the column. Thus, at any height.

$$M_u = P_u (e + \Delta)$$

and maximum moment is $$M_{u\,.\,max} = P_u \cdot e + P_u \cdot \Delta_{max} = P_u \cdot e + P_u \cdot e_a \quad \ldots(19.1)$$

A short compression member is not in danger of buckling prior to its ultimate strength based on the material properties. In contrast to this, the buckling and additional deflection effects are more pronounced in slender columns, thus reducing their ultimate strength.

19.2. SLENDERNESS LIMITS FOR COLUMNS (IS : 456 - 2000)

The dimensions (i.e. length, width and depth) of a column should be so proportioned that they fail by material failure only and not by buckling. For this, IS 456 - 2000 has given the following two recommendations.

1. The unsupported length between end restraints shall not exceed 60 times the least lateral dimension of a column.

Thus $$l \le 60\, b$$

2. If, in any plane, one end of a column is unrestrained, its unsupported length should not exceed $100\, b^2/D$.

Thus $$l \le 100\, b^2/D$$

where b = width of the cross-section

and D = depth of the cross-section measured in the plane under consideration.

19.3. METHODS OF DESIGN OF SLENDER COLUMNS

As observed in the previous article, the effect of slenderness is to cause additional moment M_a. Due to this the column is subjected to total moment $M_{ut} = M_{up} + M_a$, where M_{up} is the primary moment or initial moment (M_{ui}). The main problem with a slender column design lies in determining the $P - \Delta$ effects. The problem of a slender column, therefore, is essentially that of *structural analysis* rather than the *structural design*. When a slender column is subjected to (P_u , M_{ux} , M_{uy}), the principle of designing the column remains the same, except that the secondary moment components M_{ax} and M_{ay} are to be found first and then taken into account in addition to P_u , M_{ux} , M_{uy}. Hence a detailed *second order structural analysis* becomes necessary. Such a second order analysis is beyond the scope of this. However the design of a slender column can be carried out by the following *simplified* methods.

(*i*) The strength reduction coefficient method

(*ii*) The additional moment method

and (*iii*) The moment magnification method

The *reduction coefficient method*, given by IS 456 - 2000 is recommended for working stress design for service load and is based on allowable stresses in steel and concrete. The *additional moment method* is recommended by Indian and British Codes. However, the ACI Code recommends the use of moment magnification method.

19.4. STRENGTH REDUCTION COEFFICIENT METHOD

The strength reduction coefficient method, recommended by IS : 456-2000, *specifically for the working stress design* (WSD), can be extended to the limit state design (LSD) for slender columns subjected to axial loading only. However, this method should not be applied to columns with uniaxial or biaxial moments. According to IS 456-2000, the *permissible stresses* in concrete and steel are reduced by a strength reduction coefficient C_r given by

$$C_r = 1.25 - \frac{l_e}{48\,b} \qquad \text{...(19.2 a)}$$

or more precisely by,

$$C_r = 1.25 - \frac{l_e}{160\,r_{min}} \qquad \text{...(19.2 b)}$$

where b = least lateral dimension of the column

and r_{min} = least radius of gyration of the column.

The Explanatory hand book to the Code further recommends that the reduction factor C_r may be directly applied to the load carrying capacity estimated for a corresponding short column. However, as stated earlier, *this method should not be applied for a eccentrically loaded slender column.*

19.5. ADDITIONAL MOMENT METHOD

This method has been recommended by IS : 456-2000. According to this method, the slender column should be designed for a total moment $M_{ut} = M_u + M_a$, where M_a is the additional moment equal to $P_u \cdot e_a$ where e_a is the additional eccentricity in the appropriate direction. Thus

$$M_{utx} = P\,(e_x + e_{ax}) = M_{ux} + M_{ax} \qquad \text{...(19.3 a)}$$

and

$$M_{uty} = P_u\,(e_y + e_{ay}) = M_{uy} + M_{ay} \qquad \text{...(19.3 b)}$$

where M_{utx} = total design moment about x-axis.

M_{uty} = total design moment about y-axis.

M_{ux} = primary factored moment about x-axis, which should not be less than corresponding to minimum eccentricity $e_{x,\,min}$

M_{uy} = primary factored moment about y-axis, which should not be less than that corresponding to minimum eccentricity $e_{y,\,min}$

M_{ax} = additional moment with reference to x-axis $= P_u \cdot e_{ax}$

M_{ay} = additional moment with reference to y-axis $= P_u \cdot e_{ay}$

e_{ax} = additional eccentricity about x-axis, which is a function of the curvature of the column for bending about x-axis,

e_{ay} = additional eccentricity about y-axis, which is a function of the curvature of the column for bending about y-axis.

IS : 456 - 2000 has recommended following formulae for the *additional moments*.

$$M_{ax} = \frac{P_u D}{2000}\left(\frac{l_{ex}}{D}\right)^2 \quad \text{...(19.4 } a)$$

and

$$M_{ay} = \frac{P_u \cdot b}{2000}\left(\frac{l_{ey}}{b}\right)^2 \quad \text{....(19.4 } b)$$

In other words,

$$e_{ax} = \frac{D}{2000}\left(\frac{l_{ex}}{D}\right)^2 \quad \text{...(19.5 } a)$$

and

$$e_{ay} = \frac{b}{2000}\left(\frac{l_{ey}}{b}\right)^2 \quad \text{...(19.5 } b)$$

where l_{ex} = effective length in respect of major axis (x-x axis)

l_{ey} = effective length in respect of minor axis (y-y axis)

D = depth of the cross-section at right angles to the major axis

b = width of the member, at right angles to the minor axis.

P_u = factored axial load on the member.

Derivation of Eqs. 19.4 for M_a : The essence of the method of additional method is a simple formulation for the determination of *additional eccentricities* e_{ax} and e_{ay}. The analysis for finding e_a (i.e. e_{ax} or e_{ay}), which is a second order analysis, is called. $P - \Delta$ analysis.

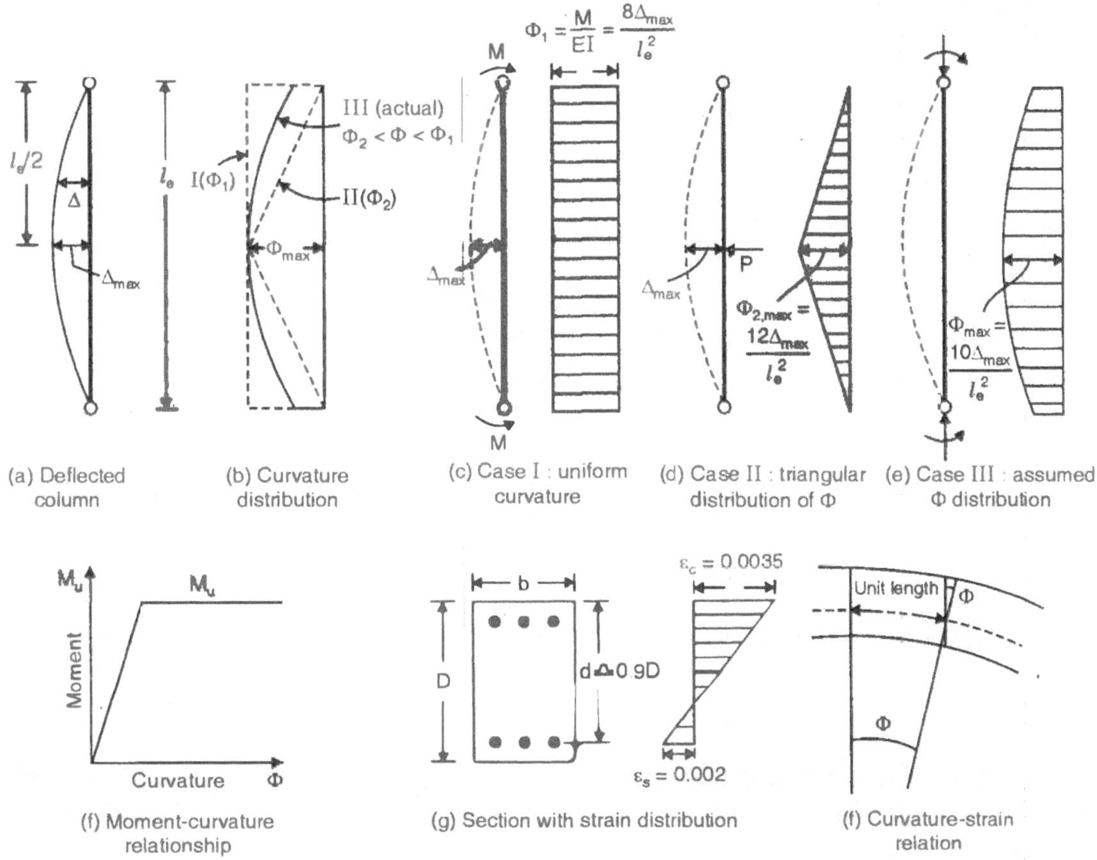

FIG. 19.2.

For this analysis, consider a braced slender column with pin jointed ends, shown in Fig. 19.2 (*a*).

The elastic deflection of a bending member is represented by

$$\frac{d^2\,\Delta}{d\,x^2} = \frac{M}{EI} = \frac{1}{R} = \Phi$$

where $1/R$ is known as curvature (Φ) defined as the change in the slope over unit length of the member, assuming it to be constant along this unit length. Integrating the above expression twice, we get

$$\Delta = \int \int \left(\frac{1}{R}\, dx \right) dx = \int \int \left(\frac{M}{EI}\, dx \right) dx$$

This shows that the deflection depends upon the distribution of curvature (Φ) or on M/EI diagram along the length of the member.

Let us assume that the moment curvature relationship is linear, as shown in Fig. 9.2 (*f*) and is not influenced by the axial load, the maximum deflection Δ_{max} can be expressed in terms of Φ_{max}.

Let us take the case of hinged column with uniformly distributed load (w) giving rise to a parabolic distribution of B.M. The maximum deflection for this can be expressed as follows :

$$\Delta_{max} = \frac{5}{384}\left(\frac{w\, l_e^2}{EI} \right) = \frac{(5)\,8}{384}\left(\frac{wl_e^2}{8} \right)\frac{l_e^2}{EI} = \frac{M_{max}}{EI} \times \frac{l_e^2}{9.6} = \frac{l_e^2}{9.6}\,\Phi_{max} \qquad \ldots(1)$$

There may be three cases of curvature distribution along the length of a bent member (Fig. 19.2 *b*).

Case I : Uniform (rectangular) distribution of curvature (Fig. 19.2 *b* and 19.2 *c*).

Case II : Triangular distribution of curvature (Fig. 19.2 *b* and 19.2 *d*)

Case III : General case of curvature distribution (Fig. 19.2 *b* and 19.2 *e*)

Let us take the first case of uniform B.M. (M) along the length of the member (Fig. 19.2 *c*). The maximum deflection is given by

$$\Delta_{max} = e_a = \frac{l_e^2}{8}\,\Phi_{max} \qquad \ldots(2)$$

Similarly taking the second case of a triangular distribution of B.M. diagram (Fig. 19.2 *d*), the maximum deflection is given by

$$\Delta_{max} = e_a = \frac{l_e^2}{12}\,\Phi_{max} \qquad \ldots(3)$$

Hence for a general case (Case III) of curvature distribution (Fig. 19.2 *b* and 19.2 *e*) the maximum deflection can be expressed as

$$\Delta_{max} = e_a = \frac{l_e^2}{10}\,\Phi_{max} \qquad \ldots(4)$$

Let us assume balanced failure condition wherein $\varepsilon_c = 0.0035$ and $\varepsilon_s = 0.002$, as shown in Fig. 19.2 (*g*). Assuming further that the additional moment is about 0.8 of the total moment, we get

$$\Phi_{max} = \frac{\varepsilon_c + \varepsilon_s}{d} \times 0.8 = \frac{0.0035 + 0.002}{0.9\, D} \times 0.8 = \frac{1}{204.5\, D} \qquad \ldots(5)$$

Combining Eq. (4) and (5), one gets

$$e_a = \Delta_{max} = \frac{l_e^2}{2045\, D} \simeq \frac{l_e^2}{2000\, D}$$

or

$$e_a = D\, \frac{(l_e/D)^2}{2000} \qquad \ldots(19.5)$$

$\therefore$

$$e_{ax} = \frac{D}{2000}\left(\frac{l_{ex}}{D}\right)^2 \qquad \ldots(19.5\ a)$$

and

$$e_{ay} = \frac{b}{2000}\left(\frac{l_{ey}}{b}\right)^2 \qquad \ldots(19.5\ b)$$

Hence

$$M_{ax} = P_u \,.\, e_{ax} = \frac{P_u\, D}{2000}\left(\frac{l_{ex}}{D}\right)^2 \qquad \ldots(19.4\ a)$$

and

$$M_{ay} = P_u \,.\, e_{ay} = \frac{P_u\, b}{2000}\left(\frac{l_{ey}}{b}\right)^2 \qquad \ldots(19.4\ b)$$

Reduction factor (k_a) for additional moment (M_a)

The expression for e_a, derived above is based on the assumption that the curvature $(\Phi)_{max}$ is corresponding to *balanced failure* where the maximum strain $\varepsilon_c = 0.0035$ in concrete and maximum tensile strain $\varepsilon_s = 0.002$ reach simultaneously. The axial load corresponding to this condition is designated as P_{ub}. If actual load P_u at failure is greater than P_{ub}, the failure mode is one of compression failure (with corresponding low e/D ratio) with strain ε_s in steel less than 0.002. Hence when $P_u > P_{ub}$, the deflection and hence the additional moment will be lesser than the values derived above. In such a case, the Code recommends that the additional moment may be multiplied by the following *reduction factor* (k_a) :

$$k_a = \frac{P_{uz} - P_u}{P_{uz} - P_b} = \frac{1 - P_u/P_{uz}}{1 - P_b/P_{uz}} \le 1 \qquad \ldots(19.6)$$

where P_u = axial load on compression member

$P_{uz} = 0.45\, f_{ck}\, A_c + 0.75\, f_y\, A_s = 0.45\, f_{ck}\, A_g + (0.75\, f_y - 0.45\, f_{ck})\, A_s$

P_b = axial load corresponding to 'balanced failure' i.e. a failure condition when maximum strain in concrete is 0.0035 and tensile strain in outer most layer in steel is 0.002.

It should be clearly noted that Eq. 19.6 is not applicable if $P_u < P_b$, i.e. $k_a = 1$ for $P_u < P_b$. The value of k_a can also be found from curves of Fig. 19.3.

Applying Eq. 19.6 to appropriate axes of bending we get

$$k_{ax} = \frac{P_{uz} - P_u}{P_{uz} - P_{bx}} \le 1 \qquad \ldots(19.6\ a)$$

and

$$k_{ay} = \frac{P_{uz} - P_u}{P_{uz} - P_{by}} \le 1 \qquad \ldots(19.6\ b)$$

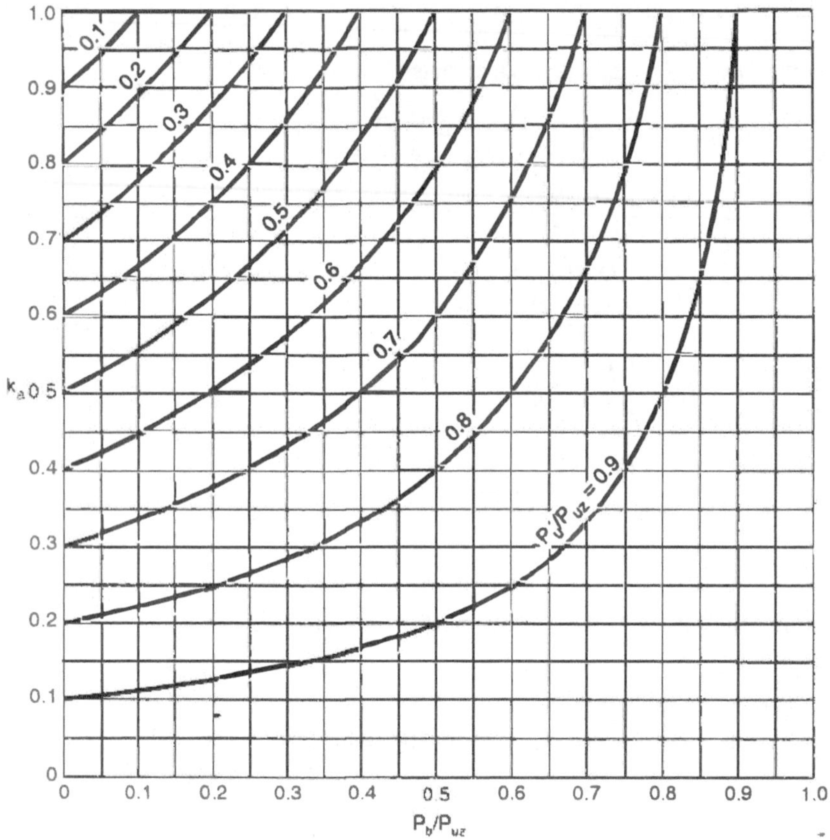

FIG. 19.3. REDUCTION FACTOR k_a FOR ADDITIONAL MOMENT IN SLENDER COLUMN

Applying the reduction factor, the value of additional moment is given by

$$M_a = k_a\, P_u \,.\, e_a \;;\; M_{ax} = k_{ax}\, P_u \,.\, e_{ax} \text{ and } M_{ay} = k_{ay}\, P_u \,.\, e_{ay} \qquad \ldots(19.7)$$

Determination of P_b

In order to determine reduction coefficient k_a, the axial load P_b is required. The value of P_b depends upon (*i*) sectional dimensions, including reinforcement ratio (p) (*ii*) arrangement of reinforcement (*iii*) cover ratio d'/D, (*iv*) grade of concrete and (*v*) grade of steel. The value of P_b can be computed from the following expressions :

(*a*) *For rectangular sections :*

$$P_b = \left(k_1 + k_2 \frac{p}{f_{ck}} \right) f_{ck}\, bD \qquad \ldots(19.8\ a)$$

(*b*) *For circular sections :*

$$P_b = \left(k_1 + k_2 \frac{p}{f_{ck}} \right) f_{ck}\, D^2 \qquad \ldots(19.8\ b)$$

In the above expressions k_1 is a dimensionless coefficient while $100\, k_2$ gives the stress in N/mm^2.

The values of k_1 and k_2 can be taken from Tables 19.1 and 19.2 respectively.

TABLE 19.1. VALUES OF k_1

Section	*Values of d'/D*			
	0.05	***0.1***	***0.15***	***0.20***
Rectangular	0.219	0.207	0.196	0.184
Circular	0.172	0.160	0.149	0.138

TABLE 19.2. VALUES OF k_2

Section	*f_y (N/mm²)*	*Values of d'/D*			
		0.05	***0.10***	***0.15***	***0.20***
1. Rectangular with equal reinforcement on opposite faces	250	– 0.045	– 0.045	– 0.045	– 0.045
	415	0.096	0.082	0.046	– 0.022
	500	0.213	0.173	0.104	– 0.001
2. Rectangular with equal reinforcement on all sides	250	0.215	0.146	0.061	– 0.001
	415	0.424	0.328	0.203	0.028
	500	0.545	0.425	0.256	0.040
3. Circular	250	0.193	0.148	0.077	– 0.020
	415	0.410	0.323	0.201	0.036
	500	0.543	0.443	0.291	0.056

In addition to the coefficients k_1 and k_2, the reinforcement percentage p is also required to determine the value of P_b and hence that of k_a. Hence one has to initially assume p and determine value of k_a and check for safety by successive trials.

Determination of total moment

The additional moment M_a (i.e. M_{ax} or M_{ay}) computed above should be added to an initial moment M_{ui} to get the total moment M_{ut}

Thus $$M_{ut} = M_{ui} + M_a$$

where M_a is the additional moment obtained after applying the reduction factor.

The value of M_{ui} has to be determined on the basis of end conditions. The Code (IS : 456-2000) has given the following recommendations for the determination of initial moment M_{ui} for a column subjected to *unequal* primary moments M_{u1} and M_{u2} at the two ends of the column.

(a) Braced column subjected to unequal primary moments

For column bent in single curvature

$$M_{ui} = 0.4\, M_{u1} + 0.6\, M_{u2} \nless 0.4\, M_{u2} \quad \text{...(19.9 a)}$$

where M_{u2} is the larger end primary moment and M_{u1} is the smaller end primary moment.

For column bent in double curvature

For column bent in double curvature, M_{u1} is assumed as negative.

Hence $$M_{ui} = 0.6\, M_{u2} - 0.4\, M_{u1} \nless 0.4\, M_{u2} \quad \text{...(19.9 b)}$$

In no case should the initial moment M_{ui} in both the cases be less than that corresponding to minimum eccentricity. Thus, $M_{ui} \nless M_{u,\,min}$

In both the cases (i.e. single curvature or double curvature),

$$M_{ut} = M_{ui} + M_a \nless M_{u2} \quad ....(19.10)$$

All the moments M_{u1}, M_{u2} and M_a are about the same axis.

(b) For unbraced column subjected to unequal primary moments

$$M_{ut1} = M_{u1} + M_a \quad ...(19.11\ a)$$

$$M_{ut2} = M_{u2} + M_a \quad ...(19.11\ b)$$

where M_{ut1} = total design moment about an axis, at an end having moment M_1

M_{ut2} = total design moment, about the same axis, at other end having moment M_2.

It is inadvisable to apply reduction factor k_a for unbraced columns.

Notes 1. A column may be considered braced in a given plane if lateral stability to the structure as a whole is provided by walls or bracing or buttressing designed to resist all lateral forces in that plane. It should otherwise be considered as unbraced. In a braced member there is no relative displacements of the ends of the member. In unbraced member, relative lateral displacement of ends of the member is not prevented.

2. Unbraced compression members, at any given level or storey, subjected to lateral load are usually constrained to deflect equally. In such cases, slenderness ratio for each column may be taken as the average for all columns acting in the same direction.

19.6. PROCEDURE FOR DESIGN OF SLENDER COLUMNS

The following procedure may be adopted for designing a slender column.

Step 1: Determine the effective length and slenderness ratio in each direction.

Step 2 (*a*) Determine initial moment (M_{ui}) from given primary end moments M_{u1} and M_{u2}, in each direction.

(*b*) Calculate e_{min} and $M_{u,min}$ in each direction

(*c*) Compare moments computed in steps (*a*) and (*b*) above and take the greater one of the two as initial moment M_{ui}, in each direction.

Step 3 (*a*) Compute additional moment (M_a) in each direction, using Eqs. 19.4.

(*b*) Compute total moment (M_{ut}) in each direction from using Eq. 19.10 or 19.11, without considering reduction factor (k_a)

(*c*) Make preliminary design for P_u and M_{ut} and find area of steel. Thus p is known.

Step 4. (*a*) Obtain P_{uz}. Also obtain P_b in each direction, for reinforcement ratio p determined above.

(*b*) Determine the value of k_a in each direction.

(*c*) Determine the modified design value of moment in each direction

$$M_{ut} = M_{ui} + k_a M_a$$

Step 5. Re-design the column for P_u and M_{ut}. If the column is slender about both the axes, design the column for biaxial bending, for (P_u . M_{uxt}) about x-axis and (P_u , M_{uyt}) about y-axis, using method explained in Chapter 18.

Note : When external moments are absent, bending moment due to minimum eccentricity should be added to additional moment about the corresponding axis.

Example 19.1. ***Determination of P_u for a slender column***

Determine the maximum factored load carrying capacity of a square column, 400 mm × 400 mm, reinforced with 8 bars of 25 mm dia., uniformly spaced along all the four sides with an effective cover of 60 mm. The column is braced against side sway and has unsupported length of 5.5 m and effective length factor of 1 about both the axes. Assume M 20 concrete and Fe 415 steel.

Solution : Given $d' = 60$ mm. Hence $d'/D = d'/b = 60/400 = 0.15$

$l = l_e = 5.5$; $\frac{l_e}{D} = \frac{5500}{400} = 13.75 < 12$. Hence column has slenderness about both the axes.

$$A_s = 8 \times \frac{\pi}{4}(25)^2 = 3926.99 \text{ mm}^2 \; ; \; p = 3926.99 \times 100/400 \times 400 = 2.454\,\%$$

$p/f_{ck} = 2.454/20 = 0.123$

We will determine P_u by trial and error method.

First trial

In the first trial, we will determine P_u, considering the column to be short and then apply the strength reduction coefficient recommended in working stress design. Final check is then made, after applying additional moments, for biaxial bending.

$$P_u = 0.4\, f_{ck}\, A_g + (0.67\, f_y - 0.4\, f_{ck})\, A_s$$

$$= [0.4 \times 20 \times 400^2 + (0.67 \times 415 - 0.4 \times 20) \times 3926.99]\, 10^{-3} = 2340.484 \text{ kN}$$

$$C_r = 1.25 - \frac{l_e}{48\, b} = 1.25 - \frac{5500}{48 \times 400} = 0.9635$$

$\therefore \quad P_u = 0.9635 \times 2340.484 \simeq 2255$ kN

Let us now apply additional moment method to check the safety of the column.

$$e_{x\,.\,min} = e_{y\,.\,min} = \frac{l}{500} + \frac{D}{30} = \frac{5500}{500} + \frac{400}{30} = 24.33 \text{ mm} > 20 \text{ mm}$$

$\therefore \quad M_{ux\,,\,min} = M_{uy\,,\,min} = M_{u\,,\,min} = P_u\,.\,e_{min} = 2255 \times 24.33 \times 10^{-3} = 54.86$ kN-m

Since there are no external moments,

$$M_{ui} = M_{u\,,\,min} = 54.86 \text{ kN-m}$$

Now $$e_{ax} = e_{ay} = e_a = \frac{D}{2000}\left(\frac{l_e}{D}\right)^2 = \frac{400}{2000}\left(\frac{5500}{400}\right)^2 = 37.81 \text{ mm}$$

Again, $$k_{ax} = k_{ay} = k_a = \frac{P_{uz} - P_u}{P_{uz} - P_b}$$

Here, $$P_{uz} = 0.45\, f_{ck}\, A_g + (0.75\, f_y - 0.45\, f_{ck})\, A_s$$

$$= [0.45 \times 20 \times 400^2 + (0.75 \times 415 - 0.45 \times 20)\, 3926.99]\, 10^{-3} = 2626.933 \text{ kN}$$

From Table 19.1, $k_1 = 0.196$ (for $d'/D = 0.15$)

From Table 19.2, $k_2 = 0.203$ (with equal A_s on all the four sides)

$$\therefore \quad P_b = \left(k_1 + k_2 \frac{p}{f_{ck}}\right) f_{ck}\, bD = \left(0.196 + 0.203 \times \frac{2.454}{20}\right) 20 \times 400^2 \times 10^{-3} = 706.91 \text{ kN}$$

$$k_a = \frac{P_{uz} - P_u}{P_{uz} - P_b} = \frac{2626.933 - 2255}{2626.933 - 706.91} \simeq 0.194$$

Hence $M_{ax} = M_{ay} = M_a = P_u \,.\, k_a \,.\, e_a = 2255 \times 0.194 \times 37.81 \times 10^{-3} = 16.54$ kN-m

$\therefore \quad M_{utx} = M_{uty} = M_{ut} = M_{ui} + M_a = 54.86 + 16.54 = 71.4$ kN-m

Hence the design moments are $M_{utx} = M_{uty} = 71.4$ kN-m

Now $\quad P_u/f_{ck}\,bD = \dfrac{2255 \times 10^3}{20 \times 400 \times 400} = 0.705$; $\dfrac{p}{f_{ck}} = 0.123$; $d'/D = 0.15$

Hence from Chart 45, we get $M_u/f_{ck}\,bD^2 = 0.058$

$\therefore \quad M_{ux1} = M_{uy1} = 0.058 \times 20 \times 400\,(400)^2 \times 10^{-6} = 74.24$ kN-m

$P_u/P_{uz} = 2255/2626.933 = 0.858 > 0.8$

Hence $\quad \alpha_n = 2$

$$\therefore \quad \left(\frac{M_{utx}}{M_{ux1}}\right)^{\alpha_n} + \left(\frac{M_{uty}}{M_{uy1}}\right)^{\alpha_n} = 2\left(\frac{71.4}{74.24}\right)^2 = 1.85 >> 1$$

Hence the column is *unsafe*. To make it safe, reduce P_u.

Second trial : Let $P_u = 1950$ kN

$\therefore \quad M_{ux,\,min} = M_{uy,\,min} = M_{u,\,min} = 1950 \times 24.33 \times 10^{-3} = 47.44$ kN-m

Hence $\quad M_{ui} = M_{u,\,min} = 47.44$ kN-m

$$k_a = \frac{P_{uz} - P_u}{P_{uz} - P_b} = \frac{2626.933 - 1950}{2626.933 - 706.91} \simeq 0.353$$

$M_{ax} = M_{ay} = M_a = P_u \,.\, k_a \,.\, e_a = 1950 \times 0.353 \times 37.81 \times 10^{-3} = 26.03$ kN-m

$\therefore \quad M_{utx} = M_{uty} = M_{ut} = M_{ui} + M_a = 47.44 + 26.03 = 73.47$ kN-m

Hence the design moments are $M_{utx} = M_{uty} = 73.47$ kN-m

$\dfrac{P_u}{f_{ck}\,b\,.\,D} = \dfrac{1950 \times 10^3}{20 \times 400 \times 400} \simeq 0.61$. Hence from Chart 45, $\dfrac{M_u}{f_{ck}\,bD^2} = 0.083$

$\therefore \quad M_{ux1} = M_{uy1} = 0.083 \times 20 \times 400\,(400)^2 \times 10^{-6} = 106.24$ kN-m

$P_u/P_{uz} = 1950/2626.933 = 0.7423$

$$\therefore \quad \alpha_n = \frac{2}{3}\left[1 + \frac{5}{2} \times 0.7423\right] = 1.904$$

$$\therefore \quad \left(\frac{M_{utx}}{M_{ux1}}\right)^{\alpha_n} + \left(\frac{M_{uty}}{M_{ux2}}\right)^{\alpha_n} = 2\left(\frac{73.47}{106.24}\right)^{1.904} = 0.991 < 1. \text{ Hence safe.}$$

Hence $\quad$ **P_u = 1950 kN**

Example 19.2. *Check for safety of axially loaded slender column*

A concrete column of 400 mm diameter, reinforced with 8 bars of 20 mm diameter, is braced and hinged at both ends, 8 m apart. Check the safety of the column if it carries a factored axial load of 1100 kN. Take concrete of M 20 grade and steel of 415 grade. The reinforcement has an effective cover of 60 mm.

Solution : Given $P_u = 1100$ kN ; $d' = 60$ mm ; $l = 8 \text{ m} = l_e$

Step 1. Determination of slenderness of column

$$l_e/D = 8000/400 = 20 > 12. \text{ Hence the column is slender}$$

$$A_s = 8\,\frac{\pi}{4}\,(20)^2 = 2513.27 \text{ mm}^2 \;;\; A_g = \frac{\pi}{4}\,(400)^2 = 125664 \text{ mm}^2$$

$$p = \frac{2513.27 \times 100}{125664} = 2\% \;;\; \frac{p}{f_{ck}} = \frac{2}{20} = 0.1 \;;\; \frac{d'}{D} = \frac{60}{400} = 0.15$$

Step 2. Determination of e_{min}, $M_{u,\,min}$ and M_{ui}

$$e_{min} = \frac{l}{500} + \frac{D}{30} = \frac{8000}{500} + \frac{400}{30} = 29.33 \text{ mm} > 20$$

$$\therefore \quad M_{u,\,min} = P_u \,.\, e_{min} = 1100 \times 29.33 \times 10^{-3} = 32.27 \text{ kN-m}$$

Since there is no external moment, $M_{ui} = M_{u,\,min} = 32.27$ kN-m

Step 3. Determination of M_a and M_{ut}

$$e_a = \frac{D}{2000}\left(\frac{l_e}{D}\right)^2 = \frac{400}{2000}(20)^2 = 80 \text{ mm}$$

$$P_{uz} = \left[\,0.45 \times 20 \times 125644 + (0.75 \times .415 - 0.45 \times 20)\ 2513.27\,\right] \times 10^{-3} = 1890.43 \text{ kN}$$

From Table 19.1 $\quad k_1 = 0.149$ for $d'/D = 0.15$

From Table 19.2, $\quad k_2 = 0.201$ for $d'/D = 0.15$

$$\therefore \quad P_b = \left(k_1 + k_2\frac{p}{f_{ck}}\right) f_{ck}\, D^2 = (0.149 + 0.201 \times 0.1)\ 20 \times 400^2 \times 10^{-3} = 541.12 \text{ kN}$$

$$k_a = \frac{P_{uz} - P_u}{P_{uz} - P_b} = \frac{1890.43 - 1100}{1890.43 - 541.12} = 0.5858$$

$$M_a = P_u \,.\, k_a \,.\, e_a = 1100 \times 0.5858 \times 80 \times 10^{-3} = 51.55 \text{ kN-m}$$

$$\therefore \quad M_{ut} = M_{ui} + M_a = 32.27 + 51.55 = 83.82 \text{ kN-m}$$

Step 4. Check for safety

$$\frac{P_u}{f_{ck}\, D^2} = \frac{1100 \times 10^3}{20\,(400)^2} = 0.344 \;;\; \frac{M_{ut}}{f_{ck}\, D^3} = \frac{83.82 \times 10^6}{20\,(400)^3} = 0.065$$

Hence from Chart 57 for $d'/D = 0.15$, $\frac{p}{f_{ck}} = 0.1$ and $\frac{P_u}{f_{ck}\, D^2} = 0.344$

we get permissible $\quad \frac{M_u}{f_{ck}\, D^3} = 0.065$.

Actual $\quad \frac{M_{ut}}{f_{ck}\, D^3} = 0.065$. Hence the column is *just safe*.

Example 19.3. ***Computation of design moments for column slender about minor axis***

A R.C. column 400 mm × 600 mm is subjected to an axial ultimate load of 3000 kN. The column is bent in single curvature about minor axis with ultimate moment $M_y = 100$ kN-m at top and $M_y = 125$ kN-m at bottom. Taking the unsupported length as 7 m and effective length as 5.6 m on both axes, compute the design moments for the column. Assume reduction factor $k_a = 1$.

Solution :

Step 1 : Test for slenderness

Given : $l_x = l_y = 7$ m ; $l_{ex} = l_{ey} = 5.6$ m.

$$\therefore \quad \frac{l_{ex}}{D} = \frac{5600}{600} = 9.33 < 12 \text{ and } \frac{l_{ey}}{b} = \frac{5600}{400} = 14 > 12$$

Hence the column is short about *x-x* axis and slender about *y-y* axis.

Step 2 : Computation of e_{min} and $M_{u, min}$

Now $$e_{x, min} = \frac{l_x}{500} + \frac{D}{30} = \frac{7000}{500} + \frac{600}{30} = 34 \text{ mm} > 20 \text{ mm}$$

$$e_{y, min} = \frac{l_y}{500} + \frac{b}{30} = \frac{7000}{500} + \frac{400}{30} = 27.33 \text{ mm} > 20 \text{ mm}$$

$$\therefore \quad M_{ux, min} = P_u \cdot e_{x, min} = 3000 \times 34 \times 10^{-3} = 102 \text{ kN-m}$$

$$M_{uy, min} = P_u \cdot e_{y, min} = 3000 \times 27.33 \times 10^{-3} = 82 \text{ kN-m.}$$

Step 3. Computation of final design moment about x-axis

Since there is no external moment in *x*-direction,

$$M_{uix} = M_{ux, min} = 102 \text{ kN-m}$$

$e_{ax} = 0$ since the column is short about *x*-axis. Hence $M_{ax} = 0$

Hence $$M_{utx} = M_{uix} + M_{ax} = 102 + 0 = 102 \text{ kN-m}$$

Step 4. Computation of final design moment about y-axis.

$$M_{uiy} = 0.6\, M_{2y} + 0.4\, M_{1y} = 0.6 \times 125 + 0.4 \times 100 = 115 \text{ kN-m.}$$

This should not be less than $M_{uy, min}$ (= 82 kN-m). Hence $M_{uiy} = 115$ kN-m

$$e_{ay} = b\,\frac{(l_{ey}/b)^2}{2000} = \frac{400\,(14)^2}{2000} = 39.2 \text{ mm}$$

$$\therefore \quad M_{ay} = P_u \cdot e_{ay} \cdot k_{ay} = 3000 \times 39.2 \times 10^{-3} \times 1 = 117.6 \text{ kN-m}$$

$$\therefore \quad M_{uty} = M_{uiy} + M_{ay} = 115 + 117.6 = 232.6 \text{ kN-m}$$

But not less than $M_{2y} = 125$ kN-m

Hence design moment $M_{uy} = M_{uty} = 232.6$ kN-m

Step 5. Final design moments

Hence design the column as axially loaded column, with biaxial bending for the following:

$$\mathbf{P_u = 3000 \text{ kN};\ M_{ux} = 102 \text{ kN-m};\ M_{uy} = 232.6 \text{ kN-m.}}$$

Example 19.4. ***Computation of design moments for column slender about both the axes***

A R.C. braced column, 300 mm × 400 mm, is subjected to $P_u = 1500$ kN and ultimate moments $M_y = 60$ kN-m at top and $M_y = 8$ kN-m at the bottom. If the column is bent in double curvature, about minor axis (y-axis) determine the design moments. Take unsupported length equal to 8 m and effective length as 6 m about both the axes. Assume $k_a = 1$.

Solution : Given $l_x = l_y = 8$ m and $l_{ex} = l_{ey} = 6$ m

Step 1 : Test for slenderness

$$l_{ex}/D = 6000/400 = 15 > 12 \ ; \ l_{ey}/b = 6000/300 = 20 > 12.$$

Hence the column is slender about both the axes.

Step 2. Computation of e_{min} and $M_{u,min}$

$$e_{x,min} = \frac{l_x}{500} + \frac{D}{30} = \frac{8000}{500} + \frac{400}{30} = 29.33 \text{ mm} > 20 \text{ mm}$$

$$e_{y,min} = \frac{l_y}{500} + \frac{b}{30} = \frac{8000}{500} + \frac{300}{30} = 26 \text{ mm} > 20 \text{ mm}$$

$$\therefore \quad M_{ux,min} = P_u \cdot e_{x,min} = 1500 \times 29.33 \times 10^{-3} \simeq 44 \text{ kN-m}$$

$$M_{uy,min} = P_u \cdot e_{y,min} = 1500 \times 26 \times 10^{-3} = 39 \text{ kN-m}$$

Step 3. Computation of initial moments, M_{ui}

Since there is no external moment in x direction, $M_{uix} = M_{ux,min} = 44$ kN-m

$$M_{uiy} = (0.6\, M_{2y} + 0.4\, M_{1y}) \nless 0.4\, M_{2y}.$$

Since the column is bent in double curvature about y-axis, $M_{2y} = 60$ and $M_{1y} = -8$

$$\therefore \quad M_{uiy} = (0.6 \times 60 - 0.4 \times 8) = 32.8 \text{ kN-m}$$

(but not less than $0.4\, M_{2y} = 0.4 \times 60 = 24$ kN-m)

Also, M_{uiy} should not be less than $M_{uy,min}$ (= 39 kN-m). Hence $M_{uiy} = 39$ kN-m

Step 4. Computation of additional eccentricities and additional moments

$$e_{ax} = \frac{D\,(l_{ex}/D)^2}{2000} = \frac{400\,(15)^2}{2000} = 45 \text{ mm}$$

$$e_{ay} = \frac{b\,(l_{ey}/b)^2}{2000} = \frac{300\,(20)^2}{2000} = 60 \text{ mm}$$

$$\therefore \quad M_{ax} = P_u \cdot e_{ax} \cdot k_a = 1500 \times 45 \times 10^{-3} \times 1 = 67.5 \text{ kN-m}$$

$$M_{ay} = P_u \cdot e_{ay} \cdot k_a = 1500 \times 60 \times 10^{-3} \times 1 = 90 \text{ kN-m}$$

Step 5. Computation of total moments

$$M_{utx} = M_{uix} + M_{ax} = 44 + 67.5 \text{ kN-m} = 111.5 \text{ kN-m}$$

Hence design moment in x-direction $= M_{ux} = 111.5$ kN-m

$$M_{uty} = M_{uiy} + M_{ay} = 39 + 90 = 129 \text{ kN-m}$$

(but not less than $M_{2y} = 60$ kN-m

Hence design moment in y-direction $= M_{uy} = 129$ kN-m

Hence design the column with axial load and biaxial moments as under :

$$\mathbf{P_u = 1500 \text{ kN} \; ; \; M_{ux} = 111.5 \text{ kN-m} \; ; \; M_{uy} = 129 \text{ kN-m}.}$$

Example 19.5. *Computation of reinforcement for an axially loaded slender column*

A braced column of M 20 grade concrete has 300 mm diameter and has 7.5 m unsupported length and 5.8 m effective length. The column is subjected to an axial load $P_u = 700$ kN. Find the area of reinforcement required, using steel of Fe 415 grade. Provide an effective cover of 60 mm.

Solution : Given $l = 7.5$ m; $l_e = 5.8$ m ; $D = 300$ mm

$$A_g = \frac{\pi}{4}(300)^2 = 70685.6 \text{ mm}^2 \; ; \; d'/D = 60/300 = 0.2$$

Step 1. Determination of slenderness of the column

$l_e/D = 5800/300 = 19.33 > 12$. Hence the column is slender.

Step 2. Determination of e_{min}, $M_{u,\,min}$ and M_{ui}

$$e_{min} = \frac{l}{500} + \frac{D}{30} = \frac{7500}{500} + \frac{300}{30} = 25 \text{ mm}$$

$$\therefore \qquad M_{u\,.\,min} = P_u \,.\, e_{min} = 700 \times 25 \times 10^{-3} = 17.5 \text{ kN-m}$$

Since there is no external B.M. $M_{ui} = M_{u\,.\,min} = 17.5$ kN-m.

Step 3. First estimation of A_s and p

The determination of A_s, and hence p, will be by trial and error since k_a and hence additional moment M_a depends upon p.

In the first estimate for p, let us adopt *equivalent axial load* for uniaxial bending as $1.2\,P_u$ and for biaxial bending as $1.4\,P_u$.

Hence $\quad 1.4\,P_u = 0.4\,f_{ck}\,A_g + (0.67\,f_y - 0.4\,f_{ck})\,A_s$

From which $\quad A_s = \dfrac{1.4 \times 700 \times 10^3 - 0.4 \times 20 \times 70685.6}{(0.67 \times 415 - 0.4 \times 20)} = 1535 \text{ mm}^2$

$$\therefore \text{ Number of 16 mm } \varphi \text{ bars } = \frac{1535}{\frac{\pi}{4}(16)^2} = 7.63$$

Hence provide 8 bars of 16 mm φ. Actual $A_s = 8\,\frac{\pi}{4}(16)^2 = 1608.5 \text{ mm}^2$

Actual $\quad p = \dfrac{1608.5 \times 100}{70685.6} = 2.276\% \quad ; \quad \dfrac{p}{f_{ck}} = \dfrac{2.276}{20} \simeq 0.114$

Step 4. Determination of e_a, k_a, M_a and M_{ut}

From Tables 19.1 and 19.2, for $d'/D = 0.2$, we get $k_1 = 0.138$ and $k_2 = 0.036$

$$\therefore \qquad P_b = \left(k_1 + k_2\frac{p}{f_{ck}}\right) f_{ck}\,D^2 = (0.138 + 0.036 \times 0.114)\,20\,(300)^2 \times 10^{-3} = 255.79 \text{ kN}$$

$$P_{uz} = [0.45 \times 20 \times 70685.6 + (0.75 \times 415 - 0.45 \times 20) \times 1608.5]\,10^{-3} = 1122.34 \text{ kN}$$

$$\therefore \qquad k_a = \frac{P_{uz} - P_u}{P_{uz} - P_b} = \frac{1122.34 - 700}{1122.34 - 255.79} = 0.4874$$

$$e_a = \frac{D}{2000}\left(\frac{l_e}{D}\right)^2 = \frac{300}{2000}(19.33)^2 = 56.07 \text{ mm}$$

$$\therefore \qquad M_a = P_u \,.\, k_a \,.\, e_a = 700 \times 0.4874 \times 56.07 \times 10^{-3} = 19.13 \text{ kN-m}$$

Hence $\quad M_{ut} = M_{ui} + M_a = 17.5 + 19.13 = 36.63$ kN-m

Step 5. Check for reinforcement ratio p of the column

$$\frac{P_u}{f_{ck}\,D^2} = \frac{700 \times 10^3}{20\,(300)^2} = 0.389 \;;\; \frac{M_{ut}}{f_{ck}\,D^3} = \frac{36.63 \times 10^6}{(20)\,(300)^3} = 0.0678 \;;\; \frac{d'}{D} = 0.2$$

Hence from Chart 58, we get $p/f_{ck} = 0.135$

$$\therefore \qquad p = 0.135 \times 20 = 2.7\% \text{ and } A_s = \frac{2.7 \times 70685.6}{100} = 1909 \text{ mm}^2$$

$\therefore$ No. of 20 mm φ bars = 1909/314.16 = 6.67. Hence provide 7 bars of 20 mm φ, thus providing $A_s = 7 \times 314.16 =$ **2199.12 mm²**

$$\therefore \quad p = \frac{2199.12}{70685.6} \times 100 = \mathbf{3.11\%}$$

Example 19.6. ***Computation of maximum ultimate moment if P_u is given***

Determine the maximum ultimate moment the column of example 19.5 will be able to carry, if P_u is reduced to 500 kN.

Solution : Given : $P_u = 500$ kN ; $p = 3.11\%$; $p/f_{ck} = 3.11/20 = 0.156$

Step 1. Determination of M_{ut}

$$\frac{P_u}{f_{ck} D^2} = \frac{500 \times 1000}{20\,(300)^2} = 0.278 \; ; \; \frac{p}{f_{ck}} = 0.156 \; ; \; \frac{d'}{D} = 0.2$$

Hence from Chart 58, we get $\frac{M_u}{f_{ck} D^3} \simeq 0.09$

$$\therefore \quad M_u = M_{ut} = 0.09 \times 20\,(300)^3 \times 10^{-6} \simeq 48.6 \text{ kN-m.}$$

Step 2. Computation of k_a and M_a

$$P_b = (0.138 + 0.036 \times 0.156)\, 20\,(300)^2 \times 10^{-3} = 258.5 \text{ kN}$$

$$P_{uz} = [0.45 \times 20 \times 70685.6 + (0.75 \times 415 - 0.45 \times 20)\, 2199.12]\, 10^{-3} = 1300.85 \text{ kN}$$

$$\therefore \quad k_a = \frac{P_{uz} - P_u}{P_{uz} - P_b} = \frac{1300.85 - 500}{1300.85 - 258.5} = 0.768 \; ; \; e_a = 56.07 \text{ mm as before}$$

$$\therefore \quad M_a = P_u \,.\, k_a\, e_a = 500 \times 0.768 \times 56.07 \times 10^{-3} = 21.54 \text{ kN-m}$$

Step 3. Determination of ultimate external moment

$M_{ui} = M_{ut} - M_a = 48.6 - 21.54 =$ **27.06 kN-m**

(Note that this is more than $M_{u,\,min} = P_u \,.\, e_{min} = 500 \times 25 \times 10^{-3} = 12.5$ kN-m)

Example 19.7. ***Design of unbraced slender column with axial load and uniaxial moment***

An unbraced column, 300 mm × 400 mm in section, is unbraced in both the principal directions. The column ends are fixed, and the column has unsupported length of 3.9 m. The column carries an ultimate axial load of 800 kN and ultimate moments of 80 kN-m and 40 kN-m about the major axis bisecting the depth, at its ends. Assuming M 20 grade concrete, Fe 415 grade steel and effective cover of 60 mm, determine the area of longitudinal reinforcement.

Solution : Given $l = 4$ m; $b = 300$ mm; $D = 400$ mm; $d' = 60$ mm

$$P_u = 500 \text{ kN} \; ; \; M_{ux1} = 40 \text{ kN-m} \; ; \; M_{ux2} = 80 \text{ kN-m}$$

Step 1. Determination of l_e, e_{min} and $M_{u,\,min}$

$$l_e = 1.2\, l = 1.2 \times 3.9 = 4.68 \text{ m}$$

$$l_e/D = 4680/400 = 11.7 < 12; \; l_e/b = 4680/300 = 15.6 > 12$$

Hence the column is short for bending about x-axis and long for bending about y-axis.

$$e_{x\,.\,min} = \frac{l}{500} + \frac{D}{30} = \frac{3900}{500} + \frac{400}{30} = 21.13 \text{ mm} > 20 \text{ mm}$$

$$e_{y\,.\,min} = \frac{l}{500} + \frac{b}{30} = \frac{3900}{500} + \frac{300}{30} = 17.8 \text{ mm} < 20 \text{ mm} \quad ; \quad \therefore \; e_{y\,.\,min} = 20 \text{ mm}$$

$$M_{ux\,.\,min} = P_u \,.\, e_{x\,.\,min} = 800 \times 21.13 \times 10^{-3} = 16.9 \text{ kN-m}$$

$$M_{uy\,.\,min} = P_u \,.\, e_{y\,.\,min} = 800 \times 20 \times 10^{-3} = 16 \text{ kN-m}$$

Step 2. Design for bending about x-axis

The column is short for bending about x-axis. The column is subjected to external ultimate B.M. of 80 kN-m and 40 kN-m. Hence the column will be designed for 80 kN-m (i.e. the greater one) and $M_{ux\,.\,min} = 16.9$ kN-m , which ever is larger.

Hence design $M_{ux} = 80$ kN-m ; Also, $d'/D = 60/400 = 0.15$

$$\therefore \quad \frac{M_{ux}}{f_{ck}\, bD^2} = \frac{80 \times 10^6}{20 \times 300\,(400)^2} = 0.0833$$

Also,
$$\frac{P_u}{f_{ck}\, bD} = \frac{800 \times 10^3}{20 \times 300 \times 400} = 0.333$$

Hence from Chart 33, corresponding to the above two values, we get

$$\frac{p}{f_{ck}} = 0.045 \quad \text{(with reinforcement on two opposite sides)}$$

$$\therefore \quad p = 0.045 \times 20 = 0.9\% \quad ; \quad A_s = \frac{0.9 \times 300 \times 400}{100} = 1080 \text{ mm}^2$$

No. of 16 mm φ bars = 1080/201.06 = 5.37

Providing 6 bars of 16 mm φ (i.e. 3 bars on each face),

Actual $A_s = 6 \times 201.06 = 1206.36 \text{ mm}^2$

Actual
$$p = \frac{1206.36 \times 100}{300 \times 400} \simeq 1 \quad \text{and} \quad \frac{p}{f_{ck}} = \frac{1}{20} = 0.05$$

Step 3. Check of the Design of bending about y-axis

The column is slender for bending about y-axis. Also $d'/D = 60/300 = 0.02$

Since there is no external B.M. about y-axis, $M_{uiy} = M_{uy\,.\,min} = 16$ kN-m

$$M_{uty} = M_{uiy} + M_{ay}\,.$$

$$e_{ay} = \frac{b}{2000}\left(\frac{l_e}{b}\right)^2 = \frac{300}{2000}(15.67)^2 = 36.83 \text{ mm}$$

$$P_{uz} = 0.45\, f_{ck}\, A_g + (0.75\, f_y - 0.45\, f_{ck})\, A_s$$

$$= [0.45 \times 20 \times 300 \times 400 + (0.75 \times 415 - 0.45 \times 20)\, 1206.36]\, 10^{-3} = 1444.62 \text{ kN}$$

From Tables 19.1 and 19.2, for $d'/D = 0.2$ and equal A_s on two sides,

$k_1 = 0.184$ and $k_2 = -0.022$

$$\therefore \quad P_b = \left(k_1 + k_2 \frac{p}{f_{ck}}\right) f_{ck}\, bD = (0.184 - 0.022 \times 0.05)\, 20 \times 300 \times 400 \times 10^{-3} = 438.96 \text{ kN}$$

$$\therefore \quad k_{ay} = \frac{P_{uz} - P_u}{P_{uz} - P_b} = \frac{1444.62 - 800}{1444.62 - 438.96} = 0.641$$

$\therefore \quad M_{ay} = P_u \,.\, k_{ay} \,.\, e_{ay} = 800 \times 0.641 \times 36.83 \times 10^{-3} = 18.89$ kN-m.

$\therefore \quad M_{uty} = M_{uiy} + M_{ay} = 16 + 18.89 = 34.89$ kN-m

Now $\dfrac{P_u}{f_{ck}\, bD} = \dfrac{800 \times 10^3}{20 \times 300 \times 400} = 0.333$, as before ; $\dfrac{p}{f_{ck}} = 0.05$ and $\dfrac{d'}{D} = \dfrac{60}{300} = 0.2$

Hence from Chart 34, we get $\dfrac{M_u}{f_{ck}\, b\, D^2} = 0.085$

$\therefore \quad M_{uy} = 0.085 \times 20 \times 400\,(300)^2 \times 10^{-6} = 61.2$ kN-m

Actual $M_{uy} = M_{uty} = 34.89$ kN-m. Hence the column is **safe.**

Example 19.8. ***Design of braced slender column with axial load and biaxial moments***

A braced column, 300 mm × 400 mm is to carry a factored axial load of 1200 kN and factored moments of 50 kN-m and 30 kN-m with respect to major and minor axis respectively, at the top end. The column is bent in double curvature in both the directions with the moments at bottom end equal to 40% of the corresponding moments at the top. Design the longitudinal reinforcement for the column, if its unsupported length is 7.2 m with an effective length ratio of 0.85 in both the directions. Use M 20 concrete and Fe 415 steel and provide 60 mm effective cover.

Solution $l = 7.2$ m. $P_u = 1200$ kN; $M_{2x} = 50$ kN-m; $M_{1x} = 0.4 \times 50 = 20$ kN-m;

$$M_{2y} = 30 \text{ kN-m};\quad M_{1y} = 0.4 \times 30 = 12 \text{ kN-m}$$

Step 1. Computation of slenderness ratio e_{min} ***and*** M_{min}

$$l_x = l_y = 7.2 \text{ m};\quad l_{ex} = l_{ey} = 0.85 \times 7.2 = 6.12 \text{ m}$$

$\therefore \quad l_{ex}/D = 6120/400 = 15.3 > 12$ and $l_{ey}/b = 6120/300 = 20.4 > 12$

Hence the column is slender for bending about both the axes.

$$e_{x,\,min} = \frac{l_x}{500} + \frac{D}{30} = \frac{7200}{500} + \frac{400}{30} = 27.73 \text{ mm} > 20 \text{ mm}$$

$$e_{y,\,min} = \frac{l_y}{500} + \frac{b}{30} = \frac{7200}{500} + \frac{300}{30} = 24.4 \text{ mm} > 20 \text{ mm}$$

$$M_{ux\,.\,min} = P_u \,.\, e_{x,\,min} = 1200 \times 27.73 \times 10^{-3} = 33.276 \text{ kN-m}$$

$$M_{uy\,.\,min} = P_u \,.\, e_{y,\,min} = 1200 \times 24.4 \times 10^{-3} = 29.28 \text{ kN-m}$$

Step 2. Computation of initial moments (M_{ui})

The column is braced and bent in double curvature and hence M_{1x} and M_{1y} are negative. Hence primary moments in x and y directions are given by

$$M_{ux} = (0.6\, M_{2x} - 0.4\, M_{1x}) = (0.6 \times 50 - 0.4 \times 20) = 22 \text{ kN-m } (\geq 0.4 \times 50 = 20 \text{ kN-m})$$

$$M_{uy} = (0.6\, M_{2y} - 0.4\, M_{1y}) = (0.6 \times 30 - 0.4 \times 12) = 13.2 \text{ kN-m } (\geq 0.4 \times 30 = 12 \text{ kN-m})$$

The initial moment in x-direction should be taken as greater of M_{ux} and $M_{ux\,.\,min}$. Similarly, the initial moment in y-direction should be taken as greater of M_{uy} and $M_{uy\,.\,min}$.

Hence $\quad M_{uix} = 33.276$ kN-m $(> M_{ux} = 22$ kN-m$)$

and $\quad M_{uiy} = 29.28$ kN-m $(> M_{uy} = 13.2$ kN-m $)$

Step 3. Computation of additional moments M_a and total moments M_{ut}

$$e_{ax} = \frac{D}{2000}\left(\frac{l_{ex}}{D}\right)^2 = \frac{400}{2000}(15.3)^2 = 46.82 \text{ mm}$$

$$e_{ay} = \frac{b}{2000}\left(\frac{l_{ey}}{b}\right)^2 = \frac{300}{2000}(20.4)^2 = 62.42 \text{ mm}$$

To start with, let us assume $k_{ax} = k_{ay} = 0.5$

$\therefore$ $$M_{ax} = P_u \,.\, e_{ax} \,.\, k_{ax} = 1200 \times 46.82 \times 0.5 \times 10^{-3} = 28.092 \text{ kN-m}$$

and $$M_{ay} = P_u \,.\, e_{ay} \,.\, k_{ay} = 1200 \times 62.42 \times 0.5 \times 10^{-3} = 37.452 \text{ kN-m}$$

$\therefore$ $$M_{utx} = M_{ux} + M_{ax} = 33.276 + 28.092 = 61.368 \text{ kN-m}$$

$$M_{uty} = M_{uiy} + M_{ay} = 29.28 + 37.452 = 66.732 \text{ kN-m}$$

Step 4. Design of trial section

Let the resultant B.M. $M_u = 1.15\sqrt{(61.368)^2 + (66.732)^2} = 104.26$ kN-m for bending about y-axis, since $M_{uty} > M_{utx}$.

Hence design the column for uniaxial bending, with $P_u = 1200$ kN and $M_u \simeq 104$ kN-m

$$\therefore \quad \frac{P_u}{f_{ck}\,bD} = \frac{1200 \times 10^3}{20 \times 400 \times 300} = 0.5 \;, \quad \frac{M_u}{f_{ck}\,bD^2} = \frac{104 \times 10^6}{20 \times 400\,(300)^2} = 0.144; \quad \frac{d'}{D} = \frac{60}{300} = 0.2$$

Providing equal reinforcement on all the four sides, we get from Chart 46 $p/f_{ck} = 0.19$. Hence $p = 0.19 \times 20 = 3.8\%$

$$\therefore \quad A_s = \frac{3.8 \times 300 \times 400}{100} = 4560 \text{ mm}^2 \text{ and}$$

$\therefore$ Number of 25 mm φ bars $= \dfrac{4560}{490.87} = 9.29.$

Hence provide 10 bars of 25 mm φ bars and arrange them uniformly on all the four sides, as shown in Fig. 19.4. Actual $A_s = 10 \times 490.87 = 4908.7$ mm^2

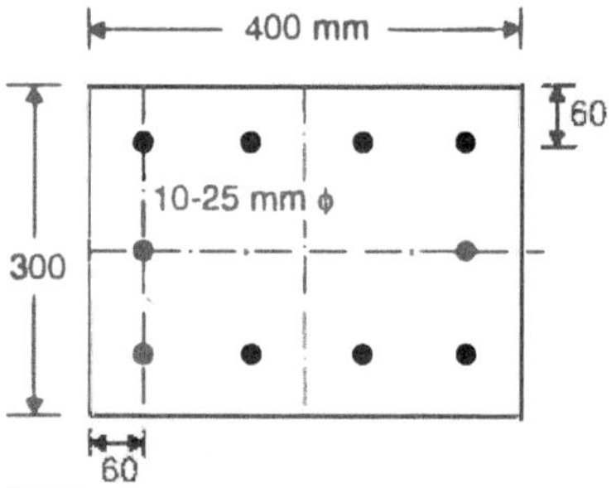

FIG. 19.4.

$$\text{Actual} \quad p = \frac{4908.7 \times 100}{300 \times 400} = 4.09\%$$

and $\quad p/f_{ck} = 4.09/20 = 0.2045.$

Provide effective cover of 60 mm.

Step 5. Computation of actual M_a and M_{ut} in each direction

$$P_{uz} = 0.45 f_{ck} A_g + (0.75 f_y - 0.45 f_{ck}) A_s$$

$$= [0.45 \times 20 \times 300 \times 400 + (0.75 \times 415 - 0.45 \times 20) \times 4908.7]\, 10^{-3} = 2563.65 \text{ kN}$$

For bending about x-axis, $d'/D = 60/400 = 0.15$. Hence from Tables 19.1 and 19.2, we get $k_{1x} = 0.196$ and $k_{2x} = 0.203$.

For bending about y-axis, $d'/D = 60/300 = 0.2$. Hence we get $k_{1y} = 0.184$ and $k_{2y} = 0.028$

$$\therefore \quad P_{bx} = \left(k_{1x} + k_{2x}\frac{p}{f_{ck}}\right) f_{ck}\,bD = (0.196 + 0.203 \times 0.2045)\, 20 \times 300 \times 400 \times 10^{-3} = 570.03 \text{ kN}$$

and $P_{by} = \left(k_{1y} + k_{2y} \frac{p}{f_{ck}} \right) f_{ck}\, bD = (0.184 + 0.028 \times 0.2045)\, 20 \times 400 \times 300 \times 10^{-3} = 455.34$ kN

$$k_{ax} = \frac{P_{uz} - P_u}{P_{uz} - P_{bx}} = \frac{2563.65 - 1200}{2563.65 - 570.03} = 0.684$$

$$k_{ay} = \frac{P_{uz} - P_u}{P_{uz} - P_{by}} = \frac{2563.65 - 1200}{2563.65 - 455.34} = 0.647$$

$$\therefore \quad M_{ax} = P_u \,.\, e_{ax} \,.\, k_{ax} = 1200 \times 46.82 \times 0.684 \times 10^{-3} = 38.43 \text{ kN-m}$$

$$M_{ay} = P_u \,.\, e_{ay} \,.\, k_{ay} = 1200 \times 62.42 \times 0.647 \times 10^{-3} = 48.46 \text{ kN-m}$$

$$\therefore \quad M_{utx} = M_{uix} + M_{ax} = 33.276 + 38.43 = 71.706 \text{ kN-m}$$

$$M_{uty} = M_{uiy} + M_{ay} = 29.28 + 48.46 = 77.74 \text{ kN-m}$$

Step 6. Check for safety under axial load with biaxial bending

$$\frac{P_u}{f_{ck}\, bD} = 0.5 \text{ as before, and } \frac{p}{f_{ck}} = 0.2045$$

For bending about x-axis, $d'/D = 0.15$.

Hence from Chart 45, we get $\dfrac{M_u}{f_{ck}\, bD^2} = 0.18$

$$\therefore \quad M_{ux1} = 0.18 \times 20 \times 300\, (400)^2 \times 10^{-6} = 172.8 \text{ kN-m}$$

Similarly, for bending about y-axis, $d'/D = 0.2$; hence from Chart 46, $M_u/f_{ck}\, bD^2 = 0.152$.

$$\therefore \quad M_{uy1} = 0.152 \times 20 \times 400\, (300)^2 \times 10^{-6} = 109.44 \text{ kN-m}$$

$$\frac{P_u}{P_{uz}} = \frac{1200}{2563.65} = 0.468, \quad \alpha_n = \frac{2}{3}\left[1 + \frac{5}{2} \times 0.468 \right] = 1.447$$

$$\therefore \quad \left(\frac{M_{utx}}{M_{ux1}} \right)^{\alpha_n} + \left(\frac{M_{uty}}{M_{uy1}} \right)^{\alpha_n} = \left(\frac{71.706}{172.8} \right)^{1.447} + \left(\frac{77.74}{109.44} \right)^{1.447} = 0.28 + 0.61 = 0.89 < 1$$

Hence the designed section is **safe.**

PROBLEMS

1. R.C. column, 300 mm × 500 mm, reinforced with 8 bars of 20 mm φ, uniformly distributed on all the faces is braced against side sway and has an unsupported length of 7 m. Taking effective length ratios equal to 0.85 in both the directions, determine the maximum factored axial load carrying capacity of the column. Use M 20 concrete and Fe 415 steel.

2. Design the longitudinal reinforcement for a braced column, 300 mm × 450 mm subject to a factored axial load of 1800 kN and factored moments of 70 kN-m and 50 kN-m with respect to major axis and minor axis respectively at the top end. Assume that the column is bent in double curvature in both the directions, with the moments at the bottom equal to 60 per cent of the corresponding moments at the top. Assume an unsupported length of 7.2 m and an effective length ratio of 0.85 in both the directions. Use M 20 concrete and Fe 415 steel.

3. Design the reinforcement for a R.C. column 300 mm × 400 mm in cross-section for the following data : $P_u = 1200$ kN; $M_{ux1} = 24$ kN-m at bottom and $M_{ux2} = 36$ kN-m at top ; $M_{uy} = 12$ kN-m at top and bottom; Unsupported length = 6 m; Effective length factor = 0.8 for x direction and 0.7 for y direction. The column is braced and bends into single curvature. Use M 20 concrete and Fe 415 steel. Provide 50 mm cover to the centre of steel.

20

Yield Line Theory and Design of Slabs

20.1. INTRODUCTION

The *yield line theory* is one of the most important developments in the analysis and design of slab systems. The yield line theory is the ultimate load theory for the design of R.C. slabs. In the case of slabs, the computation of ultimate load is quite complicated. In the yield line method, the computation of ultimate load is based on the pattern of *yield lines* that are developed in the slabs under conditions approaching collapse. The method was innovated by Ingerslav (1923) and was greatly extended and advanced by Johanssen. The most commonly used limit state or collapse method is based on Johanssen's yield line theory.

When a slab is loaded with increasing loads, the stresses in the reinforcement and the concrete increase more or less proportionately upto load level corresponding to the yield stress in the reinforcement. If the load is increased further, excessive deformations and rapid increase in strains will result. These deformations will be *elastoplastic* upto a load level called *limit load*. When limit load is reached, the slab will continue to deform without any additional load, leading to total collapse. At this stage, a pattern of cracks will form a set of lines known as *yield lines*, resulting into a mechanism leading to the total collapse of the slab. A *yield line* is defined as a line in the plane of slab across which reinforcing bars have yielded and about which excessive deformation (plastic rotations) under constant limit moment (ultimate moment) continues to occur leading to failure.

20.2. YIELD LINE PATTERNS

(*a*) **One Way Slabs** : The behaviour of one way slab regarding its ultimate moment capacity is the same as that of beam of unit width. Consider a one way slab of unit width, simply supported at the ends, and loaded with uniformly distributed load [Fig. 20.1 (*a*)]. The maximum positive B.M. will be at the mid-span of the slab, its value being equal to $+wL^2/8$. Similarly, Fig. 20.1(*b*) shows a one way slab of unit width fixed at the

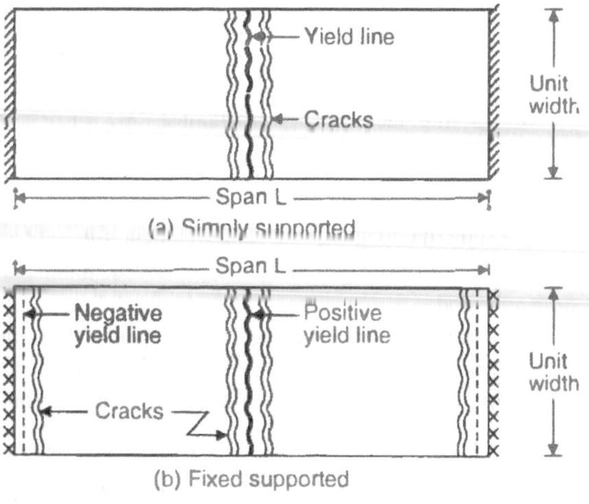

FIG. 20.1. YIELD LINES IN ONE WAY SLABS.

ends. The maximum negative B.M. will be $-wL^2/12$ at the ends and maximum mid-span positive B.M. will be $+wL^2/24$

At service loads, the slab will be uncracked, but as the load is increased, cracks will be developed at the top of slab in negative moment regions, and at bottom in the positive moment regions. As the load is further increased, the negative and/or positive steel will start yielding. At the yield load w_y, the entire width of the slab will develop cracks, and *yield lines* will be formed. In the case of simply supported slab, such an yield line will be formed at mid-span, while for slab with fixed supports, *negative yield line* will be formed at the supports and *positive yield line* will be formed at the mid-span. At the yield lines, plastification takes place, creating *'plastic hinges'*. These plastic hinges withstand plastic moments. It is to be noted that the direction of yield line is perpendicular to the direction of reinforcement.

***(b)* Two Way Slabs** : In one way slab, the direction of yield line is perpendicular to the direction of steel. But in two way slab (square or rectangular slabs) the direction of yield lines are not perpendicular to the two directions of reinforcement.

Fig. 20.2 *(a)* shows the initiation, growth and development of yield lines in square slabs. The cracks first develop along with small patches of yield of diagonals at mid-span. These cracks spread towards the corners, till yield lines are fully developed. Similarly, in a rectangular slabs, [Fig. 20.2 *(b)*], the initial crack is developed in a direction perpendicular to the short-span, at the mid-region of the slabs. Further increase in the load results in continuous growth of cracks as shown in Fig. 20.2 *(b)* *(ii)* and *(iii)*. It is seen that except for the length *AB*, the cracks are not transverse to the directions of reinforcements.

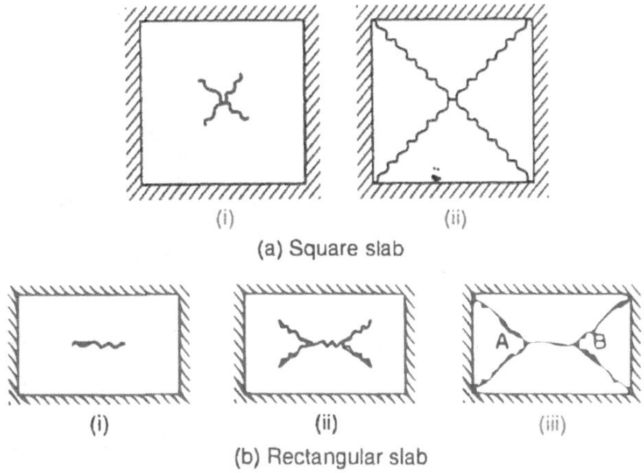

FIG. 20.2. INITIATION, GROWTH AND COMPLETION OF YIELD LINES IN SIMPLY SUPPORTED TWO WAY SLABS.

Characteristic Features of Yield Lines : The following characteristics are useful in selecting the yield line patterns for slab with various boundary conditions.

1. Yield lines are *straight lines* so that they may act as plastic hinges of a collapse mechanism.

2. Yield lines terminate at a slab boundary or at the intersection of other yield lines.

3. Yield lines act as axes of rotation for the movements of adjoining segments.

4. Each of the segments of the slab will tend to rotate in a rigid body motion. The axes of rotation generally lie along the lines of supports and pass over the columns.

5. If an edge is fixed or continuous, a yield line may form along the support.

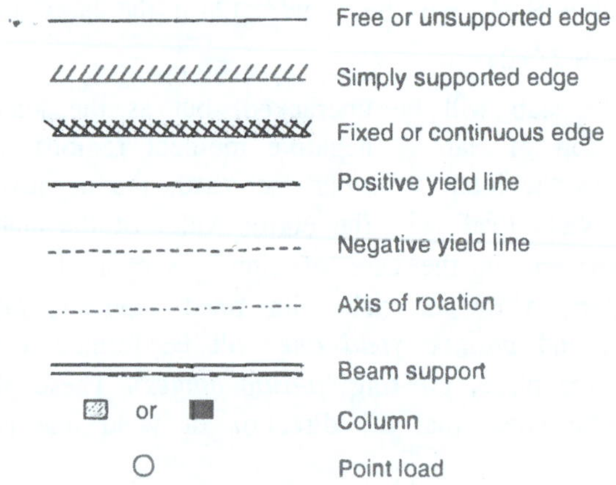

FIG. 20.3. SIGN CONVENTIONS.

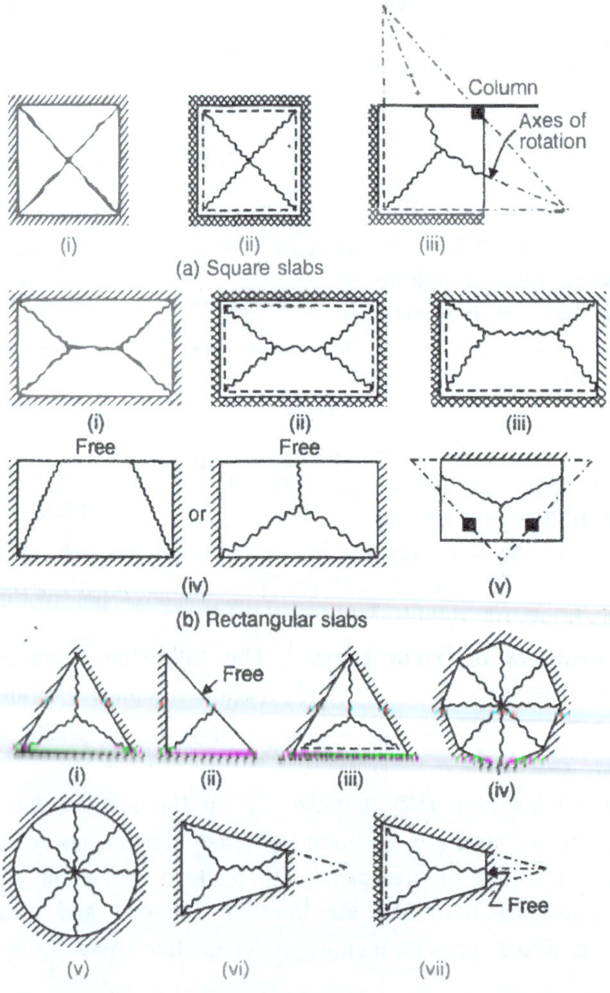

FIG. 20.4. TYPICAL YIELD LINE PATTERNS IN R.C. SLABS.

6. For mechanism to develop (*i.e.* for compatibility of deformations) yield lines or yield lines produced, pass through the intersection of the axes of rotation of adjacent slab elements.

Sign Conventions for Yield Line Patterns and Supports

Fig. 20.3 Shows generally accepted sign conventions.

Yield Line Patterns for Slabs

Fig. 20.4 shows the yield line patterns for slabs of various shapes, with various boundary conditions.

20.3. MOMENT CAPACITY ALONG AN YIELD LINE

In the case of one way slab, the yield line is perpendicular to the direction of reinforcement (Fig. 20.1). However, in the case of rectangular slabs having two-way reinforcement, the direction of the yield line may be inclined to the principal directions of reinforcement. Fig. 20.5 shows an inclined yield line AB in orthogonally reinforced slab. The normal to the yield line makes an angle α with the x-direction, *i.e.* the yield line makes an angle α with the y-direction.

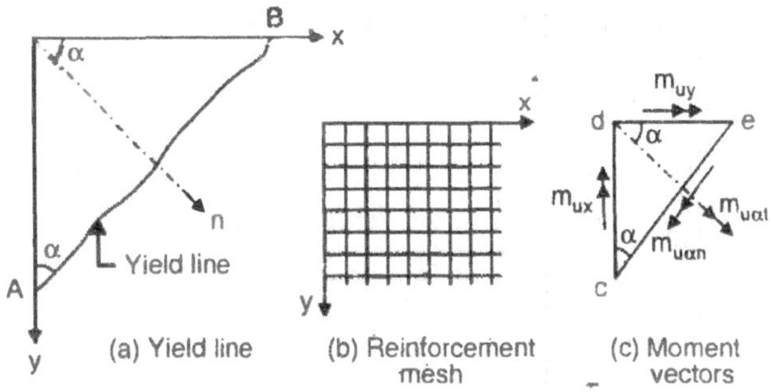

FIG. 20.5. INCLINED YIELD LINE IN ORTHOTROPICALLY REINFORCED SLAB.

Let m_{ux} = ultimate moment of resistance per unit length across a plane whose normal is in x-direction.

m_{uy} = ultimate moment of resistance per unit length, across a plane whose normal is in y-direction.

$m_{u\alpha n}$ = ultimate moment of resistance per unit length, across a plane whose normal makes an angle α with x-direction.

$m_{u\alpha t}$ = ultimate torsional moment of resistance per unit length of yield line.

Fig. 20.5 (c) shows the moment vectors. Taking moments about the side ce (parallel to the yield line),

$$m_{u\alpha n}\,(ce) = m_{ux}\,(cd)\cos\alpha + m_{uy}\,(de)\sin\alpha$$

or

$$m_{u\alpha n} = m_{ux}\left(\frac{cd}{ce}\right)\cos\alpha + m_{uy}\left(\frac{de}{ce}\right)\sin\alpha = m_{ux}\cos^2\alpha + m_{uy}\sin^2\alpha \quad \ldots(20.1)$$

If the slab is *isotropically reinforced* (*i.e.* equal reinforcement in the two directions), $m_{ux} = m_{uy} = m_u$. Hence,

$$m_{u\alpha n} = m_u(\cos^2\alpha + \sin^2\alpha) = m_u \quad \ldots(20.2)$$

Thus the ultimate moment of resistance in an isotropically reinforced slab, in any direction, is the same.

If the reinforcement in the two directions is not the same, it is said to be *orthotropically reinforced*. In such a case, let $m_{ux} = m_u$ and $m_{uy} = \mu m_u$ Then from Eq 20.2, we get

$$m_{u\alpha n} = m_u\cos^2\alpha + \mu m_u \sin^2\alpha, \quad \text{or} \quad m_{u\alpha n} = m_u(\cos^2\alpha + \mu\sin^2\alpha) \quad \ldots(20.3)$$

If $\alpha = 45°\ ,\ \cos^2 \alpha = \sin^2 \alpha = \dfrac{1}{2}$

$$\therefore \quad m_{u\alpha n} = \frac{m_u (1 + \mu)}{2} \qquad ...(20.4)$$

Again, considering the moment vectors of Fig. 20.5 (*c*) and taking moments about an axis perpendicular to *ce*, to get torsional moment per unit length of yield line as :

$$m_{u\alpha t} = (m_{ux} - m_{uy}) \sin \alpha \cos \alpha \qquad ...(20.5)$$

For an isotropically reinforced slab, $m_{ux} = m_{uy}$, and hence torsional moment is zero.

For an orthotropically reinforced slab,

$$m_{u\alpha t} = m_u (1 - \mu) \sin \alpha \cos \alpha \qquad ...(20.6)$$

20.4. ULTIMATE LOAD ON SLABS

There are two methods of determining the ultimate load capacity of reinforced concrete slabs.

1. Virtual work method. 2. Equilibrium method.

The *virtual work method* is based on the principle that the work done by the external forces in undergoing a small virtual displacement is equal to the internal work done (or energy dissipated) in rotation along the yield lines. The elastic deformations are however ignored in comparision to the plastic deformations that take place along the yield lines. In the equilibrium method, the collapse load is calculated from the equilibrium of the individual segments of a mechanism.

Both the virtual work method as well as the equilibrium method give the upper bound solution for the collapse load, *i.e.* the computed collapse load on the basis of an assumed yield line patterns is bound to be larger than the actual collapse load. Hence it is essential to examine several yield line patterns and a minimum value of collapse load has to be obtained. Test results have shown that the actual failure load of slab is greater than the one predicted by the upper bound solution of the yield line analysis, because of membrane action. Hence the results of yield line analysis can be used in design with reasonable degree of safety.

20.5. ANALYSIS BY VIRTUAL WORK METHOD

The principle of *virtual work* states that :

'*If a deformable structure in equilibrium under the action of a system of external forces is subjected to a virtual deformation compatible with its condition of support, the work done by these forces on the displacements associated with the virtual deformation is equal to the work done by the internal stresses on the strains associated with this deformation*'.

Thus, the work done during small motion of collapse mechanism is equal to work absorbed by the plastic hinges formed along the yield lines. The segments of the slab within the yield lines go through rigid body displacements with the collapse load acting on the structure. Thus,

Work done by external forces = work done by internal forces (or energy absorbed by hinges)

or

$$W_E = W_I \qquad ...(20.7)$$

Where W_E = work done by external forces

W_I = Energy absorbed by plastic hinges in undergoing corresponding rotations.

If w_u is the uniformly distributed external load, we have

$$W_E = \int\int w_u \, \delta_{x,y} \, dx \, dy = \Sigma \, W_u \, \Delta \quad \ldots(20.8)$$

where $\delta_{x,y}$ = virtual displacement at any point.

W_u = resultant load on each segment,

Δ = corresponding displacement at the centroid of load in each segment.

If $m_{u\alpha n}$ = ultimate moment across an yield line.

l_0 = length of yield line.

and θ_n = relative rotation of two adjacent plates, perpendicular to the yield line.

Then,

$$W_I = \Sigma \, m_{u\alpha n} \, \theta_n \, l_0 \quad \ldots(20.9)$$

Equating W_E and W_I, we get

$$\Sigma \, W_u \, \Delta = \Sigma \, m_{u\alpha n} \, \theta_n \, l_0 \quad \ldots(20.10)$$

It is to be noted that the work done by the twisting moment ($m_{u\alpha t}$) has not been considered in Eq. 20.9 since its effect will cancel out when the entire slab is eventually considered.

We shall now illustrate the method by considering several cases.

1. Simply Supported Square Slab : A square slab is isotropically reinforced. Hence moment of resistance $m_{u\alpha n}$ is the same in all the directions. Fig. 20.6 shows such a slab of span L, having length of diagonal equal to $\sqrt{2}\,L$. If δ is the virtual displacement at the middle, the total rotation of the diagonal segments $= 2\,\theta_n = 2\dfrac{\delta}{\frac{\sqrt{2}\,L}{2}} = \dfrac{2\sqrt{2}\,\delta}{L}$

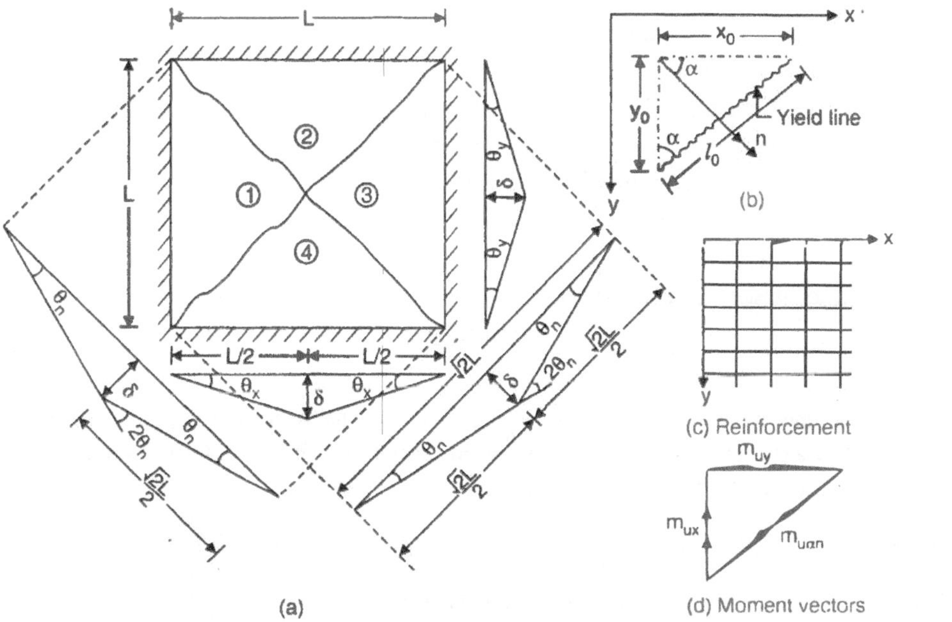

FIG. 20.6. SQUARE SLAB : FREELY SUPPORTED

Length of each diagonal yield line $= \sqrt{2}\, L$

Hence for the two diagonal yield lines,

$$W_I = \Sigma\, m_{u\alpha n}\, \theta_n\, l_0 = 2\left(m_u \frac{2\sqrt{2}\,\delta}{L} \sqrt{2}\, L \right) = 8\, m_u\, \delta \qquad ..(1)$$

There are four segments, each segment carrying equal total load $W_u = w_u \frac{1}{2} L \frac{L}{2} = w_u \frac{L^2}{4}$ acting through its centroid which undergoes a virtual displacement $\Delta = \frac{\delta}{3}$. Hence

$$W_E = \Sigma\, W_u\, \Delta = 4\left(\frac{w_u L^2}{4} \frac{\delta}{3} \right) = w_u L^2 \frac{\delta}{3} \qquad ...(2)$$

Equating W_E and W_I, we have

$$w_u L^2 \frac{\delta}{3} = 8\, m_u\, \delta \quad \text{or} \quad w_u = \frac{24\, m_u}{L^2} \qquad ...(20.11)$$

or

$$m_u = \frac{w_u L^2}{24} \qquad ...[20.11(a)]$$

Alternative Solution

The problem can be solved alternatively by combination of components of moments and rotations in the co-ordinate directions x and y. Refer Fig. 20.6 (*b*).

From Eq. 20.1, we have

$$m_{u\alpha n} = m_{ux} \cos^2 \alpha + m_{uy} \sin^2 \alpha.$$

Hence $\quad \Sigma\, m_{u\alpha n}\, \theta_n\, l_0 = \Sigma(m_{ux} \cos^2 \alpha + m_{uy} \sin^2 \alpha)\, \theta_n\, l_0$

or $\quad \Sigma\, m_{u\alpha n}\, \theta_n\, l_0 = \Sigma\, m_{ux}\, y_0 \cos \alpha\, .\theta_n + \Sigma\, m_{uy}\, x_0 \sin \alpha\, .\theta_n$

(since $\cos \alpha = y_0/l_0$ and $\sin \alpha = x_0/l_0$)

Hence $\quad \Sigma\, m_{u\alpha n}\, \theta_n\, l_0 = \Sigma\, m_{ux}\, y_0\, \theta_x + \Sigma\, m_{uy}\, x_0\, \theta_y \qquad ...(20.12)$

where θ_x = Component of θ_n in x direction.

θ_y = Component of θ_n in y dirction.

x_0 = Projected length of yield line in x direction.

y_0 = Projected length of yield line in y direction.

It is to be noted that θ_x and θ_y are the sum of the rotations of segments adjacent to the yield line under consideration.

Substituting the value of $\Sigma\, m_{u\alpha n}\, \theta_n\, l_0$ in Eq 20.10, we get

$$\Sigma\, m_{ux}\, y_0\, \theta_x + \Sigma\, m_{uy}\, x_0\, \theta_y = \Sigma\, W_u\, \Delta \qquad ....(20.13)$$

Now, for the square slab of Fig. 20.6 (*a*), we have four segments. Segment 1 and 3 rotate about y-axis and there is no rotation about x-axis. Similarly, segments 2 and 4 rotate about x-axis and there is no rotation about y-axis.

Rotation $\quad \theta_x = \theta_y = \frac{\delta}{L/2} = \frac{2\,\delta}{L}$

Also, $\quad x_0 = y_0 = L$, and $m_{ux} = m_{uy} = m_u$

$\therefore$ For segment 1, $\quad m_{ux}\, y_0\, \theta_x + m_{uy}\, x_0\, \theta_y = \left(m_u L \frac{2\,\delta}{L} \right) + 0$

For segment 2, $m_{ux}\, y_0\, \theta_x + m_{uy}\, x_0\, \theta_y = 0 + m_u\, L \dfrac{2\,\delta}{L}$

For segment 3, $m_{ux}\, y_0\, \theta_x + m_{uy}\, x_0\, \theta_y = m_u\, L \dfrac{2\,\delta}{L} + 0$

and for segment 4, $m_{ux}\, y_0\, \theta_x + \Sigma\, m_{uy}\, \theta_y\, x_0 = 0 + m_u\, L \dfrac{2\,\delta}{L}$

$\therefore \quad \Sigma\, m_{ux}\, \theta_x\, y_0 + \Sigma\, m_{uy}\, \theta_y\, x_o = 4\, m_u\, L \dfrac{2\,\delta}{L} = 8\, m_u\, \delta,$ which is the same as found earlier.

$$\Sigma\, W_u\, \Delta = 4\left(\frac{w_u\, L^2}{4}\right)\frac{\delta}{3} = w_u\, L^2\, \frac{\delta}{3} \qquad \therefore \quad 8\, m_u\, \delta = w_u\, L^2\, \frac{\delta}{3}$$

or $\quad m_u = \dfrac{w_u\, L^2}{24} \quad$ or $\quad w_u = \dfrac{24\, m_u}{L^2}$, as before.

2. Square Slab Fixed on all Edges : Fig. 20.7 shows an isotropically reinforced square slab fixed at the four supports. Due to this, negative yield lines will also be formed along the fixed edges. The rotation $\theta_x = \theta_y$ of these negative yield lines will be equal to $\dfrac{\delta}{L/2}$. The length of each negative yield line $= L$. Hence internal work done by negative yield lines AB, BC, CD and DA is

$$\Sigma\, m_{uan}\, \theta_n\, l_0 = 4\, m_u\, L \frac{\delta}{L/2} = 8\, m_u\, \delta \qquad \ldots(i)$$

Internal work done along positive yield lines along the diagonals AC and BD, as found in case 1, is

$$\Sigma\, m_{uan}\, \theta_n\, l_0 = 8\, m_u\, \delta \qquad \ldots(ii)$$

Hence $\quad W_I = 8\, m_u\, \delta + 8\, m_u\, \delta = 16\, m_u\, \delta \qquad \ldots(a)$

It is to be noted that the energy absorbed by both negative yield line as well as positive yield line has positive sign.

Again, as found in the previous case,

$$W_E = \Sigma\, W_u\, \Delta = w_u\, L^2\, \frac{\delta}{3} \qquad \ldots(b)$$

$$16\, m_u\, \delta = w_u\, L^2\, \frac{\delta}{3}$$

or $$m_u = \frac{w_u\, L^2}{48} \qquad \ldots[20.14(a)]$$

or $$w_u = \frac{48\, m_u}{L^2} \qquad \ldots(20.14)$$

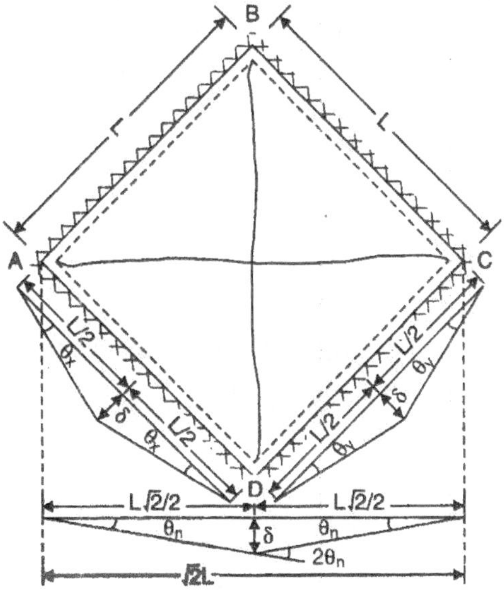

FIG. 20.7. SQUARE SLAB : FIXED EDGES.

3. Equilateral Triangular Slab Isotropically Reinforced Fig. 20.8 shows isotropically reinforced triangular slab. The ultimate moment of resistance m_u will be same in all directions. When it is simply supported along all the three edges, the three yield lines will divide the slab in three symmetrically placed sectors. Hence we need consider only one sector. The distance of point O from edge AB will be $\dfrac{L}{2\sqrt{3}}$.

For sector 1, $W_I = m_u \theta l_0 = m_u \dfrac{\delta}{L/2\sqrt{3}} L$

$= 2\sqrt{3}\, m_u \delta$ ...(1)

Also, $W_E = \dfrac{1}{2} L \times \dfrac{L}{2\sqrt{3}} \times w_u \times \dfrac{\delta}{3} = w_u \dfrac{\delta L^2}{12\sqrt{3}}$ ...(2)

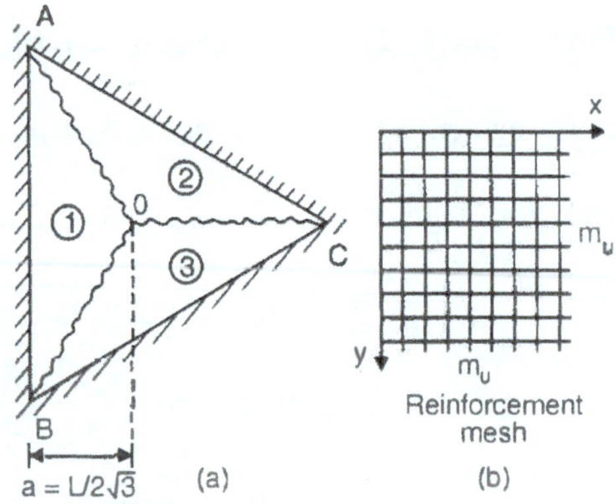

FIG. 20.8.

Equating the two, we get

$$\frac{w_u \delta L^2}{12\sqrt{3}} = 2\sqrt{3}\, m_u \delta$$

$$\therefore \quad w_u = \frac{72\, m_u}{L^2} \quad \text{...[20.15(a)]}$$

$$\text{or} \quad m_u = \frac{w_u L^2}{72} \quad \text{...(20.15)}$$

4. Generalised Triangular Slab : Let us now take the general case of a triangular slab, simply supported along two edges, third edge being free, and isotropically reinforced.

Let the side BA of length L be oriented along y-axis, and the other side BC be of length αL. Edge AC is free. Let $\angle ABC$ be β and the yield line BD make an angle ϕ with BA (or y-axis). Let the point D, having coordinates (x_d, y_d) have a downward virtual displacement δ. Let us have proper identification of rotations. Also, $m_{ux} = m_{uy} = m_u$.

For segment ABD (segment 1),

$y_{01} = y_0$

$x_{01} = x_d \quad = y_0 \tan\phi$

$\theta_{x1} = \dfrac{\delta}{x_d} = \dfrac{\delta}{y_0 \tan\phi}$

$\theta_{y1} = 0$

For segment BCD (segment 2)

$\theta_{x2} = \dfrac{\delta}{DE} = \dfrac{\delta}{y_0 \tan\beta - y_0 \tan\phi}$

$\theta_{y2} = \dfrac{\delta}{FD} = \dfrac{\delta}{y_0 - x_d \cot\beta} = \dfrac{\delta}{y_0 - y_0 \tan\phi \cot\beta}$

$y_{02} = y_0$

$x_{02} = x_d = y_0 \tan\phi$

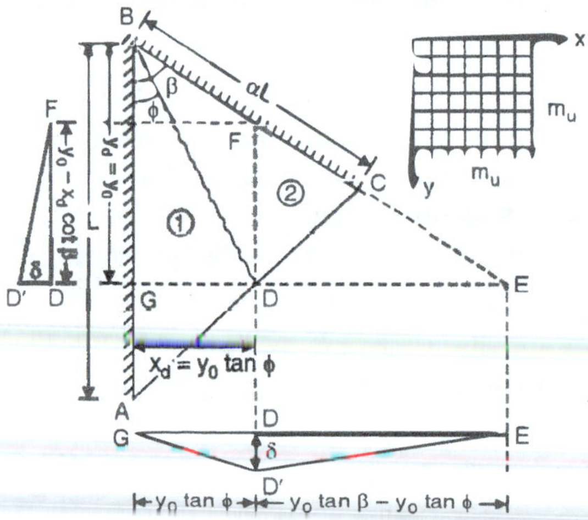

FIG. 20.9.

Now from Eq. 20.12

$W_I = \Sigma\, m_{ux}\, y_0\, \theta_x + \Sigma\, m_{uy}\, x_0\, \theta_y$

$W_I = [m_u\, y_0\, \theta_{x1} + m_u\, y_0\, \theta_{x2}] + [m_u\, x_{01}\, \theta_{y1} + m_u\, x_{02}\, \theta_{y2}]$

$$= m_u \left[y_0 \frac{\delta}{y_0 \tan\phi} + y_0 \frac{\delta}{y_0 (\tan\beta - \tan\phi)} \right] + m_u \left[y_0 \tan\phi \times \text{zero} + y_0 \tan\phi \frac{\delta}{y_0 (1 - \tan\phi \cot\beta)} \right]$$

$$= m_u \left[\frac{\delta}{\tan\phi} + \frac{\delta}{\tan\beta - \tan\phi} + 0 + \frac{\delta \tan\phi}{1 - \tan\phi \cot\beta} \right]$$

$$= m_u\,\delta\left[\cot\phi + \frac{1}{\tan\beta - \tan\phi} + \frac{\tan\phi\,\tan\beta}{\tan\beta - \tan\phi}\right]$$

or $$W_I = m_u\,\delta\left[\cot\phi + \frac{1+\tan\beta\,\tan\phi}{\tan\beta - \tan\phi}\right] = m_u\delta\,[\cot\phi + \cot(\beta-\phi)] \quad \ldots(1)$$

Also, $$W_E = \Sigma\, W_u\,\Delta = w_u \times \text{area of segments} \times \frac{1}{3}\,\delta$$

or $$W_E = w_u\,\frac{1}{2}\,L\,(\alpha\,L\sin\beta)\,\frac{\delta}{3} = w_u\,\delta\,\alpha\,L^2\,\frac{\sin\beta}{6} \quad \ldots(2)$$

Now $W_I = W_E$. $\therefore\quad m_u\,\delta\,[\cot\phi + \cot(\beta-\phi)] = w_u\,\delta\,\alpha\,L^2\,\dfrac{\sin\beta}{6}$

or $$w_u = \frac{6\,m_u}{\alpha L^2}\cdot\frac{\cot\phi + \cot(\beta-\phi)}{\sin\beta} = \frac{6\,m_u}{\alpha\,L^2}\,\frac{1}{\sin\phi\,\sin(\beta-\phi)} \quad \ldots(20.16)$$

For w_u to be minimum, $\dfrac{\partial\,w_u}{\partial\,\phi} = 0$

This gives $\cos\phi\,\sin(\beta-\phi) - \sin\phi\,\cos(\beta-\phi) = 0$

or $\tan\phi = \tan(\beta-\phi)$. $\quad\therefore\quad \phi = \beta/2$

Hence for minimum yield load, the yield line must bisect the angle β.

Substituting $\phi = \beta/2$ in Eq. 20.16, we get

$$w_u = \frac{6\,m_u}{\alpha\,L^2}\,\frac{1}{\sin\beta/2\,\sin\beta/2} = \frac{6\,m_u}{\alpha\,L^2\,\sin^2\beta/2} \quad \ldots[20.17(a)]$$

or $$m_u = \frac{w_u\,\alpha\,L^2}{6}\,\sin^2\beta/2 \quad \ldots(20.17)$$

For an equilateral triangle, $\alpha = 1$ and $\beta = 60°$.

$$m_u = \frac{w_u\,L^2}{6}\,\sin^2 30° = \frac{w_u\,L^2}{24} \quad \ldots(20.18)$$

5. Rectangular Slab, Simply Supported : Fig. 20.10 shows rectangular slab, orthotropically reinforced, and simply supported along all the four edges. Let the ultimate moment capacity in the short span direction be m_u and that in the long span direction be μm_u. The yield line pattern will be symmetrical, with point E at a distance $\beta\,L$ from AD, where β is an unknown. Let us give a virtual displacement δ to the yield line EF.

Let L = Long span

$\alpha\,L$ = Short span

θ_x = Rotation in the x-direction $= \dfrac{\delta}{\beta\,L}$;

θ_y = Rotation in the y-direction $= \dfrac{\delta}{\alpha\,L/2} = \dfrac{2\,\delta}{\alpha\,L}$

$m_{ux} = \mu\,m_u$; $\quad m_{uy} = m_u$

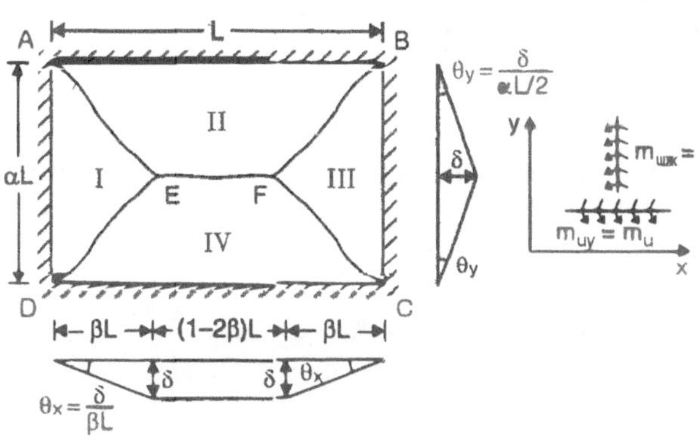

FIG. 20.10

For segment *AED* $m_{ux} y_0 \theta_x + m_{uy} x_0 \theta_y = (\mu\, m_u)(\alpha\, L)\left[\dfrac{\delta}{\beta\, L}\right] + 0$

For segment ABFE, $m_{ux} y_0 \theta_x + m_{uy} x_0 \theta_y = 0 + m_u L \dfrac{2\,\delta}{\alpha\, L}$

For segment *BCF,* $m_{ux} y_0 \theta_x + m_{uy} x_0 \theta_y = (\mu m_u)(\alpha\, L)\left(\dfrac{\delta}{\beta\, L}\right) + 0$

For segment *CDEF,* $m_{ux} y_0 \theta_x + m_{uy} x_0 \theta_y = 0 + m_u L \dfrac{2\,\delta}{\alpha\, L}$

$$\therefore \quad \Sigma\, m_{ux} y_0 \theta_x + \Sigma\, m_{uy} x_0 \theta_y = \frac{2\,\mu\, m_u\, \alpha\, L\,\delta}{\beta\, L} + 2 m_u L \frac{2\,\delta}{\alpha\, L} = 2\,\delta\left[\frac{2\, m_u}{\alpha} + \frac{\alpha\,\mu\, m_u}{\beta}\right] \quad \text{...(1)}$$

Work done by external load

For *AED,* $W_u\,\Delta = \dfrac{1}{2}\alpha\, L\,\beta\, L\, w_u \dfrac{\delta}{3} = \alpha\,\beta\, L^2 \dfrac{\delta}{6} w_u$

For *ABEF,* $W_u\,\Delta = 2\left\{\dfrac{1}{2}\beta\, L\dfrac{\alpha\, L}{2} w_u \dfrac{\delta}{3}\right\} + (1 - 2\,\beta)\, L\dfrac{\alpha\, L}{2} w_u \dfrac{\delta}{2}$

$$= \alpha\,\beta\, L^2 \frac{\delta}{6} w_u + (1 - 2\,\beta)\,\alpha\, L^2 \frac{\delta}{4} w_u$$

For *BCF* , $W_u\,\Delta = \alpha\,\beta\, L^2 \dfrac{\delta}{6} w_u.$

and for *CDEF* , $W_u\,\Delta = \alpha\,\beta\, L^2 \dfrac{\delta}{6} w_u + (1 - 2\,\beta)\,\alpha\, L^2 \dfrac{\delta}{4} w_u$

$$\therefore \quad \Sigma\, W_u\,\Delta = 2\left[\alpha\beta L^2 \frac{\delta}{6} w_u\right] + 2\left[\alpha\beta L^2 \frac{\delta}{6} w_u + (1 - 2\,\beta)\,\alpha L^2 \frac{\delta}{4} w_u\right]$$

$$= \delta\, w_u\, L^2\left[\frac{\alpha\,\beta}{3} + \frac{\alpha\,\beta}{3} + \frac{\alpha}{2} - \alpha\beta\right] = \delta\, w_u\, L^2\left[\frac{\alpha}{2} - \frac{\alpha\,\beta}{3}\right] = \delta\, w_u\, L^2 \alpha\left(\frac{3 - 2\beta}{6}\right) \quad \text{...(2)}$$

Equating internal and external work, we get

$$\alpha\,\delta\, w_u\, L^2\left(\frac{3 - 2\beta}{6}\right) = 2\,\delta\left[\frac{2 m_u}{\alpha} + \frac{\alpha\,\mu\, m_u}{\beta}\right]$$

or $\quad \alpha\, w_u\, L^2\left(\dfrac{3 - 2\,\beta}{6}\right) = 2 m_u\left(\dfrac{2}{\alpha} + \dfrac{\alpha\mu}{\beta}\right) = 2 m_u\left(\dfrac{2\beta + \alpha^2\,\mu}{\alpha\,\beta}\right)$

$$\therefore \quad m_u = \frac{\alpha\, L^2\,(3 - 2\,\beta)}{6} \times \frac{\alpha\,\beta}{2\,(2\,\beta + \alpha^2\,\mu)} \times w_u = \frac{w_u\,\alpha^2\, L^2}{12}\left[\frac{3\,\beta - 2\,\beta^2}{2\,\beta + \mu\,\alpha^2}\right] \quad \text{...(20.19)}$$

In the above equation, β is variable. For a maximum value of m_u (or minimum value of w_u), we should take the derivative with respect to β and equate it to zero.

In general, the work equation has the form

$$m_u = w_u\left[\frac{f_1\,(x_1\,,\ x_2)}{f_2\,(x_1\,,\ x_2)}\right]$$

For a maximum value of m_u , $\dfrac{\partial m_u}{\partial x_1} = 0$. By differential process,

$$\frac{\partial\, m_u}{\partial x_1} = \frac{\partial}{\partial x_1}\left[\frac{f_1\,(x_1\,,\, x_2)}{f_2\,(x_1\,,\, x_2)}\right] = \frac{f_2(x_1\,,\, x_2)\dfrac{\partial}{\partial\, x_1}\,[f_1\,(x_1\,,\, x_2)] - f_1(x_1\,,\, x_2)\dfrac{\partial}{\partial\, x_1}\,[f_2(x_1\,,\, x_2)]}{[f_2\,(x_1\,,\, x_2)]^2}$$

Equating this to zero, we get

$$f_2(x_1, x_2)\frac{\partial}{\partial x_1}[f_1(x_1, x_2)] - f_1(x_1, x_2)\frac{\partial}{\partial x_1} f_2[(x_1, x_2)] = 0$$

or
$$\frac{f_1(x_1, x_2)}{f_2(x_1, x_2)} = \frac{\frac{\partial}{\partial x_1}[f_1(x_1, x_2)]}{\frac{\partial}{\partial x_1}[f_2(x_1, x_2)]} \quad \text{...(20.20)}$$

Eq. 20.20 is a very important governing equation for maximum value of m_u. Applying this to our present case, we get

$$\frac{3\beta - 2\beta^2}{2\beta + \mu\alpha^2} = \frac{3 - 4\beta}{2}$$

or
$$6\beta - 4\beta^2 = 6\beta + 3\mu\alpha^2 - 8\beta^2 - 4\mu\beta\alpha^2$$

or
$4\beta^2 + 4\mu\alpha^2\beta - 3\mu\alpha^2 = 0$. Solution of the above quadratic equation, we get,

$$\beta = \frac{1}{2}\left[\sqrt{\mu^2\alpha^4 + 3\mu\alpha^2} - \mu\alpha^2\right] \quad \text{...(20.21)}$$

Substituting this value of β in Eq 20.19, we get

$$m_u = \frac{1}{12} w_u\alpha^2 L^2 \frac{\frac{3}{2}\left\{\sqrt{\mu^2\alpha^4 + 3\mu\alpha^2} - \mu\alpha^2\right\} - \frac{2}{4}\left(\mu^2\alpha^4 + 3\mu\alpha^2 + \mu^2\alpha^4 - 2\mu\alpha^2\sqrt{\mu^2\alpha^4 + 3\mu\alpha^2}\right)}{\frac{2}{2}\left\{\sqrt{\mu^2\alpha^4 + 3\mu\alpha^2} - \mu\alpha^2\right\} + \mu\alpha^2}$$

Simplifying the above, we get

$$m_u = \frac{w_u\alpha^2 L^2}{24}\left[\sqrt{3 + \mu\alpha^2} - \alpha\sqrt{\mu}\right]^2 \quad \text{...(20.22)}$$

Special cases :

(*i*) *For an isotropically reinforced* square slab,

$\alpha = 1$ and $\mu = 1$

$$\therefore \quad m_u = \frac{w_u\alpha^2 L^2}{24}\left[\sqrt{3 + 1} - 1\sqrt{1}\right]^2 = \frac{w_u L^2}{24} \quad \text{...(20.23)}$$

which is the same as Eq 20.11 (*a*)

(*ii*) *For an isotropically reinforced rectangular slab*,

$\mu = 1$

$$m_u = \frac{w_u\alpha^2 L^2}{24}\left[\sqrt{3 + \alpha^2} - \alpha\right]^2 \quad \text{..(20.24)}$$

6. Polygonal and Circular Slabs : Let us first consider polygonal slab, fixed along its edges, with isotropic positive moments m_u and isotropic negative moment m_u'. Let the length of each side be L, and let the perpendicular distance of each side from centre be r, which is also the radius of the inscribed circle. The positive yield lines will be along the radial lines as shown in Fig. 20.11 (*a*). The negative yield lines will be along the outer edges.

Let us give a virtual displacement δ at the apex O.

For any sector *ABO*,

$$M_I = (m_u + m_u')\, L\, \frac{\delta}{r} \qquad \ldots(1)$$

$$M_E = \frac{1}{2}\, L\, r\, \frac{\delta}{3}\, w_u \qquad \ldots(2)$$

Equating the two, we get

$$w_u\, L\, r\, \frac{\delta}{6} = (m_u + m_u')\, L\, \frac{\delta}{r}$$

or $$w_u = \frac{6\,(m_u + m_u')}{r^2} \qquad \ldots(20.25)$$

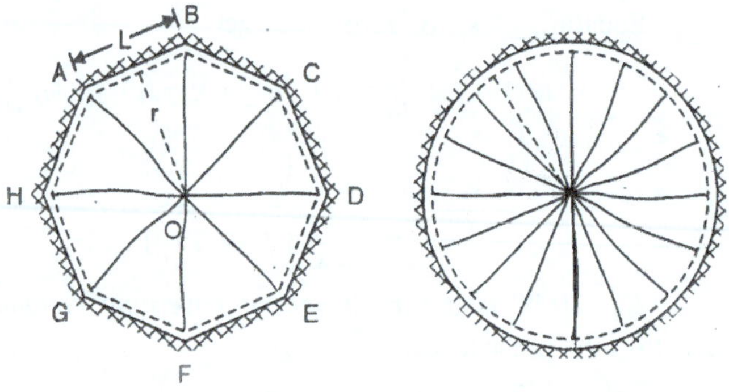

(a) Polygonal slab (b) Circular slab

FIG. 20.11

For a circular slab, there will be innumerable number of positive yield lines, such that number of sides n will tend to be infinity and L will tend to be zero. However, r will be the radius of the circle, and for any sector ABO,

$$w_u = \frac{6\,(m_u + m_u')}{r^2}$$

Thus, Equation 20.25 is applicable to both polygonal as well as circular slabs.

For circular slabs simply supported along the edge, $m_u' = 0$

$$\therefore \qquad w_u = \frac{6\, m_u}{r^2} \qquad \ldots[20.26]$$

or $$m_u = \frac{w_u\, r^2}{6} \qquad \ldots[20.26(a)]$$

For hexagonal slab, $r = \sqrt{3}\, L/2$

Hence from Eq 20.25, $w_u = \dfrac{6\,(m_u + m_u')}{(\sqrt{3}L/2)^2} = \dfrac{8\,(m_u + m_u')}{L^2}$

If the hexagonal slab is simply supported.

$$w_u = \frac{8\, m_u}{L^2} \qquad \ldots(20.28)$$

or $$m_u = \frac{w_u\, L^2}{8} \qquad \ldots[20.27(a)]$$

20.6. ANALYSIS BY EQUILIBRIUM METHOD

1. Isotropically Reinforced Square Slab : Fig. 20.12(a) shows a square slab, simply supported along the edges. m_u will be the same along all the directions. The yield lines will be along the two diagonals. Considering the equilibrium of the triangular sector ABO, and taking moments about edge AB, we get

$$m_u\, L = \frac{1}{2}\, L \cdot \frac{L}{2}\; w_u\; \frac{L}{6}$$

or $$m_u = \frac{w_u\, L^2}{24} \qquad \ldots[20.11(a)]$$

This is the same as Eq. 20.11 (a).

Again, for a fixed slab of Fig. 20.12 (*b*), positive yield lines will be along the diagonals, while the negative yield lines will be developed along the fixed edges. Considering the equilibrium of sector *ABO*, and taking moments about edge, *AB*, we get

$$m_u L + m_u L = \frac{1}{2} L \frac{L}{2} w_u \frac{L}{6}$$

or $$m_u = \frac{w_u L^2}{48} \qquad \ldots[20.14(a)]$$

This is the same as Eq. 20.14(*a*).

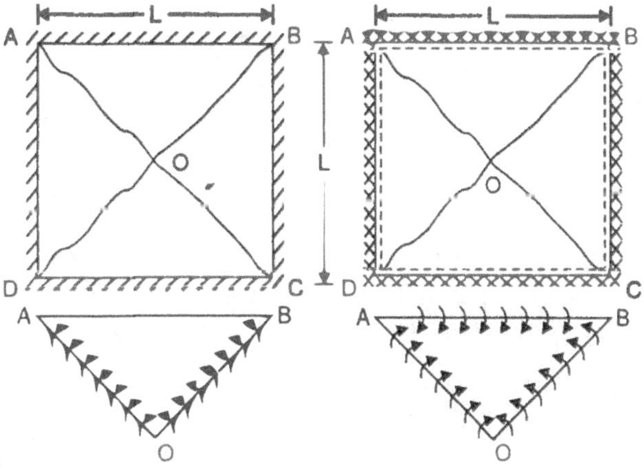

(a) Simply supported (b) Fixed

FIG. 20.12. SQUARE SLABS.

2. Orthotropically Reinforced Rectangular Slab : Fig. 20.13 (*a*) shows a rectangular slab, with length of long span equal to L and that of short edge as αL. The yield line pattern will be symmetrical with point E at a distance βL from AD, where β is unknown. Let the slab be orthotropically reinforced with $m_{uy} = m_u$ and $m_{ux} = \mu m_u$.

Consider equilibrium of sector *ABFE* [Fig. 20.13 (*b*)]. Taking moments about edge *AB*, we get.

$$m_u L = w_u \left[\left\{ (1 - 2\beta) L \frac{\alpha L}{2} \frac{\alpha L}{4} \right\} + 2 \left\{ \frac{1}{2} \beta L \frac{\alpha L}{2} \frac{\alpha L}{6} \right\} \right]$$

or $$m_u = w_u \alpha^2 L^2 \left[\frac{(1 - 2\beta)}{8} + \frac{\beta}{12} \right] = \frac{w_u \alpha^2 L^2}{24} (3 - 4\beta) \qquad \ldots(1)$$

Similarly, consider equilibrium of sector *FBC*. Taking moments about edge *BC*, we get

$$\mu m_u \alpha L = \frac{1}{2} \alpha L \beta L \frac{\beta L}{3} w_u = \frac{\alpha \beta^2 L^3}{6} w_u$$

or $$m_u = \frac{\beta^2 L^2}{6\mu} w_u \qquad \ldots(2)$$

Equating the two equilibrium equations, we get

$$\frac{w_u \alpha^2 L^2}{24} (3 - 4\beta) = \frac{\beta^2 L^2}{6\mu} w_u$$

or $$\mu \alpha^2 (3 - 4\beta) = 4\beta^2$$

or $$4\beta^2 + 4\mu\alpha^2\beta - 3\mu\alpha^2 = 0$$

$$\therefore \quad \beta = \frac{1}{2}\left[\sqrt{\mu^2 \alpha^4 + 3\mu\alpha^2} - \mu\alpha^2\right]$$

This is the same as Eq. 20.21

Substituting this value of β in Eq. (2), we get

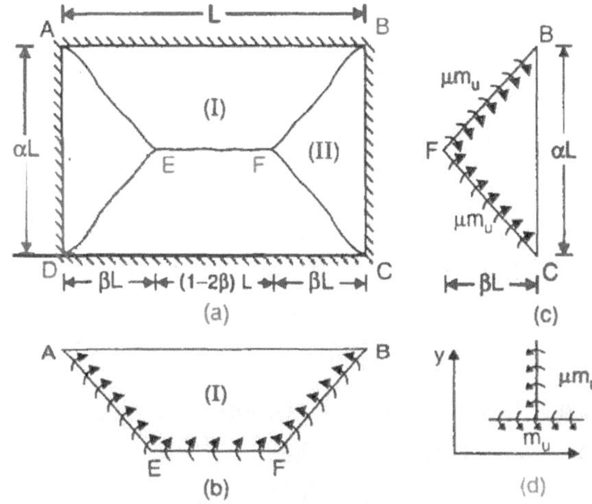

FIG. 20.13. RECTANGULAR SLAB.

$$m_u = \frac{w_u L^2}{6\mu} \times \frac{1}{4}\left[\sqrt{\mu^2\alpha^4 + 3\mu\alpha^2} - \mu\alpha^2\right]^2$$

or
$$m_u = \frac{w_u L^2}{24\mu} \times \mu\alpha^2\left[\sqrt{3+\mu\alpha^2} - \alpha\sqrt{\mu}\right]^2 = \frac{w_u \alpha^2 L^2}{24}\left[\sqrt{3+\mu\alpha^2} - \alpha\sqrt{\mu}\right]^2$$

This is the same as Eq. 20.22

3. Isotropically Reinforced Hexagonal Slab : Consider the equilibrium of sector *ABO*, and take moments about *AB*. The value of perpendicular $r = \frac{\sqrt{3}\,L}{2}$

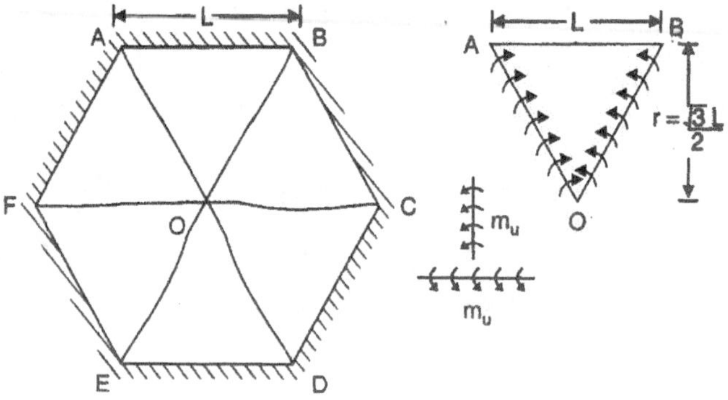

FIG. 20.14

$$\therefore \quad m_u L = \frac{1}{2} L\, r\, \frac{r}{3} w_u$$

or
$$m_u = \frac{w_u}{6} r^2 = \frac{w_u}{6}\left(\frac{\sqrt{3}\,L}{2}\right)^2$$

$$\therefore \quad m_u = \frac{w_u L^2}{8}$$

This is the same as Eq. 20.27 (*a*)

Example 20.1. *A square slab of side length 4 m is simply supported at the ends and carries a service live load of 3 kN/m². Design the slab. Use M 20 concrete and Fe 415 steel bars.*

Solution. Given : $L = 4$ m ; $f_{ck} = 20$ N/mm² ; $f_y = 415$ N/mm².

For Fe 415 steel, $\frac{x_{u,\,max}}{d} = \frac{700}{1100 + 0.87 \times 415} = 0.479$

$$\therefore \quad R_u = 0.36 f_{ck} \frac{x_{u,\,max}}{d}\left(1 - 0.416 \frac{x_{u,\,max}}{d}\right) = 0.36 \times 20 \times 0.479\,(1 - 0.416 \times 0.479)$$
$$= 2.761$$

Minimum effective depth $= \frac{\text{span}}{35} = \frac{4000}{35} = 114$ mm

Let us provide an overall depth = 130 mm.

Using 15 mm clear cover and 10 mm dia. bars, average $d = 130 - 15 - 10 = 105$ mm.

Self weight of slab = 0.13 × 25000	= 3250 N/m²
Floor finishes 50 mm thick = 0.05 × 22000	= 1100 N/m² (say)
Live load	= 3000 N/m²
Total service load	= 7350 N/m².

Load factor = 1.5. Ultimate design load $w_u = 1.5 \times 7350 = 11025$ N/m²

∴ Ultimate capacity of the slab is given by Eq. 20.11 (*a*)

$$m_u = \frac{w_u L^2}{24} = \frac{11025\,(4)^2}{24} = 7350 \text{ N-m} = 7.35 \times 10^6 \text{ N-mm}$$

The limiting or balancing moment capacity of the slab is given by

$$M_{u.lim} = R_u bd^2 = 2.761\ (1000)\ (105)^2 = 30.44 \times 10^6 \text{ N-mm}$$

Since $m_u < M_{u,lim}$, the slab is under-reinforced. The reinforcement in the slab is given by

$$M_u = m_u = 0.87\, f_y\, A_{st}\, d \left[1 - \frac{A_{st} f_y}{b\, d\, f_{ck}} \right]$$

$$\therefore \quad 7.35 \times 10^6 = 0.87 \times 415\, A_{st} \times 105 \left[1 - \frac{A_{st} \times 415}{1000 \times 105 \times 20} \right]$$

$A_{st}^2 - 5060\, A_{st} + 0.981 \times 10^6 = 0$, which gives $A_{st} = 202 \text{ mm}^2$

Alternatively, from Eq. 3.13 applicable for under-reinforced sections,

$$A_{st} = \frac{0.5 f_{ck}}{f_y} \left[1 - \sqrt{1 - \frac{4.6\, M_u}{f_{ck}\, b\, d^2}} \right] b\, d = \frac{0.5 \times 20}{415} \left[1 - \sqrt{1 - \frac{4.6 \times 7.35 \times 10^6}{20 \times 1000\ (105)^2}} \right] 1000 \times 105$$

$= 202 \text{ mm}^2$

Minimum steel area $= \frac{0.12}{100} \times 1000 \times 130 = 156 \text{ mm}^2$. Using 8 mm dia. bars, spacing $= \frac{1000 \times 50}{202}$ $= 247$ mm. Hence use 8 mm dia. bars @ 240 c/c both ways.

Example 20.2. *A reinforced concrete slab 5 m × 5 m is simply supported along the four edges and is reinforced with 10 mm dia. Fe 415 steel bars at 150 mm c/c both ways. The average effective depth of the slab is 100 mm and the overall depth of the slab is 130 mm. The slab carries a flooring of 50 mm thick having unit weight of 2.2 kN/m². Determine the maximum permissible service load, if M 20 concrete is used.*

Solution. $\quad A_{st} = \frac{1000 \times 78.54}{150} = 523.6 \text{ mm}^2/\text{m}$

Now, $\quad M_u = 0.87\, f_y\, A_{st}\, d \left[1 - \frac{A_{st} f_y}{b\, d\, f_{ck}} \right] = 0.87 \times 415 \times 523.6 \times 100 \left[1 - \frac{523.6 \times 415}{1000 \times 100 \times 20} \right]$

$= 16.85 \times 10^6$ N-mm

$M_{u\ lim} = R_u bd^2 = 2.761\ (1000)\ (100)^2 = 27.61 \times 10^6$ N-mm

Since $M_u > M_{u\ lim}$ the slab is under-reinforced.

Again, $m_u = M_u = 16.85 \times 10^6$ N-mm $= 16.85$ kN-m

But $\quad m_u = \frac{w_u L^2}{24}$ [Eq. 20.11 (*a*)]. $\quad \therefore \quad w_u = \frac{24}{L^2} m_u = \frac{24}{(5)^2} \times 16.85 = 16.176 \text{ kN/m}^2$.

$\therefore$ Service load $= 16.176/1.5 \approx 10.78$ kN/m²

Dead load of slab $= 0.130 \times 25 = 3.25$ kN/m²

Dead load of finishing $= 0.05 \times 22 = 1.10$ kN/m²

Total dead load $= 3.25 + 1.10 = 4.35$ kN/m²

$\therefore$ Permissible service load $= 10.78 - 4.35 =$ **6.43 kN/m²**.

Example 20.3. *A reinforced concrete slab, 4 m × 6 m is reinforced with 10 mm diameter bars at 150 mm spacing in the short direction and 200 mm spacing in the long direction. The slab is 100 mm thick with a average effective depth of 80 mm. If the*

yield lines are inclined at 45° to either direction of reinforcement, find the ultimate moments $m_{u\alpha n}$ and $m_{u\alpha t}$ for unit length along yield line. Use M 20 concrete and Fe 415 steel.

Solution. Refer Fig. 20.10 in which the yield lines *AE*, *DE*, *BF* and *CF* are inclined at 45° to the principal *x* and *y* directions. The reinforcement in *y*-direction is spaced at 150 mm c/c while that in *x*-direction is spaced at 200 mm c/c.

Now $M_{uy} = m_{uy}$ = moment across plane whose normal is in *y*-direction.

$$= 0.87 f_y A_{st} d \left(1 - \frac{A_{st} f_y}{bd f_{ck}}\right)$$

where $A_{st} = \frac{1000}{150} \times 78.6 = 524 \text{ mm}^2$

$$\therefore \quad m_{uy} = 0.87 \times 415 \times 524 \times 80 \left(1 - \frac{524 \times 415}{1000 \times 80 \times 20}\right) = 13.078 \times 10^6 \text{ N-mm} = 13.078 \text{ kN-m}$$

Also, $M_{ux} = m_{ux}$ = moment across plane whose normal is in *x*-direction.

$$= 0.87 f_y A_{st} d \left(1 - \frac{A_{st} f_y}{bd f_{ck}}\right)$$

where $A_{st} = \frac{1000}{200} \times 78.6 = 393 \text{ mm}^2$

$$\therefore \quad m_{ux} = 0.87 \times 415 \times 393 \times 80 \left(1 - \frac{393 \times 415}{1000 \times 80 \times 20}\right) = 10.194 \times 10^6 \text{ N-mm} = 10.194 \text{ kN-m}.$$

Now, from Eqs. 20.1 and 20.5

$$m_{u\alpha n} = m_{ux} \cos^2 \alpha + m_{uy} \sin^2 \alpha \quad ; \quad \text{and} \quad m_{u\alpha t} = (m_{ux} - m_{uy}) \sin \alpha \cos \alpha \ .$$

Here $\alpha = 45°$; $\cos \alpha = \sin \alpha = 0.707$

$$m_{u\alpha n} = 10.194\,(0.707)^2 + 13.078\,(0.707)^2 = \mathbf{11.636 \text{ kN-m}}$$

$$m_{u\alpha t} = (10.194 - 13.078)\, 0.707 \times 0.707 = \mathbf{-1.442 \text{ kN-m}}$$

(Negative sign shows that the direction of torsional moment is opposite to the one marked in Fig. 20.5 (*c*).

Example 20.4 *A rectangular slab 3.5 × 4.5 m is isotropically reinforced with 8 mm dia. bars spaced at 150 mm bothways. The average effective depth may be taken as 80 mm and the total depth of the slab is 100 mm. If Fe 415 steel and concrete of grade M 20 are used, determine the safe service live load. The dead load of floor finishing may be assumed as 1.5 kN/m².*

Solution. Given $L = 4.5$ m ; $\alpha L = 3.5$ m ; $\alpha = 3.5/4.5 = 0.7778$.

Since the slab is isotropically reinforced, $m_{ux} = m_{uy} = m_u$. Hence $\mu = 1$

$$A_{st} = \frac{1000 \times 50.27}{150} = 335 \text{ mm}^2/\text{m}.$$

The ultimate moment is given by

$$M_u = m_u = 0.87 f_y A_{st} d \left[1 - \frac{A_{st} f_y}{bd f_{ck}}\right] = 0.87 \times 415 \times 335 \times 80 \left[1 - \frac{335 \times 415}{1000 \times 80 \times 20}\right]$$

$$= 8.835 \times 10^6 \text{ N-mm} = 8.835 \text{ kN-m}$$

The yield moment of the slab is given by Eq. 20.24

$$m_u = \frac{w_u \alpha^2 L^2}{24}\left[\sqrt{3+\alpha^2} - \alpha\right]^2$$

or $$w_u = \frac{24\, m_u}{\alpha^2 L^2\left[\sqrt{3+\alpha^2} - \alpha\right]^2} = \frac{24 \times 8.835}{(0.7778)^2 \times (4.5)^2\,[\sqrt{3+(0.7778)^2} - 0.7778\,]^2}$$

$$= 13.76 \text{ kN/m}^2.$$

$\therefore$ Total serviced load = 13.76/1.5 ≈ 9.17 kN/m²

Dead load of slab = 0.1 × 25 = 2.5 kN/m²

Dead load of floor finishing = 1.5 kN/m²

$\therefore$ Service live load = 9.17 − (2.5 + 1.5) = **5.17 kN/m².**

Example 20.5. *For the slab of example 20.3. determine the safe permissible service live load. The dead load of flooring may be assumed as 1 kN/m².*

Solution. Refer Fig. 20.10

Given : $L = 6$ m. ; $\alpha L = 4$ m.

$\therefore$ $\alpha = 4/L = 4/6 = 0.667$; $D = 100$ mm ; $d = 80$ mm

As found in example 20.3.

$m_{uy} = 13.078 = m_u$; $m_{ux} = 10.194 = \mu m_u$.

$\therefore$ $\mu = 10.194/13.078 = 0.779$.

Now from Eq. 20.22

$$m_u = \frac{w_u \alpha^2 L^2}{24}\left[\sqrt{3+\mu\alpha^2} - \alpha\sqrt{\mu}\,\right]^2$$

$$w_u = \frac{24\, m_u}{[\alpha^2 L^2 \sqrt{3+\mu\,\alpha^2} - \alpha\sqrt{\mu}]^2}$$

$$= \frac{24 \times 13.078}{(0.667)^2\,(6)^2\,[\sqrt{3+0.779\,(0.667)^2} - 0.667\sqrt{0.779}\,]^2} = \frac{24 \times 13.078}{0.445 \times 36 \times 1.539}$$

$$= 12.729 \text{ kN/m}^2$$

$\therefore$ Service load = 12.729/1.5 = 8.486 kN/m²

Dead load of slab = 0.1 × 25 = 2.5 kN/m²

Dead load of flooring = 1.0 kN/m²

$\therefore$ Service live load = 8.486 − 3.5 = **4.986 kN/m².**

Example 20.6 *A rectangular slab 3.5 × 5 m in size, simply supported at the edges. The slab is expected to carry a service live load of 3 kN/m² and a floor finishing load of 1 kN/m². Use M 20 concrete and Fe 415 steel. Design the slab if (a) it is isotropically reinforced (b) If it is orthotropically reinforced with $\mu = 0.75$.*

Solution. Given : $L = 5$ m ; $\alpha L = 3.5$; $\therefore$ $\alpha = 3.5/5 = 0.7$

$$\left(\frac{\text{span}}{\text{depth}}\right) = 35. \quad \therefore \quad \text{Depth of slab} = 350/35 = 10 \text{ cm} = 100 \text{ mm}$$

Use 10 mm dia. bars both ways. Using 15 mm clear cover, assume average effective depth = 100 − 15 − 10 = 75 mm

Dead weight of slab	$= 0.1 \times 25$	$= 2.5 \text{ kN/m}^2$
Dead weight of flooring		$= 1.0 \text{ kN/m}^2$
Live load		$= 3.0 \text{ kN/m}^2$
∴ Total service load		$= 6.5 \text{ kN/m}^2$

∴ Ultimate design load $= w_u = 1.5 \times 6.5 = 9.75 \text{ kN/m}^2$

(*a*) ***Isotropically reinforced slab*** : $\mu = 1$

Now, ultimate moment on slab is given by Eq 20.22 :

$$m_u = \frac{w_u \alpha^2 L^2}{24}\left[\sqrt{3 + \mu\alpha^2} - \alpha\sqrt{\mu}\right]^2 = \frac{9.75\,(0.7)^2\,(5)^2}{24}\left[\sqrt{3 + 1(0.7)^2} - 0.7\sqrt{1}\right]^2$$

$$= 6.791 \text{ kN-m/m} = 6.791 \times 10^6 \text{ N-mm/m}$$

The reinforcement is given by

$$M_u = m_u = 0.87 f_y A_{st} d\left[1 - \frac{A_{st} f_y}{bd f_{ck}}\right]$$

or $$6.791 \times 10^6 = 0.87 \times 415 A_{st} \times 75\left[1 - \frac{A_{st} \times 415}{1000 \times 75 \times 20}\right],$$ which simplifies to

$$A_{st}^2 - 3614 A_{st} + 906459 = 0. \text{ Hence } A_{st} = 272 \text{ mm}^2$$

Using 8 mm dia. bars, spacing $= \dfrac{1000 \times 50.3}{272} = 185.5$ mm.

However provide 8 mm dia. bars @ 175 mm c/c both ways.

(*b*) ***Orthotropically reinforced slab***

$$\mu = 0.75$$

$$\therefore \quad m_u = \frac{w_u \alpha^2 L^2}{24}\left[\sqrt{3 + \mu\alpha^2} - \alpha\sqrt{\mu}\right]^2$$

$$= \frac{9.75\,(0.7)^2\,(5)^2}{24}\left[\sqrt{3 + 0.75\,(0.7)^2} - 0.7\sqrt{0.75}\right]^2$$

$$= 8.541 \text{ kN-m/m} = 8.541 \times 10^6 \text{ N-mm/m}$$

But $$m_u = 0.87 f_y A_{st} d\left[1 - \frac{A_{st} f_y}{bd f_{ck}}\right]$$

or $$8.541 \times 10^6 = 0.87 \times 415 A_{st}\,(75)\left[1 - \frac{A_{st} \times 415}{1000 \times 75 \times 20}\right]$$

or $A_{st}^2 - 3614.4 A_{st} + 1140000 = 0$, which gives $A_{st} = 350 \text{ mm}^2$.

∴ Spacing of 8 mm dia. bars in short direction $= \dfrac{1000 \times 50.3}{350} = 144$ mm

and spacing of 8 mm dia. in long direction $= 144/0.75 = 192$ mm.

Hence provide 8 mm dia. bars @ 140 mm c/c in short direction and @ 190 mm c/c in long direction.

Example 20.7. *A right angled rectangular slab is simply supported along the two edges. The length of the supported edges are 3 m and 4 m respectively. The slab is isotropically reinforced with 8 mm dia. bars spaced at 100 mm centres both ways. The* *average effective depth of the slab is 95 mm and total depth is 120 mm. Using M 20 concrete and HYSD bars (Fe 415), determine the safe permissible service live load on the slab. Assume dead load of floor finishing at 1.5 kN/m^2.*

Solution

Given : $L = 4$ m ; $\alpha L = 3$ m $\quad \therefore \quad \alpha = \frac{3}{4} = 0.75$

$$A_{st} = \frac{1000 \times 50.3}{100} = 503 \text{ mm}^2. \text{ The ultimate moment is given by}$$

$$m_u = M_u = 0.87 f_y A_{st} d \left[1 - \frac{A_{st} f_y}{bd f_{ck}} \right] = 0.87 \times 415 \times 503 \times 95 \left[1 - \frac{503 \times 415}{1000 \times 95 \times 20} \right]$$

$= 15.357 \times 10^6$ N-mm $= 15.357$ kN-m. Now, from Eq. 20.17,

$$m_u = \frac{w_u \alpha L^2}{6} \sin^2 \beta/2 \qquad \text{or} \qquad w_u = \frac{6 m_u}{\alpha L^2 \sin^2 \beta/2}$$

Here, $\beta = 90°$. $\therefore$ $w_u = \dfrac{6 \times 15.357}{0.75 (4)^2 (\sin 45°)^2} = 15.357 \text{ kN/m}^2$

$\therefore$ Total service load	$= 15.357/1.5$	$= 10.238$ kN/m^2
Dead load of slab	$= 0.12 \times 25$	$= 3$ kN/m^2
Dead load of flooring		$= 1.5$ kN/m^2
Total dead load		$= 4.5$ kN/m^2

$\therefore$ Permissible service live load $= 10.238 - 4.5 = 5.738$ kN/m^2

Example 20.8. *A hexagonal slab is simply supported along the edges with a side length of 3.5 m. It is isotropically reinforced with 10 mm dia. bars at 125 mm centres both ways, at an average effective depth of 120 mm. The overall depth of the slab is 150 mm. Calculate (i) ultimate load capacity of slab, (ii) safe permissible service live load. Use M 20 concrete and Fe 415 steel. Take the load of floor finishing as 1.25 kN/m^2.*

Solution. $L = 3.5$ m. ; $\quad A_{st} = \dfrac{1000 \times 78.54}{125} = 628.3 \text{ mm}^2$

For M 20 – Fe 415 combination, $R_u = 2.761$, as in Example 20.1

$$m_u = M_u = 0.87 f_y A_{st} d \left[1 - \frac{A_{st} f_y}{bd f_{ck}} \right] = 0.87 \times 415 \times 628.3 \times 120 \left[1 - \frac{628.3 \times 415}{1000 \times 120 \times 20} \right]$$

$= 24.264 \times 10^6$ N-mm $=$ **24.264 kN-m.**

Now, from Eq. 20.28

$$w_u = \frac{8 m_u}{L^2} = \frac{8 \times 24.264}{(3.5)^2} = 15.845 \text{ kN/m}^2$$

$\therefore$ Total service load $= 15.845/1.5 = 10.564$ kN/m^2

Dead load of slab	$= 0.15 \times 25$	$= 3.75$ kN/m^2
Load of floor finishing		$= 1.25$ kN/m^2
Total dead load		$= 5$ kN/m^2

∴ Permissible service live load $= 10.564 - 5 =$ **5.564 kN/m²**.

Example 20.9 *Design a reinforced circular slab for the following data :*

(i) Diameter of slab : 5.5 m (ii) Service live load : 4 kN/m²

(iii) Floor finishing load : 1 kN/m² (iv) Grade of concrete : M 20

(v) Grade of steel : Fe 415. The slab is simply supported along the edge.

Solution For a simply supported two way slab,

$$\frac{\text{Span}}{\text{Depth}} = 35 \quad \therefore \quad \text{Depth} = \frac{5500}{35} = 157 \text{ mm.}$$

Let us provide overall depth = 160 mm. Using 10 mm dia. bars and a clear cover of 15 mm, average effective depth = 160 − 15 − 10 = 135 mm.

Self load of slab	$= 0.16 \times 25$	$= 4$ kN/m²
Dead load of flooring		$= 1$ kN/m²
Live load		$= 4$ kN/m²
Total service load		$= 9$ kN/m².

$$w_u = 9 \times 1.5 = 13.5 \text{ kN/m}^2$$

Now, from Eq. 20.26 (*a*),

$$m_u = \frac{w_u r^2}{6} = \frac{13.5\,(5.5/2)^2}{6} = 17.016 \text{ kN-m/m} = 17.016 \times 10^6 \text{ N-mm/m.}$$

Limiting moment capacity :

$$M_{u\,lim} = R_u bd^2 = 2.761\,(1000)\,(135)^2 = 50.137 \times 10^6 \text{ N-mm.}$$

Since $m_u < M_{u\,lim}$ the slab is under reinforced.

$$\therefore \quad m_u = M_u = 0.87 f_y A_{st} d \left(1 - \frac{A_{st} f_y}{bd f_{ck}}\right)$$

or
$$17.016 \times 10^6 = 0.87 \times 415 A_{st} \times 135 \left(1 - \frac{A_{st} \times 415}{1000 \times 135 \times 20}\right)$$

or $A_{st}^2 - 6506 A_{st} + 2.271 \times 10^6 = 0$, which gives $A_{st} = 370.4$ mm²

Alternatively,

$$A_{st} = \frac{0.5 f_{ck}}{f_y}\left[1 - \sqrt{1 - \frac{4.6 M_u}{f_{ck} bd^2}}\right] bd = \frac{0.5 \times 20}{415}\left[1 - \sqrt{1 - \frac{4.6 \times 17.016 \times 10^6}{20 \times 1000\,(135)^2}}\right] \times 1000 \times 135$$

$$= 370.4 \text{ mm}^2$$

$$\text{Spacing of 10 mm dia. bars} = \frac{1000 \times 78.5}{370.4} = 212 \text{ mm}$$

However, use 10 mm dia. bars @ 200 mm centres both ways.

PROBLEMS

1. A reinforced concrete square slab, 3.5 m × 3.5 m is simply supported at the ends and is reinforced with 8 mm diameter bars spaced at 150 mm centres both ways. Determine the safe service live

load if the average effective depth of slab is 100 mm and the total thickness of slab, inclusive of flooring, is 160 mm. Use M 20 concrete and HYSD (Fe 415) bars.

2. Redesign the slab of problem 1 to carry a service load of 4 kN/m^2, inclusive of floor finishes, but exclusive of self load.

3. A rectangular slab 4 m × 5 m is simply supported at the ends. Design the slab to carry super-imposed service load of 5 kN/m^2, if the slab is to be isotropically reinforced. Use M 20 concrete and Fe 415 steel.

4. Redesign the slab of problem 3 if the slab is to be orthotropically reinforced, with $\mu = 0.8$.

5. Find the safe superimposed service load on the slab of problem 3, if it is reinforced with 8 mm diameter bars spaced at 125 mm centres both ways. The total thickness of the slab may be assumed as 125 mm.

6. A triangular slab has equal side lengths of 4.5 m is supported on two edges and is isotropically reinforced with 8 mm dia. bars of Fe 415 grade, spaced at 125 mm centres both ways. Determine (*i*) ultimate moment capacity, (*ii*) ultimate collapse load. The total thickness of slab is 125 mm, in M 20 Grade concrete.

7. A hexagonal slab, simply supported on all the sides has a side length of 3 m. Find the uniformly distributed load which would cause failure of the isotropically reinforced slab, if the ultimate moment of resistance of slab is 2.7 kN-m/m. Work from the first principles.

8. A rectangular slab 6 m × 4.5 m, simply supported at its edge is to be designed as an isotropically reinforced slab to support an uniformly distributed working load of 18 kN/m^2 which includes the self weight of the slab. Calculate the ultimate moment of resistance required for the slab section from first principles.

9. Design a circular slab of 5 m diameter, simply supported along the edge, to carry a service live load of 5 kN/m^2. Use M 20 concrete and Fe 415 steel.

21

Redistribution of Moments

21.1. INTRODUCTION : LIMIT ANALYSIS

Indian Standard Code permits the analysis of all structures by linear elastic theory to calculate internal actions produced by design loads. In lieu of rigorous elastic analysis, the Code gives a simplified analysis for frames and for continuous beams. The elastic analysis, however, does not portray the true behaviour of indeterminate structures at collapse. Elastic analysis of R.C beams gives reasonable results upto working loads only. Beyond the working loads, the elastic analysis is not applicable because of non-linearity of stress-strain curves for the materials and the cracks which develop in concrete. Contrary to elastic analysis, *limit analysis* deals with the study of strength and behaviour of member and structure at *collapse.* The limit analysis is based on *elasto-plastic theory* (commonly known as *plastic theory*) of plastic materials like steel while it is based on *ultimate load theory* (including its modified version of limit state method) for R.C. structures. However, *limit theory* primarily developed for elasto-plastic material, is not well understood for R.C. structures because R.C. element which is a combination of two materials (i.e. concrete and steel) has neither uniform strength nor large rotational capacity. Concrete is a brittle material whereas steel is a ductile material. Combination of two materials leads to an intermediate position where the rotational capacity depends as to which of the two materials is relatively more stressed. Consequently, for analysis of limit state of collapse, *redistribution of bending moments*, obtained from elastic analysis is permitted provided equilibrium between the internal forces and external loads is maintained under each combination of ultimate loads.

21.2. MOMENT CURVATURE RELATIONSHIP

When a beam is loaded beyond working load, plastic hinges are formed at certain locations. Let us first take the example of a simply supported beam, loaded with central point load W_u. Let the load be gradually increased from zero value to a certain value W. Due to this there is proportionate increase in the maximum B.M. (M_{max}) at the critical point C, with the proportionate increase in maximum stress (σ_{max}) and maximum strain (ε_{max}) at the extreme fibres of this section at C. The curvature of the member or the rotation (ϕ) of the section is proportional to maximum strain (ε_{max}), while the moment of resistance, M of the member is proportional to the maximum bending stress at the section. Hence the moment-curvature or moment-rotation ($M - \phi$) relationship of the member, shown in Fig. 21.1 (*c*) is similar to the stress-strain relationship (shown in Fig. 21.1 *b*)

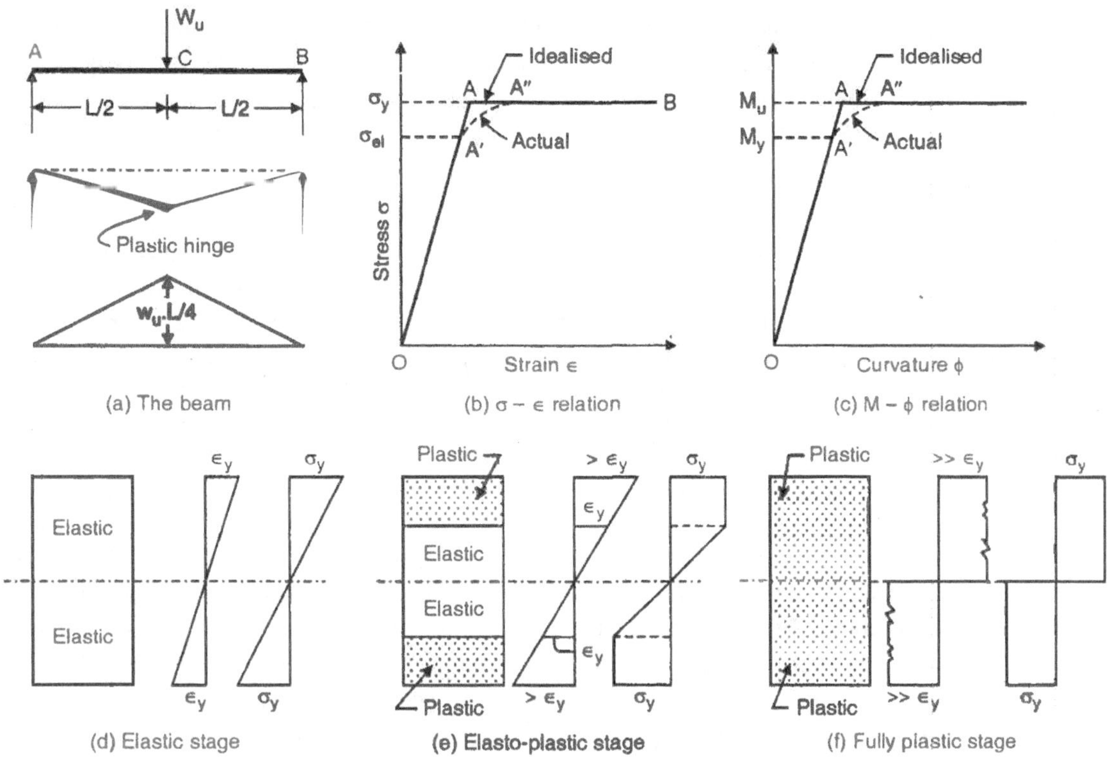

FIG. 21.1

of the member. When the maximum stress σ_{max} at the outermost fibre at *C* reaches elastic limit (σ_{el}) of the material (corresponding to point *A'* in Fig. 21.1 *b*) the M_{max} in the member becomes equal to maximum elastic moment of resistance or yield moment M_y, corresponding to point *A'* is Figs 21.1 (*c*). It should be noted that both $\sigma - \varepsilon$ curves as well as $M - \phi$ curve are linear upto point *A'* and σ_{el} is less than σ_y. This stage of the section, known as '*elastic stage*' is shown in Fig. 21.1 (*d*). With further increase in the load and hence the maximum moment, so that $M > M_y$, the outer parts of the section become *plasticised* or *partially yield*, as depicted in Fig. 21.1 (*e*) while the middle layer, known as *elastic core*, remain elastic. As the loads is increased, plasticity goes on the section from the extreme fibres towards the neutral axis. As the load is increased further, the whole section ultimately becomes *fully plastic* (Fig. 21.1 *f*) and the corresponding moment is known as *fully plastic moment* designated as M_P for ideal ductile materials like steel and M_u (i.e. ultimate moment) for R.C.C. structures. The relation between *M* and ϕ, for the range between M_y an M_u, is non-linear, as shown by dotted curve in Fig. 21.1 (*c*). The section is no more in a position to offer any additional moment of resistance with the increase in the load, and the section simply rotates at constant load about the plastic hinge so formed. Consequently, the beam turns into a *mechanism* and ultimately collapses. The load causing mechanism to form, and therefore the load causing collapse of the structure is known as *collapse load* or *ultimate load* and the corresponding maximum moment that the member can resist is known as ultimate moment of resistance M_u. It should be clearly noted that the *elastic analysis* allows a maximum value of M equal to M_y only while the *limit analysis* allows maximum value of M equal to M_u for R.C.C. structures and M_p for steel structures. The ratio of M_p/M_y is known as *shape factor* in the case of steel structures, which is the property of the section, while *the ratio of* M_u/M_y *is negligible in R.C.C. design.*

Fig. 21.1 (*c*) shows the idealised moment curvature ($M - \phi$) relationship. In real structures, the change from elastic to plastic state (*i.e.* from A' to A'') is *gradual* shown by dashed curve. However, for convenience it is assumed that the section remains elastic right upto the ultimate moment (i.e. upto point A), and then unrestricted plastic flow takes place. The error involved in this idealisation is insignificant. As stated earlier, the *limit analysis* or *the theory of plasticity* is well understood for the analysis of steel structures made of perfectly elastic and homogeneous materials and which has very large *rotational capacity* because of the ductility of steel. However this is not so in reinforced concrete which is a combination of two materials ; the R.C. beams has neither uniform strength nor does the section has large rotational capacity. Due to this reason, for analysis of limit state of collapse, of indeterminate structures (such fixed or continuous beams), redistribution of bending moments obtained from the elastic analysis under *design load* (W_u) is permitted by the Code, subject to the condition that equilibrium is maintained between the internal forces and external loads under each combination of ultimate loads.

21.3. LIMIT ANALYSIS APPLIED TO R.C. BEAM FIXED AT ENDS

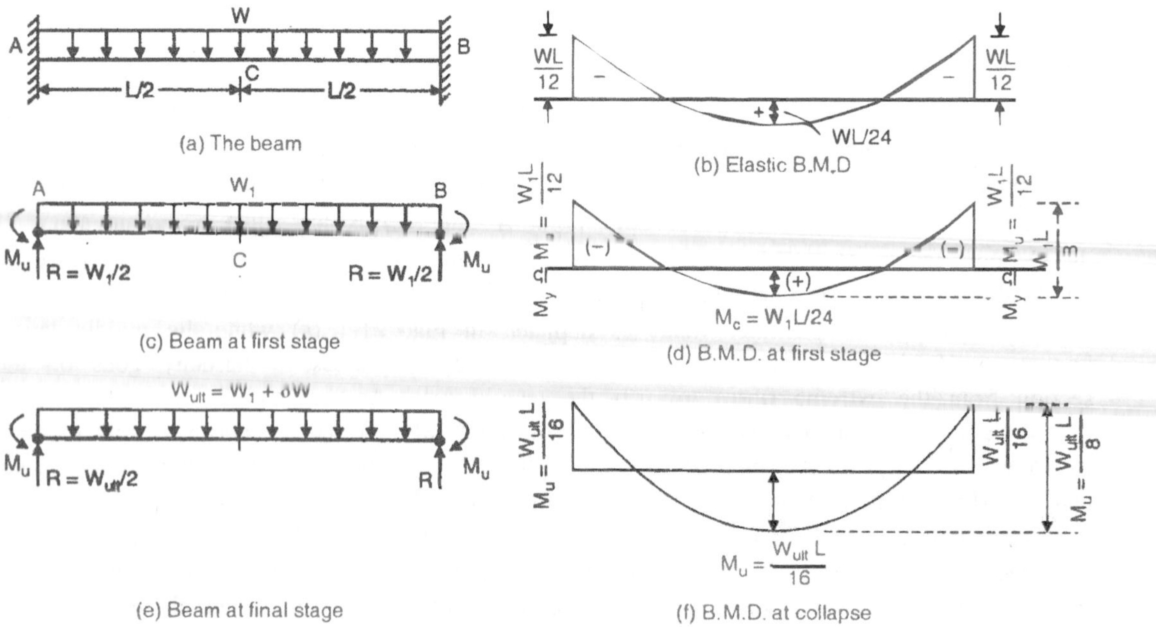

FIG. 21.2

Fig. 21.2 (*a*) shows a single span beam fixed at ends A and B and carrying total U.D.L. W. Fig. 21.2 (*b*) shows the corresponding B.M. diagram obtained by elastic analysis, where in the fixed end moments $M_A = M_B = WL/12$ and the central moment $M_c = WL/24$. Thus, the maximum B.M. will occurs at ends A and B while the B.M. at C will be half the value. As the load is increased from zero to a certain value less than W_1, where W_1 is the load causing *first yield*, the beam remains elastic. When the load becomes W_1, the moments at A and B will attain a value equal to ultimate moment of resistance M_u^* and plastic hinges will be formed at ends A and B. The behaviour of the beam will still be elastic, and it will act as a beam hinged at ends A and B, and carrying constant moment M_u at ends A and B (Fig. 21.2 *c*). This stage of formation of hinges at the ends will be called stage I, and the corresponding B.M.D. is shown in Fig. 21.2 (*d*). The beam will, however, not collapse as the mechanism has not been formed as yet. Additional load on the beam will now be carried by the span portion of the beam and *not* by supports A and B. When the load is further increased so as to reach *ultimate load* $W_{ul} = W_1 + dW$, maximum B.M. at C will reach M_u, the section at C will yield and hinge will be formed there. This will divide the beam AB into two segments AC and CB, forming a mechanism which will result in the collapse of the beam. This stage may be called stage II shown in Fig. 21.2 (*e*) ; the corresponding B.M.D. is shown in Fig. 21.2 (*f*).

From Fig. 21.2 (*f*), $$M_C = M_u = \frac{W_{ult}\,L}{8} - M_u$$

Hence $$M_u = \frac{W_{ult}\,L}{16} \quad \text{or} \quad W_{ult} = \frac{16\,M_u}{L} \qquad \ldots(i)$$

Also, from Fig. 21.2, $$M_u = \frac{W_1\,L}{12} \quad \text{or} \quad W_1 = \frac{12\,M_u}{L} \qquad \ldots(ii)$$

Hence $$\frac{W_{ult}}{W_1} = \frac{16}{12} = \frac{4}{3} \quad \text{and} \quad \delta W = W_{ult} - W_1 = \frac{W_1}{3}$$

$\therefore$ Percentage increase in load $$= \frac{W_{ult} - W_1}{W_1} \times 100 = \frac{1}{3} \times 100 = 33.3\%$$

Hence the merit of *limit analysis* lies in the fact that it gives higher load carrying capacity for statically indeterminate structure as compared to the elastic analysis. Further, we observe that redistribution of moments along the beam occurs from stage I corresponding to occurrence of first yield to stage II corresponding to the ultimate load leading to the collapse of the structure.

However, in the *limit analysis* discussed above it, is assumed that (*i*) the beam is of uniform strength, and (*ii*) the section has very large rotational capacity. Both these assumptions are not strictly true for reinforced concrete which is a combination of two different materials, one (*i.e.* concrete) being a brittle material and other (i.e. steel) being ductile material. The advantage of limit analysis for R.C. structures can be accrued only if (*i*) large rotational capacity is ensured and (*ii*) the increase in load from W_1 to W_{ult} is not too large to

*Actually, it will be equal to M_y, but the difference in M_y and M_u is extremely small and can be ignored.

affect the serviceability of the beam by way of large deflections and cracking at service loads.

21.4. REDISTRIBUTION OF MOMENTS : RECOMMENDATIONS OF IS 456 : 2000

In the application of limit analysis to beam fixed at the ends discussed in the previous article, we observe that with the formation of first set of plastic hinges at fixed ends at the end of stage I, additional load δW causes the first set of hinges to rotate, at contant moment M_u, enabling the transfer of additional moments due to $\delta \dot{W}$ to the span region, thus causing *redistribution of moments* along the span. Let us first find the magnitude or percentage of this redistribution of moments in a fixed beam.

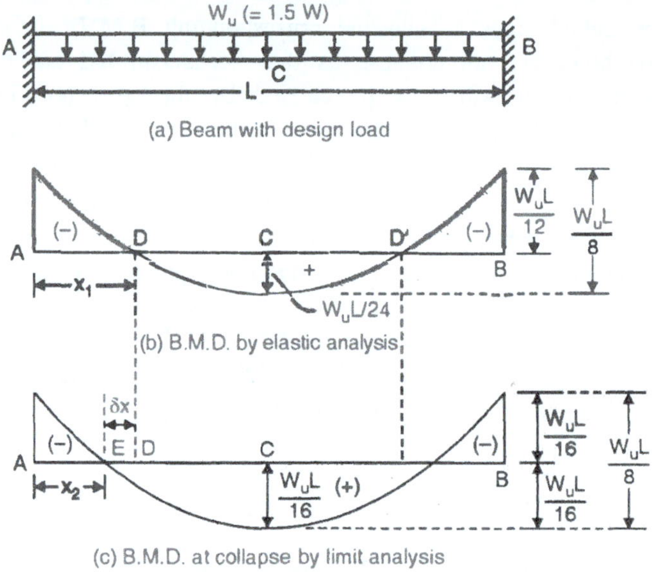

FIG. 21.3.

Fig. 21.3 (*a*) shows a fixed beam *AB* of span *L*, carrying a total uniformly distributed *design load* $W_u = 1.5\ W$, *where W* is the total working load (or service load). Fig. 21.3(*b*) shows the B.M.D. corresponding to the *elastic analysis* wherein

$$M_{eu} = \frac{M_u\, L}{12} \qquad ...(1)$$

where M_{eu} = bending moment at the fixed ends obtained by elastic analysis, under the design load W_u

Fig. 21.3 (*c*) shows the B.M.D. at collapse obtained by limit analysis, wherein

$$M_u = \frac{W_u\, L}{16} \qquad ...(2)$$

where M_u = B.M. at the ends as well as at the mid span obtained by limit analysis.

Thus we observe that the end moments are *decreased* from initial value of $W_u\, L/12$ to final value of $W_u\, L/16$, through the process of *redistribution of moments from* stage I to stage II.

$$\text{Percentage reduction in B.M.} = \frac{\frac{W_u L}{12} - \frac{W_u L}{16}}{\frac{W_u L}{12}} \times 100 = 25\%$$

The Indian Code (IS 456 : 2000), discussed below, permits a maximum reduction of 30% in order to ensure sufficient rotational capacity.

Recommendation of IS 456 : 2000

Indian Code (IS 456 : 2000) permits redistribution of moments in continuous beams and frames, obtained from elastic analysis, if the following *five conditions* are satisfied. However, where simplified analysis using coefficients is adopted, redistribution of moments shall not be done.

Condition 1 : Equilibrium between the internal forces and the external load is maintained.

Condition 2 : The ultimate moment of resistance provided at any section of a member is not less than 70 percent of the moment at that section obtained from an elastic maximum moment diagram covering all appropriate combinations of loads.

Condition 3 : The elastic moment at any section in a member due to a particular combination of load shall not be reduced by more than 30 percent of the numerically largest moment given any where by the elastic maximum moments diagram for the particular member, covering all appropriate combination of loads.

Condition 4 : At sections where the moment capacity after redistribution is less than that from the elastic maximum moment diagram, the following relationship shall be satisfied

$$\frac{x_u}{d} + \frac{\delta M}{100} \leq 0.6 \qquad \text{...(21.1)}$$

where x_u = depth of neutral axis

d = effective depth

δM = percentage reduction in moment

Condition 5 : In structure in which structural frame provides lateral stability, the reduction in moment allowed by condition 3 above, shall be restricted to 10 percent for structures over 4 storeys in height.

Comments

(*i*) In *limit state design*, the design loads are the ultimate loads. The structure should be analysed by elastic theory to calculate B.M. produced due to ultimate loads. The moments in statically indeterminate beam, obtained from such elastic analysis, may be redistributed as per conditions stated above. However during such redistribution, the equilibrium between the internal forces and external loads is maintained. This means that B.M. reduced at some sections will have to be appropriately increased at other sections.

(*ii*) As per condition 2, the design moment obtained after redistribution should not be less than 70% of the elastic moment at ultimate load. Since the ultimate load W_u is 1.5 times the working load W, moment M_{eu} obtained by elastic analysis at ultimate load W_u is 1.5 M_{ew} (where M_{ew} is the moment obtained by elastic analyes at working load W) the redistributed design moment = 0.7 M_{eu} = 0.7 (1.5 M_{ew}) = M_{ew}. *Hence the design moment after redistribution will not be less than the working moment M_{ew} at service load.*

(*iii*) As is clear from Fig. 21.3 (*b*) and (*c*), the redistribution of moments results in the shifting of point of contraflexure from D to point E, towards fixed end, by an amount δx. Thus, the portion ED is subjected to both hogging B.M. as well as sagging B.M. Due to this reason wide cracks would develop at top in portion DE under service load conditions. Hence the Code has put the restriction contained in condition 2, to avoid such cracking.

(*iv*) Rewriting Eq. 21.1, we have

$$x_{u,\,lim.} \leq \left(0.6 - \frac{dM}{100}\right) d \qquad \ldots(21.2)$$

When $dM = 30\%$, we get $x_{u,\,\text{lim}} = (0.6 - 0.3)\, d = 0.3\, d$

This is less than the balanced N.A. depth $x_{u,\,\text{max}} = 0.479\, d$ for Fe 415 steel.

However, when $dM = 10\%$, we get $x_{u,\,\text{lim}} = (0.6 - 0.1)\, d = 0.5\, d$

This is more than $x_{u,\,\text{max}}$ ($= 0.479\, d$) for Fe 415 steel.

Hence the depth of N.A. at the section under consideration, shall be the lesser of $x_{u,\,\text{lim}}$ and $x_{u,\,\text{max}}$.

From the above discussion, we note that as the percentage of redistribution of moment (dM) increases, $x_{u,\,\text{lim}}/d$ decreases and section tends to be more and more *under-reinforced*, resulting in increase in ductility and rotation capacity of the member. *It should be clearly noted that an under-reinforced section has a larger rotational capacity as compared to that of an over-reinforced section.* The Code has put a limitation of 30% on redistribution of moments to avoid large rotation and consequent excessive deflection and cracking which may affect the serviceability of the structure. Such restriction also helps to guard against crushing of the concrete at the plastic hinge.

21.5. BENDING MOMENT ENVELOPE

We have seen in the previous article that the point of contraflexure is shifted towards the support, due to redistribution of moments. Let us study its effect, in detail, with reference to the IS Code recommendations for redistribution of moments.

Fig. 21.4 (*a*) shows the fixed beam AB loaded with total uniformly distributed ultimate load (W_u). In Fig. 21.4 (*b*), the curve $a_1\, c_1\, b_1$ (designated as curve I) is the B.M.D. obtained by elastic analysis, in which support moment $Aa_1 = Bb_1 = W_u L/12$ and mid-span moment $Cc_1 = W_u L/24$. The sum of these two absolute moments is $W_u L/8$ which is the same as maximum mid-span B.M. for simply supported beam carrying the same load W_u. This moment is referred to as *free span moment*.

Curve $a_2\, c_2\, b_2$ (i.e. curve II) represents the redistributed B.M.D. obtained after numerically reducing the largest support moment by 30%. When support moments are decreased by 30%, there will be corresponding increase in the sagging moments, for the sake of static equilibrium. Hence in the redistributed B.M.D. (i.e. curve II), support moment $A\, a_2 = B\, b_2 = 0.7\, W_u L/12$ while the mid-span B.M. $Cc_2 = \dfrac{W_u L}{24} + 0.3\, \dfrac{W_u L}{12} = \dfrac{W_u L}{15}$. Hence the magnitude of the central coordinate of the basic parabola $= \dfrac{0.7\, W_u L}{12} + \dfrac{W_u L}{24} + 0.3\, \dfrac{W_u L}{12} = \dfrac{W_u L}{8} =$ free span moment.

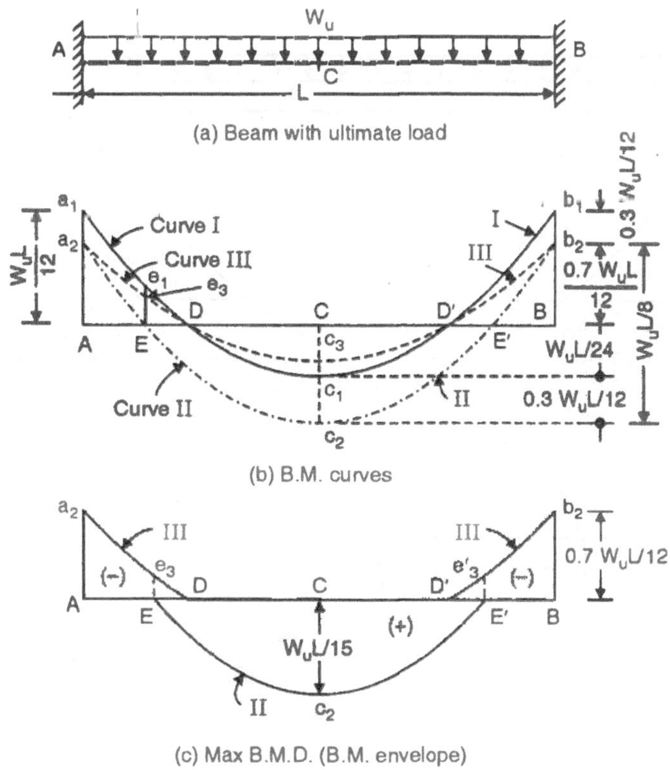

FIG. 21.4

Hence curve $a_2\,c_2\,b_2$ (i.e. curve II) is obtained by lowering curve I by $dM\ (= 30\%)$.

From the study of curve I and curve II, we observe that the point of contraflexure is shifted from point D to point E. The potion DE, which was experiencing hogging B.M. under service condition will now experience sagging B.M. after redistribution of moments, at ultimate condition. Hence the portion DE may crack. To guard against this cracking, the Code has put restriction under condition 2 that the ultimate moment of resistance *at any section* of the beam is not less than 70% of the moment at that section obtained from an elastic maximum moment diagram. After redistribuion, the B.M. at E becomes zero, giving rise to 100% reduction while as per condition 2, the B.M. at E, after redistribution should be atleast equal to $0.7\,E\,e_1$. Hence make $E\,e_3 =$ B.M. at $E = 0.7\,E\,e_1$ or $e_1\,e_3 = 0.3\,E\,e_1$. Thus curve III is obtained by reducing each ordinate of curve I by 30% to satisfy condition 2 of the Code recommendations. The resulting envelope for maximum bending moment as per IS Code recommendation is shown in Fig. 21.4 (*c*). While designing, care should be taken so that the moment of resistance any section is not less than 70% of the maximum B.M. calculated for that section by using the design loads (i.e. ultimate loads) and the elastic analysis.

Example 21.1. *Draw bending moment diagram for a beam, fixed at ends and carrying total uniformly distributed load W_u, after 20% redistribution. Determine the magnitude by which point of contraflexure is shifted.*

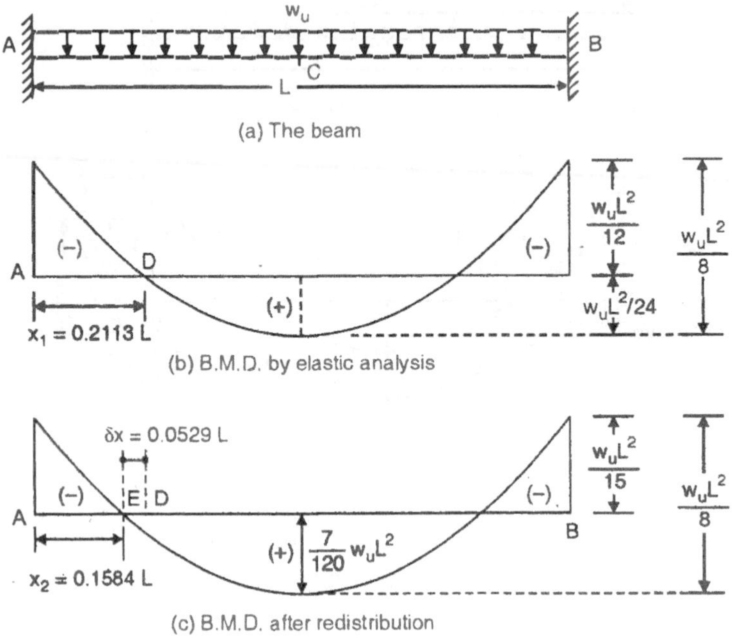

FIG. 21.5

Solution

Fig. 21.5 (*b*) shows the B.M.D. obtained by elastic analyes. The support moments are:

$$M_A = M_B = w_u L^2/12$$

Free span B.M. $= w_u L^2/8$.

At any distance x from A, $M_x = \frac{w_u L}{2} x - \frac{w_u L^2}{12} - \frac{w_u x^2}{2}$

This is zero at point D distant x given by

$$\frac{Lx}{2} - \frac{L^2}{12} - \frac{x^2}{2} = 0 \quad \text{or} \quad x^2 - Lx + L^2/6 = 0$$

which gives $x = 0.2113\,L$ or $0.7887\,L$

Hence $AD = x_1 = 0.2113\,L$

Fig. 21.5 (*c*) shows the B.M.D. after redistribution of moments. The support moments $M_A = M_B = 0.8\, w_u L^2/12 = w_u L^2/15$. When support moments are decreased by 20%, there will be corresponding increase in sagging bending moments, for the sake of static equilibrium. Hence the magnitude of mid span ordinate $M_C = \frac{w_u L^2}{24} + 0.2 \frac{w_u L^2}{12} = \frac{7\, w_u l^2}{120}$.

Alternatively, $M_C = \frac{w_u L^2}{8} - M_A = \frac{w_u L^2}{8} - \frac{w_u L^2}{15} = \frac{7\, w_u L^2}{120}$.

In order to obtain the point of contraflexure, write the equation for B.M. at any point distant x from A, noting that $M_A = M_B = 0.8\, w_u L^2/12 = w_u L^2/15$.

$\therefore \quad M_x = \frac{w_u L}{2} x - \frac{w_u x^2}{2} - \frac{w_u L^2}{15}$. Hence at point of contraflexure,

$$\frac{w_u L}{2} x - \frac{w_u x^2}{2} - \frac{w_u L^2}{15} = 0 \quad \text{or} \quad x^2 - L x + 0.1333 \, w_u L^2 = 0$$

which gives $\qquad x = 0.1584 \, L$ or $0.8416 \, L$

Hence $\qquad AE = x_2 = 0.1584 \, L$

$\therefore$ Shift in the point of contraflexure $= \delta x = x_1 - x_2 = 0.2113 \, L - 0.1584 \, L$

$= \mathbf{0.0529 \, L}$

Example 21.2. *A beam AB of 4 m span and fixed at the end, carries an uniformly distributed load of 30 kN/m at collapse. Draw the maximum bending moment diagram, as per recommendations of IS 456 : 2000.*

Solution

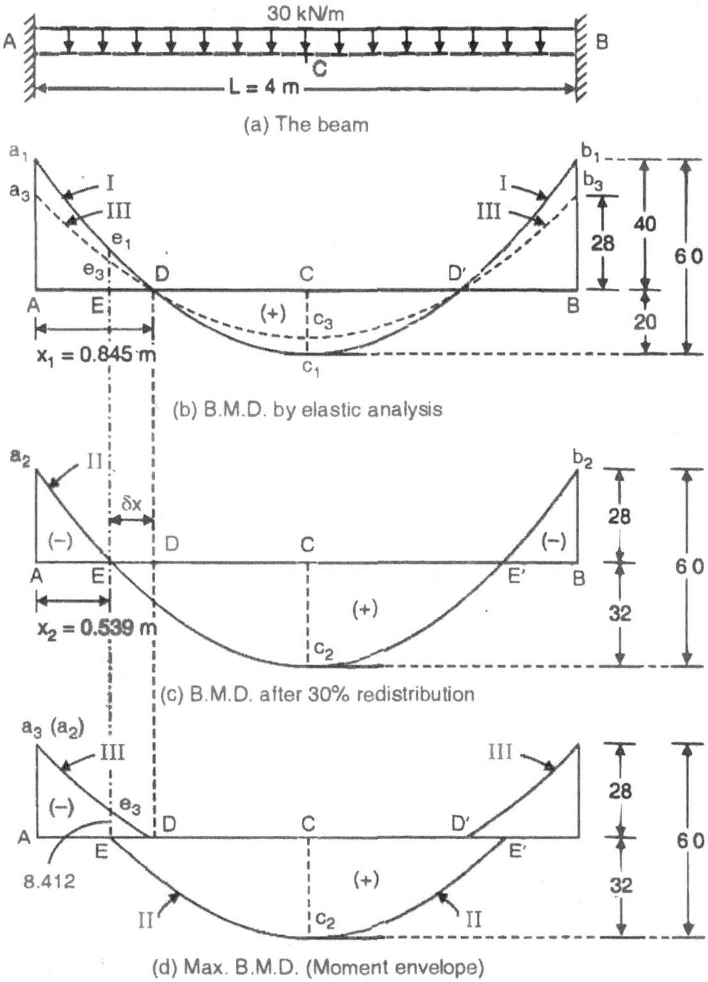

FIG. 21.6

$$M_A = M_B = -\frac{w_u L^2}{12} = -\frac{30\,(4)^2}{12} = -40 \text{ kN-m}$$

Mid span B.M. $M_C = +\dfrac{w_u L^2}{24} = +\dfrac{30\,(4)^2}{24} = +20$ kN-m

Free span B.M. $= w_u L^2/8 = 30\,(4)^2/8 = 60$ kN-m

Fig 21.6 (*b*) shows the B.M.D. by elastic analysis.

Reaction $R_A = R_B = \dfrac{w_u L}{2} = \dfrac{30 \times 4}{2} = 60$ kN

For finding the position of point of inflection D, we have $M_x = 60\,x - \dfrac{30\,x^2}{2} - 40 = 0$

or $x^2 - 4x + 2.667 = 0$ which gives $x = 0.845 \text{ m} = x_1$

Alternatively, $x = 0.2113\,L = 0.2113 \times 4 \simeq 0.845$ m

The B.M.D. obtained by elastic analysis is shown in Fig. 21.6 (*b*) in the form of curve I. Fig. 21.6 (*c*) shows the B.M.D., obtained after reduction of 30% in support moments.

Ordinate $A\,a_2 = 0.7\,A\,a_1 = 0.7 \times 40 = 28$ kN-m

$Cc_2 = Cc_1 + 0.3\,A\,a_1 = 20 + 0.3 \times 40 = 32$ kN-m

Total span moment $= Aa_2 + Cc_2 = 28 + 32 = 60$ kN-m = free span B.M.

Curve $a_2\,c_2\,b_2$ of Fig. 21.6 (*c*) is obtained by *lowering down* the curve $a_1\,c_1\,b_1$ by $0.3\,A\,a_1$ ($= 0.3 \times 40 = 12$ kN-m).

To get the position of the point of contraflexure E, write down equation for B.M. at any point distant x from A and equate it to zero.

$$\therefore \quad M_x = 60\,x - \frac{30\,x^2}{2} - 28 = 0$$

or $x^2 - 4x + 1.8667 = 0$, which gives $x = x_2 \simeq 0.539$m

$\therefore$ Shift in the point of contraflexure $= DE = x_2 - x_1 = 0.845 - 0.539 = 0.306$ m.

From Figs. 21.6 (*b*) and (*c*), we observe that B.M. at E has been reduced by 100%, while the final B.M. at E should not be less than $0.7\,e_1\,E$.

Now elastic B.M. at $E = e_1\,E = 60\,(0.539) - 15(0.539)^2 - 40 = -12.018$ kN-m.

$\therefore$ B.M. $e_3\,E = 0.7\,e_1\,E = 0.7\,(-12.018) = -8.412$ kN-m

Fig. 21.6 (*d*) shows the maximum B.M. diagram (or bending moment envelope) based on IS Code recommendations.

Example 21.3. *A beam AB of 4 m span and fixed at one end and freely supported at the other end carries an uniformly distributed load of 30 kN/m at collapse. Draw the maximum B.M. diagram as per recommendations of IS 456 : 2000.*

Solution

(*a*) ***Elastic analysis*** : Fig 21.7 (*a*) shows the beam, commonly known as propped cantilever. The elastic analysis can be done using moment distribution method, by first considering end B as fixed and then removing the fixidity at B and carrying over half the released moment at the other end A.

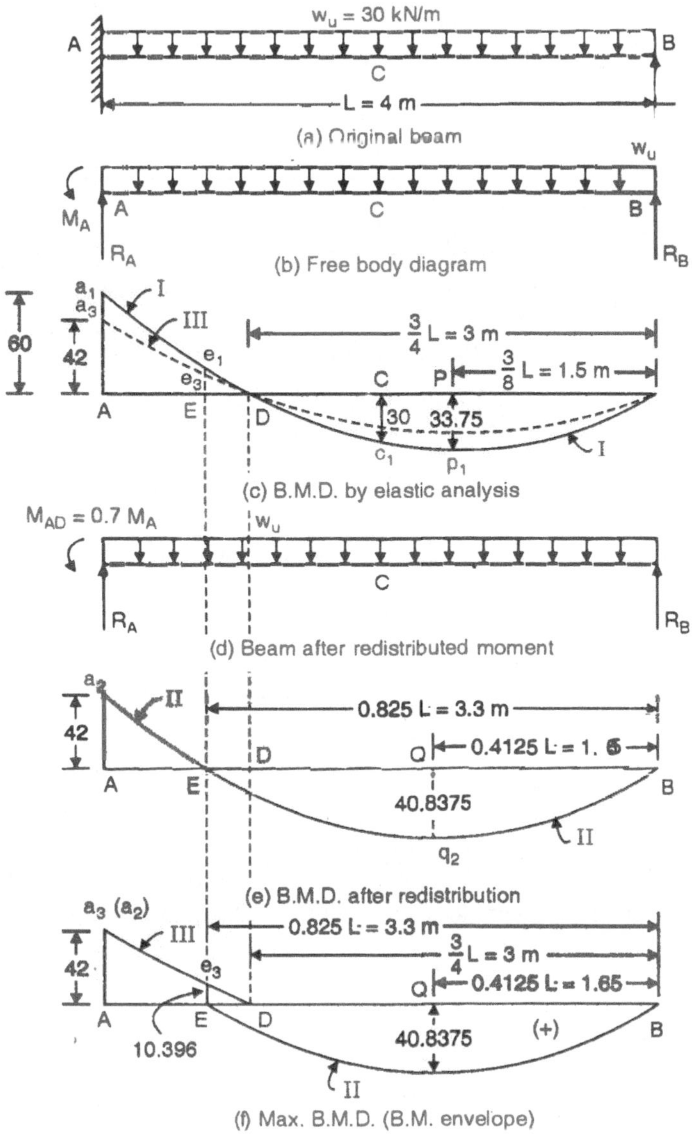

FIG. 21.7

$$M_{FAB} = -\frac{w_u L^2}{12} = -\frac{30\,(4)^2}{12} = -40 \quad \text{kN-m}$$

$$M_{FBA} = +\frac{w_u L^2}{12} = +40 \quad \text{kN-m}$$

The moment distribution can now be done is a Tabular form shown below.

Member	AB	BA
Initial FEM	$-\frac{w_u L^2}{12} = -40$	$+\frac{w_u L^2}{12} = +40$
Distribution		$-\frac{w_u L^2}{12} = -40$
Carry over	$-\frac{w_u L^2}{24} = -20$	
Final moments	$-\frac{w_u L^2}{8} = -60$	0

Thus, $M_A = -\frac{w_u L^2}{8} = -60;\ M_B = 0$

To find reactions R_A and R_B, refer Fig. 21.7 (*b*)

$$R_A = \frac{1}{L}\left(\frac{w_u L^2}{2} + \frac{w_u L^2}{8}\right) = \frac{5}{8} w_u L = \frac{5}{8} \times 30 \times 4 = 75 \text{ kN}$$

$$R_B = \frac{1}{L}\left(\frac{w_u L^2}{2} - \frac{w_u L^2}{8}\right) = \frac{3}{8} w_u L = \frac{3}{8} \times 30 \times 4 = 45 \text{ kN}$$

S.F. is zero at $x = \frac{R_B}{w_u} = \frac{3}{8} L = \frac{3}{8} \times 4 = 1.5$ m from end B.

Measuring x from B, $M_x = \frac{3}{8} w_u L x - \frac{w_u x^2}{2}$ ...(*i*)

M_{max} occurs at $x = \frac{3}{8} L$ (= 1.5 m), where S.F. is zero

Hence $$M_{max} = \frac{3}{8} w_u L\left(\frac{3}{8} L\right) - \frac{w_u}{2}\left(\frac{3}{8} L\right)^2 = \frac{9}{128} w_u L^2 = \frac{9}{128} (30)(4)^2 = 33.75 \text{ kN-m}$$

$$\text{B.M. (at } x = L/2) = M_C = \frac{w_u L^2}{8} - \frac{1}{2}\left(\frac{w_u L^2}{8}\right) = \frac{w_u L^2}{16} = \frac{30\,(4)^2}{16} = 30 \text{ kN-m}$$

Note that maximum moment does not occur at mid-span. Percentage difference between

M_{max} and $M_C = \dfrac{\frac{9}{128} w_u L^2 - \frac{8\, w_u L^2}{128}}{\frac{9}{128} w_u L^2} \times 100 = 11.1\%$. Point of contraflexure D occurs at x obtainable from Eq. (*i*). Hence $x = \frac{3}{4} L = 3$ m from B.

The B.M.D. obtained from the above elastic analysis is shown in Fig. 21.7 (*c*)

(b) Moments after redistribution (curve II)

Maximum redistribution percentage = 30%

$$\therefore \quad dM = 0.3\, M_A = 0.3 \times \frac{w_u L^2}{8} = 0.0375\, w_u L^2 = 0.0375 \times 30\,(4)^2 = 18 \text{ kN-m}$$

$\therefore$ Support moment at $A = M_{AD} = 0.7\, M_A = 0.7\,\frac{w_u L^2}{8} = 0.0875\, w_u L^2 = 42$ kN-m

With reference to redistributed moment at A for the beam shown in Fig. 21.7 (*d*),

$$R_A = \frac{1}{L}\left(\frac{w_u L^2}{2} + 0.0875\ w_u L^2\right) = 0.5875\ w_u L = 0.5875\ (30)\ (4) = 70.5\ \text{kN}$$

$$R_B = \frac{1}{L}\left(\frac{w_u L^2}{2} - 0.0875\ w_u L^2\right) = 0.4125\ w_u L = 0.4125\ (30)\ (4) = 49.5\ \text{kN}.$$

S.F. is zero at $x = \frac{R_B}{w_u} = \frac{0.4125\ w_u L}{w_u} = 0.4125\ L = 1.65$ m from B.

$$M_x = 0.4125\ w_u L x - \frac{w_u x^2}{2} \qquad \ldots(ii)$$

M_{max} (at $x = 0.4125\ L$) $= 0.4125\ w_u L\ (0.4125\ L) - \frac{w_u}{2}(0.4125\ L)^2 = 0.085\ w_u L^2 = 40.8375$ kN-m

Point of contraflexure occurs at E distant $x = 0.4125 \times 2\ L = 0.825\ L = 3.3$ m from B.

The B.M.D. after redistribution (i.e curve II) is shown in Fig. 21.7 (*e*).

(*c*) ***Curve III*** : The B.M.D. (i.e. curve III) for 0.7 of the moments obtained by elastic analysis, is shown by dotted curve in Fig. 21.7 (*c*).

Elastic B.M. at E (at $x = 0.825\ L = 3.3$ m), obtainable from Eq. (*i*) is given by

$$M_{ueE} = e_1 E = \frac{3}{8} w_u L\ (0.825\ L) - \frac{w_u}{2}(0.825\ L)^2 = -\ 0.03094\ w_u L^2 = -\ 14.80\ \text{kN-m}$$

Hence $e_3 E = 0.7\ e_1 E = 0.7\ (-\ 0.03094\ w_u L^2) = -\ 0.02166\ w_u L^2 = -\ 10.396$ kN-m

(*d*) *Final max. B.M.D. (B.M. envelope)*

The final max. B.M.D. (known as B.M. envelope) is shown in Fig. 21.7 (*f*).

Example 21.4. *Draw the maximum bending moment diagram for a fixed beam of span 4.5 m and carrying two collapse point loads of 10 kN each, acting at one third points. Follow IS Code recommendations for redistribution of moments.*

Solution

1. Elastic analysis : Using well known formulae for eccentric point load on fixed beam, we get

$$M_A = \frac{W a_1 b_1^2}{L^2} + \frac{W a_2 b_2^2}{L^2} = \frac{W\ (L/3)\ (2\ L/3)^2}{L^2} + \frac{W\ (2\ L/3)\ (L/3)^2}{L^2}$$

$$= \frac{2}{9} WL = \frac{2}{9} \times 12 \times 4.5 = 12\ \text{kN-m}$$

By symmetry $M_B = M_A = \frac{2}{9} WL = 12$ kN. Note that both M_A and M_B are of hogging nature and are negative.

$$R_A = R_B = W = 12\ \text{kN}$$

Measuring x from A, we get

$$M_x = R_A x - M_A \left| - \frac{W}{3}\left(x - \frac{L}{3}\right)\right. \quad \text{or} \quad M_x = Wx - \frac{2}{9} WL \left| - \frac{W}{3}\left(x - \frac{L}{3}\right)\right. \qquad \ldots(1)$$

This is zero, at point E distant $x = \frac{2}{9} L = 1$ m from A

At C, $$M_C = W\left(\frac{L}{3}\right) - \frac{2}{9} WL = +\frac{WL}{9} = +\frac{12 \times 4.5}{9} = 6\ \text{kN-m}$$

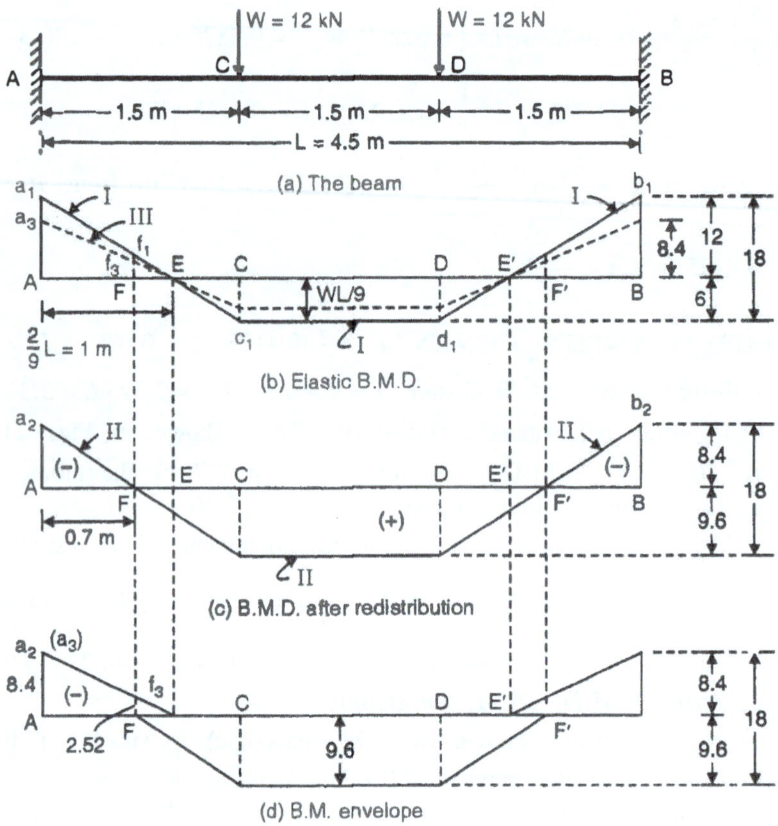

FIG. 21.8

$$\text{Sum of max. absolute hogging and sagging moments} = \left| -\frac{2}{9}WL \right| + \left| \frac{WL}{9} \right| = \frac{WL}{3} = 18 \text{ kN-m}$$

This the same as the free span B.M.

Fig. 21.8 (*b*) shows the B.M.D., (curve I) obtained by elastic analysis.

2. B.M. After 30% redistribution

Max. M_A, after 30% reduction $= -0.7 \times \frac{2}{9}WL = -0.1556\,WL = -8.4$ kN-m

Max. sagging B.M. $= \frac{WL}{9} + 0.3 \times \frac{2}{9}WL = 0.1778\,WL = +9.6$ kN-m

Sum of max. absolute hogging and sagging moments $= \left| -0.1556\,WL \right| + \left| +0.1778\,WL \right|$

$$= 0.3334\,WL = \frac{WL}{3} = 18 \text{ kN-m, as required.}$$

Also, $R_A = R_B = W = 12$ kN

$$M_x = Wx - 0.1556\,WL \left| -\frac{W}{3}\left(x - \frac{L}{3}\right) \right. \qquad \ldots(2)$$

This is zero at point F distant $x = 0.1556\,L = 0.7$ m from A.

The redistributed B.M.D (curve II) is shown in Fig. 21.8 (*c*).

3. Curve III

Curve III for 0.7 times the elastic bending moment is shown in Fig. 21.8 (*b*) by dotted line. Elastic B.M. at point F (at $x = 0.7$ m) is obtained from equation 1

Thus $$M_F = Ff_1 = 12 \times 0.7 - \frac{2}{9} \times 12 \times 45 = -3.6 \text{ kN-m}$$

Hence $$Ff_3 = 0.7\ Ff_1 = 0.7\ (-3.6) = -2.52 \text{ kN-m}$$

(*d*) **Final max B.M.D. (B.M. envelope)** : Shown in Fig. 21.8 (*d*).

21.6. CONTINUOUS BEAMS AND ONE WAY CONTINUOUS SLABS

When a single span statically indeterminate beam is loaded, the maximum bending moments occurs when the whole span is loaded fully with the design loads which consist of design dead load and design live load. However, in the case of continuous beams (and one way continuous slabs) the bending moment at any section of a given span depends not only on the type and position of loads in the same span but also on the type and position of loads in other spans. A continuous beam (or slab) is normally loaded with two types of loads : (*i*) dead load, and (*ii*) live load. The dead load (w_d) is occupied by all the spans simultaneously, but the live load (w_l) may be occupied by different sets of spans at different times. We can have the following combination of these two loads on a given span :

(*i*) ***Maximum load condition*** : *Design live load and design dead load acting simultaneously on a span.*

Thus total design load $w_{u\,,\,max} = 1.5\ (w_d + w_l)$

(*ii*) ***Minimum load condition*** : *Only dead load is acting and live load is absent.*

Thus total design load $w_{u\,,\,min.} = 1\ (w_d) = w_d$. Note that for minimum load condition, partial safety factor of only 1.0 is used for dead load.

There are two methods of analysis of continuous beams and one way continuous slabs : ***Method 1*** : Use the B.M. and S.F. coefficients given by the Code. In this case, no redistribution of moments is permitted by the Code. This method, though quicker, gives approximate results.

Method 2 : Elastic analysis done by considering various load combinations (as discussed below) and plotting the moment envelopes after carrying out redistributions as per IS Code recommendations.

Load combinations

We shall take the case of a six span continuous beam, for the purpose of illustration, to find the maximum bending moment at following three typical locations.

(*i*) Maximum sagging moment in a span

(*ii*) Maximum hogging moment at a given support

(*iii*) Maximum hogging moment in a span.

and (*iv*) Maximum sagging moment at a given support.

Case (i) : Maximum sagging moment in a span (Fig. 21.9 *a*)

For maximum sagging moment in a span (say span *CD*), the given span as well

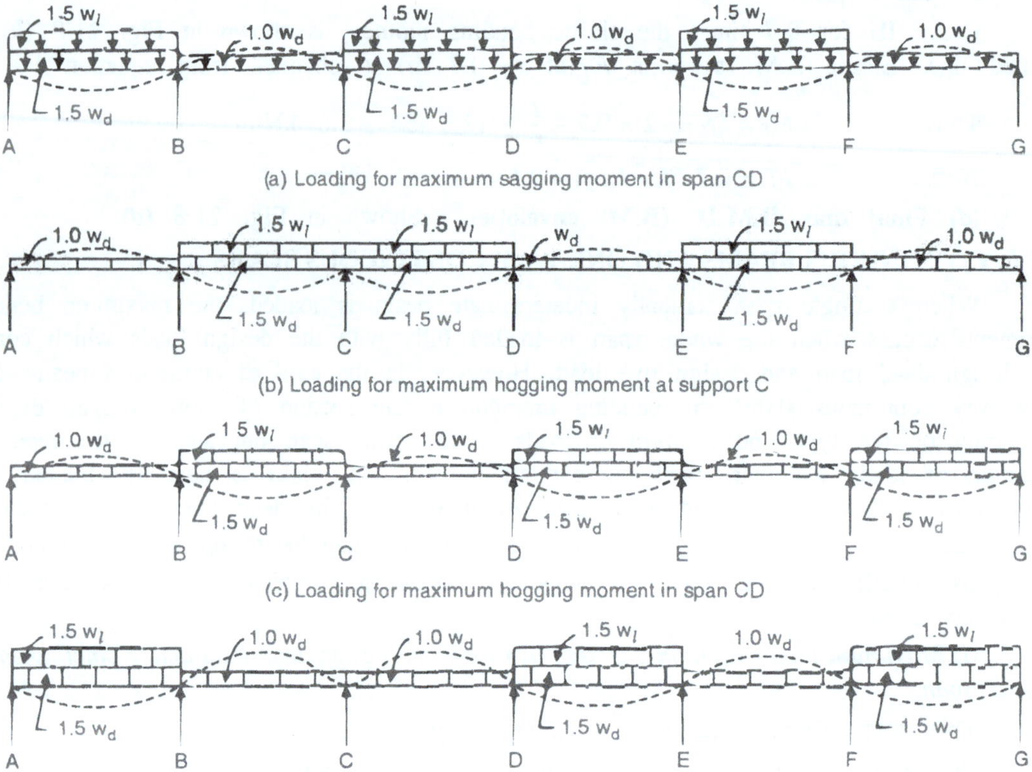

(a) Loading for maximum sagging moment in span CD

(b) Loading for maximum hogging moment at support C

(c) Loading for maximum hogging moment in span CD

(d) Loading for maximum sagging moment in support C

FIG. 21.9

as alternate spans (i.e. spans *AB* and *EF*) should be loaded under maximum load condition (i.e. $w_{u\,.\,max} = 1.5\ w_d + 1.5\ w_l$), while loads on other remaining spans (i.e. spans *BC*, *DE* and *FG*) should have minimum load $w_{u\,.\,min} = 1.0\ w_d$ as shown in Fig. 21.9 (*a*). The dotted lines show the deflected position of the beam.

Case (ii) : Maximum hogging moment at a support (Fig. 21.9 *b*)

For maximum hogging moment at a support (say support *C*), the *adjacent spans* on both sides of support *C* as well as spans alternate of these should be loaded with maximum design loads (i.e. $w_{u\,.\,max} = 1.5\ w_d + 1.5\ w_l$), while the remaining spans should be loaded with minimum load $w_{u\,.\,min} = 1.0\ w_d$, as shown in Fig. 21.9 (*b*). The dotted lines show the deflected position.

Case (iii) Maximum hogging moment in a span (Fig. 21.9 c)

For maximum hogging B.M. in a span (say span *CD*), the spans adjoining this span as well as alternate spans should be loaded with maximum load $w_{u\,.\,max} = 1.5\ w_d + 1.5\ w_l$, while the remaining spans should be loaded with minimum load $w_{u\,.\,min.} = 1.0\ w_d$, as shown in Fig. 21.9 (*c*).

Case (iv) : Maximum sagging moment at a support (Fig. 21.9 d)

For maximum sagging moment at a support (say support *C*), the adjacent spans should be loaded with minimum load $w_{u\,.\,min} = 1.0\,w_d$ while the next spans should be loaded with maximum load $w_{u\,.\,max} = 1.5\,w_d + 1.5\,w_l$. Other spans should be loaded as shown in Fig. 21.9 (*d*).

Procedure for analysis of continuous beams

Step 1 Determination of elastic moments (M_{ue}) :

Determine all possible adverse ultimate (or factored) load conditions, discussed above and obtain elastic moment (M_{ue}) diagram using elastic analysis. Hence obtain *maximum elastic moment diagram* (or elastic moment envelope) showing traces of both max. positive and max. negative bending moments at each point on the continuous beam.

Step 2: Determination of redistributed moments (M_{ud})

For *each of the above loading cases*, draw the redistributed bending moment (M_{ud}) diagram, using *desired* redistribution of moment, within permissible limits of 30% satisfying simultaneously the equilibrium condition, with the aim of *minimising* the values of maximum positive and maximum negative bending moments. Using the redistributed moment diagram, obtain the *maximum design moment envelope*.

Step 3 : Determination of $0.7\,M_{ue}$ *diagram*

As per condition 2 of IS Code recommendation the ultimate moment of resistance provided at any section of a member is not less than 70% of the moment at that section obtained from the elastic maximum moment diagram determined in step 1 above. Hence draw $0.7M_{ue}$ diagram by multiplying each ordimate of M_{ue} by 0.7 noting the fact that points of contraflexure of M_{ue} and $0.7\,M_{ue}$ diagram remain the same. Lastly, obtain the *traces* of $0.7\,M_{ue}$ diagram which lie outside the maximum design moment envelope obtained in step 2.

Example 21.5. *A two span continuous beam ABC, 10 m long, freely supported at A and C and continuous over central support B, 5 m from A, carries a characteristic dead load of 12 kN/m and characteristic live load of 18 kN/m. Plot the maximum elastic moment diagram before redistribution of moments and design moments envelope as per IS Code recommendations, using partial safety factor of 1.5 on loads.*

Solution

$w_{u\,.\,max} = 1.5\,(w_d + w_L) = 1.5\,(12 + 18) = 45$ kN/m ; $w_{u\,.\,min} = 1.0\,w_d = 1.0 \times 12 = 12$ kN/m.

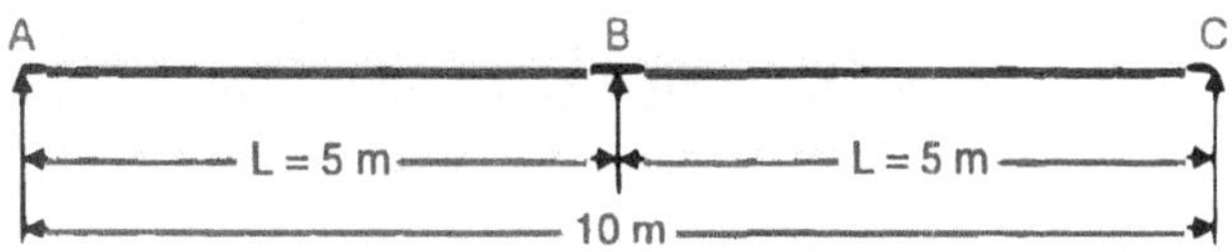

FIG. 21.10

We will consider three cases of loading as under ;

Case (*a*) : Loading for maximum hogging moment at support *B*

Case (*b*) : Loading for maximum sagging moment in *AB* : This case will incidentally given maximum hogging moment in *BC*

Case (*c*) : Loading for maximum hogging moment in *AB* : This case will incidentally give maximum sagging moment in *BC* :

Case (a) : Loading for maximum hogging moment at support *B*

As discussed in §21.6, in order to get maximum hogging B.M. at support *B*, the adjacent spans (i.e. spans *AB* and *BC*) should be loaded with maximum design loads $w_{u,\,max} = 1.5\,(w_d + w_l) = 45$ kN/m. This is shown in Fig. 21.11. The dotted lines in Fig. 21.11 show the deflected shape of the beam. Since this corresponds to the case of *support symmetry*, the support *B* may be assumed as fixed and the distribution process may be carried for half the portion only, thus getting the case of a propped cantilever.

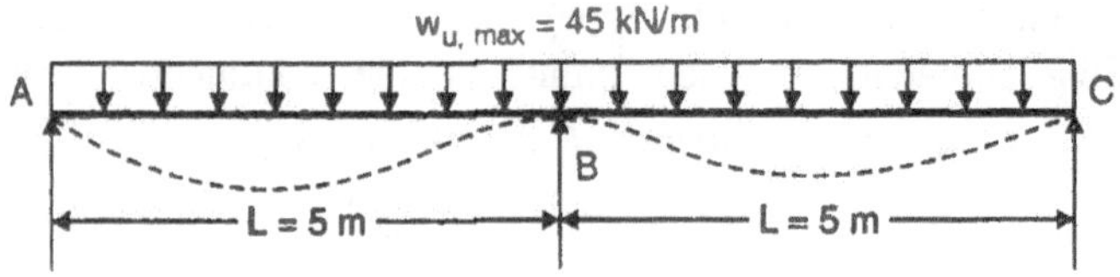

FIG. 21.11

(i) Elastic moment diagram

For a propped cantilever (example 21.3)

$$M_{ueB} = -\frac{w_u L^2}{8} = -\frac{45 \times (5)^2}{8} = -140.625 \text{ kN-m}$$

$$R_A = \frac{1}{5}\left(\frac{45 \times 5^2}{2} - 140.625\right) = 84.375 \text{ kN}$$

S.F. is zero at $x = \frac{R_A}{w_u} = \frac{84.375}{45} = 1.875$ m from *A*.

$$M_x = 84.375\,x - \frac{45\,x^2}{2} \qquad \ldots(1)$$

$$\therefore \quad M_{max} = 84.375\,(1.875) - 22.5\,(1.875)^2 = 79.10 \text{ kN-m}$$

M is zero at $x = \frac{84.375 \times 2}{45} = 3.75$ m from *A* or at 1.25 m from *B*.

Mid span moment $M_C = 84.375 \times 2.5 - 22.5\,(2.5)^2 = 70.3125$ kN-m.

At mid-span, the sum of absolute hogging and sagging moments $= \frac{140.625}{2} + 70.3125$ $= 140.625$ kN-m. This should be equal to free span moment in span $AB = w_{u\,.\,max} L^2/8$ $= 45\,(5)^2/8 = 140.625$ kN-m. Hence OK. Note that it is just a coincidence than free span B.M. is equal to max. hogging moment in the span of a symmetrical two span beam. The elastic moment diagram (M_{ue} dia.) is represented by curve a_1 (solid lines) in Fig. 21.12 (*d*).

(ii) Moment diagram (M_{ud}) after redistribution

Since the difference between max. hogging and sagging moments is quite large, we will carry-out maximum permissible redistribution @ 30%.

Reduced hogging moment at support $B = M_{udB} = -0.7 \times 140.625 = -98.4375$ kN-m.

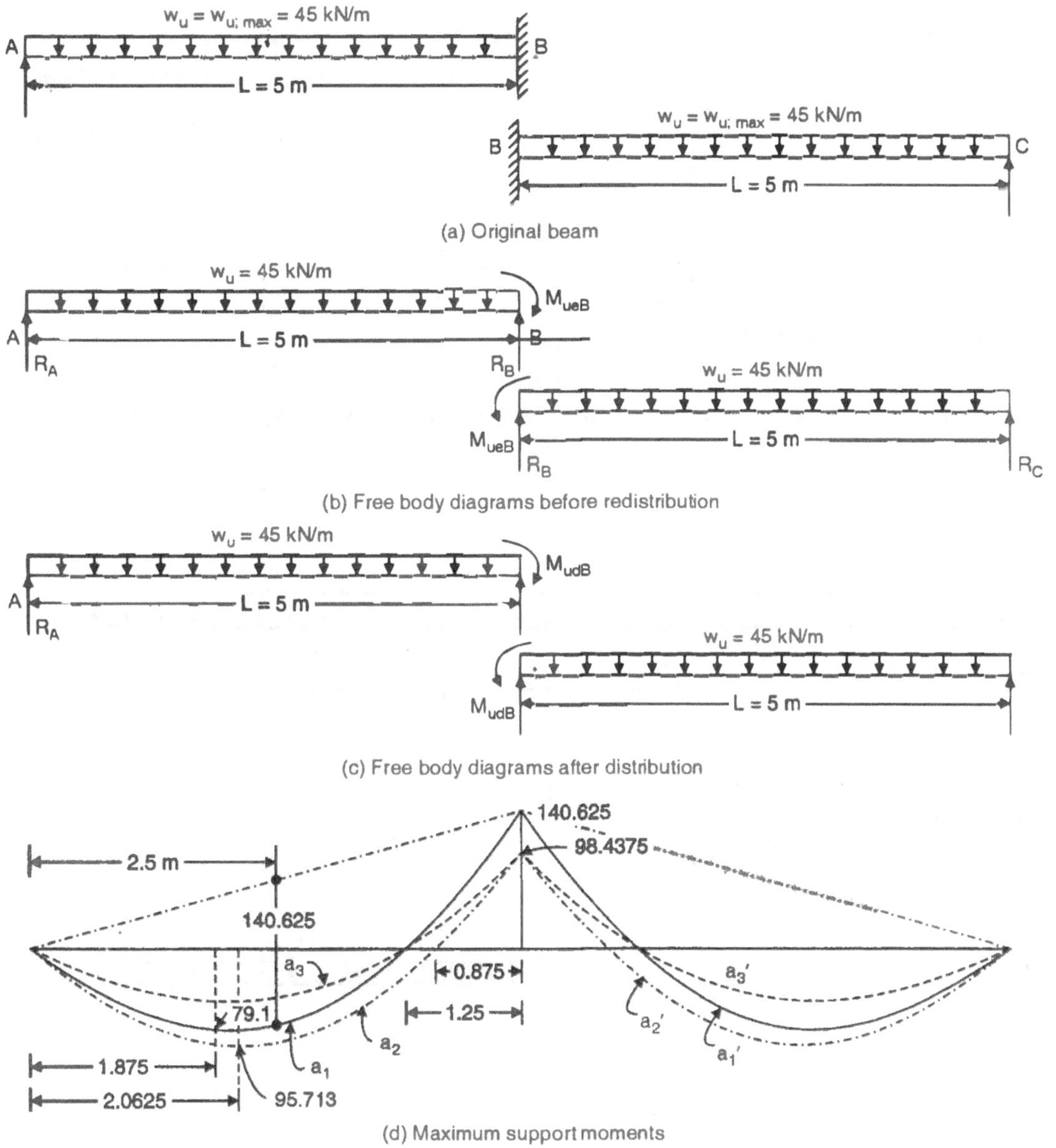

FIG. 21.12

Fig. 21.12 (*c*) shows the free body diagram of the beam *AB*, from which

$$R_A = \frac{1}{5}\left(\frac{45 \times 5^2}{2} - 98.4375\right) = 92.8125 \text{ kN}$$

S.F. is zero at $x = \dfrac{R_A}{w_u} = \dfrac{92.8125}{45} = 2.0625$ m from *A*, or 2.9375 m from *B*.

$$M_x = 92.8125\,x - 22.5\,x^2 \qquad \ldots(2)$$

$\therefore\; M_{max}$ (at $x = 2.0625$ m) $= 92.8125\,(2.0625) - 22.5\,(2.0625)^2 = 95.713$ kN-m

Mid-span moment $M_C = 92.8125\,(2.5) - 22.5\,(2.5)^2 = 91.406$ kN-m

B.M. is zero at $x = 92.8125/22.5 = 4.125$ m from A or 0.875 m from B.

Sum of absolute hogging and sagging moments at mid-span $= \dfrac{98.4375}{2} + 91.406$

$= 140.625$ kN-m = free span moment in span AB.

The B.M. diagram after redistribution is shown by chain clotted curve a_2 in Fig. 21.12 (*d*). We observe that the max. hogging and max. sagging moments are nearly equal after redistribution.

***(iii)* Moment diagram for 0.7 M_{ue}**

The curve for 0.7 M_{ue} is shown by curve a_3 drawn in dotted lines, in Fig. 21.12 (*d*).

Case (b) : Loading for maximum sagging moment in span AB

In order to get maximum sagging moment in span AB (and hence maximum hogging B.M. in BC) span AB should be loaded with $w_{u,\,max}$ (= 45 kN/m) while the span BC should be loaded with $w_{u,\,min} = 12$ kN/m, as shown Fig. 21.13 (*a*).

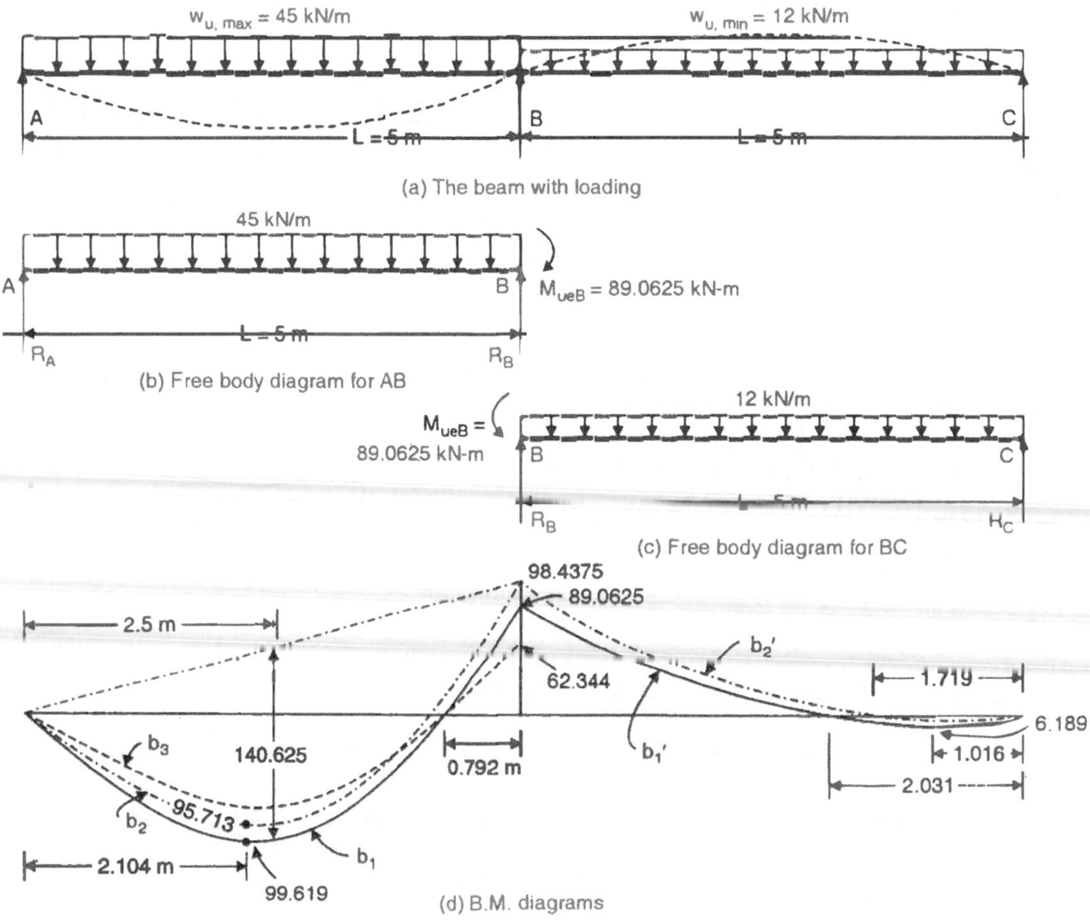

FIG. 21.13

The dotted curve shows the deflected shape of the beam.

(*i*) ***Elastic moment diagram***

$$M_{FBA} = \frac{45\,(5)^2}{8} = 140.625 \text{ kN-m} \;; \qquad M_{FBC} = \frac{12\,(5)^2}{8} = 37.5 \text{ kN-m}$$

Since both the spans *AB* and *BC* have equal section, equal span and similar end conditions at far ends *A* and *C*, the distribution factors at the joint *B* will be equal to 0.5 for each span. The moment distribution is carried out in Tabular form below.

Joint	*A*	*B*		*C*
Member	*AB*	*BA*	*BC*	*CB*
D.F.		0.5	0.5	
F.E.M. Distribution	0	+ 140.625 − 51.5625	− 37.5 − 51.5625	0
Final moments	0	+ 89.0625	− 89.0625	0

Maximum span moment in AB

Fig. 21.13 (*b*) shows the free body diagram for *AB*.

$$R_A = \frac{1}{5}\left[\frac{45\,(5)^2}{2} - 89.0625\right] = 94.6875 \text{ kN}$$

S.F. is zero at $x = \frac{R_A}{w_u} = \frac{94.6875}{45} = 2.104$ m from *A* or 2.896 m from *B*.

$$M_x = 94.6875\,x - 22.5\,x^2 \qquad \ldots(3)$$

M is zero at $x = 94.6875/22.5 = 4.208$ from *A* or 0.792 m from *B*

$$M_{max} = 94.6875\,(2.104) - 22.5\,(2.104)^2 = 99.619 \text{ kN-m}$$

Moment at mid-span $= M_C = 94.6875\,(2.5) - 22.5\,(2.5)^2 = 96.094$ kN-m

Hence sum of absolute hogging and sagging moments at mid-span

$$= \frac{89.0625}{2} + 96.094 = 140.625 \text{ kN-m} = \text{central depth of parabola.}$$

The elastic B.M.D., (curve b_1) is represented by solid lines in Fig. 21.13 (*d*)

Maximum span moment in BC

Fig. 21.13 (*c*) shows the free body diagram for *BC*.

$$R_C = \frac{1}{5}\left[\frac{12\,(5)^2}{2} - 89.0625\right] = 12.1875 \text{ kN}$$

S.F. is zero at $x = \frac{R_C}{w_u} = \frac{12.1875}{12} = 1.016$ m from *C*

$$M_x = 12.1875\,x - 6\,x^2 \qquad \ldots(4)$$

B.M. is zero at $x = 12.1875/6 = 2.031$ m from *C*

$$M_{max} = 12.1875 \times 1.016 - 6\,(1.016)^2 = 6.189 \text{ kN-m}$$

Moment at mid span $= 12.1875 \times 2.5 - 6\,(2.5)^2 = -7.031$ kN-m

$\therefore$ Central depth of parabola $= \frac{89.0625}{2} - 7.031 \simeq 37.5$ kN-m

Free span moment for span $BC = \frac{12\,(5)^2}{8} = 37.5$ kN-m

The elastic B.M.D. is represented by curve b_1' (solid lines) in Fig. 21.13 (*d*).

(*ii*) ***Moment diagram*** **(M_{ud})** ***after redistribution***

The maximum sagging moment in span *AB* is 99.619 kN-m while max. hogging moment is 89.0625 kN-m. Though maximum permissible reduction is 30%, we will *reduce* the sagging B.M. only slightly, and correspondingly *increase* the hogging B.M. at *B* so as to maintain equilibrium. From case (*a*), maximum sagging moment, after redistributions was 95.713 kN-m. Hence reduce the maximum sagging moment from 99.619 kN-m to 95.713 kN-m. The corresponding increased B.M. at the support will be 98.4375 kN-m in order to maintain static equilibrium in *AB*. The resulting B.M.D. is shown by curve b_2 (chain dotted lines) in Fig. 21.13 (*d*).

The B.M. in the span *BC* is also changed by considering its equilibrium and by noting the fact that $M_{udB} = 98.4375$ kN-m. Consider the free body diagram for *BC* shown is Fig 21.13 *c* by taking $M_{udB} = 98.4375$,

Reaction $$R_C = \frac{1}{5}\left[\frac{12\,(5)^2}{2} - 98.4375\right] = 10.3125 \text{ kN}$$

S.F. is zero distant $x = 10.3125/12 = 0.859$ m from *C*

$$M_x = 10.3125\,x - 6\,x^2 \qquad \ldots(5)$$

B.M. is zero at $x = 10.3125/6 = 1.719$ m from *C*.

$$M_{\max} = 10.3125\,(0.859) - 6\,(0.859)^2 = 4.431 \text{ kN-m}$$

The resulting B.M.D. is shown by curve b_2' (chain dotted line) in Fig. 21.13 (*d*).

(*iii*) ***Moment diagram for 0.7*** M_{ue} : Curve b_3 shows the curve for 0.7 M_{ue} for span *AB*. This moment is evidently zero at $x = 0.792$ m from *B*.

Case (*c*) Load for maximum hogging moment in *AB*

In order to get maximum hogging B.M. in span *AB* (and hence maximum sagging B.M. in *BC*), span *BC* should be loaded with $w_{u,\max}$ (= 45 kN/m) while the span *AB* should be loaded with $w_{u,\min} = 12$ kN/m. Hence the bending moment diagrams, both before and after redistribuion are mirror images of those due to loads of case (*b*). Hence these are not drawn separately.

Maximum elastic B.M. diagram (elastic moment envelope)

Elastic moment envelope is obtained by superimposing the three elastic B.M. diagrams shown by solid lines in Fig. 21.12 and 21.13. The elastic moment envelope (or maximum elastic B.M. diagram) is shown in Fig. 21.14. The max. sagging moment in span *AB* is governed by curve b_1 (case *b*). In span *BC*, the hogging moment curve (i.e. curve a_1') of case (*a*) will intersect with hogging moment curve (i.e. curve b_1') of case (*b*). Similarly, in span *AB*, hogging moment curve (i.e. curve a_1') of case (*a*) will intersect with the hogging moment curve of case (*c*).

In order to find the point of intersection of the two hogging B.M. curves (i.e. curve a_1' and b_1'), we proceed as follows.

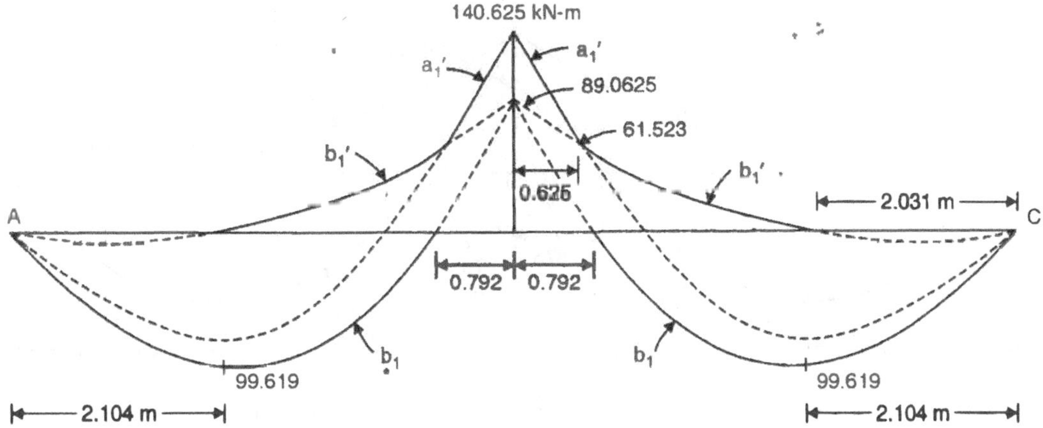

FIG. 21.14. ELASTIC MOMENT ENVELOPE

For case (*a*), refer the free body diagram of *BC* (Fig. 21.12 *b*).

Reaction $$R_B = \frac{1}{5}\left[\frac{45\,(5)^2}{2} + 140.625\right] = 140.625 \text{ kN}$$

Hogging B.M. at any distance *x* from *B*, for beam *BC* is given by

$$M_x = 140.625 + \frac{45\,x^2}{2} - 140.625\,x \qquad \ldots(i)$$

This is the equation of curve a_1'.

Similarly, for case (*b*), refer the free body diagram of *AB* (Fig. 21.13 *b*)

$$R_B = \frac{1}{5}\left[\frac{12\,(5)^2}{2} + 89.0625\right] = 47.8125 \text{ kN}$$

Hogging B.M. at any distance *x* from *B*, for beam *BC* is given by

$$M_x = 89.0625 + \frac{12\,x^2}{2} - 47.8125\,x \qquad \ldots(ii)$$

At the point of intersection, M_x given by (*i*) and (*ii*) should be equal.

$$\therefore \quad 140.625 + 22.5\,x^2 - 140.625\,x = 89.0625 + 6\,x^2 - 47.8125\,x$$

or $$x^2 - 5.625\,x + 3.125 = 0 \text{ which gives } x = 0.625 \text{ m}$$

Hence, at the point of intersection, $M_x = 89.0625 + 6\,(0.625)^2 - 47.8125\,(0.625) = 61.523$ kN-m.

Due to symmetrical loading and geometry, the point of intersection in span *AB* will also be 0.625 m from *B*, and the hogging B.M. will also be equal to 61.523 kN-m. The elastic moment envelope is shown in Fig. 21.14.

Maximum moment diagram (B.M. envelope after redistribution)

The B.M. envelope after redistribution is obtained by superimposing the three redistributed B.M. diagrams shown by chain dotted lines in Figs. 21.12 and 21.13. The resulting maximum moment diagram (or B.M. envelope after redistribution) is shown in Fig. 21.15.

The envelopes of max. hogging B.M., after redistribution, is governed by curve b_2'

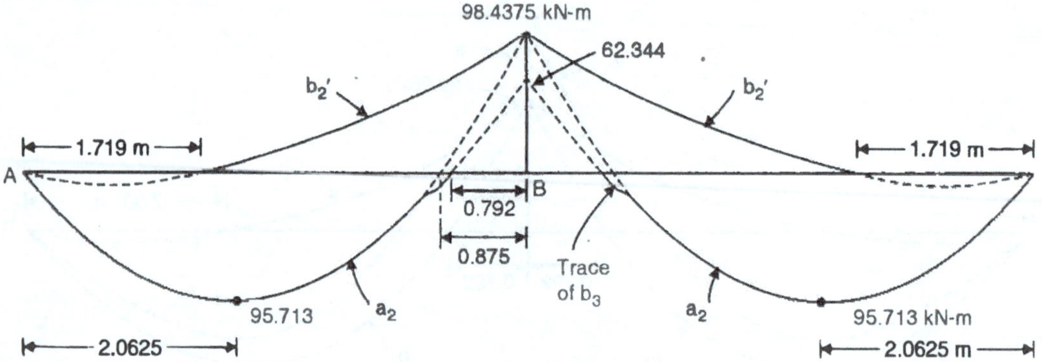

FIG. 21.15. REDISTRIBUTED MOMENT ENVELOPE

for case (*b*). The envelope of max. sagging B.M., after redistribution, is governed by curve a_2 for case (*a*) and the trace of curve b_3 drawn for $0.7\,M_{ue}$. Fig. 21.15 shows the redistributed moment envelope.

21.7. APPROXIMATE METHOD

We have seen in the previous example that the exact method of moment redistribution is quite cumbersome. The following *approximate method* gives the results quickly, though approximately, the error involved being on the safer side.

Step 1. Determine the maximum support moments. For finding maximum hogging moment at a support, it is necessary to load the adjacent spans by $w_{u\ \max} = 1.5\,(w_d + w_l)$.

Step 2. Carry out the moment redistribuion by decreasing the support moment at any one end by dM and increasing the span moment by $dM/2$. Hence if the redistribution of moments at two ends A and B of a continuous beam are dM_A and dM_B, increase the sagging B.M. in span AB by $\frac{1}{2}(dM_A + dM_B)$

Example . *To illustrate the procedure, consider the continuous beam ABC of previous example.*

By elastic analysis, $M_A = 0,\ M_C = 0$ and $M_B = -\dfrac{w_{u\ ,\ \max} L^2}{8} = -\dfrac{45\,(5)^2}{8}$

$= -\ 140.625$ kN-m, for case (*a*) loading.

$$\therefore \qquad R_A = \frac{1}{5}\left[\frac{45\,(5)^2}{2} - 140.625\right] = 84.375 \text{ kN}$$

S.F. is zero at $\qquad x = \dfrac{R_A}{w_{u\ ,\ \max}} = \dfrac{84.375}{45} = 1.875$ m from A

$$\therefore \qquad M_{\max} = 84.375\,(1.875) - \frac{45\,(1.875)^2}{2} = 79.10 \text{ kN-m}$$

Hence maximum support moment (i.e. hogging moment) = – 140.625 kN-m

and maximum span moment (i.e. sagging moment) = + 79.10 kN-m

Since the difference between these two is quite large, $dM = 30\%$ at support B. The *Moment redistribution* is carried out in a tabular form below.

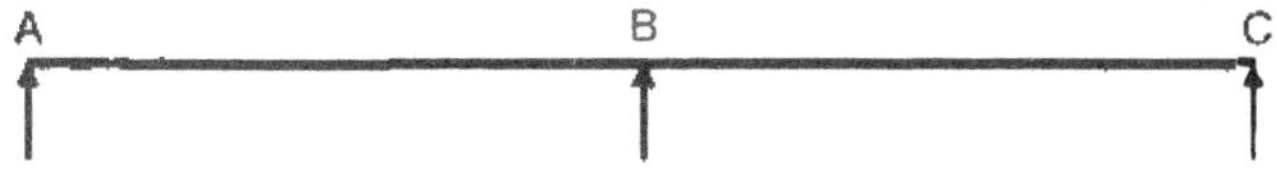

1. Initial elastic moments	0	+ 79.10	− 140.625	+ 79.10	–
2. Redistribution @ $dM = 30\%$ at B	–	–	+ 42.188	–	–
3. Increase in span moment @ $dM/2$.	–	+ 21.094	–	+ 21.094	–
4. Final moments.	–	+ 100.194	− 98.437	+ 100.194	–
5. Values by exact analysis	–	+ 95.713	− 98.437	+ 95.713	–

21.8. ANALYSIS BY MOMENT COEFFICIENTS

As per IS 456: 2000, the design bending moments can be found from the coefficients given below (Table 21.1), for beams of uniform cross-section which supports substantially uniformly distributed loads over three or more spans which do not differ by more than 15 percent of the longest.

The moment at supports where two unequal spans meet or in case where the spans are not equally loaded, the average of the two values for negative moments at the support may be taken for design.

TABLE 21.1. BENDING MOMENT COEFFICIENTS

Type of load	*Span moments*		*Support moments*	
	Near middle of end span	*At middle of interior span*	*At support next to the end support*	*At other interir supports*
1. Dead load and imposed load (fixed)	$+\frac{1}{12}$	$+\frac{1}{16}$	$-\frac{1}{10}$	$-\frac{1}{12}$
2. Imposed load (not fixed)	$+\frac{1}{10}$	$+\frac{1}{12}$	$-\frac{1}{9}$	$-\frac{1}{9}$

Note 1: For finding the B.M., the coefficient shall be multiplied by the total design load and the effective span.

Note 2: Where coefficient given in Table 21.1 are used for calculation of bending moments, *redistribution shall not be permitted.*

22

Design of Footings

22.1. INTRODUCTION

Any structure is generally considered to have two main portions (*i*) the *superstructure* and (*ii*) the *substructure*. The substructure transmits the loads of superstructure to the supporting soil and is generally termed as the *foundation*. *Footing* is that portion of the foundation which ultimately delivers the load to the soil, and is thus in contact with it. The load of the superstructure is transmitted to the foundation or substructure through either columns or walls. The object of providing foundation to a structure is to distribute the load to the soil in such a way that the maximum pressure on the soil does not exceed its permissible bearing value, and at the same time the settlement is within the permissible limits.

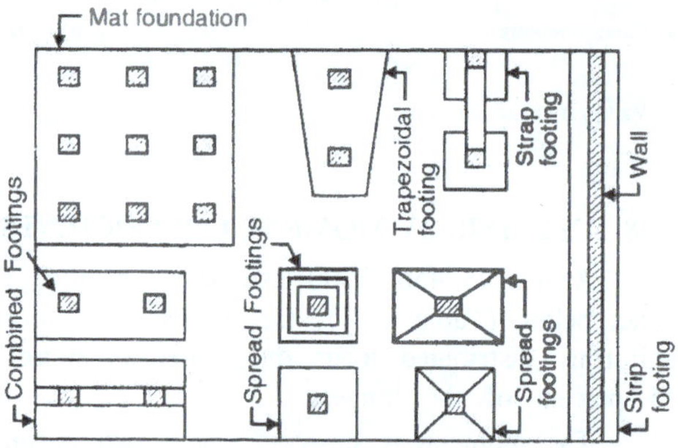

FIG. 22.1 VARIOUS TYPES OF SHALLOW FOOTINGS.

Foundations may be broadly classified under two heads : *shallow foundation and deep foundation*. According to Terzaghi, a foundation is shallow if its depth is equal to or less than its width. In the case of deep foundation, the depth is greater than the width. Apart from deep strip, rectangular or square foundations, other common forms of deep foundations are: *pier foundation, pile foundation and well foundation*. The shallow foundations are of the following types: Spread footing, strap footing, combined footing and mat or raft footing. Fig. 22.1 shows the common types of shallow foundations.

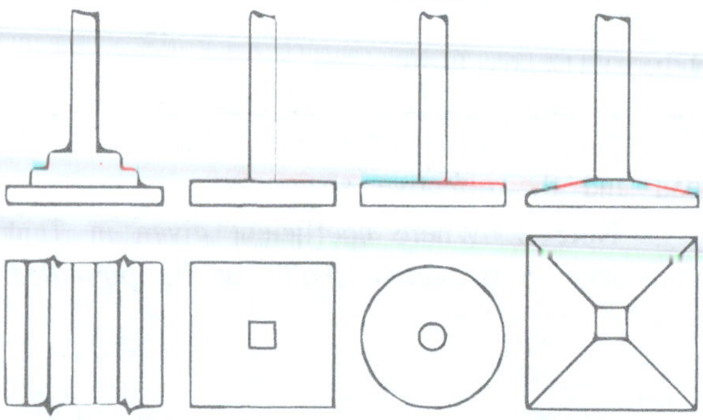

FIG. 22.2. TYPICAL SPREAD FOOTINGS.

Spread footing. A spread footing or simply footing, is a type of shallow foundation used to transmit the load of an isolated column, or that of a wall, on the subsoil. In the case of wall, the footing is *continuous* while in the case of column, it is *isolated*. Fig. 22.2 shows some common types of reinforced concrete spread footings.

Combined footing. A spread footing which supports two or more columns is termed as a *combined footing*. Such a footing is provided when the individual footings are either very near to each other, or overlap. Combined footings may either be rectangular or trapezoidal (Fig. 22.3 *a,b*).

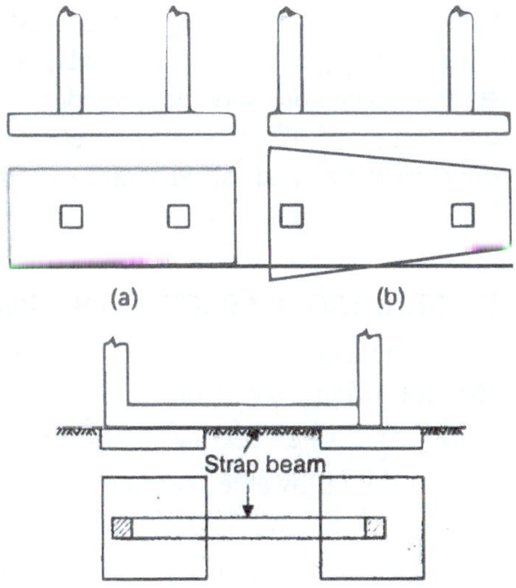

FIG. 22.3.

Strap or cantilever footing. A strap footing consists of spread footings of two columns connected by a strap beam. The strap beam does not remain in contact with soil, and thus does not transfer any pressure to the soil. Such a footing is generally used to combine the footing of outer column to the adjacent one so that the footing of the former does not extend in the adjoining property [Fig. 22.3 (*c*)].

Mat or raft foundation. A mat or raft is a combined footing that covers the entire area beneath a structure and supports all the walls and columns. When the available soil pressure is low or the building loads are heavy, the use of spread footings would cover more than one-half of the area and it may prove more economical to use mat or raft foundation.

Pile foundation. Pile foundation is a deep foundation used where the top soil is relatively weak. Piles transfer the load to a lower stratum of greater bearing capacity, by way of end bearing, or to the intermediate soil through skin friction. This is most common type of deep foundation generally used for buildings where a group of piles transfer the load of the super-structure to the sub-soil.

22.2. PRESSURE DISTRIBUTION BENEATH FOOTINGS

Both from observations as well as the analytical studies from theory of elasticity, it is known that the pressure distribution beneath footings, symmetrically loaded is not uniform. The pressure intensities depend upon the rigidity of footing, soil type, and the condition of soil. Fig. 22.4 (*a*) and (*b*) show the probable pressure distribution beneath a rigid footing on a loose cohesionless soil and cohesive soil. Fig. 22.4 (*c*) shows the usually assumed uniform pressure distribution.

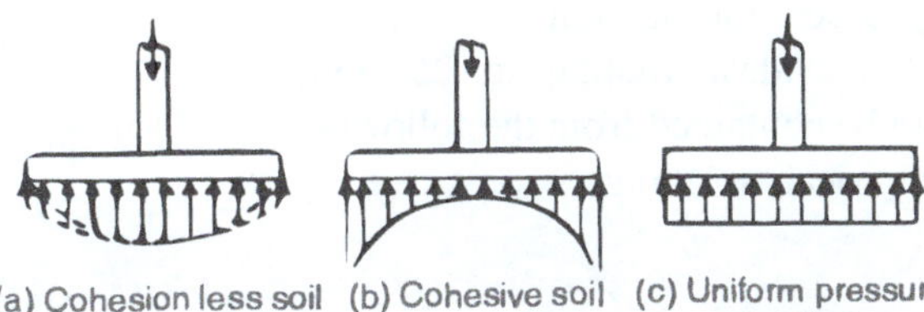

FIG. 22.4. PRESSURE DISTRIBUTION BENEATH FOOTINGS.

When a rigid footing rests on cohesionless soil, the soil grains at the outer edge have no lateral restraint, whereas in centre the soil is relatively confined resulting in a pressure distribution shown in Fig. 22.4(*a*). For the footings on cohesive soils, the edge stresses may be very large. However, the pressure distribution may be considered as *linear* as shown in Fig. 22.4(*c*), for the purpose of the design of reinforced concrete footings. Once the pressure distribution is known, the bending moment and shear force in the footing can be calculated, and the thickness of the structural member of the footing, alongwith the reinforcement etc., can be calculated using the usual principles of reinforced concrete.

22.3. BEARING CAPACITY OF SOIL AND SETTLEMENT OF FOOTINGS

In conventional design, the allowable bearing capacity should be taken as the smaller of the following two values :

(*i*) the safe bearing capacity based on ultimate capacity.

(*ii*) the allowable bearing pressure based on tolerable settlement.

The safe bearing capacity may be determined from a plate load test. Alternatively, the safe bearing capacity q_0 may be computed from the following equations based on Terzaghi's analysis:

For strip footing, $$q_0 = \frac{1}{F}[cN_c + \gamma D(N_q - 1)R_{w1} + 0.5\gamma BN_\gamma R_{w2}] + \gamma D \quad ...(22.1)$$

For square footing, $$q_0 = \frac{1}{F}[1.3\, cN_c + \gamma D(N_q - 1)R_{w1} + 0.4\gamma BN_\gamma R_{w2}] + \gamma D \quad ...(22.2)$$

For circular footing, $$q_0 = \frac{1}{F}[1.3cN_c + \gamma D(N_q - 1)R_{w1} + 0.3\gamma BN_\gamma R_{w2}] + \gamma D \quad ...(22.3)$$

where D = depth of footing

B = width of footing (strip or square) or diameter of circular footing

F = factor of safety (2 to 3)

N_c, N_q, N_γ = Bearing capacity factors for general shear failure.

For local shear failure, N_c', N_q' and N_γ' should be used. R_{w1} and R_{w2} are the water reduction factors given by (Fig. 22.5).

$$R_{w1} = 0.5\left(1 + \frac{Z_{w1}}{D}\right) \quad ...[22.4(a)]$$

$$R_{w2} = 0.5\left(1 + \frac{Z_{w2}}{B}\right) \quad ...[22.4(b)]$$

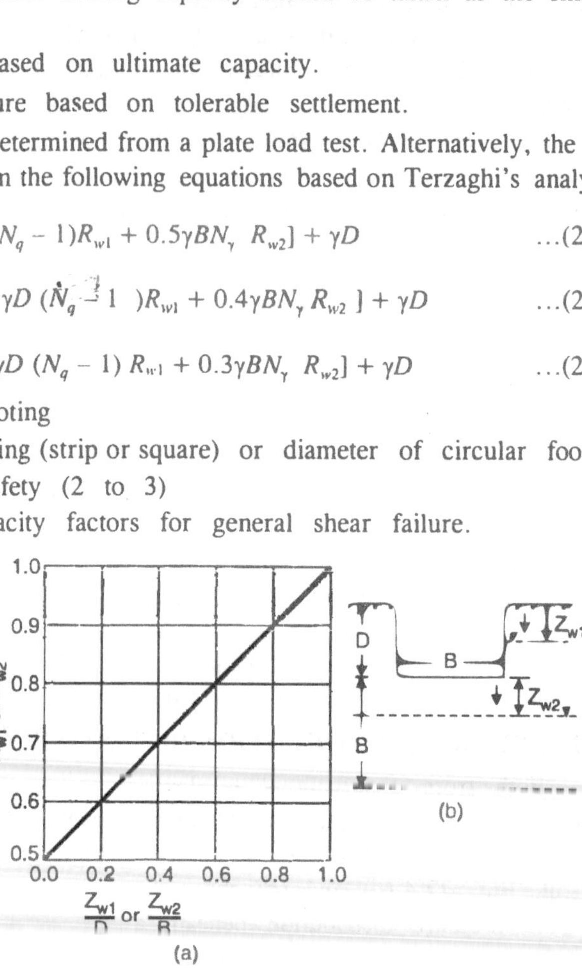

FIG. 22.5. WATER REDUCTION FACTORS.

The allowable bearing pressure q_ρ based on maximum settlement of individual footing to 25 mm, can be computed from the following empirical relation:

$$q_\rho = 34.3\,(N - 3)\left(\frac{B + 0.3}{2B}\right)^2 . R_{w2}\, R_d \quad ...(22.5)$$

where q_ρ = allowable net increase in soil pressure over existing soil pressure for settlement of 25 mm, kN/m²

N = standard penetration number with applicable corrections
B = width of footings, in metres
R_{w2} = water reduction factor
R_d = depth factor $= \left(1 + \frac{0.2\,D}{B}\right) \leq 1.20$

If the safe bearing capacity is taken to be smaller of the two values discussed above, the footings on granular soils will not suffer detrimental settlement. The total settlement of a footing in clay may be considered to consist of three components (Skempton and Bjerrum, 1957) :

$$S = S_i + S_c + S_s \qquad \ldots(22.6)$$

where S = total settlement
S_i = immediate elastic settlement
S_c = consolidation settlement
S_s = settlement due to secondary consolidation of clay.

The immediate settlement S_i is the elastic settlement, and can be computed from the following expression based on the theory of elasticity :

$$S_i = q\,B\left(\frac{1-\mu^2}{E_s}\right) I_w \qquad \ldots(22.7)$$

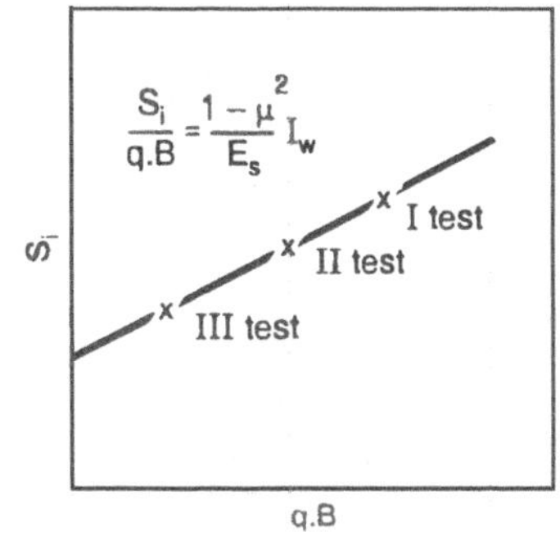

FIG. 22.6.

where q = intensity of contact pressure (kN/m²)
B = least lateral dimension of footing, metres
E_s = modulus of elasticity for soil, (kN/m²)
I_w = influence factor,
= 0.88 for rigid circular footing.
= 0.82 for rigid square footing.
= 1.06 for rigid rectangular footing with $L/B = 1.5$
= 1.70 for rigid rectangular footing width $L/B = 5$

It is difficult to determine μ and E_s for soils. However, the entire term $\frac{1-\mu^2}{E_s} I_w$ of Eq. 22.7 may be determined from plate load tests, by using two or three different size plates of the same shape. A plot between S_i and $q.B$ give a

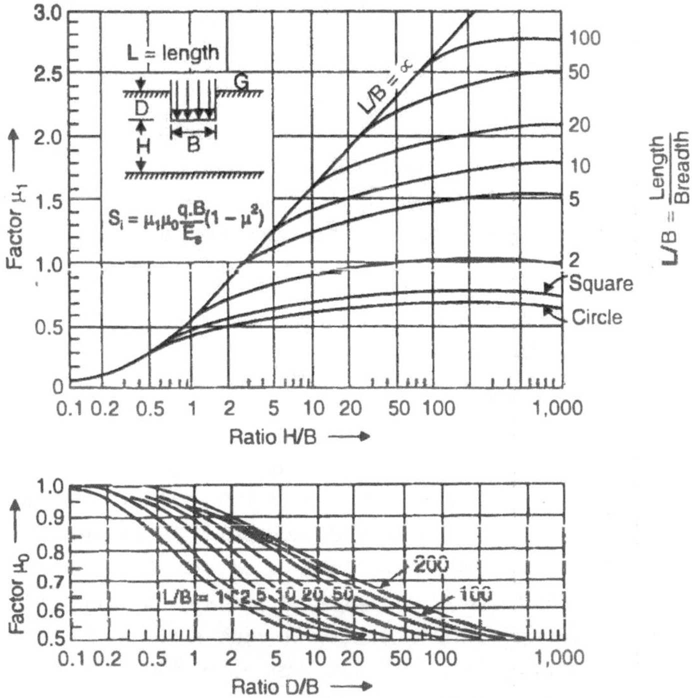

FIG. 22.7. VALUE OF μ_o AND μ_1
(AFTER JANBU, BJERRUM AND KJAERNISLI, 1966)

straight line, the slope of which is equal to $\frac{1-\mu^2}{E_s} I_w$.

Janbu, Bjerrum and Kjaernisli (1966) have proposed the following equation for computing the immediate settlement :

$$S_i = \mu_0 \; \mu_1 . \, q \, B \frac{1-\mu^2}{E_s} \qquad \ldots(22.8)$$

where the modulus of elasticity E_s is computed from the triaxial test data :

$$E_s = \frac{\sigma_1 - \sigma_3}{\Delta L / L} \qquad \ldots(22.9)$$

and μ_0 and μ_1 are taken from Fig. 22.7.

The *consolidation settlement* S_c is computed from the following expression:

$$S_c = C \frac{C_c}{1+e_0} . \, H \log_{10} \frac{\sigma_0 + \Delta\sigma}{\sigma_0} \qquad \ldots(22.10)$$

where

σ_0 = effective overburden pressure due to soil over burden

$\Delta\sigma$ = vertical stress on footing

C_c = compression index = $0.009\,(w_L - 10)$

e_0 = initial voids ratio

H = thickness of compressible layer.

C = a coefficient or correction factor (given by Fig. 22.8) depending upon the geometry of the footing and the history of loading on the clay (*i.e.*, on the pore pressure coefficient A).

w_L = Liquid limit.

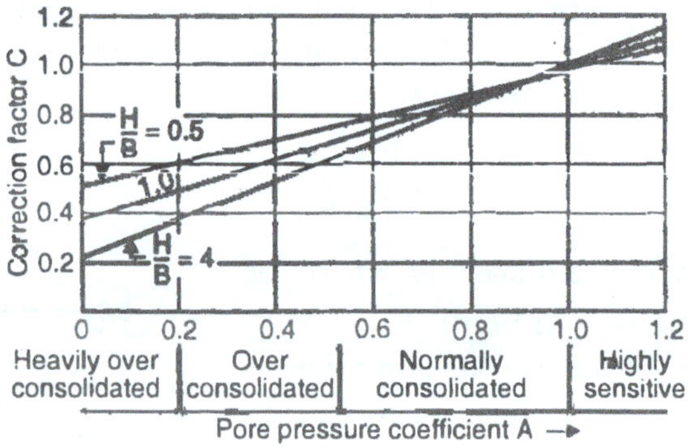

FIG. 22.8. VALUE OF FACTOR C.

22.4. INDIAN STANDARD CODE RECOMMENDATIONS FOR DESIGN OF FOOTINGS (IS : 456 - 2000)

(A) General

1. In sloped or stepped footings, the effective cross-section in compression shall be limited by the area above the neutral plane, and the angle of slope or depth and location of steps shall be such that the design requirements are satisfied at every section. Sloped and stepped footings that are designed as a unit shall be constructed to assure action as a unit.

2. *Thickness at the edge of footing*. In reinforced and plain concrete footings, the thickness at edge shall be not less than 150 mm for footings on soils nor less than 300 mm above the tops of piles for footings on piles.

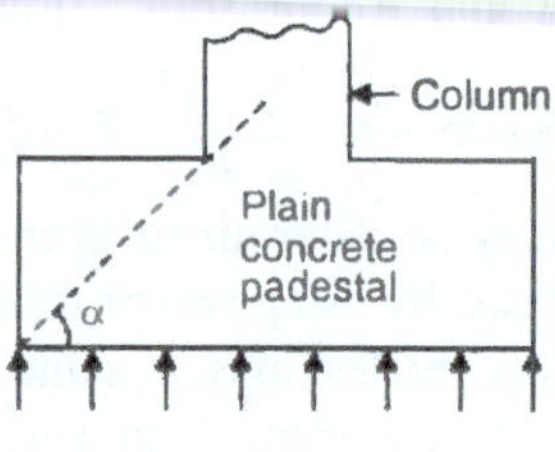

FIG. 22.9.

3. In the case of plain concrete pedestals, the angle α between plane passing through the bottom edge of the pedestal and the corresponding junction edge of the column with pedestal and the horizontal plane (Fig. 22.9) shall be governed by the expression:

$$\tan \alpha \nless 0.9 \sqrt{\frac{100\, q_0}{f_{ck}} + 1} \qquad \text{...(22.11)}$$

where q_0 = calculated maximum bearing pressure at the base of the pedestal in N/mm^2,

and f_{ck} = characteristic strength of concrete at 28 days in N/mm^2.

(B) Moments and Forces

1. In the case of footings on piles, computation for moments and shears may be based on the assumption that the reaction from any pile is concentrated at the centre of the pile.

2. For the purposes of computing stress in footings which support a round or octagonal concrete column or pedestal, the face of the column or pedestal shall be taken as the side of a square inscribed within the perimeter of the round or octagonal column or pedestal.

3. ***Bending Moment***

(*i*) The bending moment at any section shall be determined by passing through the section a vertical plane which extends completely across the footing and computing the moment of the forces acting over the entire area of the footing on one side of the said plane.

(*ii*) The greatest bending moment to be used in design of an isolated concrete footing which supports a column, pedestal or wall, shall be the moment computed in the manner prescribed in (*i*) at sections located as follows :

(*a*) At the face of the column, pedestal or wall, for footings supporting a concrete column, pedestal or wall,

(*b*) Half way between the centre line and the edge of the wall, for footings under masonry walls, and

(*c*) Half way between the face of the column or pedestal and the edge of the gussetted base, for footing under gusseted bases.

4. ***Shear and bond***

(*i*) The shear strength of footings is governed by the more severe of the following two conditions :

(*a*) The footing acting essentially as a wide beam, with a potential diagonal crack extending in a plane across the entire width, the critical section for this condition shall be assumed as a vertical section located from the face of the column, pedestal or wall at a distance equal to the effective depth of the footing in case of footings on soils, and a distance equal to half the effective depth of footing for footings on piles.

(*b*) Two-way action of the footing, with potential diagonal cracking along the surface of truncated cone or pyramid around the concentrated load : in this case, the footing shall be designed for shear in accordance with appropriate provisions discussed below (Fig 22.10).

(*ii*). In computing the external shear on any section through a footing supported on piles, the entire reaction from any pile of diameter D_p whose centre is located $D_p/2$ or more outside the section shall be assumed as producing shear on the section ; the reaction from any pile whose centre is located $D_p/2$ or more inside the section shall be assumed as producing no shear on the section. For intermediate positions of the pile centre, the portion of the pile reaction to be assumed as producing shear on the section shall be based on straight line interpolation between full value at $D_p/2$ outside the section and zero value at $D_p/2$ inside the section.

(*iii*) The critical section for checking the development length in a footing shall be assumed at the same planes as described for bending moment in *B*(3) and also at all other vertical planes where abrupt changes of section occur. If the reinforcement is curtailed, the anchorage requirements should be checked.

Thus, according to the above provision, shear stress is to be checked for (*i*) *one way action* (*i.e.*, beam shear) for which the governing section *AB* is at a distance *d* from the face of column or pedestal [Fig. 22.10(*a*)] and (*ii*) *two way action*, (*i.e.*, punching shear), for which the governing section is along the perimeter *ABCD* situated at a distance *d*/2 from the face of the column or pedestal [Fig. 22.10(*b*)].

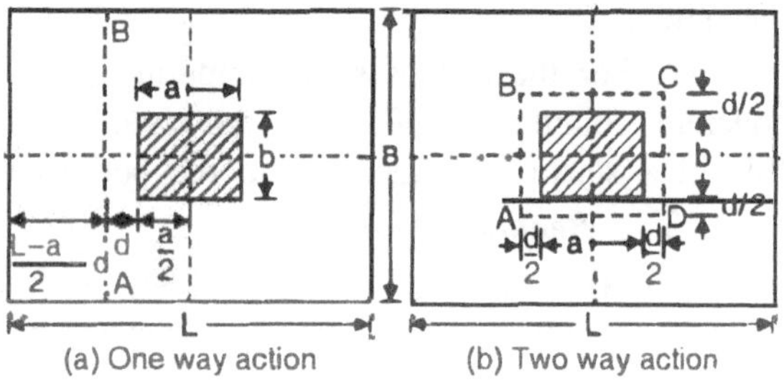

FIG. 22.10. CRITICAL SECTIONS FOR SHEAR

For the *two way action*, the calculated shear stress τ_v should satisfy the following relation

$$\tau_v \le k_s \cdot \tau_c \qquad \text{...[22.12(a)]}$$

where
$$\tau_v = \frac{F_u}{2\,[\,(a+d)+(b+d)\,]\,d} \qquad \text{...[22.12(b)]}$$

$k_s = (0.5 + \beta_c)$, but not greater than 1.0

$$\beta_c = \frac{b}{a} = \frac{\text{short side of column}}{\text{long side of column}} \qquad \text{...[22.12(c)]}$$

$$\tau_c = 0.25\sqrt{f_{ck}}\ \text{N/mm}^2 \text{ in limit state design} \qquad \text{...[22.12(d)]}$$

and F_u = net S.F. acting on the perimeter

For the *beam shear*, the nominal shear stress across *AB* should satisfy the relation

$$\tau_v \le k\,\tau_c \qquad \text{...(22.13)}$$

where τ_c = the permissible shear stress for the grade of the concrete, corresponding to the reinforcement $(100\,A_s/bd)$ as given in Table 7.1

k = factor for slabs, as given in Table 7.2.

(C) Tensile Reinforcement

The total tensile reinforcement at any section shall provide a moment of resistance at least equal to the bending moment on the section calculated in accordance with *B*(3). Total tensile reinforcement shall be distributed across the corresponding resisting section as given below.

(*i*) In one-way reinforced footing, the reinforcement shall be distributed uniformly across the full width of the footing ;

(*ii*) In two-way reinforced square footing, the reinforcement extending in each direction shall be distributed uniformly across the full width of the footing ;

(*iii*) In two-way reinforced rectangular footing, the reinforcement in long direction shall be distributed uniformly across the full width of the footing. For reinforcement in the short direction, a central band equal to the width of the footing shall be marked along the length of the footing and portion of the reinforcement determined in accordance with the equation given below shall be uniformly distributed across the central band ;

$$\frac{\text{Reinforcement in central band width}}{\text{Total reinforcement in short direction}} = \frac{2}{\beta + 1}$$

where β is the ratio of the long side to the short-side of the footing.

The remainder of the reinforcement shall be uniformly distributed in the outer portions of the footing.

(D) Transfer of load at the base of column

The compressive stress in concrete at the base of a column or pedestal shall be considered as being transferred by bearing to the top of the supporting pedestal or footing. The bearing pressure (σ_{cbr}) on the loaded area shall not exceed the permissible bearing stress in direct compression multiplied by a value equal to $\sqrt{A_1/A_2}$ but not greater than 2.

where A_1 = Supporting area for bearing, of footing, which in sloped or stepped footing may be taken as the area of the lower base of the largest frustrum of a pyramid or cone contained wholly within the footing and having for its upper base, the area actually loaded and having side slope of one vertical to two horizontal, and

A_2 = Loaded area at the column base.

For limit state method of design the permissible bearing stress shall be $0.45 f_{ck}$.

Thus, $$\sigma_{cbr} \leq 0.45 f_{ck} \sqrt{A_1/A_2} \qquad \text{....(22.14)}$$

The actual bearing pressure σ_{cbr} = column load divided by the area of column at the base.

Thus $\sigma_{cbr} = \dfrac{W}{a \times b}$ where a and b are the sides of the column.

(1) Where the permissible bearing stress on the concrete in the supporting or supported member would be exceeded, reinforcement shall be provided for developing excess force, either by extending the longitudinal bars into the supporting members or by dowels (see 3 below).

(2) Where transfer of force is accomplished by reinforcement, the development length of the reinforcement shall be sufficient to transfer the compression or tension to the supporting member.

(3) Extended longitudinal reinforcement or dowels of at least 0.5 percent of the cross-sectional area of the supported column or pedestal and a minimum of four bars shall be provided. Where the dowels are used, their diameter shall not exceed the diameter of the column bars by more than 3 mm.

(4) Column bars of diameters larger than 36 mm, in compression only can be dowelled at the footings with bars of smaller size of the necessary area. The dowel shall extend into the column, a distance equal to the development length of the column bar and into the footing, a distance equal to the development length of the dowel.

(E) Nominal cover : For footings, minimum nominal cover shall be 50 mm.

(F) Nominal reinforcement :

(*a*) The nominal reinforcement for concrete sections of thickness greater than 1 m shall be 360 mm^2 per meter length in each direction on each face. This provision does not supercede the requirements of minimum tensile reinforcement based on the depth of the section.

22.5. DESIGN OF CONTINUOUS FOOTINGS

A continuous footing is provided under walls which may be either of masonry or of concrete. For design of reinforced concrete continuous footings, it is convenient to consider the forces on unit length.

1. CONTINUOUS FOOTING UNDER MASONRY WALL

Let b = width of wall(metres).

B = width of footing

q_0 = safe bearing capacity of soil (kN/m^2)

W = load from wall, per metre run (kN/m)

W' = weight of footing per metre run.

Then, the width of footing is given by

$$B = \frac{W + W'}{q_0} \quad ...[22.15\ (a)]$$

Fig. 22.11(*b*) shows the pressure diagram. The *net* upward soil pressure p_0 on the footing will be

$$p_0 = \frac{W}{B \times 1} = \frac{W}{B} \quad ...[22.15\ (b)]$$

The pressure intensity p at the contact of wall and footing is given by

$$p = \frac{W}{b \times 1} = \frac{W}{b} \quad ...[22.15\ (c)]$$

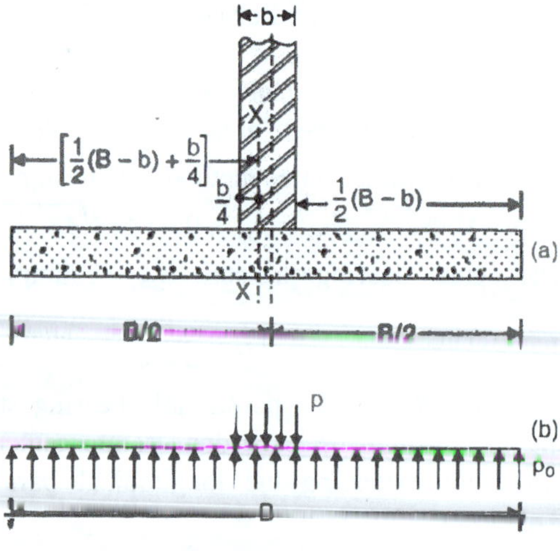

FIG. 22.11.

The entire footing can be considered to consist of two cantilevers, each projecting out by a length $\frac{1}{2}(B-b)$, and having a fixed support length $b/2$. The maximum B.M. is considered to act at a section midway between the edge of the wall and centre of the wall *i.e.*, at a distance $b/4$ from the edge of the wall. Its magnitude per metre length is thus given by

$$M = p_0 \ \frac{1}{2}\left[\frac{B-b}{2}+\frac{b}{4}\right]^2 - p \ \frac{1}{2}\left[\frac{b}{4}\right]^2 = \frac{W}{2B}\left[\frac{2B-b}{4}\right]^2 - \frac{W}{2b}\cdot\frac{b^2}{16}$$

or $$M = \frac{W}{2B}\left[\frac{4B^2+b^2-4bB}{16}\right] - \frac{Wb}{32} = \frac{W}{32}\left[4B-4b+\frac{b^2}{B}-b\right] = \frac{W}{32}\left[4B-5b+\frac{b^2}{B}\right]$$

$$M = \frac{W}{32B}\left[4B^2-5Bb+b^2\right] = \frac{W}{32B}(4B-b)(B-b) = \frac{W}{8B}(B-b)\left[B-\frac{b}{4}\right] \text{ kN-m}$$

or $$M = \frac{p_0}{8}(B-b)\left[B-\frac{b}{4}\right] \times 10^6 \text{ N-mm}$$

and $$M_u = 1.5\,M = \frac{1.5\,p_0}{8}(B-b)\left[B-\frac{b}{4}\right] \times 10^6 \text{ N-mm} \qquad \ldots(22.16)$$

$\therefore$ Effective depth $d = \sqrt{\dfrac{M_u}{1000\,R_u}}$ where $R_u = 0.36 f_{ck} \dfrac{x_{u\,.\,max}}{d}\left(1-0.416\dfrac{x_{u\,.\,max}}{d}\right)$

A minimum thickness of 150 mm should be provided at the edges.

The area of reinforcement, A_{st}, is found from Eq. 3.13 applicable for under-reinforced section.

$$A_{st} = \frac{0.5 f_{ck}}{f_y}\left[1-\sqrt{1-\frac{4.6\,M_u}{f_{ck}\,bd^2}}\right]bd \qquad \ldots(3.13)$$

In addition to this, longitudinal reinforcement is provided @ 0.15% of the area of section of the footings for mild steel bars and @ 0.12% for HYSD bars.

Check for shear. The footing slab is checked for *one way action* (*i.e.*, beam shear) for which the critical section will be at a distance of effective depth from the face of the wall. The depth of footing at that section must be sufficient to keep nominal shear stress within safe limits.

2. CONTINUOUS FOOTING UNDER CONCRETE WALL

A concrete footing under a concrete wall is cast monolithic with it. Hence the maximum B.M. will occur at the face of the wall.

As found earlier : $B = \dfrac{W+W'}{q_0}$; $\quad p = \dfrac{W}{b}$; $p_0 = \dfrac{W}{B}$

$\therefore$ Maximum B.M. per metre length is given by

$$M = p_0\ \frac{1}{2}\left[\frac{B-b}{2}\right]^2 \text{ kN-m}$$

$\therefore$ $$M_u = 1.5\,M = 1.5\,p_0\,\frac{(B-b)^2}{8} \times 10^6 \text{ N-mm} \ldots(22.17)$$

The effective depth and area of steel reinforcement are then given by

$d = \sqrt{\dfrac{M_u}{1000\,R_u}}$ and A_{st} is found from Eq. 3.13 applicable for under-reinforced section.

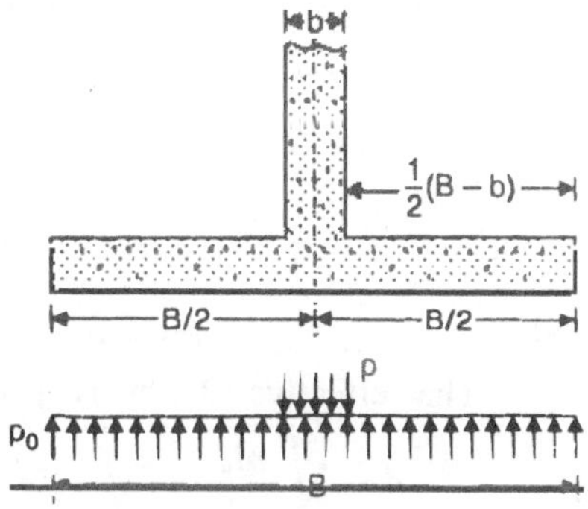

FIG. 22.12.

In the addition to this, longitudinal reinforcement is provided @ 0.15% of cross-sectional area of footing for mild steel bars and @ 0.12 % for HYSD bars. The section is checked for beam shear at a distance of effective depth from the face of the wall.

22.6. ISOLATED FOOTING OF UNIFORM DEPTH

The total area at the base of such footing is determined from the point of view of the safe bearing capacity of soil. The depth of footing is determined both from the considerations of maximum bending moments as well as from punching shear.

Let W = weight transferred from the column (kN)

W' = weight of footing

q_o = safe bearing capacity of soil (kN/m²)

A = area of footing at the base.

Then $A = \dfrac{W + W'}{q_0}$.

If p_0 = net upward pressure at footing,

then $p_0 = \dfrac{W}{A} = \dfrac{W q_0}{W + W'}$

The weight W' of the footing may be assumed to the equal to 10% W, for preliminary calculations.

1. SQUARE FOOTING

Fig. 22.13 shows a square footing of uniform thickness (depth) D.

The width B of the footing will evidently be equal to $\sqrt{\dfrac{W + W'}{q_0}}$.

The thickness is calculated both from the considerations of bending moment as well as punching shear.

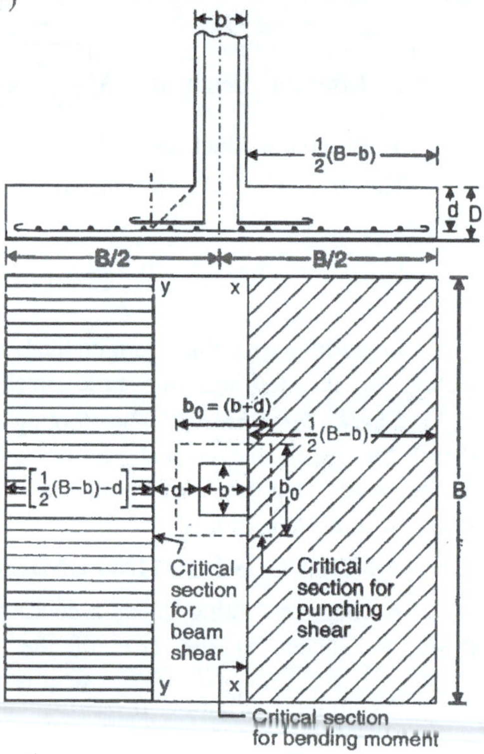

FIG. 22.13

(a) Depth for bending moment :

Since the column and footing are cast monolithic, maximum B.M. will take place about section X-X near the face of the column. The unit pressure p_0 causing bending of cantilever of length $\frac{1}{2}(B - b)$ will be equal to $\frac{W}{B^2}$. Hence,

$$M = p_0 B \left(\frac{B-b}{2}\right)\left(\frac{B-b}{4}\right) = p_0 \frac{B}{8}(B-b)^2 \text{ kN-m} = \frac{p_0 B}{8}(B-b)^2 \times 10^6 \text{ N-mm} \quad \ldots(22.18)$$

and $$M_u = 1.5\, M \quad \ldots(22.18\ a)$$

The effective depth is given by

$$d = \sqrt{\frac{M_u}{BR_u}} = \sqrt{1.5 \frac{p_0(B-b)^2\, 10^6}{8R_u}}$$

(b) Depth for shear

The footing slab is checked both for *one way action* (*i.e.* beam shear) and *two way action* (*i.e.* punching shear).

For *one way action,* the critical section *Y-Y* is situated at a distance d from the face of the column. The nominal shear stress $\tau_v \left(= V_u/B \,.\, d\right)$ is equated to the permissible shear stress $k\tau_c$ found from Table 7.1 for the grade of concrete corresponding to the reinforcement $\left(100\, A_s/bd\right)$. Thus

$$\frac{V_u}{B\,d} \leq k\,\tau_c \qquad \text{...[22.19 (a)]}$$

From this the effective depth d is found.

For *two way action,* the calculated shear stress τ_v computed from Eq. 22.12 (*b*) should be less than $k_s\ \tau_c$.

Thus, $$\frac{F_u}{4\,[b+d]\,d} \leq k_s \,.\, \tau_c$$

where $k_s = (0.5 + \beta_c) \leq 1 \;= (0.5 + 1) \leq 1 = 1$ (maximum value)

$\tau_c = 0.25\sqrt{f_{ck}}$ and $F_u =$ net S.F. acting on the periphery $= 1.5\,F = 1.5\,p_0\ [B^2 - (b+d)^2]$

Hence, $$\frac{F_u}{4(b+d)d} \leq 0.25\sqrt{f_{ck}} \qquad \text{...[22.19 (b)]}$$

The greater of the depths found from Eqs 22.19(*a*) and 22.19(*b*) should be provided. A minimum thickness of 150 mm should be provided in case d comes out to be less than 150 mm.

(*c*) **Steel reinforcement**. Steel reinforcement is provided in both the directions, in the form of mesh, for the bending moment M_u. Reinforcement is found from Eq. 3.13 applicable for under-reinforced section.

Thus $$A_{st} = \frac{0.5\,f_{ck}}{f_y}\left[1 - \sqrt{1 - \frac{4.6\,M_u}{f_{ck}\,bd^2}}\right] bd \qquad \text{...(3.13)}$$

This area is to be uniformly distributed over a width B. If A_φ is area of each bar chosen, the spacing of bars is given by

$$s = \frac{B\,.\,A_\varphi}{A_{st}}$$

(*d*) **Check for development length**. The critical section for checking the development length in the footing is assumed at the same plane as used for maximum bending moment, *i.e.* at a section along the face of the column.

If the reinforcement is curtailed, the anchorage requirements should be checked.

2. RECTANGULAR FOOTING

In the case of rectangular footing, the proportioning of the dimensions L and B should be done in such a way that the bending moments in each of the adjacent projections is equal to the moment of resistance of the footing. The C.G. of footing area should invariably coincide with the C.G. of column area. Fig. 22.14 shows a column of size $a \times b$. Sometimes, a rectangular footing may be provided even for a square column, where there are some restrictions on the maximum value of the width of the footing. The thickness of the footing is determined both from the B.M. as well as shear point of view.

(*a*) **Depth for bending moment**

Bending moment M_1 about section XX is given by

$$M_1 = p_0\ B\left(\frac{L-a}{2}\right)\left(\frac{L-a}{4}\right) = \frac{p_0\,B\,(L-a)^2}{8} \times 10^6$$

N-mm and $M_{1u} = 1.5\ M_1$ ...(22.20)

$$d = \sqrt{\frac{M_{1u}}{R_u\,B}} = \sqrt{\frac{1.5\ p_0\,(L-a)^2 \times 10^6}{8R_u}} \qquad ...(22.21)$$

Bending moment M_2 about section Y-Y is given by

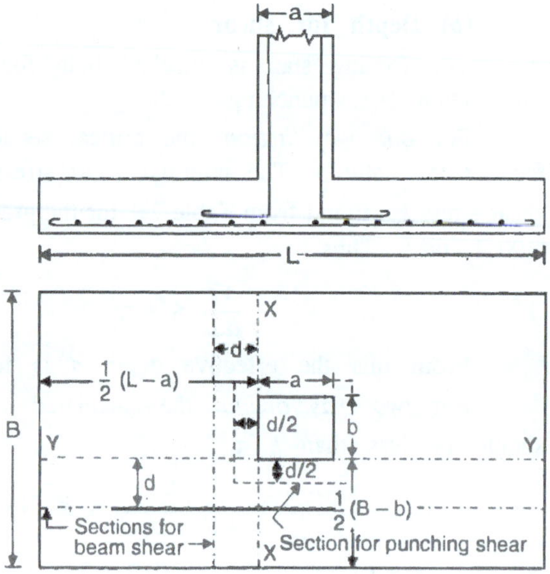

FIG. 22.14.

$$M_2 = p_0\ L\left(\frac{B-b}{2}\right)\left(\frac{B-b}{4}\right) = \frac{p_0\,L}{8}\,(B-b)^2 \times 10^6 \text{ N-mm and } M_{2u} = 1.5\ M_2 \qquad ...(22.22)$$

$$\therefore \qquad d = \sqrt{\frac{M_{2u}}{R_u\,L}} = \sqrt{\frac{1.5\ p_0 (B-b)^2 \times 10^6}{8R_u}} \qquad ...(22.23)$$

(*b*) **Depth for shear**. From shear point of view, the depth is found for both *one way action* (*i.e.* beam section) as well as *two way action* (*i.e.* punching shear action).

For one way action, the sections are situated at a distance d from each face of the column. The nominal shear stress τ_v ($= V_u/BD$ or V_u/LD) should be less than permissible shear $k\,\tau_c$ found from Table 7.1 for the grade of concrete corresponding to reinforcement ratio ($100\ A_s/bd$) To begin with, the reinforcement ratio is arbitrarily selected equal to the normal value.

$$\text{Thus} \qquad \frac{V}{Bd}\left(\text{or } \frac{V}{L\,.\,d}\right) \le k\,\tau_c \qquad ...(22.24)$$

From which effective depth is found. For *two way action*, the calculated shear stress (τ_v) should satisfy the following relation :

$$\tau_v \le k_s\,.\,\tau_c$$

where $\tau_v = \dfrac{F_u}{2\,[(a+d)+(b+d)]\,d}$; $k_s = \left(0.5 + \dfrac{b}{a}\right) \le 1$; $\tau_c = 0.25\sqrt{f_{ck}}$

and F_u = net S.F. at the periphery $= 1.5\ F = 1.5\ p_0\ [(B \times L) - (a+d)\,(b+d)]$

$$\text{Hence} \qquad \frac{F_u}{2[(a+d)+(b+d)]d} \le 0.25\ k_s\ \sqrt{f_{ck}} \qquad ...(22.25)$$

The greater of the depths found from Eq. 22.22, 22.23, 22.24 and 22.25 should be provided.

(*c*) **Steel reinforcement**: Steel reinforcement is provided in both the directions for the bending moments M_1 and M_2.

The area of steel A_{st1}, of *long bars* parallel to direction *L*, is calculated for bending moment M_{1u} as under :

$$A_{st1} = \frac{0.5 f_{ck}}{f_y}\left[1 - \sqrt{1 - \frac{4.6 M_{1u}}{f_{ck} bd^2}}\right] bd \quad \text{...[22.26(a)]}$$

This area should be uniformly distributed over a width *B*. If A_φ is the area of each bar, the spacing is given by

$$s = \frac{B A_\varphi}{A_{st1}} \quad \text{...[22.26(b)]}$$

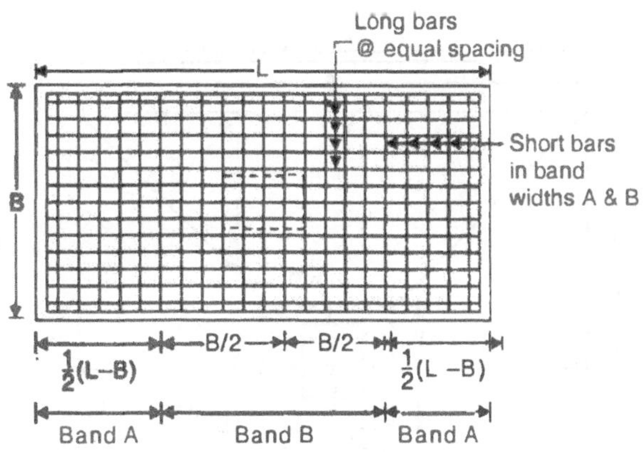

FIG. 22.15.

The area of steel A_{st2}, of short bars parallel to direction *B*, is calculated for bending moment M_{2u} as under, applicable for under reinforced section :

$$A_{st2} = \frac{0.5 f_{ck}}{f_y}\left[1 - \sqrt{1 - \frac{4.6 M_{2u}}{f_{ck} bd^2}}\right] bd \quad \text{...[22.27(a)]}$$

This area is provided in two distinct band widths : (*i*) the central band *B* of width *B*, and (*ii*) the end bands *A*, each of width $\frac{1}{2}(L - B)$.

The area $A_{st2\,(B)}$ in the central band *B* is given by

$$A_{st2\,(B)} = \frac{2 A_{st2}}{\beta + 1} \quad \text{where} \quad \beta = \frac{L}{B}. \quad \text{...[22.27(b)]}$$

The area $A_{st2(A)}$ in each of the end bands *A* is given by

$$A_{st2\,(A)} = \frac{1}{2}(A_{st2} - A_{st2(B)}) \quad \text{...[22.27(c)]}$$

The reinforcement calculated above in the three bands should be distributed uniformly. (see Fig. 22.15)

(*d*) **Check for development length.** The section should be checked for development length at each face of the column, exactly in the same manner as for square column.

22.7. ISOLATED SLOPED FOOTING

In the previous article, we have seen that bending moment, beam shear and punching shear govern the thickness or depth of footing near the column face. This depth can be reduced toward the edges of the footing where the bending moment and shear decrease rapidly. If this decrease is achieved *linearly*, we get a *sloped footing* shown in Fig. 22.16 for a square footing. However, a minimum thickness of 150 mm should be kept at the edges.

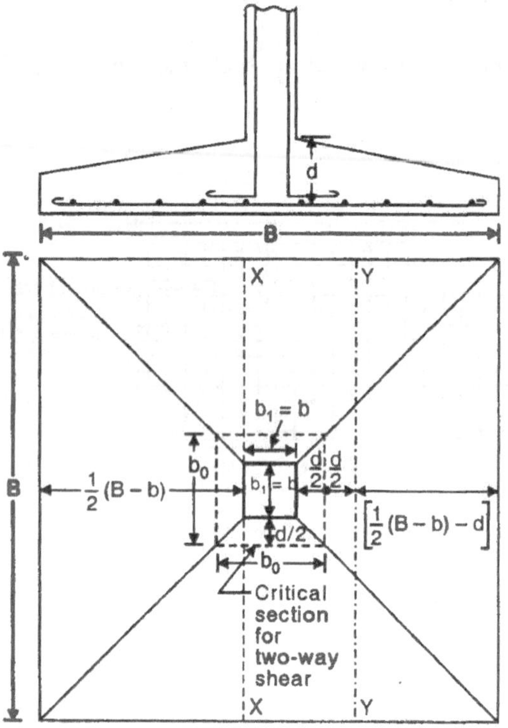

FIG. 22.16.

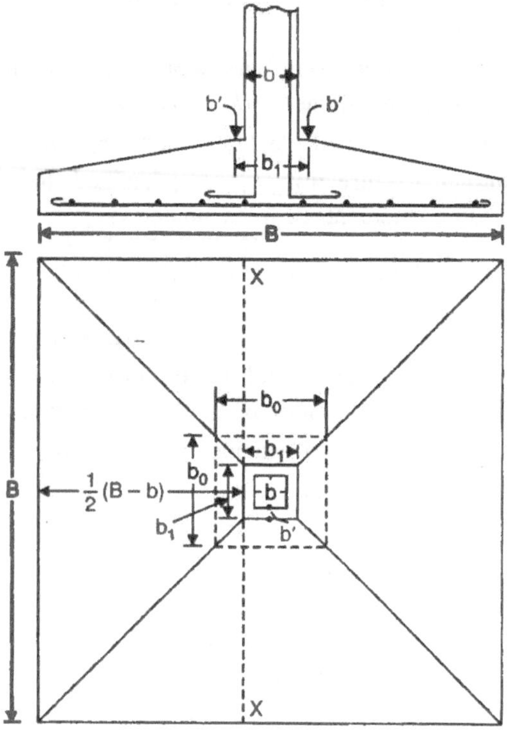

FIG. 22.17

The areas at the base of footing and the net upward pressure p_o are found exactly in the same manner as for isolated footing of uniform depth.

1. SQUARE SLOPED FOOTING

Figs. 22.16 and 22.17 show a square footing of size $B \times B$, with variable depth.

Width $\quad B = \sqrt{\dfrac{W + W'}{q_0}} \qquad$ and $\qquad p_0 = \dfrac{W}{B^2}$ (kN/m^2)

The depth at the column face is found both from the consideration of bending moment as well as punching shear.

(a) Depth for bending moment

The maximum bending moment at the section XX through the column face is given by

$$M = p_0 B\left(\frac{B-b}{2}\right)\left(\frac{B-b}{4}\right) = p_0 \ \frac{B}{8}\,(B-b)^2 \times 10^6$$

N-mm ...(22.28 *a*)

and $M_u = 1.5\ M = 1.5\ p_0\ B\ (B-b)^2 \times 10^6$ N-mm ...(22.28)

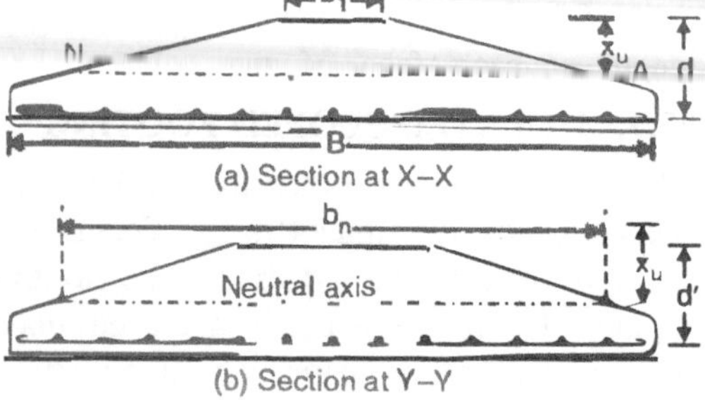

FIG. 22.18.

This should be equated to the moment of resistance of the footing section at XX. Fig. 22.18(a) shows the section of the footing at XX. The top width is b_1, the bottom width is B and the effective depth is d, where $b_1 = b$ in Fig. 22.16 and $b_1 = b + 2\,b'$ in Fig. 22.17 (*a*)

The determination of effective depth for the trapezoidal section (Fig. 22.17 *a*) can be done by one of following three methods.

(i) Method 1 : Exact method

The moment of resistance of trapezoidal section consists of the moment of resistance of the central rectangular portion plus the moment of resistance of the two triangular portions, as illustrated in § 5.7 of Chapter 5. However the method is quite cumbersome, as it involves trial and error procedure.

(ii) Method 2 : Equivalent width method

As an approximation, the trapezoidal section can be replaced by an equivalent rectangular section of width b_e given by

$$b_e = b_1 + \frac{1}{8}(B - b_1) \qquad \text{...(22.29)}$$

where $b_1 = b$ in Fig. 22.16 and $b_1 = b + 2\,b'$ in Fig. 22.17

Hence $$d = \sqrt{\frac{M_u}{R_u\, b_e}} \qquad \text{...(22.30}$$

(iii) Method 3 : Top width method

In this method, the width of the resisting section is taken equal to b_1, and the effective depth is computed from the expression

$$d = \sqrt{\frac{M_u}{R_u \,.\, b_1}} \qquad \text{...(22.31)}$$

where $b_1 = b$ in Fig. 22.16 and $b_1 = b + 2\,b'$ in Fig. 22.17

This method is quite conservative. However, this method is preferred since it gives larger value of d, giving rise to an under-reinforced section. Moreover, in most of the cases, the effective depth is governed by the criterion of *one-way shear* (as we shall see through some solved examples), it is preferred to use Eq. 22.31 for the determination of effective depth d, required for bending compression.

(*b*) Check for shear. For one way shear (beam action) the critical section is taken at a distance d from the face of column. The shear force V at the section is given by

$$V = [\tfrac{1}{2}(B - b_1) - d]\, B \times p_0 \quad \text{and} \quad V_u = 1.5\, V = 1.5\, p_0\, B\left[\frac{1}{2}(B - b_1) - d\right] \qquad \text{...(22.32)}$$

Hence nominal shear stress is given by $$\tau_v = \frac{V_u}{b_n\, d'} \qquad \text{....(22.32 a)}$$

where b_n = width of the section at N.A.

d' = effective depth of setion at that plane.

Normally, the section is designed from the considerations of *two-way shear* (punching shear), and is later checked for beam shear.

For *two way action* (punching shear), the critical section located all round, is at a distance *d/2* from the face of the column where $b_0 = (b_1 + d)$

$$\text{S.F. } F = p_0 [B^2 - b_0^2] = p_0 [B^2 - (b_1 + d)^2]$$

and
$$F_u = 1.5\, F = 1.5\, p_0 [B^2 - (b_1 + d)^2] \quad \text{...(22.33 } a)$$

$$\tau_v = \frac{F_u}{4b_0 \,.\, d_o} = \frac{F_u}{4(b_1 + d)d_0} \quad \text{...(22.33 } b)$$

where d_0 = effective depth of slab at the critical plane.

Hence $\tau_v \le k_s \,.\, \tau_c \le (0.5 + \beta_c) \,.\, \tau_c$

or
$$\frac{F_u}{4(b_1 + d)d_0} \le (0.5 + \beta_c) \times 0.25\sqrt{f_{ck}} \quad \text{...(22.33 } c)$$

where k_s or $(0.5 + \beta_c)$ has a maximum value of 1.

(*c*) **Steel reinforcement**. Steel reinforcement is provided in both the directions, in the form of mesh, for the bending moment M_u. The area of steel reinforcement is given by Eq. 3.13 applicable for under reinforced section. Thus

Thus,
$$A_{st} = \frac{0.5 f_{ck}}{f_y}\left[1 - \sqrt{1 - \frac{4.6\, M_u}{f_{ck}\, b_1\, d^2}}\right] b_1\, d \quad \text{...(22.34)}$$

Subject to minimum reinforcement given by $A_{st\,,\,\min} = \frac{0.85\, b_1\, d}{f_y}$ ...(22.34 *a*)

(*d*) **Check for development length**

The section should be checked for development length at the face of the column.

(2) **RECTANGUALAR SLOPED FOOTING**

The procedure for design of rectangular footing is similar to that of a square sloped footing, except that each of the cantilever portion is analysed separately. The reinforcements A_{st1} and A_{st2} found in the two directions are provided as explained § 22.6.

22.8. ISOLATED STEPPED FOOTING

The construction of sloped footing is sometimes difficult and when the slope of the top face of footing is more, say more than 1 vertically to 3 horizontally, it may be difficult to finish the top without having concrete slump too much. In such cases, stepped footing shown in Fig. 22.19 is an alternative.

The design is first done at the section X_1 - X_1 passing through the face of the column. The depth d is determined both on the basis of bending moment and shear.

Thus
$$M_1 = \frac{p_0\, B}{8}(B - b)^2 \times 10^6 \text{ N-mm} \quad \text{...(22.15)}$$

and
$$M_{1u} = 1.5\, M_1 = 1.5\, p_0 \frac{B}{8}(B - b)^2 \times 10^6$$

$$\therefore \quad d_1 = \sqrt{\frac{M_{1u}}{R_u\, B_1}} \quad \text{...(22.35)}$$

where B_1 is the width of the footings at the top, at section X_1 - X_1

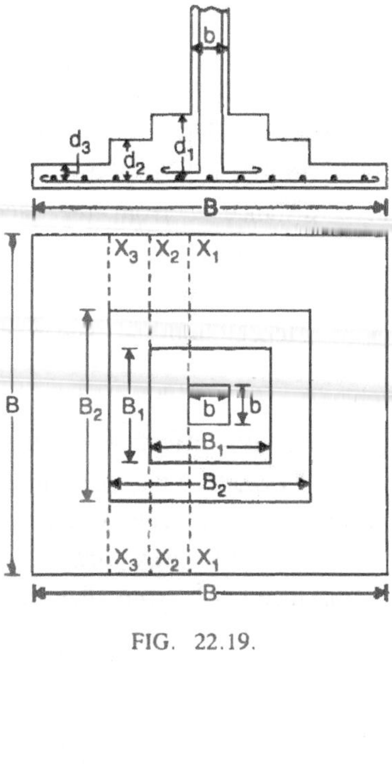

FIG. 22.19.

The depth so found is checked both for one way shear (beam shear) as well as to way shear (punching shear) as explained in § 22.6

The area of steel reinforcement is provided as usual.

Let the size of footings at its top, at the next step be $B_2 \times B_2$ and its depth d_2. The critical section for design will be at section X_2 - X_2. The curtailed depth d_2 should be safe in compression, tension, shear and development length. The bending moment at X_2 - X_2 will be the moment of the total pressure acting on the rectangular area to one side of the section. The resisting width will be B_2. Thus, from compression point of view,

$$d_2 = \sqrt{\frac{M_{2u}}{R_u B_2}}, \text{ where } M_2 \text{ is the bending moment at } X_2 - X_2 .$$

This value of d_2 is checked both for beam shear (at a section distant d_2 from X_2 - X_2) and for punching shear (at perimeter section distant $d_2/2$ from the face of the step). Similarly, depth d_3 at section X_3 - X_3 is determined.

22.9. ISOLATED FOOTING FOR CIRCULAR COLUMNS

Circular columns are usually preferred. The footing for a circular column may be either square or circular at the base.

1. SQUARE FOOTING

Fig. 22.20 show square sloped footing for a circular column. For the purposes of design, the equivalent square column of the same area as the circular column is assumed. The design of the square footing for the equivant square column is done exactly in the same manner as explained in § 22.7.

The critical section (*XX*) for bending moment will pass through the face of the equivalent square column.

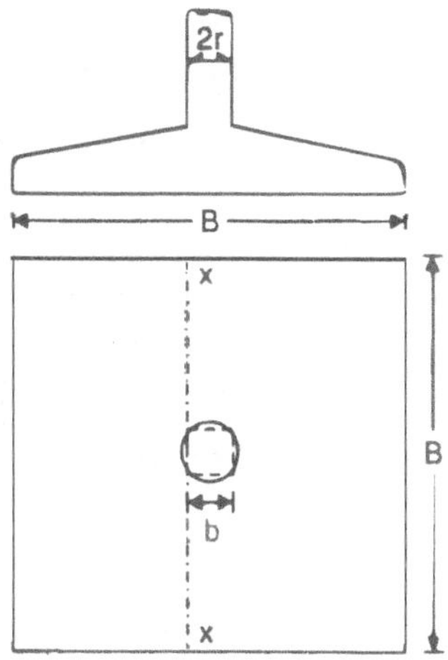

FIG. 22.20.

2. CIRCULAR FOOTING (Fig. 22.21)

Fig. 22.21 shows a column of radius *r* with circular footing of radius *R* given by

$$R = \sqrt{\frac{W + W'}{\pi q_0}} \quad ...(22.36)$$

The net upward pressure intensity of footing is given by

$$p_0 = \frac{W}{\pi R^2} \quad ...(22.37)$$

1. Depth for bending moment

The maximum bending moment in the footing is calculated at the edge of the column for one quadrant such as *PQRS* shown shaded. The distance of *C.G.* of the quadrant of a circle of radius *R* is at a distance of 0.6 *R* from the centre. Hence if $\bar{x}$ is the distance of C.G. of shaded area from the centre of column, we have

$$\bar{x} = \frac{\frac{\pi}{4} R^2 \times 0.6 R - \frac{\pi}{4} r^2 \times 0.6 r}{\frac{\pi}{4} R^2 - \frac{\pi}{4} r^2} = \frac{0.6 (R^3 - r^3)}{R^2 - r^2} = \frac{0.6 (R^2 + Rr + r^2)}{R + r} \quad ...[22.38(a)]$$

$\therefore$ Distanace of C.G. of shaded area from the face of the column is

$$\bar{x}' = \frac{0.6\,(R^2 + Rr + r^2)}{R + r} - r = \frac{0.6\,(R^2 + Rr + r^2) - Rr - r^2}{R + r} = \frac{0.6\,R^2 - 0.4Rr - 0.4\,r^2}{R + r}$$

$$= \frac{0.2}{R + r}\,[3R^2 - 2Rr - 2r^2] \qquad \ldots(22.38)$$

Upward pressure on shaded area is $= \frac{\pi}{4}(R^2 - r^2)\,p_0$

$\therefore$ Bending moment at the face of the column is

$$M = \frac{\pi}{4}(R^2 - r^2)\,p_0 \times \frac{0.2}{R + r}(3R^2 - 2Rr - 2r^2)$$

$$= \frac{p_0\,\pi}{20}(R - r)(3R^2 - 2Rr - 2r^2) \qquad \ldots(22.39)$$

and $$M_u = 1.5\,M = 1.5\,p_0\,\frac{\pi}{20}(R - r)(3R^2 - 2Rr - 2r^2) \qquad \ldots(22.39\ a)$$

Breadth b of quadrant $= PQ = \frac{\pi\,r}{2}$

Hence the effective depth is given by

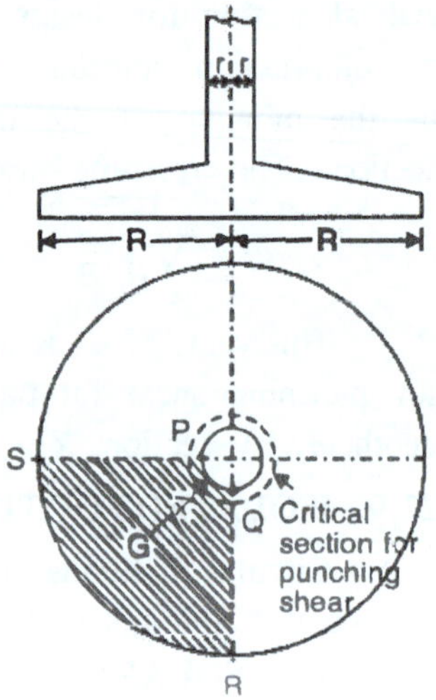

FIG. 22.21

$$d = \sqrt{\frac{M_u}{R_u\,.\,b}} \quad \text{where} \quad R_u = 0.36\,f_{ck}\,\frac{x_{u,\,max}}{d}\left(1 - 0.416\,\frac{x_{u,\,max}}{d}\right)$$

A minimum total thickness of 150 mm is provided at the outer edge of footing.

2. ***Depth for two way shear (punching shear)***

Circular footing is checked for two way shear (punching shear) for which the critical circular plane exists at a radial distance of $d/2$ from the column face, as shown in Fig. 22.21. The radius r_0 of this critical circle $= \left(r + \frac{d}{2}\right)$. Net punching shear force $F = p_0\,\pi\,(R^2 - r_0^2)$

and $$F_u = 1.5\,F = 1.5\,p_0\,\pi\,(R^2 - r_0^2)$$

If d_0 is the effective depth of slab at that plane, the shear stress

$$\tau_v = \frac{F_u}{2\,\pi\,r_0\,d_0} \quad \text{This should be less than } k_s\,.\,\tau_c. \qquad \ldots[22.40(a)]$$

$$\therefore \quad \frac{F_u}{2\,\pi\,r_0\,d_0} \le k_s\,\tau_c.$$

where $k_s = \left(0.5 + \frac{b}{a}\right) \le 1 = 1$ (since $b = a = r$). and $\tau_c = 0.25\,\sqrt{f_{ck}}$

Hence $$\frac{F_u}{2\,\pi\,r_0\,d_0} \le 0.25\,\sqrt{f_{ck}} \qquad \ldots(22.40)$$

The critiçal section for *one way shear* will occur at a radial distance d from the face of the column, *i.e.*, at AB (Fig. 22.22). The shear force V will be

$$V = p_0\,\frac{\pi}{4}\,[R^2 - (r + d)^2] \qquad \ldots[22.40(b)]$$

and $V_u = 1.5\ V = 1.5\ p_0 \frac{\pi}{4}\left[\ R^2 - (r+d)^2\ \right]$ ...[22.40(c)]

The breadth b' of the footing at critical section $= \frac{\pi}{2}(r+d)$

$$\therefore \quad \tau_v = \frac{V_u}{b'd'} = \frac{V_u}{\frac{\pi}{2}(r+d)\,d'}$$

where d' is the effective depth at the critical section.

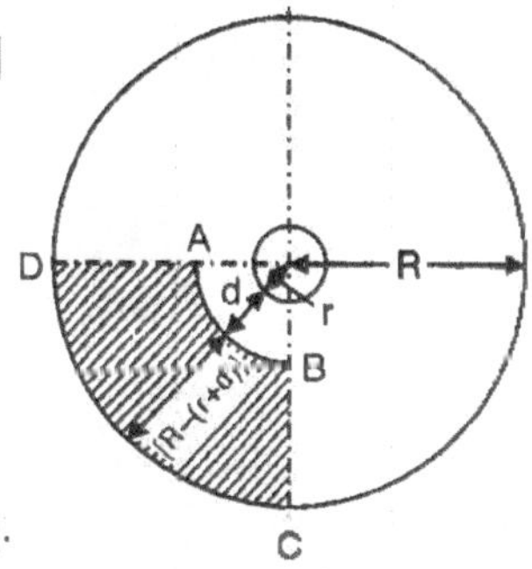

FIG. 22.22

3. ***Steel reinforcement***. Steel reinforcement is given by Eq. 3.13 applicable for under-reinforced section.

$$A_{st} = \frac{0.5\,f_{ck}}{f_y}\left[1 - \sqrt{1 - \frac{4.6\,M_u}{f_{ck}\,bd^2}}\right]bd \ ; \text{ where } b = \pi\, r/2$$

Same reinforcement (A_{st}) is provided in two directions at right angles, *i.e.*, the bars are arranged in form of square mesh centrally located under the column, the width of the square being equal to the side of square that can be inscribed in the circular footing of radius R.

4. ***Check for development length***. The section is checked for development length at the face of column.

Design Example 22.1 *Design of footing for R.C. wall*

A reinforced concrete wall 250 mm thick carries a load of 500 kN/m inclusive of its own weight. Design a reinforced concrete footing on soil having safe bearing capacity of 160 kN/m². Use M 20 concrete and Fe 415 steel.

Solution.

1. ***Design constants***. For Fe 415 steel, $f_y = 415\ \text{N/mm}^2$. For M 20 concrete $f_{ck} = 20\ \text{N/mm}^2$:

$$\therefore \quad \frac{x_{u,\max}}{d} = \frac{700}{1100 + 0.87 \times 415} = 0.479$$

and
$$R_u = 0.36\,f_{ck}\,\frac{x_{u,\max}}{d}\left(1 - 0.416\,\frac{x_{u,\max}}{d}\right)$$
$$= 0.36 \times 20 \times 0.479\,(1 - 0.416 \times 0.479) = 2.761$$

2. ***Size of footing***. $W = 500$ kN ; $q_0 = 160\ \text{kN/m}^2$

Assume the self-weight of footing equal to 10% of the super imposed load.

$$\therefore \quad W' = 50\ \text{kN/m} \ ; \quad \therefore \quad B = \frac{W + W'}{q_0} = \frac{500 + 50}{160} = 3.43\ \text{m}$$

Adopt $B = 3.5\ \text{m} = 3500\ \text{mm}$.

$\therefore$ Net upward pressure $p_0 = \frac{500}{3.5} = 142.86\ \text{kN/m}^2/\text{m}$

3. ***Determination of the thickness of the section***

(a) Thickness on the basis of bending compression

Maximum bending moment at the face of the wall is given by

$$M = p_0 \frac{(B-b)^2}{8} \times 10^6 \text{ N-mm} = \frac{142.86\,(3.5-0.25)^2}{8} \times 10^6 = 1.89 \times 10^8 \text{ N-mm}$$

$$\therefore \qquad M_u = 1.5\,M = 1.5 \times 1.89 \times 10^8 \text{ N-m} = 2.835 \times 10^8 \text{ N-mm}$$

$$d = \sqrt{\frac{M_u}{BR_u}} = \sqrt{\frac{2.835 \times 10^8}{1000 \times 2.761}} \simeq 320 \text{ mm}$$

(b) Thickness on the basis of one way shear

For one way shear, the critical section XX (Fig. 22.23) is at a distance of d from the face of the wall. Distance a of the section XX from the edge of the footing is

$$a = \tfrac{1}{2}(B-b) - \frac{d}{1000} = \tfrac{1}{2}(3.5-0.25) - \frac{d}{1000}$$

$$= 1.625 - 0.001d \quad \text{(where } d \text{ is in mm)}$$

$$\therefore\ V_u = 1.5\,p_0\,a = 1.5 \times 142.86\,(1.625 - 0.001\,d)$$

$$= 214.29\,(1.625 - 0.001d) \text{ kN/m}$$

$$= 214290\,(1.625 - 0.001\,d) \text{ N/m}$$

$$\therefore\ \tau_v = \frac{V_u}{bd} = \frac{214290\,(1.625 - 0.001\,d)}{1000\,d} \qquad \ldots(i)$$

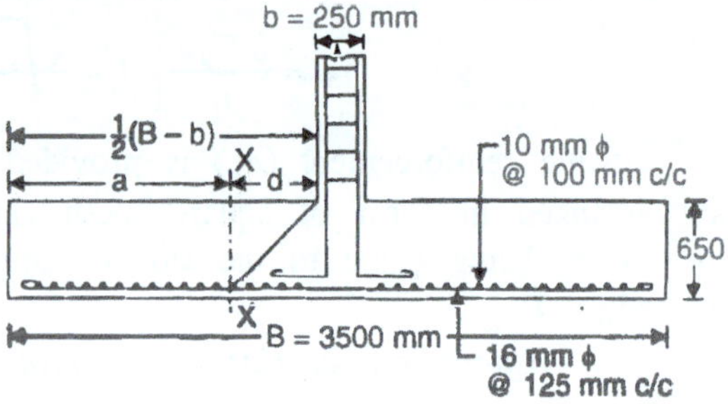

FIG. 22.23

Permissible shear stress, $= k\,\tau_c$ where $k = 1$ for $D \geq 300$ mm.

For M 20-Fe 415 combination, $p_{t\,,\,\text{lim}} = 0.414 \dfrac{f_{ck}}{f_y} \times \dfrac{x_{u\,,\,\max}}{d} = 0.414 \times \dfrac{20}{415} \times 0.479 \times 100 = 0.995\%$.

Assuming under-reinforced section with $p_t \simeq 0.3\%$, we get

$\tau_c \simeq 0.384$ for M 20 concrete, from Table 7.1.

Hence from (i), $\qquad \tau_c = 0.384 = \dfrac{214290\,(1.625 - 0.001\,d)}{1000\,d}$

From which $\qquad d \simeq 582$ mm.

Providing nominal cover of 50 mm and using 16 mm φ bars, cover to the centre of steel (i.e. effective cover) $= 50 + 16/2 = 58 \simeq 60$ mm. Hence total $D = 582 + 60 = 642$ mm. However keep $D = 650$ mm, so that available $d = 650 - 60 = 590$ mm,

4. Determination of reinforcement : Since actual d provided is much more than that required for a balanced section, we have an under-reinforced section, for which

$$A_{st} = \frac{0.5\,f_{ck}}{f_y}\left[1 - \sqrt{1 - \frac{4.6\,M_u}{f_{ck}\,bd^2}}\right]bd = \frac{0.5 \times 20}{415}\left[1 - \sqrt{1 - \frac{4.6 \times 2.835 \times 10^8}{20 \times 1000\,(590)^2}}\right] \times 1000 \times 590$$

$= 1400 \text{ mm}^2$; Using 16 mm φ bars, having $A_\varphi = 201.06$ mm

Spacing $\qquad s = \dfrac{1000\,A_\varphi}{A_{st}} = \dfrac{1000 \times 201.06}{1400} = 143$ mm

However, provide 16 mm φ bars @ 125 mm c/c, giving $A_{st} = \dfrac{201.06}{125} \times 1000 = 1608.5 \text{ mm}^2$

$\therefore$ % reinforcement, $p = \frac{1608.5}{1000 \times 590} \times 100 = 0.273\,\%$

Longitudinal reinforcement @ 0.12% $= \frac{0.12 \times 1000 \times 650}{100} = 780 \text{ mm}^2$

Using 10 mm φ bars having $A_\varphi = 78.5 \text{ mm}^2$, $s = \frac{1000 \times 78.5}{780} = 100.6$ mm

However provide 10 mm φ bars longitudingally @ 100 mm c/c.

5. ***Check for shear***. For *one way shear,* the critical section *XX* (Fig. 22.23) is at distance *d* from the face of wall. Distance of section *XX* from the edge of footing

$$= \frac{1}{2}(B - b) - d = \frac{1}{2}(3.5 - 0.25) - 0.59 = 1.035 \text{ m}$$

$$V = 1.5\, p_0 \times 1.035 = 1.5 \times 142.86 \times 1.035 = 221.79 \text{ kN/m} = 221790 \text{ N/m}$$

Hence, $\tau_v = \frac{221790}{1000 \times 590} = 0.375 \text{ N/mm}^2$

Permissible shear stress $= k\,\tau_c$ where $k = 1$ for total thickness of 650 mm (Table 7.2)

Take $\tau_c \simeq 0.375$ N/mm² (Table 7.1) for reinforcement of 0.273%. Hence $\tau_v \simeq k.\tau_c$.

Thus, the footing is just safe from shear point of view.

5. ***Check for development length***

$$L_d = \frac{\sigma_s\, \varphi}{4\, \tau_{bd}} = \frac{0.87 \times 415}{4 \times 1.6 \times 1.2}\, \varphi = 47\, \varphi = 47 \times 16 = 752 \text{ mm}$$

Providing 60 mm side cover, length of bars available $= \frac{1}{2}[B - b] - 60 = \frac{1}{2}[3500 - 250] - 60$ $= 1565$ mm, which is greater than L_d. Hence O.K.

The details of reinforcement etc. are shown in Fig 22.23.

Design Example 22.2. ***Design of footing for a brickwall***

A brick wall 300 mm thick carries a load of 180 kN/m length. Design a R.C.C. footing, if the safe bearing capacity of soil is 120 kN/m². Use M 20 concrete and Fe 415 steel.

Solution.

1. ***Design constants***. For Fe 415 steel and M 20 concrete combination, we have

$$\frac{x_{u,\max}}{d} = 0.479 \quad \text{and} \quad R_u = 2.761 \text{ as in Example 22.1}$$

2. ***Size of footing***. $W = 180$ kN ; Assume $W' = 10\%$ of $W = 18$ kN

$\therefore$ Width B of footing $= \frac{180 + 18}{120} = 1.65$ m. Adopt $B = 1.7$ m.

Net upward pressure $p_0 = 180/1.7 = 106 \text{ kN/m}^2\text{/m}$

3. ***Design of section. (a) Depth on the basis of bending compression.*** Maximum bending moment occurs at section X - X distant $b/4$ from the centre of the wall, and its magnitude is given by Eq. 22.16.

$$M = \frac{p_0}{8}(B-b)\left(B-\frac{b}{4}\right) \times 10^6 \text{ N-mm}$$

$$= \frac{106}{8}(1.7-0.3)\left(1.7-\frac{0.3}{4}\right) \times 10^6$$

$$= 30.14 \times 10^6 \text{ N-mm.}$$

$$\therefore \quad M_u = 1.5\, M = 1.5 \times 30.14 \times 10^6$$

$$= 45.216 \times 10^6 \text{ N-mm}$$

$$\therefore \quad d = \sqrt{\frac{45.216 \times 10^6}{1000 \times 2.761}} \simeq 128 \text{ mm}$$

$$D = 128 + 60 = 188 \simeq 200 \text{ mm (say)}$$

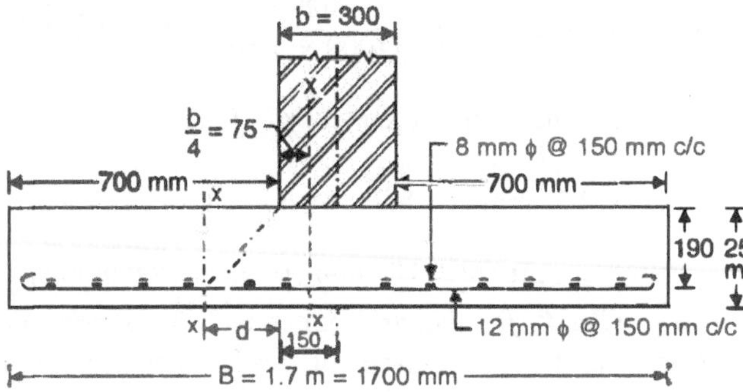

FIG. 22.24.

(b) Depth on the basis of one way shear : Assuming $p = 0.3\%$ for under-reinforced section, we get from Table 7.1, $\tau_c \approx 0.384 \text{ N/mm}^2$. Also, from Table 7.2, $k = 1.2$ for $D = 200$ mm. Hence permissible shear stress $= k\,\tau_c = 1.2 \times 0.384 = 0.461 \text{ N/mm}^2$.

The critical section *X-X* lies at a distance of *d* (mm) from the face of the wall. Hence distance *a* of *X-X* from the edge of footing is given by

$$a = \frac{1}{2}(B-b) - d = \frac{1}{2}(1.7-0.3) - 0.001\, d = 0.7 - 0.001\, d \text{ metres}$$

$$\therefore \quad V_u = 1.5 \times 106000\,(0.7 - 0.001\, d) \text{ N/m}$$

$$\therefore \quad \tau_v = \frac{V_u}{bd} = \frac{159000\,(0.7-0.001\, d)}{1000\, d} = \frac{159\,(0.7-0.001\, d)}{d}$$

Equating this to the permissible shear stress of 0.461 N/mm^2, we get

$$0.461 = \frac{159\,(0.7-0.001\, d)}{d}, \quad \text{from which we get } d = 180 \text{ mm.}$$

Providing effective cover of 60 mm, $D = 180 + 60 = 240$ mm. Provide $D = 250$ mm so that $d = 250 - 60 = 190$ mm.

4. ***Design of reinforcement***

Since actual *d* provided is more than that required for bending compression, we have an under-reinforced section for which

$$A_{st} = \frac{0.5\, f_{ck}}{f_y}\left[1-\sqrt{1-\frac{4.6\, M_u}{f_{ck}\, bd^2}}\right] bd = \frac{0.5 \times 20}{415}\left[1-\sqrt{1-\frac{4.5 \times 45.216 \times 10^6}{20 \times 1000\,(190)^2}}\right] 1000 \times 190$$

$$= 715.4 \text{ mm}^2$$

Using 12 mm φ bars, $s = \dfrac{1000 \times 113}{715.4} = 158$ mm

However, provide 12 mm φ bars @ 150 mm c/c.

Area of longitudinal reinforcement @ 0.12% area of cross-section.

$$= \frac{0.12}{100} \times 1000\, D = 1.2\, D = 1.2 \times 250 = 300 \text{ mm}^2$$

Use 8 mm φ bars having $A_\varphi = \dfrac{\pi}{4}(8)^2 = 50.26 \text{ mm}^2$.

$\therefore$ Spacing $s = \frac{1000 \times 50.26}{300} = 167$ mm.. However, provide 8 mm ϕ bars @ 150 mm c/c.

5. ***Check for development length***

$$L_d = 47\ \varphi = 47 \times 12 \approx 564 \text{ mm.}$$

Providing 60 mm side cover, length of bars available

$$= \frac{1}{2}[B - b] - 60 = \frac{1}{2}[1700 - 300] - 60 = 640 \text{ mm.} > L_d. \text{ Hence safe.}$$

Design Example 22.3. ***Isolated footing of uniform depth for R.C. Column :***

Design an isolated footing of uniform thickness of a R.C. column bearing a vertical load of 600 kN and having a base of size 500 × 500 mm. The safe bearing capacity of soil may be taken as 120 kN/m². Use M 20 is concrete and Fe 415 steel.

Solution :

1. ***Design constants.*** For Fe 415 - M 20 combination, we have

$$\frac{x_{u,\,max}}{d} = 0.479 \quad \text{and } R_u = 2.761, \quad \text{as in Example 22.1}$$

2. ***Size of footing*** . W = 600 kN. Let W' be equal to 10% W = 60 kN

$$\therefore \quad A = 660/120 = 5.5 \text{ m}^2$$

$\therefore$ Size B of square $\sqrt{5.5} = 2.345$ m. Provide a square footing of 2.4 m × 2.4 m.

Net upward pressure $p_0 = \frac{600}{2.4 \times 2.4} = 104.17$ kN/m²

3. ***Design of section. (a) Depth on the basis of bending compression*** : The maximum bending moment acts at the face of column, and its magnitude is given by Eq. 22.18.

$$M = p_0 \frac{B}{8}(B - b)^2 \times 10^6 \text{ N-mm} = 104.17 \times \frac{2.4}{8}(2.4 - 5)^2 \times 10^6 = 1.128 \times 10^8 \text{ N-mm}$$

$$\therefore \quad M_u = 1.5\ M = 1.5 \times 1.128 \times 10^8 = 1.692 \times 10^8 \text{ N-mm}$$

$$\therefore \quad d = \sqrt{\frac{M_u}{R_u B}} = \sqrt{\frac{1.692 \times 10^8}{2.761 \times 2400}} \simeq 160 \text{ mm and } D = 160 + 60 = 220 \text{ mm}$$

(b) Depth on the basis of one way shear. For *one way shear,* the critical section is located at a distance d (mm) from the face of the column (Fig 22.13), where shear force V is given by

$$V = p_0 B\left\{\frac{1}{2}(B - b) - d\right\} = 104.17 \times 2.4\left\{\frac{1}{2}(2.4 - 0.5) - 0.001\ d\right\} \times 10^3 = 250008\ [0.95 - 0.001\ d]$$

$$\therefore \quad V_u = 1.5\ V = 375012[0.95 - 0.001\ d]$$

$$\therefore \quad \tau_c = \frac{V_u}{B.\ d} = \frac{375012\ (0.95 - 0.001\ d)}{2400\ d} = \frac{156.255}{d}(0.95 - 0.001\ d) \qquad \ldots(i)$$

From Table 7.2, k = 1.16 for D = 220 mm. Also, for an under-reinforced section with p = 0.3% for M 20 concrete, $\tau_c \approx 0.384$ N/mm² from Table 7.1. Hence the permissible shear stress $= k\ \tau_c = 1.16 \times 0.384 = 0.445$ N/mm². Equating this to (*i*) we get

$$0.445 = \frac{156.225}{d}(0.95 - 0.001\ d), \text{ from which } d = 246.7 \simeq 250 \text{ mm (say)}$$

(c) Depth for two way shear/check for two way shear

Let us keep d greater one of the two, i.e. $d = 250$ mm

For *two way shear* (punching shear), the section lies at $d/2$ from the column face all round. The width b_0 of the section $= b + d = 500 + 250 = 750$ mm. Also, shear force around the section is

$$F = p_0\,[B^2 - b_0^2] = p_0\,[B^2 - (b+d)^2] = 104.17\,[(2.4)^2 - (0.75)^2] = 541.424 \text{ kN}$$

$$\therefore \quad F_u = 1.5\,F = 1.5 \times 541.424 \times 10^3 = 812136 \text{ N}$$

$$\therefore \quad \tau_v = \frac{F_u}{4\,b_0\,d} = \frac{F_u}{4\,[b+d]\,d} = \frac{812136}{4 \times 750 \times 250} = 1.083 \text{ N/mm}^2$$

Permissible shear stress $= k_s\,\tau_c$

where $k_s = (0.5 + \beta_c) = (0.5 + 1)$ with max. value of 1. Hence $k_s = 1$

Also, $\tau_c = 0.25\sqrt{f_{ck}} = 0.25\sqrt{20} = 1.118$ N/mm^2

Hence permissible shear stress $= k_s\,\tau_c = 1 \times 1.118 = 1.118$ N/mm^2. Hence safe.

Hence $d = 250$ mm. Using effective cover of 60 mm, and keeping $D = 330$ mm, effective depth $= 330 - 60 = 270$ mm in the one direction and $d = 270 - 12 = 258$ mm in other direction, with 12 mm φ bars in the two directions.

5. ***Design of steel reinforcement***

Since the actual depth d provided is more than that required for bending compression, we have an under-reinforced section, for which

$$A_{st} = \frac{0.5\,f_{ck}}{f_y}\left[1 - \sqrt{1 - \frac{4.6\,M_u}{f_{ck}\,bd^2}}\right] bd$$

$$= \frac{0.5 \times 20}{415} \times \left[1 - \sqrt{1 - \frac{4.6 \times 1.692 \times 10^8}{20 \times 2400\,(258)^2}}\right] \times 2400 \times 258 = 1944 \text{ mm}^2$$

Using 12 mm φ bars, number of bars $= 1944/113 \simeq 18$

Hence provide 18 bars of 12 mm φ bars uniformuly distributed in each direction

6. ***Check for development length***

$L_d = 47\,\Phi = 47 \times 12 \approx 564$ mm

Providing 60 mm side cover, length of bars available

$= \frac{1}{2}[B - b] - 60 = \frac{1}{2}[2400 - 500] - 60$

$= 890$ mm $> L_d$. Hence safe.

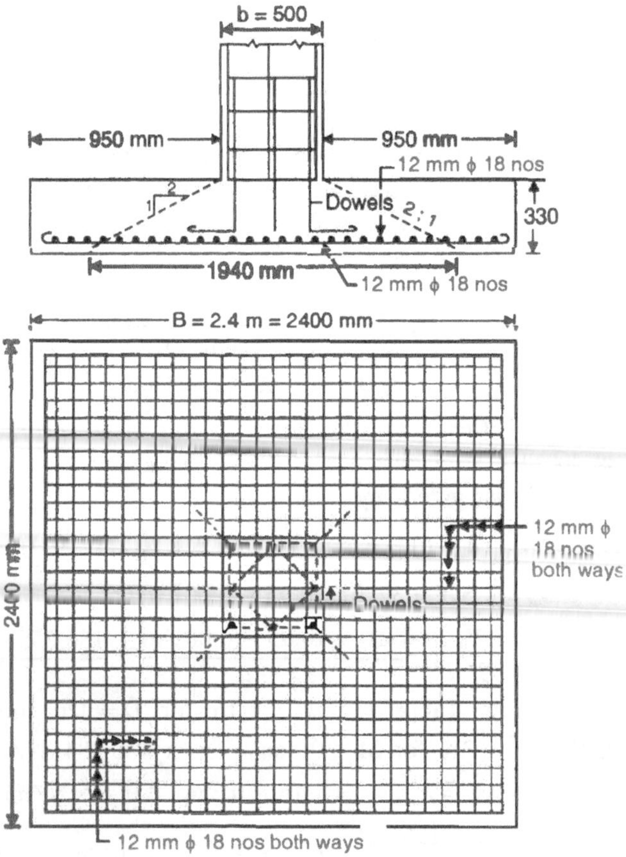

FIG. 22.25.

7. ***Transfer of load at column base***

$A_2 = 500 \times 500 = 250000 \text{ mm}^2$; $A_1 = [500 + 2(2 \times 330)]^2 = (1820)^2 = 3312400 \text{ mm}^2$

$$\therefore \quad \sqrt{\frac{A_1}{A_2}} = \sqrt{\frac{3312400}{250000}} = 3.64$$

Adopt max. value of $\sqrt{A_1/A_2} = 2$.

Hence permissible bearing stress $= 0.45 f_{ck} \sqrt{A_1/A_2} = 0.45 \times 20 \times 2 = 18 \text{ N/mm}^2$.

Actual bearing pressure $= \dfrac{1.5 \times 600000}{500 \times 500} = 3.6 \text{ N/mm}^2$,

Hence safe. Thus no separate dowel bars are required for the transfer of load. However, it is advisable to continue all the bars of the column, into the foundation.

The details of reinforcement etc. are shown in Fig 22.24.

Example 22.4. ***Isolated rectangular footing of uniform thickness for R.C. column:*** *Design a rectangular isolated footing of uniform thickness for R.C. column bearing a vertical load of 600 kN, and having a base size of 400 × 600 mm. The safe bearing capacity of soil may be taken as 120 kN/m². Use M 20 concrete and Fe 415 steel.*

Solution

1. ***Design constants.*** For M 20 – Fe 415 combination, we have

$$\frac{x_{u,\,max}}{d} = 0.479 \text{ and } R_u = 2.761, \text{ as in example 22.1}$$

2. ***Size of footing*** $W = 600$ kN. Let W' be equal to 10% $W = 60$ kN.

$$\therefore \quad A = \frac{660}{120} = 5.5 \text{ m}^2. \quad \text{Let ratio of } B \text{ to } L = \frac{40}{60} = \frac{2}{3}.$$

$$\therefore \quad \frac{2}{3} L \times L = 5.5 \quad \text{or} \quad L = 2.9 \text{ m}.$$

$$\therefore \quad B = \frac{2}{3} \times 2.9 = 1.93 \text{ m}.$$ However, provide a footing of size 2 m × 3 m.

Net upward pressure $p_0 = \dfrac{600}{2 \times 3} = 100 \text{ kN/m}^2$

3. ***Design of section.*** Refer Fig. 22.14.

(a) Design on the basis of bending compression

Bending moment M_1 about section X - X is given by

$$M_1 = \frac{p_0 B}{8}(L - a)^2 \text{ kN-m} = \frac{100 \times 2}{8}(3 - 0.6)^2 \times 10^6 \text{ N-mm} = 144 \times 10^6 \text{ N-mm}$$

$$\therefore \quad M_{1u} = 1.5 M_1 = 1.5 \times 144 \times 10^6 = 216 \times 10^6 \text{ N-mm}$$

B.M. M_2 about section Y–Y is given by

$$M_2 = \frac{p_0 L}{8}(B - b)^2 \times 10^6 \text{ N-mm} = \frac{100 \times 3}{8}(2 - 0.4)^2 \times 10^6 = 96 \times 10^6 \text{ N-mm}.$$

$$\therefore \quad M_{2u} = 1.5 M = 144 \times 10^6 \text{ N-mm} \; ; \quad \text{Thus,} \quad M_{2u} < M_{1u}.$$

$$\therefore \quad d = \sqrt{\frac{M_{1u}}{R_u B}} = \sqrt{\frac{216 \times 10^6}{2.761 \times 2000}} = 198 \text{ mm} \simeq 200 \text{ mm}$$

and $D = 200 + 60 = 260$ mm, providing 60 mm effective cover.

Provide uniform thickness for the entire footing.

(b) Depth on the basis of one way shear. For the ***beam action*** total S.F. along section AB [Fig. 22.10(*a*)] is

$$V = p_0 B\left(\frac{L}{2} - \frac{a}{2} - d\right) = p_0\ B\left(\frac{L-a}{2} - d\right) = 100 \times 2\left(\frac{3-0.6}{2} - 0.001\ d\right)10^3$$

$$= 200000\ (1.2 - 0.001\ d)\ \text{N/m}.$$

$$V_u = 1.5\ V = 300000\ (1.2 - 0.001\ d)\ \text{N/m}$$

$$\tau_v = \frac{V_u}{Bd} = \frac{300000}{2000\ d}(1.2 - 0.001\ D) = \frac{150}{d}(1.2 - 0.001\ d). \qquad \ldots(i)$$

Assuming under-reinforced section, with $p = 0.3\%$, we get $\tau_c = 0.384$ N/mm², from Table 7.1 for M 20 concrete. Also, $k = 1$ (from Table 7.2) for $D \geq 300$ mm.

$\therefore$ Permissible shear stress $= 1 \times 0.384 = 0.384$ N/mm². Equating this to (*i*), we get

$0.384 = \dfrac{150}{d}(1.2 - 0.001\ d)$, from which $d = 337$ mm $\simeq 340$ mm (say)

Hence adopt $d = 340$ mm.

(c) Check for two way shear action

For the *two way action or punching shear action* along $ABCD$ [Fig. 22.10(*b*)],

Perimeter $ABCD$ $= 2\{(a+d) + (b+d)\} = 2\{(0.6 + 0.34) + (0.4 + 0.34)\}$

$= 2\{0.94 + 0.74\} = 3.36$ m $= 3360$ mm.

Area $ABCD = 0.94 \times 0.74 = 0.6956$ m²

$\therefore$ Punching shear $= 1.5 \times 100\ [(2 \times 3) - 0.6956] = 795.66$ kN

$$\therefore \qquad \tau_v = \frac{795.66 \times 1000}{3360 \times 340} = 0.696\ \text{N/mm}^2.$$

Allowable shear stress τ_c is given by

$$\tau_c = 0.25\sqrt{f_{ck}} = 0.25\sqrt{20} \approx 1.118\ \text{N/mm}^2.$$

$k_s = (0.5 + \beta_c) = \left(0.5 + \dfrac{0.4}{0.6}\right) = 1.17$; However, adopt max. $k_s = 1$.

$$\therefore \qquad k_s\,\tau_c = 1 \times 1.118 = 1.118\ \text{N/mm}^2.$$

This is more than $\tau_v = 0.696$ N/mm². Hence safe.

Thus the effective depth $d = 340$ mm is alright. Keep $D = 420$ mm, so that using effective cover of 60 mm, $d = 420 - 60 = 360$ mm in one direction and $d = 360 - 12 = 348$ mm in the other direction, with 12 mm φ bars.

5. ***Design for reinforcement.***

Since actual d provided is more than that required for bending compression, we have an under-reinforced section. Hence area A_{st1} of long bars calculated for B.M. M_{1u} is given by

$$A_{st1} = \frac{0.5 f_{ck}}{f_y}\left[1 - \sqrt{1 - \frac{4.6\ M_{1u}}{f_{ck}\ B\ d_1^2}}\right]B\ d_1 = \frac{0.5 \times 20}{415}\left[1 - \sqrt{1 - \frac{4.6 \times 216 \times 10^6}{20 \times 2000 \times (360)^2}}\right]2000 \times 360$$

$$= 1752\ \text{mm}^2$$

Hence Number of 12 mm φ bars = 1752/113 ≏ 16

These are to be distributed uniformly in a width $B = 2$ m

Effective depth for top layer of reinforcement $= d_2 = 348$ mm

The area A_{st2} of short bars, calculated for M_{2u} is given by

$$A_{st2} = \frac{0.5 f_{ck}}{f_y}\left[1 - \sqrt{1 - \frac{4.6 M_{2u}}{f_{ck} L d_2^2}}\right] L d_2 = \frac{0.5 \times 20}{415}\left[1 - \sqrt{1 - \frac{4.6 \times 144 \times 10^6}{20 \times 3000\,(348)^2}}\right] 3000 \times 348$$

$$= 1175 \text{ mm}^2$$

This area is to be provided in two distinct band widths. Area $A_{st2(B)}$ in central band of width $B = 2$ m is given by

$$A_{st2(B)} = \frac{2A_{st2}}{\beta + 1} = \frac{2 \times 1175}{\frac{3}{2} + 1} = 940 \text{ mm}^2.$$

∴ No. of 12 mm φ bars = 940/113 ≈ 9 to be provided in central band width = 2 m.

Remaining area in each end band strip $= \frac{1}{2}(1175 - 940) = 117.5$ mm².

No. of 12 mm φ bars = 117.5/113 = 1.04. However provide minimum 3 bars in each end band of width $\frac{1}{2}(L - B) = \frac{1}{2}(3 - 2) = 0.5$ m.

6. ***Test for development length***

$L_d = 47\,\varphi = 47 \times 12 = 564$ mm.

Providing 60 mm side cover, length available $\frac{1}{2}[B - b] - 60 = \frac{1}{2}[2000 - 400] - 60 = 740$ mm, which is greater than L_d. Hence O.K.

7. ***Check for transfer of load at the base***

$A_2 = 600 \times 600 = 360000$ mm²

At a rate of spread of 2:1,

$A_1 = [600 + 2(2 \times 420)]^2 = (2280)^2$

$= 5.198 \times 10^6$ mm²

$$\therefore \quad \sqrt{\frac{A_1}{A_2}} = \sqrt{\frac{5.198 \times 10^6}{360000}} = 3.8 > 2$$

Adopt max. value of $\sqrt{A_1/A_2}$ as 2.

∴ Permissible bearing stress

$= 0.45 f_{ck} \sqrt{A_1/A_2}$

$= 0.45 \times 20 \times 2 = 18$ N/mm².

$$\text{Actual bearing stress} = \frac{1.5 \times 600000}{600 \times 600} = 2.5 \text{ N/mm}^2$$

Hence satisfactory.

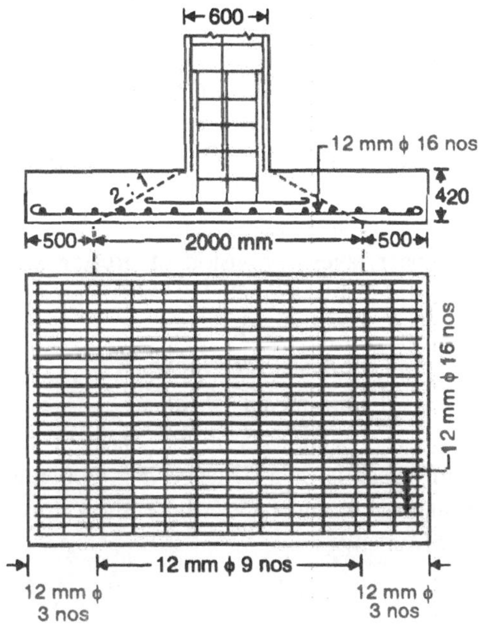

FIG. 22.26

The details of reinforcement etc. are shown in Fig. 22.26.

Example 22.5. ***Isolated square sloped footing*** *: Design an isolated square sloped footing for a column 500 mm × 500 mm, transmitting an axial load of 1200 kN. The column is reinforced with 8 bars of 20 mm diameter. The safe bearing capacity of soil is 120 tonnes/m². Use M 20 concrete and Fe 415 steel reinforcement.*

Solution.

1. ***Design constants..***

For M 20 – Fe 415 combination, we have :

$$\frac{x_{u,\max}}{d} = 0.479 \text{ and } R_u = 2.761, \text{ as in example 22.1}$$

2. ***Size of footing.*** $W = 1200$ kN. Assume $W' = 10\ \%\ W = 120$ kN

$\therefore$ Area of footing $A = \dfrac{1200 + 120}{12} = 11 \text{ m}^2$; Provide a footing of size 3.4 m × 3.4 m.

Actual upward pressure intensity $p_0 = \dfrac{1200}{3.4 \times 3.4} = 104 \text{ kN/m}^2$.

3. ***Design of footing.*** Maximum bending moment occurs at the face of the column and its magnitude is given by Eq. 22.18.

$$M = p_0 \frac{B}{8} (B - b)^2 \text{ kN-m} = \frac{104 \times 3.4}{8} (3.4 - 0.5)^2 \times 10^6 \text{ N-mm} = 372 \times 10^6 \text{ N-mm.}$$

$$\therefore \qquad M_u = 1.5\ M = 1.5 \times 372 \times 10^6 = 558 \times 10^6 \text{ N-mm}$$

The section of the footing at the columm face will be trapezoidal, as shown in Fig. 22.16. Let the effective depth at the column face be d and that at the edges be 0.2 d.

$$b_1 = b = 500 \text{ mm}$$

$$d = \sqrt{\frac{M_u}{R_u\, b_1}} = \sqrt{\frac{558 \times 10^6}{2.761 \times 500}} = 636 \text{ mm.}$$

Let us fix $D = 760$ mm. Using an effective cover of 60 mm, available $d = 760 - 60 = 700$ mm for the lower layer and $700 - 12 = 688$ mm for the upper layer. At the end, provide $D = 200$ mm, so that available $= 200 - 60 = 140$ mm for the lower layer and $140 - 12 = 128$ mm for the upper layer. Provision of greater effective depth will give rise to under-reinforced section, as desired by the Code. Also, provision of greater d will make the section more safe in shear (next step).

4. ***Check for shear***

(a) For beam shear

The section is to be checked for beam shear is at a distance $d = 700$ mm from the column face, where shear force is given by

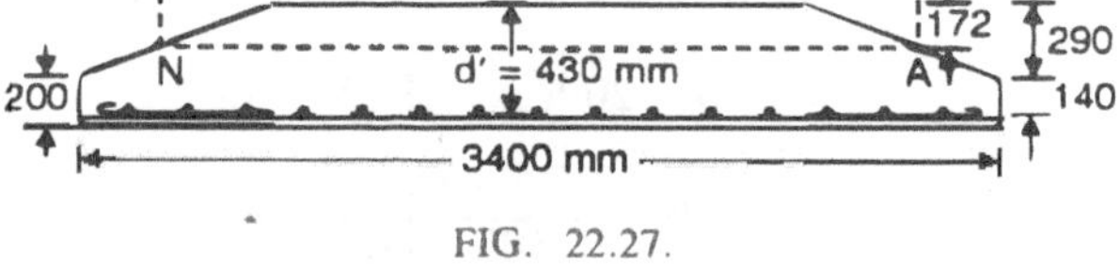

FIG. 22.27.

$$V = p_0 B \left\{\frac{1}{2}(B - b) - d\right\} = 104 \times 3.4 \left\{\frac{1}{2}(3.4 - 0.5) - 0.7\right\} = 266 \text{ kN.}$$

$$\therefore \qquad V_u = 1.5\ V = 1.5 \times 266 = 399 \text{ kN}$$

The section for footing is shown in Fig. 22.27.

$$\text{Effective depth } d' \text{ at that location} = 140 + \frac{700 - 140}{1450}(1450 - 700) = 430 \text{ mm.}$$

$$\text{Top width of section} = 500 + \frac{3400 - 500}{1450} \times 700 = 1900 \text{ mm}$$

Factor $x_{u,\,max}/d = 0.479$ for balanced section. However, for an under-reinforced section, adopt $x_u/d \simeq 0.4$ so that $x_u = 0.4\ d' = 0.4 \times 420 = 172$ mm.

$$\text{Width } b_n \text{ at N.A.} = 1900 + \frac{3400 - 1900}{290} \times 172 \simeq 2790 \text{ mm.}$$

$$\therefore \qquad \tau_v = \frac{V_u}{b_n\, d'} = \frac{399 \times 1000}{2790 \times 430} = 0.333 \text{ N/mm}^2$$

Assuming $p = 0.3\%$ for an under-reinforced section $\tau_c = 0.384$ N/mm^2, for M 20 concrete. Also, $k = 1$ for $D > 300$ mm. Thus, $\tau_v < \tau_c$. (Hence safe).

(*b*) ***Check for two way shear (punching shear)*** : [Fig. 22.10(*a*)]

Perimeter $ABCD = 2\,[(a + d) + (b + d)] = 4\,(a + d) = 4\,(500 + 700) = 4800$ mm.

Area $ABCD = 1.2 \times 1.2 = 1.44$ m^2

Punching shear $= 104\,[3.4 \times 3.4 - 1.44] = 1052.5$ kN.

$$\therefore \qquad \tau_v = 1.5 \times \frac{1052.5 \times 1000}{4800 \times 700} = 0.47 \text{ N/mm}^2.$$

Allowable shear stress τ_c is given by

$$\tau_c = 0.25\sqrt{f_{ck}} = 0.25\sqrt{20} \approx 1.118 \text{ N/mm}^2$$

$k_s = (0.5 + \beta_c) = (0.5 + 1)$. However adopt max. $k_s = 1$.

$\therefore \qquad k_s \tau_c = 1 \times 1.118 = 1.118$ N/mm^2. Hence safe.

5. ***Steel reinforcement*** : For an under-reinforced section,

$$A_{st} = \frac{0.5\, f_{ck}}{f_y}\left[1 - \sqrt{1 - \frac{4.6\, M_u}{f_{ck}\, b_1\, d^2}}\right] b_1\, d = \frac{0.5 \times 20}{415}\left[1 - \sqrt{1 - \frac{4.6 \times 558 \times 10^6}{20 \times 500\,(700)^2}}\right] 500 \times 700$$

$= 2615$ mm^2

Number of 12 mm φ bars

$= 2615/113.1 = 23.1 \simeq 24$

Hence provide 12 mm φ bars, 24 Nos. uniformly spaced in the width 3.4 m in each direction.

6. ***Check for development length***

$L_d = 47 \times 12 = 564$ mm.

Providing 60 mm side cover, length available $= \frac{1}{2}(B - b) = \frac{1}{2}(3400 - 500) - 60$ $= 1390$ mm.

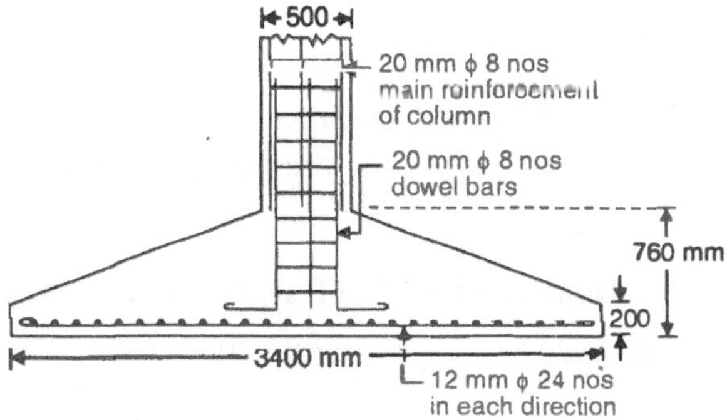

FIG. 22.28.

Hence O.K. The details of reinforcement etc. as shown in Fig. 22.27.

7. ***Check for transfer of load at the column base***

$A_2 = 500 \times 500 = 250000$ mm². At a rate of spread of 2 : 1,

$A_1 = [500 + 2(2 \times 760)]^2 = (3540)^2 = 12531600$ mm². Hence,

$$\sqrt{\frac{A_1}{A_2}} = \sqrt{\frac{12531600}{250000}} = 7.08 > 2 \text{ ; Adopt a max. value of } \sqrt{\frac{A_1}{A_2}} = 2.$$

$\therefore$ Permissible bearing stress $= 0.45 f_{ck} \sqrt{A_1/A_2} = 0.45 \times 20 \times 2 = 18$ N/mm²

Actual bearing stress $= \dfrac{1.5 \times 1200000}{500 \times 500} = 7.2$ N/mm². Hence satisfactory.

In the present case, the load transfer is there without exceeding the permissible bearing stress. However it is always a good practice to extend all the column bars into the foundation and anchored. If dowels bars are provided, development length requirements should be satisfied as required in § 22.4.

The details of reinforcement etc. are shown in Fig. 22.28.

Design Example 22.6. ***Sloped footing for circular column :*** *Design a footing for circular column 560 mm diameter, transmitting an axial load of 1200 kN. The column is reinforced with 8 bars of 10 mm diameter. The safe bearing capacity of soil is 120 kN/m². Use M 20 concrete and Fe 415 steel.*

Solution (*a*) ***Circular footing***

1. ***Design constants***. For M 20 – Fe 415 combination, we have

$x_{u,\max}/d = 0.479$ and $R_u = 2.761$

2. ***Size of footing***. $W = 1200$ kN. Assume $W' = 10\% \; W = 120$ kN

$\therefore$ Area of footing $A = \dfrac{1200 + 120}{120} = 11$ m².

Radius $R = \sqrt{A/\pi} = \sqrt{11/3.14} = 1.87$ m.

Adopt radius $R = 1.9$ m or diameter $= 3.8$ m at the base.

Radius of column $r = 280$ mm $= 0.28$ m

Net upward pressure $p_0 = \dfrac{W}{\pi R^2} = \dfrac{1200}{\pi(1.9)^2} = 106$ kN/m²

3. ***Design of footing***. Refer Fig. 22.29.

Upward force on one quadrant of footing

$$= p_0 \frac{\pi}{4}(R^2 - r^2)$$

Distance of C.G. of shaded area from the face of the column $= \dfrac{0.2}{R + r}[3R^2 - 2Rr - 2r^2]$

Bending moment at the face of the column is given by

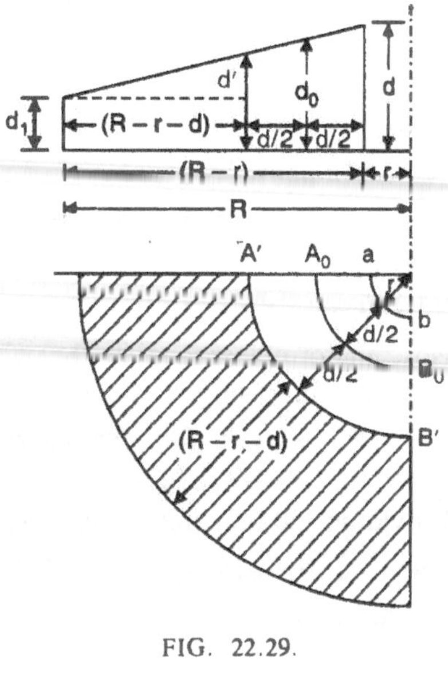

FIG. 22.29.

$$M = p_0 \frac{\pi}{4}(R^2 - r^2) \times \frac{0.2}{R+r}[3R^2 - 2Rr - 2r^2] = p_0 \frac{\pi}{20}(R - r)(3R^2 - 2Rr - 2r^2)$$

$$= \frac{106\,\pi}{20}(1.9 - 0.28)[3(1.9)^2 - 2 \times 1.9 \times 0.28 - 2(0.28)^2] \text{ kN-m}$$

$$= 2.592 \times 10^2 \text{ kN-m} = 2.592 \times 10^8 \text{ N-mm}$$

$$M_u = 1.5\,M = 1.5 \times 2.592 \times 10^8 = 3.888 \times 10^8 \text{ N-mm}$$

Breadth b of quadrant $= \frac{1}{2}\pi r = \frac{\pi}{2}(280) = 440$ mm.

$$\therefore \qquad d = \sqrt{\frac{3.888 \times 10^8}{440 \times 2.761}} \simeq 566 \text{ mm.}$$

Let us keep $D = 700$ mm, so that with 60 mm effective cover, $d = 640$ mm for outer layer and $640 - 16 = 624$ mm for inner layer.

4. Design for shear. The depth found above from bending compression should be adequate from shear point of view. Circular footings are checked for two way shear (punching shear) at a circular critical section distant $d/2$ (≈ 320 mm) away from the face of the column.

$$\therefore \qquad r_0 = r + \frac{d}{2} = 280 + 320 = 600 \text{ mm}$$

Punching shear $F_u = 1.5\,p_0\,\pi\,(R^2 - r_0^2) = 1.5 \times 106\,\pi\,[(1.9)^2 - (0.6)^2] = 1623.4$ kN

Hence from Eq. 22.40, $\frac{F_u}{2\,\pi\,r_0\,d_0} \le 0.25\sqrt{f_{ck}}$

$$\text{From which } d_0 = \frac{F_u}{2\pi r_0 \times 0.25\sqrt{f_{ck}}} = \frac{1623.4 \times 1000}{2\pi \times 600 \times 0.25\sqrt{20}} \approx 386 \text{ mm} \qquad \ldots(3)$$

Let the effective depth d_1 at the outer edge be 140 mm. Then the available depth d_0 at the critical plane, corresponding to $d = 640$ mm (fixed from bending compression) is given by (Fig. 22.29)

$$d_0 = d_1 + \frac{(d - d_1)}{R - r}\left(R - r - \frac{d}{2}\right) = 140 + \frac{640 - 140}{1900 - 280}\left[1900 - 280 - \frac{640}{2}\right] \simeq 541 \text{ mm}$$

This is more than the one found above from punching shear of view.

For *beam shear,* the critical plane $A'B'$ lies at a radial distance d from the column face (Fig. 22.29), where the effective depth d' is given by

$$d' = d_1 + \frac{(d - d_1)}{R - r}[R - r - d] = 140 + \frac{640 - 140}{1900 - 280}[1900 - 280 - 640] \simeq 442 \text{ mm}$$

Breadth b' at this section $= \frac{\pi}{2}(r + d) = \frac{\pi}{2}(280 + 640) = 1445$ mm

Shear force $V_u = 1.5\,p_0 \frac{\pi}{4}\{R^2 - (r + d)^2\} = 1.5 \times 106 \times \frac{\pi}{4}[(1.9)^2 - (0.28 + 0.64)^2] = 345.11$ kN.

$$\tau_v = \frac{V_u}{b'd'} = \frac{345.11 \times 1000}{1445 \times 442} = 0.54 \text{ N/mm}^2$$

Assuming under-reinforced section, permissible shear stress at $p = 0.3\%$ is $\tau_c = 0.384$ N/mm^2 (Table 7.1) with $k = 1$ for $D > 300$ mm. Hence $d = 640$ mm found on the basis of bending compression is inadequate.

Revised $d = \frac{0.54}{0.384} \times 640 = 900$ mm

However, keep $d = 800$ mm (and $D = 800 + 60 = 860$ mm), since V_u will be decreased and b' will be increased with increase of d from 640 mm to 800 mm. Let us, therefore, again check for one way shear.

$$d' = 140 + \frac{800 - 140}{1900 - 280}[1900 - 280 - 800] = 474 \text{ mm}; \quad b' = \frac{\pi}{2}(280 + 800) = 1696.4 \text{ mm}$$

$$V_u = 1.5 \times 106 \times \frac{\pi}{4}[(1.9)^2 - (0.28 + 0.8)^2] = 305.15 \text{ kN}$$

$$\therefore \quad \tau_v = \frac{305.15 \times 10^3}{1696.4 \times 474} = 0.379 \text{ N/mm}^2 < 0.384 \text{ N/mm}^2. \text{ Hence safe.}$$

Hence keep $D = 860$ mm ($d = 860 - 60 = 800$ mm) at the column face, and $D = 200$ mm at the ends as shown in Fig. 22.30.

5. ***Design of reinforcement***

$$A_{st} = \frac{0.5 f_{ck}}{f_y}\left[1 - \sqrt{1 - \frac{4.6 M_u}{f_{ck} bd^2}}\right] bd = \frac{0.5 \times 20}{415}\left[1 - \sqrt{1 - \frac{4.6 \times 3.888 \times 10^8}{20 \times 440 (800)^2}}\right] 440 \times 800$$

$= 1475 \text{ mm}^2$

$\therefore$ No. of 12 mm φ bars required $= 1475/113.1 \simeq 14$

Hence, provide 14 bars each, in two mutually perpendicular directions. These bars are to be accommodated in the form of square mesh, centrally located under the column. If B is the size of inscribed square, its magnitude $= \frac{D}{\sqrt{2}} = \frac{2R}{\sqrt{2}} = R\sqrt{2} = 1.9\sqrt{2} \approx 2.7$ m.

Average spacing of bars $= 2700/13 = 208$ mm c/c

Provide two additional bars in each sector beyond this square. Provide 60 mm cover to centre of steel.

6. ***Check for Development Length***

$L_d \approx 47\,\varphi = 47 \times 12 = 564$ mm

Available length $= (R - r)$ − side cover $= 1900 - 280 - 60 = 1560 \text{ mm} > L_d$ Hence safe.

The details of reinforcement etc. are shown in Fig. 22.30.

***(b)* Square footing** : For square footing, the width b of an equivalent square column is given by

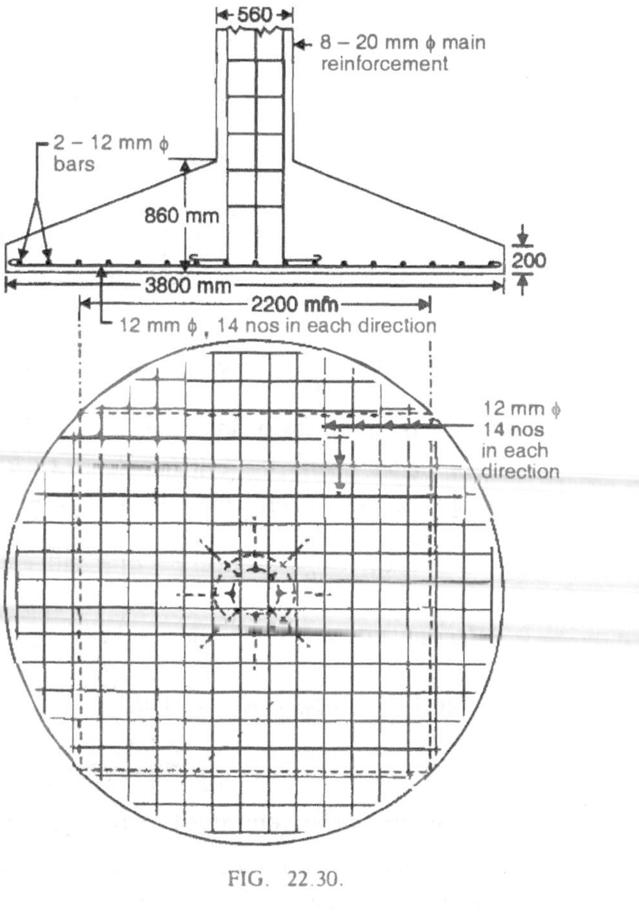

FIG. 22.30.

$b^2 = \pi r^2 \qquad \therefore \quad b = r\sqrt{\pi} = 280\sqrt{\pi} = 496 \approx 500$ mm

Thus the problem reduces to the design of a square footing for a square column of size 500 mm × 500 mm carrying an axial load of 1200 kN. The footing design will be practically the same as worked out in Example 22.6.

22.10. ISOLATED FOOTING SUBJECTED TO ECCENTRIC LOAD

When a column is subjected to axial loading only, the C.G. of column area should coincide with the C.G. of footing area, so that the soil pressure distribution below the footing is *rectangular.* However, a column may be subjected to an axial load along with certain bending moment in one plane, or it may be subjected to an eccentric load. In both the cases, the soil pressure distribution will be trapezoidal if the C.G. of footing area coincides with the C.G. of column as shown in Fig. 22.32. If however, a rectangular pressure distribution is required, the C.G. of footing area may be placed eccentric with the C.G. of the column, by a distance $e = M/W$, or equal to eccentricity of axial load W, as shown in Fig. 22.31.

(*a*) Design of unsymmetrical footing

Fig. 22.31 shows such a case in which the footing is arranged eccentric with respect to column, *i.e.* the axis *Y-Y* of footing is placed at a distance $e\left(= M/W\right)$ from the axis of the column. The pressure distribuion will be rectangular.

As before, $$B^2 = \frac{W + W'}{q_0} \qquad \ldots(i)$$

and $$p_0 = W/B^2 \qquad \ldots(ii)$$

The bending moment will be maximum at face *AB* of the column, since the cantilever distance in front of this face is more, its value being equal to

$$\frac{B}{2} + e - \frac{b}{2} = \frac{1}{2}(B - b) + e.$$

$$\therefore \; M_u = 1.5\, p_0\, B\left\{\frac{1}{2}(B - b) + e\right\} \times \frac{1}{2}\left\{\frac{1}{2}(B - b) + e\right\}$$

$$= \frac{1.5 p_0\, B}{2}\left\{\frac{1}{2}(B - b) + e\right\}^2 \qquad \ldots(22.41)..(iii)$$

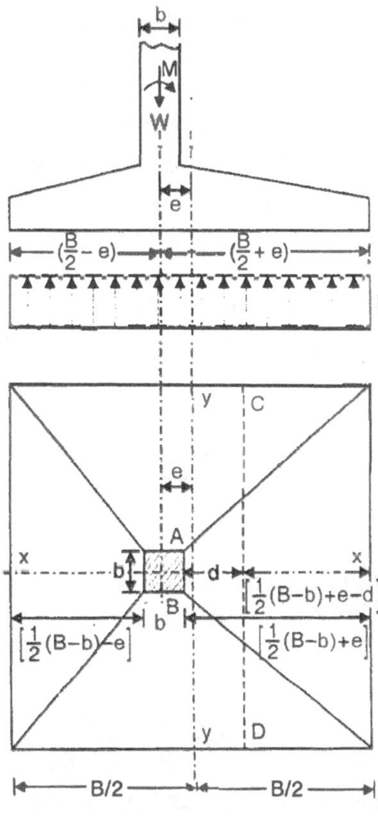

FIG. 22.31.

The section of footing at plane *AB* will be trapezoidal and its moment of resistance can be determined by the method explained in section § 22.7. Alternatively, the trapezoidal section may be replaced by an equivalent rectangular section of width $b + \frac{1}{8}(B - b)$.

Equating the moment of resistance to the external bending moment, the effective depth can be calculated.

The effective depth is also determined(or checked) from two way shear(punching shear), the critical section for which is at distance *d/2* all round the column face. The width of the section = $b_0 = (b + d)$ and the shear stress will be $\frac{F_u}{4b_0\, d_0}$ where F_u= punching shear force and d_0 is the effective depth at the section. This shear stress should be less than $k_s\, \tau_c$ where $k_s = (0.5 + \beta_c)$ but not greater than 1 and $\tau_c = 0.25\sqrt{f_{ck}}$. The shear force $F_u = 1.5\, p_0\, [B^2 - b_0^2]$.

For *one way shear*, the critical section is located at distance d from column face, where shear force V is given by

$$V = p_0\, B\, [\tfrac{1}{2}(B - b) + e - d] \qquad \text{and} \quad V_u = 1.5\, V$$

$$\therefore \qquad \tau_v = \frac{V_u}{b'\, d'}$$

where b' and d' are the width of the section at N.A. and the effective depth respectively, of the section CD (Fig. 22.31).

The shear stress τ_v should not exceed the permissible shear stress $k\, \tau_c$ (Table 7.1 and 7.2).

The steel reinforcement, parallel to XX axis is given by the following expression applicable for under-reinforced section :

$$A_{st} = \frac{0.5\, f_{ck}}{f_y}\left[1 - \sqrt{1 - \frac{4.6\, M_u}{f_{ck}\, bd^2}}\right] bd$$

The same steel reinforcement is provided in the other direction, though the bending moment will be less in the other direction because of lesser cantilever projection.

(*b*) Design of symmetrical footing

Fig. 22.32 shows a symmetrically placed footing so that C.G. of footing coincides with that of column. The pressure distribution below the footing will evidently be trapezoidal.

Let B be the size of footing. The maximum stress is equated to the safe bearing capacity q_0 of the soil, to determine the magnitude of B :

$$\frac{W + W'}{B^2} + \frac{M}{B^3/6} = q_0$$

or $\quad 6M + (W + W')\, B = q_0\, B^3 \qquad \ldots(22.42\ a)$

The above equation can be solved for B, by trial and error.

The net upward pressure distribution is given by

$$p_0 = \frac{W}{B^2} \pm \frac{M}{B^3/6} \qquad \ldots(22.42\ b)$$

so that pressure p_{01} and p_{02} at the two extremities are known. The lengths of cantilevers on all the four sides will be equal. However, the bending

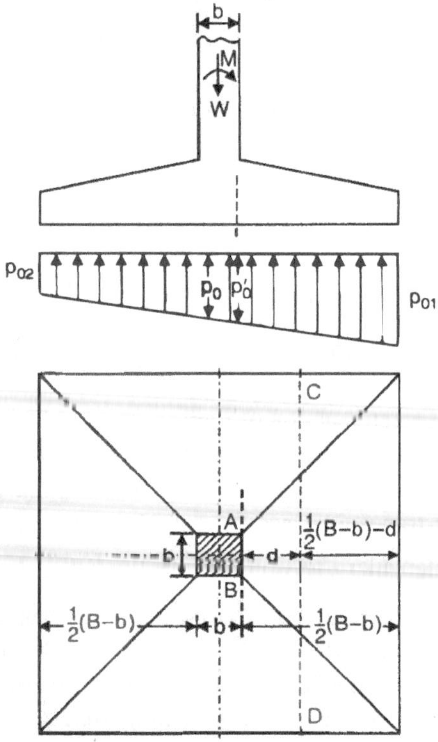

FIG. 22.32

moment will be maximum at the face AB since the pressure intensity is maximum to the right of it. The total upward force F to the right of AB is given by

$$F = B\left\{p_0'\left(\frac{B-b}{2}\right)+\frac{1}{2}(p_{01}-p_0')\left(\frac{B-b}{2}\right)\right\} = \frac{B\,(B-b)}{2}\left\{\frac{p_0'+p_{01}}{2}\right\} = \frac{B\,(B-b)}{4}(p_0'+p_{01}) \quad ..(22.43)$$

Hence B.M. M at AB is given by

$$M = B\left\{p_0'\left(\frac{B-b}{2}\right)\left(\frac{B-b}{4}\right)+\frac{1}{2}(p_{01}-p_0')\left(\frac{B-b}{2}\right)\frac{2}{3}\left(\frac{B-b}{2}\right)\right\}$$

$$= B(B-b)^2\left\{\frac{p_0'}{8}+\frac{p_{01}-p_0'}{12}\right\} = \frac{B(B-b)^2}{24}(2p_{01}+p_0') \quad ...(22.44)$$

and $\quad M_u = 1.5\,M \quad ...(22.44\ a)$

where p_0' is the pressure intensity at AB, and is given by

$$p_0' = p_{02}+\frac{p_{01}-p_{02}}{B}\left\{\frac{1}{2}(B-b)+b\right\} = p_{02}+\frac{p_{01}-p_{02}}{2B}(B+b) \quad ...(22.45)$$

The maximum bending moment can be equated to the moment of resistance of the section to get d. The effective depth should be sufficient so that punching shear stress is not exceeded. Similarly, effective depth should be sufficient so that nominal shear stress τ_v resulting from one way shear is not exceeded. Steel reinforcement at the plane of maximum B.M. is given by the following expression applicable for under-reinforced section :

$A_{st} = \dfrac{0.5 f_{ck}}{f_y}\left[1-\sqrt{1-\dfrac{4.6\,M_u}{f_{ck}\,bd^2}}\right]bd$. The reinforcement is checked for the development length.

Design Example 22.7. ***Unsymmetrical footing for column with axial load and moment:*** *Design an isolated unsymmetrical square footing for a column 500 mm × 500 mm, transmitting a load of 600 kN and a moment of 30 kN-m. The safe bearing capacity of soil is 120 kN/m². Use M 20 concrete and Fe 415 steel.*

Solution.

1. ***Design constants.*** For M 20 - Fe 415 combination

$$x_{u,\,max}/d = 0.479 \quad \text{and} \quad R_u = 2.761$$

2. ***Size of footing.*** $\quad W = 600$ kN

Assume $\quad W' = 10\%$ of $W = 60$ kN

$$\text{Area of footing} = \frac{600+60}{120} = 5.5\ \text{m}^2$$

$$\therefore \quad B = \sqrt{5.5} = 2.35\ \text{m},$$

Provide a square footing of size 2.4 m × 2.4 m.

Eccentricity $\quad e = \dfrac{M}{W} = \dfrac{30\times1000}{600} = 50$ mm

Hence place C.G. of footing 50 mm away from the axis of column as shown in Fig. 22.33. Thus the load and moment is equivalent to a load W at the C.G. of the footing.

$$p_0 = \frac{W}{B^2} = \frac{600}{2.4\times2.4} = 104\ \text{kN/m}^2$$

3. ***Design of footing***

Cantilever length to the right of AB

$= \frac{1}{2}[2400 - 500] + 50 = 1000$ mm = 1 m

Bending moment at the face AB of the column is given by

$$M = p_0 B \frac{(1)^2}{2} = 104 \times 2.4 \times \frac{1}{2} \text{ kN-m} = 124.8 \text{ kN-m}$$

$$= 1.248 \times 10^8 \text{ N-mm}$$

$$\therefore \quad M_u = 1.5\, M = 1.5 \times 1.248 \times 10^8$$

$$= 1.872 \times 10^8 \text{ N-mm}$$

Here $b_1 = b$ = top width

$$\therefore \quad d = \sqrt{\frac{M_u}{R_u\, b_1}} = \sqrt{\frac{1.872 \times 10^8}{2.761 \times 500}} = 369 \text{ mm}$$

However, keep $d = 400$ mm and $D = 400 + 60 = 460$ mm, thus providing an effective cover of 60 mm. Effective depth available for the second (inner) layer $= 400 - 12 = 388$ mm, using 12 mm φ bars. Also, keep $D = 200$ mm at the ends, so that available d at the ends is equal to $200 - 60 = 140$ mm.

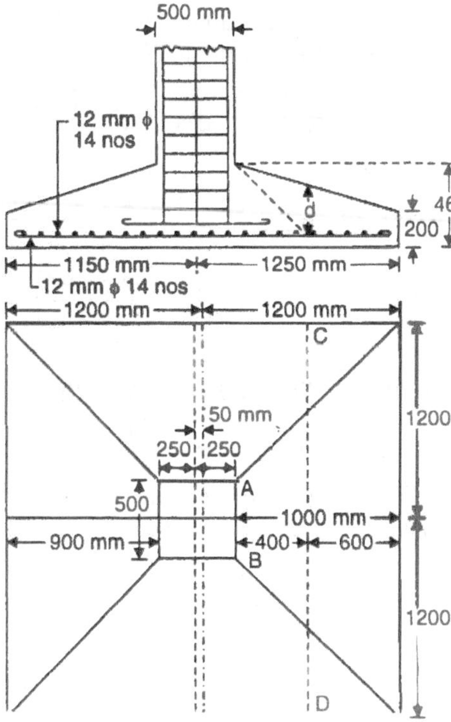

FIG. 22.33

4. ***Check for shear.***

For two way shear (punching shear) the section is situated at distance $d/2$ from the column face all round. The side b_0 of the section is $b_0 = b + d = 500 + 400 = 900$ mm.

Shear force $F_u = 1.5\, p_0\, [B^2 - b_0^2] = 1.5 \times 104\, [(2.4)^2 - (0.9)^2] = 772.2$ kN

$$\therefore \qquad \tau_v = \frac{F_u}{4\, b_0\, d_0}, \quad \text{from which } d_0 = \frac{F_u}{4\, b_0\, \tau_v}$$

Now permissible shear stress $= k_s\, \tau_c = 1 \times 0.25 \sqrt{f_{ck}} = 0.25 \sqrt{20} = 1.118$ N/mm^2

Limiting τ_v to the above permissible shear stress,

$$d_0 = \frac{772.2 \times 1000}{4 \times 900 \times 1.118} = 192 \text{ mm.}$$

If an effective depth of 140 mm (total depth = 200 mm) is provided at the outer edge, and if effective depth d = 400 mm is kept at face AB (found on the basis of bending compression), the available effective depth at $d/2$ from column face will be $= 140 + \frac{400 - 140}{1250 - 250} \times (1250 - 250 - 200) = 348$ mm, which is more than the required depth d_0 = 192 mm. Hence the section will be safe in punching shear.

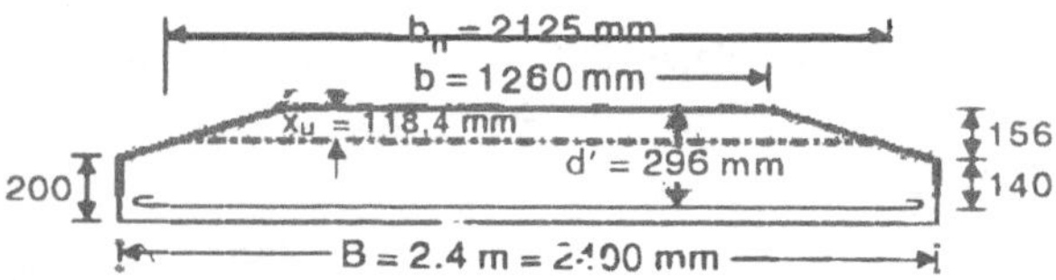

FIG. 22.34

Also, for one way shear, the critical section CD will be at a distance d = 400 mm from the column face. Cantilever length to the right of section $CD = 1000 - 400 = 600$ mm $= 0.6$ m. The shear force there is given by

$$V_u = 1.5\, p_0\, B \times 0.6 = 1.5 \times 104 \times 2.4 \times 0.6 = 224.64 \text{ kN}$$

The section will be trapezoidal in shape as shown in Fig. 22.34. The width b at the top is given by

$$b = 500 + \frac{2400 - 500}{1000} \times 400 = 1260 \text{ mm}$$

The effective depth $$d' = 140 + \frac{400 - 140}{1000} \times 600 = 296 \text{ mm}$$

For balanced section, $x_{u,\max}/d = 0.479$. Let $x_u/d \simeq 0.4$ for under reinforced section,

So that $$x_u = 0.4\, d' = 0.4 \times 296 = 118.4 \text{ mm}$$

Hence the width of section at N.A. is given by

$$b_n = 1260 + \frac{2400 - 1260}{156} \times 118.4 \simeq 2125 \text{ mm}$$

Hence $$\tau_v = \frac{V_u}{b_n\, d'} = \frac{224.64 \times 1000}{2125 \times 296} = 0.357 \text{ N/mm}^2 .$$

This is less than the permissible shear stress of 0.384 N/mm^2 at $p = 0.3\%$ (Table 7.1) corresponding to an under-reinforced section.

Hence $d = 400$ mm found on the basis of the bending compression is adequate.

5. ***Steel reinforcement*** : For an under-reinforced section.

$$A_{st} = \frac{0.5 f_{ck}}{f_y}\left[1 - \sqrt{1 - \frac{4.6\, M_u}{f_{ck}\, b_1\, d^2}}\right] b_1\, d = \frac{0.5 \times 20}{415}\left[1 - \sqrt{1 - \frac{4.6 \times 1.872 \times 10^8}{20 \times 500\, (400)^2}}\right] 500 \times 400$$

$= 1545$ mm^2

Using 12 mm φ bars $A_\varphi = 113$ mm^2. ∴ No. of bars $= 1545/113 \simeq 14$

Arrange them uniformly spaced in width of 2400 mm. Use the same reinforcement in the perpendicular direction.

6. ***Check for the development length.*** $L_d = 47\,\varphi = 47 \times 12 \approx 564$ mm

Using 60 mm side cover, available length $= 1250 - 50 - 60 = 1140$ mm $> L_d$.

Hence O.K. The details of reinforcement etc. are shown in Fig. 22.33.

Design Example 22.8. *Symmetrical sloped footing for column with axial load and moment*: *Redesign the symmetrical sloped square footing for the column of Example 22.7.*

1. ***Design constants*** : For M 20 – Fe 415 combination, we have

$$x_{u,\max}/d = 0.479 \quad \text{and} \quad R_u = 2.761$$

2. ***Size of footing.*** $W = 600$ kN. $W' = 60$ kN ; $M = 30$ kN-m

Equating the maximum soil pressure to the safe bearing capacity of soil,

$$\frac{660}{B^2} + \frac{30}{B^3/6} = 120 \qquad \text{or} \qquad 660\, B + 180 = 120\, B^3$$

Solving this by trial and error, we get $B = 2.47$ m

However, provide the footing of size 2.5 m × 2.5 m

The maximum and minimum soil pressure are given by

$$p_{01} = \frac{W}{B^2} + \frac{6M}{B^3} = \frac{600}{(2.5)^2} + \frac{6 \times 30}{(2.5)^3}$$

$$= 96 + 11.5 = 107.5 \text{ kN/m}^2.$$

$$p_{02} = \frac{W}{B^2} - \frac{6\,M}{B^3} = 96 - 11.5$$

$$= 84.5 \text{ kN/m}^2$$

Pressure intensity under the column axis is

$$p_0 = \tfrac{1}{2}\,[107.5 + 84.5] = 96 \text{ kN/m}^2$$

3. ***Design of section for bending compression***

Arrange the footing centrally under the column as shown in Fig. 22.35.

Intensity of net soil pressure below the column face is

$$p_0' = 96 + \frac{107.5 - 96}{1.25} \times 0.25$$

$$= 98.3 \text{ kN/m}^2$$

Maximum B.M. will occur at face *AB*. Cantilever length $\tfrac{1}{2}(B - b) = \tfrac{1}{2}(2.5 - 0.5) = 1$ m.

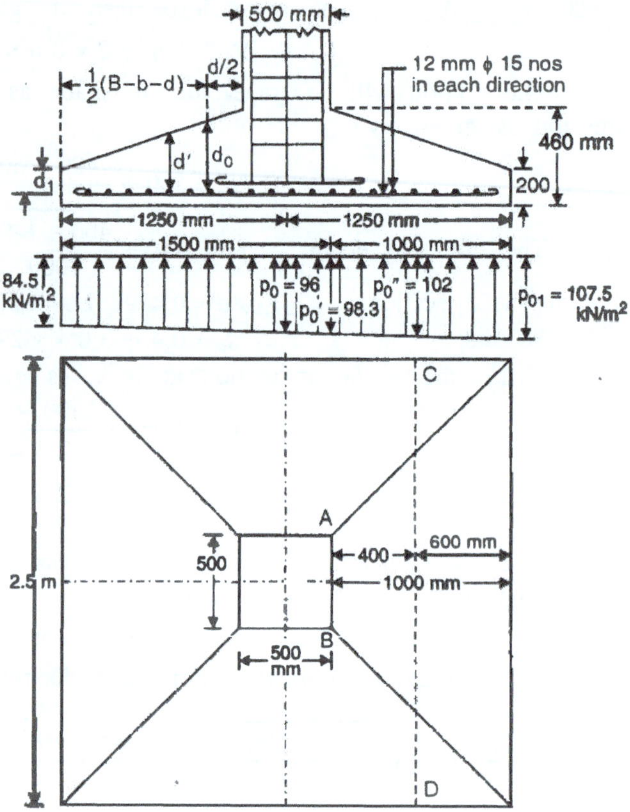

FIG. 22.35.

$$\therefore \quad \text{Total force under cantilever} = (2.5 \times 1)\left[\frac{98.3 + 107.5}{2}\right] = 257 \text{ kN}$$

$$\text{Distance of its centroid from } AB = \left(\frac{98.3 + 2 \times 107.5}{98.3 + 107.5}\right)\frac{1}{3} = 0.507 \text{ m}$$

$$\therefore \quad \text{B.M. } M = 257 \times 0.507 = 130 \text{ kN-m} = 1.3 \times 10^8 \text{ N-mm.}$$

Alternatively,

$$M = \frac{B\,(B - b)^2}{24}\,(2\,p_{01} + p_0') = \frac{2.5\,(2.5 - 0.5)^2}{24}\,[2 \times 107.5 + 98.3]$$

$$= 130 \text{ kN-m} = 1.3 \times 10^8 \text{ N-mm.}$$

$$\therefore \quad M_u = 1.5\,M = 1.5 \times 1.3 \times 10^8 = 1.95 \times 10^8 \text{ N-mm.}$$

$$d = \sqrt{\frac{M_u}{R_u\, b_1}} = \sqrt{\frac{1.95 \times 10^8}{2.761 \times 500}} = 376 \text{ mm}$$

However, keep $d = 400$ mm and $D = 400 + 60 = 460$ mm, thus providing an effective cover of 60 mm. Effective depth available for the second (or inner) layer = 400 − 12 = 388 mm, using 12 mm φ bars. Also, keep $D = 200$ mm at the ends, so that available d at the ends is equal to 200 − 60 = 140 mm.

4. ***Design for shear***. The depth found above should be safe for shear. For two way shear (punching shear), the critical plane lies at $d/2 = 200$ mm from the column face for which width $b_0 = b + d/2 = 500 + 200 = 700$ mm.

The punching shear stress $\tau_v = \dfrac{F_u}{4 b_0 d_0}$

where F_u = Punching shear $= 1.5\,[W - p_0 b_0^2] = 1.5\,[600 - 96\,(0.7)^2] = 829.44$ kN

$$d_0 = d_1 + \frac{d - d_1}{(B-b)/2}\left(\frac{B - b - d}{2}\right) \quad \text{...(see Fig. 22.35)}$$

or $$d_0 = 140 + \frac{400 - 140}{2500 - 500}(2500 - 500 - 400) = 348 \text{ mm}$$

$$\therefore \quad \tau_v = \frac{829.44 \times 1000}{4 \times 700 \times 348} = 0.851 \text{ N/mm}^2$$

Permissible shear stress $= k_s \times 0.25\sqrt{f_{ck}} = 1 \times 0.25\sqrt{20} = 1.118 \text{ N/mm}^2 > \tau_v$

Hence the thickness found from point of view of bending compression is safe.

For one way shear, the critical plane CD lies at distance d (= 400 mm) from column face. The cantilever length to the right of $CD = 1.25 - 0.25 - 0.40 = 0.6$ m.

Intensity of pressure p_0'' at CD is given by

$$p_0'' = 84.5 + \frac{107.5 - 84.5}{2.5}(2.5 - 0.60)$$

$\simeq 102 \text{ kN/m}^2$

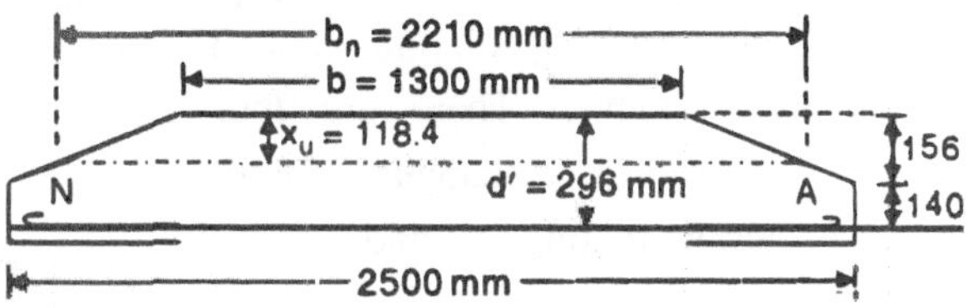

FIG. 22.36

The section at CD will be trapezoidal as shown in Fig. 22.36. The width at top is given by

$$b = 500 + \frac{2500 - 500}{1000} \times 400 = 1300 \text{ mm.}$$

$$\text{Effective depth } d' = 140 + \frac{400 - 140}{1000} \times 600 = 296 \text{ mm}$$

For balanced section, $x_{u,\,max}/d = 0.479$. Let $x_u/d \simeq 0.4$ for under-reinforced section so that $x_u = 0.4\,d' = 0.4 \times 296 = 118.4$ mm.

$\therefore$ Width of the section at N.A. is given by

$$b_n = 1300 + \frac{2500 - 1300}{156} \times 118.4 \simeq 2210 \text{ mm}$$

$\therefore$ Shear force V_u' at CD is given by

$$V_u' = 1.5 \times 2.5 \times 0.6 \times \tfrac{1}{2}\,[107.5 + 102] = 235.69 \text{ kN}$$

$$\therefore \quad \tau = \frac{V_u'}{b_n d'} = \frac{235.69 \times 1000}{2210 \times 296} = 0.36 \text{ N/mm}^2$$

Permissible shear stress at $p = 0.3\%$ (under-reinforced section) $= 0.384 \text{ N/mm}^2$ with $k = 1$ (Tables 7.1 and 7.2). Hence thickness provided from point of view of bending compression is adequate.

5. ***Design for bending tension*** : For an under-reinforced section

$$A_{st} = \frac{0.5\, f_{ck}}{f_y}\left[1 - \sqrt{1 - \frac{4.6\, M_u}{f_{ck}\, b_1\, d^2}}\right] b_1\, d = \frac{0.5 \times 20}{415}\left[1 - \sqrt{1 - \frac{4.6 \times 1.95 \times 10^8}{20 \times 500\,(400)^2}}\right] 500 \times 400$$

$= 1625\ \text{mm}^2$

$\therefore$ No. of 12 mm φ bars $= 1625/113.1 \simeq 15$

Provide the same reinforcement in the other direction.

6. ***Check for development length*** : $L_d = 47\,\varphi = 47 \times 12 = 564$ mm

Available length, providing a side a cover of 60 mm $= 1000 - 60 = 940$ mm.

Hence safe. The details of reinforcement etc. are shown in Fig 22.35.

22.11. COMBINED FOOTINGS

A combined footing supports the load of two or more adjacent columns. Such a footing is provided under the following circumstances: (*i*) when the columns are very near to each other so that their footings overlap (*ii*) when the bearing capacity of soil is less, requiring more area under individual footing (*iii*) when the end column is near a property line, so that its footing cannot be spread in that direction. A combined footing may be rectangular or trapezoidal in plan. The aim is to get uniform pressure distribution under the footing. For this, the C.G. of footing area should coincide with the C.G. of the combined loads of the two columns. There are four types of combined footings commonly used:

1. Combined rectangular footing 2. Combined trapezoidal footing
3. Strap footing, and
4. Raft footing

Rectangular footing is used when the two columns carry equal loads. If, however, one column near the property line carries heavier load, provision of trapezoidal footing becomes essential. A strap footing consists of spread footings of two columns connected by a strap beam. A raft or mat is a combined footing that covers the entire area beneath a structure and supports all the walls and columns.

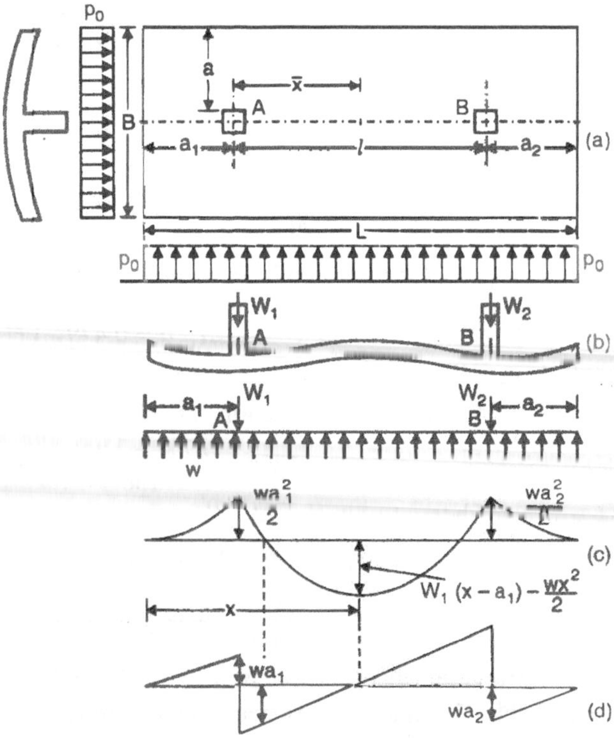

FIG. 22.37. COMBINED RECTANGULAR FOOTING

22.12. COMBINED RECTANGULAR FOOTING

1. **Proportioning of footing** : Fig. 22.37 (*a*) shows two columns, with loads W_1 and W_2 spaced at l centre to centre. If W' is the weight of the footing, and q_0 is the safe bearing capacity, the footing area is given by

$$A = \frac{W_1 + W_2 + W'}{q_0} \quad \text{...(22.46)}$$

Suitable values of length L and breadth B of the footing are chosen so that $B \times L = A$. The projections a_1 and a_2 should be so chosen that the C.G. of footing coincides with the C.G. of the two loads.

Let $\bar{x}$ = distance of C.G. of column loads from centre of column A.

Then $$a_1 + \bar{x} = \frac{L}{2} = a_2 + l - \bar{x} \quad \text{...(22.47)}$$

From which a_1 and a_2 can be determined. The net upward pressure p_0 is given by

$$p_0 = \frac{W_1 + W_2}{L.B} \quad \text{...(22.48)}$$

2. Bending pattern : The footing will bend both in longitudinal as well as transverse directions, as shown. In the longitudinal direction, the footing will have sagging bending moment in the two cantilever portions, and hogging bending moment in some length of middle portion. In the transverse direction the footing will have sagging bending moment. Near the columns, therefore, the footing will have a tendency to bend in the form of saucer. The *transverse bending will decrease at distance away from the column, where the bending will be primarily longitudinal.* Hence sections near and around the column will be subjected to heavier bending stress.

(a) *Design for longitudinal bending*

The exact analysis of such a footing, having no transverse or longitudinal beam, is very difficult since the bending pattern is not precisely known at all points. The design is based primarily on empirical practices. One of the commonly adopted practice is to consider the footing as a longitudinal beam of width B. Thus, the footing will be a beam loaded with uniformly distributed load $w = p_0 B$ per unit length and supported on two columns with reactions W_1 and W_2 as shown in Fig 22.37 (*b*). The B.M. and S.F. diagrams are shown in Fig. 22.37 (*c*) and (*d*) respectively. The maximum hogging bending moment will occur bet-

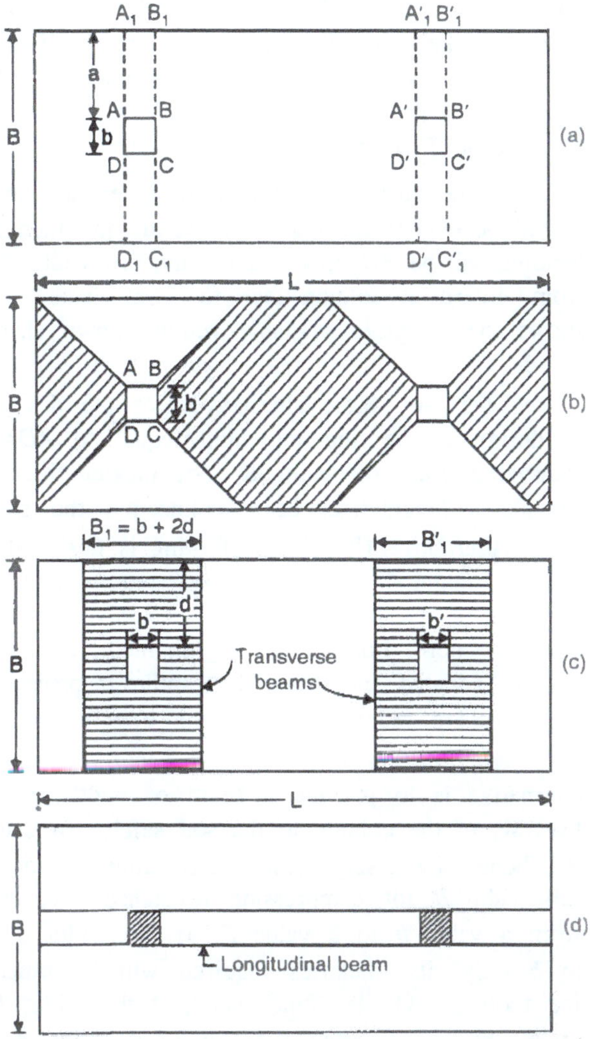

FIG. 22.38

weeen W_1 and W_2 where S.F. is zero. The maximum sagging moment at the column axis will be $\frac{w a_1^2}{2}$, while the maximum hogging bending moment will be $W_1 (x - a_1) - \frac{w x^2}{2}$, where x is the distance of the section, from the edge, where shear is zero, However, the footing will be designed for sagging moment at the *outer face* of each column, and for the maximum hogging moment. The reinforcement will be placed on the bottom face for sagging bending moment and on the top face for the hogging moment.

The above analysis is based on the assumption that full width B is available for longitudinal bending. However, in actual bending, full width B will not be available for the whole length of the footing. Fig. 22.38 (*a*) shows a critical strip $A_1 B_1 C_1 D_1$ for transverse bending.

The total shear force on the two planes $A_1 D_1$ and $B_1 C_1$ plus the upward soil reaction on the strip $A_1 B_1 C_1 D_1$ will be equal to the column load W_1. Thus, the strip $A_1 B_1 BA$ will bend up like a cantilever, with a total uniformly distributed upward force equal to half the column load. If p_0' is the net intensity of upward pressure on the strip, we have

$$p_0' = \frac{W_1}{\text{Area } A_1 B_1 C_1 D_1}$$

If the strip $A_1 B_1 B A$ is designed as cantilever with an upward pressure p_0 only, it will be over-strained and will yield, and then the effective available *width* for longitudinal bending will be AD only and not the full width B. The available width however will increase from $AD = b$ at column face to B at distance B from the column face. Fig. 22.38 (*b*) shows the *probable* available widths shown shaded, for longitudinal bending all along the length.

There are three alternatives for the design of footing for longitudinal bending : (*i*) to design the footing for width equal to $AD = b$ at the column face, (*ii*) to provide a transverse beam of width B_1 , of greater thickness then the rest of the slab, or (*iii*) to provide a longitudinal beam connecting the columns.

***Alternative* (*i*)** : If the footing is designed as longitudinal beam with a width b only, at the column face, its thickness required for compression to resist bending moment at the face of the column will be excessive and uneconomical, specially when the cantilever length is more. For the hogging moment, however, full width L may be used. This method may be adopted only when the cantilever projections a_1 or a_2 are small, so that the sagging moments are much less than the hogging moment.

Alternative (ii) : Transverse beam. In order that strip $A B B_1 A_1$ may not yield, another alternative is to provide a beam of width B_1 as shown in Fig. 22.37(*c*) to transfer the load W_1 of the column to the soil safely. In that case, the shaded portions in Fig. 22.38(*c*) will bend transversely, while the remaining portion will bend longitudinally as a beam utilizing full width B for compression resistance. The width B_1 of the transverse beam may vary from a value b to a value B. If the width of transverse beam $A_1 B_1 C_1 D_1$ is kept equal to b only, its thickness required will be much more than the rest of the thickness of the footing. On the other hand, if the width B_1 is made equal to width B, each portion under the two transverse beams of columns A and B will act as individual footings. An intermediate value of B_1, between b and B, should therefore be chosen. Failure due to shear occurs approximately along 45° lines when footings are tested to destruction. This suggests that width of the transverse beam should be taken as $B_1 = b + 2d$, but not more than B. (Fig. 22.39).

After having determined B_1, the upward soil pressure p_0' for transverse bending will be $= \frac{W_1}{B_1 \times B}$. The thickness of the transverse beam should be found for bending moment at column face equal to $(p_0' B_1) a^2/2$. This thickness will normally be greater than the thickness d for the rest of the footing. Even if both the thicknesses are equal, the reinforcement in the transverse direction should be found for the above B.M., and provided in width B_1. Similar reinforcement for width B_1' for the other column should be provided. For the rest of the portion of footing, minimum transverse reinforcement @ 0.15% of the area of cross-section of the footing (for mild steel bars) or @ 0.12% (for HYSD bars) should be provided. In the longitudinal direction, the footing will be designed for B.M. and S.F. shown Fig. 22.37(*c*) and (*d*).

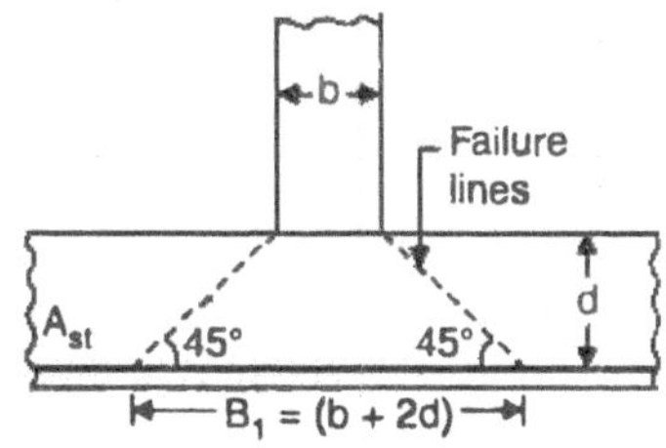

FIG. 22.39

Alternative (iii) Longitudinal beam. A third and more common alternative is to provide a longitudinal beam, all along the length, joining the two columns as shown in Fig. 22.38(*d*). In this case, the beam alone will be subjected to bending moments and shear forces shown in Fig. 22.37(*c*),(*d*) in the longitudinal direction. The slab of the footing on either side of the beam is designed as cantilever, for the transverse bending. The web of the longitudinal beam should project *below* the slab if the sagging moments in the beams are more, so that T-beam action is available in the cantilever portions of the beam. If, however, the hogging bending moment is more, the web of the beam should project at the top of the slab so that T-beam action is available for central portion.

3. Two way shear (punching shear). The depth found on the basis of bending moment should always be tested for punching shear, so that punching shear stress does not exceed the permissible value. The critical section will be at distance $d/2$ from the column face.

4. One way shear. The footing should be tested for one way shear at a distance d from the two column faces, in the respective cantilever portions, as shown in Fig. 22.40. For the central portion, two locations, F and G, should be tried for diagonal tension. At F, distant d from column face, tension crack will occur only if sagging bending moment occurs at F. Another possibility of tension crack is at G, just to the right of point of inflexion, so that there is hogging bending moment giving rise to a crack on the top face. The corresponding S.F. at G can be determined. Out of the three points E, F and G, the diagonal tensional crack may occur at a point which is *nearer* to the column,

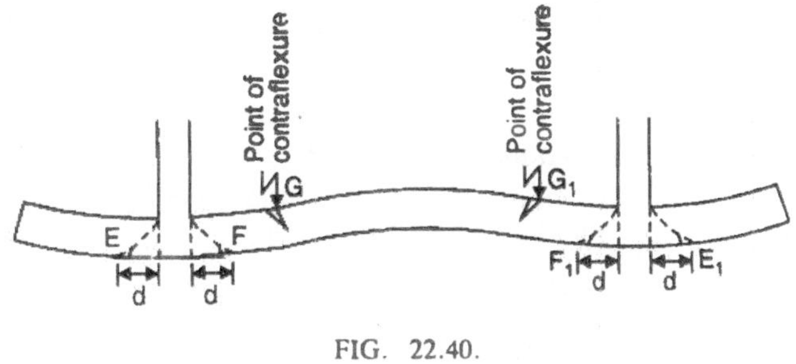

FIG. 22.40.

face, or at which S.F. is more. If the shear stress comes out to be more than permissible one, shear stirrups should be provided for the required portion. However, nominal shear stirrups should always be provided throughout the length of the beam.

5. Development length. The reinforcement should be tested for development length.

22.13. COMBINED TRAPEZOIDAL FOOTING

Combined trapezoidal footing for two columns becomes essential when the heavily loaded column is near the property line, or when there is some restriction on the total length of the footing. The widths B_2- and B_1 are so proportioned that the C.G. of footing coincides with the C.G. of column loads.

Fig. 22.41 shows two columns with loads W_1 and W_2, spaced at distance l apart. Let L be the length of the footing, and a_1 and a_2 be the cantilever projections. The widths B_1 and B_2 are unknown. Hence,

$$\frac{(B_1 + B_2)\,L}{2} = \frac{W_1 + W_2 + W'}{q_0}$$

$$B_1 + B_2 = \frac{2\,(W_1 + W_2 + W')}{q_0\,L} \quad \text{..(1)...(22.49)}$$

Distance $\bar{x}$ of C.G. of load $= \dfrac{W_2\, l}{W_1 + W_2}$

$\therefore$ Distance of C.G. of load from short edge $= a_1 + \dfrac{W_2\, l}{W_1 + W_2}$ ...(*i*)

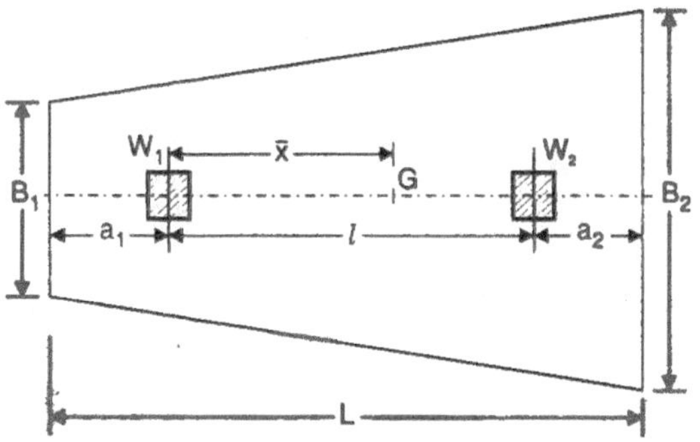

FIG 22.41.

Distance C.G. of trapezium from the short edge $= \dfrac{L}{3}\left(\dfrac{B_1 + 2\,B_2}{B_1 + B_2}\right)$ ...(*ii*)

Equating the two, $$a_1 + \frac{W_2\, l}{W_1 + W_2} = \frac{L}{3}\left(\frac{B_1 + 2\,B_2}{B_1 + B_2}\right) \quad \text{...(22.50)}$$

From Eqs. 22.49 and 22.50, the unknowns B_1 and B_2 are determined. The net upward soil pressure intensity p_0 will be uniform throughout, and its magnitude is given by

$$p_0 = \frac{W_1 + W_2}{\frac{1}{2}(B_1 + B_2)\,L} \quad \text{...(22.51)}$$

Knowing p_0, B.M. and S.F. diagrams may be drawn. The footing is designed exactly in the manner as for rectangular footing. Transverse reinforcement, for transverse bending, in provided near each column, in strip equal to width of column plus twice the effective depth of footing.

Design Example 22.9 *Combined rectangular footing : Design a combined rectangular footing for two columns A and B, carrying loads of 500 and 700 kN respectively. Column A is 300 mm × 300 mm in size and column B is 400 mm × 400 mm in size. The centre to centre spacing of the columns is 3.4 metres. The safe bearing capacity of soil may be taken as 150 N/m². Use M 20 concrete and Fe 415 steel.*

Solution.

1. ***Design constants***. For M 20 concrete and Fe 415 steel we have

$$x_{u,\max}/d = 0.479 \quad \text{and} \quad R_u = 2.761$$

2. ***Size of footing***. $W_1 = 500$ kN and $W_2 = 700$ kN

Let the weight of footing $= W' = 10\%$ of $(W_1 + W_2) = 120$ kN

$\therefore \quad A = \dfrac{500 + 700 + 120}{150} = 8.8$ m². Let the size of the footing be 1.8 m × 5 m.

The projections a_1 and a_2 should be such that C.G. of footing coincides with the C.G. of column loads. The distance $\bar{x}$ of the C.G. of column loads from centre of column A is given by

$$\bar{x} = \frac{W_2\, l}{W_1 + W_2} = \frac{700 \times 3.4}{500 + 700} \approx 2 \text{ m}$$

Now $\quad a_1 + \bar{x} = \dfrac{L}{2} \quad$ or $\quad a_1 = \dfrac{L}{2} - \bar{x} = \dfrac{5}{2} - 2 = 0.5$ m

$$a_2 = L - (l + a_1) = 5 - (3.4 + 0.5) = 1.1 \text{ m}$$

The net upward pressure p_0 is given by

$$p_0 = \frac{W_1 + W_2}{B \times L} = \frac{500 + 700}{1.8 \times 5} = \frac{400}{3} \text{ kN/m}^2$$

$\therefore$ Pressure w per metre length $= p_0\, B = \dfrac{400}{3} \times 1.8 = 240$ kN/m²

3. ***B.M. and S.F. diagrams***

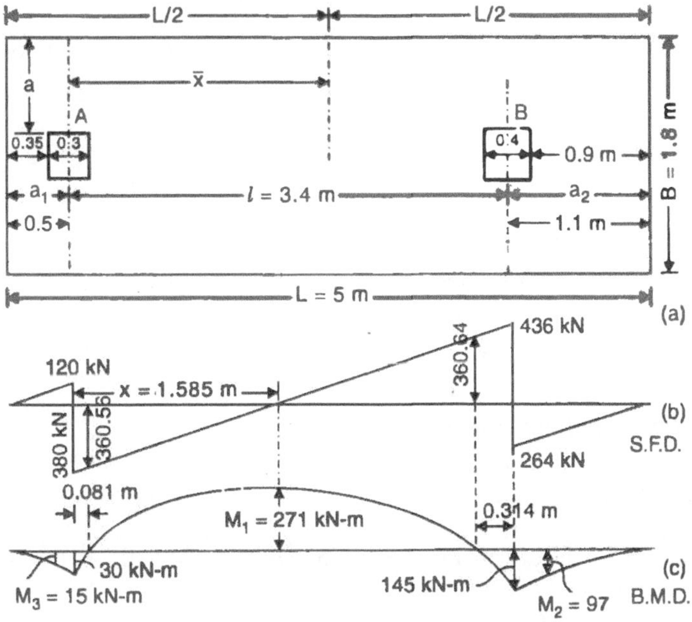

FIG. 22.42

The B.M. and S.F. diagrams in the longitudinal directions are shown in Fig. 22.42.

S.F. just to the left of A $= 240 \times 0.5 = 120$ kN

S.F. just to the right of A $= 500 - 120 = 380$ kN

S.F. just to the right of B $= 240 \times 1.1 = 264$ kN

S.F. just to the left of B $= 700 - 264 = 436$ kN

S.F. will be zero at distanace x from A, its magnitude being given by $x = \frac{380}{240} = 1.585$ m.

The maximum hogging B.M. M_1 will therefore occur at this section, its magnitude being given by

$$M_1 = (500 \times 1.585) - \frac{240}{2}(0.5 + 1.585)^2 \approx 271 \text{ kN-m} = 2.71 \times 10^8 \text{ N-mm}$$

$$\text{Sagging B.M. at } B = \frac{240\,(1.1)^2}{2} = 145 \text{ kN-m} = 1.45 \times 10^8 \text{ N-mm}$$

$$\text{Sagging B.M. at } A = \frac{240}{2}(0.5)^2 = 30 \text{ kN-m} = 0.3 \times 10^8 \text{ N-mm}$$

Sagging B.M. at the outer face of column B is

$$M_2 = \frac{240}{2}(0.9)^2 = 97 \text{ kN-m} = 0.97 \times 10^8 \text{ N-mm}$$

Sagging B.M. at the outer face of column A is

$$M_3 = \frac{240}{2}(0.35)^2 = 15 \text{ kN-m} = 0.15 \times 10^8 \text{ N-mm}$$

Let the point of contraflexure occur at x from the centre of column A.

Then $M_x = 500\,x - \frac{240}{2}(x + 0.5)^2 = 0$ which gives $x = 0.081$ or 3.086 m

Hence first point of contraflexure occurs at 0.081 m to the right of centre of column A and the second one occurs at 3 m from A, or $3.4 - 3.086 = 0.314$ m to the left of centre of column B.

S.F. at the first point of contraflexure $= 380 - (240 \times 0.081) = 360.56$ kN

S.F. at the second point of contraflexure $= 436 - (240 \times 0.314) = 360.64$ kN

4. *Effective depth from bending compression*

The effective depth d for hogging B.M. M_{1u} is given by

$$d = \sqrt{\frac{M_{1u}}{R_u\,B}} = \sqrt{\frac{1.5 \times 2.71 \times 10^8}{2.761 \times 1800}} = 286 \text{ mm} \simeq 300 \text{ mm (say)}$$

5. *Check for punching shear.*

Let us check the above depth for punching shear, for which the critical plane lies at $d/2 = 150$ mm around column B. Width b_0 of critical plane $= 400 + 300 = 700$ mm.

$$\text{Punching shear force at column } B = 1.5\left[700 - \frac{400}{3} \times (0.7)^2\right] \approx 952 \text{ kN.}$$

$$\therefore \quad \text{Shear stress} = \frac{952 \times 1000}{4 \times 700 \times 300} = 1.133 \text{ N/mm}^2.$$

Permissible shear stress $= k_s \tau_c = 1 \times 0.25 \sqrt{f_{ck}} = 1 \times 0.25 \sqrt{20} = 1.118$ N/mm^2

This is less than the actual shear stress. Hence the depth found from bending compression is inadequate.

The required depth $d = \dfrac{952 \times 1000}{4 \times 700 \times 1.118} = 305$ mm.

Using 60 mm cover to the centre of reinforcement, $D = 305 + 60 = 365$ mm.

However, provide $D = 370$ mm so that available $d = 370 - 60 = 310$ mm.

6. *Design for bending tension in longitudinal direction*

(*i*) *Reinforcement for hogging bending moment*

$$M_{1u} = 1.5\, M_1 = 1.5 \times 2.71 \times 10^8 = 4.068 \times 10^8 \text{ N-mm}$$

For under-reinforced section, area of steel reinforcement is given by

$$A_{st} = 0.5 \frac{f_{ck}}{f_y}\left[1 - \sqrt{1 - \frac{4.6\, M_{1u}}{f_{ck}\, Bd^2}}\right] Bd = \frac{0.5 \times 20}{415}\left[1 - \sqrt{1 - \frac{4.6 \times 4.068 \times 10^8}{20 \times 1800\,(310)^2}}\right] 1800 \times 310$$

$= 4336$ mm^2

Using 16 mm φ bars having $A_\varphi = \dfrac{\pi}{4}(16)^2 = 201$ mm^2

$\therefore$ No. of bars $= 4336/201 = 21.6 \simeq 22$.

Actual $p_t = \dfrac{22 \times 201.06 \times 100}{1800 \times 310} \simeq 0.79\%$

Thus provide 22 bars uniformly distributed over the width of 1.8 m over top face. At the point of contraflexure near B, shear force $F_u = 1.5 \times 360.64 = 540.96$ kN. Let us continue all the bars upto point of contraflexure, from the point of requirements of Code provisions inherent in the equation

$$\frac{M_1}{V_u} + L_0 \geq L_d$$

Actual available $A_{st} = 22 \times 201 = 4422$ mm^2

$$\therefore \quad x_u = \frac{0.87\, f_y\, A_{st}}{0.36\, f_{ck}\, B} = \frac{0.87 \times 415 \times 4422}{0.36 \times 20 \times 1800} = 123.2 \text{ mm}$$

$$\therefore \quad M_1 = 0.87\, f_y\, A_{st}\,(d - 0.416\, x_u) = 0.87 \times 415 \times 4422\,(310 - 0.416 \times 123.2)$$

$$= 4.131 \times 10^8 \text{ N-mm}$$

$$V_u = 540.96 \text{ kN} = 540.96 \times 10^3 \text{ N}$$

$L_0 = d$ or $12\,\varphi$, whichever is greater $= 310$ mm

$$L_d = 47\,\varphi = 47 \times 16 = 752 \text{ mm}$$

$$\frac{M_1}{V_u} + L_0 = \frac{4.131 \times 10^8}{540.96 \times 10^3} + 310 \simeq 1073 \text{ mm} > L_d$$

Hence Code requirements are satisfied.

However, all the 22 bars are taken up to the outer faces of both the columns. Out of these, 14 bars are stopped and 8 bars are taken straight upto either edge of footing. These bars will serve as anchorage for the stirrups.

(*ii*) ***Reinforcement for sagging B.M.*** M_{2u} (= 1.5 × 0.97 × 10⁸ = 1.455 × 10⁸ N-mm) is given by

$$A_{st2} = \frac{0.5 \times 20}{415}\left[1 - \sqrt{1 - \frac{4.6 \times 1.455 \times 10^8}{20 \times 1800\,(310)^2}}\right] 1800 \times 310 = 1371 \text{ mm}^2$$

Using 12 mm φ bars, No. of bars = 1371 / 113 = 12.2 ≏ 13.

Actual $A_{st} = 13 \times 113 = 1469 \text{ mm}^2$ and $p_t = \frac{1469}{1800 \times 361} \times 100 \simeq 0.23\%$

These No. of bars should be sufficient from development length point of view so as to satisfy the following criterion at the point of contraflexure near the inner face of column B :

$$\frac{M_1}{V_u} + L_0 \geq L_d$$

Actual $A_{st} = 13 \times 113 = 1469 \text{ mm}^2$

$$x_u = \frac{0.87 f_y A_{st}}{0.36 f_{ck} B} = \frac{0.87 \times 415 \times 1469}{0.36 \times 20 \times 1800} = 40.9 \text{ mm}$$

$$\therefore\ M_1 = 0.87 f_y A_{st} (d - 0.416\, x_u) = 0.87 \times 415 \times 1469\,(310 - 0.416 \times 40.9) = 1.55 \times 10^8 \text{ N-mm}$$

$$V_u = 1.5 \times 360.64 \times 10^3 \text{ N} = 540960 \text{ N}$$

$$L_d = 47\,\varphi = 47 \times 12 = 564 \text{ mm} \;;\; L_0 = d \text{ or } 12\,\varphi = 310 \text{ mm}$$

$$\therefore \quad \frac{M_1}{V_u} + L_0 = \frac{1.55 \times 10^8}{540960} + 310 = 596 \text{ mm} > L_d$$

Hence Code requirements are satisfied.

Hence 13 bars of 12 mm φ will be sufficient. These bars should be extended by L_0 (= 310 mm) beyond the point of contraflexure. At this point curtail 5 bars and continue 8 bars throughout the length to serve an anchor bars for stirrups.

(*iii*) *Reinforcement for sagging B.M.* M_{3u}= 1.5 × 0.15 × 10⁸ = 0.225 × 10⁸ N-mm

$$A_{st} = \frac{0.5 \times 20}{415}\left[1 - \sqrt{1 - \frac{4.6 \times 0.225 \times 10^8}{20 \times 1800\,(310)^2}}\right] 1800 \times 310 = 203 \text{ mm}^2$$

Minimum reinforcement @ 0.12% of cross-sectional area = $\frac{0.12 \times 370 \times 1800}{100} = 800 \text{ mm}^2$

∴ No. of 12 mm φ bars = 800 / 113 ≏ 8

Let us find the reinforcement from development length point of view so that the following relation is satisfied at the point of contraflexure near the inner face of column A :

$$\frac{M_1}{V_u} + L_0 \geq L_d$$

Let n = number of bars required to satisfy the above criterion

$x_{u\,.\,max}/d = 0.479$ for balanced section. However, let us assume $x_u/d = 0.3$ for an under-reinforced section, so that $x_u = 0.3\,d = 0.3 \times 310 = 93$ mm

$\therefore \quad M_1 = 0.87 f_y A_{st} (d - 0.416\, x_u) = 0.87 \times 415 \times n \times 113\,(310 - 0.416 \times 93)$

$= n \times 0.111 \times 10^8$ N-mm

V_u = S.F. at point of contraflexure $= 1.5 \times 360.56 \times 10^3 = 540.84 \times 10^3$ N

$L_0 = 12\,\varphi$ or d whichever is more = 310 mm

$L_d = 47\,\varphi = 47 \times 12 = 564$ mm

Substituting the values, we have $\dfrac{n \times 0.111 \times 10^8}{540.84 \times 10^3} + 310 = 564$

From which $n = 12.4 \simeq 13$. Continue these upto 310 mm beyond the point of contraflexure. At this point, curtail 4 bars and continue remaining 8 bars to serve as anchor bars for shear stirrups.

6. *Check for one way shear.* In the cantilever portion, test for one way shear (diagonal tension) is made at a distance $d = 0.31$ m from the column face or at $0.31 + 0.20 = 0.51$ m from the centre of column B. Shear $V_u = 1.5\,(264 - 240 \times 0.51) = 212.4$ kN.

$$\therefore \quad \tau_v = \frac{212.4 \times 1000}{1800 \times 310} = 0.38 \text{ N/mm}^2$$

Permissible shear stress $= k\,\tau_c = \tau_c = 0.34$ N/mm^2 for $p_t = 0.23\%$ (under-reinforced), obtained from Table 7.1. Hence unsafe and shear reinforcement will be necessary.

Using 8 mm φ 8-legged strirrups, having $A_{sv} = 8 \times 50.26 \simeq 402$ mm^2

$$\therefore \text{ Spacing } s_v = \frac{0.87 f_y A_{sv} d}{V_{us}} = \frac{0.87 \times 415 \times 402 \times 310}{212.4 \times 1000} = 211.8 \text{ mm} < 0.75\,d$$

(Taking $V_{us} = V_u$)

Hence provide these @ 200 mm c/c. Provide similar stirrups in the other cantilever portion, near column A.

For diagonal tension between A and B, near column B, crack can occur at the bottom face of the footing (*i.e.* for sagging B.M.) at a distance of $d = 310$ mm, or at the top of footing (*i.e.*, for the hogging B.M.) at the point of contraflexure distant 300 mm from the centre of the column, whichever is nearer. Shear force at the point of contraflexure $= 1.5 \times 364$ kN.

$$\therefore \quad \tau_v = \frac{1.5 \times 364 \times 1000}{1800 \times 310} = 0.978 \text{ N/mm}^2$$

which is more than the permissible shear stress $\tau_c = 0.57$ N/mm^2 for p_t 0.79% (Table 7.1)

Hence shear reinforcement is necessary.

Using 10 mm φ 8-legged stirrups, $A_{sv} = 8 \times 78.54 \simeq 628.3$ mm^2

$$\therefore \text{ Spacing } \quad s = \frac{0.87 f_y A_{sv} d}{V_{us}} = \frac{0.87 \times 415 \times 628.3 \times 310}{1.5 \times 364 \times 1000} = 128 \text{ mm}$$

Hence provide these @ 125 mm c/c.

The S.F. for which stress is 0.57 N/mm^2 is $= \dfrac{0.57}{0.978} \times 364 = 212$ kN

This occurs at a distance of $\frac{436 - 212}{240} = 0.93$ m from the centre of the column B or 0.73 m from the inner face of the column. Hence provide at least 7 stirrups @ 125 mm c/c, for a distance of 0.75 m from the face of the column B.

Similarly, near column A, tension crack can occur in the hogging portion at the inner face of the column since the point of contraflexure happens to fall under the column. S.F. at the inner face = 380 − 240 × 0.15 = 344 kN

$$\therefore \qquad \tau_v = \frac{1.5 \times 344 \times 1000}{1800 \times 310} = 0.927 \text{ N/mm}^2$$

against a permissible value of 0.57 N/mm^2. Hence shear reinforcement is necessary.

Hence provide 6 Nos. 10 mm φ 8-legged stirrups @ 125 mm c/c, for a distance of 625 mm from the inner face of the column. Spacing of 8 mm φ 8 lgd nominal stirrups is given by

$$s_v = \frac{2.175\, A_{sv} f_y}{b} = \frac{2.175 \times 402 \times 415}{1800} = 201.6 \text{ mm} \simeq 200 \text{ mm}$$

(subject to a maximum of 0.75 d or 300 mm, which ever is less).

Hence for the rest of the length, provide 8 mm φ 8-lgd nominal stirrups @ 200 mm c/c.

7. ***Transverse reinforcement*** : The footing will bend transversely near each column face. Projection a beyond the face of column $A = \frac{1}{2}(1800 - 300) = 750$ mm. Width B_1 of bending strip $= b + 2d = 300 + 2 \times 310 = 920$ mm.

$$\text{Net upward pressure } p_0' = \frac{W_1}{B \times B_1} = \frac{500}{1.8 \times 0.92} = 301.93 \text{ kN/m}^2$$

Maximum B.M. will occur at the face of the column.

$$M = p_0' \frac{a^2}{2} = 301.93 \frac{(0.75)^2}{2} = 81.918 \text{ kN-m} = 81.918 \times 10^6 \text{ N-mm.}$$

$$\therefore \qquad M_u = 1.5\, M = 1.5 \times 81.918 \times 10^6 \text{ N-mm} = 127.378 \times 10^6 \text{ N-mm}$$

$$d = \sqrt{\frac{127.378 \times 10^6}{2.761 \times 1000}} = 215 \text{ mm.}$$

Hence actual $d = 310$ mm provided earlier is sufficient. The transverse beam will thus be of the same thickness as the rest of the footing. Since transverse reinforcement will be provided *above* the longitudinal one, available $d = 310 - 12 = 298$ mm.

The transverse reinforcement is given by

$$A_{st} = \frac{0.5 \times 20}{415}\left[1 - \sqrt{1 - \frac{4.6 \times 127.378 \times 10^6}{20 \times 1000\,(298)^2}}\right] 1000 \times 298 = 1303 \text{ mm}^2$$

Using 12 mm φ bars, spacing is given by

$$s = \frac{1000\, A_{\varphi}}{A_{st}} = \frac{1000 \times 113}{1303} = 86 \text{ mm}$$

Provide these @ 80 mm c/c. The reinforcement provided above should have sufficient development length. $L_d \approx 47\ \varphi = 47 \times 12 \approx 564$ mm.

Using 60 mm clear cover on the sides, length of bar available = 750 – 60 = 690 mm. Hence safe.

Similarly, for column B, width B_1 for transverse beam = 400 + 2 × 310 = 1020 mm.

$$\therefore \qquad p_0' = \frac{W_2}{B \times B_1} = \frac{700}{1.8 \times 1.02} = 381.26 \text{ kN/m}^2$$

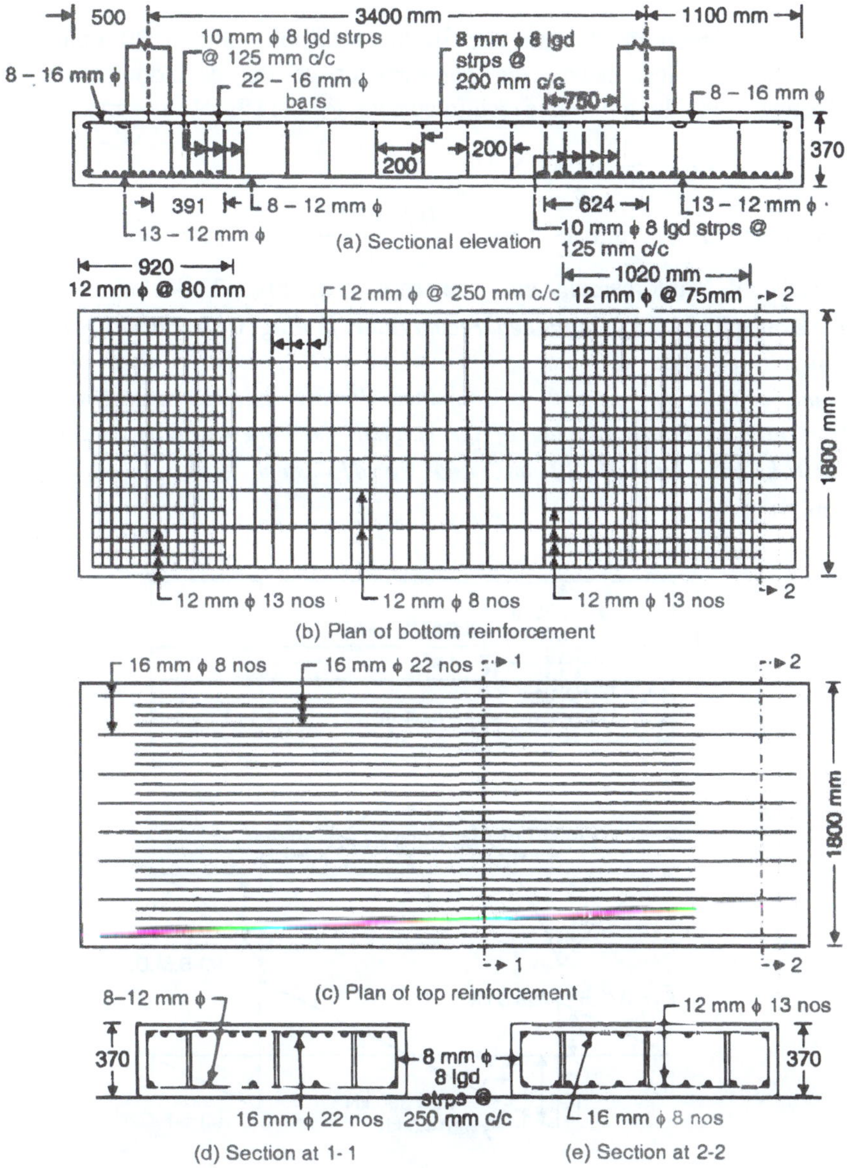

FIG. 22.43

Cantilever projection $a = \frac{1}{2}[1800 - 400] = 700 \text{ mm} = 0.7 \text{ m}$

$$\therefore \quad M = p_0' \frac{a^2}{2} = 381.26 \frac{(0.7)^2}{2} = 93.41 \quad \text{kN-m} = 93.41 \times 10^6 \quad \text{N-mm}$$

$$\therefore \quad M_u = 1.5\, M = 1.5 \times 93.41 \times 10^6 = 140.114 \times 10^6$$

$$\therefore \quad A_{st} = \frac{0.5 \times 20}{415}\left[1 - \sqrt{1 - \frac{4.6 \times 140.114 \times 10^6}{20 \times 1000\,(298)^2}}\right] 1000 \times 298 = 1450 \text{ mm}^2$$

Using 12 mm φ bars, $s = \frac{1000 \times 113}{1450} \approx 78$ mm.

Hence provide these @ 75 mm c/c in the strip of width 1020 mm. The available length of the bars will just satisfy the requirements of the development length. For the rest of the footing, provide transverse reinforcement @ 0.12% of the area of cross-section.

$$\therefore \quad A_{st} = \frac{0.12}{100}(1000 \times 370) = 444 \text{ mm}^2$$

$$\therefore \quad \text{Spacing of 12 mm φ bars} = \frac{1000 \times 113}{444} \simeq 250 \text{ mm}$$

The details of reinforcement etc. are shown in Fig. 22.43.

Design Example 22.10. ***Combined rectangular footing with central joining beam*** *Two reinforced concrete columns 400 mm × 400 mm in section carry a load of 1000 kN each, inclusive of the self-weight. Design a combined footing having central beam joining the columns. The centre to centre spacing of the columns is 4 metres. The safe bearing capacity of soil is 150 kN/m². Use M 20 mix and Fe 415 steel.*

Solution.

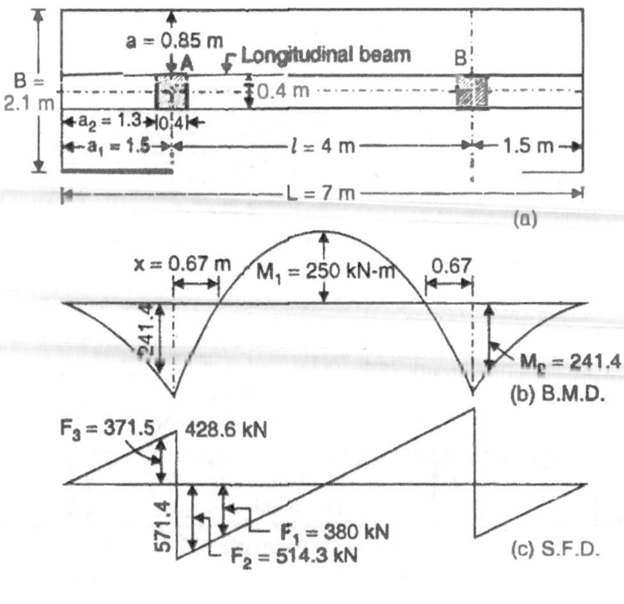

FIG. 22.44

1. ***Design constants***. For M 20 concrete, and Fe 415 steel we have the following:

$$\frac{x_{u,\max}}{d} = 0.479 \quad \text{and} \quad R_u = 2.761$$

2. ***Proportioning of footing***. $W_1 = W_2 = W = 1000$ kN

Let the weight of footing $W' = 0.1\,(1000 + 1000) = 200$ kN

$$\therefore \text{ Area of footing} = \frac{1000 + 1000 + 200}{150} = 14.65 \text{ m}^2$$

$$\text{For the beam, upward load } w \text{ per unit length} = \frac{1000 + 1000}{L} = \frac{2000}{L} \text{ kN/m}$$

Since the footing is symmetrically loaded, the projection of the beam on side of either column will be equal. Projection a_1 measured from centre of column is

$$a_1 = \frac{1}{2}(L - l) = \frac{1}{2}(L - 4) = \frac{L}{2} - 2 \text{ metres.}$$

$$\therefore \text{ Clear projection } a_2 \text{ from outer face} = \left(\frac{L}{2} - 2\right) - 0.2 = \left(\frac{L}{2} - 2.2\right) \text{ m.}$$

This projection a_2 should be such that maximum sagging bending moment at the *face* of column is equal to the maximum hogging bending moment as the middle of the beam.

$$\text{Maximum sagging B.M.} = \frac{w\,a_2^2}{2} = \frac{2000}{L}\frac{1}{2}\left(\frac{L}{2} - 2.2\right)^2 = \frac{1000}{L}\left(\frac{L}{2} - 2.2\right)^2 \qquad \ldots(i)$$

$$\text{Maximum hogging B.M.} = \frac{W\,l}{2} - \frac{w}{2}\left(\frac{L}{2}\right)^2 = 1000 \times 2 - \frac{2000}{2L}\left(\frac{L}{2}\right)^2 = 2000 - 250\,L \qquad \ldots(ii)$$

Equating the two, we get

$$\frac{1000}{L}\left(\frac{L}{2} - 2.2\right)^2 = 2000 - 250\text{ L}$$

or $\quad L^2 - 8.4\,L + 9.68 = 0.$ From which $L = 7.02$ m

Keep $\quad L = 7$ m and $B = 2.1$ m, so that $A = 7 \times 2.1 = 14.7$ m^2

$\therefore \quad a_1 = \frac{L}{2} - 2 = 3.5 - 2 = 1.5$ m ; and $\quad a_2 = \frac{L}{2} - 2.2 = 3.5 - 2.2 = 1.3$ m.

Clear projection a of the slab from the face of the column is

$$a = \frac{1}{2}(B - b) = \frac{1}{2}(2.1 - 0.4) = 0.85 \text{ m.}$$

$$\text{Net upward soil reaction} \qquad p_0 = \frac{2000}{7 \times 2.1} = 136 \text{ kN/m}^2$$

Upward load intensity on beam, $w = 2000/L = 2000/7 = 285.7$ kN/m.

3. ***Design of slab.*** Assuming width of beam equal to width of column, the cantilever projection of slab $= a = 0.85$ m.

(*a*) ***Depth from bending compression***

$$\text{B.M. } M = p_0\frac{a^2}{2} = \frac{136}{2}(0.85)^2 = 49.13 \text{ kN-m} = 49.13 \times 10^6 \text{ N-mm}$$

$$\therefore \quad M_u = 1.5\,M = 73.695 \times 10^6$$

$$d = \sqrt{\frac{M_u}{R_u\, b}} = \sqrt{\frac{73.695 \times 10^6}{2.761 \times 1000}} = 164 \text{ mm}$$

However, provide d = 240 mm and total depth = 300 mm. Reduce the total depth to 200 mm at the edges so that an effective depth of 140 mm is available there.

(b) Depth from one way shear

The critical section for one way shear (diagonal tension) is situated at a depth d (mm) from the column face. The S.F. V_u at that point is

$$V_u = 1.5\, p_0\, (a - 0.001\, d) = 1.5 \times 136 \times 10^3\, (0.85 - 0.001\, d) = 204 \times 10^3\, (0.85 - 0.001\, d) \text{ N/m}$$

The effective depth d' at that point $= 140 + \dfrac{d - 140}{850}(850 - d)$ mm (where d is in mm)

$$\therefore \quad \tau_v = \frac{V_u}{bd'} = \frac{204 \times 10^3\,(0.85 - 0.001\, d)}{1000\left[140 + \dfrac{d-140}{850}(850 - d)\right]} = \frac{173400\,(0.85 - 0.001\, d)}{119000 + (d - 140)\,(350 - d)} \qquad \ldots(i)$$

Permissible shear stress $= k\,\tau_c$ where $k = 1$ for $D \geq 300$ mm (Table 7.2)

Also, $\tau_c = 0.384$ N/mm² for $p = 0.3\%$ for under-reinforced section

$$\therefore \quad \tau_v = \tau_c = 0.384 = \frac{173400\,(0.85 - 0.001\, d)}{119000 + (990d - 119000 - d^2)} = \frac{173400\,(0.85 - 0.001\, d)}{990\, d - d^2}$$

or $\quad d^2 - 1437\, d + 380065 = 0$ from which $d \simeq 350$ mm.

Hence keep $D = 350 + 60 = 410$ mm near the face of the column, and $D = 200$ mm at the edges.

4. Determination of steel reinforcement

Since actual d provided is more than that required from bending compression, we have an under-reinforced section, for which

$$A_{st} = \frac{0.5\, f_{ck}}{f_y}\left[1 - \sqrt{1 - \frac{4.6\, M_u}{f_{ck}\, bd^2}}\right] bd = \frac{0.5 \times 20}{415}\left[1 - \sqrt{1 - \frac{4.6 \times 73.695 \times 10^6}{20 \times 1000\,(350)^2}}\right] 1000 \times 350$$

$\simeq 606$ mm²

Using 12 mm φ bars, $A_\varphi = 113$ mm². $\therefore$ Spacing $s = \dfrac{1000 \times 113}{606} = 186$ mm

However provide 12 mm φ bars @ 100 mm c/c at the bottom of the slab.

Actual $A_{st} = \dfrac{1000 \times 113}{100} = 1130$ mm² and $p = \dfrac{1130 \times 100}{1000 \times 350} = 0.32\%$ which is slightly more that the value p (= 0.3%) assumed for determining the value of τ_c.

$$L_d = 47\, \varphi = 47 \times 12 = 564 \text{ mm.}$$

Providing 60 mm side cover, available length of bar = 850 − 60 = 790 mm, which is more than L_d. Hence safe. Area of distribution reinforcement @ 0.12% of area of cross-section will be $= \dfrac{0.12}{100} \times 1000 \times 410 = 492$ mm². Using 12 mm φ bars,

Spacing $= \frac{1000 \times 113}{492} \simeq 230$ mm. However, provide 12 mm φ bars @ 200 mm c/c.

4. *B.M. and S.F. diagrams for longitudinal beam.*

The B.M. and S.F. diagrams for the beam are shown in Fig. 22.44 (*b*) and (*c*) respectively.

Sagging B.M. M_2 at column face $= \frac{w\,a_2^2}{2} - \frac{285.7}{2}(1.3)^2 - 241.4$ kN-m $= 241.4 \times 10^6$ N-mm

Hogging B.M. $M_1 = W\frac{l}{2} - \frac{w}{2}\left(\frac{L}{2}\right)^2 = 1000 \times \frac{4}{2} - \frac{285.7}{2}\left(\frac{7}{2}\right)^2 = 250$ kN-m $= 250 \times 10^6$ N-mm

(Note. Both these moments are not exactly equal since the length L adopted as 7 m is slightly less than 7.02 m found on the basis of equal values of M_1 and M_2).

The point of contraflexure occurs at x from the centre of column, its position being given by :
$$\frac{w\,(a_1 + x)^2}{2} - W x = 0$$

or
$$\frac{285.7}{2}(1.5 + x)^2 = 1000\,x. \quad \text{From which } x = 0.67 \text{ m}$$

S.F. just to the left of $A = 285.7 \times 1.5 = 428.6$ kN

S.F. just to the right of $A = 1000 - 428.6 = 571.4$ kN

S.F. at the point of contraflexure is $F_1 = 571.4 - (285.7 \times 0.67) = 380$ kN

S.F. at the inner face of column is $F_2 = 571.4 - (285.7 \times 0.2) = 514.3$ kN

S.F. at the outer face of column is $F_3 = 428.6 - (285.7 \times 0.2) = 371.5$ kN.

5. *Depth of Longitudinal beam.* (*a*) *Depth from bending compression* : Let us provide longitudinal beam projecting *below* the footing slab. In the cantilever portion, compression face is upwards and hence T-beam action will be available. In the central portion, however, T-beam action will not be available. Let us therefore, find the depth of beam for M_1 where T-beam action is not available.

Keep width b of beam = 400 mm

$$d = \sqrt{\frac{M_{2u}}{R_u\,b}} = \sqrt{\frac{1.5 \times 250 \times 10^6}{2.761 \times 400}} = 583 \text{ mm.}$$

(b) Depth from two way shear (punching shear) : The critical plane occurs at a distance of half the effective depth from the column face. In this case, the effective depth d_b of the beam is larger than the effective depth d_s (= 350 mm) of the slab. The column will punch through both these depths, along its perimeter. Let us assume that the distance of critical plane $= d_b/2$ to one side and $d_s/2$ to the other side. Hence widths of the punching shear perimeter are

$b_{01} = 400 + d_b$

and $b_{02} = 400 + d_s = 400 + 350 = 750$ mm

∴ Punching shear area of concrete available along the perimeter for the critical plane.

$= (2b_{01} \times d_s') + (2b_{02} \times d_b)$

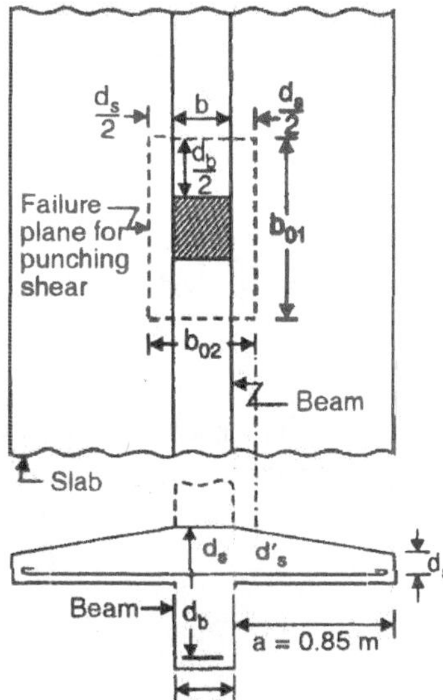

FIG. 22.45

where d_s' = depth of slab at the critical plane

$$= d_{se} + \frac{d_s - d_{se}}{a}\left[a - \frac{d_s}{2}\right] = 140 + \frac{350 - 140}{0.85}\left[0.85 - \frac{0.35}{2}\right] \simeq 306 \text{ mm}$$

Punching shear $F = 1000 \times 10^3 - 136 \times 10^3 \, [b_{01} \times b_{02}]$

$$= 1000 \times 10^3 - 136 \times 10^3 \frac{[(400 + d_b) \times 750]}{(1000)^2} = 10^6 - 102\,(400 + d_b)$$

$$\text{Punching shear stress} = \frac{F_u}{2\,[(b_{01} \times d_s') + (b_{02} \times d_b)]} = \frac{1.5\,[10^6 - 102\,(400 + d_b)]}{2\,[(400 + d_b)\,306 + 750\,d_b]}$$

$$= \frac{10^6 - 102\,(400 + d_b)}{1.333\,[122400 + 1056\,d_b]} \qquad \ldots(1)$$

Permissible shear stress $= k_s\,\tau_c$ where $k_s = (0.5 + \beta_c) \not> 1 = 1$;

$\tau_c = 0.25\sqrt{f_{ck}} = 0.25\sqrt{20} = 1.118$ N/mm². Hence permissible shear stress = 1.118 N/mm²

Equating this to (1), we get $\dfrac{10^6 - 102\,(400 + d_b)}{1.333\,[122400 + 1056\,d_b]} = 1.118$. From which $d_b = 464$ mm.

This is less than effective depth of 583 mm found from the bending compression.

Keeping 50 mm nominal cover and using 20 mm φ main bars in two layers with 20 mm space between them, with 12 mm φ stirrups, effective cover = 50 + 12 + 20 + 10 = 92 mm. However, keep effective cover = 100 mm, so that D = 583 + 100 = 683 mm. However, keep D = 720 mm and effective cover = 100 mm so that available d = 720 − 100 = 620 mm. which is more than that required from bending compression so as to get an under-reinforced section.

6. Reinforcement in longitudinal beam : For hogging B.M., the beam acts as a simple rectangular section with reinforcement at its top face. The area of steel is given by the following, applicable for under-reinforced section.

$$A_{st1} = \frac{0.5\,f_{ck}}{f_y}\left[1 - \sqrt{1 - \frac{4.6\,M_{1u}}{f_{ck}\,bd^2}}\right] bd = \frac{0.5 \times 20}{415}\left[1 - \sqrt{1 - \frac{4.6 \times 1.5 \times 250 \times 10^6}{20 \times 4000\,(620)^2}}\right] \times 400 \times 620$$

$$= 2017 \text{ mm}^2$$

No. of 20 mm φ bars = 2017/314.1 = 6.5 ≃ 7

Hence provide 7 bars of 20 mm φ. Actual $A_{st} = 7 \times 314.1 \simeq 2198$ mm²

and $$p = \frac{2198 \times 100}{4000 \times 620} = 0.886\%.$$

Let us check this reinforcement for development length at the point of contraflexure, as envisaged in the relation

$$\frac{M_1}{V_u} + L_0 \geq L_d$$

$$x_u = \frac{0.87\,f_y\,A_{st}}{0.36\,f_{ck}\,b} = \frac{0.87 \times 415 \times 2198}{0.36 \times 20 \times 400} = 275.6 \text{ mm}$$

$$M_1 = 0.87\,f_y\,A_{st}\,(d - 0.416\,x_u) = 0.87 \times 415 \times 2198\,(620 - 0.416 \times 275.6) = 401.04 \times 10^6 \text{ N-mm}$$

V_u = S.F. at the point of contraflexure $= 1.5 \times 380 \times 10^3 = 570 \times 10^3$ N

$L_0 = 12\ \varphi$ or d which ever is greater $= 620$ mm

$L_d = 47\ \varphi = 47 \times 20 = 940$ mm

$$\therefore \quad \frac{M_1}{V_u} + L_0 = \frac{401.04 \times 10^6}{570 \times 10^3} + 620 = 1323 > L_d. \text{ Hence Code requirements are satisfied.}$$

The above reinforcement (*i.e.* 7-20 mm φ bars) may be curtailed at the face of the column. However, continue 3 bars in the cantilever portion, to serve as anchor bars for shear reinforcement.

The reinforcement in the cantilever portion is to be found for the bending moment $M_{2u} = 1.5 \times 241.4$ kN-m. Here the beam will act as T-beam.

$$A_{st2} = \frac{0.5 \times 20}{415}\left[1 - \sqrt{1 - \frac{4.6 \times 1.5 \times 241.4 \times 10^6}{20 \times 400\ (620)^2}}\right] 400 \times 620 = 1930 \text{ mm}^2$$

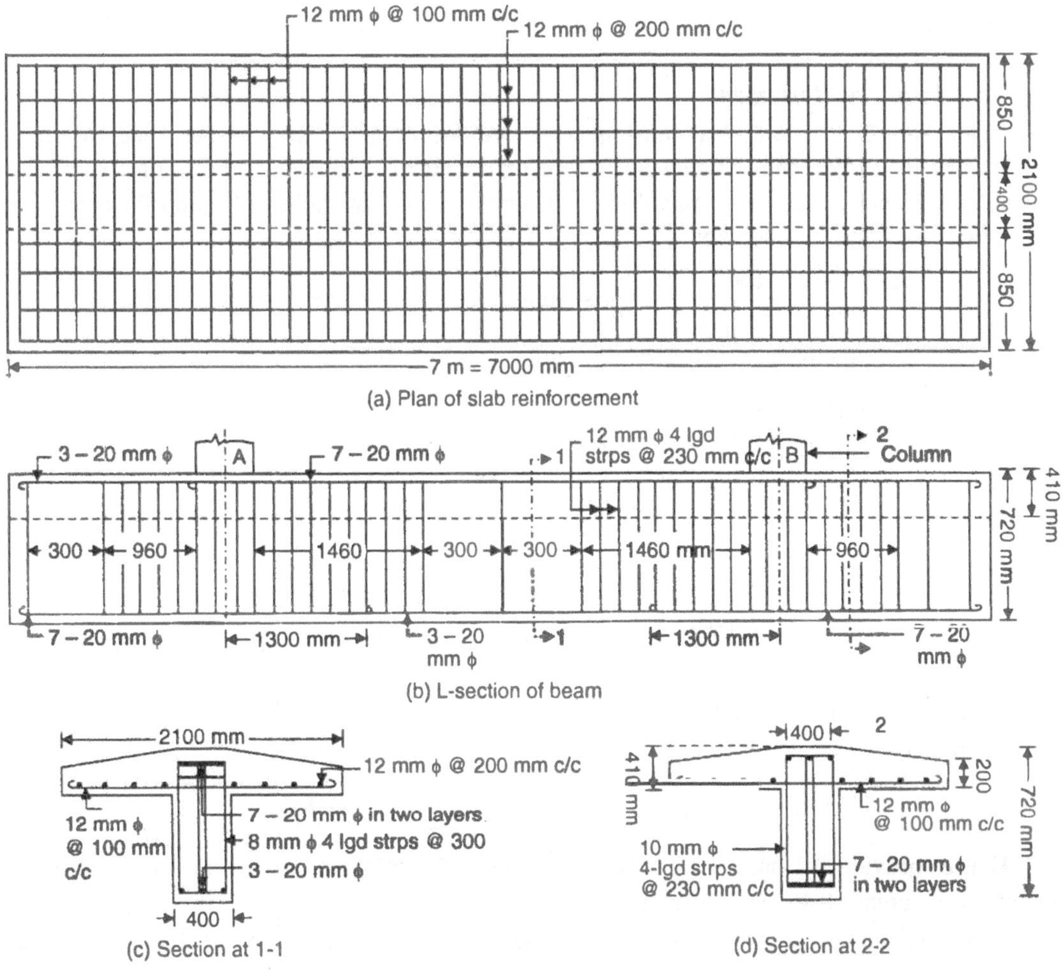

FIG. 22.46

$\therefore$ No. of 20 mm φ bars = 1930/314.1 = 6.2 ≃ 7

This reinforcement should be safe from development length (L_d) point of view, to satisfy the following equation at the point of contraflexure near the inner face of the column:

$$\frac{M_2}{V} + L_0 \geq L_d.$$

S.F. at this point of contraflexure is 1.5 × 380 kN, and it has already been established in case of A_{st1} that the above criterion is satisfied. Since the point of contraflexure is the same, and since $A_{st2} = A_{st1}$, the bars of A_{st2} will also satisfy the Code requirements.

It is to be noted that A_{st2} bars should be extended beyond the point of contraflexure by a distance $= d$ or 12 φ whichever is more (*i.e.* by 620 mm in this case) or to a distance of L_d (= 940 mm) from the centre of column (where bending moment M_2 occurs), whichever is more. Hence distance upto which A_{st2} bars are to be extended beyond centre of column = 0.67 + 0.62 ≈ 1.3 m. At this point all the bars except 3 Nos, may be discontinued. The remaining 3 bars may be taken throughout the length to serve as anchor bars for shear reinforcement.

7. *Shear reinforcement*

Refer Fig. 22.40. Diagonal tension crack will occur at *F*, only if the zone of sagging B.M. extends to d = 620 mm beyond inner face of the column. However, point of contraflexure falls earlier. Hence there is a possibility of crack at *G*, at point of contraflexure where shear force F_{1u} = 1.5 × 380 kN.

$$\therefore \quad \tau_v = \frac{1.5 \times 380 \times 1000}{620 \times 400} = 2.298 \text{ N/mm}^2. < \tau_{c,\max} \; (= 2.8 \text{ N/mm}^2, \text{ Table 7.3})$$

However, this is more than the permissible value of $\tau_c = 0.593$ N/mm² for $p = 0.886\%$. Hence shear reinforcement is required.

$$\therefore \quad V_{uc} = \tau_c bd = 0.593 \times 400 \times 620 = 146990 \text{ N}$$

$$\therefore \quad V_{us} = V_{1u} - V_{uc} = 1.5 \times 380000 - 146990 = 423010 \text{ N}$$

Using 12 mm φ 4-legged stirrups,

$$A_{sv} = 4 \times \frac{\pi}{4}(12)^2 = 452 \text{ mm}^2$$

$$\therefore \quad s_v = \frac{0.87 f_y A_{sv} d}{V_{us}} = \frac{0.87 \times 415 \times 452 \times 620}{423010} = 239 \text{ mm} \simeq 230 \text{ mm}$$

Distance upto which this is required (*i.e.*, at the point where S.F. $= V_{uc}$ = 146990 = 146.99 kN) is given by

$$x_1 = \frac{V_{2u} - V_{uc}}{1.5 \times 285.7} = \frac{1.5 \times 514.3 - 146.99}{1.5 \times 285.7} \simeq 1.46 \text{ m}$$

Hence provide 12 mm φ 4-legged stirrups, @ 230 mm c/c from the face of the column to a point 1.46 m, near each column.

In the cantilever portion, shear V_{3u} at the face of column = 1.5 × 371.5 kN. Distance x_2 of point at which shear force is equal to V_{uc} (= 146.99 kN) is given by

$$x_2 = \frac{V_{3u} - V_{uc}}{1.5 \times 285.7} = \frac{1.5 \times 371.5 - 146.99}{1.5 \times 285.7} \simeq 0.96 \text{ m}$$

Hence shear stirrups are to be provided for a distance of 0.96 m from the face of the column.

$$V_{us} = V_{3u} - V_{uc} = 1.5 \times 371.5 - 146.99 = 410.26 \text{ kN}$$

$\therefore$ Spacing of 12 mm φ 4 lgd stirrups $= \dfrac{0.87 \times 415 \times 452 \times 620}{410.20 \times 1000} = 246$ mm

However provide these @ 230 mm c/c upto 0.96 m from the face of the column.

For the remaining of length of beam, provide 8 mm φ 4-legged nominal stirrups at a spacing given by

$$s_v = \frac{2.175\, A_{sv}\, f_y}{b} = \frac{2.175 \times 201 \times 415}{400} = 453 \text{ mm but } \not> 300 \text{ mm}$$

Hence provide these @ 300 mm c/c. The details of reinforcement etc. are shown in Fig. 22.46.

Design Example 22.11 ***Combined trapezoidal footing.***

Design a combined trapezoidal footing for two columns A and B, spaced 5 metres centre to centre. Column A is 300 × 300 mm in size and transmits a load of 600 kN. Column B is 400 mm × 400 mm in size and carries a load of 900 kN. The maximum length of footing is restricted to 7 metres only. The safe bearing capacity of soil may be taken as 120 kN/m². Use M 20 mix for concrete and Fe 415 steel.

Solution.

1. ***Design constants***. For M 20 concrete and Fe 415 steel, we have :

$$\frac{x_{u,\,max}}{d} = 0.479 \quad \text{and} \quad R_u = 2.761$$

2. ***Size of footing***. Refer Fig. 22.47.

$$L = 7 \text{ m} \quad l = 5 \text{ m}; \quad \text{Keep } a_1 = a_2 = \tfrac{1}{2}(7-5) = 1 \text{ m}.$$

Let the width of footings be B_1 and B_2 as shown. The values of B_1 and B_2 should be such that C.G. of footing coincides with C.G. of the column loads.

Let weight of footing $W' = 0.1\,(W_1 + W_2) = 0.1\,(600 + 900) = 150$ kN

$$\therefore \quad \frac{(B_1 + B_2)L}{2} = \frac{W_1 + W_2 + W'}{q_0}$$

or $$B_1 + B_2 = \frac{(600 + 900 + 150)}{120} \times \frac{2}{7} = 3.93 \text{ m} \quad \ldots(1)$$

$$\text{Distance } \bar{x} \text{ of C.G. of load} = \frac{W_2\, l}{W_1 + W_2} = \frac{900 \times 5}{600 + 900} = 3 \text{ m.}$$

Distance of C.G. of load from edge $B_1 = a_1 + \bar{x} = 1 + 3 = 4$ m. ...(*i*)

Distance of C.G. of footing from edge $B_1 = \dfrac{L}{3}\left(\dfrac{B_1 + 2B_2}{B_1 + B_2}\right) = \dfrac{7}{3}\left(\dfrac{B_1 + 2B_2}{B_1 + B_2}\right)$ ...(*ii*)

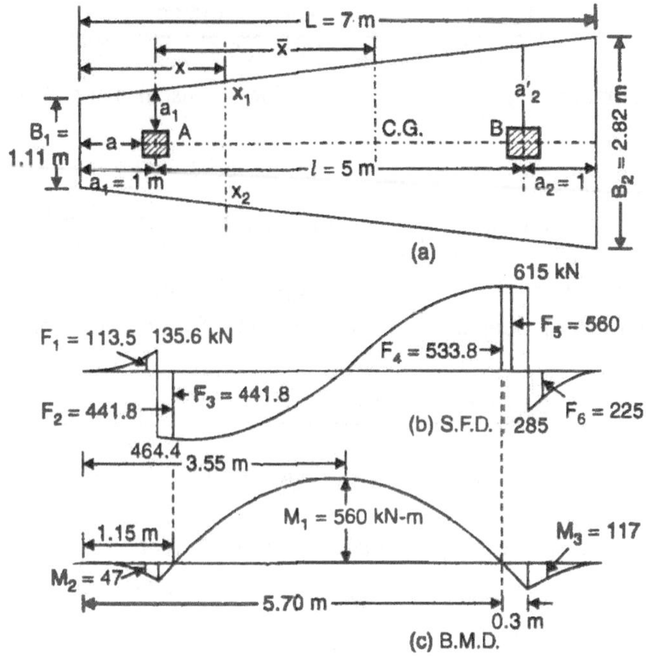

FIG. 22.47.

Equating (*i*) and (*ii*), we get $\frac{7}{3}\left(\frac{B_1 + 2B_2}{B_1 + B_2}\right) = 4$ ...(2)

From (1) and (2), we get $B_1 = 1.11$ m and $B_2 = 2.82$ m.

Net upward soil pressure intensity is given by

$$p_0 = \frac{W_1 + W_2}{\frac{1}{2}(B_1 + B_2) L} = \frac{(600 + 900)}{\frac{1}{2}(1.11 + 2.82)\, 7} \approx 110 \text{ kN/m}^2$$

3. *Bending moment and shear force*

Consider the bending of footing in the longitudinal direction. At any distance x from the face B_1, the width B_x of section X_1X_2 is given by

$$B_x = B_1 + \frac{B_2 - B_1}{L} x = 1.11 + \frac{2.82 - 1.11}{7} x = 1.11 + 0.245\, x$$

∴ Area A_x to the left of section $X_1 X_2 = \frac{1}{2}(B_1 + B_x)\, x$

$$= \frac{1}{2}(1.11 + 1.11 + 0.245x)\, x = \frac{x}{2}(2.22 + 0.245x) \quad \text{...}(i)$$

Distance of C.G. from section $X_1 X_2 = \frac{x}{3}\left[\frac{2B_1 + B_x}{B_1 + B_x}\right]$

$$= \frac{x}{3}\left[\frac{2.22 + 1.11 + 0.245x}{1.11 + 1.11 + 0.245x}\right] = \frac{x}{3}\left[\frac{3.33 + 0.245x}{2.22 + 0.245x}\right] \quad \text{...}(ii)$$

Hence total upward force to the left of X_1X_2

$$= \frac{p_o x}{2}(2.22 + 0.245x) = \frac{110}{2}x\ (2.22 + 0.245\ x) \qquad ...(1)$$

∴ S.F. at outer face of column A, where $x = a = 0.85$ m, is

$$F_1 = \frac{110}{2} \times 0.85(2.22 + 0.245 \times 0.85) = 113.5 \text{ kN.}$$

∴ S.F. at x to the right of $A = 600 - \frac{110}{2}x\ (2.22 + 0.245x)$ ...(2)

∴ S.F. at inner face of column A, where $x = 1.15$ m, is

$$F_2 = 600 - \frac{110}{2} \times 1.15\ (2.22 + 0.245 \times 1.15) = 441.8 \text{ kN.}$$

To find the position of section where S.F. is zero, put (2) to zero.

∴ $600 - \frac{110}{2}x\ [2.22 + 0.245\ x] = 0.$ From which $x = 3.55$ m-

B.M. at section X_1X_2 is given by

$$M_2 = 600\ (x - 1) - (\text{Upward force} \times \text{Distance of its C.G.})$$

$$= 600\ (x - 1) - \left\{\frac{110}{2}x\ (2.22 + 0.245\ x)\right\} \times \frac{x}{3}\left\{\frac{3.33 + 0.245\ x}{2.22 + 0.245\ x}\right\}$$

$$= 600\ (x - 1) - \frac{110}{6}x^2\ (3.33 + 0.245\ x) \qquad ...(3)$$

Maximum hogging bending moment (M_1) will occur at $x = 3.55$ m where S.F. is zero

∴ $M_1 = 600(3.55 - 1) - \frac{110}{6}(3.55)^2\ (3.33 + 0.245 \times 3.55) = 560$ kN-m

To find the value of x where M_x is zero, equate (3) to zero.

∴ $600\ (x - 1) - \frac{110}{6}x^2\ (3.33 + 0.245\ x) = 0$

Solving this by trial and error, we get, $x = 1.15$ m and 5.70 m

Thus the position of both the points of contraflexure are known.

For getting the value of S.F. (F_3) at the first point of contraflexure, put $x = 1.15$ m in Eq. (2).

∴ $F_3 = 600 - \frac{110}{2}(1.15)\left(2.22 + 0.245 \times 1.15\right) = 441.8$ kN

Similarly S.F. (F_4) at $x = 5.70$ m is

$$F_4 = 600 - \frac{110}{2}(5.7)\left(2.22 + 0.245 \times 5.7\right) = 533.8 \text{ kN (numerically)}$$

S.F. at inner face of column B, at $x = 5.8$ m is

$$F_5 = 600 - \frac{110}{2}(5.8)\left(2.22 + 0.245 \times 5.8\right) = 560 \text{ kN (numerically)}$$

S.F. at outer face of column B, at $x = 6.2$ m is

$$F_6 = 600 - \frac{110}{2}(6.2)\ (2.22 + 0.245 \times 6.2) + 900 = 225 \text{ kN}$$

Sagging bending moment (M_2) at the outer face of column A is obtained by putting $x = 0.85$ m in the second term of Eq. (3) and neglecting the first term.

$$\therefore \quad M_2 = \frac{110}{6}(0.85)^2\,(3.33 + 0.245 \times 0.85) = 47 \text{ kN-m}$$

Sagging B.M. for the cantilever portion to the right of column B is given by

$$M_x = \frac{110}{6}x^2\,(3.33 + 0.245\,x) - 600\,(x - 1) - 900\,(x - 6)$$

Putting $x = 6.2$ m, we get B.M. (M_3) at the outer face of column B :

$$M_3 = \frac{110}{6}(6.2)^2\,(3.33 + 0.245 \times 6.2) - 600(5.2) - 900(0.2) = 117 \text{ kN-m}$$

The B.M. and S.F. diagrams are shown in Fig. 22.47.

4. ***Effective depth of footing. (a) Depth from bending compression***: The effective depth is determined on the basis of hogging moment $M_{1u} = 1.5 \times 560$ kN-m which is the maximum bending moment in the footing. It occurs at a section distant $x = 3.55$ m from the edge B_1.

The width B_x at that section is given by

$$B_x = B_1 + \frac{B_2 - B_1}{L}x = 1.11 + 0.245\,x = 1.11 + 0.245 \times 3.55 = 1.87 \text{ m} = 1870 \text{ mm}$$

$$\therefore \quad d = \sqrt{\frac{M_{1u}}{R_u\,B_x}} = \sqrt{\frac{1.5 \times 560 \times 10^6}{2.761 \times 1870}} = 404 \text{ mm}$$

(b) Depth from punching shear consideration

This depth should be sufficient from punching shear point of view. Punching shear will be maximum near column B, where load $W_{2u} = 1.5 \times 900$ kN.

$$\text{Side } b_0 = b + d = (400 + d) \text{ mm}$$

Punching shear

$$F_u = 1.5\,(W_2 - p_0\,b_0^2)$$

$$= 1.5\left[900 - 110\left(\frac{400 + d}{1000}\right)^2\right] \times 10^3 = 1350000 - 0.165\,(400 + d)^2$$

$$\therefore \quad \tau_v = \frac{1350000 - 0.165\,(400 + d)^2}{4\,(400 + d)\,d}$$

Permissible shear stress $= k_s\,\tau_c = 1 \times 0.25\sqrt{f_{ck}} = 1 \times 0.25\sqrt{20} = 1.118$ N/mm^2

$$\therefore \quad \frac{1350000 - 0.165\,(400 + d)^2}{4\,(400 + d)\,d} = 1.118$$

From which $d = 367$ mm which is less than the one found from bending compression.

Hence $d = 404$ mm . Using 60 mm effective cover, $D = 404 + 60 = 464$ mm. However, keep $D = 500$ mm so that available $d = 500 - 60 = 440$ mm.

5. ***Reinforcement in the longitudinal direction***

(i) Reinforcement for hogging bending moment M_{1u} is given by the following, applicable for under-reinforced section :

$$A_{st1} = \frac{0.5 f_{ck}}{f_y}\left[1 - \sqrt{1 - \frac{4.6 M_{1u}}{f_{ck} B_x d^2}}\right] B_x \cdot d$$

$$= \frac{0.5 \times 20}{415}\left[1 - \sqrt{1 - \frac{4.6 \times 1.5 \times 560 \times 10^6}{20 \times 1870 (440)^2}}\right] \times 1870 \times 440 = 6288 \text{ mm}^2$$

Using 20 mm φ bars, $A_\varphi = \frac{\pi}{4}(20)^2 = 314.16 \text{ mm}^2$

No. of bars required $= 6288/314.16 \simeq 21$. Thus provide 21 bars at the top face of the footing. These bars should be checked for development length at the point of contraflexure, to satisfy the criteria : $\frac{M_1}{V_u} + L_0 \geq L_d$.

Out of the two points of contraflexure in the hogging B.M. range, the one near column B, has maximum shear force $F_{4u} = 1.5 \times 533.8$ kN

Actual $A_{st} = 21 \times 314.16 \simeq 6597 \text{ mm}^2$; $p = 6597 \times 100/1870 \times 440 = 0.8\%$

$$x_u = \frac{0.87 f_y A_{st}}{0.36 f_{ck} B_x} = \frac{0.87 \times 415 \times 6597}{0.36 \times 20 \times 1870} = 176.9 \text{ mm}$$

$$M_1 = 0.87 f_y A_{st} (d - 0.416 x_u) = 0.87 \times 415 \times 6597 (440 - 0.416 \times 176.9) = 872.73 \times 10^6 \text{ N-mm}$$

$$V_u = f_{4u} = 1.5 \times 533.8 \times 10^3 = 800700 \text{ N}$$

$L_0 = 12\,\varphi$ or d whichever is more $= 440$ mm

$$L_d = 47\,\varphi = 47 \times 20 = 940 \text{ mm}$$

$$\therefore \quad \frac{M_1}{V_u} + L_0 = \frac{872.53 \times 10^6}{800700} + 440 = 1530 \text{ mm}$$

which is more than L_d, provided all the 21 bars are taken upto the point of contraflexure. These bars are to be taken beyond this point for a distance of 12 φ (= 12 × 20 = 240 mm) or d (=440 mm), whichever is more. After that, curtail 13 bars and continue 8 bars straight upto either edge of footing so that these may serve as anchorage bars for shear stirrups. These 8 bars will also satisfy the requirement of minimum reinforcement @ 0.12% of the gross area.

(ii) Reinforcement for sagging bending moment M_{3u} near outer face of column B

Width of footing near the outer face of column B, where $x = 6.2$ m is

$$B_x = 1.11 + 0.245\,x = 1.11 + 0.245 \times 6.2 = 2.63 \text{ m}$$

$$A_{st3} = \frac{0.5 \times 20}{415}\left[1 - \sqrt{1 - \frac{4.6 \times 1.5 \times 117 \times 10^6}{20 \times 2630 (440)^2}}\right] 2630 \times 440 \simeq 1130 \text{ mm}^2$$

$\therefore$ Minimum reinforcement @ 0.12% $= \frac{0.12}{100} \times 2630 \times 500 = 1578 \text{ mm}^2$

Hence provide 1578 mm² area of reinforcement. Using 12 mm φ bars, $A_\varphi = 113 \text{ mm}^2$

$$\therefore \quad \text{Spacing} = \frac{B_x A_\varphi}{A_{st}} = \frac{2630 \times 113}{1578} \simeq 188 \text{ mm.}$$

However, provide 12 mm φ bars @ 180 mm c/c.

Actual $A_{st} = \dfrac{2630 \times 113}{180} \simeq 1650 \text{ mm}^2$ and actual $p = \dfrac{1650 \times 100}{2630 \times 440} = 0.14\%$

This reinforcement should be checked for development length at the point of contraflexure near the inner face column B, so as to satisfy the equation $\dfrac{M_{1u}}{V_u} + L_0 \geq L_d$.

$$x_u = \frac{0.87 f_y A_{st}}{0.36 f_{ck} B_x} = \frac{0.87 \times 415 \times 1650}{0.36 \times 20 \times 2630} = 31.5 \text{ mm}$$

$\therefore$ $M_1 = 0.87 f_y A_{st} (d - 0.416 x_u) = 0.87 \times 415 \times 1508 (440 - 0.416 \times 31.5) = 232.4 \times 10^6$ N-mm

$$V_u = 1.5 F_{4u} = 1.5 \times 533.8 \times 10^3 = 800.7 \times 10^3 \text{ N}$$

$$L_0 = 12\varphi \text{ or } d \text{ whichever is more} = 440 \text{ mm}$$

$$L_d = 47\varphi = 47 \times 12 = 564 \text{ mm}$$

$\therefore$ $\dfrac{M_{1u}}{V_u} + L_0 = \dfrac{232.4 \times 10^6}{800.7 \times 10^3} + 440 = 730 \text{ mm} > L_d$. Hence safe

Also, length of bars available in the cantilever portion, beyond the centre of column $B = 1$ m − side cover = 1000 − 60 = 940 mm > L_d. Hence safe.

(iii) Reinforcement for sagging B.M. M_{2u} near other face of column A.

Width of footing near the outer face of column A, where $x = 0.85$ m is

$B_x = 1.11 + 0.245\,x = 1.11 + 0.245 \times 0.85 \approx 1.32$ m.

$$A_{st2} = \frac{0.5 \times 20}{415}\left[1 - \sqrt{1 - \frac{4.6 \times 1.5 \times 47 \times 10^6}{20 \times 1320\,(440)^2}}\right] 1320 \times 440 = 451.3 \text{ mm}^2$$

Minimum reinforcement @ 0.12% $= \dfrac{0.12}{100} \times 1320 \times 500 = 792 \text{ mm}^2$

Hence provide 792 mm² area of reinforcement. Using 12 mm φ bars having

$A_\varphi = 113 \text{ mm}^2$, spacing $= \dfrac{B_x A_\varphi}{A_{st}} = \dfrac{1320 \times 113}{792} = 188$ mm

However, provide 12 mm φ bars @ 180 mm c/c, as near B.

Actual $A_{st} = \dfrac{1320 \times 113}{180} = 828.7 \text{ mm}^2$ and actual $p = \dfrac{828.7 \times 100}{1320 \times 440} = 0.14\%$

Since the S.F. F_{3u} at the point of contraflexure is less than S.F. F_{4u} in (*ii*) above the conditions of development length will also be satisfied.

Hence provide 12 mm φ bars @ 180 mm c/c throughout the length, at the bottom.

5. *Check for diagonal tension*

Test for diagonal tension should be made in the cantilever portion at a distance d from the column face. In the present case, $d = 440$ mm $= 0.44$ m. At 0.44 m from the outer face of column B, distance $x = 6.2 + 0.44 = 6.64$ m. At this section, shear force is extremely small and available width B_x is large. Hence shear stress will be small. Similarly, for the cantilever portion to the left of column A, shear force, and hence shear stress will be small.

For diagonal tension between A and B, near column B, crack can occur at the bottom of the footing (*i.e.*, for sagging B.M.) at a distance d = 440 mm, or at the top of footing (*i.e.* for hogging B.M.) at the point of contraflexure distant 300 mm from the centre of column B. Point of contraflexure being nearer, more S.F. will be at the point. S.F. at point of contraflexure $= F_4 = 533.8$ kN. At the point of contraflexure $x = 5.70$ m.

$$\therefore \quad B_x = 1.11 + 0.245\,x = 1.11 + 0.245 \times 5.7 \approx 2.5 \text{ m}$$

$$\therefore \quad s_v = \frac{F_{4u}}{bd} = \frac{1.5 \times 533.8 \times 10^3}{2500 \times 440} = 0.728 \text{ N/mm}^2$$

$$\frac{100\,A_s}{bd} = \frac{100\,(21 \times 314)}{2500 \times 440} \simeq 0.6\%.$$ Hence from Table 7.1, $\tau_c = 0.48$ N/mm^2

Thus, actual shear stress is more than the permissible one, and shear reinforcement is necessary.

$$V_{uc} = \tau_c\,bd = 0.48 \times 2500 \times 440 \times 10^{-3} = 528 \text{ kN}$$

$$\therefore \quad V_{us} = V_u - V_{uc} = 1.5 \times 533.8 - 528 = 272.7 \text{ kN}$$

Using 10 mm φ 8-lgd stirrups, $A_{sv} = 8 \times \frac{\pi}{4}(10)^2 = 628 \text{ mm}^2$

$$s_v = \frac{0.87\,f_y\,A_{sv}\,d}{V_{us}} = \frac{0.87 \times 415 \times 628 \times 440}{272.7 \times 100} = 365 \text{ mm}$$

Maximum spacing of nominal stirrups b given by

$$s_v = \frac{2.175\,f_y\,A_{sv}}{b} = \frac{2.175 \times 415 \times 628}{2500} = 226 \text{ mm}$$

Hence provide 10 mm φ leg stirrups @ 225 mm c/c throughout the length.

7. ***Transverse Reinforcement.*** The footing will bend transversely near column face. Projection a_2' beyond face of column $B = \frac{1}{2}(B_x - b_2)$ where B_x is the width of footing at the centre of the column B, where x = 6 m.

$$\therefore \quad B_x = 1.11 + 0.245\,x = 1.11 + 0.245 \times 6 = 2.58 \text{ m}$$

$$\therefore \quad a_2' = \frac{1}{2}(2.58 - 0.4) = 1.09 \text{ m} = 1090 \text{ mm.}$$

Width of bending strip near column $B = d + b_2 + d = 440 + 400 + 440 = 1280$ mm

∴ Average area of bending strip $= 2.58 \times 1.280 = 3.3024 \text{ m}^2$

∴ Net upward pressure $p_0' = 900/3.3024 \approx 272.5 \text{ kN/m}^2$

(This is approximately two and half times the value of average pressure p_0)

Maximum B.M. at the face of column is

$$M = p_0' \frac{(a_2')^2}{2} = \frac{272.5}{2}(1.09)^2 = 161.88 \text{ kN-m} = 161.88 \times 10^6 \text{ N-mm}$$

$$M_u = 1.5\,M = 1.5 \times 161.88 \times 10^6 = 242.82 \times 10^6 \text{ N-mm}$$

$$\therefore \quad d \text{ required} = \sqrt{\frac{242.82 \times 10^6}{2.761 \times 1000}} = 297 \text{ mm.}$$

Actual d available = 440 mm. Hence the transverse beam, of width = 1.28 m will be of the same thickness as the remaining footing. Since transverse reinforcement will be provided *above* the main reinforcement, available $d = 440 - 20 = 420$ mm (approx).

The transverse reinforcement is given by the following expression applicable for under-reinforced section.

$$A_{st} = \frac{0.5 f_{ck}}{f_y}\left[1 - \sqrt{1 - \frac{4.6 M_u}{f_{ck} bd^2}}\right] bd = \frac{0.5 \times 20}{415}\left[1 - \sqrt{1 - \frac{4.6 \times 242.82 \times 10^6}{20 \times 1000 (420)^2}}\right] 1000 \times 420$$

$= 1755 \text{ mm}^2.$

Using 16 mm φ bars, $A_\varphi = 201 \text{ mm}^2$.

$\therefore$ Spacing $s = \dfrac{1000 \times 201}{1755} = 114$ mm. Hence provide 16 mm φ bars @ 110 mm c/c.

Similarly, for column A, width B_x at the centre of column, where $x = 1$ m is,

$$B_x = 1.11 + 0.245\, x = 1.11 + 0.245 \times 1 = 1.36 \text{ m}.$$

Projection $a_1' = \frac{1}{2}(1.36 - 0.3) = 0.53$ m

Width of bending strip $= 2d + b_1 = 2 \times 440 + 300 = 1180$ mm

$\therefore$ Area bending strip $= 1.36 \times 1.18 = 1.6048 \text{ m}^2$

$\therefore$ Net upward pressure $p_0' = 600/1.6048 \simeq 374 \text{ kN/m}^2$

$\therefore$ Max. B.M. at the face of the column is

$$M = p_0 \frac{(a_1')^2}{2} = 374 \frac{(0.53)^2}{2} = 52.528 \text{ kN-m} = 52.528 \times 10^6 \text{ N-mm}.$$

$$\therefore \quad M_u = 1.5\, M = 1.5 \times 52.528 \times 10^6 = 78.79 \times 10^6 \text{ N-mm}.$$

The available depth $d = 440$ mm will be sufficient. Area of transverse reinforcement is

$$A_{st} = \frac{0.5 \times 20}{415}\left[1 - \sqrt{1 - \frac{4.6 \times 78.79 \times 10^6}{20 \times 1000 (420)^2}}\right] 1000 \times 420 = 534 \text{ mm}^2$$

Minimum reinforcement $= \dfrac{0.12}{100}(1000 \times 500) = 600 \text{ mm}^2$.

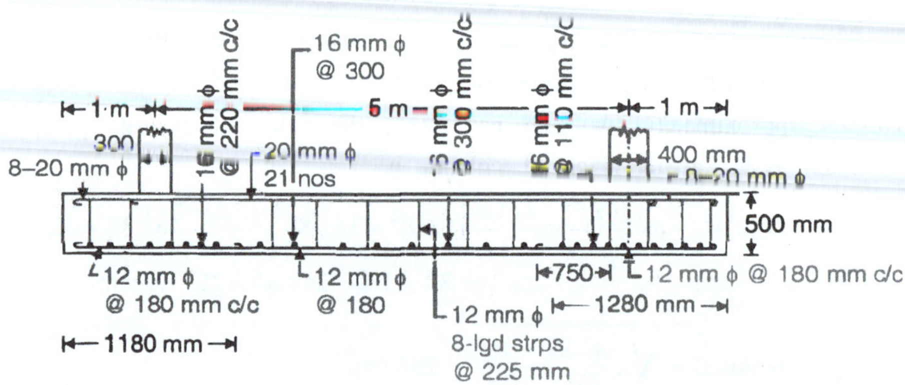

FIG. 22.48.

Using 16 mm φ bars, having $A_\varphi = 201\ mm^2$

$\therefore$ Spacing $s = \frac{1000 \times 201}{600} = 335$ mm subject to a maximum, of 300 mm

However provide 16 mm φ bars @ 220 mm c/c in the strip of width 1.18 m.

For the remaining portion, provide 16 mm φ bars @ 300 mm c/c, so that this reinforcement at any section is not less than 0.12% of area of cross-section. The details of reinforcement etc. are shown in Fig. 22.48.

Design Example 22.12. ***Trapezoidal combined footing with central connecting beam*** *Redesign Example 22.11 by providing a longitudinal beam connecting the two columns.*

Solution. For the design of trapezoidal footing with a longitudinal beam, steps 1, 2 and 3 of example 22.11 will remain the same. The bending moment and shear force diagrams drawn in Fig. 22.47(*b*) and (*c*) will be applicable for the bending of the longitudinal beam which should be designed properly to resist these bending moments and shear forces at various sections. The slab of the footings will bend transversely.

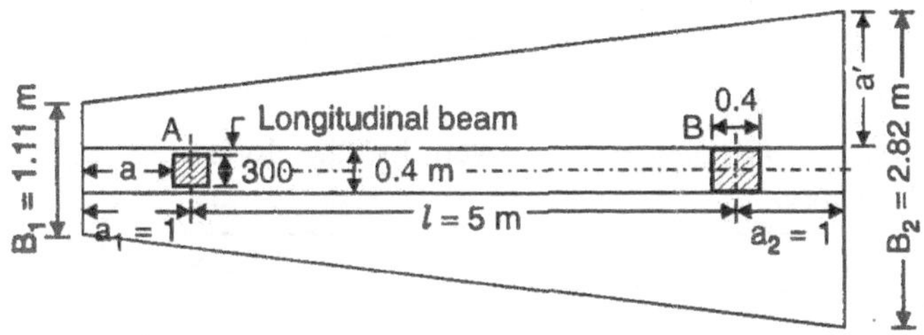

FIG. 22.49.

The salient dimensions of the footing, determined in Example 22.11 are shown in Fig. 22.49. Let the beam project above the footing. Thus, for the hogging bending moment, T-beam action will be available. Let the width of beam = 400 mm.

1. ***Design of slab***. The slab will bend as cantilever on either side of the beam. The maximum cantilever projection of the slab will be at the end near width B_2, its magnitude being given by $a' = \frac{1}{2}(2.82 - 0.4) = 1.21$ m.

Intensity of net upward pressure p_0 =110 kN/m², determined in the previous example. Maximum bending moment at the face of longitudinal beam is given by

$$M = p_0\, a'^2 = \frac{110\,(1.21)^2}{2} = 80.53 \text{ kN m} = 80.53 \times 10^6 \text{ N-mm}$$

$$\therefore \qquad M_u = 1.5\, M = 1.5 \times 80.53 \times 10^6 = 120.795 \times 10^6 \text{ N-mm}$$

$\therefore$ Thickness of slab $= \sqrt{\frac{120.795 \times 10^6}{2.761 \times 1000}} \simeq 210$ mm.

However, adopt $d = 240$ mm to get an under-reinforced section. Providing 60 mm effective cover, total $D = 240 + 60 = 300$ mm. Reduce it to 200 mm at the end, where available $d = 140$ mm.

(*b*) *Depth from one way shear*

The critical plane for probable diagonal tension crack may occur at a distance d (mm) from the beam face. The S.F. F_u' at that section $= 1.5 \times 110\,(1.21 - 0.001\,d) \times 10^3$ N/m $= 165 \times 10^3\,(1.21 - 0.001\,d)$ N/m.

The effective depth d' at that section $= 140 + \dfrac{d - 140}{1210}(1210 - d)$ mm

$$\therefore \qquad \tau_v = \frac{F_u'}{bd'} = \frac{165 \times 10^3\,(1.21 - 0.001\,d)}{1000\left[140 + \dfrac{d-140}{1210}(1210 - d)\right]} = \frac{199650\,(1.21 - 0.001\,d)}{169400 + (d - 140)(1210 - d)} \qquad \ldots(i)$$

Permissible shear stress $= k\,\tau_c$, where $k = 1$ for $D \geq 300$ mm (Table 7.2)

Also, $\tau_c = 0.384\ \text{N/mm}^2$ for $p = 0.3\%$ for under-reinforced section.

$$\therefore \qquad \tau_v = \tau_c = 0.384 = \frac{199650\,(1.21 - 0.001\,d)}{169400 + (1350\,d - 169400 - d^2)} = \frac{199650\,(1.21 - 0.001\,d)}{1350\,d - d^2}$$

or $$d^2 - 1870\,d + 629105 = 0, \text{ from which } d = 440 \text{ mm}$$

Hence keep $d = 440$ mm and $D = 440 + 60 = 500$ mm at the face of the column. Reduce $D = 200$ mm at the edges.

The area of steel reinforcement is given by the following expression applicable for an under-reinforced section.

$$A_{st} = \frac{0.5\,f_{ck}}{f_y}\left[1 - \sqrt{1 - \frac{4.6\,M_u}{f_{ck}\,bd^2}}\right]bd = \frac{0.5 \times 20}{415}\left[1 - \sqrt{1 - \frac{4.6 \times 120.795 \times 10^6}{20 \times 1000\,(440)^2}}\right] \times 1000 \times 440$$

$= 791\ \text{mm}^2$

Minimum area a steel @ 0.12% $= \dfrac{0.12}{100}(1000 \times 500) = 600\ \text{mm}^2 < A_{st} = 791\ \text{mm}^2$

Using 16 mm φ bars, spacing $= \dfrac{1000 \times 201}{791} = 254$ mm and $p = \dfrac{791 \times 100}{1000 \times 440} \simeq 0.18\%$ only.

However, provide 16 mm φ bars @ 150 mm c/c to provide $A_{st} = \dfrac{1000 \times 201}{150} = 1340\ \text{mm}^2$

and $p = \dfrac{1340 \times 100}{1000 \times 440} = 0.304\%$, which is almost equal to the one assumed for determining permissible value of τ_c ($= 0.384\ \text{N/mm}^2$). Provision for lesser reinforcement will make the slab *unsafe* against diagonal tension. Hence provide 16 mm φ bars @ 150 mm c/c at the bottom of the section.

Let us check this for development length : $L_d = 47\,\varphi = 47 \times 16 = 752$ mm.

Available length, using 60 mm side cover $= 1210 - 60 = 1150$ mm. Hence safe.

Similarly, at the end strip near column A, the cantilever projection of slab beyond the beam $= \frac{1}{2}(1.11 - 0.4) = 0.355$ m.

$$\therefore \qquad \text{B.M. } M = 110\,\frac{(0.355)^2}{2} \approx 6.93 \text{ kN-m} = 6.93 \times 10^6 \text{ N-mm}$$

and $$M_u = 1.5\,M = 1.5 \times 6.93 \times 10^6 = 10.395 \times 10^6 \text{ N-mm}$$

This is extremely small, and reinforcement found on the basis of above B.M. will be extremely small. However, for 16 mm φ bars, $L_d = 47\,\varphi = 752$ mm while available length of bars (after providing 60 mm side cover = 355 − 60 = 295 mm only. Hence maximum stress σ_s that can be developed $= \frac{295}{752} \times 415 \simeq 162$ N/mm², so that bending failure is not there.

$$A_{st} = \frac{0.5 f_{ck}}{f_s}\left[1 - \sqrt{1 - \frac{4.6\,M_u}{f_{ck}\,bd^2}}\right]bd = \frac{0.5 \times 20}{162}\left[1 - \sqrt{1 - \frac{4.6 \times 10.395 \times 10^6}{20 \times 1000\,(440)^2}}\right] \times 1000 \times 440$$

$= 169$ mm²

Also, minimum area of steel @ 0.12% = 600 mm² (found above)

Using 16 mm φ bars, $s = \frac{1000 \times 201}{600} = 325$ mm, subject to a maximum of 300 mm.

Hence the spacing of 16 mm φ bars @ 150 mm c/c at the edge near column B can be gradually increased to 300 mm at the edge near column A.

Transverse reinforcement area @ 0.12% = 600 mm².

Using 10 mm φ bars $A_\varphi = 78.5$ mm².

$$\therefore \qquad \text{Spacing} = \frac{1000 \times 78.5}{600} \simeq 180 \text{ mm c/c.}$$

2. *Depth of longitudinal beam*. Let us provide longitudinal beam of 400 mm width, projecting above the slab. In the hogging portion, therefore, T-beam action will be available. The hogging bending moment $M_{1u} = 1.5 \times 560$ kN-m is much more than the sagging bending moment. Hence the depth of longitudinal beam will be designed for moment M_{1u}, taking as a T-beam.

$$\text{Average thickness of slab} = \frac{200 + 500}{2} = 350 \text{ mm} = D_f$$

The maximum bending moment occurs at $x = 3.55$ m, where width B_x is given by

$$B_x = 1.11 + 0.245\,x = 1.11 + 0.245 \times 3.55 = 1.98 \text{ m} = 1980 \text{ mm.}$$

The width b_f of the flange will be corresponding to an isolated T-beam given by the expression

$$b_f = \frac{l_0}{\left(\frac{l_0}{b} + 4\right)} + b_w$$

where b = actual width of flange = 1980 mm

l_0 = distance between points of zero moment in the beam

= 5.70 − 1.15 = 4.55 m = 4550 mm

b_w = width of web = 400 mm.

$$\therefore \qquad b_f = \frac{4550}{\frac{4550}{1980} + 4} + 400 = 1122 \text{ mm}$$

Fixation of effective depth from bending consideration

Let $$d \simeq \frac{2}{3}\sqrt{\frac{M_{1u}}{R_u\, b_w}} = \frac{2}{3}\sqrt{\frac{1.5 \times 560 \times 10^6}{2.761 \times 400}} = 582 \text{ mm} \simeq 600 \text{ mm (say)}$$

Using 50 mm nominal cover, 25 mm φ main bars in two layers with 25 mm gap between them and 12 mm φ stirrups, effective cover $= 50 + 12 + 25 + 25/2 = 99.5$ mm.

However keep effective cover = 100 mm, so that $d = 600 + 100 = 700$ mm.

Effective depth from punching shear : Let us find the depth from punching shear point of view also. The column B will have to punch through depth d_b of the beam on two sides, and through depth d_s of the slab on the other two sides. The critical plane occurs at a distance of half the effective depth d_b of column on one side and $d_s/2$ to the other side. Hence widths of the punching shear zone are :

$$b_{01} = 400 + d_b \quad \text{and} \quad b_{02} = 400 + d_s = 400 + 440 = 840 \text{ mm}$$

∴ Punching shear area of concrete available along the perimeter of the critical plane

$$= (2b_{01} \times d_s') + (2b_{02} \times d_b)$$

where d_s' = depth of slab at the critical plane (Fig. 22.45)

$$= d_{se} + \frac{d_s - d_{se}}{a}\left[a - \frac{d_s}{2}\right] = 140 + \frac{400 - 140}{1000}\left[1000 - \frac{440}{2}\right] = 374 \text{ mm}$$

Punching shear

$$F_u = 1.5\left[900 \times 10^3 - 110 \times 10^3\left\{\frac{(1400 + d_b) \times 840}{(1000)^2}\right\}\right] = 13.5 \times 10^5 - 138.6\,(400 + d_b)$$

$$\therefore \text{ Punching shear stress} = \frac{F_u}{2\,[(b_{01} \times d_s') + (b_{02} \times d_b)]} = \frac{13.5 \times 10^5 - 138.6\,(400 + d_b)}{2\,[(400 + d_b)\,374 + 840\, d_b]}$$

Permissible shear stress $= k_s\, \tau_c$ where $k_s = (0.5 + \beta_c) \not> 1 = 1$

and $$\tau_c = 0.25\sqrt{f_{ck}} = 0.25\sqrt{20} = 1.118 \text{ N/mm}^2$$

$$\therefore \quad \frac{13.5 \times 10^5 - 138.6\,(400 + d_b)}{2\,[(400 + d_b)\,374 + 840\, d_b]} = 1.118$$

or $$\frac{1294560 - 138.6\, d_b}{149600 + 1214\, d_b} = 2.236, \text{ from which } d_b = 337 \text{ mm}$$

This is less than effective depth of 600 mm found from bending compression.

(c) ***Effective depth from consideration of diagonal tension cracks***

Refere Fig. 22.40. Diagonal tenson crack will occur at F only if the zero sagging B.M. extends to $d = 600$ mm beyond the inner face of column A (or B). However, point of contraflexure falls earlier. Hence there is a possibility of crack at the point of contraflexure, near column B where S.F. $F_4 = 533.8$ kN, and hence $F_{4u} = 1.5 \times 533.8 = 800.7$ kN

$$\therefore \quad \tau_v = \frac{F_{4u}}{bd} = \frac{F_{4u}}{400\, d}$$

Equating this to $\tau_{c,\,max} = 2.8$ N/mm² (Table 7.3), we get

$$d = \frac{F_{4u}}{400 \times 2.8} = \frac{800.7 \times 10^3}{400 \times 2.8} = 715 \text{ mm}$$

However, keep $d = 800$ mm, so that $D = 800 + 100 = 900$ mm.

3. Steel reinforcement : Steel reinforcement is to be found for design moment M_{uD} = Max. hogging B.M. $= M_{1u} = 1.5 \times 560 \times 10^6 = 840 \times 10^6$ N-mm.

Let us assume $x_u = D_f = 350$ mm

$\therefore \quad M_u = 0.36 f_{ck}\, b_f\, D_f\, (d - 0.416\, D_f) = 0.36 \times 20 \times 1106 \times 350\, (600 - 0.416 \times 350)$

$= 1266.47 \times 10^6$ N-mm

Since $M_u >> M_{uD}$, neutral axis falls inside the flange, i.e. $x_u \le D_f$

$$\therefore \quad M_{uD} = 0.87 f_y A_{st}\, d \left[1 - \frac{f_y}{f_{ck}} \frac{A_{st}}{b_f\, d}\right], \text{ from which}$$

$$A_{st} = 0.5 \frac{f_{ck}}{f_y}\left[1 - \sqrt{1 - \frac{4.6 M_{uD}}{f_{ck}\, b_f\, d^2}}\right] b_f\, d = \frac{0.5 \times 20}{415}\left[1 - \sqrt{1 - \frac{4.6 \times 840 \times 10^6}{20 \times 1122\, (800)^2}}\right] \times 1122 \times 800$$

$$= 3152 \text{ mm}^2$$

Using 25 mm φ bars having $A_\varphi \simeq 490.8$ mm², No. of bars $= 3132/491.8 \simeq 7$

Actual A_{st} provided $= 7 \times 490.8 \simeq 3436$ mm²

Let us check this for development length at the point of contraflexure, so that the following equation is satisfied.

$$\frac{M_1}{V_u} + L_0 \ge L_d$$

Now $\quad x_u = \dfrac{0.87 f_y A_{st}}{0.36 f_{ck}\, b_f} = \dfrac{0.87 \times 415 \times 3436}{0.36 \times 20 \times 1122} = 154$ mm

$\therefore \quad M_1 = 0.87 f_y A_{st}\, (d - 0.416\, x_u) = 0.87 \times 415 \times 3436\, (800 - 0.416 \times 154)$

$= 913 \times 10^6$ N-mm.

$V_u =$ S.F. F_{4u} at the point of contraflexure $= 1.5 \times 533.8$ kN $= 800.7 \times 10^3$ N

$L_0 = 12\,\varphi$ or d, whichever is more $= 800$ mm

$L_d = 47\,\varphi = 47 \times 25 = 1175$ mm

$$\therefore \quad \frac{M_1}{V_u} + L_0 = \frac{913 \times 10^6}{800.7 \times 10^3} + 800 = 1140 + 800 = 1940 \text{ mm} >> L_d. \text{ Hence safe.}$$

Continue these 7 bars upto a point distant L_0 (= 800 mm) beyond the point of contraflexure or to a point distant 800 – 300 – 200 = 300 mm from the outer face of the column *B*. Similarly, at the other side, continue these bars upto the outer face of column *A*. These bars are to be provided at the upper face of the beam, in two layers. Beyond the above two points, discontinue all bars except 3 bars which may be continued in the cantilever portions of the beam to serve as anchor bars.

The reinforcement in the cantilever portion near column *B* is to be found for B.M. $M_{3u} = 1.5 \times 117$ kN-m. Here, the beam will act as simple under-reinforced rectangular beam, for which :

$$A_{st3} = \frac{0.5 \times 20}{415}\left[1 - \sqrt{1 - \frac{4.6 \times 1.5 \times 117 \times 10^6}{20 \times 400\,(800)^2}}\right] \times 400 \times 800 = 634 \text{ mm}^2$$

$$A_{st\,,\,min.} = \frac{0.85\,bd}{f_y} = \frac{0.85 \times 400\,(800)}{415} = 656 \text{ mm}^2$$

∴ No. of 12 mm φ bars = 656/113 ≏ 6. Actual $A_{st} = 6 \times 113 = 678$ mm²

Let us check this reinforcement for development length at the point of contraflexure near column B, so as to satisfy the equation $\frac{M_1}{V_u} + L_0 \geq L_d$

Here, $$x_u = \frac{0.87\,f_y\,A_{st}}{0.36\,f_{ck}\,b} = \frac{0.87 \times 415 \times 678}{0.36 \times 20 \times 400} ≏ 85 \text{ mm}$$

$M_1 = 0.87\,f_y\,A_{st}\,(d - 0.416\,x_u) = 0.87 \times 415 \times 678\,(800 - 0.416 \times 85) = 187.18 \times 10$ N-mm

$V_u =$ S.F. at the points of contraflexure $= F_{4u} = 1.5 \times 533.8 = 800.7$ kN

$L_0 = 12\,\varphi$ or d whichever is more $= 800$ mm

$$\therefore \quad \frac{M_1}{V_u} + L_0 = \frac{187.18 \times 10^6}{800.7 \times 10^3} + 800 = 1033 \text{ mm} >> L_d. \text{ Hence O.K.}$$

However, continue all the 6 bars upto a distance of L_0 (= 800 mm) from the point of contraflexure, or upto a point distant 100 + 800 = 900 mm from the inner face of column B. Beyond this point, discontinue all bars, except 3 bars. These bars are to provided at the lower face of the beam in two layers.

Similarly, the steel reinforcement for the cantilever portion to the left of column A can be determined for $M_{2u} = 1.5 \times 47$ kN-mm.

$$A_{st} = \frac{0.5 \times 20}{415}\left[1 - \sqrt{1 - \frac{4.6 \times 1.5 \times 47 \times 10^6}{20 \times 400\,(800)^2}}\right] 400 \times 800 = 249 \text{ mm}^2$$

$$A_{st\,,\,min} = \frac{0.85\,bd}{f_y} = \frac{0.85 \times 400 \times 800}{415} = 656 \text{ mm}^2 \quad \text{Hence } n - 12\,\varphi = 656/113 ≏ 6$$

Let us check this reinforcement for development length at the point of contraflexure near column A, so as to satisfy the equation $\frac{M_1}{V_u} + L_0 \geq L_d$. Actual $A_{st} = 6 \times 113 = 678$ mm²

$$x_u = \frac{0.87\,f_y\,A_{st}}{0.36\,f_{ck}\,b} = \frac{0.87 \times 415 \times 678}{0.36 \times 20 \times 400} ≏ 85 \text{ mm}$$

∴ $M_1 = 0.87\,f_y\,A_{st}(d - 0.416\,x_u) = 0.87 \times 415 \times 678(800 - 0.416 \times 85) = 181.18 \times 10^6$ N-mm

$V_u =$ S.F. $F_{2u} = 1.5 \times 441.8 = 662.7$ kN

$L_0 = 12\,\varphi$ or d whichever is greater $= 800$ mm

$L_d = 47\,\varphi = 47 \times 12 = 564$ mm

$$\therefore \quad \frac{M_1}{V_u} + L_0 = \frac{181.18 \times 10^6}{662.7 \times 1000} + 800 = 1082 \text{ mm} >> L_d.$$

Hence the criterion is satisfied. However, continue the 6 bars of 12 mm φ, upto a point distant L_0 (= 800 mm) beyond point of contraflexure, or upto a point distant

1150 + 800 = 1950 mm from the free edge of the beam. At this point, 2 bars may be discontinued, while the remaining 3 bars may be continued throughout the length, to serve an anchor bars.

4. *Shear reinforcement*. Refer Fig. 22.40. Diagonal tension crack will occur at F only if the zero sagging B.M. extends to d = 600 mm beyond the inner face of column A (or B). However, point of contraflexure falls earlier. Hence there is a possibility of crack at the point of contraflexure near column B where S.F. $F_4 = 533.8$ kN.

$$\therefore \quad \tau_v = \frac{1.5 \times 533.8 \times 10^3}{400 \times 800} = 2.5 \text{ N/mm}^2 < \tau_{c\,,\,max} = 2.8 \text{ N/mm}^2. \text{ Hence O.K;}$$

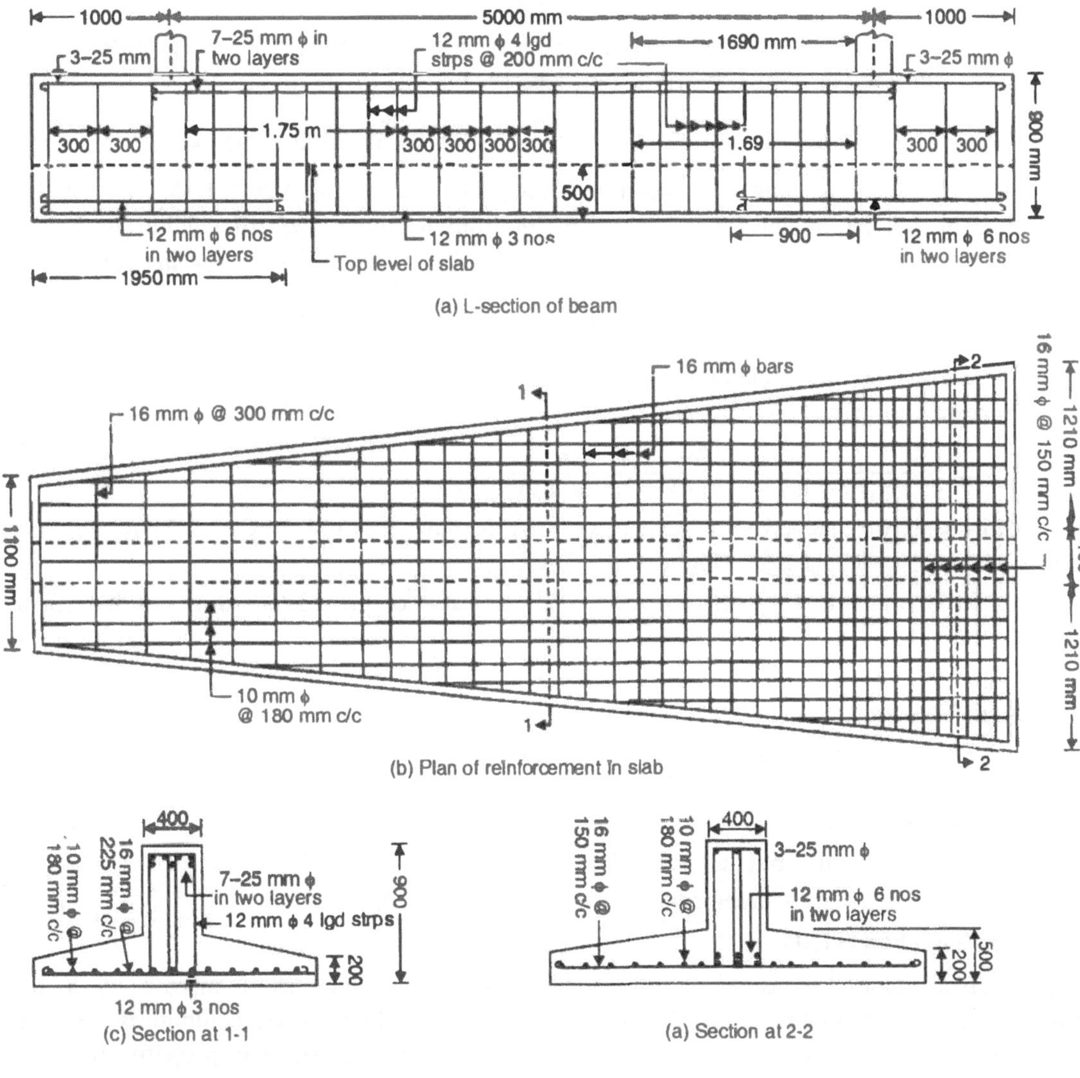

FIG. 22.50.

$$100 \frac{A_s}{bd} = 100 \times \frac{7 \times 491.8}{400 \times 800} \simeq 1.07\%$$

Hence from Table 7.1, $\tau_c \simeq 0.615$ N/mm^2. Thus $\tau_v > \tau_c$ and shear reinforcement is necessary. Hence crack will occur at the upper face of the beam if shear reinforcement is not provided.

$$V_{uc} = \tau_c . bd = 0.615 \times 400 \times 800 = 196800 \text{ N} = 196.8 \text{ kN}$$

$$V_{us} = 1.5 \times 533.8 - 196.8 = 603.9 \text{ kN}$$

Using 12 mm φ 4-lgd stirrups, $A_{sv} = 4 \times 113 = 452$ mm^2.

$$\therefore \qquad s_v = \frac{0.87 f_y . A_{sv} \, d}{V_{us}} = \frac{0.87 \times 415 \times 452 \times 800}{693.9 \times 1000} = 216 \text{ mm} \simeq 200 \text{ mm c/c (say).}$$

Distance upto which this is required (*i.e.* at a point where S.F. $= V_{uc} = 196.8$ kN) is obtained by putting $F_x = V_c = \pm 196.8/1.5$ in Eq. 2 of example 22.11.

Hence $F_x = \pm 196.8/1.5 = 600 - \frac{110}{2} x \, (2.22 + 0.245 \, x)$. Solving this, we get $x = 2.90$ m and 4.11 m. These two points are correspondingly at distances of 2.90 − 1.15 = 1.75 m from inner face of column *A* and 5.8 − 4.11 = 1.69 m from the inner face of column *B*. Between these two values of x_1, provide 8 mm φ 4-lgd stirrups at nominal spacing of

$$s_v = \frac{2.175 \, A_{sv} \, f_y}{b} = \frac{2.175 \times 201 \times 415}{400} = 453 \text{ mm, subject to a maximum of 300 mm.}$$

From the inner face of column *A* to a distance of 1.75 m, the spacing of 12 mm φ 4-lgd stirrups is given by

$$s_v = \frac{0.87 \, f_y \, A_{sv} . \, d}{1.5 \, (F_3 - V_c)} = \frac{0.87 \times 415 \times 452 \times 800}{1.5 \, (441.8 - 196.8) \, 1000} = 280 \text{ mm}$$

However, for the sake of symmetry, provide these @ 200 mm c/c. Thus, the shear reinforcement is summarised below :

1. From inner face of column *A* to 1.75 m distance, provide 12 mm φ 4-lgd stirrups @ 200 mm c/c

2. From inner face of column *B* to a distance of 1.66 mm, provide 12 mm φ 4-lgd stirrups @ 200 mm c/c.

3. For intermediate portion of beam, provide 8 mm φ 4-lgd stirrups @ 300 mm c/c.

4. For the cantilever portions, where S.F. is less than V_{cu} provide 8 mm φ 4-lgd stirrups @ 300 m c/c, but at least 2 stirrups are necessary.

The details of reinforcement etc. are shown in Fig. 22.50

22.14. STRAP FOOTING

A strap footing consists of spread footings of two columns connected by a *strap beam*. This type of footing is useful when the external column (wall column) is very near to the property line so that its footing cannot be spread beyond the property line. If a combined trapezoidal footing is provided, the bending moment and shear force in the footing are very high since the footing is continuous under both the columns. In the case of strap footing, the individual spread footings have very large area under the columns, resulting in decrease in the maximum bending moment and shear force. The beam connecting the two spread footings does not transfer any load to the soil. The individual footing areas are so arranged that the C.G. of the combined loads of the two columns pass though the combined C.G. of the two footing areas. Once this criterion is achieved, the pressure distribution below each individual footing will be uniform. The function of the strap beam is to transfer the load of heavily loaded outer column to the inner one. In doing so, the strap beam is subjected to bending moment and shear force, and it should be suitably designed to withstand these.

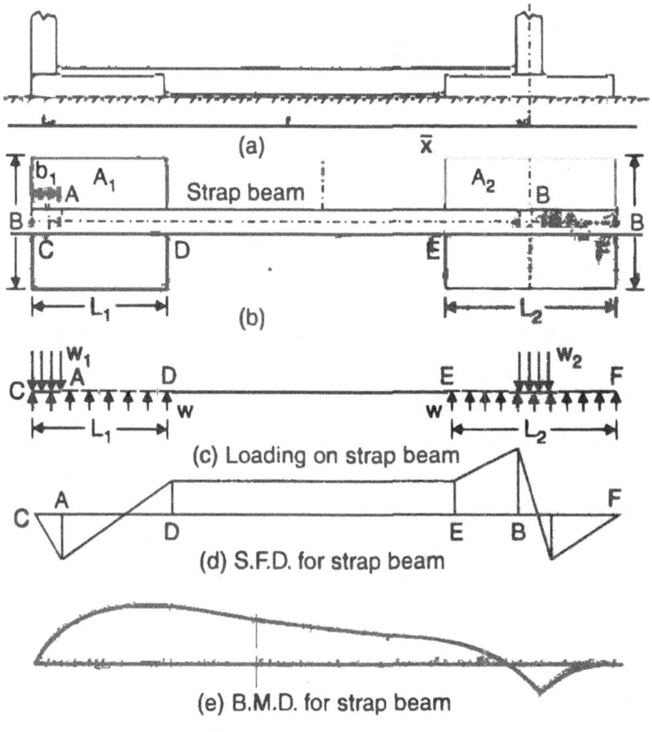

FIG. 22.51

Fig. 22.51 shows two columns A and B, transmitting axial loads W_1 and W_2 and are spaced l apart, centre to centre. Let W' be the total weight of both the individual footings. If A_1 and A_2 are the individual footing areas, and q_0 is the safe bearing capacity of soil, we have

$$A_1 + A_2 = \frac{W_1 + W_2 + W'}{q_0} \quad \text{or} \quad B(L_1 + L_2) = \frac{W_1 + W_2 + W'}{q_0} \qquad \ldots(1)$$

where B is the common width of each footings and L_1 and L_2 are the individual lengths of the footings. Length L_2 is arranged centrally under column B.

The C.G. of the resultant load $W = W_1 + W_2$, falls at $\bar{x}$ from the centre of column B, so that

$$\bar{x} = \frac{W_1\, l}{W_1 + W_2} \qquad \ldots(i)$$

Let $(b_1 \times b_1)$ and $(b_2 \times b_2)$ be the size of the columns A and B respectively. Taking moments of the footing areas about centre of column B, we get

$$\bar{x} = \frac{(B \times L_1)\left(l + \frac{1}{2} b_1 - \frac{L_1}{2}\right)}{B(L_1 + L_2)} \qquad \ldots(ii)$$

Equating (i) and (ii), we get

$$\frac{L_1\left(l + \frac{1}{2} b_1 - \frac{L_1}{2}\right)}{L_1 + L_2} = \frac{W_1\, l}{W_1 + W_2} \qquad \ldots(2)\ldots(22.52)$$

From (1) and (2), the unknowns L_1 and L_2 can be known in terms of any suitable value of B.

Net upward soil pressure $p_0 = \dfrac{W_1 + W_2}{B(L_1 + L_2)}$

Each individual footing is designed as cantilever slab, having sagging bending moment in each of the cantilever portion.

The strap beam transfers a part of load of footing A to footing B in such a way that C.G. of two loads coincide with the C.G. of the footing areas. In doing so, it is subjected to bending moment and shear force all along its length. Fig. 22.51 (*c*) shows the loading diagram on the strap beam *CDEF*.

The upward uniform load w per unit length, under CD and EF will be equal to $(p_0 \times B)$. The downward load w_1 per unit length in portion under column $A = W_1/b_1$ per unit length. Similarly downward load $w_2 = W_2/b_2$ per length. The characteristic bending moment diagram and S.F. diagram are sketched in Fig. 22.51 (*e*) and (*d*) respectively. For large portion of the strap beam, it will experience hogging bending moment, the maximum being at the section where S.F. is zero. For the cantilever portion beyond B, it will have sagging bending moment also, its maximum value being at the place where S.F. is zero.

Design Example 22.13. ***Design of strap footing*** *: Design a strap footing for two columns A and B, spaced 5 metres centre to centre. Column A, 300 mm × 300 mm in size carries a load of 600 kN and is on the property line. Column B, 400 mm × 400 mm in size, carries a load of 900 kN. The bearing capacity of soil is 120 kN/m². Use M 20 mix. and Fe 415 steel reinforcement.*

Solution.

1. *Design constants*

For M 20 mix and Fe 415 steel, we have the following:

$$x_{u\,.\,\max}/d = 0.479 \qquad \text{and} \qquad R_u = 2.761$$

2. *Size of footings*

Fig. 22.52 (*a*) shows the general arrangement of the footing.

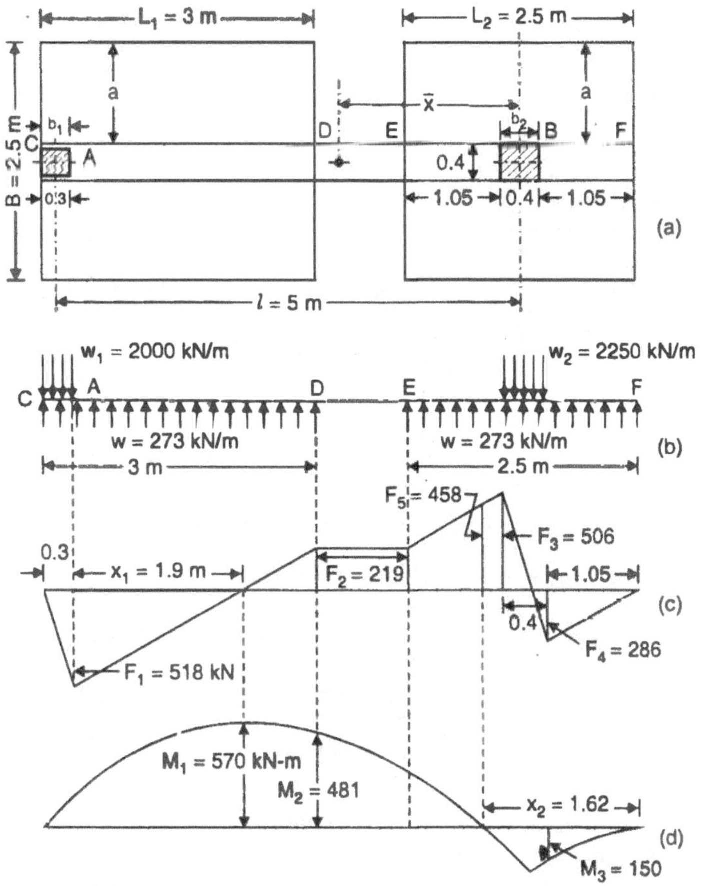

FIG. 22.52

Let the width of the two spread footings be B metres each. Length of footing under $A = L_1$ and length of footing under B be L_2 centrally arranged under B.

$$W_1 = 600 \text{ kN} \; ; \; W_2 = 900 \text{ kN}.$$

Weight of footing, $W' = 0.1\,(600 + 900) = 150$ kN. Hence,

$$B(L_1 + L_2) = \frac{600 + 900 + 150}{120} = 13.75 \text{ m}^2$$

or

$$L_1 + L_2 = \frac{13.75}{B} \qquad \ldots(1)$$

Let $\bar{x}$ = distance of C.G. of loads from centre of column B,

$$\therefore \quad \bar{x} = \frac{W_1 l}{W_1 + W_2} = \frac{600 \times 5}{600 + 900} = 2 \text{ m}.$$

If $\bar{x}$ is also the distance of C.G. of areas, from the centre of column B, we have

$$\bar{x} = \frac{B \times L_1\left(l + \frac{b_1}{2} - \frac{L_1}{2}\right)}{B(L_1 + L_2)}$$

Substituting the values of $\bar{x}$, b_1 and l_1 we get, $2 = \frac{L_1(5 + 0.15 - 0.5\, L_1)}{(L_1 + L_2)}$ ...(2)

Substituting the value of $(L_1 + L_2)$ from (1), and choosing B=2.5 m, we get

$$\frac{L_1(5.15 - 0.5\, L_1)}{13.75/2.5} = 2$$

or $L_1^{\ 2} - 10.3L_1 + 22 = 0$. From which $L_1 = 3.025$ m $\simeq$ 3 m (say).

$\therefore \quad L_2 = \frac{13.75}{2.5} - 3.0 = 2.5$ m.

Net upward pressure $p_0 = \frac{600 + 900}{2.5\,(3 + 2.5)} = 109$ kN/m^2

3. ***Design of footing slab.*** Let the width of strap beam = 400 mm

$\therefore$ Cantilever projection beyond the beam is $a = \frac{1}{2}(2.5 - 0.4) = 1.05$ m

(*a*) ***Depth on the basis of bending compression***

Max. bending moment $M = p_0 \frac{a^2}{2} = 109 \frac{(1.05)^{\,2}}{2} = 60$ kN-m $= 60 \times 10^6$ N-mm

$\therefore \quad M_u = 1.5\, M = 1.5 \times 60 \times 10^6$ N-mm $= 90 \times 10^6$ N-mm.

$\therefore \quad d = \sqrt{\frac{90 \times 10^6}{2.761 \times 1000}} = 181$ mm.

Keeping cover to centre of steel = 60 mm, $D = 181 + 60 = 241$ mm. However, keep D = 300 mm and d = 240 mm, so that effective cover available = 60 mm. Reduce to total depth to 200 mm at the edges, so that $d = 200 - 60 = 140$ mm there.

(*b*) ***Depth on the basis of one way shear***

The critical plane of one way shear is located at a distance d (mm) from the face of the beam, and its distance from the edge $= a' = a - 0.001\, d$ metres.

$$\therefore \quad V_u = 1.5\, p_0\, a' = 1.5\, p_0\,(a - 0.001\, d) = 1.5 \times 109 \times 10^3\,(1.05 - 0.001\, d)$$

$$= 163.5 \times 10^3\,(1.05 - 0.001\, d) \text{ N/m}$$

Effective depth at that plane $= d' = 140 + \frac{d - 140}{1050}(1050 - d)$ mm

$$\therefore \quad \tau_v = \frac{V_u}{bd'} = \frac{163.5 \times 10^3\,(1.05 - 0.001\, d)}{1000\left[140 + \frac{d - 140}{1050}(1050 - d)\right]} = \frac{171675\,(1.05 - 0.001\, d)}{147000 + (d - 140)(1050 - d)}$$

Permissible shear stress $= k\,\tau_c$ where $k = 1$ (for $D \geq 300$ mm) and $\tau_c = 0.384$ N/mm^2 for under-reinforced section having $p \simeq 0.3\%$ (Table 7.1)

$$\therefore \quad \frac{171675\,(1.05 - 0.001\, d)}{147000 + (d - 140)(1050 - d)} = 0.384$$

or $d^2 - 1637\,d + 469424 = 0$, from which $d = 371$ mm

However, keep $D = 450$ mm so that available $d = 450 - 60 = 390$ mm

Reduce D to 200 mm at the ends so that available $d = 200 - 60 = 140$ mm

Determination of reinforcement

$$A_{st} = \frac{0.5\,f_{ck}}{f_y}\left[1 - \sqrt{1 - \frac{4.6\,M_u}{f_{ck}\,bd^2}}\right]bd = \frac{0.5 \times 20}{415}\left[1 - \sqrt{1 - \frac{4.6 \times 90 \times 10^6}{20 \times 1000\,(390)^2}}\right] \times 1000 \times 390$$

$$= 663 \text{ mm}^2$$

$$\text{Spacing of 12 mm } \varphi \text{ bars} = \frac{1000 \times 113}{663} \simeq 170 \text{ mm c/c.}$$

However, provide these @100 mm c/c so that $A_{st} = 1000 \times 113/100 = 1130$ mm^2, so that $p = 1130 \times 100/1000 \times 390 \simeq 0.29\%$ which is quite close to the assumed value of $p = 0.3\%$.

$$\text{Distribution reinforcement} = \frac{0.12}{100} \times 1000\frac{(450 + 200)}{2} = 390 \text{ mm}^2.$$

$$\therefore \text{ Spacing of 10 mm } \varphi \text{ bars} = \frac{1000 \times 78.5}{390} \approx 200 \text{ mm c/c.}$$

4. *Bending moment and shear force diagrams for strap beam*. Upward load w per metre run, on the strap beam is given by $w = p_0 B = 109 \times 2.5 = 273$ kN/m.

$$\text{Downward load of column } A = w_1 = \frac{600}{0.3} = 2000 \text{ kN/m.}$$

$$\text{Downward load of column } B = w_2 = \frac{900}{0.4} = 2250 \text{ kN/m.}$$

The loading diagram on the strap beam is shown in Fig. 22.52(*b*).

The shear force diagram for the strap beam is shown in Fig. 22.52(*c*).

S.F. at inner face of column $A = F_1 = (2000 \times 0.3) - (273 \times 0.3) = 518$ KN.

S.F. at edge $D = F_2 = 273 \times 3 - 600 = 219$ kN.

S.F. at edge $E = F_2 = 219$ kN.

S.F. at inner edge of column $B = F_3 = 219 + 273\left\{\frac{1}{2}(2.5 - 0.4)\right\} = 506$ KN.

S.F. at outer edge of column $B = F_4 = 900 - 506 - (273 \times 0.4) = 286$ KN.

or $$= 273 \times \left\{\frac{1}{2}(2.5 - 0.4)\right\} = 286 \text{ kN}$$

In the range CD, shear force is zero at distance x_1 from the inner face of column A, its value being given by

$$x_1 = 518/273 = 1.9 \text{ m.}$$

Hogging bending moment will be maximum at this section, its values being given by

$$M_1 = (2000 \times 0.3)(1.9 + 0.15) - 273\frac{(0.3 + 1.9)^2}{2} = 570 \text{ kN-m.}$$

Hogging B.M. at edge D is given by

$$M_2 = (2000 \times 0.3)(3 - 0.15) - \frac{273(3)^2}{2} = 481 \text{ kN-m}$$

Sagging B.M. at the outer face of column B is given by

$$M_3 = \frac{273}{2}\left[\frac{1}{2}(2.5 - 0.4)\right]^2 = 150 \text{ kN-m}$$

Let the point of contraflexure occur at x_2 from point F.

$$\therefore \quad M_x = 0 = 273\left(\frac{x_2^2}{2}\right) - (2250 \times 0.4)(x_2 - 1.25), \text{ which gives } x_2 = 1.62 \text{ m}$$

Shear force at the point of contraflexure is given by

$$F_5 = (2250 \times 0.4) - (273 \times 1.62) = 458 \text{ kN}$$

Depth of strap beam : (a) Depth on the basis of bending compression

Let the width of strap beam be 400 mm. Since it projects above the footings, T-beam action will be available in range CD. Maximum bending moment is $M_1 = 570$ kN-m. However, since T-beam action is available here, the critical section will be at D, where the beam acts as rectangular beam. Bending moment at $D = M_2 = 481$ kN-m $= 481 \times 10^6$ N–mm. Hence $M_{2u} = 1.5 \times 481 \times 10^6 = 721.5 \times 10^6$ N-mm

$$\therefore \quad d = \sqrt{\frac{M_{2u}}{R_u\, b}} = \sqrt{\frac{721.5 \times 10^6}{2.761 \times 400}} \simeq 810 \text{ mm.}$$

(b) Depth on the basis of diagonal tension

$$\tau_v = \frac{V_{u,\max}}{bd} = \frac{1.5 \times 518 \times 10^3}{400\, d} = \frac{1942.5}{d} \text{ N/mm}^2$$

From Table 7.3, $\tau_{c,\max} = 2.8$ N/mm^2 for M 20 concrete.

$\therefore \; d = \dfrac{1942.5}{2.8} = 694$ mm, which is less than the one found on the basis of bending compression. Using 20 mm φ bars in 2 layers with 20 mm space between them and providing 10 mm φ strirrups with 50 mm nominal cover, effective cover = 50 + 10 + 20 + 20/2 = 90 mm. Hence $D = 810 + 90 = 900$. However, keep $D = 1000$ mm, so that $d = 1000 - 90 = 910$ mm, so as to obtain an under-reinforced section.

For the portion CD, the beam acts as T-beam, with $d = 910$ mm.

Average thickness of slab $= \frac{1}{2}(150 + 200) = 320$ mm $= D_f$

$$\text{Now} \qquad b_f = \frac{l_0}{(l_0/b) + 4} + b_w$$

where l_0 = distance between points of zero moment in beam

$= (5 + 0.15 + 1.25) - 1.62 = 4.78$ m $= 4780$ mm

b = actual width of flange $= 2.5 = 2500$ mm; b_w = width of web $= 400$ mm

$$\therefore \quad b_f = \frac{4780}{(4780/2500) + 4} + 400 = 1208 \text{ mm}$$

6. *Reinforcement in strap beam*

(c) For hogging B.M. $M_{1u} = 1.5 \times 570 \times 10^6 = 855 \times 10^6$ **N-mm** $= M_{uD}$

The beam will act as T-beam for the length CD.

Let $x_u = D_f = 325$ mm.

$\therefore$ $M_{u1} = 0.36\, f_{ck}\, b_f\, D_f\, (d - 0.416\, D_f) = 0.36 \times 20 \times 1208 \times 320\, (910 - 0.416 \times 325)$

$= 2156 \times 10^6$ N/mm

Since $M_{u1} >> M_{uD}$ ($= 855 \times 10^6$), N.A. fall inside the flange, i.e. $x_u \le D_f$

$\therefore$ $M_{uD} = 0.87\, f_y\, A_{st}\, d\left[1 - \frac{f_y}{f_{ck}}\frac{A_{st}}{b_f\, d}\right]$, from which we obtain

$$A_{st1} = \frac{0.5\, f_{ck}}{f_y}\left[1 - \sqrt{1 - \frac{4.6\, M_{uD}}{f_{ck}\, b_f\, d^2}}\right] b_f\, d = \frac{0.5 \times 20}{415}\left[1 - \sqrt{1 - \frac{4.6 \times 855 \times 10^6}{20 \times 1208\, (910)^2}}\right] \times 1208 \times 910$$

$= 2746$ mm^2

(*b*) ***For hogging B.M.*** $\mathbf{M_{2u} = 1.5 \times 481 \times 10^6 = 721.5 \times 10^6}$ **N-mm**

The beam will act as rectangular beam, since slab is not available

$$A_{st2} = \frac{0.5\, f_{ck}}{f_y}\left[1 - \sqrt{1 - \frac{4.6\, M_{2u}}{f_{ck}\, bd^2}}\right] bd = \frac{0.5 \times 20}{415}\left[1 - \sqrt{1 - \frac{4.6 \times 721.5 \times 10^6}{20 \times 400\, (910)^2}}\right] \times 400 \times 910$$

$= 2576$ mm^2

Hence provide reinforcement equal to 2746 mm^2

$\therefore$ Number of 20 mm φ bars $= 2746/314 = 8.74 \simeq 9$. Actual $A_{st} = 9 \times 314 = 2826$ mm^2

Hence provide these in two layers at the upper face of the beam.

Let us check the reinforcement for development of stresses, at point of contraflexure, so as to satisfy the criterion :

$$\frac{M_1}{V_u} + L_0 \ge L_d.$$

Assuming that the 9 bars, provided at the top face of the beam, are *available* at the point of contraflexure, we have

$$x_u = \frac{0.87\, f_y\, A_{st}}{0.36\, f_{ck}\, b} = \frac{0.87 \times 415 \times 2826}{0.36 \times 20 \times 400} = 354 \text{ mm}$$

$\therefore$ $M_1 = 0.87\, f_y\, A_{st}\, (d - 0.416\, x_u) = 0.87 \times 415 \times 2826\, (910 - 0.416 \times 354)$

$= 778.24 \times 10^6$ N-mm

V_u = S.F. at point of contraflexure $= F_{5u} = 1.5 \times 458 = 687$ kN

$L_0 = 12\, \varphi$ or d, whichever is more $= 910$ mm

$L_d = 47\, \varphi = 47 \times 20 = 940$ mm

$$\therefore \quad \frac{M_1}{V_u} + L_0 = \frac{778.24 \times 10^6}{687 \times 10^3} + 910 = 2042 \text{ mm} > L_d.$$

This suggests that it is not necessary to take all the 9 bars upto the point of contraflexure. However, we will continue all the bars upto a distance of L_0 (= 910 mm) from the point of contraflexure, or to a distance of 1620 − 910 = 710 mm from the end of the beam. After that, discontinue all the bars, except 3 bars which will serve as anchor bars for shear stirrups.

(c) ***For maximum sagging B.M.*** $\mathbf{M_{3u} = 1.5 \times 150 \times 10^6 = 225 \times 10^6}$ **N-mm.**

This occurs at the outer face of column B. Providing effective cover of 70 mm available $d = 1000 - 70 = 930$ mm.

$$A_{st3} = \frac{0.5\, f_{ck}}{f_y}\left[1 - \sqrt{1 - \frac{4.6\, M_{3u}}{f_{ck}\, bd^2}}\right] bd = \frac{0.5 \times 20}{415}\left[1 - \sqrt{1 - \frac{4.6 \times 225 \times 10^6}{20 \times 400\,(930)^2}}\right] \times 400 \times 930$$

$= 698 \text{ mm}^2$

$$A_{st,\,\min} = \frac{0.85\, bd}{f_y} = \frac{0.85 \times 400 \times 930}{415} = 762 \text{ mm}^2$$

$\therefore$ No. of 16 mm φ bars $= 762/201 = 3.79 \simeq 4$.

Hence actual $A_{st} = 4 \times 201 = 804 \text{ mm}^2$

Let us check whether this reinforcement is sufficient from the point of view of development of stress at the point of contraflexure so as to satisfy the criterion $\frac{M_1}{V_u} + L_0 \geq L_d$.

$$x_u = \frac{0.87\, f_y\, A_{st}}{0.36\, f_{ck}\, b} = \frac{0.87 \times 415 \times 804}{0.36 \times 20 \times 400} = 100.8 \text{ mm}$$

$\therefore$ $M_1 = 0.87\, f_y\, A_{st}\,(d - 0.416\, x_u) = 0.87 \times 415 \times 804\,(930 - 0.416 \times 100.8)$

$= 257.79 \times 10^6$ N-mm.

V_u = S.F. at the point of contraflexure $= 1.5 \times 458 \times 10^3 = 687 \times 10^3$ N

$L_0 = 12\,\varphi$ or d whichever is more $= 930$ mm

$L_d = 47\,\varphi = 47 \times 16 = 752$ mm.

$$\therefore \quad \frac{M_1}{V_u} + L_0 = \frac{257.79 \times 10^6}{687 \times 10^3} + 930 = 1305 \text{ mm}$$

which is more than L_d, and hence safe.

Hence provide 4-16 mm φ bars at the lower face, and continue these throughout the length.

7. ***Reinforcement for diagonal tension***

$$100\,\frac{A_s}{bd} = \frac{100 \times 9 \times 314}{400 \times 910} = 0.776\,\%.$$ Hence from Table 7.1, $\tau_c \simeq 0.565 \text{ N/mm}^2$.

Hence shear reinforcement is necessary.

$\therefore$ Shear resistance of concrete, $V_{uc} = 0.565 \times 400 \times 910 \times 10^{-3}$ N $= 205.66$ kN.

Hence shear reinforcement is necessary wherever shear force exceeds 205.66 kN At the inner face of column A, shear force $F_{1u} = 1.5 \times 518$ kN $= 777$ kN

$$\therefore \quad V_{us} = F_{1u} - V_{uc} = 777 - 205.66 = 571.34 \text{ kN}.$$

Using 10 mm φ 4-lgd stirrups, $A_{sv} = 4 \times \frac{\pi}{4}(10)^2 = 314 \text{ mm}^2$

$$\therefore \text{ Spacing} \quad s_v = \frac{0.87\, f_y\,.\, A_{sv}\,.\, d}{V_{us}} = \frac{0.87 \times 415 \times 314 \times 910}{571.34 \times 1000} = 180 \text{ mm}.$$

Distance of point where shear force is 205.66 kN, from the inner face of column $= \frac{518 - 205.66/1.5}{273} \simeq 1.4$ m. Hence this distance from end of beam $= 1.4 + 0.3 = 1.7$ m.

Hence provide 10 mm φ 4-lgd stirrups @ 180 mm c/c from end of the beam, upto a distance of 1.7 m.

At the inner face of column B, shear force $F_{3u} = 1.5 \times 506$ kN. Hence spacing of 10 mm φ 4-lgd stirrups is given by

$$s_v = \frac{0.87 \times 415 \times 314 \times 910}{(1.5 \times 506 - 205.66)\, 1000} = 186 \text{ mm}$$

However provide these stirrups @ 180 mm c/c from point E to the inner face of the column. Similarly, at the outer face of column B, shear force $F_{4u} = 1.5 \times 286 = 429$ kN.

$$s_v = \frac{0.87 \times 415 \times 314 \times 190}{(429 - 205.66)\, 1000} = 460 \text{ mm}$$

Subject to a maximum of 300 mm.

Maximum spacing of stirrups is given by

$$s_v = \frac{2.175\, A_{sv} f_y}{b} = \frac{2.175 \times 314 \times 415}{400} = 708 \text{ mm subject to a maximum of 300 mm.}$$

Hence keep $s_v = 300$ mm from the outer face of the column to the end of the beam.

At the point D, shear force $F_{2u} = 1.5 \times 219 = 328.5$ kN.

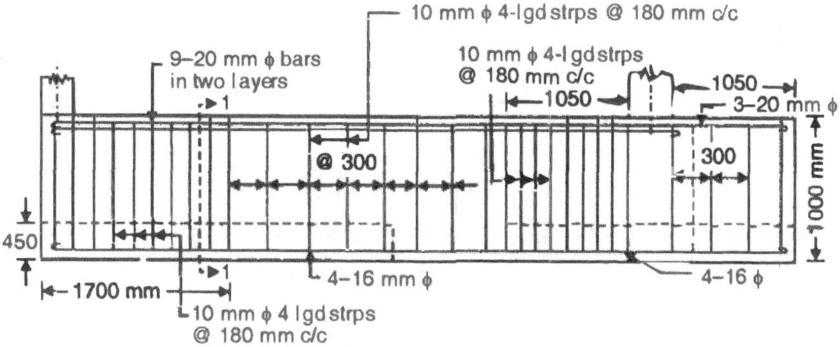

(a) L-section at strap beam

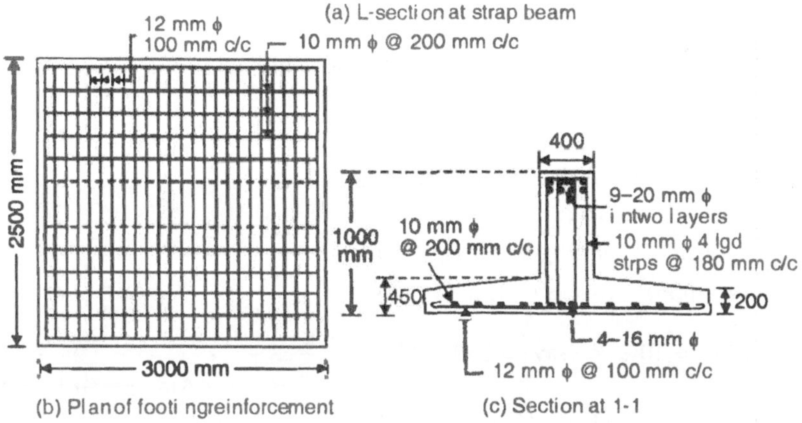

(b) Plan of footing reinforcement (c) Section at 1-1

FIG. 22.53.

$$s_v = \frac{0.87 \times 415 \times 314 \times 910}{(328.5 - 205.66)\,1000} = 840 \text{ mm, subject to a maximum of 300 mm.}$$

Hence for the remaining portion, provide stirrups of spacing of 300 mm c/c.

8. *Check for punching shear*. The beam can be checked against punching shear at columns *A* and *B*, as explained in Example 22.10. The section will be found to be safe at both the locations.

22.15. RAFT FOOTING

A raft or mat is a combined footing that covers the entire area beneath a structure and supports all the walls and columns. When the allowable soil pressure is low, or the building loads are heavy, the use of spread footings would cover more than one-half of the area and it may prove more economical to use mat or raft foundation. They are also used where the soil mass contains compressible lenses or the soil is sufficiently erratic so that the differential settlement would be difficult to control. The mat or raft tends to bridge over the erratic deposits and eliminates the differential settlement. Raft foundation is also used to reduce settlement above highly compressible soil, by making the weight of structure and raft approximately equal to the weight of the soil excavated.

Ordinarily, rafts are designed as inverted beam slab system. If the C.G. of loads coincide with the centroid of the raft, the upward load is regarded as a uniform pressure equal to the downward load divided by the area of the raft. The weight of raft is not considered in the structural design because it is assumed to be carried directly by the sub-soil. When the column loads are equal, flat slab construction can be adopted. When raft foundation is also subjected to wind load moments, and when these are considered in design, the safe stresses can be increased by $33\frac{1}{3}$ %. However, when the wind load reactions are less than $\frac{1}{3}$rd reactions due to other loads, wind is not considered in the design. Example 22.14 illustrates the design procedure.

Design Example 22.14. ***Design of raft foundation*** : *Fig. 22.54 shows the layout of the columns of a building. The outer columns are 300 × 300 mm in size and carry a load of 500 kN each. The inner columns are 400 × 400 mm in size and carry a load 800 kN each. In addition to this, each column carries a moment of 160 kN-m due to wind load on the length of the building. Design a suitable raft foundation, if the bearing capacity of soil is 100 kN/m². Use M 20 concrete and Fe 415 steel.*

Solution.

1. *Design constants*. For M 20 concrete and Fe 415 steel we have the following :
$x_{u\,.\,max}/d = 0.479$ and $R_u = 2.761$

2. *Size and general arrangement of mat*

Total weight $W = (4 \times 500) + (2 \times 800) = 3600$ kN

Self-weight of raft $W' = 10\%\ W$ (say) $= 360$ kN

∴ Total load $= 3600 + 360 = 3960$ kN

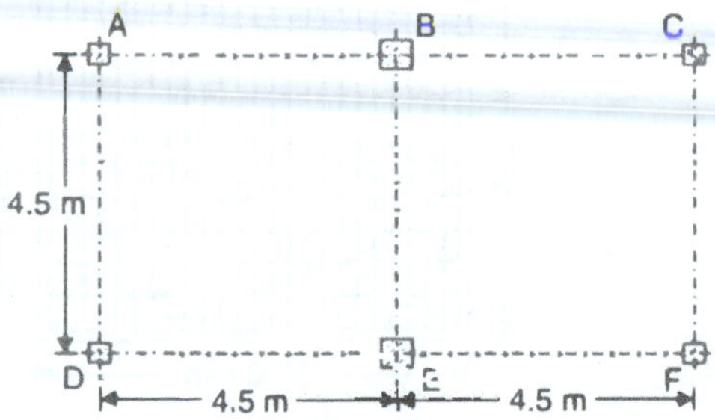

FIG. 22.54

$$\text{Total moment} = 160 \times 6 = 960 \text{ kN-m}$$

$$\therefore \quad \text{Eccentricity } e = \frac{960}{3960} = 0.242 \text{ m}$$

Let the raft project 500 mm outside the column all round, so that length of raft $= (6 \times 1.5 + 0.3 + 2 \times 0.5) = 10.3$ m, and width of raft $= (4.5 + 0.3 + 2 \times 0.5) = 5.8$ m.

Hence maximum and minimum intensities of pressures would be

$$= \frac{\Sigma W}{A}\left(1 \pm \frac{6e}{B}\right) = \frac{3960}{10.3 \times 5.8}\left(1 \pm \frac{6 \times 0.242}{5.8}\right) \simeq 83 \text{ N/m}^2 \text{ and } 5 \text{ N/m}^2$$

Since the maximum pressure is less than 100 kN/m^2, the size of the footing is satisfactory.

$$\text{Net upward soil reaction } p_0 = \frac{3600}{10.3 \times 5.8} \approx 60 \text{ kN/m}^2$$

$$\text{Upward reaction due to wind moment is given by } p_w = \pm \frac{960}{\frac{1}{6} \times 10.3\,(5.8)^2} = \pm 16.6 \text{ kN/m}^2$$

i.e., +16.6 kN/m^2 on leeward side and −16.6 kN/m^2 on windward side.

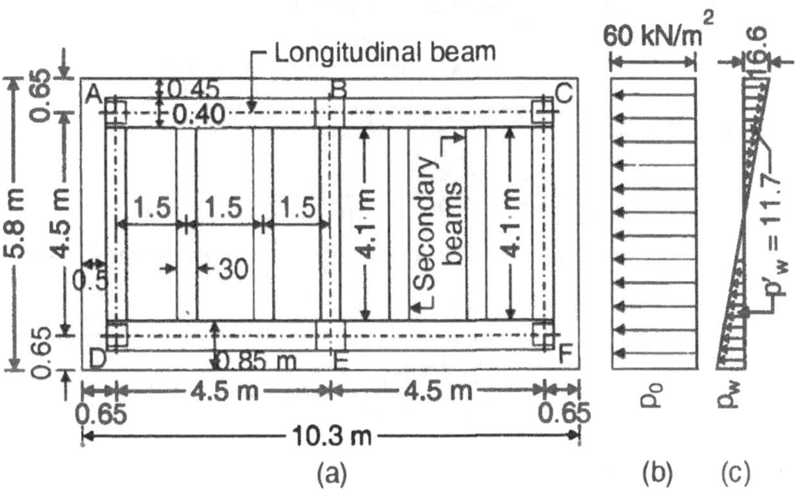

FIG. 22.55.

$$\text{Pressure intensity at inner face of longitudinal beams} = \pm 16.6 \times \frac{4.1}{5.8} = \pm 11.7 \text{ kN/m}^2$$

Let us provide two longitudinal beams in direction *AC* and *DF*, joining three columns on each side. These two beams are joined by seven secondary beams. The general arrangement of beams etc. are shown in Fig. 22.55.

3. *Design of slab*. The soil pressure distribution due to load and wind moment are shown in Fig. 22.55 (*b*) and (*c*) respectively. Since the pressure p_w due to wind moments is less than $\frac{1}{3}$rd the pressure p_0 due to dead load, the effect of wind will not be considered in the design of slab.

The slab is to be designed for the greater of the following two moments :

(*a*) Bending moment in the cantilever portion of slab.
(*b*) Bending moment in the continuous slab.

Cantilever span = 0.5 m = a

$$M_1 = p_0 \frac{a^2}{2} = \frac{60}{2}(0.5)^2 = 7.5 \text{ kN-m}; \quad M_{1u} = 1.5 \times 7.5 = 11.25 \text{ kN-m} \quad \ldots(i)$$

Shear force $F_1 = p_0\, a = 60 \times 0.5 = 30$ kN ; $F_{1u} = 1.5 \times 30 = 45$ kN ...(*ii*)

The slab is continuous over the secondary beams. Centre to centre spacing of secondary beams = 1.5 m. Taking width of secondary beams as 0.3 m, the clear span of slab, between two beams = 1.5 − 0.3 = 1.2 m. Hence bending moment in interior spans is

$$M_2 = \frac{p_0 L^2}{12} = \frac{60(1.2)^2}{12} = 7.2 \text{ kN-m} \; ; \; M_{2u} = 1.5 \times 7.2 = 10.8 \text{ kN-m} \quad \ldots(iii)$$

$$\text{S.F.} \quad F_2 = p_0 \frac{L}{2} = \frac{60(1.2)}{12} = 36 \text{ kN}; \; F_{2u} = 1.5 \times 36 = 54 \text{ kN} \quad \ldots(iv)$$

Hence design B.M. $M_{uD} = 11.25$ kN-m $= 11.25 \times 10^6$ N-mm

Design shear force $F_{uD} = 54$ kN.

(*a*) ***Depth on the basis of bending compression***

$$d = \sqrt{\frac{M_{uD}}{R_u \, . \, b}} = \sqrt{\frac{11.25 \times 10^6}{2.761 \times 1000}} = 64 \text{ mm}$$

(*b*) ***Depth on the basis of one way shear***

$$\tau_v = \frac{V_u}{bd} = \frac{F_{2u}}{bd} = \frac{54 \times 10^3}{1000 \times d} = \frac{54}{d} \text{ N/mm}^2$$

Assuming $p = 0.3\%$ for under-reinforced section, $\tau_c = 0.384$ N/mm². Also, $k \simeq 1.29$ for $D = 160$ mm.

∴ Permissible value of strear stress $= k\,\tau_c = 1.29 \times 0.384 = 0.495$ N/mm²

Hence from (*i*), we get $d = \dfrac{54}{k\,\tau_c} = \dfrac{54}{0.495} = 109 \text{ mm} \simeq 110 \text{ mm}$

Providing 60 mm effective cover, $D = 110 + 60 = 170$ mm.

Main Reinforcement : For an under-reinforced section,

$$A_{st} = \frac{0.5 f_{ck}}{f_y}\left[1 - \sqrt{1 - \frac{4.6\, M_{uD}}{f_{ck}\, bd^2}}\right] bd = \frac{0.5 \times 20}{415}\left[1 - \sqrt{1 - \frac{4.6 \times 11.25 \times 10^6}{20 \times 1000\,(110)^2}}\right] \times 1000 \times 110$$

$= 301$ mm²

However, A_{st} corresponding to $p = 0.3\%$ (assumed for choosing τ_c) is given by

$$A_{st} = \frac{pbd}{100} = \frac{0.3 \times 1000 \times 110}{100} = 330 \text{ mm}^2$$

$$\therefore \text{ Spacing of 10 mm } \varphi \text{ bars} = \frac{1000 \times 78.5}{330} = 237 \text{ mm.}$$

However, provide these @ 200 mm c/c.

$$\text{Actual } A_{st} = \frac{1000 \times 78.5}{200} = 392.5 \text{ mm}^2 \text{ and } p = \frac{392.5 \times 100}{1000 \times 110} = 0.36\%$$

Let us check the reinforcement for development length.

L_d required $= 47\ \varphi = 47 \times 10 = 470$ mm.

Providing a side cover of 50 mm, available length of each bar $= 500 - 50 = 450$ mm only .

However actual stress in bars $= \dfrac{A_{st}\ required}{A_{st}\ provided} \times f_{yd} = \dfrac{330}{392.5} f_{yd} = 0.84\ f_{yd}$

$\therefore$ Actual required $L_d = 0.84\ (47\ \varphi) = 0.84 \times 47 \times 10 = 395$ mm < 450 mm. Hence OK.

For cantilever portion, and portion under the face of the secondary beams, this reinforcement is to be provided at the bottom face of the slab since bending moment is of sagging nature, while for intermediate portion between secondary beams, the bars are to be provided on the top face, as is done in a continuous slab.

Area of transverse reinforcement $= \dfrac{0.12}{100}(1000 \times 170) = 204\ \text{mm}^2$

Using 10 mm φ bars, spacing $= \dfrac{1000 \times 78.5}{204} = 384$ mm $\simeq 380$ mm

For the portion of slab projecting from the face of the main beam, cantilever projection $= 0.45$ m.

$$\therefore \qquad M = \frac{60(0.45)^2}{2} = 6\ \text{kN-m} \ ; \ M_u = 1.5 \times 6 \times 10^6 = 9 \times 10^6\ \text{N-mm}$$

Now, actual length of each bar in the cantilever portion, using 50 mm side cover $= 450 - 50 = 400$ mm. Required $L_d = 47 \times 10 = 470$ mm.

Hence actual σ_{st} carried by each bar $= \dfrac{400}{470} \times f_y = 0.85\ f_y$ N/mm^2

Effective depth available $= 110 - 10 = 100$ mm

$$A_{st} = \frac{0.5\ f_{ck}}{0.85\ f_y}\left[1 - \sqrt{1 - \frac{4.6\ M_u}{f_{ck}\ bd^2}}\right] bd = \frac{0.5 \times 20}{0.85 \times 415}\left[1 - \sqrt{1 - \frac{4.6 \times 9 \times 10^6}{20 \times 1000\ (100)^2}}\right] \times 1000 \times 100$$

$= 311\ \text{mm}^2$

$\therefore$ Spacing of 10 mm φ bars $= \dfrac{1000 \times 78.5}{311} = 252$ mm

However for the sake of symmetry, provide these also at 200 mm c/c.

Also, the spacing of transverse reinforcement (found as 408 mm c/c) may be kept at $2 \times 200 = 400$ mm c/c, so that every alternate bar of cantilever reinforcement may be used as distribution bars.

4. ***Design of intermediate secondary beam.*** There are in all seven secondary beams spaced at 1.5 m c/c out which the two outer secondary beams and the central secondary beams, join columns. The remaining four secondary beams are intermediate beams which join the longitudinal beams. These intermediate beams are subjected to reaction from the slab beneath. Since the reaction due to wind moment is less than $\frac{1}{3}$rd the dead load reactions, it will not be considered in design. Span of the beam $= 4.5$ m.

However, the load on the beam will be the upward load acting on the slab on the clear span of 4.1 m. For the remaining poriton of the slab, load will be transferred directly to the longitudinal beams.

Load per metre run $w = p_0 \times 1.5 = 60 \times 1.5 = 90$ kN/m

The loading is shown in Fig. 22.56. End reactions transferred to each longitudinal beam

$$= R = \frac{1}{2}(4.1 \times 90) = 184.5 \text{ kN}$$

$$M = 184.5 \times \frac{4.5}{2} - \frac{90}{2}\left(\frac{4.1}{2}\right)^2 = 226 \text{ kN-m} = 226 \times 10^6 \text{ N-mm;}$$

$$M_u = 1.5 \times 226 \times 10^6 = 339 \times 10^6 \text{ N-mm}$$

$$F = R = 184.5 \text{ kN} ; F_u = 1.5 \times 184.5 = 276.75 \text{ kN}$$

Since the beam experiences hogging bending moment, it will act as T-beam. Let the width of beam $= b_w = 300$ mm. The width of flange is given by

$$b_f = \frac{l_o}{6} + b_w + 6D_f$$

where $l_0 = 4.5$ m $= 4500$ mm

$b_w = 300$ mm

$D_f = 160$ mm.

$$\therefore \quad b_f = \frac{4500}{6} + 300 + 6 \times 160 = 2010 \text{ mm}$$

subject to maximum of centre to centre spacing $= 1.5$ m $= 1500$ mm. Hence $b_f = 1500$ mm.

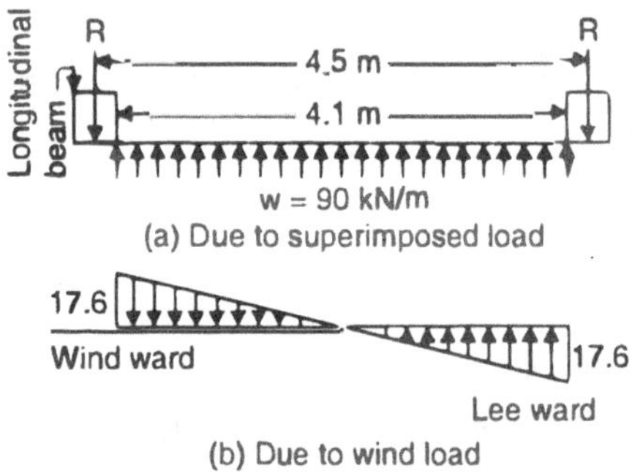

FIG. 22.56.

(*a*) ***Fixation of effective depth on the basis of B.M.***

Let $$d \simeq \frac{2}{3}\sqrt{\frac{M_u}{R_u b_w}} = \frac{2}{3}\sqrt{\frac{339 \times 10^6}{2.761 \times 300}} = 427 \text{ mm}$$

(*b*) ***Fixation of effective depth on the basis of*** $\tau_{c,max}$

$$d = \frac{F_u}{b_w \tau_{c,max}}, \text{ where } \tau_{c,max} = 2.8 \text{ N/mm}^2 \text{ for M 20 concrete}$$

$$\therefore \quad d = \frac{276.75 \times 1000}{300 \times 2.8} \simeq 330 \text{ mm}$$

Keeping nominal cover of 30 mm, using 25 mm φ bars in two layer with 25 mm space between these, and providing 10 mm φ stirrups, effective cover $= 50 + 10 + 25 + 25/2 = 97.5$ mm. However provide effective cover of 100 mm, so that keeping $D = 530$ mm, availalbe $d = 530 - 100 = 430$ mm.

Design of main reinforcement : Design moment $M_{uD} = 339 \times 10^6$ N-mm

Assume $x_u = D_f = 160$ mm

$$\therefore \quad M_{u1} = 0.36 f_{ck} b_f D_f (d - 0.416 D_f) = 0.36 \times 20 \times 1500 \times 160 (430 - 0.416 \times 160)$$

$$= 628 \times 10^6 \text{ N-mm}$$

Since $M_{u1} >> M_{uD}$, N.A. falls in the flange, i.e. $x_u < D_f$

$\therefore \quad M_u = M_{uD} = 0.87 f_y A_{st} d \left[1 - \frac{f_y}{f_{ck}} \frac{A_{st}}{b_f d} \right]$, from which

$$A_{st} = \frac{0.5 f_{ck}}{f_y} \left[1 - \sqrt{\frac{4.6 M_{uD}}{f_{ck} b_f d^2}} \right] b_f d = \frac{0.5 \times 20}{415} \left[1 - \sqrt{1 - \frac{4.6 \times 339 \times 10^6}{20 \times 1500 (430)^2}} \right] \times 1500 \times 430$$

$= 2365 \text{ mm}^2$

$\therefore$ Number of 20 mm φ bars $= 2365/314.1 \simeq 8$. Actual $A_{st} = 8 \times 314.1 \simeq 2513 \text{ mm}^2$

Near ends, bend 4 bars downwards to take negative bending moment induced due to fixidity.

At the end, let us check for development length so as to satisfy the relation

$$\frac{M_1}{V_u} + L_0 \geq L_d$$

Available $\quad A_{st} = \frac{1}{2} \times 2513 = 1256.5 \text{ mm}^2$

$$\therefore \quad x_u = \frac{0.87 f_y A_{st}}{0.36 f_{ck} b_f} = \frac{0.87 \times 415 \times 1256.5}{0.36 \times 20 \times 1500} \simeq 42 \text{ mm}$$

$$M_1 = 0.87 f_y A_{st} (d - 0.416 x_u) = 0.87 \times 415 \times 1256.5 (430 - 0.416 \times 42)$$

$$= 187.16 \times 10^6 \text{ N-mm}$$

$$V_u = 1.5 \times 184.5 = 276.75 \text{ kN} = 276.75 \times 10^3 \text{ N}$$

$$L_0 = l_s/2 - x' = 400/2 - 50 = 150 \text{ mm}$$

$$L_d = 47 \varphi = 47 \times 20 = 940 \text{ mm.}$$

$\therefore \quad \frac{M_1}{V_u} + L_0 = \frac{187.16 \times 10^6}{270.75 \times 10^3} + 150 = 841$ mm. This is less than required L_d.

Hence continue all the 8 bars upto end and bend these down wards by two 90° bends so as to make these available at the lower face of the beam. Hence available $A_{st} = 1256.5 \text{ mm}^2$ and

$$x_u = \frac{0.87 \times 415 \times 2513}{0.36 \times 20 \times 1500} = 84 \text{ mm}$$

$\therefore$ Revised $\quad M_1 = 0.87 \times 415 \times 2513 (430 - 0.416 \times 84) = 358.44 \times 10^6$ N-mm

$$\therefore \quad \frac{M_1}{V_u} + L_0 = \frac{358.44 \times 10^6}{276.75 \times 10^3} + 150 = 1445 \text{ mm} >> L_d$$

Now $\quad p = \frac{100 A_s}{b_w d} = \frac{100 \times 2513}{300 \times 430} \simeq 1.95\%$

Hence from Table 7.1, $\tau_c \simeq 1.78 \text{ N/mm}^2$; $\tau_v = \frac{276.75 \times 1000}{300 \times 430} = 2.15 \text{ N/mm}^2$

Since $\tau_v > \tau_c$, shear reinforcement is necessary.

$$\therefore \quad V_{uc} = \tau_c b_w d = 1.78 \times 300 \times 430 \times 10^{-3} = 229.62 \text{ kN}$$

$$\therefore \quad V_{us} = V_u - V_{uc} = 276.75 - 229.62 = 47.13 \text{ kN}$$

Using 8 mm φ 4 lgd stirrups, $A_{sv} = 4 \times \frac{\pi}{4}(8)^2 \simeq 201 \text{ mm}^2$

$$\therefore \quad s_v = \frac{0.87 f_y A_{sv} d}{V_{us}} = \frac{0.87 \times 415 \times 201 \times 430}{47.13 \times 10^3} \simeq 670 \text{ mm}$$

Spacing of nominal stirrups, $s_v = \frac{2.175 f_y A_{sv}}{b} = \frac{2.175 \times 415 \times 201}{300} = 604$ mm

subject to a maximum of 300 mm.

Hence provide 8 mm φ 4 lgd stirrups @ 300 c/c throughout the length of the beam.

Let us calculate the reaction transferred by the secondary beam to main beam, due to wind load reaction p_w, since this will be useful for the design of main beam.

Load due to p_w, per metre run $= \pm 1.5\, p_w = \pm 1.5 \times 11.7 = \pm 17.6$ kN/m at the face of the main beam.

$$\text{Couple due to this} = \left(\frac{1}{2} \times 17.6 \times 2.05\right)\left(\frac{2}{3} \times 4.1\right) = 49.3 \text{ kN-m}$$

$$\therefore \quad \text{End reactions} = 49.3/4.5 \approx 11 \text{ kN.}$$

$\therefore$ Net reaction at windward end $= 184.5 - 11 = 173.5$ kN

Net reaction at leeward end $= 184.5 + 11 = 195.5$ kN.

5. ***Design of central secondary beam.*** The central secondary beam *BE* joins columns *B* and *E* and hence is subjected to end moments transmitted to it by the columns. Therefore, let us consider both p_0 as well as p_w, along with the transferred end moments, to calculate the maximum bending moment and shear force. If the bending moment is greater than the intermediate secondary beam atleast by $33\frac{1}{3}\%$, then the design will change. Otherwise it will remain the same.

From Fig. 22.55 (*c*), the pressure intensity p_w' due to wind, at the face of longitudinal beam $= 11.7$ kN/m².

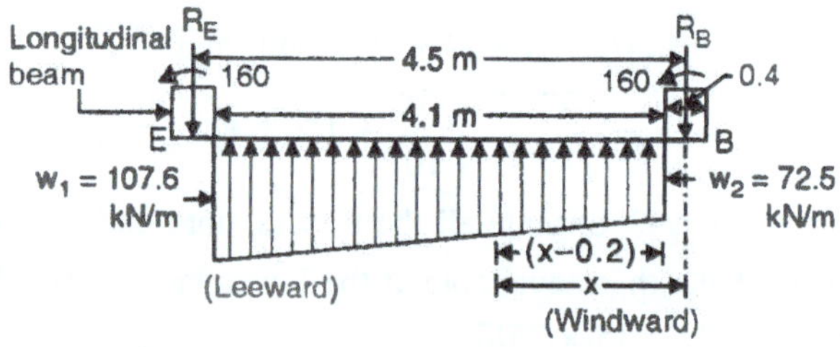

FIG. 22.57

$\therefore$ Net pressure intensity at $E' = 60 + 11.7$

$= 71.7$ kN/m².

Net pressure intensity at $B' = 60 - 11.7 = 48.3$ kN/m²

Width of slab transferring load $= 1.5$ m.

$\therefore$ Load intensity w_1 at end $B' = 71.7 \times 1.5 = 107.6$ kN/m

Load intensity w_2 at end $B' = 48.3 \times 1.5 = 72.5$ kN/m

The effective span of the beam $= 4.5$ m while it will receive load only in a length of 4.1 m. The loading on the beam is shown in Fig. 22.57

To find reaction at *E*, take moments at *B*.

$$\therefore \quad R_E = \frac{1}{4.5}\left[72.5 \times 4.1\left(\frac{4.1}{2} + 0.2\right) + \frac{107.6 - 72.5}{2} \times 4.1 \times \left\{\frac{2}{3}(4.1) + 0.2\right\} - 2 \times 160\right]$$

$= 124.5$ kN.

Similarly, taking moments E,

$$R_B = \frac{1}{4.5}\left[72.5 \times 4.1\left(\frac{4.1}{2} + 0.2\right) + \frac{107.7 - 72.5}{2} \times 4.1 \times \left\{\frac{1}{3}(4.1) + 0.2\right\} + 2 \times 160\right]$$
$$= 244.8 \text{ kN}$$

Bending moment will be maximum where S.F. is zero. Let us write down equation of shear force at x from support D.

$$F_x = 244.8 - \left[72.5(x - 0.2) + \frac{107.6 - 72.5}{4.1}(x - 0.2)\frac{1}{2}(x - 0.2)\right] \quad \ldots(i)$$

Equating this to zero, we get $x = 3.1$ m. Hence maximum bending moment is given by

$$M = (244.8 \times 3.1) - 160 - \left[\frac{72.5(2.9)^2}{2} + \left(\frac{107.7 - 72.5}{4.1}\right) \times 3.1 \times \frac{1}{2} \times 3.1 \times \frac{3.1}{3}\right]$$
$$= 251.5 \text{ kN-m} \quad \ldots(ii)$$

It will be seen that maximum bending moment and shear force found above by taking the wind effect into account are *less* than $1\frac{1}{3}$ times the corresponding values for the intermediate secondary beam. *Hence the same depth and same main reinforcement will be used.* If however, the B.M. and S.F. found above were more than $1\frac{1}{3}$ times the B.M. and S.F. for the intermediate secondary beam, the central beam could be designed for the above increased B.M. and S.F., taking higher permissible values of stresses.

However, due to partial fixidity at the ends, the beam will be subjected to sagging B.M. $= WL/24$, in addition to wind moment of 160 kN-m.

$\therefore$ Total sagging B.M. $= (124.5 + 244.8)\frac{4.5}{24} + 160 = 229.2$ kN-m.

Since the intermediate secondary beam has been designed for a B.M. of 226 kN-m, its depth $d = 430$ mm will be sufficient to take compression, but special reinforcement at the bottom face will have to be provided at the ends. However, provide the same reinforcement as the centre, *i.e.* 8 bars of 20 mm φ provided in two layers at the bottom face for the portion under the columns for a distance at least equal to $L_d = 47 \times 20 \approx 1000$ mm.

6. *Design of end secondary beam*. The end secondary beam will get less load by way of reaction from the slab, because it will receive load from slab of width $= 0.65 + \frac{1}{2}(1.5) = 1.4$ m only against 1.5 m width for central secondary beam. However, provide the same section and reinforcement as for the central secondary beam. The ratio of load on the outer beam to the central one will be approximately equal to $1.4/1.5 \approx 0.93$.

7. *Design of main beam*. The bending moments and shear forces in the main beam are to be found under the following three conditions :

(*a*) Ignoring wind effect.

(*b*) Taking wind effect : leeward beam.

(*c*) Taking wind effect ; windward beam.

(a) Ignoring wind effect. When there is no wind acting, the reaction transferred to the main beam by all the intermediate as well as central beams will be equal, while

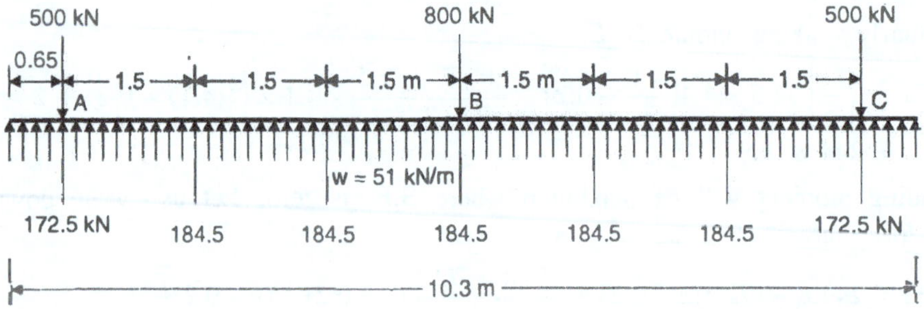

FIG. 22.58.

the reaction transferred by the end beam will be 0.93 times that transferred by intermediate beam. In addition to this, the beam will get soil reaction from slab immediately below it and the slab cantilevering from its face.

Reaction transferred by each intermediate beam = 184.5 kN

Reaction transferred by end secondary beam = 0.93 × 184.5 = 172.5 kN

Reaction transferred by slab, per metre run of beam = 60 (0.4 + 0.45) = 51 kN/m.

The loading diagram is shown in Fig. 22.58.

It should be checked that total downward load of column is equal to the total upward load. Total downward load = 500 + 800 + 500 = 1800 kN

Total upward load = (184.5 × 5) + (172.5 × 2) + (10.3 × 51) = 1793 kN

Thus, both of these are approximately equal. The discrepancy is there because of rounding off the value of p_0 to 60 kN/m^2 in step 2.

Maximum sagging bending moment will occur at B, its value being given by

$$M_1 = 51 \frac{(5.15)^2}{2} + (184.5 \times 1.5) + (184.5 \times 3.0) - (500 - 172.5)\, 4.5 = 32.8 \text{ kN-m (sagging)} \quad \ldots(1)$$

Maximum hogging bending moment occurs somewhere between A and B, where shear force is zero. This is likely to occur between the two intermediate beams at distance x from A.

$$\therefore \quad F_x = 51\,(x + 0.65) + 172.5 + 184.5 - 500 = 0$$

This gives x = 2.15 m from A or 2.8 m from the edge. Hence the maximum hogging bending moment is given by

$$M_2 = (500 - 172.5)\, 2.15 - 184.5\,(2.15 - 1.5) - \frac{51}{2}(2.8)^2 = 384.3 \text{ kN-m (hogging)} \quad \ldots(2)$$

Shear force to the right of A is $F_1 = (500 - 172.5) - (51 \times 0.65) = 294.4$ kN ...(3)

Shear force to the left of B is $F_2 = \frac{1}{2}(800 - 184.5) = 307.8$ kN ...(4)

(b) Wind effect : leeward beam. From step 4, reaction transferred by the intermediate secondary beam on the leeward end = 195.5 kN.

From step 5, reaction transferred by the central secondary beam on the leeward end = 124.5 kN. Reaction transferred by the end secondary beam, on the leeward end

$$= 0.93 \times 195.5 - \frac{160 + 160}{4.5} = 182 - 71 = 111 \text{ kN.}$$

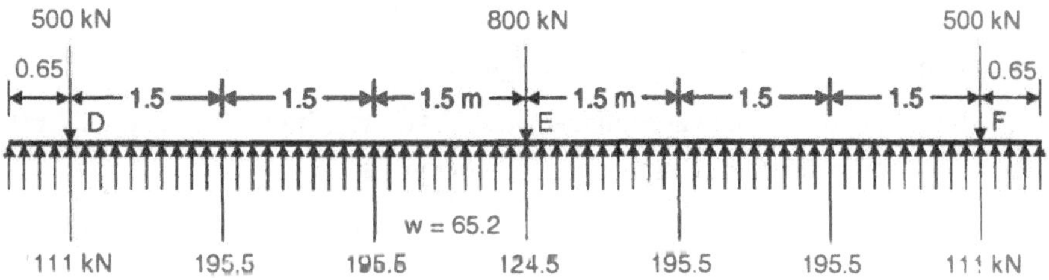

FIG. 22.59.

Total downward load = 500 + 800 + 500 = 1800 kN

∴ Upward load from slab = 1800 − (195.5 × 4 + 124.5 + 111 × 2) = 671.50 kN

Uniform upward load w = 671.5/10.3 = 65.2 kN/m

The loading on the leeward beam DF is shown in Fig. 22.59.

Maximum sagging bending moment will occur at E, its value being

$$M_3 = \frac{65.2(5.15)^2}{2} + (195.5 \times 1.5) + (195.5 \times 3) - (500 - 111)\,4.5 = -6.1 \text{ N-m } (\textit{i.e.} \text{ hogging}) \quad ...(5)$$

Maximum hogging bending moment occurs somewhere between D and E where shear force is zero. This is likely to occur between the two intermediate beams, at distance x from D.

$$F_x = 65.2(0.65 + x) + 111 + 195.5 - 500 = 0$$

This gives x = 2.32 m from D, or 2.97 m from the edge. Hence maximum hogging bending moment is given by

$$M_4 = (500 - 111) \times 2.32 - 195.5\,(2.32 - 1.5) - \frac{65.2}{2}(2.97)^2 = 454.6 \text{ kN-m (hogging)} \quad ...(6)$$

S.F. to the right of D is $F_3 = (500 - 111) - (65.2 \times 0.65) = 346.6$ kN ...(7)

S.F. to the left of E is $F_4 = \frac{1}{2}(800 - 124.5) = 337.8$ kN ...(8)

(c) *Wind effect : windward beam*. From step 4, reaction transferred by the intermediate secondary beam on the windward side = 173.5 kN

From step 5, reaction transferred by central secondary beam on the windward end = 244.8 kN.

Reaction transferred by the end secondary beam on the windward end

$$= (0.93 \times 173.5) + \frac{160 + 160}{4.5} = 232.5 \text{ kN}$$

Total downward load = 500 + 800 + 500 = 1800 kN

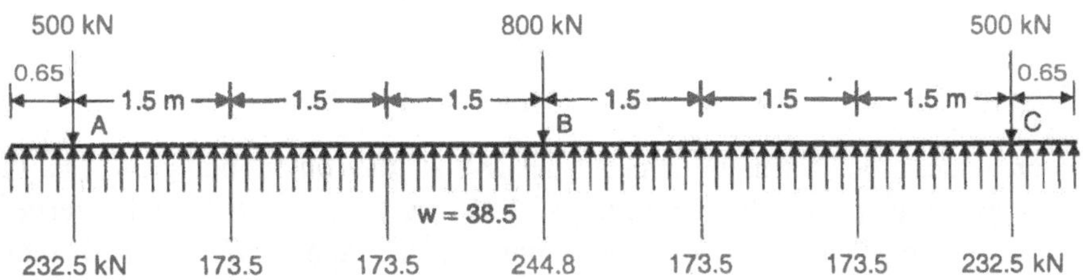

FIG. 22.60.

$\therefore$ Upward load from the slab = 1800 − (4 × 173.5 + 244.8 + 2 × 232.5) = 396.2 kN.

$\therefore$ Upward w = 396.2/10.3 ≈ 38.5 kN/m.

The loading on the windward main beam AC is shown in Fig. 22.60.

Maximum sagging bending moment will occur at B :

$$M_5 = \frac{38.5(5.15)^2}{2} + (2 \times 173.5 \times 2.25) - (500 - 232.5) \times 4.5 = 87.6 \text{ kN-m}$$

Maximum hogging bending moment will occur where S.F. is zero, some where between A and B. Let it occur at x from A between the two intermediate beams.

$$F_x = 38.5\,(x + 0.65) + 173.5 + 232.5 - 500 = 0$$

which gives x = 1.79 m from A or 2.44 m from the edge.

Hence maximum hogging bending moment is given by

$$M_6 = (500 - 232.5) \times 1.79 - 173.5\,(1.79 - 1.5) - \frac{38.5}{2}(2.44)^2 = 314 \text{ kN-m (hogging)} \quad \ldots(10)$$

S.F. to the right of A is given by $F_5 = (500 - 232.5) - (0.65 \times 38.5) = 242.5$ kN ...(11)

S.F. to the left of B is $F_6 = \frac{1}{2}(800 - 244.8) = 277.6$ kN ...(12)

Design values. From the above cases, we get the following values for design purposes:

Without wind :

Maximum sagging moment, M_1 = 32.8 kN-m ; M_{1u} = 1.5 × 32.8 = 49.2 kN-m

Maximum hogging moment, M_2 = 384.3 kN-m, M_{2u} = 1.5 × 384.3 = 576.45 kN-m

Maximum shear, F_2 = 307.8 kN; F_{2u} = 1.5 × 307.8 = 461.7 kN

With wind :

Maximum sagging moment, M_5 = 87.6 kN-m ; M_{5u} = 1.5 × 87.6 = 131.4 kN-m

Maximum hogging moment, M_4 = 454.6 kN-m ($<< \frac{4}{3} \times 384.3$)

Maximum shear, F_3 = 346.6 kN ; F_{3u} = 1.5 × 346.6 = 519.9 kN

Thus critical conditions are under 'no wind'. T-beam action will be available for maximum hogging bending moment M_{2u} = 1.5 × 384.3 = 576.45 kN-m. Let the width of rib =400 mm.

The flange width of T-beam is given by

$$b_f = \frac{l_0}{6} + b_w + 6\,D_f$$

where l_0 = distance between points of zero moments in the beam = 10.3 − (2 × 2.8) = 4.7 m = 4700 mm

b_w = 400 mm ; D_f = 160 mm

$$\therefore \quad b_f = \frac{4700}{6} + 400 + 6 \times 160 \approx 2140 \text{ mm.}$$

However, since the beam is situated nearly at the outer boundary, its half flange is restricted in its width to a value = 0.20 + 0.45 = 0.65 m = 650 mm. Hence maximum value of b_f is restricted to 2 × 650 = 1300 mm. Hence adopt b_f = 1300 mm.

(*a*) ***Depth on the basis of bending compression***

Let $d_{min} \simeq \frac{2}{3}\sqrt{\frac{M_u}{R_u\, b_w}} = \frac{2}{3}\sqrt{\frac{576.45 \times 10^6}{2.761 \times 400}} = 482$ mm.

(*b*) ***Depth on the basis of*** $\tau_{c,\,max}$:

From Table 7.3, $\tau_{c,\,max} = 2.8$ N/mm^2 for M 20 concrete

$$\therefore \qquad d = \frac{F_{u,\,max}}{b_w\, \tau_{c,\,max}} = \frac{307.8 \times 10^3}{400 \times 2.8} \simeq 275 \text{ mm.}$$

Providing 50 mm nominal cover, using 25 mm φ bars in two layers with 25 mm space between the two and using 12 mm φ stirrups, effective cover = 50 + 12 + 25 + 125 = 99.5 mm. Keep effective cover = 100 mm, so that providing $D = 650$ mm available $d = 650 - 100 = 550$ mm. A greater value of d will make the beam section as under-reinforced.

This depth will be sufficient to take sagging B.M. when the beam will act as a simple rectangular beam.

Determination of reinforcement at top : Assume $x_u = D_f = 160$ mm

$$\therefore \qquad M_{u1} = 0.36\, f_{ck}\, b_f\, D_f\,(d - 0.416\, D_f) = 0.36 \times 20 \times 1300 \times 160\,(550 - 0.416 \times 160)$$

$$= 724 \times 10^6 \text{ N-mm}$$

Since $M_{u1} > M_{uD}$ (= 576.45 × 10^6), N.A. falls in the flange, i.e. $x_u \le D_f$

$$\therefore \qquad M_u = M_{uD} = 0.87\, f_y\, A_{st}\, d\left[1 - \frac{f_y \cdot A_{st}}{f_{ck}\, b_f\, d}\right), \text{ from which}$$

$$A_{st} = \frac{0.5\, f_{ck}}{f_y}\left[1 - \sqrt{1 - \frac{4.6\, M_u}{f_{ck}\, b_f\, d^2}}\right] b_f\, d = \frac{0.5 \times 20}{415}\left[1 - \sqrt{1 - \frac{4.6 \times 576.45 \times 10^6}{20 \times 1300\, (550)^2}}\right] \times 1300 \times 550$$

$$= 3202 \text{ mm}^2$$

∴ No. of 25 mm φ bars = 3202/490.8 = 6.52 ≃ 7. Actual $A_{st} = 7 \times 490.8 = 3435.6$ mm^2

Provide these bars at the top face of the beam in two layers. This reinforcement should be safe to satisfy the development length criterion at the point of contraflexure, so as the satisfy the relation

$$\frac{M_1}{V_u} + L_0 \ge L_d.$$

$$x_u = \frac{0.87\, f_y\, A_{st}}{0.36\, f_{ck}\, b_f} = \frac{0.87 \times 415 \times 3435.6}{0.36 \times 20 \times 1300} = 132.5 \text{ mm.}$$

$$\therefore \qquad M_1 = 0.87\, f_y\, A_{st}\,(d - 0.416\, x_u) = 0.87 \times 415 \times 3435.6\,(550 - 0.416 \times 132.5)$$

$$= 613.9 \times 10^6 \text{ N-mm}$$

V_u = Shear at the point of contraflexure

Assuming that the point of contraflexure falls just to the left of B, shear force $V_u = F_{2u}$ = 1.5 × 307.8 kN = 461.7 × 10^3 N

L_0 = 12 φ or d whichever is more = 550 mm

L_d = 47 × 25 = 1175 mm.

$$\therefore \quad \frac{M_1}{V_u} + L_0 = \frac{613.9 \times 10^6}{461.7 \times 10^3} + 550 = 1879 \text{ mm. Hence safe}$$

Continue all the 8 bars throughout the length, in two layers.

Determination of reinforcement at bottom

The beam should also have reinforcement at its bottom face to resist sagging bending moment at the centre and at the ends, for a bending moment of 1.5 × 87.6 = 131.4 kN-m under wind load conditions. The reinforcement required for bending moment will be less than 2 bars of 25 mm φ . However, this reinforcement should satisfy the criterion of development length at the point of contraflexure near the centre of beam, where S.F. is maximum. Hence try 5 bars of 16 mm φ, having $A_{st} = 5 \times 201 = 1005 \text{ mm}^2$.

$$x_u = \frac{0.87 f_y A_{st}}{0.36 f_{ck} b} = \frac{0.87 \times 415 \times 1005}{0.36 \times 20 \times 400} = 126 \text{ mm}$$

$$M_1 = 0.87 f_y A_{st} (d - 0.416 x_u) = 0.87 \times 415 \times 1005 (550 - 0.416 \times 126) = 180.55 \times 10^6 \text{ N-mm}$$

V_u = S.F. at the point of contraflexure

≈ 1.5 × 307.8 kN (assuming this to be equal to S.F. at B)

L_0 = 12 φ or d, whichever is more = 550 mm.

L_d = 47 φ = 47 × 16 = 752 mm.

$$\therefore \quad \frac{M_1}{V_u} + L_0 = \frac{180.55 \times 10^6}{1.5 \times 307.8 \times 10^3} + 550 = 941 \text{ mm} > L_d. \text{ Hence OK.}$$

Hence provide 5 bars of 16 mm diameter at the bottom face, throughout the length of the beam.

Design of shear reinforcement

$$V_u = 1.5 \times 307.8 \text{ kN};$$

$$\therefore \quad \tau_v = \frac{1.5 \times 307.8 \times 1000}{400 \times 550} = 2.1 \text{ N/mm}^2 \text{ (at } B\text{).}$$

$$\frac{100 A_s}{bd} = \frac{100(7 \times 490.8)}{400 \times 550} \simeq 1.56\%$$

Hence $\tau_c \simeq 0.725 \text{ N/mm}^2$ (Table 7.1)

Since $\tau_v > \tau_c$, shear reinforcement is necessary.

$$V_{uc} = 0.725 \times 400 \times 550 \times 10^{-3} = 159.5 \text{ kN}$$

$$\therefore \quad V_{us} = 1.5 \times 307.8 - 159.5 = 302.2 \text{ kN.}$$

Using 10 mm φ 4-lgd stirrups, $A_{sv} = 314.1 \text{ mm}^2$.

$$\therefore \quad \text{Spacing} = \frac{0.87 f_y . A_{sv} . d}{V_{us}} = \frac{0.87 \times 415 \times 314.1 \times 550}{302.2 \times 1000} \approx 200 \text{ mm}$$

Hence provide the stirrups @ 200 mm c/c. for 1 m on either side of the faces of the column B.

Similarly, shear force at A = 1.5 × 294.4 kN. Hence provide 10 mm φ 4-legged stirrups @ 200 mm c/c, for a distance of 0.6 m to the right of face of column A, and 0.6 m to left of column C.

For the remaining length, provide nominal shear reinforcement at a maximum spacing of

$$s_v = \frac{2.175 \, A_{sv} f_y}{b} = \frac{2.175 \times 314.1 \times 415}{400} \simeq 700 \text{ mm.}$$

But not more than 300 mm nor more than 0.75 d (= 0.75 × 550 = 412.5 mm).

Hence provide 12 mm φ 4-lgd nominal stirrups @ 300 mm c/c for the remaining length. The details of reinforcement etc. are shown in Fig. 22.61.

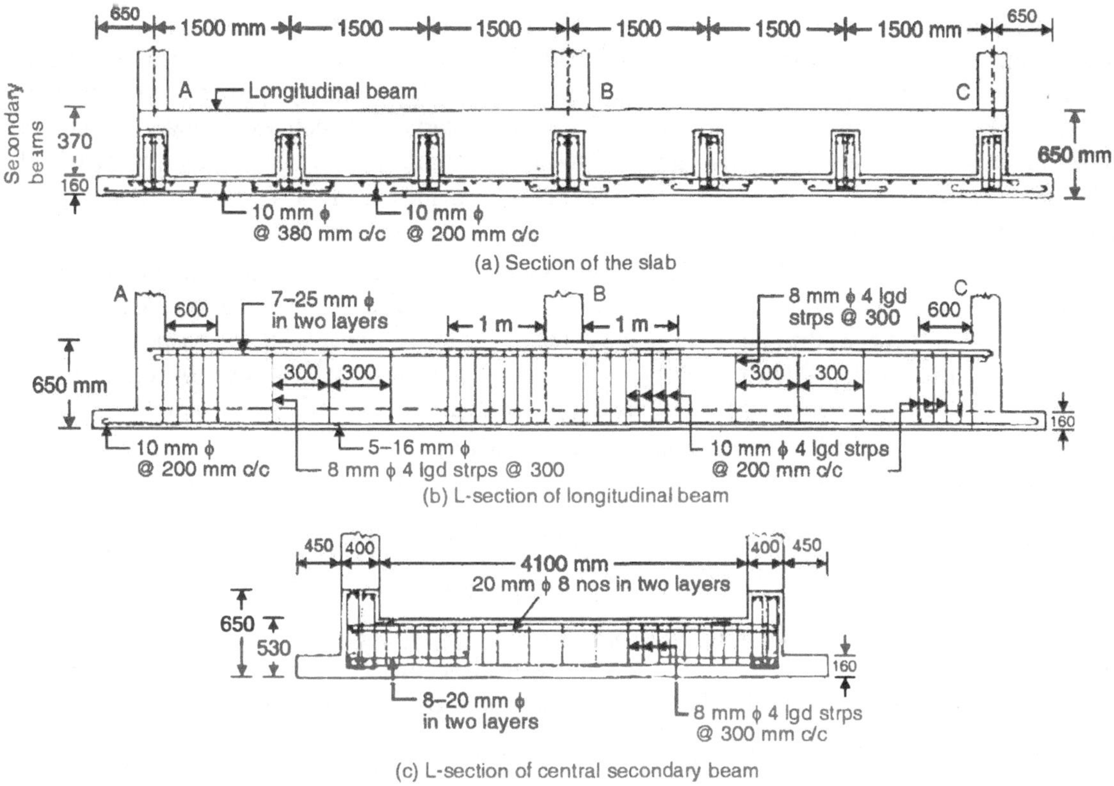

FIG. 22.61.

22.16 STRUCTURAL DESIGN OF R.C. PILE

A R.C. pile is designed for the following : (1) total load coming on it from the structure (2) handling stresses (3) driving stresses. While designing the pile as a column, it may be considered as fixed at one end and hinged at the other. The length of the pile may be taken as 2/3 rd the length embedded in firm stratum. The cross-section of pile varies with its overall length.

The main longitudinal reinforcement in a pile should not be less than the following values:

(*i*) 1.25 percent of the cross-sectional area of the pile, for piles having length upto 30 times their least width.

(*ii*) 1.5 percent of the cross-sectional area of the pile, for piles having length between 30 to 40 times its least width.

(*iii*) 2 percent of the gross cross-sectional area of the pile, for piles having length greater than 40 times its least width.

Lateral reinforcement should be provided in the form of links, of not less than 5 mm diameter wire. The volume of such lateral reinforcement should not be less than 0.2 percent of the gross volume of the pile, and their centre to centre spacing should not exceed half the least width of the pile. Near each end of pile, upto a distance of three times the least lateral dimension, lateral reinforcement should be provided @ 0.6% of the gross' volume. If the pile has to penetrate hard stratum, the lateral reinforcement at the top of the pile for a distance of three times the width, should be in the form of helix. In addition to this, the longitudinal bars should be held apart by provision of *spacers* made of 12 to 16 mm wire and provided at the interval of not exceeding 1.5 m. In order to prevent the outward displacement of the bars, 6 mm φ links should also be provided with the spacers (Fig. 22.66). The pile should be provided with a cast iron shoes at its lower end (Fig. 22.66).

Stresses during handling. Precast piles should be checked against handling stresses. When a pile is lifted by means of a derrick, it is subjected to bending moment due to its own weight. When the pile is of less than 12 metre length, it is suspended from one point at its middle. Piles longer than 12 metres are suspended at two or three points suitably spaced at its length so that handling moment is as small as possible.

(*i*) ***Pile suspended at one point.*** [Fig. 22.62(*a*)]

If w is the weight of the pile per unit length, and L is the length of the pile, it will be subjected to a maximum bending moment $= \frac{wL}{2} \times \frac{L}{4} = \frac{wL^2}{8}$ at the point of suspension.

(2) ***Pile suspended at two points.*** [Fig. 22.62(*b*)].

Let the distance of each point of suspension be x from the respective ends. The value of x should be such that maximum bending moment anywhere in the pile should be the least. This is possible when the hogging moment is equal to maximum sagging moment.

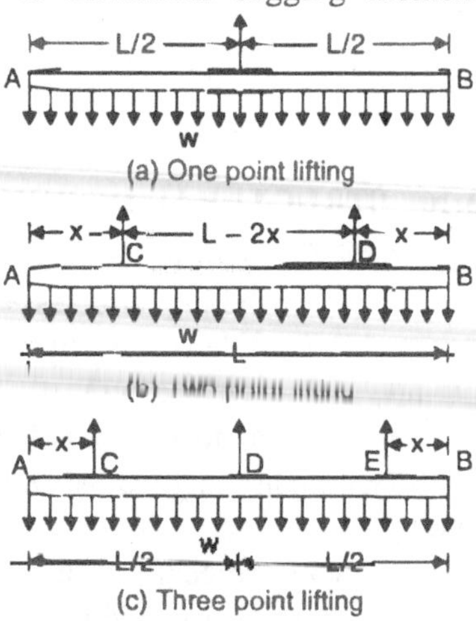

FIG. 22.62

Reaction $R = \frac{wL}{2}$ at each point of suspension.

Maximum hogging moment at the end of each cantilever

$$= \frac{w x^2}{2} \qquad \ldots(i)$$

Maximum sagging moment at the middle of the pile

$$= \frac{wL}{2}\left(\frac{L}{2} - x\right) - \frac{wL^2}{8} \qquad \ldots(ii)$$

Equating the two, we get

$$\frac{w x^2}{2} = \frac{wL}{2}\left(\frac{L}{2} - x\right) - \frac{wL^2}{8} = \frac{wL}{2}\left(\frac{L}{4} - x\right)$$

or $\quad x^2 + Lx - \frac{L^2}{4} = 0,\quad$ which gives $x = 0.206\ L$

Hence maximum B.M. $= \frac{w}{2}(0.206\ L)^2 = \frac{wL^2}{47}$

(3) ***Pile suspended at three points*** [Fig. 22.62(c)]. It can be shown that if the pile is suspended at three points, the end points should be located at $x = 0.15L$ from the corresponding ends so that B.M. is the least. The third point is located at the middle of the pile. The maximum bending moment in that case will be $wL^2/90$

Erecting the pile from one end

Fig. 22.63 shows a pile being erected from one point C, the end of the pile being resting on the ground. Let the distance of the point be x from the end. In order that bending moment on the pile is the least, x should be equal to $0.293\ L$. The maximum bending moment will be equal to $wL^2/23$

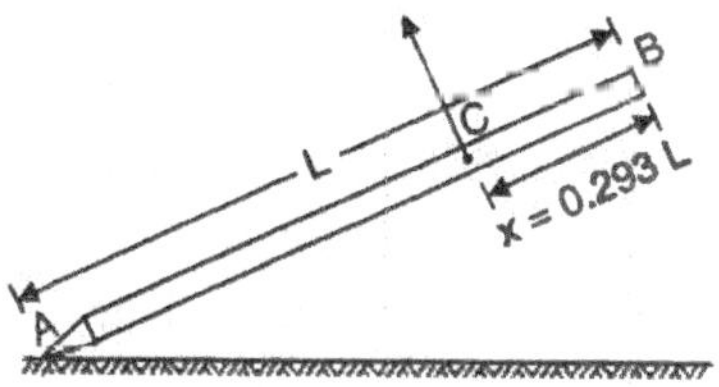

FIG. 22.63

22.17. DESIGN OF PILE CAP

When a column or pier is supported on the pile only, the column should rest centrally on the pile. However, when the column is supported on more than one pile, the piles should be connected through a rigid pile cap, to distribute the load to the individual piles. The pile cap consists of a rigid, deep, reinforced concrete slab which acts monolithically with the group of piles. The piles should be arranged symmetrically about the axis of the column so that the load from column is distributed uniformly to all the columns. The pile cap slab is provided in uniform thickness. The pile cap should be extended beyond exterior piles by 10 to 15 cm. The pile should be embedded by at least 15 cm in the pile cap, and the reinforcement in the cap should be placed at least 10 cm above the pile head. Thus, effective depth of the pile cap will be equal to the total depth minus 25 cm. The pile cap, provided over the entire area of the piles, is considered to be divided into a framework of rectangular beams, along which main reinforcement is provided. The arrangement of these beams depends upon the number of piles, and the width of beam is taken equal to width of pile.

Pile cap for three piles. Fig. 22.64 (*a*) shows the pile cap for three piles. The pile cap is considered to be composed of two beams AB and CD ; A, B and C being the three piles placed at distance L centre to centre. The column W is placed on the beam DC, at the centroid of the triangle ABC.

Let W = total load of column.

$\therefore$ Load on each pile $= W/3$

Length of beam $CD = l = \dfrac{L\sqrt{3}}{2}$, Distance of W from $D = \frac{1}{3}l = \dfrac{L}{2\sqrt{3}}$

Beam CD is supported on C and D, and is loaded with W, such that reaction at C is $W/3$. Hence maximum bending moment is

$$M_1 = \frac{W}{3} \times \frac{2l}{3} = \frac{W}{3} \times \frac{2}{3}\left(\frac{L\sqrt{3}}{2}\right) = \frac{WL}{3\sqrt{3}}$$

Reaction transferred to $D = W - \dfrac{W}{3} = \dfrac{2W}{3}$

The beam AB is of length L, and is loaded at its middle point D with a load $(2W)/3$. Hence maximum bending moment is given by

$$M_2 = \frac{W}{3}\left(\frac{L}{2}\right) = \frac{WL}{6} .$$

The reinforcement for both beams can now be computed.

Pile cap for four piles [Fig. 22.64(*b*)].

Load on each pile $= \dfrac{W}{4}$

The pile cap is considered to be composed of two beams AB and CD.

$$\therefore\ M_{max} = \frac{W}{4} \times \frac{L}{2} = \frac{WL}{8} \text{ for each beam.}$$

Secondary reinforcement. In order to prevent the outward splaying tendency of piles, secondary reinforcement should always be provided. The reinforcement is provided at the bottom of the pile cap, running round the longitudinal reinforcement projecting from the piles into the cap. It should be so bent that there is change of its direction at the head of every pile. The area of secondary reinforcement changing direction at every head of the pile should not be less than 20% of the tensile reinforcement.

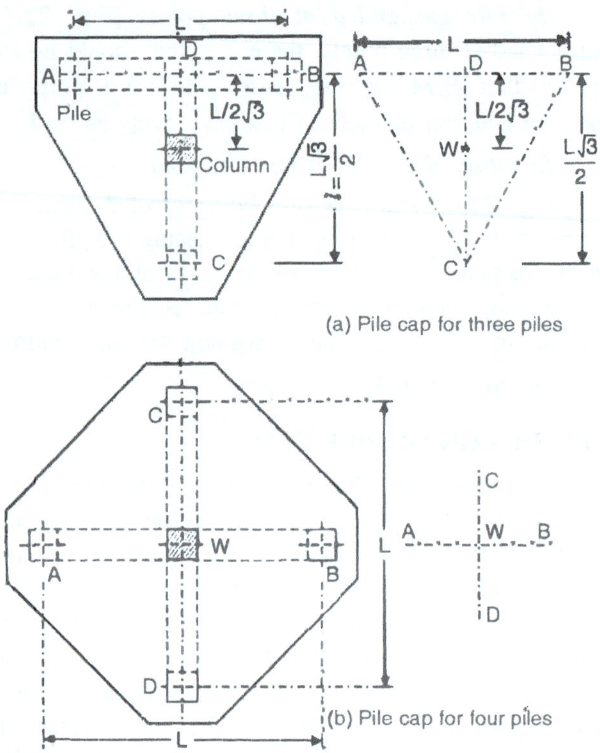

(a) Pile cap for three piles

(b) Pile cap for four piles

FIG. 22.64

Design Example 22.15. *Design a pile under a column transmitting an axial load of 800 kN. The pile is to be driven to a hard stratum available at a depth of 8 metres. Use M 20 concrete and Fe 415 steel.*

Solution.

1. *Main reinforcement*. Let the length of the pile above ground, including pile cap, etc. = 0.6 m. $\therefore$ Total length of pile = 8.6 m.

Let the size of the pile be 400 mm × 400 mm

$\dfrac{l}{D}$ ratio $= \dfrac{8.6}{0.4} = 21.5$ Since this is greater than 12, the pile behaves as a long column.

Hence reduction coefficient $C_r = 1.25 - \dfrac{l_{ef}}{48\,D} = 1.25 - \dfrac{8.6}{48 \times 0.4} \approx 0.8$

$\therefore$ Design load for a short column $= P_u = 1.5 \times 800/0.8 = 1500$ kN

Load carrying capacity of a column is given by $P_u = 0.4\, f_{ck}\, A_g + (0.67\, f_y - 0.4\, f_{ck})\, A_{sc}$

$\therefore\ P_u = 1500 \times 10^3 = 0.4 \times 20(400 \times 400) + (0.67 \times 415 - 0.4 \times 20)A_{sc}$ from which $A_{sc} = 815\ \text{mm}^2$

Since the length of pile is less than 30 times the width, minimum reinforcement @ 1.25% of gross cross-sectional area $= \frac{1.25}{100}(400 \times 400) = 2000\ mm^2$.

However, provide 4 bars of 25 mm φ giving total area of steel $= 4 \times 490 = 1960\ mm^2$ which is quite close to the minimum requirements. Provide an nominal cover of 50 mm. Using 8 mm φ stirrups, cover to the centre of reinforcement $= 50 + 8 + 25/2 = 70.5$ mm

2. ***Lateral reinforcement in the body of the pile***

Lateral reinforcement in the body of pile is provided @ 0.2% of gross volume.

$\therefore$ Volume needed per mm length $= \frac{0.2}{100}(400 \times 400 \times 1) = 320\ mm^3$.

Nominal cover = 50 mm

Using 8 mm φ ties, length of each side of tie

$= 400 - 2 \times 50 - 8 = 292$ mm

Area $A_\varphi = \frac{\pi}{4}(8)^2 = 50.3\ mm^2$.

$\therefore$ Volume of each tie $= 4 \times 292 \times 50.3 = 58750\ mm^3$

$\therefore$ Pitch $= 58750/320 \simeq 183$ mm

Maximum pitch permissible $= \frac{1}{2} \times 400 = 200$ mm.

Hence provide 8 mm φ ties @ 180 mm c/c throughout the length of the pile.

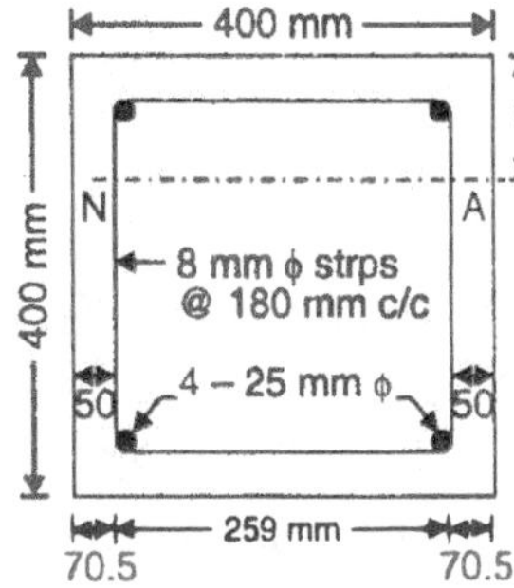

FIG. 22.65

3. ***Lateral reinforcement near pile head.*** Near pile head, special spiral reinforcement is to be provided for a length of $3 \times 400 = 1200$ mm. Volume of spiral, @ 0.6% of gross volume, per mm length is $= \frac{0.6}{100}(400 \times 400 \times 1) = 960\ mm^3$

Using 8 mm φ spiral, having $A_\varphi = 50.3\ mm^2$, pitch is given by

$$s = \frac{\text{circumference of spiral} \times A_\varphi}{960} = \frac{\pi \times 292 \times 50.3}{960} = 48 \text{ mm}$$

Provide the spiral at 45 mm pitch. Provide 6 additional bars of 16 mm φ vertically within the spiral. These spirals will be in addition to the normal ties.

4. ***Lateral reinforcement near pile end***

Volume of ties per mm length @ 0.6% of gross volume $= 960\ mm^3$.

$\therefore$ Volume of each tie $= 58750\ mm^3$. Hence, Pitch $= 58750/960 \simeq 60$ mm.

Provide ties @ 60 mm c/c in a bottom length of $3 \times 400 = 1200$ mm.

5. ***Spacer forks and links.*** Provide two pairs of 12 mm φ spacer forks with 6 mm φ links @ 1.5 m c/c along the length.

6. ***Check for moment of resistance.*** Provide three holes in the pile as follows :

(*i*) One hole at $0.293L = 0.293 \times 8.6 \approx 2.5$ m from the pile for the purpose of hoisting it.

(*ii*) One hole each from either end, at a distance of $0.206L$

$= 0.206 \times 8.6 \approx 1.75$ m for the purpose of stacking.

(*iii*) Weight of pile per meter run $= 0.4 \times 0.4 \times 1 \times 25000 = 4000$ N/m

$$M = \frac{4000\,(2.5)^2}{2} = 12500 \text{ N-mm} = 12.5 \times 10^6 \text{ N-mm}$$

$$\therefore \quad M_u = 1.5 \times 12.5 \times 10^6 = 18.75 \times 10^6 \text{ N-mm.}$$

Effective depth of pile section $= 400 - 70.5 = 329.5$ mm. Let the neutral axis be situated at x_u below one face.

$$x_{u,\,max} = 0.479\, d = 0.479 \times 329.5 = 157.8 \text{ mm.}$$

Let $x_u = 0.5\, x_{u,\,max} = 0.5 \times 157.8 \simeq 80$ mm

$$A_{sc} = 2 \times \frac{\pi}{4}(25)^2 = 981.7 \text{ mm}^2$$

$$\varepsilon_{sc} = \frac{0.0035\,(x_u - d')}{x_u} = \frac{0.0035\,(80 - 70.5)}{80} = 0.000416$$

$$\therefore \quad f_{sc} = E_s\,.\,\varepsilon_{sc} = 2 \times 10^5 \times 0.000416 = 83.1 \text{ N/mm}^2$$

$\therefore$ $C_u = 0.36\, f_{ck}\, x_u\, b + f_{sc}\, A_{sc} - 0.446\, f_{ck}\, A_{sc}$

$= 0.36 \times 20 \times 80 \times 400 + 83.1 \times 981.7 - 0.446 \times 20 \times 981.7$

$= 303222$ N

$T_u = 0.87\, f_y\, A_{st} = 0.87 \times 415 \times 981.7$

$= 354442$ N

Since $C_u < T_u$, increase the value of x_u. Let next $x_u = 85$ mm

$$\therefore \quad \varepsilon_{sc} = \frac{0.0035\,(85 - 70.5)}{85} = 0.000597$$

$\therefore$ $f_{sc} = 2 \times 10^5 \times 0.000597$

$= 119.4$ N/mm^2

$\therefore$ $C_u = 0.36 \times 20 \times 85 \times 400 + 119.4 \times 981.7 - 0.446 \times 20 \times 981.7$

$= 353258$ N $\simeq T_u$

Hence $x_u \simeq 85$ mm.

$M_{ur} = C_{cu}\,(d - 0.416\, x_u) + C_{su}\,(d - d')$

$C_{cu} = 0.36 \times 20 \times 85 \times 400 - 0.446 \times 20 \times 981.7 = 236043$ N

$C_{su} = 119.4 \times 981.7 = 117215$ N

$\therefore$ $M_{ur} = 236043\,(329.6 - 0.416 \times 85) + 117215\,(329.6 - 70.5)$

$= 69.45 \times 10^6 + 30.37 \times 10^6$

$= 99.82 \times 10^6$ N-mm

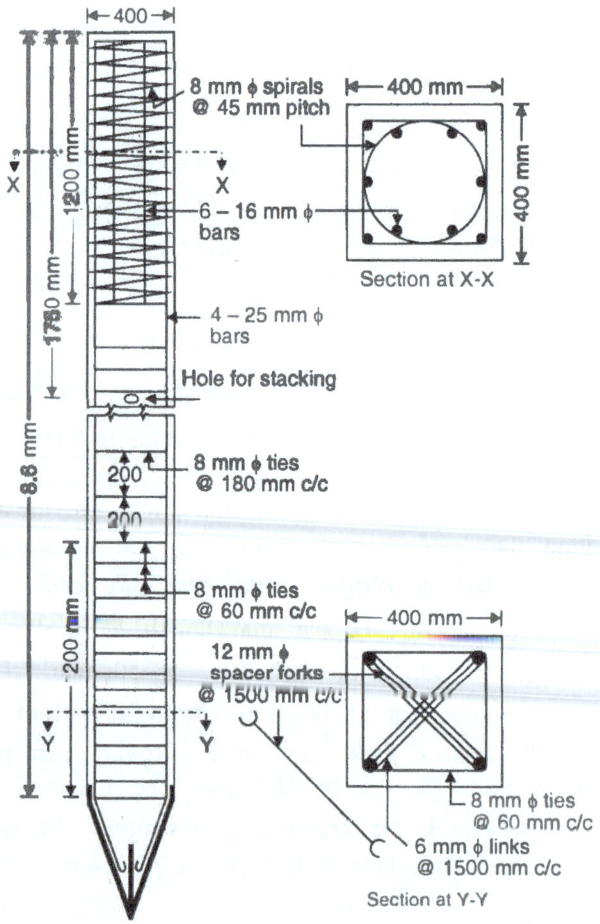

FIG. 22.66

Actual $M_{uD} = M_u = 18.75 \times 10^6$ N-mm Hence safe.

The details of reinforcement etc. is shown in Fig. 22.66.

Design Example 22.16. *A R.C. column, 400 mm × 400 mm carrying a load of 600 kN, is supported on three piles 400 mm × 400 mm in section. The centre to centre distance between the piles is 1.5 m. Design a suitable pile cap. Use M 20 concrete and Fe 415 steel.*

Solution. [Fig. 22.64 (*a*)]

1. ***Dimensions of pile cap.*** Centre to centre spacing of piles $= L = 1.5$ m.

Keep 200 mm clear projection of the cap beyond pile face. Overall length of the cap along the direction $AB = 1.5 + 0.4 + 0.4 = 2.3$ m.

$$\text{Length of beam } CD = l = \frac{L\sqrt{3}}{2} = 1.5\frac{\sqrt{3}}{2} = 1.3 \text{ m.}$$

$\therefore$ Length of cap in the direction $DC = 1.3 + 0.4 + 0.4 = 2.1$ m.

2. ***Design of beam DC***

$$\text{Load on each pile} = \frac{W}{3} = \frac{600}{3} = 200 \text{ kN}$$

Let the width of beam = width of column = 400 mm

$$\therefore \text{ B.M. due to load} = \frac{WL}{3\sqrt{3}} = \frac{600 \times 1.5}{3\sqrt{3}} = 173 \text{ kN-m} = 173 \times 10^6 \text{ N-mm} \quad \ldots(i)$$

In order to calculate the bending moment due to self weight of the beam plus weight of part of slab, let us assume total thickness of slab to be 800 mm. The self-weight of the beam is calculated on the assumption that weight of slab equal to two times the width of the beam acts with the beam.

$$\therefore \quad w = \left[\frac{3 \times 400 \times 800}{10^6}\right] \times 1 \times 25000 = 24000 \text{ N/m}$$

$$\text{Length of beam } l = \frac{L\sqrt{3}}{2} = 1.3 \text{ m.}$$

$$\therefore \quad \text{Total load} = 1.3 \times 24000 = 31200 \text{ N}$$

$$\therefore \text{ Reaction at } C = \frac{31200}{2} = 15600 \text{ N}$$

Distance of point of application of column load $= \frac{2}{3}l = \frac{2}{3} \times 1.3 = 0.87$ m.

$\therefore$ B.M. at the centre of column, due to self-weight is $= (15600 \times 0.87) - \frac{24000}{2}(0.87)^2$

$$= 4489 \text{ N-mm} = 4.489 \times 10^6 \text{ N-mm}$$

$$\therefore \quad \text{Total B.M.} = 173 \times 10^6 + 4.489 \times 10^6 \approx 177.5 \times 10^6 \text{ N-mm} \;;$$

$$M_u = 1.5\,M = 1.5 \times 177.5 \times 10^6 = 266.25 \times 10^6 \text{ N-mm}$$

$$\therefore \quad d = \sqrt{\frac{M_u}{R_u\,b}} = \sqrt{\frac{266.25 \times 10^6}{2.761 \times 400}} = 491 \text{ mm}$$

However keep $d = 600$ mm so as to get an under-reinforced section.

$$A_{st} = \frac{0.5\, f_{ck}}{f_y}\left[1 - \sqrt{1 - \frac{4.6\, M_u}{f_{ck}\, bd^2}}\right] bd = \frac{0.5 \times 20}{415}\left[1 - \sqrt{1 - \frac{4.6 \times 266.25 \times 10^6}{20 \times 400\, (600)^2}}\right] \times 400 \times 600$$

$= 1399 \text{ mm}^2$

$\therefore$ No. of 25 mm φ bars = 1399/490.8 = 2.83.

However, provide 4 bars of 25 mm φ. Actual $A_{st} = 4 \times 490.8 \simeq 1963 \text{ mm}^2$

3. *Design of beam AB*

Span $L = 1.5$ m

B.M. due to load from beam $CD = \dfrac{WL}{6} = \dfrac{600 \times 1.5}{6} = 150$ kN-m $= 150 \times 10^6$ N-mm

B.M. due to self weight $= \dfrac{24000\,(1.5)^2}{8} = 6750$ N-m $= 6.75 \times 10^6$ N–mm

$\therefore$ Total B.M. $= 150 \times 10^6 + 6.75 \times 10^6 = 156.75 \times 10^6$ N-mm

$M_u = 1.5 \times 156.75 \times 10^6 = 234.375 \times 10^6$ N-mm.

The reinforcement in direction *AB* will be placed below the reinforcement of *CD*. Hence available $d = 600 - 25 = 575$ mm

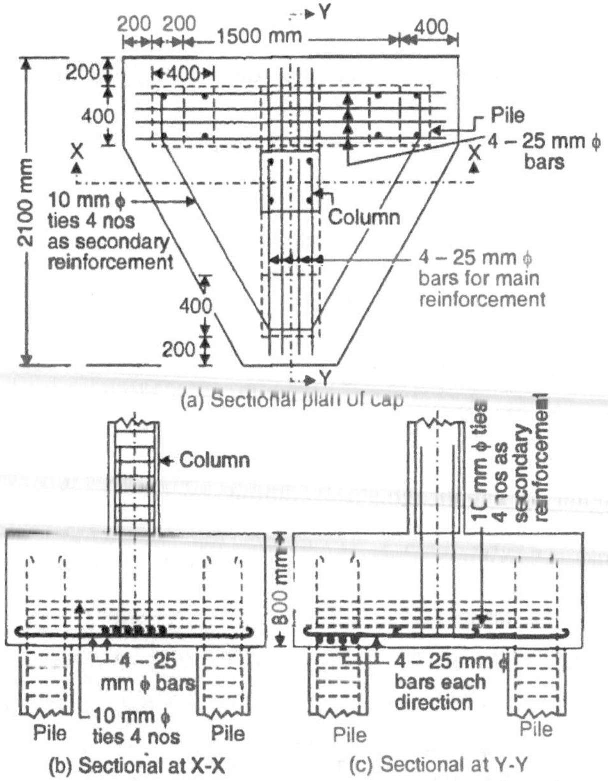

FIG. 22.67

$$A_{st} = \frac{0.5 \times 20}{415}\left[1 - \sqrt{1 - \frac{4.6 \times 234.375 \times 10^6}{20 \times 400\,(575)^2}}\right] \times 400 \times 575 = 1277 \text{ mm}^2$$

However, provide the same reinforcement, *i.e.* 4 Nos. of 25 mm φ bars. Keep total depth = 700 mm.

4. *Secondary reinforcement*

Area of secondary reinforcement running round each pile head = 0.2 × 1399 ≏ 280 mm^2

Using 10 mm φ bars, A_φ = 78.5 mm^2

∴ No. of bars $= \frac{280}{78.5} = 3.57 \approx 4.$

5. *Check for shear*

Shear is tested at a distance *d* from the beam. The dispersion lines (at 45°) transfer the load directly to the column. Hence there is no possibility of diagonal tension cracks.

The details of reinforcement etc. are shown in Fig. 22.67.

PROBLEMS

1. A steel stanchion carries 400 mm × 200 mm heavy beam section and carries a load of 1000 kN. Design a R.C.C. base for column. The safe bearing capacity of the soil may be taken as 100 kN/m^2 at a depth of 1 m below ground surface. Use M 20 concrete and Fe 415 steel.
2. A R.C. column 400 mm × 400 mm in section, carries an axial load of 750 kN. Design sloping R.C. footing for the column using M 20 concrete and Fe 415 steel.
   Take safe bearing capacity of soil = 120 kN/mm^2
3. Redesign problem 2, if the column is circular in section having diameter = 40 cm. Provide square footing.
4. Redesign problem 3, by providing circular base for the footing.
5. A rectangular R.C. column, 240 mm × 300 mm carries as axial load of 400 kN. Design a rectangular footing of uniform thickness, if the safe bearing capacity of the soil is 80 kN/m^2. Use M 20 concrete and Fe 415 steel.
6. Design as isolated stepped footing for the column of problem 2.
7. Design a sloped footing for a rectangular column 400 mm × 500 mm carrying an axial load of 800 kN. The safe bearing capacity of soil is 150 kN/m^2. Use M 20 concrete and Fe 415 steel.
8. Redesign problem 2 if the column also carries a moment of 150 kN-m along with axial load of 750 kN. The footing should be located centraily under the column.
9. Design a combined footing for two columns, each 600 mm × 600 mm, 5 metres apart, and each carrying a load of 1800 kN. The available width is restricted to 2:4 metres. The safe bearing capacity of soil may be taken as 200 kN/m^2. Use M 20 concrete and Fe 415 steel.
10. Redesign problem 9 by providing a longitudinal beam joining the column. There is no restriction on the width of the footing.
11. Design a combined footing for two R.C. columns *A* and *B*, separated by a distance of 4 m centre to centre :

Column A is 500 mm square and carries a load of 1200 kN
Column B is 600 mm square and carries a load of 1600 kN
The safe bearing capacity of a soil is 200 kN/m^2. Use M 20 concrete and Fe 415 steel.

12. Redesign problem 11 if the property line is located 400 mm from centre of column A measured outwards along the centre line of the columns and runs at right angles to the column line. No encroachment beyond the property line is permissible.

13. Design a trapezoidal combined footing for the following requirements :
Column A = 400 mm × 400 mm
Column B = 600 mm × 600 mm
Axial load on column A = 400 kN
Axial load on column B = 800 kN
Distance between centre of columns = 2.5 m
Safe bearing capacity of soil = 200 kN/ m^2
The footing is not to project more than 0.4 m beyond the outer face of the columns. Use M 20 concrete and Fe 415 steel.

14. Redesign problem 13 by providing a central longitudinal beam joining the columns.

15. Design a strap footing for the following data:
Column A, 400 mm × 400 mm carries a load of 800 kN
Column B, 600 mm × 600 mm carries a load of 1000 kN
Centre to centre distance between the two column is 8 metres.
Column A is on the property line.
Safe bearing capacity of soil = 150 kN / m^2
Use M 20 mix. and Fe 415 steel.

16. A column 600 mm × 600 mm carries an axial load of 1200 kN and is supported on three piles. The piles are driven to hard strata available at the depth of 10 metres. Using M 20 concrete, and Fe 415 steel, design the pile and the pile cap. The column is placed at the centroid of the pile group. The centre to centre distance of the piles is restricted to 2 m.

23

Design of Retaining Walls

23.1. INTRODUCTION

A *retaining wall* or retaining structure is used for maintaining the ground surfaces at different elevations on either side of it. Whenever embankments are involved in construction, retaining walls are usually necessary. In the construction of buildings having basements, retaining walls are mandatory. Similarly in bridge work, the wing walls and abutments etc. are designed as retaining walls, to resist earth pressure along with superimposed loads. The material retained or supported by a retaining wall is called *backfill* which may have its top surface horizontal or inclined. The position of the backfill lying above the horizontal plane at the elevation of the top of a wall is called the *surcharge,* and its inclination to the horizontal is called the *surcharge angle* β.

In the design of retaining walls or other retaining structures, it is necessary to compute the lateral earth pressure exerted by the retaining mass of soil. The equation of finding out the lateral earth pressure against retaining wall is one of the oldest in the Civil Engineering field. The *plastic state of stress,* when the failure is imminent, was investigated by Rankine in 1860. A lot of theoretical and experimental work has been done in this field and many theories and hypothesis have been proposed.*

23.2. TYPES OF RETAINING WALLS

Retaining walls may be classified according to their mode of resisting the earth pressure, and according to their shape. Following are some of common types of retaining walls (Fig. 23.1).

(*i*) Gravity walls
(*ii*) Cantilever retaining walls : (*a*) T-shaped (*b*) L-shaped
(*iii*) Counterfort retaining walls.
(*iv*) Buttressed walls.

A gravity retaining wall shown in Fig. 23.1(*a*) is the one in which the earth pressure exerted by the backfill is resisted by dead weight of the wall, which is either made of masonry or of mass concrete. The stress developed in the wall is very low. These walls are so proportioned that no tension is developed anywhere, and the resultant of forces remain within the middle third of the base.

* For detailed account of various earth pressure theories, the reader may refer to the Author's book 'Soil Mechanics and Foundations'

The cantilever retaining wall resists the horizontal earth pressure as well as other vertical pressures by way of bending of various components acting as cantilevers. A common form of cantilever retaining wall is the T-shaped wall shown in Fig. 23.1(*b*). The wall consists of stem *AB*, heel slab *BC* and toe slab *DB*. Each of these bend as cantilevers, about *B*. They are, therefore, reinforced on the tension face. Another form of cantilever retaining walls are the L-shaped walls shown in Fig. 23.1(*c*) and (*d*). They also resist the soil pressures by bending.

A counterfort retaining wall is shown in Fig. 23.1(*e*). The vertical stem and the heel slab are strengthened by providing counterforts at some suitable intervals. Because of provision of counterforts, the vertical stem as well as the heel slab acts as *continuous slab,* in contrast to the cantilevers of cantilever retaining wall. The toe slab however acts as cantilever bending upwards. This type of retaining wall is used when backfill of greater height is to be retained. A buttressed wall is a modification of the counterfort retaining wall in which the counterforts, called the buttresses, are provided to the other side of the backfill. However the buttresses reduce the clearance in front of the wall, and therefore these walls are not commonly used.

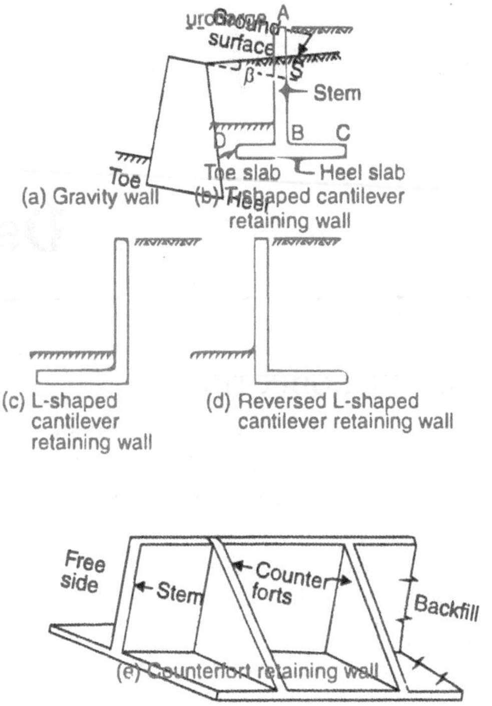

FIG. 23.1. VARIOUS TYPES OF RETAINING WALLS.

23.3. ACTIVE EARTH PRESSURE : RANKINE'S THEORY

As originally proposed, Rankine's theory of lateral earth pressure is applied to uniform cohesionless soil only. Later, it was extended to include cohesive soils, by Resal and by Bell. The theory has been extended to stratified, partially immersed and submerged soils. Following are the assumptions of Rankine Theory: (1) The soil mass is semi-infinite, homogeneous, dry and cohesionless. (2) The ground surface is plane which may be horizontal or inclined. (3) The back of the wall is vertical and smooth. In other words, there are no shearing stresses between the wall and the soil, and the stress relationship for any element adjacent to the wall is the same as for any other element for away from the wall. (4) The wall yields about the base and satisfies the deformation conditions for plastic equilibrium.

However, the retaining walls are constructed of masonry or concrete, and hence the back of the wall is never smooth. Due to this, frictional forces develop. As a consequence of Rankine's assumption of non-existence of frictional forces at the wall face, the resultant pressure must be parallel to the surface of the backfill. The existence of friction makes the resultant pressure inclined to the normal to the wall at an angle that approaches the friction angle between the soil and the wall. We shall consider the following cases of cohesionless backfill

1. Dry or moist backfill with no surcharge. 2. Submerged backfill.
3. Backfill with uniform surcharge. 4. Backfill with sloping surfaces.

1. Dry or moist backfill with no surcharge. Fig. 23.2 (*a*) shows a retaining wall of height H, having dry or moist backfill, with no surcharge. According to Rankine's theory, the intensity of *active earth pressure*, p_a, trying to move the wall away from the fill, is given by

$$p_a = K_a \gamma H \qquad ...(23.1)$$

where K_a = co-efficient of active earth pressure,

$$= \frac{1 - \sin \Phi}{1 + \sin \Phi} \qquad ...(23.2)$$

Φ = angle of internal friction, for the backfill.
γ = unit weight of soil (backfill)
H = height of retaining wall.

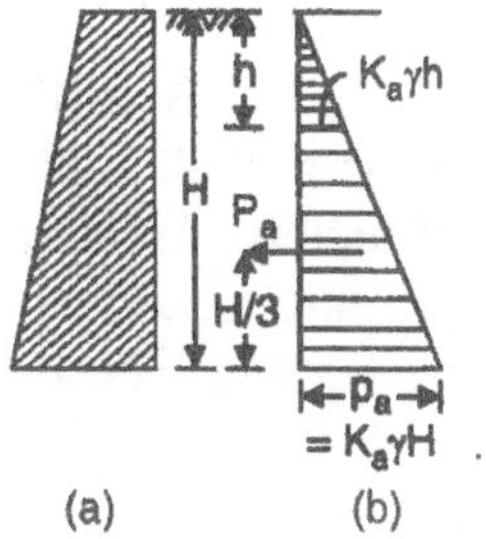

FIG. 23.2.

At any height h below the horizontal ground surface, the intensity of active earth pressure is $K_a \gamma h$. Fig. 23.2(*b*) shows the distribution of active earth pressure p_a over the retaining wall. The total active earth pressure P_a or the resultant pressure per unit length of the wall is given by

$$P_a = \frac{1}{2} K_a \cdot \gamma \cdot H^2 \qquad ...(23.3)$$

acting at $H/3$ above the base of retaining wall. If the soil is dry, γ is the dry weight of the soil, and if wet, γ is the moist weight, to be substituted in Eq. 23.1 or 23.3.

2. Submerged backfill. In this case, the sand fill behind the retaining wall is saturated with water. The lateral pressure is made up of two components.

(*i*) lateral pressure due to submerged weight γ' of soil, and

(*ii*) lateral pressure due to water.

Thus, at any depth h below the surface,

$$p_a = K_a \gamma' h + \gamma_w \cdot h \qquad ...(23.4)$$

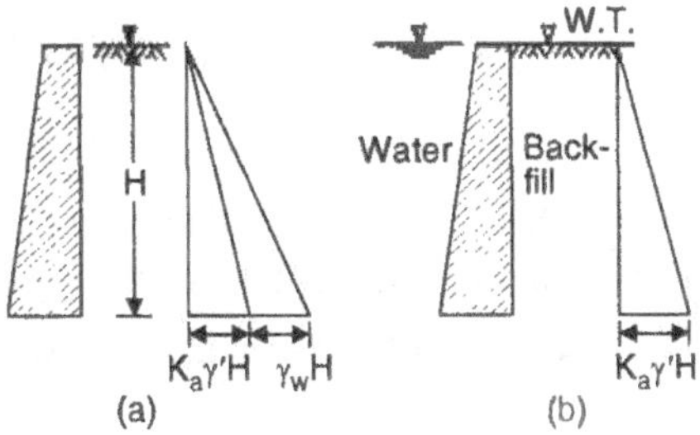

FIG. 23.3. SUBMERGED BACKFILL.

The pressure at the base of the retaining wall ($h = H$) is given by

$$p_a = K_a \gamma' H + \gamma_w H \qquad ...(23.5)$$

If the free water stands to both the sides of the wall [Fig 23.3(*b*)], the water pressure need not be considered, and the net lateral pressure given by

$$p_a = K_a \gamma' H \qquad ...(23.6)$$

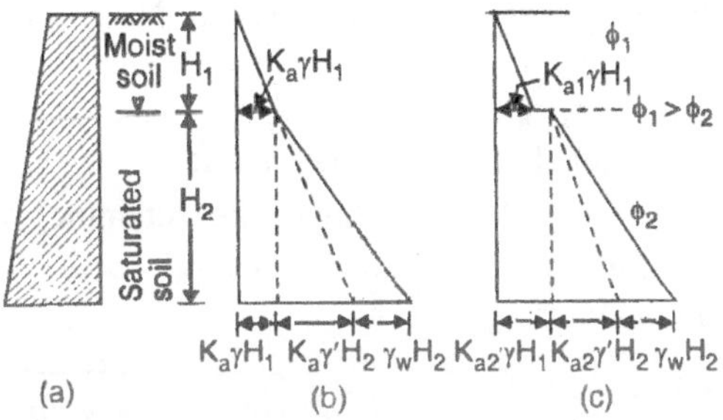

FIG. 23.4. PARTLY SUBMERGED BACKFILL.

If the backfill is partly submerged, *i.e.*, the backfill is moist to a depth H_1 below the ground level, and then it is submerged, the lateral pressure intensity at the base of the wall is given by

$$p_a = K_a \cdot \gamma H_1 + K_a \gamma' H_2 + \gamma_w \cdot H_2 \qquad \ldots(23.7)$$

The above expression is based on the assumption that the value of Φ is the same for the moist as well as submerged soil. If it is different (say Φ_1 and Φ_2 respectively), the earth pressure coefficient K_{a1} and K_{a2} for both the portions will be different. As Φ decrease, K_a increases. The lateral pressure intensity [Fig. 23.4(*c*)] at the base of wall is

$$p_a = K_{a2} \gamma H_1 + K_{a2} \gamma' H_2 + \gamma_w H_2 \qquad \ldots(23.8)$$

3. Backfill with uniform surcharge. If the backfill is horizontal and carries surcharge of uniform intensity w per unit area, the vertical pressure increment, at any depth h, will increase by w. The increase in the lateral pressure due to this will be $K_a w$. Hence the lateral pressure at any depth h is given by

$$p_a = K_a \cdot \gamma \cdot h + K_a \cdot w \qquad \ldots(23.9)$$

At the base of the wall, the pressure intensity is :

$$p_a = K_a \gamma H + K_a w \qquad \ldots[23.9(a)]$$

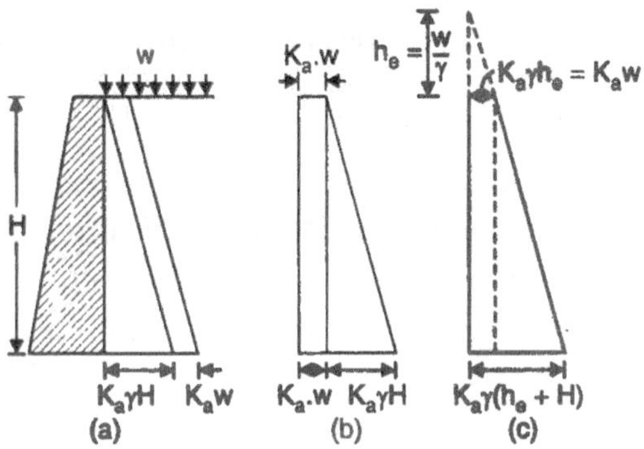

FIG. 23.5. BACK FILL WITH UNIFORM SURCHARGE

Fig. 23.5(*a*) and (*b*) show two alternative methods of plotting the lateral pressure diagram for this case. The lateral pressure increment $K_a \cdot w$ due to the surcharge is the same at every point of the back of the wall, and does not vary with height h. The height of fill h_e equivalent to the uniform surcharge intensity is given by the relation:

$$K_a \cdot \gamma \cdot h_e = K_a w \quad \text{or} \quad h_e = \frac{w}{\gamma} \qquad \ldots(23.10)$$

This means that the effect of the surcharge of intensity w is the same as that of a fill of height h_e above the ground surface [Fig. 23.5(*c*)].

4. Backfill with sloping surface. Let the sloping surface behind the wall be inclined at an angle β with the horizontal ; β is called the surcharge angle. In finding out the active earth pressure for this case by Rankine's theory, an additional assumption that the vertical and lateral stresses are *conjugate* is made. It can be shown that if the stress on a given plane at a given point is parallel to another plane, the stress on the latter plane at the same point must be parallel to the first plane.

Fig. 23.6(*a*) shows the retaining wall with a sloping backfill. The intensity of lateral earth pressure at the base of wall is given by

$$p_a = \gamma H \cdot \cos\beta \, \frac{\cos\beta - \sqrt{\cos^2\beta - \cos^2\Phi}}{\cos\beta + \sqrt{\cos^2\beta - \cos^2\Phi}}$$

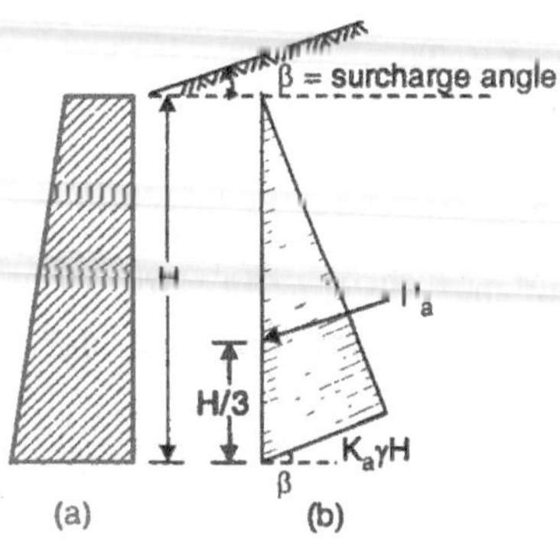

FIG. 23.6. LATERAL PRESSURE DISTRIBUTION FOR SLOPING SURCHARGE

or $$p_a = K_a \gamma H \qquad \text{...[23.11(a)]}$$

where $$K_a = \cos\beta \frac{\cos\beta - \sqrt{\cos^2\beta - \cos^2\Phi}}{\cos\beta + \sqrt{\cos^2\beta - \cos^2\Phi}} \qquad \text{...[23.11(b)]}$$

The pressure distribution is shown in Fig. 23.6(*b*). The pressure acts *parallel* to the sloping surface of the surcharge, *i.e.*, at β with the horizontal. The total active pressure P_a for the wall of height H is given by

$$P_a = \frac{1}{2} K_a \gamma H^2$$

The resultant acts at $H/3$ above the base, in direction parallel to the surcharge.

When $\beta = 0$ (*i.e.* horizontal ground surface), Eq. 23.11(*b*) reduces to

$$K_a = \frac{1 - \sin\Phi}{1 + \sin\Phi}$$ which is the same as Eq. 23.2.

23.4. PASSIVE EARTH PRESSURE

Passive earth pressure is exerted on a wall when it has a tendency to move towards the backfill. Such a condition may occur when the retaining wall supports an arch, and is subjected to arch thrust, moving it towards the fill. Another condition of passive earth pressure may be when the wall supports soil of different heights on both the sides as shown in Fig. 23.7. Due to active earth pressure from the right hand side, the wall moves to the left. The soil to the left is thus compressed and in turn exert passive earth pressure, resisting such movement.

If h is the height of fill, the intensity of passive pressure at height h is given by

$$p_p = K_p \cdot \gamma h \qquad \text{...(23.12)}$$

where K_p = coefficient of passive earth pressure

$$= \frac{1 + \sin\Phi}{1 - \sin\Phi} = \frac{1}{K_a} \qquad \text{...(23.13)}$$

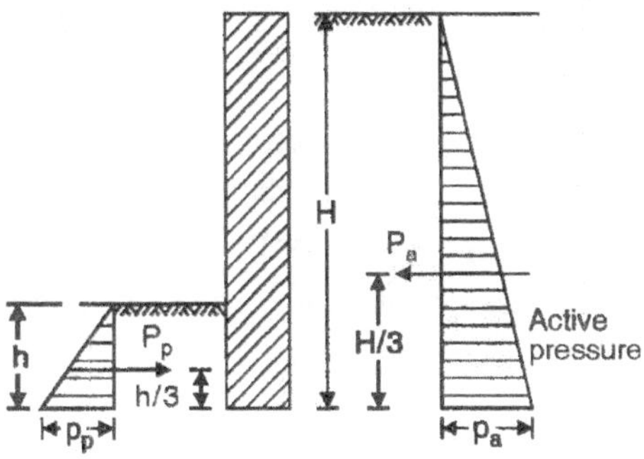

FIG. 23.7

The passive pressure distribution will thus be a triangle. The total pressure is given by

$$P_p = K_p \cdot \frac{\gamma h^2}{2} \qquad \text{...(23.14)}$$

acting at $h/3$ above base. It should be noted that K_p is very much greater than K_a. For example when $\Phi = 30°$, $K_a = \frac{1 - \sin\Phi}{1 + \sin\Phi} = \frac{1}{3}$ and $K_p = \frac{1 + \sin\Phi}{1 - \sin\Phi} = 3$.

Thus K_p, for this case, is 9 times K_a. In making calculations for active earth pressure behind retaining walls, the passive earth pressure due to other side of walls is generally neglected, since it is possible that this soil will erode away or that shearing will occur, thus eliminating the passive pressure.

If a uniform surcharge of intensity w per unit area acts over the surface of the backfill, the increase in the passive pressures will be equal to $K_p \cdot w$. The passive pressure intensity at depth h is given by $p_p = K_p\ (\gamma h + w)$ ...(23.15)

If the backfill is having its top surface inclined at an angle β, the passive pressure is given by

$$p_p = \gamma h \cos\beta \frac{\cos\beta + \sqrt{\cos^2\beta - \cos^2\Phi}}{\cos\beta - \sqrt{\cos^2\beta - \cos^2\Phi}} \quad \text{or} \quad p_p = K_p \cdot \gamma h \qquad \ldots(23.16)$$

where

$$K_p = \cos\beta \frac{\cos\beta + \sqrt{\cos^2\beta - \cos^2\Phi}}{\cos\beta - \sqrt{\cos^2\beta - \cos^2\Phi}} \qquad \ldots[23.16(b)]$$

23.5. STABILITY OF CANTILEVER RETAINING WALL

Fig. 23.8 shows a cantilever retaining wall subjected to the following forces :

(1) Weight W_1 of the stem AB. (2) Weight W_2 of the base slab DC.

(3) Weight W_3 of the column of soil supported on heel slab BC.

(4) Horizontal force P_a, equal to active earth pressure acting at $H/3$ above the base.

The following are the modes of failure of a retaining wall :

1. Overturning about the toe. 2. Sliding. 3. Failure of soil due to excessive pressure at toe or tension at the heel. 4. Bending failure of stem or base of slab or heel slab.

1. Overturning. The most hazardous mode of failure of retaining walls is due to overturning because of unbalanced moments. Thus, in Fig 23.8, the overturning moment, due to active earth pressure, at toe is

$$M_0 = P_a \frac{H}{3} = K_a \frac{\gamma H^2}{2} \cdot \frac{H}{3} = K_a \cdot \frac{\gamma H^3}{6} \qquad \ldots(23.17)$$

The resisting moment is due to the weights W_1, W_2 and W_3, neglecting the passive earth pressure and weight of soil above the toe slab.

Hence $M_R = W_1 \bar{x}_1 + W_2 \bar{x}_2 + W_3 \bar{x}_3$ ...[23.17(a)]

Hence factor of safety due to overturning (F_1) is given by

$$F_1 = \frac{M_R}{M_0} \qquad \ldots(23.18)$$

A minimum factor of safety of 2 should be used.

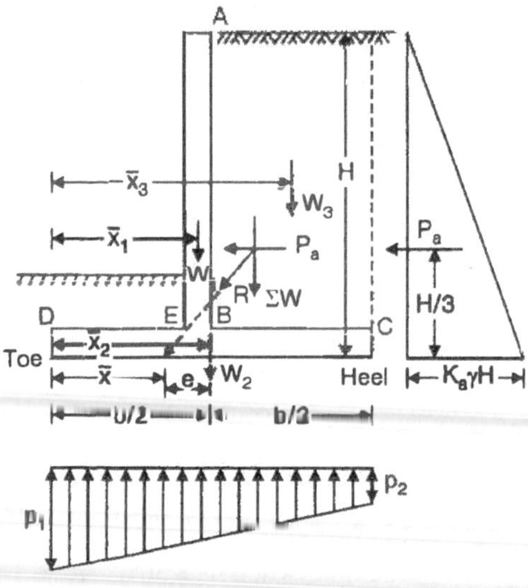

FIG. 23.8.

2. Sliding. The horizontal force P_a tends to slide the wall away from the fill. The tendency to resist this is achieved by the friction at the base.

The force of resistance, F is given by

$$F = \mu \Sigma W \qquad \ldots(23.19)$$

where μ is the coefficient of friction(or friction factor) between soil and concrete, and ΣW is the sum of vertical forces. The factor of safety F_2 due to sliding is given by

$$F_2 = \frac{\mu \Sigma W}{H} \qquad \ldots(23.20)$$

If the wall is found to be unsafe against sliding, shear key below the base [Fig. 23.9(*b*)] should be provided. Such a key develops passive pressure which resists completely the sliding tendency of the wall. A factor of safety of 1.5 must be used against sliding.

In the absence of elaborate tests, the following values of μ may be adopted:

	μ
1. Coarse grained soil without silt	0.55
2. Coarse grained soil with silt	0.45
3. Silt	0.35

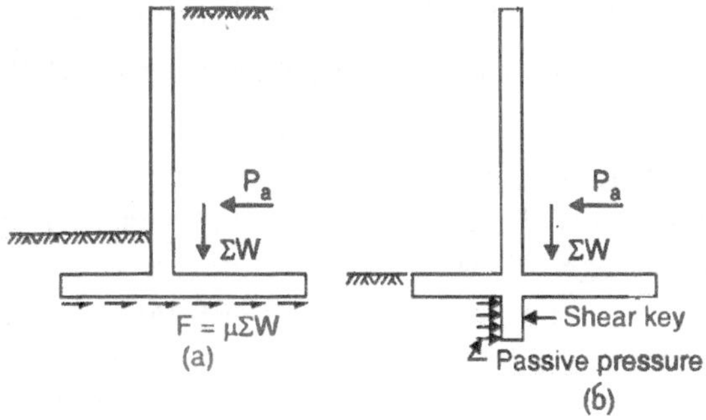

FIG. 23.9.

3. Soil pressure distribution. Fig. 23.8 shows the various forces acting on the wall. If ΣW is the sum of all vertical forces, and P_a is the horizontal active earth pressure, the resultant R will strike the base slab at a distance e (say) from the middle point of the base.

Let $$\Sigma M = W_1 \bar{x}_1 + W_2 \bar{x}_2 + W_3 \bar{x}_3 - P_a \cdot \frac{H}{3} = \text{net moment at the toe.}$$

Then $$\bar{x} = \text{distance of point of application of resultant} = \frac{\Sigma M}{\Sigma W} \quad \text{...(23.21)}$$

Hence eccentricity $$e = \frac{b}{2} - \bar{x} \quad \text{...(23.22)}$$

The pressure distribution below the base, is shown in Fig. 23.8. The intensity of soil pressure at the toe and heel is given by

$$p_1 = \frac{\Sigma W}{b}\left(1 + \frac{6\,e}{b}\right) \text{ at toe} \quad \text{...(23.23)}$$

and $$p_2 = \frac{\Sigma W}{b}\left(1 - \frac{6\,e}{b}\right) \text{ at heel} \quad \text{...(23.24)}$$

p_1 at toe should not exceed the safe bearing capacity of soil otherwise soil will fail. Similarly, p_2 at heel should be compressive. If p_2 comes to be tensile, the heel will be lifted above the soil, which is not permissible. In an extreme case, p_2 may be zero, where $e = b/6$. Hence in order that tension is not developed, the resultant should strike the base within the middle third.

4. *Bending failure*. There are three distinct parts of T-shaped cantilever retaining wall : the stem AB, the heel slab BC and toe slab DE. The stem AB will bend as cantilever, so that tensile face will be towards the backfill. The critical section will be at B, where cracks may occur at the inner face if it is not properly reinforced. The heel slab will have net pressure acting downwards, and will bend as a cantilever, having tensile face upwards. The critical section

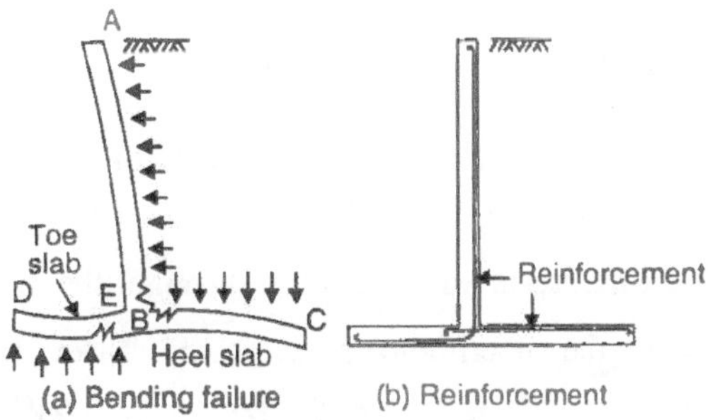

FIG. 23.10.

will be at B, where cracks may occur if it is not reinforced properly at the upper face as shown in Fig. 23.10(b). The net pressure on toe slab will act upwards, and hence it must be reinforced at the bottom face. The thickness of stem, heel slab and toe slab must be sufficient to withstand compressive stresses due to bending.

23.6. DESIGN PRINCIPLES OF CANTILEVER RETAINING WALL

The various dimensions of cantilever retaining wall should be so proportioned that the various failure criteria discussed above are taken care of. The design of a cantilever retaining wall consists of the following (*i*) fixation of base width b (*ii*) design of stem (*iii*) design of heel slab (*iv*) design of toe slab.

1. Fixation of base width (b). The base width b of the retaining wall should be so chosen that the resultant of the forces remain within middle third, and the ratio of length of the toe slab to the base width should be such that the stress p_1 at toe does not exceed the safe bearing capacity of soil. The method below gives an approximate method of finding the base width b and the width of toe slab $DE = \alpha b$.

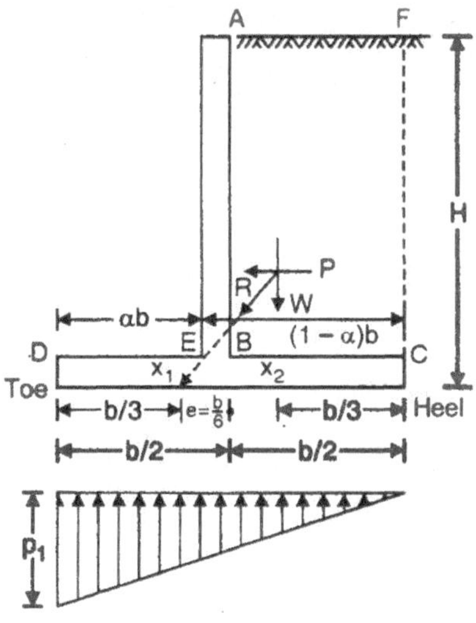

FIG. 23.11

Let us assume that the average unit weight of concrete and soil is equal to 1.1 γ where γ is the unit weight of soil. Neglecting the weight of toe slab DE, the total weight W of retaining wall plus the weight of soil in column $AFCB$ will be

$$W = (1 - \alpha)b \times H(1.1\gamma) = 1.1\gamma bH(1 - \alpha) \quad \ldots(i)$$

This acts at a distance of $\frac{1}{2}(1 - \alpha)b$ from E or C.

The horizontal earth pressure $P = K_a \dfrac{\gamma H^2}{2}$ ...(*ii*)

Let X_1 and X_2 be the outer and inner third points respectively. In order that no tension is developed at heel, the resultant should pass through the outer third point X_1. In that case, moment of resultant R will be zero, or the algebraic sum of moments of W and P about X_1 is zero. Hence, we have

$$W\left[\frac{2b}{3} - \frac{1}{2}(1 - \alpha)b\right] - P \cdot \frac{H}{3} = 0 \quad \text{or} \quad 1.1\gamma\, b\, H\,(1 - \alpha)\left\{\frac{2b}{3} - \frac{b}{2} + \frac{b\alpha}{2}\right\} - K_a \frac{\gamma H^3}{6} = 0$$

$$1.1\gamma b^2 H(1 - \alpha)(1 + 3\alpha) = K_a \gamma H^3, \quad \text{which gives } b = 0.95\, H \sqrt{\frac{K_a}{(1 - \alpha)(1 + 3\alpha)}} \quad \ldots(23.25)$$

Thus b can be determined, if α, *i.e.* the ratio of heel slab DE to the total base width is known. The ratio α can be fixed on the basis of maximum pressure under the toe. When the resultant R passes through the outer third point X_1, eccentricity $e = \dfrac{b}{2} - \dfrac{b}{3} = \dfrac{b}{6}$.

Hence the maximum pressure p_1 under the toe is

$$p_1 = \frac{W}{b}\left(1 + \frac{6e}{b}\right) = \frac{W}{b}(1 + 1) = \frac{2W}{b} = \frac{2}{b}\{1.1\gamma\, b\, H\,(1 - \alpha)\} \quad \ldots(iii)$$

The value of p_1 should not exceed the safe bending capacity q_0 of the soil. Hence,

$$q_0 = p_1 = 2.2\ \gamma H (1 - \alpha)$$

$$\therefore\ (1 - \alpha) = \frac{q_0}{2.2\gamma H}$$

or $$\alpha = 1 - \frac{q_0}{2.2\ \gamma\ H} \qquad \ldots(23.26)$$

This value of α can be substituted in Eq. 23.25 to get the value of b. Eqs. 23.25 and 23.26 are valid, though as a rough guide, only when there is no surcharge. For the case of sloping backfill with surcharge angle β, the following approximate expression may be used :

$$\alpha = 1 - \frac{q_0}{2.7\gamma H} \qquad \ldots(23.26(a))]$$

and $$b = H\sqrt{\frac{K_a \cos\beta}{(1-\alpha)(1+3\alpha)}} \qquad \ldots[23.26(a)]$$

Fig. 23.12 shows common values of b for various conditions of cantilever retaining walls.

Determination of base width from the considerations of sliding. Eq. 23.25. gives the base width from the stress considerations. Let us find the base width from sliding consideration also. The factor of safety against sliding should at least be equal to 1.5. Thus,

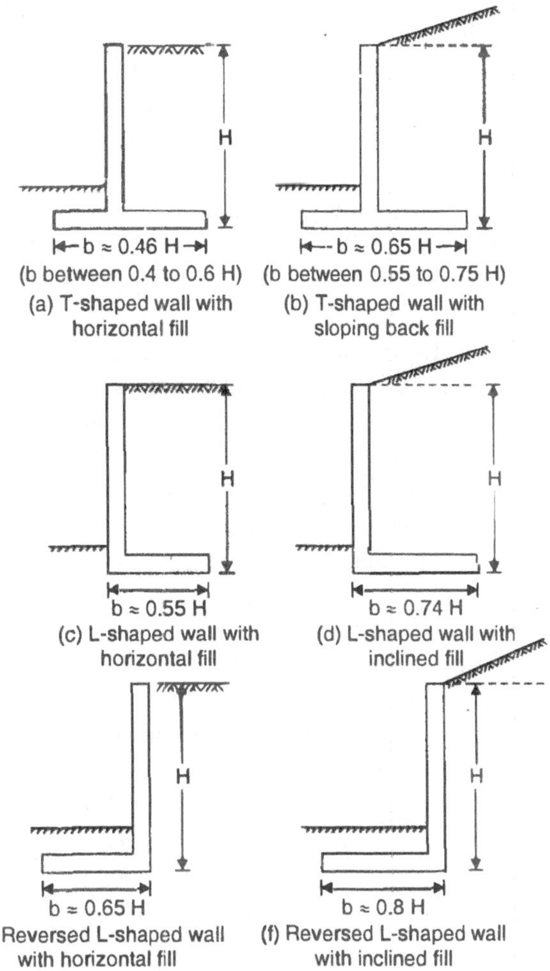

FIG. 23.12. FIXATION OF LENGTH OF BASE.

$$\text{F.S.} = \frac{\mu W}{P} = 1.5$$

Substituting the values of $W = 1.1\ \gamma b\ H\ (1 - \alpha)$ and $P = K_a \frac{\gamma\ H^2}{2}$, we get

$$\mu[1.1\gamma b\ H (1 - \alpha)] = 1.5 K_a \frac{\gamma H^2}{2}$$

or $$b \approx \frac{0.7H}{1-\alpha} \cdot \frac{K_a}{\mu} \qquad \ldots(23.27)$$

Thus knowing α from Eq. 23.26, b can be determined from Eq. 23.27.

2. Design of stem. The vertical stem AB is designed as cantilever, for triangular loading. At any section h below the top point A, the force is equal to $K_a \frac{\gamma h^2}{2}$ and its bending moment about the section is $K_a\ \gamma\ h^3/6$. The thickness at B is maximum. The minimum thickness at A should vary from 20 to 30 cm depending upon the height of the wall. Reinforcement is provided towards the inner face of stem, *i.e.* towards side of fill. The reinforcement towards the top of stem can be curtailed, since B.M. varies as h^3. Distribution reinforcement (HYSD bars) is provided @ 0.12% of the area of cross-section along the length of retaining wall at inner face. Similarly, at the outer face of the stem, temperature reinforcement (HYSD bars) is provided both in horizontal as well as in vertical direction, at the rate of 0.12% of the area of cross-section.

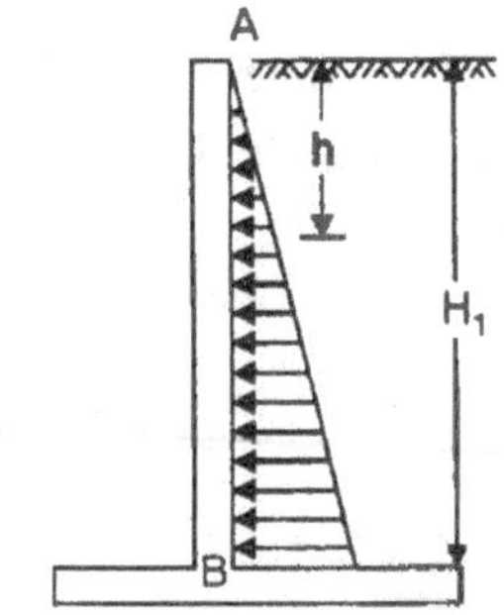

FIG. 23.13.

3. Design of heel slab. The heel is also to be designed as a cantilever. It has both downward pressure(due to weight of soil and self-weight) as well as upward pressure due to soil reaction. However, the *net* pressure is found to act *downward* and hence reinforcement is provided at the upper face BC.

4. Design of toe slab. Neglecting the weight of the soil above it, the toe slab will bend upwards as a cantilever due to upward soil reaction. Hence reinforcement is placed at the bottom face. Normally, the thickness of both toe slab and heel slab is kept the same, determined on the basis of greater of the cantilever bending moments.

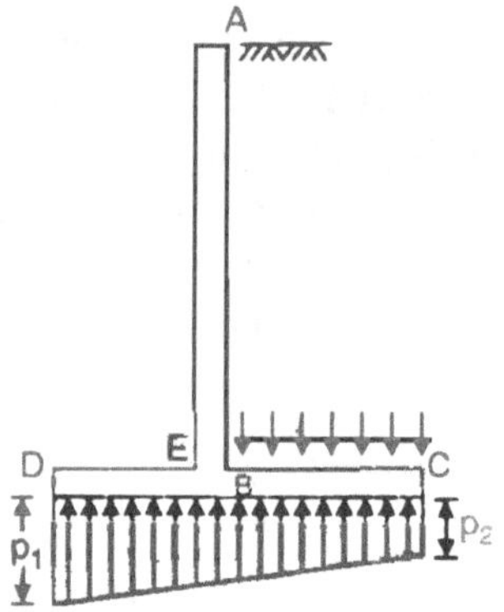

FIG. 23.14.

Depth of foundation. The height H_2 of the retaining wall, above ground level is fixed on the basis of height of the backfill to be retained. The depth of foundation y should be such that good quality of soil to bear the induced pressures is available. However, a minimum depth of foundation given below by Rankine's formula should be provided :

$$y_{min} = \frac{q_0}{\gamma}\left(\frac{1 - \sin \phi}{1 + \sin \phi}\right)^2 = \frac{q_0}{\gamma} K_a^{\ 2} \qquad \ldots(23.28)$$

where q_0 is the safe bearing capacity of soil, or equal to the maximum pressure likely to occur on soil.

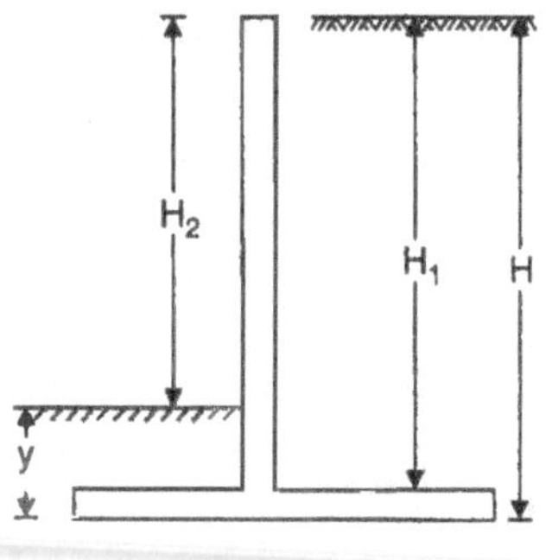

FIG. 23.15

23.7. DESIGN OF CANTILEVER RETAINING WALL WITH HORIZONTAL BACKFILL

Design Example 23.1. *Design a T-shaped cantilever retaining wall to retain earth embankment 3 m high above ground level. The unit weight of earth is 18 kN/m³ and its angle of repose is 30°. The embankment is horizontal at its top. The safe bearing capacity of soil may be taken as 100 kN/m² and the coefficient of friction between soil and concrete as 0.5. Use M 20 mix. and Fe 415 bars.*

Solution

1. ***Design constants.*** For M 20 concrete and HYSD bars we have the following:

$$\frac{x_{u\,.\,max}}{d} = 0.479 \quad \text{and} \quad R_u = 2.761$$

2. *Depth of foundation*

$$\gamma = 18 \text{ kN/m}^3 = 18000 \text{ N/m}^3$$

The minimum depth of foundation is

$$y_{min} = \frac{q_o}{\gamma}\left(\frac{1-\sin\Phi}{1+\sin\Phi}\right)^2 = \frac{100}{18}\left(\frac{1-\sin 30°}{1+\sin 30°}\right)^2 = 0.62 \text{ m.}$$

However, keep depth $= 1$ m to accommodate thickness of base wall below the ground surface. Hence height of wall above its base $= H = 3 + 1 = 4$ m.

3. *Dimensions of base*. The ratio of the length of toe slab (DE) to the base width b is given by Eq. 23.26 :

$$\alpha = 1 - \frac{q_o}{2.2\gamma H} = 1 - \frac{100}{2.2 \times 18 \times 4} = 0.365$$

Keep $\alpha = 0.37$ ...(*i*)

The width of base is given by Eq. 23.25 :

$$b = 0.95\, H\sqrt{\frac{K_a}{(1-\alpha)(1+3\alpha)}}$$

$$K_a = \frac{1-\sin\Phi}{1+\sin\Phi} = \frac{1-\sin 30°}{1+\sin 30°} = \frac{1}{3}$$

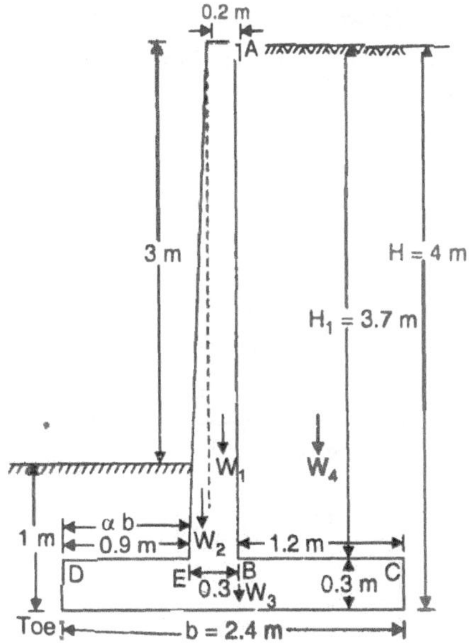

FIG. 23.16

$$b = 0.95 \times 4\sqrt{\frac{1}{3(1-0.37)(1+3\times 0.37)}} = 1.90 \text{ m.}$$

The base width from the considerations of sliding is given by Eq. 23.27.

$$b = \frac{0.7H\,K_a}{(1-\alpha)\,\mu} = \frac{0.7\times 4}{1-0.37} \times \frac{1}{3\times 0.5} = 2.96 \text{ m.}$$

This width is excessive. Normal practice is to provide b between 0.4 to 0.6 H. Taking maximum value of 0.6 H,

$$b = 0.6\,H = 0.6 \times 4 = 2.4 \text{ m.}$$

Hence provide $b = 2.4$ m. The wall will be unsafe against sliding. This will be made safe by providing a shear key at the base.

Width of toe slab $= 0.37 \times 2.4 = 0.89$ m. Provide toe slab 0.9 m long.

Let the thickness of base be $\frac{1}{12}\cdot H \approx 0.3$ m for preliminary calculations.

4. *Thickness of stem*.

Height $AB = 4 - 0.3 = 3.7$ m. Consider one meter length of retaining wall.

Maximum bending moment at $B = K_a\,\gamma\,\frac{H_1^3}{6} = \frac{1}{3}\,\frac{18}{6}(3.7)^3 = 50.65$ kN-m $= 50.65 \times 10^6$ N-mm

$$\therefore \quad M_u = 1.5 \times 50.65 \times 10^6 = 75.975 \times 10^6 \text{ N-mm}$$

Hence the effective depth is $d = \sqrt{\frac{M_u}{R_u b}} = \sqrt{\frac{75.975\times 10^6}{1000 \times 2.761}} = 166$ mm

However, keep d = 240 mm and total thickness = 300 mm so that an effective cover of 60 mm is available and nominal cover = 60 − 8 = 52 mm, using 16 mm φ bars. Reduce the total thickness to 200 mm at the top so that effective depth of 140 mm is available at the top.

5. *Stability of wall*. Fully dimensioned wall is shown in Fig. 23.16.

Let W_1 = weight of rectangular portion of stem

W_2 = weight of triangular portion of stem

W_3 = weight of base slab ; W_4 = weight of soil on heel slab.

The calculations are arranged in Table 23.1

TABLE 23.1

S. No.	*Designation*	*Force (kN)*	*Lever arm (m)*	*Moment about toe (kN-m)*
1	W_1	$1 \times 0.2 \times 3.7 \times 25 = 18.5$	1.1	20.35
2	W_2	$\frac{1}{2} \times 0.1 \times 3.7 \times 25 = 4.63$	0.97	4.49
3	W_3	$1 \times 2.4 \times 0.3 \times 25 = 18.0$	1.2	21.60
4	W_4	$1 \times 1.2 \times 3.7 \times 18 = 79.97$	1.8	143.86
		$\Sigma W = 121.05$		M_R = 190.30

The total resisting moment M_R = 190.30 kN-m ...(1)

$$\text{Earth pressure } P = K_a \gamma \frac{H^2}{2} = \frac{1}{3} \times \frac{18}{2}(4)^2 = 48 \text{ kN} \quad ...(2)$$

Overturning

$$\text{Overturning moment } M_0 = 48 \times \frac{4}{3} = 64 \text{ kN-m} \quad ...(3)$$

$$\therefore \quad \text{F.S. against overturning} = \frac{190.3}{64} = 2.97 > 2. \text{ Hence safe.}$$

Sliding

$$\text{F.S. against sliding} = \frac{\mu \Sigma W}{P} = \frac{0.5 \times 121.05}{48} = 1.26 < 1.5. \quad \text{Hence unsafe.}$$

To make it safe against sliding, we will have to provide a shear key.

***Pressure distribution*.** Net moment $\Sigma M = 190.3 - 64 = 126.3$ kN-m

∴ Distance $\bar{x}$ of the point of application of resultant, from toe is

$$\bar{x} = \frac{\Sigma M}{\Sigma W} = \frac{126.3}{121.05} = 1.04 \text{ m}$$

$$\text{Eccentricity } e = \frac{b}{2} - \bar{x} = 1.2 - 1.04 = 0.16 \text{ m. This is less than } \frac{b}{6}\left(= \frac{2.4}{6} = 0.4 \text{ m}\right)$$

$$\text{Pressure } p_1 \text{ at toe} = \frac{\Sigma W}{b}\left(1 + \frac{6e}{b}\right) = \frac{121.05}{2.4}\left(1 + \frac{6 \times 0.16}{2.4}\right) = 70.61 \text{ kN/m}^2 < 100.$$

Hence safe.

$$\text{Pressure } p_2 \text{ at heel} = \frac{\Sigma W}{b}\left(1 - \frac{6e}{b}\right) = \frac{121.05}{2.4}\left(1 - \frac{6 \times 0.16}{2.4}\right) = 30.26 \text{ kN/m}^2$$

Pressure p at the junction of stem with toe slab is

$$p = 70.61 - \frac{70.61 - 30.26}{2.4} \times 0.9 = 55.48 \quad kN/m^2$$

Pressure p' at the junction of stem with heel slab is

$$p' = 70.61 - \frac{70.61 - 30.26}{2.4} \times 1.2 = 50.44 \quad kN/m^2$$

6. *Design of toe slab*. The upward pressure distribution on the toe slab is shown in Fig. 23.17. The weight of the soil above the toe slab is neglected. Thus two forces are acting on it:

(*i*) upward soil pressure. (*ii*) downward weight of slab.

Downward weight of slab per unit area = $0.3 \times 1 \times 1 \times 25 = 7.5$ kN/m²

Hence net pressure intensities will be = 70.61 − 7.5 = 63.11 kN/m² under D and 55.48 − 7.5 = 47.98 kN/m² under E.

Total force = S.F. at $E = \frac{1}{2}(63.11 + 47.98)0.9 = 50$ kN

$$\bar{x} \text{ from } E = \left(\frac{47.98 + 2 \times 63.11}{47.98 + 63.11}\right)\frac{0.9}{3} = 0.47 \text{ m.}$$

∴ B.M. at $E = 50 \times 0.47 = 23.52$ kN-m $= 23.52 \times 10^6$ N-mm

∴ M_u at $E = 1.5 \times 23.62 \times 10^6 = 35.28 \times 10^8$ N-mm

$$\therefore \quad d = \sqrt{\frac{35.28 \times 10^6}{1000 \times 2.761}} = 114 \text{ mm}$$

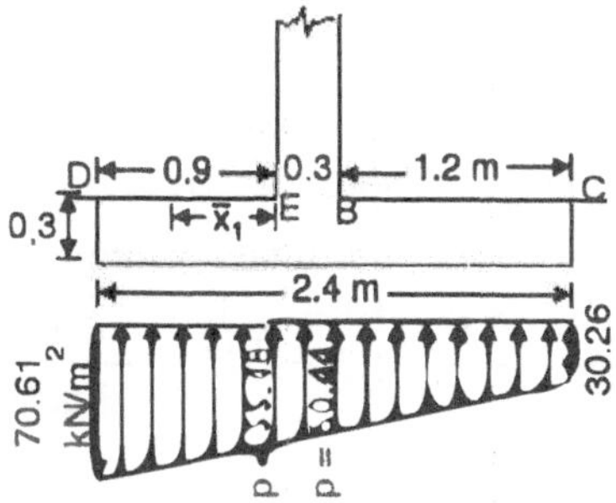

FIG. 23.17.

Let us keep total depth = 260 mm and effective depth of 200 mm so that 60 mm effective cover is available. Thickness can be reduced to 200 mm at the edge. Since actual d provided is more, we have an under-reinforced section, for which we have

$$A_{st} = \frac{0.5 f_{ck}}{f_y}\left[1 - \sqrt{1 - \frac{4.6 M_u}{f_{ck} bd^2}}\right] bd = \frac{0.5 \times 20}{415}\left[1 - \sqrt{1 - \frac{4.6 \times 35.28 \times 10^6}{20 \times 1000\,(200)^2}}\right] 1000 \times 200$$

$\simeq 517$ mm²

This reinforcement has to be provided at the bottom face. If alternate bars of stem reinforcement are bent and continued in the toe slab, area available $= \frac{1}{2} \times 1130 = 565$ mm² (see step 8). This reinforcement will consist of 12 mm φ bars @ 200 mm c/c. Let us check this reinforcement for development length.

$$L_d \approx 47\,\varphi = 47 \times 12 = 564 \text{ mm.}$$

Providing 50 mm clear side cover, actual length available = 900 − 50 = 850 mm > L_d. Hence safe.

$$\text{Distribution reinforcement} = \frac{0.12}{100} \times 1000\left[\frac{260 + 200}{2}\right] = 276 \text{ mm}^2$$

Using 8 mm φ bars, $A_\varphi = 50.3$ mm². ∴ Spacing $= \frac{1000 \times 50.3}{276} = 182$ mm

However provide these @ 180 mm c/c.

7. ***Design of heel slab***. Three forces act on it : (*i*) downward weight of soil 3.7 m high (*ii*) downward weight of heel slab (*iii*) upward soil pressure

Total weight of soil = 1.2 × 3.7 × 1 × 18 = 80 kN acting at 0.6 m from *B*.

Total weight of heel slab = 1.2 × 0.26 × 1 × 25 = 7.8 acting at 0.6 m from *B*.

Total upward soil reaction = $\frac{1}{2}(50.44 + 30.26)1.2 = 48.42$ kN

$$\text{acting at } \left(\frac{50.44 + 2 \times 30.26}{50.44 + 30.26}\right)\frac{1.2}{3} = 0.55 \text{ m from } B$$

∴ Total force = S.F. at $B = 80 + 7.8 - 48.42 = 39.38$ kN

∴ B.M. at $B = (80 \times 0.6) + (7.8 \times 0.6) - (48.42 \times 0.55) = 26.05$ kN-m

$= 26.05 \times 10^6$ N-mm ; $M_u = 1.5 \times 26.05 \times 10^6 = 39.075 \times 10^6$ N-mm

$$\therefore \quad d = \sqrt{\frac{39.075 \times 10^6}{1000 \times 2.761}} = 119 \text{ mm.}$$

However keep the same total depth (=260 mm) as that of toe slab, so that available effective depth = 200 mm. The thickness is reduced to 200 mm at the edge.

For an under-reinforced section, we have

$$A_{st} = \frac{0.5 f_{ck}}{f_y}\left[1 - \sqrt{1 - \frac{4.6 M_u}{f_{ck}\, bd^2}}\right] bd = \frac{0.5 \times 20}{415}\left[1 - \sqrt{1 - \frac{4.6 \times 39.075 \times 10^6}{20 \times 1000\,(200)^2}}\right] \times 1000 \times 200$$

$$\simeq 576 \text{ mm}^2$$

Using 10 mm φ bars, $A_\varphi = 78.53$ mm². ∴ Spacing $= \frac{1000 \times 78.53}{576} \approx 136$ mm.

However, provide these @ 130 mm c/c at the top of heel slab. Take the reinforcement into the toe slab for a distance of 47 φ = 47 × 10 = 470 ≃ 500 mm to the left of *B*,

$$\text{Distribution steel} = \frac{0.12}{100} \times 1000\left[\frac{260 + 200}{2}\right] = 276 \text{ mm}^2$$

∴ Spacing of 8 mm φ bars $= \frac{1000 \times 50.3}{276} = 182$ mm. Hence provide these @ 180 mm c/c.

$$\text{Shear stress } \tau_v = \frac{1.5 \times 39.38 \times 1000}{1000 \times 200} = 0.295 \text{ N/mm}^2$$

This is much less than the permissible shear stress even at the minimum percentage of steel (Table 7.1).

8. ***Reinforcement in the stem***. We had earlier assumed the thickness of heel slab as 0.3 m, while it has now been fixed as 0.26 m only. Hence revised $H_1 = 4 - 0.26 = 3.74$ m

$$\therefore \quad M = K_a \gamma \frac{H_1^3}{6} = \frac{1}{3} \times \frac{18}{6}(3.74)^3 = 52.31 \text{ kN-m} = 52.31 \times 10^6 \text{ N-mm}$$

$$\therefore \quad M_u = 1.5 \times 52.31 \times 10^6 = 78.465 \times 10^6 \text{ N-mm}$$

$$\therefore \quad d = \sqrt{\frac{78.465 \times 10^6}{1000 \times 2.761}} = 169 \text{ mm.}$$

However keep $d = 250$ mm so that $D = 310$ mm. Reduce the total thickness to 200 mm at the top.

For an under-reinforced section we have

$$A_{st} = \frac{0.5\, f_{ck}}{f_y}\left[1 - \sqrt{1 - \frac{4.6\, M_u}{f_{ck}\, bd^2}}\right] bd = \frac{0.5 \times 20}{415}\left[1 - \sqrt{1 - \frac{4.6 \times 78.465 \times 10^6}{20 \times 1000\,(250)^2}}\right] \times 1000 \times 250$$

$$\simeq 944 \text{ mm}^2$$

$\therefore$ Spacing of 12 mm φ bars $= \dfrac{1000 \times 113}{944} = 119$ mm. However, provide 12 mm φ @ 100 mm c/c

Actual A_{st} provided $= \dfrac{1000 \times 113}{100} = 1130 \text{ mm}^2$

Continue alternate bars in the toe slab to serve as tensile reinforcement there. Discontinue the remaining half bars after a distance of 47 φ $= 47 \times 12 \approx 564$ mm beyond B, in the toe slab.

Between A and B, some of the bars can be curtailed. Consider a section at depth h below the top of the stem. The effective depth d' at that section is

$$d' = 140 + \frac{250 - 140}{3.74} h = (140 + 29.4\, h) \text{ mm (where } h \text{ is in metres.)} \quad \ldots(1)$$

Now $A_{st} \propto \dfrac{H^3}{d}$, or $H = (A_{st}\, d)^{1/3}$; Hence $\dfrac{h}{H_1} = \left(\dfrac{A_{st}'\, d'}{A_{st}\, d}\right)^{\frac{1}{3}}$ ...(2)

where A_{st}' = reinforcement at depth h ; d' = effective depth at depth h

A_{st} = reinforcement at depth H_1 ; d = effective depth at depth H_1

If $A_{st}' = \frac{1}{2} A_{st}$, $\dfrac{A_{st}'}{A_{st}} = \dfrac{1}{2}$. $\therefore$ $\dfrac{h}{H_1} = \left(\dfrac{1}{2} \cdot \dfrac{d'}{d}\right)^{\frac{1}{3}}$

Substituting $d = 250$ mm and $d' = (140 + 29.4\ h)$, we get

$$h = H_1\left[\frac{140 + 29.4\, h}{2 \times 250}\right]^{\frac{1}{3}} = 3.74\left[\frac{140 + 29.4\, h}{500}\right]^{\frac{1}{3}} = 0.471\,[140 + 29.4\, h]^{\frac{1}{3}} \quad \ldots(3)$$

This can be solved by trial and error, noting that if the effective thickness of stem were constant, h would have been equal to $\dfrac{H_1}{(2)^{1/3}} \approx 0.79\, H_1 \approx 2.96$ m.

Solving (3) by trial, we get $h = 2.86$ m. Thus, half the bars can be curtailed at this point. However, the bars should be extended by a distance of 12 φ ($= 12 \times 12 = 144$ mm) or d (= 250 mm) whichever is more, beyond the point.

$\therefore$ $h = 2.86 - 0.25 = 2.61$ m. Hence curtail half the bars at a height 2.6 m below the top. If we wish to curtail half of the remaining bars so that remaining reinforcement is one-fourth of that provided at B, we have

$\dfrac{A_{st}'}{A_{st}} = \dfrac{1}{4}$. Hence from (2), $\dfrac{h}{H_1} = \left(\dfrac{1}{4}\dfrac{d'}{d}\right)^{\frac{1}{3}}$

or $$h = H_1\left[\frac{140 + 29.4\, h}{4 \times 250}\right]^{\frac{1}{3}} = 3.74\left[\frac{140 + 29.4\, h}{1000}\right]^{\frac{1}{3}} = 0.374\,[143 + 29.4\, h]^{\frac{1}{3}} \quad \ldots(4)$$

This can be solved by trial and error, noting that if the effective thickness of stem were constant, h would have been equal to $H_1/(4)^{\frac{1}{3}} = 0.63\, H_1 \approx 2.36$ m.

Solving (4) by trial and error, we get $h = 2.2$ m. However the bars should be extended by 250 mm beyond this. Thus, $h = 2.2 - 0.25 = 1.95$ m. Hence stop half the remaining bars by 1.95 m below the top of the stem. Continue rest of the bars to the top of the stem.

Check for shear : Shear force $= P = K_a \gamma \dfrac{H^2}{2} = \dfrac{1}{3} \times \dfrac{18}{2} (3.74)^2 = 41.96$ kN

$\therefore \quad F_u = 1.5 \times 41.96 = 62.94$ kN

$\therefore \quad \tau_v = \dfrac{62.94 \times 1000}{1000 \times 250} = 0.25 \text{ N/mm}^2 < \tau_c$. Hence safe.

Distribution and temperature reinforcement.

Average thickness of stem $= \frac{1}{2}(310 + 200) = 255$ mm

$\therefore$ Distribution reinforcement $= \dfrac{0.12}{100} \times 1000 \times 255$

$= 306 \text{ mm}^2$. Using 8 mm φ bars, $A_\varphi = 50.3 \text{ mm}^2$

$\therefore \quad$ Spacing $= \dfrac{1000 \times 50.3}{306} = 164$ mm

However provide 8 mm φ bars @ 150 mm c/c at the inner face of the wall, along its length.

For temperature reinforcement, provide 8 mm φ @ 300 mm c/c both ways, in the outer face.

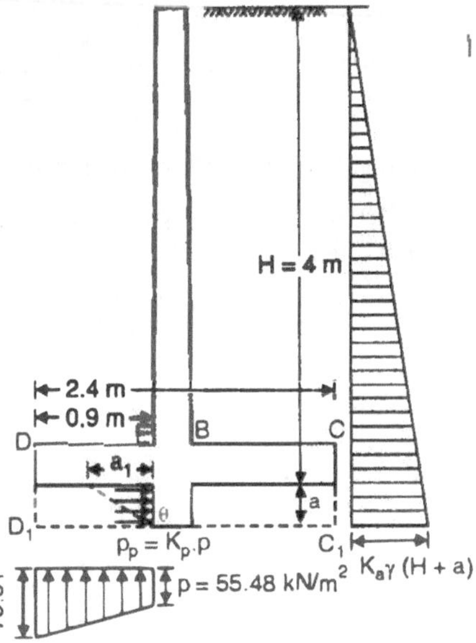

FIG. 23.18

9. *Design of shear key.* The wall is unsafe in sliding, and hence shear key will have to be provided below the stem, as shown in Fig. 23.18.

Let the depth of key $= a$. Intensity of passive pressure p_p developed in front of the key depends upon the soil pressure p in front of the key.

$\therefore \qquad p_p = K_p\, p = 3 \times 55.48 = 166.4 \text{ kN/m}^2$

$\therefore \quad$ Total passive pressure $P_p = p_p\, a = 166.4\, a$ ...(1)

Sliding force at level $D_1C_1 = \dfrac{1}{3} \times \dfrac{18}{2} (4 + a)^2$ or $P_H = 3\,(4 + a)^2$ ...(2)

Weight of soil between bottom of base and $D_1C_1 = 2.4\, a \times 18 = 43.2\, a$

$\therefore \quad \Sigma W = 121.05 + 43.2\, a$ (Refer Table 23.1)

Hence for equilibrium of the wall, permitting F.S. = 1.5 against sliding, we have

$$1.5 = \frac{\mu\, \Sigma W + P_p}{P_H} = \frac{0.5\,(121.05 + 43.2\, a) + 166.4\, a}{3\,(4 + a)^2}$$

or $\qquad a^2 - 33.8\, a + 2.54 = 0$, which gives $a \approx 0.09$ m $= 90$ mm

However, provide a minimum value of $a = 0.3$ m $= 300$ mm. Keep width of key $= 300$ mm. It should be noted that passive pressure taken into account above will be developed *only when* a length a_1 given below is available in front of the key :

$a_1 = a \tan \theta = a \tan\left(45° + \dfrac{\Phi}{2}\right) = a\sqrt{K_p}$, where $\left(45° + \dfrac{\Phi}{2}\right)$ = shearing angle of passive resistance.

$\therefore \quad a_1 = 0.3\sqrt{3} = 0.52$ m.

Actual length of the slab available $= DE = 0.9$ m. Hence satisfactory.

Now size of key

$= 300$ mm $\times$ 300 mm

$P_H = 3(4+a)^2 = 3(4+0.3)^2$

$= 55.47$ kN

$P_P = 166.4\, a = 166.4 \times 0.3$

$= 49.92$ kN. Hence,

$\Sigma W = 121.05 + 43.2\, a$

$= 121.05 + 43.2 \times 0.3 = 134.01$ kN

Actual force to be resisted by the key, at F.S.= 1.5 is

$= 1.5\, P_H - \mu\, \Sigma W$

$= 1.5 \times 55.47 - 0.5 \times 134.01$

$= 16.2$ kN

$$\therefore \quad \text{Shear stress} = \frac{16.2 \times 1000}{300 \times 1000}$$

$= 0.054$ N/mm^2 (safe)

$$\text{Bending stress} = \frac{16.2 \times 150 \times 1000}{\frac{1}{6} \times 1000\,(300)^2}$$

$= 0.16$ N/mm^2 (safe)

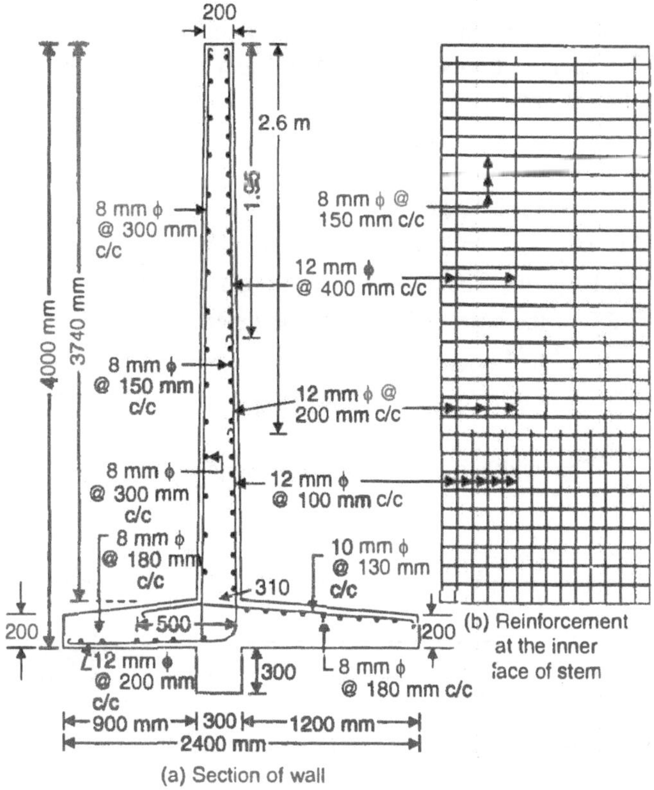

FIG. 23.19.

The details of reinforcement etc. are shown in Fig. 23.19.

23.8. DESIGN OF CANTILEVER RETAINING WALL WITH HORIZONTAL BACKFILL AND TRAFFIC LOAD

Design Example 23.2. *Design a cantilever retaining wall for a road for the following requirements:*

1. *Height of wall from the bottom of base to top of stem = 6 m*
2. *Superimposed load due to road traffic = 18 kN/m²*
3. *Unit weight of fill = 18 kN/m³*
4. *Angle of internal friction for fill material = 30°*
5. *Allowable bearing pressure on ground = 160 kN/m²*
6. *Coefficient of friction between concrete and ground = 0.4*

Also, provide a parapet wall 1 m high on the top of stem.
Use M 20 concrete and Fe 415 steel.

Solution.

1. ***Design constants***. For M 20 concrete, and Fe 415 steel reinforcement, we have the following : $x_{u\,.\,max}/d = 0.479$ and $R_u = 2.761$.

2. *Dimensions of base*. Assume that a horizontal force $Q = 2$ kN/m length of parapet wall will act because of person standing near the parapet. Due to surcharge, equivalent height of fill is given by

$$h_e = \frac{w}{\gamma} = \frac{18}{18} = 1 \text{ m}$$

Hence, in determining the values of b and α etc., we will use a height $H_2 = H + h_e = 6 + 1 = 7$ m. The ratio of the length of toe slab EF to the base width b may be determined by Eq. 23.28.

$$\alpha = 1 - \frac{160}{2.2 \times 18 \times 7} = 0.42 .$$

The width of base may be found from Eq. 23.25.

$$K_a = \frac{1 - \sin 30^\circ}{1 + \sin 30^\circ} = \frac{1}{3} . \qquad \text{Hence}$$

$$b = 0.95 \times 7 \sqrt{\frac{1}{3(1 - 0.42)(1 + 1.26)}} = 3.36 \text{ m}$$

However, keep minimum b between 0.5 to 0.6 H_2. Hence adopt $b = 4$ m.

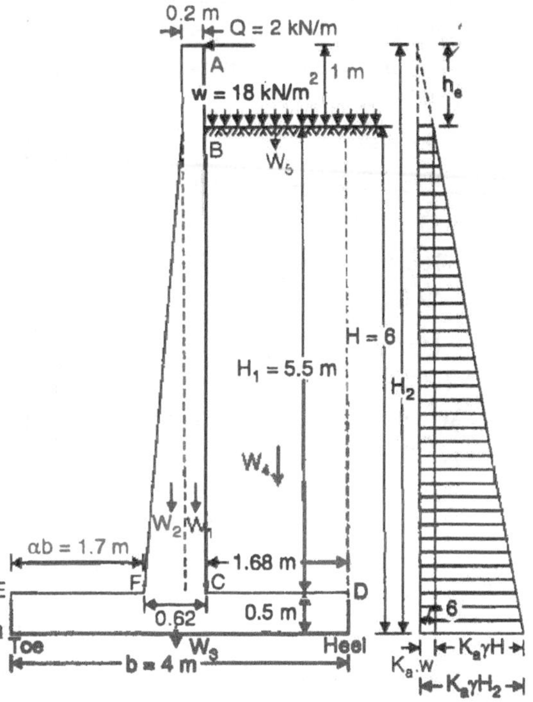

FIG. 23.20.

$\therefore$ Length of toe slab $= \alpha\, b = 0.42 \times 4 = 1.68$ m.

Keep it equal to 1.7 m. Let the thickness of base = 50 cm for preliminary calculations.

3. *Thickness of stem*. Height $H_1 = 6 - 0.5 = 5.5$ m

Due to retained soil, the earth pressure diagram will be a triangle, having an ordinate equal to $K_a \gamma h$, at h below B. Due to surcharge w, there will be a uniform horizontal pressure $K_a w = \frac{1}{3} \times 18 = 6 \text{ kN/m}^2$ throughout the height. The total bending moment at C will be due to moment of horizontal force $Q =$ 2 kN acting at A, plus moment of rectangular pressure distribution, plus moment of triangular pressure distribution.

$$\therefore \quad M = Q\,(H_1 + 1) + K_a\, w \frac{H_1^2}{2} + K_a\, \gamma \frac{H_1^3}{6} = 2\,(5.5 + 1) + \frac{1}{3} \times 18 \frac{(5.5)^2}{2} + \frac{1}{3} \times \frac{18}{6} (5.5)^3$$

$$= 13 + 90.75 + 166.35 = 270.1 \text{ kN-m} = 270.1 \times 10^6 \text{ N-mm}$$

$$M_u = 1.5 \times 270.1 \times 10^6 = 405.16 \times 10^6 \text{ N-mm}$$

$$\therefore \quad d = \sqrt{\frac{405.16 \times 10^6}{1000 \times 2.761}} = 383 \text{ mm.}$$

Provide total depth $D = 500$ mm so that using as effective cover of 60 mm, effective thickness of stem = 500 − 60 = 440 mm. Reduce the total thickness to 200 mm at B, so that effective depth at $B = 200 - 60 = 140$ mm. Continue uniform thickness of 200 from B to A.

4. ***Stability of wall.*** Fully dimensioned wall is shown in Fig. 23.20.

Length of heel slab $CD = 4 - 1.7 - 0.5 = 1.80$ m.

Let W_1 = weight of rectangular portion of stem.

W_2 = weight of triangular portion of stem.

W_3 = weight of base slab ; W_4 = weight of soil above heel slab.

W_5 = total superimposed road traffic load, over heel slab.

The calculations are arranged in Table 23.2.

TABLE 23.2

S. No.	*Designation*	*Force (kN)*	*Level arm (m)*	*Moment about toe (kN-m)*
1	W_1	$0.2 \times 6.5 \times 1 \times 25 = 32.5$	2.1	68.25
2	W_2	$\frac{1}{2} \times 0.3 \times 5.5 \times 1 \times 25 = 20.625$	1.9	39.1875
3	W_3	$0.5 \times 4 \times 1 \times 25 = 50$	2.00	100.0
4	W_4	$1.8 \times 5.5 \times 1 \times 18 = 178.2$	3.1	552.42
5	W_5	$1.8 \times 1 \times 18 = 32.4$	3.1	100.44
		$\Sigma W \approx 313.725$		M_R = 860.2975

Total earth pressure at base $= K_a \dfrac{\gamma H^2}{2} = \dfrac{1}{3} \times \dfrac{18}{2} (6)^2 = 108$ kN

Overturning moment at toe, due to earth pressure $= 108 \times \dfrac{6}{3} = 216$ kN-m

Overturning moment at toe, due to horizontal force $Q = 2 \times 7 = 14$ kN-m

Horizontal pressure due to live load $= K_a\, wH = \frac{1}{3} \times 18 \times 6 = 36$ kN

Overturning moment due to horizontal pressure caused by live load

$$= K_a\, w \frac{H^2}{2} = \frac{1}{3} \times 18 \times \frac{(6)^2}{2} = 108 \text{ kN-m}$$

$\therefore$ Total overturning moment $M_0 = 216 + 14 + 108 = 338$ kN-m

$\therefore$ F.S. against overturning $= \dfrac{860.2975}{338} = 2.54 > 2.$ Hence safe.

Total horizontal pressure $= 2 + 108 + 36 = 146$ kN $= \Sigma P$

$\therefore$ F.S. against sliding $= \dfrac{\mu \Sigma W}{\Sigma P} = \dfrac{0.4 \times 313.725}{146} = 0.86 < 1.5$

Hence unsafe. Special shear key will have to be designed to make the wall safe against sliding.

Pressure distribution. Algebraic sum of moments $= \Sigma M = M_R - M_0$
$= 860.2975 - 338 = 522.2975$ kN-m

Algebraic sum of vertical forces $= \Sigma W = 313.725$ kN

Hence distance $\bar{x}$ from toe, of the point of application of resultant is given by

$$\bar{x} = \frac{\Sigma M}{\Sigma W} = \frac{522.2975}{313.725} = 1.665 \text{ m}$$

$\therefore$ Eccentricity $e = \frac{b}{2} - \bar{x} = \frac{4}{2} - 1.665 = 0.335$ m

This is less than $\frac{b}{6} = \frac{4}{6} = 0.67$ m. Hence no tension is developed.

$$\text{Pressure } p_1 \text{ at toe} = \frac{\Sigma W}{b}\left(1 + \frac{6e}{b}\right) = \frac{313.725}{4}\left[1 + \frac{6 \times 0.335}{4}\right]$$

$$= 117.8 \text{ kN/m}^2 < 160. \quad \text{Hence safe.}$$

$$\text{Pressure } p_2 \text{ at heel} = \frac{\Sigma W}{b}\left(1 - \frac{6e}{b}\right) = \frac{313.725}{4}\left(1 - \frac{6 \times 0.335}{4}\right) = 39 \text{ kN/m}^2$$

Pressure p'' at the junction F of stem with toe slab is

$$p'' = 117.8 - \frac{117.8 - 39}{4} \times 1.7 = 84.3 \text{ kN/m}^2$$

Pressure p' at the junction C of stem with heel slab is

$$p' = 117.8 - \frac{117.3 - 39}{4} \times 2.2 = 74.7 \text{ kN/m}^2$$

5. ***Design of toe slab***. The upward pressure distribution on the toe slab is shown in Fig. 23.21. The weight of the soil above the toe slab is neglected. Thus two forces are acting on it : (*i*) upward soil pressure (*ii*) downward weight of slab.

Downward weight of slab per unit area $= 0.5 \times 1 \times 1 \times 25 = 12.5$ kN/m^2. Hence net pressure intensities will be $117.8 - 12.5 = 105.3$ kN/m^2 under E and $84.3 - 12.5 = 71.8$ kN/m^2 under F.

$$\therefore \quad \text{Total force} = \text{S.F. at } F = \tfrac{1}{2}(105.3 + 71.8) \times 1.7 = 150.535 \text{ kN}$$

$$\text{C.G. of force from } F = \bar{x}_1 = \left(\frac{71.8 + 2 \times 105.3}{71.8 + 105.3}\right)\frac{1.7}{3} = 0.904 \text{ m}$$

$$\therefore \quad \text{B.M. at } F = 150.535 \times 0.904 = 136.02 \text{ kN-m} = 136.02 \times 10^6 \text{ N-mm}$$

$$\therefore \quad M_u = 1.5 \times 136.02 \times 10^6 = 204.034 \times 10^6 \text{ N-mm}$$

$$\therefore \quad d = \sqrt{\frac{204.034 \times 10^6}{1000 \times 2.761}} = 272 \text{ mm}$$

However keep $d = 440$ mm and total depth equal to 500 mm so that 60 mm effective cover is available. This increased depth is provided to make the toe slab safe against shear. Reduce the total depth to 200 mm at the edges.

$$A_{st} = \frac{0.5 \times 20}{415}\left[1 - \sqrt{1 - \frac{4.6 \times 204.034 \times 10^6}{20 \times 1000\,(440)^2}}\right] \times 1000 \times 440 = 1374 \text{ mm}^2$$

Bars available from stem reinforcement are 20 mm φ @ 100 mm c/c giving A_{st} = 3141 mm^2 (step 7).

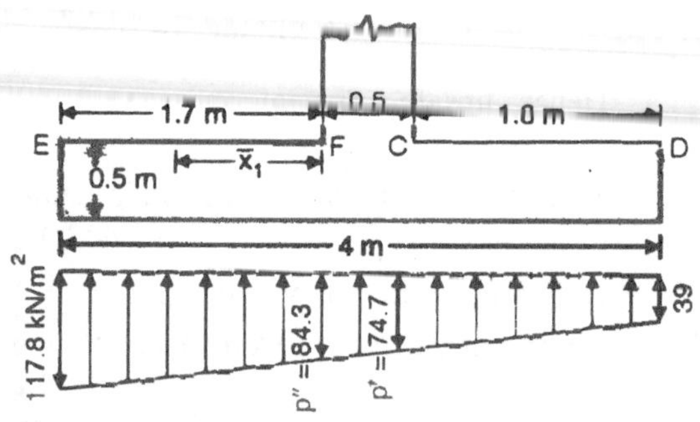

FIG. 23.21.

$$\tau_v = \frac{1.5 \times 150.535 \times 1000}{1000 \times 440} = 0.513 \ \text{N/mm}^2 \ ; \quad 100 \frac{A_s}{bd} = \frac{100 \times 3141}{1000 \times 440} \simeq 0.714 \ \%.$$

Hence from Table 7.1 $\tau_c = 0.548$ N/mm². Hence the slab is safe in shear.

$$\text{Distribution steel} = \frac{0.12}{100}\left[\frac{460 + 200}{2}\right] \times 1000 = 396 \text{ mm}^2$$

Using 8 mm φ bars, having $A_\varphi = 50.3$ mm², spacing $= \frac{1000 \times 50.3}{396} = 127$ mm

Hence provide these @ 120 mm c/c.

6. *Design of heel slab*. Four forces act on it :

(*i*) downward weight of soil 5.5 m high (*ii*) live load of traffic
(*iii*) weight of heel slab (*iv*) upward soil pressure.

∴ Weight of soil = 1.8 × 1 × 5.5 × 18 = 178.2 kN acting at 0.9 m from *C*.

Live load = 18 × 1 × 1.8 = 32.4 kN acting at 0.9 m from *C*.

Weight of heel slab = 0.5 × 1.8 × 25 = 22.5 kN acting at 0.9 m from *C*

$$\text{Upward soil reaction} = \frac{1}{2}(74.7 + 39) \times 1.8 = 102.33 \text{ kN}$$

$$\text{acting at } \left(\frac{74.7 + 2 \times 39}{74.7 + 39}\right)\frac{1.8}{3} = 0.806 \text{ m from } C$$

∴ Total force = S.F. at *C* = 178.2 + 32.4 + 22.5 − 102.33 = 130.77 kN

B.M. at *C* = (178.2 + 32.4 + 22.5) 0.9 − 102.33 × 0.806 = 127.312 kN-m
$= 127.312 \times 10^6$ N-mm

Keep the same depth as that of toe slab, *i.e.* $d = 440$ mm and total depth = 500 mm. Reduce total depth to 200 mm at edges. For an under-reinforced section, we have

$$A_{st} = \frac{0.5 f_{ck}}{f_y}\left[1 - \sqrt{1 - \frac{4.6 M_u}{f_{ck} bd^2}}\right] bd$$

$$= \frac{0.5 \times 20}{415}\left[1 - \sqrt{1 - \frac{4.6 \times 1.5 \times 127.312 \times 10^6}{20 \times 1000\,(440)^2}}\right] \times 1000 \times 440 = 1280 \text{ mm}^2$$

$$\therefore \text{ Spacing of 20 mm } \varphi \text{ bars} = \frac{1000 \times 314}{1280} = 245 \text{ mm}$$

$$\text{Also,} \quad \tau_v = \frac{1.5 \times 130.77 \times 10^3}{1000 \times 440} = 0.446 \text{ N/mm}^2$$

From Table 7.1, % reinforcement required to get $\tau_c \simeq 0.45$ N/mm² is $\simeq 0.44\%$

$$\text{This gives} \quad A_{st} = \frac{0.44\, bd}{100} = \frac{0.44 \times 1000 \times 440}{100} = 1936 \text{ mm}^2$$

∴ Spacing of 20 mm φ bars $= \frac{1000 \times 314}{1936} = 162$ mm. Hence provide 20 mm φ bars @ 160 mm c/c, giving actual $A_{st} = 1000 \times 314/160 = 1962.5$ mm²

$$\text{Distribution steel} = \frac{0.12}{100}\left[\frac{460 + 200}{2}\right] \times 1000 = 396 \text{ mm}^2$$

$\therefore$ Spacing of 8 mm φ bars $= \dfrac{1000 \times 50.3}{396} \approx 127$ mm. $\simeq$ 120 mm c/c.

7. *Reinforcement in the stem.*

Area of steel for an under-reinforced section is given by

$$A_{st} = \frac{0.5\, f_{ck}}{f_y}\left[1 - \sqrt{1 - \frac{4.6\, M_u}{f_{ck}\, bd^2}}\right] bd = \frac{0.5 \times 20}{415}\left[1 - \sqrt{1 - \frac{4.6 \times 405.16 \times 10^6}{20 \times 1000\, (440)^2}}\right] \times 1000 \times 440$$

$$= 2967 \text{ mm}^2$$

Using 20 mm φ bars, $A_\varphi = 314.1 \text{ mm}^2$. $\therefore$ $s = \dfrac{1000 \times 314.1}{2967} = 106$ mm

However provide 20 mm φ bars @ 100 mm c/c

$$\text{Actual } A_{st} \text{ provided} = \frac{1000 \times 314.1}{100} = 3141 \text{ mm}^2$$

Bend all the bars in the toe slab to serve as the reinforcement there. Due to this, sufficient bond length will be available to both the sides of point C (point of maximum bending moment).

$$\text{Total S.F. at } C = Q + K_a\, w\, H_1 + K_a\, \gamma\, \frac{H_1^2}{2} = 2 + 6 \times 5.5 + \frac{1}{3} \times \frac{18}{2}(5.5)^2 = 125.75 \text{ kN}$$

$$\therefore \qquad \tau_v = \frac{1.5 \times 125.75 \times 1000}{1000 \times 440} = 0.429 \text{ N/mm}^2$$

$$\frac{100\, A_s}{bd} = \frac{100 \times 3141}{1000 \times 440} = 0.714\ \%. \text{ Hence } \tau_c \simeq 0.55 \text{ N/mm}^2 > \tau_v \text{ Hence safe.}$$

Let us curtail reinforcement between C and B. If there were no other external force, except the earth pressure, and if the depth of stem were constant, half the bars could have been curtailed at a depth $\dfrac{H_1}{(2)^{1/3}} \simeq 0.79\, H_1$ below the point B. Because of presence of other forces, let us try at depth $= 0.65\, H_1 = 0.65 \times 5.5 = 3.575$ m below B to see whether half the bars could be curtailed there or not. Thus, depth of section below $B = h = 3.575$ m.

$$\text{B.M.} = Q\,(h + 1) + K_a\, w\, \frac{h^2}{2} + K_a\, \gamma\, \frac{h^3}{6} = 2\,(3.575 + 1) + \frac{6\,(3.575)^2}{2} + \frac{1}{3}\,\frac{18}{6}\,(3.575)^3$$

$$= 9.15 + 38.34 + 45.69 = 93.18 \text{ kN-m} = 93.18 \times 10^6 \text{ N-mm}$$

$$d = 140 + \frac{440 - 140}{5.5} \times 3.575 = 335 \text{ mm}$$

$$\therefore \qquad A_{st} = \frac{0.5 \times 20}{415}\left[1 - \sqrt{1 - \frac{4.6 \times 1.5 \times 93.18 \times 10^6}{20 \times 1000\, (335)^2}}\right] \times 1000 \times 335 = 1254 \text{ mm}^2$$

This is less than half of that provided at C. Hence half the bars can be curtailed at this depth. However, the bars should be extended by a distance of $12\,\varphi\ (= 12 \times 20 = 240$ mm) or d (= 335 mm) beyond this point, whichever is more. Hence $h = 3.575 - 0.335 = 3.24$ m. Hence curtail half the bars at a height of 3.2 m below B. If we wish to curtail half of the remaining part, let us try it at a section at depth $h = 0.65 \times 3.575 \approx 2.32$ m below B.

$$\text{B.M.} = 2\,(2.32 + 1) + \frac{6\,(2.32)^2}{2} + \frac{1}{3} \times \frac{18}{6}\,(2.32)^3$$

$$= 6.64 + 16.15 + 12.49 = 35.28 \text{ kN-m} = 35.28 \times 10^6 \text{ N-mm}$$

$$d = 140 + \frac{440 - 140}{5.5} \times 2.32 = 266 \text{ mm} \;;$$

$$A_{st} = \frac{0.5 \times 20}{415}\left[1 - \sqrt{1 - \frac{4.6 \times 1.5 \times 35.28 \times 10^6}{20 \times 1000\,(266)^2}}\,\right] \times 1000 \times 266 = 578 \text{ mm}^2$$

This is less than one fourth of that provided at C. Hence we can curtail half of the remaining bars here. However, extend the bars by a distance of $d = 266$ mm beyond this, *i.e.* curtail half of the remaining bars at a distance of $2.32 - 0.26 \simeq 2.0$ m below B. Rest of the bars will be continued upto the top of the parapet.

Distribution reinforcement. Average thickness of stem = (200 + 500)/2 = 350 mm

$\therefore$ Area distribution reinforcement $= \frac{0.12}{100}(1000 \times 350) = 420 \text{ mm}^2$

Using 10 mm φ bars, $A_\varphi = 78.5 \text{ mm}^2$ $\therefore$ Spacing $= (1000 \times 78.5)/420 = 186$ mm

$\therefore$ Hence provide 10 mm bars @ 175 mm c/c at the inner face of the wall, along its length.

Temperature reinforcement

Total area = 420 mm^2. Provide 10 mm φ bars @ 350 mm c/c both ways on the outer face.

8. ***Design of shear key.*** The wall is unsafe against sliding. Let us provide a shear key of depth a below the stem.

Let p_p be the intensity of passive pressure developed just in front of the shear key. This intensity p_p depends upon the soil pressure p just in front of shear key.

$$p_p = K_p \cdot p \quad \text{where } K_p = \frac{1 + \sin\Phi}{1 - \sin\Phi} = \frac{1}{K_a} = 3$$

and $p = 84.3 \text{ kN/m}^2$, from Fig. 23.21

Hence $p_p = 3 \times 84.3 = 252.9 \text{ kN/m}^2$

This intensity may be considered to be constant along the depth of key, though there will be little increase in p_p, because of increase in p with depth. We will, however, consider the constant value of $p_p = 252.9 \text{ kN/m}^2$.

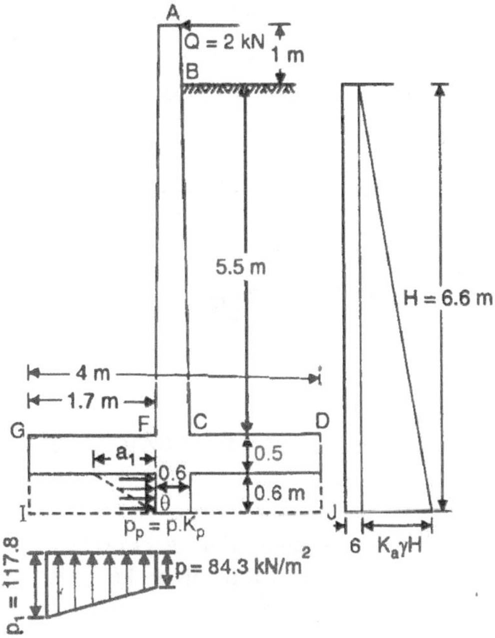

FIG. 23.22.

$\therefore$ Total passive force $P_p = p_p\,a = 252.9\,a$ kN. Keeping $a = 600$ mm,

$$P_p = 252.9 \times 0.6 = 151.74 \text{ kN} \qquad \ldots(1)$$

Total sliding force at the bottom level of key is

$$\Sigma P = 2 + (6 \times 6.6) + \frac{1}{3} \times \frac{18}{2}(6.6)^2 = 2 + 39.6 + 130.7 = 172.3 \text{ kN} \quad ...(2)$$

Weight of soil between the base and level $IJ = 0.6 \times 4 \times 1 \times 18 = 43.2$ kN (approx.)

$$\therefore \Sigma W = 313.725 + 43.2 = 356.925 \text{ kN} \quad ...(3)$$

Hence F.S. against sliding is $= \frac{(\Sigma \mu W) + P_p}{\Sigma P} = \frac{0.4 \times 356.925 + 151.74}{172.3} = 1.71$. Hence safe.

However, it should be noted that the above passive pressure will be developed *only when* a length a_1 given below is available in front of the key:

$$a_1 = a \tan \theta = a \tan\left(45 + \frac{\Phi}{2}\right)$$

$$= a\sqrt{K_p} = 0.6\sqrt{3} = 1.04 \text{ m.}$$

where $\theta = 45 + \Phi/2$ = shearing angle of passive resistance.

Actual length of toe slab available $= GF = 1.7$ m. Hence satisfactory.

Let us keep the width of key = 500 mm.

Actual force to be resisted by the key $= (1.5\, \Sigma P) - \mu\, \Sigma W$

$$= 1.5 \times 172.3 - 0.4 \times 356.925$$

$$\approx 116 \text{ kN}$$

Hence shear stress $= \frac{1.5 \times 116 \times 1000}{1000 \times 500}$

≈ 0.348 N/mm². Safe

$\therefore$ Bending stress

$$= \frac{1.5 \times 116 \times 1000 \times 350}{1/6 \times 1000\,(500)^2} = 1.04 \text{ N/mm}^2$$

Since concrete can take this much of tensile stress no special reinforcement is necessary for the key. The key is to be cast monolithically with the base.

9. *Construction Joint* A construction joint, in the form of a key, is to be provided at the junction of stem with the base slab. The width of key is kept equal to $d/4 = 440/4 = 110$ mm.

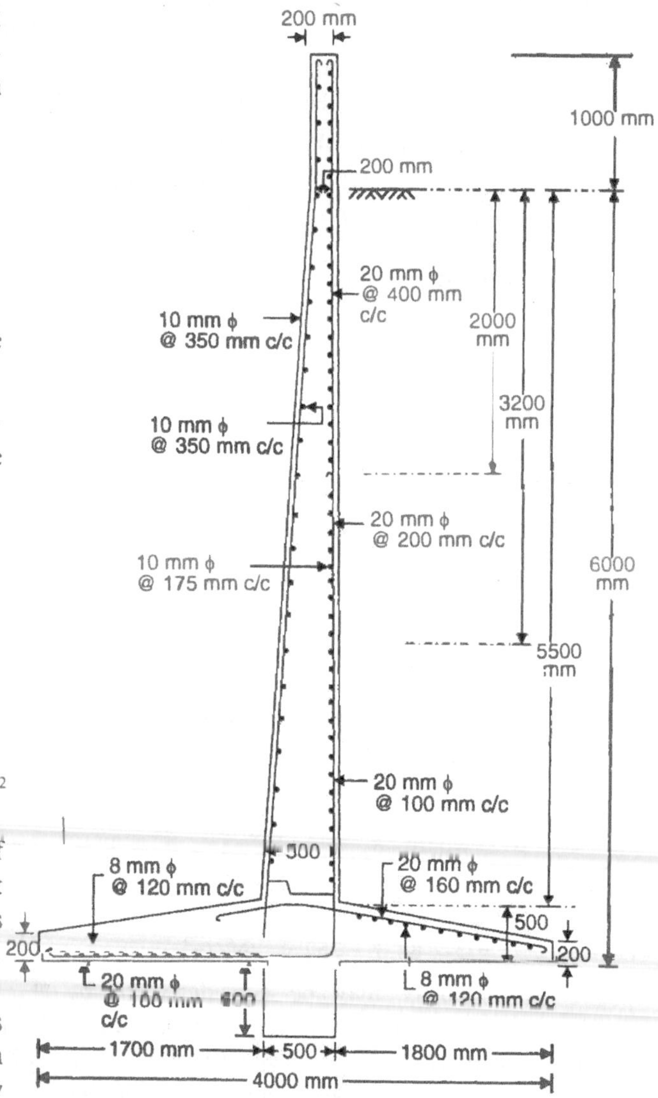

FIG. 23.23.

23.9. DESIGN OF CANTILEVER RETAINING WALL WITH SLOPING BACKFILL

Design Example 23.3. *Design a T-shaped cantilever retaining wall to retain earth embankment 3 m high above ground level. The embankment is surcharged at angle of 16°*

to the horizontal. The unit weight of earth is 18 N/m³ and its angle of repose is 30°. The safe bearing capacity may be taken as 100 kN/m² at a depth of 1 m below the ground. The coefficient of friction between concrete and soil may be taken as 0.5. Use M 20 concrete and Fe 415 steel.

Solution

1. *Design constants.* For M 20 concrete, and Fe 415 steel reinforcement, we have the following: $x_{u\,.\,max}/d = 0.479$ and $R_u = 2.761$

2. *Dimensions of base.* $\sin \beta = 0.276$; $\cos \beta = 0.961$; $\tan \beta = 0.287$;

$$\sin \phi = \sin 30° = 0.5 \quad ; \quad \cos \phi = 0.866 \; ; \qquad K_a = \cos \beta \, \frac{\cos \beta - \sqrt{\cos^2 \beta - \cos^2 \phi}}{\cos \beta + \sqrt{\cos^2 \beta - \cos^2 \phi}}$$

Substituting the values, we get $K_a = 0.38$

For a surcharged wall, the ratio of the length of the toe slab to the base width b may be fixed approximately by the expression :

$$\alpha = 1 - \frac{q_0}{2.7\,\gamma H} = 1 - \frac{100}{2.7 \times 18 \times 4} = 0.488. \quad \text{Adopt } \alpha = 0.49.$$

The base width can be determined from the expression

$$b = H \sqrt{\frac{K_a \cos \beta}{(1-\alpha)(1+3\alpha)}}$$

$$= 4 \sqrt{\frac{0.38 \times 0.961}{(1-0.49)(1+3 \times 0.49)}} = 2.15 \text{ m.}$$

From sliding point of view, b will be much more than this, as seen in example 23.1. However, a minimum base width $b = 0.6H = 2.4$ m should be adopted. Hence keep $b = 2.4$ m.

Length of toe slab $= 0.49 \times 2.4 \approx 1.2$ m.

Let the thickness of base slab be 300 mm for preliminary calculations.

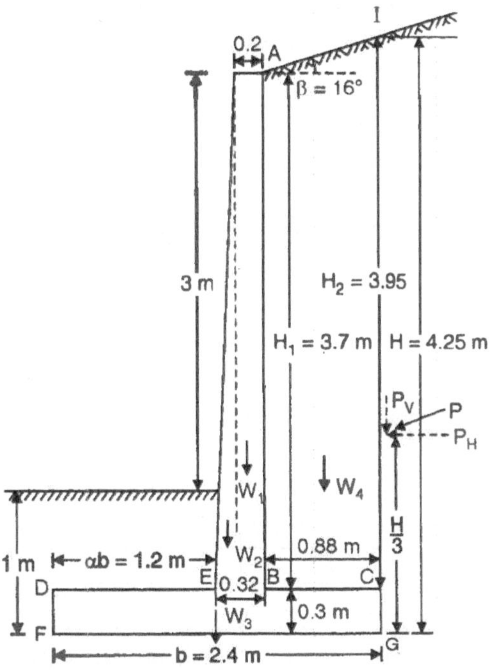

FIG. 23.24.

3. *Thickness of stem*

Height $AB = H_1 = 4 - 0.3 = 3.7$ m.; $K_a = 0.38$

Earth pressure on stem is

$$P = K_a \frac{\gamma H_1^2}{2}, \text{ acting at } 16° \text{ to horizontal}$$

$$= \frac{0.38}{2} \times 18\,(3.7)^2 = 46.82 \text{ kN}$$

Hence the horizontal earth pressure is

$$P_H = P \cos \beta = 46.82 \cos 16° \approx 45 \text{ kN}$$

$$\text{B.M. at } B = P_H \frac{H_1}{3} = 45 \times \frac{3.7}{3} = 55.5 \text{ kN-m} = 55.5 \times 10^6 \text{ N-mm}$$

$$M_u = 1.5 \times 55.5 \times 10^6 = 83.25 \times 10^6 \text{ N-mm}$$

S.F. at $B = P_H = 45$ kN ; $F_u = 1.5 \times 45 = 675$ kN

$$d = \sqrt{\frac{83.25 \times 10^6}{2.761 \times 1000}} = 174 \text{ mm.}$$

However keep $d = 260$ mm and total thickness = 320 mm so that an effective cover of 60 mm is available. Reduce total thickness to 200 mm at A, so that effective depth available there = 140 mm. Greater thickness has been provided to get an under-reinforced section, and also to make it safe against shear.

$$\tau_v = \frac{1.5 \times 45 \times 1000}{1000 \times 260} = 0.26 \text{ N/mm}^2. > \tau_c \text{ even at minimum } p = 0.15\% \text{ steel. Hence safe.}$$

4. *Stability of wall*

Length of heel slab = 2.4 − 1.2 − 0.32 = 0.88 m

Height $H_2 = H_1 + 0.88 \tan 16° = 3.7 + 0.25 = 3.95$ m

Height $H = 3.95 + 0.3 = 4.25$ m

The fully dimensioned wall is shown in Fig. 23.24.

Earth pressure $P = K_a \dfrac{\gamma H^2}{2}$, acting parallel to the surcharge $= \dfrac{0.38}{2} \times 18\,(4.25)^2 = 61.77$ kN

Its horizontal and vertical components are

$$P_H = P \cos \beta = 61.77 \times 0.961 = 59.36 \text{ kN} \; ; \; P_V = P \sin \beta = 61.77 \times 0.276 = 17.05 \text{ kN}$$

P is acting on vertical face IG, at $H/3$ and hence P_V will act along the vertical line IG (Fig. 23.24).

Let W_1 = weight of rectangular portion of stem
W_2 = weight of triangular portion of stem
W_3 = weight of base slab
W_4 = weight of soil on heel slab.

TABLE 23.3

S. N.	Designation	Force (kN)	L.A. (m)	Moment about toe (kN-m)
1	W_1	$0.2 \times 3.7 \times 25 = 18.5$	1.42	26.27
2	W_2	$\frac{1}{2} \times 0.12 \times 3.7 \times 25 = 5.55$	1.28	7.10
3	W_3	$2.4 \times 0.3 \times 25 = 18$	1.20	21.60
4	W_4	[illegible]	1.965	119.06
5	W_5	$P_V = 17.05$	2.4	40.92
		$\Sigma W = 119.69$		$M_R = 214.95$

∴ Total resisting moment $M_R = 214.95$ kN-m

$$\text{Overturning moment } M_0 = P_H \frac{H}{3} = 59.36 \times \frac{4.25}{3} = 84.1 \text{ kN-m}$$

$$\text{F.S. against overturning} = \frac{214.95}{84.1} = 2.56. \text{ Hence safe.}$$

$$\text{F.S. against sliding} = \frac{\mu \Sigma W}{P_H} = \frac{0.5 \times 119.69}{59.36} = 1 < 1.5. \text{ Hence unsafe.}$$

To make it safe against sliding, will have to provide shear key.

Pressure distribution

Net moment $\Sigma M = M_R - M_0 = 214.95 - 84.1 = 130.85$ kN ...(1)

Distance $\bar{x}$ of the point of application of resultant from toe is

$$\bar{x} = \frac{\Sigma M}{\Sigma W} = \frac{130.85}{119.69} = 1.09 \text{ m.}$$

∴ Eccentricity $e = \frac{b}{2} - \bar{x} = 1.2 - 1.09 = 0.11$ m. This is much less than $b/6$.

$$\text{Pressure } p_1 \text{ at toe} = \frac{\Sigma W}{b}\left(1 + \frac{6e}{b}\right) = \frac{119.69}{2.4}\left(1 + \frac{6 \times 0.11}{2.4}\right)$$

$$= 63.59 \text{ kN/m}^2 < 100. \text{ Hence safe.}$$

$$\text{Pressure } P_2 \text{ at heel} = \frac{119.69}{2.4}\left(1 - \frac{6 \times 0.11}{2.4}\right) = 36.16 \text{ kN/m}^2$$

Pressure p at the junction of stem with heel slab is

$$p = 63.59 - \frac{63.59 - 36.16}{2.4} \times 1.2 = 49.87 \text{ kN/m}^2$$

Presssure p' at the junction of stem with heel slab is

$$p' = 63.59 - \frac{63.59 - 36.16}{2.4} \times 1.52 = 46.22 \text{ kN/m}^2$$

5. ***Design of toe slab***. The upward pressure distribution on the base is shown in Fig. 23.25. The weight of the soil above the toe slab is neglected. Thus two forces are acting on it :

(*i*) upward soil pressure (*ii*) downward weight of the slab.

Downward weight of slab per unit area = $0.3 \times 1 \times 1 \times 25 = 7.5$ kN/m²

Hence net pressure intensities will be $= 63.59 - 7.5 = 56.09$ kN/m² under D and $49.87 - 7.5 = 42.37$ kN/m² under E.

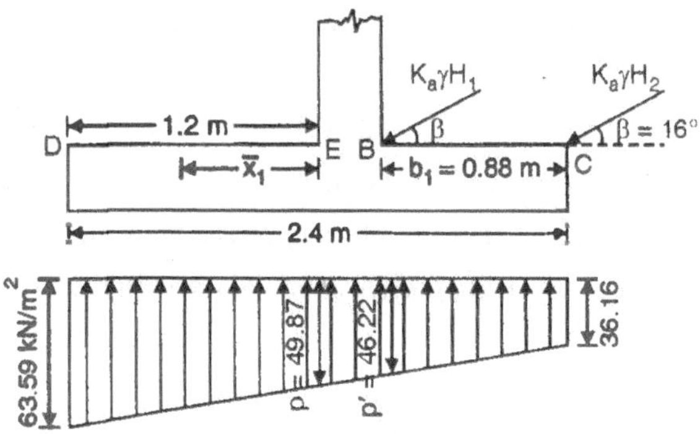

FIG. 23.25.

Total force = S.F. at $E = \frac{1}{2}(56.09 + 42.37) \times 1.2 = 59.08$ kN; $F_u = 1.5 \times 59.08 = 88.62$ kN

$$\text{C.G. of force from } E = \bar{x} = \left(\frac{42.37 + 2 \times 56.09}{42.37 + 56.09}\right)\frac{1.2}{3} = 0.63 \text{ m}$$

$$\text{B.M. at } E = 59.08 \times 0.63 = 37.09 \text{ kN-m} = 37.09 \times 10^6 \text{ N-mm;}$$

$$M_u = 1.5 \times 37.09 \times 10^6 = 55.635 \times 10^6 \text{ N-mm}$$

$$\therefore \quad d = \sqrt{\frac{55.635 \times 10^6}{1000 \times 2.761}} = 142 \text{ mm}$$

Howeer, provide minimum total depth = 300 mm so that with an effective cover of 60 mm, $d = 240$ mm. Reduce the total depth to 200 mm at the edge.

$$\tau_v = \frac{88.62 \times 1000}{1000 \times 240} = 0.37 \text{ N/mm}^2.$$

This is less than the permissible shear stress of $\tau_c = 0.38$ N/mm^2 even at $p = 0.3\%$

$$A_{st} = \frac{0.5 \times 20}{415}\left[1 - \sqrt{1 - \frac{4.6 \times 55.635 \times 10^6}{20 \times 1000\,(240)^2}}\right] \times 1000 \times 240 = 683 \text{ mm}^2$$

Hence 12 mm φ bars bent down from the stem @ 100 mm c/c will be sufficient, providing $A_{st} = 1130$ mm^2 (see step 7).

$$\text{Distribution reinforcement} = \frac{0.12}{100} \times 1000\left[\frac{300 + 200}{2}\right] = 300 \text{ mm}^2$$

Using 8 mm φ bars, $A_\varphi = 50.3$ mm^2, $\therefore \quad s = \dfrac{1000 \times 50.3}{300} = 167$ mm

However provide 8 mm φ bars @ 150 mm c/c.

6. *Design of heel slab* (Fig. 23.25). Let $b_1 = 0.88$ m = length of heel slab. Four forces act on it :

(*i*) weight of soil over the heel slab (*ii*) weight of heel slab

(*iii*) downward component of earth pressure (*iv*) upward soil pressure

$$\text{Weight of soil over heel slab} = \tfrac{1}{2}(3.7 + 3.95) \times 0.88 \times 18 = 60.59 \text{ kN}$$

$$\text{Acting at } \left(\frac{3.7 + 2 \times 3.95}{3.7 + 3.95}\right)\frac{0.88}{3} = 0.445 \text{ m from } B.$$

Weight of heel slab $= 0.3 \times 0.88 \times 1 \times 25 = 6.6$ kN acting at 0.44 m from B.

Earth pressure intensity at $B = K_a \gamma H_1$ per unit inclined area, at β to horizontal.

$\therefore$ Earth pressure at B, on horizontal unit area $= K_a \gamma H_1 \,.\, \tan\beta$

Vertical component of this, at $B = K_a \gamma H_1 \tan\beta \,.\, \sin\beta$ ...(*i*)

Similarly, vertical component of earth pressure intensity at $C = K_a \gamma H_2 \tan\beta \,.\, \sin\beta$ ...(*ii*)

Hence total force due to vertical component of earth pressure is

$$= K_a \,.\, \gamma \frac{(H_1 + H_2)}{2} \times b_1 \tan\beta \,.\, \sin\beta = \frac{0.38 \times 18}{2}(3.7 + 3.95) \times 0.88 \times 0.287 \times 0.276 = 1.82 \text{ kN}$$

$$\text{This acts at } \left(\frac{H_1 + 2H_2}{H_1 + H_2}\right)\frac{b_1}{3} = \frac{3.7 + 2 \times 3.95}{3.7 + 3.95} \times \frac{0.88}{3} = 0.445 \text{ m from } B.$$

$$\text{Upward soil pressure} = \tfrac{1}{2}(46.22 + 36.16) \times 0.88 = 36.25 \text{ kN}$$

$$\text{Acting at } \left(\frac{46.22 + 2 \times 36.16}{46.22 + 36.16}\right) \times \frac{0.88}{3} = 0.42 \text{ m from } B.$$

$\therefore$ Total force = S.F. at $B = 60.59 + 6.6 + 1.82 - 36.25 = 32.76$ kN.

$$F_u = 1.5 \times 32.76 = 49.14 \text{ kN}$$

B.M. at $B = (60.59 \times 0.445) + (6.6 \times 0.44) + (1.82 \times 0.445) - (36.25 \times 0.42) = 15.45$ kN-m

$= 15.45 \times 10^6$ N-mm ; $M_u = 1.5 \times 15.45 \times 10^6 = 23.175 \times 10^6$ N-mm

This is much less than the B.M. on the toe slab. However, we will keep the same depth, as that of toe slab, *i.e.* $d = 240$ mm and $D = 300$ mm, reducing it to 200 mm at the edges.

$$A_{st} = \frac{0.5 \times 20}{415}\left[1 - \sqrt{1 - \frac{4.6 \times 23.175 \times 10^6}{20 \times 1000\,(240)^2}}\right] 1000 \times 240 = 275 \text{ mm}^2$$

$$\text{Minimum } A_{st} = \frac{0.12}{100} \times 1000 \times \frac{1}{2}(300 + 200) = 300 \text{ mm}^2$$

$$\text{Hence } s - 8 \text{ mm } \varphi = \frac{1000 \times 50.3}{300} = 167 \text{ mm}$$

However, provide 8 mm φ bars @ 150 mm c/c. Take the reinforcement in the toe slab for a distance of $47\,\varphi = 47 \times 8 = 376 \simeq 400$ mm to the left of end B.

$$\tau_v = \frac{49.14 \times 1000}{1000 \times 240} = 0.2 \text{ N/mm}^2. \text{ Hence safe.}$$

Distribution steel @ 0.12% = 300 mm². Hence provide 8 mm φ bars @ 150 mm c/c.

7. ***Reinforcement in the stem***. Revised $H_1 = 4 - 0.3 = 3.7$ m

$$\text{S.F. at } B = \frac{0.38}{2} \times 18\,(3.7)^2 \cos 16° = 45 \text{ kN} = P_H \;;\; F_u = 1.5 \times 45 = 67.5 \text{ kN}$$

$$\therefore \quad \text{B.M. at } B = 45 \times \frac{3.7}{3} = 55.5 \text{ kN-m} = 55.5 \times 10^6 \text{ N-mm}$$

$$M_u = 1.5 \times 55.5 \times 10^6 = 83.25 \times 10^6 \text{ N-mm} \;;\; d = 260 \text{ mm},$$

$$A_{st} = \frac{0.5 \times 20}{415}\left[1 - \sqrt{1 - \frac{4.6 \times 83.25 \times 10^6}{20 \times 1000\,(260)^2}}\right] 1000 \times 260 = 961 \text{ mm}^2$$

$$\therefore \text{ Spacing of 12 mm } \varphi \text{ bars} = \frac{1000 \times 113}{961} = 117 \text{ mm}$$

However, provide 12 mm φ bars @ 100 mm c/c. Actual $A_{st} = \dfrac{1000 \times 113}{100} = 1130$ mm²

Bend these bars into the toe slab, to serve as reinforcement there. Sufficient development length is available.

Between A and B, some of the bars can be curtailed. Consider a section at a depth h below the top of the step. The effective depth d' at the section is

$$d' = 140 + \frac{260 - 140}{3.7} h = 140 + 32.4\,h \qquad \ldots(1)$$

$$\text{Now} \qquad A_{st} \propto \frac{H^3}{d} \qquad \text{or} \qquad H \propto (A_{st}\, d)^{\frac{1}{3}}$$

$$\text{Hence} \qquad \frac{h}{H_1} = \left(\frac{A_{st}'\, d'}{A_{st}\, d}\right)^{\frac{1}{3}} \qquad \ldots(2)$$

where A_{st}' = reinforcement at depth h ; A_{st} = reinforcement at depth H_1

d = effective thickness at depth H_1 .

If $\quad A_{st}' = \frac{1}{2}\ A_{st}\ ;\ \frac{A_{st}'}{A_{st}} = \frac{1}{2}$

$\therefore \quad \frac{h}{H_1} = \left(\frac{1}{2}\frac{d'}{d}\right)^{1/3}$. Substituting $d = 260$ mm and $d' = 140 + 32.4\ h$, we have :

$$h = H_1 \left(\frac{140 + 32.4\ h}{2 \times 260}\right)^{1/3} = 3.7 \left(\frac{140 + 32.4\ h}{520}\right)^{1/3} = 0.46\ (140 + 32.4\ h)^{1/3} \qquad \ldots(3)$$

This can be solved by trial and error, noting that if the effective thickness of stem were constant, h would have been equal to $H_1/(2)^{1/3} \approx 0.79\ H_1 \approx 2.92$ m. Solving (3) by trial, we get $h = 2.82$ m. Thus half the bars can be curtailed at this point. However, bars should be extended by a distance of 12φ (= 12×16) or d(= 260) beyond this point, whichever is more.

$\therefore$ $h = 2.82 - 0.26 = 2.56$ m. However, curtail half the bars at a height 2.5 m below the top. If we wish to curtail half of the remaining bars, so that reinforcement is one-fourth of that provided at B, we have

$\frac{A_{st}'}{A_{st}} = \frac{1}{4}$. Hence from (2), $\frac{h}{H_1} = \left(\frac{1}{4}\frac{d'}{d}\right)^{1/3}$

or $\quad h = H_1 \left(\frac{140 + 32.4\ h}{4 \times 260}\right)^{1/3} = 3.7 \left(\frac{140 + 32.4\ h}{1040}\right)^{1/3} = 0.365\ (140 + 32.4\ h)^{1/3}$

This can be solved by trial, noting that if the effective thickness of the stem were constant, h would have been equal to $\frac{H}{(4)^{1/2}} \approx 0.63\ H_1 = 2.32$ m.

Hence, we get $h = 2.16$ m. However, bars should be extended by d (= 260 mm) beyond this.

$\therefore \quad h = 2.16 - 0.26 = 1.90$ m

Hence stop half of the remaining bars at 1.90 m below the top of the stem. Continue rest of the bars to the top of the stem.

Distribution reinforcement

Average thickness $= \frac{1}{2}(320 + 200) = 260$ mm

$\therefore$ Distribution reinforcement

$$= \frac{0.12}{100}(1000 \times 260) = 312 \text{ mm}^2$$

Using 8 mm φ bars, $A_\varphi = 50.3$ mm^2

$\therefore \quad \text{Spacing} = \frac{1000 \times 50.3}{312} = 161$ mm

However provide 8 mm φ bars @ 150 mm c/c at the inner face of the stem.

Temperature reinforcement : Provide 8 mm φ bars @ 300 mm c/c both ways on the outer face of the stem.

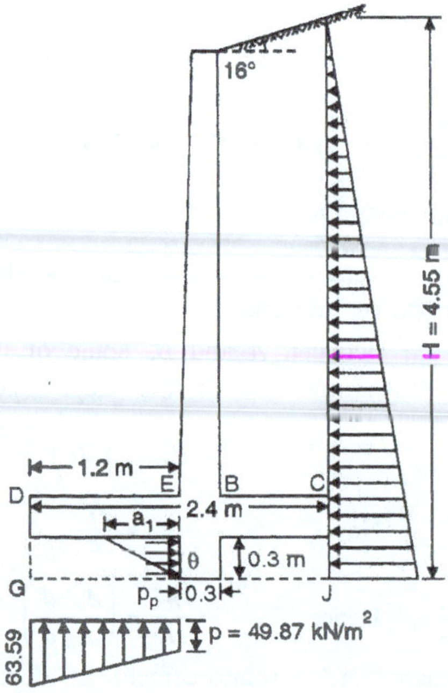

FIG 23.26.

9. *Design of shear key.* The wall is unsafe in shear. Let us provide a shear key 300 mm × 300 mm. Let p_p be the intensity of passive pressure developed just in front of the shear key. This intensity p_p depends upon the soil pressure p just in front of shear key.

$$p_p = K_p\, p \quad \text{where} \quad K_p = \frac{1}{K_a} = \frac{1}{0.38} = 2.63$$

$$\therefore \quad p_p = 2.63 \times 49.87 = 131.2 \ \text{kN/m}^2$$

This intensity may be considered to be constant along the depth of key though there will be little increase in p_p because of increase in p with depth. We will, however, consider the constant value of $p_p = 131.2$ kN/m².

$$\text{Total passive force } P_p = 131.2 \times 0.3 = 39.36 \ \text{kN} \qquad \ldots(1)$$

Let us calculate total sliding force at level GJ.

$$P_H = \frac{0.38 \times 18}{2}(4.55)^2 \cos 16° = 68.06 \ \text{kN} \qquad \ldots(2)$$

Weight of soil between the base and level $GJ = 0.3 \times 2.4 \times 18 = 12.96$ kN (approx.) ..(3)

$\therefore \ \Sigma W = 119.69 + 12.96 = 132.65$ kN (There will be some more downward weight because of increasing vertical component of earth pressure).

$\therefore$ F.S. against sliding

$$= \frac{\mu \Sigma W + P_p}{P_H}$$

$$= \frac{0.5 \times 132.65 + 39.36}{68.06}$$

$= 1.55 > 1.5$. Safe.

However, it should be noted that the above passive pressure will be developed *only* when a length a_1 given below is available in front of the key :

$$a_1 = a \tan\theta = a \tan\left(45 + \frac{\Phi}{2}\right)$$

$$= a\sqrt{K_p} = 0.3\sqrt{2.63} = 0.49 \ \text{m}.$$

Actual length of toe slab available = 1.2 m. Hence satisfactory

Now, force to be resisted by key, at F.S. of 1.5 is

$= 1.5\, P_H - \mu \Sigma W$

$= 1.5 \times 68.06 - 0.5 \times 132.65$

$= 35.8$ kN

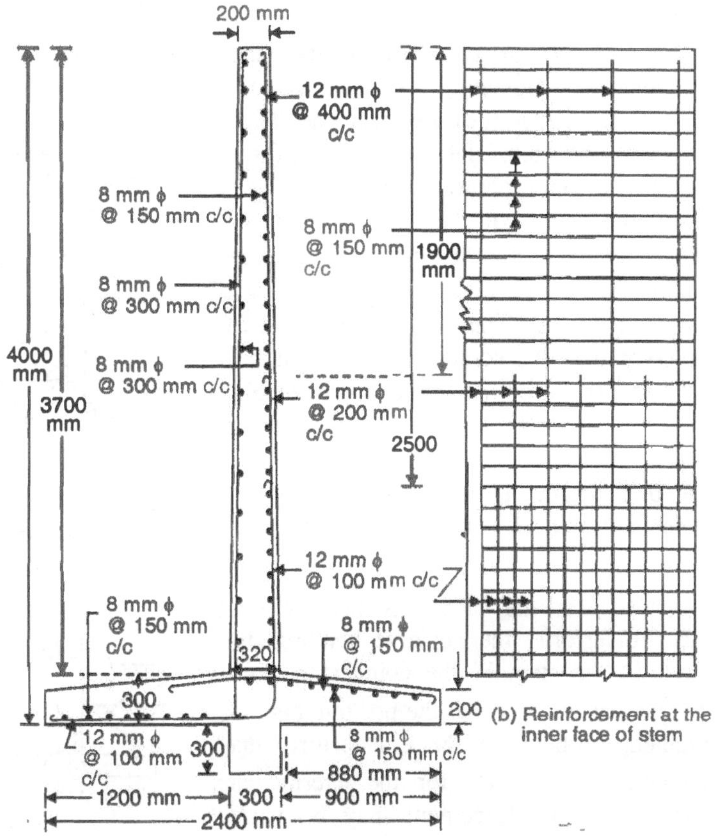

(a) Section of the wall

FIG. 23.27

$$\therefore \quad \text{Shear stress in key} = \frac{35.8 \times 10^3}{1000 \times 300} = 0.12 \ \text{N/mm}^2 \ \text{(Safe)}$$

$$\text{Bending stress} = \frac{35.8 \times 10^3 \times 150}{\frac{1}{6} \times 1000\,(300)^2} = 0.36 \ \text{N/mm}^2$$

Since concrete can take this much of tensile stress, no special reinforcement is necessary for the key. The details of reinforcement etc. are shown in Fig. 23.27.

23.10. DESIGN OF COUNTERFORT RETAINING WALL

We have seen that in case of cantilever retaining wall, the bending moment in the vertical stem varies as h^3, where h is the height of the stem. When this height exceeds 6 m, a thick section is required for the stem, and the design becomes uneconomical. In that case, counterfort retaining walls are used.

The counterforts support both the vertical stem as well as heel slab. Sometimes, counterforts are also provided over the toe slab, upto ground level, as shown in Fig. 23.28(*b*). Design principles for various component parts are discussed below in brief. The same criterion for fixing the length of base (b), and length of toe slab ($\alpha\, b$) are used as cantilever retaining wall.

1. Design of stem. Unlike the stem of cantilever retaining wall, the stem of a counterfort retaining wall acts as a continuous slab supported on counterforts. Due to the varying earth pressure over the height of stem, the stem slab deflects outwards, and hence main reinforcement is provided along the length of the retaining wall, at the outer face of the stem between the counterforts, and at the inner face near the counterforts. The reaction of the stem is taken by the counterforts to which it is firmly anchored. The maximum bending moment occurs at B, where the uniformly distributed earth pressure load is calculated for unit height. If w is the load on the stem slab, at B, per unit length

$$w = p_a \times 1 \times 1 = K_a \gamma H.$$

If l is the clear distance between the counterforts, the negative bending moment in the stem at the counterforts may be taken as $w\,l^2/12$ and the positive bending moment in between the counterforts may be taken as $w\,l^2/16$, for the intermediate panels. The reinforcement may be varied along the height of the stem.

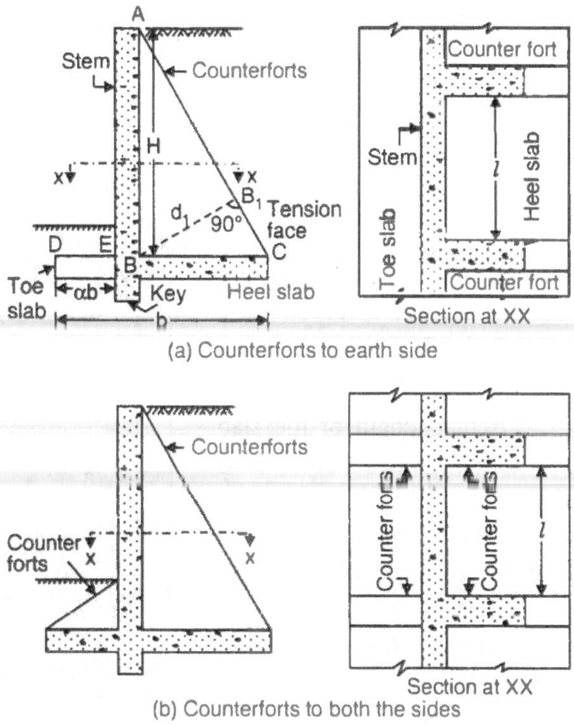

FIG. 23.28.

2. Design of heel slab. The action of heel slab is similar to that of stem. The heel slab is subjected to the downward load due to weight of soil and self-weight, and upward load due to soil reaction. The net load is found to act downwards. The maximum net pressure is found to act on a strip of unit width near edge C, since the upward soil reaction is minimum there. If p is the net downward pressure per unit area, the maximum bending moment (negative) near counterforts will be $(p\,l^2)/12$ and positive bending moment will be $(p\,l^2)/16$. The total load (reaction) from the heel slab is transferred to the counterforts, and this load helps to provide a balancing moment against its overturning. The heel slab is firmly attached to the counterforts by means of vertical ties (two legged stirrups).

3. Design of toe slab. The toe slab bends as a cantilever due to upward soil reaction as in the case of cantilever retaining wall. Due to bending of the toe as cantilever, clockwise moments are induced at E, which are shared by both vertical stem and heel slab. The distribution of this moment between the stem and heel slab are not known. These transferred edge moments will cause bending of stem as well as heel slab, in a direction normal to their usual bending. These edge moments also cause shears in both the slabs which are not precisely known. Hence the design is done empirically, because of uncertain bending behaviour. However, if counterforts are also provided over the toe slab, upto the height of soil (*i.e.* upto ground level), their behaviour becomes certain. In such a case, the reaction is transferred directly to the counterforts, without affecting the bending behaviour of stem and heel slab.

4. Design of counterforts. The counterforts take reactions both from the stem as well as the heel slab. Since the active pressure on stem acts outwards, and net pressure on heel slab acts downwards, the counterforts are subjected to tensile stresses along the outer face AC of the counterforts. The angle ABC between stem and heel slab has a tendency to increase from 90°, and this tendency is resisted by counterforts. Thus the counterfort may be considered to bend as a cantilever, fixed at BC. The counterfort acts as an inverted T-beam of varying rib depth. The maximum depth of this T-beam is at the junction B. The depth is measured perpendicular to the sloping face AB, *i.e.*, depth $d_1 = BB_1$ at B. This depth thus goes on decreasing towards A where the bending moment also decreases. The width b_1 of the counterfort is kept constant throughout its height. Main reinforcement is provided parallel to AC. The reinforcement may be varied along AC, maximum being at B_1. The faces AB and BC of the counterfort remain in compression. The compressive stresses on face AB are counter-balanced by the vertical upward reaction transferred by the slab. In addition to the main reinforcement, the counterforts are jointed firmly to the stem and heel slab by horizontal and vertical ties respectively, provided in the form of two legged stirrups. Normally, the spacing between the counterforts should not be kept less than 2 m. The spacing of the counterforts depends upon many factors, but usually ranges from $\frac{1}{3}$ to $\frac{1}{2}$ of the height of the wall. For an economical design the clear spacing l between the counterforts may be determined from the following approximate expression:

$$l = 3.5\left(\frac{H}{\gamma}\right)^{\frac{1}{4}} \qquad \ldots(23.29)$$

where H is the height of wall in m and γ is the unit weight of soil in kN/m^3.

23.11. BACK ANCHORING OF RETAINING WALL

When the height of retaining wall is much more, it becomes uneconomical to even provide counterforts. In order to reduce the section of stem etc. in the high retaining walls, the stem may be anchored at its back. The anchor practically takes all the earth pressure and B.M. and S.F. in the stem are thus greatly reduced.

Fig. 23.29 shows such a wall anchored at its back, at a height $H/3$ where the resultant earth pressure acts. Due to the provision of the anchor at this location, bending moments in the stem at B and A would be zero, and the whole of the earth pressure will be resisted by the anchor rod which is provided at some suitable spacing l centre to centre along the length of the wall. The anchor bar is attached to a concrete slab which resists the pressure by way of development of frictional forces along both the surface EF and GH. The concrete slab is located well beyond the plane of rupture.

Let A = surface area of each face of slab, in a distance l.

μ = coefficient of friction between concrete and soil.

h = height to embedment of slab $= \frac{2}{3} H$.

l = centre to centre spacing of anchor bars

A_s = area of cross-section of each anchor bars.

Horizontal sliding force $= \frac{1}{2} K_a \gamma H^2 \, l$ ...(*i*)

Resisting friction force $= 2\, A\mu.\gamma h$ ...(*ii*)

Equating the two, $2A\mu\gamma h = \frac{1}{2} K_a \gamma H^2 l$

or
$$A = \frac{K_a H^2 l}{4\mu h} = \frac{3}{8}\frac{K_a H\, l}{\mu} \qquad ...(23.30)$$

Hence width
$$b_1 = \frac{A}{l} = \frac{3K_a H}{8\,\mu} \qquad ...(23.31)$$

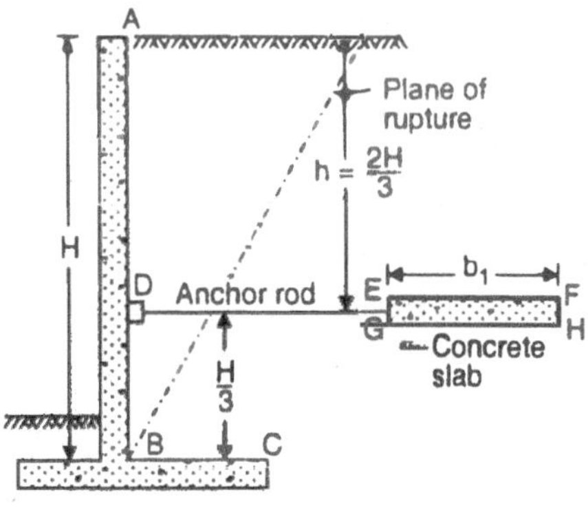

FIG. 23.29.

$$\therefore \text{ Area of rod } A_s = \frac{1.5\left(\frac{1}{2} K_a \gamma H^2 l\right)}{0.87 f_y} = \frac{K_a \gamma H^2 l}{1.16 f_y} \qquad ...(23.32)$$

If the height is more, two or more tie rods can used at different heights. The tie rod is fixed to the wall through horizontal or vertical beams cast monolithically with the stem. In case vertical beams are provided at interval l, the stem will act as a continuous slab as in the case of counterfort retaining wall. If horizontal beam is provided at height $H/3$, the portions of stem above and below it will act as cantilevers.

In some cases, such as in road or railway embankments, high retaining walls are provided to both the sides of the fill. In such a circumstance, concrete slab is not required to develop frictional force of resistance, but instead, both the walls are connected by common tie rods at suitable intervals, so that earth pressure on one is neutralised by the earth pressure on the other.

Design Example 23.4 *Design a counterfort retaining wall to retain 7 m high embankment above ground level. The foundation is to be taken 1 m deep where the safe bearing capacity of soil may be taken as 180 kN/m². The top of earth retained is horizontal, and soil* *weighs 18 kN/m³ with angle of internal friction $\Phi = 30°$. Coefficient of friction between concrete and soil may be taken as 0.5. Use M 20 concrete and Fe 415 steel.*

Solution.

1. ***Design constants*** : For M 20 concrete and Fe 415 steel reinforcement we have the following : $x_{u\,.\,max}/d = 0.479$ and $R_u = 2.761$

For the soil, $\Phi = 30°$ $\quad \therefore \quad K_a = \frac{1 - \sin 30}{1 + \sin 30} = \frac{1}{3}$ and $K_p = \frac{1}{K_a} = 3.$

2. ***Dimensions of various parts***. The ratio of length of toe slab (DE) to the base of width b may be found by the expression :

$$\alpha = 1 - \frac{q_o}{2.2\gamma H} = 1 - \frac{180}{2.2 \times 18 \times 8} = 0.43.$$

$$b = 0.95\,H\sqrt{\frac{K_a}{(1-\alpha)(1+3\alpha)}}$$

$$= 0.95 \times 8\sqrt{\frac{1}{3(1-0.43)(1 + 3 \times 0.43)}} = 3.64$$

Normal range of b is between 0.5 H to 0.6 H.

However, keep minimum b = 0.5 H = 4 m.

$\therefore$ Length of toe slab = 0.43 × 4 = 1.72 m.

Keep the length of toe slab = 1.70 m.

Taking uniform thickness of stem = 300 mm, length of heel slab $BC = 4 - 1.70 - 0.3 = 2$ m.

Let the thickness of base slab = 300 mm.

Clear spacing of counterforts is

$$l = 3.5\left(\frac{H}{\gamma}\right)^{\frac{1}{4}} = 3.5\left(\frac{8}{18}\right)^{\frac{1}{4}} = 2.86 \text{ m.}$$

Keep them at 3 m apart. Let us also provide counterforts over toe slab, upto ground level, at 3 m clear distance.

3. ***Stability of wall***. The preliminary dimensions of the wall are marked in Fig. 23.30.

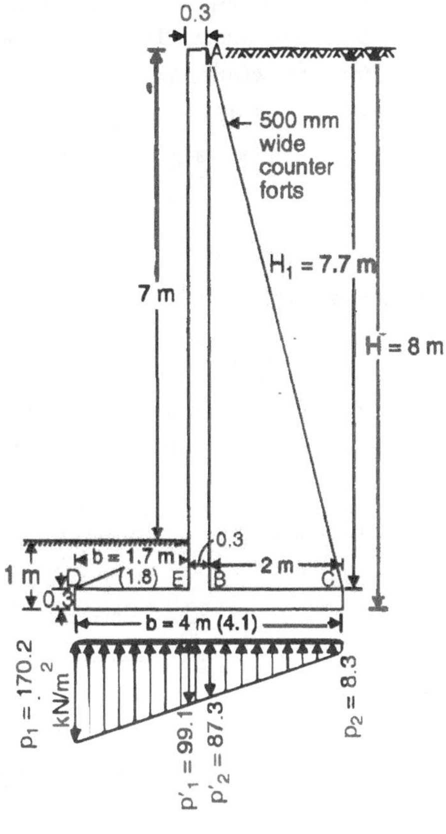

FIG. 23.30.

Let W_1 = weight of stem, per metre length.

W_2 = weight of base slab.

W_3 = weight of soil on heel slab. The calculations are arranged in Table 23.4.

TABLE 23.4

S. No.	Desig-nation	Force (kN)	L.A. (m)	Moment about toe (kN-m)
1	W_1	0.3 × 7.7 × 1 × 25 = 57.8	1.85	106.8
2	W_2	0.3 × 4 × 1 × 25 = 30	2	60.0
3	W_3	2 × 7.7 × 1 × 18 = 277.2	3	831.6
		$\Sigma W = 365$	M_R =	998.4

$\therefore$ Resisting moment $M_R = 998.4$ kN-m ...(1)

Horizontal earth pressure $P_H = K_a \gamma \dfrac{H^2}{2} = \dfrac{1}{3} \times \dfrac{1}{2} \times 18\ (8)^2 = 192$ kN ...(2)

Overturning moment $M_0 = P_H \cdot \dfrac{H}{3} = 192 \times \dfrac{8}{3} = 512$ kN-m ...(3)

$$\text{F.S. against overturning} = \frac{M_R}{M_0} = \frac{998.4}{512} = 1.95 < 2. \quad \text{Hence unsafe.}$$

$$\text{F.S. against sliding} = \frac{\mu \Sigma W}{P_H} = \frac{0.5 \times 365}{192} = 0.95$$

Hence the wall is unsafe against overturning and sliding. A shear key will have to be provided under the base. Also, width of the base will have to be increased.

Pressure distribution. Net moment $\Sigma M = M_R - M_0 = 998.4 - 512 = 486.4$ kN-m.

Distance $\bar{x}$ from toe, of the point of application of the resultant is given by

$$\bar{x} = \frac{\Sigma M}{\Sigma W} = \frac{486.4}{365} = 1.333 \text{ m} \quad \therefore \text{ Eccentricity } e = \frac{b}{2} - \bar{x} = 2 - 1.333 = 0.667 \text{ m}$$

This is exactly equal to $\dfrac{b}{6} = \dfrac{4}{6} = 0.667$ m. However no tension will be developed. Pressure p_1 under the toe is given by

$$p_1 = \frac{\Sigma W}{b}\left(1 + \frac{6e}{b}\right) = \frac{365}{4}\left(1 + \frac{6 \times 0.667}{4}\right) = 182.5 \text{ kN/m}^2 > 180 \text{ Hence unsafe.}$$

In order to make it safe, increase the length of toe slab *DE* to 1.8 m, so that total width $b = 4.1$ m. The revised computations are arranged in Table 23.4 (*a*)

TABLE 23.4 (*a*)

S. No.	Designation	Force (kN)	L.A. (m)	Moment about toe (kN-m)
1	W_1	0.3 × 7.7 × 1 × 25 = 57.8	1.95	112.6
2	W_2	0.3 × 4.1 × 1 × 25 = 30.8	2.05	63.0
3	W_3	2 × 7.7 × 1 × 18 = 277.2	3.1	859.3
		ΣW = 365.8		M_R = 1034.9

Net moment $\Sigma M = M_R - M_0 = 1034.9 - 512 = 522.9$ kN-m

$$\bar{x} = \frac{\Sigma M}{\Sigma W} = \frac{522.9}{365.8} = 1.43 \text{ m} \quad \therefore \quad e = \frac{b}{2} - \bar{x} = 2.05 - 1.43 = 0.62 \text{ m} < \frac{b}{6}$$

$$p_1 = \frac{\Sigma W}{b}\left(1 + \frac{6e}{b}\right) = \frac{365.8}{4.1}\left(1 + \frac{6 \times 0.62}{4.1}\right) = 170.2 \text{ kN/m}^2 < 180. \text{ Hence safe}$$

Pressure p_2 under the heel is given by $p_2 = \dfrac{\Sigma W}{b}\left(1 - \dfrac{6e}{b}\right) = \dfrac{365.8}{4.1}\left(1 - \dfrac{6 \times 0.62}{4.1}\right) = 8.3$ kN/m²

The pressure intensity p_1' under *E* is $p_1' = 170.2 - \dfrac{170.2 - 8.3}{4.1} \times 1.8 = 99.1$ kN/m²

The pressure intensity p_2' under *B* is $p_2' = 170.2 - \dfrac{170.2 - 8.3}{4.1} \times 2.1 = 87.3$ kN/m²

4. ***Design of heel slab***. Clear spacing betwen counterforts = 3 m. The pressure distribution below the heel slab is shown in Fig. 23.30. Consider a strip, 1 m wide, near the outer edge *C*. The upward pressure intensity is 8.3 kN/m², which is minimum at *C*.

Downward load due to weight of earth $= 7.7 \times 18 = 138.6$ kN/m², and that due to selfweight of heel slab $= 0.3 \times 25 = 7.5$ kN/m². Hence net downward intensity of load p is given by

$$p = 138.6 + 7.5 - 8.3 = 137.8 \text{ kN/m}^2$$

∴ Maximum negative bending moment in heel slab, at counterforts is

$$M_1 = \frac{pl^2}{12} = \frac{137.8(3)^2}{12} = 103.4 \text{ kN-m} = 103.4 \times 10^6 \text{ N-mm};$$

$$M_{u1} = 1.5 \times 103.4 \times 10^6 = 155.1 \times 10^6 \text{ N-mm}$$

$$\therefore \quad d = \sqrt{\frac{155.1 \times 10^6}{1000 \times 2.761}} = 237 \text{ mm}$$

$$\text{Shear force } V = \frac{137.8 \times 3}{2} = 207 \text{ kN. } ; \ V_u = 1.5 \times 207 = 310.5 \text{ kN}$$

For under-reinforced section, having $p = 0.5\%$, $\tau_c = 0.48$ N/mm². Hence depth required from shear point of view is

$$d = \frac{V_u}{\tau_c b} = \frac{310.5 \times 1000}{0.48 \times 1000} = 647 \text{ mm. This is excessive.}$$

However, keep $D = 500$ mm. Providing 60 mm effective cover, $d = 500 - 60 = 440$ mm.

$\tau_v = \frac{310.5 \times 1000}{1000 \times 440} = 0.706$ N/mm² $> \tau_c$. Hence shear reinforcement will be necessary.

Area of steel at supports is given by

$$A_{st1} = \frac{0.5 \times 20}{415}\left[1 - \sqrt{1 - \frac{4.6 \times 155.1 \times 10^6}{20 \times 1000\,(440)^2}}\right]1000 \times 440 = 1027 \text{ mm}^2$$

Providing 12 mm φ bars, $A_\varphi = 113.1$ mm². ∴ Spacing $= \frac{1000 \times 113.1}{1027} \simeq 110$ mm

Provide these @ 100 mm c/c. Actual A_{st} provided $= \frac{1000 \times 113.1}{100} = 1131$ mm²

Let us check this reinforcement for development length at the point of contraflexure, so as to satisfy the criterion : $\frac{M_1}{V_u} + L_0 \geq L_d$.

For a fixed beam or slab carrying U.D.L., the point of contraflexure is situated at a distance of 0.211 L. In our case, the slab is continuous, but we will assume the same position of point of contraflexure, *i.e.* at $0.211 \times 3 \approx 0.63$ m from the face of counterforts. Shear force at this point is given by

$$V = \frac{pL}{2}\left(\frac{l}{2} - x\right) \div \left(\frac{L}{2}\right) = p\left(\frac{l}{2} - x\right) = 137.8\left(\frac{3}{2} - 0.63\right) = 119.9 \text{ kN} ; V_u = 1.5 \times 119.9 = 179.85 \text{ kN}$$

Assuming that all the bars will be available at the point of contraflexure,

$$x_u = \frac{0.87 f_y A_{st}}{0.36 f_{ck} b} = \frac{0.87 \times 415 \times 1131}{0.36 \times 20 \times 1000} = 56.7 \text{ mm}$$

$$M_1 = 0.87 f_y A_{st}(d - 0.416 x_u) = 0.87 \times 415 \times 1131\,(440 - 0.416 \times 56.7) = 170.04 \times 10^6 \text{ N-mm}$$

$L_0 = 12\ \varphi$, or d which ever is more $= 440$ mm

$L_d = 47\ \varphi = 47 \times 12 = 564$ mm

$$\therefore \quad \frac{M_1}{V_u} + L_0 = \frac{170.04 \times 10^6}{179.85 \times 10^3} + 440 = 1385 \text{ mm} > L_d. \text{ Hence safe.}$$

Continue these bars by a distance l_0 ($= d = 440$ mm) beyond the point of contraflexure. After that, curtail half the bars, and continue the remaining half throughout the length. At the point of curtailment, length of each bar available $= 630 + 440 = 1070$ mm $> L_d$. These bars will be provided at the top face of the heel slab.

$$\text{Maximum positive B.M.} = \frac{pl^2}{16} = \frac{3}{4} M_1$$

$$\therefore \quad \text{Area of bottom steel } A_{st2} = \frac{3}{4} A_{st1} = \frac{3}{4} \times 1131 = 849 \text{ mm}^2$$

$$\text{Spacing of 12 mm } \varphi \text{ bars} = \frac{1000 \times 113.1}{849} \approx 133 \text{ mm} \simeq 130 \text{ mm}$$

$$\text{Actual} \quad A_{st} = \frac{1000 \times 113.1}{130} = 870 \text{ mm}^2$$

Let us check whether these bars satisfy the development length criterion at the point of contraflexure, inherent in the criterion : $\frac{M_1}{V_u} + L_0 \geq L_d$.

where $V_u =$ shear at point of contraflexure $= 1.5 \times 119.9$ kN distant 0.63 m from the face of support

Assuming that all the bars are available at the point of contraflexure,

$$x_u = \frac{0.87 \times 415 \times 870}{0.36 \times 20 \times 1000} = 43.6 \text{ mm};$$

$$M_1 = 0.87 \times 415 \times 870\ (440 - 0.416 \times 43.6) = 132.51 \times 10^6 \text{ N-mm}$$

$$L_0 = 440 \text{ mm and } L_d = 932 \text{ mm, as before.}$$

$$\therefore \quad \frac{M_1}{V_u} + L_0 = \frac{132.5 \times 10^6}{1.5 \times 119.9 \times 10^3} + 440 = 1176 \text{ mm This is more than } L_d. \text{ Hence safe.}$$

Thus continue all the bottom bars to a point distant L_0 (= 440 mm) from the point of contraflexure, *i.e.* upto a point distant $630 - 440 \approx 190$ mm from the centre of support. At this point, half the bars can be discontinued. Since this distance is quite small, it is better to continue these bars upto the centre of counterforts.

Reinforcement near B The c/c spacing of reinforcement near B may be increased, because p decreases due to increase in upward soil reaction. Consider a strip 1 m wide near B. Upward soil reaction at B is $= 87.3$ kN/m², as found earlier.

$\therefore$ Net downward load $p' = 138.6 + (0.5 \times 25) - 87.3 = 63.8$ kN/m²

This is about $63.8/137.8 = 0.463$ of load intensity at C.

Hence spacing of steel bars $= 100/0.463 \approx 200$ mm c/c. at the top face, near supports.

Spacing of steel bars at the bottom face, at midspan $= 133/0.463 \simeq 280$ mm c/c.

Distribution steel. Area of steel $= \frac{0.12}{100}(1000 \times 500) = 600 \text{ mm}^2$.

Using 12 mm φ bars, having $A_\varphi = 113 \text{ mm}^2$.

$$\therefore \quad \text{Spacing} = \frac{1000 \times 113}{600} \approx 180 \text{ mm c/c}$$

Shear reinforcement. Shear stress at $C = \tau_v = 0.706 \text{ N/mm}^2$. $\frac{100 A_s}{b\, d} = \frac{100 \times 1131}{1000 \times 440} = 0.257\%$

Hence from Table 7.1, $\tau_c = 0.36 \text{ N/mm}^2 < \tau_v$. Hence shear reinforcement is necessary.

$$V_{uc} = \tau_c\, bd = 0.36 \times 1000 \times 440 = 158.4 \times 10^3 \text{ N} = 158.4 \text{ kN}.$$

Consider a section distant x_1 from the face of the counterfort, where shear force is $158.4/1.5 = 105.6$ kN. The position is given by $\frac{105.6}{207} = \frac{1.5 - x_1}{1.5}$, which gives $x_1 = 0.73$ m.

Hence shear stirrups are required upto distance of 0.73 m on either side of each counterforts. The requirement is there for a strip of unit width passing through C. Let us consider a strip through B_1, distant y_1 from C, such that shear force at the counterforts is 105.6 kN.

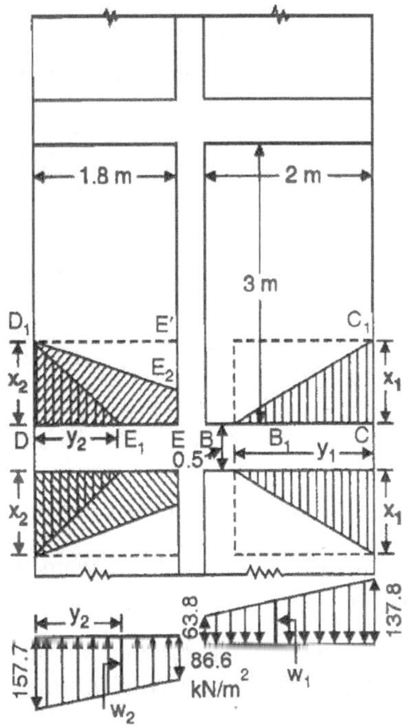

FIG 23.31

Net downward pressure at $C = 137.8 \text{ kN/m}^2$

Net downward pressure at $B = 63.8 \text{ kN/m}^2$

Let the net downward pressure at B_1 be w_1 kN/m²

$\therefore$ S.F. at face of counterforts at $B_1 = w_1 \times 3/2 = 1.5\, w_1$.

This should be equal to 105.6 kN.

$$\therefore \quad w = 105.6/1.5 = 70.4 \text{ kN/m}^2 \quad \ldots(1)$$

However, at y_1 from C,

$$w_1 = 137.8 - \frac{137.8 - 63.8}{2} \times y_1 = 137.8 - 37\, y_1 \ldots(2)$$

Equating the two, we get $137.8 - 37\, y_1 = 70.4$

From which $y_1 = 1.82$ m.

Hence shear reinforcement is required in the triangular portion on the other side of the counterforts shown hatched in Fig. 23.31. However, we will provide shear stirrups in rectangular portions $x_1 \times y_1 = 0.8 \text{ m} \times 1.82$ m on either side of counterforts. Let us provide 4-legged stirrups of 8 mm φ wire.

$$\therefore \quad A_{sv} = 4\,\frac{\pi}{4}\,(8)^2 = 201 \text{ mm}^2$$

$$\therefore \quad s_v = \frac{0.87\, f_y\, A_{sv}\, d}{V_u - V_{uc}} = \frac{0.87 \times 415 \times 201 \times 440}{1.5\,(207 - 105.6)\,1000} \simeq 200 \text{ mm}$$

Hence provide 8 m φ 4-lgd stirrups @ 200 mm c/c on either side of each counterfort.

5. ***Design of Toe slab***. Since the toe slab is also large, provide counterforts over it, upto ground level at 3 m clear distance face to face. The toe slab will thus bend like a continuous slab. Assume total depth of toe slab = 500 mm.

∴ Downward weight per unit area, due to self weight = $0.5 \times 1 \times 1 \times 25 = 12.5$ kN/m^2.

Upward pressure intensity at $D = 170.2$ kN/m^2

∴ Net upward pressure intensity at $D = 170.2 - 12.5 = 157.7$ kN/m^2.

Similarly, net upward pressure intensity at $E = 99.1 - 12.5 = 86.6$ kN/m^2

Consider strip of unit width at D. Max negative B.M. is

$$M_1 = \frac{wl^2}{12} = \frac{157.7\,(3)^2}{12} = 118.3 \text{ kN-m} = 118.3 \times 10^6 \text{ N-mm}$$

$$M_{u1} = 1.5 \times 118.3 \times 10^6 = 177.45 \times 10^6 \text{ N-mm}$$

$$d = \sqrt{\frac{177.45 \times 10^6}{2.761 \times 1000}} = 254 \text{ mm}$$

Shear force $V = \dfrac{157.7 \times 3}{2} = 236.6$ kN. ; $V_u = 1.5 \times 236.6 = 354.9$ kN

Taking stress $\tau_c = 0.38$ N/mm^2 for an under-reinforced section at $p = 0.3\%$ (Table 7.1) the depth of slab required from shear point of view is given by

$$d = \frac{V_u}{b\,.\,\tau_c} = \frac{354.9 \times 1000}{1000 \times 0.38} = 934 \text{ mm}$$

This is excessive. However, we will keep the same depth as that of heel, *i.e.*, $d = 440$ mm and $D = 500$ mm, and provide shear stirrups to take up excessive shearing stress.

Area of steel at supports, at bottom is

$$A_{st1} = \frac{0.5 \times 20}{415}\left[1 - \sqrt{1 - \frac{4.6 \times 177.45 \times 10^6}{20 \times 1000\,(440)^2}}\right] 1000 \times 440 = 1184 \text{ mm}^2$$

Using 12 mm φ bars, $A_\varphi = 113$ mm^2, $s = \dfrac{1000 \times 113}{1184} = 95.4$ mm

However, provide these @ 80 mm c/c. Actual $A_{st1} = \dfrac{1000 \times 113}{90} = 1256$ mm^2.

Let us check these from point of view of development length, at the point of contraflexure, so as to satisfy the criterion : $M_1/V_u + L_0 \geq L_d$.

The point of contraflexure occurs at a distance $x = 0.63$ m from the support. Hence S.F. at the point of contraflexure is

$$V = w\left(\frac{l}{2} - x\right) = 157.7\left(\frac{3}{2} - 0.63\right) = 137.2 \text{ kN} \; ; \; V_u = 1.5 \times 137.2 = 205.8 \text{ kN}$$

$$x_u = \frac{0.87 \times 415 \times 1256}{0.36 \times 20 \times 1000} = 63 \text{ mm}$$

∴ $M_1 = 0.87 \times 415 \times 1256\,(440 - 0.416 \times 63) = 187.6 \times 10^6$ N-mm.

$L_0 = 12\,\varphi$ or d, whichever is more = 440 mm

$L_d = 47 \times 12 = 564$ mm

$$\therefore \quad \frac{M_1}{V_u} + L_0 = \frac{187.6 \times 10^6}{1.5 \times 137.2 \times 10^3} + 440 = 1351 \text{ mm} > L_d$$

Hence satisfied. Continue these bars, at the bottom of toe slab, beyond the point of contraflexure by a distance of L_0 (= 440 mm) *i.e.*, by a distance of 630 + 440 = 1070 mm from the face of the counterforts.

$$\text{Again, positive B.M.} = \frac{3}{4} M_1 = \frac{3}{4} \times 118.3 \times 10^6 = 85 \times 10^6 \text{ N-mm}$$

$$\text{Area of bottom steel} = \frac{3}{4} A_{st1} = \frac{3}{4} \times 1184 = 888 \text{ mm}^2$$

$$\text{Spacing of 12 mm } \varphi \text{ bars} = \frac{1000 \times 113}{888} = 127 \text{ mm}$$

Hence provide these @ 120 mm c/c. Actual area of steel $= \frac{1000 \times 113}{120} = 942 \text{ mm}^2$. This steel should satisfy the following criterion at the point of contraflexure

$$\frac{M_1}{V_u} + L_0 \geq L_d$$

where V_u = S.F. at the point of contraflexure = 1.5 × 137.2 kN

Assuming that all the bars, provided at the top face, are available at the point of contraflexure.

$$x_u = \frac{0.87 \times 415 \times 942}{0.36 \times 20 \times 1000} = 47.2 \text{ mm}$$

$$M_1 = 0.87 \times 415 \times 942\,(440 - 0.416 \times 47.2) = 142.97 \times 10^6 \text{ N-mm}$$

L_0 = 12 φ or d, whichever is more = 440 mm. L_d = 47 φ = 47 × 12 = 564 mm

$$\therefore \quad \frac{M_1}{V_u} + L_0 = \frac{142.47 \times 10^6}{1.5 \times 137.2 \times 10^3} + 440 = 1134 \text{ mm}$$

This is greater than L_d. Hence the criterion is satisfied. However, these bars are to be continued for a distance of L_0(= 440 mm) beyond the point of contraflexure, their distance from the face of counterforts being only = 630 – 440 = 190 mm. Hence it is better to continue these bars upto counterfort, at the top face of the toe slab.

Reinforcement at E. At a section distant 1 m from *E*,

$$\text{Upward soil pressure} = 170.2 - \frac{170.2 - 8.3}{4.1} \times 0.8 = 138.6 \text{ kN/m}^2.$$

$\therefore$ Net upward pressure = 138.6 – 12.5 = 126.1 kN/m².

This is 126.1/157.7 ≏ 0.80 of w at D.

$\therefore$ Spacing of bottom steel = 95.4/0.80 = 119 ≏ 110 mm c/c

Spacing of top steel = 127/0.80 = 158 ≏ 150 mm c/c

Distribution steel. $A_{sd} = \frac{0.12}{100}(1000 \times 500) = 600 \text{ mm}^2$

Using 12 mm φ bars, $A_\varphi \approx 113$ mm². $\therefore$ Spacing $= \frac{1000 \times 113}{600} \approx 180$ mm

Shear reinforcement. Shear force at D = 236.6 kN; $\therefore$ $F_u = 1.5 \times 236.6 = 354.9$ kN

$$\therefore \quad \tau_v = \frac{354.9 \times 1000}{1000 \times 440} = 0.807 \text{ N/mm}^2$$

$$\frac{100\,A_s}{bd} = \frac{100 \times 1256}{1000 \times 440} \approx 0.285\%.$$ Hence from Table 7.1, $\tau_c \simeq 0.37$ N/mm^2,

Since $\tau_v > \tau_c$, shear reinforcement is necessary.

Now $V_{uc} = \tau_c \,.\, bd = 0.37 \times 1000 \times 440 = 162800$ N = 162.8 kN ; $V = 162.8/1.5 = 108.5$

Consider a section distant x_2 from the face of the counterfort (Fig. 23.31) where S.F. is 108.5 kN. Position x_2 is given by

$$\frac{108.5}{236.6} = \frac{1.5 - x_2}{1.5}, \text{ which gives } x_2 = 0.81 \text{ m.}$$

Hence shear stirrups are required upto a distance of 0.81 m on either side of counterforts. This requirement is there for a strip of unit width passing through D. Let us consider a strip through E_1, distant y_2 from D, such that shear force at the counterforts is 108.5 kN. To find the position of y_2, consider the net pressure distribution below the toe slab.

Self weight of toe slab = 12.5 kN/m^2. Hence net pressure intensity below D and E are respectively $170.2 - 12.5 = 157.7$ kN/m^2 and $99.1 - 12.5 = 86.6$ kN/m^2 respectively (Fig. 23.31). Let the net pressure intensity at E_1 be w_2 kN/m^2.

$\therefore$ S.F. at the counterforts at $E_1 = w_2 \times \frac{3}{2} = 1.5\, w_2$ kN. This should be equal to 108.5 kN.

$$\therefore \quad w_2 = 108.5/1.5 = 72.3 \text{ kN/m}^2 \qquad \ldots(1)$$

However, at y_2 from D, $w_2 = 157.7 - \frac{157.7 - 86.6}{1.8} y_2 = 157.7 - 39.5\, y_2$ ...(2)

Equating the two we get, $157.7 - 39.5\, y_2 = 72.3$

From which $y_2 = 2.16$ m. This is more than DE (= 1.8 m).

Hence shear force at E is more than 108.5 kN.

Actual S.F. at $E = 1.5 \times 86.8 = 130.2$ kN.

Consider a section distant z from the face of counterforts (Point E), where S.F. is 108.5 kN. The position of z is given by

$$\frac{108.5}{130.2} = \frac{1.5 - z}{1.5}, \text{ which gives } z = 0.25 \text{ m.}$$

Hence shear stirrups are to be provided for a region DEE_2D_1, where $EE_2 = 0.25$ m only. However, we will provide stirrups for whole of rectangular area (shown dotted), for width $DD_1 = x_2 = 0.81$ m and length DE = 1.8 m.

Let us provide 8 legged stirrups, of 8 mm φ wire.

$$A_{sv} = 8 \frac{\pi}{4}(8)^2 = 402 \text{ mm}^2$$

$$\therefore \quad \text{Spacing} = \frac{0.87 f_y \,.\, A_{sv}\, d}{V_u - V_{uc}} = \frac{0.87 \times 415 \times 402 \times 440}{1.5\,(236.6 - 108.5)\,1000} = 332 \text{ mm} \simeq 300 \text{ mm}$$

6. ***Design of stem (vertical slab)***. The stem acts as a continuous slab. Consider 1 m strip at B. The intensity of earth pressure is given by

$$p_h = K_a\gamma H_1 = \frac{1}{3} \times 18 \times 7.5 = 45 \text{ kN/m}^2 \quad \text{(where revised value of } H_1 = 8 - 0.5 = 7.5 \text{ m)}$$

$\therefore$ Negative B.M. in slab near counterforts is

$$M_1 = \frac{p_h \cdot l^2}{12} = \frac{45(3)^2}{12} = 33.75 \text{ kN-m} = 33.75 \times 10^6 \text{ N-mm}$$

$$M_{1u} = 1.5 \times 33.75 \times 10^6 = 50.625 \times 10^6 \text{ N-mm}$$

$$\therefore \quad d = \sqrt{\frac{50.625 \times 10^6}{1000 \times 2.761}} = 136 \text{ mm.}$$

Providing effective cover = 60 mm, total depth = 136 + 60 = 196 mm. However, provide total depth = 300 mm so that available $d = 300 - 60 = 240$ mm. This increased thickness will keep the shear stress within limits so that additional shear reinforcement is not required.

$$\text{Shear force} \quad V = \frac{45 \times 3}{2} = 67.5 \text{ kN} \; ; \quad V_u = 1.5 \times 67.5 \text{ kN}$$

$$\therefore \quad \tau_v = \frac{1.5 \times 67.5 \times 1000}{1000 \times 240} = 0.42 \text{ N/mm}^2$$

This is equal to $\tau_c = 0.42$ N/mm² at 0.375% reinforcement (see Table 7.1)

Area of steel near counterforts is

$$A_{st1} = \frac{0.5 f_{ck}}{f_y}\left[1 - \sqrt{1 - \frac{4.6 M_u}{f_{ck}\, bd^2}}\right] bd = \frac{0.5 \times 20}{415}\left[1 - \sqrt{1 - \frac{4.6 \times 50.625 \times 10^6}{20 \times 1000\,(240)^2}}\right] \times 1000 \times 240$$

$= 618 \text{ mm}^2$

$$\text{Reinforcement corresponding to } p = 0.375\% \text{ is } = \frac{0.375 \times 1000 \times 240}{100} = 900 \text{ mm}^2$$

$$\therefore \text{ Spacing of 12 mm } \varphi \text{ bars } = \frac{1000 \times 113}{900} = 125 \text{ mm}$$

However provide 12 mm φ bars @ 120 mm c/c.

$$\text{Actual } A_{st} = \frac{1000 \times 113}{120} \simeq 942 \text{ mm}^2 \quad \text{and} \quad \frac{100\, A_s}{bd} = \frac{100 \times 942}{1000 \times 240} = 0.39\% .$$

From Table 7.1, $\tau_c = 0.427$ N/mm² corresponding to $p = 0.39\%$

Let us check these bars for development length, near points of contraflexure, so as to satisfy the criterion : $\frac{M_1}{V_u} + L_0 \geq L_d.$

For a fixed beam or slab carrying U.D.L., the point of contraflexure is at a distance of 0.211 L = 0.63 m from the face of counterforts, S.F. at this point is given by

$$V = \frac{pL}{2}\left(\frac{l}{2} - x\right) \div \left(\frac{L}{2}\right) = p\left(\frac{l}{2} - x\right) = 67.5\left(\frac{3}{2} - 0.63\right) = 58.53 \text{ kN}$$

$$V_u = 1.5 \times 58.53 \simeq 87.8 \text{ kN}$$

Assuming that all the bars will be available at the point of contraflexure

$$x_u = \frac{0.87 \times 415 \times 942}{0.36 \times 20 \times 1000} = 47.2 \text{ mm};$$

$$\therefore \quad M_1 = 0.87 \times 415 \times 942\,(240 - 0.416 \times 47.2) = 74.95 \times 10^6 \text{ N-mm}$$

$$L_0 = 12\,\varphi \text{ or } d, \text{ whichever is more } = 240 \text{ mm}$$

$$L_d = 47 \times 12 = 564 \text{ mm}$$

$$\therefore \quad \frac{M_1}{V_u} + L_0 = \frac{74.95 \times 10^6}{87.8 \times 1000} + 240 = 1093 \text{ mm}$$

Hence safe. It is thus essential to continue all the bars upto a point distant 240 mm beyond point of contraflexure, *i.e.* upto a point 630 + 240 = 870 mm ≈ 900 mm from the face of counterforts. These bars are to be provided at the inner face of the stem slab.

Maximum positive B.M. $= \frac{3}{4} M_1 = \frac{3}{4} \times 33.75 \times 10^6 = 25.31 \times 10^6$ N-mm

$$\text{Area of steel} = \frac{3}{4} A_{st1} = \frac{3}{4} \times 942 = 706.5 \text{ mm}^2$$

and spacing of 12 mm φ bars $= \frac{1000 \times 113}{706.5} \simeq 159$ mm. Hence, provide these @ 150 mm c/c.

$$\text{Actual} \quad A_{st} = \frac{1000 \times 113}{150} = 753 \text{ mm}^2.$$

Let us check these bars for development length at the point of contraflexure, so as to satisfy the criterion : $\frac{M_1}{V_u} + L_0 \geq L_d$

Assuming that all the reinforcement is extended upto point of contraflexure

$$x_u = \frac{0.87 \times 415 \times 753}{0.36 \times 20 \times 1000} = 37.8 \text{ mm}$$

$$\therefore \quad M_1 = 0.87 \times 415 \times 753\,(240 - 0.416 \times 37.8) = 60.98 \times 10^6 \text{ N-mm}$$

$$L_0 = 12\,\varphi \text{ or } d, \text{ whichever is more} = 240 \text{ mm}; \quad L_d = 47 \times 12 = 564 \text{ mm}$$

$$V_u = 1.5 \times 58.53 \text{ kN, as before,}$$

$$\therefore \quad \frac{M_1}{V_u} + L_0 = \frac{60.98 \times 10^6}{1.5 \times 58.53 \times 10^3} + 240 = 934 \text{ mm} > L_d. \text{ Hence safe}$$

The spacing of the reinforcement at B, found above can be increased with height. The pressure p_h and hence the bending moment decreases linearly with height.

$$\therefore \quad A_{st} \propto h$$

Hence the spacing of the bars can be increased gradually to say 300 mm c/c near the top.

Distribution steel $= \frac{0.12}{100}(1000 \times 300) = 36 \text{ mm}^2$. Using 10 mm φ bars, $A_\varphi = 78.5 \text{ mm}^2$.

$$\therefore \quad \text{Spacing} = \frac{1000 \times 78.5}{360} = 218 \text{ mm. Hence provide these @ 200 mm c/c.}$$

7. ***Design of main counterfort***. Let us assume thickness of counterfort = 500 mm. The counterforts will thus be spaced @ 350 cm c/c. They will thus receive earth pressure from a width of 3.5 m and downward reaction from the heel slab for a width of 3.5 m.

At any section at depth h below the top A, the earth pressure acting on each counterfort will be $= \frac{1}{3} \times 18 \times h \times 3.5 = 21\,h$ kN/m ...(1)

Similarly, net downward pressure on heel at C is $= 7.5 \times 18 + 0.5 \times 25 - 8.3 = 139.2$ kN/m^2

and that at B is $= 7.5 \times 18 + 0.5 \times 25 - 87.3 = 60.2$ kN/m^2

Hence reaction transferred to each counterfort will be

At C, $139.2 \times 3.5 = 487.2$ kN/m. At B, $60.2 \times 3.5 = 210.7$ kN/m.

The variations of horizontal and vertical forces on the counterfort are shown in Fig. 23.32.

The critical section for the counterfort will be at F, since below this, enormous depth will be available to resist bending.

Pressure intensity at $h = 7$ m is $= 21 \times 7 = 147$ kN/m

Shear force at $F = \frac{1}{2} \times 147 \times 7 = 514.5$ kN

$$\text{B.M. } M = 514.5 \times \frac{7}{3} = 1200.5 \text{ kN-m} = 1200.5 \times 10^6 \text{ N-mm.}$$

$$M_u = 1.5 \times 1200.5 \times 10^6 = 1800.75 \times 10^6 \text{ N-mm}$$

The counterfort acts as a T-beam. However, even as a rectangular beam, depth required is

$$d = \sqrt{\frac{1800.75 \times 10^6}{500 \times 2.761}} \approx 1142 \text{ mm}$$

$\therefore$ Total depth $= 2.762 + 60 = 1202$ mm. However, to make it safe against shear, keep $D = 1720$ mm so that $d = 1720 - 60 = 1660$ mm.

Angle θ of face AC is given by

$\tan\theta = 2/7.5 = 0.267$. $\therefore$ $\theta = 14° 56'$

$\sin\theta = 0.2577$; $\cos\theta = 0.9662$

Depth $F_1G_1 = AF_1 \sin\theta = 7 \times 0.2577$

$= 1.8$ m $= 1800$ mm

$\therefore$ Depth $FG = 1800 + 300 = 2100$ mm.

This is much more than required.

Assuming that steel reinforcement is provided in two layers with 20 mm space between them and providing a nominal cover of 50 mm and 20 mm φ main bars, the effective depth may be assumed approximately to be equal to $2100 - (50 + 12 + 20 + 10) = 2008$ mm.

$$\therefore\ A_{st} = \frac{0.5 \times 20}{415}\left[1 - \sqrt{\frac{4.6 \times 1800.75 \times 10^6}{20 \times 500\,(2008)^2}}\right] \times 500 \times 2008 = 2628 \text{ mm}^2$$

Using 20 mm φ bars, $A_\varphi = 314$ mm^2

No. of bars $= 2628/314 \approx 9$. Provide 10 bars in two layers.

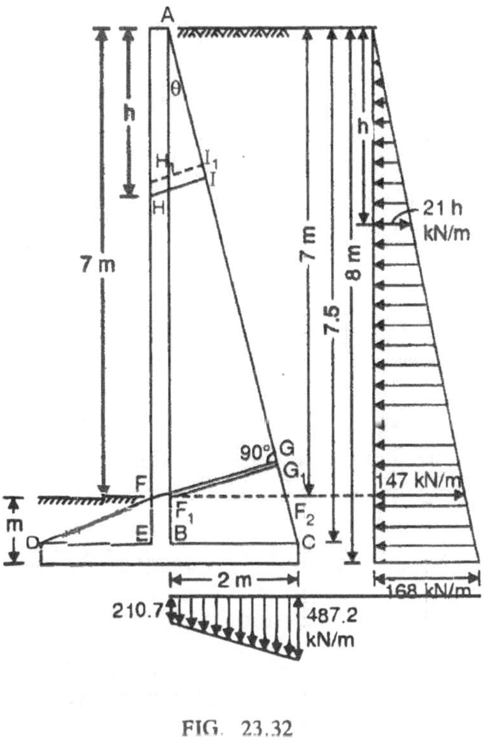

FIG. 23.32

$$\text{Effective S.F.} = Q - \frac{M}{d'} \tan\theta, \quad \text{where } d' = \frac{d}{\cos\theta} = \frac{2008}{0.9662} = 2078 \text{ mm}$$

$$\therefore \text{ Effective S.F.} = Q - \frac{M}{d'} \tan\theta = 514.5 \times 10^3 - \frac{1200.5 \times 10^6}{2078} \times 0.267 = 360250 \text{ N}$$

$$V_u = 1.5 \times 360250 \text{ kN}$$

$$\therefore \quad \tau_v = \frac{1.5 \times 360250}{500 \times 2078} = 0.52 \text{ N/mm}^2$$

$$100 \frac{A_s}{bd} = \frac{100\,(10 \times 314)}{500 \times 2078} = 0.3\%. \quad \therefore \quad \tau_c \approx 0.38 \text{ N/mm}^2.$$

Thus the shear stress τ_v is more than permissible shear stress τ_c. However, the vertical and horizontal ties provided in the counterforts will bear the excess shear stress.

The height h where half of the reinforcement can be curtailed will be equal to $\sqrt{H} = \sqrt{8} \simeq 2.8$ m below A, *i.e.* at point H. To locate the position of point of curtailment on AC, draw HI parallel to FG. Thus, half the bars can be curtailed at I. However, these should be extended by a distance $12\,\varphi = 240$ mm beyond I, *i.e.* extended upto I_1. The location of H, corresponding to I_1 can be located by drawing line $I_1 H_1$ parallel to FG. It should be noted that $I_1 G$ should not be less than $47\,\varphi = 940$ mm. Similarly, other bars can be curtailed, if desired.

Design of horizontal ties. The vertical stem slab has a tendency to separate out from the counterfort, and hence should be tied to it by horizontal ties. At any depth h below the top, force causing separation $= \frac{1}{3} \times 18\,h \times 3 = 18\,h$ kN/m.

At $h = 7$ m, force $= 18 \times 7 = 126$ kN/m.; $F_u = 1.5 \times 126 = 189$ kN

$$\therefore \quad \text{Steel area required} = \frac{189 \times 1000}{0.87 \times 415} = 524 \text{ mm}^2 \text{ per metre height.}$$

Using 10 mm φ 2 legged ties, $A_\varphi = 2\,\frac{\pi}{4}(10)^2 = 157$ mm^2

$$\therefore \quad \text{Spacing} = \frac{1000 \times 157}{524} \approx 299 \text{ mm c/c. However, provide these at 250 mm c/c.}$$

This spacing can be increased gradually to 300 mm c/c towards top.

Design of vertical ties. Similar to the stem slab, heel slab has also tendency to separate out from the counterfort, due to net downward force, unless tied properly by vertical ties. The downward force at C will be $487.2 \times \frac{3}{3.5} = 417.6$ kN/m and $210.7 \times \frac{3}{3.5} = 180.6$ kN/m at B (see Fig. 23.32). Near end C, the heel slab is tied to the counterforts with the help of main reinforcement of counterforts.

$$\therefore \quad \text{Steel area at } C = \frac{1.5 \times 417.6 \times 1000}{0.87 \times 415} = 1735 \text{ mm}^2$$

Using 12 mm φ 2 legged ties, $A_\varphi = 2 \times \frac{\pi}{4}(12)^2 = 226.2$ mm^2

$$\therefore \quad \text{Spacing of ties} = \frac{1000 \times 226.2}{1735} \approx 130 \text{ mm c/c.}$$

$$\text{Steel area at } B = \frac{1.5 \times 180.6 \times 1000}{0.87 \times 415} = 751 \text{ mm}^2$$

$$\therefore \quad \text{Spacing of ties} = \frac{1000 \times 226.2}{751} \approx 300 \text{ mm c/c.}$$

Thus the spacing of vertical ties can be increased gradually from 130 mm c/c at C to 300 mm c/c at B.

8. ***Design of front counterforts.***

Refer Fig. 23.30. The upward pressure intensity varies from 170.2 kN/m^2 at D to 99.1 kN/m^2 at E. Downward weight of 500 mm thick toe slab = 0.5 × 25 = 12.5 kN/m^2. Hence net w at D = 170.2 − 12.5 = 157.7 kN/m^2 and at E = 99.1 − 12.5 = 86.6 kN/m^2. The centre to centre spacing of counterforts, 500 mm wide is 3.5 m. Hence upward force transmitted to counterfort at D = 157.7 × 3.5 = 552 kN/m and that at E = 86.6 × 3.5 = 303.1 kN/m.

The counterfort will bend up as cantilever about face FE. Hence DF will be in compression while $D_1 E_1$ will be in tension, and main reinforcement will be provided at bottom face $D_1 E_1$.

Total upward force

$$= \frac{1}{2}[552 + 303.1] \times 1.8 = 770 \text{ kN acting at}$$

$$\bar{x} = \frac{303.1 + 2 \times 552}{303.1 + 552} \times \frac{1.8}{3} = 0.99 \text{ m from } E$$

$$\text{B.M.} = 770 \times 0.99 = 760 \text{ kN-m} = 760 \times 10^6 \text{ N-mm}$$

$$M_u = 1.5 \times 760 \times 10^6 = 1140 \times 10^6 \text{ N-mm}$$

$$\therefore \quad d = \sqrt{\frac{1140 \times 10^6}{500 \times 2.761}} \simeq 910 \text{ mm}$$

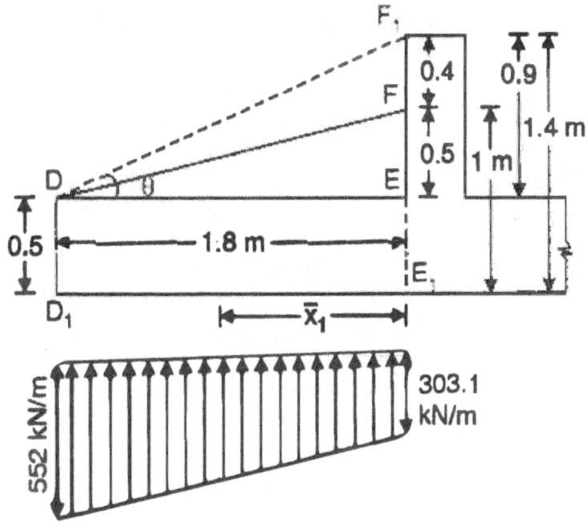

FIG. 23.33.

However, to make it safe against shear, provide a total depth of 1400 mm, so that with an effective cover of 80 mm, available d = 1400 − 80 = 1320 mm.

Thus, project the counterfort 400 mm above ground level, to point F_1 as shown.

$$A_{st} = \frac{0.5 \times 20}{415}\left[1 - \sqrt{1 - \frac{4.6 \times 1140 \times 10^6}{20 \times 500\,(1320)^2}}\right] 5000 \times 1320 = 2607 \text{ mm}^2$$

$\therefore$ No. of 25 mm φ bars = 2607/491 = 5.3 ≏ 6. Provide these in one layer. These bars should be continued by a distance of 47 φ = 47 × 25 ≈ 1175 mm beyond E.

$$\text{Net shear force} = F - \frac{M}{d}\tan\theta. \quad \text{From Fig 23.33,} \quad \tan\theta = \frac{0.9}{1.8}$$

$$\therefore \quad V = 770 - \frac{760}{1.3} \times \frac{0.9}{1.8} = 477.7 \text{ kN} \quad ; \quad V_u = 1.5 \times 477.7 = 716.54 \text{ kN}$$

and
$$\tau_v = \frac{716.54 \times 1000}{500 \times 1320} = 1.086 \text{ N/mm}^2 < \tau_{c,\,max}.$$

$$100\frac{A_s}{bd} = \frac{100\,(6 \times 491)}{500 \times 1320} = 0.45\%$$

Hence from Table, 7.1 $\tau_c = 0.456$ N/mm^2. Since $\tau_v > \tau_c$, shear reinforcement is required.

Using 12 mm φ 2-legged stirrups, $A_{sv} = 2 \times \frac{\pi}{4}(12)^2 = 226$ mm^2.

$$V_{uc} = \tau_c . bd = 0.456 \times 500 \times 1320 = 300960 \text{ N} = 300.96 \times 10^3 \text{ N}$$

$$\therefore \quad V_{us} = V_u - V_{uc} = 716.54 \times 10^3 - 300.96 \times 10^3 = 415.58 \times 10^3 \text{ N}$$

$$\therefore \quad s_v = \frac{0.87 f_y . A_{sv} . d}{V_{us}} = \frac{0.87 \times 415 \times 226 \times 1320}{415.58 \times 10^3} = 259$$

However, provide these @ 250 mm c/c. Provide 2-16 mm Φ holding bars at the top.

9. ***Fixing effects in stem, toe and heel slab***

At the junction of stem, toe and heel slabs fixing moments are induced, which are at right angles to their normal direction of bending. These moments are not determinate, but normal reinforcement given below may be provided :

(*i*) In stem @ 0.8 × 0.3 = 0.24% of cross section, to be provided at the inner face, in vertical direction, for a length = 47 φ.

$$\therefore \quad A_{st} = \frac{0.24}{100} \times 1000 \times 300 = 720 \text{ mm}^2$$

Using 10 mm φ bars, $A_{\varphi} = 78.5$ mm^2

$$\therefore \quad \text{Spacing} = \frac{1000 \times 78.5}{720} \approx 100 \text{ mm c/c.}$$

Length of embedment in stem, above heel slab ≈ 470 mm.

(*ii*) In toe slab @ 0.12% to be provided at the lower face

$$A_{st} = \frac{0.12}{100}(1000 \times 500) = 600 \text{ mm}^2$$

Spacing of 10 mm φ bars $= \frac{1000 \times 78.5}{600} \approx 125$ mm

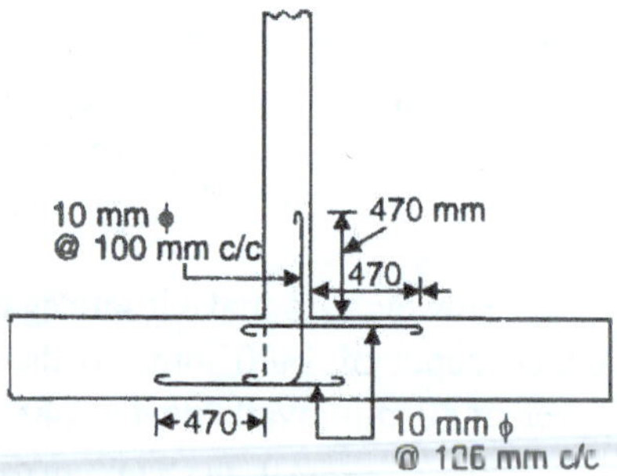

FIG. 23.34

Length of embedment in toe slab = 470 mm.

(*iii*) In heel, @ 0.12% to be provided at the upper face.

∴ Provide 10 mm bars @ 125 c/c to be embeded by a distance of 470 mm.

Each of the above reinforcement should be anchored properly in the adjoining slab, as shown in Fig. 23.34.

10. ***Design of shear key***. The wall is unsafe in sliding, and hence shear key will have to be provided, as shown in Fig. 23.35. Let the depth of key = *a*.

Intensity of passive pressure p_p developed in front of the key depends upon the soil pressure p in front of key.

$$p_p = K_p\, p = 3 \times 99.1 = 297.3 \quad \text{kN/m}^2$$

∴ Total passive pressure $P_p = p_p . a = 297.3\, a$ ...(1)

Sliding force at level D_1C_1 $= \frac{1}{3} \times \frac{18}{2}(8+a)^2$

or $P_H = 3(8+a)^2$ ...(2)

Weight of soil between bottom of base and $D_1 C_1 \approx 4.1 \times a \times 18 = 73.8\ a$ kg ...(3)

$\therefore\ \Sigma W = 365.8 + 73.8\ a$ (Refer Table 23.4*a*)

Hence for equilibrium of wall, permitting F.S. = 1.5 against sliding, we have

$$1.5 = \frac{\mu \Sigma W + P_P}{P_H} = \frac{0.5(365.8 + 73.8\,a) + 297.3\,a}{3(8+a)^2}$$

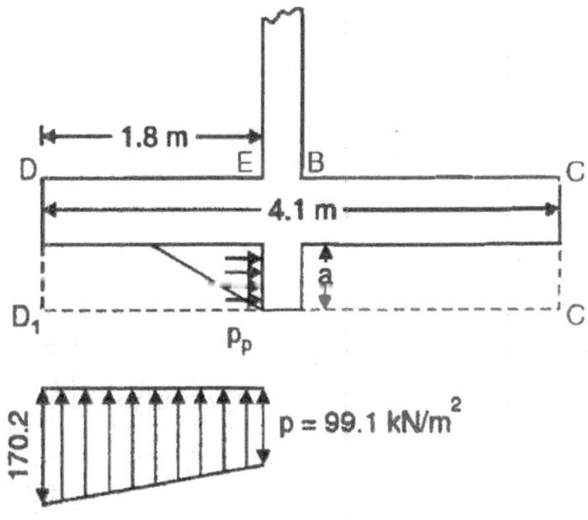

FIG. 23.35

or $a^2 - 57a + 23.3 = 0$ which gives $a = 0.40$ m

Hence keep a = 400 mm and width of key = 400 mm.

Actual force to be resisted by key $= 1.5P_H - \mu \Sigma W$

$$= 1.5 \times 3\,(8+0.4)^2 - 0.5(365.8 + 73.8 \times 0.4) = 119.8 \text{ kN}$$

$$\therefore \quad \text{Shear stress} = \frac{119.8 \times 1000}{400 \times 1000} \approx 0.3 \text{ N/mm}^2 \text{ (safe)}$$

$$\text{Bending stress} = \frac{119.8 \times 10^3 \times 200}{\frac{1}{6} \times 1000(400)^2} = 0.9 \text{ N/mm}^2 \text{ (safe)}$$

The details of reinforcement etc. are shown in Fig. 23.36.

PROBLEMS

1. Explain the methods of designing vertical stem, toe slab and heel slab of a T-shaped cantilever retaining wall. What will be the changes in the design if counterforts are provided at regular interval towards the side of backfil ?
2. Explain the method designing a shear key for a retaining wall.
3. Design a T-shaped cantilever retaining wall for the following data :
   (*i*) Height of wall above ground, 4.5 m
   (*ii*) Depth of foundation, 1.5 m
   (*iii*) Unit weight of earth fill, 17 kN/m^3
   (*iv*) Angle of internal friction, 20°
   (*v*) Coefficient of friction between soil and concrete, 0.45
   (*vi*) Safe bearing capacity of soil 130 kN/m^2. Use M 20 mix and Fe 415 steel.
4. Redesign problem 3 if the top surface of backfill is inclined at an angle of 15° with horizontal.
5. Redesign problem 3 if the embankment carries a load of 20 kN/m^2 over the backfill. Also provide a R.C. parapet 1 m high over the top of the stem of the retaining wall.
6. Design a suitable counterfort retaining wall for the data of problem 3.
7. Design a counterfort retaining wall for the following data :
   (*i*) Height of wall above ground, 8 m.
   (*ii*) Depth of foundation, 1.5 m
   (*iii*) Safe bearing capacity, 200 kN/m^2

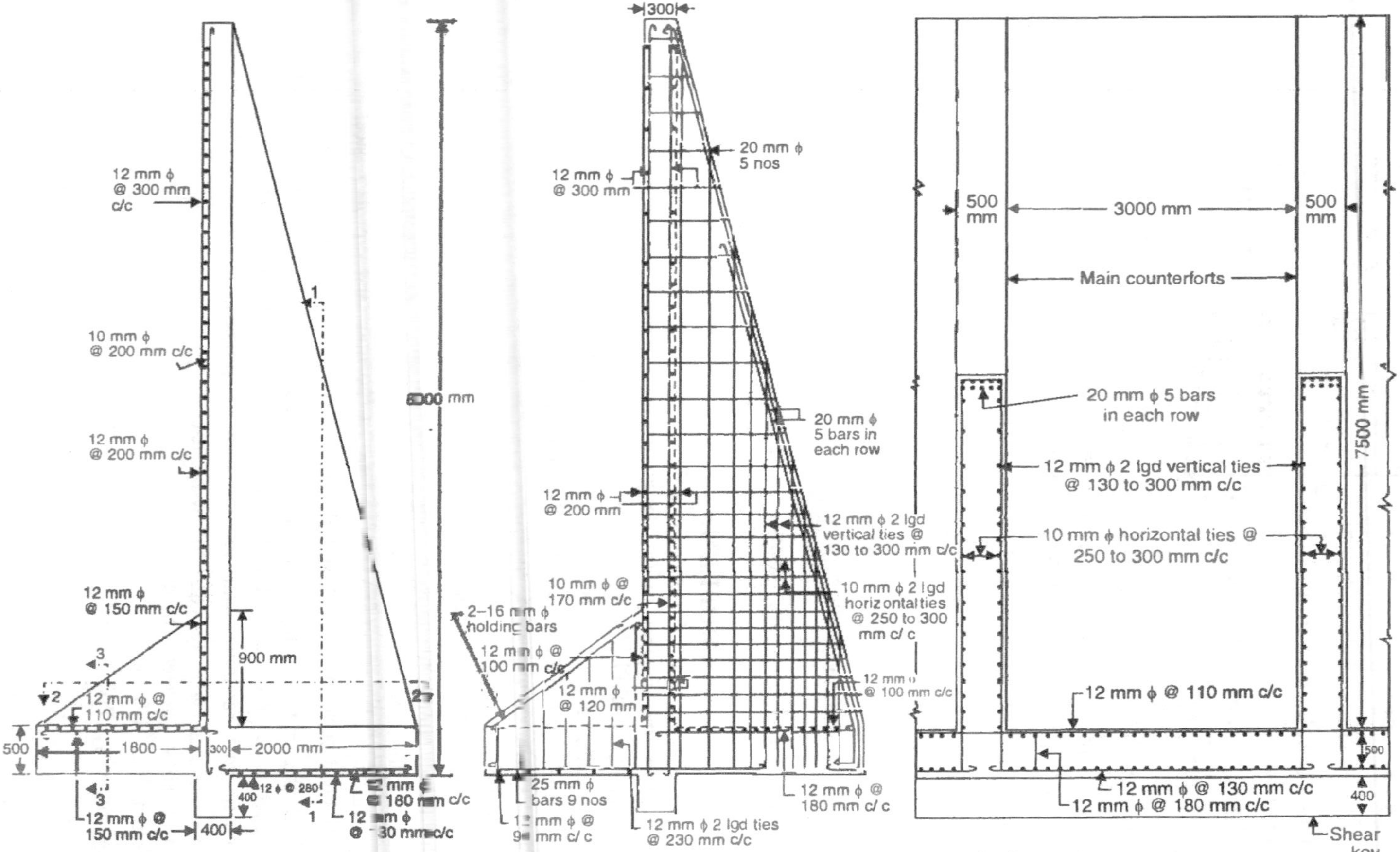

(a) Cross-section midway between counterforts

(b) Cross-section through counterforts

(c) Sectional elevation through section 1-1

FIG. 23.36

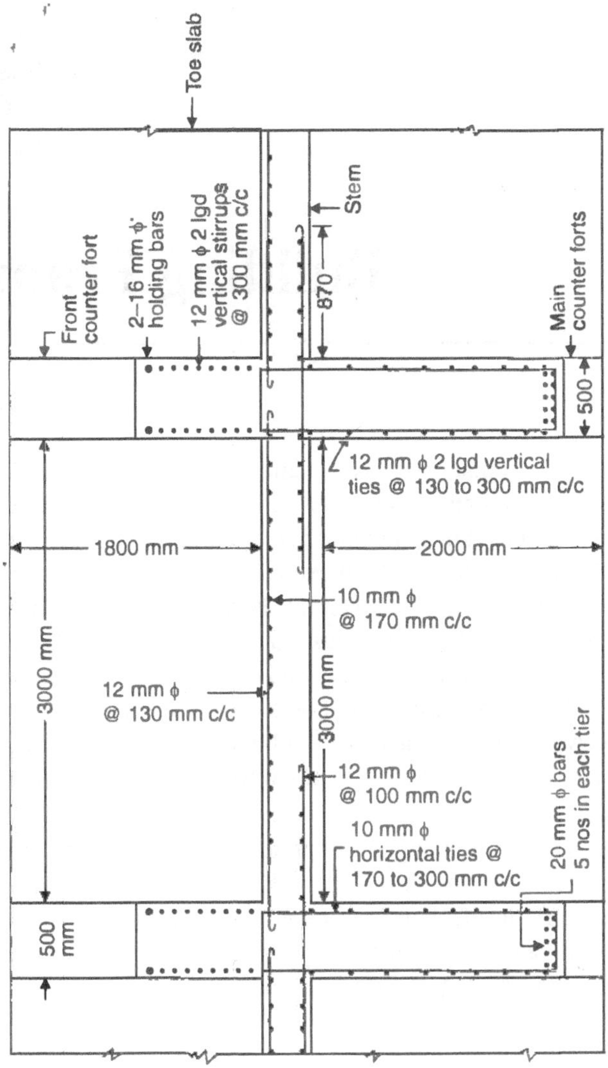

(d) Sectional plan through 2-2

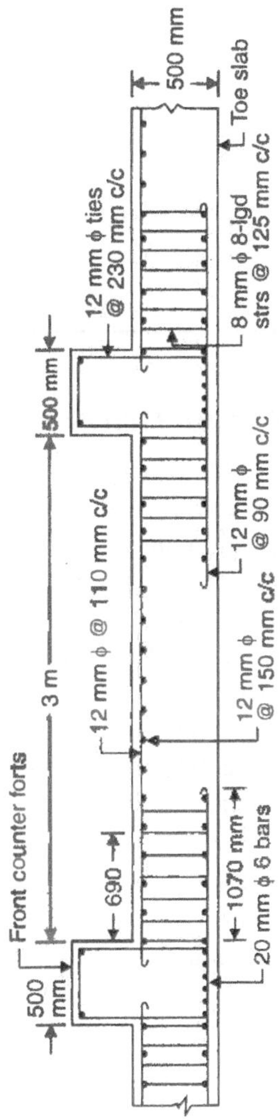

(e) Sectional elevation of toe slab and front counterforts at section 3-3

FIG. 18.36 (CONTD.)

(*iv*) Unit weight of earthfill, 18 kN/m^3

(*v*) Surcharge angle 18°

(*vi*) Angle of internal friction for backfill, 30°

(*vii*) Face to face spacing of front counterforts, 2 m.

(*viii*) Face to face spacing of front counterforts, provided upon ground level, 2 m.

(*ix*) Coefficient of friction between soil and concrete, 0.55.

Use M 20 mix and Fe 415 steel reinforcement

24

Building Frames

24.1. INTRODUCTION

A building frame may contain a number of bays, and may have several storeys. A multi-storeyed, multi-panelled frame is a complicated statically indeterminate structure. It consists of a number of beams and columns built monolithically, forming a net work. The floors and the walls are supported on beams which transmit the loads to the columns. A building frame is subjected to both the vertical as well as horizontal loads. The vertical loads consists of the dead weight of structural components such as beams, slabs, columns etc., and live load. The horizontal loads consist of the wind forces and the earth quake forces. The ability of multi-storey-buildings to resist wind and other lateral forces depends upon the rigidity of connections between the beams and columns. When the connections of beams and columns are fully rigid, the structure as a whole is capable of resisting the lateral forces acting on the structure.

In ordinary reinforced concrete skeleton buildings, a continuous beam is rigidly connected with columns. Due to this, the moments in the beam depend not only upon the number and length of spans composing the beam itself, but also upon the rigidity of columns with which it is connected. The bending moment at the end of any one span of the continuous beam cannot be transferred to the beam in the next span without subjecting the columns to bending. Instead of transmitting the bending moment in full to the beam in the next span, part of the moment is transferred to the columns above and below the beam, and the balance to the beam. Due to this, the effect of loading on one span upon the other spans is much lower than in ordinary continuous beams which are not connected to the columns.

24.2. SUBSTITUTE FRAMES

The analysis of a multi storeyed multi-panelled building frame is very cumbersome, since the frame contains a number of continuous beams and columns. As stated in the previous article, the effect of loading on the span upon other spans is much smaller. The moments in any beam or column are mainly due to the loads on spans very close to it. Loads on distant spans do not have appreciable effect. Due to this, a simple method of analysis, accurate enough for practical purpose, is used by analysing a small portion of the frame, called '*substitute frame*' rather than analysis of the whole frame.

It has been found by exact analysis that the moments carried from floor to floor, through columns, are very small in comparison to the beam moments. In other words, the moments in one floor have negligible effect of the moments of the floors above and below it. Therefore, a substitute frame consists of one floor, connected above and below with their far end either hinged or fixed or restrained. Fig. 24.1(*a*) shows a building frame consisting of five storeys and three bays. Fig. 24.1(*b*) shows the substitute frame for determining bending moment in the second floor. Generally, it is sufficient to consider two adjacent spans on each side of joint considered. The substitute frame gives the results which are safe for all practical purposes.

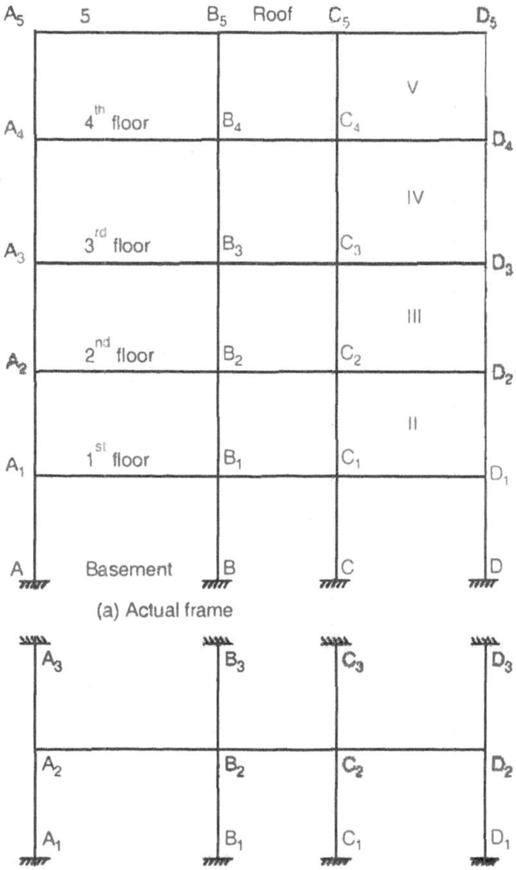

(a) Actual frame

(b) Substitute frame

FIG. 24.1

Types of Substitute Frames : Under ordinary conditions, the following four types of substitute structures are considered sufficient:

(*a*) three-span structure with two storey columns

(*b*) substitute frame for wall columns

(*c*) substitute frame for two panel wide building

(*d*) substitute frame for one panel wide building.

Fig. 24.2(*a*) shows the most general substitute frame consisting of *three span, two-storey substitute structure* with irregular spacing of columns. Fig. 24.2(*b*) shows the substitute frame for finding the bending moments in wall columns. This consists of three spans and three two-storey columns, one of which is the wall column. Fig. 24.2(*c*) shows the substitute frame for structures two panels wide. Fig.24.2(*d*) shows the substitute frame for one span multi-storey frame.

End conditions for substitute frames

The restraining effect of any one member, upon other members forming a joint depends also upon the condition existing on the other end of the restraining member. The other end may have three conditions: (*i*) free to turn (*i.e.* hinged),(*ii*) partially restrained, or (*iii*) rigidly fixed. The restraining effect is largest for the rigidly fixed conditions of the end and smallest for free end. It should be noted that the restraining effect of a fixed member is one third larger than the restraining effect if it were free to turn. The rigidity of any member is expressed by the ratio I/l where I is its moment of inertial and l is its length (for beam) or height (for column). If the loaded member has rigidity I/l, and the restraining member has rigidity I_1/l_1 then this restraining member is considered as fixed at the other end if $\frac{I_1}{l_1} \div \frac{I}{l}$ is equal to or greater than 10. The end of a member is considered as partly restrained when it runs into another joint composed of several members, a condition which is often found in concrete skeleton structure. No restraint exists if $\frac{I_1}{l_1} \div \frac{I}{l} = 0$. The outer ends of the member of the substitute frame are sometimes taken as hinged (except for columns fixed in the ground). This gives severset condition for a particular reaction under investigation. The moments obtained by assuming the ends hinged gives the moments nearest to the value obtained from full frame analysis and compensates to some extent for the error caused due to neglecting loads on members of distant span.

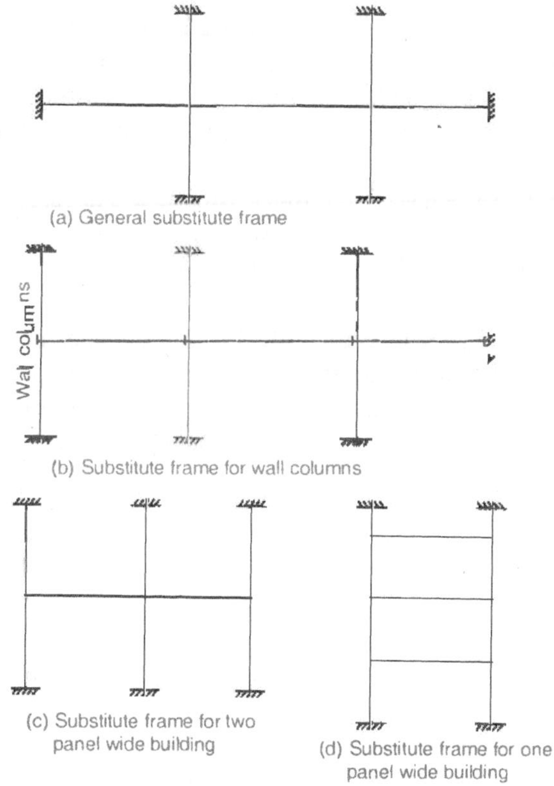

FIG. 24.2. VARIOUS TYPES OF SUBSTITUTE FRAMES.

24.3. ANALYSIS FOR VERTICAL LOADS

A building frame is a three dimensional structure consisting of a number of bays in two directions at right angles to each other. A building structure may be assumed to be consisting of two sets of plane frames crossing each other at right angles. The vertical members (*i.e.*, columns) are common to both these sets of frames. Each set of frames are analysed separately. Since moments in the vertical members occur in two planes, the stresses in columns should be found for moments acting in two planes simultaneously and the corresponding vertical loads.

(*a*) **Maximum bending moments in beams**. The magnitude of bending moments in beams and columns respectively depend upon their relative rigidity . *Generally*, the beams are made of the same dimensions in all floors, while the dimensions of column vary from storey to storey. Columns have smallest dimensions at the top and largest dimensions at the bottom. Due to this reason, the ratio of the rigidity of the beam to that of the column is larger in the upper floors than in the lower floors. The positive bending moments in the beams increase with decrease of the rigidity of the columns, while the negative B.M. in them increase with the increase in the rigidity of the columns. Due to this, the positive B.M. are largest in the upper storeys where the columns are least rigid and the negative bending moments are maximum in the lower storeys where the columns are rigid.

In order to find the maximum moment in a given span of the beam, the substitute frame is so. selected that span under investigation forms the centre span. This substitute frame may be moved from floor to floor. However, since the beams in all floors are made of the same dimensions and provided with same amount of steel, only one substitute frame may be sufficient when placed in a position in the structure for which the bending moments are the largest. The beam should be loaded with live loads as follows for maximum effects:

(*i*) ***For maximum positive B.M.*** at the mid-point C of a span AB, the loads should be placed on the span and on alternative spans, as shown in Fig. 24.3(*a*).

(*ii*) ***For maximum negative B.M.*** at the mid-point C of a span AB, the span AB should be unloaded while load should be placed on spans adjacent to the span under consideration, as shown in Fig. 24.3(*b*).

(*iii*) ***For maximum negative B.M.*** at the support A, loads should be placed on the two spans adjacent to the support, as shown in Fig.24.3(*c*).

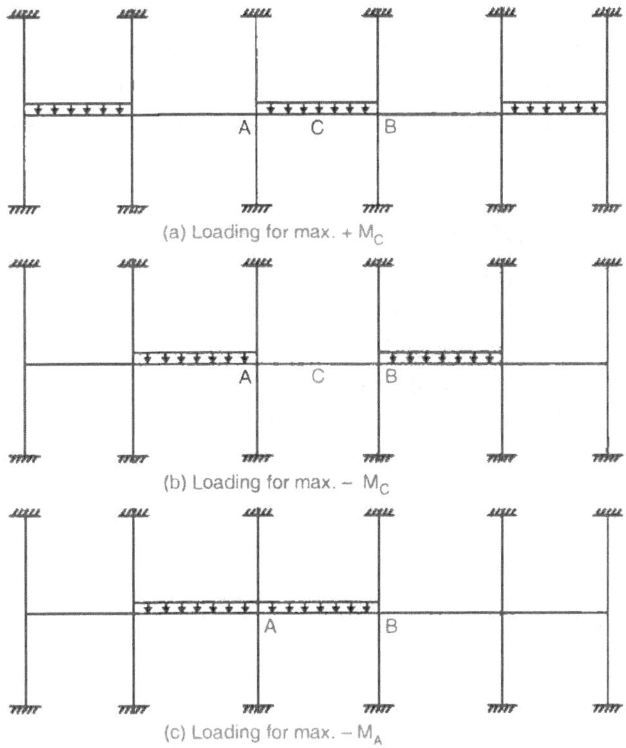

FIG. 24.3. LOADING FOR BENDING MOMENTS IN BEAMS.

When the spans of beams are not equal, substitute frames should be selected in which the largest span forms the centre span and also frames in which the smallest span forms the centre span. Several trial computations may be necessary to get the frame for which the bending moments are maximum. To get bending moment in the wall columns and wall beams, substitute frame shown in Fig. 24.2(*b*) should be used.

The bending moments due to dead loads are found separately. The bending moments for dead and live loads are then added, and the beam is designed.

(*b*) Maximum bending moment in columns

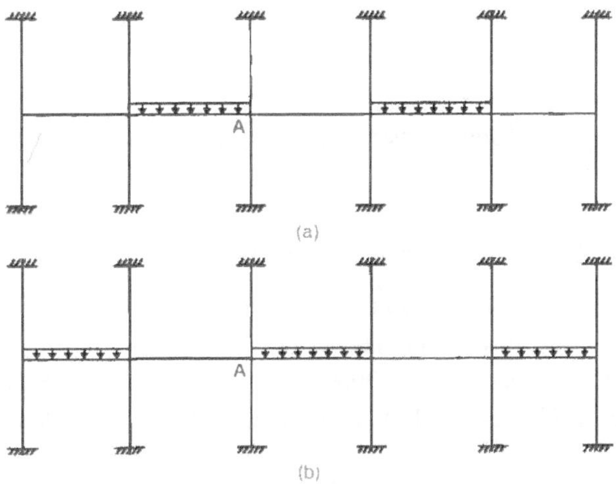

FIG. 24.4. LOADING FOR MAX. B.M. AT COLUMN A.

The bending moments in columns increase with increase in their rigidity. Hence they are largest in the lower storeys, and smallest in the upper storeys. The maximum compressive stresses in columns is found by combining maximum vertical loads with maximum bending moments. The maximum tensile stresses in columns is found by combining the maximum bending moment with minimum vertical loads. Though the bending moment is smallest in the upper floors, its effect is much larger since the dimensions of the columns are the smallest there and also the vertical loads are much smaller than in lower storeys. Also the possibility of tensile stresses in columns is much larger in upper storeys than in lower storeys.

The maximum moments is columns occur when alternative spans are loaded as shown in Fig. 24.4(*a*),(*b*). The corresponding axial loads are found. The column is designed to resist the stresses provided by every combination of axial load and the corresponding moment.

24.4. METHODS OF COMPUTING B.M.

The bending moments in the beams and columns of a substitute frame may be computed by the following methods :

1. Slope-deflection method.
2. Moment distribution method.
3. Building frame formulae.
4. Kani's method.

The slope deflection method results in too many equations to be solved simultaneously. Hence moment distribution method, using two cycles is used. Taylor, Thomson and Smulski recommend the use of building frame formulae which they have developed using slope deflection equations.

Example 24.1. *Fig. 24.5 shows an intermediate frame of a multistoreyed building. The frames are spaced at 4 metres centre to centre. Analyse the frame taking live load of 4000 N/m² and dead load as 3000 N/m², 3250 N/m² and 2750 N/m² for the panels AB, BC and CD respectively. The self weight of the beams may be taken as under:*

Beams of 7 m span : 5000 N/m

Beams of 5 m span : 3500 N/m

Beam of 3.5 m span : 2500 N/m.

The relative stiffness of the members are marked on the figure itself.

Solution

1. ***Substitute frame*** : Let us analyse the second floor *ABCD*. The substitute frame is shown in Fig. 24.6, assuming the far ends of the columns fixed. Other floors can be analysed in a similar manner.

2. ***Loading and fixed end moments***

Since the frames are spaced @ 4 m c/c, the live loads transferred from the floors will be 4000 × 4 = 16000 N/m.

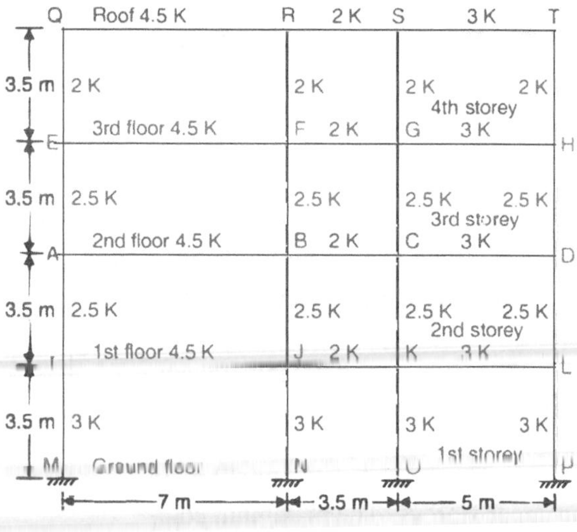

FIG. 24.5.

The total dead load on a beam will be equal to the dead load from the floors plus the dead load due to the self-weight of the beam. Thus, the total dead load on beam *AB*, per metre run = (3000 × 4) + 5000 = 17000 N/m. Dead loads for other beams are tabulated in Table 24.1. The fixed end moment is calculated from the following expressions :

$$M_F = \pm \frac{w L^2}{12}$$ where w is the U.D.L. and L is span of the beam. Clockwise moments are taken as positive and anticlockwise moments are taken as negative.

The fixed end moments due to dead load, and due to dead and live load combined are tabulated in Table 24.1.

3. ***Distribution factors*** : The distribution factors at a joint depends upon the relative stiffness of the members meeting at the joints. These are tabulated in Table 24.2.

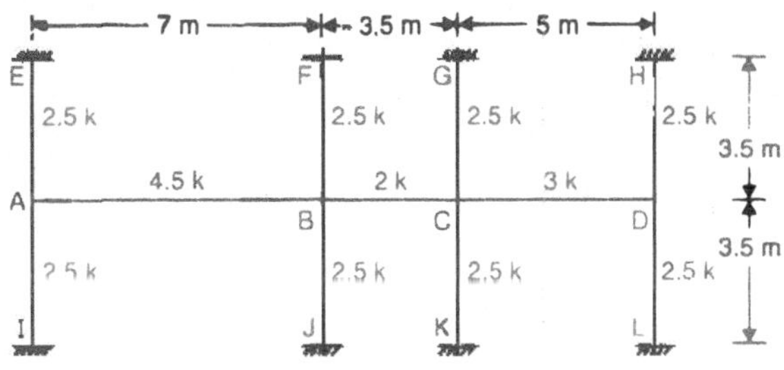

FIG. 24.6.

TABLE 24.1 LOADING AND FIXED END MOMENTS

Member	*Dead load (N/m)*	*Live load (N/m)*	*F.E.M. due to dead load (N-m)*	*F.E.M. due to dead and live load combined (N-m)*
AB	17000	16000	69420	134750
BC	15500	16000	15820	32160
CD	14500	16000	30210	63540

TABLE 24.2 DISTRIBUTION FACTORS

Joint	*members*	*Relative stiffness*	*Sum*	*Distribution factors*
	AE	2.5 *K*		0.263
A	*AI*	2.5 *K*	9.5 *K*	0.263
	AB	4.5 *K*		0.474
	BA	4.5 *K*		0.392
	BF	2.5 *K*		0.217
B	*BC*	2.0 *K*	11.5 *K*	0.174
	BJ	2.5 *K*		0.217
	CB	2.0 *K*		0.200
	CG	2.5 *K*		0.250
C	*CD*	3.0 *K*	10 *K*	0.300
	CK	2.5 *K*		0.250
	DC	3.0 *K*		0.375
D	*DH*	2.5 *K*	8 *K*	0.3125
	DL	2.5 *K*		0.3125

(*A*) MAXIMUM NEGATIVE B.M. AT SUPPORTS

4. Maximum negative B.M. at joint A : The condition of loading to obtain max. negative B.M. at joint *A* is as follows: Live load on *AB* only, while the dead load is on *AB* and *CD*. The effect of load on other spans is neglected. The moment distribution is carried out in Table 24.3. The distribution is done at joint *B* and the carry over effect (*i.e.* half the moment) is transferred to joint *A*. After adding the total moments at *A*, distribution is done at *A*. The distribution at joint *B* has not been recorded in the Table 24.3.

5. ***Maximum negative B.M. at joint B*** : The loading conditions are : Live load on *AB* and *BC*, while dead load on whole of *ABCD*. The Moment distribution is carried out in Table 24.4. In the first cycle, joints *A* and *C* are balanced, and half the moments are carried over to joint *B* for the beams *BA* and *BC* respectively. In the second cycle, joint *B* is balanced and final moments are found. Thus, in the first cycle, unbalanced moment at *C* is + 1950, the distributed moment to *CB* will be 1950 × 0.2 ≈ − 400, the

TABLE 24.3 MOMENT DISTRIBUTION FOR – VE B.M. AT A

Joint	A	B		C		D
Member	*AB*	*BA*	*BC*	*CB*	*CD*	*DC*
D.F.	0.474	0.392	0.174	0.20	0.30	0.375
1. F.E.M. due to D.L.			– 15820	+15820		
2. F.E.M. due to total load	– 134750	+134750				
3. Distribution at *B* and carry over to *A*	– 23310					
4. Add (2) and (3)	– 158060					
5. Distribution	+74920					
6. Total sum of (4) and (5)	– 83140					

carry over moment at $B = -400/2 = -200$. Similarly, the unbalanced moment at A is –134750, the distributed moment of $AB = +134750 \times 0.474 = +63880$, the carried over moment to $B = +31940$. The total moments at BA and BD will be $+166690$ and -32260, leaving an unbalanced moment of $+134330$. The distributed moments to BA and BC will be $-134330 \times 0.392 = -52660$ and $-134330 \times 0.174 = -23370$ respectively.

TABLE 24.4 MOMENT DISTRIBUTION FOR – VE B.M. AT *B*

Joint	A	B		C		D
Member	*AB*	*BA*	*BC*	*CB*	*CD*	*DC*
D.F.	0.474	0.392	0.174	0.20	0.30	0.375
1. F.E.M. due to D.L.					+30210	
2. F.E.M. due to total load	– 134750	+134750	– 32160	+32160		
3. Distribution at *A* and *C* and carry over to *B*		+31940	– 200			
4. Add (2) and (3)		+166690	– 32360			
5. Distribution		– 52660	– 23370			
6. Total sum of (4) and (5)		+114030	– 55730			

6. *Maximum negative B.M. at C*

The conditions of loadings are : Live load on BC and CD, and dead load on $ABCD$. In the first cycle, joints B and D are balanced and effects are carried over to C. In the second cycle, joint C is balanced, as shown in Table 24.5.

7. *Maximum negative B.M. at D*

The conditions for loadings are : Live load on CD and dead load on $ABCD$. In the first cycle, joint C is balanced and its effect is carried over to D. In the second cycle, joint D is balanced as shown in Table 24.6.

TABLE 24.5 MOMENT DISTRIBUTION FOR – VE B.M. AT *C*

Joint	***A***	***B***		***C***		***D***
Member	***AB***	***BA***	***BC***	***CB***	***CD***	***DC***
D.F.	0.474	0.392	0.174	0.20	0.30	0.375
1. F.E.M. due to D.L.	– 69420	+69420				
2. F.E.M. due to total load			– 32160	+32160	– 63540	+63540
3. Distribution at *B* and *D* carry over to *C*				– 3240	– 11910	
4. Add (2) and (3)				+28920	– 75450	
5. Distribution				+9310	+13960	
6. Total sum of (4) and (5)				+38230	– 61490	

TABLE 24.6 MOMENT DISTRIBUTION FOR – VE B.M. AT *D*

Joint	***A***	***B***		***C***		***D***
Member	***AB***	***BA***	***BC***	***CB***	***CD***	***DC***
D.F.	0.474	0.392	0.174	0.20	0.30	0.375
1. F.E.M. due to D.L.			– 15820	+15820		
2. F.E.M. due to total load					– 63540	+63540
3. Distribution at *C* and carry over to *D*						+7160
4. Add (2) and (3)						+70700
5. Distribution						– 26510
6. Total sum of (4) and (5)						+44190

(*B*) MAXIMUM POSITIVE B.M. AT MID-SPANS

8. ***Maximum +ve B.M. in mid-span of AB*** : The conditions of loadings are : Live load on *AB* and *CD*, and Dead load on *ABCD*. In the first cycle, distribution is performed at joints *A*, *B* and C. Half of these distributed moments are carried over to the opposite ends, *i.e.* from *A* to *B* and *B* to *A*, and from *C* to *B*. In the second cycle, distribution is performed at *A* and *B* as illustrated in Table 24.7. Thus the end moments at *A* and *B* for beam *AB* are – 83140 and + 105680 respectively. The free B.M. at

$$\text{mid-span of } AB = \frac{w L^2}{8} = \frac{(17000 + 16000)(7)^2}{8} = 202120 \text{ N-m}$$

$$\therefore \text{ Net B.M. at centre of } AB = 202120 - \frac{(83140 + 105680)}{2} = 107710 \text{ N-m}$$

9. ***Maximum +ve B.M. in mid-span of BC*** : The conditions for loadings are : Live load on *BC* and dead load on *ABCD*. In the first cycle, moments are distributed at *A*, *B*, *C* and *D*. These distributed moments are carried over from *A* to *B*, from *B* to *C*, from *C* to *D* and from *D* to *C*. Finally, the moment are distributed at joints *B* and *C*, as shown in Table 24.8.

TABLE 24.7 MOMENT DISTRIBUTION FOR + VE B.M. AT *AB*

Joint	*A*	*B*		*C*		*D*
Member	*AB*	*BA*	*BC*	*CB*	*CD*	*DC*
D.F.	0.474	0.392	0.174	0.20	0.30	0.375
1. F.E.M. due to D.L.			− 15820	+15820		
2. F.E.M. due to total load	− 134750	+134750			− 63540	+63540
3. Distribution at *A*, *B* and *C*	+63870	− 46620	+4770			
4. Carry over	− 23310	+31940		+9540		
5. Distribution at *A* and *B*	+11050	− 14390				
6. Total moments (sum of 2, 3, 4, 5)	− 83140	+105680				

TABLE 24.8 MOMENT DISTRIBUTION FOR − VE B.M. AT *BC*

Joint	*A*	*B*		*C*		*D*
Member	*AB*	*BA*	*BC*	*CB*	*CD*	*DC*
D.F.	0.474	0.392	0.174	0.20	0.30	0.375
1. F.E.M. due to D.L.	− 69420	+69420			− 30210	+30210
2. F.E.M. due to total load			− 32160	+32160		
3. Distribution at *A*, *B*, *C* and *D*	+32900		− 6480	− 390		− 11330
4. Carry over		+16450	− 200	− 3240	− 5660	
5. Distribution at *B* and *C*			− 2830	+1780		
6. Total moments			− 41670	+30310		

Thus, the end moments at *B* and *C* are − 41670 and + 34190 respectively. The free B.M. at the centre of span *BC* is $= \frac{w L^2}{8} = \frac{(15500 + 16000)}{8} (3.5)^2 = 48230$

$$\text{Net B.M. at centre of } BC = 48230 - \frac{41670 + 30310}{2} = 12240 \text{ N-m}$$

10. ***Maximum +ve B.M. in mid-sapn of CD*** : Conditions of loadings are : Live load on *CD* and *AB* and dead load on *ABCD*. In the first cycle, moment distribution is done at joints *B*, *C* and *D*, and half the distributed moments are carried over to the opposite ends, i.e. from *D* to *C* and *C* to *D*, and from *B* to *C*. In the second cycle, distribution is performed at *C* and *D*, as illustrated in Table 24.9.

Thus the end moments at *C* and *D* are − 54450 and + 44200 respectively. The free B.M. at the centre of span *CD*

$$= \frac{w L^2}{8} = \frac{(14500 + 16000) \times 5^2}{8} = 95310 \text{ N-m}$$

$$\therefore \quad \text{Net B.M. at the centre of } CD = 95310 - \frac{54450 + 44200}{2} = 45980 \text{ N-m}$$

TABLE 24.9 MOMENT DISTRIBUTION FOR – VE B.M. AT *CD*

Joint	*A*	*B*		*C*		*D*
Member	***AB***	***BA***	***BC***	***CB***	***CD***	***DC***
D.F.	0.474	0.392	0.174	0.20	0.30	0.375
1. F.E.M. due to D.L.			– 15820	+15820		
2. F.E.M. due to total load	– 134750	+134750			– 63540	+63540
3. Distribution at *B*, *C* and *D*			– 20690		+14320	– 23820
4. Carry over				– 10350	– 11910	+7160
5. Distribution at *C* and *D*					+ 6680	– 2680
6. Total moments					– 54450	+44200

(*C*) MAXIMUM NEGATIVE B.M. AT CENTRE OF SPANS

11. ***Maximum Negative B.M. at the centre of span BC***

The condition for loadings are : Live loads on *AB* and *CD*, and dead load on *ABCD*. In the first cycle, moment distribution is carried out at all the four joints *A*, *B*, *C* and *D*. These moments are then carried over to joints *B* and *C* from joints *A* and *B*, as well as between themselves. The second distribution is carried out at joints *B* and *C*, as shown in Table 24.10.

TABLE 24.10 MOMENT DISTRIBUTION FOR – VE B.M. AT CENTRE OF SPAN BC

Joint	*A*	*B*		*C*		*D*
Member	***AB***	***BA***	***BC***	***CB***	***CD***	***DC***
D.F.	0.474	0.392	0.174	0.20	0.30	0.375
1. F.E.M. due to D.L.			– 15820	+15820		
2. F.E.M. due to total load	– 134750	+ 134750			– 63540	+63540
3. Distribution at *A*, *B*, *C* and *D*	+63880	– 46620	– 20700	+9540	+14320	– 23820
4. Carry over to *B* and *C*		+31940	+4770	– 10350	– 11910	
5. Distribution at *B* and *D*			– 6380	+4450		
6. Final moments			– 38130	+19460		

Thus the end moment at *B* and *C* are –38130 and +19460 respectively. Free B.M. at the centre of span *BC* = Dead load intensity $\times \frac{L^2}{8} = \frac{15500\,(3.5)^2}{8} = 23730$

$$\text{Net B.M. at centre of } BC = 23730 - \frac{38130 + 19460}{2} = -5065.$$

12. ***Maximum Negative B.M. at centre of spans AB and CD*** : Since spans *AB* and *CD* are large, free B.M. at their mid-span will be large. It will be seen that the net B.M. at the centre of these spans will either be positive, or will be negative but of negligibly small magnitude. Due to this reason, these spans are not being investigated for max. negative B.M. However, conditions for max. negative B.M. at the centre of span *AB* will be when live load is on *BC* and dead load is on *ABCD*. Similarly, the loading condition for max. negative B.M. at centre of *CD* will be when span *BC* is loaded with live load, and dead load is on *ABCD*.

(*D*) BENDING MOMENTS IN COLUMNS : For maximum B.M. in columns, alternate spans should be loaded with live load, while the whole floor is loaded with dead load. The two possible load conditions are shown in Fig. 24.4. For the present case, the loadings will be as shown in para (13) and (14) below. See Tables 24.11 and 24.12.

13. ***Max. B.M. in columns*** : The loading conditions are : live load on *AB* and *CD* and dead load on *ABCD*. The moment distribution is carried out as illustrated in Table 24.11. In the first cycle, distribution is done at all the four joints *A*, *B*, *C* and *D*. The carry over moments are then transferred to the appropriate points. These carried-over moments are added to the original F.E.M. to get new moments at each joint. These new moments are distributed to the columns meeting at the joints.

TABLE 24.11 BENDING MOMENT IN COLUMNS

Joint	*A*	*B*		*C*		*D*
Column D.F.						
(*a*) Just above floor	0.263	0.217		0.25		0.3125
(*b*) Just below floor	0.263	0.217		0.25		0.3125
Horizontal members	***AB***	***BA***	***BC***	***CB***	***CD***	***DC***
D.F.	0.474	0.392	0.174	0.20	0.30	0.375
1. F.EM. due to D.L.			− 15820	+15820		
2. F.E.M. due to total load	− 134750	+134750			− 63540	+63540
3. Distribution	+63880	− 46620	− 20700	+9540	+14320	− 23820
4. Carry over	− 23310	+31940	+4770	− 10350	− 11910	+7160
5. New moments (total 1, 2, 4)	− 158060	+166690	−11050	+5470	− 75450	+70700
6. Distribution to Columns						
(*i*) just above floor	+41570	− 33770		+17500		− 22160
(*ii*) just below floor	+41570	− 33770		+17500		− 22160

TABLE 24.12 BENDING MOMENT IN COLUMNS

Joint	*A*	*B*		*C*		*D*
Column D.F.						
(*a*) Just above floor	0.263	0.217		0.25		0.3125
(*b*) Just below floor	0.263	0.217		0.25		0.3125
Horizontal members	***AB***	***BA***	***BC***	***CB***	***CD***	***DC***
D.F.	0.474	0.392	0.174	0.20	0.30	0.375
1. F.EM. due to D.L.	− 69420	+69420			− 30210	+30210
2. F.E.M. due to total load			− 32160	+32160		
3. Distribution	+32900	− 14600	− 6480	− 400	− 580	− 11320
4. Carry over	− 7300	+16450	− 200	− 3240	− 5660	− 290
5. New moments (total 1, 2, 4)	− 76720	+85870	− 32360	+28920	− 35870	+29920
6. Distribution to Columns						
(*i*) just above floor	+20180	− 11610		+1740		− 9350
(*ii*) just below floor	+20180	− 11610		+1740		− 9350

14. ***Max. B.M. in columns: Alternative loading*** : The loading conditions are: live load on *BC* and dead load on *ABCD*. The moment distribution is carried out as illustrated in Table 24.12. In the first cycle, distribution is done at all the four joints. The carry over moments are then transferred to appropriate points. These carried over moments are added to the original F.E.M. to get new moments at each joint. These new moments are distributed to the columns meeting at the joints.

24.5. ANALYSIS OF FRAMES SUBJECTED TO HORIZONTAL FORCES

A building frame is subjected to horizontal forces due to wind pressure and seismic effects. These horizontal forces cause axial forces in columns and bending moment in all the members of the frame. As stated earlier, a building frame is a highly indeterminate structure. The degree of indeterminacy of a building bent (Fig. 24.7) is found by providing a cut near mid-span of each beam. Each cut beam will thus have three unknown reaction components: moment (*M*), shear (*F*) and axial thrust (*H*). Each column with its cut beams will act as a cantilever, which is in a statically determinate structure. Thus, if n is the number of beams in a bent, the degree of indeterminancy will be $3n$. For the building bent shown in Fig. 24.7, there are eight beams and hence the bent is statically indeterminate upto 24th degree. An ordinary 20 storey building with 20 storeys and 5 stacks of columns has 80 beams, thus having the degree of indeterminacy of 240.

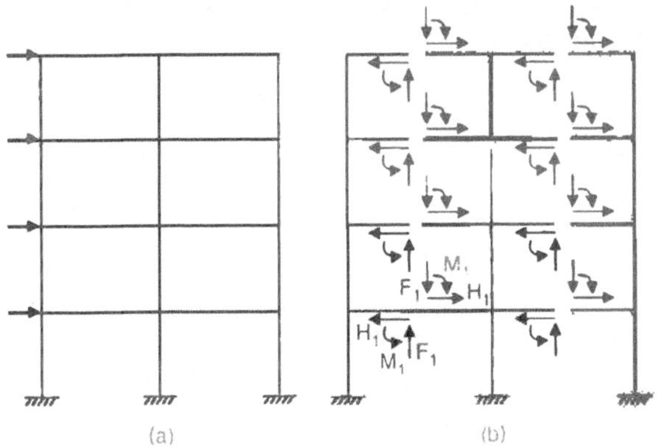

FIG. 24.7

Due to this reason, suitable assumptions are made so that the frame subjected to horizontal forces can be analysed by using simple principles of mechanics. Following approximate methods are commonly used for the analysis of building frames subjected to lateral forces:

1. Portal method. 2. Cantilever method. 3. Factor method.

24.6. PORTAL METHOD

For the purposes of analysis it is assumed that the horizontal forces are acting on the joints. The portal method is based on the following two important assumptions:

(*i*) the points of contraflexure in all the members lie at their mid-points.

and (*ii*) horizontal shear taken by each interior column is double the horizontal shear taken by each of exterior column.

Fig. 24.8 shows a three storey building frame with three spans. Let P_1, P_2, P_3 be the external horizontal forces acting at the joints of the wall columns. Under the action of the horizontal forces, the frame will deflect. The point of contraflexure will lie at the middle of each member. Only horizontal shears will act at these points of contraflexure, since B.M. will be zero at these points.

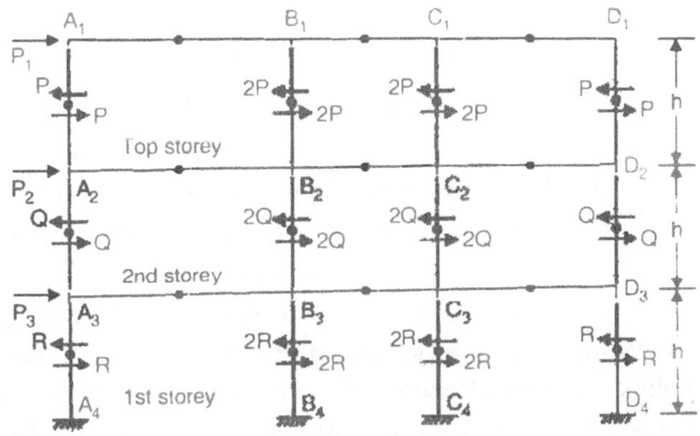

FIG. 24.8

Consider the top storey having vertical members $A_1 A_2$, $B_1 B_2$, $C_1 C_2$ and $D_1 D_2$. The horizontal shear for the outer columns A_1A_2 and D_1D_2 will be

P each while that for the inner columns B_1B_2 and C_1C_2 will be $2P$ each, as marked. The value of P is given by

$$P_1 = P + 2\,P + 2\,P + P$$

$$\therefore \qquad P = \frac{1}{6} P_1$$

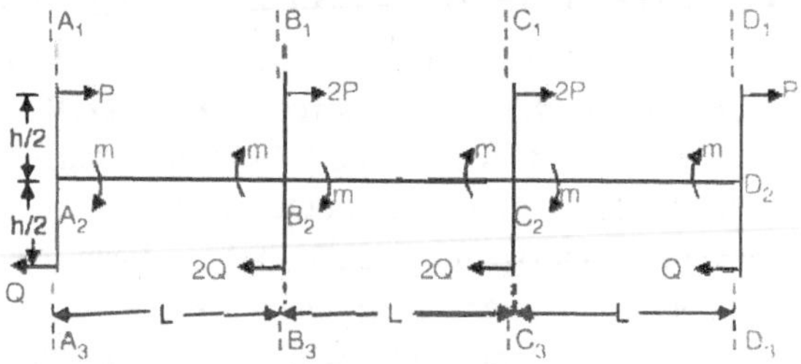

FIG. 24.9

Similarly, consider the second storey, where the exterior columns A_2A_3 and D_2D_3 have shear Q. The value of shear Q is found by :

$$P_1 + P_2 = Q + 2\,Q + 2\,Q + Q \qquad \therefore \qquad Q = \frac{1}{6}(P_1 + P_2)$$

Similarly, for the bottom storey, the shear R is given by

$$P_1 + P_2 + P_3 = R + 2\,R + 2\,R + R \qquad \therefore \qquad R = \frac{1}{6}(P_1 + P_2 + P_3)$$

Knowing the horizontal shears at the points of contraflexure, the bending moments in the columns can be easily found. Let us consider the floor $A_2\,B_2\,C_2\,D_2$ between third and second storey. The shear acting at the point of contraflexure are as shown in Fig.24.9. The joint A_2 is subjected a clockwise moment of $P\,h/2$ at A_2 in column $A_1\,A_2$, and to a clockwise moment equal to $Q\,h/2$ at A_2 in column $A_2\,A_3$. The beam $A_2\,B_2$ is thus required to resist a clockwise moment of $m = (P + Q)\,h/2$ at A_2. Similarly, at joint B, there will be a clockwise moment equal to $(2\,P + 2\,Q)\,h/2$. But there are two beams to resist this. Hence clockwise moment in each beam will be $(P + Q)\,h/2$. Thus the ends of each beam receive the same clokwise moment of $(P + Q)\,h/2$, with the result that points of contraflexure will lie in the middle of the beams.

The moment m acting at each end of the beams $A_2\,B_2$, $B_2\,C_2$, $C_2\,D_2$, give rise to vertical reactions in columns. If L is the span of these beams, each beam will impose an upward pull of $2m/L$ on windward column and a push of $2m/L$ on leeward column connected to the beam, for each span. The vertical reactions will neutralize for any intermediate column, provided span of beams on either side are equal. Only the end columns will experience vertical reactions. The windward column will have an upward pull of $2m/L$ and the leeward column will have a downward push of $2m/L$. The method of analysis is illustrated in Example 24.2.

24.7. CANTILEVER METHOD

The cantilever method. is based on the following assumptions:

(i) Ponts of contraflexure in each member lies at its mid-span or mid-height.

(ii) The direct stresses (axial stress) in the columns, due to horizontal forces, are directly proportional to their distance from the centroidal vertical axis of the frame.

Fig. 24.10. (a) shows a building frame subjected to horizontal forces. Fig 24.10(b) shows the top storey, up to the points of contraflexure of columns. The reactions at the points of contraflexure will be direct and shear forces only. Let V_1, V_2, V_3 and V_4 be the axial forces in the columns AE, BF, CG and DH, having areas of cross-sections a_1, a_2, a_3 and a_4 respectively.

From statics, we have

$$P = H_1 + H_2 + H_3 + H_4 \quad \ldots(i)$$

From assumption (2), we have

$$\frac{V_1/a_1}{x_1} = \frac{V_2/a_2}{x_2} = \frac{V_3/a_3}{x_3} = \frac{V_4/a_4}{x_4} \quad \ldots(ii)$$

where x_1, x_2, x_3 and x_4 are the centroidal distances of the columns from the vertical centroidal axis of the frame.

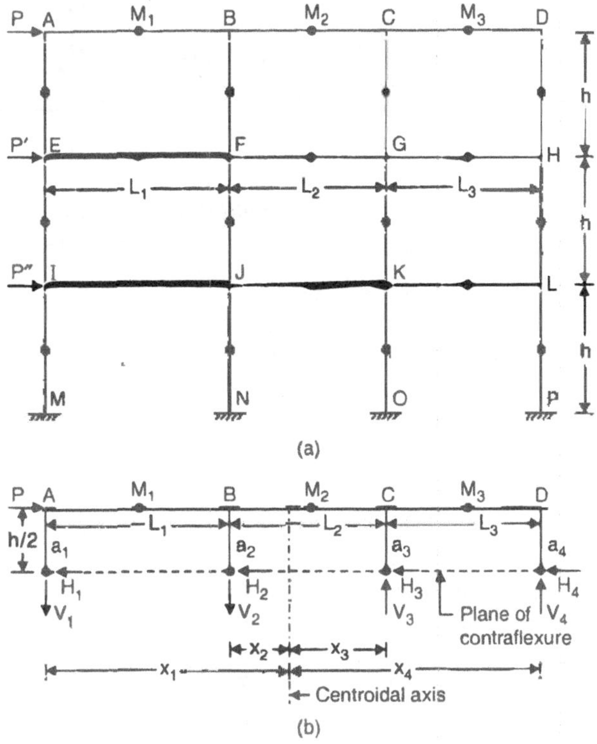

FIG. 24.10

By taking moments about the point of intersection of the vertical centroidal axis and the top beam, we get

$$(H_1 + H_2 + H_3 + H_4)\frac{h}{2} = V_1 x_1 + V_2 x_2 + V_3 x_3 + V_4 x_4$$

or

$$V_1 x_1 + V_2 x_2 + V_3 x_3 + V_4 x_4 = \frac{P h}{2} \quad \ldots(iii)$$

From (*ii*) and (*iii*), axial forces V_1, V_2, V_3 and V_4 can be determined.

In order to determine H_1, take moments about the point of contraflexure M_1 in beam *AB* [Fig. 24.11(*a*)].

$$H_1 \frac{h}{2} = V_1 \cdot \frac{L_1}{2} \qquad \therefore \qquad H_1 = \frac{V_1 L_1}{h} \quad \ldots(a)$$

Similarly, taking moments about point of contraflexure M_2 in beam *BC*,

$$H_1 \cdot \frac{h}{2} + H_2 \cdot \frac{h}{2} = V_1\left(L_1 + \frac{L_2}{2}\right) + V_2 \frac{L_2}{2}$$

$$\therefore (H_1 + H_2) = \frac{2\left[V_1 L_1 + (V_1 + V_2)\dfrac{L_2}{2}\right]}{h} \quad \ldots(b)$$

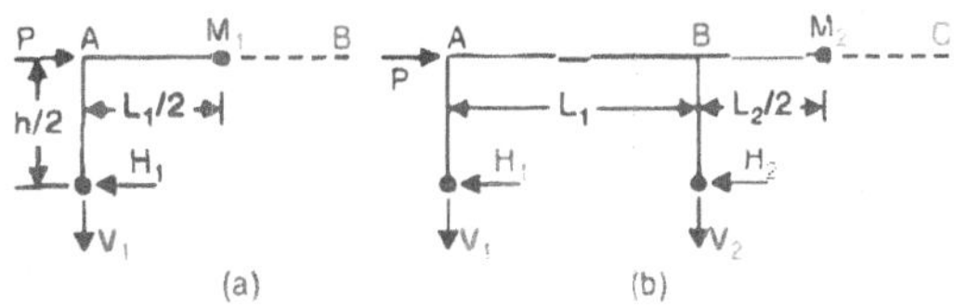

FIG. 24.11

Since H_1 is known from (*a*), H_2 can be determined. In a similar manner, H_3 and H_4 can be determined.

Problem 24.2. *Analyse the building frame, subjected to horizontal forces, as shown in Fig. 24.12. Use portal method.*

Solution

1. ***Horizontal shears*** : Let the horizontal shears in the exterior columns be P and in the interior columns be $2P$ for the top storey. Similarly, for the bottom storey, let the shears be R and $2R$ for the exterior and interior columns. For the top storey, we have $\quad P + 2P + 2P + P = 120$

$$\therefore \quad P = \frac{120}{6} = 20 \text{ kN}$$

For the bottom storey, we have

$$R + 2R + 2R + R = 120 + 180$$

$$\therefore \quad R = \frac{300}{6} = 50 \text{ kN}$$

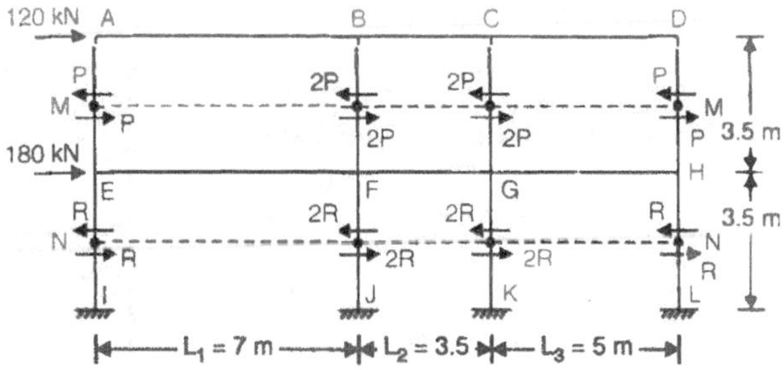

FIG. 24.12.

2. *Moments at the ends of columns*

For the top storey,

$$M_{EA} = M_{AE} = M_{HD} = M_{HD}$$

$$= P \times \frac{h}{2} = 20 \times \frac{3.5}{2} = 35 \text{ kN-m}$$

$$M_{FB} = M_{BF} = M_{GC} = M_{CG} = 2P \times \frac{h}{2} = 20 \times 3.5 = 70 \text{ kN-m}$$

For the bottom storey. $\quad M_{IE} = M_{EI} = M_{LH} = M_{HL} = R \times \frac{h}{2} = 50 \times \frac{3.5}{2} = 87.5 \text{ kN-m}$

$$M_{JF} = M_{FJ} = M_{KG} = M_{GK} = 2R \times \frac{h}{2} = 50 \times 3.5 = 175 \text{ kN-m}$$

3. *Moments at the ends of beams*

First floor beams : $\quad m_{EF} = M_{EA} + M_{EI} = 35 + 87.5 = 122.5 \text{ kN-m}$

Similarly, $m_{FE} = m_{FG} = m_{GF} = m_{GH} = m_{HG} = 122.5$ kN-m, since the point of contraflexure lies at the middle of each span.

In general, $\quad m = (P + R)\frac{h}{2} = (20 + 50) \times \frac{3.5}{2} = 122.5$

Roof beams : $\quad m_{AB} = m_{BA} = m_{BC} = m_{CB} = m_{CD} = m_{DC} = P \cdot \frac{h}{2} = 20 \times \frac{3.5}{2} = 35 \text{ kN-m}$

4. *Shear in beams* : Since no external vertical force is acting on the beam, shear F is given by

$$F = \frac{m_1 + m_2}{L}$$

where m_1 and m_2 are the moments at ends of the beams of span L.

Thus, $\quad F_{EF} = \frac{122.5 + 122.5}{7} = 35 \text{ kN} \uparrow \qquad F_{FE} = 35 \downarrow$

$$F_{FG} = F_{GF} = \frac{122.5 + 122.5}{3.5} = 70 \text{ kN} \qquad F_{GH} = F_{HG} = \frac{122.5 + 122.5}{5} = 49 \text{ kN}$$

$$F_{AB} = F_{BA} = \frac{35 + 35}{7} = 10 \text{ kN} \qquad F_{BC} = F_{CB} = \frac{35 + 35}{3.5} = 20 \text{ kN}$$

$$F_{CD} = F_{DC} = \frac{35 + 35}{5} = 14 \text{ kN}$$

5. *Axial force in columns.* The axial force in the columns will be as under :

Column AE = shear in beam $AB = 10 \uparrow$

Column EI = axial force in AE + shear in $EF = 10 + 35 = 45 \uparrow$

Column DH = shear in beam $DC = 14$ KN $\downarrow$

Column HL = axial force in DH + shear in $HG = 14 + 49 = 63 \downarrow$

Since the spans are not equal, interior columns will also have axial forces.

$$\text{Column } BF = F_{BA} - F_{BC} = 10 - 20 = -10 \quad (i.e.\ \uparrow)$$

$$\text{Column } FJ = (-10) + (F_{FE} - F_{FG}) = (-10) + (35 - 70) = -45 \quad (i.e.\ \uparrow)$$

Alternatively, axial force in $BF = \frac{2\,m'}{L_1} - \frac{2\,m'}{L_2} = \frac{2 \times 35}{7} - \frac{2 \times 35}{3.5} = -10$ kN

and axial force in column $FJ = (-10) + \left(\frac{2\,m}{L_1} - \frac{2\,m}{L_2}\right) = (-10) + \left(\frac{2 \times 122.5}{7} - \frac{2 \times 122.5}{3.5}\right)$

$$= -45 \ (\uparrow)$$

Axial force in column $CG = \frac{2m'}{L_2} - \frac{2\,m'}{L_3} = \frac{2 \times 35}{3.5} - \frac{2 \times 35}{5} = 6 \ (\downarrow)$

Axial force in column $GK = 6 + \left(\frac{2\,m}{L_2} - \frac{2\,m}{L_3}\right) = 6 + \left(\frac{2 \times 122.5}{3.5} - \frac{2 \times 122.5}{5}\right) = 27 \downarrow$

Check. Total axial force at the base = $-45(\uparrow) - 45(\uparrow) + 27(\downarrow) + 63(\downarrow) =$ zero.

Example 24.3. *Re-analyse the frame of example 24.2 by cantilever method, assuming that all the columns have the same area of cross-section.*

Solution. 1. ***Location of centroidal axis of the columns***

Let the centroidal axis be at a distance of $\bar{x}$ from the wind- ward columns *AEI*. Taking the moment of areas of the columns about *AEI*, we get

$$\bar{x} = \frac{(2 \times 0) + (2 \times 7) + (2 \times 10.5) + (2 \times 15.5)}{8} = 8.25 \text{ m. Hence,}$$

$x_1 = 8.25\ (= \bar{x})$ m

$x_2 = 8.25 - 7 = 1.25$ m

$x_3 = 3.5 - 1.25 = 2.25$ m

$x_4 = (7+3.5+5) - 8.25 = 7.25$ m.

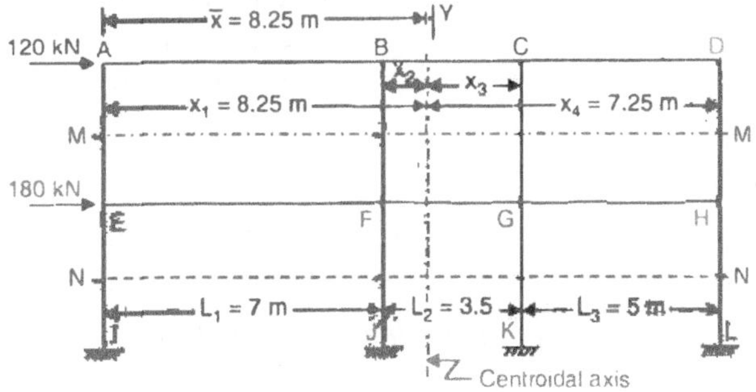

FIG. 24.13.

2. ***Axial forces in columns of first storey***

Let the axial force in column $EI = V_1 = V$.

Since the areas are equal, the axial forces in other columns will be in proportion to their distances from the centroidal axis.

$$\frac{V_1}{x_1} = \frac{V_2}{x_2} = \frac{V_3}{x_3} = \frac{V_4}{x_4}$$

$$V_2 = V\,\frac{x_2}{x_1} = \frac{1.25}{8.25}\,V$$

$$= 0.1515\ V\ (\downarrow)$$

$$V_3 = V\,\frac{x_3}{x_1} = \frac{2.25}{8.25}\,V = 0.2727\ V\ (\uparrow)$$

$$V_4 = V\,\frac{x_4}{x_1} = \frac{7.25}{8.25}\,V$$

$$= 0.8788\ V\ (\uparrow)$$

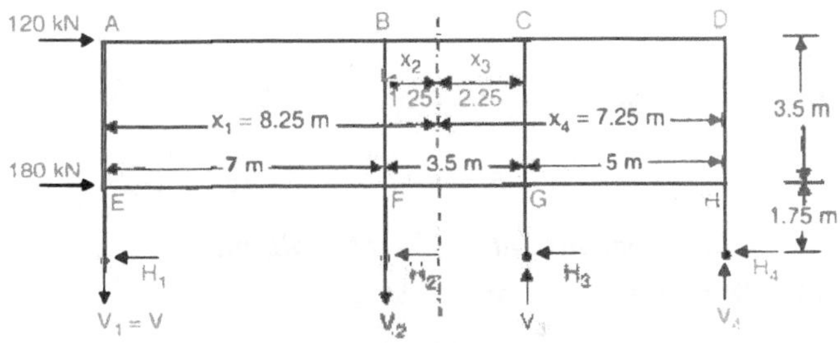

FIG. 24.14.

Taking moments of all forces about the point of contraflexure N of the leeward column, we get, $(120 \times 5.25) + (180 \times 1.75) - (V \times 15.5) - (0.1515\ V \times 8.5) + (0.2727\ V \times 5) = 0$

which gives $V = V_1 = 61.267$ kN $(\downarrow)$ $\quad \therefore \quad V_2 = 0.1515 \times 61.267 = 9.282 \ (\downarrow)$

$V_3 = 0.2727 \times 61.267 = 16.707 \ (\uparrow) \qquad V_4 = 0.8788 \times 61.267 = 53.842 \ (\uparrow)$

Check $\Sigma V = 61.267 + 9.282 - 16.707 - 53.842 = 0$

3. *Axial forces in the columns of second storey*

Let $V_1' = V' =$ axial force in column AE

$\therefore \quad V_2' = 0.1515\ V'$; $V_3' = 0.2727\ V'$ and $V_4' = 0.8788\ V'$

Taking moments about point of contraflexure M, we get

$(120 \times 1.75) - (V' \times 15.5) - (0.1515\ V' \times 8.5) + (0.2727\ V' \times 5) = 0$

From which $V' = V_1' = 13.615 \ (\downarrow)$

$V_2' = 0.1515 \times 13.615 = 2.063 \ (\downarrow)$

$V_3' = 0.2727 \times 13.615 = 3.713 \ (\uparrow)$

$V_4' = 0.8788 \times 13.615 = 11.965 \ (\uparrow)$

Check

$\Sigma V = 13.615 + 2.603 - 3.713 - 11.965 = 0$

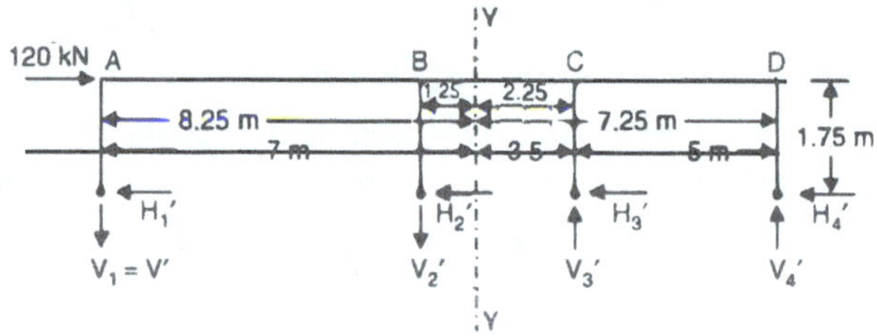

FIG. 24.15.

4. *Shears at ends of beams* : The shears at the ends of beams can be determined from the axial forces in the columns at various joints. Let us assume downward force as negative.

Joint E : $F_{EF} = V_1' - V_1 = 13.615 - 61.267 = -47.652$

Joint F : $F_{FG} = -47.652 + 2.063 - 9.282 = -54.871$

Joint G : $F_{GH} = -54.871 - 3.713 + 16.707 = -41.877$

Joint A : $F_{AB} = -13.615$

Joint B : $F_{BC} = -13.615 - 2.063 = -15.678$

Joint C : $F_{CD} = -15.678 + 3.713 = -11.965$

5. *Moments at the ends of beams* : Since there is a point of contraflexure at the middle of each beam, the moment at the end of beams is found by multiplying shear at the end by half the length of the beam.

(*a*) *First floor* : $M_{EF} = M_{FE} = F_{FE} \times \dfrac{L_1}{2} = 47.652 \times \dfrac{7}{2} = 166.8$

$M_{FG} = M_{GF} = 54.871 \times \dfrac{3.5}{2} = 96.0$; $\quad M_{GH} = M_{HG} = 41.877 \times \dfrac{5}{2} = 104.7$

(*b*) *Second floor* : $M_{AB} = M_{BA} = 13.615 \times \dfrac{7}{2} = 47.6$

$M_{BC} = M_{CB} = 15.678 \times \dfrac{3.5}{2} = 27.4$; $\quad M_{CD} = M_{DC} = 11.965 \times \dfrac{5}{2} = 29.9.$

6. *Moments at ends of column*

(*a*) *Top storey* : $M_{AE} = M_{AB} = 47.6$ kN-m

Since there is point of contraflexure at the middle of column AE, $M_{EA} = 47.6$

$M_{BF} = M_{BA} + M_{BC} = 47.6 + 27.4 = 75$ kN-m $\quad \therefore \quad M_{FB} = 75$ kN-m

$$M_{CG} = M_{CB} + M_{CD} = 27.4 + 29.9 = 57.3 \qquad \therefore \quad M_{GC} = 57.5$$

$$M_{DH} = M_{DC} = 29.9 \text{ kN-m} \qquad \therefore \quad M_{HD} = 29.9 \text{ kN-m}$$

(*b*) *Bottom storey* $\quad M_{EI} + M_{EA} = M_{EF}$

$\therefore \quad M_{EI} = M_{EF} - M_{EA} = 166.8 - 47.6 = 119.2$

$\therefore \quad M_{IE} = 119.2$ kN-m

$M_{FJ} + M_{FB} = M_{FE} + M_{FG}$

$\therefore \quad M_{FJ} = 166.8 + 96 - 75 = 187.8$ kN-m

Hence $\quad M_{JF} = 187.8$ kN-m

$M_{GK} = M_{GF} + M_{GH} + M_{GC}$

$\therefore \quad M_{GK} = 96 + 104.7 - 57.5 = 143.2$

$\therefore \quad M_{KG} = 143.2$ kN-m

$M_{HL} + M_{HD} = M_{GH}$

$\therefore \quad M_{HL} = 104.7 - 29.9 = 74.8$

$\therefore \quad M_{LH} = 74.8.$

Alternatively, the moment at the column ends can be found by first determining horizontal shears (H) at the point of contraflexure, and multiplying these by half the height of the column. Thus, $\quad M_{AE} = H_1' \times \frac{h}{2} \; ; \; M_{BF} = H_2' \times \frac{h}{2}$ etc.

Similarly, $\quad M_{EI} = H_1 \times \frac{h}{2} \; ; \; M_{FJ} = H_2 \times \frac{h}{2}$ etc.

The method of determining horizontal shears have been explained in § 24.7.

For example, $\quad H_1' = \frac{V_1' L_1}{h} = 13.615 \times \frac{7}{3.5} = 27.23$

$$H_2' = \frac{2\left[V_1' L_1 + (V_1' + V_2')\frac{L_2}{2}\right]}{h} - H_1' = \frac{2\left[13.615 \times 7 + (13.615 + 2.063) \times \frac{3.5}{2}\right]}{3.5} - 27.23$$

$$= 42.908$$

$$\therefore \quad M_{AE} = H_1' \times \frac{h}{2} = 27.23 \times \frac{3.5}{2} = 47.65.$$

$M_{BF} = H_2' \times \frac{h}{2} = 42.908 \times \frac{3.5}{2} = 75$, which is the same as found earlier.

24.8. FACTOR METHOD

This method is more accurate than either the portal method or cantilever method, and is more useful when the moments of inertia of various members [*i.e.* of columns and beams] are different. Both cantilever method as well as portal method assume uniform moments of inertia of members. These methods therefore depend on some stress assumptions, thus limiting the analysis to be based on equations of statics only. The factor method is based on assumptions regarding the elastic action of the structure. For analysis by factor method, the relative stiffness $K\,(=I/L)$ for each member of the frame or structure should be known. The procedure consists of the following steps :

Step 1. Calculate the *girder factor* g for each joint from the following expression

$$g = \frac{\Sigma K_c}{\Sigma K} \quad ....(24.1)$$

where ΣK_c = Sum of relative stiffnesses of all column members at the the joint considered

ΣK = Sum of relative stiffnesses of all the members at the joint considered.

These values of girder factor g are entered in a tabular form as shown in Table 24.13. The values are entered at the end of each girder meeting at that joint.

Step **2.** Calculate the *column factor* c for all joints from the following expression:

$$c = 1 - g \quad ....(24.2)$$

where g = girder factor of the joint.

The values of column factors are entered in Table 24.13 at the end of each column at the joint.

For columns which are fixed at the base, the column factor c is taken as 1.00.

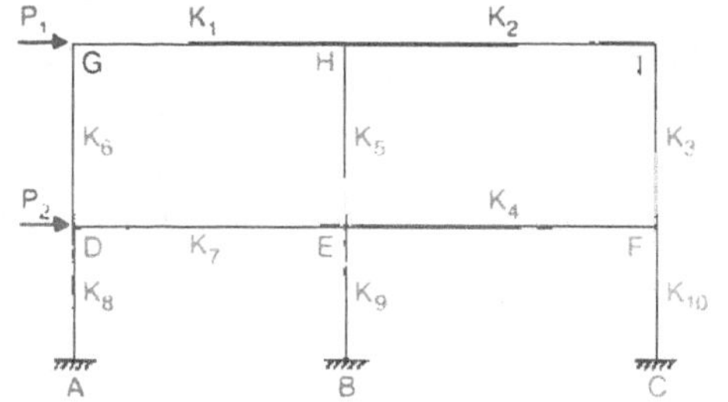

FIG. 24.16.

Fig. 24.16 shows a simple frame with two storeys and two bays, used for illustration purpose. The relative stiffnesses k_1, k_2, k_3, ... k_{10} of all the ten members are entered on/near each member. Table 24.13 is used for computation of *column factor* (c), *girder factor*(g), *column moment factor* (C) *and girder moment factor* (G).

TABLE 24.13

Joint	*Member*	*Column/girder factors c or g*	*Half value of factor at opposite end of member*	*(3) + (4)*	*K= I/L*	*Column/ girder moment factor C or G* = (5) × (6)
(1)	*(2)*	*(3)*	*(4)*	*(5)*	*(6)*	*(7)*
D	DG	c_1	$c_2/2$	$c_1 + c_2/2$	K_6	
	DA	c_1	$c/2$ from AD	$c_1 + (c/2)_{AD}$	K_8	
	DE	g_1	$g/2$ from ED	$g_1 + (g/2)_{ED}$	K_7	
G	GD	c_2	$c_1/2$	$c_2 + c_1/2$	K_6	
	GH	g_2	$g/2$ from HG	$g_2 + (g/2)_{HG}$	K_1	

Note : Refer Table 24.14 for further illustration.

Step **3.** As shown in Table 24.13, in the first column of the table, the name of all the joints are entered. The 2nd column contains all the members at each joint. In 3rd column, the corresponding girder or column factors are entered against each girder or column. In column 4 of the table, half the values of the column factor/girder factor of opposite end of the members are entered. For example, if c_1 = column factor of member DG, it is entered in column 3 opposite DG, while column factor c_2 of member GD is entered in column 3 opposite member GD. Hence in column 4, half the column factor of opposite end i.e. $c_2/2$ is entered opposite member DG. Similarly, for member GD, a value of $c_1/2$ is entered opposite it for the same reason. So in this way column no. 4 is entered. The values in column no. (3) and (4) of Table 24.13 for each member are added and entered in column no. 5. In column no. 6, the relative stiffness values $K = I/L$ for each member is entered.

Step **4.** The sum of columns (3) and (4), which is entered in column no. (5) is multiplied by relative stiffness of respective members (which are entered in column no. 6). This product is termed as *column moment factor C* for columns and *girder moment factor G* for girders. This is entered in column No. (7) of Table 24.13.

The column moment factor C, gives the relative values of moments at the ends of columns for each storey in which the column occurs. The sum of column end moments is equal to the horizontal shear on that storey multiplied by the storey height. Hence the column moment factors (C) are converted into end moments for columns by direct proportion for each storey. Similarly, the girder moment factor G, gives the relative values of moment at ends of each girder for the joint. The sum of girder end moment at each joint is equal to the sum of end moments in the columns at the joint. Hence the girder moment factors are converted into end moments for girders by direct proportion for each storey.

***Step* (5) *Calculation of column moments* :**

(a) Total column moments (A) for each storey is found by the relation

$$A = \frac{H \times h}{\Sigma C} \qquad \ldots(24.3)$$

where A = Total column moment for each storey

H = Total horizontal force above the storey considered

h = height of the storey considered.

ΣC = Sum of column end moment factors of that storey

Thus, for each storey, different column moments A_1, A_2 ... etc. are calculated.

(b) The column moment factor C of each member is multiplied by the total column moment (A) of that storey in which the column occurs. For example, in Fig. 24.16 if we want to find the column moment M_{GD} of column GD, we have

$$M_{GD} = A_1 \times C_{GD} \qquad \ldots.(24.4)$$

where A_1 = Total column moments of first storey (Eq. 24.3) and

C_{GD} = Column moment factor for column GD.

Similarly, moment M_{HE} in column HE of first storey is $M_{HE} = A_1 \times C_{HE}$

Also for column DA of ground storey, $M_{DA} = A_0 \times C_{DA}$

where A_0 = total column moments of ground storey.

Step* (6) *Calculations of Girder/beam moments

(a) For calculation of girder/beam moments, a constant B is found for each joint.

$$B = \frac{\text{Sum of column moments at the joint}}{\text{Sum of the girder moment factors at that joint}}$$

(b) This constant B is multiplied by the girder moment factor(G) to obtain the girder moments. For example (Fig. 24.16),

$$M_{DE} = B_D \times G_{DE} \;; \qquad M_{HI} = B_H \times G_{HI}$$

$$M_{IH} = B_I \times G_{IH} \quad \text{and so on.}$$

Here, B_D, B_H, B_I are the constants for joints D, H and I respectively.

The factor method of analysing the building frame has been illustrated in Example 24.4

Example 24.4. *Analyse the frame as shown in Fig. 24.17 by factor method. Sketch the B.M.D. The relative k values are written on the members.*

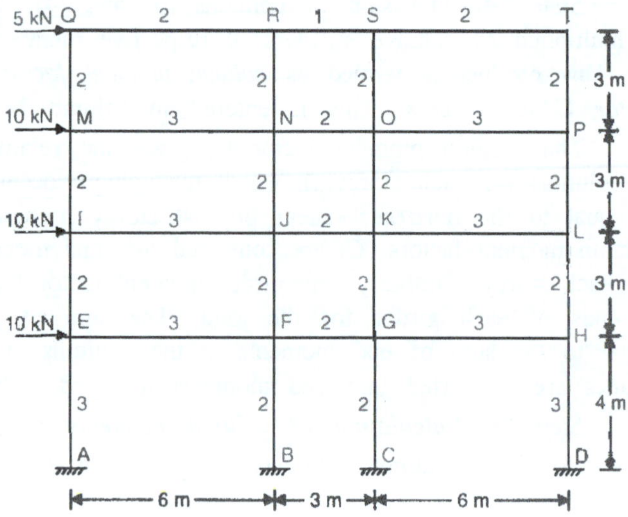

FIG 24.17

***Step* 1.** Calculate the girder factor '*g*' at all joints by Eq. 24.1.

$$g = \frac{\Sigma K_C}{\Sigma K}$$

where ΣK_C = Sum of relative Stiffness of columns at that joint

ΣK = Sum of relative Stiffness of all the members at that joint.

Joint Q $\quad g_Q = \dfrac{K_{QM}}{K_{QM} + K_{QR}} = \dfrac{2}{2+2} = 0.5$

Joint R $\quad g_R = \dfrac{K_{RN}}{K_{RN} + K_{RQ} + K_{RS}} = \dfrac{2}{2+2+1} = 0.4$

Joint S $\quad g_S = \dfrac{K_{SO}}{K_{SO} + K_{SR} + K_{ST}} = \dfrac{2}{2+1+2} = 0.4$

Joint T $\quad g_T = \dfrac{K_{TP}}{K_{TP} + K_{TS}} = \dfrac{2}{2+2} = 0.5$

Joint M $\quad g_M = \dfrac{K_{MI} + K_{MQ}}{K_{MI} + K_{MQ} + K_{MN}} = \dfrac{2+2}{2+2+3} = 0.57$

Joint N $\quad g_N = \dfrac{K_{NJ} + K_{NR}}{K_{NJ} + K_{NR} + K_{NM} + K_{NO}} = \dfrac{2+2}{2+2+3+2} = 0.44$

Joint O $\quad g_O = \dfrac{K_{OK} + K_{OS}}{K_{OK} + K_{OS} + K_{ON} + K_{OP}} = \dfrac{2+2}{2+2+2+3} = 0.57$

Joint P $\quad g_P = \dfrac{K_{PL} + K_{PT}}{K_{PL} + K_{PT} + K_{PO}} = \dfrac{2+2}{2+2+3} = 0.57$

Joint I $\quad g_I = \dfrac{K_{IE} + K_{IM}}{K_{IE} + K_{IM} + K_{IJ}} = \dfrac{2+2}{2+2+3} = 0.57$

Joint J $\quad g_J = \dfrac{K_{JF} + K_{JN}}{K_{JF} + K_{JN} + K_{JI} + K_{JK}} = \dfrac{2+2}{2+2+3+2} = 0.44$

Joint K $\quad g_K = \dfrac{K_{KO} + K_{KG}}{K_{KO} + K_{KG} + K_{KJ} + K_{KL}} = \dfrac{2+2}{2+2+2+3} = 0.44$

Joint L $\quad g_L = \dfrac{K_{LH} + K_{LP}}{K_{LH} + K_{LP} + K_{LK}} = \dfrac{2+2}{2+2+3} = 0.57$

Joint E $\qquad g_E = \dfrac{K_{EA} + K_{EI}}{K_{EA} + K_{EI} + K_{EF}} = \dfrac{3+2}{3+2+3} = 0.63$

Joint F $\qquad g_F = \dfrac{K_{FB} + K_{FJ}}{K_{FB} + F_{FJ} + K_{FE} + K_{FG}} = \dfrac{2+2}{2+2+3+2} = 0.44$

Joint G $\qquad g_G = \dfrac{K_{GC} + K_{GK}}{K_{GC} + K_{GK} + K_{GF} + K_{GH}} = \dfrac{2+2}{2+2+2+3} = 0.44$

Joint H $\qquad g_H = \dfrac{K_{HD} + K_{HL}}{K_{HD} + K_{HL} + K_{HG}} = \dfrac{3+2}{3+2+3} = 0.63.$

These values of girder factors are written at the ends of girders, beams meeting at each joint as shown in col. 3, Table 24.14.

Step 2. Calculate the column factor c at the joints by the relation

$$c = 1 - g \quad \text{where } g = \text{girder factor at that joint} \qquad \ldots(24.2)$$

Joint Q $\qquad c_Q = 1 - g_Q = 1 - 0.5 = 0.5$

Joint R $\qquad c_R = 1 - g_R = 1 - 0.4 = 0.6$

Joint S $\qquad c_s = 1 - g_S = 1 - 0.4 = 0.6$

Joint T $\qquad c_T = 1 - g_T = 1 - 0.5 = 0.5$

Joint M $\qquad c_M = 1 - g_M = 1 - 0.57 = 0.43$

Joint N $\qquad c_N = 1 - g_N = 1 - 0.44 = 0.56$

Joint O $\qquad c_O = 1 - g_O = 1 - 0.44 = 0.56$

Joint P $\qquad c_P = 1 - g_P = 1 - 0.57 = 0.43$

Joint I $\qquad c_I = 1 - g_I = 1 - 0.57 = 0.43$

Joint J $\qquad c_J = 1 - g_J = 1 - 0.44 = 0.56$

Joint K $\qquad c_K = 1 - g_K = 1 - 0.44 = 0.56$

Joint L $\qquad c_L = 1 - g_L = 1 - 0.57 = 0.43$

Joint E $\qquad c_E = 1 - g_E = 1 - 0.63 = 0.37$

Joint F $\qquad c_F = 1 - g_F = 1 - 0.44 = 0.56$

Joint G $\qquad c_G = 1 - g_G = 1 - 0.44 = 0.56$

Joint H $\qquad c_H = 1 - g_H = 1 - 0.63 = 0.37$

Joint A $\qquad c_A = 1.00$
Joint B $\qquad c_B = 1.00$
Joint C $\qquad c_C = 1.00$
Joint D $\qquad c_D = 1.00$

For columns fixed at the base, column factor is taken as 1.00

These values of column factors are written at the end of columns meeting at the joint, and have been entered in column 3 of Table 24.14.

Step 3. Half the values of column/girder factors of the opposite ends are entered in column 4. The values in col. 3 and col. 4 are added and entered in col. 5 of Table 24.14.

Step 4. Enter the values of relative stiffnesses of all members in col.6. These values of stiffnesses are multiplied by the values of column 5, to get the values of girder/column moment factors (*G* or *C*), and are entered in col. 7 of Table 24.14.

TABLE 24.14

Joint	*Member*	*Girder/ Column factor (c or g)*	*Half values of the factors from opposite end*	*(3)+(4)*	*Relative stiffness (K=I/L)*	*Girder/ column Moment factor (C or G) =(5) × (6)*
1	2	3	4	5	6	7
Q	*QR*	0.5	0.2	0.7	2	1.4
	QM	0.5	0.21	0.71	2	1.42
R	*RQ*	0.4	0.25	0.65	2	1.30
	RS	0.4	0.2	0.60	1	0.60
	RN	0.6	0.28	0.88	2	1.76
S	*SR*	0.4	0.2	0.6	1	0.6
	ST	0.4	0.25	0.65	2	1.3
	SO	0.6	0.28	0.88	2	1.76
T	*TS*	0.5	0.2	0.7	2	1.4
	TP	0.5	0.21	0.71	2	1.42
M	*MN*	0.57	0.22	0.79	3	2.37
	MI	0.43	0.21	0.64	2	1.28
	MQ	0.43	0.25	0.68	2	1.36
N	*NM*	0.44	0.28	0.72	3	2.16
	NO	0.44	0.22	0.66	2	1.32
	NR	0.56	0.3	0.86	2	1.72
	NJ	0.56	0.28	0.84	2	1.68
O	*ON*	0.44	0.22	0.66	2	1.32
	OP	0.44	0.28	0.72	3	2.16
	OS	0.56	0.30	0.86	2	1.72
	OK	0.56	0.28	0.84	2	1.68
P	*PO*	0.57	0.22	0.79	3	2.37
	PL	0.43	0.21	0.64	2	1.28
	PT	0.43	0.25	0.68	2	1.36
I	*IJ*	0.57	0.22	0.79	3	2.37
	IM	0.43	0.21	0.64	2	1.28
	IE	0.43	0.18	0.61	2	1.22
J	*JI*	0.44	0.28	0.72	3	2.16
	JK	0.44	0.22	0.66	2	1.32
	JN	0.56	0.28	0.84	2	1.68
	JF	0.56	0.28	0.84	2	1.68
K	*KJ*	0.44	0.22	0.66	2	1.32
	KL	0.44	0.28	0.72	3	2.16
	KO	0.56	0.28	0.84	2	1.68
	KG	0.56	0.28	0.84	2	1.68
L	*LK*	0.57	0.22	0.79	3	2.37
	LP	0.43	0.21	0.64	2	1.28
	LH	0.43	0.18	0.61	2	1.22
E	*EF*	0.63	0.22	0.85	3	2.55
	EI	0.37	0.21	0.58	2	1.16
	EA	0.37	0.5	0.87	3	2.61
F	*FE*	0.44	0.31	0.75	3	2.25
	FG	0.44	0.22	0.66	2	1.32
	FB	0.56	0.5	1.06	2	2.12
	FJ	0.56	0.28	0.84	2	1.68

1	2	3	4	5	6	7
G	*GF* *GH* *GK* *GC*	0.44 0.44 0.56 0.56	0.22 0.31 0.28 0.50	0.66 0.75 0.84 1.06	2 3 2 2	1.32 2.25 1.68 2.12
H	*HG* *HL* *HD*	0.63 0.37 0.37	0.22 0.21 0.50	0.85 0.58 0.87	3 2 3	2.55 1.16 2.61
A	*AE*	1.00	0.18	1.18	3	3.54
B	*BF*	1.00	0.28	1.28	2	2.56
C	*CG*	1.00	0.28	1.28	2	2.56
D	*DH*	1.00	0.18	1.18	3	3.54

Thus Table 24.14 is completed.

Step 5. Calculation column moments : Total columm moments for each storey

$$A = \frac{H \times h}{\Sigma C}$$

where H = Total horizontal force above the storey considered

h = height of the storey considered

ΣC = sum of *column end moment factors* of that storey

Let A_3 = Total column moments of third/top storey

A_2 = Total column moments of second storey

A_1 = Total column moments of first storey

A_0 = Total column moments of ground storey.

$$\therefore \quad A_3 = \frac{5 \times 3}{C_{QM} + C_{MQ} + C_{RN} + C_{NR} + C_{SO} + C_{OS} + C_{TP} + C_{PT}}$$

$$= \frac{15}{1.42 + 1.36 + 1.76 + 1.72 + 1.76 + 1.72 + 1.42 + 13.6} = 1.2 \text{ kN-m}$$

$$A_2 = \frac{(10 + 5) \times 3}{C_{MI} + C_{IM} + C_{NJ} + C_{JN} + C_{OK} + C_{KO} + C_{PL} + C_{LP}}$$

$$= \frac{45}{1.28 + 1.28 + 1.68 + 1.68 + 1.68 + 1.68 + 1.28 + 1.28} = 3.80 \text{ kN-m}$$

$$A_1 = \frac{(10 + 10 + 5) \times 3}{C_{IE} + C_{EI} + C_{JF} + C_{FJ} + C_{KG} + C_{GK} + C_{LH} + C_{HL}}$$

$$= \frac{75}{1.22 + 1.16 + 1.68 + 1.68 + 1.68 + 1.68 + 1.22 + 1.16} = 6.422 \text{ kN-m}$$

$$A_0 = \frac{(10 + 10 + 10 + 5) \times 4}{C_{EA} + C_{AE} + C_{FB} + C_{BF} + C_{GC} + C_{CG} + C_{HD} + C_{DH}}$$

$$= \frac{140}{2.61 + 3.54 + 2.12 + 2.56 + 2.12 + 2.56 + 2.61 + 3.54} = 6.464 \text{ kN-m.}$$

Column Moments :

Top Storey : $A_3 = 1.2$ kN-m

$$M_{QM} = A_3 \times C_{QM} = 1.2 \times 1.42 = 1.704 \text{ kN-m}$$

$$M_{MQ} = A_3 \times C_{MQ} = 1.2 \times 1.36 = 1.632 \text{ kN-m}$$
$$M_{RN} = A_3 \times C_{RN} = 1.2 \times 1.76 = 2.112 \text{ kN-m}$$
$$M_{NR} = A_3 \times C_{NR} = 1.2 \times 1.72 = 2.064 \text{ kN-m}$$
$$M_{SO} = A_3 \times C_{SO} = 1.2 \times 1.76 = 2.112 \text{ kN-m}$$
$$M_{OS} = A_3 \times C_{OS} = 1.2 \times 1.72 = 2.064 \text{ kN-m}$$
$$M_{TP} = A_3 \times C_{TP} = 1.2 \times 1.42 = 1.704 \text{ kN-m}$$
$$M_{PT} = A_3 \times C_{PT} = 1.2 \times 1.36 = 1.632 \text{ kN-m}$$

Second Storey : $A_2 = 3.80$ kN-m

$$M_{MI} = 3.80 \times 1.28 = 4.864 \text{ kN-m}$$
$$M_{IM} = 3.80 \times 1.28 = 4.864 \text{ kN-m}$$
$$M_{NJ} = 3.80 \times 1.68 = 6.384 \text{ kN-m}$$
$$M_{JN} = 3.80 \times 1.68 = 6.384 \text{ kN-m}$$
$$M_{OK} = 3.80 \times 1.68 = 6.384 \text{ kN-m}$$
$$M_{KO} = 3.80 \times 1.68 = 6.384 \text{ kN-m}$$
$$M_{PL} = 3.80 \times 1.28 = 4.864 \text{ kN-m}$$
$$M_{LP} = 3.80 \times 1.28 = 4.864 \text{ kN-m}$$

First Storey : $A_1 = 6.422$ kN-m

$$M_{IE} = 6.422 \times 1.22 = 7.834 \text{ kN-m}$$
$$M_{EI} = 6.422 \times 1.16 = 7.449 \text{ kN-m}$$
$$M_{JF} = 6.422 \times 1.68 = 10.789 \text{ kN-m}$$
$$M_{FJ} = 6.422 \times 1.68 = 10.789 \text{ kN-m}$$
$$M_{KG} = 6.422 \times 1.68 = 10.789 \text{ kN-m}$$
$$M_{GK} = 6.422 \times 1.68 = 10.789 \text{ kN-m}$$
$$M_{LH} = 6.422 \times 1.22 = 7.834 \text{ kN-m}$$
$$M_{HL} = 6.422 \times 1.16 = 7.449 \text{ kN-m}$$

Ground Storey : $A_0 = 6.464$ kN-m

$$M_{EA} = 6.464 \times 2.61 = 16.871 \text{ kN-m}$$
$$M_{AE} = 6.464 \times 3.54 = 22.882 \text{ kN-m}$$
$$M_{FB} = 6.464 \times 2.12 = 13.703 \text{ kN-m}$$
$$M_{BF} = 6.464 \times 2.56 = 16.547 \text{ kN-m}$$
$$M_{GC} = 6.464 \times 2.12 = 13.703 \text{ kN-m}$$
$$M_{CG} = 6.464 \times 2.56 = 16.547 \text{ kN-m}$$
$$M_{HD} = 6.464 \times 2.61 = 16.871 \text{ kN-m}$$
$$M_{DH} = 6.464 \times 3.54 = 22.882 \text{ kN-m}$$

***Step 6* : Calculation of beam moments**

(*a*) Constant $B = \dfrac{\text{Sum of column moments at the joint}}{\text{Sum of girder moment factors at that joint}}$

Joint Q $$B_Q = \frac{M_{QM}}{G_{QR}} = \frac{1.704}{1.4} = 1.217$$

Joint R $$B_R = \frac{M_{RN}}{G_{RQ} + G_{RS}} = \frac{2.112}{1.30 + 0.60} = 1.111$$

Joint S $$B_S = \frac{M_{SO}}{G_{SR} + G_{ST}} = \frac{2.112}{0.6 + 1.3} = 1.111$$

Joint T $$B_T = \frac{M_{TP}}{G_{TS}} = \frac{1.704}{1.4} = 1.217$$

Joint M $$B_M = \frac{M_{MQ} + M_{MI}}{G_{MN}} = \frac{1.632 + 4.864}{2.37} = 2.741$$

Joint N $$B_N = \frac{M_{NR} + M_{NJ}}{G_{NM} + G_{NO}} = \frac{2.064 + 6.384}{2.16 + 1.32} = 2.427$$

Joint O $$B_O = \frac{M_{OS} + M_{OK}}{G_{ON} + G_{OP}} = \frac{2.064 + 6.384}{1.32 + 2.16} = 2.427$$

Joint P $$B_P = \frac{M_{PT} + M_{PL}}{G_{PO}} = \frac{1.632 + 4.864}{2.37} = 2.741$$

Joint I $$B_I = \frac{M_{IE} + M_{IM}}{G_{IJ}} = \frac{7.834 + 4.864}{2.37} = 5.357$$

Joint J $$B_J = \frac{M_{JN} + M_{JF}}{G_{JI} + G_{JK}} = \frac{6.384 + 10.789}{2.16 + 1.32} = 4.934$$

Joint K $$B_K = \frac{M_{KO} + M_{KG}}{G_{KJ} + G_{KL}} = \frac{6.384 + 10.789}{1.32 + 2.16} = 4.934$$

Joint L $$B_L = \frac{M_{LP} + M_{LH}}{G_{LK}} = \frac{4.864 + 7.834}{2.37} = 5.357$$

Joint E $$B_E = \frac{M_{EI} + M_{EA}}{G_{EF}} = \frac{7.449 + 16.871}{2.55} = 9.537$$

Joint F $$B_F = \frac{M_{FJ} + M_{FB}}{G_{FE} + G_{FG}} = \frac{10.789 + 13.703}{2.25 + 1.32} = 6.86$$

Joint G $$B_G = \frac{M_{GK} + M_{GC}}{G_{GF} + G_{GH}} = \frac{10.789 + 13.703}{1.32 + 2.25} = 6.86$$

Joint H $$B_H = \frac{M_{HD} + M_{HL}}{G_{HG}} = \frac{16.871 + 7.449}{2.55} = 9.537$$

(b) Beam Moments

Beam moments = $B \times$ Girder Moment Factor.

$M_{QR} = B_Q \times G_{QR} = 1.217 \times 1.40 = 1.704$ kN-m

$M_{RQ} = B_R \times G_{RQ} = 1.111 \times 1.30 = 1.444$ kN-m

$M_{RS} = B_R \times G_{RS} = 1.111 \times 0.60 = 0.667$ kN-m

$M_{SR} = B_S \times G_{SR} = 1.111 \times 0.60 = 0.667$ kN-m

$M_{ST} = B_S \times G_{ST} = 1.111 \times 1.30 = 1.444$ kN-m

$M_{TS} = B_T \times G_{TS} = 1.217 \times 1.40 = 1.704$ kN-m

$M_{MN} = B_M \times G_{MN} = 2.741 \times 2.37 = 6.496$ kN-m

$M_{NM} = B_N \times G_{NM} = 2.427 \times 2.16 = 5.242$ kN-m

$M_{NO} = B_N \times G_{NO} = 2.427 \times 1.32 = 3.203$ kN-m

$M_{ON} = B_O \times G_{ON} = 2.427 \times 1.32 = 3.203$ kN-m

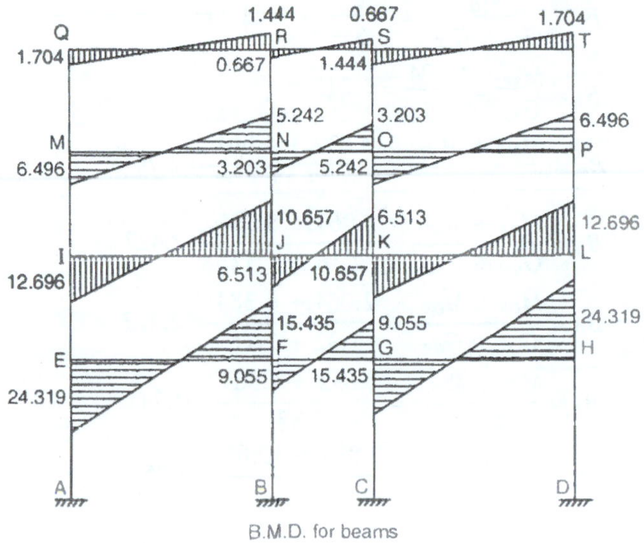

FIG. 24.18.

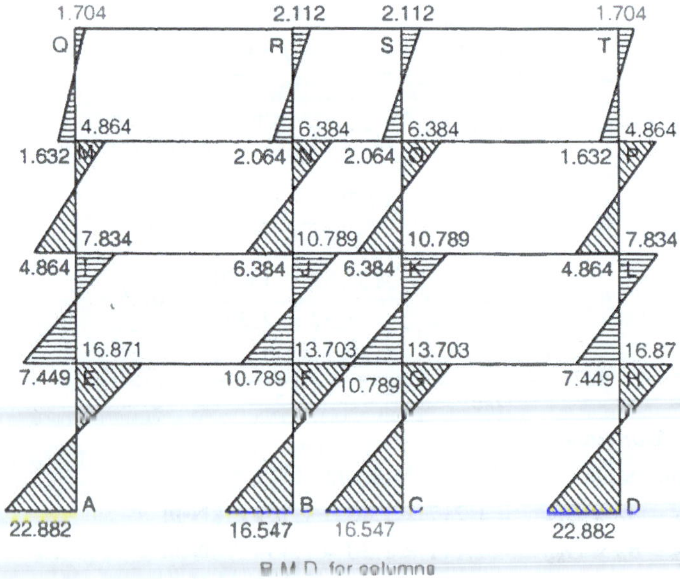

FIG. 24.19

$$M_{OP} = B_O \times G_{OP} \quad = 2.427 \times 2.16 = 5.242 \quad \text{kN-m}$$
$$M_{PO} = B_P \times G_{PO} \quad = 2.741 \times 2.37 = 6.496 \quad \text{kN-m}$$
$$M_{IJ} = B_I \times G_{IJ} \quad = 5.357 \times 2.37 = 12.696 \quad \text{kN-m}$$
$$M_{JI} = B_J \times G_{JI} \quad = 4.934 \times 2.16 = 10.657 \quad \text{kN-m}$$
$$M_{JK} = B_J \times G_{JK} \quad = 4.934 \times 1.32 = 6.513 \quad \text{kN-m}$$
$$M_{KJ} = B_K \times G_{KJ} \quad = 4.934 \times 1.32 = 6.513 \quad \text{kN-m}$$

$M_{KL} = B_K \times G_{KL} \quad = 4.934 \times 2.16 = 10.657$ kN-m

$M_{LK} = B_L \times G_{LK} \quad = 5.357 \times 2.37 = 12.696$ kN-m

$M_{EF} = B_E \times G_{EF} \quad = 9.537 \times 2.55 = 24.319$ kN-m

$M_{FE} = B_F \times G_{FE} \quad = 6.86 \times 2.25 = 15.435$ kN-m

$M_{FG} = B_F \times G_{FG} \quad = 6.86 \times 1.32 = 9.055$ kN-m

$M_{GF} = B_G \times G_{GF} \quad = 6.86 \times 1.32 = 9.055$ kN-m

$M_{GH} = B_G \times G_{GH} \quad = 6.86 \times 2.25 = 15.435$ kN-m

$M_{HG} = B_H \times G_{HG} \quad = 9.537 \times 2.55 = 24.319$ kN-m

Thus the moments in all the columns and girder are found. Fig. 24.18 shows the B.M. diagram for girder/beams while Fig. 24.19 shows the B.M. diagram for columns.

PROBLEMS

1. What do you understand by a substitute frame? How do you select it ? Discuss in brief the method of analysis.
2. Explain the portal method for analysing a building frame subjected to horizontal forces.
3. Explain the cantilever method for analysing a building frame subjected to horizontal forces.
4. A two-span intermediate frame of a multi-storeyed building is shown in Fig. 24.20. The frames are spaced at 5 m intervals. The dead load and live load per metre run of the beam may be taken as 15 kN/m and 20 kN/m respectively. Analyse the frame using twocycle method of moment distribution.

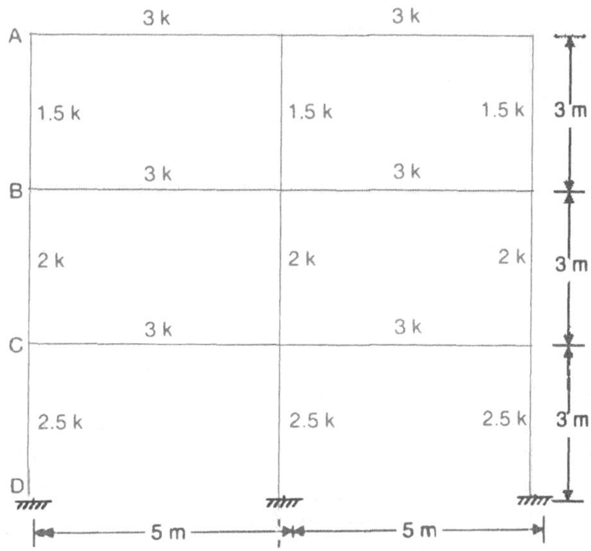

FIG. 24.20.

5. If wind loads of 15 kN, and 30 kN are acting at joint *A*, *B* and *C* respectively, analyse the frame (Fig. 24.20) by (*a*) portal method, (*b*) cantilever method. Assume that all the columns have equal area of cross-section for the purpose of analysis.

25

Prestressed Concrete

25.1. INTRODUCTION

In ordinary reinforced concrete, consisting of concrete and mild steel as basic components, the compressive streses are borne by concrete while tensile stresses are borne entirely by steel. The concrete surrounding steel reinforcement does not take part in resisting the external forces/moments since concrete is considered weak in tension. It simply acts as a bonding material. Thus only that portion of concrete, which lies above the neutral axis, is considered to be useful in resisting the external forces. This results in heavy sections. In the case of *prestressed concrete,* comprising of concrete and high tensile steel as basic components, both steel and concrete are stressed prior to the application of external loads. If such induced *pre-stress* in concrete is of compressive nature, it will balance the tensile stress produced in concrete surrounding steel due to external loads. Due to this, the whole of the concrete can participate in resisting the external forces.

According to ACI Committee on Pre-stressed Concrete, *pre-stressed concrete is the one in which there have been introduced internal stresses of such magnitude and distribution that the stresses resulting from given external loadings are counteracted to a desired degree.* In reinforced concrete member, the prestress is commonly introduced by tensioning the steel reinforcement. Thus, prestressing is the intentional creation of permanent stresses in a structure or assembly, for the purpose of improving its behaviour and strength under various service conditions (Lin, 55).

In prestressed concrete, compression is induced prior to loading in the zones where external loads would normally cause tensile stresses. Simple reinforced concrete has two major disadvantages.

(1) It has weak crack-resistance in consequence of the low tensile strength—all the reinforced concrete beams and other bending members have cracks at working loads. Due to this, corrosion of both the reinforcement and concrete takes place.

(2) It is impossible to use high tensile steel as reinforcement for simple reinforced concrete. If high tensile steel is used, the high stresses in the tensile reinforcement will result in such wide cracks in the tensile zone of concrete that the load carrying capacity of the members will practically be lost. A prestressed concrete construction on the other hand, has no cracks at working loads and has offered the possibility of employing high-tensile steel as reinforcement.

The earlier attempts of prestressing were made immediately after the development of reinforced concrete. P. Jackson (1886) of U.S.A. obtained patents for pretensioning steel tie rods in artificial stones and concrete arches to serve as floor slabs. K. Doring (1888) of Germany suggested pretensioning of wires in reinforced concrete floor structures. However, in all the earlier attempts, low tensile steel was used as prestressing material with the result that low pretension was lost in *shrinkage* and *creep* in concrete. If an ordinary mild steel bar is prestressed to a working stress of 140 N/mm^2, the resulting strain in it will be

$$\varepsilon = \frac{140}{2.1 \times 10^5} = 0.00066.$$

The permanent negative strain due to both shrinkage and creep is of order of the 0.0008. Since this permanent strain is greater than the initial strain in the mild steel caused by tensioning it, the prestress induced in mild steel would soon disappear, leaving the member simply reinforced.

To avoid this trouble, C.R. Steiner (1908) of U.S.A. recommended the tightening of reinforcing rods after some shrinkage and creep of concrete had taken place. R.E. Dill (1925) of Nebraska used high strength steel bars coated to prevent bond with concrete. The bars were tensioned and anchored to concrete by means of nuts, after the concrete had set. For economical reasons, these methods could not be applied to an appreciable extent. It was E. Freyssinet (1928) of France who made the first specific contribution of prestressing as it is today. He used high strength steel wires for prestressing. These wires has an ultimate strength in tension of 1750 N/mm^2 and yield point of about 1260 N/mm^2. If these are prestressed to 1050 N/mm^2 , the resulting strain will be

$$\varepsilon = \frac{1050}{2.1 \times 10^5} = 0.005.$$

Allowing for a total loss of 0.0008 due to shrinkage and creep a net strain of 0.005 − 0.0008 = 0.0042 will remain, giving rise to an effective or final prestress in steel equal to

$$f_s = 2.1 \times 10^5 \times 0\ 0042 = 882 \text{ N/mm}^2$$

According to Freyssinet, to prestress a structure is artificially to create either prior to or simultaneously with the application of external loads such permanent stresses that in combination with the stress due to external load, total stresses will every where and for all states of loads envisaged, be within limits of stress that the material can support indefinitely. In his earlier attempts, he tried pretensioning in which steel was bonded to concrete without end anchorages. However, in 1939, he developed conical wedges for end anchorages and designed double acting jacks which tensioned the wires and then thrust the male cones into the female cones for anchoring them. After this successful work, prestressed concrete was very widely developed in many countries and many famous workers like Guyon, Kani, Leonhardt, Magnel and others, started working on this. G. Magnel (1940) of Belgium developed the Magnel system, wherein two wires were stretched at a time and anchored with a simple metal wedge at each end. In the U.S.S.R., investigations on prestressed concrete were initiated by V.V. Michailov and then by A.A. Gvosdev, S.A. Dmitriev and others.

25.2. BASIC CONCEPTS

As stated earlier, *prestressing is the initial application of stresses, controlled in magnitude and direction, to a structural member to counteract the undesirable stresses due to working loads*. Prestressing is commonly introduced by tensioning the steel reinforcement. According to Lin, three different concepts may be applied to explain and analyse the basic behaviour of prestressed concrete. These concepts are as follows : (1) Stress concept, (2) Strength concept and (3) Balanced load concept.

1. Stress concept : Prestressing to transform concrete into an elastic material. This concept is credited to Eugene Freyssinet who visualised prestressed concrete as essentially concrete which is transformed from a brittle material into an elastic one by precompression given to it. If an ordinary concrete, whether plain or reinforced, is subjected to only compressive stresses, it behaves as a perfect elastic material because no tension cracks are there. But if it is subjected to flexural stresses, some portion of it will be in tension resulting in tension cracks ; the material under such circumstance no longer remains elastic. In prestressed concrete, on the other hand, concrete is visualised as being subjected to two system of forces : internal prestress, which is compressive and external load causing tensile stresses. The tensile stresses caused due to external load are counterbalanced by the compressive stress due to prestress, with the result that final stress in the extreme fibre is either compressive or zero. Due to absence of final tensile stress, no tension cracks would be there in concrete, and it will thus be transformed from brittle to elasic material. To elaborate this point, let us consider two cases: (*a*) concentric tendon (steel reinforcement) and (*b*) eccentric tendon.

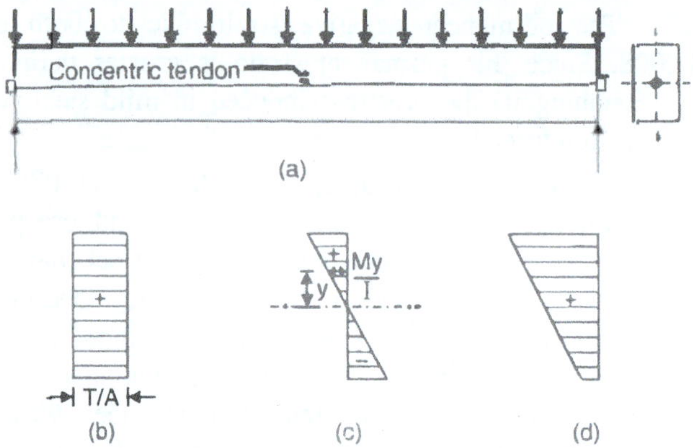

FIG. 25.1. CONCENTRICALLY PRESTRESSED SECTION.

Fig. 25.1(*a*) shows a concentrically prestressed concrete beam. Due to prestress force T in tendon, a uniform compressive stress $= T/A$ will be induced in concrete. If the beam is subjected to a moment M due to external load, inclusive of its own weight, the stress at any point will be $\frac{M}{I}y$ where y is the distance of the point from the centriodal axis and I is the moment of inertia of the section. The final stress at any point is given by

$$f = \frac{T}{A} \pm \frac{M\,y}{I} \qquad \text{...(25.1)}$$

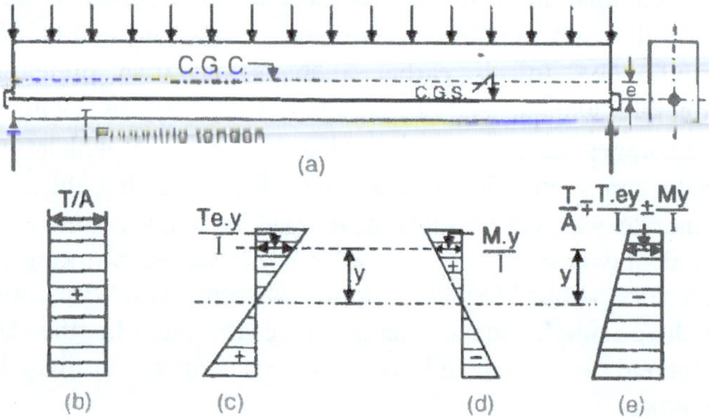

FIG. 25.2. ECCENTRICALLY PRESTRESSED SECTION.

Fig. 25.2(*a*) shows an eccentrically prestressed beam with external loading. Due to prestress force T in tendon, applied eccentrically, the moment produced due to prestress will be $T.e$. Hence the stress f' due to prestress at any point will be

$$f' = \frac{T}{A} + \frac{T \cdot e \cdot y}{I} \qquad \ldots[25.2(a)]$$

If M is the external moment, the final stress at any point is given by

$$f = \frac{T}{A} + \frac{T \cdot e \cdot y}{I} \pm \frac{M \cdot y}{I} \qquad \ldots(25.2)$$

The first concept is further illustrated through Example 25.1

2. Strength concept : Prestressing for combination of high strength steel and concrete In reinforced concrete, steel takes tension while concrete takes compression and the couple formed by the resultant compressive force C and tensile force T (where $C = T$) resists the external moment [Fig. 25.3(*a*)].

The same concept can be applied for a prestressed concrete section. The prestressed concrete is considered as a combination of steel and concrete, with steel taking T (passing through the tendon) and concrete taking compression C [passing through the C.G. of the stress distribution shown shaded in Fig. 25.3(*a*)], so that the two materials form a resisting couple to resist the external moment. This concept has been well utilised to determine the ultimate strength of prestressed concrete beams. This concept has been illustrated in Example 25.2.

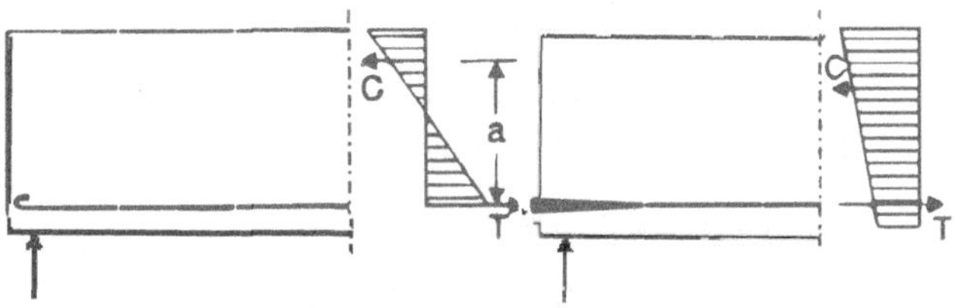

(a) R.C. section (b) Prestressed concrete

FIG. 25.3.

3. Load Balancing concept : Prestressing to achieve load balancing. This concept was developed by Lin, in which a flexural member is transformed into a member under direct stress. In the overall design of a prestressed concrete structure, the effect of prestressing is essentially viewed as the balancing of gravity loads (both self-weight as well as external loads) so that members under bending will not be subjected to flexural stresses under a given load conditions. It will be clear in Example 25.1 that by placing the tendon eccentrically, moment can be resisted. Flexural members, such as beam, slabs etc. are normally subjected to bending moments which are not uniform along their length. In the case of simply supported beam, moment is maximum at the mid-span and zero at the ends. Hence the eccentricity e should also vary along the span. For this, either a curved (parabolic) tendon [Fig. 25.4(*a*)] or inclined tendon [Fig. 25.5(*a*)] is used.

For the application of the third concept, concrete is taken as free body and the tendon is replaced by forces acting on the concrete as shown in Fig. 25.4(*b*). Fig. 25.4(*a*) shows the parabolic profile of tendon with central sag h with respect to the C.G. of the section. Consider two points P and Q at x_1 and x_2 from mid-span. The cable is under uniform tensile force T. If θ_1 and θ_2 are the inclinations of the cable with horizontal, at P and Q, we have, from Fig. 25.4(*d*)

Upward force $= T(\sin\theta_2 - \sin\theta_1)$

If $h < L$, θ_1 and θ_2 will be small, with the result that $\sin\theta \approx \tan\theta$

$\therefore$ Upward force $= T(\tan\theta_2 - \tan\theta_1)$ ...(1)

The equation of the parabola [Fig. 25.4(*c*)] is

$$y = \frac{4h}{L^2}\left(x^2 - \frac{L^2}{4}\right) \quad ...(2)$$

$$\therefore \quad \frac{dy}{dx} = \frac{8h}{L^2}x.$$

$$\therefore \quad \left(\frac{dy}{dx}\right)_1 = \tan\theta_1 = \frac{8h}{L^2}x_1$$

$$\left(\frac{dy}{dx}\right)_2 = \tan\theta_2 = \frac{8h}{L^2}x_2$$

$\therefore$ Upward force on PQ

$$= T\frac{8h}{L^2}(x_2 - x_1).$$

If $(x_2 - x_1)$ is taken as unity, upward focrce p per unit length is

$$p = T\frac{8h}{L^2} \quad ...(25.3)$$

From Eq.(2), $\dfrac{d^2y}{dx^2} = \dfrac{8h}{L^2}$

Hence $$p = T\frac{d^2y}{dx^2} \quad ....[25.3(a)]$$

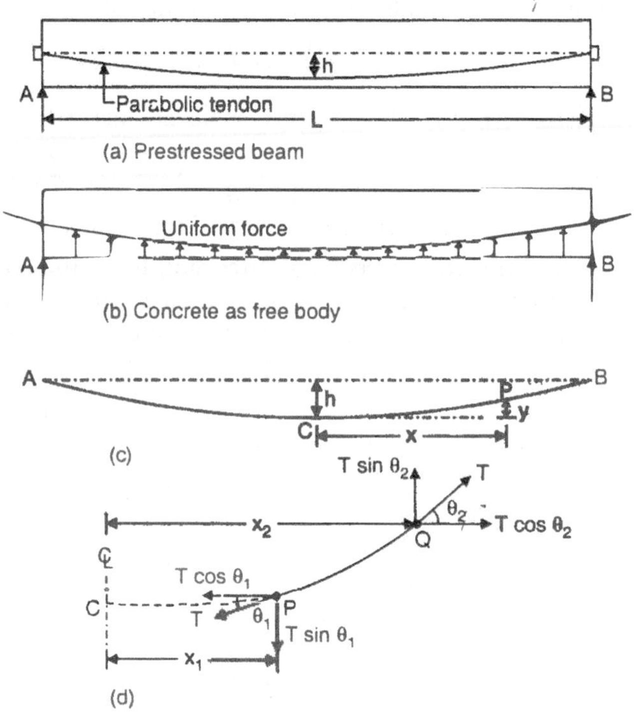

FIG. 25.4. PRESTRESSED BEAM WITH PARABOLIC TENDON.

Thus, the parabolic cable supports a uniformly distributed load p given by the above equations. If the external U.D.L. w is equal to p, transverse load on the beam is balanced and the beam is subjected to only the axial force T, which produces uniform stress in concrete $= f = T/A$. If the external load w is greater than p, the mid-span moment M for the remaining force $(w - p)$ can be computed and fibre stresses due to that moment would be My/I. The resulting stresses will then be equal to $\dfrac{T}{A} \pm \dfrac{My}{I}$

Fig. 25.5(*a*) shows a beam with bent tendon while Fig. 25.5(*b*) shows the freebody of concrete with tendon replaced by forces. In the figure, C.G.C. stands for centre of gravity line for concrete while C.G.S. stands for C.G. line of steel cable. The balancing force V is given by

$$V = 2\,T\sin\theta \quad ....[25.4(a)]$$

If this V exactly counterbalanes a concentrated load W applied at midspan, the beam is not subjected to any transverse load (neglecting the weight of the beam).

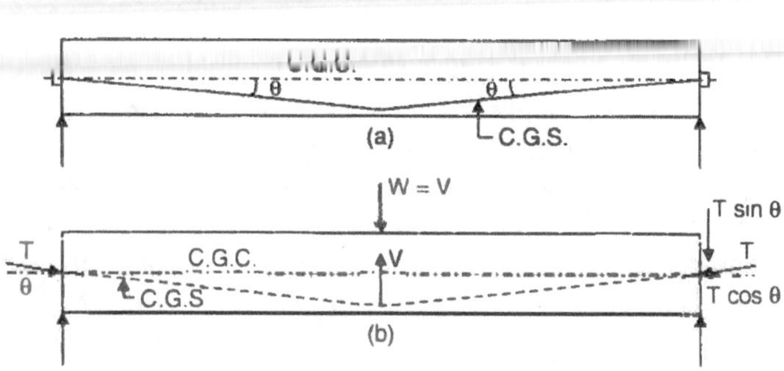

FIG. 25.5.

The uniform compressive stress in beam is given by

$$f = \frac{T \cos \theta}{A_c} \approx \frac{T}{A_c} \quad \text{(if } \theta \text{ is small)} \qquad \ldots[25.4(b)]$$

While this 'load balancing concept' often represents the simplest approach to prestressed design and analysis, its advantage over the other two concepts is not significant for statically determinate structures. However, the method offers tremendous advantages when dealing with statically indeterminate structures.

Example 25.1. *Stress concept. : A simply supported prestressed concrete beam of rectangular cross-section 400 mm × 600 mm, is loaded with a total uniformly distributed load of 256 kN over a span of 6 m. Sketch the distribution of stresses at mid-span and end sections if the prestressing force is 1920 kN and the tendon is (a) concentric, (b) eccentric, located at 200 mm above the bottom fibre.*

Solution

(*a*) ***Concentric tendon***

$A = 400 \times 600 = 240000 \text{ mm}^2$

$I = \frac{1}{12}(400)(600)^3 = 72 \times 10^8 \text{ mm}^4$

f_1 due to prestressing force

$= \frac{1920 \times 1000}{240000} = 8 \text{ N/mm}^2.$

At the end section, B.M. is zero, and hence the section will be subjected to uniform compressive stress of 8 N/mm^2 throughout its depth.

Midspan moment $= \frac{W\,L}{8} = \frac{256 \times 6}{8}$

$= 192$ kN-m $= 192 \times 10^6$ N-mm

$\therefore$ Extreme fibre stresses $= f_b$

$= \pm \frac{192 \times 10^6}{72 \times 10^8}(300) = \pm 8 \text{ N/mm}^2$

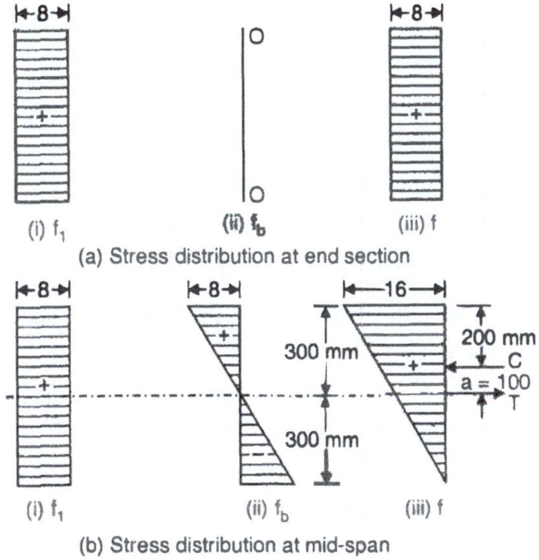

FIG. 25.6. STRESS DISTRIBUTIONS DUE TO CONCENTRIC TENDON.

Hence the final stresses at extreme fibres

$$f = f_1 \pm f_b = 8 \pm 8 = 16 \text{ or } 0.$$

The stress distributions at end and mid-span are shown in Fig. 25.6(*a*).

It is to be noted in Fig. 25.6(*b*) that for final stress diagram, $C = \frac{1}{2} \times 16 \times 24000 = 1920000$ N = 1920 kN. The compressive force C acts at a distacne of 100 mm from T, giving rise to a couple $= 100 \times C = 100\,T = 100 \times 1920 = 192000$ kN-mm = 192 kN-m = applied moment.

(*b*) ***Eccentric tendon.*** Eccentricity ; $e = 300 - 200 = 100$ mm.

The stresses due to eccentric prestress force is

$$f' = \frac{T}{A} \mp \frac{T\,.\,e}{I} y \quad ; \qquad f' = f_1 \mp f_2, \text{ where}$$

$$f_1 = \frac{1920 \times 1000}{240000} = 8 \text{ N/mm}^2$$

$$f_2 = \mp \frac{1920 \times 1000 \times 100}{72 \times 10^8} \times 300$$

$$= \mp 8 \text{ N/mm}^2$$

f_2 will be tensile at the top and compressive at the bottom fibres.

At the end section, where B.M. is zero, the final stresses f will be 8 ∓ 8 *i.e.* zero at top fibre and 16 N/mm² at the bottom fibre, as shown in Fig. 25.7(*b*). In the final stress distribution, line of action of C and T coincide, giving rise to zero moment.

At the mid-span, the bending stress f_b is equal to ± 8 N/mm², as found earlier. Hence the final stress f is given by $f = f_1 \mp f_2 \pm f_b = 8 \mp 8 \pm 8$. Hence f at top fibre = 8 N/m m² ; f at bottom fibre= 8 N/mm². The stress distribution is shown in Fig. 25.6(*c*).

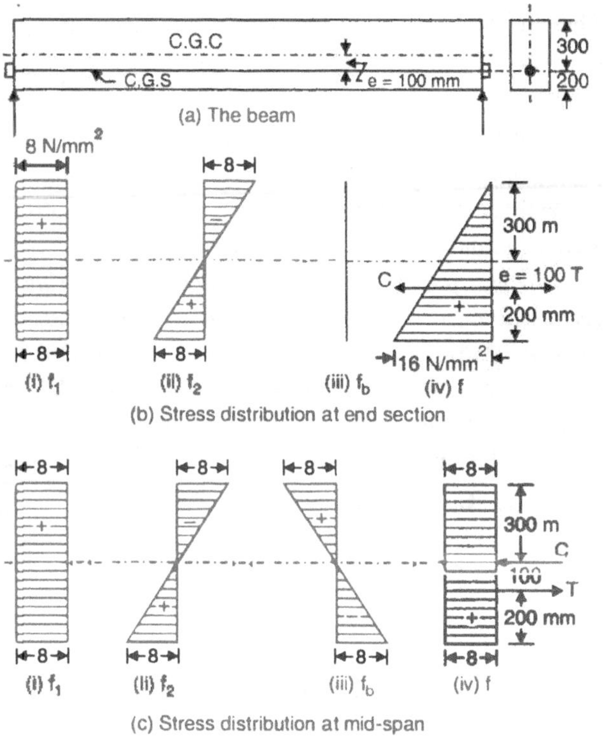

FIG. 25.7. STRESS DISTRIBUTION DUETO ECCENTRIC TENDON

It will be seen that by shifting the position of the tendon, the extreme fibre stress in concrete is reduced from 16 N/mm² to 8 N/mm²

Example 25.2. *Strength Concept. Find the extreme fibre stresses at mid-span of the beam of example 25.1 by applying the principle of internal resisting couple.*

Solution : External B.M. $= M = \frac{W\,L}{8} = \frac{256 \times 6}{8} = 192$ kN-m $= 192 \times 10^6$ N-mm

(a) Concentric tendon : Internal couple is furnished by forces $C = T = 1920$ kN. The lever arm a given by

$$T \times a = M$$

or $$a = \frac{M}{T} = \frac{192 \times 10^6}{1920 \times 1000} = 100 \text{ mm}$$

[Note : See Fig. 25.6(*b*) (*iii*) also.]

Since T acts at 300 mm above bottom fibre, distance of C from bottom fibre = 300 + 100 = 400 mm. Thus C is located. It acts at a distance of $e' = 400 - 300 = 100$ mm from the c.g.c., as shown in Fig. 25.8(*a*).

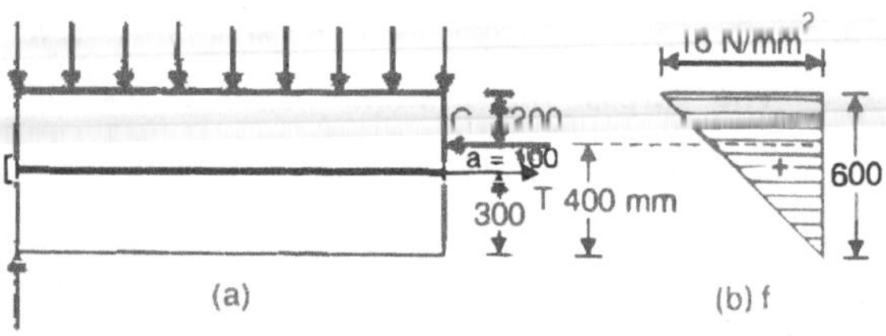

FIG. 25.8.

Hence extreme fibre stresses are given by

$$f = \frac{C}{A} \pm \frac{C \cdot e'}{I} \cdot y \;=\; \frac{1920 \times 1000}{240000} \pm \frac{(1920 \times 1000)\,100}{72 \times 10^8}\,(300)$$

$= 16 \text{ N/mm}^2$ at top fibre and 0 N/mm^2 at bottom fibre.

The stress distribution is sketched in Fig. 25.8(*b*), which is the same as that shown in Fig. 25.6(*b*)(*iii*).

(*b*) *Eccentric tendon*

$C = T = 1920$ kN

Also, $T \times a = M$

$$a = \frac{M}{T} = \frac{192 \times 10^6}{1920 \times 1000} = 100 \text{ mm}$$

[See Fig. 25.7 *c* (*iii*) also.]

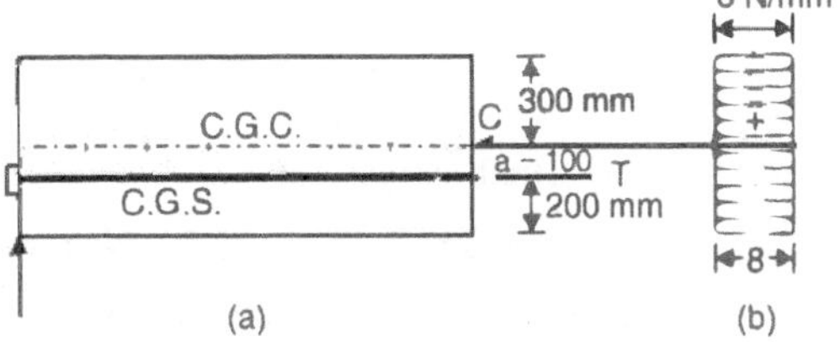

FIG. 25.9.

Since *T acts at 200 mm above bottom fibre, distance of C* from bottom fibre = 200 + *a* = 200 + 100 = 300 mm. Thus *C* is located. It acts at a distance of $e' = 300 - 300$ = zero cm from c.g.c., as shown in Fig. 25.9(*a*). Hence only uniform stress will be induced, having magnitude $f = \frac{C}{A} = \frac{1920 \times 1000}{240000} = 8 \text{ N/mm}^2$, as shown in Fig. 25.9(*b*).

Example 25.3. *Load Balancing Concept.* *The beam of Example 25.1 is pre-stressed by a parabolic tendon with a prestressing force of 1920 kN. The tendon has a sag of 100 mm at mid-span as shown in Fig. 25.10(a). Find the extreme fibre stresses by load balancing concept, if it is subjected to (a) total U.D.L. of 256 kN, (b) total U.D.L. of 360 kN.*

Solution

$h = 0.1$ m ; $A = 400 \times 600 = 240000 \text{ mm}^2$; $I = 72 \times 10^8 \text{ mm}^4$

The upward uniform force from the tendon on the concrete is

$$p = T \cdot \frac{8h}{L^2} \qquad \ldots(25.3)$$

$$= \frac{1920 \times 8 \times 0.1}{6 \times 6} = 42.67 \text{ kN/m.}$$

(*a*) *External U.D.L.*

$= w = 256/6 = 42.67$ kN/m.

Thus the external load is prefectly balanced. Hence there will be no moment, and the beam will be subjected to a uniform compressive stress.

$$= f = \frac{1920 \times 1000}{400 \times 600} = 8 \text{ N/mm}^2$$

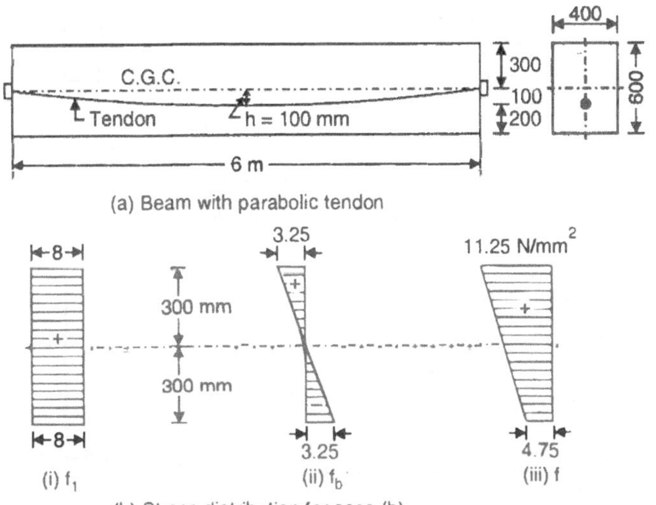

FIG. 25.10.

(b) *External U.D.L.* $= w = \dfrac{360}{6} = 60$ kN/m.

$$w' = w - p = 60 - 42.67 = 17.33 \text{ kN/m.}$$

$$M = \frac{w' L^2}{8} = \frac{17.33\,(6)^2}{8} \approx 78 \text{ kN-m.} = 78 \times 10^6 \text{ N-mm.}$$

$$\therefore \quad f_b = \pm \frac{78 \times 10^6 \times 300}{72 \times 10^8} = \pm 3.25 \text{ N/mm}^2. \;;\; f_1 = \frac{T}{A} = \frac{1920 \times 1000}{240000} = 8 \text{ N/mm}^2$$

Hence final stress are

$f = f_1 \pm f_b = 8 \pm 3.25 = 11.25$ N/mm² at top fibre and 4.75 N/mm² at bottom fibre.

Example 25.4. *Load Balancing Concept.* *Solve Example 25.3 if the parabolic profile of the tendon has an eccentricity of 100 mm at ends and 200 mm at mid-span, as shown in Fig. 25.11. The beam carries a U.D.L. of 60 kN/m.*

Solution The upward uniform force from tendon on the concrete is

$$p = T \cdot \frac{8h}{L^2}, \text{ where}$$

h = sag of centre of tendon with respect to ends = 200 − 100 = 100 mm = 0.1 m. Hence,

$$p = \frac{1920 \times 8 \times 0.1}{6 \times 6} = 42.67 \text{ kN/m.}$$

Effective vertical load

$$= w' = w - p = 60 - 42.67$$
$$= 17.33 \text{ kN/m.}$$

$$M = \frac{w' L^2}{8} = \frac{17.33\,(6)^2}{8}$$
$$= 78 \text{ kN-m} = 78 \times 10^6 \text{ N-mm.}$$

$$\therefore f_b = \pm \frac{78 \times 10^6 \times 300}{72 \times 10^8}$$
$$= \pm 3.25 \text{ N/mm}^2.$$

The uniform prestress force has an eccentricity of 100 mm at ends. Due to this, the entire beam will be subjected to uniform compressive stress $f_1 = \dfrac{T}{A}$ and bending stress $f_2 = \mp \dfrac{T \cdot e}{I} \cdot y$. Hence,

$$f_1 = \frac{1920 \times 1000}{240000} = 8 \text{ N/mm}^2 \text{ (comp.)}$$

$$f_2 = \mp \frac{1920000 \times 100 \times 300}{72 \times 10^8}$$

$$= \mp 8 \text{ N/mm}^2$$

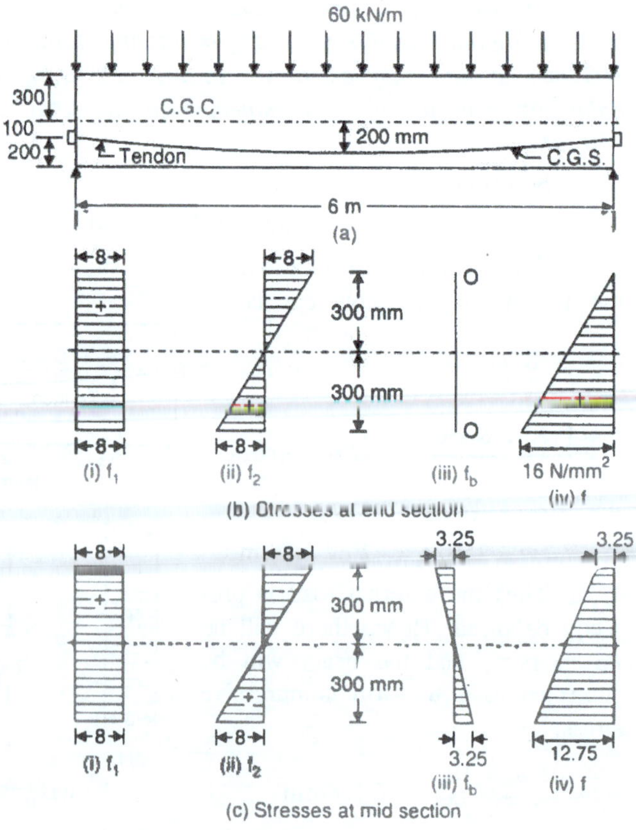

FIG. 25.11.

Hence the final stresses at end section are

$$f = f_1 \mp f_2 = 8 \mp 8 = \text{zero at top fibre}$$

$= 16\ \text{N/mm}^2$ at bottom fibre. Final stresses at mid-span are

$$f = f_1 \mp f_2 \pm f_b = 8 \mp 8 \pm 3.25 = 3.25\ \text{N/mm}^2 \text{ at top fibre (comp.)}$$

$= 12.75\ \text{N/mm}^2$ at bottom fibre (comp.)

Example 25.5. ***Load Balancing Concept.*** *Solve Example 25.4 if the eccentricity of tendon is − 100 mm at ends and +100 mm at mid-span, with straight line variation as shown in Fig. 25.12. The beam carries a U.D.L. of 60 kN/m.*

Solution Slope of the tendon being small,

$$\sin\theta \approx \tan\theta = \frac{20}{300} = \frac{1}{15}$$

The vertical component of the tendon force at the middle of the section is

$$V = 2\,T\sin\theta = 2 \times 1920000 \times 1/15 = 256000 \text{ N.}$$

Horizontal component $= T\cos\theta \approx T$.

At the end section, eccentricity $e = 100$ mm, above the C.G.C.

$$f = f_1 \pm f_2 = \frac{T}{A} \pm \frac{T.e.y}{I}$$

$$= \frac{1920000}{240000} \pm \frac{1920000 \times 100 \times 300}{72 \times 10^8}$$

$= 8 \pm 8$

$= 16\ \text{N/mm}^2$ top fibre

$=$ zero at bottom fibre.

At the mid-section, B.M. due to live load is

$$M_l = \frac{wL^2}{8} = \frac{60\,(6)^2}{8} = 270 \text{ kN-m}$$

$= 270 \times 10^6$ N-mm.

B.M. due to V is

$$M_T = -\frac{VL}{4} = -\frac{256000 \times 6000}{4}$$

$= -384 \times 10^6$ N-mm.

$\therefore$ Net B.M. $M = (270 - 384)\,10^6$

$= -114 \times 10^6$ N-mm

(*i.e.* hogging)

$$f_b = \mp\frac{My}{I} = \mp\frac{114 \times 10^6 \times 300}{72 \times 10^8}$$

$= \mp 4.75\ \text{N/mm}^2$.

Hence final stresses at mid-span section are

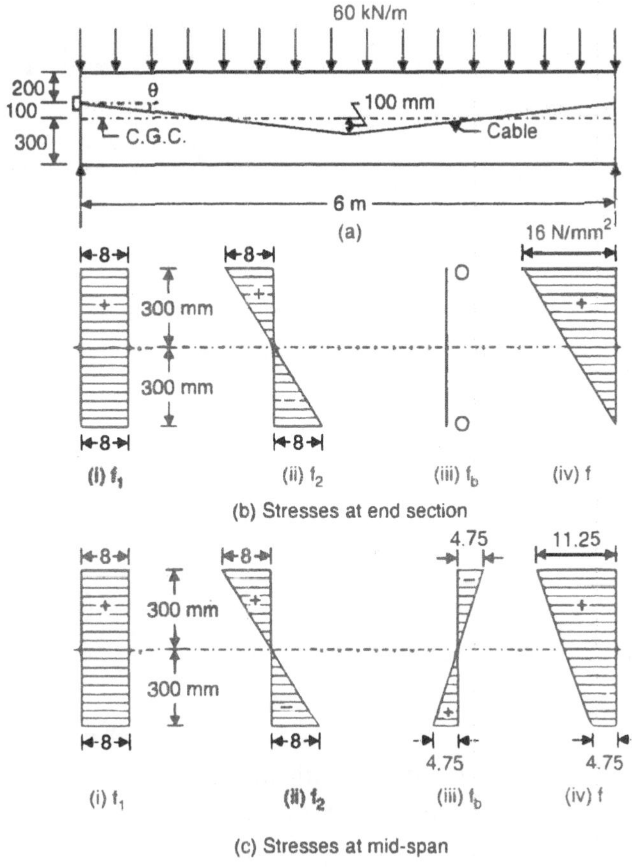

FIG. 25.12.

$$f = f_1 \pm f_2 \mp f_b = 8 \pm 8 \mp 4.75 = 11.25 \text{ N/mm}^2 \text{ (comp.) at top fibre and}$$

$$= 4.75 \text{ N/mm}^2 \text{ (comp.) at bottom fibre.}$$

Example 25.6. ***Load Balancing method.***

Solve Example 12.4 if the eccentricity of tendon is +100 mm at ends and +200 mm at mid-span, with straight line variation as shown in Fig. 25.13. The beam carries U.D.L. of 60 kN/m.

Solution : $\sin\theta \approx \tan\theta = \dfrac{10}{300} = \dfrac{1}{30}.$

$$V = 2\,T\sin\theta = 2 \times 1920000 \times \frac{1}{30} = 128000 \text{ N}$$

Horizontal component $= T\cos\theta \approx T$

At the end section, $e = 100$ mm below C.G.C.

$$\therefore \quad f = f_1 \mp f_2 = \frac{T}{A} \mp \frac{T \cdot e \cdot y}{I} = \frac{1920000}{240000} \mp \frac{1920000 \times 100 \times 300}{72 \times 10^8} = 8 \mp 8$$

$$= \text{zero at top fibre} = 16 \text{ N/mm}^2 \text{ at bottom fibre.}$$

Moment at mid-span, due to live load is

$$M_l = \frac{wL^2}{8} = \frac{60\,(6)^2}{8}$$

$$= 270 \text{ kN-m} = 270 \times 10^6 \text{ N-mm}$$

$$M_T = -\frac{VL}{4} = -\frac{128000 \times 6000}{4}$$

$$= -192 \times 10^6 \text{ N-mm}$$

$$\therefore \quad \text{Net } M = (270 - 192)\,10^6$$

$$= 78 \times 10^6 \text{ N-mm (sagging)}$$

$$f_b = \pm\frac{78 \times 10^6 \times 300}{72 \times 10^8}$$

$$= \pm 3.25 \text{ N/mm}^2$$

$\therefore$ Final stresses at mid-span section are

$$f = f_1 \mp f_2 \pm f_b = 8 \mp 8 \pm 3.25$$

$= 3.25$ N/mm² (comp.) at top fibre

$= 12.75$ N/mm² (comp.) at bottom fibre.

Compare the stress distribution with that of example 25.4.

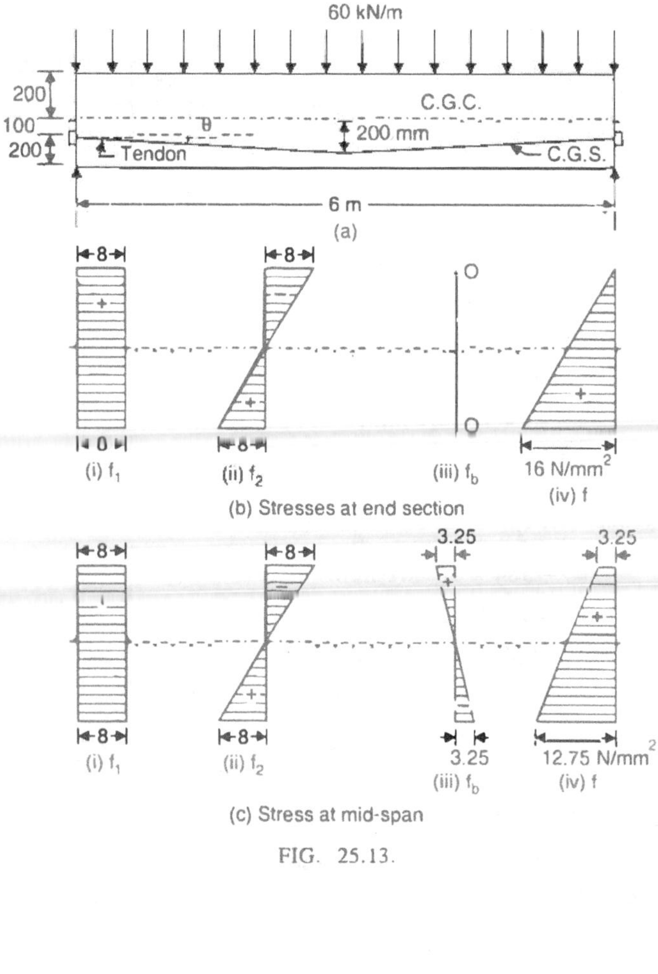

FIG. 25.13.

25.3. CLASSIFICATION AND TYPES OF PRESTRESSING

Prestressed concrete may be classified into a number of ways depending upon design, construction, method of applying prestressing and purpose of structure. Some of these are given below.

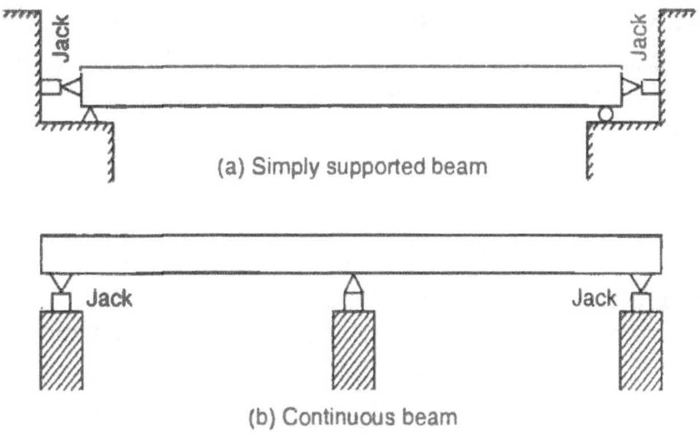

FIG. 25.14.

1. External or Internal Prestressing System : The prestress in a concrete structure is maintained either by adjusting its external reactions, or by a built-up internal thrust using high tensile steel. The *external prestressing,* though not common, is achieved by external reactions introduced through different support conditions. Fig. 25.14(a) shows a simply supported beam, externally prestressed by jacking against abutments. Fig. 25.14(*b*) shows a continuous beam prestressed externally by jacking the appropriate supports. External prestressing system is not common because it requires greater degree of accuracy in planning, maintenance and execution. *Internal prestressing* obtained by tensioning the steel reinforcement is the most common because greater degree of accuracy can be achieved in execution.

2. Linear or Circular Prestressing : Linear prestressing is applied in straight members such as beams, slabs, piles, electric poles etc. Circular prestressing is applied to circular structure such as tanks, pipes, bunkers, silos where the prestressing tendons are wound around in circles. However, the prestressing tendons in the linear prestressing need not be necessarily straight. They can be either bent or even curved, but they are not wound in round circles around the structure. The *Preload system* has been developed in America for prestressing circular structures.

3. Pretensioning and Post Tensioning System : Internal prestressing has two systems: pretensioning and post-tensioning, according to whether the steel is tensioned before or after casting the concrete.

(a) Pre-tensioning system (fully bonded construction) : It is that method of prestressing in which the tendons are prestressed before the concrete is placed. The tendons passing through the mould are temporarily anchored against some abutments. The tension is maintained when the concrete is placed. When the concrete is sufficiently hard, the ends of tendons are slowly released from anchorages, thereby transferring the prestress from steel to concrete *through bond.* Pretensioning is applied to relatively simple units precast in factories. In mass production, the wires may be stretched the full length of the casting shop (say 100 to 150 m) between anchorages rigidly held in the ground ; the wires pass through a whole series of 'moulds' as shown in Fig. 25.15. Each mould is mounted on rollers and spaced with small end gaps between adjacent moulds to allow the wires to be cut later and the individual

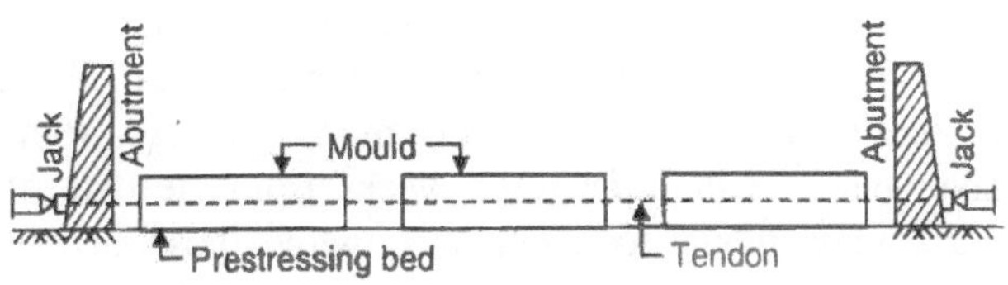

FIG. 25.15. PRESTRESSING BED

units freed. When the concrete in the mould has set, the wires are released from end anchorages, thereby transferring the prestress to concrete through bond. Since tendon force is transferred from steel to concrete through bond, it is desirable to use small diameter wires. The diameter of wires often used is 2 to 3 mm. They are sometimes twisted or notched to improve bond. When wires are released, their ends loose their stress (within bond length) and recover to their original diameter. This increase in diameter provides mechanical anchorage. The *bond length or transmission length* varies from 84 Φ (for 2 mm wires) to 120 Φ (for 5 mm wire), within which the stress in the wire increases from zero at free end to the maximum value at the end of bond length. If the profile of the tendon is curved, necessary profile fixing arrangements are needed.

(b) Post-tensioing system (end anchored construction) : In the post-tensioning system, the member is concreted and a duct is formed in the member either with a tube or with a metal sheath. When the concrete is sufficiently hard, tendon consisting of either a bar, cable or strand is threaded through the duct, tensioned and anchored to the concrete member at its ends. Thus the tendon force is transferred to the member through necessary anchorage wedges or similar blocks at the end of the member. In some cases the wires or cables are assembled and lie freely in tubes or ducts carefully located in the form work before the concrete is cast.

The cable ducts can be cast in the members to any desired curved paths, and by sweeping the cables up at the ends of the beams the eccentricity can be reduced as required to suit the tailing off of the bending moments ; the upward component of the cable force acts also to assist in carrying the end shear force. The anchoring of the wires to the concrete itself needs more care than the anchoring to bulk heads done in a pretensioning system. Many popular post-tensioning systems have been developed under patented propritories, notable among these being the Freyssinet system, Magnel system, Hobling system, Leonhardts system, Lee McCall system and the Ryerson BBRV system.

After applying the prestressing force to the tendons through jack and end anchorages, the ducts may be filled completely with cement grout injected under pressure. Such a system is known as fully bonded system. The wires in this system are protected against rusting. They are also doubly secured, primarily by mechanically anchorage devices and secondly can by bond by the grouting. It has been claimed that the anchorage can be safely released and re-used after grout has set ; however, invariably they are retained and concreted over for protection against corrosion. If the duct is not grouted, an unbonded system is obtained, in which the protection for the tendons from corrosion is provided by galvanising, greasing or some other means. In some cases, the bonded tendon may be purposely unbonded along certain portions of their length.

In post-tensioned structure, complications arise from friction of wires in the ducts, at the anchorage and in the jack itself at the time of prestressing. In pretensioned structure, a major loss is the elastic deformation of the concrete when prestress is first applied. Pretensioned work also suffers more severely from effects of shrinkage, about half of which takes place in the first three weeks from the date of casting. The total losses in post-tensioned work may amount to about 15% while those in pre-tensioned work may be as high as 25%.

4. Full Prestressing or Partial Prestressing : The concrete is said to be fully prestressed if there are no tensile stresses in it under working load. However, if some tensile stresses are produced in the concrete under working load conditions, it is said to be partially prestressed. In partially prestressed concrete, additional mild steel bars are frequently provided to reinforce the portion under tension.

25.4. PRESTRESSING SYSTEMS : END ANCHORAGES

A *prestressing system* comprises essentially a method of stressing the steel combined with a method of anchoring it to the concrete. A number of different systems and their patents for tensioning and anchoring the tendons exist. The practicing engineer who simply wants to design some structures of prestressed concrete, is free to specify and design for any system without studying the intrigues of patent rights. In fact, the royalty of the patent owner is indirectly included in the bid price for the supplying of prestressing steel and anchorages, which sometimes also includes the furnishing of equipment for prestressing and some technical supervision for jacking. Table 25.1 gives various types of linear prestressing systems. Some prominent and more common systems are discussed here.

1. **Pre-Tensioning Systems and End Anchorages** : The simplest pre-stressing system consists of two bulkheads anchored against the ends of a stressing bed. The tendons are pulled between the two bulk heads. In some cases, the bulk heads may be independent. In either case the bulk head and the bed must be strong enough to resist the force of pre-stress and its eccentricity. A prestressing bed is generally costly, but its use has two advantages:

(*i*) intermediate bulkheads can be inserted in the bed to use and tension shorter wires for casting shorter units and (*ii*) it can support vertical reactions, due to which prestressing of bent cables can be done.

For mass production of pre-tensioned members, Hoyer system is used, which is the extension of the above method. In this system, the wires are stretched the full length of the casting stop (say 100 to 150 m) between anchorages rigidly held in the ground. The wires pass through a whole series of moulds, as shown in Fig. 25.15. Each mould is mounted on rollers and spaced with small ends gaps between adjacent moulds to allow the wires to be cut later and the individual units freed. When the concrete in the mould has set, the wires are released from end anchorages, thereby transferring the prestress to concrete through bond.

The anchoring devices for holding pre-tensioning wires to the bulkheads are made on the *wedge and friction* principle. One common device consists of a split cone wedge, which is made from a tapered conical pin. The tapered conical pin is drilled axially and tapped, and then cut in half longitudinally to form pair of wedges [Fig. 25.16(*a*)]. The anchoring block also has a conical hole, in which the tapered conical pin holds the wires. These grips can be used for single wires as well as for twisted wire strands.

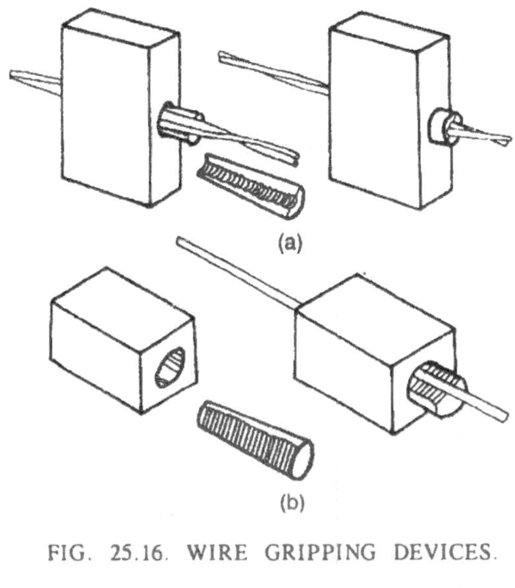

FIG. 25.16. WIRE GRIPPING DEVICES.

TABLE 26.1 LINEAR PRESTRESSING SYSTEMS

Type (1)	*Classification* (2)	*Description* (3)		*Name of system* (4)	*Country of origin* (5)
Pre-tensioning	*Method of stressing*	(*i*) Against buttresses or stressing beds		Hoyer	Germany
		(*ii*) Against central steel tube		Shorer Chalos	U.S. France
		(*iii*) Continuous stressing against moulds		Continuous wire winding	U.S.S.R.
		(*iv*) Electric current to heat steel		Electro-thermal	U.S.S.R.
	Method of anchoring	During Pre-stressing	(*i*) Wire	Various wedges	
			(*ii*) Strands	Strandwise	U.S.
		For transfer of prestress	(*i*) Bond, for strands and small wires		Europe, U.S.
			(*ii*) Corrugated clips, for big wires	Dorland	U.S.
Post-tensioning	*Methods of stressing*	(*i*) Steel against concrete		Most systems	Germany / U.S.
		(*ii*) Concrete against conerete		Leonhardt/Billner	
		(*iii*) Expanding cement			France
		(*iv*) Electrical Prestressing		Lossier	U.S.
		(*v*) Bending steel beams		Biliner Preflex	Belgium
	Methods of anchoring	(*i*) *Wires*, by frictional grips		Freyssinet Magnel Morandi Holzmann Preload	France Belgium Italy Germany U.S.
		(*ii*) *Wires*, by bearing		B.B.R.V. Prescon	Switzerland U.S.
		(*iii*) *Wires*, by loops and combination of methods		Texas, P.I. Biller Monier bau	U.S. U.S. Germany
		(*iv*) *Bars*, by bearing and by grips		Leoba Leonhardt Lee Mc Call Stress-steel	Germany Germany England U.S.
		(*v*) *Strands*, by bearing		Stress rods Dywidag Wet's	U.S. Germany Belgium
		(*vi*) *Strands*, by friction grips		Bakker Roebling Wayss and Freytag CCL Freyssinet Anderson Atlas	Holland U.S. Germany England U.S. U.S. U.S.

In some cases, the pin is not drilled, but is cut in half longitudinally and the flat surface is machined and serrated, as shown in Fig. 25.16(*a*). In some cases, quick release grips, which are more complicated and costly, are used–specially when wires are to be held in tension only for short periods

As stated earlier, the dependence on bond to transmit prestress between steel and concrete necessitates the use of small wires to ensure good anchorage. When the wires are relased, their ends lose their stress and recover to their original diameter, thus providing mechanical anchorage. If the *bond length or transfer length* (Fig. 25.17) is insufficient, specially when cracks occur near the end of a beam, the bond may be broken and the wires may slip. A more reliable method is to add mechanical end anchorages to the pre-tensioned wires. For example, *Dorland anchorage,* consisting of clips, can be gripped to the tendon under high pressure, and the edges of the clips can then be welded together at several points. If such mechanical anchorages are used, tendons of greater diameter can be permitted.

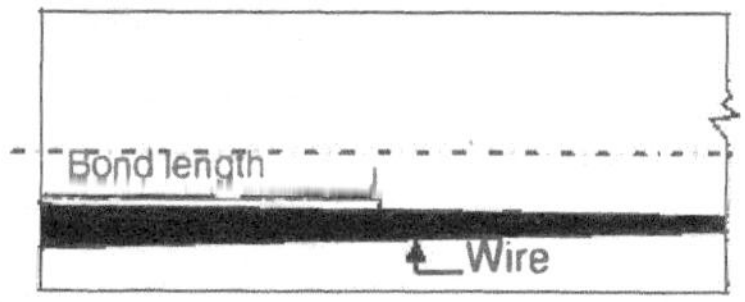

FIG. 25.17

2. Post-Tensioning Systems : Various post-tensioning systems can be classified under four groups : (*a*) mechanical prestressing, using jacks, (*b*) electrical prestressing by application of heat, (*c*) chemical prestressing by use of expanding cement, and (*d*) miscellaneous methods.

(A) MECHANICAL PRESTRESSING : This is the most common method, in which the stretching of wire is done by means of hydraulic jacks, screw jacks or by a system of levers. For large prestress, hydraulic jacks are used having capacity between 30 to 1000 kN. Systems of jacking vary from pulling one or two wires upto several hundred wires at a time. For example, the Magnel system employs a hydraulic jack which pulls two wires at a time while the Freyssinet double acting jack can pull up to 18 wires or 12 strands at a time. Pressure gauges for jacks are calibrated either to read the pressure on the piston or to read directly the amount of tension applied to the tendon. Usually, the elongation of steel is measured to check against the gauge readings. When several tendons in a member are to be tensioned in succession, care should be taken to pull them in the proper order so that no serious eccentric loading will result during the tensioning process.

Before the pull in the jack is released, the wires are anchored in their extended position against the end of concrete, by some wedging device. There are three basic principles or methods by which steel wires are anchored to concrete after stretching :

(*i*) By method of wedge action, producing a friction grip on the wires ;

(*ii*) By method of direct bearing from rivet or bolt heads formed at the end of wires; and (*iii*) By method of looping the wires around concrete.

Variations on the wire arrangement and jacking and anchorage devices constitute the difference between various systems. A number of stressing systems are now available, each varying detail and protected by patent rights. Some of the common systems are described below, in brief.

(*i*) **Freyssinet system.** This system was the first method of post-tensioning to be developed. The method, developed by French engineer Freyssinet is one of the most widely used method of prestressing. It uses high tensile steel wires of 5 mm dia (0.196") in units of 8, 10, 12 and 18, and 7 mm (0.276") arranged in 12 wires units. The wire units, commonly known as cables are encased in flexible tube or sheething of 32 gauge metal sheet, with a helical spring inside the wires. The spring keeps a proper spacing between the wires and forms a channel for cement grout. The wires are anchored by being held between two reinforced concrete cones which fit one inside the other, as shown in Fig. 25.18. The female part of the anchorage is a conical steel wound lining heavily reinforced with high tensile steel spirals to resist bursting force. The male cone is of mesh reinforced concrete, fluted to space evenly the requisite number of wires. A central tube pass axially through the male cone. This tube permits the grout to be injected through it. All the wires are stressed simultaneously by means of Freyssinet double acting jack, which can pull upto 18 wires at a time. Wires are wedged around the jack casting and are stretched by means of main ram that reacts against the embeded anchorage [Fig. 25.19(*a*)]. When the required tension is reached the inner piston pushes the plug into the anchorage to secure the wires. The pressure on the main ram and that on the inner piston are then released gradually and the jack is removed. Since the anchoring unit is buried flush with the face of the concrete, it helps to transmit the reaction of the jack as well as the prestress of the concrete. After the completion of prestressing, grout is injected through the hole at the centre of male cone. Fig. 25.19(*b*) shows the concreted face.

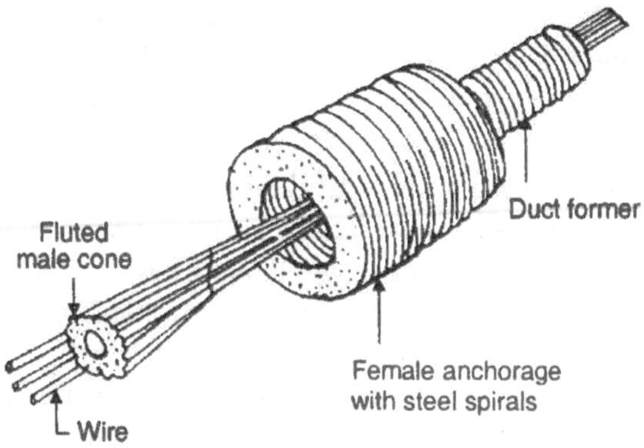

FIG. 25.18. FREYSSINET SYSTEM.

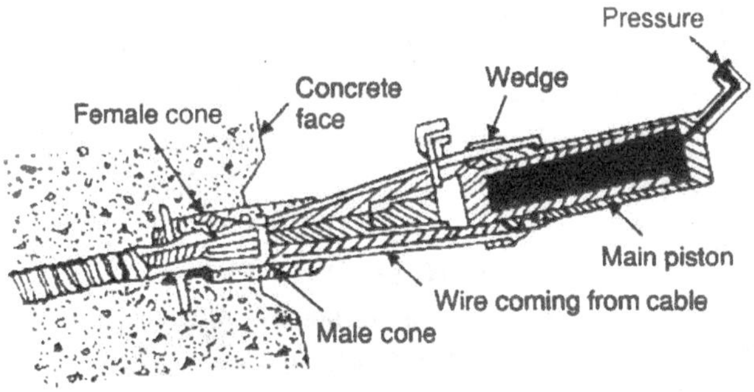

FIG. 25.19(*a*) PRESTRESSING JACK

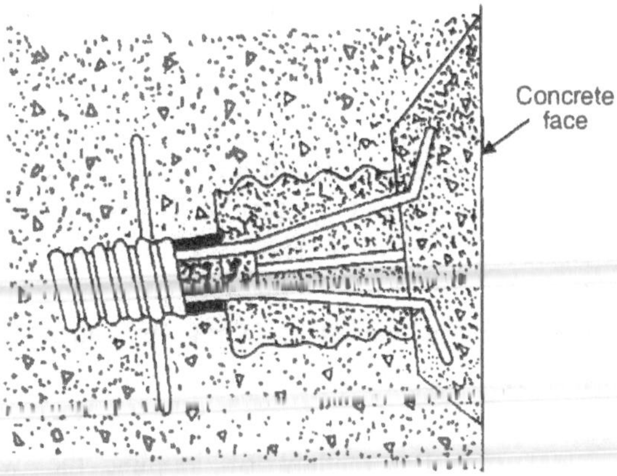

FIG. 25.19(*b*). FINAL CONCRETED FACE.

Table 25.2 gives the details about standard size wires.

TABLE 25.2

Types of cables	*8 × 5 mm*	*10 × 5 mm*	*12 × 5 mm*	*18 × 5 mm*
Steel area	1.57 cm^2	1.96 cm^2	2.35 cm^2	3.53 cm^2
Weight	1.2 kg/m	1.5 kg/m	1.8 kg/m	2.7 kg/m
O.D. metal hose	28.6 mm	28.6 mm	31.5 mm	41.3 mm
I.D. metal hose	25.4 mm	25.4 mm	28.6 mm	38.1 mm

(*ii*) **Magnel Blaton system.** This system was developed by Prof. Magnel and contractors Blaton, both of Belgium. The unit of stretching is two wires at a time, as against 8 to 18 wires in the Freyssinet system. The cable consists of any even number of wires upto 64, using high tensile steels wires of 5 mm (0.2") dia. or 7 mm (0.276") dia. The anchorage device consists of steel sandwich plates having grooves to hold the wires and wedges which also are grooved. Each wire is separated from neighbouring wire in the same layer and those in the adjacent layers by a space of 0.75 mm. Each layer holds four wires. Sixteen such layers can be used, making a total of 64 wires. The wire pattern is maintained throughout the length of cable by means of grills (spacers) spaced at regular interval of 1.5 to 2.5 m along it. These grills offer little frictional resistance to the wires which are free to move relative to each other during tensioning. The steel plate wedges are about 25.4 mm thick, with two wedge shaped grooves in their upper and lower faces. The sandwich plates are placed in the form of a bank, one above the other against a distribution plate and held thus by means of temporary bolted clamp during tensioning operations.

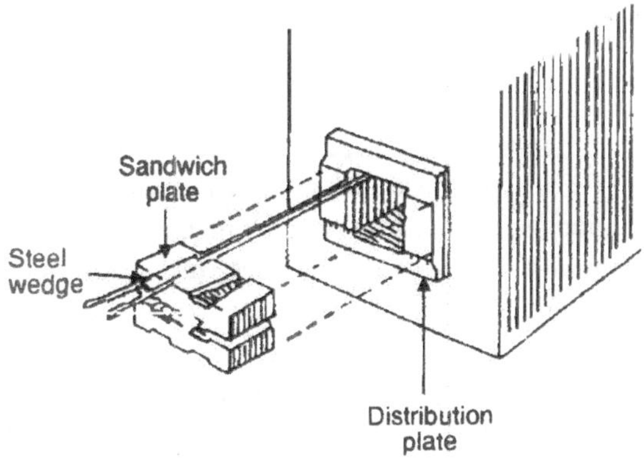

FIG. 25.20. MAGNEL BLATON SYSTEM.

Regular ducts are cast at suitable places along the length of the beam by introducing rubber cores in the mould. The rubber cores have an initial diameter of 3 cm. Steel tubes are introduced in the holes, to make the core rigid. The rubber cores are pulled out after 6 to 8 hours of concreting. Even after 1 or 2 days, the cores have been pulled out without any difficulty. Ducts are thus formed; the wires are introduced in these ducts only at the time of prestressing. Thus, the Magnel Blaton System results in the saving of cost of sheathing used in Freyssinet system. The distribution plates are either cast into the member at the extremity of the duct or placed on thin mortar beds after the concrete has hardened, prior to assembling the anchorage.

TABLE 25.3 (*a*) DETAILS FOR 5 mm WIRE SYSTEM

No. of wires in cable	*End plate*		*Duct hole*	
	Length (cm)	*Width (cm)*	*Width (cm)*	*Depth (cm)*
16	13	13	5.5	5.5
24	16	16	5.5	7.5
32	20	17	5.5	10.0
40	23	18	5.5	12.5
48	26	19	5.5	15.0
56	29	20	5.5	17.5
64	32	21	5.5	20.0

TABLE 25.3 (b) DETAILS FOR 7 mm WIRE SYSTEM

No. of wires in cable	*End plate*		*Duct hole*	
	Length (cm)	*Width (cm)*	*Width (cm)*	*Depth (cm)*
16	15	20	6.5	6
24	18.5	26	6.5	9
32	24	26	6.5	13
40	29	27	6.5	15
48	34	28	6.5	18
56	37	29	6.5	19
64	39	30	6.5	22

For stressing the wires, hydraulically operated stressing jacks are used, with arrangement of holding two wires at a time. The jack takes its support on the end plate of suitable size. When tensioning is completed, the wedge for two wires is driven home. When stressing of one cable is completed, it is invariable grouted and the anchorage is covered with concrete for protection.

Table 25.3 gives the dimensions of the end plate and duct hole in the Magnel Blaton System.

Fig. 25.21 shows the details of the sandwich plate and the wedge. Table 25.4 gives the values of width *a*, length *e* and other dimensions of the plate and the wedge.

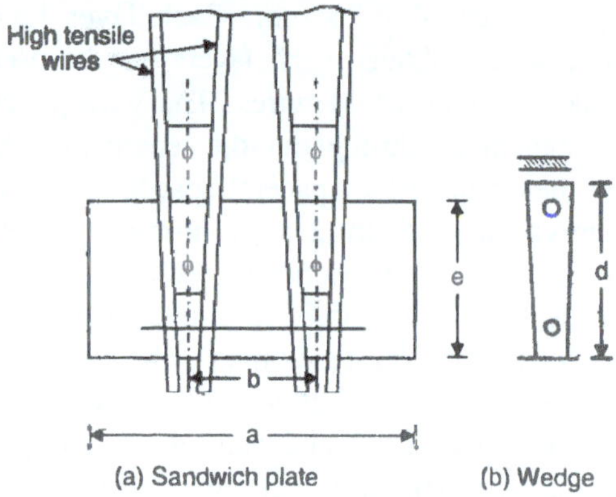

FIG. 25.21. SANDWICH PLATE AND WEDGE.

TABLE 25.4

Dia. of wire	*Dimensions (cm)*				
	a	*b*	*c*	*d*	*e*
5 mm	11.4	4.0	2.1	6.0	5.4
7 mm	16.0	5.0	2.5	10.4	8.9

(*iii*) **Gifford Udall system.** Gifford Udall system is one of the most widely used systems, using 7 mm wires. The cable consists of parallel wires which are separated by means of circular spacers placed at one metre c/c. The diameter of spacer is 39 mm for cable upto 8 wires and of 51 mm for cable upto 12 wires. The system uses two types of anchorage systems: (*a*) plate anchorage (*b*) tube anchorage.

(*a*) ***Plate anchorage.*** Fig. 25.22 shows the plate anchor system originally used by Gifford Udall system. It essentially consists of a bearing plate, a thrust ring anchor grips and a steel helix. The wires are stressed and anchored individually by small wedge type Udall grips seating against a bearing plate. The thrust ring is cast into

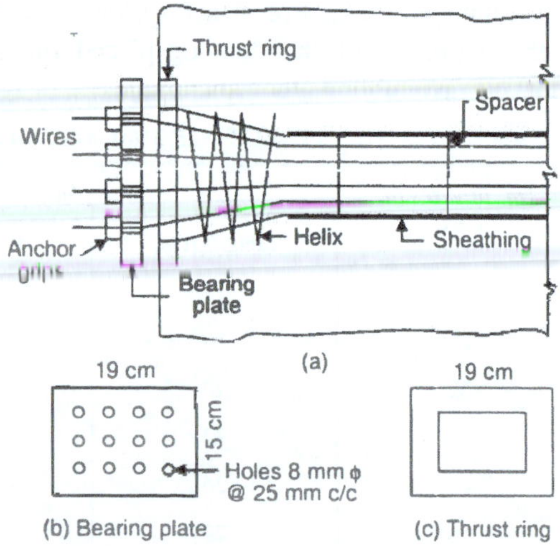

FIG. 25.22. PLATE ANCHORAGE WITH GRIPS

the concrete. The bearing plate locates against the thrust ring. The end of the dust is encircled by a helix. The anchor grips consist of an outer barrel and two semi circular wedges which fit in the tappered hole of the barrel. The wire is gripped between the serratted parallel faces of the wedge.

(*b*) ***Tube anchorage***. This is the most recent anchorage system in which the wedges fit directly into the tappered recesses in the bearing plate which is thick. Thus the need for separate grips is eliminated (Fig. 25.23). The bearing plate includes a threaded hole for grouting purposes. The bearing plate seats against a *tube unit* incorporating the thrust ring and helix in a single element which bolts into the form work and is cast into the concrete. The tube unit provides the necessary spreading out of the wire in the cable so as to pass through the hole of the bearing plate. This type of anchorage is available for 8 wire and 12 wire cables. The jack pulls one wire at a time.

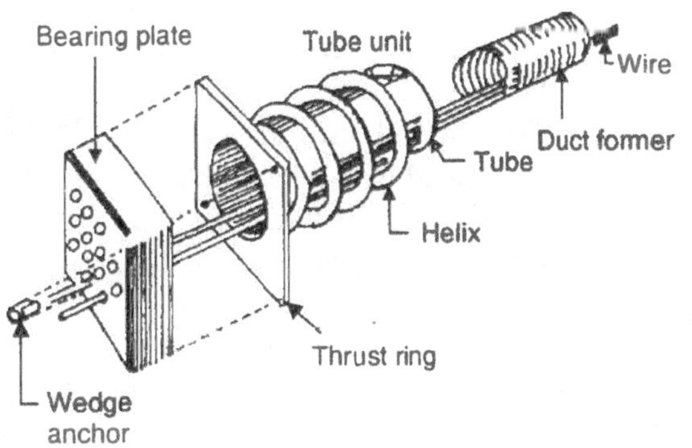

FIG. 25.23. GIFFORD UDALL TUBE ANCHORAGE SYSTEM.

(*iv*) **P.S.C. Mono-wire system**: In this system, wires are tensioned individually. The system uses *collect sleeves* wedging in conical holes. These sleeves are of single piece serrated internally to give positive grip to wire passing through them. A steel trunked guide leads each wire from its cable position to its point of anchorage through a gentle curvature (Fig. 25.24). The mono-wire anchorage are available for 1-, 2-, 4-, 8-, and 12 wire cables of size upto 7 mm. For 8- and 12-wire cable, a high strength plastic spacer is used to make up the cable and separate the wires for stressing.

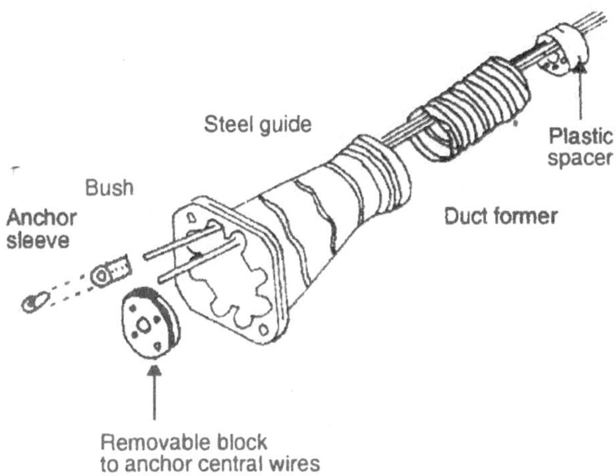

FIG. 25.24 P.S.C. MONO-WIRE SYSTEM.

(*v*) **Lee McCall system**

Lee-McCall system uses high-tensile alloy steel bars as the prestressing tendons, in the place of high-tensile steel wires used in other systems. *Macalloy steel* used in the system is an open-hearth Silico-Manganese steel, hot rolled into bars and subsequently processed to give the required physical properties. The diameter of the rods may very from 12 to 28 mm. The rods are threaded at the ends. Holes are made in the member by means of rubber core. When the concrete has set, bars are introduced in the hole, after removing the core. After the desired stretching, a nut is tightened at its end to prevent its return to original length. Fig. 25.25 shows the details of the system. The bars can be either bonded or unbonded to the concrete.

(*B*) ELECTRIC PRESTRESSING

This is essentially a post-tensioning method in which bars are stretched by means of heating these using electric current. Thus the use of jack is eliminated altogether. The method uses smooth reinforcing bars coated with thermoplastic material such as sulphur, before embeding them in concrete. After the concrete has set, a low voltage high amperage current is passed through the bar to heat it upto 170° C. The sulphur melts and allows steel to expand. The nuts on the protruding ends of the bars are then tightened against heavy washers. On cooling, the sulphur solidifies and restores the bond. The method did not find commerical application with high tensile steel because a much higher temperature would be required for its prestressing.

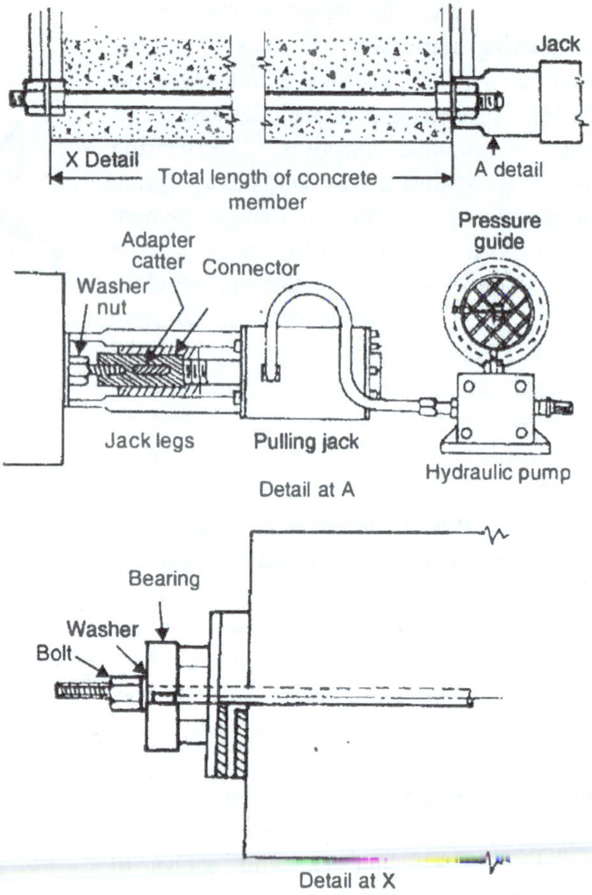

FIG. 25.25. LEE-MCCALL SYSTEM.

(*a*) Chemical prestressing or self-stressing : In this method, *self stressing cement* is used that expand chemically after setting and during hardening. When steel is embeded in such a concrete, it is elongated by the expansion of concrete, and thus gets prestressed. The steel, in turn, produces compressive prestress in concrete. If the concrete made with the expanding cement is un-restrained, the amount of expansion produced by chemical reaction between cement and water could be between 3 to 5%, resulting in disintegration of concrete by itself. However, if the concrete is restrained internally (with steel) or externally (by other means), the expansion would be controlled, resulting in compressive prestress of the concrete. Such cement has been successfully applied for many interesting projects in France. Extensive applications of expansive cement may not be too far in future.

25.5. LOSSES OF PRESTRESS

The prestressing force applied to the member does not remain constant, but gradually reduces. The amount of force which gets reduced is known as *loss of prestress* and may vary from 15 to 20%. The losses in prestress may be due to the following reasons :

(*i*) loss due to elastic deformation of concrete when prestress is applied to it. (*ii*) loss due to creep of concrete, (*iii*) loss due to shrinkage of concrete, (*iv*) loss due to relaxation of steel, (*v*) loss due to deformation and slip of anchorage and (*vi*) loss due to friction between tendon and concrete.

1. Loss due to elastic deformation of concrete : This loss takes place only in pretensioned beams. When prestress is transferred to concrete, elastic stress and strains are induced in it. Due to this the concrete member gets shortened, alongwith shortening of steel, thereby reducing the prestress in steel. Let us first consider a uniformly prestressed beam.

Let P_i = initial prestress force.

A_e = area of tranformed section $= A_c + m\,A_{st}$

A_{st} = area of tensioned steel.

Unit shortening $\delta = \dfrac{P_i}{A_c\,E_c + A_{st}\,E_s}$

$\therefore$ Loss of prestress in steel is

$$\Delta f_{el} = E_s\,\delta = \frac{E_s\,P_i}{A_c\,E_c + A_{st}\,E_s} = \frac{m\,P_i}{A_c + m\,A_{st}} = \frac{m\,P_i}{A_e} = m\,\frac{f_{si}\,A_{st}}{A_e} \quad \ldots(25.5)$$

where f_{si} = stress in steel at the time of initial tension.

The loss due to elastic deformation due to concrete may range from 3 to 6% in pre-tensioned beams. Let us now consider a non-uniformly prestressed beam in which the prestressing force (P_i) has been applied at an eccentricity e. The stress in concrete at any height y from the centroidal axis is

$$f_{ci} = \frac{P_i}{A_e} \pm \frac{P_i\,.\,e}{I_e}\,y$$

The effective prestress in steel will be

$$f_{se} = f_{si} - m\,f_{ci} = f_{si} - m\left(\frac{P_i}{A_e} + \frac{P_i\,.\,e}{I_e}\,y_t\right)$$

where y_t = distance of tensile steel from centriodal axis.

Hence the loss of prestress in steel is

$$\Delta f_{el} = f_{si} - f_{se} = m\left[\frac{P_i}{A_e} + \frac{P_i\,.\,e}{I_e}\,y_t\right] = m\left[\frac{f_{si}\,A_{st}}{A_e} + \frac{f_{si}\,A_{st}}{I_e}\,.\,e\,.\,y_t\right] \quad \ldots(25.6)$$

Values of Moduli of Elasticity : The value of the modulus of elasticity of steel (E_s) used for the design of prestress concrete should preferably by determined by tests on samples of steel to be used for construction. Where it is not possible to ascertain the modulus of elasticity by test or from the manufacturer of steel, the following values may be adopted :

Type of Steel	E_s (N/mm^2)
Plain-drawn wires (cold drawn or cold worked)	2.1×10^5
High tensile steel bars (rolled or heat treated)	2.0×10^5

The modulus of elasticity of concrete may be assumed to have the values given by the following expression :

$$E_c = 5700\sqrt{f_{ck}}\ (\text{N/mm}^2) \qquad \ldots(25.7)$$

where f_{ck} is the characteristic cube strength of concrete.

Post-Tensioned Beams : In the case of post tensioned beam with a single concentric tendon, as the tendon is stretched and anchored against concrete, all the elastic strain has occurred in concrete when the jacking force is reached in tendon. Since the force in the cable is measured after the elastic shortening of the concrete has taken place, no loss in prestress due to this need be accounted for. However, in actual practice there are more than one tendon. If the tendons are stressed in succession, the prestress is applied gradually. Due to this, the tendon that is stressed first will suffer the maximum loss while the one that is tensioned last will not suffer any loss. Accurate calculation for losses is complicated, but for all practical purposes, it may be assumed that average loss of each cable is equal to half the loss in first cable.

In order to counteract elastic loss all the tendons are anchord at an *overstress* equal to the average elastic loss in practice. The average loss in post-tensioned beam may be of the order 40%.

2. Loss due to creep of concrete : Creep in concrete is defined as its time-dependent deformation resulting from the presence of stress. It is the plastic flow of concrete under compression. Creep strain varies with the intensity of stress, and is about two to three times the elastic strain. The rate of creep is high initially, and then decreases as time increases. The creep increases with higher water-cement ratio and with a lower aggregate cement ratio. It varies inversely with strength of concrete. It also depends upon the humidity of surrounding atmosphere and the strength at the time of loading. It is known that the failure of early efforts at prestressing was attributed largely to the lack of knowledge concerning creep in concrete. It is still one of the main source of loss, and a serious one, if the prestress in the steel is low and the compression in concrete is high. Pre-tensioned members have more loss than post-tensioned ones, because transfer of prestress usually takes place earlier in pre-tensioned members.

The ratio of the final creep strain to the elastic strain in concrete is defined as *creep coefficient* (C_c) the values of which are given in Table 25.5 for different atmospheric conditions. Creep strain (ε_{cr}) is given by

$$\varepsilon_{cr} = (C_c - 1)\,\varepsilon_{el} \qquad \ldots(25.8)$$

where ε_{el} = elastic strain in concrete.

TABLE 25.5 CREEP COEFFICIENT C_c

Condition of exposure	*Final creep coefficient* (C_c)
In water	0.5 K_c to K_c
In very humid atmosphere	1.5 K_c to 2.0 K_c
In average humidity	2.0 K_c to 3.0 K_c
In dry atmosphere (*i.e.* in dry internal rooms)	2.5 K_c to 4.0 K_c

The value of K_c depends upon the ratio of cube strength (f_{cr}) of concrete at transfer of prestress on 15 cm cube and cube strength (f_{cu}) of concrete at 28 days where Portland cement is used and 7 days where rapid hardening cement is used. Its value may be taken from Table 25.6

TABLE 25.6 VALUES OF K_c

f_{cr} / f_{cu} *at the time of stressing*	K_c
0.50	2.20
0.84	1.50
1.00	1.00
1.08	0.75
1.30	0.50

The loss of prestress due to creep of concrete under load should be determined for all the permanently applied loads including the prestress. The creep loss due to live load stresses, erection stresses and other stresses of short duration may be ignored. The loss of prestress due to creep of concrete is obtained as the product of modulus of elasticity of prestress steel (E_s) and the integrated ultimate creep strain of concrete fibre along the line of centre of gravity of the prestressing steel over its entire length. The total creep strain during any specific period should be assumed, for all practical purpose, to be the creep strain due to a sustained stress equal to the average of the stresses at the beginning and end of the period.

As per IS : 1343-1980, the ultimate creep strain may be estimated from the following values of creep-coefficient (that is, ultimate creep strain/elastic strain at the age of loading).

Age of Loading	*Creep coefficient*
7 days	*2.2*
28 days	*1.6*
1 year	*1.0*

The above ultimate creep strain does not include the elastic strain. For the calculation of deformation at some stage before the total creep is reached, it may be assumed that about half the total creep takes place in the first month after the loading and that about three quarters of the total creep takes place in the first six months after loading.

The loss in prestress due to creep may range from 5 to 10%.

3. Loss due to shrinkage of concrete : Shrinkage in concrete is its contraction due to drying and chemical changes. It depends upon the quantity of water, type of aggregates used in the mix and surrounding atmospheric conditions. If minimum shrinkage is desired, the water cement ratio and the proportion of cement paste should be kept to a minimum. Aggregates of larger size, well graded for minimum void, need a smaller amount of cement paste, and shrinkage will be smaller. Harder and denser aggregates of low absorptions and high modulus of elasticity will exhibit small shrinkage. Shrinkage is relatively small for cements high in tricalcium silicate and low in the alkalies and the oxides of sodium and potassium. Shrinkage is also time dependent, though bulk of the shrinkage takes place in the early days of the hardening of concrete. In pre-tensioned work, wires are subjected to some compression due to creep of concrete, before the release of wires. The concrete around the wires is subjected to tension as the anchored wires are not free to shorten its length. The concrete thus tends to slip in the early stages and develop fine cracks.

When the wires are released, these cracks close, causing a loss of prestress. After that, the time dependent shrinkage continue to shorten the length of member, though at a much reduced rate. In post-tensioning work, shrinkage takes place unhampered and hence loss of shrinkage is relatively small. As per IS : 1343-1980, the approximate value of shrinkage strain for design is assumed as follows :

$$\text{For pre-tensioning} = 0.0003$$

$$\text{For post-tensioning} = \frac{0.0002}{\log_{10}(t+2)} \qquad \ldots(25.9)$$

where t = age of concrete at transfer, in days.

The value of shrinkage strain for design of post-tensioned concrete may be increased by 50 percent in dry atmospheric conditions, subject to a maximum value of 0.0003. Thus if ε_{sh} is the shrinkage strain, the loss of prestress ΔF_{sh} due to shrinkage is given by

$$\Delta f_{sh} = \varepsilon_{sh} \cdot E_s \qquad \ldots(25.10)$$

The loss of prestress due to shrinkage may be of the order of 4 to 6% for pre-tensioned work and 3 to 4% for post-tensioned work.

4. Loss due to relaxation of steel. When the stress in steel is more than half the yield stress, there is creep of steel also. It is sometimes measured by the amount of lengthening when maintained under a constant stress for a period of time. Relaxation varies with steel of different compositions and treatments. The percentage of creep increases with increasing stress, and when a steel is under low stress, the creep is negligible. While creep in steel is a function of time, there is evidence to show that under the ordinary working stress for high-tensile steel, creep takes place mostly during of the first few days.

The magnitude of the loss of prestress due to relaxation of steel may be taken to vary from 2 to 8 percent to the average initial stress depending on the quantity of the steel and the order of the intitial prestress. Temporary overstressing for the purpose of the reducing the loss of prestress due to relaxation of steel may be permitted provided such overstressing does not exceed 85% of the ultimate tensile strength (f_{su}) of the prestressing steel.

5. **Loss due to deformation and slip of anchorage.** This loss is applicable in post-tensioning system in which, after tensioning a tendon, the jack is released to transfer prestress to concrete by bearing of the anchorage. During the transfer, the anchorage fixures deform under stress, and the friction wedges used in several system of prestressing slip slightly before wires are firmly gripped. The amount of slippage (δ_s) depends on the type of wedge and stress in the wires, and an average value of 2 to 3 mm may be taken. For heavy strands, the slip may be 5 mm. Generally, manufacturers supply the value of such slip. In the case of Freyssinet cones, the slip is 6 mm for 5 mm of wires and 9 mm for 7 mm of wire cables. If the release of strain is assumed to be uniform throughout the length of wire, the loss of prestress is given by

$$\Delta f_{slip} = E_s \frac{\delta_s}{L} \qquad \ldots(25.11)$$

where L is the length of wire. It is necessary to note that this loss of prestress depends upon the length of the tendons, being high when a tendon is short and low when a tendon is long.

Due to friction forces between the piston and cylinder etc. of the jack, there may be some loss of stress in jack itself. Due to this, the actual force in the wire will be less than the one registered by the gauge of the jack. The value of this loss depends upon the type of jack. For Freyssinet jack, it is about 36 kN.

Any loss of prestress which may occur due to slip of wires during anchoring or due to the strain of anchorage may be allowed for in the design. Loss due to slip in anchorage may be allowed for in the design. Loss due to slip in anchorage is of special importance with short members and the necessary additional elongation should be provided for at the time of tensioning to compensate for this loss.

6. Loss due to friction. During tensioning operations in post-tensioned members, there will be considerable movement or sliding of tendon relative to the surrounding. Since the tendon is in direct contact with the duct or with spacers provided, the friction will cause a reduction in the prestressing force as the distance from the jack increases. The frictional loss can be conveniently considered in two parts :

(*a*) the curvature effect and (*b*) length or wobble effect.

(*a*) ***Curvature effect.*** The loss of prestress due to curavature effect results from the intended curvature of the tendons. The loss depends upon :

(*i*) coefficient of friction between the contact materials.

(*ii*) Stress in the tendon and (*iii*) total change in the angle.

Fig. 25.26 shows a small length dx of a prestressing tendon whose centroid follows the arc of a circle of radius R. The change $d\alpha$, as the tendon goes around the length dx is

$$d\alpha = \frac{dx}{R} \qquad \ldots(i)$$

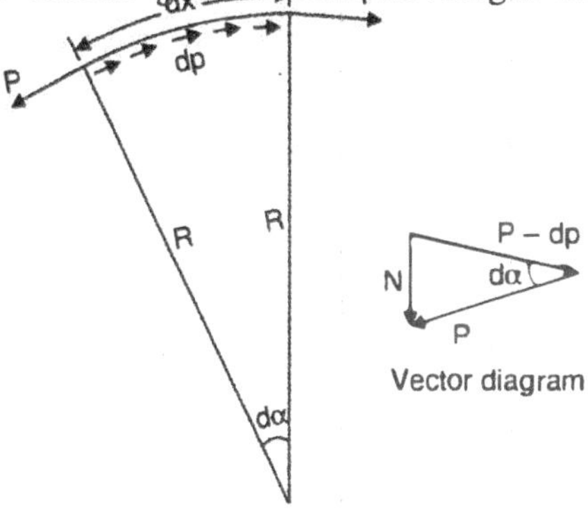

FIG 25.26.

The normal component of pressure produced by stress P bending around an angle $d\alpha$ is given from the vector diagram

$$N \approx P \,.\, d\alpha = P \,.\, \frac{dx}{R} \qquad \ldots(ii)$$

Hence frictional loss dP around the length dx is

$$dP = -\mu\, N = -\mu\, P d\alpha$$

where μ is the coefficient of friction.

(The minus sign indicates that P decreases as α increases.)

$$\frac{dP}{P} = -\mu\, d\alpha \qquad \ldots(iii)$$

If P_1 is the force at one end, and P_2 is force at the other end of the curved cable, subtending an angle α we get, by integrating Eq. (*iii*)

$$\Big[\log_e P\Big]_{P_1}^{P_2} = -\Big[\mu\,\alpha\Big]_0^{\alpha} = -\mu\,\alpha$$

or

$$P_2 = P_1 \,.\, e^{-\mu\alpha} \qquad \ldots(25.12)$$

The loss of force $(P_1 - P_2)$ is called *loss due to curvature.*

Eq. 25.12 may also be written in the alternative from :

$$P_2 = P_1 . e^{-\mu L/R} \quad \text{...[25.12(a)]}$$

since $\alpha = L/R$ for a length L of constant radius R.

For tendons with a succession of curves of varying radii, it is necessary to apply this formula to the different sections in order to obtain the total loss.

(b) Wobble or length effect. In post-tensioning there is another loss known as *length, wobble* or *wave loss.* This is due to the cable touching the duct or surrounding concrete. This is mainly caused by the deviation of the duct from its intended depth. The length effect is the amount of friction that would be encountered if the tendon is a straight one, that is one that is not purposely bent or curved. This loss is dependent on the length and stress of tendon and coefficient of friction between the contact materials, and the workmanship and method used in aligning and obtaining the duct. The prestressing force P_2 at the other end can be computed by substituting a factor $K_f L$ in the place of $\mu\alpha$ in Eq. 25.12 :

$$P_2 = P_1 e^{-K_f L} \quad \text{...(25.13)}$$

Combining both the effects, we have

$$P_2 = P_1 e^{-(\mu \alpha + K_f L)} = P_1 e^{-\left[\frac{\mu L}{R} + K_f L\right]} \quad \text{...(25.14)}$$

If the total difference in tension in the tendon at the start and that at end of of the curve is not excessive (say not more than 15 to 20%) the following approximate formula may be used :

$$\left(\frac{P_2 - P_1}{P_1}\right) = -\left(K_f + \frac{\mu}{R}\right) L \quad \text{...(25.15)}$$

or

$$\frac{P_1 - P_2}{P_1} = (K_f L + \mu\, \alpha) \quad \text{...[25.15(a)]}$$

Strain loss,

$$\varepsilon_f = \frac{P_1 - P_2}{E_s . A_{st}} \quad \text{...[25.15(b)]}$$

As per IS : 1343-1980, for straight or moderately curved structures with curved or straight cables, the value of prestressing force P_x at a distance x from the tensioning end may be determined by the formula, $P_x = P_0\, e^{-(\mu\, \alpha + k\, .\, x)}$ ...[25.14(b)]

where P_0 = Prestressing force in the prestressed steel at the tensioning end.

For small values of $(\mu\, \alpha + kx)$, the above equation may be reduced to the form

$$P_x = P_0 (1 - \mu . \alpha - k . x)$$

where α = the cumulative angle in radians through which the tangent to the cable profile has turned between any two point under consideration.

k = coefficient of wave effect, varying from 15×10^{-4} to 50×10^{-4} per metre.

μ = coefficient of friction in curve, to be taken as follows :

= 0.55 for steel moving on smooth curve,

= 0.30 for steel moving fixed to duct,

= 0.25 for steel moving on lead.

Methods of reducing friction loss : The friction loss can be reduced by the the following methods :

1. Passing the cables through metal tubes at bends

2. Making the bends through as small an angle as possible.

3. Making the radius of curvature at bends as large as possible, but not lesser than 800 times the diameter of the wire.

4. Avoiding double curvatures and using catenary curves.

5. Supporting the cables at closer intervals.

6. Using lubricants of unbonded concrete. For bonded concrete, water-soluble oils have been successfully employed to reduce the friction while tensioning. The lubricant is flushed off with water afterwards.

7. Prestressing the wires from both the ends. This involves more work in the field, but is often resorted to when the tendons are long or when the angles of bending are large.

8. Over-tensioning the wire. When friction loss in not excessive the amount of overtension may be equal to maximum frictional loss. The expected extra elongation can be computed to serve as a check. This extra pull required for overcoming the loss due to slip in anchorage or that required for minimising the creep in steel. The extra pull should be the greatest of the three required values. At this pull, the desired prestress will be created at the centre of the cable and the tension will gradually increase towards the pulling ends. If the wire is released now, part of stress will be lost at the end due to slip in anchorage. This will result in lessening the strain throughtout the length of wire, and the resulting movement of wire toward the centre will take place. Due to this, the friction force will act in the opposite direction, without letting the stress at the centre be reduced. Thus, the effect of overtensioning with a subsequent release-back is to put the frictional difference in the reverse direction. However, when the frictional loss is very high, it cannot be totally overcome by over tensioning, since maximum amount of tensioning is limited by strength or yield point on the tendon. In such a case, overtensioning may be limited to 10% only and the remaining loss must be allowed for in the design.

Total Amount of Losses : The amount of losses to be deducted will differ depending upon the definition of the term *initial prestress* and other terms defined below :

Apparent jacking stress . It is the force shown on the gauge of the jack.

Jacking stress. It is the actual pull in the wire near the jack before the anchorage. This stress is always less than the apparent jacking stress.

Initial prestress. The jacking stress minus the anchorage loss will be the stress at anchorage, after relase, and is frequently called the initial prestress.

Effctive prestress or design prestress. It is equal to the initial prestress minus the losses.

If the jacking stress minus the anchorage loss is taken as the initial prestress as defind above, the losses to be deducted will consist of elastic shortening, creep and shrinkage in concrete and creep in steel. For points away from the jacking end, the effect of friction must be considered in addition.

Magnitude of losses are generally expressed in the following four ways :

(*i*) *In unit strains.* Losses like creep, shrinkage and elastic shortening of concrete are more conveniently expressed by unit strains.

(*ii*) *In total strains* . Generally, anchorage losses are expressed in terms of total strain.

(*iii*) *In unit stresses.* If E_s is known, all losses expressed in strains can be converted into unit stresses in steel.

(*iv*) *In per cent of prestress.* This is more common form, specially for losses due to creep in steel and friction. Other losses expressed in unit stresses can be converted into percentage of initial prestress.

For average steel and concrete properties, cured under average air conditions, the following values may be taken as representative of average losses :

TABLE 25.7

Loss due to	*Pre-tensioning %*	*Post-tensioning %*
1. Elastic shortening and bending of concrete	3	1
2. Creep of concrete	6	5
3. Shrinkage of concrete	7	6
4. Creep in steel	2	3
Total loss	18	15

The above table is based on the assumption that overtensioning has been applied to reduce creep in steel and to overcome friction and anchorage losses.

If the frictional loss is excessive, that part which has not been balanced by overtensioning should be accounted for separately, in the design.

When the average prestress (P_0/A_c) is high (say about 7 N/mm^2) the above losses should be increased to 25% for pretensioning and 20% for post-tensioning.

When the average prestress is low (say about 1.75 N/mm^2) the above total losses may be reduced to 14% for pre-tensioning and 12% for post-tensioning.

Accurate computation of various losses is possible only after a preliminary design of a member is made. However, the preliminary design must take into account the losses. Due to this, some losses are assumed, and for this guidance may be taken from the above table.

25.6. COMPUTATION OF ELONGATION OF TENDONS

Pre-tensioning results in the elongation of tendons. The computation of this elongation is important, so that it can be compared with the measured elongation, thus serving as a check on the accuracy of the gauge readings or on the magnitude of frictional loss along the length of the tendon. We shall consider two cases.

Case 1. Friction along tendon neglected : If the friction along the tendon is neglected, it will be subjected to uniform tension P along its entire length, resulting in uniform stress.

Thus,
$$\Delta_s = \frac{P\,L}{A_{st}\,.\,E_s} \qquad ...(25.16)$$

This formula is valid only if the prestress does not exceed the proportional limit. If, however, it exceeds the proportional limit, it is necessary to refer to the stress-section diagram for the corresponding strain.

Case 2. Considering friction along tendon : Let L be the total length of the cable and α be the total change in the angle. At a distance x from the pulling end, the change in angle $\alpha_x = \dfrac{x \cdot \alpha}{L}$

The pull at distance x from the support is given by Eq. 25.14 (*b*).

$$P_x = P_1 \cdot e^{-(\mu \cdot \alpha_x + K_f \cdot x)} = P_1 \cdot e^{-x\left(\frac{\alpha}{L}\mu + K_f\right)}$$

$\therefore$ Extension in length $\quad dx = \dfrac{P_1 \cdot e^{-x\left(\frac{\alpha}{L}\mu + K_f\right)}}{A_{st} \cdot E_s} \cdot dx$

Hence total extension of steel is $\Delta s = \dfrac{P_1}{A_{st} \cdot E_s} \displaystyle\int_0^L e^{-x\left(\frac{\alpha}{L}\mu + K_f\right)} \cdot dx$

$$\Delta s = \frac{P_1 L}{-A_{st} E_s (\alpha \mu + K_f L)}\left[e^{-(\alpha \mu + K_f L)} - 1\right] = \frac{P_1 L}{A_{st} E_s (\alpha \mu + K_f L)}\left[1 - \frac{P_2}{P_1}\right] \quad ...(25.17)$$

$$\frac{\text{Actual extension}}{\text{Extension of frictionless cable}} = \frac{1 - \dfrac{P_2}{P_1}}{(\mu \alpha + K_f L)}$$

Example 25.7. *A pretensioned prestress concrete beam of 9 m span has a cross-section of 400 mm × 800 mm, and is prestressed with 2400 kN at transfer. The cable has cross-sectional area of 2000 mm² of steel and has a parabolic profile with maximum eccentricity of 120 mm at the middle of span. Determine the loss of prestress, given that $E_s = 2.1 \times 10^5$ N/mm². Use M 30 concrete. Assume minimum ultimate tensile strength of prestressing steel as 1500 N/mm².*

Solution

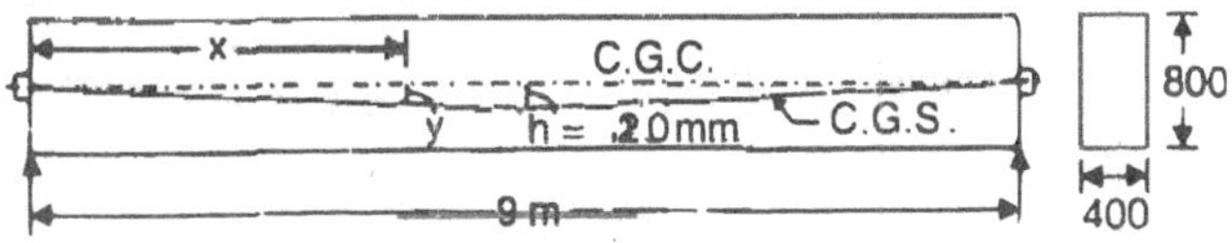

FIG. 25.27

Prestress at transfer,

$$f_s = \frac{2400 \times 1000}{2000} = 1200 \text{ N/mm}^2$$

$$E_c = 5700\sqrt{f_{ck}} = 5700\sqrt{30}$$

$$= 31220 \text{ N/mm}^2$$

Strain in steel $\quad \epsilon_s = \dfrac{f_s}{E_s} = \dfrac{1200}{2.1 \times 10^5} = 57.1 \times 10^{-4}$

(*i*) ***Loss due to elastic shortening of concrete.*** If the cable were straight and parallel to the centroidal axes, at a constant eccentricity y, Eq. 25.6 could be used for computing the elastic loss. However, in the present case, the cable is parabolic. Let us derive an expression fro strain in concrete (E_c) at the level of steel.

The equation for parabola is $y = \dfrac{4\,hx\,(L - x)}{L^2}$

where y is the eccentricity of cable at distance x from the end of the span, and h is the maximum eccentricity of the mid-span.

Hence compressive deformation in concrete, at steel level, due to prestressing force is

$$\delta_{cx} = -\frac{1}{E_c}\left(\frac{P_i}{A_e} + \frac{P_i \cdot y^2}{I_e}\right)$$

If $\in_{el}$ is the mean elastic strain, we have

$$\in_{el} = \frac{1}{LE_c}\int_0^L \left(\frac{P_i}{A_e} + \frac{P_i \cdot y^2}{I_e}\right) dx$$

Substituting the value of y, we get

$$\in_{el} = \frac{P_i}{L\,E_c}\int_0^L \left[\frac{1}{A_e} + \frac{16\,h^2\,x^2\,(L-x)^2}{I_e \cdot L^4}\right] dx = \frac{P_i}{LE_c}\int_0^L \left[\frac{1}{A_e} + \frac{16\,h^2}{I_e \cdot L^4}\left\{x^2\,L^2 + x^4 - 2\,Lx^3\right\}\right] dx$$

$$= \frac{P_i}{LE_c}\left[\frac{x}{A_e} + \frac{16\,h^2}{I_e L^4}\left\{\frac{x^3L^2}{3} + \frac{x^5}{5} - \frac{2\,L\,x^4}{4}\right\}\right]_0^L$$

or
$$\in_{el} = \frac{P_i}{E_c}\left[\frac{1}{A_e} + \frac{8}{15}\frac{h^2}{I_e}\right] \qquad \text{...(25.18)}$$

Since the losses are of small nature, A_e can be approximately taken equal to A_c and I_e can be taken equal to I_c without any appreciable error.

$\therefore$ $$A_e \approx A_c = 400 \times 800 = 32 \times 10^4 \text{ mm}^2; \quad I_e \approx I_c = \frac{1}{12}(400)(800)^3 = 1.7 \times 10^{10}$$

Substituting the values in Eq. 25.18

$$\in_{el} = \frac{2400 \times 1000}{E_c}\left[\frac{1}{32 \times 10^4} + \frac{8}{15}\;\frac{120 \times 120}{1.71 \times 10^{10}}\right] = \frac{8.58}{E_c}$$

An approximate value of strain can be obtained by taking the average of strains at maximum and minimum strains points. Minimum compressive strain occures at the ends while maximum compressive strain occurs at mid span.

$$\text{Strain at end section} = \frac{P_i}{A\,E_c} = \frac{2400 \times 1000}{32 \times 10^4\,E_c} = \frac{7.5}{E_c}$$

$$\text{Strain at mid span} = \frac{1}{E_c}\left[\frac{P_i}{A} + \frac{P\,h^2}{I}\right] = \frac{2400 \times 1000}{E_c}\left[\frac{1}{32 \times 10^4} + \frac{120 \times 120}{1.71 \times 10^{10}}\right] = \frac{9.53}{E_c}$$

$$\therefore \text{ Approximate mean strain } \in_{el} = \frac{(7.5 + 9.53)}{2\,E_c} = \frac{8.515}{E_c}$$

This is nearly the same as found by exact expression. Hence the approximate method can be used. For the present case,

$$\in_{el} = \frac{8.58}{E_c} = \frac{8.58}{31220} = 2.75 \times 10^{-4}$$

$$\text{Percentage loss} = \frac{2.75 \times 10^{-4}}{60 \times 10^{-4}} \times 100 = \mathbf{4.58\%}$$

$\therefore$ Loss of prestress $= 0.0458 \times 1200 = 55$ N/mm^2

(*ii*) ***Loss due to shrinkage of concrete*** : As per I.S. Code, shrinkage strain = 0.0003 for pretensioned members.

$$\% \text{ Loss} = \frac{3 \times 10^{-4}}{60 \times 10^{-4}} \times 100 = \mathbf{5\%} \quad ; \quad \text{Loss of prestress} = 0.05 \times 1200 = 60 \text{ N/mm}^2$$

(*iii*) ***Loss due to creep of concrete.*** As per IS 1343-1980 the creep coefficient at transfer of prestress at 28 days of curing of concrete may be taken as 1.6 .

$$\therefore \quad \text{Creep strain} = C\,\epsilon_{el} = 1.6 \times 2.75 \times 10^{-4} = 4.4 \times 10^{-4}$$

$$\text{Percentage loss} = \frac{4.4 \times 10^{-4}}{60 \times 10^{-4}} \times 100 = 7.33\%; \quad \text{Loss of prestress} = 0.0733 \times 1200 = 88 \text{ N/mm}^2$$

(*iv*) ***Loss due to relaxation of steel*** : As per IS : 1343-1980, the characteristic strength of prestressing steel (f_p) shall be assumed as the minimum ultimate tensile stress.

$$\therefore \quad f_p = 1500 \text{ N/mm}^2$$

$$\text{Ratio of intial stress in steel to } f_p = \frac{1200}{1500} = 0.8$$

Thus, initial stress in steel $= 0.8 f_p$. Hence from Table 25.13, the relaxation loss $= 90$ N/mm^2

$$\text{Per cent loss} = \frac{90}{1200} \times 100 = 6.5\% \quad ; \quad \text{Loss of strain} = 60 \times 10^{-4} \times 0.075 = 4.5 \times 10^{-4}$$

(*v*) ***Anchorage loss***

Let us assume loss of prestress due to anchorage take-up and friction of spacers and end block as equivalent to 2% of strain loss,

Loss of strain $= 0.02 \times 60 \times 10^{-4} = 1.2 \times 10^{-4}$; Loss of prestress $= 0.02 \times 1200 = 24$ N/mm^2

The various losses are tabulated below :

TABLE 25.8 (*a*) SUMMARY OF LOSSES

Loss due to	*Loss of strain*	*Percentage Loss*	*Loss of prestress (N/mm²)*
1. Elastic shortening	2.75×10^{-4}	4.58	55
2. Shrinkage	3.00×10^{-4}	5.00	60
3. Creep in concrete	4.4×10^{-4}	7.33	88
4. Creep in steel	4.50×10^{-4}	7.50	90
5. Anchorage etc.	1.20×10^{-4}	2.00	24
Total	15.85×10^{-4}	26.41	317

Thus total loss is 26.41%

$$\eta = \frac{\epsilon_s - \text{loss}}{\epsilon_s} = \frac{P_e}{P_i} = (1 - 0.2641) = 0.7369 \quad \text{and} \quad P = \eta P_i = 0.7369 \times 1200 = 833 \text{ N/mm}^2$$

Total loss of prestress = 1200 − 883 = **317 N/mm²**

Example 25.8. *A post-tensioned prestress concrete beam of 30 m span is subjected to a transfer prestress force of 2500 KN at 28 day's strength. The profile of the cable is parabolic with maximum eccentricity of 200 mm at midspan. Determine the loss of prestress,* *and the jacking force required if jacking is done from both ends of the beam. The beam has a cross-section of 500 mm × 800 mm, and is prestressed with 9 cables, each cable consisting of 12 wires of 5 mm diameter. Take $E_s = 2.1 \times 10^5$ N/mm² and $E_c = 3.5 \times 10^4$ N/mm². One cable is tensioned at a time.*

Solution $A_{st} = A_p = 9 \times 12 \times \frac{\pi}{4}(5)^2 = 2121 \text{ mm}^2;$

$$f_s = \frac{2500 \times 100}{2121} = 1178.7 \text{ N/mm}^2$$

Elastic strain in steel at transfer $\in_s = \frac{f_s}{E_s} = \frac{1178.7}{2.1 \times 10^5} = 56.12 \times 10^{-4}$

(*i*) ***Loss due to elastic shortening*** : We have seen in Example 25.7 that this loss can be found by taking the average of elastic strains at the ends and the middle of the beam.

Area of concrete $A_c = 500 \times 800 = 40 \times 10^4 \text{ mm}^2$; $I = \frac{1}{12} \times 500(800)^3 = 213.33 \times 10^8 \text{ mm}^4$

Elastic strain at the end section $= \frac{P_i}{A_c E_c} = \frac{2500 \times 1000}{40 \times 10^4 \times 3.5 \times 10^4} = 1.79 \times 10^4$

Elastic strain at mid span section.

$$= \frac{P_i}{A E_c} + \frac{P_i \cdot h^2}{I E_c} = \frac{P_i}{E_c}\left[\frac{1}{A} + \frac{h^2}{I}\right] = \frac{2500 \times 1000}{3.5 \times 10^4}\left[\frac{1}{40 \times 10^4} + \frac{(200)^2}{213.33 \times 10^8}\right] = 3.13 \times 10^{-4}$$

∴ Mean elastic strain $= \frac{1}{2}(1.79 + 3.13)\, 10^{-4} = 2.46 \times 10^{-4}$

Since tendons are tensioned one after the other, the first tendon will loose stress due to elastic shortening of subsequent eight tendons. Hence loss of strain in first tendon

$$= \frac{8}{9} \times 2.46 \times 10^{-4} = 2.18 \times 10^{-4}$$

Average elastic loss for all tendons is given by

$$\in_{el} = \frac{1}{2}(2.18 \times 10^{-4}) = 1.09 \times 10^{-4}$$

∴ Percentage loss $= \frac{1.09 \times 10^{-4}}{56.12 \times 10^{-4}} \times 100 =$ **1.94%**

(*ii*) ***Loss due to shrinkage*** : As per IS : 1343-1980, the shrinkage strain is given by

$$\in_{sh} = \frac{0.0002}{\log_{10}(t+2)} = \frac{0.0002}{\log_{10}(28+2)} = 1.35 \times 10^{-4}$$

Percentage loss $= \frac{1.35 \times 10^{-4}}{56.12 \times 10^{-4}} \times 100 =$ **2.41** %

(*iii*) ***Loss due to creep in concrete*** : As per IS : 1343-1980, the creep coefficient at transfer of prestress at 28 days of curing of concrete may be taken as 1.6

∴ $\in_{cr} = (C \cdot \in_{el}) = (1.6) \times 2.46 \times 10^{-4} = 3.936 \times 10^{-4}$

Percentage loss $= \frac{3.936 \times 10^{-4}}{56.12 \times 10^{-4}} \times 100 =$ **7.01** %

(*iv*) ***Loss due to relaxation of steel*** : As per IS : 1343-1980 , the characterstic strength (f_p) of prestressing steel shall be assumed as the minimum ultimate tensile strain.

From Table 25.11, ultimate tensile strain for 5 mm wires = 1600 N/mm^2

$$\therefore \quad f_p = 1600 \text{ N/mm}^2$$

Ratio of initial stress to characteristic strength = 1178.7/1600 = 0.74

∴ Initial stress in steel = 0.74 f_p. Hence from Table 25.73,

The relaxation loss = 78 N/mm^2

Percent loss $= \frac{78}{1178.7} \times 100 =$ **6.62** %; Loss of strain $= 56.12 \times 10^{-4} \times 0.0662 = 3.7 \times 10^{-4}$

(*v*) ***Loss due to anchorage take up*** : Let 2.5 mm be the anchorage take-up at each jacking

end. $\quad$ Loss of strain $= \frac{2.5}{L/2} = \frac{2.5}{15 \times 1000} = 1.67 \times 10^{-4}$

$$\text{Percentage loss} = \frac{1.67 \times 10^{-4}}{56.12 \times 10^{-4}} \times 1000 = \mathbf{2.98\%}$$

(*vii*) ***Friction loss*** : Let us assume $\mu = 0.3$ and $k = 15 \times 10^{-4}$ per metre.

Since jacking is done from both ends, effective length of cable

$L_e = L/2 = 15$ m. The cable profile is given by

$$y = \frac{4\,hx\,(L-x)}{L^2}$$, where x is measured from the free end.

$$\therefore \quad y = \frac{4 \times 0.2\, x\,(30-x)}{900} = \frac{8}{9000}(30\,x - x^2)$$

$$\therefore \quad \left(\frac{dy}{dx}\right)_{x=0} = \frac{8}{9000}(30) = 0.0267 \text{ radians} = \alpha$$

$$P_2 = P_1 \,.\, e^{-(\mu\alpha + k\,.\,L_e)} \text{ Hence } \quad \frac{P_2}{P_1} = e^{-(\mu\,.\,\alpha + k\,.\,L_e)}$$

or $$\frac{P_1 - P_2}{P_1} = 1 - e^{-(\mu\alpha + k\,.\,L_e)} = 1 - e^{-(0.3 \times 0.0267 + 0.0015 \times 15)} = 0.03.$$

Alternatively, from approximate method, $P_2 = P_1\,(1 - \mu\alpha + k\,L_e)$

or $$\frac{P_1 - P_2}{P_1} = \mu\alpha + k\,L_e = 0.3 \times 0.0267 + 0.0015 \times 15 = 0.0305$$

Loss of prestress = 0.03 P_1 = 3%

$$\therefore \quad \text{Strain loss } \epsilon_f = \frac{P_1 - P_2}{E_s\,.\,A_{st}} = \frac{0.03\,P_1}{E_s\,.\,A_{st}} = \frac{0.03 \times 2500 \times 1000}{2.1 \times 10^5 \times 2131} = 1.68 \times 10^{-4}$$

It is assumed that the friction loss will be neutralised by applying additional force at the jack. It will therefore not be counted in the total loss.

Jacking force = Transfer force + Friction loss = $P_1 + 0.03\,P_1 = 1.03\,P_1$

The loss of prestress due to various causes are tabulated below.

TABLE 25.8 (*b*) SUMMARY OF LOSSES

Loss due to	*Loss of strain*	*Percent loss*
(*i*) Elastic shortening	1.09×10^{-4}	1.94
(*ii*) Shrinkage	1.35×10^{-4}	2.41
(*iii*) Creep in concrete	3.936×10^{-4}	7.01
(*iv*) Creep in steel	3.70×10^{-4}	6.62
(*v*) Anchorage take up	1.67×10^{-4}	2.98
Total	11.746×10^{-4}	20.96

Percent loss in prestress = 20.96 ; Loss of prestress = $0.2096 \times 1178.7 = 248$ N/mm^2

Effective prestress force $P_e = (1 - 0.2096)\, P_1 = 0.7904 \times 2500 =$ **1976 kN.**

Example 25.9. *If the beam of Example 25.8 carries a superimposed load of 20 kN/m, find the total lossses.*

Solution B.M. $M = \dfrac{wL^2}{8} = \dfrac{20000(30)^2}{8} \times 1000 = 2.25 \times 10^9$ N-mm.

Strain in concrete at the steel level at midspan, due to B.M.

$$= -\frac{M\,.\,h}{I\,E_c}\text{ , where } h \text{ is the eccentricity of steel at midspan.}$$

$$= -\frac{2.25 \times 10^9 \times 200}{213.33 \times 10^8 \times 3.5 \times 10^4} = -6.03 \times 10^{-4}$$

This strain is negative, *i.e.* there is a gain in strain instead of loss.

$$\text{Percentage gain} = \frac{6.03 \times 10^{-4}}{56.12 \times 10^{-4}} \times 100 = 10.74\%$$

Hence final loss of strain = $(11.746 - 6.03)\, 10^{-4} = 5.716 \times 10^{-4}$

Percentage loss = 20.96 − 10.74 = **10.22%**

Example 25.10. *Fig. 25.28 shows a prestressed concrete beam, continuous over two spans. The curved tendon is to be tensioned from both ends. Compute the percentage loss of prestresss due to friction, from one end to the centre of beam (i.e. from A to E). Take the coefficient of friction between the cable and duct as 0.45, and the coefficient of wobble effect as 0.0015 per metre.*

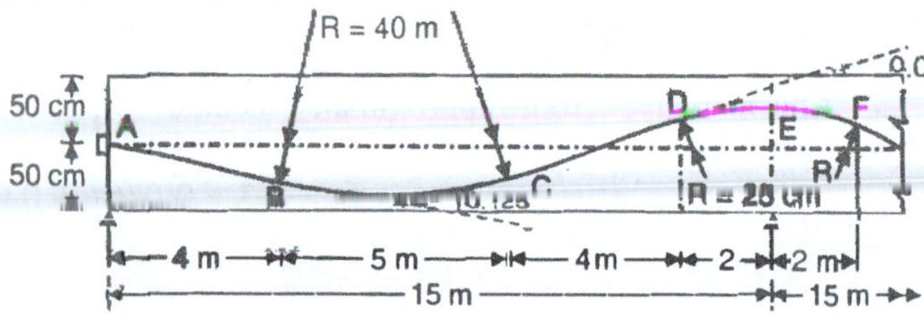

FIG. 25.28

Solution

(*a*) ***Approximate solution :***

Change in angle for curve $BC = \hat{L}/R = 5/40 = 0.125$.

Change in angle for curve $DE = \hat{L}/R \approx 2/25 = 0.08$

$$\left(\frac{P_1 - P_2}{P_1}\right) = kL + \mu\,\alpha = 0.0015 \times 15 + 0.45\,(0.125 + 0.08) = 0.1148$$

∴ Loss due to friction = $P_1 - P_2 = 0.1148\ P_1 = \mathbf{11.48\%}$

(*b*) ***Exact solution*** : In this method, the cable is divided into segments *AB* (straight) *BC* (curved), *CD* (straight) and *DE* (curved), and stress at the end of each segment is computed, using Eq. 25.14 (*b*). The calculations are tabulated below.

TABLE 25.9

Segment	x	$k\ .\ x$	α	$\mu\ \alpha$	$kx + \mu\ \alpha$	$e^{-(k.x+\mu\alpha)}$	*Stress at end of segments*
AB	4	0.006	0	0	0.006	0.9940	0.9940 P_1
BC	5	0.0075	0.125	0.0563	0.0638	0.9382	0.9326 P_1
CD	4	0.006	0	0	0.006	0.9940	0.9270 P_1
DE	2	0.003	0.08	0.036	0.039	0.9618	0.8915 P_1

∴ Total frictional loss = $(1 - 0.8915)\ P_1 = 0.1085\ P_1 = \mathbf{10.85\%}$

Example 25.11. *Fig. 25.29 shows a prestressed concrete beam, in which the cable is pulled from both the ends, with a pulling stress of 1200 N/mm² at each end. Assuming* $\mu = 0.40$ *and* $k = 0.0015$ *per metre, compute the extension at each end. Take* $E_s = 2.1 \times 10^5\ N/mm^2$

Solution The mid-point *C* will have least tension.

For portion AB

$\mu = L/R \approx 10/50 = 0.20$ radians; $k\ .\ x + \mu\ \alpha = (0.0015 \times 10) + (0.4 \times 0.20) = 0.095$

Stress at $B = 1200\ e^{-0.095} = 1091\ N/mm^2$

$$\therefore \quad \Delta_{ab} = \frac{f_1\ x}{E_s\ (kx + \mu\ \alpha)}\left[1 - \frac{f_2}{f_1}\right] = \frac{1200\ (10 \times 1000)}{2.1 \times 10^5 \times 0.095}\left[1 - \frac{1091}{1200}\right] = 54.6 \text{ mm}$$

For portion BC

$\alpha = 0$. Hence,

$k\ .\ x + \mu\ \alpha = 0.0015 \times 10 = 0.015$

Stress at $B = f_1 = 1091\ N/mm^2$

∴ Stress at $C = f_2 = 1091\ e^{-0.015}$

$= 1075\ N/mm^2$

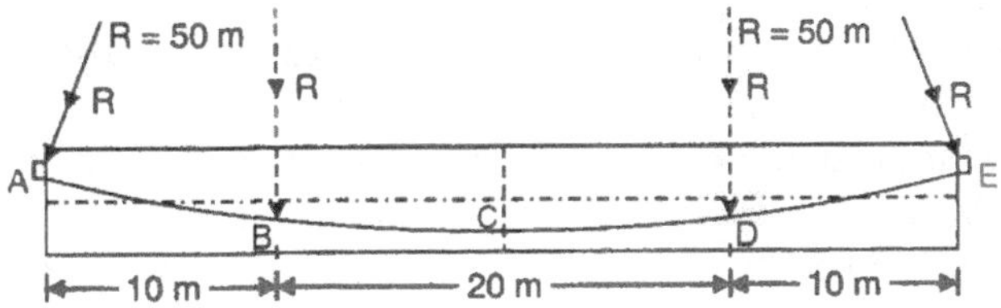

FIG. 25.29.

$$\therefore \quad \Delta_{bc} = \frac{1091\ (10 \times 1000)}{2.1 \times 10^5 \times 0.015}\left[1 - \frac{1075}{1091}\right] = 50.8 \text{ mm}$$

∴ Total extension = 54.6 + 50.8 = **105.4 mm.**

25.7. PROPERTIES OF MATERIALS

In prestressed cement concrete, the principal materials used are: (*i*) high tensile steel and (*ii*) high strength concrete.

1. High Tensile Steel : As stated earilier, ordinary mild steel wires are not useful because no prestress will be left in wires after losses (specially the shrinkage and creep losses) have taken place. The initial tensile strain in steel is reduced after tensioning by as much as 15 to 20 % (or even more) due to various losses. The reduction of strain must be small compared to initial strain otherwise most of the prestress will be lost. Hence the first principal requirement of a steel for use in prestressing is high *ultimate tensile strength.* The second principal requirement is high ultimate elongation. High tensile steel have ultimate strength capacity as high as 2100 N/mm^{2}and the use of such steel will provide considerable amount of effective prestressing force even after losses in prestress. Mild and hard steel used in reinforced concrete construction have a yield limit of 200 to 300 N/mm^2 only.

High tensile steel is produced by *alloying* . Carbon is extremely economical element for alloying, since it is cheap and easy to handle. Other alloys include manganese and silicon. High tensile steel contains about 3/4 per cent of carbon against only about 0.2% in mild steel. Approximate chemical composition of the high tensile steel wires is as follows:

Carbon	: 0.6 to 0.85 %	Sulphur	: 0.055%
Manganese	: 0.7 to 1.0%	Silicon	: 0.1 to 0.35 %
Phosphorus	: 0.05%		

High tensile steel is hot rolled, and then given heat treatment or cold treatment by rolling or drawing. The most common method for increasing the tensile strength of steel for prestressing is by cold drawing high tensile bars through a series of dyes. The process of cold drawing tends to realign the crystals and the strength of the wires is increased by each drawing so that the strength of wire increases as its diameter decreases. The ductility of wires is some what decreased as a result of cold drawing, which is a disadvantage.

Another method of increasing the tensile strength is to obtain the wire as *as-drawn* wires. These wires have low proportional limit, and in order to increase the proportional limit, the wires are subjected to some type of stress relieving processes. Fig. 25.30 shows the stress-strain curves for mild steel, high tensile rod and high tensile wires. Unlike mild steel, high tensile steel does not have any well defined yield point. While the ultimate strength of high tensile steel can be easily determined by testing, its elastic limit or its yield point cannot be so simply ascertained. The yield point of high tensile steel is generally defined at 0.2% set, *i.e.* at 0.2% of permanent inelastic deformation. Thus in Fig. 25.30, *OA* is the straight portion of the curve for high tensile wire, while *P* is the so selected yield point on the curve portion. If the specimen is loaded upto *P* and then unloaded, the unloading curve will follow line *PQ* which is practically parallel to the straight portion *OA* of the loading curve. *OQ* is then the permanent strain which is equal to 0.2%. The yield stress, defined as the stress which produces a residual strain equal to 0.2% on unloading, is also termed as 0.2 *percent proof stress*. The approximate values of yield point and proportional limit for various types of high steel wires are given in Table 25.10. On an average, 0.2% proof stress of wires is about 80% of their ultimate strength. The working stress in high tensile steel is taken equal to 80% of its 0.2% proof stress. Also, 0.2% proof stress should not be less than 80% of the minimum specified ultimate strength.

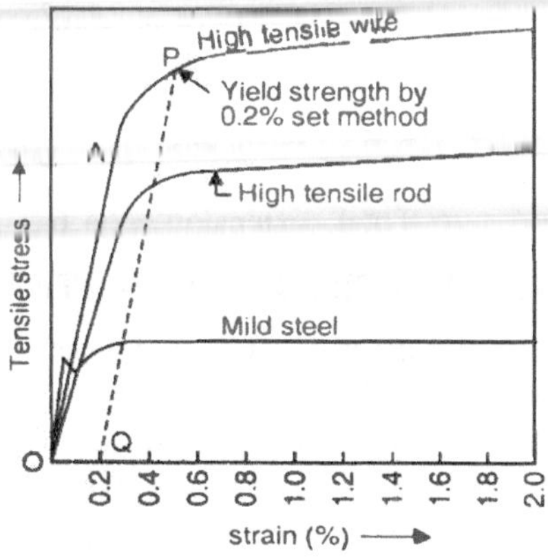

FIG. 25.30.

TABLE 25.10 APPROXIMATE YIELD STRESS AND PROPORTIONAL LIMITS

Wire	Yield point	Proportional limits
1. Wires as drawn	0.75 f_{su}	0.35 f_{su}
2. Prestretched	0.85 f_{su}	0.55 f_{su}
3. Temperature treated wires	0.87 f_{su}	0.70 f_{su}
4. Strands-stress relieved	0.90 f_{su}	0.75 f_{su}

The modulus of elasticity for cold drawn wires and heat treated wires may be taken as 2.1×10^5 N/mm^2 and 2.0×10^5 N/mm^2 respectively. In most of the high tensile wires the creep is negligible upto 0.45 f_{su} and it is about 3% at 0.5 f_{su} and 0.55 f_{su} stress level.

Wires are supplied in drums. In order to have least permanent set due to winding on drums, the drum diameter are made as big as possible, with a minimum of 1.5 to 2 m. Most wires have slight permanent set when unwound, and require some straightening. Small diameter wires possess higher unit strength and furnish better bond, which is often vital in pre-tensioning. However, large diameter are preferred for pre-tensioning to save labour and anchorage costs. The prestressing steel is also available in the form of strands which are obtained by twisting wires together. Mostly, seven-wire strands are used, having a centre wire slightly larger then the outer six wires which enclose it tightly in a helix with a uniform pitch between 12 to 16 times the nominal diameter of the strand. After stranding, all strands are subjected to a stress-relieving continuous heat treatment to produce the prescribed mechanical properties.

The high tensile steel wire used for prestressed concrete should conform to Table 25.11 regarding minimum ultimate tensile strength.

TABLE 25.11 MINIMUM ULTIMATE TENSILE STRENGTH OF TENSILE STEEL WIRES

Nominal dia. of wires (mm)	Min. ultimate tensile strength (f_{su}) (N/mm^2)
8.0	1500
7.0	1500
5.00	1600
4.00	1750
3.00	1900
2.50	2050
2.00	2200
1.50	2350

The proof stress shall not be less than 80% of the minimum ultimate tensile strength. The elongation at rupture shall not be less than 2% over a gauge of 200 mm. The wire shall be free from rust, scale and other similar deleterious matter liable to affect adversely proper tensioning or its bond with concrete. Slight rust may be permitted provided it is not loose. High tensile steel wires shall be supplied in coils having a sufficiently large diameter for the wire to layout straight. Cables used as prestressing steel shall consists of a group of single wires arranged in parallel or stranded formation. For the purposes of tests, the strength of a cable shall be taken as the multiple of the strength of a single or stranded wires forming the cable.

High tensile alloy steel bars shall have a minimum ultimate tensile strength (f_{su}) of 950 N/mm^2. Proof stress (f_{su}) of the bars shall not be less than 80% of its ultimate tensile strength. The bars as supplied shall be free from end jagged edges. Coupling units and similar fixtures used in conjunction with the wires or bars shall have an ultimate tensile strength of not less than the individual strength of the wires or bars being joined.

2. High Strength Concrete : Because of the use of high tensile steel in pre-stressed concrete construction, the concrete has to be of good quality and of high strength. A good, well compacted dense concrete has less elastic strain, and has less shrinkage plastic flow, thus reducing the loss of prestress considerably. It is cheaper comparatively to prepare a high strength concrete as its cost does not increase in the same proportion as its strength does. High quality concrete is also essential to bear the high concentration of stresses under the end anchorages. Ordinary concrete would require bigger size of anchorages. In some cases, high grade concrete can be very useful from the point of view of crack resistance of the members as also resistance against corrosion. The durability of the members will also be higher too. Also, concrete of high compressive strength offers high resistance in tension and shear, as well as in bond and bearing, and is desirable for prestressed concrete structures whose various portions are under higher stresses than ordinary reinforced concrete. Not only should high strength concrete be specified for prestressed work, but such strength should be more closely attained in the field than for reinforced concrete. Indian Standard Code recommends a minimum cube strength of 40 N/mm^2 for pre-tensioned system.

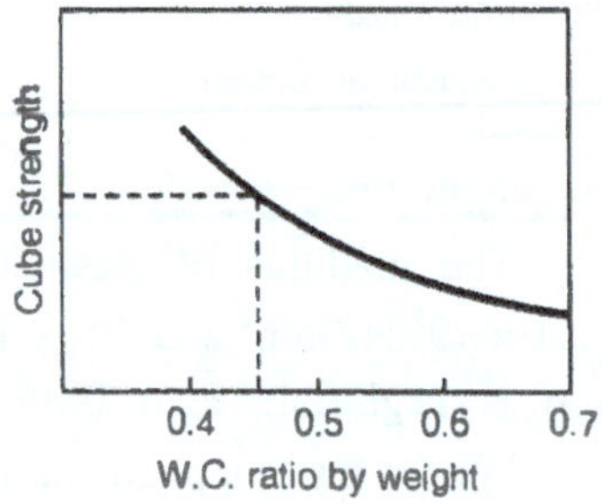

FIG. 25.31.

The strength of concrete decreases with increase in water-cement ratio as is evident from Fig. 25.31. To attain a strength of 36.5 N/mm^2, the water cement ratio should not be higher than 0.45 by weight. However, if water cement ratio is less than 0.4, the workability will decrease and consequently compact and high density concrete would not be obtained. High workability with less water-cement ratio would require a higher precentage of cement and well graded aggregate. Since excessive cement tends to increase shrinkage, a lower cement factor is desirable. To this end, good vibration is advised. Slumps of 12 to 25 cm are sometimes used with controlled vibration. Workability can also be increased by proper admixtures. Air entrainment of 3 to 5% improves workability and reduced bleeding.

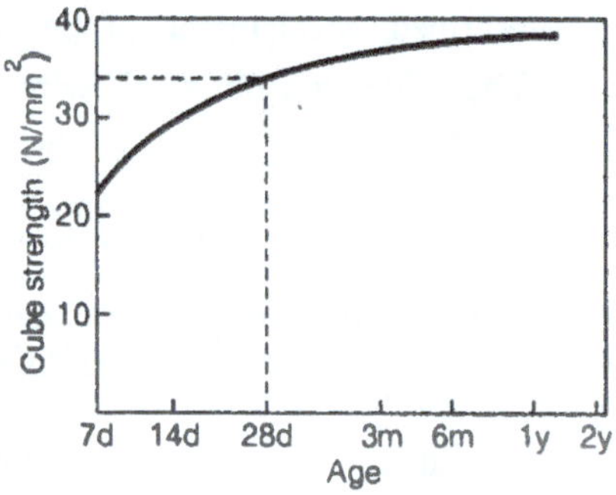

FIG. 25.32.

The high compressive strength mentioned above is of 15 cm cube at 28 days. Actually, the strength of concrete increases with age, as is evident from Fig. 25.32.

Fig. 25.33 shows the stress strain curve for concrete. The stress-strain relation can be expressed by the following relations : $f_c = E_c \cdot \varepsilon \left(1 - \dfrac{\varepsilon}{2\,\varepsilon_u}\right)$ ...(25.19)

where ε_u = ulimate strain = 0.003.

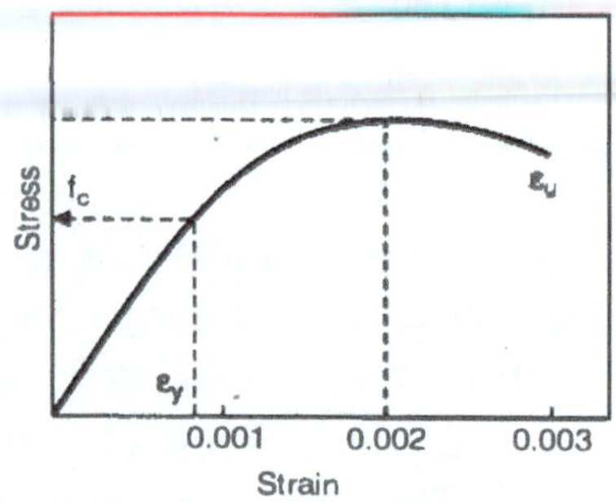

FIG. 25.33

The modulus of elasticity E_s used in the above equation is an average one. It is well known that the modulus of elasticity of concrete decreases with increase in stress. Since the compression on the section of beam varies from top to bottom, each fibre of concrete will have different E_c, though the difference will not be much. Also, modulus of elasticity of concrete during a compression operation is lower than during decompression and there is a residual strain after every decompression. If these cycles of compression and decompression are repeated, E_c tends to equalise and behaviour of concrete becomes almost elastic. An average modulus of elasticity can conveniently be calculated from the deflection measurements of a prestressed concrete beam.

3. Cement Grout : In the case of bonded post-tensioning, cement grouting is done which also serves to protect steel against corrosion. Conduits can be made of aluminium, steel, tin or other metal sheathing or tubes. It is also possible to form the duct by withdrawing steel tubing or rod before concrete hardens. Entry of the grout into the cable way is provided by means of holes in the anchorage heads and cones. Where the space between the wires is large, such as in Magnel system, a 1:1 cement-sand mix is used with a water cement ratio of 0.5 by volume. In other systems, such as in Freyssinet or Strescon cable, neat cement paste with 0.5 water-cement ratio is injected. On a big job, saving of cement is important. In that case fine sand of about 0.4 mm grain size may be used, with water : cement : sand proportion of about 1:1.3:0.7 by volume. The grout pressure may be between 0.5 to 0.7 N/mm^2. The injection can be applied at one end of the member until it is forced out at the other end ; the end is then plugged and pressure is again applied at the injecting end to compact the grout.

25.8. MERITS AND DEMERITS OF PRESTRESSED CONCRETE

Prestressed concrete has the following merits :

1. Since the technique of prestressing eliminates cracking of concrete under all stage of loading, the entire section of the structures takes part in resisting the external load. In contrast to this, in the reinforced concrete, only portion of concrete above the neutral axis is effective.

2. Since concrete does not crack, the possibility of steel to rust and concrete to deteriorate is minimised.

3. Absence of cracks results in higher capacity of the structure to bear reversal of stresses, impact, vibration and shock.

4. In prestressed concrete beams, dead loads are practically neutralised. The reactions required are therefore, much smaller than required in reinforced concrete. The reduced dead weight of structure results in saving in the cost of foundations.

The neutralisation of dead weight is of importance in large bridges.

5. The use of curved tendons and the precompression of concrete helps to resist shear.

6. The quantity of steel required for prestressing about $\frac{1}{3}$ of that required for reinforced concrete, though the steel for the former should have high tensile strength.

7. In prestressed concrete, precast blocks and elements can be assumed and used as one unit. This saves in the cost of shuttering and centring for large structures.

8. With the advent of prestressed concrete, it has been possible now to construct large size liquid retaining structures not economical to build otherwise. Such structures have low cost and are preferably safe against cracking and consequent leakage.

9. Prestressed concrete can be used with advantage in all those structures where tension develops, such as tie and suspender of a bow string girder, railway sleepers, electric poles, upstream face of gravity dam etc.

10. Prestressed concrete beams have usually low deflection.

However, prestressed concrete construction has the following demerits :

1. It requires high quality dense concrete of high strength. Perfect quality control in production, placement and compaction is required.

2. It requires high tensile steel, which is 2.5 to 3.5 times costlier than mild steel.

3. It requires complicated tensioning equipment and anchoring devices, which are usually covered under patented rights.

4. Construction requires perfcet supervision at all stages of construction.

25.9. BASIC ASSUMPTIONS

The design of prestressed concrete members is based on the following assumptions:

1. A plane section before bending remains plane after bending.

2. Within the range of working stresses both concrete and steel behave elastically notwithstanding the small amount of creep which occurs in both materials under sustained loading.

3. Up to the limit of the modulus of rupture of concrete, any change in the loading produces a change of stress in the concrete only, the sole function of the prestressing tendon being to impart and maintain the prestress in the concrete notwithstanding the comparatively small changes in the stress of steel due to changes in strain of concrete surrounding it.

25.10. ANALYSIS OF BEAMS FOR FLEXURE

1. Concentrically Prestressed Section : If a horizontal force P is applied at the centroid of the concrete section, the stress in concrete is uniform across the section and is given by

$$f = \frac{P}{A}$$

The release of resistance from bulkheads is equivalent to the application of an opposite force P_i to the member. Using transformed section method,

$$f_c = \frac{P_i}{A_c + m\,A_{st}} = \frac{P_i}{A_e} \qquad \text{...(25.20)}$$

where A_e is the area of equivalent or transformed section of concrete.

The stress induced in steel is

$$\Delta f_s = m\,f_c = \frac{m\,P_i}{A_c + m\,A_{st}} = \frac{m\,P_i}{A_e} \qquad \text{...(25.21)}$$

where Δf_s represents immediate reduction of the prestress in the steel as a result of the transfer.

In practice, however, the loss in prestress is computed by the following approximate expression:

$$\Delta f_s = \frac{m P_i}{A_c} \qquad \ldots[25.21(a)]$$

2. Eccentrically Prestressed Section : Let the prestressed force be applied at an eccentricity e. Using elastic theory, the fibre stress at any point is given by

$$f = \frac{P}{A} \pm \frac{P \cdot e \cdot y}{I}$$

Using transformed section method, and taking $P = P_i =$ initial prestressing force,

$$f_c = \frac{P_i}{A_e} \pm \frac{P_i \cdot e \cdot y}{I_e} \qquad \ldots(25.22)$$

where I_e is the equivalent moment of inertia of transformed section.

However, in practice, the gross or net concrete section is used, and either inital or reduced prestress (P_e) is used :

$$f_c = \frac{P}{A_e} \pm \frac{P e \cdot y}{I_e} \qquad \ldots[25.22(a)]$$

where P is either P_i or P_e.

3. Stresses due to Live Load : Let the beam be subjected to a B.M. M, either due to weight of beams, or due to externally applied loads, or due to both. The resulting stress in concrete is given by

$$f_c = \frac{P}{A} \pm \frac{P \cdot e\, y}{I} \pm \frac{M \cdot y}{I} = \frac{P}{A}\left(1 \pm \frac{e \cdot y}{r^2}\right) \pm \frac{M\, y}{I} \qquad \ldots(25.23)$$

where $r =$ radius of gyration $= \sqrt{I/A}$.

In the above expression, $P = P_i$ if M is due to dead load only, while $P = P$ if M is both due to dead and live loads.

For a pretensioned beam, steel is always bonded to concrete before M is applied. Hence y and I are computed on the basis of transformed section. As an approximation, either gross or net section of concrete may be used without any appreciable error. In the post-tensioned bonded construction, transformed section should be used. If however, weight of the beam is applied before bonding takes place, net section should be used. For post-tensioned unbonded beam, net section is to be used.

Example 25.12. *Fig. 25.34 shows the midspan cross-section of a prestressed post-tensioned beam of 24 m span. There are 6 cables each of 12 wires of 5 mm diameter, and are stressed with initial prestress of 1150 N/mm^2. Assuming an effective prestress of 1000 N/mm^2 after all losses have taken place, compute the extreme fibre stresses in concrete. The profile of cable is parabolic with zero ecccentricity at the ends. The beam carries a live load of 9 kN/m in addition to its own weight. Take unit weight of concrete as 24 kN/m^3.*

Solution. Area of prestressing steel

$$= 6 \times 12\left(\frac{\pi}{4} \times 25\right) = 1413.7 \text{ mm}^2$$

$$P_i = 1150 \times 1413.7 = 1625770 \text{ N}$$

$$P_e = 1000 \times 1413.7 = 1413700 \text{ N}$$

$$A_e = (1000 \times 150) + (150 \times 750) + (400 \times 400)$$

$$= 150000 + 112500 + 160000 = 422500 \text{ mm}^2$$

$$y_1 = \frac{(150000 \times 75) + (112500 \times 525) + (16000 \times 1100)}{422500}$$

$$= 583 \text{ mm.}$$

$$y_2 = \frac{(150000 \times 1225) + (112500 \times 775) + (160000 \times 200)}{422500}$$

$$= 717 \text{ mm.}$$

Check : $y_1 + y_2 = 583 + 717 = 1300$ mm.

Eccentricity $e = y_2 - 200 = 717 - 200 = 517$ mm

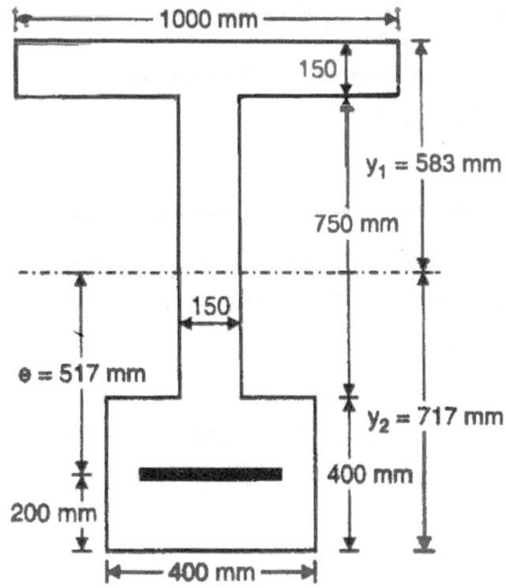

FIG. 25.34

$$I = \left[\frac{1}{12} 1000\,(150)^3 + 150000(583 - 75)^2\right] + \left[\frac{1}{12} \times 150\,(750)^3 + 112500\,(583 - 150 - 375)^2\right]$$

$$+ \left[\frac{1}{12} \times 400\,(400)^3 + 160000\,(717 - 200)^2\right] = 8.954 \times 10^{10} \text{ mm}^4$$

Dead load of beam $= (422500 \times 10^{-6}) \times 1 \times 24000 = 10140$ N/m.

Dead load B.M. $M_d = \dfrac{10140 \times 24^2}{8} \times 1000 = 730 \times 10^6$ N-mm.

Live load B.M. $M_l = \dfrac{9000 \times 24^2}{8} \times 1000 = 648 \times 10^6$ N-mm.

$\therefore$ Total $M = M_d + M_l = 1378 \times 10^6$ N-mm.

Initial Condition : Due to moment caused by eccentrically applied pre-stress, the top fibres will be subjected to tensile stress while the bottom fibres will be subjected to compressive stresses. However, due to dead weight of beam, top fibres will be subjected to compressive stresses. Using plus sign for compressive stress and minus sign for tensile stresses, we get

$$f = \frac{P_i}{A_c} \mp \frac{P_i \cdot e \cdot y}{I} \pm \frac{M_d \cdot y}{I}$$

$$P_i / A_e = 1625770 / 422500 = 3.848; \quad \frac{P_i \cdot e}{I} = \frac{1625770 \times 517}{8.954 \times 10^{10}} = 93.87 \times 10^{-4}$$

$$\frac{M_d}{I} = \frac{730 \times 10^6}{8.954 \times 10^{10}} = 81.53 \times 10^{-4}$$

f_1 (topfibre) $= 3.848 - 93.87 \times 10^{-4} \times 583 + 81.53 \times 10^{-4} \times 583 = \mathbf{3.13\ N/mm^2}$.

f_2 (bottom fibre) $= 3.848 + 93.87 \times 10^{-4} \times 717 - 81.53 \times 10^{-4} \times 717 = \mathbf{4.73\ N/mm^2}$.

Final Condition $\quad f = \dfrac{P_e}{A_c} \mp \dfrac{P_e \cdot e \cdot y}{I} \pm \dfrac{M \cdot y}{I}$; $P_e / A_c = 1413700/4225000 = 3.346$

$$\frac{P_e \cdot e}{I} = \frac{1413700 \times 517}{8.954 \times 10^{10}} = 81.63 \times 10^{4} \; ; \; \frac{M}{I} = \frac{1378 \times 10^{6}}{8.954 \times 10^{10}} = 153.9 \times 10^{-4}$$

$\therefore$ f_1 (top fibre) $= 3.346 - 81.63 \times 10^{-4} \times 583 + 153.9 \times 10^{-4} \times 583 =$ **7.56 N/mm²**

f_2 (bottom fibre) $= 3.346 + 81.63 \times 10^{-4} \times 717 - 153.9 \times 10^{-4} \times 717 =$ **– 1.84 N/mm²**.

Example 25.13. *A prestressed concrete beam of inverted T-section has dimensions as shown in Fig. 25.35 and is simply supported over a span of 16 m. The beam is post-tensioned with 3 Freyssinet cables, each containing 12 wires of 7 mm dia. placed as shown, at the midspan. If the initial prestress is 1000 N/mm², calculate maximum uniformly distributed load if the maximum compressive stress in concrete is limited to 14 N/mm² and tensile stress is limited to 1 N/mm². Assume loss of prestress =15%.*

Solution $A_{st} = A_p = 3 \times 12 \times \frac{\pi}{4} (7)^2 = 1385.4 \text{ mm}^2$

$P_i = 1385.4 \times 1000 = 1385400 \text{ N}; \; P_e = 0.85 \times 1385400 = 1177600 \text{ N}$

$A_c = (300 \times 900) + (300 \times 600) = 270000 + 180000 = 450000 \text{ mm}^2$

$$y_2 = \frac{(180000 \times 150) + (270000 \times 750)}{450000} = 510 \text{ mm.}$$

$$y_1 = \frac{(270000 \times 450) + (180000 \times 1050)}{450000} = 690 \text{ mm.}$$

$e = 510 - 150 = 360$ mm.

$$I = \left[\frac{1}{12} \times 600 (300)^3 + 180000 \times (510 - 150)^2\right] \times \left[\frac{1}{12} \times 300 (900)^3 + 270000 \times (690 - 450)^2\right]$$

$= 5.8455 \times 10^{10} \text{ mm}^4$.

Dead load $w_d = 450000 \times 10^{-6} \times 24000 = 10800$ N/m.

Compressive stress due to $P_e = 1177600/450000$
$= 2.62 \text{ N/mm}^2$

Stress due to eccentricity of prestress $= \mp \frac{P_e \cdot e}{I} y$

$$f_1 \text{ (top fibre)} = -\frac{1177600 \times 360}{5.8455 \times 10^{10}} \times 690 = -5 \text{ N/mm}^2$$

$$f_2 \text{ (bottom fibre)} = +\frac{1177600 \times 360}{5.8455 \times 10^{10}} \times 510 = +3.7 \text{ N/mm}^2$$

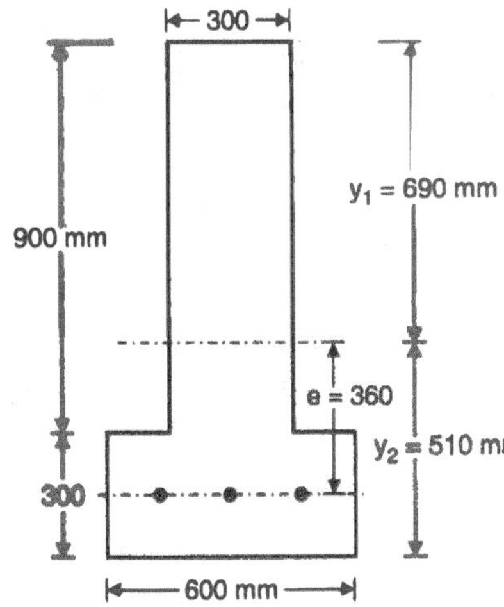

FIG. 25.35.

Total stress at top $= 2.62 - 5 = -2.38 \text{ N/mm}^2$.

Total stress at bottom $= 2.62 + 3.7 = +6.32 \text{ N/mm}^2$.

Due to super-imposed load (inclusive of self load), there will be compressive stress at top and tensile stress at bottom. In order to limit the compressive stress to 14 N/mm², the stress due to super-imposed load $= 14 - (-2.38) = 16.38$ N/mm² at top. Similarly, in order to limit the tensile stress to 1 N/mm², the stress due to supper-imposed load $= 1 + 6.32 = 7.32$ N/mm². Let M_1 and M_2 be the respective moments to cause these.

$\therefore \qquad M_1 = \dfrac{16.38 \times 5.8455 \times 10^{10}}{690} = 13.38 \times 10^8$ N-mm

and $\qquad M_2 = \dfrac{7.32 \times 5.8455 \times 10^{10}}{510} = 8.39 \times 10^8$ N-mm

$\therefore$ Minimum B.M $= 8.39 \times 10^8$ N-mm.

$$M = \frac{w\,L^2}{8} = \frac{w \times 256 \times 1000}{8} = 32000\ w \text{ N-mm}$$

$w = \dfrac{8.39 \times 10^8}{32000} = 2.62 \times 10^4$ N/m $= 26.2$ kN/m; Dead load $= 10.8$ kN/m.

$\therefore$ External load $= 26.2 - 10.8 =$ **15.4** kN/m.

25.11. KERN DISTANCES AND EFFICIENCY OF SECTION

Let us consider an unsymmetrical beam section subjected to a prestressing force P_e acting at an eccentricity e, with respect to the C.G. of concrete section as shown in Fig. 25.36(*a*). Fig.25.36(*b*) shows the stress distribution due to combined effect of eccentric prestress and dead load moment M_d. Let the resultant compressive force in $C\ (= P_e)$ act at distance of k_b' from the C.G. of concrete area.

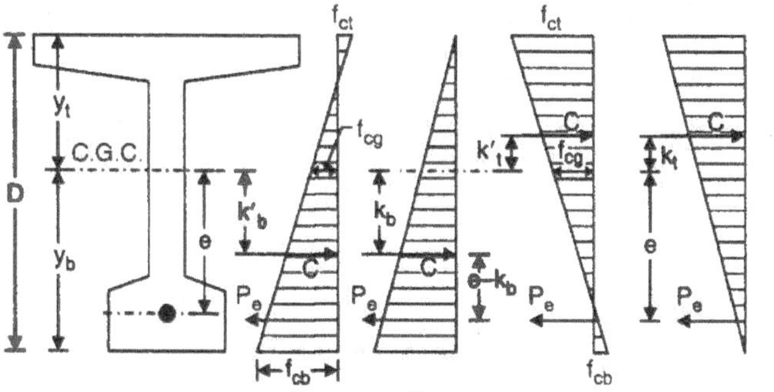

(a) Section (b) Stress distribution due to P_e and M_d (c) Stress distribution due to P_e and M_T

FIG. 25.36

If $f_{cg}\left(=\dfrac{C}{A}\right)$ is the stress ordinate at C.G. of the section, the stress at the top (f_{ct}) is given by

$$f_{ct} = f_{cg} - \frac{C\,.\,k_b'\,.\,y_t}{I} = f_{cg} - \frac{C\,.\,k_b'\,.\,y_t}{A\,r^2} = f_{cg} - f_{cg}\,.\,\frac{k_b'\,.\,y_t}{r^2} = f_{cg}\left(1 - \frac{k_b'\,y_t}{r^2}\right)$$

(where r is the radius of gyration for the section).

$$k_b' = \frac{r^2}{y_t}\left(1 - \frac{f_{ct}}{f_{cg}}\right) \qquad \text{...[25.24(a)]}$$

If no tension is allowed at top, f_{ct} will be zero, and the stress distribution will be triangular, as shown in Fig. 25.36(*b*)(*ii*). In that case, the distance k_b of point of application of C is given by

$$k_b = \frac{r^2}{y_t} = \frac{I}{A\,y_t} \qquad \ldots(25.24)$$

The distance k_b is known as *bottom kern distance.* It is the distance from the C.G.C. where the resultant compressive force acts, resulting in zero stress at the top.

Let us now consider the effect of total moment M consisting of dead load moment M_d and live load moment M_L. Fig. 25.36(*c*)(*i*) shows the final stress distribution. Let the resultant compressive force C act at a distance k_t' from C.G. of concrete. If f_{cg} is the stress in concrete at C.G., the stress at bottom (f_{cb}) is given by

$$f_{cb} = f_{cg} - \frac{C\,.\,k_t'\,y_b}{I} = f_{cg} - \frac{C\,k_t'\,.\,y_b}{A\,r^2} = f_{cg} - \frac{f_{cg}\,k_t'\,.\,y_b}{r^2} = f_{cg}\left(1 - \frac{k_t'\,y_b}{r^2}\right)$$

$$k_t' = \frac{r^2}{y_b}\left(1 - \frac{f_{cb}}{f_{cg}}\right) \qquad \ldots[25.25(a)]$$

If no tension is allowed at bottom, f_{cb} will be zero, and the stress distribution will be triangular, as shown in Fig. 25.36(*c*)(*ii*). In that case, the distance k_t of point of application of C is given by

$$k_t = \frac{r^2}{y_b} = \frac{I}{A\,y_b} \qquad \ldots(25.25)$$

The distance k_t is known as *top kern distance.* Let us consider Fig. 25.36(*b*)(*ii*). The lever arm between C and P_e is $e - k_b$. Equating the internal moment to the external moment, we get

$$M_d = P_e\,(e - k_b) \qquad \ldots(25.26)$$

Similarly, consider Fig. 25.36(*c*)(*ii*), where lever arm between C and P_e is $(k_t + e)$. Equating the external moment to the internal moment, we get

$$M_T = (M_d + M_l) = P_e\,(k_t + e) \qquad \ldots(25.27)$$

or

$$M_l = P_e\,(k_t + e) - M_d = P_e\,(k_t + e) - P_e\,(e - k_b)$$

$\therefore$

$$M_l = P_e\,(k_t + k_b) \qquad \ldots(25.28)$$

Equation 25.28 is very important, suggesting that greater the sum of the two kern distances $(k_t + k_b)$, greater will be the capacity of the beam to carry the external moment. If D is the depth of the beam the *efficiency factor* ρ on crack resistance may be expressed as

$$\rho = \frac{k_t + k_b}{D} = \frac{\dfrac{r^2}{y_b} + \dfrac{r^2}{y_t}}{y_b + y_t}$$

$$\text{or } \rho = \frac{r^2}{y_t\,.\,y_b} = \frac{I}{A\,y_t\,y_b} \qquad \ldots(25.29)$$

For a rectangular section of width B and depth D,

$$\rho = \frac{\frac{1}{12}\,BD^3}{B\,D\left(\dfrac{D}{2} \times \dfrac{D}{2}\right)} = \frac{1}{3}$$

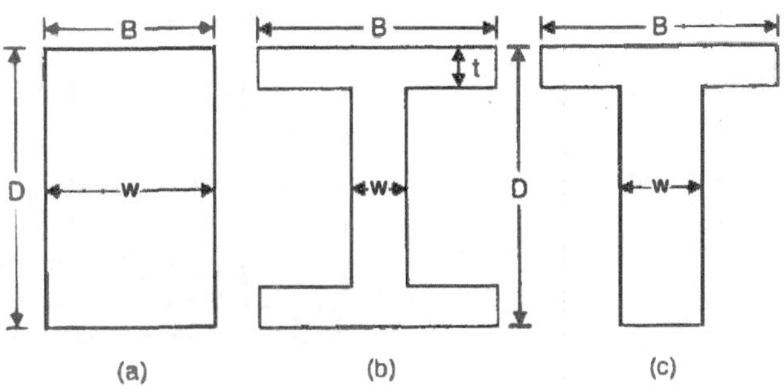

FIG. 25.37.

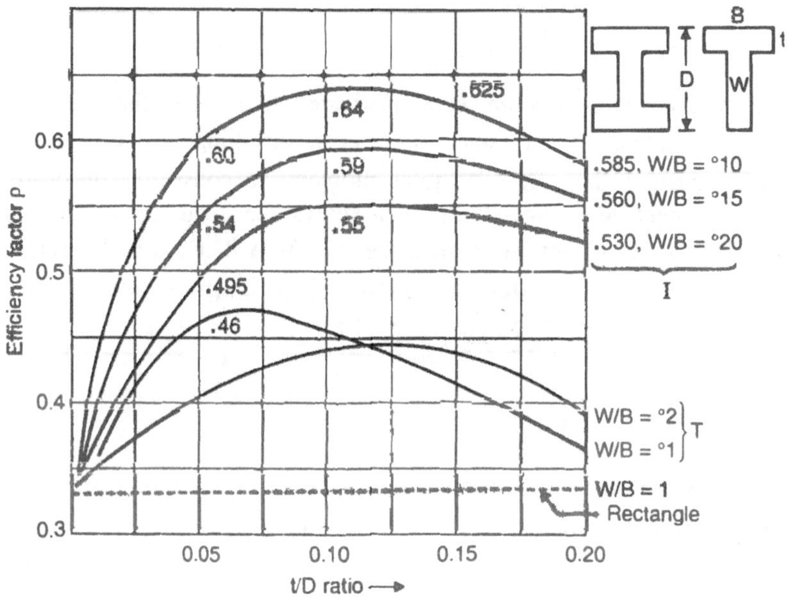

FIG. 25.38. EFFICIENCY OF SECTIONS.

For a symmetrical I-section,

$$\rho = \frac{I}{A\, y_t\, y_b} = \frac{4\, I}{A\, D^2} = \frac{1}{3}\, \frac{B\, D^3 - (B - W)\,(D - 2\,t)^3}{D^2\,[B\, D - (B - W)\,(D - 2\,t)]} = \frac{1}{3}\left[\frac{1 - (1 - W/B)(1 - 2\,t/D)^3}{1 - (1 - W/B)(1 - 2\,t/D)}\right] \quad \ldots(25.30)$$

Similar expressions can be derived for T and other sections. Fig. 25.38 shows the efficiency curves for I, T and rectangular sections for various W/B and t/D ratios. It is found that the rectangular section ($B/W = 1$) has a constant $\rho = 0.333$. Efficiency of I-section is maximum for $t/D \approx 0.1$. Also, smaller the W/B ratio, greater the efficiency. A rectangular section gives minimum efficiency on crack resistance.

Example 25.14. *Fig. 25.39 shows four-sections each having equal area of cross section. Determine the kern distance and efficiency of each section.*

Solution (*a*) ***Rectangular section***

$A = 30 \times 90 = 2700 \text{ cm}^2$; $y_t = y_b = 45 \text{ cm} = y$ $I = \frac{1}{12} \times 30\,(90)^3 = 1822500 \text{ cm}^4$

$$\therefore k_t = k_b = \frac{I}{A\,y} = \frac{1822500}{2700 \times 45}$$

$= 15$ cm and $\rho = \dfrac{I}{A\, y_t \,.\, y_b}$

$$= \frac{k_t}{y} = \frac{15}{45} = \frac{1}{3} = \mathbf{0.3333}$$

(*b*) ***Symmetrical I-section***

$A = 2(60 \times 15) + (60 \times 15)$

$= 2700 \text{ cm}^2$

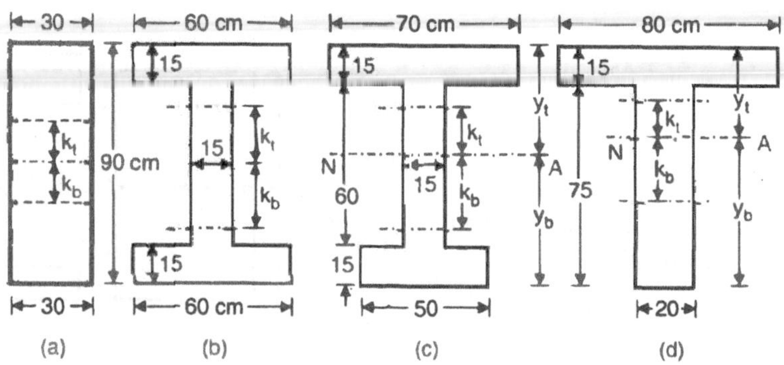

FIG. 25.39.

$$I = \frac{1}{12} \times 60\,(90)^3 - \frac{1}{12}\,(60-15)(90-30)^3 = 2835000 \text{ cm}^4$$

$$y_t = y_b = 45 \text{ cm} = y \quad ; \qquad k_t = k_b = \frac{I}{A\,y} = \frac{2835000}{2700 \times 45} = 23.33 \text{ cm}$$

$$\rho = \frac{I}{A\,y_t\,y_b} = \frac{I}{A\,y^2} = \frac{k_t}{y} = \frac{23.33}{45} = \mathbf{0.5185.}$$

(*c*) ***Unsymmetrical* I-*section* [Fig. 25.39(*c*)]**

$$A = (70 \times 15) \times (50 \times 15) + (60 \times 15) = 2700 \text{ cm}^2$$

$$y_t = \frac{(70 \times 15 \times 7.5) + (60 \times 15 \times 45) + (50 \times 15 \times 82.5)}{2700} = 40.83 \text{ cm}$$

$$y_b = 90 - 40.83 = 49.17 \text{ cm}$$

$$I = \left[\frac{1}{12} \times 70 \times 15^3 + 1050\,(40.83 - 7.5)^2\right] \times \left[\frac{1}{12} \times 15 \times 60^3 + 60 \times 15\,(45 - 40.83)^2\right] + \left[\frac{1}{12} \times 50 \times 15^2 + 750\,(49.17 - 7.5)^2\right] = 2788124 \text{ cm}^4$$

$$k_t = \frac{I}{A\,y_b} = \frac{2788214}{2700 \times 49.17} = \mathbf{21.00 \text{ cm}}. \quad ; \qquad k_b = \frac{I}{A\,y_t} = \frac{2788214}{2700 \times 40.83} = \mathbf{25.29 \text{ cm}}.$$

$$\rho = \frac{k_t + k_b}{D} = \frac{21.00 + 25.29}{90} = \mathbf{0.514}$$

Alternatively, $$\rho = \frac{I}{A\,y_t\,y_b} = \frac{2788124}{2700 \times 40.83 \times 49.17} = 0.514$$

(*d*) **T-*section* [Fig. 25.39(*d*)]**

$$A = (80 \times 15) + (75 \times 20) = 2700 \text{ cm}^2$$

$$y_t = \frac{(80 \times 15 \times 7.5) + (75 \times 20 \times 52.5)}{2700} = 32.5$$

$$y_b = 90 - 32.5 = 57.5 \text{ cm}.$$

$$I = \left[\frac{1}{12} \times 80 \times 15^3 + 80 \times 15\,(32.5 - 7.5)^2\right] + \left[\frac{1}{12} \times 20 \times 75^3 + 75 \times 20\,(57.5 - 37.5)^2\right]$$

$$= 2075625 \text{ cm}^4$$

$$k_t = \frac{I}{A\,y_b} = \frac{2075625}{2700 \times 57.5} = \mathbf{13.37 \text{ cm}} \quad ; \qquad k_b = \frac{I}{A\,y_t} = \frac{2075265}{2700 \times 32.5} = \mathbf{23.65 \text{ cm}}$$

$$\rho = \frac{k_t + k_b}{D} = \frac{13.37 + 23.65}{90} = \mathbf{0.41}$$

Alternatively, $$\rho = \frac{I}{A\,y_t\,y_b} = \frac{2075625}{2700 \times 32.5 \times 57.5} = 0.41$$

It will thus be seen that unsymmetrical I-section has the maximum efficiency while the rectangular section has minimum efficiency.

25.12. DESIGN OF SECTIONS FOR FLEXURE : MAGNEL'S METHOD

Notations. In the notations given below, suffix t stands for top fibre, while suffix b stands for the bottom fibre of the concrete section. Suffix i stands for initial stress condition, suffix d corresponds to dead loads while l corresponds for live load conditions.

P_i = Initial prestressing force.
P_e = Effective prestressing force $= \lambda P_i$.
λ = Loss factor or proportion of prestressing that remains permanently.
A = Sectional area.
I = Moment of inertia of the section.
r = Radius of gyration.
e = Eccentricity of pre-stressing force.
M_d = Moment due to dead weight.
M_l = Moment due to external or superimposed load.
M_T = Moment due to total load $= M_d + M_l$.
f_{yp} = Stress at any level y due to prestressing force.
f_{td} = Tensile stress in concrete at top, due to initial dead load only.
f_{bd} = Compressive stress in concrete at bottom, due to initial dead load only.
f_{ti} = Total stress at top, under initial stress condition.
f_{bi} = Total stress at bottom under initial stress condition.
f_{tl} = Compressive stress at top due to external load.
f_{bl} = Tensile stress at bottom due to external load.
f_{tT} = Total stress at top (compressive) under final stress conditions.
f_{bT} = Total stress at bottom (tensile) under final stress condition.
p_{ci} = Permissible compressive stress in concrete, under initial stress condition.
p_{ti} = Permissible tensile stress in concrete, under initial stress condition.
p_c = Permissible compressive stress in concrete under final stress condition.
p_t = Permissible tensile stress in concrete under final stress condtion.
y_t = Distance of top fibre from C.G.C.
y_b = Distance of bottom fibre from C.G.C.
k_t = Distance of top kern level.
k_b = Distance of bottom kern level.

Initial stress conditions : Fig. 25.40(*a*) shows a concrete section, prestressed with a force p_i acting at eccentricity e. Fig. 25.40(*b*) shows the stresses induced by the eccentric prestress force. The stress at any distance y is given by

$$f_{yp} = \frac{P_i}{A} \mp \frac{P_i \cdot e \cdot y}{I} = \frac{P_i}{A} \mp \frac{P_i \cdot e \cdot y}{A\, r^2} = \frac{P_i}{A}\left(1 \mp \frac{e\, y}{r^2}\right),$$ with minus sign for upper fibres and plus sign for lower fibres.

Fig. 25.40(*c*) shows the stress distribution due to dead load moment M_d, which will cause compressive stress (f_{td}) at top and tensile stress (f_{bd}) at bottom. The stress at any level is given by $f_{bi} = \pm \dfrac{M_d \cdot y}{I}$

Fig. 25.40(*d*) shows the total stress due to eccentric prestress force and the dead load moment. The stresses for top and bottom fibres, due to this stress conditions are as follows : Tensile stress at top fibre:

$$f_{ti} = \frac{P_i}{A}\left(-1+\frac{e\,y_t}{r^2}\right) - \frac{M_d\,y_t}{I} \quad \ldots(a)$$

Compressive stress at bottom fibre :

$$f_{bi} = \frac{P_i}{A}\left(1+\frac{e\,y_b}{r^2}\right) - \frac{M_d\,y_b}{I} \quad \ldots(b)$$

By inspection of Fig. 25.40(*d*), there is possibility of tensile stress at the top. *The stress should be lesser than the permissible tensile stress* (p_{ti}). Similarly, maximum compressive stress at bottom (f_{bi}) should be less than the permissible compressive stress (p_c) for concrete. Hence we have two condition equations for initial stress conditions

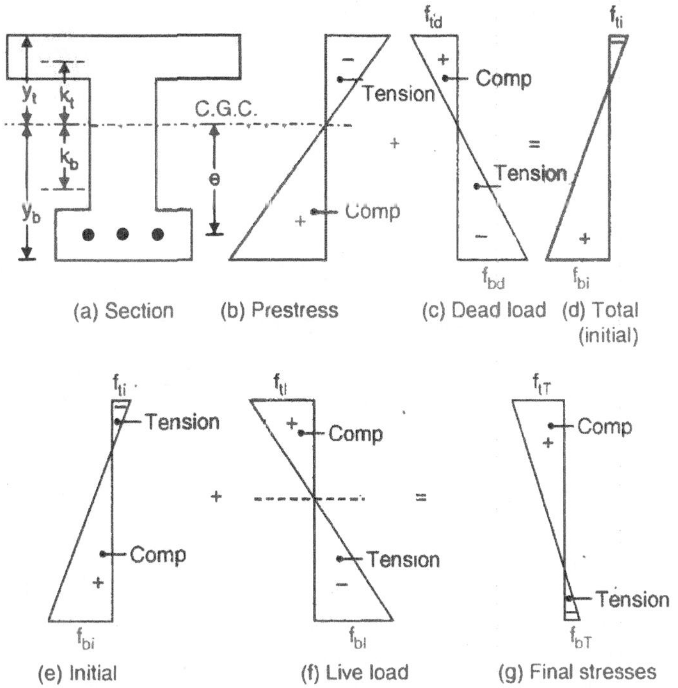

FIG. 25.40. INITIAL AND FINAL STRESS CONDITIONS

$$\frac{P_i}{A}\left(-1+\frac{ey_t}{r^2}\right) - \frac{M_d\,y_t}{I} \le p_{ti} \quad \ldots(25.31)$$

$$\frac{P_i}{A}\left(1+\frac{e\,y_b}{r^2}\right) - \frac{M_d\,y_b}{I} \le p_{ci} \quad \ldots(25.32)$$

Final stress conditions : Final stress conditions are included when external moment (M_l) is applied to the section. Fig. 25.40 (*f*) shows the stress distribution due to M_l the stress being compressive at top (f_{tl}) and tensile at bottom (f_{bl}). Fig. 25.40(*g*) shows the final stress distribution due to eccentric prestress, dead load moment M_d and live load moment M_l. The stress at top fibre (f_{tT}) will be compressive while stress at bottom fibre (f_{bT}) is likely to be tensile. Both these stresses should be larger than the permissible values. If the effective prestress $P_e = \lambda\, P_i$, we get the following two condition equations :

Compressive stress at top, $f_{tT} \le p_c$

or
$$-\frac{\lambda\, P_i}{A}\left(-1+\frac{e\,y_t}{r^2}\right) + \frac{M_d\,y_t}{I} + \frac{M_l\,.\,y_t}{I} \le p_c \quad \ldots(25.33)$$

Similarly tensile stress at bottom, $f_{bT} \le p_t$

or
$$-\frac{\lambda\, P_i}{A}\left(1+\frac{e\,y_b}{r^2}\right) + \frac{M_d\,y_b}{I} + \frac{M_l\,y_b}{I} \le p_t \quad \ldots(25.34)$$

In the above four equations, there are four unknowns : A, I, P_i and e. The values of these unknowns should be such that these satisfy all the above equations simultaneously. Magnel suggested a graphical solution. But before that, let us find the section modulus.

Adding Eq. 25.31 and 25.33, we get

$$\frac{P_i}{A}(1-\lambda)\left(-1+\frac{e\,y_t}{r^2}\right)+\frac{M_l\,y_t}{I}\le p_c+p_{ti} \qquad \text{or} \qquad \frac{I}{y_t}\ge\frac{M_l}{(p_c+p_{ti})-\frac{P_i}{A}(1-\lambda)\left(-1+\frac{e\,y_t}{r^2}\right)}$$

Let $P_i / A =$ average compressive stress $= f_c$

$$\frac{I}{y_t}\ (\text{or } Z_t)\ge\frac{M_l}{(p_c+p_{ti})-f_c(1-\lambda)\left(-1+\frac{e\,y_t}{r^2}\right)} \qquad \ldots(25.35)$$

Similarly, adding Eq. 25.32 and 25.34, we get

$$\frac{P_i}{A}(1-\lambda)\left(1+\frac{e\,y_b}{r^2}\right)+\frac{M_l\,y_b}{I}\le p_{ci}+p_t$$

or

$$\frac{I}{y_b}\ (\text{or } Z_b)\ge\frac{M_l}{(p_{ci}+p_t)-f_c(1-\lambda)\left(1+\frac{e\,y_b}{r^2}\right)} \qquad \ldots(25.36)$$

Equations 25.35 and 25.36 gives the section moduli. It is clear from the above equations that *the section modulus is independent of dead load moments*. The prestressed beam is known as 'carry itself' beam, since its section does not have to be increased to carry its weight, as is required for reinforced concrete beams. Hence the section of prestressed beam is much smaller than the ordinary reinforced concrete beam. This conclusion is not valid if the dead load moment is very large in comparision to M_l. For obtaining dimensions approximately, the following values may be adopted, for use in Eq. 25.35 and 25.36.

$\lambda = 0.85$ (*i.e.* loss equal to 15%)

$\frac{P_i}{A} = f_c = \frac{1}{2}p_c$ or $\frac{1}{2}p_{ci}$ as the case be.

$\frac{e\,y_t}{r^2} = \frac{e\,y_b}{r^2} = 2$ in beams which have more or less uniform section.

For sections with heavy top flange, $\frac{e\,y_t}{r^2}$ may be taken as 2 and $\frac{e\,y_b}{r^2}$ equal to 3.

For a section to be most economical, the sectional area (A) should be the least for a required value $\frac{I}{y_b}$. In other words $\frac{I}{A\,y_b}\left(\text{or } \frac{r^2}{y_b}\right)$ should be as large as possible.

But $r^2/y_b = k_t =$ distance of upper kern point from the centroid, and its maximum value can be only y_t. For this extreme condition

$$\frac{r^2}{y_b} = y_t \qquad \text{or} \qquad \frac{r^2}{y_b\,y_t} = 1.$$

When r^2/y_b is less that y_t, for a general case, the factor $r^2/y_b y_t$ ($=\rho$) is a measure of the efficiency of the section (see Eq. § 25.29 also). For rectangular section, $r^2/y_b\,y_t = 1/3$ (see § 25.11). For I-section, $r^2/y_b\,y_t \ge 0.5$ shows an economical section, while $r^2/y_b\,y_t \le 0.45$ gives a heavy section. However, it is usually not possible to raise ρ higher than 0.55 without making the web thinner than desirable. After having determined the section moduli

(I/y_t and I/y_b) the other quantities to be found are P_i and e. In order that these values satisfy Eqs. 25.31 to 25.34, Magnel suggested the following graphical procedure:

(a) Equation 25.31 (Line T₁)

$$\frac{P_i}{A}\left(\frac{e\,y_t}{r^2}-1\right)\le\frac{M_d}{Z_t}+p_{ti}$$

or
$$\frac{A}{P_i}\ge\frac{\dfrac{e\,y_t}{r^2}-1}{\dfrac{M_d}{Z_t}+p_{ti}}\qquad\ldots[25.31(a)]$$

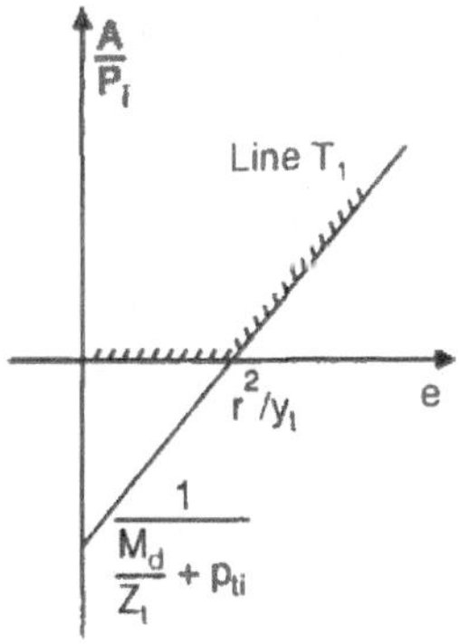

FIG. 25.41

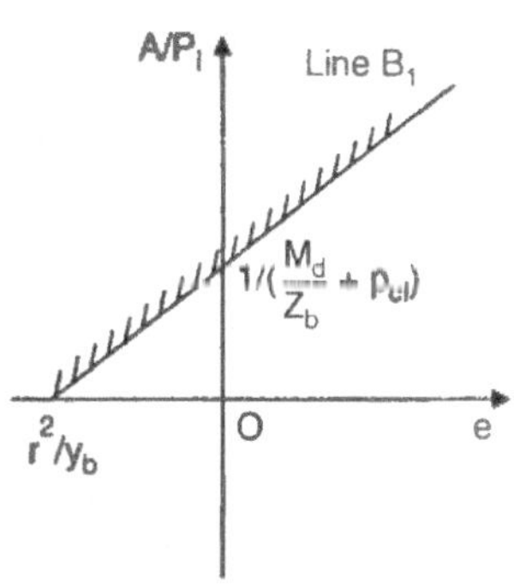

FIG. 25.42

If a graph is plotted between A/P_i and e, a straight line will be obtained as shown in Fig. 25.41. For $A/P_i = 0$, $e \le r^2/y_t$.

Similarly, for $e = 0$, $\dfrac{A}{P_i}\ge\dfrac{1}{\dfrac{M_d}{Z_t}+p_{ti}}$

These give the values of co-ordinates at points of crossing the line with the axis. Any value of A/P_i and e lying in the shaded area will satisfy this equation.

(b) Equation 25.32 (Line B₁)

$$\frac{P_i}{A}\left(1+\frac{e\,y_b}{r^2}\right)\le\frac{M_d}{Z_b}+p_{ct}\quad\text{or}\quad\frac{A}{P_i}\ge\frac{\left(1+\dfrac{e\,y_b}{r^2}\right)}{\dfrac{M_d}{Z_b}+p_{ct}}\qquad\ldots[25.32\ (a)]$$

Fig. 25.42 shows the plot of Eq. 25.32(a). Any value of A/P_i and e lying in the shaded area will satisfy Eq. 25.32(a).

(c) Equation 25.33 (Line T_2)

$$\frac{\lambda P_i}{A}\left(1-\frac{e\,y_t}{r^2}\right)\le p_c-\frac{M_T}{Z_t}$$

or
$$\frac{A}{P_i}\ge\frac{\lambda\left(1-\dfrac{e\,y_t}{r^2}\right)}{\left[p_c-\dfrac{M_T}{Z_t}\right]}\qquad\ldots[25.33(a)]$$

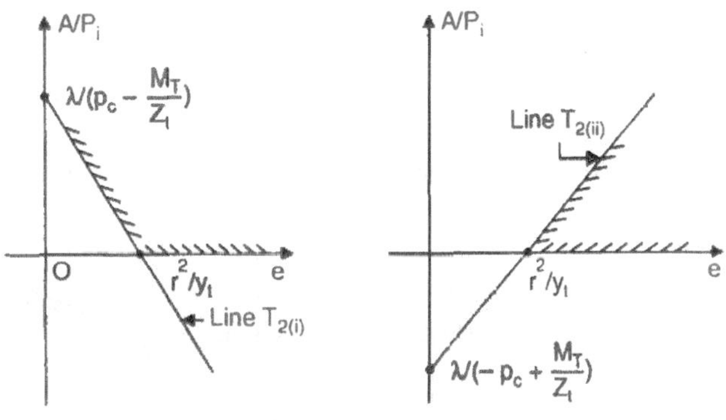

(a) (b)

FIG. 25.43

Two conditions, may arise depending upon the relative magnitudes of p_c and M_T/Z_t

Case (i) $p_c \ge M_T/Z_t$: If e more than r^2/y_t, equation 25.33(a) will be satisfied for all values of P_i. If e is less than r^2/y_t, the prestress force will be at point higher than the lower kern, causing compressive stress at top, instead of the usual tensile stress.

or
$$\frac{A}{P_i} \geq \frac{\lambda\left(1 - \frac{e\,y_t}{r^2}\right)}{\left[p_c - \frac{M_T}{Z_t}\right]} \qquad \ldots[25.33(a)(i)]$$

This is represented by line T_2 (i) in Fig. 25.43(a)

Case (ii) $p_c \leq M_T/Z_t$: If e is less than r^2/y_t Eq. 25.33 (a) will never be satisfied. For $e > r^2/y_t$.

$$\frac{A}{P_i} \leq \frac{\lambda\left(-1 + \frac{e\,y_t}{r^2}\right)}{\left[-p_c + \frac{M_T}{Z_t}\right]} \qquad \ldots[25.33\ (a)(ii)]$$

This is represented by line T_2 (ii) in Fig. 25.43(b)

Equation 25.34 (Line B_2)

$$\frac{\lambda\,P_i}{A}\left(1 + \frac{e\,y_b}{r^2}\right) \leq \frac{M_T}{Z_b} - p_t$$

or
$$\frac{A}{P_i} \geq \frac{\lambda\left(\frac{e\,y_b}{r^2} + 1\right)}{\left[\frac{M_T}{Z_b} - p_t\right]} \qquad \ldots[25.34\ (a)]$$

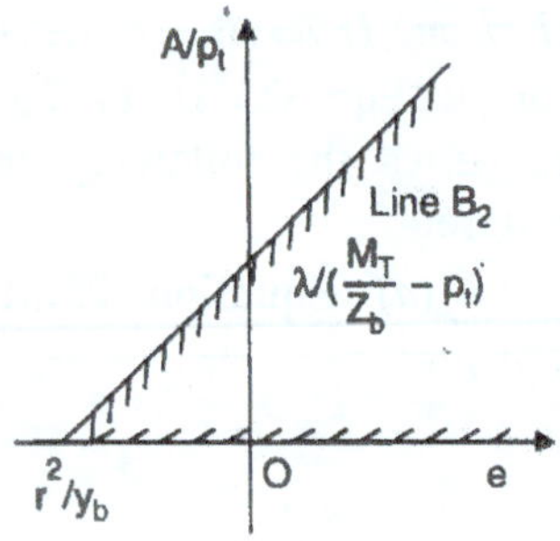

FIG. 25.44

Fig. 25.44 shows the straight line B_2 represented by the above equation.

Values of A/P_i and e lying in the shaded area will satisfy Eq. 25.34(c).

Combined effect : Figs. 25.45(a) and (b) show the four equations plotted together. The shaded area common to all the four lines give the magnitude of A/P_i and e that would satisfy all the four equations. In both the diagrams, point R and R' indicate minimum value of e, while points S and S' indicate minimum values of P_i. Hence for design purposes P_i and e should be selected with respect to points S and S'. The values of e corresponding to S and S' are large. If these values of e are not available, the maximum values of e available should be used.

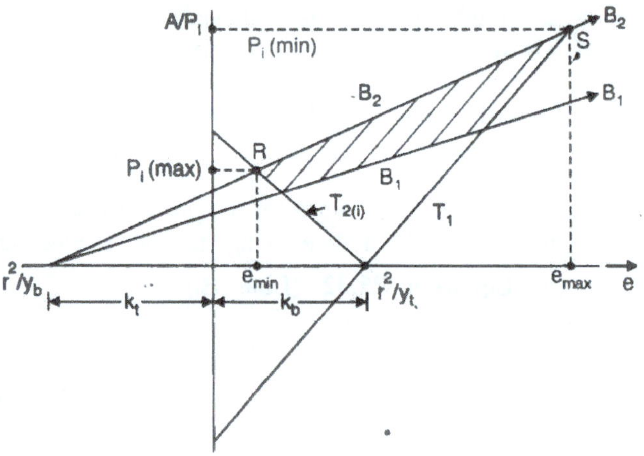

(a) Lines B_1, B_2, T_1 and $T_{2(i)}$

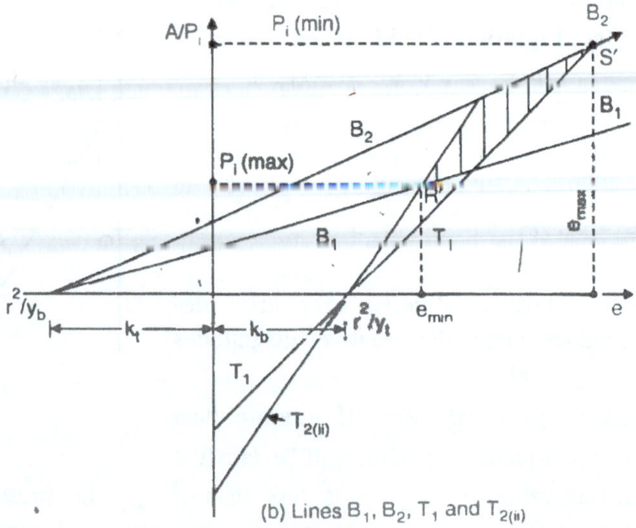

(b) Lines B_1, B_2, T_1 and $T_{2(ii)}$

FIG. 25.45

For a simply suported beam, both M_d and M_l go on decreasing towards the support, and become zero at the supports. Since P_i remains constant, e should be changed from section to section, giving rise to a curved cable profile.

Cable zone : It is clear from the above that the C.G. of the cable should remain within some limits so that the stresses are within the allowable limits.

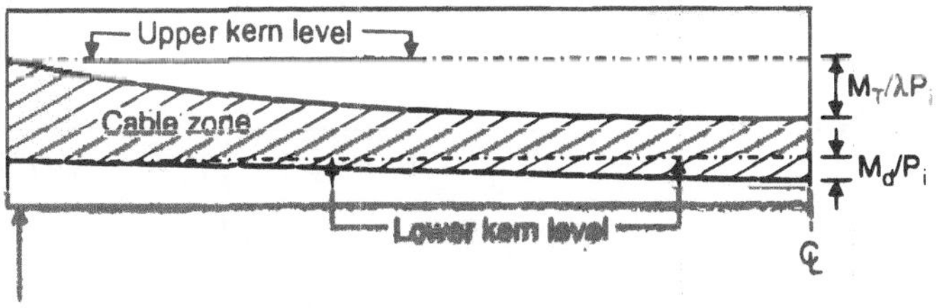

FIG. 25.46. CABLE ZONE.

Consider Eq. 25.31 corresponding to initial prestress and dead load moment M_d . If the minimum stress is equal to zero, we have

$$\frac{P_i}{A}\left(1-\frac{ey_t}{r^2}\right)+\frac{M_d\,y_t}{I}=0$$

Putting $I = Ar^2$, $\quad P_i\left(-1+\frac{e\,y_t}{r^2}\right)=\frac{M_d\,y_t}{r^2}\quad$ or $\quad e=\frac{r^2}{y_t}+\frac{M_d}{P_i}$

or
$$e = k_b + \frac{M_d}{P_i} \qquad \text{...[25.37 (a)]}$$

Eq. 25.37(a) suggests that for no tension to be developed, the cable should be placed at M_d/P_i below the lower kern point.

Again, consider Eq. 25.35 corresponding to eccentric prestress M_d and M_l. Minimum stress will occur at the bottom fibre. If this minimum stress is to be zero, we get

$$\frac{\lambda P_i}{A}\left(1+\frac{e\,y_b}{r^2}\right)-\frac{M_d\,y_b}{Ar^2}-\frac{M_l\,y_b}{Ar^2}=0 \quad \text{or} \quad \lambda P_i\left(1+\frac{e\,y_b}{r^2}\right)=\frac{M_T\,y_b}{r^2}$$

$$e=\frac{M_T}{\lambda P_i}-\frac{r^2}{y_b}=\frac{M_T}{\lambda P_i}-k_t \qquad \text{...(25.37)}$$

This suggests that the cable should be plotted at $M_T/\lambda P_i$ below the upper kern point.

Both Eqs. 25.37(a) and 25.37(b) give the limiting positions of the cable for the section. If these positions are found for all sections, and curves are drawn as shown in Fig. 25.46, the zone (shown shaded) between the two curves will be the safe zone in which the resultant cable axis should lie. This zone is known as the ***cable zone***. In actual practice, several cables may be used. It is not necessary to locate each cable within the shaded area. The individual cables may be placed anywhere in the cross-section, provided that the resultant cable line (*i.e.* c.g.s.) lies within the cable zone.

25.13. RECTANGULAR SECTION

A rectangular section of width B and depth D is symmetrical about the C.G. of the section, thus having $y_t = y_b = D/2$, and $Z_b = Z_t = Z = BD^2/6$. Hence Eqs. 25.35 and 25.36, giving the values of section modulus are modified as follows, taking $\frac{e\,y_t}{r^2}=\frac{e\,y_b}{r^2}=2$ and $f_c=\frac{1}{2}p_c$.

$$Z_t = Z = \frac{M_l}{(p_c + p_{ti}) - \frac{1}{2} p_c (1 - \lambda)(-1 + 2)} \qquad \text{....[25.35(a)]}$$

or

$$Z = \frac{B\,D^2}{6} = \frac{M_l}{\frac{p_c}{2}(\lambda + 1) + p_{ti}} \qquad \text{....[25.35(b)]}$$

Similarly,

$$Z_b = Z = \frac{M_l}{(p_{ci} + p_t) - \frac{1}{2} p_c (1 - \lambda)(1 + 2)} \qquad \text{...[25.36(a)]}$$

Taking $p_{ci} = p_c$,

$$Z = \frac{B\,D^2}{6} = \frac{M_l}{\frac{p_c}{2}(3\,\lambda - 1) + p_t} \qquad \text{...[25.36(b)]}$$

It will be seen that Eq. 25.36(*b*) gives larger section modulus, should be adopted for design. If λ is assumed to be 0.85, Eq. 25.36 (*b*) simplifies to :

$$Z = \frac{M_l}{0.775\ p_c + p_t} \qquad \text{...[25.36(c)]}$$

Similarly, Eq. 25.35(*b*) simplifies to :

$$Z = \frac{M_l}{0.925\ p_c + p_{ti}} \qquad \text{...[25.35(c)]}$$

For a rectangular section, the depth of the beam should be between 1/15 to 1/20 th of span. From the buckling point of view, the width of the beam should not be less than 1/30 th of the length between effective restraints or 1/4 th of the depth of the beam.

After having obtained the section modulus, *B* and *D*, the other unknowns to be determined are P_i and *e*.

Consider Eqs. 25.31 (Line T_1) and 25.34 (Line B_2), to get the minimum value of prestressing force (max. value of $1/P_i$). From Eq. 25.31, we get

$$\frac{P_i}{A}\left(e - \frac{r^2}{y_t}\right)\frac{y_t}{r^2} = M_d \frac{y_t}{A\,.\,r^2} + p_{ti} \quad \text{or} \quad P_i\left(e - \frac{r^2}{y_t}\right) = M_d + p_{ti}\frac{A\,r^2}{y_t}$$

But

$$\frac{r^2}{y_t} = \frac{r^2}{y_b} = \frac{r^2}{y} = \frac{I}{A\,.\,y} = \frac{Z}{A} = \frac{D}{6} \quad \text{and} \quad \frac{A\,r^2}{y} = \frac{I}{y} = Z$$

$$\therefore \quad P_i\left(e - \frac{D}{6}\right) = M_d + p_{ti}\,Z \qquad \text{...(a)}$$

Similarly, from Eq. 25.34

$$-\frac{\lambda\,P_i}{A}\left(e + \frac{r^2}{y_b}\right)\frac{y_b}{r^2} = p_t - \frac{M_d\,y_b}{A\,r^2} - \frac{M_l\,y_b}{A\,r^2} \quad \text{or} \quad \lambda\,P_i\left(e + \frac{D}{6}\right) = (M_d + M_l) - p_t\,.\,Z \qquad \text{...(b)}$$

Dividing (*a*) and (*b*), we get

$$\frac{\left(e - \frac{D}{6}\right)}{\left(e + \frac{D}{6}\right)} = \frac{\lambda\,(M_d + p_{ti}\,Z)}{(M_d + M_l) - p_t\,Z} \qquad \text{...(25.38)}$$

From this equation, *e* can be found. If p_t and p_{ti} are taken as zero, this equation further simplifies to :

$$\frac{e-\frac{D}{6}}{e+\frac{D}{6}}=\frac{\lambda\, M_d}{M_T} \qquad \text{...[25.38(a)]}$$

Knowing e, value of P_i can be obtained from Eq. 25.34 (line B_2) simplified as follows:

$$-\frac{\lambda\, P_i}{A}\left(e+\frac{r^2}{y_b}\right)\frac{y_b}{r^2}+\frac{M_d\, y_b}{I}+\frac{M_l\, y_b}{I}=p_t$$

$$\lambda\, P_i\left(e+\frac{D}{6}\right)=(M_d+M_l)-p_t\, Z \quad \text{or} \quad P_i=\frac{(M_d+M_l)-p_t\, Z}{\lambda\left(e+\frac{D}{6}\right)} \qquad \text{...(25.39)}$$

If, however, available eccentricity is less than the one found from Eq. 25.38(a), the corresponding value of P_i is obtained by putting the available value of e in Eq. 25.34 (Eq. 25.39) or Eq. 25.33 whichever gives higher value. Eq. 25.33 simplifies as follows:

$$P_i=\frac{(M_d+M_l)-p_c\, Z}{\lambda\left(e-\frac{D}{6}\right)} \qquad \text{...[25.39(a)]}$$

Critical span for slabs and rectangular beams : Critical span (L_c) is that span of the beam upto which the dead load does not affect the design of the section, and only live load need be considered as is evident from Eq. 25.36(c). Let us derive the expression for the same.

Fig. 25.47(a) shows the rectangular section with $k_t = k_b = D/6$ while Fig. 25.47(b) shows the stress distribution due to dead load. For stress to be zero at the top fibre (*i.e.* for triangular stress distribution, the resultant compressive force C must pass through the lower kern point, so that

$$M_d = C\,(e-k_b) = P_i\,(e-k_b) \qquad \text{(See Eq. 25.26 also)} \qquad \text{...(a)}$$

Now let ΔM be increase in moment, due to external load, and Δf_t be the increase in compressive stress at top fibre due to this.

Then $\Delta M = \Delta f_t\, Z$

If $\Delta M = M_l$ and $\Delta f_t = p_c$ = allowable compressive stress at top [Fig. 25.47(c)], we have $M_l = p_c \,.\, Z$ ...(b)

Now for a live load of w_l and critical span L_c,

$$M_l=\frac{w_l\, L_c^2}{8}, \quad \text{and} \quad Z=\frac{B\, D^2}{6}$$

$$\therefore \quad \frac{w_l\, L_c^2}{8}=p_c\,\frac{B\, D^2}{6}$$

$$\therefore \quad \frac{D}{L_c}=\sqrt{\frac{3}{4}\frac{w_l}{p_c}} \qquad \text{...(25.40)}$$

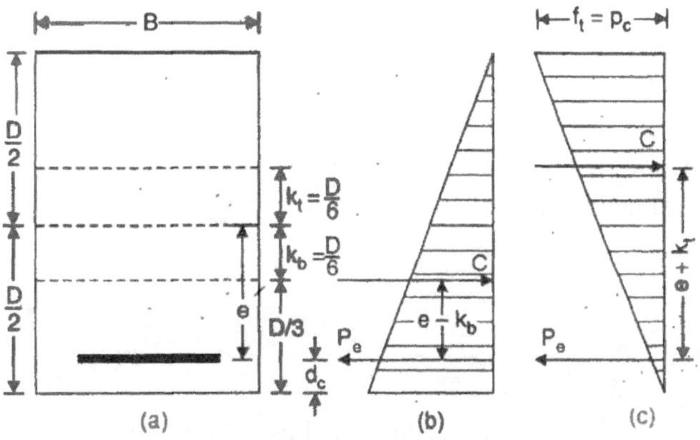

FIG. 25.47

Fig. 25.47(c) shows the final stress distribution due to P_e, M_d and M_l. For the triangular stress distribution,

the resultant compressive force must act at the upper kern, thus getting a lever arm of $(e + k_b)$ for the internal momnet.

$$\therefore \quad M_d + M_l = P_e\,(e + k_t) \qquad \text{...(see Eq. 25.27 also)}$$

Subtracting Eq. (*a*) from this, we get

$$M_l = P_e\,(k_b + k_t) \qquad \text{...(see Eq. 25.27 also)}$$

$$\frac{M_d}{M_l} = \frac{w_d}{w_l} = \frac{e - k_b}{k_b + k_t} \qquad \text{...(25.41)}$$

But $k_b = k_t = D/6$. If the distance of centre of steel from bottom fibre is d_c, we have $e = 0.5\,D - d_c$. Also, $w_d = 0.025\,BD$ N/m^2 where B and D are in mm.

$$\frac{0.025\,B\,D}{w_l} = \frac{0.5\,D - d_c - D/6}{D/3} \quad \text{or} \quad D = \frac{w_l}{0.025\,B}\left(1 - \frac{3\,d_c}{D}\right)$$

Equating this to Eq. 25.40, we get $\quad \dfrac{L_c}{2}\sqrt{3\,\dfrac{w_l}{p_c}} = \dfrac{w_l}{0.025\,B}\left(1 - 3\,\dfrac{d_c}{D}\right)$

or

$$L_c = \frac{1}{0.0125\,B}\sqrt{\frac{w_l \cdot p_c}{3}}\left(1 - 3\,\frac{d_c}{D}\right) \qquad \text{...(25.42)}$$

In a slab, $B = 1000$ mm $\quad \therefore \quad L_c = \dfrac{1}{1.25}\sqrt{\dfrac{w_l\,p_c}{3}}\left(1 - \dfrac{3\,d_c}{D}\right) \qquad$ ...[25.42(*a*)]

If the actual span (L) is less than L_c given above, the dead load is fully carried. *The critical span increases with eccentricity e* of the prestressing force, and attempts should be made to provide maximum possible eccentricity.

25.14. I-SECTION

(*a*) Determination of section modulus : Let us start with the case of an unsymmetrical I-section. The section moduli are given by Eqs. 25.35 and 25.36.

$$Z_t = \frac{M_l}{(p_c + p_{tt}) - f_c\,(1 - \lambda)\left(-1 + \dfrac{e\,y_t}{r^2}\right)} \qquad \text{...(25.35)}$$

$$Z_b = \frac{M_l}{(p_{ct} + p_t) - f_c\,(1 - \lambda)\left(1 + \dfrac{e\,y_b}{r^2}\right)} \qquad \text{...(25.36)}$$

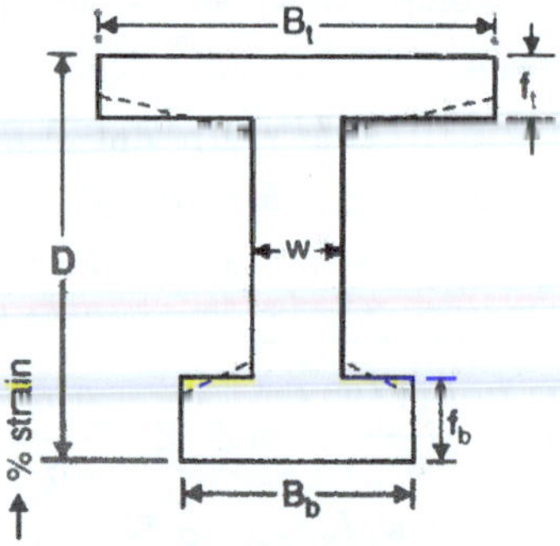

FIG. 25.48.

Taking $\quad f_c = \dfrac{p_c}{2}, \ \dfrac{e\,y_c}{r^2} = \dfrac{e\,y_b}{r^2} = 2$ (to start with),

$p_{ct} = p_c$ and $p_{tt} = p_t$, we get

$$Z_t = \frac{M_l}{\dfrac{p_c}{2}(\lambda + 1) + p_t} \quad \text{...[25.35 (A)]}; \quad Z_b = \frac{M_l}{\dfrac{p_c}{2}(3\,\lambda - 1) + p_t} \quad \text{...[25.36 (A)]}$$

By the inspection of Eqs. 25.35(*A*) and 25.36(*A*), it is clear that Z_b is more that Z_t, *i.e.* the bottom flange should be heavier (or wider) from the top flange. However, from buckling point of view, top flange is kept wider than the bottom flange. Bottom flange is kept thicker for accommodating prestressing cables. For the prestressing force to be minimum, eccentricity should be maximum. Eccentricity can be increased by providing heavier top flange so that C.G. is shifted up. All these factors are to be kept in mind while selecting the section. Taking, $\lambda = 0.85$,

$$Z_t = \frac{M_l}{0.925\ p_c + p_t} \quad \text{...[25.35(B)]} \quad \text{and} \quad Z_b = \frac{M_l}{0.775\ p_c + p_t} \quad \text{...[25.36(B)]}$$

When the dead load is greater, the depth of the section has to be increased beyond the value given by Eq. 25.36(*B*).

Fig. 25.48 shows an unsymmetrical I-section, with wider top flange, and a thicker bottom flange. Based on experience and practical considerations, Guyon has suggested the following proportions. However, the flanges are splayed for easiness of pouring concrete and to avoid sharp corners, as shown by dotted lines.

(*i*) ***Depth D***: If the dead moment (M_d) is 20% or more of total moment (M_T),

$$D = k\sqrt{M_T} \qquad \text{...[25.43 (a)]}$$

where k varies from 10 to 13 and M_T is in tonne-cm.

If M_d less than 20% of M_T, $D = k\sqrt{M_l}$ ...[25.43 (*b*)]

where M_l is the live load B.M.

According to Guyon : $D = \dfrac{L}{25} + 10$ cm (for spans 18 to 36 m) ...[25.43 (*c*)]

and $D = \dfrac{L}{20}$ (for spans 36 to 45 m) ...[25.43 (*d*)]

(*ii*) ***Web thickness (w)*** : The thickness should be adequate to accommodate the cables with sufficient cover. For this the details about the size of duct for various system should be known. A minimum cover of 30 mm or the size of cable whichever is more is provided. For example, if Freyssinet cable consisting of 12 wires of 5 mm wires are provided, the external diameter of sheathing required is 34 mm. Hence minimum $w = 3 \times 3.34 = 10.2$ cm for a single cable.

From other considerations,

For $D < 75$ cm, $w = \dfrac{D}{7}$ to $\dfrac{D}{8}$ ...[25.44 (*a*)]

For $D > 75$ cm, $w = \dfrac{D}{40} + 10$ cm ...[25.44 (*b*)]

(*iii*) ***Top flange***

$$B_t = 0.6\ D \text{ to } 0.8\ D \qquad \text{...(25.45)}$$

Minimum width of flange may be taken equal to $L/30$.

The depth of flange may be taken equal to 0.1 to 0.2 times overall depth.

(*iv*) ***Bottom flange*** : The thickness and width of the bottom flange are decided on the basis of accommodating all the cables at the section of maximum moment. The minimum clear distance between cables should not be less then 40 mm or the minimum dimension of the cable or 6.5 mm in excess of the larger size of aggregate used, whichever is greater.

(*b*) **Determination of e and P_i** : After having determined section modulus and other dimensions of the section, the next step to is determine the eccentricity e of the prestressing force. For minimum prestressing force (or maximum value of A/P_i) consider Eq. 25.31 (Line T_1) and Eq. 25.34 (Line B_2).

$$P_i\left[-\frac{1}{A}+\frac{e}{Z_t}\right] \le \frac{M_d}{Z_t}+p_{ti} \quad ...[25.31(b)] \; ; \text{ and } \quad \lambda P_i\left[\frac{1}{A}+\frac{e}{Z_b}\right] \le \frac{M_T}{Z_b}-p_t \quad ...[25.34\ (b)]$$

$$\therefore \qquad \frac{\dfrac{e}{Z_t}-\dfrac{1}{A}}{\dfrac{e}{Z_b}+\dfrac{1}{A}} \le \frac{\lambda\left(\dfrac{M_d}{Z_t}+p_{ti}\right)}{\left(\dfrac{M_T}{Z_b}-p_t\right)} \qquad ...(25.46)$$

From the above equation, maximum allowable e can be found. If this eccentricity is available, P_i can be found from Eq. 25.34, which simplifies as follows :

$$P_i = \frac{M_T - p_t Z_b}{\lambda\,(e + k_t)} \qquad ...(25.47)$$

If $p_t = 0$, we get $P_i = \dfrac{M_T}{\lambda\,(e+k_t)}$, which is the same as Eq. 25.27.

If, however, eccentricity e available is lesser than that given by Eq. 25.46, the corresponding value of P_i is obtained by putting the available value of e in Eq. 25.34 or 25.33 whichever gives higher value. Eq. 25.33 simplifies as follows :

$$P_i = \frac{M_T - p_c Z_t}{\lambda\,(e - k_b)} \qquad ...(25.48)$$

Alternatively, e and P_i can be obtained by plotting lines T_1 (Eq. 25.31), B_1 (Eq. 25.32), T_2 (Eq. 25.33) and B_2 (Eq. 25.34), as explained earlier.

Design Example 25.15. ***Design of prestressed concrete slab.*** ; *Design a prestressed concrete slab over a span of 16 metres to carry a superimposed load of 15 kN/m². Assume permissible stresses in concrete of 12 N/mm² and no tension. Also, assume that the initial losses in prestress in steel will amount to 15 per cent. The live load will be brought to bear after the ducts are grouted. The initial prestress in the wires will be 1100 N/mm².*

Solution

1. *Determination of the Section*

Consider one metre width of slab. $p_c = p_{ci} = 12$ N/mm². $p_t = p_{ti} = 0$.

The section modulus is given by Eq. 25.37,

$$Z = \frac{B\,D^2}{6} = \frac{M_l}{\dfrac{p_c}{2}(3\lambda - 1) + p_t} \quad \text{Taking } \lambda = 0.85, \text{ this reduces to}$$

$$\frac{B\,D^2}{6} = \frac{M_l}{0.775\,p_c + p_t} \qquad ...[25.36\ (c)]$$

$$M_l = \frac{15000\,(16)^2}{8} \times 1000 = 480 \times 10^6 \text{ N-mm}$$

$$p_c = 12 \text{ N/mm}^2 \; ; \; p_t = 0 \; ; \; B = 1000 \text{ mm}$$

$$\therefore \quad \frac{1000\, D^2}{6} = \frac{480 \times 10^6}{0.775 \times 12}$$

$$D = \sqrt{\frac{480 \times 10^6 \times 6}{0.775 \times 12 \times 1000}} = 556 \text{ mm} \; ; \quad \text{Adopt} \quad D = 560 \text{ mm}$$

$$\text{Dead load} = 0.56 \times 1 \times 1 \times 25000 = 14000 \text{ N/m}$$

$$\therefore \quad M_d = \frac{14000\,(16)^2}{8} \times 1000 = 448 \times 10^6 \text{ N-mm}$$

Total moment $\quad M_T = M_d + M_l = (480 + 448)\, 10^6 = 928 \times 10^6$ N-mm

Section modulus $\quad Z = \frac{1}{6} B\, D^2 = \frac{1}{6} \times 1000\,(560)^2 = 52.267 \times 10^6 \text{ mm}^3$

2. ***Determination of Eccentricity,*** e : The eccentricity e is given by Eq. 25.38

$$\frac{e - \frac{D}{6}}{e + \frac{D}{6}} = \frac{\lambda\,(M_d + p_{ti}\, Z)}{(M_d + M_l) - p_t\, Z} \quad \ldots(25.38)$$

Taking $p_t = p_{ti} = 0$, we get

$$\frac{e - \frac{D}{6}}{e + \frac{D}{6}} = \frac{\lambda\, M_d}{M_T} \quad \text{or} \quad \frac{e - \frac{560}{6}}{e + \frac{560}{6}} = \frac{0.85 \times 448 \times 10^6}{928 \times 10^6} = 0.4103$$

$$e - \frac{560}{6} = 0.4103 \left(e + \frac{560}{6} \right) ; \quad \text{From which } e = 223.3 \text{ mm.}$$

This is the maximum permissible value of e to get the minimum value of P_i. Let us assume that the size of the cable duct is such that this much eccentricity will be available.

3. ***Determination of Prestressing Force*** P_i : The prestressing force is given by Eq. 25.39

$$P_i = \frac{(M_d + M_l) - p_t \,.\, Z}{\lambda \left(e + \frac{D}{6} \right)} \quad \ldots(25.39)$$

Taking $\quad p_t = 0, \quad P_i = \dfrac{M_T}{\lambda \left(e + \frac{D}{6} \right)} = \dfrac{928 \times 10^6}{0.85 \left(223.3 + \frac{560}{6} \right)} = 3.44 \times 10^6$ N.

4. ***Design of Cables*** : Permissible prestress in cable $= 1100$ N/mm².

$$\therefore \quad \text{Area of steel} = \frac{3.44 \times 10^6}{1100} = 3127 \text{ mm}^2.$$

Using 7 mm ϕ wires, number of wires required per metre width of slab $= \dfrac{3127}{\frac{\pi}{4}(7)^2} \approx 81$ wires.

Adopting Magnel system, and providing 16 wires in a cable, the width of duct hole will be 65 mm and depth will be 60 mm. The duct spacing $= \dfrac{16}{81} \times 1000 \approx 200$ mm c/c.

Hence use ducts @ 200 mm c/c. Area of steel in each duct = $16 \times \frac{\pi}{4}(7)^2 = 615.8$ mm^2

5. ***Check for Stresses before Grouting Ducts*** : Take a unit of slab 200 mm wide. There will be one duct of size 65 × 60 mm in the unit. Let the N.A. be at $\bar{y}$ from the top fibre.

Net area $= (200 \times 560) - (65 \times 60) = 112000 - 3900 = 108100$ mm^2

$$\bar{y} = \frac{(112000 \times 280) - (3900 \times 503.3)}{108100} = 271.9 \text{ mm}$$

$$\therefore \quad \bar{x} = 280 - 271.9 = 8.1 \text{ mm}$$

$$I_{NA} = \left[\frac{1}{12} \times 200\,(560)^3 + 112000(8.1)^2\right] - \left[\frac{1}{12}\,65\,(60)^3 + 3900(8.1 + 223.3)^2\right]$$

$$= 2934.28 \times 10^6 - 210.0 \times 10^6 = 2724.28 \times 10^6 \text{ mm}^4$$

$$Z_t = \frac{2724 \times 10^6}{280 - 8.1} = 10.019 \times 10^6 \text{ mm}^3$$

$$Z_b = \frac{2724.28 \times 10^6}{280 + 3.1} = 9.456 \times 10^6 \text{ mm}^3$$

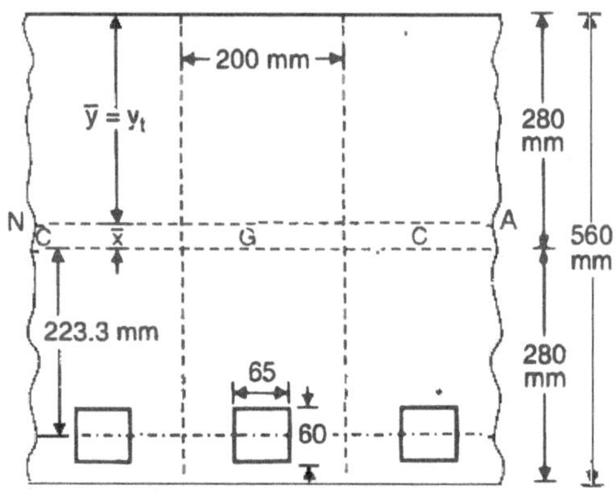

FIG. 25.49

∴ Stresses due to dead load are

$$f_{td} = +\frac{M_d'}{Z_t} = \frac{448 \times 10^6}{10.019 \times 10^6}\left(\frac{200}{1000}\right)$$

$$= +8.94 \text{ N/mm}^2 \quad (i.e.\ \text{compression})$$

$$f_{bd} = -\frac{M_d'}{Z_b} = \frac{448 \times 10^6}{9.456 \times 10^6}\left(\frac{200}{1000}\right)$$

$$= -9.48 \text{ N/mm}^2 \quad (i.e.\ \text{tensile})$$

Stresses due to prestressing force are

Top fibre : $\frac{P_i}{A} - \frac{P_i \cdot e}{Z_t} = 3.44 \times 10^6\left[\frac{1}{108100} - \frac{223.3}{10.019 \times 10^6}\right] \times \frac{200}{1000} = -8.97 \text{ N/mm}^2$

Bottom fibre : $\frac{P_i}{A} + \frac{P_i \cdot e}{Z_b} = 3.44 \times 10^6\left[\frac{1}{108100} + \frac{223.3}{9.456 \times 10^6}\right] \times \frac{200}{1000} = +22.61 \text{ N/mm}^2$

Hence the total stresses for the initial condition are

$$f_{ti} = 8.94 - 8.97 = 0.03 \text{ N/mm}^2 \ ; \qquad f_{bi} = -9.48 + 22.61 = 13.13 \text{ N/mm}^2$$

It is seen that maximum stresses are slightly exceeded but accepted since it diminishes as soon as any losses in cable pull materializes.

6. ***Check for Stresses after Grouting*** : Since live load is applied after grouting, section moduli with steel are used. Let us use $m = 7$.

Equivalent area $= (200 \times 560) + (7 - 1)\,615.8 = 112000 + 3695 = 115695$ mm^2

$$\bar{y} = \frac{(11200 \times 280) + 3695 \times 503.3}{115695} = 287 \text{ mm} \ ; \ \bar{x} = 287 - 280 = 7 \text{ mm.}$$

$$I_e = \left[\frac{1}{12} \times 200(560)^3 + 112000(7)^2\right] + [0 + 3695(223.3 - 7)^2]$$
$$= 2932.42 \times 10^6 + 172.87 \times 10^6 \text{ mm}^4 = 3105.29 \times 10^6 \text{ mm}^4$$

$$Z_t = \frac{3105.29}{280 + 7} = 10.82 \times 10^6 \text{ mm}^3 \; ; \qquad Z_b = \frac{3105.29}{280 - 7} = 11.375 \times 10^6 \text{ mm}^3$$

$$\frac{\lambda P_i}{A_e} = \frac{0.85 \times 3.44 \times 10^6}{115695} = 25.27 \text{ N/mm}^2$$

$$M_T' = 928 \times 10^6 \left(\frac{200}{1000}\right) = 185.6 \times 10^6 \text{ N-mm per 200 mm width}$$

Hence the final stresses are as follows :

$$f_t = \left[25.27 - \frac{0.85 \times 3.44 \times 10^6 \times 223.3}{10.82 \times 10^6}\right]\frac{200}{1000} + \frac{185.6 \times 10^6}{10.82 \times 10^6} = 10.14 \text{ N/mm}^2$$

$$f_b = \left[25.27 + \frac{0.85 \times 3.44 \times 10^6 \times 223.3}{11.375 \times 10^6}\right]\frac{200}{1000} - \frac{185.6 \times 10^6}{11.375 \times 10^6} = 0.08 \text{ N/mm}^2$$

The stresses are within safe limit.

Alternative solution for determining P_i and e : P_i and e can be alternatively found by solving Eqs. 25.31 to 25.34 graphically. Consider one metre width of slab. Thus, $B = 1000$ mm and $D = 560$ mm as found in step 1.

$$A = B \times D = 560000 \text{ mm}^2. \; ; \qquad Z = \frac{1}{6} BD^2 = \frac{1}{6} \times 1000\,(560)^2 = 52267000 \text{ mm}^2$$

$$\frac{y_t}{r^2} = \frac{y_b}{r^2} = \frac{A\,y}{I} = \frac{A}{Z} = \frac{560000}{52267000} = 0.01071; \qquad \frac{r^2}{y} = \frac{1}{0.01071} = 93.33 \text{ mm}$$

(*a*) ***Eq. 25.31*** **(Line T_1)** :
$$\frac{P_i}{A}\left(-1 + \frac{e\,y_t}{r^2}\right) - \frac{M_d}{Z_t} \le p_{ti}$$

or
$$\frac{P_i}{A}(-1 + 0.01071\,e) \le \frac{448 \times 10^6}{52267000} \le 8.571 \quad \text{or} \quad \frac{A}{P_i} \ge \frac{-1 + 0.01071\,e}{8.571} \qquad \ldots(T_1)$$

Hence for
$$e = 0, \; \frac{A}{P_i} = -\frac{1}{8.571} = -11.67 \times 10^{-2}$$

For
$$\frac{A}{P_i} = 0, \; e = +\frac{1}{0.01071} = +93.33$$

(*b*) ***Eq. 25.32 (Line B_1)*** :
$$\frac{P_i}{A}\left(1 + \frac{e\,y_b}{r^2}\right) \le p_{ci} + \frac{M_d}{Z_b}$$

or
$$\frac{P_i}{A}(1 + 0.01071\,e) \le 12 + \frac{448 \times 10^6}{52267000} \le 20.571; \quad \text{or} \quad \frac{A}{P_i} \ge \frac{1 + 0.01071\,e}{20.571} \qquad \ldots(B_1)$$

For
$$e = 0 \; ; \; \frac{A}{P_i} = \frac{1}{20.571} = 4.86 \times 10^{-2}$$

For
$$\frac{A}{P_i} = 0, \; e = -\frac{1}{0.01071} = -93.33$$

(*c*) ***Eq. 25.33*** **(Line T_2)** :
$$-\lambda\frac{P_i}{A}\left(-1 + \frac{e\,y_t}{r^2}\right) + \frac{M_T}{Z_t} \le p_c$$

or $$= -\frac{0.85\,P_i}{A}(-1 + 0.01071\,e) \le 12 - \frac{928 \times 10^6}{5226700} \le -5.75$$

or $$\frac{A}{P_i} \ge \frac{-0.85 + 0.009104\,e}{5.75} \quad \ldots (T_2)$$

For $$e = 0\,,\ \frac{A}{P_i} = -\frac{0.85}{5.75} = -14.78 \times 10^{-2}$$

For $$\frac{A}{P_i} = 0,\quad e = \frac{0.85}{0.009104} = 93.33 \text{ mm}$$

(*d*) ***Eq. 25.34 (Line*** B_2 **)**

$$-\frac{\lambda P_i}{A}\left(1 + \frac{e\,y_b}{r^2}\right) + \frac{M_T}{Z_b} \le p_t \quad \text{or} \quad -0.85\,\frac{P_i}{A}(1 + 0.01071\,e) \le -\frac{928 \times 10^6}{52267000} \le -17.75$$

or $$\frac{A}{P_i} \ge \frac{0.85\,(1 + 0.01071\,e)}{17.75} \ge \frac{0.85 + 0.009104\,e}{17.75} \quad \ldots (B_2)$$

For $e = 0$,

$$\frac{A}{P_i} = +\frac{0.85}{17.75} = 4.789 \times 10^{-2}$$

For $\frac{A}{P_i} = 0$,

$$e = -\frac{0.85}{0.009104} = -93.33.$$

Fig. 25.50 shows the plot of all the four lines. The eccentricity e is found corresponding to the intersection of lines T_1 and B_2

From Fig. 25.50

$$e = 224 \text{ mm and } \frac{A}{P_i} \times 10^2 = 16.4$$

$$\therefore P_i = \frac{A \times 10^2}{16.4} = \frac{560000 \times 10^2}{16.4}$$
$$= 3.415 \times 10^6 \text{ N.}$$

Both these values are very near to the one found by analytical soultion. However, graphical solution is cumbersome and does not give very accurate results. Hence analytical solution is always preferred.

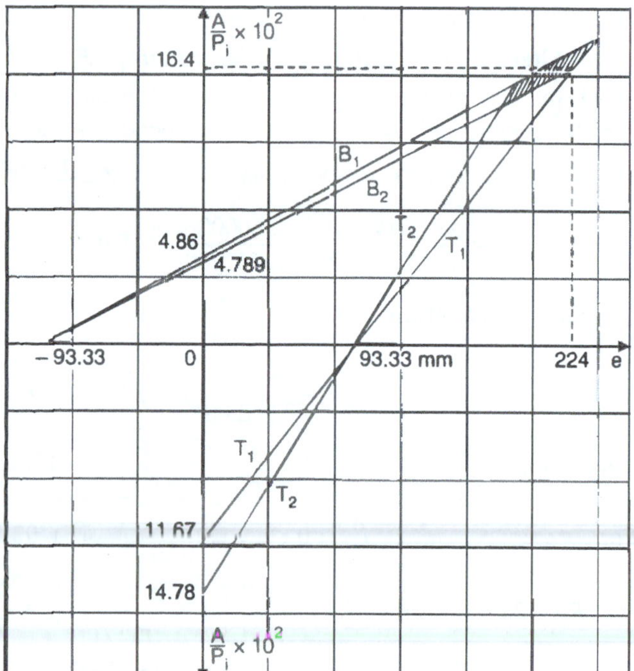

FIG. 25.50

Design Example 25.16. *Rectangular beam*

A rectangular beam of prestressed concrete is required to support a dead load moment of 15 × 10⁶ N-mm (inclusive of its own weight) and a live load moment of 40 × 10⁶ N-mm at its mid-section. Determine the initial prestressing force and its eccentricity at the mid-span section. Take the following values :

Allowable initial compressive stress $= 17\ N/mm^2$

Allowable final compressive stress $= 14\ N/mm^2$

Allowable initial or final tensile stress $= 1\ N/mm^2$.

Ultimate stress in steel $= 1500\ N/mm^2$. *Assume losses 15%.*

Solution

1. ***Section Modulus*** : Given $p_{ci} = 17\ N/mm^2$; $p_c = 14\ N/mm^2$; $p_{ti} = p_t = 1\ N/mm^2$

As the dead load moment is less than live load moment, the section modulus will be governed by Eqs. 25.35(*a*) and 25.36(*a*) :

$$Z_t = Z = \frac{40 \times 10^6}{(14+1) - \frac{14}{2}(1-0.85) \times 1} = 2.8674 \times 10^6\ mm^3$$

and
$$Z_b = Z = \frac{40 \times 10^6}{(17+1) - \frac{14}{2}(1-0.85) \times 3} = 2.6936 \times 10^6\ mm^3$$

Hence adopt
$$Z = \frac{1}{6} B D^2 = 2.8674 \times 10^6\ mm^3$$

Taking $B = 150$ mm, $D = \sqrt{\frac{2.8674 \times 10^6 \times 6}{150}} = 338.7$ mm.; Adopt $D = 340$ mm

Actual
$$Z = \frac{1}{6} \times 150(340)^2 = 2.89 \times 10^6\ mm^3; \quad A = 150 \times 340 = 51000\ mm^2$$

$$y_t = y_b = y = 170\ mm$$

$$\frac{r^2}{y} = \frac{I}{Ay} = \frac{Z}{A} = \frac{D}{6} = 56.7\ mm. ; \qquad \frac{y}{r^2} = \frac{1}{56.7} = 0.01765.$$

2. ***Determination of e*** : The maximum value of *e* is given by solving (*i.e.* dividing) Eqs. 25.31 (Line T_1) and 25.34 (Line B_2). The solution is given by Eq. 25.38.

$$\frac{e - D/6}{e + D/6} = \frac{\lambda(M_d + p_{ti} \cdot Z)}{(M_d + M_l) - p_t Z} \qquad \ldots(25.38)$$

$$\frac{e - 56.7}{e + 56.7} = \frac{0.85(15 \times 10^6 + 1 \times 2.89 \times 10^6)}{(15+40)\,10^6 - 1 \times 2.89 \times 10^6} = 0.2918.$$ From which $e = 103.4$ mm.

Let us assume average size of duct = 60 mm. Providing 40 mm clear cover to the duct, available $e = \frac{340}{2} - \frac{60}{2} - 40 = 100$ mm. Hence adopt $e =$ **100 mm.**

3. ***Determination of P_i*** : Since the available value of *e* is lesser than that given by Eq. 25.38, P_i is obtained either from Eq. 25.34 (simplified to Eq. 25.39) or by Eq. 25.33 [simplified to Eq. 25.39(*a*)]. From Eq. 25.39 (Eq. 25.34) : Line B_2

$$P_i = \frac{(M_d + M_l) - p_t \cdot Z}{\lambda\left(e + \frac{D}{6}\right)} = \frac{(15+40)\,10^6 - 1 \times 2.89 \times 10^6}{0.85\,(100 + 56.7)} = 391230\ N$$

From Eq. 25.39(*a*) (Eq. 25.33) : Line T_2

$$P_i = \frac{(M_d + M_l) - p_c \, . \, Z}{\lambda\left(e - \frac{D}{6}\right)} = \frac{(15+40)\,10^6 - 14 \times 2.89 \times 10^6}{0.85\,(100 - 56.7)} = 395050 \text{ N}$$

Hence adopt $P_i = \mathbf{395050}$ **N.**

4. ***Alternative solution for e and*** P_i : Alternatively, e and P_i can be obtained graphically from plotting Eqs. 25.31 to 25.34.

(*a*) ***Line*** T_1 **[*Eq.* 25.31(*a*)]**

$$\frac{A}{P_i} \geq \frac{\frac{ey}{r^2} - 1}{\frac{M_d}{Z} + p_{ti}} \quad \text{or} \quad \frac{A}{P_i} \geq \frac{0.01765\,e - 1}{\frac{15 \times 10^6}{2.89 \times 10^6} + 1} \geq \frac{0.01765\,e - 1}{6.19} \qquad \text{...(Line } T_1)$$

For $e = 0, \ \frac{A}{P_i} = -\frac{1}{6.19} = -16.15 \times 10^{-2}$; For $\frac{A}{P_i} = 0, \ e = \frac{1}{0.01765} = 56.7$ mm

(*b*) ***Line*** B_1 **[*Eq.* 25.32(*a*)]**

$$\frac{A}{P_i} \geq \frac{1 + \frac{ey}{r^2}}{\frac{M_d}{Z} + p_{ci}} \quad \text{or} \quad \frac{A}{P_i} \geq \frac{1 + 0.01765\,e}{\frac{15 \times 10^6}{2.89 \times 10^6} + 17} \geq \frac{1 + 0.01765\,e}{22.19} \qquad \text{....(Line } B_1)$$

For $e = 0, \ \frac{A}{P_i} = -\frac{1}{22.19} = 4.51 \times 10^{-2}$; For $\frac{A}{P_i} = 0, \ e = \frac{1}{0.01765} = -56.7$ mm

(*c*) ***Line*** [T_2 **[*Eq.* 25.33(*a*)]**

$$\frac{A}{P_i} \geq \frac{\lambda\left(1 - \frac{e\,y}{r^2}\right)}{p_c - \frac{M_T}{Z}} \quad \text{or} \quad \frac{A}{P_i} \geq \frac{0.85(1 - 0.01765\,e)}{14 - \frac{(15+40)10^6}{2.89 \times 10^6}} \geq \frac{0.015\,e - 0.85}{5.03} \qquad \text{...(Line } T_2)$$

For $e = 0, \ \frac{A}{P_i} = -\frac{0.85}{5.03} = -16.89 \times 10^{-2}$; For $\frac{A}{P_i} = 0, \ e = \frac{0.85}{0.015} = +56.7$ mm

(*d*) ***Line*** B_2 **[*Eq.* 25.34(*a*)]**

$$\frac{A}{P_i} \geq \frac{\lambda\left(\frac{ey}{r^2} + 1\right)}{\frac{M_T}{Z} \ p_t} \quad \text{or} \quad \frac{A}{P_i} \geq \frac{0.85\,(0.01765\,e + 1)}{\frac{(15+40)\,10^6}{2.89 \times 10^6} - 1} \geq \frac{0.015\,e + 0.85}{18.031} \qquad \text{...(Line } B_2)$$

For $e \approx 0$, $\quad \frac{A}{P_i} = 0.85/18.031 = 4.71 \times 10^{-2}$.

For $\frac{A}{P_i} = 0$, $\quad e = -0.85/0.015 = -56.7$ mm

Fig. 25.51 shows the plotted positions of all the four lines. Lines B_1 and B_2 nearly coincide, though B_2 is slightly higher than B_1. The maximum value of e, at the intersection of B_2 and T_1 comes out to be 104 mm. However, available $e = 100$ mm, corresponding to which value of A/P_i comes out to be 12.90×10^{-2}. Hence,

$$P_i = \frac{A}{12.90 \times 10^{-2}} = \frac{51000}{12.90 \times 10^{-2}}$$

≈ 395350 N.

5. ***Design of reinforcement***

Allowable stress in steel after losses

$= 0.6 \times 1500 = 900 \text{ N/mm}^2$

Allowable stress in steel before losses

$= 900/0.85 \approx 1059 \text{ N/mm}^2$

Initial prestressing force

$= P_i = 395050$

Use 7 mm wires. Area of steel

$= 395050/1059 = 373 \text{ mm}^2$

No. of 5 mm Φ bars

$$= \frac{373}{\frac{\pi}{4}(5)^2} \approx 19$$

Hence provide two cables of 10 wires each and use the Freyssinet system.

Actual $A_{st} = 20 \times \frac{\pi}{4}(5)^2 = 392.7 \text{ mm}^2$

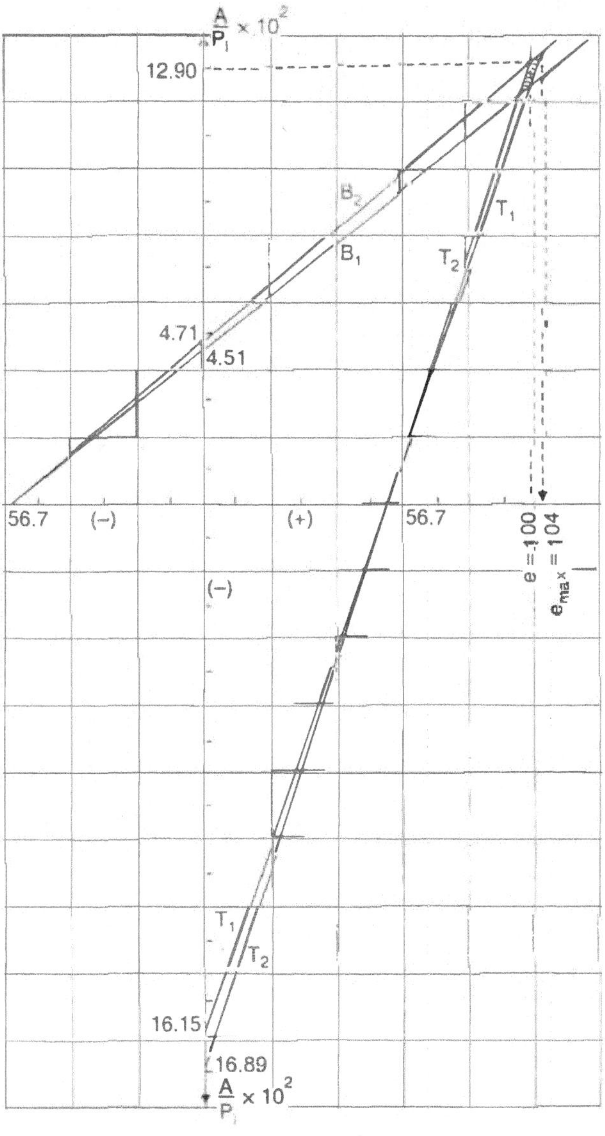

FIG. 25.51.

25.15. ALTERNATIVE DESIGN PROCEDURE

The design procedure discussed in § 25.12 is cumbersome, since it involves simultaneous solution of four condition equations, as suggested by Magnel. A simplified design procedure is given below. The procedure takes into account two stress conditions, after the losses have taken place :

(1) Stresses due to P and M_d (2) Stresses due to P, M_d and M_l

It is justified to neglect the initial stress conditions due to P_i and M_d since higher compressive stress (p_{ci}) is normally permitted.

(*a*) **Determination of section modulus** : Fig. 25.52 (*b*), (*c*) and (*d*) show the stress distribution due to P, $P.e.$ and M_d , while Fig. 25.52(*e*) shows the combined effect due to these (first condition). Let the top fibre have tensile stress f_{t1} and bottom fibre has compressive stress f_{c1}. Fig. 25.52(*h*) shows the final stress distribution due to P, M_d and M_l in which the top fibre has compressive stress f_{c2} and bottom fibre has tensile stress f_{t2}.

***For the top fibre* :**

From Fig. 25.52(*e*), max. tensile stress is given by

$$f_{t1} = -\frac{P}{A} + \frac{P \cdot e}{Z_t} - \frac{M_d}{Z_t} \qquad \ldots(i)$$

From Fig. 25.52(*h*), max. compressive stress is given by

$$f_{c2} = \frac{P}{A} - \frac{P \cdot e}{Z_t} + \frac{M_d}{Z_t} + \frac{M_l}{Z_t} \qquad \ldots(ii)$$

Adding the two, we get

$$f_{t1} + f_{c2} = \frac{M_l}{Z_t}$$

or
$$f_{c2} = \frac{M_l}{Z_t} - f_{t1} \qquad \ldots[25.49(a)]$$

For the bottom fibre

Max. compressvie stress is given by [Fig. 25.52(*e*)]

$$f_{c1} = \frac{P}{A} + \frac{P \cdot e}{Z_t} - \frac{M_d}{Z_b} \qquad \ldots(iii)$$

Max. tensile stress is given by [Fig. 25.52(*h*)]

$$f_{t2} = -\frac{P}{A} - \frac{P \cdot e}{Z_b} + \frac{M_d}{Z_b} + \frac{M_l}{Z_b} \ldots(iv)$$

Adding the two, we get

$$f_{c1} + f_{t2} = \frac{M_l}{Z_b}$$

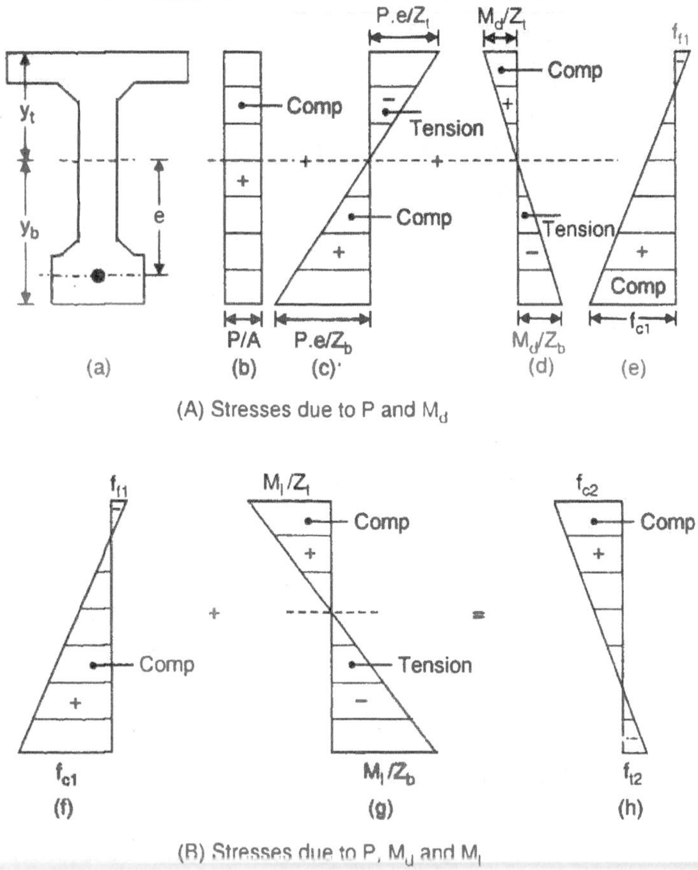

FIG. 25.52.

or
$$f_{c1} = \frac{M_l}{Z_b} - f_{t2} \qquad \ldots[25.49(b)]$$

From Eqs. 25.49(*a*) and (*b*), it is evident that if f_{t1} and f_{t2} are limited to equal permissible value of p_t (often zero), the compressive fibre stress will be maximum at the top or bottom, depending upon whether Z_t or Z_b is smaller. Let maximum permissible compressive stress be p_c. Then,

$$p_c = \frac{M_l}{Z_{min}} - p_t \qquad \ldots(25.49)$$

From the above equation, the section modulus can be found.

(b) Determination of Prestress Force (P) : For a given prestress force, the cable eccentricities must be properly determined so as to limit tensile stress to permissible value p_t. Let e_1 and e_2 be the cable eccentricities to limit the tensile fibre stresses under condition 1 [Fig. 25.52(*e*)] and condition 2 [Fig. 25.52(*h*)], to permissible value p_t.

$$\therefore \qquad f_{t1} = -\frac{P}{A} + \frac{P \cdot e_1}{Z_t} - \frac{M_d}{Z_t} = p_t \qquad \ldots(i)$$

Re-arranging the terms, we get

$$e_1 = \frac{M_d}{P} + \frac{I}{y_t}\left[\frac{p_t}{P} + \frac{1}{A}\right] \qquad \ldots[25.50(a)]$$

and

$$f_{t2} = \frac{M_d}{Z_b} + \frac{M_l}{Z_b} - \frac{P}{A} - \frac{P \cdot e_2}{Z_b} = p_t \qquad \ldots(iv)$$

Re-arranging the terms we get

$$e_2 = \frac{M_d + M_l}{P} - \frac{I}{y_b}\left[\frac{p_t}{P} + \frac{1}{A}\right] \qquad \ldots[25.50(b)]$$

It will be seen from Eq. 25.50(*a*) that an increase of eccentricity above e_1 will give a tensile stress in excess of p_t in the *top* fibre. Similarly, from Eq. 25.50(*b*), a decrease in eccentricity below e_2 will give a tensile stress in excess of p_t in the *bottom* fibre. In other words, the cable should be laid along a zone (known as cable zone), if the permissible stress p_t is not to be exceeded. The cable zone is shown shaded in Fig. 25.53(*a*) and (*b*).

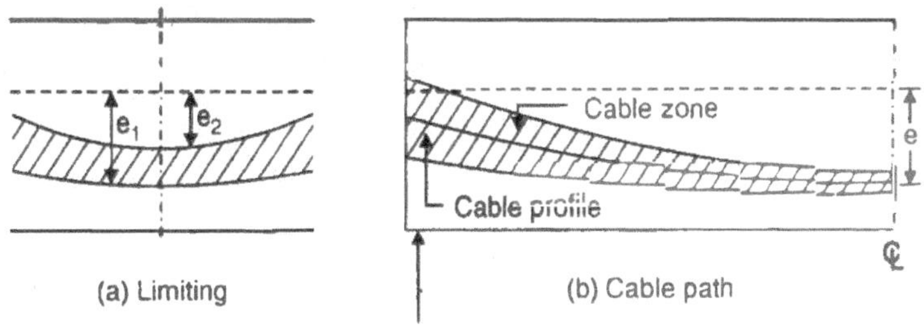

FIG. 25.53. CABLE PATH.

For the sake of economy, the cable zone should be limited to a minimum. Equating e_1 and e_2 at the critical section (where M_l is maximum), we get

$$\text{or} \qquad \frac{M_l}{P} - \frac{I}{y_b}\left[\frac{p_t}{P} + \frac{1}{A}\right] = \frac{I}{y_t}\left[\frac{p_t}{P} + \frac{1}{A}\right] \qquad \text{or} \qquad \frac{M_l}{P} \cdot \frac{y_b \cdot y_t}{I\left(\frac{p_t}{P} + \frac{1}{A}\right)} = y_b + y_t = D$$

$$\text{or} \qquad \left(p_t + \frac{P}{A}\right) = \frac{y_b \cdot y_t \, M_l}{ID}$$

From which

$$P = A\left[\frac{y_b \, y_t}{ID} M_l - p_t\right] \qquad \ldots(25.51)$$

(c) Determination of eccentricity : Re-arranging Eq. 25.51,

$$\frac{I}{y_t} = \frac{M_l \cdot y_b}{D\left(\frac{P}{A} + p_t\right)} = \frac{M_l \cdot y_b}{PD\left(\frac{1}{A} + \frac{p_t}{P}\right)}$$

Substituting this in equation 25.50(*a*) and noting that $e_1 = e_2 = e$ at the critical section, we get

$$e = \frac{M_d}{P} + \frac{M_l \cdot y_b}{P \cdot D} = \frac{1}{P}\left[M_d + \frac{M_l \cdot y_b}{D} \right] \qquad \ldots(25.52)$$

If the available eccentricity is less than the one given by Eq. 25.52, the available value of e should be re-substituted in Eq. 25.52 to get the revised prestressing force. Alternatively, a large section may be provided and both P and e may be determined.

Thus the position of the cable at the critical section is determined. It will not, in general, be possible to achieve zero cable-zone (*i.e.* $e_1 = e_2 = e$) elsewhere. Thus the designer can fix the path of the cable at his discretion, keeping in mind the friction losses and the end-shear requirements.

Knowing P and estimating the losses (or assuming the losses), the initial prestressing force (P_i) can be determined.

Thus. $$P_i = P + \text{Losses} = \frac{P}{\lambda}$$

The above method is extremely useful in designing unsymmetrical sections, specially T-section and unsymmetrical I-section.

Design Example 25.17. *Rectangular Beam*

Redesign the rectangular beam of example 25.16. Check the stresses in the section.

Solution. Given : $p_c = 14\ \text{N/mm}^2$; $p_t = 1\ \text{N/mm}^2$

1. ***Determination of section modulus*** : The minimum section modulus is given Eq. 25.49:

$$p_c = \frac{M_l}{Z_{min}} - p_t \quad \text{or} \quad Z_{min} = \frac{M_l}{p_c + p_t} \quad \text{or} \quad \frac{B\,D^2}{6} = \frac{40 \times 10^6}{14 + 1} = 2666667\ \text{mm}^3$$

Let $B = 150$ mm. $\therefore$ $D = \sqrt{\dfrac{6 \times 2666667}{150}} = 327$ mm. Adopt $D = 330$ mm

Actual $Z = \frac{1}{6} \times 150\,(330)^2 = 2722500\ \text{mm}^3 \approx 2.722 \times 10^6$

$y_t = y_b = y = 165$ mm ; $I = 2722500 \times 165 = 449.21 \times 10^6$

$A = 150 \times 330 = 49500\ \text{mm}^2$

2. ***Determination of tendon force*** : From Eq.. 25.51

$$P = A\left[\frac{y_b \cdot y_t}{ID} M_l - p_t\right] = A\left[\frac{M_l}{2Z} - p_t\right] = 49500\left[\frac{40 \times 10^6}{2 \times 2.722 \times 10^6} - 1\right] = 314140\ \text{N.}$$

3. ***Determination of eccentricity e*** : From Eq. 25.52

$$e = \frac{1}{P}\left[M_d + \frac{M_l \cdot y_b}{D}\right] = \frac{1}{314140}\left[15 \times 10^6 + \frac{40 \times 10^6 \times 165}{330}\right] = 111.4\ \text{mm}$$

Let us assume average size of duct = 60 mm. Providing 40 mm clear cover to the duct, available $e = \frac{330}{2} - \frac{60}{2} - 40 = 95$ mm.. This is much less than the required one. If we adopt this value, corresponding P will be high, giving rise to high initial compressive stress.

Hence let us increase the depth from 330 to 350 mm.

Available $e = \dfrac{350}{2} - \dfrac{60}{2} - 40 = 105$ mm

$\therefore$ $y_t = y_b = \dfrac{D}{2} = \dfrac{350}{2} = 175$ mm ; $A = 150 \times 350 = 52500$ mm^2

$Z = \frac{1}{6} \times 150\,(350)^2 = 3.0625 \times 10^6$ mm^3 ; $I = 3.0625 \times 10^6 \times 175 = 535.94 \times 10^6$ mm^4

Revised $P = \dfrac{1}{105}\left[15 \times 10^6 + \dfrac{40 \times 10^6 \times 175}{350}\right] = 333333$ N

Assuming losses @ 15%, we get : $P_i = \dfrac{333333}{0.85} \approx$ **392157 N**

4. *Check for stresses :Initial condition*

$$f = \frac{P_i}{A} \mp \frac{P_i \cdot e}{Z} \pm \frac{M_d}{Z} = \frac{392157}{52500} \mp \frac{392157 \times 105}{3.0625 \times 10^6} \pm \frac{15 \times 10^6}{3.0625 \times 10^6} = 7.47 \mp 13.45 \pm 4.90$$

$\therefore$ $f_{(top)} = -1.08$ N/mm^2 (tensile) > 1 ; $f_{(bottom)} = 16.02$ N/mm^2 < 17

Hence the stresses are within safe limits, except that the tensile stress is slightly higher.

Final condition

$$f = \frac{P}{A} \mp \frac{P \cdot e}{Z} \pm \frac{M_d}{Z} \pm \frac{M_l}{Z} = \frac{333333}{52500} \mp \frac{333333 \times 105}{3.0625 \times 10^6} \pm \frac{(15 + 400)\,10^4}{3.0625 \times 10^6} = 6.35 \mp 11.43 \pm 17.96$$

$\therefore$ $f_{(top)} = 12.88$ N/mm^2 < 14(safe) ; $f_{(bottom)} = -0.18$ N/mm^2 < 1 (safe)

25.16. SHEAR AND DIAGONAL TENSION

In the case of prestressed concrete, the whole section of the beam is effective since the section remains uncracked. Hence the calculation of shear resistance of a prestressed member at safe working condition is more straight forward than for ordinary R.C. sections. The shear stres at any plane of a pre-stressed concrete section is given by the usual formula:

$$q = \frac{Q}{I\,b} A_a \bar{y}_a \qquad \ldots(25.53)$$

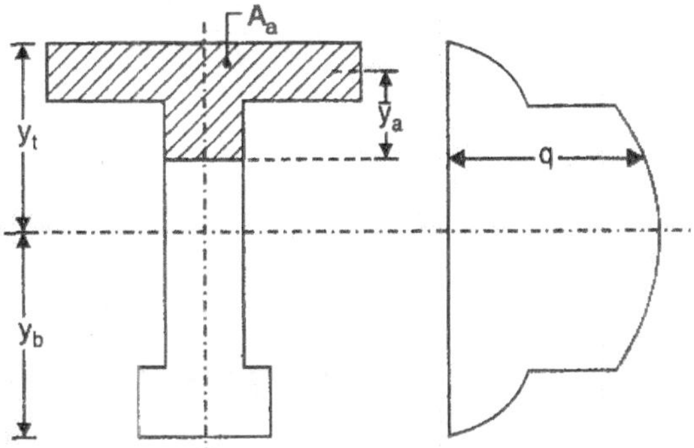

FIG. 25.54.

where

Q = net shear force.

I = moment of inertia of the section.

b = width of the section

A_a and $\bar{y}_a$ arc as marked in Fig. 25.54.

The shear force at any section is decreased by the vertical component (V) of the cable tension, and the net shear force Q used in Eq. 25.53 will be equal to shear force due to dead and live loads minus V. The up-swept cable has thus the balancing effect, and this effect can be determined if we first decide the cable slope. It is also necessary to check for shear stresses by combining dead load S.F. with V.

Determination of cable slope (shear cable) : The available latitude of cable eccentricity were determined in Eq. 25.50(a) and (b) from considerations of bending. At the end section, where B.M. is zero, the cable eccentricity can be $\pm\frac{D}{6}$ without producing tensile stress. Let us now determine the cable path from considerations of shear. Let,

Q_d = S.F. due to dead load ; Q_l = S.F. due to live load.

∴ Q = Net shear force ;

V = Vertical component of prestressing force = $P \tan\theta$.

When live load and dead load act, net S.F.= $Q_d + Q_l - V$

When only dead load acts, net S.F.= $V - Q_d$ (*i.e.* there is reversal of direction of S.F.)

From design considerations, the cable slope should be such that at the ends, where the S.F. is maximum, the net shear when dead load is acting is equal to the net shear when both dead and live loads are acting.

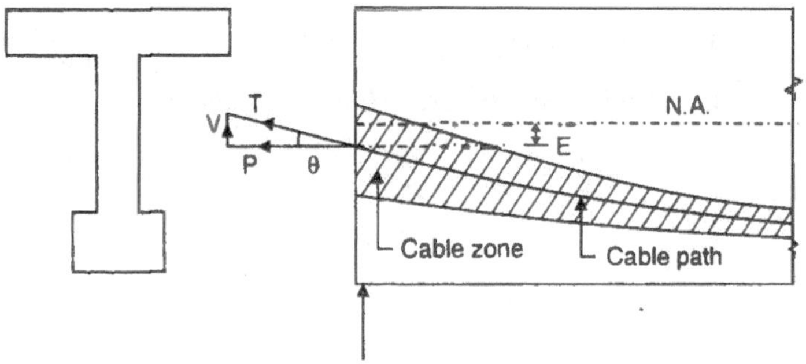

FIG. 25.55

$$Q_d + Q_l - V = V - Q_d \quad \text{or} \quad V = Q_d + \frac{Q_l}{2} \qquad \text{...(25.54)}$$

Substituting $V = P \tan\theta = T \sin\theta$ we get

$$\tan\theta = \frac{1}{P}\left(Q_d + \frac{Q_l}{2}\right) \quad \text{...(25.55(a)]} \quad \text{or} \quad \sin\theta = \frac{1}{T}\left(Q_d + \frac{Q_l}{2}\right) \qquad \text{...[25.55(b)]}$$

Let us assume that the conditions for worst shear occur with the same loading that cause worst moment conditions. If e is the cable eccentricity, and x is the distance along the member.

$$\tan\theta = \frac{de}{dx} = \frac{V}{P} \;; \qquad V = P\frac{de}{dx} \qquad \text{...(25.56)}$$

For simple beam theory, $Q = \frac{dM}{dx}$

Hence from Eq. 25.54, $V = \frac{d}{dx}\left(M_d + \frac{M_l}{2}\right)$ ...(25.57)

Equating Eqs. 25.56 and 25.57 and intergrating

$$e = \frac{1}{P}\left(M_d + \frac{M_l}{2}\right) + E \qquad \text{...(25.58)}$$

where E is some constant of integration.

The above equation is the equation for shear cables, which is similar to Eq. 25.52 found from considerations of bending, except for the addition of constant E. For rectangular and other symmetrical section, where $y/D = 1/2$, both the equations are identical with value of $E = 0$. For non-symmetrical section, the value of E, as marked in Fig. 25.55, can be determined by substituting the values of e, M_d and M_l as calculated for bending at the critical section.

Determination of principal tensile stresses : Thus the cable path (*i.e.* inclination θ), V and net shear Q are known. The shear stress q can then be computed from Eq. 25.53 and it may then be combined with the bending stress to compute the principal tension, using the following expression ;

$$\sigma_t = \frac{1}{2}\sqrt{f_c^2 + 4q^2} - f_c \quad \ldots(25.59)$$

where σ_t = principle tensile stress

f_c = bending stress at the point considered.

q = shear stress at the point considered.

From study of Eq. 25.59 as well Fig. 25.56, it is clear that the principal tension (diagonaly tension) will be less than the shear stress of q. This is one of the great advantages of prestressing techniques. The elimination of principal tension, or its reduction to a small value enables the adoption of a thin web. The safe value of diagonal tension in prestressed concrete may be taken as equal to $0.4\sqrt{f_{su}}$. Under ulimate load conditions, it should not exceed $1.9\sqrt{f_{cu}}$.

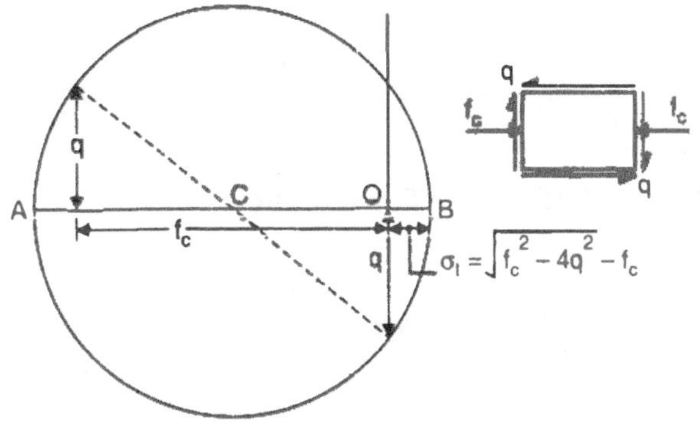

FIG. 25.56. MOHR CIRCLE.

If the diagonal tension is more than the permissible one, shear stirrups are provided to bear the total effective S.F. Q_r allowing usual working stress in steel. Some times, *vertical prestressing* is done to neutralise the vertical component of principal tension. Generally, the I-section of a beam is thickened into a rectangular section, at the ends, to accommodate the anchorage. This further reduces the shear stresses. Hence diagonal tension is seldom a problem in prestressed concrete beams.

Guyon's recommendations : Guyon recommends that even though the principal tension may be less than the permissible value, nominal shear reinforcement should be provided. The *pitch* of the vertical stirrups may be determined from the following expression :

$$p = \frac{Z \cot\phi\, A_w f_w}{Q_d + Q_l - T\sin\theta} \quad \ldots(25.60)$$

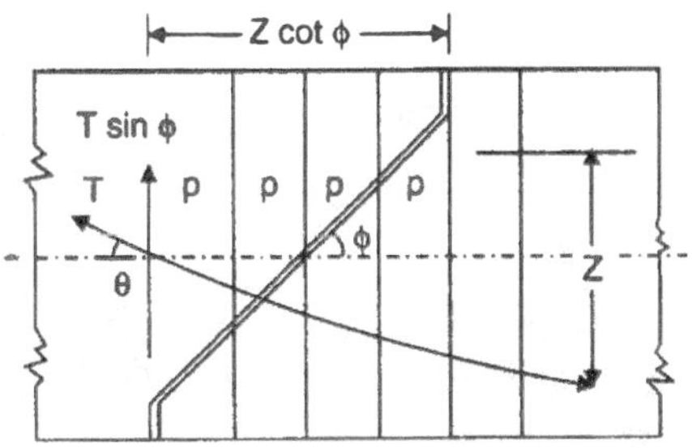

FIG. 25.57.

where p = pitch of the stirrups.

$\phi \approx \tan^{-1}\left(\frac{2q}{f}\right)$

q = shear stress in concrete

f_c = compression stress in concrete

A_w = area of stirrup

f_w = yield stress in shear reinforcement

Z = lever arm= $I/A_a \cdot \bar{y}_a$

$A_a \cdot \bar{y}_a$ = first moment of area, above the level of the point at which shear stress is to be calculated, about the N.A.

I = moment of inertia of the section.

The above expression is based on the assumption of the formation of a full crack, as shown in Fig. 25.57. Guyon recommended the following values of tan ϕ and f_w

σ_t or p	$tan\ \phi$	f_w
$\sigma_t < p_c/20$	$2\ q/f_c$	210 N/mm^2
$\sigma_t > p_c/20$	$2\ q/f_c$	140 N/mm^2
$q > f_c/2$	1	140 N/mm^2

where σ_t = principal tensile stress in concrete

p_c = permissible compressive stress in concrete

f_c = actual compressive stress in concrete.

25.17. STRESSES AT ANCHORAGE

The anchorage devices in post-tensioned beams bear against the ends of the concrete, resulting in very high local bearing stresses. The stresses in the immediate vicinity of the anchorage are different from those at sections farther away. At the ends, the prestressing force is transmitted over a small area ; the force then disperses itself till it covers the whole section. The length of the beam between the end and the section where the full section shares the force is known as *lead-in-zone* or *anchorage zone*. It is generally believed that the length of the zone is equal to the depth of the beam.

Fig. 25.58(a) shows a prestressing force P applied to the section AB, through a small area. The stress trajectories transfer the force from bearing face AB across the lead-in-zone to face CD, where the stress $p = P/BD$. These lines of forces (or stress trajectories) can be considered to be acting as individual struts which are first convex and then concave towards the centroidal axis. Because of curvatures of the stress trajectories, transverse stresses are induced as shown in Fig. 25.58(b) and (c). These stresses are compressive near the face AB, and tensile at a greater distance from it. Fig. 25.58(d) shows the isobars of transverse stresses, known from photoelastic studies. The distribution of stresses near anchorage is discontinuous,

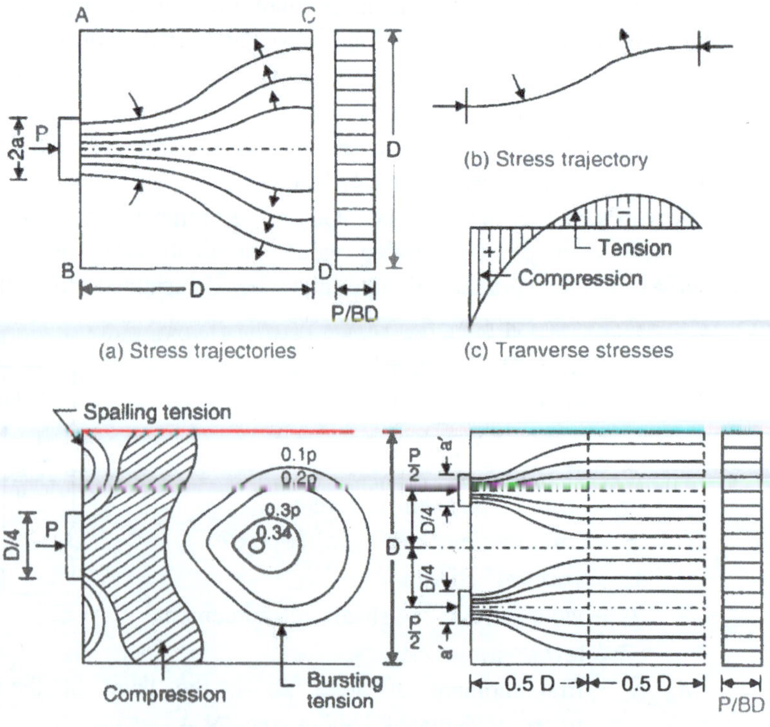

(a) Stress trajectories

(b) Stress trajectory

(c) Tranverse stresses

(c) Isobars of transverse stresses

(d) Stress trajectories

FIG 25.58. ANCHORAGE STRESSES.

and besides longitudinal compressive stresses, transverse tensile stresses, normal to the beam axis also occur. The transverse tensile stresses as high as 0.3 to 0.5 p may occur deep inside the concrete ; this zone is known as *bursting zone*. Also, the corners of the end section are found to be subjected to high spalling tensile stresses the tensile stress may be as high as 0.68 p. It is, therefore necessary to provide one layer of reinforcement very near to the end (just behind the anchorage), to take care of *spalling zone* and another layer of reinforcement away from the anchorage to take care of bursting zone.

Fig. 25.58(d) shows the advantage of distributing the total prestressing force P over a large number of small units. When the force on each bearing surface is $P/2$, the tensile stresses act only on half the length (0.5 D) and the resultant forces are halved.

A variety of analysis of the problem of stresses at anchorage have been developed. However, the method suggested by Prof. Magnel is given here.

Magnel's Method : Magnel assumes that the local bearing stresses are dissipated over a length of the beam equal to its depth. Within the anchorage zone, any elemental unit is acted on by three stresses (i) horizontal stress f_x , (ii) vertical stress f_y and (iii) shear q, shown in Fig. 25.59(a).

1. ***Horizontal stress*** f_x : Consider Fig. 25.59(b) in which it is assumed that the compression from the anchorage device is assumed to be spread out at 45°, cutting the top and bottom faces at H and G respectively. Upto vertical section FG, the stressed zone is symmetrical about the cable, while beyond this the stressed zone has greater depth above the cable and small depth below the cable. Thus, line $EKLM$ is the *line of centroid* of the stressed zone, and can be easily located. At any vertical section RS, of height h ($i.e.$ $RS = h$), the cable eccentricity e_v is thus known. At any point, distant y from the centroid, the magnitude to horizontal stress f_x is given by

$$f_x = \frac{P}{b\,.\,h} \pm \frac{P\,.\,e_v\,.\,y}{I_v} \quad \ldots(25.61)$$

where b = width of beam section;

I_v = second moment of area of zone under stress.

Thus horizontal stress distribution across any vertical plane RS can be found.

2. ***Shear stress*** **(q)** : The distribution of shear stress, at any horizontal plane XX [Fig. 25.59(c)] is given by the following expression :

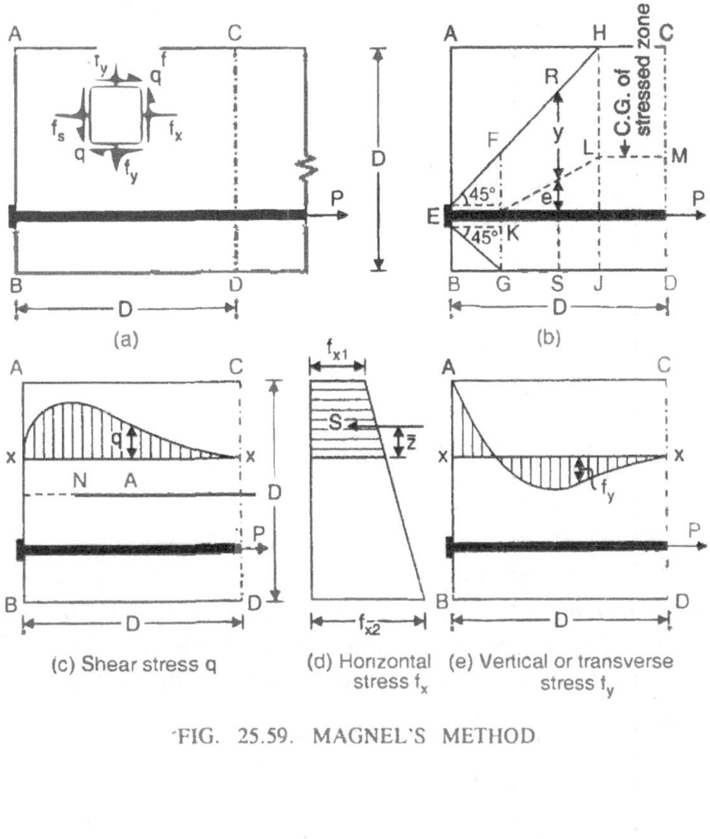

FIG. 25.59. MAGNEL'S METHOD

$$q = K_q \frac{S}{b\,D} \qquad \ldots(25.62)$$

where

K_q = *shear stress factor*, which depends upon the distance of the point from the end section *AB*, and can be taken from Fig. 25.60.

S = Summation of all horizontal stresses (Σf_x) above plane *XX* [Fig. 25.59(*d*)].

In determining *S*, the amount of opposing cable force should be deduced if it occurs above the plane *XX*.

3. ***Vertical stress (or transverse stress)*** f_y

Fig. 25.59(*e*) shows the distribution of vertical stress on any horizontal plane *XX*. Magnel gives the following expression for f_y :

$$f_y = K_y \cdot \frac{M}{b\,D^2} \qquad \ldots(25.63)$$

where

K_y = *vertical stress factor*, to be taken from Fig. 25.60

$M = S \times \bar{z}$

$\bar{z}$ = distance of centroid *S* above *XX*.

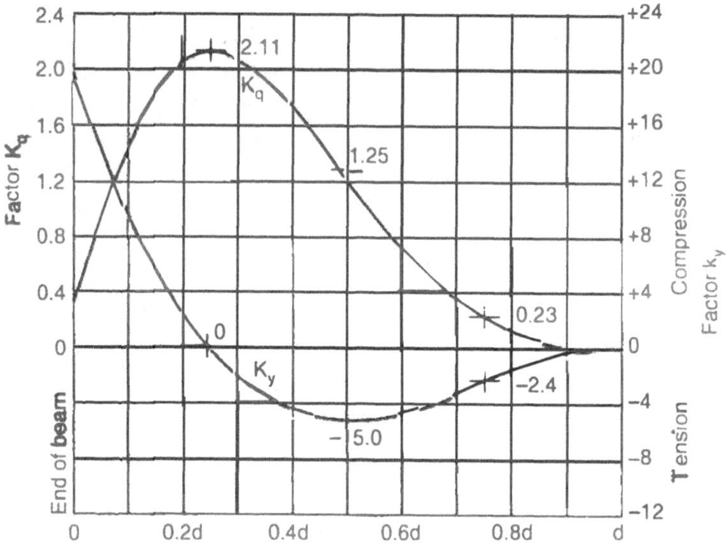

FIG. 25.60. STRESS FACTORS.

4. ***Principal stresses***

After having determined f_x , f_y and q, the principal stress can be found from the following expression :

$$\sigma = \frac{1}{2}(f_x + f_y) \pm \frac{1}{2}\sqrt{(f_x - f_y)^2 + 4\,q^2}$$

and
$$\tan 2\theta = \frac{2\,q}{f_x - f_y}$$

In case the principal tensile stress is more than the allowable, mild steel reinforcement is provided in from of a mat or spirals.

Empirical Method : The empirical method gives the approximate location and the magnitude of maximum transverse tensile stress. Fig. 25.61 (*a*) shows the loaded area of side $2\,a_1$. If the loaded area is circular, the side of the equivalent square of the same area may be found. The surrounding prism has a side $2a$, which is the smallest dimension of the section, either in width or in depth direction. It has been observed that positions of zero and maximum transverse stresses area independent on the ratio a_1/a, through its magnitude is dependent on this ratio.

Fig. 25.61(*b*) shows the distribution of the transverse tensile stress. It increases from *zero* at 0.2*a* to a maximum at 0.5 *a*, and then again decreases to zero at 2*a* from the end section. The maximum transverse tensile stress is given by

$$f_{y\,(max)} = f_c\left[0.98 - 0.825\,\frac{a_1}{a}\right] \; ; \quad \text{where} \quad f_c = \frac{P}{(2\,a_1)^2} \qquad \ldots(25.64)$$

Total tensile force,

$$F_t = P\left[0.48 - 0.4\left(\frac{a_1}{a}\right)\right] \qquad \ldots(25.65)$$

Reinforcement is provided in the form of mats, helices or links, in the region 0.2 *a* to 1.0 *a*. The area of steel is equal to $F_t/1400$ cm^2, where F_t is the total tensile force in kg. If the concrete can take some tensile stress p_t ,

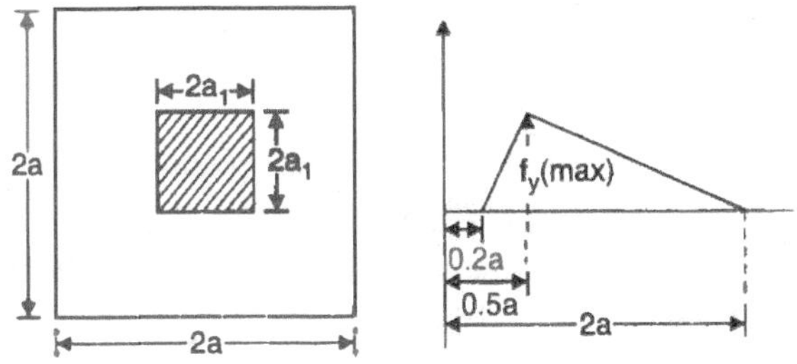

FIG. 25.61.

$$\text{Corrected } F_t = F_t\left[1 - \left(\frac{p_t}{f_{y\,(max)}}\right)^2\right] \qquad \ldots(25.66)$$

Generally, a rectangular mesh of bars of 6 to 10 mm Φ is provided. In addition to this, a light spiral of 6 mm Φ at a pitch of 5 cm is provided round individual anchorages.

25.18. INDIAN STANDARD CODE RECOMMENDATIONS (IS : 1343-1980)

1. Concrete

TABLE 25.12 GRADES OF CONCRETE

Grade Designation	*Specified characteristic compressive strength at 28 days f_{ck} (N/mm^2)*
M 30	30
M 35	35
M 40	40
M 45	45
M 50	50
M 55	55
M 60	60

(*a*) **Grades of concrete** : The concrete used shall be in grades designated in Table 25.12.

For pre-tensioned prestressed concrete, the grade of concrete shall be not less than M 40. Where it can be shown that a member will not receive its full design stress within a period of 28 days after the casting of the member, the characteristic compressive strength given in Table 25.12 may be increased by multiplying by factors given below :

Minimum Age of Member when Full Design Stress is Expected (*months*)	*Age Factor*
1	1.00
3	1.10
6	1.15
12	1.20

(*b*) **Tensile strength of concrete** : When the designer wishes to use an estimate of flexural strength from the compressive strength, the following formula may be used:

$$f_{cr} = 0.7\sqrt{f_{ck}}\ \text{N/mm}^2 \qquad \ldots(25.67)$$

where f_{cr} = flexural strength in N/mm^2

f_{ck} = characteristic compressive strength of concrete (N/mm^2)

(*c*) **Modulus of elasticity** : In absence of test data, the modulus of elasticity for structural concrete may be assumed as follows :

$$E_c = 5700\sqrt{f_{ck}}\,. \qquad \ldots(25.68)$$

where E_c = short-term static modulus of elasticity, in N/mm^2.

(*d*) **Shrinkage strain** : The shrinkage of concrete depends upon the constituents of concrete, size of the member and environmental conditions. For a given environment, the shrinkage of concrete is most influenced by the total amount of water present in the concrete at the time of mixing and, to a lesser extent by the cement content. In the absence of test data, the approximate value of shrinkage strain for design shall be assumed as follows:

$$\text{For pre-tensioning} = 0.0003 \ ; \quad \text{For post-tensioning} = \frac{0.0002}{\log_{10}(t+2)}$$

where t = age of concrete at transfer, in days.

The value of shrinkage strain for design of post-tensioned concrete may be increased by 50 per cent in dry atmospheric conditions, subjected to a maximum value of 0.0003. For the calculation of deformation of concrete at some stage before the maximum shrinkage is reached, it may be assumed that half of the shrinkage takes place during the first month and that about three-quarters of the shrinkage takes place in the first six months after commencement of drying.

(*e*) ***Creep of concrete*** : Creep of concrete depends, in addition to the factors listed in (*d*) above, on the stress in concrete, age at loading and the duration of loading. As long as the stress in concrete does not exceed one third of the characteristic compressive strength, creep may be assumed to be proportional to the stress.

In the absence of experimental data and detailed information of the effect of variables, the ultimate creep strain may be estimated from the following values of crep coefficient (that is, ultimate creep strain/elastic strain at the age of loading).

Age of loading	*Creep Coefficient*
7 days	2.2
28 days	1.6
1 year	1.1

The ultimate creep strain estimated as per above clause does not include the elastic strain.

For the calculation of deformation at some stage before the total creep is reached, it may be assumed that about half the total creep takes place in the first month after loading and that about three quarters of the total creep takes place in the first six months after loading.

2. Prestressing steel

(*a*) ***Type of steel and modulus of elasticity***

The prestressing steel shall be any one of the following types:

Type of Steel	*Modulus of elasticity* $E(kN/mm^2)$
(*i*) Plain hard-drawn wire conforming to IS : 1785 (Part I)-1966 and IS 1785 (Part II)-1967	210
(*ii*) Cold drawn indented wire conforming to IS : 6003-1970	210
(*iii*) High tensile steel bar conforming to IS : 2090-1962	200
(*iv*) Uncoated stress relieved strand conforming to IS : 6006-1900	195

(*b*) ***Untensioned steel*** : Reinforcement used as untensioned steel shall be any of the following:

(*i*) Mild steel and medium tensile steel bars conforming to IS : 432 (Part I)-1966.

(*ii*) Hot rolled deformed bars conforming to IS : 1139-1966.

(*iii*) Cold twisted bars conforming to IS : 1786-1979.

(*iv*) Hard drawn steel wire fabric conforming to IS : 1566-1967

(*c*) ***Characteristic strength*** (f_p) : For prestressing steel, the characterisitic strength shall be assumed as the minimum ultimate tensile stress. For steel bars, $f_p = 1400\ N/mm^2$. For steel wires, $f_p = 1500$, 1600, 1650 and 1700 N/mm^2 depending upon the diameter of wires (higher value for thinner wire). Table 25.11 gives the minimum ultimate tensile strength of tensile steel wires.

(*d*) ***Cover*** : In pre-tensioned work, the cover of concrete measured from the outside of the pretressing tendon shall be atleast 20 mm. In post-tensioned work, where cables and large-sized bars are used, the maximum clear cover from sheathing/duct shall be at least 30 mm or the size of the cable or bar whichever is bigger. Where prestressed concrete members are located in aggressive environment, the cover specified above shall be increased by 10 mm.

(*e*) ***Spacing*** : In the case of single wires used in pre-tension system, the minimum clear spacing shall not be less than greater of the following :

(*i*) 3 times diameter of wire, and (*ii*) $1\frac{1}{3}$ times the maximum size of aggregate. In the case of cables or large bars, the minimum clear spacing (measured between sheathings/ducts, wherever used) shall not be less than greater of the following :

(*i*) 40 mm

(*ii*) Maximum size of cable or bar

(*iii*) 5 mm plus maximum size of aggregate.

(*f*) ***Grouped cables*** : Cables or ducts may by grouped together in group of not more than four, as shown in Fig. 25.62. The minimum clear spacing between groups of cables or ducts of grouped cables shall be greater of the following :

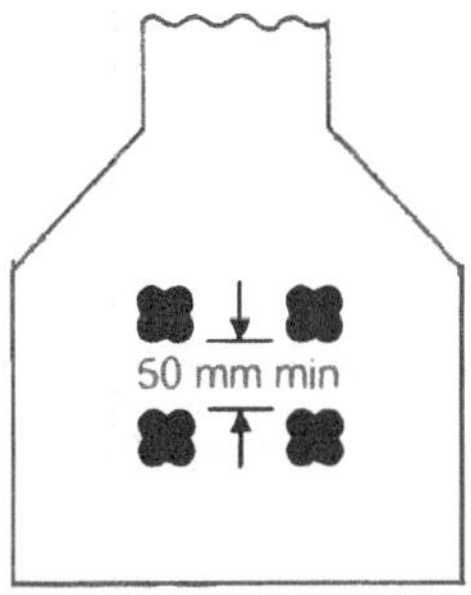

FIG. 25.62. SPACING OF GROUPS OF CABLES

(*i*) 40 mm and (*ii*) 5 mm plus maximum size of aggregates.

The vertical distance between groups shall not be less than 50 mm (See Fig. 25.62).

(*g*) ***Minimum longitudinal reinforcement*** : (*i*) A minimum longitudinal reinforcement of 0.2 per cent of the total concrete area shall be provided in all cases except in the case of pretensioned units of small sections. This reinforcement may be reduced to 0.15 per cent in the case of high yield strength deformed reinforcement. The percentage of steel provided, both tensioned and untensioned taken together, should be sufficient so that when the concrete in the precompressed tensile zone cracks, the steel is in a position to take up the additional tensile stress transferred on to it by the cracking of the adjacent fibres of concrete and a sudden failure is avoided.

(*ii*) When the depth of web exceeds 50 cm, longitudinal distribution reinforcement not less than 0.5 per cent of the area of the web shall be provided on each face. The spacing of the individual bars of such reinforcement shall not exceed 20 cm.

(*h*) ***Transverse reinforcement*** : (*i*) The amount and spacing of transverse reinforcement shall be governed by shear and torsion requirements. It is however, desirable to provide transverse reinforcement in the web when the web is thin and cables are located in the web.

(*ii*) In case of all members subjected to dynamic loading, webs shall be provided with transverse reinforcement not less than 0.3 per cent of the sectional area of the web in plan. This percentage of reinforcement may be reduced to 0.2 percent in members where the depth of the web is not more than four times the thickness of the web. These values may be reduced to 0.2 and 0.15 percent respectively when high strength reinforcement is used.

(*iii*) In case of members not subjected to dynamic loading, reinforcement shall be provided when the depth of the web is more than 4 times the thickness. Such reinforcement shall not be less than 0.1 percent of the sectional area of the web in plan. The reinforcement shall be spaced at a distance not greater than the clear depth of web and the size of such reinforcement shall be as small as possible.

(*iv*) Reinforcement in the form of links or helix shall be provided perpendicular to the line of heavy compression or shock loading to resist the induced tensile stresses.

(*i*) ***Deductions for Prestressing Tendons***

In calculating areas, centroid and moment of inertia of cross-section, deduction for prestressing tendons shall be made as follows :

(*i*) In the case of pre-tensioned members, where the prestressing tendons are single wires distributed on the cross-section or strands of wires of relatively small cross-sectional area, allowances for the prestressing tendons need not be made. Where allowance is made, it shall be on the basis of $(m - 1)$ times the area of the prestressing tendons, m being the modular ratio.

(*ii*) In the case of post-tensioned members, deductions shall invariably be made for prestressing tendons, cable ducts or sheaths and such other openings whether they are formed longitudinally or transversely. These deductions need not however, be made for determining the effect of loads applied after the ducts, sheaths or openings have been grouped or filled with concrete. Where such deductions are not made, a transformed area equivalent to $(m - 1)$ times the area of the prestressing tendon shall be taken in calculation.

Note : Modular ratio m shall be calculated as E_s/E_c. Wherever necessary, creep effects shall also be taken into consideration.

3. Minimum Initial Prestress : At the time of initial tensioning, the maximum tensile stress f_{pi} (f_{si}) immediately behind the anchorages shall not exceed 80 per cent of the ultimate tensile strength of the wire or bar or strand.

4. Losses in Prestress : While assessing the stresses in concrete and steel during tensioning operations and later in service, due regard shall be paid to all losses and variations in stress resulting from creep of concrete, shrinkage of concrete, relaxation of steel, shortening (elastic deformation) of concrete at transfer, and friction and slip of anchorages. In computing the losses in prestress when untensioned reinforcement is present, the effect of tensile stresses developed by the untensioned reinforcement due to shrinkage and creep shall be considered.

a) ***Loss of prestress due to creep of concrete*** : The loss of prestress due to creep of concrete under load shall be determined for all the permanently applied loads including the prestress.

The creep loss due to live load stresses, erection stress and other stresses or short duration may be ignored. The loss of prestress due to creep of concrete is obtained as the product of modulus of elasticity of the prestressing steel and the ultimate creep strain of the concrete fibre integrated along the line of centre of gravity of the prestressing steel over its entire length. The total creep strain during any specific period shall be assumed for all practical purposes, to be creep strain due to sustained stress equal to the average of the stresses at the beginning and end of the period.

(*b*) ***Loss of prestress due to shrinkage of concrete*** : The loss of prestress due to shrinkage of concrete shall be the product of the modulus of elasticity of steel and the shrinkage strain of concrete.

(*c*) ***Loss of prestress due to relaxation of steel*** : The relaxation losses in prestressing steels vary with type of steel, initial prestress, age and temperature, and therefore shall be determined from experiments. When experimental values are not available, the relaxation losses may be assumed as given in Table 25.13.

TABLE 25.13 RELAXATION LOSSES FOR PRESTRESSING STEEL (At 1000 H at 27° C)

Initial stress	*Relaxation loss (N/mm²)*
0.5 f_p	0
0.6 f_p	35
0.7 f_p	70
0.8 f_p	90

Note : *f_p is the characteristic strength of prestressing steel.* : For tendons at higher temperatures or subjected to large lateral loads, greater relaxation losses as specified by engineer-in-charge shall be allowed for. No reduction in the the value of relaxation losses should be made for a tendon with a load equal to or greater than the relevant jacking force that has been applied for a short time prior to the anchoring of the tendon.

(*d*) ***Loss of prestress due to shortening of concrete*** : This loss is proportional to the modular ratio and initial prestress in the concrete and shall be calculated as below, assuming that the tendons are located at their centroid :

(*i*) For pretensioing, the loss of prestress in the tendons at transfer shall be calculated on a modular ratio basis using the stress in the adjacent concrete.

(*ii*) For members with post-tensioned tendons which are not stressed simultaneously, there is progressive loss of prestress during transfer, due to the gradual application of the prestressing forces. This loss of prestress should be calculated on the basis of half the product of the stress in the concrete adjacent to the tendons averaged along their lengths and the modular ratio. Alternatively, the loss of prestress may be exactly computed on the sequence of tensioning.

(*e*) ***Loss of prestress due to slip anchorage*** : Any loss of prestress which may occur due to slip of wires during anchoring or due to the strain of anchorage shall be allowed for in the design. Loss due to slip in anchorage is of special importance with short member and the necessary additional elongation should be provided for at the time of tensioning to compensate for this loss.

(*f*) ***Loss of prestress due to friction*** : The design shall take into consideration all losses in prestress that may occur during tensioning due to friction between the prestressing tendons and the surrounding concrete or any flexure attached to the steel or concrete.

For straight or moderately curved structures with curved or straight cables, the value of prestressing force P_x at a distance x metres from tensioning end and acting in the direction of the tangent to the curve of the cable shall be calculated as below :

$$P_x = P_0 \cdot e^{-(\mu\alpha + kx)} \qquad \ldots(25.69)$$

where P_0 = Prestressing force in the prestressed steel at the tensioning end acting in the direction of the tangent to the curve of the cable.

α = cumulative angle in radians through which the tangent to the cable profile has turned between any two points under consideration.

μ = coefficient of friction in curve ; unless otherwise proved by tests, μ may be taken as :

0.55 for steel moving on smooth concrete.
0.30 for steel moving on steel fixed to duct, and
0.25 for steel moving on lead.

k = coefficient of wave effect varying from 15×10^{-4} to 50×10^{-4} per metre.

Note 1 : Expansion of equation for P_x for small values of $(\mu\alpha + kx)$ may be $P_x = P_0 (1 - \mu\alpha - k \cdot x)$

Note 2 : In circular constructions, where circumferential tendons are tensioned by jacks, values of μ for calculating friction may be taken as :

0.45 for steel moving in smooth concrete,
0.25 for steel moving on steel bearers fixed to the concrete, and
0.10 for steel moving on steel rollers.

Note 3 : The effect of reverse friction shall be taken into consideration in such cases where the initial tension applied to a prestressing tendon is partially released and action of friction in the reverse direction causes an alteration in the direction of stress along the length of the tendon.

5. **Bearing Stress** : (*i*) On the areas immediately behind external anchorages the permissible unit bearing stress on the concrete, after accounting for all losses due to relaxation

of steel, elastic shortening, creep of concrete, slip and/or seating of anchorages etc., should not exceed $0.48 f_{ci} \sqrt{\frac{A_{br}}{A_{pun}}}$ or $0.8 f_{ci}$ whichever is smaller, where, f_{ci} is the cube strength at transfer, A_{br} is the bearing area and A_{pun} is the punching area.

(*ii*) During tensioning, the allowable bearing stress specified in (*i*) may be increased by 25 per cent, provided that this temporary value does not exceed f_{ci}.

(*iii*) The bearing stress specified in (*i*) and (*ii*) for permanent and temporary bearing stress may be increased suitabily if adequate hoop reinforcement is provided at anchorages.

(*iv*) When the anchorages are embedded in concrete, the bearing stress shall be investigated after accounting for the surface friction between the anchorage and the concrete.

(*v*) The effective punching area shall generally be the contact area of the anchorage devices which, if circular in shape, shall be replaced by a square of equivalent area. The bearing area shall be the maximum area of that portion of the member which is geometrically similar and concentric to the effective punching area.

(*vi*) Where a number of anchorages are used, the bearing area A_{br} shall not overlap. Where there is already a compressive stress prevailing over the bearing area, as in the case of anchorage placed in the body of a structure, the total stress shall not exceed the limiting values specified in (*i*), (*ii*) and (*iii*). For stage stressing of cables, the adjacent unstressed anchorages shall be neglected while determining the bearing area.

6. Structural Design : Limit State Method

(*a*) ***Limit state design*** : The structural design shall be based on limit state concept. In this method of design, the structure shall be designed to withstand safely all loads liable to act on it throughout the life ; it shall also satisfy the serviceability requirements, such as limitation on serviceability requirements before failure occurs is called as *'limit state'*. The aim of the design is to achieve acceptable probabilities that the structure will not become unfit for use for which it is intended, that is, it will not reach a limit state.

The design should be based on *characteristic values* for material strength and applied loads. The *design values* are derived from the characteristic values through the use of partial safety factors, one for the material strengths and the other for loads. The values of partial safety factors for material strength (γ_m) should be taken as 1.5 for concrete and 1.15 for steel. It should be noted that γ_m values are already incorporated in the equation and tables given in IS : 1343-1980. The partial safety factors (γ_f) for loads shall be the same as given in the previous chapter on limit state design.

(*b*) ***Limit state of collapse*** : The limit state of collapse of the structure or part of the structure could be assessed from rupture of one or more critical sections and from buckling due to elastic or plastic instability (including the effects of sway where appropriate) or overturning. The resistance to bending, shear, torsion and axial load at every section shall not be less than appropriate value at that section produced by the probable most unfavourable combination of loads on the structure using the appropriate partial safety factor.

(*c*) ***Limit state of serviceability : Deflection.***

The deflection shall be limited to the following :

(*i*) The final deflection, due to all loads including the effects of temperature, creep and shrinkage and measured from the as-cast level of the supports of floors, roofs and all other horizontal members, should not normally exceed span / 250.

(*ii*) The deflection including the effects of temperature creep and shrinkage occurring after erection of partitions and the application of finishes should not normally exceed span / 350 or 20 mm whichever is less.

(*iii*) If finishes are to be applied to prestressed concrete members, the total upward deflection should not exceed span/300, unless uniformity of camber between adjacent units can be ensured.

(*d*) ***Limit state of serviceability : Cracking***

Cracking of concrete shall not affect the appearance of durability of structure. The criteria of limit state of cracking for the *three types* of prestressed concrete members shall be as follows :

(*i*) ***For type 1*** : No tensile stress.

(*ii*) ***For type 2*** : Tensile stresses are allowed but no visible cracking.

(*iii*) ***For type 3*** : Cracking is allowed, but should not affect the appearance or durability of the structure ; the acceptable limits of cracking would vary with the type of structure and environment and will vary between wide limits and the prediction of absolute maximum width is not possible.

Note : For design of type 3 members, as a guide, the following paragraph may be regarded as reasonable limits : The flexural tensile stress at any section of the structure, both at transfer and under the most unfavourable combination of design loads, shall satisfy the criteria for the corresponding type of structure.

(*e*) ***Limit state of serviceability : Maximum compression*** : The compressive stresses both at transfer and under design loads shall be limited to the values given in paragraph follows, for all types of structures.

7. ***Limit State of Collapse : Flexure***

(*a*) ***Assumption*** : Design for the limit-state of collapse in flexure shall be based on the following assumptions :

(*i*) Plane sections normal to the axis remain plane after bending.

(*ii*) The maximum strain in concrete at the outermost compression fibre is taken as 0.0035 in bending.

(*iii*) The relationship between the compressive stress distribution in concrete and the strain in concrete may be assumed to be rectangle, trapezoid, parabola or any other shape which results in prediction of strength in substantial agreement with the result of tests. An acceptable stress-strain curve is given in Figs. 2.3 and 3.1. For design purposes the compressive strength of concrete on the structure shall be assumed to be 0.67 times the characteristic strength. The partial safety factor $\gamma_m = 1.5$ shall be applied in addition to this

For the stress block shown in Fig 3.1, the parameters may be adopted as follows:

Area of stress block $= 0.36 \, f_{ck} \, . \, x_u$

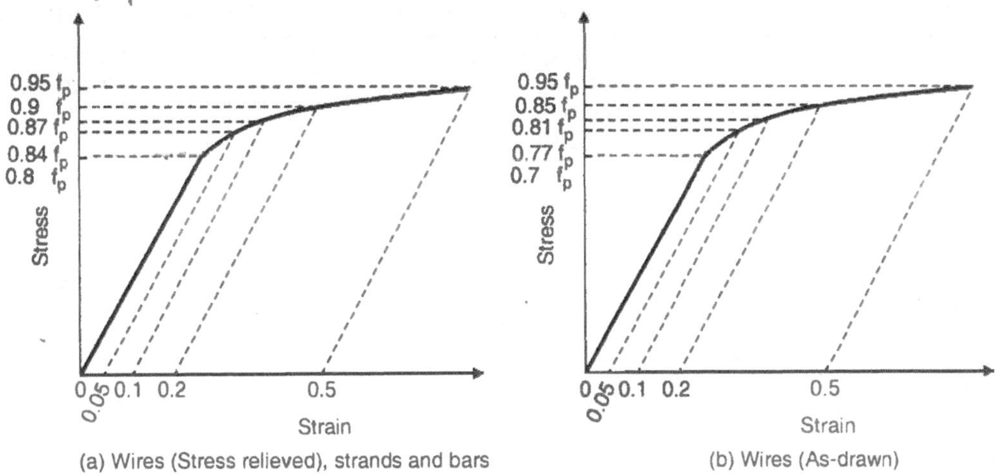

FIG. 25.63. STRESS-STRAIN CURVES FOR STEEL

Depth of centre of compressive force from the extreme fibre in compression $= 0.42\, x_u$

(*iv*) The tensile strength of the concrete is ignored.

(*v*) The stresses in bonded prestressing tendons, whether initially tensioned or untensioned, and in additional reinforcement are derived from the representative stress-strain curve for the type of steel used given the manufactures of typical curves given in Fig. 25.63 for prestressing tendons and in Figs. 2.4, 2.5 and 2.6 for reinforcement. For design purposes, the partial safety factors γ_m equal to 1.15 shall be applied.

(*iv*) If tendons are unbonded in post-tensioned members, the stress in tendons may be obtained from a rigorous analysis or from tests.

(*b*) ***Design Formulae***

(*i*) The moment of resistance of rectangular section or T-section in which neutral axis lies within the flanges be obtained as follows:

$$M = f_{pu} \cdot A_p(d - 0.42\, x_u) \qquad \ldots(25.70)$$

where M = moment of resistance of the section,

f_{pu} = ultimate tensile stress in the tendons,

f_p = characteristic strength of prestressing steel,

A_p = area of pretensioning tendons,

d = effective depth, and

x_u = neutral axis depth.

For pretensioned members and for post-tensioned members with effective bond between the concrete and tendons values of f_{pu} and x_u are given in Table 25.14 .The effective prestress after all losses should not be less than $0.45\, f_p$ where f_p is the characteristic strength of prestressing steel.

TABLE 25.14 CONDITIONS AT THE ULTIMATE LIMIT STATE FOR RECTANGULAR BEAMS WITH PRE-TENSIONED TENDONS OR WITH POST-TENSIONED TENDONS HAVING EFFECTIVE BOND

$\frac{A_s f_p}{bd\, f_{ck}}$	*Stress in tendon as a proportion of the design strength $f_{pu}/0.87 f_p$*		*Ratio of the depth of N.A. to that of the centroid of the tendon in the tension zone x_u/d*	
	Pre-tensioning	*Post-tensioning with effective bond*	*Pre-tensioning*	*Post-tensioning with effecive bond*
0.025	1.0	1.0	0.054	0.054
0.05	1.0	1.0	0.109	0.109
0.10	1.0	1.0	0.217	0.217
0.15	1.0	1.0	0.326	0.316
0.20	1.0	0.95	0.435	0.414
0.25	1.0	0.90	0.542	0.488
0.30	1.0	0.85	0.655	0.558
0.40	0.9	0.75	0.783	0.653

The post-tensioned members with unbonded tendons, the values of f_{pu} and x_u are given in Table 25.15.

TABLE 25.15 CONDITIONS AT THE ULTIMATE LIMIT STATE FOR POST-TENSIONED RECTANGULAR BEAMS HAVING UNBONDED TENDONS

$\frac{A_p f_p}{bd \cdot f_{ck}}$	*Stress in tendon as a proportion to the effective pressure f_{pu}/f_p for values of l/d $\left(\frac{\text{Effective span}}{\text{Effective depth}}\right)$*			*Ratio of depth of N.A. to that of the centroid of the tendon in the tension zone, x_u/d for values of l/d $\left(\frac{\text{Effective span}}{\text{Effective depth}}\right)$*		
	30	*20*	*10*	*30*	*20*	*10*
0.025	1.23	1.34	1.45	0.10	0.10	0.10
0.05	1.21	1.32	1.45	0.16	0.16	0.18
0.10	1.18	1.26	1.45	0.30	0.32	0.36
0.15	1.14	1.20	1.36	0.44	0.46	0.52
0.20	1.11	1.16	1.47	0.56	0.58	0.64

(*ii*) For flanged section in which the N.A. lies outside the flange, the moment of resistance shall be determined using assumption in 7(*a*) above.

8. ***Limit State of Collapse : Compression*** : Prestressed concrete compression members in framed structures, where the mean stress in concrete section imposed by tendons is less than 2.5 N/mm², may be analysed as reinforced concrete compression members in accordance with IS : 456 ; in other cases, special literature may be referred to.

9. ***Limit State of Collapse : Tension***. Tensile strength of tension member shall be based on the design strength (0.87 times characteristic strength of prestressing tendons) and the strength developed by any additional reinforcement. The additional reinforcement may usually be assumed to be acting at its design stress (0.87 times characteristic strength of reinforcement); in special cases it may be necessary to check the stress in the reinforcement using strain compatibility.

10 ***Limit State of Collapse : Shear*** : The ultimate shear resistance of concrete along V_c should be considered at both sections uncracked and cracked in flexure, and lesser value taken, and if necessary, shear reinforcement provided.

(*a*) ***Section uncracked in flexure*** : The ultimate shear resistance of a section uncracked in flexure, $V_c = V_{co}$ is given by

$$V_{co} = 0.67\, b\, D \sqrt{f_t^2 + 0.8\, f_{cp}\, f_t} \qquad \text{...(25.71)}$$

where b = breadth of the members which for T, I and L beams should be replaced by breadth of rib b_w.

D = overall depth of the member.

f_t = maximum principal stress given by $0.24\sqrt{f_{ck}}$ taken as positive, where f_{ck} is the characteristic compressive strength of concrete.

f_{cp} = compressive stress at centroidal axis due to prestress taken as positive.

In flanged members where the centroidal axis occurs in the flange, the principal tensile stress should be limited to $0.24\sqrt{f_{ck}}$ at the intersection of the flanged web; in this calculation, 0.8 of the stress due to prestress at this intersection may be used, in calculating V_{co}. For a section uncracked in flexure and with inclined tendons or vertical prestress, the component of prestressing force normal to the longitudinal axis of the member may be added to V_{co}.

(*b*) ***Section cracked in flexure*** : The ultimate shear resistance of a section cracked in flexure $V_c = V_{cr}$ is given by :

$$V_{cr} = \left(1 - 0.55\frac{f_{pe}}{f_p}\right)\xi_c \,.\, bd + M_0 \frac{V}{M} \qquad \text{...(25.72)}$$

where f_{pe} = effective prestress after losses have occurred, which shall not be put greater than $0.6\, f_p$

f_p = characteristic strength of prestressing steel.

ξ_c = ultimate shear stress capacity of concrete, obtained from Table 25.16.

b = breadth of members, which, for flanged sections shall be taken as the breadth of the web, b_w.

d = distance from extreme compression-fibre to the centroid of the tendons at the section considered.

M_0 = moment necessary to produce zero stress in concrete at the depth given by

$$M_0 = 0.8\, f_{pt}\, I/y \qquad \text{...(25.73)}$$

where f_{pt} = is the stress due to prestress only at depth d and distance y from the centroid of the concrete section which has second moment of area I, and

V and M = shear force and bending moment respectively at the section considered due to ultimate loads. ; V_{cr} should be taken not less than $0.1\, bd\sqrt{f_{ck}}$

TABLE 25.16 DESIGN SHEAR STRENGTH OF CONCRETE ξ_c N/mm²

$100\frac{A_p}{bd}$	Concrete grade		
	M 30	*M 35*	*M 40 and above*
0.20	0.37	0.37	0.38
0.50	0.50	0.50	0.51
0.75	0.59	0.59	0.60
1.00	0.66	0.67	0.68
1.25	0.71	0.73	0.74
1.50	0.76	0.78	0.79
1.75	0.80	0.82	0.84
2.00	0.84	0.86	0.88
2.25	0.88	0.90	0.92
2.50	0.91	0.93	0.95
2.75	0.94	0.96	0,98
3.00	0.96	0.99	1.01

Note : A_p is the area of prestressing tendon.

The value of V_{cr} calculated at a particular section may be assumed to be constant for a distance equal to $d / 2$, measured in the direction of increasing moment, from that particular section.

For a section cracked in flexure and with inclined tendons, the component of prestressing force normal to the longitudinal axis of the member should be ignored.

(*c*) ***Shear reinforcement***

(*i*) When V, the shear force due to the ultimate loads, is less than V_c the shear force which can be carried by the concrete, minimum shear reinforcement should be provided in the form of stirrups such that

$$\frac{A_{sv}}{b \cdot s_v} \geq \frac{0.4}{0.87 f_y} \qquad \ldots(25.74)$$

where A_{sv} = total cross-sectional area of stirrups legs effective in shear.

b = breadth of the members which for T, I and L beams should be taken as the breadth of the rib, b_w

s_v = stirrup spacing along the length of the member.

f_y = characteristic strength of stirrrup reinforcement which shall not be taken greater than 415 N/mm^2.

However, shear reinforcement need not be provided in the following cases : (*i*) where V is less than 0.5 V_c , and (*ii*) in members of minor importance.

(*ii*) When V exceeds V_c shear reinforcement shall be provided such that

$$\frac{A_{sv}}{s_v} = \frac{V - V_c}{0.87 f_y \cdot d_t} \qquad \ldots(25.75)$$

In rectangular beams, at both corners in tensile zone, a stirrup should pass around a longitudinal bar, a tendon or a group of tendons t having a diameter not less than the diameter of the stirrup. The depth d_t is then taken as the depth from extreme compression fibre either to the longitudinal bars or to the centriod of the tendons which ever is greater.

The spacing of stirrpus along a member should not exceed 0.75 d_t nor 4 times the web thickness for flanged members. When V exceeds 1.8 V_c the maximum spacing should be reduced to 0.5 d_t. The lateral spacing of the individual legs of the stirrups provided at a cross-section should not exceed 0.75 d_t.

(*iii*) In no circumstances should the shear force V, due to ultimate loads, exceed the approprate values given in Table 25.17 multiplied by *b.d.*

TABLE 25.17 MAXIMUM SHEAR STRESS

Grade of Concrete	*M 30*	*M 35*	*M 40*	*M 45*	*M 50*	*M 55 and over*
Maximum shear stress (N/mm^2)	3.5	3.7	4.0	4.3	4.6	4.8

11. ***Limit State of Serviceability : Deflection***

(*a*) ***Type 1 and type 2 members.***

The *instantaneous deflection* (short-term deflection) due to design loads may be calculated using elastic analysis based on uncracked section and modulus of elasticity of concrete.

The total *long term deflection* due to the prestressing force, dead load and any sustained imposed load may be calculated using elastic analysis taking into account the effects of cracking and of creep and shrinkage. Due allowance shall be made for the loss of prestress after the period considered. The deflection should comply with limits given in clause 6(*c*) above.

(*b*) ***Type 3 members.*** Where the permanent load is less than or equal to 25 per cent of the design imposed load, the deflection may be calculated as in (*i*) above. When the permanent load is more than 25 per cent of the design imposed load, the vertical deflection limits of beams and the design imposed load, the vertical deflection limits of beams and slabs may generally be assumed to be satisfied provided that the span to effective depth ratio are not greater than values obtained as below :

For values of span to effective depth ratios for spans upto 10 m:

Cantilever : 7 ; Simply supported : 20 ; Continuous ; 26

For spans above 10 m, the above values may be multiplied by 10/span in metres except for cantilever in which case deflection calculation should be made.

12. ***Limit State of Serviceability: Cracking***

For the members made up of precast units, no tension shall be allowed at any stage at mortar or concrete joints. For a member which is free of joints, the tensile stress shall not exceed the values specified below for the 3 types of members:

(*a*) **Type 1.** No tensile stress.

(*b*) **Type 2.** The tensile stress shall not exceed 3 N/mm^2. However, where part of service loads is temporary in nature, this value may be exceeded by 1.5 N/mm^2 provided under the per-mancent component of the service load the stress remains compressive.

(*c*) **Type 3.** For type 3 members in which cracking is permitted, it may be assumed that the concrete section is uncracked, and that hypothetical tensile stresses exist at the maximum size of cracks. The hypothetical tensile stresses for use in these calculations for member with either pre-tensioned or post tensioned tendons are given in Table 25.18. The values in Table 25.18 shall be multiplied by the depth factors obtained from Fig. 25.64.

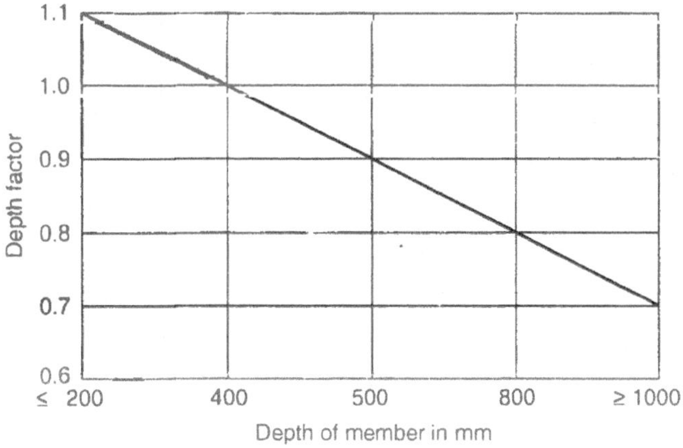

FIG. 25.64. DEPTH FACTORS FOR TENSILE STRESSES FOR TYPES 3 MEMBERS.

TABLE 25.18 HYPOTHETICAL FLEXURAL TENSILE STRESSES FOR TYPE 3 MEMBERS

Type of Tendons	*Limiting crack width (mm)*	*Stress of Concrete of Grade*				
		M 30	*M 35*	*M 40*	*M 45*	*M50 & above*
(*i*) Pre tensioned tendons	0.1	—	—	4.1	4.4	4.8
	0.2	—	—	5.0	5.4	5.8
(*ii*) Grouted post tensioned tendons	0.1	3.2	3.6	4.1	4.4	4.8
	0.2	3.8	4.4	5.0	5.4	5.8
(*iii*) Pre tensioned tendons distributed in the tensile zone and positioned close to the tension face of concrete.	0.1	—	—	5.3	5.8	6.3
	0.2	—	—	6.3	6.8	7.3

Note : When additional reinforcement is distributed within the tension zone and positioned close to the tension face of concrete, the hypothetical tensile stresses may be increased by an amount which is proportioned to the cross-sectional areas of the addtional reinforcement expressed as a percentage of the cross-sectional area of the concrete. For 1 per cent of additional reinforcement, the stress may be increased by 4 N/mm^2 for members with pre-tensioned and grouted post-tensioned tendons and by 3 N/mm^2 for other members. For other percentages of additional reinforcement the stresses may be increased in proportion excepting that total hypothetical tensile stress shall not exceed 0.25 times the characteristic compressive strength of concrete.

13. ***Limit State of Serviceability : Maximum Compression***

(*a*) ***Maximum Stress Under Service Conditions***

(*i*) *Compressive stress in flexure.* The maximum permissible compressive stress, prestress and service loads after deduction of the full losses in the specified prestress shall be determined by a straight line relation as shown in Fig. 25.65 but different stress limits shall apply to the concrete of the structure depending on whether if falls in a part of the section where the compressive stresses are not likely to increase in service (zone I) or in part of the section where the compressive stresses are likely to increase in service (zone II).

For zone I , the straight line relation of permissible stress shall be determined by straight line joining a point given by a permissible stress of 0.41 f_{ck} for concrete of grade M 30 to another point given by a permissible stress of 0.35 f_{ck} of concrete of grade M 60.

For zone II, the determining points of the graph shall be reduced to 0.34 f_{ck} and 0.27 f_{ck} respectively.

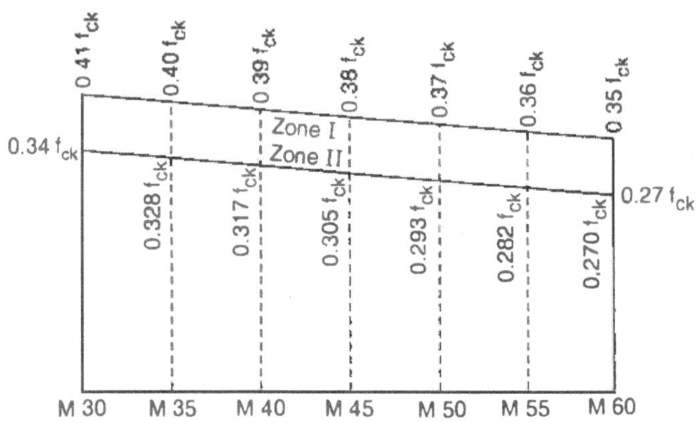

FIG. 25.65. COMPUTATION OF MAXIMUM PERMISSIBLE COMPRESSIVE STRESSES ON FLEXURE DUE TO FINAL PRESTRESS.

(*ii*) *Stress in direct compression.* Except in part immediately behined the anchorage, the maximum stress in direct compression shall be limited to 0.8 times the permissible stress obtained in (*i*) above.

(*b*) ***Maximum Stress at Transfer***

(*i*) *Compressive stress in flexure.* The maximum permissible compressive stress due to bending and direct force at the time of transfer of prestress shall be determined from a graph in which a straight line joins a point given by 0.54 f_{ci} for a concrete of grade M 30 to a second point giving a permissible stress of 0.37 f_{ci} for concrete of grade M 60, [Fig. 25.66(*a*)] ; f_{ci} being cube strength of concrete at transfer which in no case shall be less than half the corresponding characteristic compressive strength of concrete. These values apply to post-tensioned work ; for pre-tensioned work the variation represented by Fig. 25.66(*b*) will apply.

Note. The strength of concrete at the time of transfer, f_{ci} shall be established by tests carried out on cubes at the age of the concrete at transfer for bridges and such other major structures

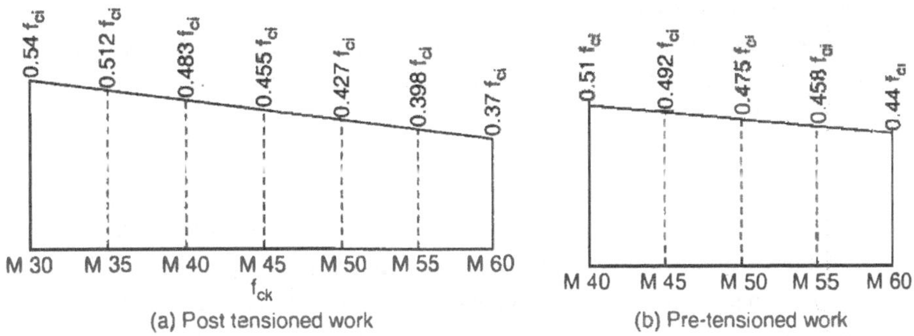

FIG. 25.66. COMPUTATION OF MAXIMUM PERMISSIBLE COMPRESSIVE STRESS IN FLEXURE AT TRANSFER

and in other cases, if more convenient, from straight line graph, joining the characteristic compressive strength of concrete and cube strength at 7 days. The transfer of prestress shall be made only after the concrete has attained a strength of atleast half the characteristic compressive strength of concrete.

(*ii*) *Stress in direct compression.* Except in the parts immediately behind the anchorages, the maximum stress in direct compression shall be limited to 0.8 times the permissible stress obtained in (*i*) above.

25.19. PROCEDURE FOR LIMIT STATE DESIGN

As mentioned in § 25.18, there are three types of prestressed concrete members : Type 1, Type 2 and Type 3.

Type 1 members are those in which no tensile stresses are permitted. The common examples of such members are the water retaining structures, nuclear structures, bridge girders etc. The compressive stress in concrete as well as tensile stresses in steel are also limited to specified values.

Type 2 members are those in which though tensile stress are allowed to certain permissible value but no visible cracking is allowed. Small bridge girders and certain building elements fall under this category.

Type 3 members are those in which cracking is allowed ; however such cracking should not effect appearance or durability of the structure. Office and residential buildings etc. fall under these category. The flexural tensile stress at any section of the structure both at transfer and under most unfavourable combination of design loads shall satisfy the criteria for corresponding type of structure.

In general, a prestressed concrete structure is designed for the following limiting states:

1. **Collapse Limit State (or Strength limit state)**

(*a*) Limit state for flexure
(*b*) Limit state for compression
(*c*) Limit state for tension
(*d*) Limit state for shear
(*e*) Limit state for torsion
(*f*) Limit state for bursting and bond.

2. **Serviceability Limit States**

(*a*) Limit state for allowable stresses both for service conditions as well as transfer condition.
(*b*) Limit state for deflection.
(*c*) Limit state for cracking.

3. **Limit State of Durability** :

Out of these, the *collapse limit state* (or strength limit state) are *common* to all the types of structures. The *serviceability limit state* are same for type 1 and type 2 structures but different for type 3 structures.

The partial safety factors for material strength are 1.5 for concrete and 1.15 for steel. The partial safety factor γ_f for loads shall be the same as discussed in the chapter for limit state design. The design is carried out in the following steps.

Step 1. Determination of design bending moment at collapse : The superimposed dead load and live loads are generally given. The self weight of the beams is estimated on the basis of assumed dimensions of the beam. For this purpose, the width *b* and depth *d* of rectangular beams, simply supported at the ends, may be taken as *L*/30 and *L*/15 respectively. The design bending moment at collapse (M_c) is then computed, after applying partial safety factor for loads. $\gamma_f = 1.5$.

Step 2. Design for limit state of flexure : Fig. 25.67(*c*) shows stress block, having depth of N.A. equal to x_u. The stress block has rectangular portion upto depth of $\frac{3}{7}x_u$ and then parabolic for the remaining depth of $\frac{4}{7}x_u$. As proved in the chapter 33, the design compressive force C_u in concrete is equal to $0.36 f_{ck} \cdot b\, x_u$ which acts at a depth $= 0.416\, x_u \simeq 0.42\, x_u$ below the top fibre.

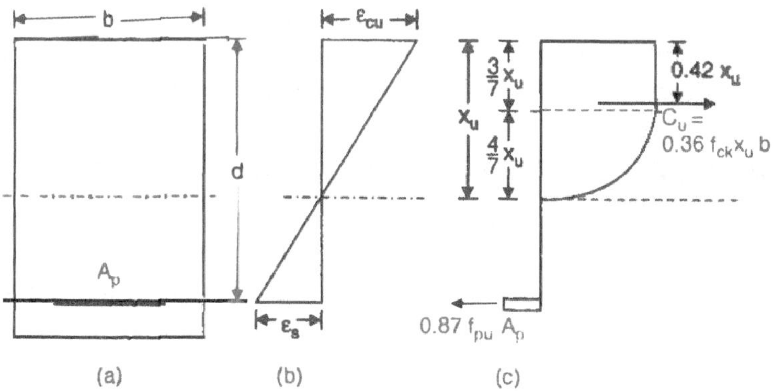

FIG. 25.67. COMPUTATION OF MAXIMUM PERMISSIBLE COMPRESSIVE STRESSES ON FLEXURE DUE TO FINAL PRESTRESS.

The neutral axis can be located from the strain compatibility [Fig. 25.67(*b*)].

$$\frac{\varepsilon_{cu}}{\varepsilon_{cu} + \varepsilon_s} = \frac{x_u}{d} \quad \text{or} \quad x_u = \frac{\varepsilon_{cu}}{\varepsilon_{cu} + \varepsilon_s}\, d$$

where ε_{cu} = strain in concrete at limit state = 0.0035

ε_s = compatible strain in steel $= \varepsilon_y - \varepsilon_{sc}$

ε_{sc} = effective prestrain in steel

ε_y = total strain in steel

$$= 0.0020 + \frac{0.87 f_p}{E_s} \quad \text{(IS Code)} \qquad ...(25.76)$$

Normally accepted value of $\varepsilon_{sc} = 0.004$. Taking $E_s = 2 \times 10^5$ N/mm^2 and substituting the values in Eq. 25.76, we get

$$x_u = \frac{0.0035\, d}{0.0035 + \left(0.0020 + \frac{0.87 f_p}{\varepsilon_s}\right) - 0.004} = \frac{0.0035\, d}{0.0015 + \frac{0.87 f_p}{2 \times 10^5}} = \frac{35\, d}{15 + 0.0435 f_p} \qquad ...(25.77)$$

Now the design moment caused by the compressive force about the centre of steel is given by

$$M_u = C_u (d - 0.42 x_u)$$

Substituting the values of C_u and x_u, we get

$$M_u = 0.36 f_{ck} . b \frac{35 d}{15 + 0.0435 f_p} \left(d - 0.42 \times \frac{35 d}{15 + 0.0435 f_p} \right)$$

or $$M_u = f_{ck} b d^2 \frac{12.6}{15 + 0.0435 f_c} \left(1 - \frac{14.6}{15 + 0.0435 f_p} \right)$$

or $$M_u = R_u b d^2 . f_{ck} \qquad \ldots(25.78)$$

where $$R_u = \frac{12.6}{15 + 0.0435 f_p} \left(1 - \frac{14.6}{15 + 0.0435 f_p} \right) \qquad \ldots[25.78(a)]$$

The area of pretensioned steel can be obtained by equating the compressive and tensile forces : $C_u = T_u$ or $0.36 f_{ck} . b x_u = 0.87 f_p . A_p$

where A_p = area of pretensioned steel; f_p = characteristic strength of pretensioned steel.

$$\therefore \qquad A_p = \frac{0.36}{0.87} \frac{f_{ck}}{f_p} . b x_u$$

Substituting the value of x_u, we get

$$A_p = \frac{0.36}{0.87} \frac{f_{ck}}{f_p} . b \frac{35 d}{15 + 0.0435 f_c} = \frac{14.49 b d}{15 + 0.0435 f_p} \frac{f_{ck}}{f_p}$$

or $$\frac{A_p}{b d} = p_0 \frac{f_{ck}}{f_p} \qquad \ldots(25.79)$$

where $$p_0 = \frac{14.49}{15 + 0.0435 p_c} = \frac{A_p f_p}{b d f_{ck}} \qquad \ldots[25.79(a)]$$

Lever arm factor $$j = 1 - 0.42 \frac{x_u}{d} = 1 - 0.42 \times \frac{35 d}{(15 + 0.0435 f_p) d} = 1 - \frac{14.6}{15 + 0.0435 f_p} \qquad \ldots(25.80)$$

The design coefficients $\frac{x_u}{d}$, R_u, j and p_0 depend upon the characteristic strength f_p of pretensioned steel. Their values for various values of f_p are given in Table 25.19.

TABLE 25.19 DESIGN COEFFICIENTS

f_p (N/mm^2)	$\frac{x_u}{d}$	j	R_u	p_0
1400	0.461	0.808	0.134	0.191
1500	0.436	0.818	0.128	0.181
1600	0.414	0.827	0.123	0.171
1700	0.393	0.836	0.118	0.163
1800	0.384	0.840	0.116	0.159
1900	0.358	0.850	0.110	0.148

Eq. 25.79 is applicable for balanced sections. Type 1 and Type 2 members are likely to be over-reinforced, while Type 3 members are often under-reinforced in which the steel reaches the yield stress before the crushing in concrete is initiated. The condition is represented by

$$\frac{A_p}{b\,d} < p_0 \frac{f_{ck}}{f_p} \qquad \ldots[25.79(b)]$$

For over-reinforced section, we have

$$\frac{A_p}{b\,d} \geq p_0 \frac{f_{ck}}{f_p} \qquad \ldots[25.79(c)]$$

The amount of prestressing steel in a section is limited by the following equation:

$$\frac{A_p}{b\,d} \leq 0.24 \frac{f_{ck}}{f_p} \qquad \ldots(25.81)$$

For under-reinforced section, the moment of resistance is given by

$$M_u = f_{pu} \cdot A_p\,(d - 0.42\,x_u) \qquad \ldots(25.82)$$

where f_{pu} = ultimate tensile stress in tendons $= \dfrac{\alpha \cdot f_p}{1.15} = \alpha\,(0.87)\,f_p$

$\alpha = \dfrac{f_{pu}}{0.87\,f_p}$ = coefficient in the range of 1 to 0.9 depending upon steel ratio $\left(\dfrac{A_p}{b\,d} \cdot \dfrac{f_p}{f_{ck}}\right)$ and is given in Table 25.14.

For over-reinforced section, ultimate moment of resistance is limited to that one for balanced section. Thus, the area of steel in excess of the balanced proportion should be neglected.

For flanged section, if the N.A. lies in the flange (i.e. if $x_u \leq D_f$),

$$M_u = R_u\,b_f \cdot d^2\,f_{ck} \qquad \ldots(25.83)$$

where b_f = width of flange and D_f = thickness of flange.

If $x_u \geq D_f$, the ultimate moment of resistance is given by

$$M_u = R_u\,b_w\,d^2\,f_{ck} + 0.45\,(b_f - b_w)\,D_f\,(d - 0.5\,D_f)\,f_{ck} \qquad \ldots(25.84)$$

The area of steel is given by balancing the compressive and tensile forces :

$$0.87\,A_p \cdot f_{pu} = 0.36\,b_w\,x_u\,f_{ck} + 0.45\,(b_f - b_w)\,D_f \cdot f_{ck}$$

or

$$\mathbf{A_p = 1.15\left[0.36\,b_w\,x_u + 0.45\,(b_f - b_w)\,D_f\right]\frac{f_{ck}}{f_{pu}}} \qquad \ldots(25.85)$$

where

$$x_u = \frac{35\,d}{15 + 0.0435\,f_p} \qquad \ldots(25.77)$$

In Eq. 25.85, b_f and D_f are considered to be known, while b_w is suitably assumed. Hence d and A_p can be found.

Thus, in the *second step*, the depth d and area of steel A_p can be determined from the relation $M_u \geq M_c$

where M_u is given by Eq. 25.78 for rectangular section and Eq. 25.84 for flanged section.

Determine ultimate prestressing force from the relation

$$P_u = \gamma_{fp} \cdot A_p \cdot f_p \qquad \ldots.(25.86)$$

where γ_{fp} can be taken between 1.0 to 1.20.

Step 3. Design for limit state strength at transfer condition: At the transfer state, the prestressing force may be considered as an external load, along with the dead load.

The ultimate tendon force P_u is given by Eq. 25.86, in which is γ_{fp} is the partial safety factor which may be taken equal to 1.0 to 1.20.

If M_{dy} is the dead load moment due to self weight of beam, the tensile force produced in the tendons is given by

$$F_{dg} = \frac{\gamma_{fd} \cdot M_{dg}}{d - 0.42\, x_u} = \frac{\gamma_{fd} M_{dg}}{j\, d} \qquad \ldots(25.87)$$

where $\gamma_{fd} =$ partial safety factor for dead load which may be taken equal to 0.9 since the dead load moment is of opposite nature than the moment due to prestressing force.

Net compressive force to be resisted by concrete section is

$$F_u = P_u - F_{dg} \qquad \ldots(25.88)$$

The area of concrete has to resist this force, with the centre of compresion assumed at the centroid of prestressing steel, its value being given by

$$A_{ct} \geq \frac{\gamma_m F_u}{k_p f_{ci}} \qquad \ldots(25.89)$$

In the above expression, $\gamma_m =$ partial safety factor for concrete strength = 1.5 ; $f_{ci} =$ concrete cube strength at the time of transfer of prestress and $k_p =$ ratio of the compressive strength of concrete in flexure to that of the cube and taken equal to $\frac{2}{3}$ and A_{ct} is the area of concrete with its centroid at the prestressing steel level.

$$A_{ct} \geq \frac{1.5\, F_u}{\frac{2}{3} f_{ci}} \geq 2.25 \frac{F_u}{f_{ci}} \qquad \ldots(25.90)$$

For rectangular beams, the area of concrete centred at the steel level is

$$A_{ct} = 2\,(D - d)\, b \;; \qquad 2\,(D - d)\, b \geq 2.25 \frac{F_u}{f_{ci}}$$

or

$$(D - d)\, b \geq 1.125 \frac{F_u}{f_{ci}} \qquad \ldots(25.91)$$

The above equation is applicable both for rectangular as well as T-beams.

For I-beams (Flanged beam) such as shown in Fig. 25.68,

$A_{ct} = A_{ctw} + A_{ctb}$, where,

A_{ctw} = area of web which contributes to the concrete action $= 2\,(D - d)\, b_w$

A_{ctb} = area of bottom flange overhanges $= (b_b - b_w)\, D_b$

$$A_{ct} = 2\,(D - d)\, b_w + (b_b - b_w)\, D_b \qquad \ldots(25.92)$$

Substituting in Eq. 25.90, we get

$$2\,(D - d)\, b_w + (b_b - b_w)\, D_b \geq 2.25 \frac{F_u}{f_{ci}}$$

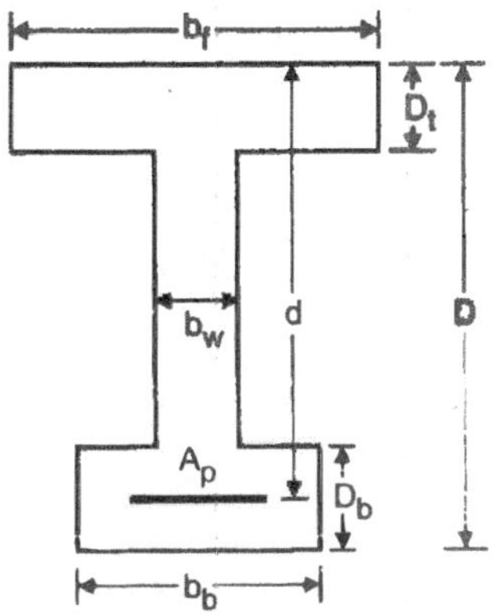

FIG. 25.68

or $$(D-d)\,b_w+\tfrac{1}{2}(b_b-b_w)\,D_b \geq 1.125\,\frac{F_u}{f_{ci}} \qquad \text{...(25.93)}$$

From the above relation, the width of bottom flange (b_b) can be found for assumed value (D_b) of its thickness.

Step 4. Design for limit state of collapse in shear : The section for critial shear failure is at a distance *d* from the effective support. The shear force *V* at this section is computed after applying partial safety factor γ_f.

The ultimate shear resistance of section uncracked in flexure is given by Eq. 25.94

$$V_c = 0.67\,b\,D\sqrt{f_t^2+0.8\,f_{cp}\cdot f_t} \qquad \text{...(25.94)}$$

If the tendon has inclination θ at the section, the component $P_c\,.\,\theta$ of cable force in transverse direction should be *added* to V_c, where

$$P_c = \frac{A_p\cdot f_{pu}}{\gamma_{fp}} \qquad \text{...(25.95)}$$

where γ_{fp} = partial safety factor for prestressing force, the value of which is normally taken as 1.67.

(*a*) In no case should the S.F. *V* due to ultimate load exceed the appropriate values given in Table 25.17 multiplied by *b.d.*

(*b*) If $V < V_c$, minimum reinforcement given by Eq. 25.74 should be provided.

(*c*) If $V > V_c$, shear reinforcement should be provided in accordance with Eq. 25.75:

$$\frac{A_{sv}}{s_v} = \frac{V-V_c}{0.87\,f_y\,d_t} \qquad \text{...(25.75)}$$

(*d*) No shear reinforcement is necessary if *V* is less than 0.5 V_c or if the member is of minor importance.

Step 5. Determination of prestressing steel area

For the case of *rectangular section*, we have, from Eq. 25.82

$$A_p = \frac{M_u}{(d-0.42\,x_u)\,f_{pu}} = \frac{M_u}{j\,d\,f_{pu}} \qquad \text{...[25.82(a)]}$$

For balanced flanged sections A_p is given by Eq. 25.85

$$A_p = 1.15\,[0.36\,b_w\,x_u + 0.45\,(b_f-b_w)\,D_f]\,\frac{f_{ck}}{f_{pu}}$$

Step 6. Design for limit state of serviceability : allowable stresses

The initial prestressing force is computed from the relation

$$P_i = A_p\cdot f_{pi}$$

where f_{pi}= maximum tensile stress immediately behined the anchorages. As per IS : 1343-1980, this should not exceed 0.8 times ultimate tensile strength of the wire = 0.8 f_p.

$$\therefore \quad P_i = 0.8\,A_p\,f_p \qquad \text{...(25.96)}$$

Note : $f_{pi\,(max)}$ is equal to 0.8 f_p . Generally, f_{pi} is choosen between 0.7 f_p to 0.8 f_p.

The losses of prestress are estimated, generally to a value equal to 0.15 $f_p\,.\,A_p$. Subtracting these losses from P_i , we get effective prestressing force given by

$$P_e = P_i - 0.15 f_p A_p = 0.8 A_p f_p - 0.15 A_p f_p = 0.65 A_p f_p \quad ...(25.97)$$

Determine the stresses in concrete at transfer and working load condition and check whether these are within the allowable limits as specified by IS : 1343-1980.

Step 7. ***Check for limit state of serviceability : deflection*** : Compute (*i*) deflection due to creep, (*ii*) deflection at initial stage, (*iii*) deflection at working load condition. The total deflection including that due to creep should be less than the permissible value.

Example 25.18. ***Design of Rectangular Beam.***

Design a bonded post-tensioned prestressed concrete beam of type 2 simply supported over an effective span of 9 m. The beam has to be of rectangular cross-section. It carries superimposed dead and live loads of 6 and 12 kN/m respectively. Use grade M 35 concrete. Take unit weight of concrete as 24 kN/m³ .

Solution. *Given data* : $L = 9$ m

Superimposed dead load $w_d = 6$ kN/m. Superimposed live load $w_l = 12$ kN/m

$$f_{ck} = 35 \text{ N/mm}^2 \ ; \ f_{ci} = 30 \text{ N/mm}^2$$

The allowable stresses at the time of transfer are :

$$\sigma_{cci} = \left[0.54 - \frac{0.54 - 0.37}{6} \times 1 \right] f_{ci} = 0.512 \times 30 = 15.35 \text{ N/mm}^2 \quad \text{[Fig. 25.66 (a)]}$$

$$\sigma_{cti} = 3 \text{ N/mm}^2 .$$

The allowable stresses under service conditions are

$$\sigma_{cce} = \left[0.34 - \frac{0.34 - 0.27}{6} \times 1 \right] f_{ck} = 0.328 \times 35 = 11.49 \text{ N/mm}^2 \quad \text{(Fig. 25.65 zone II)}$$

and $\sigma_{cte} = 3 \text{ N/mm}^2$.

Step 1. ***Determination of design bending moment***

For the purpose of computing self weight of the beam, let

$$b = \frac{L}{30} = \frac{9000}{30} = 300 \text{ mm} \ ; \qquad D = \frac{L}{15} = \frac{9000}{15} = 600 \text{ mm}$$

∴ Self weight $w_{dg} = (300 \times 600 \times 10^{-6}) \times 1 \times 24000 \approx 4300 \text{ N/m} = 4.3$ kN/m

∴ Characteristic load $w_k = w_{dg} + w_d + w_l = 4.3 + 6 + 12 = 22.3$ kN/m

γ_f for loads = 1.5. ∴ Design bending moment at collapse is

$$M_c = \frac{\gamma_f w_k L^2}{8} = \frac{1.5 \times 22.3 \times 1000}{8} (9)^2 \times 1000 \text{ N-mm} = 337.2 \times 10^6 \text{ N-mm}$$

Step 2. ***Design for limit state of flexure*** : Let the characteristic strength of steel wires (7 mm Φ) be 1500 N/mm² (See Table 25.11). From Eq. 25.77,

$$\frac{x_u}{d} = \frac{35}{15 + 0.0435 \times 1500} = 0.436$$

From Eq. 25.80, $$j = 1 - \frac{14.6}{15 + 0.0435 \times 1500} = 0.818$$

From Eq. 25.78 (*a*) $$R_u = \frac{12.6}{15 + 0.0435 \times 1500} \left(1 - \frac{14.6}{15 + 0.0435 \times 1500} \right) = 0.128 \text{ N/mm}^2.$$

From Eq. 25.79(*a*) $$p_0 = \frac{14.49}{15 + 0.0435 \times 1500} = 0.181$$

(Alternatively, all the above four values could be taken from Table 25.19).

The ultimate moment of resistance of the beam is given by $M_u = R_u \cdot b\, d^2 f_{ck}$

Hence the design criterion is

$$\therefore \quad M_u \geq M_c$$

$$\therefore \quad R_u\, b\, d^2 f_{ck} \geq M_c$$

or $$d = \sqrt{\frac{M_c}{R_u \cdot b \cdot f_{ck}}}$$

Choosing $b = L/30 = 300$ mm,

$$d = \sqrt{\frac{337.2 \times 10^6}{0.128 \times 300 \times 35}} \approx 500 \text{ mm}$$

The area of pre-tensioned steel is estimated from Eq. 25.79

$$A_p = p_0\, b\, d \cdot \frac{f_{ck}}{f_p} = 0.181 \times 300 \times 500 \times \frac{35}{1500} = 633.5 \text{ mm}^2$$

Limit of ultimate prestressing force is given be Eq. 25.86 :

$$P_u = \gamma_{fp} \cdot A_p \cdot f_p$$

Choosing $\gamma_{fp} = 1.2$, $P_u = 1.2 \times 633.5 \times 1500 = 1.14 \times 10^6$ N

Step 3. ***Design for limit state of strength at transfer condition***

Self weight $W_{dg} = 4300$ N/m

$$\therefore \quad M_{dg} = \frac{4300 \times (9)^2}{8} \times 100 = 43.54 \times 10^6 \text{ N-mm}$$

Axial tension caused by the above moment is given by Eq. 25.87 :

$$F_{dg} = \frac{\gamma_{fd}\, M_{dg}}{d - 0.42\, x_u} = \frac{\gamma_{fd} \cdot M_{dg}}{j \cdot d}$$

Taking $\gamma_{fd} = 0.9$, and $j = 0.818$,

$$F_{dg} = \frac{0.9 \times 43.54 \times 10^6}{0.818 \times 500} = 95800 \text{ N}$$

Hence from Eq. 25.88,

$$F_u = P_u - F_{dg} = 1.14 \times 10^6 - 0.096 \times 10^6 = 1.044 \times 10^6 \text{ N}$$

For rectangular beams, the design conditions is given by Eq. 25.91 :

$$(D - d)\, b \geq 1.125 \frac{F_u}{f_{ci}}$$

Taking $f_{ci} = 30$ N/mm^2, we get

$$D = \frac{1.125}{b} \frac{F_u}{f_{ci}} + d = \frac{1.125}{300} \times \frac{1.044 \times 10^6}{30} + 500 = 130.5 + 500 = 630.5 \text{ mm}$$

The assumed value of $D = 600$ mm.

The increase in the self weight due to increased value of D will be only marginal. Hence effective depth can be increased marginally.

Thus, choose the following dimensions :

$$D = 650 \text{ mm} \;;\; b = 300 \text{ mm} \;;\; d = 510 \text{ mm}.$$

$$w_{dg} = 0.3 \times 0.65 \times 24000 = 4680 \text{ N} = 4.68 \text{ kN}$$

$$w = w_{dg} + w_d + w_L = 4.68 + 6 + 12 = 22.68 \text{ kN/m}.$$

$$M_c = \gamma_f \frac{w L^2}{8} = \frac{1.5 \times 22.68 (9)^2}{8} \times 10^6 \text{ N-mm} = 334.5 \times 10^6 \text{ N-mm}.$$

The ultimate moment of resistance of the beam is

$$M_u = R_u\, b\, d^2 f_{ck} = 0.128 (300) (510)^2 \times 35 = 349.6 \times 10^6 \text{ N-mm}.$$

Since, $M_u > M_c$, the section is safe against its limit state of collapse in flexure.

Step 4. *Design for limit state of collapse in shear*

S.F. at distance d from the support is

$$V = \gamma_f \left(\frac{w L}{2} - w\, d \right) = 1.5 \left(\frac{22.68 \times 9}{2} - 22.68 \times 0.51 \right) = 135.74 \text{ kN} = 135740 \text{ N}$$

Let us provide the prestressing tendons parabolic in shape, with say $h = 0$ at end section, and maximum sag at the centre $= h = d - 0.5\, D = 510 - 0.5 \times 650 = 185$ mm.

The slope θ of the cable at distance d from the support is

$$\theta = \frac{4\, h}{L^2} (L - 2\, d) = \frac{4 \times 0.185}{(9)^2} (9 - 2 \times 0.51) = 0.073 \text{ radians}$$

$$P_c = \frac{A_p \cdot f_{pu}}{\gamma_{fp}} \quad \text{(Eq. 25.95)}$$

Taking $f_{pu} = f_p = 1500 \text{ N/mm}^2$ and $\gamma_{fp} = 1.67$, $P_c = \dfrac{633.5 \times 1500}{1.67} = 569000$ N

$$P_c \,.\, \theta = 569000 \times 0.073 \approx 41500 \text{ N}$$

The ultimate shear resistance of uncracked section is given by Eq. 25.71.

$$V_{co} = 0.76\, b\, D \sqrt{f_t^2 + 0.8 f_{cp} \cdot f_t}$$

where $f_t = 0.24 \sqrt{f_{ck}} = 0.24 \sqrt{35} = 1.42 \text{ N/mm}^2$

$$f_{cp} = \frac{P_c}{A} = \frac{569000}{300 \times 650} = 2.92 \text{ N/mm}^2$$

$$\therefore \quad V_{co} = 0.67 \times 300 \times 650 \sqrt{(1.42)^2 + 0.8 (1.42)(2.92)} = 301730 \text{ N}$$

$$\therefore \quad V_c = V_{co} + P_c \,.\, \theta = 301730 + 41500 = 343230 \text{ N}$$

Since the shear resistance V_c is more than twice the actual ultimate S.F., *no shear reinforcement is necessary.*

Step 5. *Determination of prestressing steel area*

The area of steel is given by Eq. 25.82, in which $f_{pu} = f_p$

$$\therefore \quad A_p = \frac{M_u}{j\, d\, f_p} = \frac{349.6 \times 10^6}{0.818 \times 510 \times 1500} = 558.7 \text{ mm}^2$$

Using 7 mm dia. wires, having $A_\Phi = \frac{\pi}{4}(7)^2 = 38.48 \text{ mm}^2$. No. of wires $= \frac{558.7}{38.48} = 14.5$

Provide 15 wires of 7 mm Φ . Actual $A_p = 15 \times 38.48 = 577.3 \text{ mm}^2$

Step 6. ***Design for limit state of serviceability : allowable stresses***

The initial prestressing force is given by $= P_i = A_p \cdot f_{pi}$

Choosing $f_{pi} = 0.75 f_p$, we have

$$P_i = A_p(0.75 f_p) = 577.3 \times (\quad \times 1500 \approx 649500 \text{ N}$$

Assuming losses $= 0.15 f_p A_p = 0.15 \times 1500 \times 577.3 \approx 130000$ N

$$P_e = 649500 - 130000 = 519500 \text{ N}$$

The properties of the section, neglecting steel area, are as follows :

$$A = 300 \times 650 = 195000 \text{ mm}^2 \; ; \; Z = \frac{300(650)^2}{6} = 21.125 \times 10^6 \text{ mm}^3.$$

Eccentricity $e = 185$ mm at midspan.

(*a*) ***At the transfer condition***

$$f = \frac{P_i}{A} \mp \frac{P_i \cdot e}{Z} \pm \frac{M_{dg}}{Z}$$

where $M_{dg} = \frac{w_{dg} \cdot L^2}{8} = \frac{4680(9)^2}{8} \times 1000 = 47.38 \times 10^6$ N-mm

$$\frac{P_i}{A} = \frac{649500}{195000} = 3.33 \text{ N/mm}^2 \; ; \quad \frac{P_i \cdot e}{Z} = \frac{649500 \times 185}{21.125 \times 10^6} = 5.69 \text{ N/mm}^2$$

$$\frac{M_{dg}}{Z} = \frac{47.38 \times 10^6}{21.125 \times 10^6} = 2.24 \text{ N/mm}^2$$

$\therefore$ f_1 (top fibre) $= 3.33 - 5.69 + 2.24 = -0.12 \text{ N/mm}^2$ (i.e. tensile)

This is less than permissible tensile stress of 3 N/mm^2

f_2 (bottom fibre) $= 3.33 + 5.69 - 2.24 = 6.78 \text{ N/mm}^2$

This is less than permissible value of 15.35 N/mm^2

(*b*) ***At the service load condition***

$$f = \frac{P_e}{A} \mp \frac{P_e \cdot e}{Z} \pm \frac{M}{Z}$$

where $M = \frac{w L^2}{8} = \frac{22.68(9)^2}{8} \times 10^6 = 229.6 \times 10^6$ N-mm; $P_e = 519500$ N

$$\therefore \quad \frac{P_e}{A} = \frac{519500}{195000} = 2.66 \text{ N/mm}^2 \; ; \quad \frac{P_e \cdot e}{Z} = \frac{519500 \times 185}{21.125 \times 10^6} = 4.55 \text{ N/mm}^2$$

$$\frac{M}{Z} = \frac{229.6 \times 10^6}{21.125 \times 10^6} = 10.87 \text{ N/mm}^2$$

$\therefore$ f_1 (top fibre) $= 2.66 - 4.55 + 10.87 = 8.98 \text{ N/mm}^2$ (Safe)

f_2 (bottom fibre) $= 2.66 + 4.55 - 10.87 = -3.66 \text{ N/mm}^2$ (*i.e.*, tensile)

This is slightly more than the permissible tensile stress of 3 N/mm^2. In order to bring it back to the permissible value, let us increase the prestressing force. The net stress due to prestressing force = 2.66 + 4.55 = 7.21 against requirement of 10.87 − 3 = 7.87.

Hence increase the prestressing force by ratio 7.87/7.21 = 1.09.

∴ No. of wires = 15 × 1.09 = 16.4

Let us provide 17 wires of 7 mm Φ. ∴ $A_p = 17 \times 38.48 = 654.16$ mm^2

$$P_i = A_p\,(0.75\,f_p) = 654.16 \times 0.75 \times 1500 = 735930 \text{ N}$$

$$\text{Losses} = 0.15\,f_p\,A_p = 0.15 \times 1500 \times 654.16 = 147186 \text{ N}$$

$$\therefore \quad P_e = 735930 - 147186 = 588744 \text{ N.}$$

The raised stresses will be as under ;

(*a*) *At transfer condition*

$$f = \frac{P_i}{A} \mp \frac{P_i\,.\,e}{Z} + \frac{M_{dg}}{Z}\ ; \qquad \frac{P_i}{A} = \frac{735930}{195000} = 3.77 \text{ N/mm}^2$$

$$\frac{P_i\,.\,e}{Z} = \frac{735930 \times 185}{21.125 \times 10^6} = 6.44 \text{ N/mm}^2 \qquad \frac{M_{dg}}{Z} = \frac{47.38 \times 10^6}{21.125 \times 10^6} = 2.24 \text{ N/mm}^2$$

$$\therefore \quad f_1 \text{ (top fibre)} = 3.77 - 6.44 + 2.24 = -0.43 \text{ N/mm}^2 \text{ (safe)}$$

$$f_2 \text{ (bottom fibre)} = 3.77 + 6.44 - 2.24 = 7.97 \text{ N/mm}^2 \text{ (safe)}$$

(*b*) *At service load condition*

$$f = \frac{P_e}{A} \mp \frac{P_e\,.\,e}{Z} \pm \frac{M}{Z}\ ; \qquad \frac{P_e}{A} = \frac{588744}{195000} = 3.02 \text{ N/mm}^2$$

$$\frac{P_e\,.\,e}{Z} = \frac{588744 \times 185}{21.125 \times 10^6} = 5.16 \text{ N/mm}^2\ ; \qquad \frac{M}{Z} = \frac{229.6 \times 10}{21.125 \times 10^6} = 10.87 \text{ N/mm}^2$$

$$f_1 \text{ (top fibre)} = 3.02 - 5.16 + 10.87 = 8.73 \text{ N/mm}^2 \text{ (safe)}$$

$$f_2 \text{ (bottom fibre)} = 3.02 + 5.16 - 10.87 = -2.69 \text{ N/mm}^2 \text{ (safe)}$$

Since A_p has been increased, the section should be re-checked for limit state of strength at transfer condition (step 3).

Revised $\quad P_u = \gamma_{fp}\,.\,A_p\,.\,f_p = 1.2 \times 654.16 \times 1500 = 1.178 \times 10^6$ N.

$$F_{dg} = \frac{\gamma_{fd}\,.\,M_{dg}}{j\,d} = \frac{0.9 \times 47.38 \times 10^6}{0.818 \times 510} = 102050 \text{ N} = 0.102 \times 10^6 \text{ N}$$

$$\therefore \quad F_u = P_u - F_{dg} = 1.178 \times 10^6 - 0.102 \times 10^6 = 1.076 \times 10^6$$

The criterion for safety is

$$(D - d)\,b \geq 1.125\,\frac{F_u}{f_{ci}} \qquad \text{...(Eq. 25.91)}$$

$$\therefore \quad (650 - 510)\,300 \geq 1.125\,\frac{1.076 \times 10^6}{30}; \quad 42000 \geq 40350. \quad \text{Hence satisfied.}$$

Step 7. ***Check for limit state of serviceability : deflection***

$$I = \frac{b\,D^3}{12} = \frac{300\,(650)^3}{12} = 6866 \times 10^6 \text{ mm}^4; \quad E_c = 5700\sqrt{f_{ck}} = 5700\sqrt{35} = 33722 \text{ N/mm}^2$$

The equivalent distributed load of cables at the initial and transfer stage are as under.

$$w_{pi} = \frac{8\,P_i\,.\,e}{L^2} = \frac{8 \times 735930 \times 0.185}{(9)^2} = 13450 \text{ N/m} = 13.45 \text{ kN/m}.$$

$$w_{pe} = \frac{8\,P_e\,.\,e}{L^2} = \frac{8 \times 588744 \times 0.185}{(9)^2} = 10760 \text{ N/m} = 10.76 \text{ kN/m}.$$

The effective distributed load on the beam at the transfer and service load conditions are:

$$w_i = w_{dg} - w_{pi} = 4.68 - 13.45 = -\ 8.77 \text{ kN/m} = -\ 8770$$

$$w_e = w - w_{pe} = 22.68 - 10.76 = 11.92 \text{ kN/m} = 11920 \text{ N/m}$$

The effective permanent load is given by

$$w_{pd} = w_{dg} + w_d - w_{pe} = 4.68 + 6 - 10.76 = -\ 0.08 \text{ kN/mm}^2 = -\ 80 \text{ N/mm}^2$$

The deflections at various stages will be as follows :

(*i*) ***Deflection at initial stage***

$$\delta_i = \frac{5\,w_i\,L^4}{384\,E_c\,I} = \frac{5\,(8770)\,(9)^4\,(1000)^3}{384 \times 32722 \times 6866 \times 10^6} = -\ 3.24 \text{ mm } (i.e. \text{ upwards})$$

Allowable upward deflection $= \frac{L}{300} = \frac{9000}{300} = 30$ mm. Hence safe.

(*ii*) ***Deflection at working load condition***

$$\delta_e = \frac{5 w_e\,L^4}{384\,E_c\,I} = \frac{5\,(11920)\,(9)^4\,(1000)^3}{384 \times 32722 \times 6866 \times 10^6} = 4.4 \text{ mm}.$$

(*iii*) ***Deflection due to effective permanent load***

$$\delta_{pd} = \frac{5\,w_{pd}\,L^4}{384\,E_c\,I} = \frac{5\,(-\ 80)\,(9)^4\,(1000)^3}{384 \times 3372 \times 6866 \times 10^6}$$

$$= -\ 0.03 \text{ mm}$$

(*iv*) ***Total deflection, including effect of creep***

Net creep coefficient = 1.6

$$\delta_t = \delta_e + (\delta_{pd} \times \text{creep coefficient}) = 4.4 + (-\ 0.03)\,(1.6) = 4.35 \text{ mm}$$

Allowable downward deflection $= \frac{L}{250} = \frac{9000}{250} = 36$ mm. Hence safe.

PROBLEMS

1. A simply supported prestressed concrete beam of rectangular cross-section 300 mm × 500 mm is loaded with a total uniformly distributed load of 200 kN over a span of 5 m. Sketch the distribution of stresses at the mid-span and end sections if the prestressing force is 1500 kN and the tendon is eccentric, located at 150 mm above the bottom fibre.

2. Find the extreme fibre stresses at mid-span of the beam of problem 1 by applying the principle of internal resisting couple.

3. The beam of problem 1 is prestressed by a parabolic tendon with a prestressing force of 1500 kN. The tendon has a sag of 100 mm at the mid-span. Find the extreme fibre stress by load balancing concept if it is subjected to (a) total U.D.L. of 200 kN, (b) total U.D.L. of 300 kN.

4. A pretensioned prestress concrete beam of 7 m span has a cross-section of 300 mm × 500 mm, and is prestressed with 1500 kN force at transfer. The cable has cross-sectional area of 1500 mm^2 of steel, and has a parabolic profile with a maximum eccentricity of 1500 mm at the mid-span. Determine the loss of prestress, given that $E_s = 2.1 \times 10^5$ N/mm^2 and $E_c = 3 \times 10^4$ N/mm^2

5. A post-tensioned prestressed concrete beam of 24 m span is subjected to a transfer prestress force of 2000 kN at 25 days strength. The profile of the cable is parabolic wth maximum eccentricity of 160 mm at the mid-span. Determine the losses of prestress and the jacking force required if jacking is done from both the ends of the beam. The beam has a cross-sectional area of 450 mm × 700 mm and is prestressed with 8 cables, each cable consisting of 12 wire of 5 mm dia. One cable is tensioned at a time. Take $E_s = 2.1 \times 10^5$ N/mm^2 and $E_c = 3.5 \times 10^4$ N/mm^2.

6. (*a*) Write notes on (*i*) kern distance (*ii*) efficiency factor (*b*). Find the kern distance, and compare the efficiency factors of (*i*) rectangular section, (*ii*) symmetrical I-section, (*iii*), T-section, if all the three have the same cross-sectional area of 2800 cm^2, and same depth of 70 cm. Assume suitable thickeness of web and flange.

7. Design a prestressed concrete slab over a span of 12 metres to carry a super-imposed load of 15 kN/m^2. Assume permissible stresses in concrete of 12 N/mm^2 and no tension. Also assume that the initial losses in prestress in steel will amount to 15%. The live load will be brought to bear after the ducts are grouted. The initial prestress in the wires is to be 1100 N/mm^2 .

8. A rectangular beam of prestressed concrete is required to support a dead load moment of 12×10^6 N-mm (inclusive of its own weight) and a live load moment of 30×10^6 N-mm at its mid-section. Determine the initial prestressing force and its eccentricity at the mid-span section. Adopt the following values:

   Allowable initial compressive stress = 16 N/mm^2

   Allowable final compressive stress = 13 N/mm^2

   Allowable initial or final tension = 1 N/mm^2

   Ultimate stress in steel = 1500 N/mm^2

   Assume losses as 15%.

26

Design of Miscellaneous Structures

26.1. INTRODUCTION

Following structural elements, commonly encountered is buildings, bridges, water tanks and other structures, need special care in analysis and design :

1. Deep beams
2. Slender beams
3. Concrete walls
4. Corbels, brackets and nibs
5. Grid or waffle slabs
6. Beams curved in plan.

In this chapter, we shall consider the analysis and design of the first four structural elements.

26.2. DEEP BEAMS

Beams having large depths in comparison of its effective span are known as deep beams. Such beams are encountered in the following situations.

(*i*) *Foundation beams* transferring concentrated column loads to the supporting soil.

(*ii*) *Transfer girders* or *wall-beams* at an intermediate floor level of a building where some columns are required to be stopped. Such beams of depth equal to a storey height transfer the loads of the stopped columns to the adjoining columns.

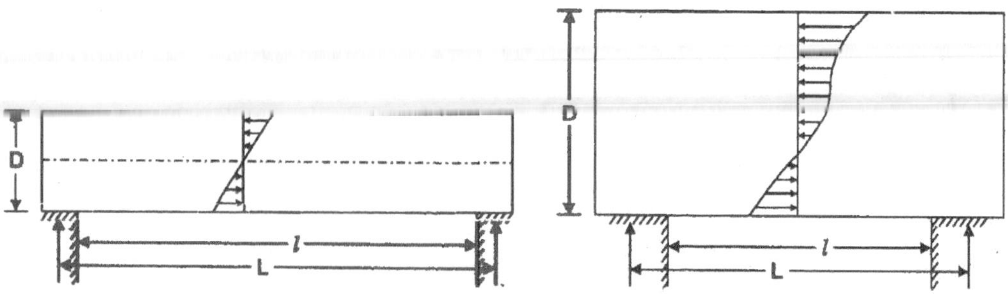

(a) Shallow beam (v) Deep beam

FIG. 26.1.

(*iii*) *Bunker walls* or *tank walls* acting as deep beams supported on columns at the corners or intermediate points.

In the case of shallow beams, we have the basic assumption that plane sections remain plane after bending and hence the distribution of bending stress across the depth is linear (Fig. 26.1 *a*). However in the case of deep beams, there is non-linear stress distribution along the depth of the beam (Fig. 26.1 *b*).

26.2.1. Definition

As per IS 456 : 2000, a beam shall be deemed to be a deep beam when the ratio of the effective span (L) to the overall depth (D), *i.e.* L/D ratio is less than

(*i*) 2.0 for simply supported beams, and

(*ii*) 2.5 for a continuous beam.

where L = effective span of the beam, taken as centre to centre distance between supports or 1.15 times the clear span whichever is smaller

D = over all depth of the beam.

A deep beam complying with the requirements of 26.2.2 and 26.2.3 shall be deemed to satisfy the provision for shear.

26.2.2. Lever arm

As shown in Fig. 26.1 (*b*), the stress distribution along the depth of the beam is non-linear. Due to this, the *lever arm* between the compressive and tensile forces cannot be easily determined. Therefore, IS 456 : 2000 gives the following *empirical expressions* for lever arm (z):

(*a*) ***For simply supported beams***

$$z = 0.2\,(L + 2D), \quad \text{where } 1 \le \frac{L}{D} \le 2 \qquad \text{...(26.1 } a)$$

or

$$z = 0.6\,L \quad \text{when } \frac{L}{D} < 1 \qquad \text{...(26.1 } b)$$

(*b*) ***For continuous beams***

$$z = 0.2\,(L + 1.5\,D) \quad \text{where } 1 \le \frac{L}{D} \le 2.5 \qquad \text{...(26.2 } a)$$

or

$$z = 0.5\,L \quad \text{where } \frac{L}{D} < 1 \qquad \text{...(26 2 } b)$$

26.2.3. Reinforcement

For a given B.M. M, the area of tensile steel reinforcement is given by

$$A_{st} = \frac{M}{\sigma_{st} \,.\, z} \qquad \text{...(26.3)}$$

where σ_{st} = permissible stress for the reinforcement (Table A-3)

z = lever arm given by Eq. 26.1 or 26.2.

26.2.3.1. Positive Moment Reinforcement

In the case of deep beams, tension steel is to be extended over the whole length of the span and it should be well anchored in the supports. This is necessary as arch-action takes place and tension steel acts as an arch tie. The depth of tension steel zone is given by

$$D' = (0.25\,D - 0.05\,L) \qquad \ldots(26.4)$$

As per IS 456 : 2000, the tensile reinforcement required to resist positive bending moment in any span of a deep beam shall :

(*a*) extend without curtailment between the supports.

(*b*) be embedded beyond the face of the support so that at the face of the support it shall have a development length not less than 0.8 L_d, where L_d is the development length for the design stress in the reinforcement, and

(*c*) be placed within a tension zone of depth equal to 0.25 D – 0.5 L adjacent to the tension face of the beam.

26.2.3.2. Negative moment reinforcement

The tensile reinforcement required to resist negative B.M. over a support should satisfy the following requirements :

(*a*) *Termination of negative moment reinforcement* :

(*i*) It shall be permissible to terminate not more than half the negative moment reinforcement at a distance of 0.5 D from the face of the support.

(2) The remaining half shall extend the full span.

(*b*) *Distribution of negative moment reinforcement* : When the ratio of clear span to overall depth is in the rage 1.0 to 2.5, tensile reinforcement over the support of a deep beam shall be placed in two zones comprising of the following :

(1) a zone of depth 0.2 D from the top tension fibre which shall contain an area A_{st1} of the tension steel given by

$$A_{st1} = 0.5\left(\frac{l}{D} - 0.5\right)A_{st} \qquad \ldots(26.5)$$

where l = clear span of the beam

A_{st} = total area of negative moment reinforcement

(2) A zone measuring 0.3 D on either side of the mid-depth of the beam, which shall contain the remainder of the tension steel (i.e. $A_{st2} = A_{st} - A_{st1}$) evenly distributed.

For span to depth ratio less than unity, the steel shall be evenly distributed over a depth of 0.8 D measured from the top tension fibre at the support.

26.2.3.3. Vertical reinforcement

If forces are applied to a deep beam in such a way that hanging action is required, bars or suspension stirrups shall be provided to carry all the forces concerned.

26.2.3.4. Side face reinforcement

Side face reinforcement, both in the vertical as well as horizontal directions, is extremely important in the case of deep beams, to provide adequate resistance to the effects of shrinkage and temperature. As per IS 456 : 2000, side face reinforcement shall comply with requirements of minimum reinforcement of walls (§ 26.4.5). Table 26.1 gives the minimum area of side face reinforcement as per IS 456 : 2000. The spacing of each of these reinforcements shall be not more than three times the width of the beam or 450 mm whichever is smaller.

TABLE 26.1. SIDE FACE REINFORCEMENT IN DEEF BEAMS (IS 456 : 2000)

S.N.	*Type of reinforcing steel*	*Percentage of side face reinforcement of gross area of concrete*	
		Vertical reinforcement	*Horizontal reinforcement*
1.	Deformed bars of dia. ≤ 16 mm and $f \geq 415$ N/mm^2	0.12	0.2
2.	Bars of other types	0.15	0.25
3.	Welded wire fabric not larger than 16 mm in dia.	0.12	0.20

Note Deep beams provided with reinforcements discussed above are deemed to satisfy the provision for shear. *Hence no separate shear reinforcement is provided in deep beams.*

Example 26.1. ***Design of simply supported deep beam***

Design a deep beam 300 mm wide and 4 m deep, simply supported over a clear span of 6 m. The beam carries a live load of 160 kN/m at the service state and is supported on walls of 600 mm thick on each end. Use M 20 concrete and Fe 415 steel having permissible tensile stress of 230 N/mm².

Solution : Given $b = 300$ mm $D = 4000$ mm ; clear span $l = 6000$ mm

1. Determination of positive moment reinforcement

Self weight of beam, $w_d = 0.3 \times 4 \times 1 \times 25 = 30$ kN/m

Live load $w_l = 160$ kN/m

∴ Total U.D.L., $w = w_d + w_l = 30 + 160 = 190$ kN/m.

Effective span of beam $L = (6000 + 600)$ or (1.15×6000) which ever is small

$$\therefore \quad M = M_{\max} = \frac{wL^2}{8} = \frac{190\,(6.6)^2)}{8} = 1034.55 \text{ kN-m} = 1034.55 \times 10^6 \text{ N-mm}$$

$\sigma_{st} = 230$ N/mm² for Fe 415 steel.

$L/D = 6.6/4 = 1.65$

∴ Lever arm $z = 0.2\,(L + 2D)$ for $1 < L/D < 2 = 0.2\,(6600 + 2 \times 4000) = 2920$ mm

Hence $$A_{st} = \frac{M}{\sigma_{st} \,.\, z} = \frac{1034.55 \times 10^6}{230 \times 2920} = 1540.5 \text{ mm}^2$$

∴ No. of 16 mm φ bars $= 1540.5/201.06 = 7.66$

Hence provide 8 bars of 16 mm φ bars, giving a total area $A_{st} = 8 \times 201.6 \simeq 1608.5$ mm². This reinforcement has to be provided within a depth $D' = 0.25\,D - 0.05\,L = 0.25 \times 4000 - 0.05 \times 6600 = 670$ mm from the bottom fibre of the beam. Provide these 8 bars in four layers, with two bars in each layer.

Each bar of tension reinforcement has to be embedded into the support by a length $0.8\,L_d = \frac{0.8 \times \sigma_{st}}{4\,\tau_{bd}}\,\varphi = \frac{0.8 \times 230}{4\,(1.6 \times 0.8)} \times 16 = 575$ mm. This is achieved by extending each bar into the support by maximum possible length and then providing 90° bend which project along the width of the beam, thus giving an anchorage length much greater than 575 mm.

2. *Determination of side face reinforcement*

(*a*) *Vertical reinforcement* (A_{sv})

$$A_{sv} = 0.12\% \text{ of gross-sectional area } = \frac{0.12}{100} \times b \times 1 \text{ m length}$$

$$= 0.0012 \times 300 \times 1000 = 360 \text{ mm}^2$$

$\therefore$ A_{sv} on each face $= 360/2 = 180 \text{ mm}^2$

$\therefore$ Spacing of vertical reinforcement, $s_v = \dfrac{1000\, A_{\varphi}}{A_{sv}}$

Using 8 mm φ bars, $s_v = \dfrac{1000 \times 50.27}{180} = 279.2$ mm c/c

However, provide vertical reinforcement in the form of 8 mm φ bars @ 250 mm c/c on each face, giving total $A_{sv} = 2\,(1000 \times 50.27/250) = 420 \text{ mm}^2$ per metre length of the beam. This spacing is less than 450 mm or 3 times the width of the beam.

(*b*) *Horizontal reinforcement* (A_{sh})

$$A_{sh} = 0.2\% \text{ of gross-sectional area } = \frac{0.2}{100} \times b \times 1 \text{ m depth}$$

$$= 0.002 \times 300 \times 1000 = 600 \text{ mm}^2$$

Hence A_{sh} on each face $= 600/2 = 300 \text{ mm}^2$

$\therefore$ Spacing of 8 mm φ bars, $s_h = \dfrac{1000 \times 50.27}{300} = 167.5$ mm

However, provide 8 mm φ bars horizontally at 150 mm c/c on each face, giving total area $A_{sh} = 2\,(1000 \times 50.27/150) = 670.3 \text{ mm}^2$.

3. *Details of reinforcement* : Shown in Fig. 26.2.

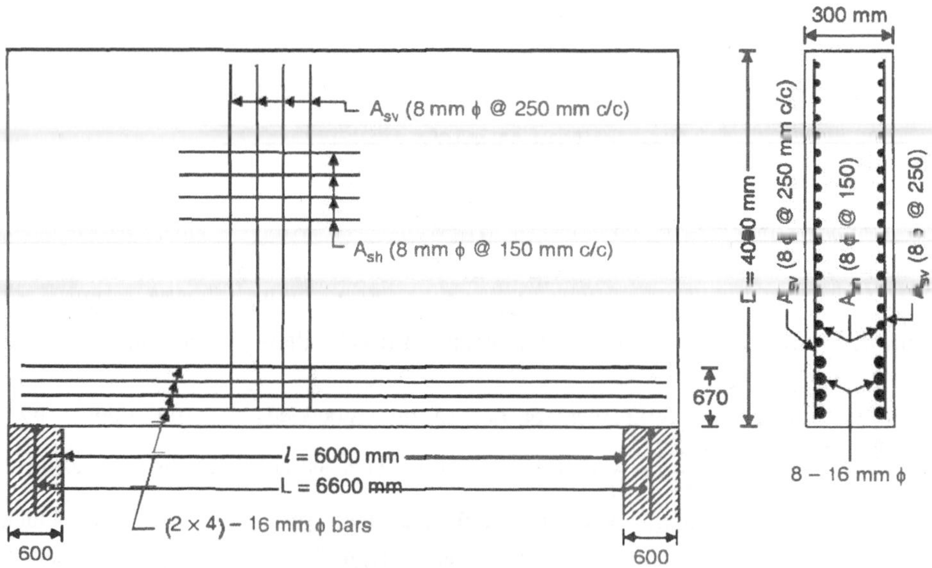

FIG. 26.2.

Example 26.2. ***Design of continuous deep beam***

A beam 3500 mm deep and 250 mm wide continuous over three spans, as shown in Fig. 26.3, carries uniformly distributed service load of 160 kN/m. Design the beam, using M 20 concrete and Fe 415 steel.

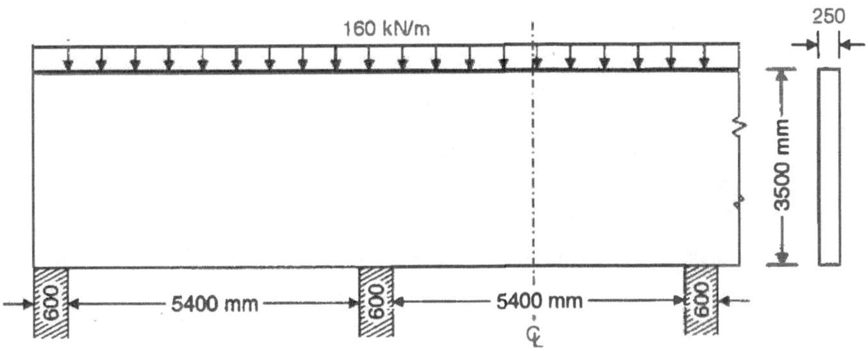

FIG. 26.3.

Solution : Given : $b = 250$ mm ; $D = 350$ mm

1. Effective span, lever arm and maximum moments

Effective span, $L = (5400 + 600)$ or (1.15×5400), which ever is smaller

$= 6000$ mm

$L/D = 6000/3500 = 1.714$

Lever arm, $z = 0.2\,(L + 2\,D)$ for $1 < L/D < 2.5$

$= 0.2\,(6000 + 2 \times 3500) = 2600$ mm

w_d = dead load of beam $= 0.25 \times 3.5 \times 1 \times 25 = 21.875$ kN/m

w_l = live load on beam $= 160$ kN/m

$\therefore$ w = total U.D.L. $= w_d + w_l = 21.875 + 160 = 181.875$ kN/m

Max. positive B.M., $M_p = \dfrac{wL^2}{24} = \dfrac{181.875}{24}(6)^2 = 272.8125$ kN-m

Max. negative B.M. $M_n = \dfrac{wL^2}{12} = \dfrac{181.875\,(6)^2}{12} = 545.625$ kN-m

2. Positive moment reinforcement (A_{stp}) : For Fe 415 steel, $\sigma_{st} = 230\ \text{N/mm}^2$. Also, for M 20 - Fe 415 combination, $\tau_{bd} = 1.6 \times 0.8 = 1.28\ \text{N/mm}^2$.

$$A_{stp} = \frac{M_p}{\sigma_{st}\,.\,z} = \frac{272.8125 \times 10^6}{230 \times 2600} = 456.2\ \text{mm}^2$$

No. of 10 mm φ bars $= 456.2/78.54 = 5.81$. However, provide 6 - 10 mm φ bars in three layers (*i.e.* 2 bars in each layer) giving total $A_{stp} = 471.2\ \text{mm}^2$, within a depth $D' = 0.25\,D - 0.05\,L = 0.25 \times 3500 - 0.05 \times 6000 = 575$ mm from the bottom fibre of the beam.

These bars are to be embedded into the support for a minimum length of

$$0.8\ L_d = \frac{0.8\ \sigma_{st}}{4\ \tau_{bd}}\ \varphi = \frac{0.8 \times 230}{4 \times (1.6 \times 0.8)} \times 10 = 359.4 \text{ mm} \simeq 400 \text{ mm}.$$

This embedment can be achieved by extending each bar into the support by maximum possible length and then providing 90° bend along the width of the beam.

3. Negative moment reinforcement (A_{stn})

$$A_{stn} = \frac{M_n}{\sigma_{st}\ .\ z} = \frac{545.625 \times 10^6}{230 \times 2600} = 912.4 \text{ mm}^2$$

This negative reinforcement is to be provided in two zones ; top zone of depth 0.2 D = 0.2 × 3500 = 700 mm from the top fibre of the beam and middle zone of 0.3 D = 0.3 × 3500 = 1050 mm on either side of mid depth of the beam.

$$\text{Reinforcement in top zone} = A_{stn1} = 0.5\left(\frac{L}{D} - 0.5\right) A_{stn}$$

$$= 0.5\left(\frac{6000}{3500} - 0.5\right) \times 912.4 \simeq 554 \text{ mm}^2$$

Reinforcement in the middle zone $= A_{stn2} = A_{stn} - A_{stn1} = 912.4 - 554 = 358.4$ mm^2.

Hence provide 8-10 mm φ bars (giving $A_{stn1} = 628.3$ mm^2) in the top zone in the form of 4-layers of 2 bars each. Also provide 8 - 8 mm φ bars (giving $A_{stn2} = 402.1$ mm^2) in the two middle zones, in the forms of 2 layers of 2 bars in each middle zone, uniformly spaced.

Each bar of negative moment reinforcement is to be embedded into the support for a minimum length of $0.8\ L_d \simeq 400$ mm, as done for positive moment reinforcement.

4. Side face reinforcement

(a) Vertical reinforcement (A_{sv})

$$A_{sv} = 0.12\% \text{ of gross-sectional area} = \frac{0.12}{100} \times b \times 1 \text{ m length}$$

$= 0.0012 \times 250 \times 1000 = 300$ mm^2 per metre length of the beam

∴ A_{sv} on each face $= 300/2 = 150$ mm^2

$$\therefore \text{ Spacing of 8 mm } \varphi \text{ bars, } s_v = \frac{1000 \times 50.270}{150} = 335 \text{ mm}$$

However, provide these @ 300 mm c/c, giving total $A_{sv} = 2\ (1000 \times 50.27)/300 = 335$ mm^2

(b) Horizontal reinforcement (A_{sh})

$$A_{sh} = 0.2\% \text{ of gross-sectional area} = \frac{0.2}{100} \times b \times 1 \text{ m depth}$$

$= 0.002 \times 250 \times 1000 = 500$ mm^2 per metre depth of the beam.

Provide 8 mm φ bars @ 200 mm c/c on *each face* giving total $A_{sh} = 2\ (1000 \times 50.270)/200$ = 502.7 mm^2

5. Details of reinforcement : Fig. 26.4.

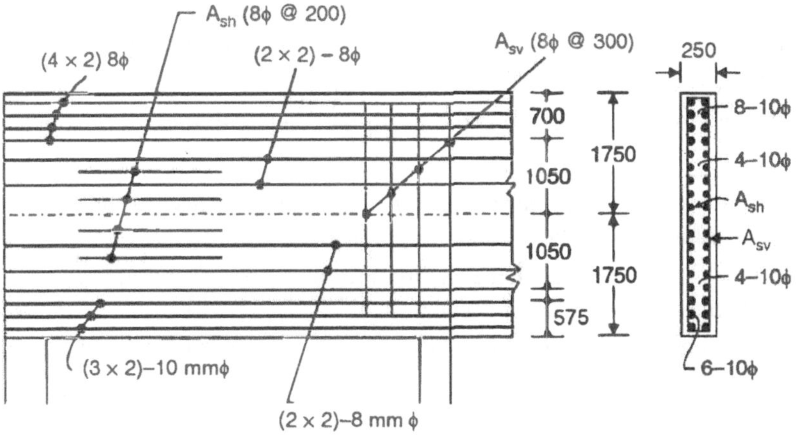

FIG. 26.4.

26.3. SLENDER BEAMS

A beam is said to be slender when the ratio of its length between lateral restraints (L) to its width (b) is equal to or more than a certain value say 60 for a simply supported or continuous beam and 25 for a cantilever beam. A slender beam is subjected to lateral buckling. In order to prevent lateral buckling, it is essential to provide lateral restraint to the *compression edge* of the beam. In the case of a down-turned beam (i.e. T-beam or L-beam), the slab at the top automatically provides lateral restraint to the compression edge of the beam at the top. However, such a restraint is not available in the case of an *inverted beam* since the slab is attached to the bottom (or tension zone) and the top compression edge is unrestrained.

As per IS 456 : 2000, a simply supported or continuous beam shall be so proportioned that the clear distance between the lateral restraints does not exceed $60\,b$ or $250\,b^2/d$ whichever is less, where d is the effective depth of the beam and b is the breadth of the compression face midway between the lateral restraints. For a cantilever, the clear distance from the free end of the cantilever to the lateral restraint shall not exceed $25\,b$ or $100\,b^2/d$ whichever is less.

Thus, for a simply supported or continuous rectangular beam of 300 mm width and 500 mm depth, it is essential to provide lateral restraints at clear distance of 60×300 mm (= 18 m) or $250\,(300)^2/500$ mm (= 45 m), whichever is less, i.e. at a clear distance of 18 m. Similarly, if this beam is used as a cantilever, the clear distance between lateral restraints should not exceed 25×300 mm (= 7.5 m), to prevent lateral bukling of the compression zone of the beam.

26.4. CONCRETE WALLS (IS 456 : 2000)

26.4.1. General : The design of R.C. walls subjected to direct compression or combined flexure and direct compression should be done in accordance with the provision of the Code discussed in Chapters 16 and 17 respectively, provided vertical reinforcement is provided on each face. Braced walls subjected to only vertical compression may be designed as per *empirical procedure* given in §26.4.2. The guide lines or design of walls subjected to horizontal and vertical loads are given in § 26.4.3. *The minimum thickness of walls shall be 100 mm.*

26.4.2. Empirical method for walls subjected to inplane vertical loads

26.4.2.1. Braced walls

Walls shall be assumed to be braced if they are laterally supported by a structure in which all the following apply :

(*a*) Walls or vertical braced elements are arranged in two directions so as to provide lateral stability to the structure as a whole.

(*b*) Lateral forces are resisted in shear in the planes of these walls or by braced elements.

(*c*) Floors and roof systems are designed to transfer lateral forces.

(*d*) Connections between the walls and the lateral supports are designed to resist a horizontal force not less than :

(1) the simple static reactions to the total applied horizontal forces at the level of lateral support; and

(2) 2.5 percent of the total vertical load that the wall is designed to carry at the level of lateral support.

26.4.2.2. Eccentricity of vertical load

The design of a wall shall take account of the actual eccentricity of the vertical force subject to a minimum value of 0.05 t, where t is the thickness of the wall.

The vertical load transmitted to a wall by a discontinuous concrete floor or roof shall be assumed to act at one third the depth of the bearing area measured from the span face of the wall. Where there is an in-situ concrete floor continuous over the wall, the load shall be assumed to act at the centre of the wall.

The resultant eccentricity of the total vertical load on a braced wall at any level between horizontal lateral supports, shall be calculated on the assumption that the resultant eccentricity of all the vertical loads above the upper support is zero.

26.4.2.3. Maximum effective height to thickness ratio

The ratio of effective height (H_{we}) to thickness, *i.e.* H_{we}/t shall not exceed 30.

26.4.2.4. Effective height

The effective height of a braced wall shall be taken as follows :

(*a*) Where restrained against rotation at both ends by

(1) Floors : $0.75\ H_w$

or (2) Intersecting walls or similar members whichever is lesser : $0.7\ L_1$

(*b*) Where not restrained against rotation at both ends by

(1) Floors : $1.0\ H_w$

or (2) Intersecting walls or similar members whichever is the lesser : $1.0\ L_1$

where H_w = the unsupported height of the wall

L_1 = the horizontal distance between centres of lateral restraint.

26.4.2.5. Design axial strength of wall

The design axial strength P_{uw} per unit length of a braced wall in compression may be calculated from the equation

$$P_{uw} = 0.3\ (t - 1.2\ e - 2\ e_a)\ f_{ck} \qquad \text{...(26.6)}$$

where e = eccentricity of load measured at right angles to the plane of the wall determined in accordance with 26.4.2.2, and

e_a = additional eccentricity due to slenderness effect, taken as $H_{we}/2500\ t$

26.4.3. Walls subjected to combined horizontal and vertical forces

26.4.3.1. When horizontal forces are in the plane of the wall, it may be designed for vertical forces in accordance with § 26.4.2. and horizontal shear in accordance with § 26.4.4. In plane bending may be neglected in case a horizontal cross-section of the wall is always under compression due to combined effect of horizontal and vertical loads.

26.4.3.2. Walls subjected to horizontal forces perpendicular to the wall and for which the design axial load does not exceed $0.04\, f_{ck}\, A_g$ shall be designed as slabs where A_g is the gross area of the section.

26.4.4. Design for horizontal shear

26.4.4.1. Critical section for shear

The critical section for maximum shear shall be taken at a distance from the base of $0.5\, L_w$ or $0.5\, H_w$ whichever is less.

26.4.4.2. Nominal shear stress

The nominal shear stress, τ_{vw} in walls shall be obtained as follows.

$$\tau_{vw} = \frac{V_u}{td} \qquad \text{...(26.7)}$$

where V_u = shear force due to design loads

t = wall thickness

$d = 0.8 \times L_w$, where L_w is the length of the wall.

Under no circumstance shall the nominal shear stress τ_{vw} in walls exceed $0.17\, f_{ck}$ in limit state method.

26.4.4.3. Design shear strength of concrete

The design shear strength of concrete in walls (τ_{cw}), without shear reinforcement shall be taken as below :

(*a*) For $H_w/L_w \le 1$,

$$\tau_{cw} = \left(3.0 - \frac{H_w}{L_w}\right) K_1 \sqrt{f_{ck}} \qquad \text{...(26.8)}$$

where $K_1 = 0.2$ in limit state method.

(*b*) For $H_w/L_w > 1$, lesser of the values calculated from (*a*) above and from

$$\tau_{cw} = K_2 \sqrt{f_{ck}}\ \frac{(H_w/L_w + 1)}{(H_w/L_w - 1)} \qquad \text{...(26.9)}$$

where $K_2 = 0.045$ in limit state method, but shall be not less than $K_3 \sqrt{f_{ck}}$ in any case, where $K_3 = 0.15$ in limit state method.

26.4.4.4. Design of shear reinforcement

Shear reinforcement shall be provided to carry a shear equal to $V_u - \tau_{cw} \cdot t\,(0.8\,L_w)$. The strength of shear reinforcement shall be calculated as per § 7.11, with A_{sv} defined as below.

$$A_{sv} = p_w\,(0.8\,L_w)\,t \qquad \ldots(26.10)$$

where p_w is determined as follows.

(*a*) For walls where $H_w/L_w \le 1$, p_w shall be the lesser of the ratios of either the vertical reinforcement area or the horizontal reinforcement area to the cross-sectional area of the wall in the respective direction.

(*b*) For walls where $H_w/L_w > 1$, p_w shall be the ratio of the horizontal reinforcement area to the cross-sectional area of the wall per vertical metre.

26.4.5. Minimum requirements for reinforcement in walls

The reinforcement for walls shall be provided as below :

(*a*) the minimum ratio of vertical reinforcement to the gross area shall be
   (1) 0.0012 for the deformed bars not larger than 16 mm in diameter, and with a characteristic strength of 415 N/mm^2 or greater
   (2) 0.0015 for other types of bars
   (3) 0.0012 for welded wire fabric not larger than 16 mm in diameter.

(*b*) Vertical reinforcement shall be spaced not farther apart than three times the wall thickness, nor 450 mm.

(*c*) The minimum ratio of horizontal reinforcement to gross concrete area shall be.
   (1) 0.0020 for deformed bars not larger than 16 mm in diameter and with a characteristic strength of 415 N/mm^2 or greater.
   (2) 0.0025 for other types of bars.
   (3) 0.0020 for welded wire fabric not larger than 16 mm in diameter.

(*d*) Horizontal reinforcement shall be spaced not farther apart than three times the wall thickness nor 450 mm.

Note 1 : The minimum reinforcement may not always be sufficient to provide adequate resistance to the effects of shrinkage and temperature.

Note 2 : The above mentioned minimum vertical and horizontal reinforcements are the same as given in Table 26.1.

26.4.5.1. For walls having thickness more than 200 mm, the vertical and horizontal reinforcement shall be provided in two grids, one near each face of the wall.

26.4.5.2. Vertical reinforcement need not be enclosed by transverse reinforcement as given for column, if the vertical reinforcement is not greater than 0.01 times the gross sectional area or where the vertical reinforcement is not required for compression.

26.5. CORBELS, BRACKETS AND NIBS

26.5.1. General

A *corbel* (Fig. 26.5 *a*) is a short cantilever brought out of a R.C. column to support gantry girders, precast beams or steel trusses where some space is required to accommodate rain water gutter. In contrast to this, a *nib* is a continuous slab type bracket, brought out of R.C.C. wall, to support slab or any other prefabricated structural component all along its length (Fig. 26.5 *b*)

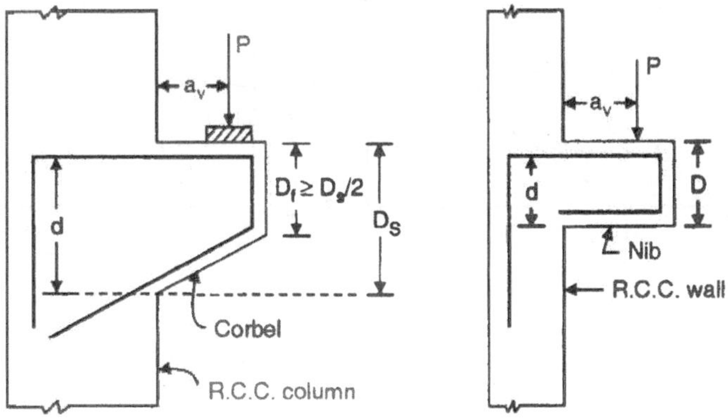

(a) Corbel (b) Nib

FIG. 26.5. CORBEL AND NIB.

Structurally, corbel is a short cantilever whose a_v/d values is less than 1.0 and whose depth D_f at the end face is not less than half the depth D_s at the support. When this criterion is satisfied, the design of the corbel is governed mostly by shear. However, if $a_v > d$, it is termed as a *bracket* which is designed for flexure and shear as per usual procedures.

Results of laboratory tests show that when a_v/d ratio is low, the load transfer to the support can be assumed to take place more by 'strut' action than by simple bending. Under these circumstances, *'enhanced shear strength'* may be taken into account by increasing the design shear strength (τ_c) of concrete to $2\,d\,\tau_c/a_v$ provided that the design stress at the face of the support remains less than $\tau_{c\,.\,max}$.

Thus, enhanced design shear strength $\tau_{ce} = \tau_c \dfrac{2d}{a_v}$ (But not greater than $\tau_{c\,.\,max}$) ...(7.21)

The depth of corbel at the face of the support is determined on the basis of this enhanced shear strength.

26.5.2. Analysis by 'strut and tie' system : Corbel action

Fig. 26.6 shows a corbel having depth D_s at the support, depth D_f at the end and subjected to a design vertical load P at a distance of a_v from the column face. The external vertical force P is resisted by 'strut action'. The corbel is in equilibrium under the action of horizontal force T in steel reinforcement and the inclined force F_c from concrete compression, with a potential crack (*ae*) along the face of the supporting member.

Fig. 26.6 (*b*) shows the simplified stress block as per BS 8110, with uniform compressive stress of $0.4\,f_{ck}$ in concrete ; the compressive force C acts at $0.45\,x_u$ from the extreme compression fibre, where x_u is the depth of N.A. The lever arm $z = d - 0.45\,x_u$. The forces

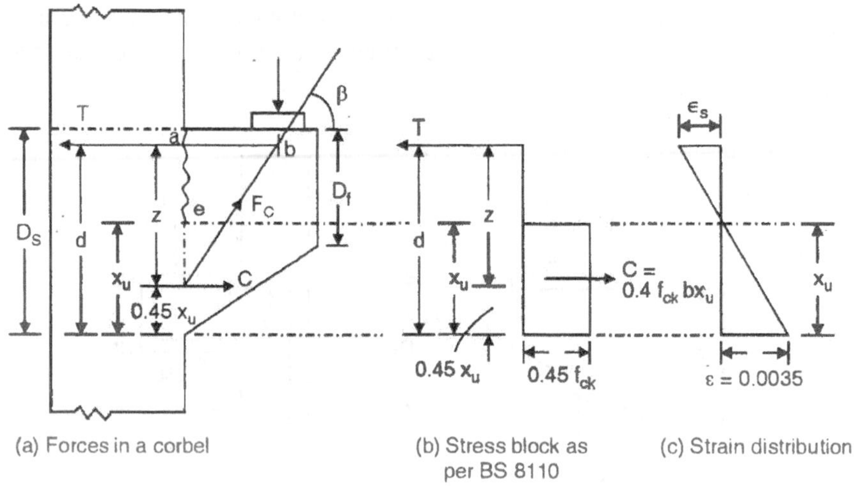

FIG. 26.6. ANALYSIS BY STRUT AND TIE SYSTEM

T and F_c can be calculated on the assumption that F_c acts at a point distant $0.45\, x_u$ from the extreme compression fibre of concrete.

From the triangle of forces, we have

$$\frac{T}{P} = \frac{a_v}{z} \quad \text{or} \quad T = P\,(a_v/z) \qquad \ldots(26.11)$$

and

$$F_C = \frac{P}{z}\sqrt{a_v^2 + z^2} \qquad \ldots(26.12)$$

The value of compressive force F_c can be found from the strength of concrete, noting that the area perpendicular to the force F_c over which the compression is built up is $x \cos \beta$.

Hence $$F_c = 0.4\, f_{ck}\, b\,(x_u \cos \beta) \qquad \ldots(i)$$

where $$\cos \beta = \frac{a_v}{\sqrt{a_v^2 + z^2}} \qquad \ldots(ii)$$

Also $$z = d - 0.45\, x_u \quad \text{or} \quad x_u = 2.22\,(d - z) \qquad \ldots(iii)$$

Substituting the values of $\cos \beta$ and z in (i), we get

$$F_c = 0.889\, f_{ck}\, b\,(d - z)\frac{a_v}{\sqrt{a_v^2 + z^2}} \qquad \ldots(26.13)$$

Equating Eqs. 26.12 and 26.13, we get

$$0.889\, f_{ck}\, bd\left(1 - \frac{z}{d}\right) a_v\, z = P\,(a_v^2 + z^2)$$

Putting $$\frac{P}{0.889\, f_{ck}\, bd} = m \quad \text{and} \quad \frac{a_v}{d} = n, \text{ we get}$$

$$\left(\frac{z}{d}\right)^2 - \left(\frac{n}{m+n}\right)\frac{z}{d} + \left(\frac{m}{m+n}\right)n^2 = 0 \qquad \ldots(26.14)$$

This is quadratic equation in terms of z/d, the value of which can be found from its solution, for given value of $m\ (= P_u/0.889 f_{ck} bd)$ and $n\ (= a_v/d)$. Thus the value of lever arm z is known. Hence the value of x_u and T can be computed.

26.5.3. Computation of area of tension steel

The area of tension steel is computed by the relation.

$$A_{st} = \frac{T}{f_s} \qquad \ldots(26.15)$$

where f_s is the stress in steel, corresponding to strain ε_s in steel either from stress strain curve or from Table 2.3. The strain ε_s in steel, corresponding to a maximum strain of 0.0035 in concrete is given by

$$\varepsilon_s = \frac{0.0035\ (d - x_u)}{x_u} \qquad \ldots(26.16)$$

It is observed in most of the cases that the value of ε_s is very small and hence f_c will also be small. Due to this reason, it is better to use mild steel bars in preference to HYSD bars. Also mild steel bars have the added advantage of ease with which they can be bent and welded.

Also, the value of x_u should be less than $x_{u\,,\,max}$. If not, compression steel of at least 0.1 percent should be provided to satisfy the limit conditions. Secondly the value of T should not be less than 0.5 P_u. It there is any horizontal force (H) acting on the bearing, it should be added to T, for computation of A_{st}.

Thus, $$A_{st} = \frac{T + H}{f_s} \qquad \ldots(26.15\ a)$$

26.5.4. Computation of horizontal shear steel (A_{sh})

Extensive tests have shown that premature diagonal tension failure will not occur in corbels if closed stirrups or ties parallel to main tensile steel is provided. The area A_{sh} of such steel should not be less than 0.5 A_{st} provided to resist T. This shear reinforcement should be placed in the upper two-third effective depth of the corbel at the support, and should be properly anchored (see Fig. 26.7).

26.5.5. Provision of compression steel (A_{sc})

Compression steel is not theoretically required. However the practice illustrated in SP 34 is (Fig. 26.7) is follows :

1. In the first arrangement, shown in Fig. 26.7 (*a*) for main bars greater than 16 mm φ, the main tension bars at the front face of the corbel are welded to transverse bars, thus providing *nominal* compression steel to anchor the horizontal stirrups.

2. In the second arrangement, shown in Fig. 26.7 (*b*) for main bars of diameter equal to 16 mm φ or lesser, the main tension bars are bent back to form hoops to act as compression steel.

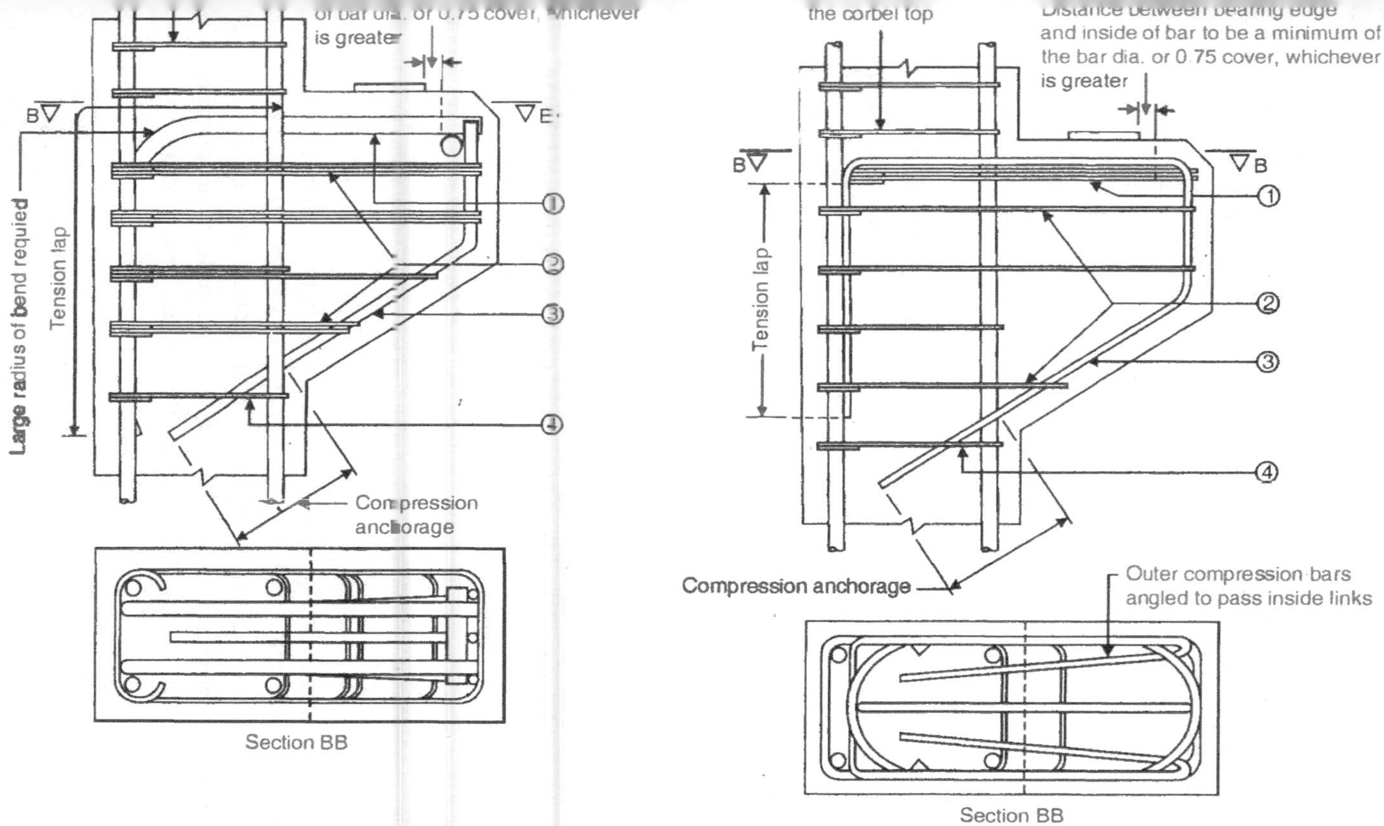

(a) DETAILS FOR $\Phi < 16$ mm

(b) DETAILS FOR $\Phi > 16$ mm

1. MAIN TENSILE BARS
2. HORIZONTAL LINKS. TOTAL AREA OF WHICH SHOULD NOT BE LESS THAN 0.25 OF AREA OF MAIN TENSILE REINFORCEMENT.
3. COMPRESSION BARS. TOTAL AREA OF WHICH SHOULD NOT BE LESS THAN 1000 mm2 /m WIDTH OF CORBEL
4. EXTRA BINDERS.

FIG. 26.7. DETAILS OF REINFORCEMENT IN CORBELS

26.5.6. Recommendations of IS 456 : 2000

1. General conditions

A corbel is a short cantilever projection which supports a load bearing member and where :

(*a*) The distance a_v between the line of reaction to the supported load and root of the corbel is less than d (the effective depth of the root of the corbel), and

(*b*) the depth of the outer edge of the contact area of the supported load is not less than one-half of the depth at the root of the corbel.

The depth of the corbel at the face of the support is determined on the basis of enhanced shear strength of concrete.

2. Design

2.1. Simplifying assumptions

The concrete and reinforcement may be assumed to act as elements of a simple strut-and-tie system, with the following guide lines.

(*a*) The magnitude of the resistance provided to horizontal force should be not less than one half of the design vertical load on the corbel.

(*b*) compatibility of strains between the strut and tie at the corbel root should be ensured.

It should be noted that the horizontal link requirement described in para 2.3 below will esure satisfactory serviceability performance.

2.2. Reinforcement anchorage

At the front face of the corbel, the reinforcement should be anchored either by :

(*a*) *welding to a transverse bar of equal strength* — in this case the bearing area of the load should stop short of the face of the support by a distance equal to the cover of the tie reinforcement, or

(*b*) *bending back the bars to form a loop* — in this case the bearing area of the load should not project beyond the straight portion of the bars forming the main tension reinforcement.

2.3. Shear reinforcement

Shear reinforcement should be provided in the form of horizontal links distributed in the upper two-third of the effective depth of the root of the corbel; this reinforcement should be not less than one-half of the area of the main tension reinforcement and should be adequately anchored.

2.4. Additional horizontal force

Additional reinforcement connected to the supported member should be provided to transmit this force in its entirety.

26.5.7. Procedure for design of corbels

The design of corbels is done in the following steps.

1. Fixation of dimensions of bearing plate

In order that ultimate bearing pressure on concrete is not exceeded, fix the dimensions of the bearing plate or bearing area taking the following values of bearing pressure :

(*a*)	Bearing with no padding material	:	$0.4\,f_{ck}$
(*b*)	Bearing in cement mortar	:	$0.6\,f_{ck}$
(*c*)	Bearing on steel plate cast into member	:	$0.8\,f_{ck}$ (BS)
			$0.9\,f_{ck}$ (IS)

2. *Fixation of dimensions of corbel* : Fix the enhanced value of shear (τ_{ce}) as given by Eq. 7.21. To start with, assume τ_{ce} *nearer* to $\tau_{c\,,\,max}$. Compute the effective depth of corbel at the root on the basis of vertical shear :

$$d = \frac{P}{\tau_{ce}\,.\,b}\,, \qquad \text{where } \tau_{ce} \le \tau_{c\,,\,max}$$

Hence $$D_s = d + \frac{\varphi}{2} + \text{nominal cover. Round off this value of } D_s.$$

Compute the total depth at the face of the corbel : $D_f \ge 0.5\,D_s$

Check the dimension for 'corbel action' wherein the value of a_v/d is less than 1.0. However, it is *preferable* to keep a_v/d ratio less than 0.6.

3. *Determination of z and x_u*

Knowing *b*, *d* and a_v, determine the value of z/d (and hence that of *z*), using Eq. 26.14 which is a quadratic equation. Knowing *z*, compute $x_u = 2.22\,(d - z)$ and check that this is less than $x_{u\,,\,max}$.

4. *Computation of tensile force T*

Compute $$T = P\left(\frac{a_v}{z}\right) \qquad \text{...(26.11)}$$

As per BS Code, *T* should be taken at least equal to 0.5 *P*.

5. *Computation of tensile steel (A_{st})*

Compute $$\varepsilon_s = 0.0035\left(\frac{d - x_u}{x_u}\right) \qquad \text{...(26.16)}$$

Find stress f_s in tensile steel either from the stress-strain curve for the steel reinforcement or from Table 2.3.

Compute $$A_{st} = \frac{T}{f_s}$$

However, if there is any additional horizontal force (*H*) acting on the bearing plate, compute $A_{st} = \dfrac{T + H}{f_s}$

The value of A_{st} found above should neither be less than 0.004 *bd* nor more than 0.013 *bd*. If the value of A_{st} comes out to be be more than 0.013 *bd*, increase the depth (*d*) and redesign.

6. *Computation of horizontal shear steel*

$$A_{sh} = 0.5\,A_{st}$$

Provide these in the form of horizontal hoops in the upper two third part of the total depth of the corbel at the support.

7. Check for vertical shear

Since the exact value of A_{st} is known, find τ_c from Table 7.1. Compute enhanced shear stress τ_{ce} from Eq. 7.21. Compute the actual shear stress (τ_v) at the section :

$$\tau_v = \frac{P}{bd}$$

Check that $$\tau_v \le \tau_{ce} \le \tau_{c,\,max}$$

If τ_v comes out to be greater than τ_{ce}, redesign the section by taking higher value of d.

8. Detailing reinforcement : Detail the reinforcement as per standard practice as given in SP 34 and shown in Fig. 26.7.

Example 26.3. Design of a Corbel

A corbel, attached to a 250 mm × 250 mm R.C.C. column, carries a factored load of 400 kN at a distance of 170 mm from the face of the column. Design the corbel using M 25 concrete.

Solution : Given $P = P_u = 400$ kN; $a_v = 175$ mm

Dimensions of bearing plate

Let the length of plate = width of column = 250 mm

Using steel plate, bearing strength $= 0.8 f_{ck} = 0.8 \times 25 = 20$ N/mm²

$$\therefore \quad \text{Width of plate} = \frac{400 \times 10^3}{250 \times 20} = 80 \text{ mm}$$

However, since the corbel is an isolated member, increase the width to 90 mm. Hence provide 90 mm × 250 mm plate.

Computation of dimensions of corbel

Length of corbel = width of column = 250 mm.

For M 25 concrete, $\tau_{c\,.\,max} = 3.1$ N/mm².

Hence take enhanced shear stress $\tau_{ce} = 2.8$ N/mm².

$$\therefore \quad d = \frac{P}{\tau_{ce}\,.\,b} = \frac{400 \times 10^3}{2.8 \times 250} = 571 \text{ mm}$$

$$D_s = d + \varphi/2 + \text{nominal cover.}$$

Providing 40 mm nominal cover and using 16 mm φ bars,

$$D_s = 571 + 16/2 + 40 = 619 \simeq 620 \text{ mm (say)}$$

Actual $$d = 620 - 16/2 - 40 = 572 \text{ mm}$$

Depth at the face, $D_f \ge 0.5\, D_s = 310$ mm. Provide $D_f = 320$ mm.

Available $\dfrac{a_v}{d} = \dfrac{170}{572} = 0.297 < 0.6$. Hence strut action will be available

3. Determination of z and x_u

$$m = \frac{P}{0.889 f_{ck}\, bd} = \frac{400 \times 10^3}{0.889 \times 25 \times 250 \times 572} = 0.126$$

$$n = \frac{a_v}{d} = 0.297$$

Substituting in Eq. 26.14, we get

$$\left[\left(\frac{z}{d}\right)^2 - \frac{0.297}{0.126 + 0.297}\left(\frac{z}{d}\right) + \frac{0.126}{0.126 + 0.297}(0.297)^2\right] = 0$$

$$\left(\frac{z}{d}\right)^2 - 0.702\left(\frac{z}{d}\right) + 0.0263 = 0$$

The solution of the above gives $\frac{z}{d} = 0.662$

$\therefore$ $z = 0.662 \times 572 = 378.83$ mm

Also, $x_u = 2.22\ (d - z) = 2.22\ (572 - 378.83) = 429.2$ mm

For mild steel, $x_{u\,,\,max} = 0.531\ d = 0.531 \times 572 = 303.7$ mm

Since $x_u > x_{u\,,\,max}$, adequate steel should be provided in compression also. However, the support steel for main reinforcement and horizontal shear steel can satisfy this condition.

4. Computation of tensile force (T)

$$T = P\left(\frac{a_v}{z}\right) = 400 \times 10^3\left(\frac{170}{378.83}\right) = 179500 \text{ N.}$$

However, $T_{min} = \frac{1}{2} P = \frac{1}{2} \times 400 \times 10^3 = 200000$ N.

Hence adopt $T = 200000$ N.

5. Computation of tensile steel (A_{st})

$$\varepsilon_s = 0.0035\left(\frac{d - x_u}{x_u}\right) = 0.0035\left(\frac{572 - 429.2}{429.2}\right) = 0.00116$$

Yield strain for mild steel $= 0.87\,f_y/E_s = 0.87 \times 250/2 \times 10^5 = 0.0010875$

Hence the mild steel bars reach yield strain.

$\therefore$ Stress in steel, $f_s = 0.87\,f_y = 0.87 \times 250 = 217.5$ N/mm^2

$\therefore$ Area of mild steel bar $= \frac{T}{f_s} = \frac{T}{0.87\,f_y} = \frac{200000}{217.5} \simeq 920$ mm^2

$\therefore$ No. of 16 mm φ bars $= 920/201.06 \simeq 4.6$

However, if Fe 415 bars are used, they do not reach minimum yield strain of 0.00144.

$$f_s = \varepsilon_s E_s = 0.00116 \times 2 \times 10^5 = 232 \text{ N/mm}^2$$

$\therefore$ Area of HYSD bars $= \frac{T}{f_s} = \frac{200000}{232} = 862$ mm^2

$\therefore$ Number of 16 mm φ bar $= 862/201.06 = 4.3$

Hence provide 5 – 16 mm φ mild steel bars.

Actual $A_{st} = 5 \times 201.06 = 1005.3$ mm^2

$$\therefore \quad \frac{100\,A_{st}}{bd} = \frac{100 \times 1005.3}{250 \times 572} = 0.703$$

This is greater than 0.4% but lesser than 1.3 %. Hence OK.

6. Computation of horizontal steel

$$A_{sh} = 0.5\, A_{st} = 0.5 \times 1005.3 \simeq 503 \text{ mm}^2$$

Providing 10 mm φ 2-lgd links, number of links $= \dfrac{503}{2 \times 78.54} = 3.2$

Hence provide 4 links of 10 mm φ mild steel bars, each two legged, in the upper two third depth.

$$\therefore \quad \text{Spacing} = \frac{2}{3} \times \frac{572}{4} = 95.3 \text{ mm.}$$

7. Check for vertical shear

From Table 7.1, τ_c (for 0.703% steel) $\simeq 0.55$ N/mm^2

$$\therefore \quad \tau_{ce} = \tau_c \cdot \frac{2d}{a_v} = 0.55 \times \frac{2 \times 572}{170} = 3.7 \text{ N/mm}^2$$

subject to a maximum equal to $\tau_{c\,.\,max} = 3.1$ N/mm^2

$$\text{Hence} \quad \tau_{ce} = 3.1 \text{ N/mm}^2$$

$\therefore$ Shear capacity of concrete $= \tau_{ce}\, bd = 3.1 \times 250 \times 572 \times 10^{-3} = 443.3$ kN > 400 kN

Hence OK.

Actual shear stress, $\quad \tau_v = \dfrac{P}{bd} = \dfrac{400 \times 10^3}{250 \times 572} \simeq 2.8 \text{ N/mm}^2 < \tau_{ce}.\ (= 3.1 \text{ N/mm}^2)$

Hence safe.

27

Detailing of Earthquake Resistant R.C. Structures

27.1. INTRODUCTION

Earthquake causes ground motions in random fashion, both horizontally and vertically, in all directions radiating from the epicentre. Consequently, structures founded in the ground vibrate, inducing *inertial forces* on them. It is therefore essential to ensure (*i*) stability, (*ii*) strength and (*iii*) serviceability with acceptable levels of safety, by way of suitable design and detailing. The characteristics (*i.e.* intensity, duration etc.) of seismic ground vibrations expected at any location depends upon the magnitude of the earthquake, its depth of focus, distance from the epicentre, characteristics of the path through which seismic waves travel, and the soil strata on which the structure stands. The random earthquake ground motions, which cause the structure to vibrate, can be resolved in any three mutually perpendicular directions. *The prominent direction of ground vibration is usually horizontal.*

Earthquake-generated vertical inertial forces are to be considered in design unless directed and proven by specimen calculations to be not significant. Vertical accelerations should be considered in structures with large spans, those in which stability is a criterion for design, or for overall stability analysis of structures. Reduction in gravity force due to vertical component of ground motion can be particularly detrimental in cases of prestressed horizontal members and of cantilevered members. Hence special attention should be paid to the effect of vertical component of the ground motion on prestressed or cantilevered beams, girders and slabs.

27.2. DESIGN APPROACH

IS 1893 (Part 1): 2002 (*Criteria for Earthquake Resistant Design of Structures*) adopts the following three criteria for fixing the level of design seismic loading :

1. Structure should possess at least a minimum strength to with stand *minor earthquakes* (< DBE) which occur frequently, *without damage*. Here DBE stands for '*design basis earthquake*' defined as the earthquake which can reasonably be expected to occur at least once during the life of the structure. Thus, a structure is expected to resist frequently occurring minor shocks within its elastic range of stresses.

2. Structure should be able to resist *moderate earthquake* (DBE) without significant structural damage though some non-structural damage may occur.

3. Structure should withstand a *major earthquake* (or *maximum considered earthquake*, MCE) without collapse, but with some structural and non-structural damage.

27.3. LATERAL LOAD RESISTING SYSTEMS : SEISMIC SYSTEMS

During an earthquake of given intensity, the magnitude of forces induced in a structure mainly depends on (*i*) damping (*ii*) ductility and (*iii*) energy dissipation capacity of the structure. The induced seismic forces on the structure can be reduced by *enhancing* ductility and energy dissipation capacity in the structure, thus reducing the probability of collapse. In general, *ductility of a structure, or its members, is the capacity to undergo large elastic deformations (beyond the initial yield deformations) without significant loss of strength or stiffness*. A ductile material is the one that can undergo large strains while resisting loads. According to Blume, a structure must have both strength as well as ductility for satisfactory performance during an earthquake. The main structural elements and their connections should be designed to have a ductile failure. This will enable the structure to avoid sudden collapse. Since reinforced concrete is relatively less ductile in compression and shear, dissipation of seismic energy is best achieved by *flexural yielding*.

In order to resist lateral forces induced during an earthquake, we may have the following systems.

1. Moment resisting space frames
   (*a*) ordinary moment resisting frame
   (*b*) ductile moment resisting frame
2. Shear walls
3. Dual system consisting of ductile moment resisting space frame and ductile shear (or flexural) wall.

An *ordinary moment resisting frame* is a space frame capable of carrying all vertical and horizontal loads, by developing bending moments in the members and at joints, but not meeting the special detailing requirements for ductile behaviour. In contrast to this, *special moment resisting frame or ductile moment resisting frame* is a moment resisting frame *detailed* to provide ductile behaviour and comply with the requirements given in IS 4326 (*Earthquake resistant design and construction of buildings—Code of practice*) or IS 13920 (*Ductile detailing of reinforced concretes structures subjected to seismic forces—Code of practice*) or SP 6(6). A frame of continuous construction comprising flexural members and columns designed and detailed to accommodate reversible lateral displacements after the formation of plastic hinges (without decrease in strength) is known as *ductile moment resisting frame*. Horizontal forces at any floor or roof level is transmitted to the foundation (ground) by using the strength rigidity and ductility of a moment resisting space frame. A space frame will survive a major earthquake only if it can yield without essential loss of lateral resistance or vertical load capacity. The energy dissipation, ductility and structural response (deformation) of space frames depend upon type of members, connections (joints), and materials of construction used.

A *shear wall* (or *flexural wall*) is a wall designed to resist lateral forces in its own plane. Shear walls are reinforced concrete walls cantilevering vertically from the base (*i.e.* foundations) designed and detailed to be ductile so as to resist seismic forces and to dissipate energy through *flexural yielding* at one or more plastic hinges. Shear walls should extend from the foundations either to the top of the building or to a lesser height as required from the design consideration. Studies show that shear walls of height about 85 percent of total height of building are advantageous. A shear wall building is normally quite rigid as compared to a framed structure.

Buildings with *dual system* consist of ductile shear walls and ductile moment resisting frames such that :

(*a*) The two systems are designed to resist the total design lateral force in proportion to their lateral stiffness considering the interaction of dual system at all floor levels, and

(*b*) The moment resisting frames are designed to independently resist atleast 25 percent of the design *base shear.*

27.4. BEHAVIOUR OF CONCRETE AND REINFORCEMENT UNDER CYCLIC LOADING

(*a*) Brittle and ductile behaviour of materials

As stated earlier, the term *ductility* implies the ability of a material to sustain significant deformation prior to collapse. In contrast to this, a brittle material or structure is the one that fails suddenly upon attaining the maximum load. Mild steel reinforcement is a ductile material while plain concrete is a brittle material. Fig. 27.1 (*a*) shows the ductile P–Δ behaviour of reinforcing steel while Fig. 27.1 (*b*) shows the brittle P–Δ behaviour of plain concrete.

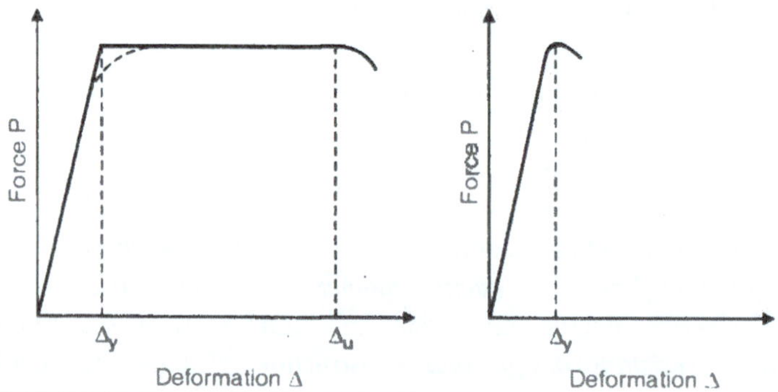

(a) Ductile behaviour (b) Brittle behaviour

FIG. 27.1. DUCTILE AND BRITTLE P–Δ BEHAVIOUR

(*b*) Behaviour of plain concrete under repeated compressive loading

Plain concrete comes under the category of *brittle materials*. Fig. 27.2 shows the effect of repeated *compressive* loading on a specimen of plain concrete. During the first cycle of loading the stress-strain curve is the same as obtained from static loading test. However, when the specimen is unloaded and reloaded, the maximum stress attained and the slope of the stress-strain curve decreases. With repeated loading and unloading, the slope of the curve decrease and the maximum stress also decreases. Hence it is concluded that the stress-strain relationship for plain concrete, subjected to repeated compressive loading is *cycle dependent*. Also, it is be noted that plain concrete cannot be subjected to repeated tensile loading since its tensile strength is negligible.

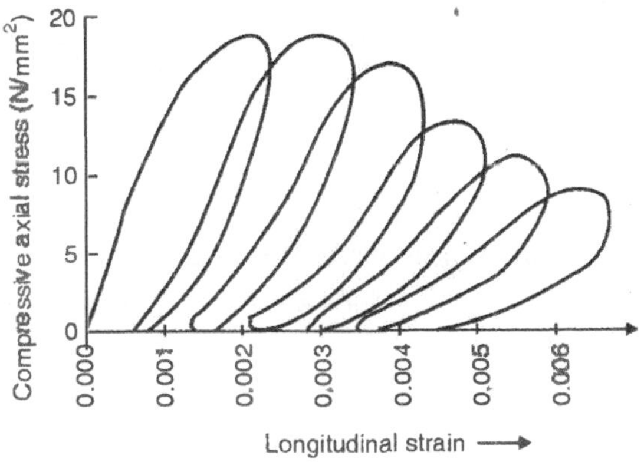

FIG. 27.2. BEHAVIOUR OF PLAIN CONCRETE SPECIMEN UNDER REPEATED COMPRESSIVE LOADING.

(C) Behaviour of reinforcing steel under repeated reversed loading

When a reinforcing steel bar is subjected to tensile/compressive loading the stress-strain curve (*OAB*) for the first cycle is similar to that for static loading.

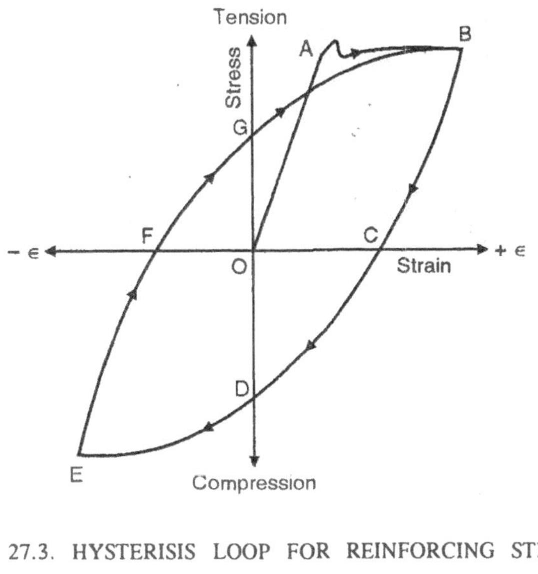

FIG. 27.3. HYSTERISIS LOOP FOR REINFORCING STEEL.

However, after the specimen has reached the yield level (upto point *B*), and the direction of loading is reversed (i.e. during unloading), the unloading curve *BC* is not straight but is curvilinear. This shape of the curve *BC*, during unloading, is popularly known as *Bauschinger effect*. After point *C*, the cycle is continued by applying compressive load (curve *CDE*), unloading (curve *EF*), and formally applying tensile loading (curve *FGB*). The whole curve so obtained during one complete cycle of loading (both tensile as well as compressive) and unloading, is known as *hysterisis loop*. In subsequent cycles, practically the same path is repeated, indicating that the stress-strain curve for steel reinforcement subjected repeated reversed loading is independent of number of cycles. Also, the reinforcing steel has much more ductility than plain concrete. The ultimate strain in mild steel is of the order of 25% where as ultimate strain of concrete is of the order of 0.3% only.

(*D*) Behaviour of reinforced concrete under repeated reversed loading

As seen earlier, the specimen of plain concrete can not be subjected to repeated tensile loading because of its poor tensile strength while the specimen of reinforcing steel can be subjected to both tensile as well as tensile loading. However, a specimen of reinforced concrete can be subjected to repeated compressive as well as tensile loading.

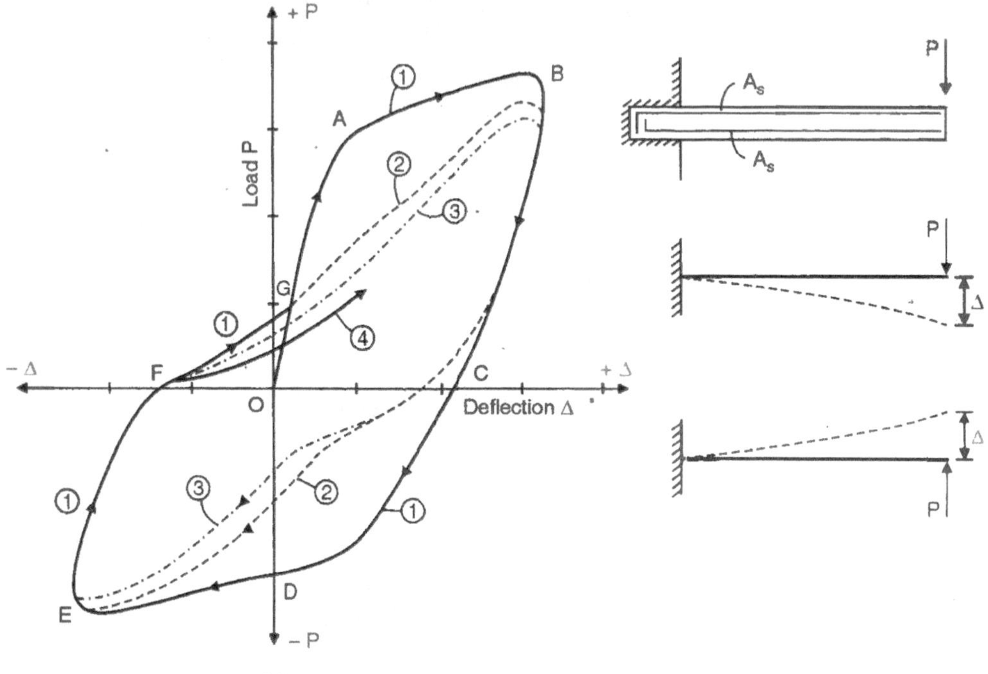

(a) (b)

FIG. 27.4. HYSTERISIS BEHAVIOUR OF A DOUBLY REINFORCED CANTILEVER

Let us take the effect of repeated reversed loading on a R.C. cantilever beam, equally reinforced on both the faces. Fig. 27.4 (*a*) shows the resulting load deflection curves obtained during *reversed cyclic loading*. During first cycle of loading and unloading in both the directions, curve *OABCDEFG* is obtained. On repeated cycles of reversed loading and unloading, the slope of the curves are successively decreased, indicating the decrease in stiffness of the beam with the number of cycles. In addition to this, the curves tend to *pinch-in* near zero load. These two effects of reversed loading cycles are known as *stiffness degradation* and *pinching-in* effects. Various factors affecting such non-linear behaviour of reinforced concrete are (*i*) cracking of concrete, (*ii*) strain hardening and Bauschinger effect in reinforcing steel, (*iii*) effectiveness of bond and anchorage between concrete and reinforcing steel, and (*iv*) presence of high shear.

27.5. SIGNIFICANCE OF DUCTILITY IN SEISMIC DESIGN

As defined earlier, ductility, of a structure, or its members, is the capacity to undergo large inelastic deformation (beyond the initial yield deformation), without significant loss of strength or stiffness. In general, ductility denotes those properties of materials, system or structures which have large energy absorption capacities in the *inelastic range*. The capability of a structure to absorb energy, within acceptable deformations and without failure is a very desirable characteristic of any earthquake resistant design. When a ductile structure is subjected to overloading it will tend to deform inelastically and in doing so, will redistribute, the excess load to elastic parts of the structure. If a structure is ductile, it can be expected to adapt to unexpected overloads, load reversals, impact and structural moments due to foundation settlement and volume changes. Secondly, if the structure is ductile, its occupants will have sufficient warning of the impending failure, thus reducing the probability of loss of life in the event of collapse.

Quantitative measures of ductility

Qualitative definition/meaning of ductility was given in the above para. However quantitatively, the *member ductility* or *structural ductility is defined as the ratio of the maximum deformation to the corresponding yield deformation*. Thus, the quantitative measure of the ductility is with reference to a load-deformation response—such response being horizontal, leading to deformation increasing at nearly constant load. Thus, the ductility μ is defined by the following relationship :

$$\mu = \frac{\text{Ultimate deformation}}{\text{deformation at first yield}} = \frac{\delta_u}{\delta_y} \qquad \ldots(27.1)$$

In the above relationship, the *deformation* may be of the following forms in the case of a beam member (*i*) curvature (*ii*) deflection or displacement (*iii*) rotation. Thus, we have curvature ductility, deflection (or displacement) ductility and rotational ductility.

Curvature ductility $$\mu_c = \frac{\phi_u}{\phi_y} \qquad \ldots(27.1\ a)$$

Deflection or displacement ductility $$\mu_d = \frac{\Delta_u}{\Delta_y} \qquad \ldots(27.1\ b)$$

and Rotation ductility, $$\mu_r = \frac{\theta_u}{\theta_y} \qquad \ldots(27.1\ c)$$

Curvature ductility (μ_c) : Ductile member section response

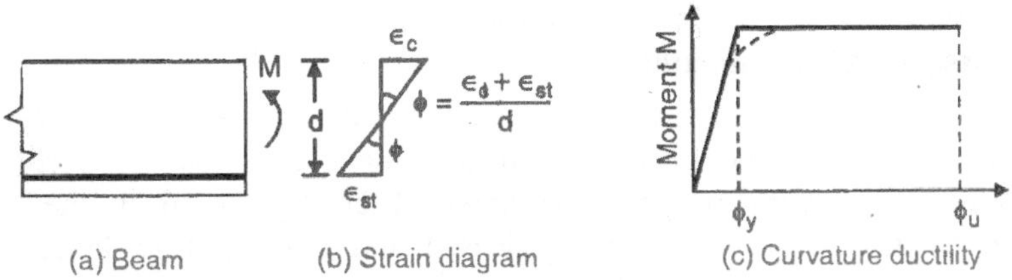

FIG. 27.5. CURVATURE DUCTILITY

When an under-reinforced beam section is subjected to increasing B.M. the curvature at any stage is given by $\phi = (\varepsilon_c + \varepsilon_{st})/d$. Fig. 27.5 (*c*) shows idealised $M - \phi$ behaviour, where ϕ_y is the curvature at the yield point and ϕ_u is the maximum curvature at ultimate stage. The curvature ductility (μ_c) is given by

$$\mu_c = \frac{\phi_u}{\phi_y} \qquad \ldots(27.1\ a)$$

Thus the curvature ductility is a property of the beam cross-section. The curvature ductility can be *increased* by (*i*) *decrease* in percentage tensile steel (*ii*) *increase in percentage compressive steel* (*iii*) *decrease* in tensile stress of steel (*iv*) *increase* in compressive strength of concrete, though higher grades of concrete have lower ultimate ε_c (*v*) increase in transverse reinforcement and (*vi*) increase in compression flange area in flanged beams. As per IS Codes, a curvature ductility of at least 5 is considered to be adequate for reinforced concrete.

Displacement and rotation ductilities

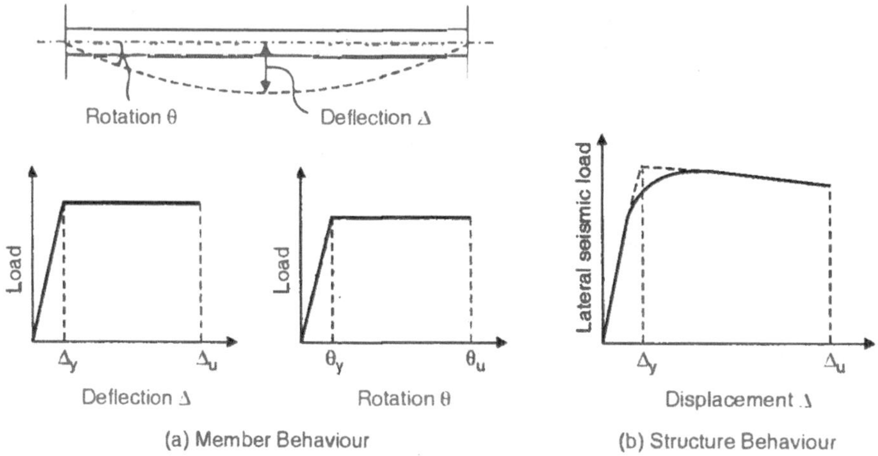

FIG. 27.6. DEFLECTION, ROTATION AND DISPLACEMENT DUCTILITIES

Fig. 27.6 (*a*) illustrates to member behaviour with respect to deflection and rotation ductilities defined by Eqs. 27.1 (*a*) and (*c*) respectively. Fig. 27.6 (*b*) illustrates displacement ductility of a structure as a whole, subjected to lateral seismic load. Such a ductility is achieved by ensuring ductile member section responses (as indicated by the M-ϕ relation of Fig. 27.5 *c*), thus ensuring the formation of adequate number of plastic hinges at appropriate locations, under extreme lateral seismic forces.

Energy dissipation by ductility

During an earthquake, a structure is subjected to several cycles of reversed cyclic loading. Fig. 27.7 (*a*) shows the linear load displacement behaviour of a structure modelled as a single degree of freedom system. The potential energy stored in the structure during the load (shown by the shaded area) gets released and converted to kinetic energy as the structure return to its zero-load position.

In contrast to this, if the structure responds in an elastoplastic (or ductile) manner, developing fully plastic behaviour at load level F' (less than F of Fig. 27.7 *a*), the corresponding

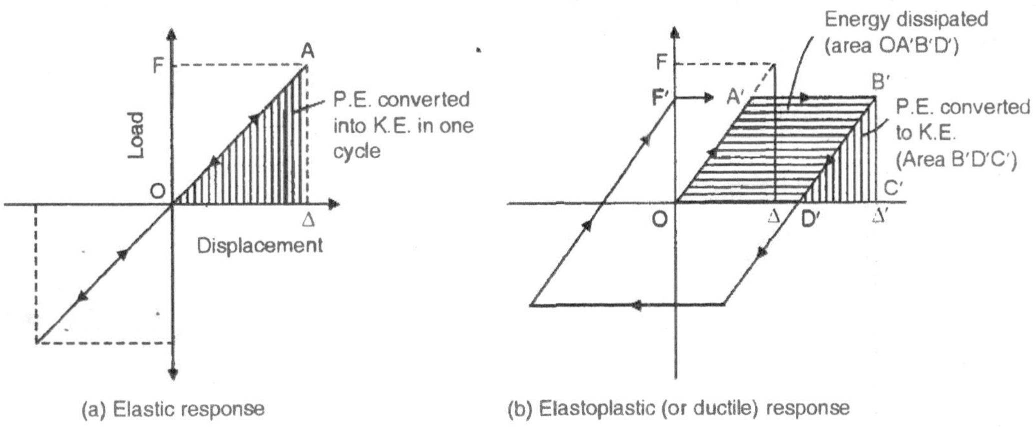

(a) Elastic response

(b) Elastoplastic (or ductile) response

FIG. 27.7. ENERGY DISSIPATION BY DUCTILITY

load-displacement behaviour is depicted in Fig. 27.7 (*b*), wherein the maximum deflection Δ' is greater than the deflection Δ of elastic behaviour shown in Fig. 27.7 (*a*). When the structure returns to zero load position, the portion of potential energy (P.E.) that gets converted into kinetic energy (K.E.) is limited to the triangular area $B'\ D'\ C'$. The remainder of the input energy (given by area $OA'B'D'$) gets dissipated by plastic hinge. Hence an elastoplastic (or ductile) response has the following characteristics. (*i*) a large part of energy gets dissipated by plastic hinge (*ii*) the induced force is less, and (*iii*) the maximum deflection is more. Thus, we conclude that the *ductility helps in reducing the induced forces and in dissipating part of the input energy ; however it demands larger deformations to be accommodated by the structure.*

27.6. DISSIPATION OF SEISMIC ENERGY BY FLEXURAL YIELDING

Reinforced concrete is relatively less ductile in compression and shear. Hence dissipation of seismic energy is best achieved by *flexural yielding* (i.e by formation of plastic hinges). Hence the structure should not be weak in compression and shear, in relation to flexure. The desired ductile (or inelastic) response is obtained by formation of plastic hinges in the members of a structure composed of ductile moment resisting frames and /or shear (or flexural) walls. These plastic hinges may be formed either in the beams or in the columns of a ductile moment resisting frame. However, it is desirable to design the frame in such a way that plastic hinges are formed in the beams rather than in columns. Such a design philosophy is aimed at due of the following reasons.

1. The plastic hinges formed in the beams have larger rotation capacity than those formed in columns.

2. The beam mechanisms, so obtained by formation of hinges in the beams, have larger energy absorptive capacity because of larger number of hinges, each with greater rotation capacity.

3. Only *'localised failure'* occurs by eventual collapse of a beam while *'global failure'* occurs due to collapse of a column.

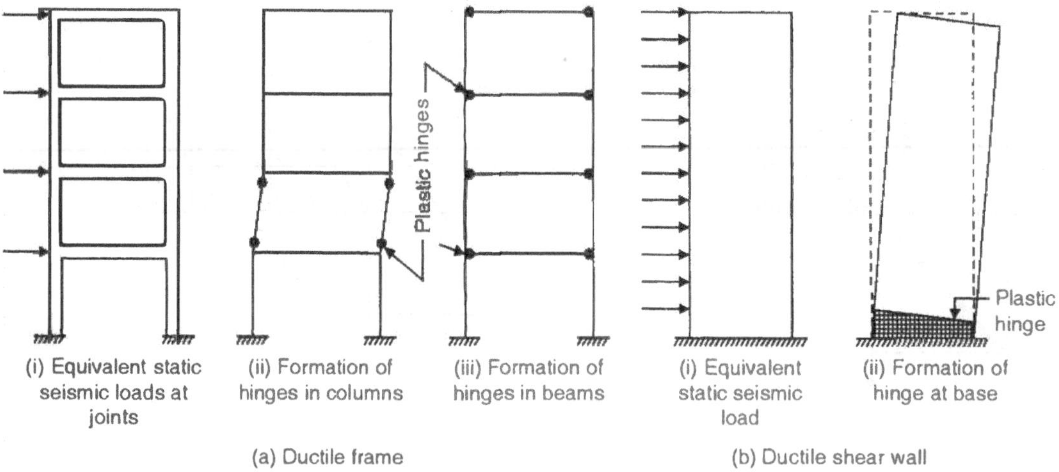

FIG. 27.8. PLASTIC HINGES IN A DUCTILE STRUCTURE

4. It is easier to repair beams in the event of residual deformation and damage, in comparison to straightening/ repairing of columns.

27.7. GENERAL OBJECTIVES OF DESIGN FOR DUCTILITY

The objective of the special design and detailing provision laid down in IS 13920 (*Code of practice for ductile detailing of reinforced concrete structures subjected to seismic forces*) is to ensure adequate toughness and ductility with ability to undergo large inelastic reversible deformations, for individual members such as beams, columns and walls, and to prevent other nonductile types of failure. In order to maintain overall ductile behaviour of structure, with minimal damage, it is essential to achieve combinations of (*i*) relatively strong foundations and weak super-structure (*ii*) each member relatively stronger in shear than in flexure, and (*iii*) relatively strong columns with beams with little over-strength. From stability point of view, the structural system should be so designed as to ensure that formation of plastic hinges at suitable locations may at worst result in the failure of individual elements, but will not lead to *instability* or *progressive collapse*. This calls for building-in redundancy into the structural system. Redundancy helps in the development of alternative load paths, thereby helping redistribution of forces, dissipation of energy and avoidance of progressive collapse. Also the structure must have sufficient *stiffness* to limit the lateral deflection or drift.

To ensure sufficient ductility and adequate stiffness, the designer should pay attention to detailing of reinforcement, bar cut offs, splicing and joint details. Following are main design considerations to be followed in providing ductility.

1. The structural layout should be simple and regular, avoiding offsets of beams and columns or offsets of columns from floor to floor. Changes in stiffness should be gradual from floor to floor.

2. The amount of tensile reinforcement in beams should be restricted.

3. More compression reinforcement should be provided; the compression reinforcement should be enclosed by stirrups to prevent it from buckling.

4. Relatively low grade steel (such as mild steel reinforcement of Fe 250 grade) should be preferred. Lower grade steel has clearly defined and longer yield plateau and hence the plastic hinges formed will have larger rotation capacities leading to greater energy dissipation. Lower the grade of steel, the higher is the ratio of ultimate tensile strength (f_u) to the yield strength (f_y). A high ratio of f_u/f_y is desirable, as it results in an increased length of plastic hinges along the member axis, and thereby an increased plastic rotation capacity. However, use of low grade steel will necessitate larger sections for flexural members. Hence IS Codes permit the steel of grade Fe 415 but prohibits the steel grade higher than Fe 415.

5. Adequate stirrups should be provided to ensure that shear failure does not precede flexural failure. This will prevent a non-ductile shear failure before the fully reversible flexural strength of a beam member has been developed.

6. Beams and columns in a R.C. frame should be so designed that inelasticity is confined to beams only and the column remain elastic. This is ensured by providing the sum of the moment capacities of the column for the design axial loads at a beam column joint greater than the moment capacities of the beams along each principal plane.

7. Closed stirrups or spirals should be used to confine concrete at section of maximum moments. Such sections include upper and lower ends of columns and within beam-column joints which do not have beams on all sides. If axial load exceed 0.4 times the balanced axial load, a spiral column is preferred.

8. Splices and bar anchorages must be adequate to prevent bond failures.

9. Beam-column connections should be made monolithic.

10. Indian Code limits the minimum grade of concrete to M 20. However, higher grade of concrete results in lower ultimate compressive strain (ε_{cu}) resulting in reduction in ductility. The ACI and Canadian Codes limit the maximum cylinder strength of low density concrete for use in earth quake resistant design to 30 MPa.

General specifications laid down in IS 13920-1993

1. The design and constructions of reinforced concrete building shall be governed by the provision of IS 456, except as modified by the provisions of IS 13920.

2. For all buildings which are more than 3 storeys in height, the minimum grade of concrete shall be M 20 ($f_{ck} = 20\ N/mm^2$).

3. Steel reinforcement of grade Fe 415 or less only shall be used.

However, high strength deformed steel bars, produced by the thermo-mechanical treatment (TMT) process, of grades Fe 500 and Fe 550 having elongation more than 14.5 percent may also be used for the reinforcement.

27.8. DUCTILE DETAILING OF FLEXURAL MEMBERS (IS 13920 : 1993)

1. General

These requirements apply to frame members resisting earthquake induced forces and designed to resist flexure. These members shall satisfy the following requirements.

1.1. The factored axial stress on the member under earthquake loading shall not exceed $0.1\ f_{ck}$.

1.2. The member shall preferably have a width-to-depth ratio of more than 0.3.

1.3. The width of the member shall not be less than 200 mm.

1.4. The depth D of the member shall preferably be not more than 1/4 of the clear span.

2. Longitudinal Reinforcement

2.1. (*a*) The top as well as bottom reinforcement shall consist of a least two bars throughout the member length.

(*b*) The tension steel ratio on any face, at any section, shall not be less than

$$\rho_{min} = 0.24 \sqrt{f_{ck}} / f_y \quad \text{...(27.2. } a)$$

where f_{ck} and f_y are MPa (N/mm^2).

2.2. The maximum steel ratio on any face at any section, shall not exceed

$$\rho_{max} = 0.025 \quad \text{...(27.2 } b)$$

2.3. The positive steel at a joint face must be at least equal to half the negative steel at that face.

2.4. The steel provided at each of the top and bottom face of the member at any section along its length shall be at least equal to one-fourth of the maximum negative moment steel provided at the face of either joint. It may be clarified that redistribution of moments permitted in IS 456 will be used only for vertical load moments and not for lateral load moments.

2.5. In an external joint, both the top and the bottom bars of the beam shall be provided with anchorage length, beyond the inner face of the column, equal to the development length in tension plus 10 times the bar diameter minus the allowance for 90 degree bend(s) (*see* Fig. 27.9). In an internal joint, both face bars of the beam shall be taken continuously through the column.

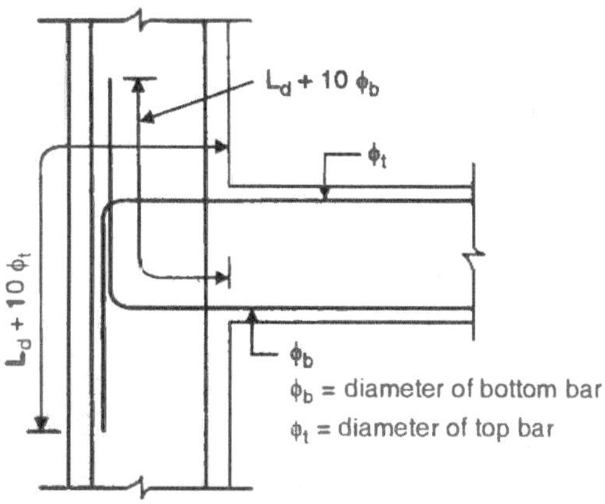

FIG. 27.9. ANCHORAGE OF BEAM BARS IN AN EXTERNAL JOINT

2.6. The longitudinal bars shall be spliced, only if hoops are provided over the entire splice length at a spacing not exceeding 150 mm (see Fig. 27.10). The lap length shall not be less than the bar development length in tension. Lap splices shall not provided (*a*) within a joint, (*b*) within a distance of 2 *d* from joint face, and (*c*) within a quarter length of the member where flexural yielding may generally occur under the effect of earthquake forces. Not more than 50 percent of the bars shall be spliced at one section.

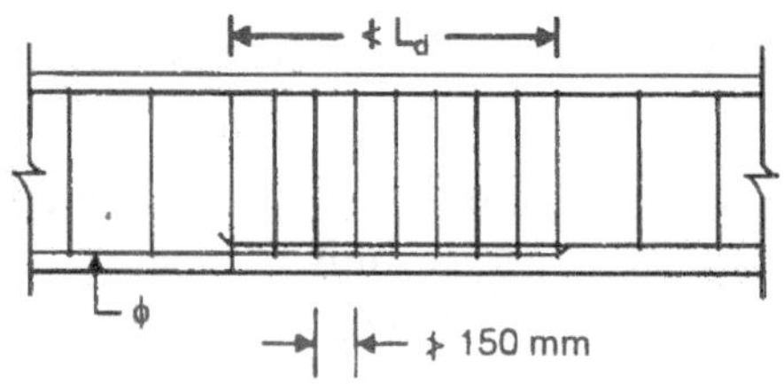

FIG. 27.10. LAP SPLICE IN BEAM

2.7. Use of welded splices and mechanical connections may also be made, as per IS 456. However, not more than half the reinforcement shall be spliced at a section where flexural yielding may take place. The location of splices shall be governed by para 2.6.

3. Web Reinforcement

3.1. Web reinforcement shall consist of vertical hoops. A vertical hoop is a closed stirrup having a 135° hook with a 10 diameter extension (but not < 75 mm) at each end that is embedded in the confined core (see Fig. 27.11 *a*). In compelling circumstances, it may also be made up of two pieces of reinforcement; a U-stirrup with a 135° hook and a 10 diameter extension (but not < 75 mm) at each end, embedded in the confined core and a cross-tie (see Fig. 27.11 *b*). A cross-tie is a bar having a 135° hook with a 10 diameter extension (but not < 75 mm) at each end. The hooks shall engage peripheral longitudinal bars.

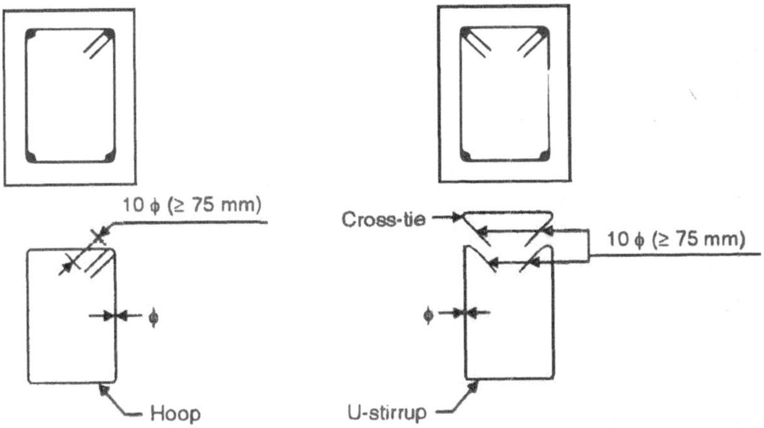

(a) (b)

FIG. 27.11. BEAM WEB REINFORCEMENT

3.2. The minimum diameter of the bar forming a hoop shall be 6 mm. However, in beams with clear span exceeding 5 m, the minimum bar diameter shall be 8 mm.

3.3. The shear force to be resisted by the vertical hoops shall be the maximum of :

(*a*) calculated factored shear force as per analysis, and

(*b*) shear force due to formation of plastic hinges at both ends of the beam plus the factored gravity load on the span. This is given by (see Fig. 27.12)

(*i*) *for sway to right* :
$$V_{u,a} = V_a^{D+L} - 1.4\left[\frac{M_{u,lim}^{As} + M_{u,lim}^{Bh}}{L_{AB}}\right] \qquad \text{...(27.3 } a)$$

and
$$V_{u,b} = V_b^{D+L} + 1.4\left[\frac{M_{u,lim}^{As} + M_{u,lim}^{Bh}}{L_{AB}}\right] \qquad \text{...(27.3 } b)$$

(*ii*) *for sway to left* :
$$V_{us} = V_a^{D+L} + 1.4\left[\frac{M_{u,lim}^{Ah} + M_{u,lim}^{Bs}}{L_{AB}}\right] \qquad \text{...(27.3 } c)$$

and
$$V_{u,b} = V_b^{D+L} - 1.4\left[\frac{M_{u,lim}^{Ah} + M_{u,lim}^{Bs}}{L_{AB}}\right] \qquad \text{...(27.3 } d)$$

where $V_a^{D+L} = V_b^{D+L} = \dfrac{1.2\,(D+L)}{2}$ ...(27.4)

where $M^{As}_{u,\,lim}$, $M^{Ah}_{u,\,lim}$, $M^{Bs}_{u,\,lim}$ and $M^{Bh}_{u,\,lim}$ are the sagging and hogging moments of resistance of the beam section at ends *A* and *B*, respectively. These are to be calculated as per IS 456. L_{AB} is clear span of beam. V_a^{D+L} and V_b^{D+L} are the shears at ends *A* and *B*, respectively, due to vertical loads with a partial safety factory of 1.2 on loads (Fig. 27.12 *a*). The design shear at end *A* shall be the larger of the two values of $V_{u,a}$ computed above. Similarly, the design shear at end *B* shall be the larger of the two values of $V_{u,b}$ computed above.

3.4. The contribution of bent up bars and inclined hoops to shear resistance of the section shall not be considered.

3.5. The spacing of hoops over a length of 2 *d* at either end of a beam shall not exceed (*a*) *d*/4, and (*b*) 8 times the diameter of the smallest longitudinal bar; however, it need not be less than 100 mm (see Fig. 27.13). The first hoop shall be at a distance not exceeding 50 mm from the joint face. Vertical hoops at the same spacing as above, shall also be provided over a length equal to 2 *d* on either side of a section where flexural yielding may occur under the effect of earthquake forces. Elsewhere, the beam shall have vertical hoops at a spacing not exceeding *d*/2.

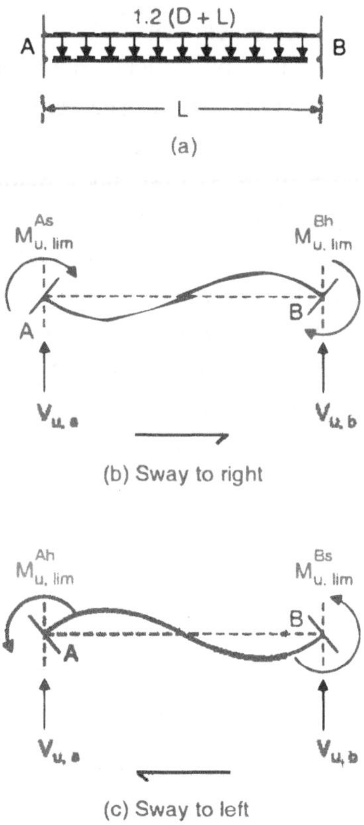

FIG. 27.12. CALCULATION OF DESIGN SHEAR FORCE FOR BEAM

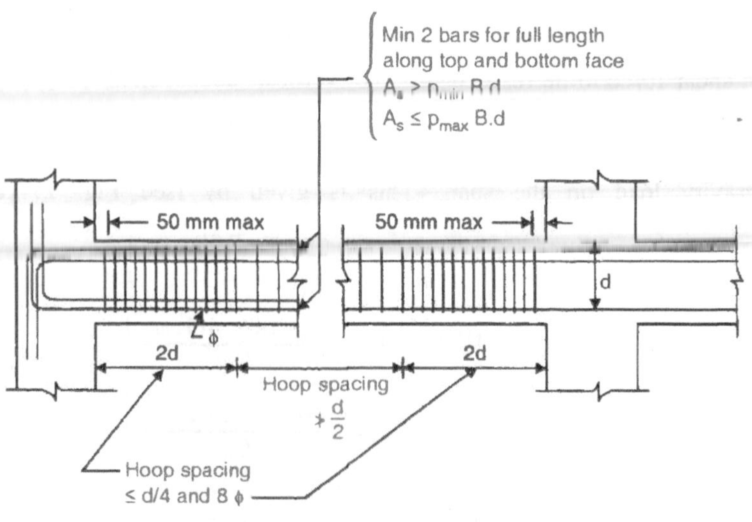

FIG. 27.13. DETAILS OF BEAM REINFORCEMENT

27.9. DUCTILE DETAILING FOR COLUMNS AND FRAME MEMBERS SUBJECTED TO BENDING AND AXIAL LOAD (IS 13920 : 1993)

1. General

1.1. These requirements apply to frame members which have a factored axial stress in excess of $0.1 f_{ck}$ under the effect of earthquake forces.

1.2. The minimum dimension of the member shall not be less than 200 mm. However, in frames which have beams with centre to centre span exceeding 5 m or columns of unsupported length exceeding 4 m, the shortest dimension of the column shall not be less than 300 mm.

1.3. The ratio of the shortest cross sectional dimension to the perpendicular dimension shall preferably not be less than 0.4.

2. Longitudinal Reinforcement

2.1. Lap splices shall be provided only in the central half of the member length. It should be proportioned as a tension splice. Hoops shall be provided over the entire splice length at spacing not exceeding 150 mm centre to centre, Not more than 50 percent of the bars shall be spliced at one section.

2.2. Any area of a column that extends more than 100 mm beyond the confined core due to architectural requirements, shall be detailed in the following manner. In case the contribution of this area to strength has been considered, then it will have the minimum longitudinal and transverse reinforcement as per this Code.

However, if this area has been treated as non-structural, the minimum reinforcement requirements shall be governed by IS 456 provision for minimum longitudinal and transverse reinforcement (see Fig. 27.14)

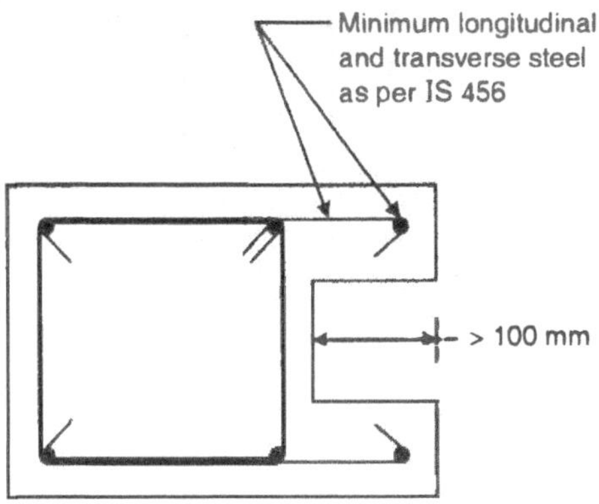

FIG. 27.14. REINFORCEMENT REQUIREMENT FOR COLUMN WITH MORE THAN 100 mm PROJECTION BEYOND CORE

3. Transverse Reinforcement

3.1. Transverse reinforcement for circular columns shall consist of spiral or circular hoops. In rectangular columns, rectangular hoops may be used. A rectangular hoop is a closed stirrups, having a 135° hook with a 10 diameter extension (but not < 75 mm) at each end, that is embedded in the confined core (see Fig. 27.15 *a*)

3.2. The parallel legs of rectangular hoop shall be spaced not more than 300 mm centre to centre. If the length of any side of the hoop exceeds 300 mm, a cross-tie shall be provided (Fig. 27.15 *b*). Alternatively, a pair of overlapping hoops may be provided within the column (see Fig. 27.15 *c*). The hoops shall engage peripheral longitudinal bars.

3.3. The spacing of hoops shall not exceed half the least lateral dimension of the column, except where special confining reinforcement is provided as per para 4 below.

3.4. The design shear force for columns shall be the maximum of :

(*a*) calculated factored shear force as per analysis, and

(*b*) a factored shear force given by

$$V_u = 1.4 \left[\frac{M_{u\,.\,lim}^{bL} + M_{u\,.\,lim}^{bR}}{h_{st}} \right] \qquad \ldots(27.5)$$

where $M_{u\,,\,lim}^{bL}$ and $M_{u\,,\,lim}^{bR}$ are moment of resistance, of opposite sign, of beams framing into the column from opposite faces (see Fig. 27.16) and h_{st} is the storey height. The beam moment capacity is to be calculated as per IS 456.

4. Special Confining Reinforcement

This requirement shall be met with, unless a larger amount of transverse reinforcement is required from shear strength considerations.

4.1. Special confining reinforcement shall be provided over a length l_0 from each joint face, towards midspan, and on either side of any section, where flexural yielding may occur under the effect of earthquake forces (see Fig. 27.17). The length 'l_0' shall

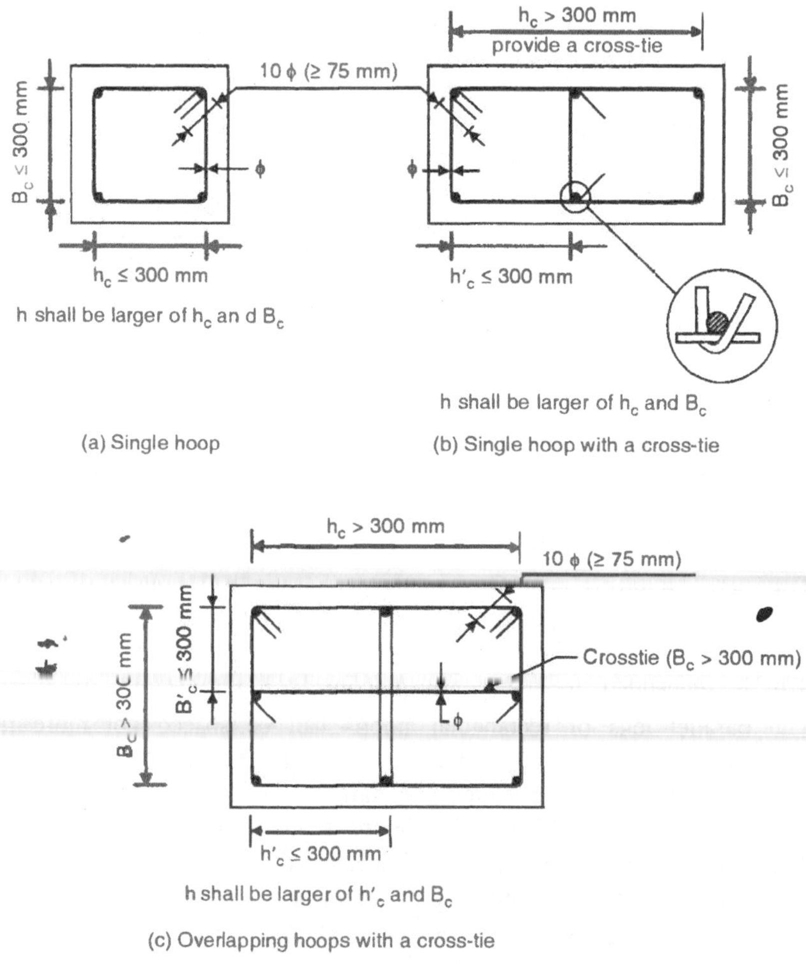

FIG. 27.15. TRANSVERSE REINFORCEMENT IN COLUMN

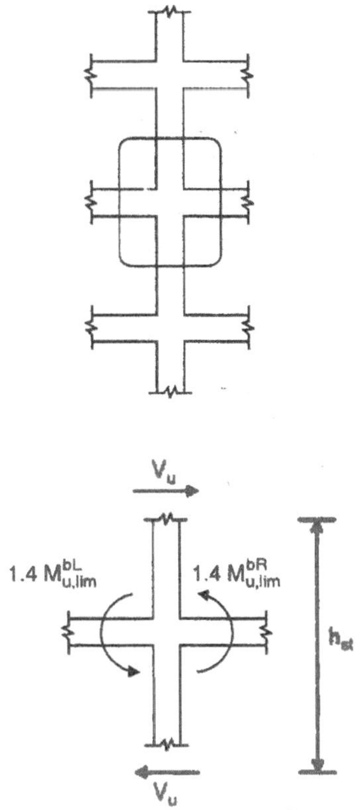

FIG. 27.16. CALCULATION OF DESIGN SHEAR FORCE FOR COLUMN

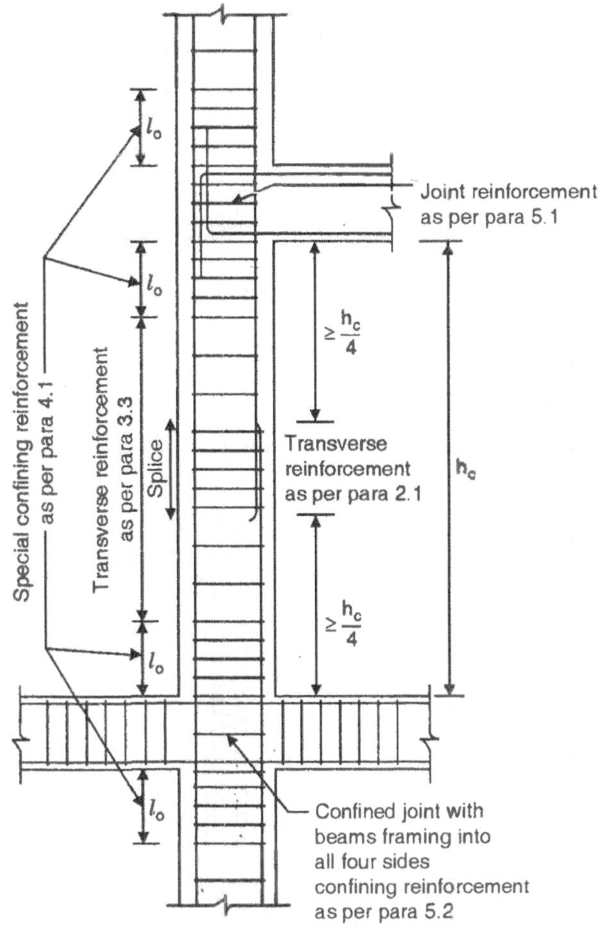

FIG. 27.17.

not be less than (*a*) larger lateral dimension of the member at the section where yielding occurs, (*b*) 1/6 of clear span of the member, and (*c*) 450 mm.

4.2. When a column terminates into a footing or mat, special confining reinforcement shall extend at least 300 mm into the footing or mat (see Fig. 27.18).

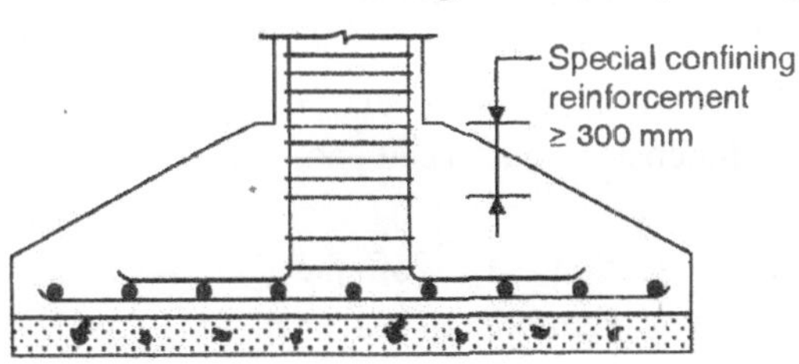

FIG. 27.18. PROVISION OF SPECIAL CONFINING REINFORCEMENT IN FOOTINGS

4.3. When the calculated point of contra-flexure, under the effect of gravity and earthquake loads, is not within the middle half of the member clear height, special confining reinforcement shall be provided over the full height of the column.

4.4. Columns supporting reactions from discontinued stiff members, such as walls, shall be provided with special confining reinforcement over their full height (see Fig. 27.19). This reinforcement shall also be placed above the discontinuity for at least the development length of the largest longitudinal bar in the column. Where the column is supported on supported on a wall, this reinforcement shall be provided over the full height of the column; it shall also be provided below the discontinuity for the same development length.

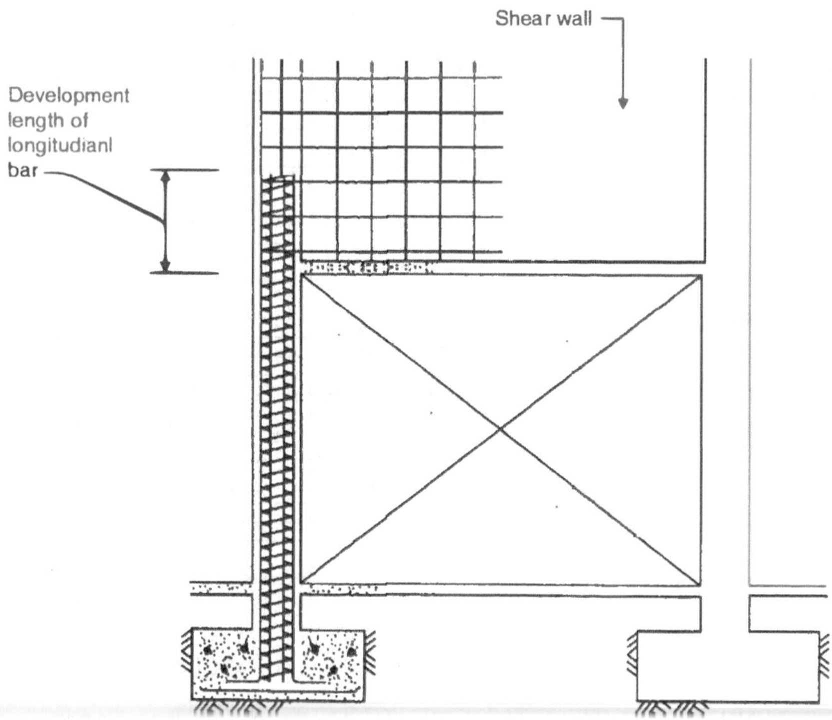

FIG. 27.19. SPECIAL CONFINING REINFORCEMENT REQUIREMENT FOR COLUMNS UNDER DISCONTINUED WALLS

4.5. Special confining reinforcement shall be provided over the full height of a column which has significant variation in stiffness along its height. This variation in stiffness may result due to the presence of bracing, a mezzanine floor or a R.C.C. wall on either side of the column that extends only over a part of the column height (see Fig. 27.20).

4.6. The spacing of hoops used as special confining reinforcement shall not exceed 1/4 of minimum member dimension but need not be less than 75 mm nor more than 100 mm.

4.7. The area of cross section, A_{sh}, of the bar forming circular hoops or spiral, to be used as special confining reinforcement, shall not be less than :

$$A_{sh} = 0.09\, sD_k \frac{f_{ck}}{f_y}\left[\frac{A_g}{A_k} - 1.0\right] \quad \ldots(27.6)$$

where A_{sh} = area of the bar cross section.

s = pitch of spiral or spacing of hoops,

D_k = diameter of core measured to the outside of the spiral or hoop.

f_{ck} = characteristic compressive strength of concrete cube,

f_y = yield stress of steel (of circular hoop or spiral).

A_g = gross area of the column cross section, and

A_k = area of the concrete core $= \frac{\pi}{4} D_k^2$

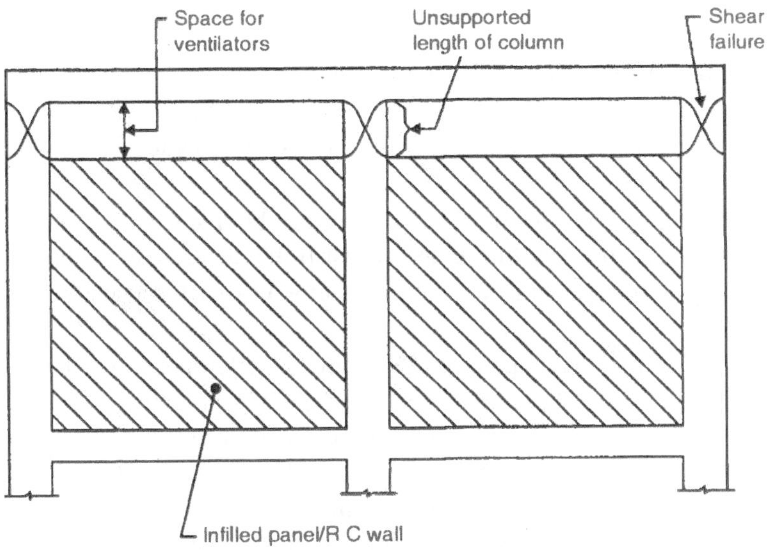

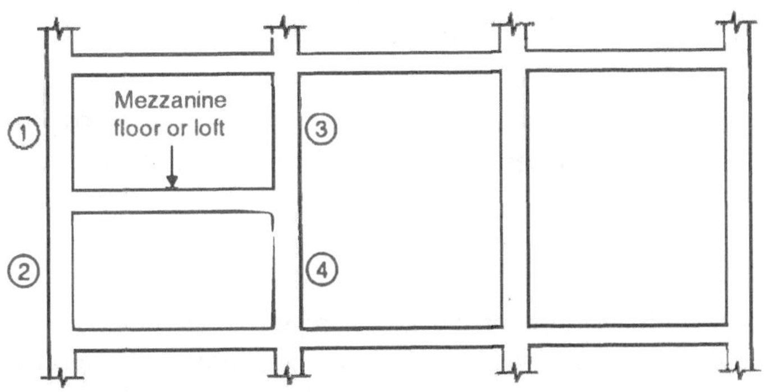

(1), (2), (3) and (4) relatively stiff columns — They attract large seismic shear force

FIG. 27.20. COLUMNS WITH VARYING STIFFNESS

Example Consider a column of diameter 300 mm. Let the grade of concrete be M 20, and that of steel Fe 415, for longitudinal and confining reinforcement. The spacing of circular hoops, s shall not exceed the smaller of (a) 1/4 of minimum member dimension = $1/4 \times 300$ = 75 mm, and (b) 100 mm. Therefore $s = 75$ mm. Assuming 40 mm clear cover to the longitudinal reinforcement and circular hoops of diameter 8 mm, $D_k = 300 - 2 \times 40 + 2 \times 8 = 236$ mm. Thus, the area of cross section of the bar forming circular hoop works out to be 47.28 mm^2. This is less than the cross sectional area of 8 mm bar (50.27 mm^2). Thus, circular hoops of diameter 8 mm at a spacing of 75 centre of centre will be adequate.

4.8. The area of cross section, A_{sh}, of the bar forming rectangular hoop, to be used as special confining reinforcement shall not be less than

$$A_{sh} = 0.18\, sh \frac{f_{ck}}{f_y}\left[\frac{A_g}{A_k} - 1.0\right] \qquad \ldots(27.7)$$

where h = longer dimension of the rectangular confining hoop measured to its outer face. It shall not exceed 300 mm (see Fig. 27.15), and

A_k = area of confined concrete core in the rectangular hoop measured to its outside dimensions.

Note : The dimension 'h' of the hoop could be reduced by introducing cross-ties, as shown in Fig. 27.15 (b). In this case, A_k shall be measured as the overall core area, regardless of the hoop arrangement. The hooks of crossties shall engage peripheral longitudinal bars.

Example . Consider a column of 650 mm × 500 mm. Let the grade of concrete be M 20 and that of steel Fe 415, for the longitudinal and confining reinforcement. Assuming clear cover of 40 mm to the longitudinal reinforcement and rectangular hoops of diameter 10 mm, the size of the core is 590 mm × 440 mm. As both these dimensions are greater than 300 mm, either a pair of overlapping hoops or a single hoop with crossties, in both directions, will have to be provided. Thus, the dimension 'h' will be the larger of (i) 590/2 = 295 mm, and (ii) 440/2 = 220 mm. The spacing of hoops s, shall not exceed the smaller of (a) 1/4 of minimum member dimensions = $1/4 \times 500 = 125$ mm, and (b) 100 mm. Thus, $s = 100$ mm The area of cross section of the bar forming rectangular hoop works out to be 64.47 mm^2. This is less than the area of cross section of 10 mm bar (78.54 mm^2). Thus, 10 mm diameter rectangular hoops at 100 mm c/c will be adequate. Similar calculations indicate that as an alternative, one could also provide 8 mm diameter rectangular hoops at 70 mm c/c.

5. Joints of Frames

5.1. The special confining reinforcement as required at the end of column shall be provided through the joint as well, unless the joint is confined as specified by para. 5.2.

5.2. A joint which has beams framing into all vertical faces of it and where each beam width it at least 3/4 of the column width, may be provided with half the special confining reinforcement required at the end of the column. The spacing of hoops shall not exceed 150 mm.

27.10. DUCTILE SHEAR (OR FLEXURAL) WALLS

As stated earlier, ductile shear walls (also known as flexural walls), which form part of the lateral load resisting system, are vertical members, cantilevering vertically from the foundations. Shear walls are subjected to bending moment, shear and axial load. It is relatively a thin and deep flexural member, subjected to substantial axial forces. Hence it is designed as axially loaded cantilever beam capable of forming reversible plastic hinges, usually at the base, with sufficient rotation capacity. A shear wall building is normally quite rigid compared to a framed structure.

Recommendations of IS 13920 : 1993

1. General Requirements

1.1. The requirements of this sections apply to the shear walls, which are part of the lateral force resisting system of the structure.

1.2. The thickness of any part of the wall shall preferably, not be less than 150 mm.

1.3. The effective flange width, to be used in the design of flanged wall sections, shall be assumed to extend beyond the face of the web for a distance which shall be the smaller of (*a*) half the distance to an adjacent shear wall web, and (*b*) 1/10th of the total wall height.

1.4. Shear walls shall be provided with reinforcement in the longitudinal and transverse directions in the plane of the wall. The minimum reinforcement ratio shall be 0.0025 of the gross area in each direction. This reinforcement shall be distributed uniformly across the cross section of the wall.

1.5. If the factored shear stress in the wall exceeds $0.2\sqrt{f_{ck}}$ or if the wall thickness exceeds 200 mm, reinforcement shall be provided in two curtains, each having bars running in the longitudinal and transverse directions in the plane of the wall.

1.6. The diameter of the bars to be used in any part of the wall shall not exceed 1/10th of the thickness of that part.

1.7. The maximum spacing of reinforcement in either direction shall not exceed the smaller of $l_w/5$, $3\,t_w$, and 450 mm; where l_w is the horizontal length of the wall, and t_w is the thickness of the wall web.

2. Shear Strength

2.1. The nominal shear stress, τ_v, shall be calculated as :

$$\tau_v = \frac{V_u}{t_w\, d_w} \qquad \ldots(27.8)$$

where V_u = factored shear force,

t_w = thickness of the web, and

d_w = effective depth of wall section.

This may be taken as $0.8\,l_w$ for rectangular sections.

2.2. The design shear strength of concrete, τ_c, shall be calculated as per IS 456.

2.3. The nominal shear stress in the wall, τ_v shall not exceed $\tau_{c,\,max}$ as per IS 456 : 2000.

2.4. When τ_v is less than τ_c , shear reinforcement shall be provided in accordance with para 1.4, 1.5 and 1.7 above.

2.5. When τ_v is greater than τ_c, the area of horizontal shear reinforcement, A_h, to be provided within a vertical spacing, s_v is given by

$$V_{us} = \frac{0.87 f_y A_h d_w}{s_v} \qquad \text{...(27.9)}$$

where $V_{us} = (V_u - \tau_c t_w d_w)$, is the shear force to be resisted by the horizontal reinforcement. However, the amount of horizontal reinforcement provided shall not be less than the minimum, as per para 1.4 above.

2.6. The vertical reinforcement, that is uniformly distributed in the wall, shall not be less than the horizontal reinforcement calculated as per para 2.5.

3. Flexural Strength

3.1. The moment of resistance, M_{uv}, of the wall section may be calculated as for columns subjected to combined bending and axial load as per IS 456. The moment of resistance of slender rectangular shear wall section with uniformly distributed vertical reinforcement is given in para 10.

3.2. The cracked flexural strength of the wall section should be greater than its uncracked flexural strength.

3.3. In walls that do not have boundary elements, vertical reinforcement shall be concentrated at the ends of the wall. Each concentration shall consist of a minimum of 4 bars of 12 mm diameter arranged in at least 2 layers.

4. Boundary Elements

Boundary elements are portions along the wall edges that are strengthened by longitudinal and transverse reinforcement. Though they may have the same thickness as that of the wall web it is advantageous to provide them with greater thickness.

4.1. Where the extreme fibre compressive stress in the wall due to factored gravity loads plus factored earthquake force exceeds $0.2 f_{ck}$, boundary elements shall be provided along the vertical boundaries of walls. The boundary elements may be discontinued where the calculated compressive stress becomes less than $0.15 f_{ck}$. The compressive stress shall be calculated using a linearly elastic model and gross section properties.

4.2. A boundary element shall have adequate axial load carrying capacity, assuming short column action, so as to enable it to carry an axial compression equal to the sum of factored gravity load on it and the additional compressive load induced by the seismic force. The latter may be calculated as :

$$\frac{M_u - M_{uv}}{C_w}$$

where M_u = factored design moment on the entire wall section.

M_{uv} = moment of resistance provided by distributed vertical reinforcement across the wall section, and

C_w = centre to centre distance between the boundary elements along the two vertical edges of the wall.

4.3. If the gravity load adds to the strength of the wall, its load factor shall be taken as 0.8.

4.4. The percentage of vertical reinforcement in the boundary elements shall not less than 0.8 percent, nor greater than 6 percent. In order to avoid congestion, the practical upper limit would be 4 percent.

4.5. Boundary elements, where required, as per para 4.1, shall be provided throughout their height, with special confining reinforcement, as per para 4 of § 27.9.

4.6. Boundary elements need not be provided if the entire wall section is provided with special confining reinforcement, as par para 4 of §27.9.

5. Coupled Shear Walls

5.1. Coupled shear walls shall be connected by ductile coupling beams. If the earthquake induced shear stress in the coupling beam exceeds

$$\frac{0.1\, l_s \sqrt{f_{ck}}}{D}$$

where l_s is the clear span of the coupling beam and D is its overall depth, the entire earthquake induced shear and flexure shall, preferably, be resisted by diagonal reinforcement.

5.2. The area of reinforcement to be provided along each diagonal in a diagonally reinforced coupling beam shall be :

$$A_{sd} = \frac{V_u}{1.74\, f_y . \sin \alpha} \qquad \ldots(27.10)$$

where V_u is the factored shear force, and α is the angle made by the diagonal reinforcement with the horizontal. At least 4 bars of 8 mm diameter shall be provided along each diagonal. The reinforcement along each diagonal shall be enclosed by special confining reinforcement, as per para 4 of § 27.9. The pitch of spiral or spacing of ties shall not exceed 100 mm.

5.3. The diagonal or horizontal bars of a coupling beam shall be anchored in the adjacent walls with an anchorage length of 1.5 times the development length in tension.

6. Opening in Walls

6.1. The shear strength of a wall with openings should be checked along critical planes that pass through openings.

6.2. Reinforcement shall be provided along the edges of openings in walls The area of the vertical and horizontal bars should be such as to equal that of the respective interrupted bars. The vertical bars should extended for the full storey height. The horizontal bars should be provided with development length in tension beyond the sides of the opening.

7. Discontinuous Walls

Columns supporting discontinuous walls shall be provided with special confining reinforcement, as per para 4.4 of § 27.9.

8. Construction Joints

The vertical reinforcement ratio across a horizontal construction joint shall not be less than

$$\frac{0.92}{f_y}\left(\tau_v - \frac{P_u}{A_g}\right)$$

where τ_v is the factored shear stress at the joint, P_u is the factored axial force (positive for compression), and A_g is the gross cross sectional area of the joint.

9. Development, Splice and Anchorage Requirement

9.1. Horizontal reinforcement shall be anchored near the edges of the wall or in the confined core of the boundary elements.

9.2. Splicing of vertical flexural reinforcement should be avoided, as far as possible, in regions where yielding may take place. This zone of flexural yielding may be considered to extend for a distance of l_w above the base of the wall or one sixth of the wall height, whichever is more. However, this distance need not be greater than 2 l_w. Not more than one third of this vertical reinforcement shall be spliced at such a section. Splices in adjacent bars should be staggered by a minimum of 600 mm.

9.3. Lateral ties shall be provided around lapped spliced bars that are larger than 16 mm in diameter. The diameter of the tie shall not be less than one fourth that of the spliced bar nor less than 6 mm. The spacing of ties shall not exceed 150 mm centre to centre.

9.4. Welded splices and mechanical connections shall conform to IS 456. However, not more than half the reinforcement shall be spliced at a section, where flexural yielding may take place.

10. Moment of Resistance of Rectangular Shear Wall Sections

10.1. The moment of resistance of a slender rectangular shear wall section with uniformly distributed vertical reinforcement may be estimated as follows :

(*a*) For $x_u/l_w < x_u^*/l_w$,

$$\frac{M_{uv}}{f_{ck}\, t_w\, l_w^2} = \Phi\left[\left(1+\frac{\lambda}{\Phi}\right)\left(\frac{1}{2}-0.415\,\frac{x_u}{l_w}\right)-\left(\frac{x_u}{l_w}\right)^2\left(0.168+\frac{\beta^2}{3}\right)\right] \quad \text{...(27.11)}$$

where

$$\frac{x_u}{l_w} = \left(\frac{\Phi+\lambda}{2\,\Phi+0.36}\right) \quad \text{...(27.12)}$$

$$\frac{x_u^*}{l_w} = \left(\frac{0.0035}{0.0035+0.87\,f_y/E_s}\right) \quad \text{...(27.13)}$$

$$\Phi = \left(\frac{0.87\,f_y\,\rho}{f_{ck}}\right) \quad \text{...(27.14)}$$

$$\lambda = \left(\frac{P_u}{f_{ck}\, t_w\, l_w}\right) \quad \text{...(27.15)}$$

ρ = vertical reinforcement ratio $= A_{st}/(t_w\, l_w)$,

A_{st} = area of uniformly distributed vertical reinforcement,

$\beta = 0.87\,f_y/(0.0035\,E_s)$

E_s = elastic modulus of steel, and

P_u = axial compression on wall.

(*b*) For $x_u^*/l_w < x_u/l_w < 1.0$

$$\frac{M_{uv}}{f_{ck}\, t_w\, l_w^2} = \alpha_1 \left(\frac{x_u}{l_w}\right) - \alpha_2 \left(\frac{x_u}{l_w}\right)^2 - \alpha_3 - \frac{\lambda}{2} \qquad \text{...(27.16)}$$

where

$$\alpha_1 = \left[0.36 + \Phi\left(1 - \frac{\beta}{2} - \frac{1}{2\beta}\right)\right] \qquad \text{...(27.17)}$$

$$\alpha_2 = \left[0.15 + \frac{\Phi}{2}\left(1 - \beta - \frac{\beta^2}{2} - \frac{1}{3\beta}\right)\right] \;;\; \text{and} \;\; \alpha_3 = \frac{\Phi}{6\beta}\left(\frac{1}{(x_u/l_w)} - 3\right) \qquad \text{...(27.18)}$$

The values of x_u/l_w to be used in this equation, should be calculated from the quadratic equation.

$$\alpha_1 \left(\frac{x_u}{l_w}\right)^2 + \alpha_4 \left(\frac{x_u}{l_w}\right) - \alpha_5 = 0 \qquad \text{...(27.19)}$$

where

$$\alpha_4 = \left(\frac{\Phi}{\beta} - \lambda\right) \;;\; \text{and} \;\; \alpha_5 = \left(\frac{\Phi}{2\beta}\right) \qquad \text{...(27.20)}$$

These equations were derived, assuming a rectangular wall section of depth l_w and thickness t_w that is subjected to combined uni-axial bending and axial compression. The vertical reinforcement is represented by an equivalent steel plate along the length of the section. The stress-strain curve assumed for concrete is as per IS 456. whereas that for steel is assumed to be bi-linear. Two equations are given for calculating the flexural strength of the section. Their use depends on whether the section fails in flexural tension or in flexural compression.

Appendix-A

Working Stress Design Method

A-1. SINGLY REINFORCED BEAM

A-1.1. General

Fig. A-1(*a*) shows a singly reinforced concrete beam section. At very small loads, both concrete and steel resist tension. However, as the load increases, the concrete at the bottom of the section will reach a tensile stress at which cracks occur, and thus the steel bars alone resist the tension. Figs. A-1 (*b*), (*c*) and (*d*) show the general character of distribution of compressive stress in concrete as the load increases. Since the stress-strain diagram for concrete [Fig.A-1(*f*)] is not linear, the stresses in the concrete do not increase in direct proportion to increase in strain. Therefore, the stress diagram is slightly curved. The straight line theory assumes that the stress distribution in the concrete at working stresses of approximately half the ultimate compressive strength of concrete (σ_{cu}) is like that of Fig. A-1 (*c*) but the slight curvature in the compression diagram is neglected.

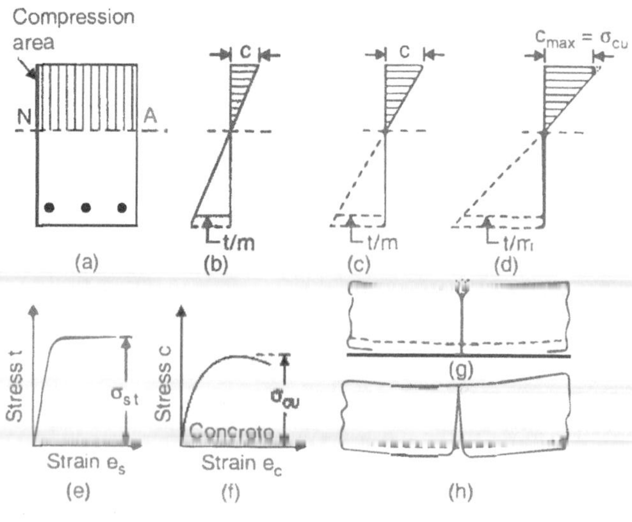

FIG. A-1

When the loads increase further so that maximum compressive stress occurs at the top of the section, the concrete flows plastically. The corresponding stress diagram in concrete is shown in Fig. A-1 (*d*), where σ_{cu} represents the ultimate compressive strength of concrete. At such a stage, the beam may continue to deform without offering proportionately increased resistance, and a small increase in load may cause complete failure of beam, without sufficient warning. Such compression failure is indicated in Fig. A-1 (*g*). However, if the tensile stress in steel reaches its ultimate value first, the beam would "pop open" [Fig A-1(*h*)] and the beam will fail in tension or by local crushing of concrete at top of the concrete. However, upto the elastic limit of steel, the reinforcing bars resist tension in proportion to their strain.

A-1.2. Equivalent Area

(*i*) ***Uncracked section.*** Fig. A-2 (*a*) shows a R.C. beam section having A_{st} = area of steel reinforcement, a_c = area of concrete and A = total area of cross section. If the load on the beam is *small* so that the tensile stress set-up in the concrete below the neutral axis is smaller than the permissible, the concrete area below the N.A. will not crack. In that case, $e_s = e_c$ at the centre of the reinforcement.

$$\therefore \quad \frac{t}{E_s} = \frac{c'}{E_c}$$

or $$t = \frac{E_s}{E_c} c' = m \cdot c' \quad \text{...(A-1)}$$

or $$c' = \frac{t}{m} \quad \text{...[A-1(a)]}$$

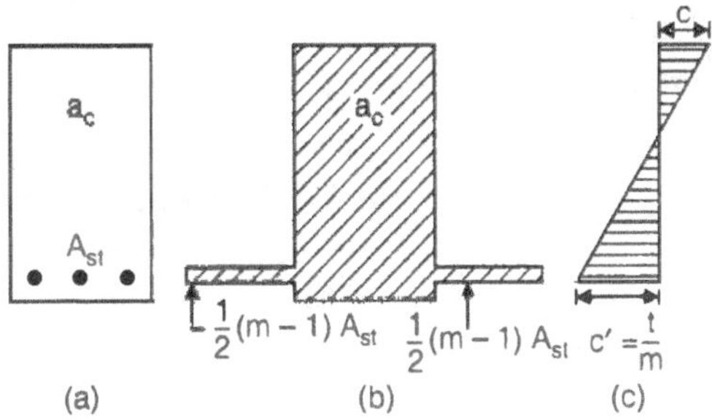

FIG. A-2. EQUIVALENT AREA : UNCRACKED SECTION.

where

e_s = strain in steel

e_c = strain in concrete surrounding steel

t = stress in steel

c' = stress in concrete surrounding steel, assuming that concrete is not cracked.

c = compressive stress in concrete at the extreme fibre.

m = modular ratio = E_s/E_c

Eq. A-1 suggests that stress in steel is m times the stress in concrete, or conversely, the stress in concrete is $1/m$ times the stress in steel. Fig A-2(*c*) shows the stress distribution in the beam section consisting of an equivalent concrete section of Fig. A-2(*b*). If a_c' is the additional area of concrete equivalent to steel area, we have

Load carried by steel = Load carried by equivalent concrete

or $$A_{st} \cdot t = a_c' c' \quad \text{or} \quad a_c' = \frac{t}{c'} \cdot A_{st} = m A_{st} \text{ (since } t/c' = m) \quad \text{...(A-2)}$$

However, net increase in the equivalent concrete area a_{ce} (uncracked) is

$$a_{ce} = a_c' - A_{st} = m A_{st} - A_{st} = (m-1) A_{st} \quad \text{...[A-2(a)]}$$

(*ii*) ***Cracked Section.***. If the load on the beam is increased, the tensile stress in the concrete below the N.A. will increase and will crack it. Even under normal loading conditions, the tensile stresses set up in concrete will be more than the permissible one. Hence it is assumed that the concrete area in tension zone will be *ineffective*. In that case, the compressive stress will be taken up by the concrete above N.A. and the tensile stresses will be taken up by the steel reinforcement. Fig. A-3 (*c*) shows the stress distribution in the *equivalent concrete section* shown in Fig. A-3(*b*). Since concrete is assumed to have cracked, the equivalent area of concrete will be such that the load carried by

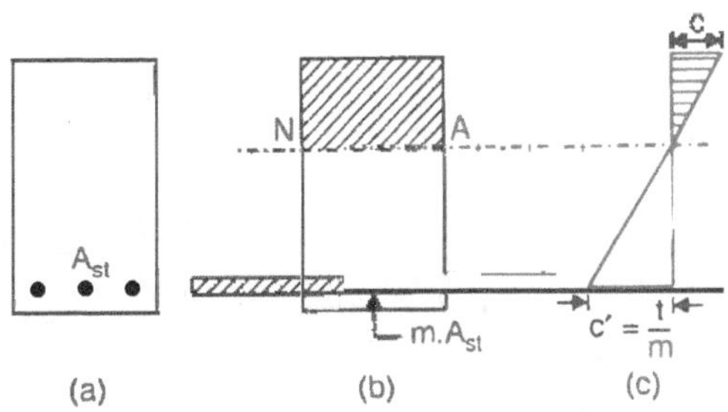

FIG. A-3. EQUIVALENT AREA : CRACKED SECTION

steel is the same as the load carried by an equivalent concrete area a_{ce} (strong enough to carry that tensile load),

i.e. $$A_{st} \, . \, t = a_{ce} \, . \, c' \quad \text{or} \quad a_{ce} = \frac{t}{c'} A_{st} = m \, A_{st} \qquad \text{...(A-3)}$$

where c' is the tensile stress in the *equivalent concrete area.*

The difference between the stress diagrams of Fig. A-2(*c*) and A-3(*c*) must be clearly borne in mind. In Fig. A-2(*c*), the tensile stress distribution is all over the portion of the concrete below the N.A., while in Fig. A-3 (*c*), the tensile stress $c' (= t/m)$ is in the equivalent concrete area concentrated at the level of steel reinforcement.

A-1.3. Neutral Axis of Beam Section

In order to determine the position of the neutral axis in a reinforced concrete beam section, we shall consider two cases: (*i*) when the stresses developed in the concrete section are known and (*ii*) when the dimensions of the section are known but stresses are not known. Fig. A-4 (*a*) shows the reinforced concrete beam section having steel reinforcement placed in the tension zone. Such a section is known as *singly reinforced section.*

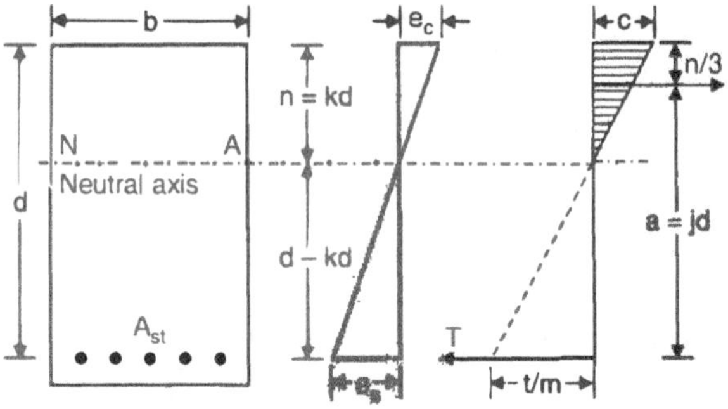

(a) Section (b) Strain diagram (c) Stress diagram

FIG. A-4. SINGLY REINFORCED SECTION

CASE (*i*) STRESSES IN CONCRETE AND STEEL KNOWN

Let c = compressive stress in the extreme fibre of concrete

t = tensile stress in steel reinforcement ; b = breadth of beam

d = depth to the centre of reinforcement (known as the *effective depth)*

n = kd = depth of N.A. below the top of the beam

k = neutral axis depth factor = n/d

A_{st} = area of tensile reinforcement,

Since there is no resultant force across the section,

Total Compression = Total Tension.

Neglecting the tensile stress in concrete, we get $\frac{1}{2} c \, . \, b \, . \, kd = t \, . \, A_{st}$

or $$\frac{A_{st}}{b \, . \, d} = k \, . \, \frac{c}{2t} \qquad \text{...(A-4)}$$

If t is the stress in steel, the stress c' in the equivalent concrete area will be equal to t/m as proved in Eq. A-1. Hence from the diagram [Fig. A-4 (*c*)], we get

$$\frac{c}{k \, . \, d} = \frac{t}{m} \cdot \frac{1}{d - kd} = \frac{t}{md\,(1-k)}$$

$$\frac{c}{k} = \frac{t}{m\,(1-k)} \quad \text{or} \quad mc - mck = kt$$

From which $$k = \frac{mc}{mc + t} = \frac{1}{1 + \frac{1}{m}\frac{t}{c}} = \frac{1}{1 + \frac{r}{m}} \qquad \text{...(A-5)}$$

where r = stress ratio = t/c

Eq. A-5 gives the location of the neutral axis when the stresses in concrete and steel are known. If the stress ratio $r\ (= t/c)$ and the modular ratio m are known, N.A. can be located from Eq. A-5. Fig. A-5 shows the graphical representation of Eq. A-5 for various values of stress ratio and modular ratio. Let,

σ_{st} = Permissible tensile stress in reinforcement (*i.e.* permissible value of t) given in Table A-3

σ_{cbc} = Permissible compressive stress in concrete due to bending (*i.e.* permissible value of c), given in Table A-1

k_c = Critical neutral axis depth factor

n_c = Depth of critical neutral axis, corresponding to permissible values of stresses in steel and concrete. Then

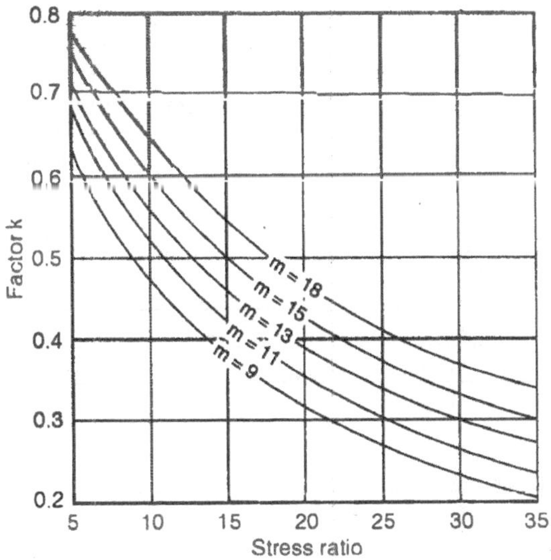

FIG. A-5. DETERMINATION OF k

$$k_c = \frac{m\,\sigma_{cbc}}{m\,\sigma_{cbc} + \sigma_{st}} \qquad \text{...[A-5 (a)]}$$

$$n_c = k_c \,.\, d = \frac{m\,\sigma_{cbc}}{m\sigma_{cbc} + \sigma_{st}} \,.\, d = \frac{1}{1 + \frac{r_c}{m}} d \qquad \text{...[A-5 (b)]}$$

where r_c = stress ratio = σ_{st}/σ_{cbc}

The section in which the stresses in concrete and steel reach their corresponding permissible values simultaneously is called a *balanced or critical section* (See § A-1.5).

Note. As per IS : 456 – 2000, the modular ratio is given by

$$m = \frac{280}{3\,\sigma_{cbc}} \quad \text{or} \quad m \,.\, \sigma_{cbc} = 280/3 = 93.33$$

Table A-2 gives the values of modular ratio m for various grades of concrete. Hence Eq. A-5 reduces to the following form

$$k_c = \frac{m\,\sigma_{cbc}}{m\sigma_{cbc} + \sigma_{st}} = \frac{93.33}{93.33 + \sigma_{st}} = \frac{1}{1 + 0.0107\,\sigma_{st}} \qquad \text{...[A-5(c)]}$$

This suggests that value of k_c does not depend upon the grade of the concrete. It depends only on the permissible stress in reinforcement See Table A-4.

The area of steel for the section is given by Eq. A-4. If p is the *percentage steel*, we have

$$p = \frac{A_{st}}{b\,.\,d} \times 100 \;; \quad \text{But} \quad \frac{A_{st}}{b\,.\,d} = \frac{k\,.\,c}{2t} \quad \text{(from Eq. A-4)}$$

$$\therefore \quad p = \frac{k\,.\,c}{2t} \times 100 = 50\,\frac{c}{t} \cdot \frac{mc}{mc + t} = \frac{50\,mc^2}{t\,(mc + t)} \qquad \text{...(A-6)}$$

Expressed in terms of the *permissible values* of the stresses, the above equation may be written as :

$$p_c = \frac{50\, m\sigma_{cbc}^2}{\sigma_{st}\,(m\,\sigma_{cbc} + \sigma_{st})} \qquad \ldots[\text{A-6 } (a)]$$

TABLE A-1. PERMISSIBLE STRESSES IN CONCRETE (IS : 456–2000)

Grade of concrete	*Permissible stress in compression (N/mm²)*		*Permissible stress in Bond (Average) for plain bars in tention (N/mm²)* τ_{bd}
	Bending (σ_{cbc})	*Direct* (σ_{cc})	
M 10	3.0	2.5	—
M 15	5.0	4.0	0.6
M 20	7.0	5.0	0.8
M 25	8.5	6.0	0.9
M 30	10.0	8.0	1.0
M 35	11.5	9.0	1.1
M 40	13.0	10.0	1.2
M 45	14.5	11.0	1.3
M 50	16.0	12.0	1.4

TABLE A-2 MODULAR RATIO

Grade of concrete	*M 10*	*M 15*	*M 20*	*M 25*	*M 30*	*M 35*	*M 40*
Modular ratio m	31 (31.11)	19 (18.67)	13 (13.33)	11 (10.98)	9 (9.33)	8 (8.11)	7 (7.18)

TABLE A-3 PERMISSIBLE STRESSES IN STEEL REINFORCEMENT

Type of stress in steel reinforcement	*Permissible stresses*		
	(a) Mild steel bars conforming to Grade I of IS : 432 (Part I)	*(b) Medium tensile steel conforming to IS : 432 (Part I)*	*(c) High yield strength deformed bars (HYSD bars) conforming to IS : 1786 (Grade Fe 415)*
(1)	*(2)*	*(3)*	*(4)*
1. Tension (σ_{st} or σ_{sv}) (a) up to and including 20 mm (b) over 20 mm	 140 N/mm² 130 N/mm²	Half the guaranteed yield stress subject to a maximum of 190 N/mm²	 230 N/mm² 230 N/mm²
2. Compression in column bars (σ_{sc})	130 N/mm²	130 N/mm²	190 N/mm²
3. Compression in bars in a beam or a slab where the compressive resistance of the concrete is taken in to account.	The calculated compressive stress in the surrounding concrete multiplied by 1.5 times the modular ratio or σ_{sc} whichever is lower.		
4. Compression in bars in a beam or a slab where compressive resistance of the concrete is taken into account. (a) Upto and including 20 mm. (b) Over 20 mm.	 140 N/mm² 130 N/mm²	Half the guaranteed yield stress subject to a maximum of 190 N/mm²	 190 N/mm² 190 N/mm²

CASE (*ii*) DIMENSIONS OF THE BEAM COMPLETELY KNOWN

Let us now take the case when the dimensions of the section, including the reinforcement are known, but the stresses in concrete and steel are *not known*. Refer Fig. A-4. The neutral axis of a homogeneous beam is the horizontal line that passes through the centre of gravity of the cross-section. Hence equating the moment of area in compression to the moment of the *equivalent area* in tension about the N.A., we get

$$b \cdot kd \cdot \frac{kd}{2} = mA_{st}(d - kd)$$

or $k^2 \cdot \dfrac{b\,d^2}{2} + m \cdot kd\, A_{st} - m \cdot d \cdot A_{st} = 0$ or $k^2 + 2k \cdot m \cdot \dfrac{A_{st}}{b\,d} - 2m\dfrac{A_{st}}{b\,d} = 0$

Puting $\dfrac{A_{st}}{b\,d} = p'$ = reinforcement ratio, we get

$$k^2 + 2k\,mp' - 2mp' = 0,$$

From which, $k = \dfrac{-2mp' \pm \sqrt{4m^2p'^2 + 8mp'}}{2} = \sqrt{2mp' + (mp')^2} - mp'.$ ...(A-7)

Thus if p' is known (*i.e.* if the dimensions b and d, and the area of steel A_{st} are known) the neutral axis depth factor k can be calculated from Eq. A-7 and the N.A. can be located.

A-1.4. Moment of Resistance

Since there is no resultant force across the section, the total compressive force acting at the centre of gravity of the compressive forces is equal to the total tensile force acting at the centre of gravity of the steel reinforcement. Hence the moment of resistance of the reinforced concrete beam section is equal to the moment of the couple consisting of the compressive force and the tensile force. The total compressive force $C = \frac{1}{2} c \cdot k \cdot d \cdot b \; (= \frac{1}{2}\sigma_{cbc} \cdot k_c d \cdot b)$ and acts at a distance of $kd/3$ from the top of the section. Similarly, the total tensile force $T = t \cdot A_{st}$ $(= \sigma_{st} A_{st})$ and acts at a distance d from the top of the section. If jd is the lever arm (*i.e.* the distance between the compressive force and the tensile force), we have

Lever arm $a = jd = d - \dfrac{kd}{3} = d\left(1 - \dfrac{k}{3}\right)$

or $j = 1 - \dfrac{k}{3}$ or $j_c = 1 - \dfrac{k_c}{3}$ ...(A-8)

Eq. A-8 gives the *lever arm factor* j in terms of the neutral axis depth factor k.

Now, moment of resistance M_r is given by

M_r = force (either compressive or tensile) × lever arm

∴ $M_r = \frac{1}{2} c \cdot kd \cdot b\,(jd) = (\frac{1}{2} cjk)\, bd^2$ ...[A-9(*a*)]

or $M_r = Rbd^2$ ...(A-9)

where $R = \frac{1}{2} cjk$ or $R_c = \frac{1}{2}\sigma_{cbc} \cdot j_c k_c$ ...(A-10)

Table A-4 gives the values of design constants k_c, j_c, R_c, and p_c for balanced design.

TABLE A-4. VALUES OF DESIGN CONSTANTS

Grade of concrete →		M 15 *	M 20	M 25	M 30	M 35	M 40
Modular ratio m →		18.67	13.33	10.98	9.33	8.11	7.18
σ_{cbc} N/mm^2 →		5.0	7.0	8.5	10.0	11.5	13.0
$m\sigma_{cbc}$		93.33	93.33	93.33	93.33	93.33	93.33
(a) σ_{st} = 140 N/mm^2 (Fe 250)	k_c	0.400	0.400	0.400	0.400	0.400	0.400
	j_c	0.867	0.867	0.867	0.867	0.867	0.867
	R_c	0.867	1.214	1.474	1.734	1.994	2.254
	p_c (%)	0.714	1.000	1.214	1.429	1.643	1.857
(b) σ_{st} = 190 N/mm^2	k_c	0.329	0.329	0.329	0.329	0.329	0.329
	j_c	0.890	0.890	0.890	0.890	0.890	0.890
	R_c	0.732	1.025	1.244	1.464	1.684	1.903
	p_c (%)	0.433	0.606	0.736	0.866	0.997	1.127
(c) σ_{st} = 230 N/mm^2 (Fe 415)	k_c	0.289	0.289	0.289	0.289	0.289	0.289
	j_c	0.904	0.904	0.904	0.904	0.904	0.904
	R_c	0.653	0.914	1.110	1.306	1.502	1.698
	p_c (%)	0.314	0.440	0.534	0.628	0.722	0.816
(d) σ_{st} = 275 N/mm^2 (Fe 500)	k_c	0.253	0.253	0.253	0.253	0.253	0.253
	j_c	0.916	0.916	0.916	0.916	0.916	0.916
	R_c	0.579	0.811	0.985	1.159	1.332	1.506
	p_c (%)	0.230	0.322	0.391	0.460	0.530	0.599

* **As per IS 456: 2000, the minimum grade of concrete shall be not less than M 20 in R.C.C. work.**

From Eq. A-9, the dimensions of the beam section can be determined to develop a given moment of resistance. *It should be noted that for a given type of concrete and steel reinforcement, factor R_c is a constant and does not depend upon the beam dimensions.*

Also $$M_r = t \cdot A_{st} \cdot jd = \sigma_{st} \cdot A_{st} \cdot j_c\, d \qquad \text{...(A-11)}$$

Hence $$A_{st} = \frac{M_r}{t \cdot jd} = \frac{M}{t \cdot jd} = \frac{M}{\sigma_{st} \cdot j_c\, d} \qquad \text{...(A-12)}$$

where M is the external bending moment and σ_{st} is permissible stress in steel reinforcement.

Eq A-12 is used for determining the cross sectional area of steel reinforcement for given moment of resistance and effective depth.

If the section of a beam, including the steel reinforcement, is completely known, its moment of resistance is calculated both by Eqs. A-9(*a*) and A-11, using maximum permissible values of stresses c (= σ_{cbc}) and t (= σ_{st}). Thus the values of M_r will be obtained and if the two values do not agree, it will be concluded that both concrete and steel do not develop their allowable stresses simultaneously. The moment of resistance of the beam will then be equal to the lesser of the two values.

A-1.5. Balanced, Under-reinforced and Over-reinforced Section

(*a*) **Balanced Section**. In a beam section, if the area of steel reinforcement A_{st} is of such magnitude that the permissible stresses c (= σ_{cbc}) and t (= σ_{st}) in concrete and steel respectively, are developed simultaneously, the section is known as the balanced section, *critical* *section* or *economical section*. For such a section (Fig A-6), the neutral axis depth factor k_c is determined by Eq. A-5 (*a*), using $c\ (=\sigma_{cbc})$ and $t\ (=\sigma_{st})$ as the maximum permissible stress in the two materials and depth n_c of the critical N.A. is taken equal to k_c times d. For such a balanced section, the moment of resistance obtained from the compressive force will be equal to the moment of resistance obtained from the tensile force.

The percent area of reinforcement for a balanced section is obtained from Eq. A-6 (*a*):

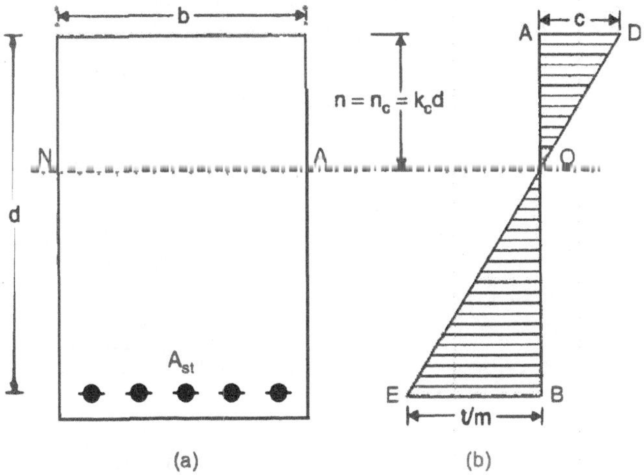

FIG. A-6. BALANCED SECTION.

$$p_c = \frac{1}{2} k_c \cdot \frac{\sigma_{cbc}}{\sigma_{st}} \times 100 \qquad \text{...[A-6 (a)]}$$

or $$p_c = \frac{50\, k_c}{r_c}$$ where r_c is the stress ratio

For example, if $c = \sigma_{cbc} = 5$ N/mm^2, $t = \sigma_{st} = 140$ N/mm^2 and $m = 19$, we have

$$k_c = \frac{1}{1 + \frac{1}{m}\frac{t}{c}} = \frac{1}{1 + \frac{140}{19 \times 5}} = 0.404 \quad \text{and} \quad p_c = \frac{1}{2} \times 0.404 \times \frac{5}{140} \times 100 = 0.72\%$$

Thus for a reinforced concrete having the above properties if the reinforcement provided is 0.72% of the cross-section ($b \times d$), a balanced section will be obtained. The N.A. of a balanced section is called the *critical neutral axis*..

(*b*) **Under-reinforced Section**. An under-reinforced section is the one in which the percentage steel provided is *less* than that given in Eq A-6 and therefore full strength of concrete in compression is not developed. The actual neutral axis of such a section will falls *above* the critical neutral axis of a balanced section.

Fig A-7 (*a*) shows the under-reinforced section in which the actual N.A. is above the critical N.A. Let $c = \sigma_{cbc}$ be the permissible stress in concrete and $t = \sigma_{st}$ be the permissible stress in steel. If full permissible compressive stress $c = \sigma_{cbc} = AD_1$ is permitted to be developed in concrete, the corresponding stress t_1 in steel

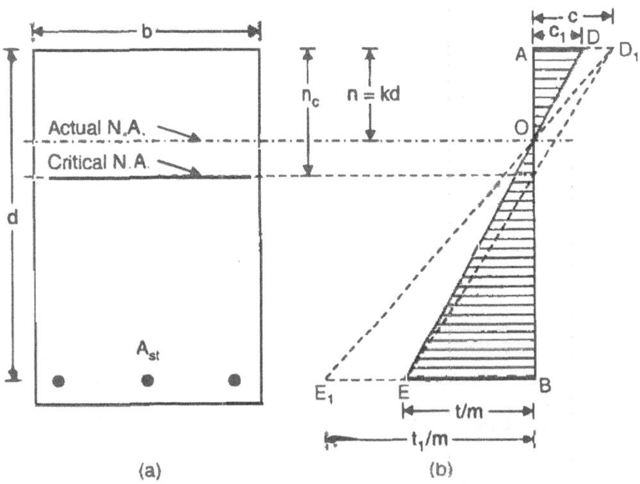

FIG. A-7. UNDER-REINFORCED SECTION.

will be such that $BE_1 = t_1/m$. Evidently, since $t_1 > t$, full compressive stress c cannot develop in concrete. Instead, a compressive stress $c_1 = AD$ is developed of such a magnitude that $BE = t/m = \sigma_{st}/m$ is obtained. Thus, in a under-reinforced concrete, the concrete is not fully stressed to its permissible value when stress in steel reaches its maximum value of $t = \sigma_{st}$. The moment of resistance of an under-reinforced section is, therefore, computed on the basis of the tensile force in steel :

$$M_r = t\,A_{st}\,.\,jd = \sigma_{st}\,.\,A_{st}\,.\,jd \qquad \text{...(Eq. A-11)}$$

(*c*) **Over-reinforced Section**. In an over-reinforced section, the reinforcement provided is more than critical one and therefore the actual N.A. of such a section falls below the critical N.A. of a balanced section.

Fig A-8 (*a*) shows a over-reinforced section in which the actual N.A. falls below the critical one. If stress in steel is permitted to be equal to maximum permissible value $t = (\sigma_{st})$ the corresponding stress in concrete will be equal to $AD_1 = c_1$ which is greater than permissible value $c = (\sigma_{cbc})$ in concrete. Hence the stress distribution will be governed by line DOE (and not $D_1\,O\,E_1$), giving rise to a stress t_1 in steel which is lesser than t. Thus, in a over-reinforced section, steel reinforcement is not fully stressed to its permissible value and the moment of resistance is determined on the basis of compressive force developed in concrete :

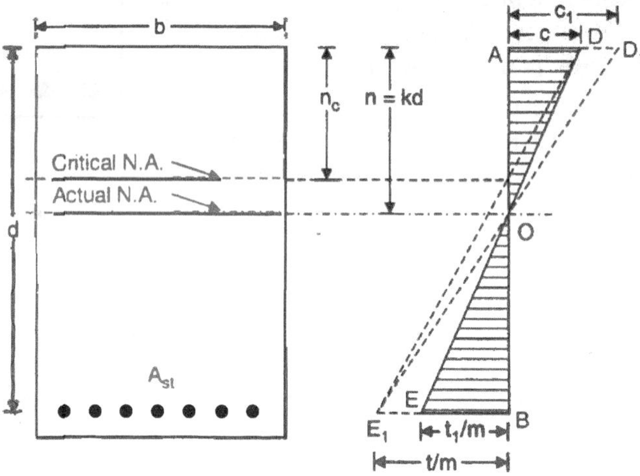

FIG. A-8. OVER-REINFORCED SECTION.

$$M_r = \frac{1}{2}c\,.\,kd\,.\,b\,.\,jd = \frac{1}{2}c\,.\,j\,.\,k\,.\,bd^2 = \frac{1}{2}\sigma_{cbc}\,.\,j\,.\,k\,.\,bd^2 \qquad \text{...[A-9(a)]}$$

or $$M_r = \frac{1}{2}c\,.\,n\,.\left(d - \frac{n}{3}\right)b = \frac{1}{2}\sigma_{cbc}\,.\,n\left(d - \frac{n}{3}\right)b \qquad \text{...[A-9(b)]}$$

where n = depth of the neutral axis

Note : It should be noted that compression failure is *sudden*, and therefore not desirable. *The Code recommends that if n (= kd) is found to be greater than the limiting value* *(n_c), the beam should be redesigned.*

Example A-1. *For a balanced rectangular section (b × d) of a singly reinforced beam, determine (i) depth of neutral axis (ii) moment of resistance and (iii) percentage of steel using M 20 concrete and Fe 415 steel.*

If b = 200 mm and d = 300 mm, determine the numerical values of n, M_r and A_{st}.

Solution. We have $c = \sigma_{cbc} = 7$ N/mm², $m = 13.33$ and $t = \sigma_{st} = 230$ N/mm²

(*i*) For balanced section, $k_c = \dfrac{mc}{mc + t} = \dfrac{13.33 \times 7}{13.33 \times 7 + 230} = 0.289$; $\therefore\ n_c = k_c\,d = \mathbf{0.289\,d}$

(*ii*) From Eq. A-9(*a*), $M_r = \frac{1}{2} cjk \cdot b\,d^2 = R_c \cdot b\,d^2$

where $j_c = 1 - \frac{k_c}{3} = 1 - \frac{0.289}{3} = 0.904$

$\therefore$ $R_c = \frac{1}{2}\,\sigma_{cbc}\,j_c\,k_c = \frac{1}{2} \times 7 \times 0.904 \times 0.289 = 0.914$

$\therefore$ $M_r = R_c\,bd^2 = \mathbf{0.914\,bd^2}$ (where M_r is in N-mm units)

(*iii*) From Eq. A-6 (*a*), percentage of steel is given by

$$p_c = \frac{50 \times 13.33\,(7)^2}{230\,(13.33 \times 7 + 230)} = 0.440\%$$

Alternatively, $p_c = \frac{0.289 \times 7}{2 \times 230} \times 100 = 0.440\%$, from Eq. A-6 (*a*).

For the given numerical values,

$$n_c = 0.289 \times 300 = \mathbf{86.7\ mm}$$

$$M_r = 0.914\,(200)\,(300)^2 = 16.452 \times 10^6 \text{ N-mm} = \mathbf{16.452\ kN\text{-}m}$$

$$A_{st} = p_c \cdot \frac{b\,d}{100} = \frac{0.44 \times 200 \times 300}{100} = \mathbf{264\ mm^2}$$

Example A-2. *Determine the moment of resistance of a singly reinforced beam 160 mm wide and 300 mm deep to the centre of reinforcement, if the stresses in steel and concrete are not to exceed 140 N/mm² and 5 N/mm². The reinforcement consists of 4 bars of 16 mm diameter. Take m=18. If the above beam is used over an effective span of 5 m, find the maximum load the beam can carry, inclusive of its own weight.*

Solution. $A_{st} = 4 \times \frac{\pi}{4}\,(16)^2 = 804\ \text{mm}^2$

$$c = \sigma_{cbc} = 5\ \text{N/mm}^2\ ;\ t = \sigma_{st} = 140\ \text{N/mm}^2\ ;\ m = 18.$$

Equating the moment of area in compression to the moment of the equivalent area in tension, about the N.A., we get

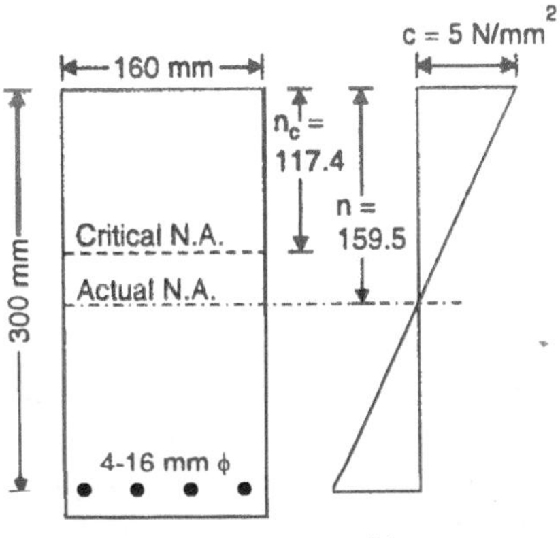

FIG. A-9

$$b \times n \times \frac{n}{2} = m\,A_{st}\,(d - n)$$

or $$\frac{160 \times n^2}{2} = 18 \times 804\,(300 - n)$$

or $n^2 + 181\,n - 54287 = 0$

From which $n = 159.5$ mm

If n_c is the depth of critical neutral axis, we have, from Eq. A-5

$$n_c = k_c \cdot d = \frac{18 \times 5}{(18 \times 5) + 140} \times 300 = 117.4\ \text{mm}$$

Since the depth of actual neutral axis is more than the critical one, the section is over reinforced. Thus, concrete reaches its maximum stress earlier to steel. Hence the moment of resistance is found on the basis of compressive force developed in concrete.

$$\text{Lever arm} = d - \frac{n}{3} = 300 - \frac{159.5}{3} = 246.8 \text{ mm}$$

$$\therefore \quad M_r = \frac{1}{2} c \,.\, n \,.\, b \,.\left(d - \frac{n}{3}\right) = \frac{1}{2} \times 5 \times 159.5 \times 160\,(246.8) \times 10^{-6} = 15.75 \quad \text{kN-m}$$

Let w =uniformly distributed load in kN/m inclusive of the self-weight of the beam.

$$\text{Maximum B.M.} = \frac{w L^2}{8} = \frac{w \times 25}{8} \text{ kN-m}$$

Equating this to the moment of resistance of the beam we get

$$w = \frac{15.75 \times 8}{25} = \mathbf{5.04 \ kN/m.}$$

Example A-3. *The cross-section of a simply supported reinforced beam is 200 mm wide and 300 mm deep to the centre of the reinforcement which consists of 3 bars of 16 mm dia. Determine from the first principles the depth of N.A. and the maximum stress in concrete when steel is stressed to 120 N/mm². Take m=19.*

Solution. Here, the dimensions are fully known.

$$A_{st} = 3 \times \frac{\pi}{4} (16)^2 = 603.2 \text{ mm}^2.$$

Let the depth of N.A. be n. Equating the moment of the compressive area to the moment of equivalent area of steel, about N.A., we get

$$b \,.\, n \,.\, \frac{n}{2} = m A_{st} (d - n)$$

or $\quad 100 n^2 = 19 \times 603.2 (300 - n)$

or $\quad n^2 + 114.6 n - 34382 = 0$

From which $n = \mathbf{136.8 \ mm}$

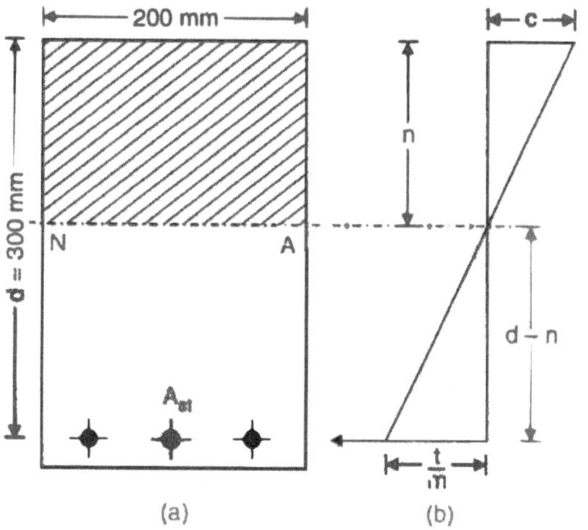

FIG. A-10

Fig. A-10 (*b*) shows the stress diagram.

Stress in steel = t = 120 N/mm² (Given)

$$\therefore \quad \frac{t}{m} = \frac{120}{19} = 6.316$$

From the stress diagram $\quad \frac{c}{n} = \frac{t/m}{d-n} \quad$ or $\quad \frac{c}{136.8} = \frac{6.316}{300 - 136.8} = 0.0387$

From which $\quad c = \mathbf{5.29 \ N/mm^2}$

Example A-4. *A rectangular, singly reinforced beam, 300 mm wide and 500 mm effective depth is used as a simply supported beam over an effective span of 6 m. The reinforcement consists of 4 bars of 20 mm diameter. If the beam carries a load of 12 kN/m, inclusive of the self-weight, determine the stresses developed in concrete and steel. Take m=19.*

Solution Let n = depth of N.A.; $A_{st} = 4 \times \frac{\pi}{4} (20)^2 = 1256.6 \text{ mm}^2.$

Equating the moments of two areas about N.A. we get

$$b \times n \times \frac{n}{2} = m\, A_{st}\,(d-n) \quad \text{or} \quad 300 \times \frac{n^2}{2} = 19 \times 1256.6\,(500-n)$$

or $\quad n^2 + 159.2\,n - 79587 = 0 \quad$ From which $n = 213.5$ mm

Lever arm, $\quad a = d - \frac{n}{3} = 500 - \frac{213.5}{3} = 428.8$ mm

$$\text{Maximum B.M.} = \frac{w\,l^2}{8} = \frac{12(6)^2}{8} = 54 \text{ kN-m} = 54 \times 10^6 \text{ N-mm}$$

Let c be the compressive stress in concrete.

$$\therefore \quad M_r = \tfrac{1}{2}\, c.n.b \times a = \tfrac{1}{2}\, c \times 300 \times 213.5 \times 428.8 = 13.732 \times 10^6\, c \text{ N-mm}$$

Equating M_r to the external B.M. we get

$$13.732 \times 10^6\, c = 54 \times 10^6 \quad \text{or} \quad c = \frac{54}{13.732} = \mathbf{3.93\ N/mm^2}$$

If t is the corresponding stress in steel, we get

$$t = \frac{m\,c}{n}(d-n) = \frac{19 \times 3.93}{213.5}(500 - 213.5) = \mathbf{100.2\ N/mm^2}$$

Alternatively, $\quad M_r = A_{st}\,.\,t\,a = 1256.6 \times t \times 428.8$

$$\therefore \quad 1256.6 \times 428.8\, t = 54 \times 10^6 \quad \text{or} \quad t = \frac{54 \times 10^6}{1256.6 \times 428.8} = 100.2 \text{ N/mm}^2$$

Example A-5. *Design a reinforced concrete beam subjected to a bending moment of 20 kN-m. Use M 20 concrete, and Fe 415 reinforcement. Keep the width of the beam equal to half the effective depth.*

Solution. For M 20 concrete, $c = \sigma_{cbc} = 7$ N/mm² and $m = 13.33$. For Fe 415 steel, $\sigma_{st} = 230$ N/mm². For a balanced section,

$$k_c = \frac{m\sigma_{cbc}}{m\sigma_{cbc} + \sigma_{st}} = \frac{13.33 \times 7}{13.33 \times 7 + 230} = 0{\cdot}289;\; j_c = 1 - \frac{k_c}{3} = 1 - \frac{0.289}{3} = 0.904$$

$$R_c = \tfrac{1}{2}\, c\,.\,j_c\,.\,k_c = \tfrac{1}{2} \times 7 \times 0.904 \times 0.289 = 0.914$$

Now $\quad M_r = R_c\,.\,b\,d^2 = \frac{R_c\,d^3}{2} = \frac{0.914}{2}\,d^3 = 0.457\,d^3$ N-mm

Given B.M. $= M = 20$ kN-m $= 20 \times 10^6$ N-mm

$$\therefore \quad 0.457\,d^3 = 20 \times 10^6$$

or $$d = \left(\frac{20 \times 10^6}{0.457}\right)^{\frac{1}{3}} \approx \mathbf{352.4\ mm}$$

$$b = \tfrac{1}{2}\,d = \mathbf{176.2\ mm.}$$

Area of steel is given by $\quad A_{st} = \frac{M}{\sigma_{st}\,j_c\,d} = \frac{20 \times 10^6}{230 \times 0.904 \times 352.4} = \mathbf{273\ mm^2}$

Thus provide a beam of 176.2 mm width and 352.4 mm effective depth, with area of tensile reinforcement equal to 273 mm².

Example A-6. *A reinforced concrete beam 200 mm × 400 mm effective depth is used over an effective span of 5 m. It is subjected to a uniformly distributed load of 5 kN/m inclusive of its own weight. Find the necessary steel reinforcement at the centre of the span. Take allowable stresses in steel and concrete as 130 N/mm² and 4 N/mm² respectively, and m=16.*

Solution. Given : $c = \sigma_{cbc} = 4$; $t = \sigma_{st} = 130$ and $m = 16$.

$$M = \frac{w\,l^2}{8} = \frac{5 \times 25}{8} \text{ kN-m} = 15.625 \text{ kN-m} = 15.625 \times 10^6 \text{ N-mm}$$

$$M_r = R\,.\,b\,d^2 = 15{\cdot}625 \times 10^6 \quad \text{or} \quad R = \frac{15{\cdot}625 \times 10^6}{200\,(400)^2} = 0{\cdot}488$$

Hence $\quad M_r = 0{\cdot}488\; b\,d^2 \qquad \ldots(i)$

For the balanced section, $\quad k_c = \dfrac{m\,c}{m\,c + t} = \dfrac{16 \times 4}{16 \times 4 + 130} = 0{\cdot}33.$

$$j_c = 1 - \frac{k_c}{3} = 1 - 0{\cdot}11 = 0{\cdot}89 \; ; \; \therefore \; R_c = \frac{1}{2}\,c\,j_c\,k_c = \frac{1}{2} \times 4 \times 0.89 \times 0.33 = 0{\cdot}587$$

$\therefore \qquad M_r = 0{\cdot}587\; b\,d^2 \qquad \ldots(ii)$

The moment of resistance of the given beam has to be less than the moment of resistance of the critical section. Hence the steel reinforcement will be corresponding to an under-reinforced beam. For such a section, the stress in steel will reach the maximum value of 130 N/mm².

Let n = depth of N.A. of the actual section. Then the corresponding stress in concrete is given by

$$\frac{c}{t/m} = \frac{n}{d - n} \quad \text{or} \quad c = \frac{t}{m}\,.\,\frac{n}{d-n} = \frac{130}{16} \times \frac{n}{400 - n} \qquad \ldots(iii)$$

Now $\quad M_r = \dfrac{1}{2}\,c\,.\,n\,.\,b\left(d - \dfrac{n}{3}\right)$

$$\therefore \quad 15{\cdot}625 \times 10^6 = \frac{1}{2}\left\{\frac{130}{16} \times \frac{n}{400 - n}\right\} n \times 200 \left(400 - \frac{n}{3}\right)$$

or $\quad n^2 \dfrac{(1200 - n)}{400 - n} = 57700$

This is a cubic equation in n. Solving it by trial and error, we get, $n = 122$ mm

Hence from (iii) $\quad c = \dfrac{130}{16} \times \dfrac{122}{400 - 122} = 3{\cdot}566$ N/mm²

Now, Total compression = Total tension, $\quad \therefore \quad \dfrac{1}{2}\,cnb = A_{st}\,.\,t$

or $\quad A_{st} = \dfrac{1}{2}\,\dfrac{cnb}{t} = \dfrac{1}{2}\,\dfrac{3{\cdot}566 \times 122 \times 200}{130} =$ **334.6 mm²**

Check. For the actual section, $\quad k = \dfrac{m\,c}{m\,c + t} = \dfrac{16 \times 3{\cdot}566}{16 \times 3{\cdot}566 + 130} = 0{\cdot}305.$

$$j = 1 - \frac{0{\cdot}305}{3} = 0{\cdot}8983 \; ; \; R = \frac{1}{2}\,c\,j\,k = \frac{1}{2} \times 3{\cdot}566 \times 0{\cdot}8983 \times 0{\cdot}305 = 0{\cdot}4885$$

$$M_r = 0{\cdot}4855 \times 200 \times (400)^2 \simeq 15{\cdot}63 \times 10^6 \text{ N-mm}$$

Also $M_r = A_{st}\, t\, j\, d = 334{\cdot}6 \times 130 \times 0.8983 \times 400 \simeq 15{\cdot}63 \times 10^6$ N-mm.

Hence O.K.

Example A-7. *A reinforced concrete beam 200 mm × 400 mm effective depth is used over an effective span of 5 m. It is subjected to a uniformly distributed load of 7 kN/m inclusive of its own weight. Find the necessary steel reinforcement at the centre of the span. Take allowable stresses in steel and concrete as 130 N/mm² and 4 N/mm² respectively and m=16.*

Solution. This example is exactly the same as the previous example except that w has been increased to 7 kN/m.

$$c = \sigma_{cbc} = 4 \text{ N/mm}^2 \;;\; t = \sigma_{st} = 130 \text{ N/mm}^2 \text{ and } m = 16.$$

$$M = \frac{w L^2}{8} = \frac{7 \times 25}{8} = 21{\cdot}875 \text{ kN-m} = 21{\cdot}875 \times 10^6 \text{ N-mm}$$

$$\therefore \quad R = \frac{M_r}{b\,d^2} = \frac{M}{b\,d^2} = \frac{21{\cdot}875 \times 10^6}{200 \times (400)^2} = 0{\cdot}6836$$

Hence $\quad M_r = 0{\cdot}6836\, b\, d^2 \qquad \ldots(i)$

For the balanced section

$$k_c = \frac{m\,c}{m\,c + t} = \frac{16 \times 4}{16 \times 4 + 130} = 0{\cdot}33. \;;\quad j_c = 1 - \frac{0{\cdot}33}{3} = 1 - 0{\cdot}11 = 0{\cdot}89.$$

$$R_c = \frac{1}{2}\, c\, j_c\, k_c = \frac{1}{2} \times 4 \times 0.89 \times 0.33 = 0{\cdot}5874 \;;\; M_r = 0{\cdot}5874\, b\, d^2 \qquad ..(ii)$$

Since the moment of resistance of the given beam is greater than the moment of resistance of the critical section, the steel reinforcement will be corresponding to an over-reinforced beam. For such a section, the stress in concrete will reach the maximum value of 4 N/mm².

Now, $\quad M_r = \frac{1}{2}\, c\, n\, b\left(d - \frac{n}{3}\right)$ or $21{\cdot}875 \times 10^6 = \frac{1}{2} \times 4\, n \times 200\left(400 - \frac{n}{3}\right)$

$\therefore \quad n^2 - 1200\, n + 164100 = 0.$ From which, $n = 1042$ or $157{\cdot}5$ mm

Hence $\quad n = 157{\cdot}5$ mm

(**Note.** For critical section, $n_c = k\,d = 0{\cdot}33 \times 400 = 132$ mm). The stress in steel is given by

$$t = \left(\frac{d - n}{n}\right) m\, c = \frac{400 - 157{\cdot}5}{157{\cdot}5} \times 16 \times 4 = 98.5 \text{ N/mm}^2$$

Now, total tension = total compression

$$\therefore \quad t\,.\,A_{st} = \frac{1}{2}\, cn\,.\,b$$

or $$A_{st} = \frac{1}{2}\,\frac{c\,n\,b}{t} = \frac{1}{2}\,\frac{4 \times 157{\cdot}5 \times 200}{98{\cdot}5} = \mathbf{639.6\ mm^2}$$

Check. $M = A_{st}\,.\,t\,j\,d = A_{st}\,.\,t\left(d - \frac{n}{3}\right) = 639{\cdot}6 \times 98{\cdot}5\left[400 - \frac{157.5}{3}\right] = 21.89 \times 10^6$ N-mm

Note. In the above example of an over-reinforced section, the stress in steel is only 98.5 N/mm² which is very much less than the permissible value of 130 N/mm². Thus the strength of steel is not fully utilized. Such a design is undesirable and uneconomical. In circumstances where the dimensions of the beam are limited, and where the section has to develop greater moment of resistance than that of the balanced section, it is always desirable to design the section as *doubly reinforced* (see § A-2). In doubly reinforced section, steel reinforcement is placed in the compression zone also, which increases its moment of resistance.

Hence a doubly reinforced section is always preferred over an over-reinforced section.

A-2. DOUBLY-REINFORCED BEAM

A-2.1. Location of Neutral Axis

Fig. A-11 shows a doubly reinforced section. The theory of doubly reinforced section is based on the same assumptions as for singly reinforced section. The neutral axis of a doubly reinforced section can be found by finding the centre of gravity of the combined section consisting of concrete in compression only and steel in compression and tension. Let,

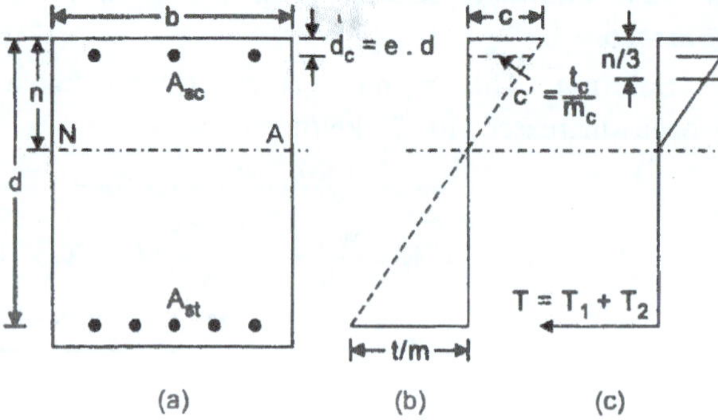

FIG. A-11. DOUBLY REINFORCED SECTION.

b = breadth of the beam.

d = effective depth of the beam

d_c = depth of the centre of the compressive steel $= e.d$

e = compressive steel depth factor $= d_c / d$

c = maximum stress in concrete

t = maximum stress in steel

c' = stress in concrete surrounding compressive steel

t_c = stress in compressive steel $= m_c \,.\, c' = 1.5\, m \,.\, c'$

A_{st} = area of tensile steel

A_{sc} = area of compressive steel

m_c = modular ratio for compression zone $= 1.5\, m$

From Fig. A-11(*b*) we get : $\frac{c}{c'} = \frac{kd}{kd - ed} = \frac{k}{k - e}$ or $c' = \frac{k - e}{k} c$

Hence stress in steel $t_c = m_c\, c' = m_c \frac{k - e}{k} c = 1.5\, m \frac{k - e}{k} \,.\, c$

If maximum stresses in concrete and tensile steel are known, the N.A. can be located from stress diagram [Fig. A-11(*b*)] exactly in the same manner as in singly reinforced section:

$$\frac{c}{t/m} = \frac{n}{d - n} = \frac{kd}{d - kd} = \frac{k}{1 - k} \quad \text{or} \quad k = \frac{mc}{mc + t} = \frac{m\,\sigma_{cbc}}{m\,\sigma_{cbc} + \sigma_{st}}$$

where σ_{cbc} and σ_{st} are the permissible stresses in concrete and steel respectively. Thus the presence of steel in compression zone does not affect the depth of N.A.

Again neglecting the concrete in the tensile zone and equating the moment of the compressive area about N.A. to the moment of the tensile area about N.A. we get

$$\frac{bn^2}{2} + m_c \,.\, A_{sc}\,(n - d_c) - A_{sc}\,(n - d_c) = m\, A_{st}\,(d - n)$$

or $$\frac{bn^2}{2} + (m_c - 1) A_{sc} (n - d_c) = m A_{st} (d - n) \quad \text{...[A-13(a)}$$

or $$\frac{bn^2}{2} + (1.5m - 1) A_{sc} (n - d_c) = m A_{sc} (d - n) \quad \text{...(A-13)}$$

where $(m_c - 1) A_{sc}$ or $(1.5 m - 1) A_{sc}$ is the concrete area equivalent to compressive steel. In the above equation, all the terms except n is known. Hence the N.A. can be located. The above equation can also be expressed in terms of the neutral axis depth factor k and compresssive steel depth factor e as follows :

$$\frac{b}{2} k^2 d^2 + (m_c - 1) A_{sc} (k - e) d = m A_{st} (1 - k) d$$

or $$\frac{b k^2 d}{2} + (m_c - 1) A_{sc} (k - e) = m A_{st} (1 - k) \quad \text{[A-14]}$$

A-2.2. Moment of Resistance

The moment of resistance of a doubly reinforced section can be determined by taking the moment of the compressive force about the centre of the tensile reinforcement. Let,

M = total bending moment to be resisted by the section
= moment of resistance of the section $= M_1 + M_2$

M_1 = moment that can be developed by the balanced section, without compressive reinforcement
= moment of the compressive force in concrete about the centre of tensile reinforcement.

M_2 = moment, in excess of M_1, that is to be provided by the compression steel
= moment of the compressive force in steel, about the centre of the tensile reinforcement.

A_{st} = total tensile reinforcement $= A_{st1} + A_{st2}$.

A_{st1} = tensile steel for the balanced section, corresponding to the moment M_1.

A_{st2} = additional tensile steel necessary to develop the remaining moment M_2

Refer Fig. A-11. Taking the moments of the compressive forces about the centre of the tensile steel, we get

$$M = b\, n \frac{c}{2}\left(d - \frac{n}{3}\right) + (m_c - 1) A_{sc} \,.\, c' (d - d_c) \quad \text{...(A-15)}$$

where c may be equal to or less than σ_{cbc}. See note 1. Expressed in terms of k and e,

we get $$M = b\, kd \frac{c}{2}\left(d - \frac{kd}{3}\right) + (m_c - 1) A_{sc} \,.\, c' (d - ed)$$

or $$M = \frac{1}{2} jck \,.\, bd^2 + (m_c - 1) A_{sc} \,.\, c' d (1 - e)$$

or $$M = R_c\, b\, d^2 + c' d (m_c - 1) A_{sc} (1 - e) \quad \text{...[A-15(a)]}$$

$$\downarrow \qquad\qquad \downarrow$$

or $$M = M_1 + M_2 \quad \text{...[A-15(b)]}$$

Now $$A_{st1} = \frac{M_1}{t \times \text{lever arm}} = \frac{M_1}{t\, j\, d} = \frac{M_1}{\sigma_{st} \,.\, jd} \quad \text{...(A-16)}$$

and $$A_{st2} = \frac{M_2}{t \times \text{lever arm}} = \frac{M_2}{t\,(d - d_c)} \quad \text{...(A-17)}$$

(where t may be equal to or less than σ_{st}. See Note 3 below)

The additional tensile area A_{st2} is actually required to balance the compressive steel. Hence A_{st2} can also alternatively be determined by equating the moment of the equivalent concrete area of the compressive steel to the moment of equivalent concrete area of the tensile steel A_{st2} about N.A.. Thus,

$$(m_c - 1)\, A_{sc}(n - d_c) = m\, A_{st2}\,(d - n)$$

or
$$A_{st2} = A_{sc}\,\frac{m_c - 1}{m}\left(\frac{n - d_c}{d - n}\right) = A_{sc}\,\frac{m_c - 1}{m}\left(\frac{k - e}{1 - k}\right) \qquad \text{...(A-18)}$$

The total area A_{st} of tensile steel will be equal to $A_{st1} + A_{st2}$.

Note 1. The value of c in Eq A-15 will be equal to the permissible value of stress σ_{cbc} in concrete if the total area of steel A_{st} is so provided that it balances both the compressive force in concrete as well as compressive force in steel. However, there are *two cases* in which the value of c in Eq. A-15 may be less than the working or permissible stress. *Case (i)* : In this case the tensile steel may be insufficient even to balance the compressive concrete, with the result that neutral axis will be above the critical one and the stress in steel will reach its permissible value first. *Case (ii)* : In this case the compressive steel A_{sc} may be more than the amount required to balance the additional tensile reinforcement A_{st2} (Eq A-18) with the result that the N.A. will again be above the critical N.A. and the stress in tensile steel will reach its maximum permissible value first. In *both these cases*, the value of c has to be first determined from the known value of t from the following :

$$c = \frac{t}{m}\,\frac{n}{d - n} = \frac{\sigma_{st}}{m}\cdot\frac{n}{d - n}$$

Except in these two cases, the value of c to be substituted in Eq. A-15 will be equal to its permissible value σ_{cbc}.

Note 2. The stress in concrete (c') surrounding the compression steel can also be found from the following relation easily derived from the stress diagram [Fig. A-11(*b*)] :

$$\frac{c'}{c} = \frac{n - d_c}{n} \quad \text{or} \quad c' = c\,\frac{n - d_c}{n} = c\,\frac{k - e}{k} \qquad \text{...(A-19)}$$

In this equation, the value of c should be determined as explained in note 1.

Note 3. If the additional tensile reinforcement A_{st2} required to balance the compressive reinforcement cannot be provided in the same tier or layer as that of A_{st1}, the value of stress t, to be used in Eq. A-17, will be slightly less than the permissible value σ_{st} and will have to be first determined from the stress diagram.

A-2.3. Steel Beam Theory

We have seen that the total moment of resistance of a doubly reinforced beam consists of two terms : the moment of resistance M_1 of the compression concrete and the corresponding tensile steel (A_{st1}) to balance it, and the moment of resistance M_2 of the compression steel (A_{sc}) and the remaining tensile steel (A_{st2}) to balance it. In the steel beam theory, the concrete is completely neglected and the moment of resistance is taken equal to the moment of steel couple, taking the permissible value of stresses in compressive steel equal to permissible value in tensile steel, σ_{st}.

$$M = A_{st} \cdot t\,(d - d_c) = A_{st} \cdot \sigma_{st}\,(d - d_c) \qquad \text{...(A-20)}$$

Thus, in the steel beam theory, the concrete serves only as a web and the compressive and tensile steels as the flanges of an imaginary steel joist having equal area of both the flanges. However, there seems to be no justification in neglecting the compressive area of concrete and permitting the stress in compressive steel equal to stress in tensile steel, as the stress in concrete would pass the limit of the safe working stress. However, the method can be used to find the approximate value of the moment of resistance of the section specially when the area of compression steel is equal to or more than the area of the tensile steel and when these areas are within 2 to 3 percent of the cross-sectional area.

Example A-8. *A beam section, 300 mm wide and 560 mm deep is reinforced with 4 bars of 25 mm diameter in the tensile zone and 4 bars of 12 mm diameter in the compression zone. The cover to the centre of both the reinforcement is 40 mm. Determine the moment of resistance of the section, if M 20 concrete and HYSD bars are used.*

Solution : $\sigma_{cbc} = 7\ \text{N/mm}^2$: $\sigma_{st} = 230\ \text{N/mm}^2$; $m = 13.33$; $m_c = 13.33 \times 1.5 \simeq 20$

$$A_{sc} = 4 \times \frac{\pi}{4}(12)^2 = 452.4\ \text{mm}^2$$

and $$A_{st} = 4 \times \frac{\pi}{4}(25)^2 = 1963.5\ \text{mm}^2$$

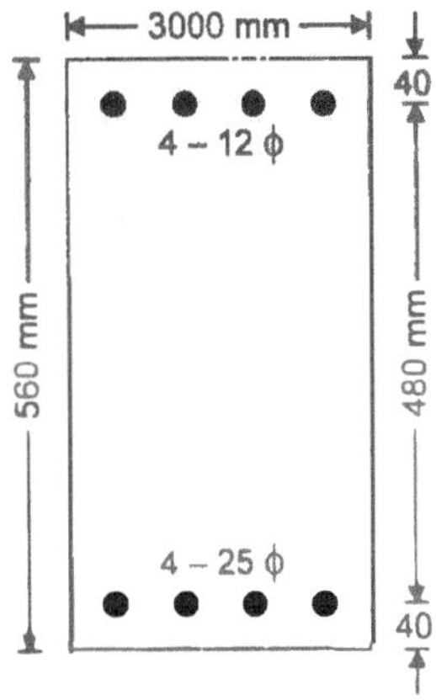

FIG. A-12

From Eq. A-13

$$\frac{300}{2} n^2 + (20 - 1) \times 452.4\,(n - 40) = 13.33 \times 1963.5\,(520 - n)$$

or $$n^2 + 231.8\,n - 93027 = 0$$

From which $n = 210.4$ mm. The critical N.A. is given by

$$n_c = \frac{m\sigma_{cbc}}{m\sigma_{cbc} + \sigma_{st}}\, d = \frac{13.33 \times 7}{13.33 \times 7 + 230} \times 520 = 150.1\ \text{mm}$$

Since the actual N.A. is below the critical N.A., the streses in concrete will reach its maximum permissible value first.

$$\therefore \quad c = \sigma_{cbc} = 7\ \text{N/mm}^2$$

$$c' = c\,\frac{n - d_c}{n} = 7 \times \frac{210.4 - 40}{210.4} = 5.669\ \text{N/mm}^2$$

$$M_r = 300 \times 210.4 \times \frac{7}{2}\left(520 - \frac{210.4}{3}\right) + (20 - 1) \times 452.4 \times 5.669\,(520 - 40)$$

$$= 122.77 \times 10^6\ \text{N-mm}$$

$$= \mathbf{122.77\ kN\text{-}m}$$

Note $t_c = m_c\,c' = 20 \times 5.669 = 113.34\ \text{N/mm}^2$ which is less than the permissible value σ_{sc} ($= 190\ \text{N/mm}^2$).

Example A-9. *A doubly reinforced rectangular beam of breadth b and effective depth d has tensile reinforcement equal to 1.2 percent and compressive reinforcement also 1.2 percent. The compressive reinforcement is at 0.1d below the top. Determine the moment* *of resistance of the section if the permissible streseses in concrete and steel are 5 N/mm² and 140 N/mm² respectively, and m = 19. What will be the moment of resistance, by the steel beam theory ?*

Solution. Given : $\sigma_{cbc} = 5\ \text{N/mm}^2$; $\sigma_{st} = 140\ \text{N/mm}^2$; $m = 19$

$$\frac{bn^2}{2} + (1.5 \times 19 - 1)\frac{1.2}{100} bd\,(n - 0.1\,d) = 19 \times \frac{1.2}{100} bd\,(d - n)$$

or $$n^2 + 1.116\,n\,d - 0.522\,d^2 = 0 \text{ which gives } n = 0.355\,d.$$

$$n_c = \frac{m\sigma_{cbc}}{m\sigma_{cbc} + \sigma_{st}}\,d = \frac{19 \times 5}{19 \times 5 + 140}\,d = 0.404\,d$$

The actual N.A. is above the critical one ; hence the stress in tensile steel reaches its maximum value of 140 N/mm² before the stress in concrete can reach the value of 5 N/mm². Corresponding to $t = 140\ \text{N/mm}^2$, c and c' can be found as under :

$$c = \frac{t}{m} \cdot \frac{n}{d-n} = \frac{140}{19} \times \frac{0.355\,d}{0.645\,d} = 4.06\ \text{N/mm}^2$$

$$c' = \frac{t}{m} \cdot \frac{n - d_c}{d-n} = \frac{140}{19}\left(\frac{0.355 - 0.1}{0.645}\right) = 2.91\ \text{N/mm}^2.$$

$$M_r = b \times 0.355\,d \times \frac{4.06}{2}\left(d - \frac{0.355\,d}{3}\right) + (1.5 \times 19 - 1) \times \frac{1.2\,bd}{100} \times 2.91\,(d - 0.1\,d)$$

$= 0.653\,bd^2 + 0.864\,bd^2 = \mathbf{1.517\,bd^2}$ **N-mm** (where b and d are in mm)

Using the *steel beam theory*

Eq. A-15 $$M_r = A_{st} \cdot \sigma_{st}\,(d - d_c) = \frac{1.2\,bd}{100} \times 140\,(d - 0.1\,d) = \mathbf{1.512\ bd^2}\ \textbf{N-mm}$$

Thus it is seen that the moment of the resistance found by the approximate theory is nearly the same as that calculated by the exact theory, the difference being 0.3%.

Example A-10. *A doubly reinforced concrete beam is 400 mm wide and 600 mm deep to the centre of tensile reinforcement. The compression reinforcement consists of 4 bars of 16 mm diameter, and is placed with its centre at a depth of 40 mm from the top The tensile reinforcement consists of 4 bars of 20 mm diameter. The section is subjected to a bending moment of 100 kN-m. Detemine the stresses in concrete and steel. Take m =16.*

Solution. $$A_{sc} = 4 \times \frac{\pi}{4}(16)^2 = 805\ \text{mm}^2$$

$$A_{st} = 4 \times \frac{\pi}{4}(20)^2 = 1250\ \text{mm}^2$$

Taking moment of equivalent areas, about N.A.

$$\frac{bn^2}{2} + (m_c - 1)A_{sc}\,(n - a_c) = m\,A_{st}\,(d - n)$$

or $$\frac{400}{2}n^2 + (1.5 \times 16 - 1)\,805(n - 40) = 16 \times 1250\,(600 - n)$$

or $$n^2 + 192.6\,n - 63700 = 0$$

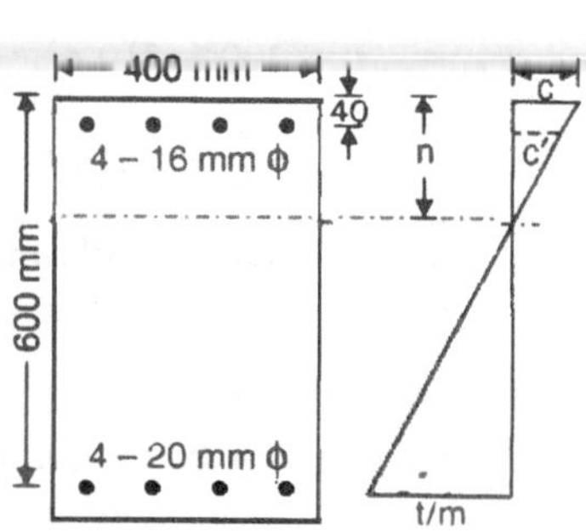

FIG. A-13.

From which $n = 173.8$ mm.

$$c' = \frac{n - d_c}{n} c = \frac{173.8 - 40}{173.8} c = 0.77\ c$$

Now, $M_r = b\,n \cdot \frac{c}{2}\left(d - \frac{n}{3}\right) + (m_c - 1) A_{sc} \cdot c'(d - d_c)$

$$100 \times 10^6 = 400 \times 173.8\ \frac{c}{2}\left(600 - \frac{173.8}{3}\right) + (1.5 \times 16 - 1)\ 805\ \times 0.77c\ (600 - 40)$$

From which $c =$ **3.73 N/mm²** ; $c' = 0.77\ c = 0.77 \times 3.73 = 2.87$ N/mm²

The stress in compressive steel $= 1.5\ mc' = 1.5 \times 16 \times 2.87 =$ **68.9 N/mm²**

Stress in tensile steel is $t = \frac{cm\ (d - n)}{n} = \frac{3.73 \times 16\ (600 - 173.8)}{173.8} =$ **146.3 N/mm²**

A 3. T AND L-BEAMS

A-3.1. Position of Neutral Axis

The neutral axis in a T-beam section may fall in two alternative locations : (*i*) within the flange [Fig. A-14(*a*)], or (*ii*) outside the flange [Fig. A-14 (*b*)]. Case (*i*) may happen when the slab is comparatively thicker. In that case, the N.A. may be located exactly in the same manner that is adopted in singly reinforced beam :

$$\frac{b_f n^2}{2} = m\ A_{st}\ (d - n) \quad \text{...(A-21)}$$

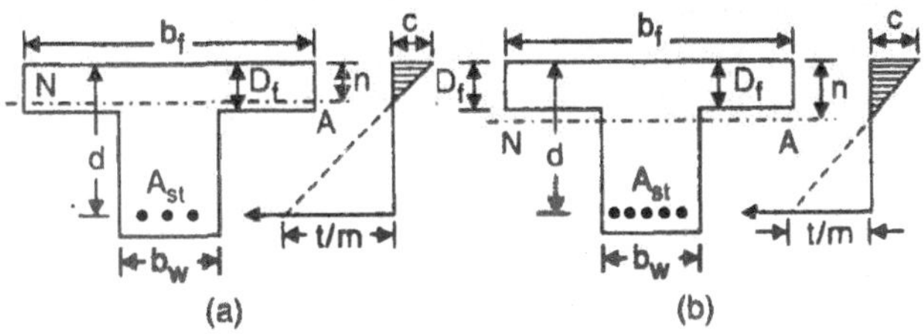

FIG. A-14. LOCATION OF NEUTRAL AXIS.

If, however, the N.A. falls exactly at the junction of flange and rib, the above equation reduces to :

$$\frac{b_f D_f^2}{2} = m\ A_{st}\ (d - D_f) \quad \text{...(A-22)}$$

In the second position, which is the most common one, the N.A. falls outside the flange. In that case, N.A. can be located by equating the moments of compressive and tensile areas about the N.A. :

$$b_f \cdot D_f\left(n - \frac{D_f}{2}\right) + \frac{b_w(n - D_f)^2}{2} = m\ A_{st}\ (d - n) \quad \text{...(A-23)}$$

In the above equation, the term $\frac{b_w\ (n - D_f)^2}{2}$ is usually very small and it is customary to reject this term. In that case, the above expression reduces to :

$$b_f D_f\left(n - \frac{D_f}{2}\right) = mA_{st}\ (d - n) \quad \text{...(A-24)}$$

Thus, if the dimensions of the T-beam section are completely known, the N.A. can be located either by Eq. A-21 or by Eq. A-24. If by application of Eq. A-21, n comes out to be more than D_f , calculations should be repeated by using Eq. A-24. Similarly, if Eq. A-24 is used first and n comes out to be less than D_f calculations should be repeated by using Eq. A-21.

If, however, the maximum stresses in concrete ($c = \sigma_{cbc}$) and steel ($t = \sigma_{st}$) are *known*, the N.A. is located by using the stress diagram :

$$\frac{n}{d-n} = \frac{c}{t/m} \quad \text{...(A-25)}$$

A-3.2. Lever Arm and Moment of Resistance

(*a*) Lever Arm : Fig. A-15 (*b*) shows the stress diagram for a T-beam section in which the N.A. falls outside the flange, which is more general case.

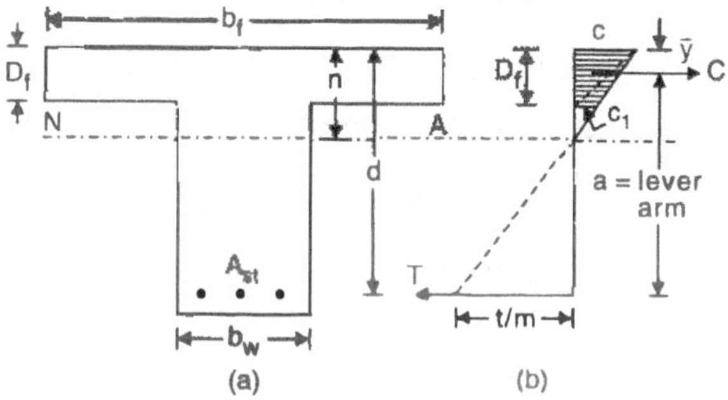

FIG. A-15. LEVER ARM.

Let c be the stress in the concrete at the top (extreme fibre) and c_1 be the stress in the concrete at the junction of flange with the rib. Since the compression area lying between the junction and the N.A. is very small, the compressive force of this area is *neglected*. Therefore, the shaded area in Fig. A-15(*b*) shows the net compressive stress diagram, the centroid of which is at a depth $\bar{y}$ (say) below the top. The value of $\bar{y}$ can be found by dividing the trapezoidal stress diagram into two triangles by a diagonal (shown dotted) and taking moments at the top :

$$\bar{y} = \frac{\left[c \cdot \frac{D_f}{2} \cdot \frac{D_f}{3}\right] + \left[c_1 \cdot \frac{D_f}{2} \cdot \frac{2}{3} D_f\right]}{\left[c \cdot \frac{D_f}{2}\right] + \left[c_1 \frac{D_f}{2}\right]} = \left[\frac{c + 2c_1}{c + c_1}\right] \frac{D_f}{3} \quad \text{...(A-26)}$$

But from stress diagram, $\frac{c_1}{n - D_f} = \frac{c}{n}$ or $c_1 = \frac{n - D_f}{n} c$ ...(A-27)

Substituting this value of c_1 in Eq. A-26, we get

$$\bar{y} = \frac{1 + 2\left\{\frac{n - D_f}{n}\right\}}{1 + \left\{\frac{n - D_f}{n}\right\}} \times \frac{D_f}{3} = \frac{3n - 2D_f}{2n - D_f} \times \frac{D_f}{3} \quad \text{...[A-27(a)]}$$

$$\text{Lever arm } a = d - \bar{y} = d - \frac{D_f}{3}\left\{\frac{1 + 2\left\{\frac{n - D_f}{n}\right\}}{1 + \frac{n - D_f}{n}}\right\} = d - \frac{D_f}{3}\left\{\frac{3n - 2D_f}{2n - D_f}\right\} = d - \frac{D_f}{3}\left\{\frac{3}{2} - \frac{D_f}{2(2n - D_f)}\right\}$$

or

$$a = d - \frac{D_f}{2} + \frac{D_f^2}{6(2n - D_f)} \quad \text{...(A-28)}$$

If the actual N.A. is above the critical N.A., the section is under-reinforced, the stress (t) in steel reaches its maximum permissible value σ_{st} first and the stress c in concrete should be found from Eq. A-25 first, before substituting Eq. A-26. If, however, the actual

N.A. falls below the critical N.A., the section is over-reinforced and the stress (c) in concrete reaches its maximum permissible value (σ_{cbc}) first.

(*b*) **Moment of Resistance** : The moment of resistance of the T-beam is found by multiplying the total compression and the lever arm.

If the N.A. lies in the flange , $M_r = b_f \cdot n \cdot \frac{c}{2}\left(d - \frac{n}{3}\right)$ ...[A-29(*a*)]

If the N.A. lies in the web, the moment of resistance is given by

$$M_r = b_f \cdot D_f \cdot \frac{c + c_1}{2} \times a \text{ (Neglecting the compression of the rib)} \quad ...[A\text{-}29(b)]$$

Hence $$M_r = \frac{b_f \cdot D_f \cdot c}{2}\left[1 + \frac{n - D_f}{n}\right]\left[d - \frac{D_f}{2} + \frac{D_f^2}{6(2n - D_f)}\right] \quad ...(A\text{-}29)$$

Also, $$M_r = t \cdot A_{st} \cdot a = \sigma_{st} \cdot A_{st} \cdot a. \quad ...(A\text{-}30)$$

A-3.3. Moment of Resistance Taking Compression of Rib into Account

When the compression of the rib is taken into account, the neutral axis can be located by :

$$b_f \cdot D_f \cdot \left(n - \frac{D_f}{2}\right) + \frac{b_w}{2}(n - D_f)^2 = m\, A_{st}\,(d - n) \quad ...(A\text{-}31)$$

The moment of resistance of the beam can be found by two methods :

Method 1. In this method, the distance $\bar{y}$ of the C.G. of compressive force is found first and then lever arm is determined. The moment of resistance is then found by multiplying the total compressive force by the lever arm.

Thus, taking moments at the top of the beam :

$$\bar{y} = \frac{b_f \cdot D_f \cdot \frac{c + c_1}{2} \times \frac{D_f}{3}\left[\frac{2c_1 + c}{c_1 + c}\right] + b_w (n - D_f) \times \frac{c_1}{2}\left[D_f + \frac{n - D_f}{3}\right]}{b_f \cdot D_f \cdot \frac{c + c_1}{2} + b_w (n - D_f)\frac{c_1}{2}} \quad ...(A\text{-}32)$$

where $$c_1 = \frac{n - D_f}{n} \cdot c$$

$\therefore$ Lever arm $a = d - \bar{y}$. Hence,

$$M_r = \left\{b_f \cdot D_f \cdot \frac{c + c_1}{2} + b_w (n - D_f)\frac{c_1}{2}\right\} a \quad ...(A\text{-}33)$$

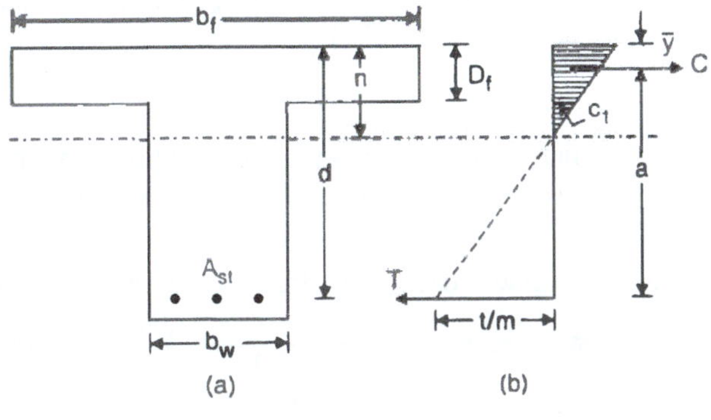

FIG. A-16.

Method 2. The moment of resistance of the beam is the sum of the moment of resistance due to force in flange and the moment of resistance due to force in rib. Thus

$$M_r = M_{r1} + M_{r2}$$

where M_{r1} = moment of resistance of the force in flange

or $$M_{r1} = \frac{b_f \cdot D_f \cdot c}{2}\left[1 + \frac{n - D_f}{n}\right]\left[d - \frac{D_f}{2} + \frac{D_f^2}{6(2n - D_f)}\right] \quad ...(A\text{-}34)$$

and M_{r2} = moment of resistance of the force in rib.

$$= b_w (n - D_f) \times \frac{c_1}{2}\left[d - n + \frac{2}{3}(n - D_f)\right]$$

$$= \frac{b_w}{2}(n - D_f)\left\{\frac{n - D_f}{n} c\right\}\left[d - \frac{n}{3} - \frac{2}{3} D_f\right]$$

$$= \frac{c \, . \, b_w}{2n}(n - D_f)^2\left[d - \frac{n + 2D_f}{3}\right] \qquad \text{...(A-35)}$$

A-3.4. Depth of Balanced Section of T-beam

Let us find the depth of balanced section of T-beam, corresponding to external moment M. The critical neutral axis depth factor k is given by :

$$k = \frac{mc}{mc + t} = \frac{m\sigma_{cbc}}{m\sigma_{cbc} + \sigma_{st}} \qquad \text{...}(i)$$

The depth of balanced section of T-beam is determined on the assumption that the compression of the rib is negligible and can be ignored. Hence the compressive force (C) of beam is equal to the compressive force in flange, given by

$$C = b_f \, . \, D_f \, . \, \frac{1}{2}(c + c_1)$$

or
$$C = b_f \, . \, D_f \, . \, \frac{1}{2}\left[c + \frac{n - D_f}{n} c\right] = \frac{b_f \, . \, c \, . \, D_f}{2}\left[\frac{2n - D_f}{n}\right] = \frac{b_f \, . \, c \, . \, D_f}{2}\left[\frac{2kd - D_f}{kd}\right] \qquad \text{...}(ii)$$

$$\text{Lever arm} = a = d - \frac{D_f}{2} + \frac{D_f^2}{6\,(2n - D_f)} \qquad \text{...(Eq. A-28)} \qquad \text{...}(iii)$$

$$M = M_r = \text{compressive force } C \times \text{lever arm } a$$

or
$$M = \frac{b_f \, . \, c \, . \, D_f}{2}\left[\frac{2kd - D_f}{kd}\right]\left[d - \frac{D_f}{2} + \frac{D_f^2}{6(2\,kd - D_f)}\right]$$

or
$$M = \frac{b_f \, . \, c \, . \, D_f}{12\,kd}\,[\,12\,k\,d^2 - 6\,d\,.\,D_f - 6\,k\,d\,D_f + 3\,D_f^2 + D_f^2\,]$$

or
$$M = \frac{b_f \, . \, c \, . \, D_f}{12\,kd}\,[12\,k\,d^2 - 6\,d\,D_f\,(1 + k) + 4\,D_f^2]$$

or
$$M = b_f \, . \, c \, . \, D_f\left[d - \frac{D_f\,(1 + k)}{2k} + \frac{D_f^2}{3\,k\,d}\right] \qquad \text{...(A-36)}$$

In the above expression, all quantities except d are known. Hence d can be calculated from the solution of the above quadratic equation.

Alternatively, d can be computed by *assuming* the lever arm a equal to 0.9 d, though its actual value may vary from 0.88 d to 0.95 d. Then :

$$M = M_r = C \times a = \frac{b_f \, . \, c \, . \, D_f}{2}\left[\frac{2kd - D_f}{k \, . \, d}\right] \times 0.9\,d \qquad \text{...[A-37}(a)]$$

or
$$M = 0.45\,b_f \, . \, c \, . \, D_f\left[\frac{2kd - D_f}{k}\right] \qquad \text{...(A-37)}$$

This is a linear equation in terms of the unknown d.

A-3.5. Economical Depth of T-beam

The balanced depth found above is not economical. This is because of the fact that in T-beam large compression area is provided by the flange and consequently balanced depth is comparatively reduced and area of steel is increased. The cost of steel is several times more than the cost of concrete and thus overall cost of T-beam, having balanced depth, is increased. To find an expression for the economical depth of beam, we will have to take into account the *cost-ratio* of steel and concrete. Let,

R_s = cost of steel, per cubic mm = $r \cdot R_c$

R_c = cost of concrete, per cubic mm

r = cost ratio $= \dfrac{\text{cost of steel}}{\text{cost of concrete}}$ per unit volume $= \dfrac{R_s}{R_c}$

d_c = depth of cover, measured below the centre of the tensile steel.

The volume of concrete provided by the flange is *fixed,* and is, therefore, not included in the cost calculations.

Volume of concrete, per mm length, in the web $= b_w(d - D_f + d_c)$

$\therefore$ Cost of concrete in the rib, per mm length $= b_w(d - D_f + d_c)R_c$

Similarly, cost of steel, per mm length of beam $= A_{st} R_s = r \cdot R_c \cdot A_{st}$

But $A_{st} = \dfrac{M}{tjd}$, where $t = \sigma_{st}$ = permissible stress in steel

$\therefore$ Total cost (Q) of the beam, per mm length is given by

$$Q = R_c \cdot b_w(d - D_f + d_c) + r \cdot R_c \cdot \frac{M}{tjd} = R_c \cdot \left[b_w(d - D_f + d_c) + \frac{r \cdot M}{tjd} \right]$$

In the above expression, b_w, D_f and d_c are fixed dimensions while depth d is variable. For the cost to be ***minimum***,

$$\frac{dQ}{dd} = 0 = R_c \left[b_w - \frac{r\,M}{tjd^2} \right]$$

$\therefore$ $$d^2 = \frac{r \cdot M}{tjb_w}, \text{ where } t = \sigma_{st}$$

or $$d = \sqrt{\frac{r \cdot M}{tj\,b_w}} = \sqrt{\frac{r \cdot M}{\sigma_{st}\,j\,b_w}} \qquad \text{...(A-38)}$$

Taking $t = \sigma_{st} = 140$ N/mm^2 and $j = 0.9$, the above expression reduces to

or $$d = \sqrt{\frac{r \cdot M}{126\,b_w}} \qquad \text{...[A-38(a)]}$$

Similarly, taking $\sigma_{st} = 230$ N/mm^2 and $j = 0.9$, we have

$$d = \sqrt{\frac{r \cdot M}{207\,b_w}} \qquad \text{...[A-38(b)]}$$

In the above expressions, the cost of shuttering has not been taken into account. For deeper beams, the cost of shuttering may be appreciable. In that case the depth of beam found from Eq. A-38 may be further reduced by 10 to 15% to account for the additional cost of shuttering.

The width of web (b_w) is kept a certain minimum, so as to accommodate tensile steel. Sometimes, steel may be provided in two or three layers, but this reduces the lever arm. Suitable value of b_w is also selected on the consideration of shear.

Example A-11. *An isolated T-beam, simply supported over a span of 6 m has following dimensions : Width of flange 750 mm, thickness of flange 125 mm ; over all depth 400 mm ; width of web 260 mm ; effective cover to tensile reinforcement, 40 mm. The beam is reinforced with 4 bars of 20 mm dia. Determine the moment of resistance of the beam if (a) mild steel bars are used (b) Fe 415 steel bars are used. Take* $\sigma_{cbc} = 5$ *N/mm²* *and* $m=19$ *in each case.*

Solution. The section of the T-beam, is shown in Fig. A-17

Assume that N.A. lies in flange.

Area of steel $= A_{st} = 4 \times \frac{\pi}{4}(20)^2 = 1250.6 \text{ mm}^2$

For an isolated T-beam, effective flange width

$$b_f = \frac{l_0}{\frac{l_0}{b}+4} + b_w = \frac{6000}{\frac{6000}{750}+4} + 260 = 760 \text{ mm}$$

Available $b = 750$ mm. $\therefore\ b_f = b = 750$ mm

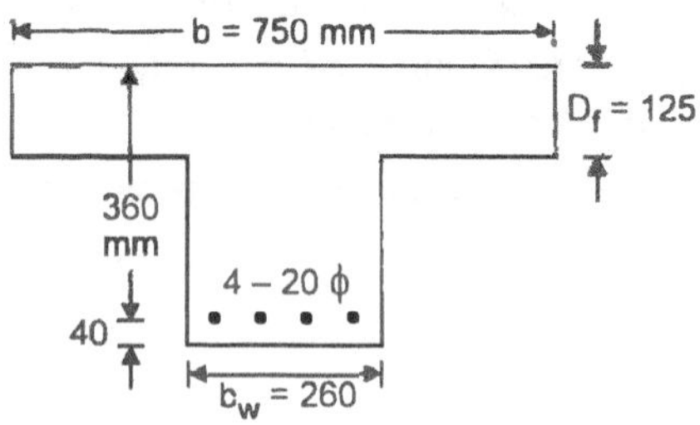

FIG. A-17

Equating the moments of equivalent areas about the N.A., we get

$$750 \times \frac{n^2}{2} = 19 \times 1250.6\,(360 - n)$$

or $$n^2 = 22920 - 63.67\,n. \text{ From which, } n = 122.9 \text{ mm}$$

This is less than D_f (=125 mm). Hence the N.A. just lies inside the flange.

(a) Mild steel reinforcement.

$$n_c = \frac{d}{1+\frac{\sigma_{st}}{m\sigma_{cbc}}} = \frac{360}{1+\frac{140}{19\times 5}} = 145.5 \text{ mm}$$

Since the actual N.A. falls *above* the critical N.A., the section is *under-reinforced* and the stress in steel reaches its maximum value first. Thus, $t = \sigma_{st} = 140$ N/mm². The corresponding stress in concrete is given by

$$c = \frac{t}{m}\frac{n}{d-n} = \frac{140}{19} \times \frac{122.9}{360 - 122.9} = 3.82 \text{ N/mm}^2$$

The moment of resistance of the beam is given by

$$M_r = b_f \cdot n \cdot \frac{c}{2}\left(d - \frac{n}{3}\right) = 750 \times 122.9 \times \frac{3.82}{2}\left(360 - \frac{122.9}{3}\right)$$

$$= 56.17 \times 10^6 \text{ N-mm} = \mathbf{56.17\ kN\text{-}m.}$$

(*b*) Fe 415 Steel.

$$n_c = \frac{d}{1+\frac{\sigma_{st}}{m\sigma_{cbc}}} = \frac{360}{1+\frac{230}{19\times 5}} = 105.2 \text{ mm}$$

Since the actual N.A. falls *below* the critical N.A., the section is *over-reinforced* and the stress in concrete reaches its maximum value first. Thus $c = \sigma_{cbc} = 5$ N/mm². The corresponding stress in steel is given by

$$t = \frac{m \cdot c\,(d-n)}{n} = \frac{19 \times 5\,(360 - 122.9)}{122.9} = 183.3 \quad \text{N/mm}^2$$

The moment of resistance of the beams is given by

$$M_r = b_f \cdot n \cdot \frac{c}{2}\left(d - \frac{n}{3}\right) = 750 \times 122.9 \times \frac{5}{2}\left(360 - \frac{122.9}{3}\right)$$

$$= 73.52 \times 10^6 \text{ N-mm} = \mathbf{73.52 \ kN\text{-}m}\ .$$

Example A-12. *An isolated T-beam has a flange width of 1000 mm, flange thickness of 80 mm and effective depth of 400 mm. The rib is 240 mm wide and reinforced with 5 bars of 20 mm diameter. Determine the moment of resistance of the section, if the permissible stresses in concrete and steel are 5.2 N/mm² and 140 N/mm² and m = 18. Neglect the compressive force in the web. The beam is simply supported over a span of 4 m.*

Solution. The effective width

$$b_f = \frac{l_0}{\left(\frac{l_0}{b}\right) + 4} + b_w$$

$$= \frac{4000}{\left(\frac{4000}{1000}\right) + 4} + 240 = 740 \text{ mm}$$

Thus the effective width of flange is very much less than the actual flange width.

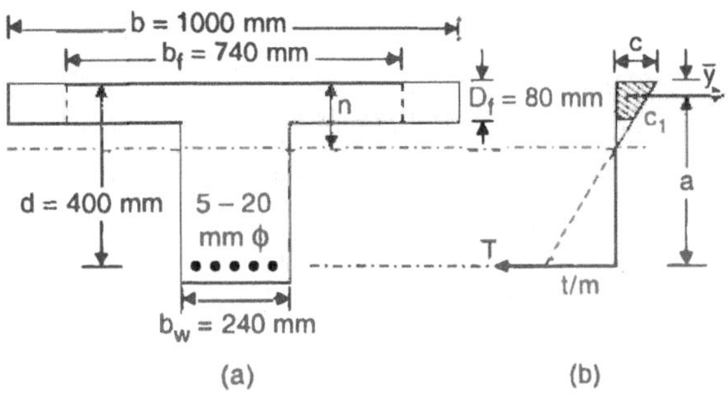

FIG. A-18.

Area of steel $A_{st} = 5\,\frac{\pi}{4}\,(20)^2 = 1570 \text{ mm}^2$

Equating the moments of equivalent areas about N.A. and assuming that N.A. falls in the flange, we get

$$b_f \cdot \frac{n^2}{2} = m \cdot A_{st}\,(d-n) \quad \text{or} \quad 740 \times \frac{n^2}{2} = 18 \times 1570\,(400-n)$$

or $\quad n^2 + 76.4n - 30550 = 0$ From which $n = 141$ mm $> D_f$. Hence N.A. falls in the web.

Again, equating the moments of equivalent areas, and neglecting the compression of the web, we get from Eq. A-24 :

or $$740 \times 80\left(n - \frac{80}{2}\right) = 18 \times 1570\,(400-n)$$

or $$1.477\,n = 230.9. \text{ From which } n = 156.4 \text{ mm}$$

Taking $\sigma_{cbc} = 5.2$ N/mm², $\sigma_{st} = 140$ N/mm² and $m = 18$,

$$n_c = \frac{d}{1 + \frac{\sigma_{st}}{m\sigma_{cbc}}} = \frac{400}{1 + \frac{140}{18 \times 5.2}} = 160.3 \text{ mm}$$

Since the actual N.A. falls above the critical N.A., the stress in steel reaches the maximum value first. Hence $t = \sigma_{st} = 140$ N/mm². The corresponding stress in concrete at the outer fibre is given by

$$c = \frac{t}{m}\frac{n}{d-n} = \frac{140}{18} \times \frac{156.4}{400 - 156.4} = 4.99 \quad \text{N/mm}^2$$

$$c_1 = \frac{n - D_f}{n} c = \frac{156.4 - 80}{156.4} \times 4.99 = 2.44 \quad \text{N/mm}^2$$

Hence
$$\bar{y} = \left[\frac{c + 2c_1}{c + c_1}\right]\frac{D_f}{3} = \left[\frac{4.99 + 4.88}{4.99 + 2.44}\right]\frac{80}{3} = 35.4 \text{ mm.}$$

Lever arm
$$a = d - \bar{y} = 400 - 35.4 = 364.6 \quad \text{mm}$$

$$M_r = b_f D_f \frac{c + c_1}{2} \times a = 740 \times 80 \, \frac{4.99 + 2.44}{2} \times 364.6$$

$$= 80.1 \times 10^6 \text{ N-mm} = \mathbf{80.1 \quad kN\text{-}m}$$

Alternatively, $M_r = \sigma_{st} A_{st} \,.\, a = 140 \times 1570 \times 364.6 = 80.1 \times 10^6$ N-mm = 80.1 kN-m

Example A-13. *Solve Example A-12 taking web compression into account.*

Solution. The N.A. will lie in the web at a distance n below the top of the flange. Equating the moments of equivalent areas about the N.A., we have from Eq. A-23.

$$740 \times 80(n - 40) + 120\,(n - 80)^2 = 18 \times 1570\,(400 - n)$$

or $n^2 + 568.8n - 107533 = 0$

From which $n = 149.7$ mm.

As before $n_c = 160.3$ mm.

The actual N.A. falls above the critical one and hence the section is under-reinforced. The stress in steel reaches its maximum value first. Thus, $t = \sigma_{st} = 140$ N/mm². The corresponding value of stress in concrete at the outer fibre is given by

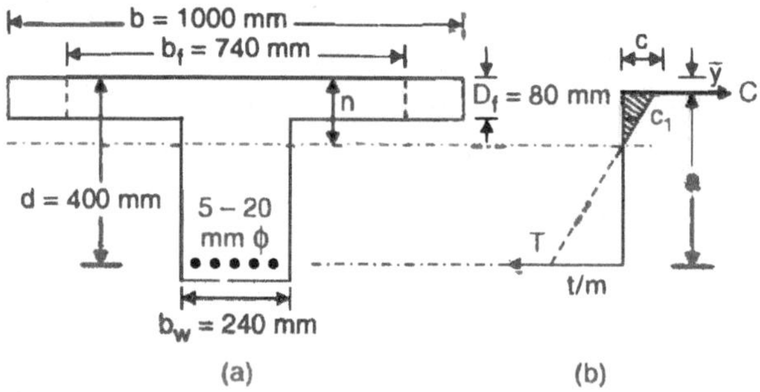

FIG. A-19.

$$c = \frac{t}{m} \cdot \frac{n}{d-n} = \frac{140}{18} \times \frac{149.7}{400 - 149.7} = 4.65 \quad \text{N/mm}^2$$

$$c_1 = \frac{n - D_f}{n} \,.\, c = \frac{149.7 - 80}{149.7} \times 4.65 = 2.16 \quad \text{N/mm}^2$$

$$\bar{y} = \frac{b_f \,.\, D_f \,.\, \dfrac{c + c_1}{2} \times \dfrac{D_f}{3}\left\{\dfrac{2c_1 + c}{c_1 + c}\right\} + b_w\,(n - D_f) \times \dfrac{c_1}{2}\left\{D_f + \dfrac{n - D_f}{3}\right\}}{b_f \,.\, D_f \,.\, \dfrac{c + c_1}{2} + b_w\,(n - D_f)\,\dfrac{c_1}{2}}$$

$$= \left[740 \times 80 \left\{\frac{4.65 + 2.16}{2}\right\} \times \frac{80}{3}\left\{\frac{4.32 + 4.65}{2.16 + 4.65}\right\} + 240\left\{149.7 - 80\right\} \times \frac{2.16}{2}\left\{80 + \frac{149.7 - 80}{3}\right\}\right]$$

$$\div \left[740 \times 80 \left\{\frac{4.65 + 2.16}{2}\right\} + 240\left\{149.7 - 80\right\}\frac{2.16}{2}\right] = 40.8 \quad \text{mm}$$

∴ L.A. $= a = d - \bar{y} = 400 - 40.8 = 359.2$ mm

$$M_r = \left\{ b_f \, . \, D_f \frac{c + c_1}{2} + b_w (n - D_f) \frac{c_1}{2} \right\} a$$

$$= \left\{ 740 \times 80 \left(\frac{4.65 + 2.16}{2} \right) + 240 \; (149.7 - 80) \frac{2.16}{2} \right\} 359.2 = 79 \times 10^6 \text{ N-mm } = \mathbf{79} \text{ kN-m}$$

Alternatively, $M_r = \sigma_{st} \, . \, A_{st} \, . \, a = 140 \times 1570 \times 359.2 = 79 \times 10^6$ N-mm = 79 kN-m.

A-3.6. Doubly Reinforced T-Beams

Sometimes, the dimensions of the web are restricted from architectural or other considerations. If the beam is expected to carry more bending moment than it can take as a balanced section, compression reinforcement is provided to increase its moment of resistance. Such a beam is known as doubly reinforced T-beam. Let :

A_{sc} = area of compressive reinforcement

d_c = cover for the compressive reinforcement

c' = compressive stress in concrete surrounding compressive steel.

The design of such a beam is done in the following steps :

1. Find the moment of resistance M_1 of the T-beam as the balanced section.

$$M_1 = b_f \, . \, D_f \, . \, \frac{c}{2} \left[1 + \frac{kd - D_f}{kd} \right] \left[d - \frac{D_f}{2} + \frac{D_f^2}{6 \, (2kd - D_f)} \right] \quad \text{...(A-39)}$$

2. Calculate the moment of resistance (M_2) to be provided by the compression steel.

$M_2 = M - M_1$, where M = total B.M.

But $M_2 = (m_c - 1) \, A_{sc} \times c' \, (d - d_c)$ where $c' = \frac{kd - d_c}{kd} \, . \, c$ and $m_c = 1.5 \, m$

$$\therefore \quad M_2 = (m_c - 1) \, A_{sc} \times c \times \frac{kd - d_c}{kd} (d - d_c) \quad \text{...(A-40)}$$

Thus, A_{sc} is known from this equation.

3. Find the tensile reinforcement as under:

$$A_{st1} = \frac{M_1}{\sigma_{st} \times a} \quad \text{...(A-41)}$$

where $a = d - \frac{D_f}{2} + \frac{D_f^2}{6 \, (2kd - D_f)}$

$$A_{st2} = \frac{M_2}{\sigma_{st}(d - d_c)} \quad \text{...(A-42)}$$

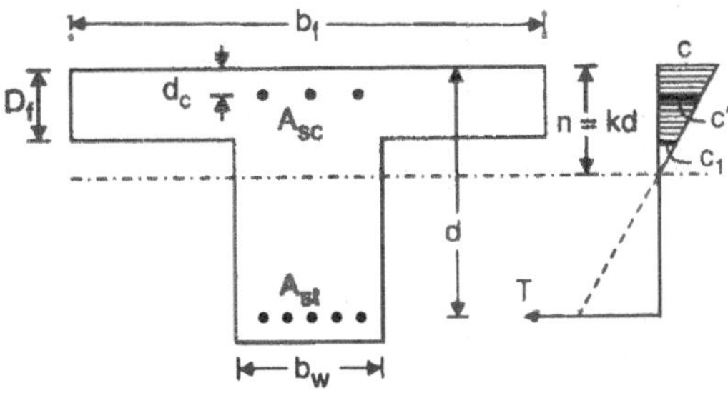

FIG. A-20

Total area of steel $A_{st} = A_{st1} + A_{st2}$ ...(A-43)

For a doubly reinforced T-beam of given dimensions, the neutral axis can be located by equating the moments of equivalent areas about the neutral axis. Thus :

$$b_f \, . \, D_f \left(n - \frac{D_f}{2} \right) + (m_c - 1) \, A_{sc}(n - d_c) = m \, A_{st} \, (d - n) \quad \text{...(A-44)}$$

(Assuming the N.A. to lie in the web)

The moment of resistance of such a beam, neglecting the compressive force in the rib, is given by

$$M_r = b_f \, . \, D_f \left(\frac{c + c_1}{2} \right) \times a + (m_c - 1) \, A_{sc} \times c \times \frac{n - d_c}{n} (d - d_c) \quad \text{...(A-45)}$$

Taking a equal to $d - \frac{D_f}{2} + \frac{D_f^2}{6\,(2n - D_f)}$, the above expression reduces to :

$$M_r = b_f \,.\, D_f \,.\, c\left[\frac{2n - D_f}{2n}\right]\left[d - \frac{D_f}{2} + \frac{D_f^2}{6\,(2n - D_f)}\right] + (m_c - 1)\, A_{sc} \,.\, \frac{c(n - d_c)}{n}\,(d - d_c) \quad \text{...(A-46)}$$

where $c = \sigma_{cbc}$ = permissible compressive stress in concrete.

Example A-14. *An isolated T-beam carries a uniformly distributed load of 40 kN/m run, inclusive of its own weight, over an effective span of 6 m. The beam has the following dimensions : width of flange : 800 mm ; thickness of flange : 100 mm; effective depth of the beam : 480 mm and width of rib 300 mm. Determine the necessary areas of tensile and compressive reinforcement. Take $\sigma_{cbc} = 5$ Nmm^2, $\sigma_{st} = 140$ N/mm^2 and $m = 18$.*

Solution. Given $\sigma_{cbc} = 5$ N/mm^2 ; $t = \sigma_{st} = 140$ N/mm^2.

$$b_f = \frac{l_0}{\left(\frac{l_0}{b}\right) + 4} + b_w = \frac{6000}{\left(\frac{6000}{800}\right) + 4} + 300 = 822 \text{ mm.}$$

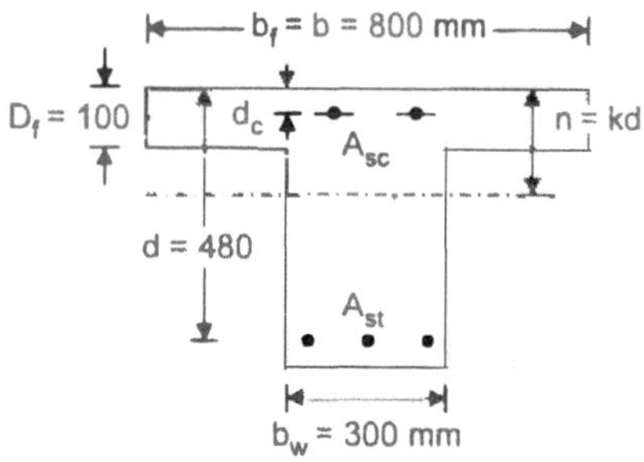

FIG. A-21.

But this cannot be more than $b(=800$ mm).

Hence $b_f = 800$ mm ;

$$M = \frac{40 \times 36}{8} = 180 \text{ kN-m} = 180 \times 10^6 \text{ N-mm.}$$

For balanced section,

$$n = kd = \frac{mc}{mc + t}\, d = \frac{18 \times 5}{18 \times 5 + 140} \times 480 = 187.8 \text{ mm.}$$

The moment of resistance of the balanced section is

$$M_1 = b_f \,.\, D_f \frac{c}{2}\left[\frac{2kd - D_f}{kd}\right]\left[d - \frac{D_f}{2} + \frac{D_f^2}{6(2\,kd - D_f)}\right]$$

$$= 800 \times 100 \times \frac{5}{2}\left[\frac{2 \times 187.8 - 100}{187.8}\right] \times \left[480 - 50 + \frac{100 \times 100}{6\,(2 \times 187.8 - 100)}\right]$$

$$= 293504 \times [436] = 128 \times 10^6 \text{ N-mm} \;;\; \text{where } 436 = a = \text{L.A.}$$

$$A_{st1} = \frac{M_1}{t \times a} = \frac{128 \times 10^6}{140 \times 436} = 2097 \text{ mm}^2$$

$$M_2 = M - M_1 = (180 - 128)\, 10^6 = 52 \times 10^6 \text{ N-mm}$$

$$\therefore \quad 52 \times 10^6 = (m_c - 1)\, A_{sc} \times c\left[\frac{kd - d_c}{kd}\right](d - d_c), \text{ Assuming } d_c = 40 \text{ mm,}$$

$$A_{sc} = \frac{(52 \times 10^6) \times (187.8)}{(1.5 \times 18 - 1) \times 5(187.8 - 40)(480 - 40)} = 1155 \text{ mm}^2, \text{ and}$$

$$A_{st2} = \frac{M_2}{t\,(d - d_c)} = \frac{52 \times 10^6}{140(480 - 40)} = 844 \text{ mm}^2$$

$\therefore$ Total tensile steel $A_{st} = 2097 + 844 =$ **2941** mm^2 and Comp. steel $A_{sc} =$ **1155** mm^2.

Example A-15. *Fig. A-22 shows the section of a doubly reinforced T-beam section. The tensile reinforcement consist of 5 bars of 24 mm diameter while the compressive reinforcement* *consists of 3 bars of 20 mm diameter. Taking $\sigma_{cbc} = 5$ N/mm², $\sigma_{st} = 140$ N/mm² and $m = 18$, determine the moment of resistance of the section (a) neglecting compression of web (b) taking compression of web into account. The beam is used over an effective span of 5 m.*

Solution. The effective width of flange is given by

$$b_f = \frac{l_0}{(l_0/b)+4} + b_w = \frac{5000}{(5000/1200)+4} + 300 = 912 \text{ mm}$$

(a) Neglecting Web Compression :

Let the N.A. lie in the web at a distance n below the top of the flange. Taking the moment of equivalent areas about N.A. we get

$$b_f D_f\left(n - \frac{D_f}{2}\right) + (m_c - 1)\, A_{sc}\,(n - d_c)$$

$$= m\, A_{st}(d - n) \qquad \text{where}$$

$$A_{sc} = 3\,\frac{\pi}{4}\,(20)^2 = 945 \text{ mm}^2 \text{ and}$$

$$A_{st} = 5\,\frac{\pi}{4}\,(24)^2 = 2260 \text{ mm}^2$$

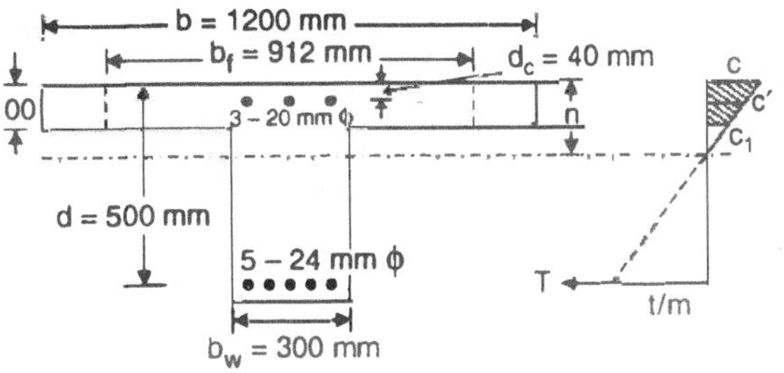

FIG. A-22.

$$\therefore\; 912 \times 100(n - 50) + (1.5 \times 18 - 1)$$

$$\times\, 945(n - 40) = 18 \times 2260\,(500 - n)$$

or $\qquad n - 50 + 0.269\, n - 10.78 = 223 - 0.446\, n$

from which $n = 165.5$ mm. Depth of critical N.A. is given by

$$n_c = \frac{m\sigma_{cbc}}{m\sigma_{cbc} + \sigma_{st}}\, d = \frac{18 \times 5}{18 \times 5 + 140} \times 500 = 196 \text{ mm}$$

Since the actual N.A. falls above the critical N.A., the section is under-reinforced and the stress in steel will reach its maximum value first. Hence $t = \sigma_{st} = 140$ N/mm².

$$\therefore \qquad c = \frac{t}{m}\,\frac{n}{d - n} = \frac{140}{18} \times \frac{165.5}{500 - 165.5} = 3.848 \quad \text{N/mm}^2$$

The moment of resistance is given by

$$M_r = b_f \,.\, D_f \frac{c}{2}\left[1 + \frac{n - D_f}{n}\right]\left[d - \frac{D_f}{2} + \frac{D_f^2}{6(2n - D_f)}\right] + (m_c - 1)\, A_{sc} \,.\, c \times \frac{n - d_c}{n}\,(d - d_c)$$

$$= 912 \times 100 \times \frac{3.848}{2}\left[1 + \frac{165.5 - 100}{165.5}\right] \times \left[500 - 50 + \frac{(100)^2}{6(2 \times 165.5 - 100)}\right]$$

$$+ (1.5 \times 18 - 1) \times 945 \times 3.848\, \frac{165.5 - 40}{165.5}\,(500 - 40)$$

$= 111.98 \times 10^6 + 32.98 \times 10^6 = 144.96 \times 10^6$ N-mm = **144.96 kN-m**

(b) ***Taking Compressive Force of Web into Account***

$$b_f \,.\, D_f\left(n - \frac{D_f}{2}\right) + \frac{b_w(n - D_f)^2}{2} + (m_c - 1)\, A_{sc}\,(n - d_c) = m\, A_{st}(d - n)$$

or $\qquad 912 \times 100(n - 50) + 150(n - 100)^2 + (1.5 \times 18 - 1)945(n - 40) = 18 \times 2260\,(500 - n)$

or $\qquad 608n - 30400 + n^2 + 10000 - 200n + 164n - 6552 = 135600 - 271n$

or $n^2 + 843\,n - 162\,552 = 0$. From which $n = 161.8$ mm. As found earlier, $n_c = 196$ mm.

The section is, therefore, under-reinforced, and $t = \sigma_{st} = 140$ N/mm².

The corresponding stresses c, c' and c_1 are as follows :

$$c = \frac{t}{m} \cdot \frac{n}{d-n} = \frac{140}{18} \times \frac{161.8}{500 - 161.8} = 3.72 \text{ N/mm}^2.$$

$$c' = \frac{t}{m} \cdot \frac{n - d_c}{d-n} = \frac{140}{18} \times \frac{161.8 - 40}{500 - 161.8} = 2.80 \text{ N/mm}^2.$$

$$c_1 = \frac{t}{m} \cdot \frac{n - D_f}{d-n} = \frac{140}{18} \times \frac{161.8 - 100}{500 - 161.8} = 1.42 \text{ N/mm}^2.$$

The moment of resistance of the beam will be equal to the sum of the moment of resistance due to (*i*) compressive force of flange (*ii*) compressive force of rib, and (*iii*) compressive force of steel.

or $$M_r = M_{r1} + M_{r2} + M_{r3}$$

Now $$M_{r1} = \text{M.R. due to compression in flange} = b_f \cdot D_f \left(\frac{c + c_1}{2}\right)(d - \bar{y})$$

where $$\bar{y} = \left(\frac{c + 2c_1}{c + c_1}\right)\frac{D_f}{3} = \left(\frac{3.72 + 2.84}{3.72 + 1.42}\right)\frac{100}{3} = 42.54 \text{ mm}$$

$$\therefore \quad M_{r1} = 912 \times 100 \left(\frac{3.72 + 1.42}{2}\right)(500 - 42.54) = 107.22 \times 10^6 \text{ N-mm}$$

$$M_{r2} = \text{M.R. due to compressive force in rib} = b_w (n - D_f)\frac{c_1}{2}\left[d - D_f - \frac{1}{3}(n - D_f)\right]$$

$$= 300(161.8 - 100)\frac{1.42}{2}\left[500 - 100 - \frac{1}{3}(161.8 - 100)\right] = 4.99 \times 10^6 \text{ N-mm}$$

$$M_{r3} = \text{M.R. due to compressive force in steel} = (m_c - 1)\,A_{sc} \cdot c'\,(d - d_c)$$

$$= (1.5 \times 18 - 1)\,945 \times 2.80\,(500 - 40) = 31.65 \times 10^6 \text{ N-mm}$$

$\therefore$ Total $M = (107.22 + 4.99 + 31.65)\,10^6 = 143.86 \times 10^6$ N-mm $=$ **143.86 kN-m.**

A-4. SHEAR AND BOND

A-4.1. Shear Stress in R.C. Beams

Fig. A-23(*b*) shows two sections *mn* and m_1n_1 distant *dx* apart, of a R.C. beam. Let the bending moment at *mn* be M and that at $m_1\,n_1$ be $M + \delta M$. Due to this variation of bending moment in a length δx, the compressive forces in concrete at these two sections will be C and $C + \delta C$, and the tensile forces in steel will be T and $T + \delta T$ as shown. Let us assume that the shear force V remains constant in the length *dx*.

Let us assume that concrete does not take any tension. Consider any plane between N.A. and centre of steel reinforcement. Let q be the intensity of shear at the plane.. Hence total horizontal shear $= q.b.\,\delta x$ at that layer. Also, the total horizontal force that tends to slide this layer past the adjacent one is equal to $(T + \delta T) - (T) = \delta T$. Hence equating the two, we get

$$q \cdot b \cdot \delta x = \delta T \qquad \text{or} \qquad q = \frac{\delta T}{b \cdot \delta x} \qquad \ldots(i)$$

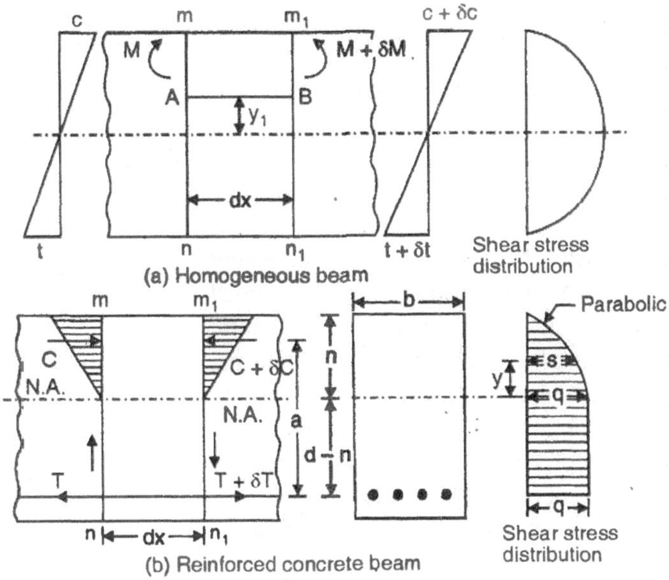

FIG. A-23. SHEAR STRESSES IN BEAMS

In order to get the value of q, it is first essential to determine the value of δT. For equilibrium $\Sigma M = 0$, Hence,

$$(T + \delta T)\, jd = Tjd + V \,.\, \delta x$$

or

$$\delta T = \frac{V \,.\, \delta x}{jd} \qquad \ldots(ii)$$

Substituting in (i), we get

$$q = \frac{V}{bjd} \qquad \ldots\text{(A-47)}$$

This value of q is evidently the same for any layer between the N.A. and the centre of steel. Thus the shear stress distribution below the N.A. is rectangular. The shear stress distribution above the N.A. is parabolic, as in the case of homogeneous beams. The intensity of shear stress s at any layer distant y above the N.A. is given by

$$s = \frac{V}{Ib} \int_y^n b \,.\, y \,.\, dy = \frac{V}{2I}(n^2 - y^2) \qquad \ldots\text{(A-48)}$$

where I is the moment of inertia of the R.C. beam section about the N.A. and can be found the following expression:

$$I = \frac{1}{3} b \,.\, n^3 + m\, A_{st}\, (d - n)^2 \qquad \ldots\text{(A-49)}$$

The complete shear stress distribution diagram is shown in Fig. A-23(b). The intensity of maximum shear (q) can also be found by equating the external shear force to the shearing resistance of the beam, as under :

$$V = (\text{ Area of shear stress diagram }) \times b = \left\{ \left(\frac{2}{3} qn \right) + q\,(d - n) \right\} b$$

$$= \left(\frac{2}{3} qn + qd - qn \right) b = qb \left(d - \frac{n}{3} \right) = q \,.\, bjd$$

Hence $$q = \frac{V}{b \cdot jd} \qquad \ldots\text{(A-47)}$$

Eq. A-47, obtained from the stress based approach, gives the value of maximum shear stress q at N.A. and for all layers below N.A. in a cracked reinforced beam. Though in pure flexure, we neglect the concrete area below N.A., the same concrete between the cracks is needed for shear transfer between steel reinforcement and the compression zone. Also the flexural cracks (assumed vertical in pure flexure) are not vertical in the shear-flexural zone under consideration. We have also neglected the shear deformations, which are really responsible for the mobilization of aggregate interlock at the cracks and dowel forces in the reinforcement. Due to these reasons, the *stress-based approach* followed above in the derivation of Eq. A-47 may not represent the real behaviour of the beam in shear and flexure. Hence Eq. A-47 , representing conventional shear stress formula $q = V/bjd$, has now been abandoned. Indian Standard Code (IS : 456 - 2000) recommends to use *nominal shear stress* given by the expression.

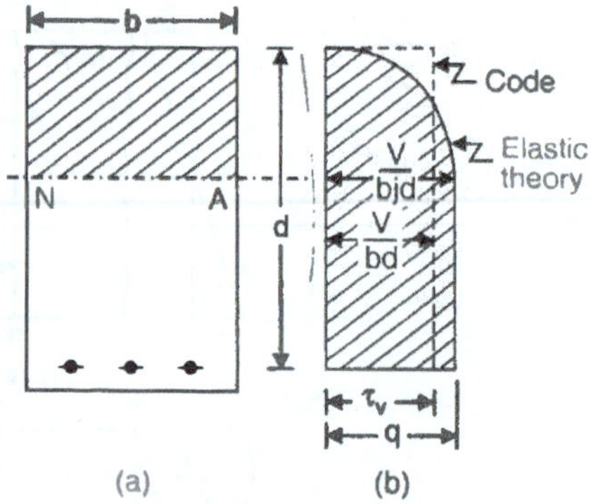

FIG. A-24.

$$\tau_v = \frac{V}{bd} \qquad \ldots\text{(A-50)}$$

where τ_v = nominal shear stress and V = shear force at the section, due to design loads.

Thus in Fig. A-24(*b*), the maximum shear stress q $(= V/bjd)$ obtained from elastic theory, is greater than the *nominal shear stress* (or average shear stress) τ_v suggested by the Code.

A-4.2. Indian Standard Code Recommendations (IS : 456 – 2000)

1. Nominal Shear Stress : Indian Standard Code (IS : 456 – 2000) uses a concept of *nominal shear stress* (τ_v) rather than *shear stress* (q) given by Eq. A-47. The nominal shear stress in beams or slabs of uniform depth shall be calculated by the following equation:

$$\tau_v = \frac{V}{bd} \qquad \ldots\text{(A-50)}$$

where V = shear force due to design loads

b = breadth of the member, which for flanged section shall be taken as the breadth of web (b_r)

and d = effective depth

In the case of beams of varying depth, the equation shall be modified as :

$$\tau_v = \frac{V \pm \dfrac{M}{d}\tan\beta}{b \cdot d} \qquad \ldots\text{[A-50(a)]}$$

where τ_v, V, b and d have the same meanings as above and β is the angle between the top and the bottom edges of the beam. The negative sign in the formula applies when the bending moment M increases numerically in the same direction as the effective depth increases, and the positive sign when the moment decreases numerically in this direction.

(2) Design Shear Strength of Concrete

(a) Beams without Shear Reinforcement : The permissible shear stress (τ_c) in concrete in beams without shear reinforcement is given in Table A-5. The permissible shear stress depends upon two factors : grade of concrete and percentage reinforcement ($100\, A_s/bd$).

TABLES A-5. PERMISSIBLE SHEAR STRESS (τ_c) IN CONCRETE (IS : 456 – 2000)

$\frac{100 A_s}{bd}$	*Permissible shear stress in concrete τ_c N/mm^2 for grade of concrete*					
	M 15	*M 20*	*M 25*	*M 30*	*M 35*	*M 40*
≤ 0.15	0.18	0.18	0.19	0.20	0.20	0.20
0.25	0.22	0.22	0.23	0.23	0.23	0.23
0.50	0.29	0.30	0.31	0.31	0.31	0.32
0.75	0.34	0.35	0.36	0.37	0.37	0.38
1.00	0.37	0.39	0.40	0.41	0.42	0.42
1.25	0.40	0.42	0.44	0.45	0.45	0.46
1.50	0.42	0.45	0.46	0.48	0.49	0.49
1.75	0.44	0.47	0.49	0.50	0.52	0.52
2.00	0.44	0.49	0.51	0.53	0.54	0.55
2.25	0.44	0.51	0.53	0.55	0.56	0.57
2.50	0.44	0.51	0.55	0.57	0.58	0.60
2.75	0.44	0.51	0.56	0.58	0.60	0.62
3.00 and above	0.44	0.51	0.57	0.60	0.62	0.63

(b) Solid slabs : For solid slab; the permissible shear strees in concrete shall be $k \cdot \tau_c$ where k has the value given in Table A-6.

TABLE A-6. FACTOR k

Overall depth of slab (mm)	300 or more	275	250	225	200	175	150 or less
k	1.00	1.05	1.10	1.15	1.20	1.25	1.30

The above requirement does not apply to flat slabs.

Note. A_s is that area of longitudinal tensile reinforcement which continues at least one effective depth beyond the section being considered except at supports where the full area of tension reinforcement may be used.

(c) Shear strength of members under axial compression

For members subjected to axial compression, P, the permissible shear stress in concrete τ_c, given in Table A-5, shall be multiplied by the following factor :

$$\delta = 1 + \frac{5P}{A_g f_{ck}} \quad \text{(but not exceeding 1.5)} \qquad \text{...(A-51)}$$

where P = axial compressive force in N

A_g = gross area of concrete section in mm^2

and f_{ck} = characteristic compressive strength of concrete

(d) Beams with Shear Reinforcement : When shear reinforcement is provided the nominal shear stress τ_v in beam shall not exceed $\tau_{c.max}$ given in Table A-7.

TABLE A-7. MAXIMUM SHEAR STRESS $\tau_{c.max}$ N/mm^2 (IS : 456–2000)

Concrete grade	M 15	M 20	M 25	M 30	M 35	M 40
τ_{cmax} *(N/mm^2)*	1.6	1.8	1.9	2.2	2.3	2.5

For slabs, τ_v shall not exceed half the value of given in Table A-7.

(3) Minimum Shear Reinforcement

When τ_v is less than τ_c given in Table A-5, minimum shear reinforcement shall be provided as explained below :

Minimum shear reinforcement in the form of stirrups shall be provided such that

$$\frac{A_{sv}}{b \cdot s_v} \geq \frac{0.4}{0.87 f_y} \qquad \text{...(A-52)}$$

where A_{sv} = total cross-sectional area of stirrup legs effective in shear.

s_v = stirrup spacing along the length of the member

b = breadth of beam or breadth of the web of flanged beam and,

f_y = characteristic strength of the stirrup reinforcement, in N/mm^2, which shall not be taken greater than 415 N/mm^2. The values of f_y for three types of steel reinforcement are given in Table 1.29.

However, in members of minor structural importance such as lintels or where the maximum shear stress calculated is less than half the permissible value, this provision need not be complied with.

Eq. A-52 can be rearranged in the following form to give the maximum spacing of stirrups:

$$s_v \leq \frac{0.87 f_y A_{sv}}{0.4 b} \leq \frac{2.175 A_{sv} \cdot f_y}{b} \qquad \text{...[A-52(a)]}$$

(4) Design of Shear Reinforcement

When τ_v exceeds τ_c given in Table A-5, shear reinforcement shall be provided in any of the following forms :

(*i*) Vertical stirrups

(*ii*) Bent-up bars along with stirrups and (*iii*) Inclined stirrups.

Where bent up bars are provided, their contribution towards shear resistance shall not be more than half that of the total shear reinforcement. Shear reinforcement shall be provided to carry a shear equal to $V_s = V - \tau_c\ bd$. The strength of shear reinforcement V_s shall be calculated as below :

(i) For vertical stirrups

$$V_s = \frac{\sigma_{sv} \cdot A_{sv} \cdot d}{s_v} \qquad \text{...(A-53)}$$

(ii) For inclined stirrups or a series of bars bent-up at different cross section :

$$V_s = \frac{\sigma_{sv} \cdot A_{sv} \cdot d}{s_v}(\sin\alpha + \cos\alpha) \qquad \text{...(A-54)}$$

(iii) For single bar or single group of parallel bars, all bent up at the same cross-section

$$V_s = \sigma_{sv} \cdot A_{sv} \cdot \sin\alpha \qquad \text{...(A-55)}$$

where: A_{sv} = total cross-sectional area of stirrups legs or bent up bars within a distance s_v

s_v = spacing of the stirrups or bent up bars along the length of the member

τ_c = design shear strength of the concrete.

b = breadth of the member which for flanged beams, shall be taken as breadth of the web b_w.

σ_{sv} = permissible tensile stress in shear reinforcement which shall not be taken greater than 230 N/mm^2.

α = angle between the inclined stirrups or bent up bar and the axis of the member, not less than 45° and

d = effective depth.

Note : Where more than one type of shear reinforcement is used to reinforce the same portion of the beam, the total shear resistance shall be computed as the sum of the resistance of the various types separately. The area of stirrups shall not be less than the minimum specified in para 3 above.

(5) Maximum Spacing of Shear Reinforcement

The maximum spacing of shear reinforcement measured along the axis of the member shall not exceed 0.75 d for vertical stirrups and d for inclined stirrups at 45°, where d is the effective depth of the section under consideration. In no case shall the spacing exceed 300 mm.

(6) Summary

(*i*) If $\tau_v \le \frac{1}{2}\tau_c$ or $V \le \frac{1}{2} V_c$: No shear reinforcement is really needed.

(*ii*) If $\frac{1}{2} V_c \le V \le V_c$: Provide nominal reinforcement given by Eq. A-52..

However, the maximum spacing is restricted to 0.75 d or 300 mm, whichever is less.

(*iii*) If $V_c \le V \le V_{cmax}$ Design the transverse reinforcement for the net S.F.

$$F = (V - V_c).$$

If ΣV_s is the sum of the resistances of various types of shear reinforcements (Eqs. A-53, A-54, A-55) then ΣV_s should be equal to or greater than $(V - V_c)$.

However, the maximum spacing (s_v) is restricted to 0.75 d or 300 mm, whichever is less. Also the contribution of bent-up bars should not be more than half of the total shear reinforcement and the area of stirrups should not be less than the one provided by Eq. A-52.

(*iv*) If $V \ge V_{c.max}$, redesign the web of the section such that $V_{c.max}$ becomes equal to or more than external shear V.

Example A-16. *A reinforced concrete beam 250 mm wide and 400 mm effective depth is subjected to a shear force of 95 kN at the supports. The tensile reinforcement at the support is 0.5 percent. Find the spacing of 12 mm diameter 2-legged stirrups to resist the shearing stress at supports, for M 15 concrete. Use (a) mild steel reinforcement (b) HYSD bars. Also, design the minimum shear reinforcement at the mid-span of the beam.*

Solution. Given $p = 100\, A_{st}/bd = 0.5\%$

Hence from Table A-5, for M 15 concrete, we get, $\tau_c = 0.29$ N/mm^2

Also from Table A-7, maximum shear stress, $\tau_{c.max} = 1.6$ N/mm^2

$$\tau_v = \frac{V}{bd} = \frac{95 \times 1000}{250 \times 400} = 0.95 \text{ N/mm}^2$$

This is greater than τ_c but less than $\tau_{c,max}$.

The shear reinforcement is provided to carry a shear V_s equal to $V - \tau_c \,.\, bd$.

$\therefore \qquad V_s = V - \tau_c\, bd = 95000 - 0.29 \times 250 \times 400 = 66000 \quad N$

(*a*) ***Mild steel bars*** (Fe 250) $\sigma_{sv} = 140$ N/mm^2

The spacing of stirrups at supports is given by Eq. A-53, rearranged as follows :

$$s_v = \frac{\sigma_{sv}\, d\, A_{sv}}{V_s} = \frac{140 \times 400}{66000} A_{sv} = 0.848 \; A_{sv} \text{ mm.}$$

Using 12 mm φ bars two legged stirrups, $A_{sv} = 2\,\frac{\pi}{4}(12)^2 = 226.2$ mm^2

$\therefore \qquad s_v = 0.848 \times 226.2 = 191.9$ mm

Hence provide stirrups at spacing of 190 mm *c/c* at the supports.

For nominal (minimum) shear reinforcement at mid span, the spacing at 12 mm φ 2-lgd stirrups is given by Eq. A-52(*a*).

$$s_v = \frac{2.175\, A_{sv} f_y}{b} = \frac{2.175 \times 226.2 \times 250}{250} = 492 \text{ mm.}$$

However, max. spacing = 0.75 *d* or 300 mm which ever is less = 0.75 × 400 = 300 mm.

Thus, use of 12 mm dia stirrups will be uneconomical. Using 10 mm dia bars for stirrups, $A_{sv} = 2 \times \frac{\pi}{4}(10)^2 = 157.1$ mm^2

$$\therefore \qquad s_v = \frac{2.175\, A_{sv} \cdot f_y}{b} = \frac{2.175 \times 157.1 \times 250}{250} = 341 \text{ mm.}$$

Subject to a maximum of 0.75 *d* = 0.75 × 400 = 300 mm. Hence use 10 mm dia 2-lgd. stirrups at 300 mm spacing at the location where nominal reinforcement is required.

(*b*) ***HYSD bars***

$$s_v = \frac{\sigma_{sv} \cdot d \cdot A_{sv}}{V_s} = \frac{230 \times 400}{66000} A_{sv} = 1.3939\, A_{sv} \text{ mm.}$$

Using 12 mm φ bars two legged stirrups, $A_{sv} = 2 \times \frac{\pi}{4}(12)^2 = 226.2$ mm^2

$\therefore \qquad s_v = 1.3939 \times 226.2 = 315.3$ mm

subject to a maximum of 0.75 *d* = 0.75 × 400 = 300 mm.

Hence provide the stirrups @ 300 mm c/c.

For nominal (minimum) shear reinforcement, let us use 8 mm dia bars.

$A_{sv} = 2 \times \frac{\pi}{4}(8)^2 = 100.53$ mm^2. The spacing is given by Eq. A-52 (*a*)

$$s_v = \frac{2.175\, A_{sv} \cdot f_y}{b} = \frac{2.175 \times 100.53 \times 415}{250} = 362 \text{ mm.}$$

subject to a maximum spacing equal to 0.75*d* = 300 mm

Hence provide 8 mm dia. 2 lgd stirrups at 300 mm spacing at the location where nominal reinforcement is required.

A-4.3. IS Code on Bond and Anchorage Requirements (IS : 456 – 2000)

(*a*) Permissible Stress in Bond

The permissible stress in bond (average) for plain bars in tension are given in Table A-8.

TABLE A-8. PERMISSIBLE BOND STRESS τ_{bd} (IS : 456–2000)

Grade of concrete	M 10	M 15	M 20	M 25	M 30	M 35	M 40	M 45	M 50
Permissible stress in bond τ_{bd} (N/mm^2)	–	0.6	0.8	0.9	1.0	1.1	1.2	1.3	1.4

Note 1. The bond stress given shall be increased by 25 percent for bars in compression.

Note 2. In the case of deformed bars conforming to IS : 1139–1966 and IS : 1786–1979, the bond stress given above may be increased by 60%.

(*b*) Development of Stress in Reinforcement

The calculated tension or compression in any bar at any section shall be developed on each side of the section by the appropriate development length of end anchorage or by a combination thereof.

Development length of bars. The development length L_d is given by

$$L_d = \frac{\varphi \cdot \sigma_s}{4\tau_{bd}} = k_d\, \varphi \qquad \text{...(A-56)}$$

where φ = nominal diameter of the bar

σ_s = stress in bar at the section considered at design load, and

τ_{bd} = design bond stress given in Table A-8

k_d = development length factor $= \sigma_s/4\,\tau_{bd}$

The values of L_d for plain mild steel bars ($\sigma_s = \sigma_{st} = 140\ N/mm^2$) and high yield strength deformed bars (HYSD bars having $\sigma_s = \sigma_{st} = 230\ N/mm^2$), for various grades of concrete are summarised in Table A-9.

TABLE A-9. DEVELOPMENT LENGTH IN TENSION

Grade of Concrete	*Plain m.s. bars*		*HYSD bars*	
	τ_{bd} (N/mm^2)	$k_d = L_d/\varphi$	τ_{bd} (N/mm^2)	$k_d = L_d/\varphi$
M 15	0.6	58	0.96	60
M 20	0.8	44	1.28	45
M 25	0.9	39	1.44	40
M 30	1.0	35	1.60	36
M 35	1.1	32	1.76	33
M 40	1.2	29	1.92	30
M 45	1.3	27	2.08	28
M 50	1.4	25	2.24	26

When the bars are in compression, the bond stress given in Table A-9 shall be increased by 25%. Hence the development length in compression is given by

$$L_{dc} = \frac{\varphi\, \sigma_{sc}}{5\, \tau_{bd}} \qquad \text{...[A-56(a)]}$$

Note 1. The development length includes anchorage values of hooks in tension reinforcement.

Note 2. For bars of sections other than circular, the development length should be sufficient to develop the stress in the bar by bond.

Note 3: When the actual reinforcement provided is more than that theoretically required so that actual stress (σ_s) is steel is less than full resistance, the development length requircal may be reduced by the relation :

$$\textbf{\textit{Reduced development length}}\ L_{dr} = L_d\left(\frac{A_{st}\ required}{A_{st}\ provided}\right) \quad \ldots\text{(A-57)}$$

This principle is used in the design of footing and other shear bending mombers where bond is critical. By providing more steel, the bond requirements are satisfied.

A-5. AXIALLY LOADED COLUMNS

A-5.1. Load Carrying Capacity of Short Columns

(*a*) **Short columns with lateral ties.** According to the classical *elastic theory* or the *compatible strain theory*, when a reinforced concrete column is loaded, both concrete and steel have equal strains during initial stage of loading, because they are well bonded. Hence if P is the load on the column, the loads carried by steel bars (P_s) and that carried by surrounding concrete (P_c) bears the relation

$$P = P_c + P_s \quad \ldots(i)$$

From compatibility requirements, $\varepsilon_s = \varepsilon_c$

$$\therefore \quad \frac{P_s}{E_s \cdot A_{sc}} = \frac{P_c}{E_c \cdot A_c} \quad \text{or} \quad \frac{\sigma_{sc}}{E_s} = \frac{\sigma_{cc}}{E_c}$$

or $$\sigma_{sc} = \frac{E_s}{E_c} \cdot \sigma_{cc} = m \cdot \sigma_{cc} \text{ where } m \text{ is the modular ratio.} \quad \ldots(ii)$$

Hence from (i) $$P = \sigma_{cc} \cdot A_c + \sigma_{sc} \cdot A_{sc}$$

or $$P = \sigma_{cc} \cdot A_c + m\sigma_{cc} \cdot A_{sc} \quad \ldots\text{(A-58)}$$

where A_{sc} = Area of steel reinforcement; A_c = Area of concrete

σ_{sc} = Compressive stress in steel; σ_{cc} = Compressive stress in concrete.

It has been found that the steel reinforcement is restrained against lateral expansion, due to concrete surrounding it, reducing the problem to be a plain strain problem rather than a plain stress problem. It is therefore essential to moderate the compatible stress in steel by restrained Poisson's effect, using modified Poisson's ratio $m_c = 1.5\ m$. Thus stress in steel is given by : $\sigma_{sc} = m_c \cdot \sigma_{cc} = 1.5\ m\ \sigma_{cc}$ and equation A-58 reduces to the form

$$P = \sigma_{cc} \cdot A_c + 1.5\ m\ \sigma_{cc} \cdot A_{sc} \quad \ldots\text{(A-59)}$$

The above theoretical approach, based on strain compatibility, has been proved to be highly unrealistic due to creep and shrinkage behaviour of concrete. The above formula therefore gives very conservative results. It is therefore recommended to use *equilibrium approach* based on allowable stresses in the two materials. Such an equilibrium approach, however, violates the compatibility of strains. According to this approach, the load carrying capacity of a short column, axially loaded and provided with lateral ties, is given by

$$P = \sigma_{cc} \cdot A_c + \sigma_{sc} \cdot A_{sc} \quad \ldots\text{(A-60)}$$

where P = permissible axial load

σ_{sc} = permissible compressive stress in column bars

σ_{cc} = permissible compressive stress in concrete

IS : 456-2000 recommends Eq. A-60 for the determination of permissible axial load P. The minimum eccentricity mentioned in § 16.6 may be deemed to be incoporated in the above equation.

(*b*) **Short column with helical reinforcement**. For a column having longitudinal re-inforcement tied with spirals (*i.e.* helical reinforcement), the load carrying capacity is taken as 1.05 times the strength of similar member with lateral ties, provided the requirements laid down in § 16.8 are satisfied.

(*c*) **Composite columns**. The allowable load P on a composite column, consisting of a structural steel or cast-iron column thoroughly encased in concrete reinforced with both longitudinal and spiral reinforcement is given by the following expression :

$$P = \sigma_{cc} \cdot A_c + \sigma_{sc} \cdot A_{sc} + \sigma_{mc} \cdot A_m \qquad \text{...(A-61)}$$

where A_c = net area of concrete section

A_{sc} = cross-sectional area of longitudinal bar reinforcement

σ_{mc} = allowable unit stress in metal core, not to exceed 125 N/mm^2 for a steel core, or 70 N/mm^2 for a cast iron core

A_m = Cross-sectional area of steel or cast iron core.

A-5.2. Long Columns.

The maximum permissible stress in a reinforced concrete column or part thereof having a ratio of effective column length to least lateral dimension above 12 shall not exceed that which results from the multiplication of the appropriate maximum permissible stresses (in concrete and steel) by the coefficient C_r given by the following formula :

$$C_r = 1.25 - (l_{ef}/48b)$$

where C_r = reduction coefficient ; l_{ef} = effective length of column, and

b = least lateral dimension of column ; for column with helical reinforcement, b is the diameter of the core.

For more exact calculations, the maximum permissible stresses in a reinforced concrete column or part thereof having a ratio of effective column length to least lateral radius of gyration above 40 shall not exceed these which result from the multiplication of the appropriate maximum permissible stresses in concrete and steel, by the coefficient C_r given by the following formula : $C_r = 1.25 - \dfrac{l_{ef}}{160\, i_{min}}$ where i_{min} is the least radius of gyration.

A-5.3. Design Procedure

A R.C. column may be designed in the following steps :

1. Depending upon the grade of concrete to be used, determine the permissible stresses in concrete, longitudinal bars and ties.

2. Find the super-imposed load the column is required to carry. To this, add dead weight (assumed) of the column to get the total load the column has to carry at its base.

3. Assume some suitable value of reinforcement A_{sc} say between 0.8 to 2% of gross area A of column. Determine the area A from the following expression :

$$P = \sigma_{cc} \cdot A_c + \sigma_{sc} \cdot A_{sc} \qquad \text{or} \qquad P = \sigma_{cc}(A_g - p \cdot A_g) + \sigma_{sc} \cdot pA_g$$

or $$A_g = \frac{P}{\sigma_{cc}(1-p) + p \cdot \sigma_{sc}} \qquad \text{...(A-62)}$$

where A_g = gross area of column. and p = ratio of steel to total area $= A_{sc}/A_g$.

4. After having known the area A_g, determine the dimensions of the column. If it is a square of side b, then $b = \sqrt{A_g}$. If a circular column of diameter D is to be used, D (= b) will be equal to $\sqrt{4A_g/\pi}$.

5. For the given end conditions, determine the effective length l_{ef} of the column, and hence calculate l_{ef}/b ratio to find whether it is short column or long column.

6. If $l_{ef}/b < 12$, it will be designed as a short column for which dimensions have already been found in steps 3 and 4. Determine the area of steel A_{sc} and distribute the bars suitably around the periphery of the column, keeping suitable cover.

7. If $l_{ef}/b > 12$, it will be designed as a long column for which the reduction factor C_r is determined from the expression $C_r = 1.25 - \frac{l_{ef}}{48\, b}$

Calculate the design load P' for an equivalent short column :

$$P' = P/C_r \qquad \text{...(A-63)}$$

Using the revised value of P', re-calculate the area A_g and hence determine the size of the column. Also, calculate the area of steel $A_{sc} = pA_g$.

8. Find the diameter of bars used as ties, and determine its pitch as per rules discussed in the previous article.

Example A-17. *(a) A reinforced concrete column 4 m long (effective) and 400 mm in diameter is reinforced with 8 bars of 20 mm diameter. Find safe load of columns can carry. Take M 20 concrete and Fe 415 steel reinforcement. The column carries lateral ties. (b) If the effective length is increased to 8 m, what will be the safe load of column can carry ?*

Solution. (*a*) $l_{ef} = 4000$ mm ; $b = D = 400$ mm

Hence $$\frac{l_{ef}}{b} = \frac{4000}{400} = 10 < 12. \text{ Thus the column is short.}$$

$$P = \sigma_{cc} \cdot A_c + \sigma_{sc} \cdot A_{sc}$$

where $$A_{sc} = 8\,\frac{\pi}{4}(20)^2 = 2513.3 \text{ mm}^2 \;;\; A_g = \frac{\pi}{4}(400)^2 = 125664 \text{ mm}^2$$

$$A_c = 125664 - 2513.3 \simeq 123151 \text{ mm}^2.$$

For M 20 concrete, $\sigma_{cc} = 5$ N/mm²; For Fe 415 steel, $\sigma_{sc} = 190$ N/mm²

∴ $$P = (5 \times 123151) + (190 \times 2513.3) = 1093281 \text{ N} \approx \mathbf{1093.281 \text{ kN}}$$

(*b*) $l_{ef} = 8000$ mm ; $D = 400$ mm

$\frac{l_{ef}}{b} = \frac{8000}{400} = 20 > 12$. Hence the column is a long column for which the reduction factor is given by

$$C_r = 12.5 - \frac{l_{ef}}{48b} = 1.25 - \frac{8000}{48 \times 400} = 0.8333$$

$$P = 0.8333 \times 1093281 \simeq 911067 \text{ N} \simeq 911.067 \text{ kN}.$$